The Rhizobiaceae

The Rhizobiaceae

Molecular Biology of Model Plant-Associated Bacteria

Edited by

Herman P. Spaink
Leiden University,
Institute of Molecular Plant Sciences,
Clusius Laboratory,
Leiden, The Netherlands

Adam Kondorosi
CNRS,
Institut des Sciences Végétales,
Gif sur Yvette,
Cedex, France

and

Paul J.J. Hooykaas
Leiden University,
Institute of Molecular Plant Sciences,
Clusius Laboratory,
Leiden, The Netherlands

SPRINGER-SCIENCE+BUSINESS MEDIA, B.V.

A C.I.P. Catalogue record for this book is available from the Library of Congress

ISBN 978-0-7923-5180-1   ISBN 978-94-011-5060-6 (eBook)
DOI 10.1007/978-94-011-5060-6

The camera ready text was prepared by
Emma Groot, Jacob Keylaan 4, 2343 HT Oegstgeest, The Netherlands

Printed on acid-free paper

Contents

Preface

The *Rhizobiaceae*, Molecular Biology of Model Plant-Associated Bacteria.

This book gives a comprehensive overview on our present molecular biological knowledge about the *Rhizobiaceae*, which currently can be called the best-studied family of soil bacteria. For many centuries they have attracted the attention of scientists because of their capacity to associate with plants and as a consequence also to specifically modify plant development. Some of these associations are beneficial for the plant, as is the case for the *Rhizobiaceae* subgroups collectively called rhizobia, which are able to fix nitrogen in a symbiosis with the plant hosts. This symbiosis results in the formation of root or stem nodules, as illustrated on the front cover. In contrast, several *Rhizobiaceae* subgroups can negatively affect plant development and evoke plant diseases. Examples are *Agrobacterium tumefaciens* and *A. rhizogenes* which induce the formation of crown galls or hairy roots on the stems of their host plants, respectively (bottom panels on front cover). In addition to the obvious importance of studies on the *Rhizobiaceae* for agronomy, this research field has resulted in the discovery of many fundamental scientific principles of general interest, which are highlighted in this book.

To mention three examples: (i) the discovery of DNA transfer of *A. tumefaciens* to plant cells, resulting in their genetic colonization by the bacterium, (ii) the discovery of the lipo-chitin oligosaccharides as mitogenic signal molecules, (iii) the discovery that the establishment of a symbiotic association is based on a molecular dialogue between the partners, consisting of the transmittal of specific signal molecules that activate gene sets or genetic programs in the partner. These principles have been shown to be of importance or used as paradigms for the study of very unrelated organisms, such as yeast and vertebrates or associations of various organisms. Molecular genetic studies on the interaction between the plant hosts and the bacterial partners have resulted in novel tools for studies of plant development and a general understanding of the signal molecules which are exchanged and their subsequent transduction pathways as summarized in the color plate C1. In addition to these signal molecules, many other factors play a role in establishing a successful association with the host plants. In this book these factors are discussed by specialists in the fields of chemistry, biochemistry, genetics and biology. However, the need to summarize all available data was not taken as an excuse to make the information less accessible to non-specialist readers. In contrast, the book is well suited for educational purposes as well, for instance in second or third year graduate courses in microbiology or plant-microbe interactions. The reason for this is that the field is presently well enough advanced to make possible a complete integration with the general textbook knowledge on biology and chemistry of microbes. The editors would like to thank the many authors of this book for preparing their outstanding contribution. We are also grateful to Mrs. Emma Groot for doing the type editing and preparing the layout of the book.

Herman P. Spaink
Adam Kondorosi
Paul J.J. Hooykaas

Color Plates

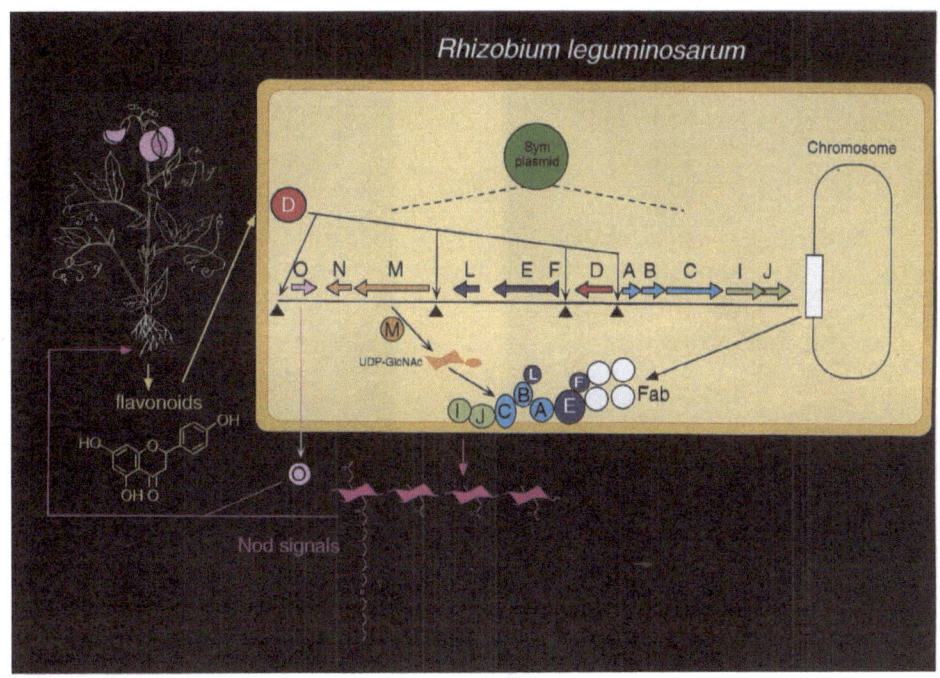

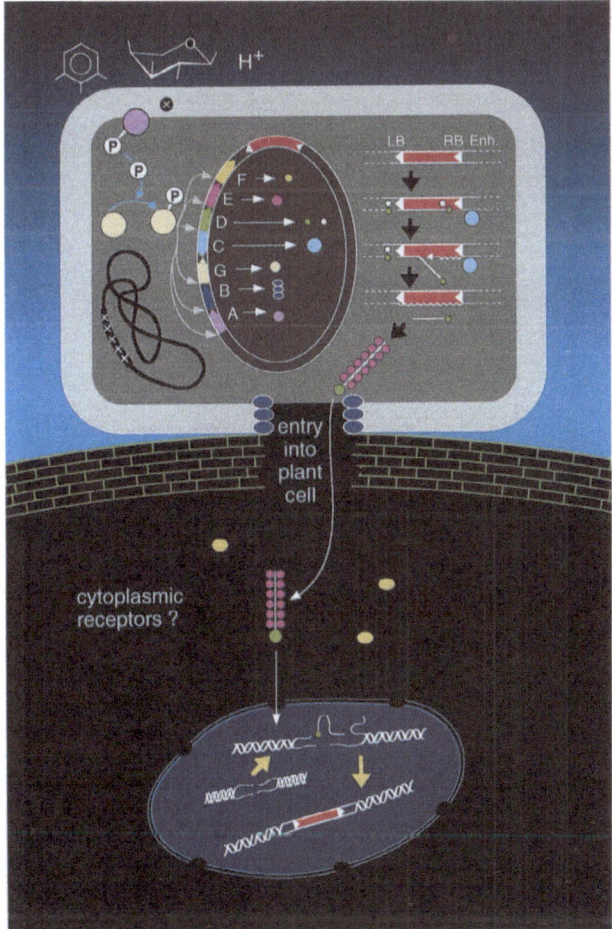

C1. Examples of molecular interactions between members of the *Rhizobiaceae* and their plant hosts.

Upper panel: Signal exchange between rhizobia and leguminous plants. Two steps of signal exchange which play a major role in the root nodulation and infection process are illustrated using as a typical example the bacterial strain *Rhizobium leguminosarum* biovar *viciae* 248. These steps comprise the induction of nodulation genes by plant flavonoids and the subsequent production of nodulation factors such as the lipo-oligosaccharides and secreted proteins. Although many variations exist in the bacterial genes involved in the illustrated steps of signal exchange within the rhizobial sub-group (e.g. see chapters 19 and 20), the basic principles and their role in determining host-specificity are general for all studied rhizobia. Drawing made by Dr. H.P. Spaink.

Bottom panel: The molecular mechanism of plant transformation by *Agrobacterium*. This mechanism includes the induction of the bacterial virulence system by a two-component regulatory system which is responsive to specific chemical signals from the plant. Virulence proteins then mediate the production of an oncogenic single stranded DNA molecule and its transfer into plant cells. Integration of this transferred DNA (T-DNA) into the plant genome and its continued expression leads to the transformation of the plant cells leading to a crown gall tumor. Drawing made by Dr. H.R.M. Schlaman.

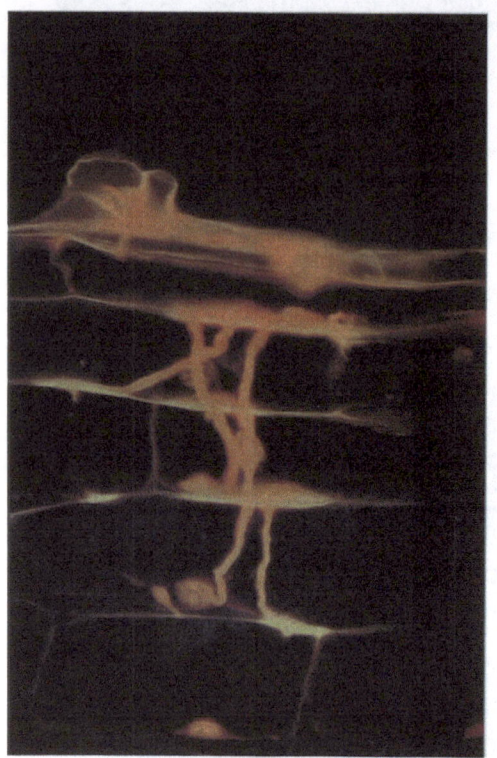
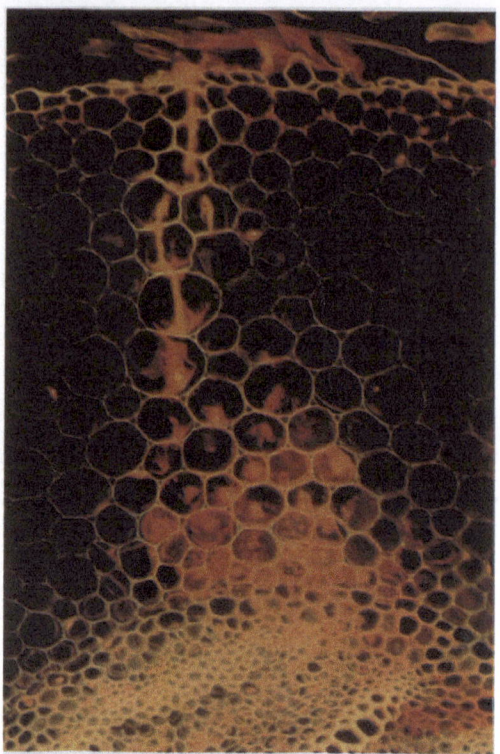
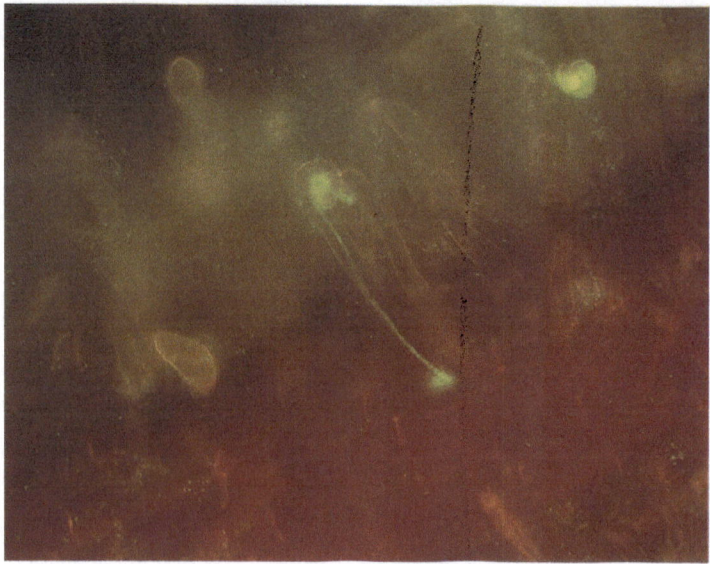

C2. Examples of the rhizobial infection process. Shown are infection threads in pea (panels A and B) and vetch (panel C) plants. In the case of pea, the infection threads are stained with acridine orange. In the case of vetch, use has been made of rhizobia which contain the gene encoding the green fluorescent protein (GFP) derived of the jelly-fish *Aequoria victoria*. The GFP protein can be detected using fluorescence microscopy (with this particular variety of GFP excitation light of 480 nm is used and fluorescence is detected at 510 nm). The use of the GFP protein is a versatile tool for non-destructive detection of gene expression and various other processes in living systems (see for example Quaedvlieg *et al.* Plant Mol. Biol. 37: 715-727, 1998). The rhizobial infection process is described in detail in chapters 18 and 22. Photographs were made by Dr. K. Libbenga, Leiden University (panels A, and B) and Dr. H.P. Spaink (panel C).

C3 (Fig. 2, Chapter 18). Examples of different types of nodules induced by rhizobia. (A) Indeterminate nodules induced on the root of a pea plant. (B) Determinate nodules induced on the stem of *Sesbania rostrata*. Nodules are induced on dormant root primordia of the stem and therefore the process is comparable to root nodulation. (C) A close-up of indeterminate root nodules of *Vicia hirsuta*. (D) A close-up of determinate root nodules induced in *Lotus preslii*. The picture in panel B is courtesy of Dr. M.Holsters (Gent University, Belgium).

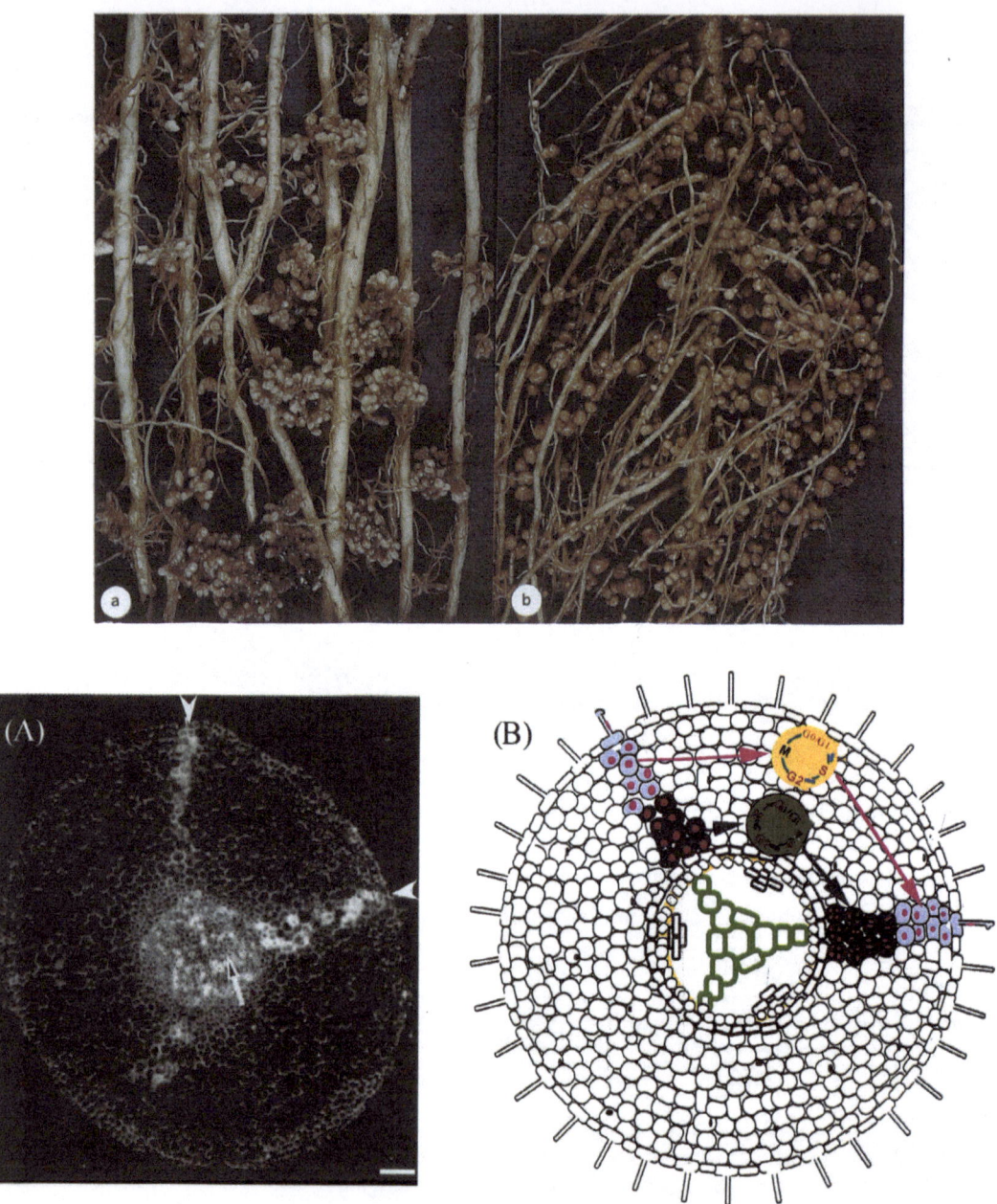

C4 (Top: Fig. 3, Chapter 26; Bottom: Fig 3, Chapter 21). Top: Root nodules on (a) alfalfa (*Medicago sativa* L.) and (b) birdsfoot trefoil (*Lotus corniculatus* L.). Bottom: Activation of the cortex during the induction of an indeterminate nodule. (A) Dark-field micrograph showing a pea root cross-section that originates from a pea root one day after inoculation with *R. l.* bv. *viciae*. This section was hybridized with a histone *H4* gene probe. In front of the infection sites indicated by arrowheads, *H4* transcripts are localized in narrow rows of cortical cells. Hybridization signal is represented by the silver grains. Infection sites are opposite to the protoxylem poles (arrow). Bar = 50 1m. (B) Schematic drawing of a pea root cross-section showing the reactivation of cortical cells after application of Nod factors or inoculation with rhizobia. The outer cortical cells shown in light grey, reenter the cell cycle, proceeding from the G0/G1 phase to the S phase, and become arrested in the G2, as indicated in the cell cycle in the light grey circle. However, inner cortical cells as shown in dark grey, progress all the way through the cell cycle, as indicated by the cell cycle in the dark grey circle, dividing and forming the nodule primordium. The cells that are activated are opposite the protoxylem poles of the root as shown in thick black.

C5. Top: Crown gall tumors on the stem of a potato plant induced by *Agrobacterium tumefaciens*. Bottom: Hairy roots on the stem of a kalanchoe plant induced by *A.rhizogenes*. The bacteria infect wounded parts of the plant and cause overgrowths by the transfer of oncogenes to the plant. An overview on the role of the *Agrobacterium* oncogenes is given in chapter 12. Photographs are courtesy of Dr. P.J.J. Hooykaas.

Steps of T-DNA transfer

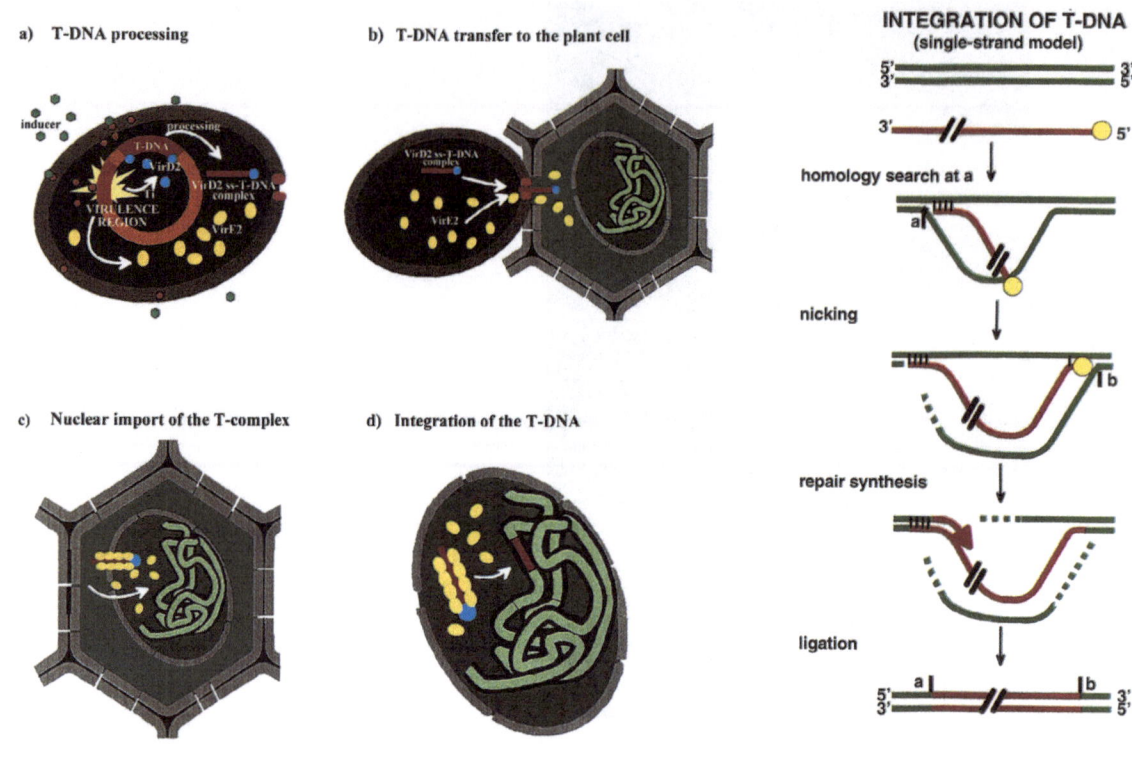

a) T-DNA processing

b) T-DNA transfer to the plant cell

c) Nuclear import of the T-complex

d) Integration of the T-DNA

INTEGRATION OF T-DNA
(single-strand model)

homology search at a

nicking

repair synthesis

ligation

The pattern of T-DNA integration

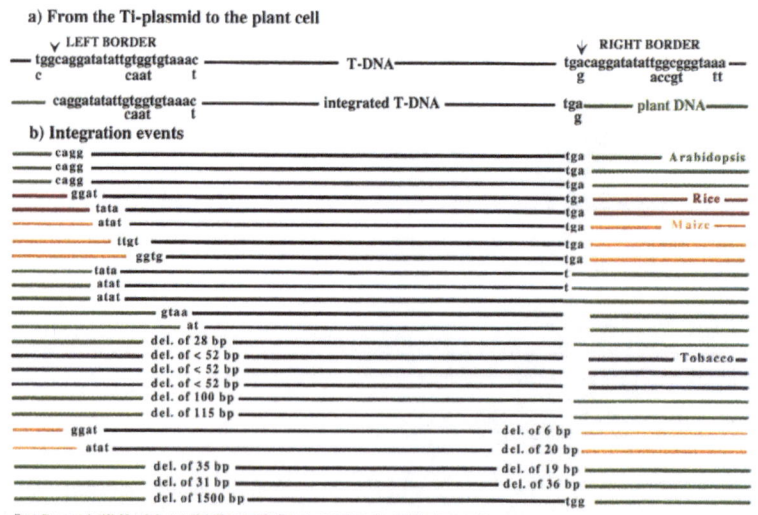

a) From the Ti-plasmid to the plant cell

b) Integration events

From Gheysen et al., 1991; Mayerhofer et al., 1991; Hiei et al., 1994; Tinland et al., 1995; Rossi et al., 1996; Ohba et al., 1995.

C6. Top left: Steps of T-DNA transfer (Figure 1, Chapter 15) Top right: Integration of T-DNA (Fig. 2, Chapter 15); Bottom: The pattern of T-DNA integration (Fig. 3, Chapter 15). As described in chapter 15, the process of T-DNA (red) transfer by *Agrobacterium tumefaciens* (brown) is mediated by *vir* genes located in the virulence region of the Ti plasmid (orange). The VirE2 (yellow) and VirD2 (blue) proteins play an important role in the T-DNA integration process inside the plant cells. T-DNA integration has been studied in detail in the above listed references (full references in Chapter 15).

Horizontal gene transfer from *Agrobacterium*, mediated by the virulence transport system

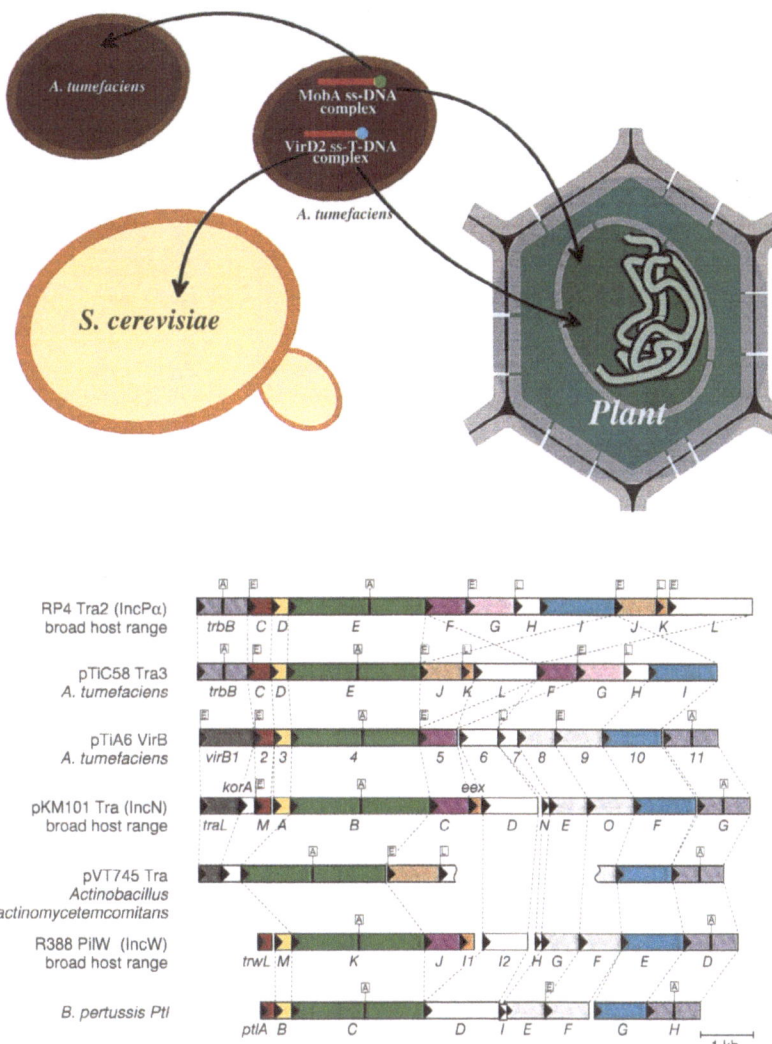

C7. Top: Horizontal gene transfer from *Agrobacterium*, mediated by the virulence transport system (Fig. 4, Chapter 15). The single stranded T-DNA (red) linked to the MobA (green) or VirD2 (blue) proteins can be transferred from *A.tumefaciens* (brown) to plant cells (green) or *Saccharomyces cerevisiae* cells (yellow). Bottom: Conserved gene organization among certain bacterial conjugation systems, the *Bordetella* toxin export system and the virulence system of *Agrobacterium* (Fig. 1, Chapter 14). Genes encoding similar products are connected by broken lines or are shown in the same color. Tags marked with E represent signal cleavage sites for *E. coli* signal peptidase Lep. Tags marked with L indicate lipoprotein signatures at the N terminus of the respective protein. Conserved nucleotide binding motifs of type A are marked by tags with A. In the Ptl operon genes D and I overlap slightly, indicated by a staggered arrangement of the bars. References/GenBank accession numbers: IncPα RP4 Tra2 ((64)/ M93696); pTiC58 Tra3 ((1)/U43674 and U43675); pTiA6 VirB ((109)/J03216); pKM101 ((87)/U09868); pVT745 (D. Galli and D. Leblanc, personal communication); R388 Pil_w ((15, 92)/X81123); *B. pertussis* Ptl ((112)/L10720). The conjugative transfer region of the large *Rhizobium* plasmid pNGR234 ((38)/U00090) is similar in gene organization to the Tra region of the *Agrobacterium* Ti plasmids, and it is not included in the Fig. for simplicity.

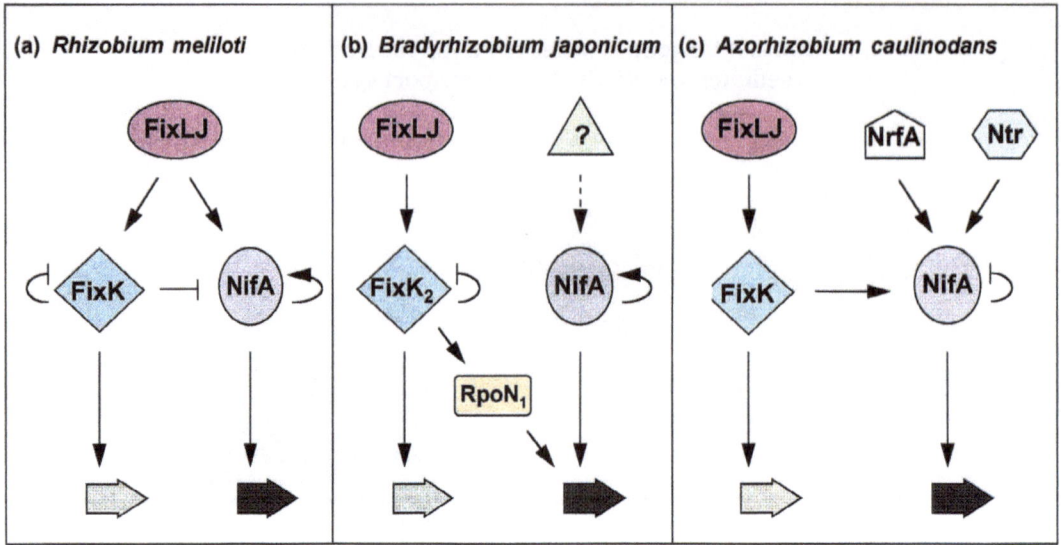

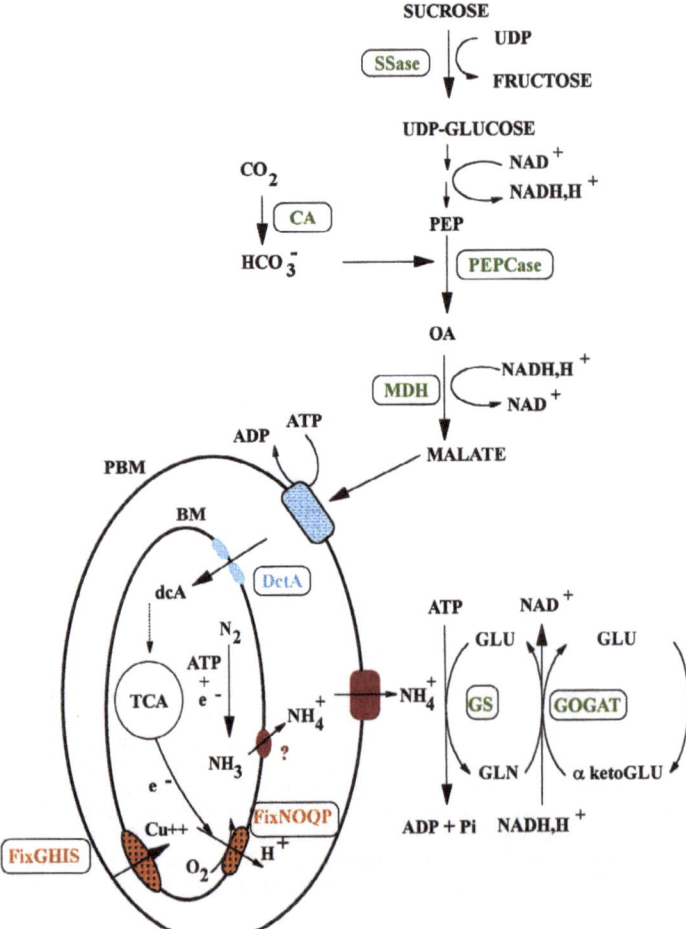

C8 (Top: Fig. 5, Chapter 23; Bottom: Fig. 3, Chapter 23). Top: Regulators and regulatory patways in rhizobia (Courtesy Dr H.M. Fischer, reproduced with permission from reference Fischer et al. 1996). Bottom: Schematic representation of metabolite exchange between a nitrogen-fixing bacteroid and the plant cell. PBM: peribacteroid membrane. BM: bacteroid membrane.

Chapter 1

Molecular Evolutionary Systematics of the *Rhizobiaceae*

Peter van Berkum and Bertrand D. Eardly

I. Introduction

The plant family Fabaceae (the Legume family), subdivided into three subfamilies, Mimosoideae, Ceasalpinioideae, and Papilionoideae, contains 674 genera (Gunn *et al.*, 1992) with an estimated 16,000 to 19,000 species (Allen and Allen, 1980). The Fabaceae have worldwide distribution and their economic importance is second only to the Poaceae (the Grass family). A unique characteristic of members of the Fabaceae is centered around the formation of root nodules, which were observed as early as the 16th century and by the 19th century were considered of diagnostic value for taxonomic identification. Contrary to early beliefs not all members of the Fabaceae nodulate and a majority of the different species have not been systematically examined for nodulation. Few of the species within the more primitive subfamily Ceasalpinioideae (23% of those examined) nodulate, while nodulation is predominant among species within the subfamilies Mimosoideae and Papilionoideae (90% and 97% of those examined, respectively) - thought to have evolved from the Ceasalpinioideae (de Faria *et al.*, 1984; de Faria *et al.*, 1987; de Faria *et al.*, 1989; Polhill *et al.*, 1981).

Nodulation and the subsequent development of nitrogen-fixing symbioses in the Fabaceae are responses to infection by bacteria of several genera in the family *Rhizobiaceae* (Conn, 1938). These plant-microbe symbioses are significant in nature as sources of nitrogen influx, while for man they represent an economical and efficient means of plant cultivation to restore the environment and for the production of crops. However, not all the genera in the family *Rhizobiaceae* are beneficial. Several species within one of the genera cause plant disease, which is evident by the presence of crown galls or hairy roots on plants. Yet another genus in the family *Rhizobiaceae* occur in leaf nodules of higher plants in the genera Myrsinaceae and Rubiaceae. Whether infection of leaf hypertrophies by *Phyllobacterium*

benefits or harms the plant is unclear. The ability to produce hypertrophies on plants is considered to be the primary characteristic of species within the family *Rhizobiaceae* (Knösel, 1984).

The study of bacteria was initiated within the fields of hygiene, medicine, fermentation, and food technology. Bacteriology as a science originated from botany and supplied the basis for microbial systematics, which is a practical need to describe and reliably identify microbes. Systematics is the study of methods and principles of classification and nomenclature; which in bacteria relies upon phenotypic (characterization by differences in morphology and physiology) as well as genetic characterization (by variations in DNA and gene products). The results of such characterization studies are classification schemes presumed to reflect natural relationships. Comparative analyses of DNA and gene products (molecular systematics) can be used to develop a phylogenetic classification scheme. Taxonomy is the science and practice of classification of distinct groups and is useful for identification purposes.

The central goal of molecular evolutionary systematics is the reconstruction of evolutionary history through the study of patterns of molecular genetic diversity in natural populations. In its search for evolutionary relationships among organisms, molecular systematics encompasses the field of taxonomy, which focuses on the classification and naming of species. The evolutionary history (or phylogeny) of a group of organisms is traditionally diagramed as a hierarchical tree to reflect putative evolutionary relationships. Normally, each successive group in a taxonomic hierarchy represents a cluster of related taxa. A taxon is said to be monophyletic if it includes only the descendants of a single ancestral population. If members of a given taxon are derived from two or more distinct ancestral populations, then the taxon is said to be polyphyletic. Although there is some disagreement among bacterial systematists as to the definition of a species, there is a general consensus that a bacterial species should be monophyletic, or at least approximately so. This means that most of the DNA of the members of a given species should be derived from a single common ancestor.

II. Historical Perspective

Although Beijerinck was the first to culture a bacterium from root nodules, naming it *Bacillus radicicola*, the taxonomy of these symbionts and the accepted number of species was in dispute for about 30 years (Fred *et al*, 1932). Frank named the root-nodule bacterium *Rhizobium leguminosarum*. Five species, *R. leguminosarum*, *R. trifolii*, *R. phaseoli*, *R. meliloti*, *R. japonicum*, and *R. lupini* were recognized by 1929 (Fred *et al.*, 1932). Speciation was primarily based on the formation of nodules upon the roots of certain legumes (the cross-inoculation groups) and not others.

By 1944 the designation of new species of *Rhizobium* on the basis of cross-inoculation was discontinued because many unexplainable, incongruous plant and bacteria reactions cast doubt on the validity of the concept (Burton, 1979; Wilson, 1944). The focus became speed of growth, acid production, serology, DNA base ratios, numerical taxonomy, DNA hybridization, and phage susceptibility (Burton, 1979). Yet, the taxonomic classification of the legume symbionts proposed as early as 1929 was formerly adopted in the approved list of bacterial names in 1980 (Skerman *et al*, 1980). Bacteria inducing grown-galls were first isolated in 1907 and were named *Bacterium tumefaciens* (Smith and Townsend, 1907). Conn (1942) concluded that the soil bacterium classified as *Alcaligenes radiobacter*, first isolated in 1902 (Beijerinck and van Delden, 1902) was very similar to the legume nodule bacteria and closely related to the crown-gall producing plant pathogens then classified as *Phytomonas*. Based upon similarity in morphology and physiology, Conn (1942) proposed a new genus name, *Agrobacterium*, and included both saprophyte and pathogens in the type species *A. tumefaciens*. Also placed in the genus were the soil saprophyte *A. radiobacter* and *A. rhizogenes*, the cause of hairy root. The approved list of bacterial names (Skerman, 1980) also includes a fourth species, *A. rubi* (Starr and Weiss, 1943), which was isolated from cane galls on *Rubus* species and was originally named *Phytomonas rubi* (Hildebrand, 1940).

The approved list of bacterial names (Skerman *et al.*, 1980) does not contain the genus *Phyllobacterium*, members of which are suspected to induce hypertrophies on the leaves of plant species of Myrsinaceae and Rubiaceae. However, Knösel (1984) revived the name in Bergey's Manual of Systematic Bacteriology, proposing the tentative classification of two species, *P. myrsinacearum* and *P. rubiacearum*, within the *Rhizobiaceae* based upon results of the DNA/rRNA hybridization studies of Gillis and de Ley (1980).

In this chapter we shall mainly consider the developments in the study of the molecular evolutionary systematics of the *Rhizobiaceae* since the publishing of the approved list of names by Skerman *et al.*, (1980). An excellent review of the population genetics and phylogeographic distribution of the *Rhizobiaceae* has recently been published (Martinez-Romero and Caballero-Mellado, 1996), therefore, these subjects will not be extensively examined in this chapter.

III. Approaches for Determining Taxonomic Relationships

Classification, nomenclature, and identification are three interrelated areas of taxonomy. Classification is the arranging of organisms into taxonomic groups based upon similarities. Nomenclature is the assignment of names to the taxonomic group according to international rules. Identification is the process of determining whether a new isolate belongs to one of the established and named groups. The approach in many higher organisms is partly based upon the accessibility of a fossil record permitting a phylogenetic classification, which is distinguishable from a classification based entirely on phenotypic characters. Early attempts at bacterial classification relied heavily on morphological characters and the pure culture technique. Although fossil records of microorganisms have been uncovered, they are not useful in the construction of prokaryotic phylogeny. Phylogenetic information of bacteria has increased over the last two decades primarily through the use of methods for the measurement of genetic relatedness. Therefore, a brief overview of these methods is required to provide perspective to interpret the molecular evolutionary systematics of the *Rhizobiaceae*.

IV. Methods for Estimating Phylogenetic Relatedness

IV.A. MULTILOCUS ENZYME ELECTROPHORESIS (MLEE)

It is possible to estimate genetic relatedness between individuals by comparing allelic variation among several highly conserved genes. In multilocus enzyme electrophoresis, the genetic relatedness between strains is estimated by comparing electrophoretic mobilities of several metabolically important, water-soluble enzymes. One advantage of this approach is that it relies on polymorphism at several well-defined loci collectively spanning several thousand nucleotides around the genome. However, only differences in charged amino acids are normally recognized by this method, and it does not reveal information on fine-scale genetic recombination. Regardless, unlike most other methods commonly employed in bacterial systematics, MLEE provides information on the organization of genetic variation in a species. Such information can serve as a basis for inferring the genetic structure in natural populations, and it is also valuable for assessing the evolutionary roles of Darwinian selection, random drift, mutation, and horizontal gene exchange (Selander et al., 1994).

General applications of the method are described by Selander et al., (1986), and methods specifically applicable to Rhizobium are described by Eardly et al., (1994). MLEE has been used in numerous studies examining species of Rhizobium and Bradyrhizobium (Bottomley et al., 1994; Demezas et al., 1991; Demezas et al., 1995; Eardly et al., 1990; Eardly et al., 1995; Engvild et al., 1990; Gordon et al., 1995; Harrison et al., 1989; Leung, et al., 1994; Martinez-Romero et al., 1991; Nour et al., 1994; Pinero et al., 1988; Segovia et al., 1991; Souza et al., 1994; Souza et al., 1992; Strain et al., 1994; Strain et al., 1995; Sullivan et al., 1996; van Berkum et al., 1995; Young, 1985; Young et al., 1987; Young and Wexler, 1988). We are unaware of any studies in which this method has been used in the study of Agrobacterium and Phyllobacterium spp. The results of the rhizobial MLEE studies through 1995 have been summarized by Martinez-Romero and Caballero-Mellado (1996); in general, their data indicate that species of Rhizobium and Bradyrhizobium are extremely diverse in comparison to human bacterial pathogens, and like many human pathogens, rhizobia appear to have a clonal population structure, where genetic recombination is rare. However, it should be noted that the appearance of clonality in some named species (e.g. S. meliloti) also may be due to the fact that they are polyphyletic (Eardly et al., 1990; Maynard Smith et al., 1993).

IV.B. ANALYSIS OF FINGERPRINT PATTERNS GENERATED WITH THE POLYMERASE CHAIN REACTION (PCR)™

DNA fragments of different molecular sizes are generated from template DNA with a thermo-cycler using either arbitrary primers or primers to conserved repeat sequences present in bacterial genomes. The approach relies upon genetic variation among genomes for the relative locations of the targets of the primers, which results in amplification products varying in molecular size. Random primers have been used to determine variation among legume symbionts (Dooley et al., 1993; Selenska-Pobell et al., 1995; van Rossum et al., 1995) by generating Random Amplified Polymorphic DNA (RAPD). However, analyses with primers for conserved repeat sequences present in bacterial genomes (Veraslovic et al, 1991) has been more extensively used (de Bruijn, 1992; Judd et al., 1993; Leung et al., 1994; Nick and Lindstrom, 1994; Selenska-Pobell et al., 1995; van Berkum et al., 1994; Veraslovic et al., 1991; Veraslovic et al., 1994). These include pairs of primers for amplification of DNA regions between Repetitive Extragenic Palindromic (REP) and Enterobacterial Repetitive Intergenic Consensus (ERIC) sequences and a single primer for "BOX" sequences (Hulton et al., 1991; Martin et al., 1992; Stern et al., 1984). Fingerprint patterns using PCR also can be obtained by using primers to amplify the spacer region between the SSU and LSU rRNA genes (Jensen and Straus, 1993; Jensen et al., 1993). The method was shown to be useful for distinguishing 28 species of bacteria belonging to eight genera, but members within the Rhizobiaceae were not included in the analysis. Amplification products are separated according to molecular size by using horizontal agarose gel electrophoresis. The gels are photographed for quantitative comparisons of the lanes according to the presence of products or their absence across lanes. This can be achieved manually or by scanning the image into a computer and by using specialized software packages. The resultant binary matrix is used to produce a dendrogram and to measure the fit of the dendrogram to the data (Verasolvic *et al.*, 1994). Within rhizobial species REP and ERIC PCR results are well correlated with genetic distance measurements made with MLEE (de Bruijn, 1992; Leung *et al.*, 1994) and, thus offer an alternative or additional approach for the measurement of genetic diversity.

A significant limitation of the PCR techniques using either random primers or primers targeted to repeat sequences is the identity of each product across the different genome templates. Products of measured or perceived identical molecular size need not represent the identical region in each genome, but could be the same size by mere coincidence. Similarly, the absence of a product is not informative whether the targets were closer together, further apart, or whether one or both targets were absent. Therefore, dendrograms constructed with the PCR data need not reflect accurately the genetic relationships among genomes and the technique is likely to be useful only for determining the variation within species. In the case of amplification of the spacer region between the SSU and LSU rRNA genes, the amplification products include single stranded and heteroduplex DNA (Jensen and Straus, 1993). Since the heteroduplex structures contain significant regions of single stranded DNA, they are difficult to use for identification purposes which relies on high reproducibility of patterns. In addition, the presence of heteroduplex DNA structures composed of single stranded sequences from two distinct loci may affect results of mapping. Thus, it would be advisable to ensure that such heteroduplex structures are not present in the PCR products or are removed prior to electrophoretic separation (Jensen and Straus, 1993).

IV.C. ANALYSIS OF FINGERPRINT PATTERNS GENERATED BY DNA RESTRICTION DIGESTION

There exist two major approaches in determining genetic relationships based upon restriction site analysis, regular Restriction Fragment Length Polymorphism (RFLP) and PCR-RFLP. For RFLP analysis, total DNA digests are examined for fingerprint patterns with specific probes after standard Southern hybridizations. In the case of PCR-RFLP, specific regions of the genome are amplified and fingerprint patterns are obtained after restriction digestion of the amplification products. The banding patterns across different enzyme digests are used to estimate the genetic diversity. However, instead of a computation based upon simple

matching coefficients as is the case with the fingerprint analysis with PCR, the data can be used to estimate sequence divergence from the proportion of shared restriction fragments (Nei and Li, 1979). PCR-RFLP of the SSU rRNA gene has been used in the analysis of legume symbionts (Laguerre *et al*, 1994). The spacer region between the SSU and LSU rRNA genes has been used to map variations among the rhizobia nodulating chickpea (Nour *et al.*, 1994a; Nour *et al.*, 1994b; Nour *et al.*, 1995).

There are several limitations to both types of RFLP analysis. Both analyses estimate sequence divergence across very small regions of the genomes, which may not be representative of the sequence divergence across the entire genomes. In the case of the Southern hybridization approach, the use of arbitrary clones as probes may result in poor and no hybridization signals with DNA of more distant lineages. Also, a variation in gene copy number may influence the estimate of sequence divergence. In the case of PCR-RFLP, gene amplification is dependent upon the presence of conserved nucleotide sequences at the 5' and 3' ends of genes. Estimates of sequence divergence from fingerprint patterns of enzyme digests may be influenced by the presence of polymorphic insertion sequences.

IV.D. MEASUREMENT OF DNA RELATEDNESS BY DNA REASSOCIATION

The measurement of genomic relatedness relies upon the double-stranded (native) nature of DNA and the property that the complementary strands dissociate at high temperature and reassociate at lower temperature. RNA is single-stranded and can pair with a complementary strand of DNA. DNA and RNA homology experiments attempt to answer whether the DNA or RNA of two organisms have a base sequence sufficiently similar to allow the formation of DNA heteroduplexes or heterologous DNA-RNA hybrids. DNA homology values are measurements of similarity in which the entire genome of one organism is compared with that of another. Ribosomal (rRNA) and transfer (tRNA) are coded by small fractions of the genomes and, therefore, homology values are specific for these regions. DNA homology values are used to detect similarities between closely related organisms, while RNA homologies detect similarities between more distantly related organisms (Johnson, 1984).

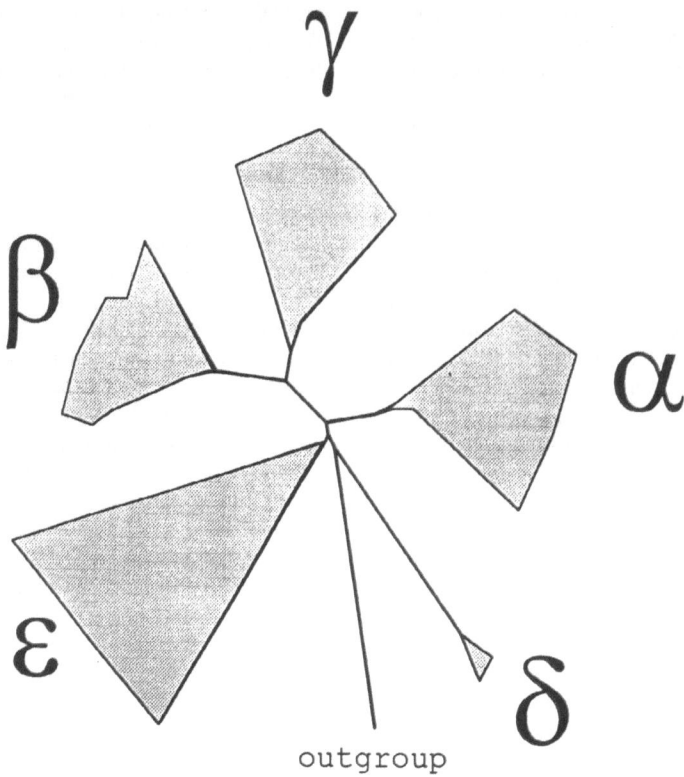

Figure 1. Phylogenetic tree of the purple bacteria estimated from aligned sequences of the small subunit rRNA gene. Jukes Cantor distances were derived from the aligned sequences to construct an optimal unrooted tree using the Neighbor-Joining method. Sequences representing each of the subdivisions indicated in the tree which were used in the phylogenetic analysis were: for the α-subdivision, *Azospirillum brasilense* (Z29617), *Bradyrhizobium japonicum* (U69638), *Paracoccus denitrificans*(X69159), and *Rickettsia rickettsii* (M21293); for the β-subdivision, *Alcaligenes eutrophus* (M32021), *Azoarcus indigens* (L15531), *Bordetella pertussis* (U04950), *Burkholderia pickettii* (X67042), *Nitrosovibrio tenuis* (M96404), and *Spirillum volutans* (M34131); for the γ-subdivision, *Legionella jamestowniensis* (X73409), *Oceanospirillum linum* (M22365), *Proteus vulgaris* (X07652), and *Vibrio cholerae* (X74694); for the δ-subdivision, *Cystobacter fuscus* (M94276), *Myxococcus xanthus* (M34114); and for the ε-subdivision, *Fusobacterium russii* (M58681), and *Helicobacter pylori* (U01330). The outgroup used in the analysis was *Streptomyces griseus* (M76388). The sequences were aligned using the PILEUP program in the Wisconsin package of the Genetics Computer Group (Madison, Wis.) and the aligned sequences were analyzed using the DNADIST and NEIGHBOR programs of the software package PHYLIP version 3.5c (Felsenstein, 1993) and the tree was constructed using DRAWTREE

Several methods that rely upon labeling of one of the nucleic acids have been developed and have been reviewed (Grimont, 1988; Johnson, 1984). Although each of the methods gives similar results, the meaning of the homology values is subject to interpretation. Johnson (1984) has proposed that within species DNA homology values range from 60% and 100%, where values between 60% and 70% are of subspecies within the species of the reference organism. Homology values of between 20% and 60% between the DNA of the test and reference organisms represent species closely related to the reference species. However, the interpretation of the exact limits in DNA homology values is at the discretion of the individual researcher.

Although DNA homology provides a credible estimate of genetic relatedness, there are some major problems with the approach. One drawback is that multiple pairwise comparisons are required making large studies laborious. Full distance matrices cannot be obtained and there are large ranges of DNA homology values for the same and closely related species. A major problem is that the technique is not sensitive enough to detect a limited number of shared short sequences, which result from genetic exchange (Maynard Smith, 1995). At moderate levels of DNA divergence heteroduplexes are not formed and the levels of hybridization decline to background level. The contributions of plasmid and chromosomal DNA cannot be distinguished. Nevertheless, DNA homology is considered by most investigators to be one of the better techniques for determining whether two organisms belong to the same species. Further support for the value of DNA hybridization analysis is that results have been reported to correlate with those of MLEE analysis (Leung *et al.*, 1994, Sullivan *et al.*, 1996). Although the identification and classification of new species must be supported by low DNA homology values with previously described reference species, it is important to interpret values of DNA homology in the context of other phylogenetic information (Maynard Smith, 1995).

IV.E. MEASUREMENT OF GENETIC RELATEDNESS BY GENE SEQUENCING METHODS.

IV.E.1. The 16S or Small Subunit Ribosomal RNA (rRNA) Gene

The most dramatic progress in the construction of microbial phylogeny is based upon sequencing analysis of the ribosomal genes (Figure 1). In particular, sequencing the SSU rRNA gene has profoundly affected the understanding of the relationships among the bacteria (Olsen *et al*, 1994). Based upon SSU rRNA nucleotide sequences, the *Rhizobiaceae* are part of the α-subdivision of the purple bacteria or the α-proteobacteria (Maidak *et al*, 1994). The α-subdivision of the purple bacteria is not exclusive to the *Rhizobiaceae*, this branch is shared by at least 24 different genera (Young, 1996). Members of the family *Rhizobiaceae* are divided into several genera that do not form a cluster distinctly separated from other genera in the α-subdivision (Figure 2).

The SSU rRNA gene is useful for estimating evolutionary relationships among bacteria because it is slowly evolving, and the gene product is both universally essential and functionally conserved. Although these are useful attributes, the reliability of using the SSU rRNA gene alone as an approach for estimating phylogenies of bacteria has been brought into question, particularly for closely related taxa. Stackebrandt and Goebel (1994) investigated the role of SSU rRNA similarity in species definition in comparison with estimates of genetic relatedness made with DNA-DNA reassociation analyses. A

correlation plot between SSU rRNA similarity and values of DNA-DNA reassociation across species of *Fibrobacter*, *Bacillus*, *Enterobacter*, and *Serratia* was not linear. A similar correlation plot with data from members of the *Rhizobiaceae* also is not linear (Figure 3). The conclusion reached by Stackebrandt and Goebel (1994) was that for a pragmatic species definition the SSU rRNA nucleotide sequence data are most reliable to estimate relatedness when these genes share less than 97% similarity. They also concluded that DNA-DNA hybridization provides a more reliable estimate of relatedness when the SSU rRNA genes are more than 97% similar. For example, different species of *Bacillus* defined by low DNA-DNA reassociation values had identical SSU rRNA gene sequences (Fox *et al.*, 1992). Therefore, Stackebrandt and Goebel (1994) indicated that additional information in the form of DNA-DNA reassociation values is required to define species limits when SSU rRNA genes share more than 97% sequence similarity.

IV.E.2. The 23S or Large Subunit (LSU) rRNA Gene

The topology of a phylogenetic tree reflecting the relationships among the five subdivisions of the proteobacteria based upon LSU rRNA gene sequences was reported to be similar to that of a tree based upon SSU rRNA gene sequences (Ludwig *et al.*, 1995). However, the LSU rRNA gene has not been extensively used to estimate the genetic relationships among the *Rhizobiaceae*, but there are several dramatic differences which may be helpful for classification and identification purposes. The 5' half of the LSU rRNA gene in bacteria may contain highly variable stem-loop structures or intervening (IV) sequences, which are subject to *in vivo* fragmentation (Burgin *et al*, 1990; Evguenieva-Hackenberg and Selenska-Pobell, 1995; Kordes *et al.*, 1994; Lessie, 1965; Mackay *et al.*, 1979; Marrs and Kaplan, 1970; Schuch and Loening, 1975; Selenska-Pobell and Evguenieva-Hackenberg 1995; Winkler, 1979) when the rRNA precursor molecules are processed (King *et al.*, 1986). Central fragmentation in the LSU rRNA in *Rhizobium* and *Agrobacterium* species cleaving the molecule into approximately two 1.3 kb fragments has been reported (Evguenieva-Hackenberg and Selenska-Pobell, 1995; Selenska-Pobell and Evguenieva-Hackenberg 1995). In addition to the possibility of a central fragmentation, the LSU rRNA in most *Rhizobiaceae* is fragmented at the 5' end

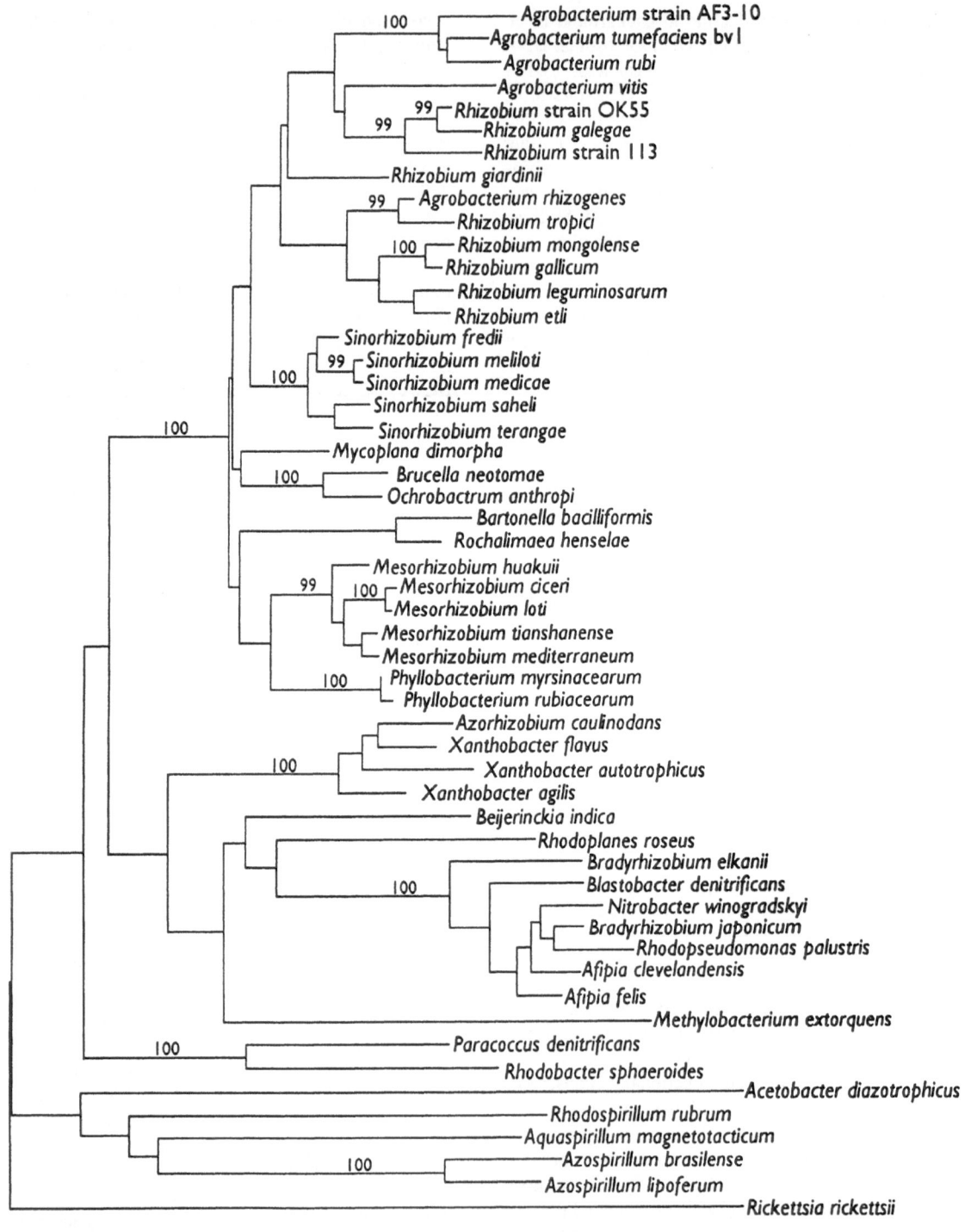

1% nucleotide difference

Figure 2. Phylogenetic relationship of the *Rhizobiaceae* within the α-subdivision of the Purple Bacteria based upon aligned sequences of the small subunit ribosomal RNA genes. Jukes-Cantor distances were derived from the aligned sequences to construct an optimal unrooted tree using the Neighbor-Joining method. Five hundred replicate trees were generated in a bootstrap analysis to derive a majority rule consensus tree. The levels of support equal to or exceeding 99% for the presence of nodes are indicated in the NJ tree. Sequences representing the genera indicated in the tree which were used in the phylogenetic analysis were, *Acetobacter diazotrophicus* (X75618), *Afipia clevelandensis* (M69186), *A. felis* (M65248), *Agrobacterium rhizogenes* (D14501), *A. rubi* (D14503), *A. tumefaciens* bv1 (D14500), *A. vitis* (D14502), *Agrobacterium* sp. strain 3-10 (Z30542), *Aquaspirillum magnetotacticum* (M58171), *Azorhizobium caulinodans* (X94200), *Azospirillum brasilense* (Z29617), *A. lipoferum* (Z29619), *Bartonella bacilliformis* (M65249), *Beijerinckia indica* (M59060), *Blastobacter denitrificans* (S46917), *Bradyrhizobium elkanii* (U35000), *B. japonicum* (U69638), *Brucella neotomae* (L26167), *Mesorhizobium ciceri* (U07934), *M. huakuii* (D13431), *M. loti* (X67229), *M. mediterraneum* (L38825), *M. tianshanense* (not submitted), *Methylobacterium extorquens* (D32224), *Mycoplana dimorpha* (D12786), *Nitrobacter winogradskyi* (L35507), *Ochrobactrum anthropi* (D12794), *Paracoccus denitrificans* (X69159), *Phyllobacterium myrsinacearum* (D127890), *P. rubiacearum* (D12790), *Rhizobium etli* (U28916), *R. gallicum* (U86343), *R. giardinii* (U86344), *R. leguminosarum* bv *viciae* (U29386), *R. galegae* (X67226), *R. mongolense* (U89817), *R. tropici* (D11344), *Rhizobium* sp. strain OK55 (D14510), *Rhizobium* sp. strain 113 (D14512), *Rhodobacter sphaeroides* (D16425), *Rhodoplanes roseus* (D25313), *Rhodopseudomonas palustris* (D25312), *Rhodospirillum rubrum* (D30778), *Rochalimaea henselae* (M73229), *Rickettsia rickettsii* (M21293), *Sinorhizobium fredii* (X67231), *S. medicae* (L39882), *S. meliloti* (X67222), *S. saheli* (X68390), *S. terangae* (X68387), *Xanthobacter agilis* (X94198), *X. autotrophicus* (X94201), and *X. flavus* (X94199). The sequences were aligned using the PILEUP program in the Wisconsin package of the Genetics Computer Group (Madison, Wis.) and the aligned sequences were analyzed using the the the Molecular Evolutionary Genetics Analysis (MEGA) package version 1.01 (Kumar *et al.*, 1993)

because of an IV sequence starting at nucleotide position 130 (Evguenieva-Hackenberg and Selenska-Pobell, 1995; Selenska-Pobell and Evguenieva-Hackenberg 1995).

V. The Genera of the *Rhizobiaceae*

Several fundamental morphological and physiological characters separate *Rhizobium* and *Bradyrhizobium* (Jordan, 1984) and their classification as separate genera is indisputable (summarized in Elkan, 1992). Similarly, the stem-nodulating symbiont of *Sesbania rostrata*, *Azorhizobium caulinodans*, is distinct from the genera *Rhizobium* and *Bradyrhizobium* (Dreyfus *et al.*, 1988). Up until very recently, the only other recognized genera within the *Rhizobiaceae* were, *Agrobacterium*, and *Phyllobacterium*. However, it has become evident that there are three genetic groups within *Rhizobium*, and suggestions for their separation into different genera have been made. The separation of the genus *Sinorhizobium* (De Lajudie *et al.*, 1994) is well supported by SSU rRNA sequences. The additional separation of the third branch also is well supported by SSU rRNA sequencing data. The genus name *Mesorhizobium* has been proposed and has been formally adopted (Jarvis *et al.*, 1997). The SSU rRNA gene nucleotide sequence similarities among the genera of the *Rhizobiaceae* vary from 88.0% to 97.0% (Table 1). The genetic information justifying separation of the seven genera of the *Rhizobiaceae* seems clear with the exception of one species of *Rhizobium*. With the

possible exception of strains SIN-1, OK55, and 113 isolated from the nodules of *Sesbania aculeata* and *Mimosa invisa* (Rana and Krishnan, 1995; Sawada *et al.*, 1993), the phylogenetic position of *R. galegae* appears to be distant from other members of the legume symbionts. Based upon phylogenetic position, the closest relative is *A. vitis* (Sawada *et al*, 1993), which has initiated the debate whether *R. galegae* should be reclassified as an *Agrobacterium* species (De Lajudie *et al.*, 1994; Lindström *et al*, 1996; Young, 1996). However, a comparison of the SSU rRNA nucleotide similarities of *R. galegae* with the different species of *Agrobacterium* and with the different species of *Rhizobium* indicates that *R. galegae* is as distant from *Agrobacterium* as it is from *Rhizobium* (Table 2). Since the SSU rRNA nucleotide similarities of *R. galegae* and the *Agrobacterium* species are similar to the range of the SSU rRNA nucleotide similarities with the different genera of the *Rhizobiaceae*, there seems no justification to reclassify *R. galegae* as a species of *Agrobacterium*, even though several published phylogenetic positions for *R. galegae* would tend to place it with *Agrobacterium*. It might be prudent to wait until other species in this group have been identified, and their relationships have been substantiated by sequence data from multiple loci. The additional information may be useful for deciding whether *R. galegae* is in a genus separate from *Agrobacterium* or *Rhizobium* as suggested by Young and Haukka (1996), should remain within the genus *Rhizobium* or should be transferred to the genus *Agrobacterium*.

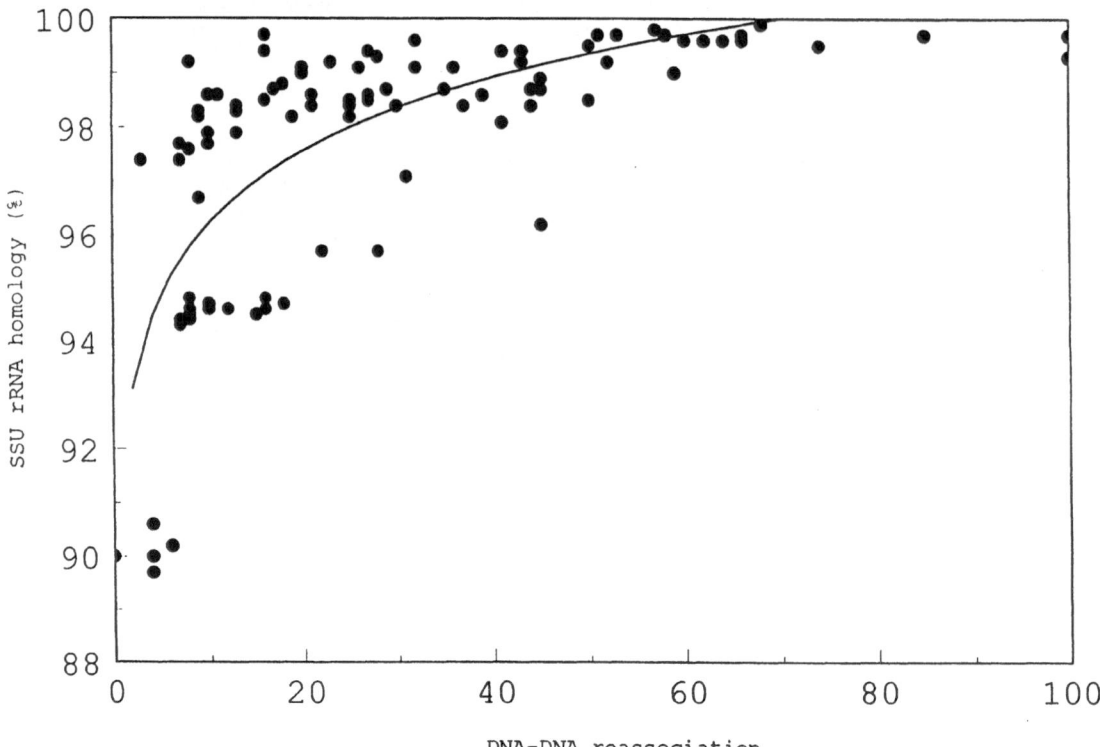

Figure 3. Comparison of small subunit rRNA homology and DNA-DNA reassociation values among members of the *Rhizobiaceae*. Data used in the figure were compiled from DeLadjudie *et al.*, 1994; Hollis *et al.*, 1981; Lindström, 1989; Sullivan *et al.*, 1996; van Berkum *et al.*, 1996; Wedlock and Jarvis, 1986 and van Berkum *et al.*, 1997

V.A. DESCRIPTION OF THE SPECIES WITHIN THE GENERA OF THE *RHIZOBIACEAE*

V.A.1. Agrobacterium (Conn, 1942)

Originally *Agrobacterium* species were classified according to their phytopathogenicity and also were divided based upon the production of 3-ketolactose. Non-pathogenic species were *A. radiobacter*, those causing hairy root were classified as *A. rhizogenes*, *A. rubi* strains cause cane gall on *Rubus* species, and *A. tumefaciens* was the cause of crown gall (Skerman *et al.*, 1980). However, *A. radiobacter* and *A. tumefaciens* cannot be distinguished other than by phytopathogenicity (Holmes, 1988). Evidence for separation of *A. tumefaciens*, *A. rhizogenes*, and *A. rubi* into separate species was by classical numerical taxonomic information (Holmes and

Roberts, 1981), which also identified an unnamed "yellow group" of *Agrobacterium*. Holmes and Roberts (1981) classification ignored phytopathogenicity as a taxonomic character because production of crown galls or hairy roots are encoded for by plasmid-borne genes (Holmes, 1988). The tumor-inducing plasmid (pTi) and the root-inducing plasmid (pRi) may be acquired or lost and lead to instability of species classifications based upon phytopathogenicity. Classification of *Agrobacterium* is further confused by suggestions in the past of profound changes in nomenclature including the assignment of the genus *Rhizobium* to include all the agrobacteria (Moffett and Colwell, 1968). Although for the most part *Agrobacterium* is recognized as a genus separate from *Rhizobium* today, several strains identified as *Agrobacterium* are suggested to share characteristics more in common with *Rhizobium* than with *Agrobacterium*.

	Mesorhizobium	*Sinorhizobium*	*Agrobacterium*	*Rhizobium*	*Azorhizobium*	*Bradyrhizobium*
Phyllobacterium	97.0	96.0	93.6	94.4	92.0	88.5
Mesorhizobium		95.9	93.6	94.4	91.4	88.1
Sinorhizobium			95.3	96.0	91.2	88.3
Agrobacterium				94.5	90.6	88.0
Rhizobium					91.4	88.9
Azorhizobium						90.3

Table 1. Small Subunit rRNA nucleotide sequence similarities among the genera of the *Rhizobiaceae*. The SSU rRNA sequences used were those of *Phyllobacterium myrsinacearum* (IAM 13584; D12789), *Mesorhizobium loti* (NZP 2213; X67229), *Sinorhizobium meliloti* (USDA 1002; X67222), *Agrobacterium tumefaciens* biovar 1 (NCPPB2437; D14500), *Rhizobium leguminosarum* bv *viceae* (USDA 2370; U29386), *Azorhizobium caulinodans* (ORS571; X67221) and *Bradyrhizobium japonicum* (USDA 6; U69638)

Holmes (1988) indicated the existence of five genetically different groups of *Agrobacterium*, strains belonging to clusters 1 and 2 as defined by Kersters *et al.*, (1973), *A. rubi*, strain NCPPB 1650, and strain NCPPB 1771, all of which share about 15% DNA homology. Strains of cluster 1 are genetically heterogeneous and consist of at least seven subgroups with DNA homology values with each other ranging from 45 to 50% (Kersters and Delay, 1984), which had not been named because of insufficient number of phenotypic characters for their distinction (Holmes, 1988), but which does include biovar 3 strains which infect grapevine (Kersters and De Ley, 1984). It was suggested that the "yellow group" of *Agrobacterium* did not belong to the genus *Agrobacterium* (Kersters and Delay, 1984). Differences among isolates can be detected using PCR-RFLP of chromosomal and plasmid regions (Ponsonnet and Nesme, 1994).

V.A.2. Agrobacterium tumefaciens (Smith and Townsend, 1907).

The type strain (ATCC 23308, NCPPB 2437) belongs to biovar 1 and causes crown galls. Strains belonging to this species are separated from other species of *Agrobacterium* according to phenotypic characterization (Holmes and Roberts, 1981), low values of DNA-DNA reassociation (Kersters and De Ley, 1984; Ophell and Kerr, 1990) and SSU rRNA nucleotide sequences (Sawada *et al*, 1993; Willems and Collins, 1993). The species includes rhizogenic strains, NCPPB 2303 is representative, and used to

include strains infecting grapevines (biovar 3) of which NCPBB 2562 is representative.

V.A.3. Agrobacterium radiobacter (Beijerinck and van Delden, 1902).

The type strain of *A. radiobacter* (ATCC 19358) also belongs to biovar 1. Because the type strain of *A. radiobacter* has 80% to 87% DNA homology with the type strain of *A. tumefaciens* (De Ley *et al.*, 1973; Ophell and Kerr, 1990; Sawada *et al.*, 1993), Sawada *et al.*, (1993) argued that both type strains belong to the same species and proposed *A. radiobacter* as the type species over *A. tumefaciens*. This has precipitated in a request for a judicial opinion whether *A. tumefaciens* or *A. radiobacter* should be the type species for *Agrobacterium* (Bouzar, 1994; Oyaizu and Sawada, 1994). Biovar 2 is represented by strain NCPPB 2407. No biovar 3 strains belonging to the *A. radiobacter* classification were identified.

V.A.4. Agrobacterium rhizogenes (Riker et al., 1930).

The type strain, ATCC 11325, belongs to biovar 2 and is distinguishable from the other species of *Agrobacterium* by DNA-DNA reassociation values (Kersters and De Ley, 1984; Ophell and Kerr, 1990), and by SSU rRNA gene nucleotide sequence (Sawada *et al*, 1993; Willems and Collins, 1993). The SSU rRNA gene nucleotide sequences of several unclassified strains, K-Ag-3 and Ch-Ag-4, are 99.7 and 100% similar to that of the type strain,

	A. vitis	R. leguminosarum	R. galegae
A. tumefaciens	96.5	96.2	96.8
A. rhizogenes	95.8	98.6	96.2
A.vitis		94.8	96.7
A. rubi	96.8	95.1	96.8
R. leguminosarum	94.8		96.4
R. galegae	96.7	96.4	

Table 2. Small subunit rRNA nucleotide sequence similarities among species of *Agrobacterium, Rhizobium leguminosarum* and *R. galegae.* The data was obtained from Sawada *et al.,* 1993

respectively. However, these two strains are different from *A. rhizogenes* phenotypically (Sawada and Ieki, 1992) and can be distinguished from the type strain by DNA relatedness values and separate speciation has been suggested (Sawada *et al.,* 1993). Another issue needing resolution in the future is whether or not there is sufficient evidence to reclassify *A. rhizogenes* and the strains K-Ag-3 and Ch-Ag-4 to the genus *Rhizobium* because of their high SSU rRNA sequence similarity with *R. tropici* (Young, 1996), which ranges from 99.6 to 100% (Sawada *et al.,* 1993). The focus of the controversy is whether the SSU nucleotide sequence similarities between these agrobacteria and *R. tropici* alone are sufficient evidence for the reclassification. This is not a problem unique to the agrobacteria, but also is relevant to *Phyllobacterium, Azorhizobium,* and *Bradyrhizobium.*

V.A.5. Agrobacterium vitis (Ophell and Kerr, 1990).

These isolates from grapevine were previously classified as biovar 3 and were included in one of the heterogenous groups of *A. tumefaciens* (Kersters and De Ley, 1984). The type strain, NCPPB 3554 can be distinguished from the other species of *Agrobacterium* phenotypically by a number of biochemical differences and serological affinity. Low levels of DNA binding (7 to 47%) with the other species of *Agrobacterium* (Ophell and Kerr, 1990) and SSU rRNA gene nucleotide sequence similarities ranging from 96.8 to 95.7% (Sawada *et al,* 1993) further support separate speciation of *A. vitis.*

V.A.6. Agrobacterium rubi (Hildebrand, 1940).

These agrobacteria were isolated from the cane galls of *Rubus* spp., but the host range is not limited to

these plants. The type strain, ATCC 13335 (NCPBB 1854), is distinguishable by low values of DNA-DNA reassociation with other species of *Agrobacterium* (Kersters and Delay, 1984; Ophell and Kerr, 1990) and SSU rRNA gene nucleotide sequence similarities ranging from 99.7 to 95.3% (Sawada *et al,* 1993; Willems and Collins, 1993). Only two additional strains are considered to belong to this species.

V.A.7. Unclassified Species of Agrobacterium.

Strains NCPBB 1650, NCPPB 1771, K-Ag-3, Ch-Ag-4 and AF3-10 are genetically and phenotypically distinct from the type strains of the different species of *Agrobacterium*. NCPPB 1650 and NCPPB 1771 were initially classified as *A. rubi* (Holmes and Roberts, 1980) based upon results of numerical taxonomy analysis and were referred to as aberrant strains of *A. tumefaciens* by Kersters and De Ley (1984). The strains K-Ag-3, Ch-Ag-4, and AF3-10 originated from Kiwifruit, Cherry, and *Ficus benjamina*, respectively (Bouzar *et al.,* 1995; Sawada and Ieki, 1992). The strains NCPBB 1650, NCPPB 1771, K-Ag-3 and Ch-Ag-4 were shown to be unusual in their phenotypic characteristics (Sawada and Ieki, 1992) and DNA relatedness levels (Sawada *et al.,* 1993). Strain AF3-10 was reportedly different from the type species in fatty acid content, carbon substrate utilization profiles, and unusual opine metabolism (Bouzar *et al.,* 1995). The strains NCPBB 1650 and NCPPB 1771 are distantly related from each other as well as to the type strains of *Agrobacterium* based upon DNA-DNA reassociation determinations (Kersters and De Ley, 1984). Martinez-Romero (1994) reported that K-Ag-3 and Ch-Ag-4 phenotypically are indistinguishable from *R. tropici* in culture, but that a DNA-DNA reassociation value of 24% distinguished Ch-Ag-4 from *R. tropici* type A (Martinez-Romero, 1994). Sawada *et al.*, (1993) concluded that the phylogenetic information available was not sufficiently reliable to combine *Agrobacterium* and *Rhizobium* species into one genus. SSU rRNA gene nucleotide sequences of strains K-Ag-3 and Ch-Ag-4 are 99.7 and 100% similar to the type strain of *A. rhizogenes*, respectively. The SSU rRNA nucleotide sequences of strain NCPBB 1650 and the type strain of *A. rubi* are 99.7% similar (Sawada *et al*, 1993). However, Sawada *et al.*, (1993) indicate that these three strains of *Agrobacterium* should be assigned to new species because of their phenotypic and genetic differences from the type strains. Strain AF3-10 also has been suggested to represent a new species of *Agrobacterium* (Bouzar *et al.*, 1995). The evidence for this is based upon SSU rRNA gene nucleotide sequence similarities between AF3-10 and NCPBB 1650, *A. tumefaciens*, *A. vitis*, and *A. rhizogenes* of 98.1, 98.0, 97.8, 95.3, and 94.3%, respectively; DNA-DNA reassociation values were not reported.

V.A.8. Azorhizobium (Dreyfus et al., 1988).

Sesbania rostrata is one of the few existing stem-nodulating legumes. Isolates from the stem nodules were shown to be different from other legume-nodulating *Rhizobiaceae* in their ability to grow with dinitrogen as sole source of nitrogen (Dreyfus *et al.*, 1983). Molecular systematic data clearly demonstrated that *Azorhizobium* as a genus is distinct from other members of the *Rhizobiaceae* (Dreyfus *et al.*, 1988; Sawada *et al.*, 1993; Willems and Collins, 1993). Only one species, *A. caulinodans*, has been described and ORS 571 has been identified as the type strain (Dreyfus *et al.*, 1988). Based upon DNA-DNA reassociation results Rinaudo *et al.*, (1991) reported that *Azorhizobium* could be divided into two genomic species, but refrained from naming the second species because they were unable to identify phenotypic characters which differentiated the two.

Azorhizobium and *Xanthobacter* have very similar rRNA cistrons and have been classified on the same rRNA subbranch (Dreyfus *et al.*, 1988). The close genetic relationship between *Azorhizobium* and *Xanthobacter* is supported by up to 98.2% sequence similarities of the SSU rRNA genes (Rainey and Wiegel, 1996). Dreyfus *et al.*, (1988) conceded that the level of similarity they described between the rRNA cistrons of *Azorhizobium* and *Xanthobacter* did not reflect their separate genus status, but argued that several morphological and biochemical

differences justified such a separation. In contrast, Rainey and Wiegel (1996) acknowledged that the molecular systematic data obtained with *Xanthobacter* and *Azorhizobium* could be interpreted several different ways and draw as one of their conclusions that the two genera together with *Aquabacter* should be combined into the one genus, *Xanthobacter*.

V.A.9. Bradyrhizobium (Jordan, 1982).

Bradyrhizobium are clearly differentiated from the other legume symbionts because they grow slowly and produce an alkaline reaction and no serum zone in Litmus milk. This genus represents an extremely heterogeneous group within which, with possibly the exception of *Phyllobacterium*, the investigation of the taxonomic relationships has not advanced as rapidly as it has with the other lineages of the *Rhizobiaceae*. The symbionts of soybean (formerly *Rhizobium japonicum*) and those that were classified as the cowpea-miscellany rhizobia are included in this group. The rRNA cistrons of *Rhizobium* and *Bradyrhizobium* resemble each other less than do the rRNA cistrons of *Rhizobium* and the other genera included in the family *Rhizobiaceae* (Jarvis *et al.*, 1986), this is supported by data on SSU rRNA gene nucleotide sequence similarities (Table 1). Jarvis *et al.*, (1986) proposed a photosynthetic ancestry for *Bradyrhizobium* because of close relationship of the rRNA cistrons of *Bradyrhizobium* and of *Rhodopseudomonas palustris*. Conclusions derived from the analysis of diverse bradyrhizobial isolates from the stem-nodulating legume *Aeschynomene indica* are in agreement with the suggested photosynthetic ancestry of *Bradyrhizobium* (van Berkum *et al.*, 1995). Even though the group is heterogeneous there are currently only two recognized species of *Bradyrhizobium*, *B. japonicum* and *B. elkanii*; both originated from soybean and a suggestion for a third species of soybean symbionts, *B. liaoningense*.

V.A.10. The Soybean Symbionts, Bradyrhizobium japonicum (Jordan, 1982) and B. elkanii (Kuykendall et al., 1992).

Because of their slow growth the scientific investigation of legume symbionts has focused on *Rhizobium* rather than on *Bradyrhizobium*. However, the symbionts of soybean have been an exception to this generalization because of the importance of this crop to agriculture. The slow-growing soybean symbionts have been classified into 17 distinct serogroups (Date and Decker, 1965). However, the serological classification of the soybean symbionts appears to be more relevant to those originating from the USA than to those from South East Asia (Thompson *et al.*, 1991). Hollis *et al.*, (1981) identified three DNA homology groups among 28 strains, only one of which originated from SouthEast Asia. DNA homology groups I and Ia are closely related to each other, but DNA homology group II is more divergent. Generally, there is correspondence between serological reaction and DNA homology grouping, serogroups 6, 38, 123 and serogroups 62 and 110 belong to groups I and Ia, respectively, while serogroups 31, 46, 76, and 94 belong to group II. DNA homologies for the other serogroups have not been reported. The divergence between strains belonging to DNA homology groups I and II also is evident from results of RFLP (Kuykendall *et al.*, 1992; Stanley *et al.*, 1985) and MLEE (Bottomly *et al.*, 1994) analyses. Based upon these differences and variations in antibiotic resistance patterns, EPS, fatty acids, and hemoproteins, Kuykendall *et al.*, (1992) proposed the separation of DNA homology group II from *B. japonicum* as the separate species *B. elkanii*. However, fatty acid-methyl ester analysis of strains of American origin and those from South-East Asia have indicated the existence of a fourth grouping besides *B. japonicum* I, Ia, and *B. elkanii* (Graham *et al.*, 1995). SSU rRNA gene nucleotide sequence analyses with the slow-growing legume symbionts have indicated the presence of two lineages within the genus *Bradyrhizobium*. One of the branches centers around *Rhodopseudomonas palustris* and contains *B. japonicum*, as well as members of the bacterial genera *Afipia*, *Nitrobacter* and *Blastobacter* (Figure 2). The pigmented bradyrhizobial strains isolated from *Aeschynomene indica* also are closely related to *R. palustris* (Wong *et al*, 1994; Young *et al.*, 1991) based upon SSU rRNA gene sequencing analysis. A distinct second lineage, supported by high bootstrap values, contains *B. elkanii* and strains of *Bradyrhizobium* isolated from various tropical legumes.

The close genetic relationship between *R. palustris*, *B. japonicum*, *Afipia* species, *Nitrobacter* species and *Blastobacter* (Figure 2) is analogous to the close relationships between *Agrobacterium* with *Rhizobium*, *Phyllobacterium* with *Mesorhizobium*, and *Azorhizobium* with *Xanthobacter*. However, the situation with *Bradyrhizobium* is more complex because of the distinct separate lineage of *B. elkanii* and several other bradyrhizobial strains originating from several tropical region legumes. Relative phylogenetic position or close SSU rRNA sequence similarity among some members within *Rhizobium*

and *Agrobacterium* has initiated a debate for reclassification of those members to the other genus. The reasons to reclassify some members of *Agrobacterium* and *Rhizobium* also would apply to the lineage centered around *Bradyrhizobium*. In the event of reclassification, consideration must also be given to the possibility of maintaining *B. elkanii* as a separate genus. Of course, no reclassification should be considered until the phylogenies are evaluated with independent analyses of other gene loci.

V.A.11. Bradyrhizobium liaoningense (Xu et al., 1995), an Additional Species Nodulating Soybean.

Extra-slowly growing (ESG) symbionts of soybean occur in the soils of Liaoning, Heilongjiang, Shanxi, Hubei, and Anhui Provinces, the People's Republic of China. Seventeen isolates were distinguishable from *B. japonicum* by carbon substrate utilization, antibiotic resistance patterns, tolerance to NaCl, growth at elevated temperatures, and serology with the exception of one isolate. DNA homology ranged from 9% to 38% with *B. japonicum* and from 7% to 30% with *B. elkanii* (Xu *et al.*, 1995). Full-length SSU rRNA gene sequences were not reported.

V.A.12. The Third Genus of Rhizobium Now Reclassified as Mesorhizobium (Jarvis et al., 1997).

The third separate genus of rhizobia shares fast growth rate and acid production as characteristics with the other two lineages formerly classified together as *Rhizobium*. In the case of two species, flagellation is polar or sub-polar compared to peritrichous in the other two genera of fast-growers. Another distinguishing characteristic is the location of the symbiotic genes, which are integrated in the genome (Cadahia *et al.*, 1986; Chua *et al.*, 1985; Pankhurst *et al.*, 1983; Pankhurst *et al.*, 1986; Ruiz-Argüeso *et al.*, 1988) and not borne on plasmids. The recognition of this group as a separate genus is warranted since they are clearly distinct based upon DNA homology classification (Crow *et al*, 1981) and phylogeny determined from SSU rRNA sequences (Nour *et al*, 1995; Sawada *et al*, 1993; Willems and Collins, 1993). A change in the name for the genus to *Mesorhizobium* has been suggested to distinguish members from the other two genera (Jarvis *et al.*, 1997). The name reflects that some of the bacteria of this genus have growth rates slower than *Rhizobium* and *Sinorhizobium* but faster than *Bradyrhizobium*, which the prefix "Meso" is meant to indicate. However, slower growth is not necessarily a characteristic shared among all the rhizobia of this genus. For instance Martinez de Drets (1972) reported similar growth rates among *S. melioti*, *R. leguminosarum*, and *Lotus* rhizobia. Also, Jordan (1984) has classified the type species as fast-growing and acid producing. The growth rate among isolates of chickpea range from fast to slow (Ruiz-Argüeso *et al.*, 1988).

Currently there are reports naming five species within this genus, *M. loti*, *M. huakuii*. *M. ciceri*, *M. mediterraneum*, and *M. tianshanense*. NZP 2213 originating from *L. tenuis* represents the type strain for the genus. Several additional species may exist which in one case have been referred to as cluster U and originate from several different legumes growing in New Zealand, Senegal, and Brazil (De Ladjudie *et al.*, 1994) and in the other two cases have been referred to as separate genomic species originating from *Cicer arietinum* (Nour *et al*, 1995) and from *Lotus corniculatus* (Sullivan *et al.*, 1996).

V.A.13. Mesorhizobium loti (Jarvis et al., 1982).

This was the first additional species of nitrogen fixing legume symbionts after publication of the list of approved bacterial names (Skerman *et al.*, 1980). Jordan (1984) indicated that many of the phenotypic and biochemical characteristics of strains nodulating *Lotus* are in common with the other fast-growers. Distinguishing features are that the symbiotic genes are integrated into the chromosome and each cell has one polar or sub-polar flagellum. The type strain nodulates most *Lotus* species, however, symbioses with some of these species are ineffective for nitrogen fixation. Furthermore, symbiotic effectiveness across *Lotus* species varies with the fast-growing strain used as inoculant (Pankhurst and Jones, 1979). The type strain, NZP 2213, can be distinguished from the type strains of other genera by results of DNA-DNA hybridization and SSU rRNA nucleotide sequence analyses (Jarvis *et al*, 1982; Willems and Collins, 1993)

Two widely divergent groups of *Lotus*-nodulating bacteria were recognized before molecular methods were used for classification purposes. Characters which separated the two groups were bacteroid ultrastructure (Craig *et al.*, 1973; internal antigens (Pankhurst, 1979; Vincent and Humphrey, 1970), extracellular polysaccharide composition (Bailey *et al.*, 1971), gel electrophoresis of soluble proteins

(Roberts *et al.*, 1980), growth rate and acid production on yeast extract media (Abdel-Ghaffar and Jensen, 1966; Brockwell *et al.*, 1966; Norris, 1965), sensitivity to antibiotics (Pankhurst, 1977), phage relationships (Patel, 1976), and plant cross-inoculation (Abdel-Ghaffar and Jensen, 1966; Crow *et al.*, 1981, Erdman and Means, 1949; Jensen, 1967; Jensen and Hansen, 1968; Pankhurst, 1977). The divergence between the fast- and slow-growing *Lotus*-nodulating rhizobia was supported by results of DNA homology experiments, based upon which Crow *et al.*, (1981) classified the *Lotus* symbionts into two groups, 4 and 6, which had less than 10% DNA homology. Similarly, Jarvis *et al.*, (1986) reported that the rRNAs of the slow-growing Lotus rhizobia are more closely related to those of *Bradyrhizobium* than *Rhizobium*. SSU rRNA nucleotide sequences of the slow-growing *Lotus* rhizobia have not been reported.

Besides the obvious wide differences between the fast and slow-growing *Lotus* rhizobia, divergence also is apparent among the fast-growing rhizobia that nodulate *Lotus* species. Crow *et al.*, (1981) reported DNA homology variability within the strains of group 4 originating from *Lotus* species. The strains CC811 from *L. corniculatus* and CC809a from *L. maroccanus* had mean relative hybridization values of only 49%. Differences in nitrogen-fixing effectiveness by NZP 2213 and NZP 2037 were reported across 9 species of *Lotus* (Pankhurst and Jones, 1979). These two strains also differ by DNA homology (De Lajudie *et al*, 1994). DNA homology values of NZP 2213 and other *Lotus*-nodulating strains, R88B and R8CS, and non-symbiotic rhizobial isolates originating from a field planted with *Lotus corniculatus* in New Zealand were less than 60%. The inoculant strain used for *Lotus corniculatus* in New Zealand, ICMP3153, was originally isolated from a plant growing in Ireland (J. Sullivan personal communication) and was reported to be of a genotype more similar to *R. huakuii* than other strains recovered from *Lotus* (Sullivan *et al.*, 1996). The published information on the genetic variation among the isolates obtained from *Lotus* species indicates that this host genus is nodulated by a wide range of species representing at least two genera.

V.A.14. Mesorhizobium huakuii (Chen et al., 1991).

The plant genus *Astragalus* contains 1,500 to 2,000 species, but few are important agriculturally (Allen and Allen, 1980). Several species are important as a source for the gum tragacanth, and others are indicator plants for selenium and uranium. Yet another group of *Astragalus* species produces a toxic principle, which could affect animal husbandry. In Asia, *A. sinicus* is important as a green manure and forage crop. An analysis of nine rhizobial isolates of *A. sinicus* originating from Hubei and Nanjing, China, indicated that they belonged to a distinct DNA homology group (Chen *et al.*, 1991). In the description of *M. huakuii*, NZP 2213 was used as a reference strain and speciation was indicated because DNA homology values were between 25 and 35%. The type strain for *M. huakuii*, CCBAU 2609, was shown to belong to the same genus as *M. loti* because of similarity in their SSU rRNA nucleotide sequences (Sawada *et al*, 1993, Willems and Collins, 1993). However, according to Chen *et al.*, (1991), *M. huakuii* differs from the other type strains which belong to the same genus because the genes required for symbiosis are located on plasmids. Rhizobial isolates originating from nodules of *A. sinicus* growing in Japan also were identified as *M. huakuii* based upon phenotypic characterization (Murooka *et al.*, 1993). A distinguishing feature of the Japanese isolates was that they and not the type strain nodulated the local variety of *A. sinicus* cv. Japan. Based upon these results, Murooka *et al.*, (1993) suggested that their isolates belonged to a new biovar, *M. huakuii* biovar *renge*. However, data to indicate the molecular evolutionary relationships between *M. huakuii* biovar *renge* and the *M. huakuii* type strain were not reported.

V.A.15. Mesorhizobium ciceri (Nour et al., 1994b) and Mesorhizobium mediterraneum (Nour et al., 1995), Chickpea-Nodulating Species.

Since the reported speciation of *M. loti* and *M. huakuii* several species have been described which nodulate *Cicer arietinum* or Chickpea. One of these descriptions reports the classification of *M. ciceri* (Nour *et al.*, 1994b; Nour *et al.*, 1995). The type strain for *M. ciceri*, UPM-Ca7, shares 44% and 19% DNA homology with the type strains for *M. loti* and *M. huakuii*, respectively (Nour *et al.*, 1994b). SSU rRNA gene sequence similarity of *M. ciceri* with that of *M. loti* and *M. huakuii* was 99.8% and 98.2%, respectively.

The other Chickpea-nodulating species was classified as *M. mediterraneum* (Nour *et al.*, 1995). The type strain, UPM-Ca36, had 31%, 29% and 42%

DNA homology with the type strains of *M. loti*, *M. huakuii* and *M. ciceri*, respectively (Nour *et al.*, 1994b, Nour *et al.*, 1995). The nucleotide similarities of the SSU rRNA genes were 98.5, 99.2, and 98.6%, respectively (Nour *et al.*, 1995). Unfortunately, the characterization to describe both *M. ciceri* and *M. mediterraneum* did not include an evaluation of the DNA homology of these two type strains with the strains used by Crow *et al.*, (1981) to define DNA homology group 4. Such an examination may have provided additional insight into the genetic relationship between the rhizobia of *Lotus* and those of chickpea since several of the chickpea strains used in the study of Crow *et al* (1981) exhibited up to 88% DNA homology with one of their reference *Lotus* strains.

V.A.16. Mesorhizobium tianshanense (Chen et al., 1995).

Chen *et al.*, (1995) described the characterization of isolates originating from seven legume species growing in Xinjiang Province, the People's Republic of China. The seventeen isolates were grouped together using the results of 148 phenotypic characters. The strain, CCBAU 3306 (A-1BS), was chosen as the type strain for *M. tianshanense*. The DNA homology values between the type strain and the other isolates classified as *M. tianshanense* ranged from 75 to 100%. The DNA homology values between the type strain for *M. tianshanense* and the type strains for other species were generally low with the exception of the type strain for *M. huakuii*, which shared 64% DNA homology with CCBAU 3306. Complete SSU rRNA gene nucleotide sequences were not reported in the description of *M. tianshanense*. However, a 260-bp partial SSU rRNA gene nucleotide sequence of CCBAU 3306 is identical with that of the type strain for *M. ciceri* (Chen *et al.*, 1995). A later report indicated that the DNA homology between the type strains of *M. tianshanense* and *M. huakuii* was considerably less (13.9%) and that the SSU rRNA gene sequence similarities between CCBAU 3306 and the type strains of *M. loti* and *M. huakuii* were 97.4 and 97.9%, respectively (Tan *et al.*, 1997). However, there are 6 nucleotide differences within the 260-bp region of the full-length sequence corresponding with the previously reported partial SSU rRNA gene sequence. Also, Jarvis *et al.*, (1997) indicated a closer relationship between *M. tianshanense* and the other species within this genus based upon an independent determination of the full-length SSU rRNA gene sequence of CCBAU 3306.

V.A.17. Phyllobacterium (Knösel, 1962).

Very little molecular evolutionary information is available for the genus *Phyllobacterium*. Besides occurring on the surfaces and in hypertrophies of the leaves of some tropical plants, phyllobacteria have been isolated from the roots of sugarbeet (Lambert *et al.*, 1990). The isolates were identified as *Phyllobacterium* by phenotypic characterization and protein electrophoretic patterns. The SSU rRNA nucleotide sequences of the two type species are very similar (Yanagi and Yamasoto, 1993), but to our knowledge DNA reassociation data is not available to demonstrate that the two strains are separate species. SSU rRNA nucleotide sequence homology with the genus *Mesorhizobium* is approximately 97.0% (Table 1) and both genera are phylogentically related. It would appear that the molecular evolutionary systematic information with ·*Phyllobacterium* and *R. galegae* is analogous with respect to potential reclassification of genus status. Therefore, for the consistency it will be necessary to evaluate *Phyllobacterium* as a genus separate from *Mesorhizobium* when considering *R. galegae* as a separate genus from *Agrobacterium*.

V.A.18. Rhizobium (Frank, 1889).

Currently there are four species recognized within the genus *Rhizobium*. Also, there have been suggestions for three additional species within this genus, *R. gallicum*, *R. hainanense*, and *R. mongolense*. The type strain *R. leguminosarum* (USDA 2370, ATCC 10004) is also the type strain for the genus *Rhizobium*. The other two recognized species are *R. etli* (type strain CFN 42) and *R. tropici* (type strain CIAT 899). The SSU rRNA gene nucleotide sequences of the three type strains are very similar ranging in homology from 97.8% to 99.0% (van Berkum *et al.*, 1996; Willems and Collins, 1993). Some evidence has been reported which would indicate that the SSU rRNA genes between *R. leguminosarum* and *R. etli* may undergo recombination (Eardly *et al.*, 1995; Eardly *et al.*, 1996). Values of DNA-DNA reassociation (Martinez-Romero *et al.*, 1991; van Berkum *et al.*, 1996) would support the separation of *Rhizobium* into these three species.

V.A.19. Rhizobium leguminosarum (Frank, 1889).

Originally this name was used to describe the symbionts of the legume genus *Vicia* distinguishing them from symbionts of the genus *Trifolium* (*R. trifolii*) and from symbionts of the genus *Phaseolus* (*R. phaseoli*). However, the majority of the symbionts of the latter two plant genera were reclassified as *R. leguminosarum* based upon extensive phenotypic and genetic characterization (Jordan, 1984). Infrasubspecific ranks (biovar) have been given to the former species to distinguish their plant affinities, which is a character coded for by plasmid-borne genes. The infrasubspecific ranks in use are biovar *viciae* for nodulation of the legume tribe *Viciae*, biovar *trifolii* for nodulation of the legume genus *Trifolium*, and biovar *phaseoli* for nodulation of the legume genus *Phaseolus*.

V.A.20. Rhizobium tropici (Martinez-Romero et al., 1991)

According to the revised classification of *R. phaseoli*, the bean rhizobial strains became *R. leguminosarum* biovar *phaseoli* (Jordan, 1984). However, there was evidence of genetic heterogeneity among rhizobial isolates of bean even before the reclassification of *R. phaseoli* to *R. leguminosarum* (Crow *et al.*, 1981; Jarvis *et al*, 1980). Bean strains originating from Mexico and South America were identified as a heterogeneous complex of strongly differentiated phylogenetic lineages, indicating that recognition of several species was warranted (Piñero *et al.*, 1988). One of the deep lineages was subsequently classified as *R. tropici* (Martinez-Romero *et al.*, 1991), but within this species two subgroups, A and B, have been identified. It may seem that these two subgroups are sufficiently different to warrant independent species status, but the additional evidence to support such a separation has not been reported. CIAT 899, a member of subgroup B, is the type strain for *R. tropici*. Based upon SSU rRNA gene sequences, Willems and Collins (1993) showed that the phylogenetic position of *R. tropici* was distinct from that of *R. leguminosarum*. All bean strains other than *R. tropici* were provisionally classified as *R. leguminosarum*.

V.A.21. Rhizobium etli (Segovia et al., 1993).

The distinctiveness of the SSU rRNA gene sequence of *R. etli* was first recognized in a partial 16S rRNA analysis of field isolates capable of nodulating both alfalfa and beans (Eardly *et al*, 1992). Based upon this information these groups of isolates were provisionally referred to as *Rhizobium* sp. Type I strains rather than *R. leguminosarum* bv *phaseoli*

(Eardly *et al.*, 1992). Subsequently, Segovia *et al.*, (1993) provided further evidence that this group was distinguishable both genetically and phenotypically and proposed the species name *R. etli*. Full length SSU rRNA sequencing and DNA-DNA reassociation analyses have confirmed the independent species status of *R. etli* (van Berkum *et al.*, 1996).

V.A.22. Rhizobium galegae (Lindström, 1989).

The rhizobia of *Galega orientalis*, and *G. officinalis* initially were classified as *R. leguminosarum* or to have taxonomic affiliation with the rhizobia of the proposed genus *Mesorhizobium* (Hauke-Pacewic-zowa, 1952; Proctor, 1963). However, these rhizobia appeared to be quite distinct from species within these two genera based upon phenotypic and bio-chemical analyses (Lindström *et al*, 1983, Lindström and Lehtomäki, 1988; Lipsanen and Lindström, 1989). In addition, DNA homology, similarity of ribosomal cistrons, and SSU rRNA sequencing analysis supported the separate speciation of *R. galegae* (Jarvis *et al.*, 1986; Lindström *et al.*, 1983; Wedlock and Jarvis, 1986; Willems and Collins, 1993). The phylogenetic position of the type strain, HAMBI 540, appears to be distant from other members of the legume symbionts. Based upon currently constructed phylogenetic trees the closest relative is *A. vitis* (Sawada *et al*, 1993).

V.A.23. Rhizobium gallicum and Rhizobium giardinii, Ttwo Additional Species Isolated from Nodules of Phaseolus vulgaris (Amarger et al., 1997).

Among a collection of bean isolates obtained from various locations in France, two groups were differentiated from *R. leguminosarum*, *R. etli*, and *R. tropici* by RFLP analysis using several DNA probes (Geniaux *et al.*, 1993; Laguerre *et al.*, 1993b). Subsequently, these two groups were assigned to new genomic species based upon results of DNA-DNA hybridization and partial SSU rRNA gene sequencing analyses (Laguerre *et al.*, 1993a). The type strains for each of these two genomic species are R602sp and H152 for *R. gallicum* and *R. giardinii*, respectively and their separate speciation is further supported by their phenotypic differentiation and their SSU rRNA gene sequences (Amarger *et al.*, 1997). However, the phylogenetic relationship between *R. giardinii* and the typical members of the genus *Rhizobium* is distant since

SSU rRNA gene sequences are only from 95.5 to 96.2% similar. Although the phylogenetic relationship of *R. giardinii* is closer to those of *R. galegae* and *Agrobacterium* species, Amarger *et al.*, (1997) concluded that their SSU rRNA nucleotide sequence similarities were too low to classify *R. giardinii* within either of these two genera.

V.A.24. Rhizobium hainanense (Chen et al., 1997).

Goa *et al.*(1994) described the phenotypic characteristics of 63 rhizobial isolates obtained from nodules of 21 legume genera growing in Hainan Province, a tropical region of the People's Republic of China. Among these there were 5 divergent fast-growing isolates which had originated from nodules of *Desmodium* and *Centrosema*. The type strain, CCBAU 57015 (166), was placed within the genus *Rhizobium* based upon results of SSU rRNA gene sequence analysis (Chen *et al.*, 1997). Separate speciation was supported with low values of DNA homology between the type strain of *R. hainanense* and those of *R. leguminosarum*, *R. etli*, *R. tropici*, and *A. rhizogenes*.

V.A.25. Rhizobium mongolense (van Berkum et al., 1997)

Medicago ruthenica, a plant species native to Inner Mongolia, is a potential new forage crop and also may be a source of genes for the genetic improvement of cultivated alfalfa (*M. sativa* L.) for tolerance to stress. However, the symbionts of *M. ruthenica* belong to the genus *Rhizobium* and not to the genus *Sinorhizobium*, which is more commonly associated *Medicago* (van Berkum *et al.*, 1997). The type strain for *R. mongolense* is USDA 1844 and its separate speciation is supported by low values of DNA-DNA hybridization with the type strains of the other species, differentiation based upon the SSU rRNA gene sequence, and by distinctive multilocus electrophoretic types. Within this species van Berkum *et al.*, (1997) reported a difficulty in placing one strain with a chimeric genotype, which they indicated could reflect a wider problem in rhizobial systematics.

V.A.26. "Rhizobium lupini" (Eckhardt et al., 1931; Schroeter, 1886).

Very little molecular evolutionary information has been reported with rhizobia which nodulate lupines, although USDA 3051 (USDA strain 3C2e1 deposited with the American Type Culture Collection as ATCC 10319) is recognized as the type strain of this "species". However, the scant information available based upon an analysis of ATCC 10319 places this type strain firmly within the genus *Bradyrhizobium*. The reason for this is because SSU rRNA sequencing with DSM 30140 (ATCC 10319 or USDA 3051) indicates a phylogenetic relationship within *Bradyrhizobium* (Ludwig *et al.*, 1995). At first "*R. lupini*" was suspected to be *B. japonicum* because Hollis *et al.*, (1981) reported up to 99% DNA homology values between ATCC 10319 and the type strain of *B. japonicum*. Subsequently, the culture preserved as ATCC 10319, supplied by the American Type Culture Collection, was shown to be contaminated with *B. japonicum*. A second report indicated that the DNA homology between ATCC 10319 and the type strain of *B. japonicum* was 35% (Scholla *et al.*, 1990), which was confirmed by us with USDA 3051. Similarly, three lupine isolates originating from Mexico and the type strain of *B. japonicum* were reported to exhibit 37% DNA homology or less (Barrera *et al.*, 1997). Although the nitrogen-fixing symbionts of lupine would appear to be a separate species of *Bradyrhizobium*, there still is a potential for a separate mesorhizobial species as well because Crow *et al.*, (1981) reported that the strain "L.densif85" originating from *Lupinus densiflorus* was associated with DNA homology group 4 as defined by the *Lotus* strains CC809a and CC811.

V.A.27. Sinorhizobium (De Lajudie et al., 1994).

The existence of distinct genetic groups of *Rhizobium* was evident from extensive DNA-DNA reassociation analyses and one genetically cohesive group (group 3) was identified which contained the rhizobia nodulating *Medicago* and *Trigonella* spp. (Crow *et al.*, 1981). Wedlock and Jarvis (1986) reported that the reference strain (SU47) used by Crow *et al.*, (1981) to identify members belonging group 3 and fast-growing soybean isolates originating from China were closely related since DNA homology values ranged from 36 to 44%. By comparison, DNA homology values of SU47 with *R. leguminosarum* and *R. galegae* were an average of 14 and 21%, respectively. In contrast, Chen *et al.*, (1988) proposed that the fast-growing soybean isolates originating from China belonged to a separate genus, *Sinorhizobium*, based upon

numerical taxonomic information. The proposal was supported by a distinct separation of the fast-growing soybean rhizobia from representatives of other genera and species of the *Rhizobiaceae*. However, the reported close genetic relationship between the rRNA cistrons of group 3 and fast-growing soybean isolates (Jarvis *et al.*, 1986) and the sequence similarities among the SSU rRNA genes (Jarvis *et al*, 1992; Willems and Collins, 1993) contradicted the phenotypic relationships reported by Chen *et al.*, (1988). Consequently, the molecular evolutionary systematic data did not support the separation of the fast-growing soybean isolates into a new genus and their recognition as a separate genus was discouraged. Subsequently, the new genus name, *Sinorhizobium*, was revived by the rules of nomenclatural priority to assign separate genus status to the members belonging to group 3, the fast-growing soybean isolates originating from China, and two new species isolated from tropical trees growing in Senegal (De Lajudie *et al.*, 1994). Currently, the genus *Sinorhizobium* consists of four recognized species, *S. meliloti* (formerly *R. meliloti*, the group 3 strains), *S. fredii* (formerly *R. fredii*, the fast-growing soybean isolates from China), *S. saheli*, and *S. terangae* as well as a more recently suggested new species *S. medicae*, which nodulates alfalfa and several annual medics.

V.A.28. Sinorhizobium meliloti (Dangeard, 1926).

Rhizobium meliloti was a recognized species included among the approved bacterial names (Skerman *et al.*, 1980). The species was described by Jordan (1984) and was separated from the other species of *Rhizobium* by differences in phenotype and host plant affinity. Its separate species status within *Sinorhizobium* is supported by DNA-DNA reassociation values and SSU rRNA nucleotide sequence information. Two subgroups (A and B) within this species were identified by MLEE analysis (Eardly *et al.*, 1990). It may be concluded from the magnitude of the genotypic differences between these groups that they may represent separate species. However, this conclusion may be inconsistent with the multilocus enzyme electrophoretic profile data of strain CC2013, which may be evidence for recombination between the two subgroups (Eardly *et al.*, 1990). The type strain for the species (USDA 1002, ATCC 9930) is a member of subgroup A.

V.A.29. Sinorhizobium medicae (Rome et al., 1996b).

Seventy three isolates originating from two French sites sown with *Medicago trunculata* were separated into two subgroups based upon PCR-RFLP analyses and relative DNA homology values (Rome *et al.*, 1996a). One of these two subgroups was identified with *S. meliloti* subgroup B strains M3 and M104. Subsequently, a subset of 11 isolates representing the two subgroups was shown to be distinguishable by 9 phenotypic characters and also by the ability to fix nitrogen with *M. polymorpha*. The SSU rRNA gene sequence similarities of the type strain for *S. medicae*, A321, and the type strain of *S. meliloti* is 99.7% (Rome *et al.*, 1996b). DNA homology between strain A321 and RCR 2011, chosen to represent *S. meliloti*, is 60% (Rome *et al.*, 1996a).

V.A.30. Sinorhizobium fredii (Scholla and Elkan, 1984).

Fast-growing isolates from soybean nodules originating from China were identified on four separate occasions (Chen *et al.*, 1988; Dowdle and Bohlool, 1985; Keyser *et al.*, 1982; Lin *et al.*, 1987). The species (type strain USDA 205, ATCC 35243) is differentiated from the other legume symbionts by DNA-DNA reassociation, plant specificity, growth rate, antibiotic resistance profiles and serology. SSU rRNA nucleotide sequence homology and DNA reassociation values indicate that *S. fredii* and *S. meliloti* are closely related (Jarvis *et al.*, 1992; Wedlock and Jarvis, 1986; Willems and Collin, 1993). The fast-growing wide-host range tropical rhizobial strain NGR 234, isolated from *Lablab purpureus*, is closely related to *S. fredii* based upon serology, *nodD* genes, partial SSU rRNA nucleotide sequences and FAME analysis (Jarvis *et al.*, 1992, Jarvis and Tighe, 1994). However, confirmation of NGR 234 as a member of *S. fredii* is necessary with DNA-DNA reassociation data. The eleven initial isolates of *S. fredii* can be divided into two subgroups, *S. fredii* chemovar fredii and *S. fredii* chemovar sinensis (Scholla and Elkan, 1984; Wedlock and Jarvis, 1986).

V.A.31. Sinorhizobium saheli (De Ladjudie et al., 1994).

These are symbionts of *Sesbania* growing in the Sahel area of Africa, but also nodulate *Acacia seyal*, *Leucaena leucocephala*, and *Neptunia oleracea*. The type strain is ORS 609 and can be differentiated from other species of *Sinorhizobium* and related genera by results of DNA-DNA reassociation, DNA-RNA hybridization, by protein fingerprint patterns using SDS-PAGE, and by SSU rRNA sequencing analysis.

V.A.32. Sinorhizobium terangae (De Ladjudie et al., 1994).

These are symbionts of various *Sesbania* and *Acacia* species growing in Senegal, Western Africa. Host range is similar to that of *S. saheli*. The type strain is ORS 1009 and can be differentiated from other species of *Sinorhizobium* and related genera by results of DNA-DNA reassociation, DNA-RNA hybridization, by protein fingerprint patterns using SDS-PAGE, and by SSU rRNA sequencing analysis.

VI. Possible Alternative Speciation Criteria

The named species within the *Rhizobiaceae* examined in this review have each been shown to be distinctive by polyphasic analysis, including colonial and cultural characteristics, with DNA-DNA hybridization and SSU rRNA sequence analysis being the standard criteria by which they are differentiated (Graham *et al.*, 1991). However, some bacterial population geneticists have suggested that these methods may not always provide accurate estimates of evolutionary relationships. For example Maynard Smith (1995) has suggested that the use of DNA-DNA hybridization to define bacterial species is arbitrary and without merit, because it does not recognize real distinctions but rather imposes arbitrary divisions upon a continuum. He supports the latter assertion by the observation that the frequency of genetic exchange in bacteria seems to vary more or less continuously with genetic distance (*e.g.* Roberts and Cohan, 1993).

Questions on the range and extent of genetic exchange in bacteria have also brought into question the use of SSU rRNA sequence data to distinguish closely related bacterial species, whose SSU rRNA sequences may only differ by a few nucleotide bases (Fox *et al.*, 1992). It could be argued that evolutionary forces such as recombination and Darwinian selection should be considered more critically when using SSU rRNA polymorphism to extrapolate organismal phylogenies. Like most bacteria examined so far, rhizobial species carry multiple rRNA operons (Honeycutt *et al.*, 1993; Kundig *et al.*,

1995), which could facilitate recombination between divergent alleles (Anderson and Roth, 1981; Hill *et al.*, 1977, Petes and Hill, 1988). It has generally been assumed that there is a strong negative selection against rRNA recombinants in most normal situations because of the decreased translational efficiency associated with recombinant rRNA alleles (Hill *et al.*, 1977; Niebel *et al.*, 1987; Peters and Hill, 1988). Nevertheless there are reports suggesting that the SSU rRNA genes in strains of *Aeromonas* (Sneath, 1993), *Rhizobium* (Eardly *et al.*, 1995; Eardly *et al*, 1996; Sullivan *et al.*, 1996) and *Escherichia* (Cilia, 1996) may have a history of recombination. In their studies of *rrn* polymorphism in a single strain of *E. coli*, Cilia *et al.*, (1996) reported that direct SSU rRNA sequencing (by PCR) reveals only an average sample of the heterogeneity present. They also pointed-out that previous gene conversion may bias the diversity observed. Also, contrary to the assumption that there will be a negative selection against recombinant rRNA alleles, Condon *et al.*, (1995) reported that the heterogeneous *rrn* operons in *E. coli* may actually provide an adaptive advantage under certain environmental conditions. Furthermore, it has been shown that certain polymorphic nucleotide positions in SSU and LSU rRNAs are subject to other environmental selection pressures such as antibiotics (Aagard *et al.*, 1994; Cundliffe, 1990; Fourmy *et al.*, 1996). SSU rRNA sequences will probably continue to be important for estimating phylogenetic relationships among bacteria, but by themselves their use may be inappropriate for confidently resolving species below the genus level.

Dykhuizen and Green (1991) have suggested that perhaps a better approach for defining species limits in bacteria, might be to apply a biological species concept, where a species is defined as a group of strains that share a common gene pool (by lateral transfer and recombination). There is evidence from both MLEE allele-assortment analyses and DNA sequence analyses that chromosomal alleles may be recombined at a low, but nevertheless detectable frequency in natural populations of both *Rhizobium* (Demezas *et al.*, 1995; Eardly *et al.*, 1990; Gordon *et al.*, 1995; Maynard Smith *et al.*, 1993; Souza *et al.*, 1994; Sullivan *et al.*, 1995; Sullivan *et al.*, 1996) and *Bradyrhizobium* (Bottomley *et al.*, 1994). In practice this could be determined by the comparison of the sequences of multiple loci in a group of strains thought to belong to the same species. If the results indicate that phylogenetic relationships within the group are generally the same for each locus exam-

ined, in other words if the phylogenies of different genes (loci) are congruent across the group of strains, then it would be concluded that the genes (and the organisms carrying them) have a history of genetic isolation, and thus may be considered distinct biological species. Alternatively, if lateral gene transfer and recombination within the group has been relatively common, then phylogenies for a given gene would be scrambled, depending on the recombinational history of the particular locus under examination. In this case, a multilocus sequence analysis would reveal incongruent phylogenies for the individual genes suggesting that the members of the group are conspecific. Cohan (1994) has suggested that perhaps a more universally applicable criterion for a biological species concept in bacteria might be permanent neutral sequence divergence. Maynard Smith (1995) has argued that the two practical disadvantages of this general approach are that entities of very different kinds would be recognized as species in different taxa (because of differing rates of recombination in each), and that massive amounts of sequence data would be required. However, it is anticipated that rapid progress in sequencing technology will facilitate identification of species by providing increased capability to identify common gene pools within populations, and also by providing historical evidence of mosaicism on the genomic level.

VII. References

Aangard, C., Phan, H., Trevisanato, S. and Garrett, R.A. (1994). J. Bacteriol. 176, 7744-7747.

Abdel-Ghaffar, A.S. and Jensen, H.L. (1966). Arch. Mikrobiol. 54, 393-405.

Allen, O.N. and Allen, E.K. (1980). The Leguminosae, A source book of characteristics, uses, and nodulation. The University of Wisconsin Press, Madison, WI.

Amarger, N., Macheret, V. and Laguerre, G. (1997). Int. J. Syst. Bacteriol. 47, 996-1006.

Anderson, P. and Roth, J.R. (1981). Proc. Natl. Acad. Sci. USA 78, 3113-3117.

Bailey, R.W., Greenwood, R.M. and Craig, A. (1971). J. Gen Microbiol. 65, 315-324.

Barrera, L.L., Trujillo, M.E., Goodfellow, M., García, F.J., Hernández-Lucas, I., Dávila, G., van Berkum, P. and Martinez-Romero, E. (1997). Int. J. Syst. Bacteriol. 47, 1086-1091.

Beijerinck, M.W. and van Delden, A. (1902). Centbl. Bakt. (etc.) II 9, 3-43.

Bottomley, P.J., Cheng, H. and Strain, S.R. (1994). Appl. Environ. Microbiol. 60, 1754-1761.

Bouzar, H. (1994). Int. J. Syst. Bacteriol. 44, 373-374.

Bouzar, H., Chilton, W.S Nesme, X., Dessaux, Y., Vaudequin, V., Petit, A., Jones, J.B. and Hodge, N.C. (1994). Appl. Environ. Microbiol. 61, 65-73.

Brockwell, J., Asuo, S.E. and Rea, G.A. (1966). J. Aust. Inst. Agric. Sci. 32, 295-297.

Burgin, A.B., Parodos, K., Lane, D.J. and Pace, N.R. (1990). Cell 60, 405-414.

Burton, J. C. (1979). Rhizobium species, in H. J. Peppler and D. Perlman (eds.), Microbial Technology 2nd ed., vol. 1, Microbial Processes, Academic Press Inc., NY.

Cadahia, E., Leyva, A. and Ruiz-Argüso, T. (1986). Arch. Microbiol. 146, 239-244.

Chen, W.X., Li, G.S., Qi, Y.L., Wang, E.T., Yuan, H.L. and Li, J.L. (1991). Int. J. Syst. Bacteriol. 41, 275-280.

Chen, W.X., Wang, E.T., Wang, S., Li, Y., Chen, X. and Li, Y. (1995). Int. J. Syst. Bacteriol. 45, 153-159.

Chen, W.X., Yan, G.H. and Li, J.L. (1988). Int. J. Syst. Bacteriol. 38, 392-397.

Chen, W., Tan, Z., Gao, J. and Wang, E. (1997). Int. J. Syst. Bacteriol. 47, 870-873.

Chua, K-Y, Pankhurst, C.E., MacDonald, P.E., Hopcroft, D.H., Jarvis, B.D.W. and Scott, B.D. (1985). J. Bacteriol. 162, 335-343.

Cilia, V., Lafay, B. and Christen, R. (1996). Mol. Biol. Evol. 13, 451-461.

Cohan, F.M. (1994). Trends Ecol. Evol. 9, 175-180.

Condon, C., Liveris, D., Squires, C., Schwartz, I. and Squires, C.L. (1995). J. Bacteriol. 177, 4152-4156.

Conn, H.J. (1938). J. Bacteriol. 36, 320-321.

Conn, H.J. (1942). J. Bacteriol. 44, 353-360.

Craig, A.S., Greenwood, R.M. and Williamson, K.I. (1973). Arch Mikrobiol. 89, 23-32.

Crow, V.L., Jarvis, B.D.W. and Greenwood, R.M. (1981). Int. J. Syst. Bacteriol. 31, 152-172.

Cundliffe, E. (1990). p. 479-490, in W.E. Hill, Moore, P.B. Dahlberg, A., Schlessinger, D., Garret, R.A. and Warner, J.R. (eds.), The ribosome structure, function, and evolution. American Society for Microbiology, Wasgington, D.C.

Dangeard, P.A. (1926). Botaniste (Paris) 16, 1-275.

Date, R.A. and Decker, A.M. (1965). Can J. Microbiol. 11, 1-8.

de Bruijn, F.J. (1992). Appl. Environ. Microbiol. 58, 2180-2187.

de Faria, S.M., Franco, A.A. de Jesus, R.M., de Menandro, M., Baitello, J.B., Mucci, E.S.F., Döbereiner, J. and Sprent, J.I. (1984). New Phytol. 98, 317-328.

de Faria, S.M., de Lima, H.C., Franco, A.A., Mucci, E.S.F. and Sprent, J.I. (1987). Plant and Soil 99, 347-356.

de Faria, S.M., Lewis, G.P., Sprent, J.I. and Sutherland, J.M. 1989. New Phytol. 111, 607-619.

De Lajudie, P., Willems, A., Pot, B., Dewettinck, D., Maestrojuan, G., Neyra, M., Collins, M.D., Dreyfus, B., Kersters, K. and Gillis, M. (1994). Int. J. Syst. Bacteriol. 44, 715-733.

De Ley, J., Tijgat, R., De Smedt, J. and Michiels, M. (1973). J. Gen. Microbiol. 78, 241-252.

Demezas, D.H., Reardon, T.B. Strain, S.R., Watson, J.M. and Gibson, A.H. (1995). Mol. Ecol. 4, 209-220.

Demezas, D.H., Reardon, T.B. Watson, J.M. and Gibson, A.H. (1991). Appl. Environ. Microbiol. 57, 3489-3495.

Dooley, J.J., Harrison, S.P., Mytton, L.J., Dye, M., Cresswell, A. and Skot, L. (1993). Can. J. Microbiol. 39, 665-673.

Dowdle, S.F. and Bohlool, B.B. (1985). Appl. Environ. Microbiol. 50, 1171-1176.

Dreyfus, B., Garcia, J.L. and Gillis, M. (1988). Int. J. Syst. Bacteriol. 38, 89-98.

Dykhuizen, D.E. and Green, L. (1991). J. Bacteriol. 173, 7257-7268.

Eardly, B.D. (1994). p. 557-573, in, R. W. Weaver, J. S. Angle, and P. J. Bottomley (eds.), Methods of Soil Analysis, Part 2-Microbiological and Biochemical Properties. Soil Science Society of America, Inc., Madison, WI.

Eardly , B.D., Materon, L.A., Smith, N.H., Johnson, D.A., Rumbaugh, M.D. and Selander, R.K. (1990). Appl. Environ. Microbiol. 56, 187-194.

Eardly, B.D., Wang, F-S. and van Berkum, P. (1996). Plant Soil 186, 69-74.

Eardly, B.D., Wang, F-S., Whittam, T.S. and Selander, R.K. (1995). Appl. Environ. Microbiol. 58, 1809-1815.

Eardly, B.D., Young, J.P.W. and Selander, R.K. (1992). Appl. Environ. Microbiol. 58, 1809-1815.

Eckhardt, M.M., Baldwin, I.R. and Fred, E.B. (1931). J. Bacteriol. 21, 273-285.

Elkan, G.H. (1992). Can J. Microbiol. 38, 446-450.

Engvild, K.C., Jensen, E.S. and Skot, L. (1990). Plant Soil 128, 283-286.

Erdman, L.W. and Means, U.M. (1949). Soil. Sci. Soc. Proc. 14, 170-175.

Evguenieva-Hackenberg, E. and Selenska-Pobell, S. (1995). Lett. Appl. Microbiol. 21, 402-405.

Felsenstein, J. (1993). PHYLIP (phylogenetic inference package). Version 3.5c. Department of Genetics, University of Washington, Seattle.

Fourmy, D., Recht, M.I., Blanchard, S.C. and Puglisi, J.D. (1996). Science 274, 1367-1371.

Fox, G.E., Wisotzkey, J.D. and Jurtshuk Jr., P. (1992). Int J. Syst. Bacteriol. 42, 166-170.

Frank, B. (1889). Ber. Deut. Bot. Ges. 7, 332-346.

Fred, E.B., Baldwin, I.L. and McCoy, E. (1932). Root nodule bacteria and leguminous plants. University of Wisconsin Studies in Sciences, Number 5. University of Wisconsin Press, Madison.

Geniaux, E., Laguerre, G. and Amarger, N. (1993). Mol. Ecol. 2, 295-302.

Gillis, M. and De Ley, J. (1980). Int. J. Syst. Bacteriol. 30, 7-27.

Gordon, D.M., Wexler, M., Reardon, T.B. and Murphy, P.J. (1995). Soil Biol. Biochem. 27, 491-499.

Graham, P.H., Sadowsky, M.J., Keyser, H.H., Barnet, Y.M., Bradley, R.S., Cooper, J.E., De Ley, J., Jarvis, B.D.W., Roslycky, E.B., Strijdom, W.B. and Young, J.P.W. (1991). Int. J. Syst. Bacteriol. 41, 582-587.

Graham, P.H., Sadowsky, M.J., Tighe, S.W., Thompson, J.A., Date, R.A., Howieson, J.G. and Thomas, R. (1995). Can. J. Microbiol. 41, 1038-1042.

Grimont, P.A.D. (1988). Can J. Microbiol. 34, 541-546.

Gunn, C.R., Wiersema, J.H., Ritchie, C.A. and Kirkbride, J.H. (1992). U. S. Department of Agriculture, Technical Bulletin No. 1796.

Harrison, S.P., Jones, D.G. and Young, J.P.W. (1989). J. Gen Microbiol. 135, 1061-1069.

Hauke-Pacewiczowa, T. (1952). Acta Microbiol. Pol. 1, 37-39.

Hildebrand, E. M. (1940). J. Agric. Sci. 61, 685-698.

Hill, C.W., Grafstrom, R.H., Harnish, B.W. and Hillman, B.S. (1977). J. Mol. Biol. 116, 407-428.

Hollis, A.B., Kloos, W.E. and Elkan, G.E. (1981). J. Gen Microbiol. 123, 215-222.

Holmes, B. (1988). Acta Hort. 225, 47-52.

Holmes, B. and Roberts, P. (1980). J. Appl. Bacteriol. 50, 443-467.

Honeycutt, R.J., NcClelland, M. and Sobral, B.W.S. (1993). J. Bacteriol. 175, 6945-6952.

Hulton, C.S., Higgins, C.F. and Sharp, P.M. 1991. Mol. Microbiol. 5, 825-834.

Jarvis, B.D.W., Dick, A.G. and Greenwood, R.M. (1980). Int. J. Syst. Bacteriol. 30, 42-52.

Jarvis, B.D.W., Downer, H.L. and Young, J.P.W. (1992). Int. J. Syst. Bacteriol. 42, 93-96.

Jarvis, B.D.W., Gillis, M. and de Ley, J. (1986). Int. J. Syst. Bacteriol. 36, 129-138.

Jarvis, B.D.W., Pankhurst, C.E. and Patel, J.J. (1982). Int. J. Syst. Bacteriol. 32, 378-380.

Jarvis, B.D.W., van Berkum, P., Chen, W.X., Nour, S.M., Fernandez, M.P., Cleyet-Marel, J.C. and Gillis, M. (1997). J. Syst. Bacteriol. 47, 895-898.

Jensen, H.L. (1967). Arch. Mikrobiol. 59, 174-179.

Jensen, H.L. and Hansen, A. (1968). Acta Agric. Scan. 18, 135-142.

Jensen, M.A. and Straus, N. (1993). PCR Methods Appl. 3, 186-194.

Jensen, M.A., Webster, J.A. and Straus, N. (1993). Appl. Environ. Microbiol. 59, 945-952.

Johnson, L.I. (1984). p. 8-11, in Krieg, N.R. and Holt, J.G. (eds.) Bergey's Manual of Systematic Bacteriology, vol. 1, Williams & Wilkins, Baltimore.

Jordan, D.C. (1982). Int. J. Syst. Bacteriol. 32, 136-139

Jordan, D.C. (1984). p. 234-244, in Krieg, N.R. and Holt, J.G. (eds.) Bergey's Manual of Systematic Bacteriology, vol. 1, Williams & Wilkins, Baltimore.

Judd, A.K., Schneider, M., Sadowsky, M.J. and de Bruijn, F.J. (1993). Appl. Environ. Microbiol. 59, 1702-1708.

Kersters, K. and De Ley, J. (1984). p. 244-254, in Krieg, N.R. and Holt, J.G. (eds.) Bergey's Manual of Systematic Bacteriology, vol. 1, Williams & Wilkins, Baltimore.

Kersters, K., de Ley, J., Sneath, P.A.H. and Sackin, M. (1973). J. Gen. Microbiol. 78, 227-239.

Keyser, H.H., Bohlool, B.B., Hu, T.S. and Weber, D.F. (1982). Science 215, 1631-1632.

King, T.C., Sirdeskmukh, R. and Schlessinger, D. (1986). Microbiol. Rev. 50, 428-451.

Knösel, D. (1962). Zentralbl. Bakteriol. Parasitenkd. Infeektionskr. Hyg. II Abt. 116, 79-100.

Knösel, D.H. (1984). p. 254-256, in Krieg, N.R. and Holt, J.G. (eds.) Bergey's Manual of Systematic Bacteriology, vol. 1, Williams & Wilkins, Baltimore.

Kordes, E., Jock, S., Fritsch, J., Bosch, F. and Klug, G. (1994). J. Bacteriol. 176, 1121-1127.

Kumar, S., Tamura, K. and Nei, M. (1993). MEGA, Molecular Evolutionary Gentics Analysis, version 1.01. The Pennsylvania State University, University Park, PA 16802.

Kündig, C., Beck, C., Hennecke, H. and Göttfert, M. (1995). J. Bacteriol. 177, 5151-5154.

Kuykendall, L.D., Saxena, B., Devine, T.E. and Udell, S.E. (1992). Can. J. Microbiol 38, 501-505.

Laguerre, G., Allard, M-R., Revoy, F. and Amarger, N. (1994). Appl. Environ. Microbiol. 60, 56-63.

Laguerre, G., Fernandez, M.P., Edel, V., Normand, P. and Amarger, N. (1993a). Int. J. Syst. Bacteriol. 43, 761-767.

Laguerre, G., Geniaux, E., Mazurier, S.I., Rodriguez-Casartelli, R. and Amarger, N. (1993b). Can. J. Microbiol. 39, 412-419.

Lambert, B., Joos, H., Dierickx, S., Vantomme, R., Swings, J., Kersters, K. and van Montagu, M. (1990). Appl. Environ. Microbiol. 56, 1093-1102.

Lessie, T.G. (1965). J. Gen. Microbiol. 39, 311-320.

Leung, K., Strain, S.R., de Bruijn, F.J. and Bottomley, P.J. (1994). Appl. Environ. Microbiol. 60, 416-426.

Lin, J., Walsh, K.B., Johnson, D.A., Canvin, D.T., Shujin, W. and Layzell, D.B. (1987). Plant Soil 99, 441-446.

Lindström, K. (1989). Int J. Syst. Bacteriol. 39, 365-367.

Lindström, K., Jarvis, B.D.W., Lindström, P.E. and Patel, J.J. (1983). Can. J. Microbiol. 29, 781-789.

Lindström, K. and Lehtomäki, S. (1988). FEMS Microbiol. Lett. 50, 277-287.

Lindström, K., van Berkum, P., Gillis, M., Martinez, E., Novikova, N. and Jarvis, B. (1996). p. 807-810, in Tikhonovich, I.A., Provorov, N.A., Romanov, N.I. and Newton, W.E. (eds.), Nitrogen Fixation, Fundamentals and Applications, Kluwer Academic Publishers, Dordrecht.

Lipsanen, P., and Lindström, K. (1989). FEMS Microbiol. Lett. 58, 323-328.

Ludwig, W., Rossello-Mora, R., Aznar, R., Klugbauer, S., Springs, S., Reetz, K., Beimfohr, C., Brockmann, E., Kirchhof, G., Dorn, S., Bachleitner, M., Klugbauer, N., Springers, N., Lane, D., Nietupsky, R., Weizenegger, M. and Schleifer, K.H. (1995). Appl. Microbial. 18, 164-188.

Mackay, M.L. Zablen, B., Woese, C.R. and Doolittle, W.F. (1979). Arch. Microbiol. 123, 165-172.

Maidak, B.L., Larsen, N., McCaughey, M.J., Overbeek, R., Olsen, G.J., Fogel, K., Blandy, J. and Woese, C.R. (1994). Nucl. Acids Res. 22, 3485-3487.

Marrs, B. and Kaplan, S. (1970). J. Mol. Biol. 49, 297-317.

Martin, R., Humbert, O., Camara, M., Guenzi, E., Walker, J., Mitchell, T., Andrew, P., Prudhomme, M., Alloing, G., Hakenbeck, R., Morrison, D.A., Boulnois, G.J. and Claverys, J-P. (1992). Nucl. Acids Res. 20, 3479-3483.

Martinez-de Drets, G. and Arias, A. (1972). J. Bacteriol. 109, 467-470.

Martinez-Romero, E. (1994). Plant and Soil 161, 11-20.

Martinez-Romero, E. and Caballero-Mellado, J. (1996). Crit. Rev. Plant Sci. 15, 113-140.

Martinez-Romero, E., Segovia, L., Mercante, F.M., Franco, A.A., Graham, P. and Pardo, M.A. (1991). Int. J. Syst. Bacteriol. 41, 417-426.

Maynard Smith, J. (1995). P. 1-12, in, Baumberg, S., Young, J.P.W., Wellington, E.M.H. and Saunders, J.R. (eds.), Population Genetics of Bacteria, Cambridge University Press, Cambridge.

Maynard Smith, J., Smith, N.H., O'Rourke, M. and Spratt, B.G. (1993). Proc. Natl. Acad. Sci. USA 90, 4384-4388.

Moffett, M. L. and Colwell, R.R. (1968). J. Gen. Microbiol. 51, 245-266.

Murooka, Y., Xu, Y., Sanada, K., Araki, M., Morinaga, T. and Yokota, A. (1993). J. Ferment. Bioeng. 76, 38-44.

Nei, M. and Li, W.H. (1979). Proc. Natl. Acad. Sci. USA 76, 5269-5273.

Nick, G. and Lindström, K. (1994). Syst. Appl. Microbiol. 17, 265-273.

Neibel, H., Dorsch, M. and Stackebrandt, E. 1987. J. Gen Microbiol. 133, 2401-2409.

Norris, D.O. (1965). Plant Soil 22, 143-166.

Nour, S.M., Cleyet-Marel, J-C., Beck, D., Effosse, A. and Fernandez, M.P. (1994a). Can J. Microbiol. 40, 345-354.

Nour, S.M., Cleyet-Marel, J-C., Normand, P. and Fernandez, M.P. (1995). Int. J. Syst. Bacteriol. 45, 640-648.

Nour, S.M., Fernandez, M.P., Normand, P. and Cleyet-Marel, J-C. (1994b). Int. J. Syst. Bacteriol. 44, 511-522.

Olsen, G.J., Woese, C.R. and Overbeek, R. (1994). J. Bacteriol. 176, 1-6.

Ophell, K. and Kerr, A. (1990). Int. J. Syst. Bacteriol. 40, 236-241.

Oyaizu, H. and Sawada, H. (1994). Int. J. Syst. Bacteriol. 44, 374.

Pankhurst, C.E. (1977). Can. J. Microbiol. 23, 1026-1033.

Pankhurst, C.E. (1979). Microbios. 24, 19-28.

Pankhurst, C.E., Broughton, W.J. and Weineke, U. (1983). J. Gen Microbiol. 129, 2535-2543.

Pankhurst, C.E. and Jones, W.T. (1979) J. Exp. Bot. 30, 1095-1107.

Pankhurst, C.E., MacDonald, P.E. and Reeves, J.M. (1986). J. Gen. Microbiol. 132, 2321-2328.

Patel, J.J. (1976). Can. J. Microbiol. 22, 204-212.

Petes, T.D. and Hill, C.W. 1988. Annu. Rev. Genet. 22, 147-168.

Piñero, D., Martinez, E. and Selander, R.K. (1988). Appl. Environ. Microbiol. 54, 2825-2832.

Polhill, R.M., Raven, P.H. and Stirton, C.H. (1981). p.1-26, in Polhill, R.M. and Raven, P.H. (eds.), Advances in Legume Systematics, Royal Botanic Gardens, Kew, England.

Ponsonnet, C. and Nesme, X. (1994). Arch. Microbiol. 161, 300-309.

Proctor. M.H. (1963). N. Z. J. Bot. 1, 419-425.

Rainey, F.A. and Wiegel, J. (1996). Int. J. Syst. Bacteriol. 46, 607-610.

Rana, D. and Krishnan, H.B. (1995). FEMS Microbiol. Lett. 134, 19-25.

Riker, A.J., Banfield, W.M., Wright, W.H., Keitt, G.W. and Sagen, H.E. (1930). J. Agr. Res. 41, 507-540.

Rinaudo, G., Orenga, S., Fernandez, M.P., Meugnier, H. and Bardin, R. (1991). Int. J. Syst. Bacteriol. 41, 114-120.

Roberts, G.P., Leps, W.T., Silver, L.E. and Brill, W.J. (1980). Appl. Environ. Microbiol. 39, 414-422.

Roberts, M.S. and Cohan, F.M. (1993). Genetics 134, 396-401.

Rome, S., Brunel, B., Normand, P., Fernandez, M. and Cleyet-Marel, J-C. (1996a). Arch. Microbiol. 165, 285-288.

Rome, S., Fernandez, M., Brunel, B., Normand, P. and Cleyet-Marel, J-C. (1996b). Int. J. Syst. Bacteriol. 46, 972-980.

Ruiz-Argüeso, T., Cadahia, E. and Leyva, A. (1988). p. 109-119, in Beck, D.P. and Materon, L.A. (eds), Nitrogen Fixation by Legumes in Mediterranean Agriculture, ICARDA.

Sawada, H. and Ieki, H. (1992). Ann. Phytopath. Soc. Japan 58, 37-45.

Sawada, H., Ieki, H., Oyaizu, H. and Matsumoto, S. (1993). Int. J. Syst. Bacteriol. 43, 694-702.

Scholla, M.H. and Elkan, G.H. (1984). Int. J. Syst. Bacteriol. 34, 484-486.

Scholla, M.H., Moorefield, J.A. and Elkan, G. 1990. Syst. Appl. Microbiol. 13, 288-294.

Schroeter, J. (1886). p 1-814, in Cohn, F. (ed.) Kryptogamenflora von Schlesien, Band 3, Heft 3, Pilze J. U. Kern's Verlag, Breslau.

Schuch, W. and Loening, U.E. (1975). Biochem. J. 149, 17-22.

Segovia, L., Piñero, D., Palacios, R. and Martinez-Romero, E. (1991). Appl. Environ. Microbiol. 57, 426-433.

Segovia, L., Young, J.P.W. and Martinez-Romero, E. (1993). Int. J. Syst. Bacteriol. 43, 374-377.

Selander, R.K., Li, J., Boyd, E.F., Wang, F-S. and Nelson, K. (1994). p. 17-49, in Priest, F.G., Ramos-Cormenzana, A. and Tindal, B.J. (eds.), Bacterial Diversity and Systematics, Plenum, New York.

Selenska-Pobell, S. and Evguenieva-Hackenberg, E. (1995). J. Bacteriol. 177, 6993-6998.

Selenska-Pobell, S., Gigova, L. and Petrova, N. (1995). J. Appl. Bacteriol. 79, 425-431.

Skerman, V.B.D., McGowan, V. and Sneath, P.H.A. (1980). Int. J. Syst. Bacteriol. 30, 225-420.

Smith, E.F. and Townsend, C.O. (1907). Science 25, 671-673.

Sneath, P.H.A. (1993). Int. J. Syst. Bacteriol. 43, 626-629.

Souza, V., Guiarte, L., Avila, G., Cappello, R., Gallardo, C., Montoya, J. and Piñero, D. (1994). Appl. Environ. Microbiol. 60, 1260-1268.

Souza, V., Nguten, T.T., Hudson, R.R., Piñero, D. and Lenski, R.E. (1992). Proc. Natl. Acad. Sci. USA 89, 8389-8393.

Stackebrandt, E. and Goebel, R.M. (1994). Int. J. Syst. Bacteriol. 44, 846-849.

Stanley, J., Brown, G.G. and Verma, D.P. (1985). J. Bacteriol. 163, 148-154.

Starr, M.P. and Weiss, J.E. (1943). Phytopathology 33, 314-318.

Stern, M.J., Ames, G.F-L., Smith, N.H., Robinson, E.C. and Higgins, C.F. 1984. Cell 37, 1015-1026.

Strain, S.R., Leung, K., Whittam, T., de Bruijn, F. and Bottomley, P.J. (1994). Appl. Environ. Microbiol. 60, 2772-2778.

Strain, S.R., Whittam, T.S. and Bottomley, P.J. (1995). Mol. Ecol. 4, 105-114.

Sullivan, J., Eardly, B.D., van Berkum, P. and Ronson, C.W. (1996). Appl. Environ. Microbiol. 62, 2818-2825.

Sullivan, J., Patrick, H.N., Lowther, W.L., Scott, D.B. and Ronson, C.W. (1995). Proc. Natl. Acad. Sci. USA 92, 8985-8989.

Tan, Z., Xu, X., Wang, E., Gao, J., Martinez-Romero, E. and Chen, W. (1997). Int. J. Syst. Bacteriol. 47, 874-879.

Thompson, J.A., Bhromsiri, A., Shutsrirung, A. and Lillakan, S. (1991). Plant and Soil 135, 53-65.

van Berkum, P., Beyene, D. and Eardly, B.D. (1996). Int. J. Bacteriol. 46, 240-244.

van Berkum, P., Beyene, D., Bao, G., Campbell, A.T. and Eardly, B.D. 1997. Int. J. Syst. Bacteriol. (In press).

van Berkum, P., Navarro, R.B. and Vargas, A.A.T. (1994). Appl. Environ. Microbiol. 60, 554-561.

van Berkum, P., Tully, R.E. and Keister, D.L. (1995). Appl. Environ. Microbiol. 61, 623-629.

van Rossum, D., Schuurmans, F.P., Gillis, M., Muyotcha, A., van Verseveld, H.W., Stouthamer, A.H. and Boogerd, F.C. (1995). Appl. Environ. Microbiol. 61, 1599-1609.

Veraslovic, J., Koeth, T. and Lupski, J.R. (1991). Nucleic Acids Res. 24, 6823-6831.

Veraslovic, J., Schneider, M., de Bruijn, F.J. and Lupski, J.R. (1994). Methods. Mol. Cell. Biol. 5, 25-40.

Wedlock, D.N. and Jarvis, B.D.W. (1986). Int J. Syst. Bacteriol. 36, 550-558.

Willems, A. and Collins, M.D. (1993). Int. J. Syst. Bacteriol. 43, 305-313.

Wilson, J.K. (1944). Soil Sci. 58, 61-69.

Winkler, M.E. (1979). J. Bacteriol. 139, 842-849.

Wong, F.Y.K., Stackebrandt, E., Ladha, J.K., Fleischman, D.E., Date, A. and Fuerst, J.A. (1994). Appl. Environ. Microbiol. 60, 940-946.

Xu, L.M., Ge, C., Cui, Z., Li, J. and Fan, H. (1995). Int. J. Syst. Bacteriol. 45, 706-711.

Yanagi, M. and Yamasoto, K. (1993). FEMS Microbiol. Lett. 107, 115-120.

Young, J.P.W. (1985). J. Gen Microbiol. 131, 2399-2408.

Young, J.P.W. (1996). Plant and Soil 186, 45-52.

Young, J.P.W., Demetriou, L. and Apte, R.G. (1987). Appl. Environ. Microbiol. 53, 397-402.

Young, J.P.W., Downer, H.L. and Eardly, B.D. (1991). J. Bacteriol. 173, 2271-2277.

Young, J.P.W, and Haukka, K.E. 1996. New Phytol. 133, 87-94.

Young, J.P.W. and Wexler, M. (1988). J. Gen Microbiol. 134, 2731-2739.

Chapter 2

General Genetic Knowledge

Michael F. Hynes and Turlough M. Finan

I. Introduction: A Strange "Family"!

This chapter is about the "general genetic knowledge" of the family *Rhizobiaceae*. Since this "family" is not a true family in any phylogenetic sense, it is very difficult to generalise about the genetics of the group (see Chapter 1). The phylogeny of its members will be discussed elsewhere, but suffice it to say here that the "family" has traditionally been defined in terms of the ability of the bacteria belonging to it to induce the formation of certain types of growths (*e.g.* nodules, tumours) on plants (Jordan, 1982). Since it has usually been because of this ability to interact with plants that individual members of the "family" have been studied, much information available about the genetics of these organisms is biased towards study of genes involved in symbiotic or pathogenic interactions. It is our intent in this chapter to review

the basic genetics of some of the better-studied members of the *Rhizobiaceae* with an emphasis on tools available for genetic study. Where useful information regarding more "exotic" bacteria is available it will be introduced. The discussion will centre on *Rhizobium leguminosarum*, *Rhizobium* (syn. *Sinorhizobium*) *meliloti*, *Agrobacterium* spp. and *Bradyrhizobium japonicum*.

Genetic studies of *Rhizobium*, at least, are not new. As long ago as 1954, studies on transformation were carried out (Balassa, 1954; 1957). Strong evidence for conjugation systems was presented in the 1960s (Higashi, 1967) and mutants of various types have been available for quite some time. Work on the genetics of *Agrobacterium* and the rhizobia accelerated in the 1970s and early 80s and resulted in the development of a number of techniques (gene replacement, transposon delivery systems *etc.*) which have been of immense benefit to the study of bacteria from many diverse genera. Studies of these organisms also began revealing interesting details about genome arrangement in bacteria, and were among the first to suggest that the *Escherichia coli* paradigm (one circular chromosome) was not universal among the prokaryotes. Thus, although much of the research into genetics of *Agrobacterium* and the rhizobia was driven by interest in plant transformation and exploitation of nitrogen fixation, it has resulted in many important discoveries about the nature of bacterial genomes, and in the development of tools and techniques with almost universal applications.

II. Genetic Maps and Genome Arrangement

Although there was early evidence for conjugation systems in the rhizobia (Higashi, 1967; Heumann, 1968) some scepticism was expressed about these results, particularly about the authenticity of the "*R. lupini*" strains used by Heumann and colleagues (Beringer, 1980). The first systematic approaches to construction of chromosomal maps of *Rhizobium* and *Agrobacterium* took advantage of the ability of broad host-range plasmids such as RP4 and its derivatives to mobilise chromosomes (Beringer and Hopwood, 1976, Beringer *et al.*, 1978 a and b, Hooykaas *et al.*, 1982, Kondorosi *et al.*, 1977, 1980, Meade *et al.*, 1977, Megias *et al.*, 1982, Zurkowski and Lorkiewicz, 1978). The majority of markers used in such studies were auxotrophic mutations (usually chemically induced) and spontaneous mutations to antibiotic resistance. Fine mapping with transducing phage has also been carried out (Casadesus and Olivares, 1979).

Some of the important information from the studies cited above included the fact that they showed circular linkage maps for chromosomes of biovars of *R. leguminosarum*, *S. meliloti*, and *A. tumefaciens*. The broad host range plasmid RP4 (or RK2) , and its more efficient derivative R68.45 (Haas and Holloway, 1976) were used for chromosome mobilisation and gave at least a satisfactory frequency of gene transfer. And finally, comparison of maps generated in this way demonstrated a similar arrangement ("colinearity") of analogous genes between all three species (Kondorosi *et al.*, 1980, Hooykaas *et al.*, 1982). However, some of the impetus for doing further work in this area was lost when it was discovered that extrachromosomal genes

largely determined the interaction of the bacteria with plant hosts, and by the time interest in genome mapping in the rhizobia and *Agrobacterium* was revived, better, more rapid techniques such as generation of physical maps using pulsed field electrophoresis, and creating artificial "Hfr" strains using Tn5-Mob (Simon, 1984) or analogous transposons were available (*e.g.* Tn5-*oriT*, Yakobson and Guiney, 1984). However, genetic mapping by conjugation using a set of Tn5-Mob insertions (Klein *et al.*, 1992, Glazebrook *et al.*, 1992, Osteras *et al.*, 1989) and by transduction using linked Tn5 insertions (Charles and Finan, 1990; 1991) have been recently demonstrated to be viable means of rapidly establishing chromosomal or plasmid maps, and of determining the location of new mutations.

Pulsed field gel electrophoresis allows the generation of physical maps of bacterial genomes based on restriction digests using enzymes that cut relatively rarely in the genome. These physical maps are then refined using enzymes which cut more frequently, and ideally, the physical map is correlated with pre-existing genetic maps by hybridisation to probes representing cloned genes of known location, or Tn5 inserts of known location. The two members of the *Rhizobiaceae* for which the most work has been done so far are *S. meliloti* and *B. japonicum*. For the former, it was shown, after suitable enzymes for genomic digestion had been identified (Sobral *et al.*, 1991a), that the three principal components of the genome could be separated by pulsed field electrophoresis (Sobral *et al.*, 1991b), and physical maps of all three components (the chromosome and two megaplasmids) could be established (Honeycutt *et al.*, 1993). The physical map of pRmeSU47b (also called pExo) correlated well with the genetic map of this replicon (Charles and Finan, 1991) and the estimated size of the two megaplasmids agreed well with previous reports using transduction linkage (Charles and Finan, 1991) and electron microscopy (Burkhardt *et al.*, 1987). As more markers are added to the physical map, it should become an increasingly useful tool, especially if it is linked to an ordered cosmid library.

B. japonicum has also been mapped physically, and had its genome size determined by summation of the sizes of restriction fragments analysed by pulsed field electrophoresis (Sobral *et al.*, 1991a; Kündig *et al.*, 1993). The physical map generated by Kündig *et al.*, was used to generate a genetic map by insertion of cassettes carrying rare-cutting restriction enzymes into several of 79 genes of known function, and, in many cases, sequence. These constructs were recombined into the genome, thus creating unique new recognition sites for these enzymes. The resulting map shows the position of many of the genes involved in symbiotic interactions, as well as other genes such as *groEL* and cytochrome and respiratory chain genes. There is however a distinct lack of auxotrophic and other markers on the map, due perhaps to the lack of work in this area in *Bradyrhizobium* (although some Tn5 auxotrophs have been produced (Hom *et al.*, 1984)) and the absence of a previously existing genetic map created by conventional techniques. These mapping studies confirm the existence of one single circular chromosome as the only major component of the genome of *B. japonicum*.

Pulsed field analysis of the DNA of *Agrobacterium tumefaciens* strain C58 has revealed an interesting new twist; whereas previous mapping studies had placed all markers on a single circular linkage map of this strain (Hooykaas *et al.*, 1982), Allardet-Servent *et al.*, (1993) demonstrated the presence of a large linear component in the DNA of this strain. The presence of ribosomal RNA genes on this linear 2100 kb replicon, as well as genes for phosphomannose isomerase and proline metabolism, suggests that this replicon may be a second (linear) chromosome, in addition to the 3000 kb chromosome to which other genes have been mapped. Presumably none of the markers with which Hooykaas *et al.*, (1982) worked mapped to this linear replicon, otherwise the construction of a circular map would have been impossible. Complete physical maps of *Agrobacterium* are not yet available. It is also not known whether this arrangement is unique to strain C58 or is the norm in *Agrobacterium*. Taking the above observations together with some preliminary *Ceu*-I mapping of several *R. leguminosarum* strains which we have carried out (C. Yost, M. Hynes, unpublished observations), the picture which we get of the organisation and size of the genomes of the better studied members of the *Rhizobiaceae* is

summarised in Table 1. It can be noted here that *B. japonicum* has the largest genome of the family, and that it is among the largest prokaryote genomes studied so far, along with some of the myxococci (Kündig *et al.*, 1993). All members of the family have genomes larger than that of *E. coli*, and compartmentalisation of the genome is a common feature in all but *Bradyrhizobium*. It is tempting to speculate that the linear component in *A. tumefaciens* C58 corresponds to the coding capacity that is represented by large plasmids or megaplasmids in *S. meliloti* and *R. leguminosarum*. There is no evidence yet for this, other than the close resemblance of the genetic maps of the circular chromosomes of all three organisms (Hooykaas *et al.*, 1982). The important role of plasmids in these organisms will be discussed at greater length below, but it should be mentioned that complete restriction maps of several Ti and Ri plasmids (Holsters *et al.*, 1980, Depicker *et al.*, 1980, Pomponi *et al.*, 1983, Jouanin *et al.*, 1986) as well as several nodulation plasmids in *Rhizobium* (Prakash *et al.*, 1981, 1982a,b; Girard *et al.*, 1991, Perret *et al.*, 1991) have been obtained and have been useful in a number of studies.

II.A. GENE STRUCTURE

The general features of rhizobial genes are similar to those observed for other bacteria. Gene regions are generally identified via hybridization or complementation experiments. Following DNA sequencing, open reading frames (ORFs) are routinely identified employing methods based on codon usage bias and high % G+C at the third nucleotide position within codons (Gribskov *et al.*, 1984). While such methods are very powerful for ORF detection, Ramseier and Göttfert (1991) have shown that *nod* and NifA regulated genes from *B. japonicum* have a less biased codon usage and

Species	Chromosome	Plasmids	Total Genome Size
S. meliloti 1021	3540 kb circular	1340 kb, 1700 kb	6600 kb
B. japonicum	8700 kb circular		8700 kb
A. tumefaciens C58	3000 kb circular 2100 kb linear	450 kb 200 kb	5750 kb
R. leguminosarum	3200 kb circular	at least four (total of ca. 2400 kb)	5600 kb

Table 1. Genome compostion of members of the *Rhizobiaceae*. Additional plasmids are present in many strains

lower G + C content than other genes from this organism. *B. japonicum* NifA regulated genes are thus difficult to detect when employing conventional codon frequency tables.

In this section we wished to review briefly our knowledge of the promoters of genes we believe to be transcribed by RNA polymerase holoenzyme incorporating the major vegetative sigma factor, SigA. Despite the fact that over four hundred genes have been sequenced from various members of the *Rhizobiaceae*, the transcriptional start sites of only a few genes have been determined. Rhizobial genes are generally believed to be poorly transcribed in *E. coli* (Bae and Stauffer, 1991; Fisher *et al.*, 1987; Luka *et al.*, 1996). However, we note, that this it is not always the case (*e.g. recA*, Better and Helinski, 1983). Using an *in vitro* system, Fisher *et al.*, (1987) showed that an *S. meliloti* RNA polymerase preparation which recognized and transcribed accurately from the constitutive *nodD* promoter also initiated and terminated accurately at the *E. coli trp* promoter-leader region. On the other hand *E. coli* RNA polymerase failed to initiate transcription from the *nodD* promoter. The deduced animo acid sequence of the major vegetative sigma factor SigA (encoded by *rpoD*) from *A. tumefaciens*, *S. meliloti*, and *R. etli* have been recently reported (Luka *et al.*, 1996, Rushing and Long, 1995, Segal and Ron, 1993). Analysis of these sequences revealed that while the proteins are very similar to each other, they differ from the *E. coli* SigA protein at four subregions within the first of the four recognized domains within SigA. Luka *et al.*, (1996) suggest that the observed differences in domain 1 may account for the altered promoter recognition and these authors also point out interesting similarities between the *Rhizobiaceae* SigA sequences and that from *Caulobacter crescentus*. *C. crescentus* genes are poorly transcribed in *E. coli*, and a consensus promoter sequence derived from a compilation of twelve sequences was recently published (Malakoot and Ely, 1995). The -35 region contained 5'-TTGACGS-3' while the -10 region contained 5'-GCTANAWC-3 (Malakoot *et al.*, 1995).

The sequence requirements of the *S. meliloti trpE(G)* promoter have been examined by Bae and Stauffer (1991). Mutations which increased the similarity of the -10 region to that of the *E. coli* consensus sequence, rendered the *trpE(G)* promoter active in *E. coli*, and most dramatically a C → T transition at the -9 position increased transcription in *E. coli* over one thousand fold. The strength of this promoter in *S. meliloti* was affected by single nucleotide changes

within the putative -10, -35 regions, but not in the -10, -35 spacer region. Comparison of the *trpE(G)* (Bae *et al.*, 1989) *nodD* (Fisher *et al.*, 1987), *hemA-P1* and *hemA-P2* (Leong *et al.*, 1985) allowed the authors to identify a highly conserved trimer near position -35 ('5-TTG-3') but no clearly conserved sequence was found in the -10 region. Subsequently, the transcriptional start sites of the *S. meliloti ntrA* and *pckA* genes have been determined (Albright *et al.*, 1989; Østeras *et al.*, 1995) and both these genes contain a '5-TTG-3' trimer in the -35 region.

The transcriptional start site of *ropA* and two of its homologs from *R. leguminosarum* biovars have been determined, but comparison of the promoter regions of these genes did not allow resolution of a consensus sequence (de Maagd *et al.*, 1992, Roest *et al.*, 1995). Segal and Ron, (1995) aligned the promoter regions of several *groESL* and *dnaK* operons from α-purple proteobacteria (including *A. tumefaciens*, *S. meliloti* and *B. japonicum*) and tentatively identified a promoter sequence which may constitute a heat shock promoter consensus sequence for these bacteria (-35 '5-CTTG-3').

With reference to rRNA (rrn) gene regions within the *Rhizobiaceae*, Kündig *et al.*, (1995) have shown that in *Bradyrhizobium japonicum*, this region has an organization which is frequently found in bacteria, '5-rrs (16S rRNA)-*ileT* (tRNA^{ile})-*alaT* (tRNA^{Ala})-rrl (23S rRNA)-rrf (5S rRNA)-3'. These authors also presented DNA hybridization data, which suggested that in general, slow growing rhizobia have a single *rrn* copy whereas faster growing rhizobia appear to have three *rrn* copies.

II.B. PLASMIDS

As can be noted above, many of the members of this family are likely to contain large plasmids as an important component of their genomes. It became apparent quite early in the molecular characterisation of the "fast growing" branch of the family (*Rhizobium*, *Sinorhizobium*, *Agrobacterium* and related species) that large plasmids played an important role in the interaction of the bacteria with plants. Most of the genes involved in tumour production in *Agrobacterium tumefaciens* and *A. rhizogenes* were demonstrated to be plasmid encoded (Van Larebeke *et al.*, 1974, White *et al.*, 1980a,b; Holsters *et al.*, 1980, Hooykaas *et al.*, 1977, Costantino *et al.*, 1980), as are most of the genes involved in nodulation and nitrogen fixation in *R. leguminosarum* (all three biovars), *R. etli*,

S. meliloti, and, in fact all "fast growing" rhizobia examined in detail so far with the exception of *R. loti*. Interestingly, though, in *R. loti* (syn *Mesorhizobium loti*), there is strong evidence that although the *nod* and *nif* genes do not appear to be on a plasmid *per se*, they are located on a mobile piece of DNA which can be transferred by conjugation, and which may well have a plasmid-like intermediary phase (Sullivan *et al.*, 1995).

The techniques used to establish the location of genes involved in symbiosis ranged from hybridisation using *nif* gene probes (*e.g.* Hombrecher *et al.*, 1981, Banfalvi *et al.*, 1981, Rosenberg *et al.*, 1981), to plasmid transfer to other (non-nodulating) species (Hooykaas *et al.*, 1981, Hirsch, 1979, Hirsch *et al.*, 1980, Zurkowski and Lorkiewicz, 1979) to plasmid curing (Zurkowski, 1982). However, there are many other plasmids which do not affect plant interactions (Merlo and Nester, 1977, Hynes *et al.*, 1989, Hynes and McGregor, 1990). It is not the intent of this chapter to review the genetics of nodulation, nitrogen fixation, or tumour formation as these will be dealt with in other chapters. Thus our discussion of plasmids will focus on methods of study, evolutionary considerations of large plasmids, and other roles that these plasmids may play in the life of their bacterial hosts.

Firstly, it must be considered that these plasmids (in both rhizobia and *Agrobacterium*) were originally referred to as "large" because of their significantly greater size than plasmids such as the F plasmid. Original estimates of their sizes turned out to be much lower than the actual size and we now know that plasmids as large as 1700 kb can be found in some members of the family (see above for *S. meliloti*). The term "megaplasmids" was coined to refer to some of these plasmids (Rosenberg *et al.*, 1982) and has been used (somewhat loosely) to refer to large plasmids ever since. We suggest that this term be restricted to plasmids of approximately 1000 kb (i.e. 1 megabase) in size and not be used to refer to smaller (*e.g.* 300 kb) plasmids. There has been some speculation that these megaplasmids should be regarded as "mini-chromosomes", but we feel that they have a number of important characteristics, which identify them as plasmids, including the following:

1) No evidence has yet been presented that these replicons carry genes <u>absolutely</u> essential for the survival of the host cell.

2) All megaplasmids we have studied can be transferred to other genera, where they will replicate stably with no apparent ill effect on the new host -- this is not to be expected of a chromosome.

3) Strains cured of these plasmids, or carrying deletions, which encompass well over 50 % of a plasmid, can be generated.

Some of these results and the techniques used to generate them, will be discussed in more detail below.

II.C. PLASMID ISOLATION

Large plasmids, because they are closer to the size of the chromosome than smaller plasmids such as cloning vectors, are much more difficult to isolate and detect by physical means. For members of the *Rhizobiaceae* various different techniques have been used to visualise and isolate plasmids. Many of these are based on using alkaline conditions to denature and eliminate chromosomal DNA (Ledeboer *et al.*, 1976, Casse *et al.*, 1979, Kado and Liu, 1981, Hirsch *et al.*, 1980). All of these techniques work, and when used on a preparative scale, can result in isolation of large amounts of plasmid DNA. However, with plasmids larger than 200-300 kb, they tend to be unreliable, and it is quite common for larger plasmids (400 kb or greater in size) to go completely undetected using these systems. A technique developed by Schwinghamer (1980) seems to give gentler lysis of cells and allow preparation of intact megaplasmids when used judiciously (Burkhardt *et al.*, 1987). However, the DNA yields are very low, and the procedure is not suited for screening large numbers of strains for the presence of plasmids.

Perhaps the most convenient technique for examining extremely large plasmids is that developed by Eckhardt (1978). This relies on lysis of cells in the wells of an agarose gel, and since there is no physical stress (*e.g.* shearing) on the lysate after lysis, extremely large replicons are released intact, in covalently closed circular form, into the well and will migrate into the gel. The Eckhardt technique was first modified for use in *S. meliloti* (Rosenberg *et al.*, 1981, Banfalvi *et al.*, 1981, Rosenberg *et al.*, 1982), and in these studies was still used as a vertical gel technique. A more convenient modification involving use of horizontal agarose gels was developed by Reinhard Simon (Universität Bielefeld), and is the basis of several published descriptions (Hynes *et al.*, 1985, Hynes *et al.*, 1989, Hynes and McGregor 1990). A highly detailed protocol is available from M. Hynes on request. In our hands, this technique reliably allows visualisation of replicons ranging in size from a few kb to over 3000 kb (the chromosomes of *S. meliloti* and *A. tumefaciens* - the latter presumably the circular one, as linear DNA of that size will not migrate into a normal agarose gel). Although intact DNA cannot be recovered from these gels, it is possible to use the modified Eckhardt technique in a preparative fashion, either by using low melting agarose, or one of a number of other techniques for isolating DNA from gels (*e.g.* GeneClean, Prep-a-Gene). Sufficient DNA yields for labelling one plasmid by nick translation or use of random primers, or for making a DNA library of one plasmid can easily be obtained. It is also possible to cut the bands out of an "Eckhardt" gel and subject the DNA in the band to enzymatic digestion while it is still in the agarose, and then run this DNA on a pulsed field gel (K. Sanderson and M. Hynes, unpublished observations). This procedure could greatly facilitate pulsed field mapping of individual plasmids.

II.D. CONJUGATION AND INCOMPATIBILITY GROUPS

Although it is often assumed that most large plasmids are capable of self-transfer by conjugation, this does not necessarily appear to be the case for the majority of *Rhizobium* and *Agrobacterium* plasmids. There have been reports of conjugative plasmids, such as pRL1JI, pRL3JI, and pRL4JI (Hirsch 1979) and pRleVF39a and pRleVF39b (Hynes *et al.*, 1988) in *R. leguminosarum*. However as there are usually from four to eight plasmids in these strains, and the majority are not transferred by conjugation (*e.g.* Hynes *et al.*, 1988), it must be assumed that either the plasmids are not conjugative, or that they require special conditions to induce transfer. This latter scenario is quite possible as *Agrobacterium* Ti plasmid transfer is inducible by the presence of opines (reviewed elsewhere in this volume). There may thus be special signal molecules, which induce transfer by conjugation of several plasmids in *Rhizobium* and *Agrobacterium*. We have tested plant root and seed exudates from peas for their ability to stimulate the transfer of Tn5-tagged, apparently non-conjugative plasmids in three different *R. leguminosarum* strains, but have yet to see any transfer at a detectable frequency (Hynes, unpublished). However, these experiments should perhaps be repeated in situ on a plant root system. There is some evidence that the megaplasmids of *S. meliloti* are conjugative plasmids, albeit at a very

low frequency (Charles, Ph.D. thesis McMaster U., Banfalvi *et al.*, 1985). There also appears to be a link between rare plasmids in *R. leguminosarum* which are self-transmissible at a high frequency, and susceptibility to "small" bacteriocins as well as suppression of synthesis of "small" bacteriocins (Wijfelman *et al.*, 1983). These "small" bacteriocins have subsequently been shown to be acyl homoserine lactone molecules of the *luxI* class of inducers (Gray *et al.*, 1996, Schripsema *et al.*, 1996) Just why they are growth inhibitory to certain strains carrying conjugative plasmids remains unclear.

The host range of all *Rhizobium* plasmids seems to include *Agrobacterium*, as all plasmids which have been tested can also replicate in this alternate host, and in fact plasmid-free strains of *Agrobacterium* (Hynes *et al.*, 1985, Rosenberg and Huguet, 1984) are useful hosts for the study and manipulation of *Rhizobium* plasmids. The evidence also suggests that *Agrobacterium* plasmids will replicate in *Rhizobium* (Hooykaas *et al.*, 1977, O'Connell *et al.*, 1987) and both an *Agrobacterium* nopaline Ti plasmid derivative and a Sym plasmid from *R. leguminosarum* have been shown to replicate and induce crown galls and root nodules, respectively, in *Phyllobacterium* (van Veen *et al.*, 1988). In fact it seems likely that *Rhizobium* and *Agrobacterium* and perhaps other soil bacteria share a common pool of plasmids.

Incompatibility between *Rhizobium* and *Agrobacterium* plasmids has been reported (Rosenberg and Huguet, 1984, Hynes *et al.*, 1985, O'Connell *et al.*, 1987), and origins of replication of several of these plasmids are related (Turner and Young, 1995, Turner *et al.*, 1996). In general, studies of plasmids have been hampered by the fact that most strains contain several plasmids, and that introduction of incompatible plasmids frequently seems to result in plasmid rearrangements rather than elimination of a given plasmid.

Given the difficulties of generating data on incompatibility groups by experiments involving plasmid transfer, it seems that a more logical approach might be to characterise incompatibility groups by hybridisations using cloned origins of replication. Several origins have been cloned, including one from a *S. meliloti* megaplasmid (Margolin and Long, 1993), one from a *S. meliloti* non-symbiotic plasmid (Mercado-Blanco and Olivares, 1994a; 1994b), one from *Rhizobium* "hedysarum" (Mozo *et al.*, 1990) and several from *R. leguminosarum* (Turner and Young, 1995; Turner *et al.*, 1996). Several origins have been isolated from plasmids of *A. tumefaciens* and *A rhizogenes*. (Gallie *et al.*, 1984; 1985a; Gallie and Kado, 1988; Vilaine and Casse-Delbart, 1987, Tabata *et al.*, 1989, Gerard *et al.*, 1992). Turner *et al.*, (1996) have shown that a certain class of *repC* gene is conserved between many plasmids in *Rhizobium, Sinorhizobium* and *Agrobacterium*. Close homology at this locus seems to indicate incompatibility, whereas more divergent sequences at these loci are found between compatible plasmids. Ideally, origins of replication of a large number of different plasmids (*e.g.* all the plasmids in one strain of *R. leguminosarum* such as JB300 or VF39) in the *Rhizobiaceae* should be cloned and used as probes to investigate incompatibility on a wider scale. It is worth noting that these large plasmids may have more than one potential origin of replication; the replication origin cloned from the pExo megaplasmid of *S. meliloti* is clearly not the only replication origin on this megaplasmid (Margolin and Long, 1993).

II.E. TOOLS FOR GENETIC MANIPULATION OF PLASMIDS

Since the properties that plasmids in the *Rhizobiaceae* seem to share are that they are large, stable, and in general not self-transmissible, it has not always been easy to determine their functions. In fact the majority of plasmids found in various strains must still be regarded as cryptic.

Over the last decade several important tools for the study of these plasmids have been developed. One of the first and most useful of the these was the transposon Tn5-Mob (Simon 1984), which has the origin of transfer of broad host range plasmid RP4 and some adjacent DNA (collectively known as the Mob site) cloned into a transposon Tn5 derivative. Any replicon in which Tn5-Mob is located can be transferred to other strains by using RP4 as a helper plasmid. Chromosomal insertions of Tn5-Mob effectively make artificial "Hfr" strains (Simon, 1984) whereas plasmid insertions will make plasmids transferable by conjugation. It is frequently possible to identify some of the functions of a plasmid by transferring it to other strains (Hynes *et al.*, 1986, Brewin *et al.*, 1980, Hirsch *et al.*, 1980, Hooykaas *et al.*, 1981). However, there are some limitations to this method. If similar plasmids exist in all strains of a species, it may not be possible to establish their functions by transfer from one strain to another. Similarly, if a host of another genus is used (*e.g.* a plasmid-free *Agrobacterium* strain), the

phenotypes encoded by certain plasmids may not be revealed because they are in fact chromosomally encoded in the new host.

One way of establishing the function of a plasmid relatively quickly and effortlessly is to cure the plasmid, or eliminate it from a cell. Although there have been several reports of curing of nodulation plasmids from rhizobia (Zurkowski, 1982) and of tumour inducing plasmids from *Agrobacterium* (*e.g.* Lin and Kado, 1977) using conventional methods such as ethidium bromide or heat treatment, in general we have found that plasmids in the *Rhizobiaceae* are extremely stable and difficult to eliminate. For this reason Tn5-Mob derivatives which carry a conditionally lethal gene, the *sacB* gene from *B. subtilis*, were constructed (Hynes *et al.*, 1989). These transposons can be used to tag a plasmid, then select for its loss by growth on sucrose (Hynes *et al.*, 1989, Hynes and McGregor 1990, Baldani *et al.*, 1992). Derivatives specifying resistance to neomycin, tetracycline and gentamicin are available (Hynes *et al.*, 1989; Quandt and Hynes, manuscript submitted) and have been widely used to cure plasmids in *R. leguminosarum, S. meliloti, R. huakuii, R. galegae* and *Agrobacterium*. Similar constructs have been designed and used by Brom *et al.*, (1992) to cure plasmids from *R. etli*. Although some plasmids are remarkably resistant to curing using this technique, it has usually been possible to obtain derivatives missing (or carrying extremely large deletions in) each plasmid. In fact it has been possible to generate strains missing several plasmids, and one *R. leguminosarum* bv. *trifolii* strain with all of its plasmids cured has been obtained by repeated application of this technique (Moënne-Loccoz *et al.*, 1995). Some of the interesting conclusions that can be drawn from these curing studies include the fact that, in *R. leguminosarum* and *R. etli* at least, several plasmids are required for a functional symbiosis with the appropriate host plant, and that the larger plasmids are required for optimal growth. In fact curing or deleting these plasmids severely debilitates the strain (Hynes and McGregor, 1990, Brom *et al.*, 1992, Moënne-Loccoz and Weaver, 1995a,b). Experiments in which multiple plasmids are cured, then reintroduced as transposon tagged derivatives into the cured strains require multiple antibiotic resistance markers, so we have constructed a series of Tn5-Mob derivatives carrying a wide variety of markers, as well as complementary RP4 derivatives which carry a similar selection of markers such as resistance to mercury, chloramphenicol, gentamicin, and spectinomycin (Quandt and Hynes, manuscript submitted).

For extremely large plasmids such as the megaplasmids of *S. meliloti*, we have found that "one step" curing using the *sacB* system does not work, but by generation of deletions, re-application of the transposon tagging procedure, and re-selection on sucrose, we have been able to produce extremely large deletions (Hynes *et al.*, 1989) and obtain apparently cured derivatives (Hynes and Yost, unpublished).

Another approach to characterising large replicons was that taken by Charles and Finan (1990, 1991), who constructed a map of the 1700 kb pRmeSU47b megaplasmid which consisted of Tn5-derivative insertions linked by transduction. Large defined deletions were then generated via homologous recombination between the IS50 elements of two insertions, which flanked various 160-600 kb regions of the megaplasmid. The detection of these deletions was facilitated through the use of Tn5-derivatives, which constitutively expressed the *lacZ* gene and thus allowed a blue/white screen. While some such deletions were obtained in strains in which the two insertions were in cis, the frequency of IS50-directed recombination was enhanced by transducing the transposon insertion at one site into a recipient strain carrying an insertion at a distant site.

II.F. PLASMID ENCODED PHENOTYPES

As the characterisation of plasmids in *Agrobacterium* and *Rhizobium* proceeds, using some of the techniques described above, as well as some purely serendipitous discoveries that certain genes are plasmid-located, a picture of a role of plasmids in these bacteria which transcends the traditional view that plasmids are "accessory" genetic elements is emerging. While the plasmids do not appear to be absolutely essential to growth of the bacteria in the laboratory, they do contribute substantially to the ability of the bacterium to survive in its natural habitat, to compete with other bacteria, and to interact with plants (Moenne Loccoz and Weaver 1995a,b, 1996). Aside from the genes directly involved in plant interactions (*nod, nif* and *fix* in *Rhizobium*, *vir* and T-DNA in *Agrobacterium*), a large number of other plasmid encoded phenotypes have been elucidated in the *Rhizobiaceae*. These include a large number of catabolic genes (Table 2), dicarboxylate transport genes (Finan *et al.*, 1988,

Watson *et al.*, 1988) exopolysaccaride genes (Hynes *et al.*, 1986, Finan *et al.*, 1986), thiamine synthesis genes (Finan *et al.*, 1986), LPS genes (Hynes and McGregor, 1990; Brom *et al.*, 1992, Baldani *et al.*, 1992), phosphate transport and phosphonate utilization (Bardin *et al.*, 1996), bacteriocin production genes (Hirsch *et al.*, 1980, Wijffelman *et al.*, 1983) melanin production genes (Lamb *et al.*, 1982, Hynes *et al.*, 1988, Mercado-Blanco *et al.*, 1993) to name but a few. However, given that plasmids can make up as much as 40 % of the genome of these bacteria, and that there can be ten or more large plasmids in some strains, it is obvious that many additional important plasmid encoded loci remain to be found. While the above discussion has focused on *Agrobacterium* and the fast growing rhizobia, it should be pointed out that plasmids are relatively common in *B. japonicum* (*e.g.* Gross *et al.*, 1979, Luyindula *et al.*, 1975) but that they do not appear to play any significant role in interactions with plants and that their function remains unclear. *Rhizobium* plasmids have been reviewed recently by Mercado-Blanco and Toro (1996) and Garcia de Los Santos *et al.*, (1996).

An exciting new development in the study of plasmids in the *Rhizobiaceae* has been the release of the complete sequence of the nodulation plasmid from *Rhizobium* broad host-range strain NGR234 (Freiberg *et al.*, 1997). This plasmid is over 536 kb in size, and the sequence reveals a number of interesting features. These include the fact that about 18% of the plasmid is made up of IS-like elements, putative integrons, and other mosaic sequences, suggesting, as does the variability in GC content, that the evolutionary history of this replicon is that of a mosaic. As would be expected, many of the *nod, nif,* and *fix* genes which have been indentified in other rhizobia are located on this plasmid, an important exception being the *fixLJ, fixK, fixNOQP,* and *fixGHIS* genes, which are apparently chromosomally located in this strain. There are a number of open reading frames whose homologies suggest involvement in protein secretion and polysaccharide synthesis, which may suggest an involvement of these two processes in establishment of symbiosis. However, it is also possible that these ORFs may have nothing to do with symbiotic nitrogen fixation, as, for example, protein secretion systems could be involved in bacteriocin export, or export of extracellular enzymes which have significance in the rhizosphere or soil, but not necessarily during nodule development. The sequence of the NRG234 pSym reveals that the origin of replication of the

Carbon sources	Plasmids	Strains	Reference
Opines (various)	various Ti and Ri plasmids	*Agrobacterium tumefaciens, A. rhizogenes*	Chapter 9
Ornithine	pRmGR4b pRleVF39f	*S. meliloti R. leguminosarum*	Soto *et al.,* (1994) Hynes unpublished
Rhizopines	pRme220-3a	*S. meliloti*	Saint *et al.,* (1993)
Trigonelline	pRme2011a	*S. meliloti*	Boivin *et al.,* (1990)
Stachydrine	pRme2011a	*S. meliloti*	Goldman *et al.,* (1994)
Callystegine	pRme41a	*S. meliloti*	Tepfer *et al.,* (1988)
Rhamnose	pRleW14-2c pRleVF39e	*R. leguminosarum*	Baldani *et al.,* (1992) Oresnik *et al.,* (submitted)*
Sorbitol	pRleW14-2c pRleVF39e	*R. leguminosarum*	Baldani *et al.,* (1992) Oresnik *et al.,* (submitted)*
Adonitol	pRleW14-2b pRleW8-7c pRleVF39d	*R. leguminosarum*	Baldani *et al.,* (1992) Oresnik *et al.,* (submitted)*
Glycerol	pRleW14-2a pRleW8-7b pRleVF39c	*R. leguminosarum*	Baldani *et al.,* (1992) Oresnik *et al.,* (submitted)*
Raffinose	pRmeSu47b	*S. meliloti*	Charles *et al.,* (1991)
Melibiose	pRmeSu47b	*S. meliloti*	Charles *et al.,* (1991)
Dulcitol	pRmeSu47b	*S. meliloti*	Charles *et al.,* (1991)
Arabinose	pRleW11-9b	*R. leguminosarum*	Baldani *et al.,* (1992)
Protocatechuate	pRmeSU47b	*S. meliloti*	Charles *et al.,* (1991)
Catechol	pRleW14-2d pAMG1	*R. leguminosarum Rhizobium* sp.	Baldani *et al.,* (1992) Gajendiran and Mahavedan (1990)

Table 2. Examples of catabolic genes on plasmids
(*Work from M. Hynes laboratory)

plasmid is highly similar to the origins of several *Rhizobium* and *Agrobacterium* plasmids. Sequences homologous to transfer genes from *Agrobacterium* Ti plasmids were also found, suggesting that the NGR plasmid is conjugative, and this was confirmed experimentally. In addition to the sequence, the authors also provided a transcriptional analysis of one 137 kb segment of the plasmid. RNA probes were prepared from broth cultures, cultures subjected to known *nod* gene inducers, and from bacteroids. This type of analysis appears to be a viable strategy for finding new symbiotically important genes. The analysis of ORFs found next to newly discovered *nod* boxes could also provide much new information.

The sequencing of the entire NGR nodulation plasmid is an important accomplishment, and will give added incentive to complete sequencing of various replicons in other members of the *Rhizobiaceae*. However, there is obviously an enormous amount of work, which remains to be done, as the biological roles of all the various potential genes remain to be determined. It is quite likely that many of the genes will have nothing whatsoever to do with plant-microbe interactions. It is also possible, given how much of plasmids like the megaplasmids of *S. meliloti* can be deleted with out much apparent effect (Charles and Finan, 1991, Hynes *et al.,* 1989), that mutations in many of the ORFs will have effects which are subtle and very difficult to study.

III. Gene Transfer and Recombination

III.A. TRANSDUCTION

Transfer of DNA from one bacterium to another via a phage is referred to as transduction. The restricted host range of individual phage has resulted in several reports describing general transducing phages for different strains of *S. meliloti* (Casadesus and Olivares, 1979a; Finan *et al.*, 1984; Kowalski, 1967; Martin and Long, 1984; Sik *et al.*, 1980). A general transducing phage, RL38, has been isolated for *R. leguminosarum* and the host range of phage RL38 appears to include many strains of the different biovars of *R. leguminosarum* (Buchanan-Wollaston, 1979); this phage also infects some *R. etli* strains (Hynes and Oresnik, unpublished). To our knowledge transduction has not been reported in *Agrobacterium* or in *Bradyrhizobium*. The *R. leguminosarum* and *S. meliloti* transducing phage have proven to be very useful for the routine construction of strains, including those carrying multiple mutations. Transductions employing lysates made from pooled colonies carrying random transposon insertions can be used to isolate transposon insertions, which are genetically linked to any allele. In practical terms, on average, one need only screen one hundred *S. meliloti* transductants in order to find a transposon insertion which is linked to the allele under study (Finan *et al.*, 1984). For example, a genetic analysis of spontaneous mutations, which suppressed the symbiotic phenotype of Fix- ndvF mutants, was made possible through the isolation of linked insertions (Oresnik *et al.*, 1994). Transduction from pools or libraries of thousands of such random insertions was the technique employed in the construction of the genetic map of the complete 1700 kb pExo megaplasmid of *S. meliloti*. Cotransduction maps for parts of the *R. leguminosarum* and *S. meliloti* chromosomes have also been published (Casadesus and Olivares, 1979, Glazebrook *et al.*, 1992).

Orosz and coworkers in Hungary have characterized a λ-like specialized transducing phage, 16-3, which lysogenes *S. meliloti* Rm41 (Svab *et al.*, 1978). They have constructed a physical and genetic map of this temperate phage, and following the identification and sequencing of both the host and phage attachment sites, they constructed plasmids vectors which allow the stable integration, in single copy, of genes into Rm41 (Hermesz *et al.*, 1992, Dorgai *et al.*, 1993). Regrettably the narrow host range of 16-3 precludes the use of this system in diverse *S. meliloti* strains.

III.B. TRANSFORMATION

The uptake and stable incorporation of DNA into the bacterial cells is referred to as transformation. There are many reports of transformation in various members of the *Rhizobiacae* (*e.g.* Balassa, 1954; Drozanska and Lorkiewicz, 1978; Courtois *et al.*, 1988; Raina and Modi, 1971, O'Gara and Dunican, 1973; Selveraj and Iver, 1981). However this technique is seldom used in the genetic manipulation of these bacteria. The latter may in part be due to difficulties in obtaining reproducible results (Courtois *et al.*, 1988), and also the ease with which plasmids can be introduced into these bacteria by conjugation. More recently, the electroporation of DNA into *Agrobacterium* spp. and *Bradyrhizobium japonicum* has been reported (Charles *et al.*, 1994; Hattermann and Stacey, 1990; Shen and Ford, 1989). Electroporation is now routinely used to introduce plasmids into agrobacteria, and Charles *et al.*, 1994 have demonstrated that it can be used to transfer chromosomal mutations directly from one strain to another.

III.C. CONJUGATION

Conjugation, or the cell contact dependent transfer of DNA from one bacterium to another, is the most common genetic technique used in the studying the *Rhizobiacae*. In practice, donor, recipient and (for triparental matings), a strain carrying a mobilizing plasmid are mixed and pipetted onto the surface of an agar medium. After overnight incubation the mating mixture is plated onto appropriate selective media to obtain the desired recombinants. Conjugation is used to: a) transfer clone banks and other recombinant clones from *E. coli*, b) introduce transposons from *E. coli* into the *Rhizobiaceae*, c) construct genetic maps of particular strains, d) transfer symbiotic, tumorigenic and other plasmids between strains.

Many of the broad host range plasmids of the incompatibility groups incP, incW and incQ replicate in these bacteria. The most popular cloning vectors are derivatives of the incP plasmid RK2 and are described under section "Vectors and other genetic tools" in this review. All these derivatives contain the origin of transfer (*oriT*) of RK2, however they lack the ca. 20 kb RK2 region containing the *tra* genes. The *tra* genes were cloned into a narrow host range ColE1 based plasmid to give the plasmid pRK2013 (Figurski and Helinski, 1979) and this plasmid is widely employed to mobilize various vector plasmids from *E. coli* into Gram negative bacteria. As an alternative to this triparental mating procedure, researchers first introduce their plasmid clone directly into the *E. coli* mobilizing strains SM10 or S17-1, which contain the RK2 *tra* genes integrated into the chromosome (Simon *et al.*, 1983, 1986).

III.D. RESTRICTION AND MODIFICATION SYSTEMS

Given that so much genetic work has been done on the *Rhizobiaceae*, it is remarkable how little information has been published about restriction and modification systems in these bacteria. *A priori*, since protection from phage infection is the ostensible reason for the existence of R/M systems, one would expect them to be relatively common. The modification of *Rhizobium* phage DNA suggests that most strains have some sort of restriction barrier (Finan *et al.*, 1984; Martin and Long, 1984; Swinton *et al.*, 1985). A search of the computer data base kept by New England Biolabs (address http://www.neb.com/rebase/index.html) revealed that there are seven entries for *Rhizobium* (several species including *S. meliloti*, *R. leguminosarum* and "*R. lupini*") and nine for *Agrobacterium*. The majority are enzymes which are unpublished and frequently the target specificity is unknown. There are also isoschizomers of *SalI*, *NaeI*, *ClaI*, *BclI*, *BamHI* and *EcoRII* in this group. The only commercially available enzyme from the *Rhizobiaceae* is *AgeI*, from *Agrobacterium gelatinovorum*, a marine isolate (Mizuno *et al.*, 1990). Some of the references given in the REBASE are to patent applications, but some publications are also given such as a reference to an *EcoRII* like enzyme from *Agrobacterium* (Le Bon *et al.*, 1978) and to *RleAI*, a novel enzyme from *R. leguminosarum* which recognises an asymmetric site (Vesely *et al.*, 1990). A *PstI* like enzyme has also been found encoded on plasmid pRleVF39b in *R. leguminosarum* strain VF39 (Rochepeau *et al.*, 1997). It is quite likely that R/M systems are in fact widespread in the *Rhizobiaceae*, but since most of the original work was done using conjugation, such

systems did not constitute a significant barrier to gene transfer, and therefore were not thoroughly investigated. No restriction enzymes have yet been reported for either *Bradyrhizobium* or *Azorhizobium*. There are reports where R/M systems appear to affect phage host range, and the frequencies of gene transfer via transduction and electroporation (Buchanan-Wollaston, 1979; Hattermann and Stacey, 1990; Selvaraj and Iyer, 1981). For example, *S. meliloti* phage φM12h1 shows reduced plating efficiency on strain Rm41 or strain SU47 depending on which strain the phage was grown on. Moreover treating Rm41 cells for 3 hours at 42°C increased both the plating efficiency and transduction frequency 100 fold when using phage grown on SU47 suggesting that the heat treatment inactivated the restriction system (Williams *et al.*, 1989).

III.E. GENETIC RECOMBINATION

Genetic recombination between homologous genes obviously takes place in the *Rhizobiaceae*, otherwise none of the many studies mentioned in this chapter (transduction, mapping by conjugation, gene replacement) would have been possible. However, the mechanisms of genetic recombination have not been extensively studied. Recombination deficient mutants and/or *recA* genes have been isolated from *Agrobacterium* (Klapwijk *et al.*, 1979; Farrand *et al.*, 1989) *S. meliloti* (Better and Helinski, 1983; Selbitschka *et al.*, 1991) *R. etli* (Martinez-Salazar *et al.*, 1991; Michiels *et al.*, 1991), and *R. leguminosarum* (Selbitschka *et al.*, 1991). The *recA* mutants are highly deficient in DNA recombination and sensitive to UV light. The *recA* genes complement *E.coli recA* mutations. The regulation of *recA* genes from *Agrobacterium* and several rhizobia and their induction by DNA damage in various hosts have been studied (Riera *et al.*, 1994). One of the enigmas surrounding recombination in the *Rhizobiaceae*, or at least *Agrobacterium* and the fast growing rhizobia, is how large amounts of DNA reiteration are tolerated without extensive rearrangements of the genome. Since the amount of DNA reiteration was initially pointed out by Flores *et al.*, (1987), it has now become well established that several genes are reiterated, both on plasmids and on chromosomes, in these genera. There are also multiple copies of some genes (*groESL*, *rpoN*) in *B. japonicum* (Kündig *et al.*, 1993) but the overall degree of reiteration seems smaller in this species. One can speculate that either

a) rearrangements occur regularly, but are selected against because they yield unstable or inefficient genomes, or b) the recombination systems are not particularly efficient. There is certainly some evidence for a) as amplification and deletion of various segments of the *R. etli* genomes has been shown to occur (Romero *et al.*, 1995), and to be *recA* mediated. The situation here could be analogous to that in enteric bacteria, where recombination between different copies of the *rrn* operons (one of the few reiterated loci in *Salmonella* spp. and *E.coli*) can take place, but only certain genomic configurations are stable, and these tend to be highly conserved (Sanderson, 1976; Krawiec and Riley, 1990). However, we have also found evidence that recombination frequencies in *R. leguminosarum* are low, and that extensive homologous sequences are required before recombination takes place at a detectable frequency (Quandt and Hynes, 1993; Hynes, unpublished observations). In some gene replacement experiments, flanking sequences have to be over 1kb in order to get recombination (Quandt and Hynes, 1993), and we have found that when there are two copies of the same gene (*e.g.* *fixNOQP*, *fixGHIS*), one will not replace the other (Mitsch 1995), implying that less than perfect homology is not adequate. Similarly, when a clone carrying a Tn5 insert in genes for adonitol catabolism was introduced on a suicide vector (pJQ200, Quandt and Hynes 1993) into several strains of *R. leguminsarum*, gene replacement was found to occur only in the strain from which the clone was isolated, whereas in other strains recombination occurred at a frequency far below that of Tn5 transposition and was essentially undetectable (Faas 1995). It may be, therefore, that despite the reiteration found in the genomes of *Agrobacterium* and *Rhizobium*, the nature of the recombination system, and selection pressures in favour of certain gene orders and genome configurations co-operate to give a relatively stable genome.

III.F. INSERTION SEQUENCES AND TRANSPOSONS

One of the circumstances which may contribute to the amount of reiterated DNA in genomes of the *Rhizobiaceae* is the presence of mobile genetic elements such as insertion sequences and transposons, as well as other repetitive sequences and IS-like elements. A large number of IS elements have been isolated from both *Agrobacterium* and *Rhizobium* species, and many of these have been sequenced completely, including: four from *R. leguminosarum*, six from *S. meliloti*, one from "*R. lupini*" and 13 from *Agrobacterium* spp. (source GenBank, see Table 3). There are also several instances of IS-like elements and repetitive sequences in the rhizobia (Østeras *et al.*, 1995; Kaluza *et al.*, 1985; Ramseier and Göttfert, 1991). The IS elements have been found in a variety of ways: several were stumbled upon when they caused frequent mutations at specific loci (Ruvkun *et al.*, 1982; Dusha *et al.*, 1987) whereas others were found by sequencing or hybridisations. There have also been some systematic attempts to isolate IS elements using vectors specifically designed for that purpose. The first of these was a vector carrying *sacB* which was used by Gay *et al.*, (1985) to isolate an IS element from *Agrobacterium*. Simon *et al.*, (1991) used three different systems (*pheS*, *sacB*, *rpsL*), all of which work on the principle of inactivation of a gene whose inactivation is readily screened for, to isolate IS elements from *S. meliloti* and *R. leguminosarum*. A total of 17 IS elements was recovered, 16 from *S. meliloti* strains and one from a *R. leguminosarum* strain. Only two of these appeared to be duplicates, and only one (ISRm2011-1) corresponded to a previously isolated IS (ISRmI). This study was particularly interesting as the identity of the IS elements recovered varied according to the system used. Hybridisation patterns using the various IS elements as probes showed that they could be used for strain fingerprinting. Some elements were present in all, or nearly all strains of *S. meliloti* examined (*e.g.* Rm2011-2 [all], Rm2011-1 [all but one], Rm2011-3 [all but two]) whereas others were much more restricted in their distribution. Copy numbers varied from one per strain to 15, depending on both the element and strain examined. This set of vectors has since been used by several other groups to isolate further transposable elements (Mazurier *et al.*, 1996; Ulrich and Pühler, 1994). Several studies on population distribution of IS elements of the *Rhizobiaceae* indicate that they can be quite widespread and occur in several different genera (Mazurier *et al.*, 1996; Selbitschka *et al.*, 1995; Wheatcroft and Watson, 1988). Given the large number of different IS elements and IS-like elements found in the *Rhizobiaceae*, it can be assumed that they have played an important role in the evolution of the genomes of these bacteria.

Element	Size bp	Accession #	Reference
Agrobacterium			
IS136	1648	X04282	Vanderleyden *et al.*, (1986)
IS426	1319	X56562	Watson and Wheatcroft (1991)
IS1312	1317	U19148	Deng *et al.*, (1995)
IS292	2494	L29283	Ponsonnet *et al.*, (1995)
IS66	2608	M10204	Machida *et al.*, (1984)
IS870	1152	Z18270	Fournier *et al.*, (1989)
IS866	2716	M25805	Bonnard *et al.*, (1989)
IS869	863	X53945	Paulus *et al.*, (1991)
IS1131	2888	M82888	Wabiko (1992)
IS427	1330	M55562	de Meirsman *et al.*, (1990)
S. meliloti			
ISRm1	1319	X56563	Watson and Wheatcroft (1991)
ISRm2	----	M15786	Dusha *et al.*, (1987)
ISRm3	1298	M60971	Wheatcroft and Laberge (1991)
ISRm4	991	X65471	Soto *et al.*, (1992)
ISRm5	1340	U08627	Laberge *et al.*, (1995)
ISRm6	1269	X95567	Zekri and Toro (1996)
ISRm2011-2	1073	U22370	Selbitschka *et al.*, (1995)
R. leguminosarum			
ISR12	932	Z37965	Mazurier *et al.*, (1996)
ISRle39a	892	X99520	Rochepeau *et al.*, (1997)
ISR11	2495	L19650	O'Brien *et al.*, (1993)
Tn163	4605	L14931	Ulrich and Pühler (1994)
"R. lupini" Rhizobium sp.			
ISR1	1260	X06616	Priefer *et al.*, (1989)
B. japonicum			
RsRjalpha9	1126	X02581, M11150	Kaluza *et al.*, (1985)
RsRjbeta9	*ca.* 950	M11149	Kaluza *et al.*, (1985)
R. fredii			
Isrf1	1001	Y08939	Vinardell *et al.*, (1996)

Table 3. Transposable elements from *Rhizobiaceae* in DNA sequence database.
Note: The recently released sequence of the nodulation plasmid of broad host range strain NGR234 has revealed that 18% of the DNA of this plasmid shows homology to insertion sequences, tranposases, and mosaic sequences (Freiberg *et al.*, 1997). The accession number of the complete sequence is U00090. We have not included any of the potential insertion elements and transposons in the above table, because none of them have been given IS designations yet. It is also possible that some of these potential IS elements are non-functional.

IV. Vectors and other Genetic Tools

The cornerstone for much of the work on genetics of the various members of the *Rhizobiaceae* was the discovery that certain broad host range resistance plasmids, such as RP4 (same as RK2 and RP1), could be transferred into rhizobia and *Agrobacterium*, and would replicate in these hosts (Beringer 1974). This enabled the linkage mapping described above to be carried out, and also was the basis for the design of a whole array of vectors, which could be mobilised into most Gram negative bacteria by conjugation (see above). These mobilisable vectors fall into two classes: 1) those that will replicate in various members of the *Rhizobiaceae*, and are useful for complementing mutations, and investigating the effects of multiple copies of genes in different hosts, and 2) those which do not replicate in the target organism, but are useful as delivery systems for transposons, and as "suicide vectors" for gene replacement.

IV.A. VECTORS WHICH REPLICATE IN TARGET HOSTS

The first vectors designed for use in rhizobia and *Agrobacterium* were smaller tetracycline resistant derivatives of the plasmid RK2. Plasmid pRK290 (Ditta *et al.*, 1980) is an example of such a vector and it and its derivatives (*e.g.* Ditta *et al.*, 1985) have been widely used in the *Rhizobiaceae*. However, since one of the principal uses of a replicating vector is to clone genes by complementation of mutants from a genomic library, broad host range cosmid vectors were considered more practical for some experiments. The first such vector, pLAFR1 (Friedman *et al.*, 1982) was derived from pRK290, and was used in the first cloning of nodulation genes from *S. meliloti* (Long *et al.*, 1982). Subsequent improvements to this vector have included the addition of multiple cloning sites and the *E.coli lacZ* α-fragment for blue-white selection (Staskawicz *et al.*, 1987; Jones and Gutterson, 1987), and the *lacZ* gene in vectors such as pMP220 and pMP190 to monitor gene expression (Spaink *et al.*, 1987). All of these vectors function well in most Gram negative bacteria including all of the *Rhizobiaceae*.

However, in some experiments there is a need for smaller vectors (pRK290 and pLAFR1 are very large, *ca.* 20 kb), other antibiotic resistances, and in some cases vectors which are compatible with RK2-derived vectors so that the effects of several genes can be studied simultaneously. For this reason several other sets of vectors, based on different replicons have been employed in the *Rhizobiaceae*. Gallie *et al.*, (1985b) designed cosmid and plasmid vectors based on the pSa replicon (incW) which function well in *Agrobacterium* and *Rhizobium*. RSF1010 (incQ) derived vectors suitable for use in *Rhizobium* and *Agrobacterium* have also been designed (Priefer *et al.*, 1985; Labes *et al.*, 1990).

The set of vectors described by Labes *et al.*, (1990) allows construction of gene fusions and inducible expression of cloned genes. Taken together these vectors provide several alternatives for cloning experiments where complementation of function in rhizobia or *Agrobacterium* is the ultimate goal. However, there is room for some improvements as restriction site availability is not always optimal, and a wider choice of antibiotic resistance markers would be desirable. As RSF1010 and its vector derivatives do not appear to replicate in *B. japonicum*, additional vectors compatible with pRK290 would be of some use in this bacterium.

IV.B. SUICIDE VECTORS

Vectors that do not replicate in the target bacterium have, as mentioned above, two principal uses: delivery of transposons, and gene replacement experiments. The original procedures for doing these two things in *Rhizobium* and *Agrobacterium* were effective, but unwieldy. Original transposon mutagenesis schemes used broad host range plasmids (such as RP4 and pPHIJI) carrying insertions of bacteriophage Mu, as this had been shown to destabilise these plasmids (Boucher *et al.*, 1977; Van Vliet *et al.*, 1978). The use of these constructs in transposon mutagenesis provided some very important mutants (Beringer *et al.*, 1978b; Hernalsteens *et al.*, 1978; Meade *et al.*, 1982) but there were certain problems created by the use of Mu (deletions, independent Mu transposition) which added unnecessary levels of complication to the experiments. The first gene replacement or marker replacement experiments in *Rhizobium* were done by Ruvkun and Ausubel (1981) using incompatible plasmids R751 or pPH1JI (Hirsch and Beringer, 1984) to select for recombinants in which a cloned gene on the broad host-range vector (pRK290) had replaced the equivalent locus on the *S. meliloti* genome. This technique functions very well, and is still quite widely used, but the necessity for identifying the recombinants in two separate steps takes extra time.

A solution to the problems of both transposon delivery and more convenient suicide vectors was to take small plasmids with a narrow host range, such as pBR322 and pACYC184, and adapt them to use in rhizobia by providing them with functions which would make them mobilisable by RP4. This approach was employed very successfully by Reinhard Simon and colleagues (Simon *et al.*, 1983;

Simon *et al.*, 1986) who constructed a complete set of such vectors which have been very widely used in nearly all Gram-negative bacteria. They function efficiently in transposon delivery, and can be used for one-step gene replacement. They are mobilised at high frequencies into *Rhizobium*, *Bradyrhizobium*, *Agrobacterium* and other bacteria, but will not replicate in these hosts. One complication which has been encountered, however, is that these vectors, when used for Tn5 mutagenesis in *Azorhizobium caulinodans* appear to integrate into the genome of this host, thus causing "vector integration mutagenesis" rather than proper transposon mutagenesis (Donald *et al.*, 1985). A further refinement to this type of vector for gene replacement is to incorporate a conditionally lethal gene such as *sacB*, which facilitates selection for double crossover (gene replacement) events (Quandt and Hynes, 1993). Additional delivery vectors for transposon mutagenesis include the mobilising plasmid pRK2013 carrying Tn5 derivatives (Finan *et al.*, 1985) and plasmids composed of a narrow host range p15A replication origin together with incN *tra* genes and Tn5, Tn1 or Tn9 (Selvaraj and Iyer, 1983).

IV.C. TRANSPOSONS

Transposon mutagenesis has played a central role in the evolution of genetics of the *Rhizobiaceae*. Because many of the attributes originally studied involve interaction with a plant host (*e.g.* nodulation) it was exceptionally important to obtain stable, non-reverting mutations which could easily be traced. Transposon insertions provide such mutations. Some of the transposons which have been used include Tn5 (Beringer *et al.*, 1978b, Meade *et al.*, 1982), Tn7 (Hernalsteens *et al.*, 1978), Tn1 (Casadesus *et al.*, 1980) Tn904 (Klapwijk *et al.*, 1980), Tn1831 (Pees *et al.*, 1986) and Tn10 (Hynes *et al.*, 1986). All of these transposons work, but some have limitations which make them less useful than others, such as non-randomness of transposition, creation of deletions, and unsuitable antibiotic resistance markers. The most widely and most successfully used transposon has been Tn5. It seems to transpose randomly in all hosts, and provides stable non-reverting mutants. The antibiotic markers neomycin, streptomycin, and bleomycin coded for by wild-type Tn5 are generally useful in the *Rhizobiaceae*, but new derivatives with other markers such as gentamicin, spectinomycin, and

tetracycline resistance are also available (Simon *et al.*, 1983; De Vos *et al.*, 1986; Hirsch *et al.*, 1986). Some other interesting and very useful Tn5 derivatives include Tn5-Mob and the *sacB* derivatives described above (Simon, 1984; Hynes *et al.*, 1989), and various derivatives which allow creation of gene fusions (transcriptional fusions to *lacZ*, luciferase and a promoterless Neomycin resistance gene), as well as "transposable promoters" which have been described by Simon *et al.*, (1989). Tn*phoA*, a Tn5 derivative which generates translational fusions to the *E. coli* alkaline phosphatase gene (Manoil and Beckwith 1985) has been used to good effect in *Rhizobium* (Long *et al.*, 1988; Yarosh *et al.*, 1989). Also of note is the Tn3 derivative Tn3HoHo1 (Stachel *et al.*, 1985) which has been widely used for generating *lacZ* fusions in *Agrobacterium* and *Rhizobium*.

IV.D. FUTURE NEEDS

The genetic systems for the analysis of rhizobia and *Agrobacterium* are highly developed as can be seen above. Nonetheless there is still progress that could be made in developing techniques and tools for the genetic analysis of these and other related bacteria. One notable missing feature is the availability of transducing phage for *Bradyrhizobium* spp. and *Agrobacterium*. While transducing phage are not a necessity, as their use can be circumvented by cloning mutations of interest, and reintroducing them into wild-type or other strains by gene replacement, transduction is quicker, easier and much cheaper. There is also a need for additional selective marker genes, as, in complex constructs, the available markers (such as gentamicin, neomycin and tetracycline) quickly become exhausted.

The determination of the complete nucleotide sequence of more than 30 microbial genomes is currently underway. It is thus clear that we are entering an era where the complete sequencing of microbial genomes is commonplace. We expect and hope that the complete genome sequence of several species within the *Rhizobiaceae* will also be determined over the next several years. Freiberg and coworkers have now completed the nucleotide sequence of the 500 kb pSym plasmid from the broad host range *Rhizobium* strain NGR234 (Freiberg *et al.*, 1996; 1997). The analysis of this sequence has already revealed some interesting features of the plasmids, and given rise to important hypotheses about genome evolution in this strain, and about additional genes potentially involved in symbiosis. As more sequences of other plasmids, and of complete rhizobial genomes are added to the data bases, the usefulness of such information in predicting possible functions of uncharacterised open reading frames is likely to increase at a rapid rate.

One anticipates that the *Rhizobiaceae* genomes will contain all of the house keeping genes involved in transcription, translation, replication, cell envelope synthesis etc. which should be common to many bacteria. We would expect to find additional genes involved in N_2-fixation and root nodule formation. It will also be fascinating to determine what additional genes are present and whether particular gene clusters are found in soil bacteria. Genetic studies have had little impact in the general area of competition and persistence of bacteria in soil. It will be interesting to see whether genome sequencing will lead to advances in this poorly understood area.

V. References

Albright, L., Ronson, C., Nixon, T. and Ausubel, F. (1989) J. Bacteriol. 171, 1832-1941.

Allardet-Servent, A., Michaux-Charachon, S., Jumas-Bilak, E., Karayan, L. and Ramuz, M. (1993) J. Bacteriol. 175, 7869-7874.

Bae, Y.M., Holmgren, E. and Crawford, I.P. (1989) J. Bacteriol. 171, 3471-3478.

Bae, Y.M. and Stauffer, G.V. (1991) J. Bacteriol. 173, 5831-5836.

Balassa, R. (1954) Acta Microbiol. Acad. Sci. Hung. 2, 51-78.

Balassa, R. (1957) Acta Microbiol. Acad. Sci. Hung. 4, 77-84.

Baldani, J.I., Weaver, R.W., Hynes, M.F. and Eardly, B.D. (1992) Appl. Environ. Microbiol. 58, 2308-2314.

Banfalvi, Z., Sakanyan, V., Koncz, C., Kiss, A., Dusha. I. and Kondorosi, A. (1981) Mol. Gen. Genet. 184, 318-325.

Banfalvi, Z., Kondorosi, E. and Kondorosi, A. (1985) Plasmid 13, 129-138.

Bardin, S., Dan, S., Osteras, M. and Finan, T.M. (1996) J. Bacteriol.178, 4540-4547.

Barnett, M.J., Rushing, B.G., Fisher, R.F. and Long, S.R. (1996) J. Bacteriol. 178, 1782-1787.

Beringer, J.E. (1974) J. Gen. Microbiol. 84, 188-198.

Beringer, J.E. (1980) J. Gen. Microbiol. 116, 1-7.

Beringer, J.E., Beynon, J.L., Buchanan-Wollaston, A.V. and Johnston, A.W.B. (1978b) Nature 276, 633-634.

Beringer, J.E., Hoggan, S.A. and Johnston, A.W.B. (1978a) J. Gen. Microbiol. 104, 201-207.

Beringer, J.E. and Hopwood, D.A. (1976) Nature 264, 291-293.

Better, M. and Helinski, D.R. (1983) J. Bacteriol. 155, 311-316.

Beynon, J.L., Beringer, J.E. and Johnston, A.W.B. (1980) J. Gen. Microbiol. 120, 421-429.

Boivin, C., Camut, S., Malpica, C., Truchet, G. and Rosenberg, C. (1990) The Plant Cell 2, 1157-1170.

Bonnard, G., Vincent, F. and Otten, L. (1989) Plasmid 22, 70-81.

Boucher, C., Bergeron, B., Barate de Bertalmio, M. and Dénarié, J. (1977) J. Gen. Microbiol. 98, 253-263.

Brewin, N., DeJong, T., Phillips, D. and Johnston, A. (1980) Nature 288, 77-79.

Brom, S., Garcia de los Santos, A., Stepkowsky, T., Flores, M., Davila, G., Romero, D. and Palacios, R. (1992) J. Bacteriol. 174, 5183-5189.

Buchanan-Wollaston, V. (1979) J. Gen. Microbiol. 112, 135-142.

Burkardt, B., Schillik, D. and Pühler, A. (1987) Plasmid 17, 13-25.

Casadesus, J., Ianez, E. and Olivares, J. (1980) Mol. Gen. Genet. 180, 405-410.

Casadesus, J. and Olivares, J. (1979) Mol. Gen. Genet. 174, 59-66.

Casadesus, J. and Olivares, J. (1979) J. Bacteriol. 139, 316-317.

Casse, F., Boucher, C., Julliot, J.S., Michel. M. and Dénarié, J. (1979) J. Gen. Microbiol. 113, 229-242.

Charles, T.C., Doty, S.L. and Nester, E. (1994) Applied and Envir. Microbiol. 60, 4192-4194.

Charles, T.C. and Finan, T.M. (1990) J. Bacteriol. 172, 2469-2476.

Charles, T.C. and Finan, T. (1991) Genetics 127, 5-20.

Costantino, P., Hooykaas, P.J.J., den Dulk-Ras, H. and Schilperoort, R.A. (1980) Gene 11, 79-87.

Courtois, J., Courtois, B. and J. Guillaume. (1988) J. Bacteriol. 170, 5925-5927.

De Maagd, R., Mulders, I., Canter Cremers, H. and Lugtenberg, B. (1992) J. Bacteriol. 174, 214-222.

de Meirsman, C., van Soom, C., Verreth, C., Van Gool, A. and Vanderleyden, J. (1990) Plasmid 24, 227-234.

Deng,W., Gordon, M.P. and Nester, E.W. (1995) J. Bacteriol. 177, 2554-2559.

Depicker, A., De Wilde, M., De Vos, G., De Vos, R., Van Montagu, M. and Schell, J. (1980) Plasmid 3, 193-211.

De Vos, G.F., Walker, G.C. and Signer, E.R. (1986) Mol. Gen. Genet. 204, 485-491.

Ditta, G., Schmidhauser, T., Yakobson, E., Lu, P., Liang, X.W., Finlay, D.R., Guiney, D. and Helinski, D.R. (1985) Plasmid 13, 149-153.

Ditta, G., Stanfield, S., Corbin, D. and Helinski, S.R. (1980) Proc. Natl. Acad. Sci. 77, 7347-7351.

Donald, R.G., Raymond, C.K. and Ludwig, R.A. (1985) J. Bacteriol., 162, 317-323.

Dorgai, L., Papp, I., Papp, P., Kalman, M. and Orosz, L. (1993) Nuc. Acid Res. 21, 1671.

Drozanska, D. and Lorkiewicz, Z. (1978) Acta Microbiol. Pol. 27, 81-88.

Dusha, I., Kovalenko, S., Banfalvi, Z. and Kondorosi, A. (1987) Mol. Gen. Genet. 169, 1403-1409.

Eckhardt, T. (1978) Plasmid 1, 584-588.

Faas, L.A. (1995) Thesis, University of Calgary, Calgary, Alberta, Canada.

Farrand, S.K., O'Morchoe, S.P. and McCutchan, J. (1989) J. Bacteriol. 171, 5314-5321.

Figurski, D.H. and Helinski, D.R. (1979) Proc. Natl. Acad. Sci. 76, 1648-1652.

Finan, T., Oresnik, I. and Bottacin, A. (1988) J. Bacteriol. 170, 3396-3403.

Finan, T.M., Hartwieg, E., LeMieux, K., Bergman, K., Walker, G.C. and Signer, E.R. (1984) J. Bacteriol. 159, 120-124.

Finan, T.M., Hirsch, A., Leigh, J., Johansen, E., Kuldau, G., Deegan, S., Walker, G. and Signer, E. (1985) Cell 40, 869-877.

Finan, T.M., Kunkel, B., De Vos, G.F. and Signer, E.R. (1986) J. Bacteriol. 167, 66-72.

Fisher, R.F., Brierley, H.L., Mulligan, J.T. and Long, S.R. (1987) J. Biological Chem. 262, 6849-6855.

Flores, M., Gonzalez, V., Brom, S., Martinez, E., Pinero, D., Romero, D., Davila, G. and Palacios, R. (1987) J. Bacteriol. 169, 5782-5788.

Fournier, P., Paulus, F. and Otten, L. (1993) J. Bacteriol. 175, 3151-3160.

Freiberg, C., Fellay, R., Bairoch, A., Broughton, W.J., Rosenthal, A. and Perret, X. (1997) Nature 387, 394-401.

Frieberg, C., Perret, X., Broughton, W. and Rosenthal, A. (1996) Genome Research 6, 590-600.

Friedman, A.M., Long, S.R., Brown, S.E., Buikema, W.J. and Ausubel, F.M. (1982) Gene 18, 289-296.

Gajendiran, N. and Mahadevan, A. (1990) FEMS Microbiol. Ecol. 73, 125-130.

Gallie, D.R., Hagiya, M. and Kado, C.I. (1985a) J. Bacteriol 161, 1034-1041.

Gallie, D.R. and Kado, C.I. (1988) J. Bacteriol. 170, 3170-76.

Gallie, D.R., Novak, S. and Kado, C.I. (1985b) Plasmid 14, 171-175.

Gallie, D.R., Zaitlin, D., Perry, K.L. and Kado, C.I. (1984) J. Bacteriol. 157, 739-745.

Garcia de los Santos, A., Brom, S. and Romero, D. (1996) World J. Microbiol. Biotech. 12, 119-125.

Gay, P., Le Coq, D., Steinmetz, M., Berkelman, T. and Kado, C.I. (1985) J. Bacteriol. 164, 918-921.

Gerard, J.C., Canaday, J., Szegedi, E., de la Salle, H. and Otten, L. (1992) Plasmid 28, 146-156.

Girard, M. de L., Flores, M., Brom, S., Romero, D., Palacios, R. and Davila, G. (1991) J. Bacteriol. 173, 2411-2419.

Glazebrook, J., Meiri, G. and Walker, G.C. (1992) Molec. Plant. Microbe. Interac. 5, 223-227.

Goldman, A., Lecoeur, L., Message, B., Delarue, M., Schoonejans, E. and Tepfer, D. (1994) FEMS Microbiol. Lett. 115, 305-312.

Gray, K.M., Pearson, J.P., Downie, J.A., Boboye, B.E. and Greenberg, E.P. (1996) J. Bacteriol. 178, 372-376.

Gribskov, M., Devereux, J. and Burgess, R. (1984) Nuc. Acids Res. 12, 539-549.

Gross, D.C., Vidaver, A.K. and Klucas, R.V. (1979) J. Gen. Microbiol. 114, 257-266.

Haas, D. and Holloway, B. (1976) Mol. Gen. Genet. 144, 243-251.

Hattermann, D. and Stacey, G. (1990) Appl. Environ. Microbiol. 56, 833-836.

Hermesz, E., Olasz, F., Dorgai, L. and Orosz, L. (1992) Gene 119, 9-15.

Hernalsteens, J.P., de Greve, H., Van Montagu, M. and Schell, J. (1978) Plasmid 1, 218-225.

Heumann, W. (1968) Mol. Gen. Genet. 102, 132-144.

Higashi, S. (1967) J. Gen. Appl. Microbiol. 13, 391-403.

Hirsch, P.R. (1979) J. Gen. Microbiol. 113, 219-228.

Hirsch, P.R. and Beringer, J.E. (1984) Plasmid 12, 139-141.

Hirsch, P.R., Van Montagu, M., Johnston, A.W.B., Brewin, N.J. and Schell, J. (1980) J. Gen. Microbiol. 120, 403-412.

Hirsch, P.R., Wang, C.L. and Woodward, M.J. (1986) Gene 48, 203-209.

Holsters, M., Silva, B., Van Vliet, F., Genetello, C., De Block, M., Dhaese, P., Depicker, A., Inzé, D., Engler, G., Villaroel, R., Van Montagu, M. and Schell, J. (1980) Plasmid 3, 212-230.

Hom, S., Uratsu, S. and Hong, F. (1984) J. Bacteriol. 159, 335-340.

Hombrecher, G., Brewin, N.J. and Johnston, A.W.B. (1981) Mol. Gen. Genet. 182, 133-136.

Honeycutt, R.J., McClelland, M. and Sobral, B.W.S. (1993) J. Bacteriol. 175, 6945-6952.

Hooykaas, P.J.J., Kalpwijk, P.M., Nuti, M.P., Schilperoort, R.A. and Rorsch, A. (1977) J. Gen. Microbiol. 98, 477-484.

Hooykaas, P.J.J., Peerbolte, R., Regensburg-Tuïnk, A.J.G., De Vries, P. and Schilperoort, R.A. (1982) Mol. Gen. Genet. 188, 12-17.

Hooykaas, P.J.J., Van Brussel, A.A.N., den Dulk-Ras, H., van Slogteren, G.M.S. and Schilperoort, R.A. (1981) Nature 291, 351-353.

Hynes, M.F., Simon, R. and Pühler, A. (1985) Plasmid 13, 99-105.

Hynes, M.F., Simon, R., Müller, P., Niehaus, K., Labes, M. and Pühler, A. (1986) Mol. Gen. Genet. 202, 356-362.

Hynes, M.F., Brucksch, K. and Priefer, U. (1988) Arch. Microbiol. 150, 326-332.

Hynes, M.F., Quandt, J., O'Connell, M.P. and Pühler, A. (1989) Gene 78, 111-120.

Hynes, M.F. and McGregor, N.F. (1990) Molec. Microbiol. 4, 567-571.

Jones, J.D.G. and Gutterson, N. (1987) Gene 61, 299-306.

Jouanin, L., Tourneur, J., Tourneur, C. and Casse-Delbart, F. (1986) Plasmid 16, 124-134.

Jordan, D. C. (1984) Bergey's manual of Systematic Bacteriology. Eds. Williams and Williams Pub Baltimore/London Vol. 1 pgs. 234-256.

Kado, C.I. and Liu, S.T. (1981) J. Bacteriol. 145, 1365-1373.

Kaluza, K., Hahn, M. and Hennecke, H. (1985) J. Bacteriol. 162, 535-542.

Kiss, G.B. and Kalman, Z. (1982) J. Bacteriol. 150, 465-470.

Klapwijk, P.M., van Beelen, P. and Schilperoort, R.A. (1979) Mol. Gen. Genet. 173, 171-175.

Klapwijk, P.M., Van Breukelen, J., Korevaar, K., Ooms, G. and Schilperoort, R.A. (1980) J. Bacteriol. 141, 129-136.

Klein, S., Lohman, K., Clover, R., Walker, G. and Signer, E. (1992) J. Bacteriol. 174, 324-326.

Kondorosi, A., Kiss, G.B., Forrai, T., Vincze, E. and Banfalvi, Z. (1977) Nature 268, 525-527.

Kondorosi, A., Kondorosi, E., Pankhurst, C.E., Broughton, W.J. and Banfalvi, Z. (1982) Mol. Gen. Genet. 188, 433-439.

Kondorosi, A., Vincze, E., Johnston, A.W.B. and Beringer, J.E. (1980) Mol. Gen. Genet. 178, 403-408.

Kowalski, M. (1967) Acta Microbiol. Polonica 16, 7-12.

Krawiec, S. and Riley, M. (1990) Microbiol. Rev. 54, 502-539.

Krishnan, H.B. and Pueppke, S.G. (1991) Mol. Plant Microb. Interact. 4, 521-529.

Krishnan, H.B. and Pueppke, S.G. (1993) Appl. Environ. Microbiol. 59, 150-155.

Kündig, C., Beck,C., Hennecke, H., Göttfert, M. (1995) J. Bacteriol. 177, 5151-5154.

Kündig, C., Hennecke, H., Göttfert, M. (1993) J. Bacteriol. 175, 613-622.

Kuykendall, L.D. (1979) Appl. Environm. Microbiol. 37, 862-866.

Laberge, S., Middleton, A.T. and Wheatcroft, R. (1995) J. Bacteriol. 177, 3133-3142.

Labes, M., Puhler, A. and Simon, R. (1990) Gene 89, 37-46.

Lamb, J.W., Hombrecher, G. and Johnston, A.W.B. (1982) Mol. Gen. Genet. 186, 449-452.

LeBon, J.M., Kado, C.I., Rosenthal, L.J. and Chirikjian, J.G. (1978) Proc. Natl. Acad. Sci. U.S.A. 75, 4097-4101.

Ledeboer, A.M., Krol, A.J.M., Dons, J.J.M., Spier, F., Schilperoort, R.A., Zaenen, I., Van Larebeke, N. and Schell, J. (1976) Nucl. Acid Res. 3, 449-463.

Leong, S.A., Williams, P.H. and Ditta, G. S. (1985) Nuc. Acids Res. 13, 5965-5976.

Lin, B.C. and Kado, C.I. (1977) Can. J. Microbiol. 23, 1554-1561.

Long, S., McCune, S. and Walker, G.C. (1988) J. Bacteriol. 170, 4257-4265.

Long, S.R., Buikema, W.J. and Ausubel, F.M. (1982) Nature 298, 485-488.

Luka, S., Patriarca, E.J., Riccio, A., Iaccarino, M. and Defez, R. (1996) J. Bacteriol. 178, 7138-7143.

Luyindula, H., Tshitenge, G., Lurquin, P. and Ledoux, L. (1975) Arch. Int. Physiol. Biochim. 83, 199-200.

Machida, Y., Sakurai, M., Kiyokawa, S., Ubasawa, A., Suzuki, Y. and Ikeda, J.E. (1984) Proc. Natl. Acad. Sci. U.S.A. 81, 7495-7499.

Malakooti, J. and Ely, B. (1995) J. Bacteriol. 177, 6854-6860.

Malakooti, J., Wang, S.P. and Ely, B. (1995) J. Bacteriol. 177, 4372-4376.

Manoil, C. and Beckwith, J. (1985) Proc. Natl. Acad. Sci. USA 82, 8129-8133.

Margolin, W. and Long, S.R. (1993) J. Bacteriol. 175, 6553-6561.

Martin, M.O. and Long, S.R. (1984) J. Bacteriol. 159, 125-129.

Martinez-Salazar, J.M., Romero, D., Girard, M.L. and Davila, G. (1991) J. Bacteriol. 173, 3035-3040.

Mazurier, S.I., Rigottier-Gois, L. and Amarger, N. (1996) Appl. Environ. Microbiol. 62, 685-693.

Meade, H.M. and Signer, E.R. (1977) Proc. Natl. Acad. Sci. USA 74, 2076-2078.

Meade, H.M., Long, S.R., Ruvkun, G.B., Brown, S.E. and Ausubel, F.M. (1982) J. Bacteriol. 149, 114-122.

Megias, M., Caviedes, M., Palomares, A. and Perez-Silva, J. (1982) J. Bacteriol. 149, 59-64.

Mercado-Blanco, J., Garcia, F., Fernandez-Lopez, M. and Olivares, J. (1993) J. Bacteriol. 175, 5403-5410.

Mercado-Blanco, J. and Olivares, J. (1994a) Gene 139, 133-134.

Mercado-Blanco, J. and Olivares, J. (1994b) Plasmid 32, 75-79.

Mercado-Blanco, J. and Toro, N. (1996) Mol. Genet. Plant-Microbe Inter. 9, 535-545.

Merlo, D.J. and Nester, E.W. (1977) J. Bacteriol. 129, 76-80.

Michiels, J., Vande Broek, A. and Vanderleyden, J. (1991) Mol. Gen. Genet. 228, 486-490.

Mitsch, M.J. (1995) Thesis, University of Calgary, Calgary, Alberta, Canada.

Mizuno, H., Suzuki, T., Akagawa, M., Yamasato, K. and Yamada, Y. (1990) Agric. Biol. Chem. 54, 1797-1802.

Moënne-Loccoz, Y., Baldani, J.I. and Weaver, R.W. (1995) Letters Appl. Microbiol. 20, 175-179.

Moënne-Loccoz, Y., Sen, D., Krause, E.S. and Weaver, R.W. (1994) Agron. J. 86, 117-121.

Moënne-Loccoz, Y. and Weaver, R.W. (1996) Appl. Soil Ecol. 3, 137-148.

Moënne-Loccoz, Y. and Weaver, R.W. (1995a) Soil Biol. Biochem. 27, 1001-1004.

Moënne-Loccoz, Y. and Weaver, R.W. (1995b) FEMS Microbiol. Ecol. 18, 139-144.

Mozo, T., Cabrera, E. and Ruiz-Argueso, T. (1990) Plasmid 23, 201-215.

O'Brien, J.K., Ryan, A.J. and O'Connell, M. (1993) Accession L19650, Unpublished.

O'Connell, M.P., Hynes, M.F. and Pühler, A. (1987) Plasmid 18, 156-163.

O'Gara, F. and Dunican, L. (1973) J. Bacteriol. 116, 1177-1180.

Oresnik, I., Charles, T. and Finan, T. (1994) Genetics 136, 1233-1243.

Østeras, M., Stanley, J. and Broughton, W. (1995) J. Bacteriol. 177, 5485-5494.

Østeras, M., Stanley, J., Broughton, W. and Dowling, D. (1989) Mol. Gen. Genet. 220, 157-160.

Paulus, F., Canaday, J., Vincent, F., Bonnard, G., Kares, C. and Otten, L. (1991) Plant Mol. Biol. 16, 601-614.

Paulus, F., Huss, B., Tinland, B., Hermann, A., Canaday, J. and Otten, L. (1991) Accession M63056 J03693, Unpublished.

Pees, E., Wijffelman, C., Mulders, I., van Brussel, A.A.N. and Lugtenberg, B.J.J. (1986) FEMS Microbiol. Lett. 33, 165-171.

Perret, X., Broughton, W.J. and Brenner, S. (1991) Proc. Natl. Acad. Sci. USA 88, 1923-1927.

Pomponi, M., Spano, L., Sabbadini, M.G. and Costantino, P. (1983) Plasmid 10, 119-129.

Ponsonnet, C., Normand, P., Pilate,G. and Nesme, X. (1995) Microbiology 141, 853-861.

Prakash, R.K., Schilperoort, R.A. and Nuti, M.P. (1981) J. Bacteriol. 145, 1129-1136.

Prakash, R.K., Van Brussel, A.A.N., Quint, A., Mennes, A.M. and Schilperoort, R.A. (1982b) Plasmid 7, 281-286.

Prakash, R.K., van Veen, R.J.M. and Schilperoort, R.A. (1982a) Plasmid 7, 271-280.

Priefer, U.B., Kalinowski, J., Ruger, B., Heumann, W. and Puhler, A. (1989) Plasmid 2, 120-128.

Priefer, U.B., Simon, R. and Pühler, A. (1985) J. Bacteriol. 163, 324-330.

Quandt, J. and Hynes, M.F. (1993) Gene 127, 15-21.

Raina, J.L. and Modi V.V. (1971) J. Gen. Microbiol. 65, 161-165.

Ramseier, T.M., Göttfert, M. (1991) Arch. Microbiol. 156, 270-276.

Riera, J., Fernandez de Henestrosa, A.R., Garriga, X., Tapias, A. and Barbe, J. (1994) Mol. Gen. Genet. 245, 523-527.

Rochepeau, P., Selinger, L.B. and Hynes, M.F. (1997) Mol. Gen. Genet. 256, 387-396.

Roest, H., Bloemendaal, C., Wijffelman, C. and Lugtenberg, B. (1992) J. Bacteriol. 177, 4985-4991.

Romero, D., Davila, G. and Palacios, R. (1997) De Bruijn, F., J. Lupski, and Weinstock G.(eds.) Chapman and Hall USA in press.

Romero, D., Martinez-Salazar, J., Girard, L., Brom, S., Davilla, G., Palacios, R., Flores, M. and Rodriguez, C. (1995) J. Bacteriol. 177, 973-980.

Rosenberg, C., Boistard, P., Dénarié, J. and Casse-Delbart, F. (1981) Mol. Gen. Genet. 184, 326-333.

Rosenberg, C., Casse-Delbart, F., Dusha, I., David, M. and Boucher, C. (1982) J. Bacteriol. 150, 402-406.

Rosenberg, C. and Huguet, T. (1984) Mol. Gen. Genet. 196, 533-536.

Rushing, B.G. and Long, S.R. (1995) J. Bacteriol. 177, 6952-6957.

Ruvkun, G.B. and Ausubel, F.M. (1981) Nature 289, 85-88.

Ruvkun, G.B., Long, S.R., Meade, H.M., Van den Bos, R.C. and Ausubel, F.M. (1982) J. Mol. Appl. Genet. 1, 405-418.

Saint, C., Wexler, M., Murphy, P., Tempe, J., Tate, M. and Murphy, P. (1993) J. Bacteriol. 175, 5205-5215.

Sanderson, K.E. (1976) Ann. Rev. Microbiol. 30, 327-349.

Schripsema, J.K., Rudder, E.E., van Vliet, T.B., Lankhorst, P.P., de Vroom, E. and van Brussel, A.A.N. (1996) J. Bacteriol. 178, 366-371.

Schwinghamer, E.A. (1980) FEMS Microbiology Letters 7, 157-162.

Segal, G. and Ron, E. (1993) J. Bacteriol. 175, 3026-3030.

Selbitschka, W., Arnold, W., Jording, D., Kosier, B., Toro, N. and Puhler, A. (1995) Gene 163, 59-64.

Selbitschka, W., Arnold, W., Priefer, U.B., Rottschäfer, T., Schmidt, M., Simon, R. and Pühler, A. (1991) Mol. Gen. Genet. 229, 86-95.

Selvaraj, G. and Iyer, V.N. (1981) Gene 15, 279-283.

Selvaraj, G. and Iyer, V.N. (1983) J. Bacteriol. 156, 1292-1230.

Shen, W. and Forde, B. (1989) Nuc. Acids Res. 17, 8385.

Sik, T., Horvath, J. and Chatterjee, S. (1980) Mol. Gen. Genet. 178, 511-516.

Simon, R. (1984) Mol. Gen. Genet. 196, 413-420.

Simon, R., Hötte, B., Klauke, B. and Kosier, B. (1991) J. Bacteriol. 173, 1502-1508.

Simon, R., O'Connell, M., Labes, M. and Pühler, A. (1986) Methods in Enzymology 118, 640-659.

Simon, R., Priefer, U. and Pühler, A. (1983) Biotechnology 1, 784-791.

Simon, R., Quandt, J. and Klipp, W. (1989) Gene 80, 161-169.

Sobral, B.W.S., Honeycutt, R.J. and Atherly, A.G. (1991a) J. Bacteriol. 173, 704-709.

Sobral, B.W.S., Honeycutt, R.J., Atherly, A.G. and McClelland, M. (1991b) J. Bacteriol. 173, 5173-5180.

Soto, M.J., Zorzano, A., Garcia-Rodriquez, F.M., Mercado-Blanco, J., Lopez-Lara, I.M., Olivares, J. and Toro, N. (1994) Mol. Plant Microbe Interact. 7, 703-707.

Soto, M.J., Zorzano, A., Olivares, J. and Toro, N. (1992) Gene 120, 125-126.

Soto, M.J., Zorzano, A., Olivares, J. and Toro, N. (1992) Plant Mol. Biol. 20, 307-309.

Spaink, H. P., Okker, R., Wijffelman, C., Pees, E. and Lugtenberg, B. (1987) Plant Mol. Biol. 9, 27-39.

Stachel, S., An, G., Flores, C. and Nester, E.W. (1985) EMBO J. 4, 891-898.

Staskawicz, B., Dahlbeck, D., Keen, N. and Napoli, C. (1987) J. Bacteriol. 169, 5789-5794.

Sullivan, J.T., Patrick H.N., Lowther, W.L, Scott, D.B. and Ronson, C.W. (1995) Proc. Natl. Acad. Sci. 92, 8985-8989.

Svab, Z., Kondorosi, A. and Orosz, L. (1978) J. Gen. Microbiol. 106, 321-327.

Swinton, D., Hattman, S., Benzinger, R., Buchanan-Wollaston, V. and Beringer, J. (1985) FEBS Lett. 184, 294-298.

Tabata, S., Hooykaas, P.J.J. and Oka, A. (1989) J. Bacteriol. 171, 1665-1672.

Tepfer, D., Goldmann, A., Pamboukdjian, N., Maille, M., Lepingle, A., Chevalier, D., Dénarié, J. and Rosenberg, C. (1988) J. Bacteriol. 170, 1153-1161.

Toro, N. and Olivares, J. (1986) Mol. Gen. Genet. 202, 331-335.

Turner, S.L., Rigottier-Gois, L., Power, R.S., Amarger, N. and Young, J.P.W. (1996) Microbiology 142, 1705-1713.

Turner, S.L. and Young, J.P.W. (1995) FEMS Microbiol. Lett. 133, 53-58.

Ulrich, A. and Puhler, A. (1994) Mol. Gen. Genet. 242, 505-516.

Vanderleyden, J., Desair, J. and De Meirsman, C. (1991) DNA Seq. 2, 163-172.

Van Larebeke, N., Engler, G., Holsters, M., Van den Elsacker, S., Zaenen, I., Schilperoort, R.A. and Schell, J. (1974) Nature 242, 171-172.

Van Veen, R., den Dulk-Ras, H., Bisseling, T., Schilperoort, R. and Hooykaas, P. (1988) Mol. Plant-Microbe Inter. 1, 231-234.

Van Vliet, F., Silva, B., Van Montagu, M. and Schell, J. (1978) Plasmid 1, 446-455.

Vesely, Z., Müller, A., Scmitz, G.G., Kaluza, K., Jarsch, M. and Kessler, C. (1990) Gene 95, 129-131.

Vilaine, F. and Casse-Delbart, F. (1987) Gene 55, 105-114.

Vinardell, J.M., Ollero, F.J. and Ruiz-Sainz, J.E. (1996) Accession Y08939, Unpublished.

Wabiko, H.(1992) Gene 114, 229-233.

Watson, R.J., Chan, Y.-K., Wheatcroft, R., Yang, A-F. and Han, S. (1988) J. Bacteriol. 170, 927-934.

Watson, R.J. and Wheatcroft,R. (1991) Accession X56562, Unpublished.

Watson, R.J. and Wheatcroft,R. (1991) DNA Seq. 2, 163-172.

Wheatcroft, R. and Laberge, S. (1991) J. Bacteriol. 173, 2530-2538.

Wheatcroft, R. and Watson, R.J. (1988) J. Gen. Microbiol. 134, 113-121.

White, F.F. and Nester, E.W. (1980) J. Bacteriol. 141, 1134-1141.

White, F.F. and Nester, E.W. (1980) J. Bacteriol. 144, 710-720.

Wijffelman, C.A., Pees, E., Van Brussel, A.A.N. and Hooykaas, P.J.J. (1983) Mol. Gen. Genet. 192, 171-176.

Williams, M., Klein, S. and Signer, E. (1989) Appl. Environ. Microbiol. 55, 3229-3230.

Yakobson, E. and Guiney, D. (1984) J. Bacteriol. 160, 451-453.

Yarosh, O., Charles, T. and Finan, T. (1989) Mol. Micro. 3, 813-823.

Zekri, S. and Toro, N. (1996) Accession X95567, Unpublished.

Zurkowski, W. (1981) Mol. Gen. Genet. 181, 522-524.

Zurkowski, W. (1982) J. Bacteriol. 150, 999-1007.

Zurkowski, W. and Lorkiewicz, Z. (1978) Acta Microbiol. Polonica 27, 309-319.

Zurkowski, W, and Lorkiewicz, Z. (1979) Arch. Microbiol. 123, 195-201.

Chapter 3

Outer Membrane Proteins

Ben J.J. Lugtenberg

I. The Cell Envelope of Gram-Negative Bacteria

The membrane of Gram-negative bacteria consists of three layers: (i) the cytoplasmic or inner membrane, (ii) the peptidoglycan layer and (iii) the outer membrane (Figure 1). The cytoplasmic membrane consists of a phospholipid bilayer in which proteins are present which can play a role in solute transport, generation of energy and synthesis of cell envelope components. The peptidoglycan layer consists of building blocks composed of N-acetylglucosamine, N-acetylmuramic acid and a few amino acids (usually L-alanine, D-glutamic acid, *meso*-diaminopimelic acid and D-alanine). It determines the cell shape and prevents lysis of the cytoplasmic membrane, and therefore of the cell, under hypotonic conditions. The outer membrane of *Enterobacteriaceae* protects the cell content against detergents, exogenous enzymes such as proteases and phospholipases and hydrophobic and large molecular weight antibiotics. This is the result of its specific structure. In the outer membrane the inner leaflet of the lipid bilayer consists of phospholipids whereas the outer leaflet consists of the fatty acids of the lipid A part of LPS (lipopolysaccharide). The phosphate and negatively charged KDO (keto-deoxyoctonate) residues of the inner core of the LPS molecule interact with divalent cations, thereby forming a virtually impermeable layer. Pore proteins form aqueous channels through the outer membrane which are permeable for hydrophilic molecules with a molecular mass of up to 600 to 700 daltons (assuming the molecules are more or less global). Such pores allow a fast permeation of nutrients, and reduce or prevent entry of hydrophobic and large molecular weight toxin and antibiotic molecules (reviewed in Lugtenberg and van Alphen, 1983).

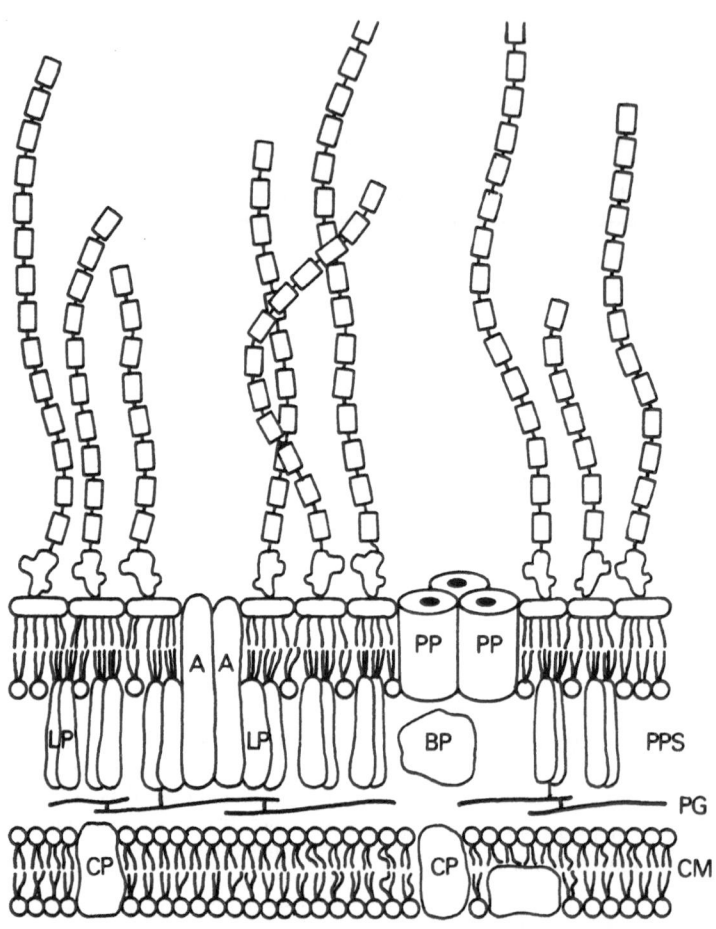

Figure 1. Molecular organization of the cell envelope of *Enterobacteriaceae*. The CM (cytoplasmic membrane) consists of a phospholipid bilayer and proteins, including CPs (carrier proteins) involved in transport of nutrients. It is surrounded by a single molecule of PG (peptidoglycan), which consists of amino sugar chains crosslinked to each other by peptide bridges. PG determines the cell shape and protects the cell from lysis by mechanically preventing expansion of the cytosol. Without the PG layer the cytosol - due to its high osmolarity - would take up water from the surrounding medium, resulting in explosion of the CM. The PPS is a compartment between the outer membrane and the CM. It contains specific proteins such as BPs (binding proteins) involved in transport of nutrients, and can contain enzymes such as alkaline phosphatase and β-lactamase. The outer membrane consists of phospholipids, LPSs (lipopolysaccharides) and proteins. Phospholipids and LPSs are exclusively found in the inner and outer leaflets, respectively, of the lipid bilayer. LPS molecules are tightly bound to each other by divalent cations. The major outer membrane proteins are PPs (pore proteins, which are aqueous channels through which nutrients can diffuse, A (OmpA protein, a transmembrane molecule which links the outer membrane non-covalently to the PG), and LP (Braun's lipoprotein, which is inserted in the inner leaflet of the bilayer by three fatty acids and of which the protein part is located in the PPS). One third of the LP molecules is covalently linked to the PG. Reprinted from: B. Lugtenberg and L. van Alphen (1983) *Bioch. Biophys. Acta* 737:51-115. With kind permission of Elsevier Science-NL, Sara Burgerhartstr. 25, 1055 KV Amsterdam, The Netherlands

The permeability properties of the outer membrane of *Rhizobiaceae* and of its pore proteins have never been studied. Several publications have indicated that the outer membrane of bacteroids differs from that of bacteria (Planqué *et al.*, 1979; van Brussel *et al.*, 1977; Bal and Wong, 1982; Bal *et al.*, 1985). In addition to pore proteins, the outer membrane of *Escherichia coli* contains other proteins such as Braun's lipoprotein (of which part of the molecules is covalently linked to peptidoglycan) and the OmpA protein. Both these proteins contribute to the rigidity of the outer membrane-peptidoglycan complex (Lugtenberg and van Alphen, 1983).

Bacteriophages use receptors in the outer membrane for recognition. These usually consist of LPS or protein-LPS complexes. In the latter case, *in vitro* experiments with OmpA protein indicate that LPS is required to keep the protein in a structure which has bacteriophage receptor activity (van Alphen *et al.*, 1977).

Flagella are used for swimming. The major part of the flagellum, the flagellar body, consists of monomers of the protein flagellin. Flagella are anchored in the cell envelope by ring structures connected to all three cell envelope layers. Fimbriae, used for attachment, usually consist of protein monomers but in *Rhizobium* (Smit *et al.*, 1987) and *Agrobacterium* (Matthysse *et al.*, 1981) they appear to consist of cellulose.

II. The Outer Membrane of *Rhizobiaceae*

In the mid-nineteeneighties attempts were undertaken to isolate the outer membrane of *Rhizobiaceae* for the following reasons. (i) Interactions of the bacterium's cell surface with the host plant can be expected in several stages of nodulation. (ii) *Rhizobium* has an unusually high calcium ion requirement for growth (Humphrey and Vincent, 1962; Humphrey and Vincent, 1965; Vincent and Humphrey, 1968). (iii) Fractionation of the bacterium will allow the sub-cellular localization of *nod* gene products. The localization of a protein may be helpful in elucidating its function and exclude other functions. (iv) In order to study the molecular basis of differentiation of bacterium into bacteroid it is first required to establish the differences in molecular terms.

II.A. SUB-CELLULAR FRACTIONA-TION OF CELLS OF *RHIZOBIACEAE*

Fractionation of Gram-negative bacteria is usually based on the following steps (i) plasmolysis of the cells, (ii) permeabilization of the outer membrane by EDTA treatment which allows lysozyme to degrade the peptidoglycan layer, (iii) osmotic shock, (iv) mechanical disruption of the cells and (v) differential centrifugation (Osborn *et al.*, 1972). De Maagd and Lugtenberg (1986) have described the sub-cellular fractionation of *Rhizobium leguminosarum* biovar *viciae* 248 cells into cytosol, cytoplasmic membrane, periplasm and outer membrane. In addition to being present in one of these fractions, a protein can also be secreted into the medium. The percentages of protein in these fractions have been estimated as 45, 19, 16, 19 and 1, respectively. Each fraction contains unique proteins as judged by SDS-PAGE. Initially no secreted proteins were detected (de Maagd and Lugtenberg, 1986). The outer and cytoplasmic membranes appeared to have buoyant densities of 1.243 and 1.137 g/cm^3, respectively, data well in accordance with those of other Gram-negative bacteria. This method for membrane separation worked also well for *R. leguminosarum* bv. *viciae* strain RBL-1 and for *Rhizobium leguminosarum* bv. *trifolii* strain RBL2020 (de Maagd and Lugtenberg, 1986). This cell fractionation method has been used for the localisation of the 24 kDa Rhi protein in the cytosol (de Maagd *et al.*, 1988b) and, after minor modifications, for the localization of the VirA protein in the cytoplasmic membrane (Melchers *et al.*, 1987).

II.B. PRELIMINARY CHARACTERI-ZATION OF OUTER MEMBRANE PROTEINS

Protein-containing fractions were incubated for 10 minutes in 2% SDS-1mM dithiothreitol and analyzed using SDS-PAGE (Lugtenberg *et al.*, 1975). The isolated cytoplasmic membrane fraction of *Rhizobium leguminosarum* bv. *viciae* strain 248 can be separated in over 20 protein bands, ranging in apparent molecular weight from 14 kDa to approximately 100 kDa. The outer membrane protein pattern is dominated by a group of approximately eight bands with apparent MWs ranging from 36 to 45 kDa (de Maagd and Lugtenberg, 1986) (Table 1).

II.C. IMMUNOLOGICAL CHARACTERI-ZATION OF OUTER MEMBRANE ANTIGENS

Polyclonal rabbit antiserum raised against whole cells of *Rhizobium leguminosarum* bv. *viciae* strain 248 reacts in Western blots with three outer membrane antigens (de Maagd *et al.,* 1989a). The band at position 19 kDa has been identified as LPS whereas the 28 and 34 kDa bands turned out to be components of flagella. These results are consistent with results from classical serological analyses which indicate that rhizobia show O- (somatic)-type and H-type reactions with homologous antisera (Graham, 1969).

The antiserum raised against whole cells hardly reacts with outer membrane proteins in Western blots. In order to obtain antibodies, which do react with these proteins in Western blots, a different strategy was used. After a primary injection with a native cell envelope preparation, booster injections were given with a presumably (partially) denatured 2% SDS-insoluble peptidoglycan/outer membrane fraction. Since LPS and flagellin are not detectable in such fractions, whereas outer membrane proteins are abundant, a good reaction with outer membrane proteins in Western blots was expected. Indeed, this polyclonal antiserum recognized a large number of antigens in a Western blot of a cell envelope preparation (de Maagd *et al.,* 1989a). To study a possible immunological relationship between outer membrane proteins, Mabs (monoclonal antibodies) were produced in a similar way. Of eight mouse Mabs, three reacted with LPS. Surprisingly, the other five Mabs all react with more than one outer membrane protein band. A set of proteins reacting with the same Mab is therefore refered to as an antigen group (Table 1). Epitopes recognized by Mab's 3, 16, 24, 38 and probably 40 are accessible in whole cells whereas the epitopes recognized by Mab's 8, 20 and 37 are either only present after denaturation by SDS or are too deeply buried in the outer membrane to be detected in any other way than after solubilization of the outer membrane.

II.D. COMPARISON OF OUTER MEMBRANE ANTIGENS IN FREE-LIVING BACTERIA AND BACTEROIDS

The protein profile of bacteroid cell envelopes is strikingly different from that of free-living rhizobia.

This difference is also reflected in the antigens as analyzed by Western blotting. In bacteroid samples antigen I is present whereas reaction with antigen IV and a number of minor protein antigens is still observed but seems to be somewhat less intense than with the bacterial cell envelope components. The reaction with antigen groups II and III is approximately 10-fold decreased. It is not clear whether the remaining activity is real or due to a contamination of the preparation with free-living bacteria (de Maagd *et al.,* 1989a).

Using anti-LPS monoclonals it was also concluded that either the relative amounts of the two LPS species are lower in bacteroid cell envelopes or that bacteroid LPS is altered. These results are consistent with earlier findings which show a decreased or changed LPS content of the bacteriod cell envelope (Planqué *et al.,* 1979; van Brussel *et al.,* 1977; Chapters 7 and 22).

II.E. ACCESSIBILITY OF OUTER MEMBRANE PROTEIN ANTIGENS IN WHOLE CELLS

Whereas the LPS antigens recognized by Mab's 3, 16 and 24 are exposed at the cell surface and are accessible for immunoglobulins, the outer membrane protein antigens recognized by Mab 38 (group III) and Mab 40 (group I) are not accessible in whole cells. It appeared that antigen accessibility can be modulated by changes in LPS and Ca^{2+} content of the cell envelope (de Maagd *et al.,* 1989a).

Rhizobium has a high requirement for Ca^{2+} for normal growth (Vincent, 1962). Ca^{2+} is concentrated in the cell wall (Humphrey and Vincent, 1962; Vincent and Humphrey, 1963). Deficiency of Ca^{2+} results in abnormal cell shape (Vincent, 1962), increased fragility (Humphrey and Vincent, 1965) and changes in antigenic properties (Humphrey and Vincent, 1965; Vincent and Humphrey, 1968). In Trypton-Yeast (TY) medium the presence of calcium chloride (7 mM) is needed for prolonged growth of strain 248. After transfer of TY/Ca^{2+} grown cells to TY medium, further growth results in decreased relative amounts of LPS which is paralleled by the appearance of LPS in the growth medium. Under these conditions the cells also react with Mab 38, which recognizes group III proteins. The suggestion that this protein is shielded by the O-antigen part of LPS is further supported by the observation that cells of a mutant strain which completely lacks the LPS species with low electrophoretic mobility, despite

Apparent Mr of Antigen (kDa)	Reaction with Antibody Type	Identity, Characteristics
17	Mab40	Antigen group I
19	Polyclonals, Mab3, Mab16, Mab24	O-antigen-containing LPS
22, 24, 26[a)]	Mab8	Antigen group IIa. Primary protein encoded by *rop*B with a calculated Mr of 20.3 kDa for the mature protein. Synthesized with a 23 amino acid signal peptide. Down-regulated in bacteroids
28	Polyclonal	Flagellar component
34	Polyclonal	Flagellar component
36-39[a)b)]	Mab37, Mab20, Mab38	Low Mr family of approximately four antigen group III proteins. Primary protein encoded by *rop*A with a calculated Mr of 31.486 kDa for the mature protein. Synthesized with a 22 amino acid signal peptide. Down-regulated in bacteroids. Involved in receptor activity for various bacteriophages
40-45[a)b)]	Mab37, Mab20, Mab38	High Mr family of approximately four antigen group III proteins. Primary protein encoded by *rop*A2 with calculated Mr of 36.028 kDa for the mature protein. Synthesized with a 22 amino acid signal peptide. Down-regulated in bacteroids
48	Weak reaction with Mab8	Therefore designated as antigen group IIb
74	Polyclonal	Antigen group IV
Antigen smear[a)]	Mab37	Oligomers op antigen group III

Table 1. Cell surface antigens of *R. leguminosarum* bv. *viciae* strain 248
[a)]Most protein molecules of group II and III are covalently bound to peptidoglycan. Families of such proteins form small ladders of approximately 3 to 4 bands which are supposed to represent the unsubstituted monomer and monomers with an increasing number of peptidoglycan subunits, resulting from degradation of the peptidoglycan by lysozyme.
[b)]Group III proteins are shielded from exogenously added antibodies by the O-antigen part of LPS. They form Ca^{2+}-stabilized oligomers which resist complete denaturation in 2% SDS at 100 EC. Likely to have porin functions.

growth in Ca^{2+}-supplemented TY, appear to be agglutinated by Mab 38. Thus, the O-antigen part of LPS shields antigen 38 from reacting with immunoglobulin. It is assumed that Ca^{2+} plays a role in maintaining an outer membrane structure responsible for shielding antigen 38 by the O-antigen of LPS (de Maagd *et al.*, 1989a).

II.F. COMPARISON OF OUTER MEMBRANE PROTEINS OF VARIOUS MEMBERS OF THE *RHIZOBIACEAE*

Comparison of the membrane protein patterns of eighteen *R. leguminosarum* strains of biovars *viciae*, *trifolii* and *phaseoli* with those of six *Sinorhizobium meliloti* and two *Agrobacterium tumefaciens* strains

revealed the following (de Maagd *et al.*, 1988a). (i) Outer membrane proteins I, II and IV, as identified using antisera, are present in all 18 *R. leguminosarum* strains and have only minor differences in electrophoretic mobility. Proteins of antigen group III show large strain-dependent differences in the number of bands and in their electrophoretic mobilities. No biovar-specific features were observed (de Maagd *et al.*, 1988a). Since no differences in outer membrane profiles between wild type strain 248 and a Sym plasmid-cured derivative were observed (de Maagd *et al.*, 1988b) these differences are not due to the Sym plasmid. (ii) In *S. meliloti* and *A. tumefaciens* epitopes of protein antigen groups I and IV are present, whereas those of protein group II are absent. The profiles of the presumable group III proteins deviate from those of *R. leguminosarum* strains. Reaction of Mab 37 with *S. meliloti* antigens was weak whereas a strong cross-reaction was observed with *A. tumefaciens*. These results confirm the already established distinction between *R. leguminosarum* and *S. meliloti* as well as between *Rhizobium* and *Agrobacterium* (de Maagd et al., 1988a).

II.G. OUTER MEMBRANE PROTEIN OLIGOMERS AND INTERACTION OF OUTER MEMBRANE PROTEINS WITH PEPTIDOGLYCAN AND PHAGES

In contrast to the situation with *Enterobacteriaceae* (Lugtenberg and van Alphen, 1983), a substantial amount of *Rhizobium* outer membrane protein can only be visualized in SDS-PAGE after lysozyme treatment of the isolated cell envelopes, suggesting a very strong, possibly covalent, interaction of these proteins with peptidoglycan. These proteins belong to the two distinct groups of immunologically related proteins, groups II and III. Most members of both groups can be radioactively labelled by growing cells in the presence of (^{3}H)*N*-acetylglucosamine. Therefore it was proposed that variation in apparent molecular weight of the antigens within each group is caused by varying numbers of peptidoglycan subunit residues on only two or three different outer membrane proteins (de Maagd et al., 1989b). This phenomenon has also been described for Braun's lipoprotein of *E. coli* (Wensink and Witholt, 1981). Another novel feature of *Rhizobium* is that group III outer membrane proteins form oligomers stabilized by divalent cations, of which Ca^{2+} is the best, which resist complete denaturation in 2% SDS at 100 EC (de Maagd et al., 1989b).

When *ropA* (rop for *Rhizobium* outer membrane protein) was expressed under the *lac* promoter in *E. coli*, RopA was found to form Ca^{2+}-stabilized outer membrane oligomers, indicating that both outer membrane localization and the formation of Ca^{2+}-stabilized oligomers do not depend on some unique property of *R. leguminosarum*. In contrast, the high extent of peptidoglycan-linkage of protein IIIa does not occur in *E. coli* and is apparently unique for *R. leguminosarum*. This suggests either that *E. coli* peptidoglycan is so different from that of *R. leguminosarum* that linkage of the protein is not possible, or that *E. coli* lacks a protein-peptidoglycan linking mechanism that is present in *R. leguminosarum* (de Maagd et al., 1992).

Strain RBL 5523-ZA4 was obtained as a mutant of strain RBL5523 that is resistant to the virulent phages K2, K5 and RL38. In comparison with the wild type, the mutants lack the fastest moving series of proteins of group III antigens (de Maagd et al., 1989b). These results indicate that the missing subgroup III proteins form transmembrane Ca^{2+}-stabilized oligomers which are covalently bound to peptidoglycan and function, possibly stabilized by LPS which shields it from antibody molecules, as the receptor for the mentioned phages.

II.H. INFLUENCE OF SYM PLASMID AND OF INDUCTION OF *NOD* GENES ON OUTER MEMBRANE PROTEIN PROFILE OF *RHIZOBIUM*

The presence of the Sym plasmid pRL1JI in *R. leguminosarum* bv. *viciae* strain 248 has no apparent influence of the outer membrane protein profile. In contrast, the cytosolic fraction of only Sym$^+$ cells contains a 24 kDa protein band identified as the Rhi protein (de Maagd et al., 1988b).

Induction of the *nod* genes of strain 248 did not yield many new proteins, which could be related to *nod* gene products. Only a 50 kDa protein, whose expression is dependent on both *nodD* and a *nod* gene inducer and which later was identified as NodO (de Maagd et al., 1989d), was observed in the secreted fraction.

The cell fractionation method developed by de Maagd and Lugtenberg (1986) has been used as the basis of localisation of *nod* gene products in combination with antibodies. In this way Schlaman et al. have localized NodD (1989) and NodI (1990) in the cytosol fraction whereas NodA (1992) was found in both the cytosolic and cytoplasmic membrane fractions in *R. leguminosarum* biovar *viciae*. Although NodC has been reported to be localized in the outer membrane (John et al, 1988), more recent results suggest it to be localized exclusively in the cytosolic membrane (Barny and Downie, 1993).

II.I. MOLECULAR CHANGES IN OUTER MEMBRANE PROTEINS DURING THE DEVELOPMENT OF BACTERIUM TO BACTEROID

Down-regulation of antigen groups II and III is a general phenomenon in the rhizobial bacteroid differentiation process. Therefore elucidation of the molecular mechanism behind these outer membrane protein changes is likely to contribute to the elucidation of the mechanism of bacteroid development. The decrease in antigen groups II and III was also found in *R. leguminosarum* bv. *viciae nifA* and *nifK* mutants and in the non-fixing pea mutant FN1 inoculated with wild type strain 248. This indicates that nitrogen fixation is not a prerequisite for the decrease in the levels of these antigens (Roest *et al.*, 1995a).

The expression of outer membrane protein *ropA*, which encodes the low Mr part of the two protein families, which constitute outer membrane protein antigen group III, during nodule development, was followed using immuno-electron microscopy and *in situ* hybridization. The first approach, now using vetch nodule sections, confirmed that RopA protein decreases. Moreover, it showed that this occurs in the nodule after release of the bacteria from the infection thread, during the transition in morphology from that of the free-living bacterium into that of the bacteroid (de Maagd *et al.*, 1994). Detection of *ropA* m-RNA in sections of pea nodules by *in situ* hybridization revealed an abrupt decrease in m-RNA level from one cell layer to the next layer, exactly at the transition from pre-infection zone II to interzone II-III. This decrease coincided with a sudden increase in *nifH* mRNA levels. Expression of *nifH* starts abruptly and the first cell layer in which *nifH* is expressed is the first cell layer in which *ropA* expression abruptly decreases. This suggests that these processes might be regulated through a similar mechanism. Although the decrease in *ropA* mRNA and the appearance of *nif* mRNA are spatially correlated, *ropA* down-regulation could be uncoupled from *nif* gene activation by using a strain that induces non-nitrogen fixing nodules on pea but does develop into bacteroids (de Maagd *et al.*, 1994). It should be noted that *ropA2*, which encodes the high Mr family of outer membrane protein antigen group III, is 91.8% identical to *ropA* at the nucleotide level (Roest *et al.*, 1995c). Therefore it is likely that the above-described experiments describe not only the fate of RopA but also that of all group III antigens.

Attempts to mimick down-regulation of group III proteins *in vitro* by incubating cells under conditions, which may be present in the nodule, such as low oxygen and succinate as the carbon source, only yielded the observation that high Ca^{2+} concentrations repress *ropA* expression. It is not clear whether a sharp change in Ca^{2+} concentration could occur from one cell layer to the next (Roest *et al.*, 1995c).

III. Genetics of Outer Membrane Proteins

III.A. *ROPA*

RopA is a 36 kDa outer membrane protein which constitutes the low Mr part of antigen group III of *R. leguminosarum* bv. *viciae* strain 248, whose levels are severely decreased during symbiosis. *ropA* encodes a 366 amino acids protein with a predicted molecular mass of 39 kDa. Sequencing of the N-terminus of RopA showed that *ropA* is the structural gene of this protein. After processing of a 22 amino acid signal peptide, mature RopA is predicted to have a molecular weight of 37 kDa, comparable to the apparent molecular mass (de Maagd *et al.*, 1992).

A *ropA* homologous gene was isolated from *R. leguminosarum* bv. *trifolii* strain LPR 5020. Except for one amino acid, the N-terminal 51 amino acids are identical to RopA of strain 248 (Roest *et al.*, 1995c).

III.B. *ROPA2* (ROEST *ET AL.*, 1995C)

In an attempt to isolate the structural gene(s) encoding the high Mr part of antigen group III, a *ropA* homologous DNA fragment was isolated which appeared to be required for the synthesis of the high Mr part of antigen group III. At the nucleotide level it is 91.8% identical to *ropA*. It encodes a putative protein of 38.1 kDa with a predicted signal peptide of 22 amino acids. The resulting molecular weight for the mature protein of 36.1 kDa is in reasonable agreement with the 40 kDa observed for the expected unsubstituted protein in SDS-PAGE. The putative signal sequence of this protein has the same length as that of RopA. The protein's deduced amino acid sequence is almost 70% identical to that of RopA and also contains the C-terminal phenylalanine often observed in outer membrane proteins. These results indicate that the gene designated as *ropA2* is the second structural gene of antigen group III. Its expression results in a family of approximately four protein bands.

III.C. *ROPA* AND *ROPA2* PROBABLY ENCODE PORINS

ropA and *ropA2* have a 54 to 62% amino acid similarity with proteins Omp2a and Omp2b, respectively, of the facultative intracellular bacterium *Brucella abortus* (Ficht *et al.*, 1988). The latter genes are believed to encode porins (Ficht *et al.*, 1989). Therefore and because of their Mr and the fact that their expression levels are high, it is very likely that also *ropA* and *ropA2* encode porins. It appeared that after infection of their animal host cells the expression of *B. abortus omp2b* is severely depleted (Roest *et al.*, 1995c), a phenomenon similar to the down-regulation of both *ropA* and *ropA2* in bacteroids.

III.D. *ROP*B (ROEST *ET AL.*, 1995B)

This structural gene of outer membrane protein antigen group IIA was isolated by expression in *R. meliloti* LPR2120 using Mab8. The open reading frame encodes a protein with a calculated molecular mass of 22.5 kDa. After subtraction of the predicted 23 amino acid signal sequence, a Mr of 20.3 kDa was calculated for the mature protein, which corresponds well with the 22 kDa estimated from SDS-PAGE. *ropB* under its own promoter is not expressed in *E. coli*. RopB contains a C-terminal phenylalamine which has been shown to be required for efficient translocation of outer membrane proteins in other organisms. Membrane spanning β-sheets of no more than 10 amino acids that have an amphipathic character and that are characteristic for outer membrane proteins are predicted to represent the major part of its secondary structure. Using the criterion of the membrane spanning amphipatic β-sheets together with other criteria governing the folding of outer membrane proteins (Tommassen, 1988) a model of the topology of RopB was predicted (Roest *et al.*, 1995b). *ropB* is well-conserved in *Rhizobium* and *A. tumefaciens* strains. No homology was visible with *B. japonicum*, *P. fluorescens* and *E. coli* (Roest *et al.*, 1995b).

III.E. REGULATION OF EXPRESSION OF *ROPA*, *ROPA2*, AND *ROPB*

The levels of the proteins which together constitute antigen group III of *R. leguminosarum* bv. *viciae* strain 248 are severely decreased during symbiosis. *ropA* and *ropA2* are the structural genes for the low and high Mr groups, respectively, of these proteins. The *ropA* promoter shows no homology to known promoter sequences. It contains two potential binding sites for integration host factor protein. The *ropA* promoter is not active in *E. coli* (de Maagd *et*

al., 1992). *ropA* and *ropA2*, together encoding group III antigens, have in common (i) a probably functional IHF (integration host factor) binding site, and (ii) the absence of a clear consensus sequence for the binding of RNA polymerase. No other regulatory consensus sequence was found (Roest *et al.*, 1995c). No homology of the promoter of *ropB*, which encodes antigen group II, with other known promoter sequences was found either.

Antigen group IIA proteins, encoded by *ropB*, are also severely decreased during symbiosis. Despite testing a large series of circumstances, no condition was found under which a reduction of the Mab8 epitope to bacteroid levels was observed. No clear -10 and -35 promoter sequence for RNA polymerase were found.

Considering the observation that down-regulation of antigen groups II and III during symbiosis is a general phenomenon in rhizobia (Roest *et al.*, 1995a), a consensus in the promoters was more or less expected. However, recent experiments indicate that the decrease in antigen group III is not observed in bacteroids isolated from nodules of the determinate nodulating host plant *Phaseolus vulgaris* (Roest *et al.*, 1995c). These results indicate that the regulation of antigen groups II and III can be different and that they therefore are subject to independent regulatory mechanisms. They also show that the host plant plays a role in down-regulation.

IV. References

Bal, A.K. and Wong, P.P. (1982) Can. J. Microbiol. 28, 890-896.

Bal, A.K., Sen, D. and Weaver, R.W. (1985) Current Microbiol. 12, 353-354.

Barny, M.A. and Downie, J.A. (1993) Mol.Plant-Microbe Int. 6 (5):669-672.

de Maagd, R.A. and Lugtenberg, B.J.J. (1986) J. Bacteriol. 167, 1083-1085.

de Maagd, R.A., van Rossum, C. and Lugtenberg, B.J.J. (1988) J. Bacteriol. 170, 3782-3785.

de Maagd, R.A., Wijffelman, C.A., Pees, E. and Lugtenberg, B.J.J. (1988) J. Bacteriol. 170, 4424-4427.

de Maagd R.A., de Rijk, R., Mulders, I.H.M. and Lugtenberg, B.J.J. (1989) J. Bacteriol. 171, 1136-1142.

de Maagd, R.A., Wientjes, F.B.,and Lugtenberg B.J.J. (1989) J. Bacteriol. 171, 3989-3995.

de Maagd, R.A., Spaink, H.P., Pees, E., Mulders, I.H.M., Wijfjes, A., Wijffelman, C.A., Okker, R.J.H. and Lugtenberg, B.J.J. (1989a) J. Bacteriol. 171, 1151-1157.

de Maagd, R.A., Wijfjes, A.H.M., Spaink, H.P., Ruiz-Sainz, J.E., Wijffelman, C.A., Okker, R.J.H. and Lugtenberg, B.J.J. (1989b) J. Bacteriol. 171, 6764-6770.

de Maagd, R.A., Mulders, I.H.M., Canter Cremers, H.C.J. and Lugtenberg, B.J.J. (1992) J. Bacteriol. 174, 214-221.

de Maagd, R.A., Yang, W-C., Goosen-de Roo, L., Mulders, I.H.M., Roest, H.P., Spaink, H.P., Bisseling, T. and Lugtenberg, B.J.J. (1994) Mol. Plant-Microbe Int. 7, 276-281.

Ficht, T.A., Bearden, S.W., Sowa, B.A. and Adams, L.G. (1988) Infect. Immun. 56, 2036-2046.

Ficht, T.A., Bearden, S.W., Sowa, B.A. and Adams, L.G. (1989) Infect. Immun. 57, 3281-3291.

Graham, P.H. (1969) Analytical Serology of Bacteria, J. Wiley and Sons, N.Y., pp. 353-378.

Humphrey, B.A. and Vincent, J.M. (1962) J. Gen. Microbiol.29, 557-561.

Humphrey, B.A. and Vincent, J.M. (1965) J. Gen. Microbiol. 41, 109-118.

John,M., Schmidt, J., Wieneke,U., Krussman, H.D. and Schell, J. (1988) EMBO J. 7:583-588.

Lugtenberg, B.J.J., Meyers, J., Peters, R., van der Hoek, P. and van Alphen, L. (1975) FEBS Lett. 58, 254-258.

Lugtenberg, B.J.J. and Peters, R. (1976) Biochim. Biophys. Acta 441, 38-47.

Lugtenberg, B.J.J. and van Alphen, L. (1983) Biochem. Biophys. Acta 737, 51-115.

Matthysse, A.G., v. Holmes, K.V. and Gurlitz, R.H.G. (1981) J. Bacteriol. 145, 583-595.

Melchers, L.S., Thompson, D.V., Idler, K.B., Neuteboom, S.T.C., de Maagd, R.A., Schilperoort, R.A. and Hooykaas, P.J.J. (1987) Plant Molec. Biol. 9, 635-645.

Osborn, M.J., Bander, J.E., Parisi, E. and Carson, J., (1972) J.Biol. Chem. 247, 3962-3972.

Planqué, K., van Nierop, J.J., Burgers, A. and Wilkinson, S.G. (1979) J. Gen. Microbiol. 110, 151-159.

Roest, H.P., Goosen-de Roo, L., Wijffelman, C.A., de Maagd, R.A. and Lugtenberg, B.J.J. (1995) Mol. Plant-Microbe Int. 8, 14-22.

Roest, H.P., Mulders, I.H.M., Wijffelman, C.A. and Lugtenberg, B.J.J. (1995) Mol Plant-Microbe Int. 8, 576-583.

Roest, H.P., Bloemendaal, C-J.P., Wijffelman C.A. and Lugtenberg, B.J.J. (1995) J. Bacteriol. 177, 4985-4991.

Schlaman, H.R.M., Spaink, H.P., Okker, R.J.H. and Lugtenberg, B.J.J. (1989) J. Bacteriol. 171, 4686-4693.

Schlaman, H.R.M., Okker, R.J.H. and Lugtenberg, B.J.J. (1990) J. Bacteriol. 172, 5486-5489.

Schlaman, H.R.M., Okker, R.J.H. and Lugtenberg, B.J.J. (1992) J. Bacteriol. 174, 5177-5182.

Smit, G., Kijne, J.W. and Lugtenberg, B.J.J. (1987) J. Bacteriol. 169, 4294-4301.

Tommassen, J. (1988) in J.A.F. Op den Kamp (ed.), NATO ASI Series, "Membrane Biogenesis", Springer-Verlag, Berlin, pp. 352-373.

van Alphen, L., Havekes, L. and Lugtenberg, B. (1977) FEBS Lett. 75, 285-290.

van Brussel, A.A.N., Planqué, K. and Quispel, A. (1977) J. Gen. Microbiol. 101, 51-56.

Vincent, J.M. (1962) J. Gen.Microbiol. 28, 658-663.

Vincent, J.M. and Humphrey, B.A. (1963) Nature (London) 199, 149-150.

Vincent, J.M. and Humphrey, J. (1968) J. Gen. Microbiol. 54, 397-405.

Wensink, J. and Witholt, B. (1981) Eur. J. Biochem. 113, 349-357.

Chapter 4

Phospholipids and Alternative Membrane Lipids

Otto Geiger

I. Introduction

The cell envelope of Gram-negative bacteria, such as the *Rhizobiaceae*, is built up of three distinctive layers. These layers are the inner or cytoplasmic membrane, the peptidoglycan sacculus, and the outer membrane. Both membranes contain membrane-forming lipids, mainly phospholipids, and proteins (Chapter 3). Outer membranes contain as additional component lipopolysaccharides. In this chapter the composition, function, and organization of phospholipids and other membrane lipids of members of the *Rhizobiaceae* will be discussed.

The phospholipid composition of the *Rhizobiaceae* is more complex than that of *Escherichia coli*. In addition to phosphatidylglycerol, cardiolipin (also called diphosphatidylglycerol), and phosphatidyl-ethanolamine also methylated derivatives of phosphatidylethanolamine, among them phosphatidylcholine, are found. Phospholipids are subject to lipid turnover. Other integral membrane lipids have been reported in only some members of the *Rhizobiaceae*. Sulfolipids are found in symbiotic species of the *Rhizobiaceae*. Hopanoids, structural relatives of steroids, are found only in representatives of the genus *Bradyrhizobium*.

The structures of fatty acids, the major building blocks of phospholipids, are very diverse as well. This complex situation in fatty acid structures is reflected by complex channelling of biosynthetic systems by various different acyl carrier proteins.

II. Lipid Components

The study of membrane lipids of *Escherichia coli* has contributed greatly to our understanding of the synthesis and functions of membrane lipids. *E. coli* contains three major phospholipids: phosphatidyl-ethanolamine (PE), phosphatidylglycerol (PG), and cardiolipin (CL). PE, as the major phospholipid of *E. coli*, constitutes about 75% of the total phospholipids (Cronan and Rock, 1996). The relative amounts of PG and CL depend on the growth phase of the cultures; PG is most abundant in log-phase cells, whereas CL accumulates in stationary-phase cells. The phospholipids, neutral lipids, and lipoproteins of *E. coli* contain similar fatty acids, consisting of the saturated fatty acids palmitic (hexadecanoic) acid and myristic (tetradecanoic) acid and the monounsaturated fatty acids palmitoleic (*cis*-9-hexadecenoic) acid and *cis*-vaccenic

(*cis*-11-octadecenoic) acid. Traces of lauric (dodecanoic), stearic (octadecanoic), and *cis*-7-tetradecenoic acids are also present (Cronan and Rock, 1996). Studies on the lipid composition of cytoplasmic and outer membranes of *E. coli* showed that the outer membrane is enriched in both saturated fatty acids and PE if compared with the cytoplasmic membrane (Lugtenberg and Peters, 1976). Like other Gram-negative bacteria, the *Rhizobiaceae* possess besides a cytoplasmic membrane an outer membrane. Cell envelopes of *R. leguminosarum* could be separated into cytoplasmic membrane and outer membrane fractions (Maagd and Lugtenberg, 1986). A similar fractionation of cytoplasmic membrane and outer membrane of *S. meliloti* demonstrated that most phospholipids were present in similar amounts in both membranes. Only cardiolipin was reduced in outer membranes (Hubac *et al.*, 1992).

Lipid synthesis in *E. coli* is divided into two parts: (i) synthesis of the fatty acids which are responsible for the characteristic hydrophobicity of lipids and (ii) attachment of completed fatty acids to *sn*-glycerol-3-phosphate followed by the addition and modification of the polar head groups to yield the major cellular phospholipids.

Medium chain intermediates of the fatty acid synthetic pathway are required for the synthesis of the coenzymes biotin and lipoic acid (de Moll, 1996) and fatty acid synthesis intermediates are also used for the synthesis of the acylated homoserine lactones involved in cell density-dependent signalling (Fuqua *et al.*, 1996; Moré *et al.*, 1996) (Chapter 7).

II.A. LIPID COMPONENTS IN *RHIZOBIACEAE*

Granules of polyhydroxybutyrate (PHB) are often present in rhizobia, and can account for up to 98% of lipid extracts. Although phospholipids can comprise over 90 % of the other lipids in rhizobia (Bunn and Elkan, 1971) and agrobacteria (Kaneshiro and Marr, 1962) substantial amounts of neutral lipids are sometimes encountered. These include the ubiquinone Q-10 (in all three genera) (Collins and Jones, 1981; Yamada *et al.*, 1982), α-tocopherolquinone (reported for rhizobia) (Hughes and Tove, 1982), fatty acids and acylglycerols (Das *et al.*, 1979; Gerson and Patel, 1975; Manasse and Corpe, 1967; Miller and Tremblay, 1983), fatty acid methyl esters, fatty alcohols and hydrocarbons (Gerson and Patel, 1975; Miller and Tremblay, 1983).

III. Genetic Analysis of Lipid Metabolism

Genetic analysis of lipid metabolism has turned out to be difficult because only few auxotrophic mutants were isolated in which the genetic defect could be complemented by a lipid present in the growth medium. Auxotrophic mutants that require fatty acids (Silbert, 1967) or sn-glycerol-3-phosphate (Bell, 1974) are described.

A second type of mutant selection is by tritium suicide (Rock and Cronan, 1985). Large quantities of highly ^{3}H-labeled lipid precursors are incorporated into mutagenized cells at the nonpermissive growth temperature (usually 42°C) and then the labeled cells are stored at -80°C. During storage, those cells competent in lipid synthesis are irradiated by the disintegration of the incorporated radioactive precursor, and they die, whereas those cells mutationally defective in lipid synthesis survive due to their lack of incorporation of the radioactive precursor.

A very useful method has been developed by Raetz (Raetz, 1975) to screen colonies of mutagenized cells, either by performing a given lipid biosynthetic enzyme assay on isolated bacterial colonies or by screening such colonies for the inability to synthesize a given lipid. The use of such brute force colony autoradiography schemes has proven extremely valuable in isolating phospholipid mutants. This method was recently also successfully employed for the isolation of mutants from *Rhizobium meliloti* deficient in phospholipid *N*-methyltransferase (Figure 1) (de Rudder *et al.*, 1997).

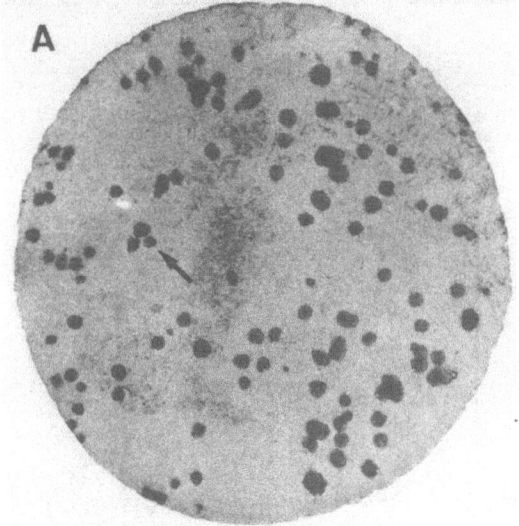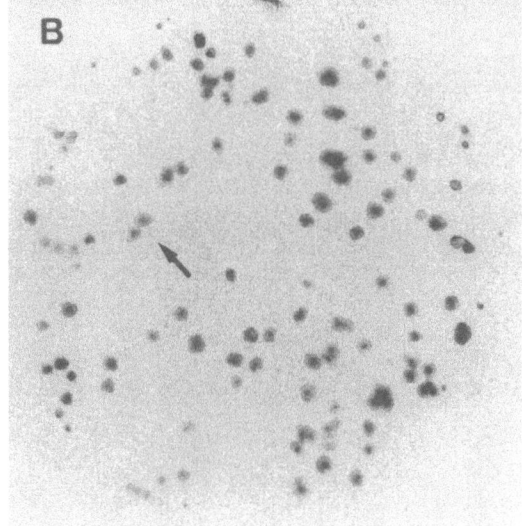

Figure 1. Isolation of *S. meliloti* mutants deficient in phospholipid *N*-methyltransferase by colony autoradiography (de Rudder *et al.*, 1997).

Mutagenized colonies growing on the agar surface in a Petri dish are blotted onto a sheet of filter paper. The paper is then immersed in a solution which lyses the colonies (membranes and proteins of the lysed cells remain bound to the paper) and is subsequently immersed in a solution containing the precursor and buffer necessary for the enzyme reaction. A water-soluble substrate (*S*-adenosylmethionine) of the reaction is included in a radioactively labeled form. The immobilized lysed cells are allowed to incorporate the radioactive precursor for a short time; then the filter is washed repeatedly with a solution in which lipids are insoluble but in which water-soluble compounds remain in solution. The washed paper is stained with a specific protein stain, dried, and exposed to an X-ray film. When the resulting autoradiogram (B) is compared with the stained filter paper (A), colonies that lack enzymatic activity (arrow) can be identified. The mutant colony is then isolated from the original agar plate.

IV. Fatty Acids in *Rhizobiaceae*

The fatty acid profiles of all species in this family which have been examined so far show that, like in *E. coli*, palmitic (16:0), palmitoleic (16:1(9c)), and *cis*-vaccenic (18:1(11c)) acid are the predominant fatty acids in the *Rhizobiaceae* (Table 1). However, the relative amount of *cis*-vaccenic acid and fatty acids derived from it (see below) exceeds 50% of the total fatty acids in most members of the *Rhizobiaceae* and therefore is much higher than in *E. coli*. One fatty acid, derived from *cis*-vaccenic acid (18:1(11c)), is lactobacillic acid (*cis*-11,12-methyleneoctadecanoic acid; 19:0 cyclo) and it contains a cyclopropane ring instead of a *cis* double bond. Other derivatives, among them 11-methyl-*cis*-vaccenic acid (11-Me-VA), can be understood as ring-opened products of lactobacillic acid (Figure 2) (Gerson *et al.*, 1975). These novel fatty acids have not yet been reported for agrobacteria (Gerson and Patel, 1975; Gerson *et al.*, 1975; Mac Kenzie *et al.*, 1979). Fatty acid composition of different classes of lipids has been determined for *Agrobacterium*

tumefaciens (Hildebrand and Law, 1964), *R. meliloti* (Miller and Tremblay, 1983) and two strains of rhizobia infective on *Lotus pedunculatus* (Gerson and Patel, 1975). Comparisons of fatty acid compositions have also been made for the free-living and the symbiont (bacteroid) forms of rhizobia (Gerson and Patel, 1975; Miller and Tremblay, 1983).

A recent study compared fatty acid profiles of 123 strains of the genus *Rhizobium* (Jarvis and Tighe, 1994) and tried to identify the fatty acids based on their retention times on a gas chromatograph with the aid of a computerized Microbial Identification System software. They showed that all species contained substances having retention times identical to fatty acid methyl esters of 16:0, 17:0, 18:0, and 19cycloω9c fatty acids. However, further characterization by mass spectrometry in order to confirm the proposed structures was not performed.

In some species of *Rhizobium* characteristic substances were present or absent. Only *R. tropici* and *M. loti* lacked substances with retention times like 16:1ω7c, 16:0 3-OH, and 20:3ω6,9,12c. Only

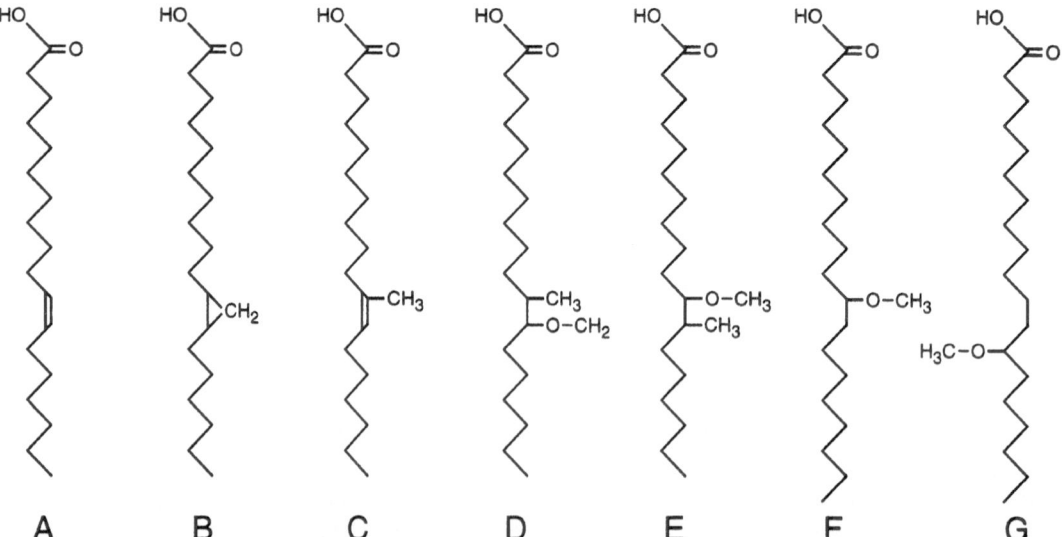

Figure 2. Unusual fatty acids derived from *cis*-vaccenate.
Shown are *cis*-vaccenate (A), lactobacillic acid (B) (Gerson *et al.*, 1975) and expected products of ring opening with OH and subsequent methylation: 11-methylvaccenate (11-methyloctadec-11-enoate; 11-Me-VA) (C), 12-methoxy-11-methylstearate (D), 11-methoxy-12-methylstearate (E), 11-methoxynonadecanoate (F), and 13-methoxynonadecanoate (G).

R. leguminosarum contained a substance with retention time like 15:0 2-OH. Only *M. loti* produced substances with retention times like 12:0 3-OH, 13:0 iso 3-OH, 18:1ω9c. Fatty acids found only in *R. tropici* were 15:0 iso 3-OH, 17:0 iso 3-OH, and 20:2ω6,9c. A comparison of gas chromatographic fatty acid methyl ester elution profiles was used to show that the strains could be divided into clusters corresponding to the six species (*R. leguminosarum, S. meliloti, R. galegae, S. fredii, M. loti,* and *R. tropici*). Fatty acid profiles for each species were described and allow a correct identification of unknown rhizobial species in 95% of the cases (Jarvis and Tighe, 1994).

IV.A. HYDROXY FATTY ACIDS

In the course of normal fatty acid biosynthesis also 3-hydroxy fatty acids are formed. In *Enterobacteriaceae* 3-hydroxy fatty acids, and especially 3-hydroxymyristic acid are common acyl substitutions of the lipid A part of the LPS. Besides 3-hydroxy fatty acids, the *Rhizobiaceae* contain also long (ω-1)-hydroxy fatty acids. An unusually long 27-hydroxyoctacosanoic acid was originally identified as a major structural fatty acyl component of the lipopolysaccharide of *R. leguminosarum* bv. *trifolii* (Hollingsworth and Carlson, 1989). Numerous representatives of the α-2 subgroup of the proteobacteria contain long (ω-1)-hydroxy fatty acids in their lipopolysaccharides (Bhat *et al.*, 1991a). Among these organisms are pathogens like the bacterial agent causing "cat-scratch disease" in humans (English *et al.*, 1988). The cat scratch bacterium seems to be devoid of a murein cell wall, but it still contains 27-hydroxyoctacosanoic acid in its LPS. It is thought that this unusually long 27-hydroxyoctacosanoic acid penetrates the whole lipid bilayer of the outer membrane of Gram-negative bacteria and that it might be covalently linked through its (ω-1)-hydroxy group to other cell constituents on the inside of the outer membrane thereby increasing the stability of this bacterium. An enzymatic system involved in the transfer of 27-hydroxyoctacosanoic acid to (Kdo)$_2$-lipid IV$_A$ has been described (Brozek *et al.*, 1996) and a detailed understanding of the enzyme systems involved in the biosynthesis and transfer of the 27-hydroxyoctacosanoic acid might allow the development of specific inhibitors of these enzymes. Such inhibitors might work as antibiotics against bacterial pathogens, like the "cat-scratch agent",

containing 27-hydroxyoctacosanoic acid in their LPS.

Interestingly, in *S. meliloti* long-chain (ω-1)-hydroxy fatty acids are not only found in the LPS but also in specific lipo-chitin oligosaccharide signal molecules (Demont *et al.*, 1994). The length of the fatty acids found in the rhizobial signal molecules vary between C18 and C26 and are therefore shorter than the (ω-1)-hydroxy fatty acids found in the LPS. In *S. meliloti* the synthesis of lipo-chitin oligosaccharide signal molecules is controlled by the regulatory genes, *nodD1, nodD2, nodD3,* and *syrM.* Only the activation of *nodD3* or *syrM* leads to the synthesis of lipo-chitin oligosaccharides that are substituted with (ω-1)-hydroxy fatty acids (Demont *et al.*, 1994). These results suggest that *nodD* genes or their products influence the specificity of the NodA-dependent acyl transferase reaction during lipo-chitin oligosaccharide biosynthesis (Chapter 21).

V. Fatty Acid Biosynthesis

Biosynthesis and transfer of fatty acids is of major importance during the formation of biological membranes, storage lipids, lipoproteins, or certain amphiphilic signal molecules (*i.e.* lipo-chitin oligosaccharides in rhizobia or *N*-acyl homoserine lactone autoinducers in Gram-negative bacteria).

In higher animals, fatty acid biosynthesis occurs at a multienzyme complex. All enzymatic activities involved are localized on a single, multifunctional polypeptide (Goodridge, 1991). In contrast, in bacteria (Cronan and Rock, 1996; Gouill *et al.*, 1993; Hale *et al.*, 1987; Shen and Byers, 1996) and in the plastids of green plants (Slabas and Fawcett, 1992) fatty acid biosynthesis is achieved through the catalytic activity of a number of monofunctional enzymes. Fatty acid biosynthesis proceeds in two stages, initiation and cyclic elongation (Cronan and Rock, 1996).

Fatty acid biosynthesis in *Rhizobiaceae* has hardly been investigated. However, the biosynthetic pathways demonstrated in *E. coli* are generally thought to be present in the *Rhizobiaceae* as well. In recent years lipid metabolism of *Rhodobacter sphaeroides*, like the *Rhizobiaceae* a member of the subgroup of α-proteobacteria, has been studied in some detail and might indicate more precisely how lipid metabolism proceeds in the *Rhizobiaceae*.

Organism	Growth phase (Source of acids)	Non-hydroxy fatty acids (% of total fatty acids)									References
		14:0	16:0	16:1	18:0	18:1	17:cy	19:cy	11-Me-VA	Others	
Rhizobium meliloti	E(TC)	2	12	2	+	46	0	14	7	17	Mac Kenzie *et al.*, 1979; Miller and Tremblay, 1983
Rhizobium leguminosarum bv. *viciae*	E(TC)	1	9	2	+	50	0	9	18	11	Mac Kenzie *et al.*, 1979
Rhizobium leguminosarum bv. *trifolii*	E(TC)	1	8	2	+	51	0	13	17	8	Mac Kenzie *et al.*, 1979
Rhizobium leguminosarum bv. *phaseoli*	E(TC)	1	6	1	+	63	0	9	15	5	Mac Kenzie *et al.*, 1979
Bradyrhizobium japonicum	E(TC)	2	11	3	+	62	0	8	3	11	Mac Kenzie *et al.*, 1979; Bunn *et al.*, 1970
Bradyrhizobium sp. (*Lupinus*)	E(TC)	2	13	8	+	66	0	3	3	5	Mac Kenzie *et al.*, 1979
Bradyrhizobium sp. (cowpea)	E(TC)	1	10	4	+	77	0	3	2	3	Mac Kenzie *et al.*, 1979
Bradyrhizobium sp. (*Lotus pedunculatus*)	ND(C-PL)	ND	13	1	7	46	ND	3	10	20	Gerson and Patel, 1975
Agrobacterium tumefaciens	ND(C-REL)	1	22	6	0	30	0	41	0	0	Das *et al.*, 1979; Kaneshiro and Marr, 1962
Agrobacterium stellulatum	E(C-REL)	Tr	4	1	1	93	0	0	0	Tr	Oliver and Colwell, 1973a

Table 1. Fatty acid composition in members of the *Rhizobiaceae*
E, exponential phase; TC, whole cells; C-PL, cellular phospholipids; C-REL, cellular extractable lipids; ND, no data; Tr, trace

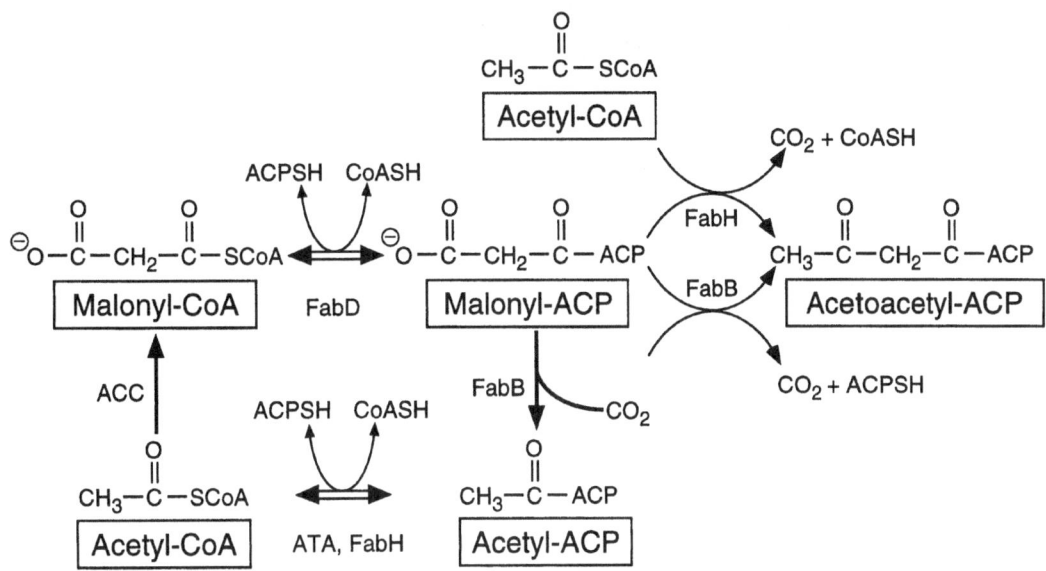

Figure 3. Pathways for the initiation of fatty acid biosynthesis.

V.A. FATTY ACID BIOSYNTHESIS IN *ESCHERICHIA COLI*

V.A.1. Initiation (Figure 3)
The precursors of fatty acid biosynthesis are derived from the acetyl coenzyme A (acetyl-CoA) pool. The first committed step is the conversion of acetyl-CoA to malonyl-CoA by acetyl-CoA carboxylase (ACC). ACC consists of two different protein subcomplexes with different enzymatic activities. In a first step the biotin carboxyl carrier protein (BCCP or AccB) is carboxylated by biotin carboxylase (AccC). The carboxyltransferase, composed of AccA and AccD, transfers the carboxy residue from carboxy-BCCP to acetyl-CoA thereby forming malonyl-CoA. Malonyl-ACP is formed from malonyl-CoA by malonyl-CoA:ACP transacylase (FabD). Acetyl-ACP can either be formed by decarboxylation of malonyl-ACP catalyzed by one of the condensing enzymes (β-ketoayl-ACP synthase I) or from acetyl-CoA. Either acetyl-CoA:ACP transacylase or β-ketoayl-ACP synthase III can catalyze the exchange of ACP for CoA on acetyl-CoA.

V.A.2. Elongation
Elongation of fatty acids by C2 units goes through a cycle of reactions, each cycle involving a condensation, a first reduction, a dehydration and a second reduction step (Figure 4).

The initial condensation yielding acetoacetyl-ACP, a β-ketide, can be performed by either one of three enzymes in *E. coli*. The condensing enzyme β-ketoacyl-ACP synthase III (FabH) uses malonyl-ACP and acetyl-CoA to form acetoacetyl-ACP. The condensation of malonyl-ACP and acetyl-ACP is catalyzed by β-ketoacyl-ACP synthase I (FabB) or the β-ketoacyl-ACP synthase II (FabF), to yield acetoacetyl-ACP. The release of CO_2 is driving this reaction. The three condensing enzymes show different preferences with regard to their acyl substrates. Whereas FabH hardly can use anything else than acetyl-CoA, medium- and long-chain acyl-ACP derivatives are condensed by FabB or FabF.

The β-ketoacyl-ACP is reduced by an NADPH-dependent β-ketoacyl-ACP reductase (FabG) forming the β-D-hydroxyacyl-ACP.

A water molecule is removed by either one of the β-hydroxyacyl-ACP dehydratases FabA or FabZ to form a *trans*-enoyl-ACP (Heath and Rock, 1996).

Enoyl reductase (FabI) can use either NADPH or NADH when reducing enoyl-ACP to a saturated acyl-ACP (Bergler *et al.*, 1996).

The resulting acyl-ACP can in turn serve as the substrate for another condensation reaction. Repeated elongations yield palmitic acid in its bound form as pamitoyl-ACP (used in PL biosynthesis) but also β-hydroxymyristoyl-ACP (used in lipid A biosynthesis) if the dehydration and second reduction step are not performed.

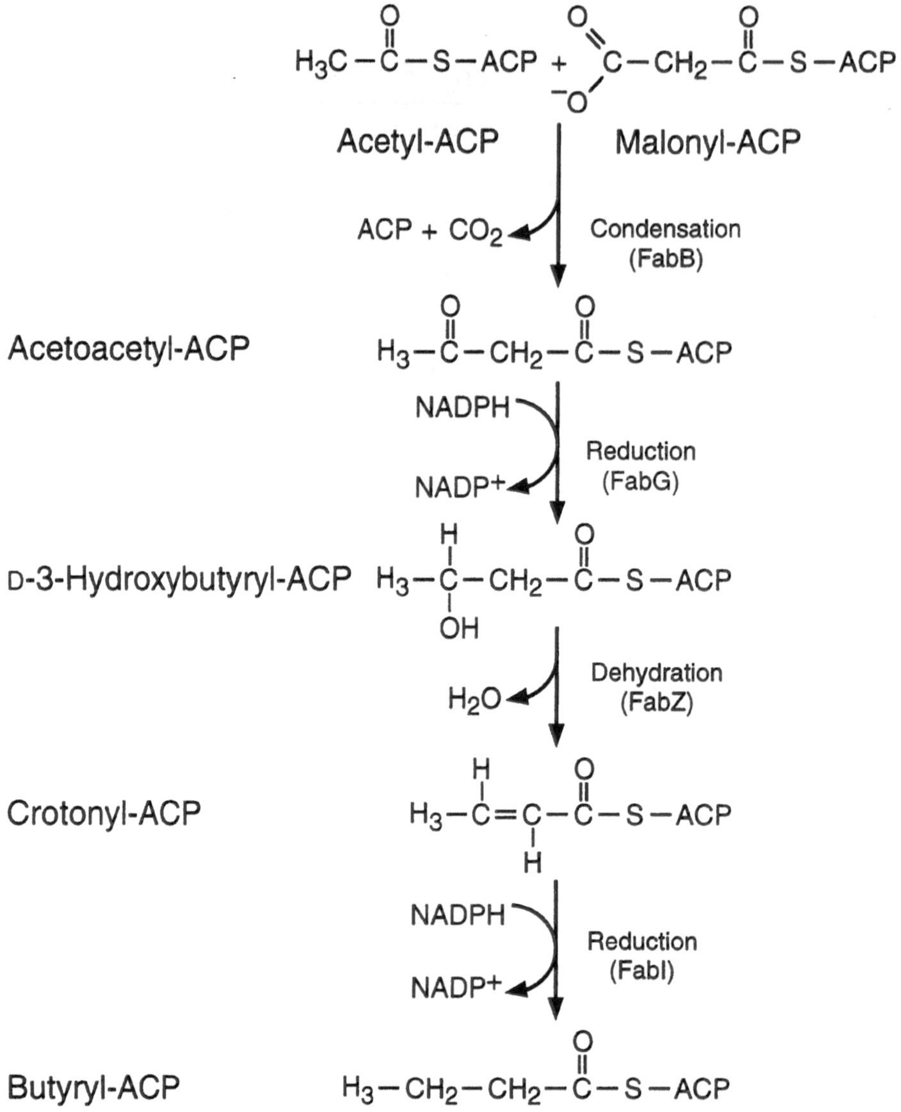

Figure 4. Elongation cycle of fatty acid biosynthesis.

V.A.3. Biosynthesis of Unsaturated Fatty Acids (Figure 5)

The β-hydroxyacyl-ACP dehydratase FabA, not only can remove water during fatty acid elongation, but is also able to catalyze the isomerization of *trans*-2-decenoyl-ACP to *cis*-3-decenoyl-ACP (Bloch, 1971). For the elongation of *cis*-3-decenoyl-ACP to palmitoleyl-ACP FabB is essential. It is involved in condensing *cis*-3-decenoyl-ACP and malonyl-ACP (Heath and Rock, 1996).

FabF catalyzes the condensation of palmitoleyl-ACP and malonyl-ACP thereby initiating the elongation to *cis*-vaccenoyl-ACP. The enzyme activity of FabF is regulated by temperature in an unusual way: The activity is high at low temperatures (25°C) and much less at higher temperatures (37°C). Consequently much more *cis*-vaccenic acid (and thereby a higher percentage of unsaturated fatty acids) is formed at lower temperatures. FabF alone is responsible for thermal regulation of fatty acid synthesis in *E. coli*.

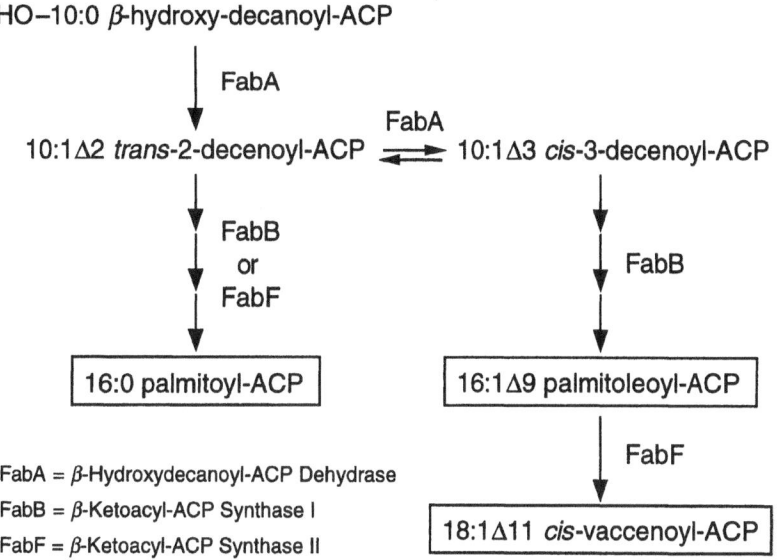

HO–10:0 β-hydroxy-decanoyl-ACP

FabA

10:1Δ2 *trans*-2-decenoyl-ACP ⇌ 10:1Δ3 *cis*-3-decenoyl-ACP
FabA

FabB
or
FabF

FabB

16:0 palmitoyl-ACP

16:1Δ9 palmitoleoyl-ACP

FabF

FabA = β-Hydroxydecanoyl-ACP Dehydrase
FabB = β-Ketoacyl-ACP Synthase I
FabF = β-Ketoacyl-ACP Synthase II

18:1Δ11 *cis*-vaccenoyl-ACP

Figure 5. Biosynthesis of *cis*-vaccenate.

V.A.4. Biosynthesis of Cyclopropane Fatty Acids

The synthesis of cyclopropane fatty acids (CFAs) occurs postsynthetically when the substrate unsaturated fatty acids are already esterified into membrane-localized phospholipid molecules. CFA synthase (Cfa) catalyzes the methylation of *cis* double bonds, the methylene donor being *S*-adenosylmethionine (SAM). Mutants that completely lack CFA synthase activity exist. They grow and survive normally under virtually all conditions (Grogan and Cronan, 1984). The only exception to this finding is that *cfa* mutant strains are more sensitive to freeze-thaw treatment than are isogenic wild type strains (Grogan and Cronan, 1986). The bulk of CFA synthesis occurs when cultures enter the stationary phase of growth but is reduced again in stationary phase cultures (Wang and Cronan, 1994).

V.B. BIOSYNTHESIS OF β-KETIDES (POLYKETIDES)

Fatty acid biosynthesis is a specialized case of the more general scheme of polyketide synthesis as it is found in *Streptomyces* (Cane, 1994; Hopwood and Sherman, 1990). Polyketides are formed by repeated condensation of C2 units resulting in multi-β-keto compounds. The β-ketide antibiotics are prominent members and their biosynthesis often involves a cyclization of β-ketides (*i.e.* tetracycline).

V.C. ASPECTS OF FATTY ACID BIOSYNTHESIS IN *RHIZOBIACEAE*

Rhizobium meliloti show a strong growth requirement for biotin the cofactor for many carboxylases (Streit *et al.*, 1996). Carbon dioxide is a required substrate for growth of *S. meliloti* (Lowe and Evans, 1962). That need probably reflects the activity of carboxylase enzymes in *Rhizobium*, including propionyl-CoA carboxylase and acetyl-CoA carboxylase which were detected in *R. meliloti* (de Hertogh *et al.*, 1964), and pyruvate carboxylase which was studied in *R. trifolii* (Ronson and Primerose, 1979).

V.C.1. Acyl Carrier Proteins in Rhizobia

Acyl Carrier Protein (ACP) of *Escherichia coli* is a small (molecular weight 8860), acidic (isoelectric point 3.8) protein which carries a 4'-phosphopantetheine prosthetic group (Figure 6) linked via phosphodiester bond to Ser-36. ACP can carry fatty acids, which are then linked via its sulfhydryl group in the form of a thioester (Jackowski *et al.*, 1991).

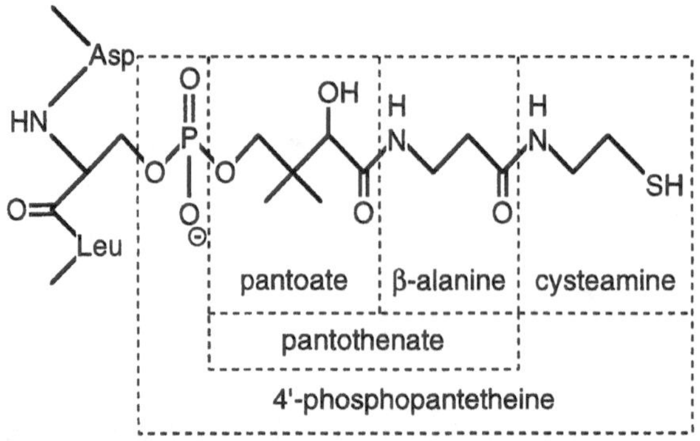

Figure 6. Prosthetic group of Acyl Carrier Proteins: 4'-phosphopantetheine.

ACPs of bacteria are not only involved in the synthesis of fatty acyl chains but also in their transfer during phospholipid, lipid A (Jackowski *et al.*, 1991), or hemolysin (Issartel *et al.*, 1991) biosynthesis. In *E. coli* all these functions seem to be realized by a single ACP which is produced constitutively and which comprises about 0.25% of the total soluble proteins of a cell (Rock and Cronan, 1979). This constitutive ACP is absolutely essential for the organism and is expressed from the *acpP* gene (Rawlings and Cronan, 1992). So far at least 14 proteins (Aas, AcpS, FabA, FabB, FabD, FabF, FabG, FabH, FabI, FabZ, LpxA, MdoH, PlsB, PlsC) are known in *E. coli* to interact with the AcpP protein during their enzymatic reaction.

Some bacteria have in addition to their constitutive ACP an inducible ACP which is involved in the biosynthesis of special β-ketides (Bibb *et al.*, 1989; Crosby *et al.*, 1995; Khosla *et al.*, 1992; Khosla *et al.*, 1993; Revill *et al.*, 1996; Sherman *et al.*, 1989). In rhizobia the inducible ACP is a protein needed for the formation of nitrogen-fixing nodules on their respective host plant, the nodulation protein NodF (Geiger *et al.*, 1991a). Together with NodE this protein is involved in the synthesis of polyunsaturated fatty acids (Geiger *et al.*, 1994a).

Surprisingly, recent evidence suggests that, besides the constitutive AcpP and the inducible NodF, there are at least two additional ACPs in rhizobia. The first one was identified by its ORF in the complementation unit I of the *fix-23* locus in *S. meliloti* (Petrovics *et al.*, 1993). The second novel ACP is involved in the transfer of the unusually long 27-hydroxyoctacosanoic acid to a sugar backbone

during lipid A biosynthesis in *R. leguminosarum* (Brozek *et al.*, 1996).

Constitutive AcpP from S. meliloti. Acyl carrier proteins involved in the synthesis of essential lipids are expressed at a constitutive level. The cell seems not to be able to accept even minor changes in expression levels (Rawlings and Cronan, 1992). After the initial characterization of the constitutive acyl carrier protein (AcpP) from *E. coli* numerous constitutive acyl carrier proteins from bacteria (Cooper *et al.*, 1987; Froelich *et al.*, 1990; Morbidoni *et al.*, 1996; Revill and Leadlay, 1991; Shen and Byers, 1996) and plants (Høj and Svendsen, 1983; Post-Beittenmiller *et al.*, 1989; Simoni *et al.*, 1967; Souciet and Weil, 1992) have been characterized, among them the constitutive acyl carrier protein (AcpP) from *S. meliloti* (Platt *et al.*, 1990). Like *E. coli* AcpP, AcpP from *S. meliloti* can be acylated by an ACP acylase isolated from *E. coli* (Platt *et al.*, 1990). Interestingly however, the rhizobial AcpP cannot substitute for *E. coli* AcpP as a cofactor in a transglycosylation function (Platt *et al.*, 1990) needed for the synthesis of membrane-derived oligosaccharides (MDO) in *E. coli* (Therisod and Kennedy, 1987; Therisod *et al.*, 1986).

Nodulation Protein NodF from Rhizobium is an Acyl Carrier Protein. In the symbiotic interaction of *Rhizobium* with its plant host, which results in nitrogen-fixing root nodules, flavonoids secreted from the plant root induce rhizobial nodulation (*nod*) genes. As a result the rhizobial *nod* gene products synthesize signal molecules (lipo-chitin oligosaccharides) which are able to cause the formation of nodulation-related phenomena (Chapter 21).

Some rhizobia (*S. meliloti*, *R. leguminosarum* bvs. *viciae* and *trifolii*) produce lipo-chitin oligosaccharides harbouring polyunsaturated fatty acyl chains (Lerouge *et al.*, 1990; Schultze *et al.*, 1992; Spaink *et al.*, 1995; Spaink *et al.*, 1991). The polyunsaturated fatty acyl chains in lipo-chitin oligosaccharides of *R. meliloti* are *trans*-2, *cis*-9-hexadecadienoic acid (Lerouge *et al.*, 1990) or *trans*-2, *trans*-4, *cis*-9-hexadecatrienoic acid (Schultze *et al.*, 1992). In lipo-chitin oligosaccharides of *R. leguminosarum* bv. *viciae* so far only one type of polyunsaturated fatty acyl chain (*trans*-2, *trans*-4, *trans*-6, *cis*-11-octadecatetraenoic acid) has been found (Firmin *et al.*, 1993; Spaink *et al.*, 1991), whereas lipo-chitin oligosaccharides of *R. leguminosarum* bv. *trifolii* can contain polyunsaturated C18 (C18:3) or C20 (C20:3 or C20:4) fatty acyl residues (Spaink *et al.*, 1995). These rhizobia are characterized by a narrow host range and by nodulating mainly hosts which form indeterminate nodules (Chapter 18).

Relatively early it was recognized that the host-specific nodulation genes *nodF* and *nodE* have limited but recognizable homologies to acyl carrier protein and to β-ketoacyl synthase of *E. coli*, respectively (Shearman *et al.*, 1986). So far, the *nodFE* genes from *S. meliloti* (Debellé and Sharma, 1986; Horvath *et al.*, 1986), *R. leguminosarum* bv. *viciae* (Shearman *et al.*, 1986), *R. leguminosarum* bv. *trifolii* (Schofield and Watson, 1986), and *Rhizobium* sp. strain N33 (Cloutier *et al.*, 1997) have been sequenced. Recently it was reported that lipo-chitin oligosaccharides from *R. huakuii* and *R. galegae* contain polyunsaturated fatty acids (Debelle *et al.*, 1997). This finding implies that *R. huakuii* and *R. galegae* also contain a set of *nodFE* genes.

Mutants lacking *nodF* or *nodE* are unable to produce lipo-chitin oligosaccharides substituted with polyunsaturated fatty acyl chains. When *nodFE* of *R. leguminosarum* was introduced into a *nodFE*-deficient strain of *S. meliloti* lipo-chitin oligosaccharides substituted with polyunsaturated C18 fatty acids, the chain length normally found in *R. leguminosarum* for polyunsaturated fatty acids, were formed (Demont *et al.*, 1993). However, if the *nodFE* from *R. leguminosarum* bv. *viciae* are expressed in *S. meliloti* not only the *trans*-2, *trans*-4, *trans*-6, *cis*-11-octadecatetraenoic acid is found as a substituent of the lipo-chitin oligosaccharides but

also *trans*-2, *trans*-4, *cis*-11-octadecatrienoic acid, and *trans*-2, *cis*-11-octadecadienoic acid (Demont *et al.*, 1993).

Studies on the biosynthesis of polyunsaturated fatty acids in *R. leguminosarum* show that after the induction of *nodFE* genes, even in the absence of the *nodABC* genes, the *trans*-2, *trans*-4, *trans*-6, *cis*-11-octadecatetraenoic acid, which has a characteristic absorbance maximum of 303 nm, is synthesized (Geiger *et al.*, 1994a). This shows that the biosynthesis of the unusual fatty acid is not dependent on the synthesis of lipo-chitin oligosaccharides. This finding also suggests that the biosynthesis of the polyunsaturated fatty acid is completed before it is linked to the sugar backbone of the lipo-chitin oligosaccharides (Geiger *et al.*, 1994a).

In an attempt to identify the lipid fraction with which the unusual C18:4 fatty acid is associated, it was found that it is linked to the *sn*-2 position of the phospholipids (Geiger *et al.*, 1994a). In addition, the phospholipids contain other *nodFE*-derived fatty acids, a C18:3 *trans*-4, *trans*-6, *cis*-11-octadecatrienoic acid which has an absorption maximum at 225 nm, and a C18:2 octadecadienoic acid which no specific absorption maximum (Geiger *et al.*, 1997). Even when lipo-chitin oligosaccharide signals are produced in a wild type *Rhizobium* cell, a fraction of all those unusual fatty acids is still bound to all major phospholipids (Geiger *et al.*, 1997; Geiger *et al.*, 1994a). Neither the *trans*-4, *trans*-6, *cis*-11-octadecatrienoic acid nor the octadecadienoic acid without a specific absorption have been observed so far in lipo-chitin oligosaccharides (Spaink *et al.*, 1991). Although phospholipids with *nodFE*-dependent polyunsaturated fatty acyl residues showed no activity in bioassays normally used to test lipo-chitin oligosaccharide nodulation factors (Geiger *et al.*, 1994b) a biological role for such phospholipids cannot be excluded.

As the spectrum of *nodFE*-dependent polyunsaturated fatty acyl residues is not identical in phospholipids and lipo-chitin oligosaccharides these data suggest that NodA, the presumptive *N*-acyltransferase responsible for the assembly of the fatty acyl chain to the sugar backbone of the lipo-chitin oligosaccharides, does not transfer just all fatty acids synthesized by the action of NodFE to the lipo-chitin oligosaccharides. Rather it selectively transfers α-β unsaturated fatty acids. For this and other reasons doubts have emerged about the notion of *nodA* being a common *nod* gene and it was shown that NodA could function as a host-specific

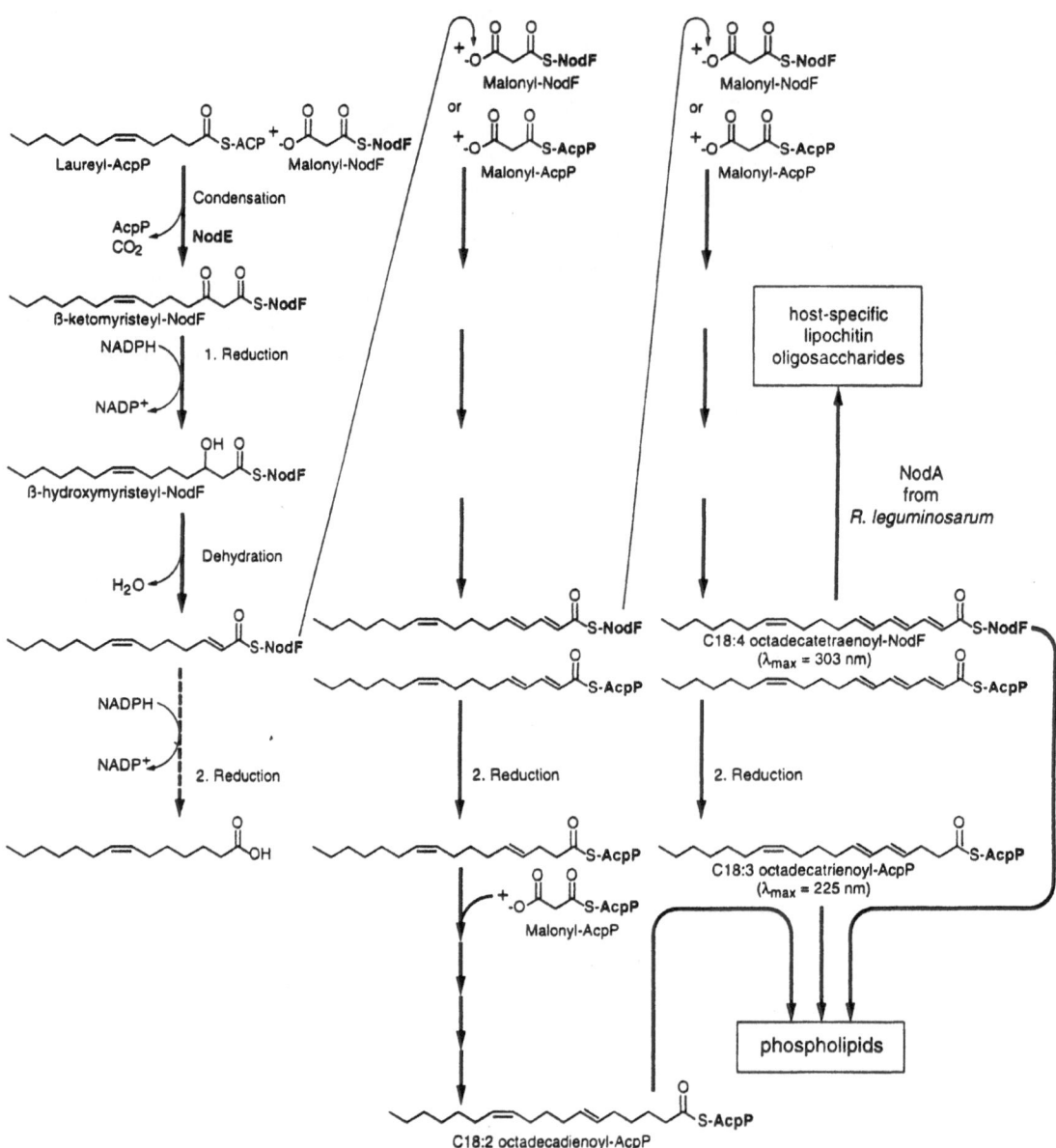

Figure 7. Biochemical function of NodF (Geiger *et al.*, 1997).

determinant of the transfer of of fatty acids in Nod factor biosynthesis (Debellé *et al.*, 1996; Ritsema *et al.*, 1996; Roche *et al.*, 1996). A model presented by Geiger *et al.* (Figure 7) might explain how the selectivity for α-β unsaturated fatty acids is actually achieved through specific interaction between NodA and NodF (Geiger *et al.*, 1997).

The NodF protein from *R. leguminosarum* bv. *viciae* has been characterized in more detail. NodF carries a 4'-phosphopantetheine prosthetic group which is characteristic for acyl carrier proteins (Geiger *et al.*, 1991a). When the active site serine residue 45 was exchanged for a threonine, the altered NodF S45T protein could no longer be substituted with 4'-phosphopantetheine and at the same time it was neither active in its biochemical function nor in its biological function (Ritsema *et al.*, 1994). Recent NMR investigations of the structural properties of the NodF protein show that NodF has three well-formed helices which fold in a parallel-antiparallel fashion and a prosthetic group attachment site near the beginning of the second helix (Ghose *et al.*, 1996). Therefore despite the limited sequence homology with AcpP from *E. coli* the structural homology is realized also in the third dimension. In order to understand the structural basis of the specialized function of NodF, chimeric genes containing part of the *nodF* gene and part of the *E. coli acpP* gene were constructed. Interestingly, a hybrid ACP consisting of the N-terminal part of AcpP and the C-terminal part of NodF is able to replace NodF in nodulation (Ritsema *et al.*, 1997).

RkpF protein from Rhizobium meliloti.

Bacterial exopolysaccharides (EPS) and lipopolysaccharides (LPS) are known to play important roles in plant-bacterium interactions. The *fix-23* locus of *Rhizobium meliloti* strain 41 can compensate for exopolysaccharide-deficient (*exo*) mutations during symbiotic nodule development. This locus is involved in the production of a novel polysaccharide, called K-like antigen, that is rich in 3-deoxy-D-*manno*-2-octulosonic acid (Kdo) but different from classical LPS (Reuhs *et al.*, 1993) (Chapter 7). One of the four *fix-23* complementation units (unit I) contains 6 open reading frames (ORFs) coding for protein products that show considerable similarity and similar organization to those of the rat fatty acid synthase multifunctional enzyme domains. The sixth ORF was proposed to code for a novel Acyl Carrier Protein (ACP) (Petrovics *et al.*, 1993) and is now named RkpF (Kiss *et al.*, 1997; Reuhs, 1996).

Recently the *rkpF* gene has been cloned and its derived protein product (RkpF) was overexpressed in *E. coli* (Epple *et al.*, 1997). After purification, the RkpF protein migrates as a single band with an apparent molecular weight of 5000 on sodium dodecyl sulfate-polyacrylamide gel electrophoresis. The RkpF protein can be effectively labeled *in vivo* with radioactive β-alanine added to the growth medium. If homogenous RkpF protein is incubated with radiolabeled coenzyme A (CoA) in the presence of purified holo-ACP synthase (AcpS) from *E. coli* (Lambalot and Walsh, 1995) an *in vitro* transfer of 4'-phosphopantetheine to the RkpF protein can be observed. The conversion from apo-RkpF protein to holo-RkpF protein seems to go along with a major conformational change of the protein structure, as the holo-RkpF protein runs significantly faster on native polyacrylamide gel electrophoresis than the apo-RkpF protein. The quantitative conversion of the apo-RkpF protein in the presence of AcpS and CoA to holo-RkpF protein can thus be followed easily using electrophoretic analysis under native conditions (Epple *et al.*, 1997).

Therefore in *S. meliloti*, in addition to the constitutive AcpP and the flavonoid-inducible nodulation protein NodF, there is at least one more 4'-phosphopantetheine-carrying small, acidic protein RkpF, presumably functioning as a novel ACP in the biosynthesis of an yet unknown β-ketide.

AcpXL protein from Rhizobium leguminosarum.

Lipid A, the hydrophobic anchor of lipopolysaccharides in the outer membranes of Gram-negative bacteria is acylated with 27-hydroxyoctacosanoic acid in all members of the *Rhizobiaceae* (Bhat *et al.*, 1991a; Bhat *et al.*, 1991b). Brozek *et al.* (1996) describe an *in vitro* system, consisting of a membrane enzyme and a cytosolic acyl donor from *R. leguminosarum*, that transfers 27-hydroxyoctacosanoic acid to (Kdo)$_2$-lipid IV$_A$, a precursor during lipid A biosynthesis. The cytosolic acyl donor was purified and amino-terminal sequencing revealed an amino acid match with a partially sequenced gene (*orf**) of *R. leguminosarum* (Colonna-Romano *et al.*, 1990). When the entire *orf** gene was sequenced, it was found to encode a protein, now named AcpXL, of 92 amino acids and a consensus sequence, LGXDSLD, for the attachment of 4'-phosphopantetheine. Mass spectrometry of the purified AcpXL revealed the presence of 27-hydroxyoctacosanoic acid in the major species. Other mass peaks indicated AcpXL acylation with ω-1-hydroxylated fatty acyl residues (C18-C26). The latter fatty acyl residues are incorporated in *nodD3*-dependent lipo-chitin oligosaccharides during their biosynthesis in *S. meliloti* (Demont *et al.*, 1994). One therefore might expect AcpXL being present in *S. meliloti* and other 27-hydroxy-octacosanoic acid-containing bacteria as well.

Organism	Growth phase (Source of lipids)	Phospholipid									References
		PE	PG	CL	PC	DMPE	MMPE	PS	PI	Others	
Rhizobium meliloti	S(C)	22	11	11	27	4	22	0	0	ND	Thompson *et al.*, 1983
Rhizobium leguminosarum bv. *viciae*	ND(C)	20	33	22	25	+	0	0	0	0	Faizova *et al.*, 1971
Bradyrhizobium japonicum (*Lotus pedunculatus*)	ND(C)	40	0	16	31	0	0	13	0	0	Bunn and Elkan, 1971
Agrobacterium tumefaciens	S(C)	18	13	19	28	0	16	0	0	6	Randle *et al.*, 1969
Agrobacterium stellulatum	E(C)	50	27	3	4	0	0	0	0	16	Oliver and Colwell, 1973b

Table 2. Phospholipid composition in members of the *Rhizobiaceae*
E, exponential phase; S, stationary phase; C, whole cell; +, present; ND, no data.

VI. Phospholipids in *Rhizobiaceae*

Like in *E. coli* the major phospholipids in the *Rhizobiaceae* are phosphatidylglycerol (PG), cardiolipin (CL) and phosphatidylethanolamine (PE). However, the *Rhizobiaceae* also contain significant amounts of methylated derivatives of PE: monomethyl phosphatidylethanolamine (MMPE), dimethyl phosphatidylethanolamine (DMPE), and phosphatidylcholine (PC) (Table 2).

A reinvestigation of the phospholipid composition of *Bradyrhizobium* (Miller *et al.*, 1990) confirms that, in contrast to earlier reports (Bunn and Elkan, 1971; Bunn *et al.*, 1970), PG is present as a major phospholipid whereas neither phosphatidylserine (PS) nor phosphatidylinositol (PI) could be detected (Miller *et al.*, 1990).

Either one or both of MMPE and DMPE may also be formed: this applies to *Agrobacterium rhizogenes* and *A. radiobacter* as well as *A. tumefaciens* (Goldfine and Ellis, 1964). There are conflicting reports (Das *et al.*, 1979; Randle *et al.*, 1969) about whether PC accumulates in stationary-phase cells of the latter species.

VII. Phospholipid (PL) Biosynthesis

Like for fatty acid biosynthesis a thorough investigation of phospholipid biosynthesis pathways was done in *E. coli*.

VII.A. BIOSYNTHESIS OF GLYCEROL-3-PHOSPHATE

Glycerol-3-phosphate forms the "backbone" of all phospholipid molecules. It can be synthesized by two different pathways:

1. During growth with glycerol as the sole carbon source, the enzymes of the glycerol catabolic (*glp*) operon are induced. One of the induced enzymes, a glycerol kinase, is able to phosphorylate glycerol and form glycerol-3-phosphate.

2. During growth on carbon sources other than glycerol, glycerol-3-phosphate is made by direct reduction of the glycolytic intermediate dihydroxyacetone phosphate (DHAP) with NADH. The enzyme catalyzing this reaction is called biosynthetic glycerol-3-phosphate dehydrogenase (GpsA) (Bell, 1974).

All further biosynthesis of phospholipids is performed with membrane-bound enzymes.

VII.B. ACYLATION OF GLYCEROL-3-PHOSPHATE TO PHOSPHATIDIC ACID (FIGURE 8)

Glycerol-3-phosphate acyltransferase (PlsB) catalyzes the first acylation of position 1 of glycerol-3-phosphate. The second fatty acid is added by another enzyme, 1-acyl-glycerol-3-phosphate acyltransferase (PlsC), to form phosphatidic acid (PA). There is a tendency that most fatty acids at the 1-position are saturated whereas most at the 2-position are unsaturated. The acylation specificity however, is not absolute and can be altered by the supply of acyl donors.

VII.C. DIVERSIFICATION OF POLAR HEAD GROUPS (FIGURE 9)

VII.C.1. Activation of Phosphatidic Acid to CDP-Diglyceride

The conversion of phosphatidic acid to CDP-diglyceride (CDP-diacylglycerol) is catalyzed by CDP-diglyceride synthase (CdsA) (Figure 9).

VII.C.2. Biosynthesis of Phosphatidyl-glycerol (PG)

Phosphatidylglycerol phosphate synthase (PgsA) transfers glycerol-3-phosphate to CDP-diglyceride under the release of CMP thereby producing phosphatidylglycerol phosphate (PGP). Also the gene encoding PGP synthase (*pgsA*) in *Rhodobacter sphaeroides* has been characterized (Dryden and Dowhan, 1996).

There are at least two enzymes with PGP phosphatase activity (PgpA, PgpB) in *E. coli* releasing inorganic phosphate from PGP to form PG. Insertional inactivation of PGP synthase is lethal (Heacock and Dowhan, 1987). There are many important cellular functions that are affected by reduced PG and/or CL content of the membrane. PG is required for protein translocation across the membrane, and SecA is the critical component affected (Kusters *et al.*, 1991), although SecA-independent translocation is also impaired (Kusters *et al.*, 1994). Acidic phospholipids are also required for channel activity of bacterial colicins A and N as well as for the interaction of antibiotics with the membrane (Van der Groot *et al.*, 1993).

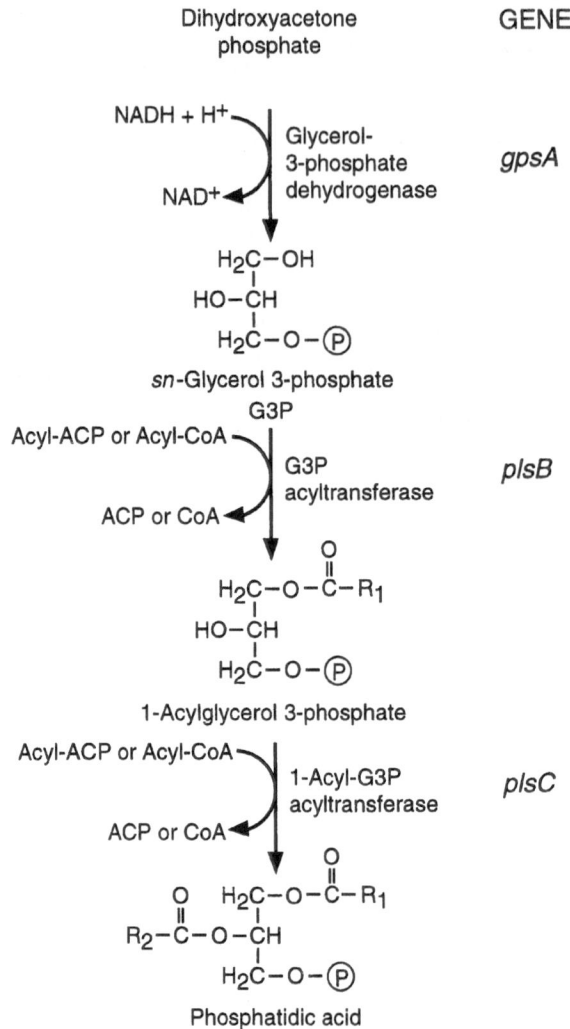

Figure 8. Biosynthesis of phosphatidic acid.

VII.C.3. Biosynthesis of Cardiolipin (CL)

CL synthase (Cls) condenses two PG molecules to yield CL and free glycerol in a transesterification reaction (Pluschke *et al.*, 1978). *E. coli* is able to survive the disruption of the *cls* gene, although the cells grow at a slower rate and to a lower density than the corresponding wild type cells, indicating that CL may confer a growth or survival function. CL accumulation and CL synthase activity increase as the cells enter the stationary phase of growth, and CL is the most stable membrane phospholipid during prolonged incubation in stationary phase. The *cls*

null mutants lose viability in stationary phase, supporting the idea that CL is important for long-term survival under non-growing conditions (Hiraoka *et al.*, 1993).

VII.C.4. Biosynthesis of Phosphatidylethanolamine (PE)

The first step in the synthesis of PE is the condensation of CDP-diacylglycerol with serine to form phosphatidylserine (PS) catalyzed by PS synthase (Pss). The second step in the formation of

PE is the decarboxylation of PS, catalyzed by PS decarboxylase (Psd).

The inability to synthesize PE appears to be lethal. Surprisingly, the lethality can be phenotypically suppressed by the addition of divalent cations to the growth medium, although this is accompanied by pertubations in the function of permeases (Bogdanov and Dowhan, 1995), electron transport (Mileykovskaya and Dowhan, 1993), as well as in motility and chemotaxis (Shi *et al.*, 1993).

VII.C.5. Biosynthesis of Phosphatidyl-choline (PC)

In eukaryotes, phosphatidylcholine (lecithin) is the major membrane-forming phospholipid. Physiological effects of phosphatidylcholine are well known. It is considered to strengthen the explicit memory (Ladd *et al.*, 1993) and it functions as a lung surfactant (Possmeyer, 1989). The elucidation of molecular mechanisms in which PC might be involved is just beginning. Interestingly, PC is thought to be involved in cellular signal chains causing the establishment of cancer. In one of the steps during mitogenic signal transduction from Ras to Raf protein, PC is cleaved by a specific phospholipase C releasing the signal molecule diacylglycerol (Cai *et al.*, 1993).

The biosynthesis of PC in eukaryotes is well studied. It can be synthesized either by the CDP-choline pathway, in which phosphocholine is transferred from CDP-choline to diacylglycerol (Kennedy and Weiss, 1956), or by the methylation pathway (Vance and Ridgway, 1988). In the methylation pathway PC is formed by three successive methylations of phosphatidylethanol-amine via the intermediates monomethyl phosphatidylethanolamine and dimethyl phospha-tidylethanolamine using the methyl donor S-adenosyl-L-methionine.

The initial work on phosphatidylcholine in bacteria was carried out by Law et al. (Law *et al.*, 1963), who showed that *A. tumefaciens* incorporated L-[^{14}C-methyl]methionine into lipids, which on hydrolysis yielded radioactive N-methylethanol-amine, N,N-dimethylethanolamine, and choline. These results were the first indication that bacteria synthesized phosphatidylcholine by a trans-methylation pathway, a pathway that had been characterized previously in a liver microsomal enzyme system (Bremer and Greenberg, 1961). The occurrence of phosphatidylcholine in bacteria however, is thought to be rather limited (Goldfine and Ellis, 1964). Since no enzymatic activities of the

CDP-choline pathway have been detected in *A. tumefaciens* (Sherr and Law, 1965), it is thought that bacteria have only the methylation pathway of phosphatidylcholine biosynthesis.

A phosphatidylethanolamine N-methyltransferase from *A. tumefaciens* was purified and its enzymatic properties were characterized (Kaneshiro and Law, 1964). The fact that a soluble enzyme was capable of catalyzing the first transmethylation and a particulate system could catalyze the formation of all three methylated phospholipids led the authors to conclude that *A. tumefaciens* must possess two independent enzymes involved in the methylation of PE. In mammalian cells the PE-N-methylation pathway has been shown to involve two distinct enzymes. One is the enzyme that catalyzes the first methylation of PE to MMPE, whereas the other one catalyzes the two-step methylation from the latter molecule to PC. The participation of two enzymes in the pathway was also demonstrated by experiments using mutant strains of *Neurospora crassa* and *Saccharomyces cerevisiae* (Yamashita *et al.*, 1982). However, in *Zymomonas mobilis* (Tahara *et al.*, 1986; Tahara *et al.*, 1987) only one enzyme seems to be responsible for all three methylation steps and in *Rhodobacter sphaeroides* there is only one structural gene coding for a phospholipid N-methyltransferase, able to catalyze all three methylation steps (Arondel *et al.*, 1993).

Mutants deficient in phospholipid N-methyltransferase had been isolated from *Rhodobacter sphaeroides* (Arondel *et al.*, 1993) and from *Zymomonas mobilis* (Tahara *et al.*, 1994) and in both cases no phosphatidylcholine was formed. Surprisingly, PC-deficient mutants of *Rhodobacter* or *Zymomonas* seem to be fully functional in their vegetative functions and no associated phenotypes have been recognized.

Recently, also from *S. meliloti* 1021 mutants, deficient in phospholipid N-methyltransferase, have been isolated (de Rudder *et al.*, 1997). These mutants are deficient in the mono and dimethylated derivatives of PE, indicating that the methylation pathway is indeed not functioning. Surprisingly however, these mutants still do synthesize PC, apparently by a second still unknown pathway. Rhizobia, unlike all other known bacterial systems, therefore have, like eukaryotes two pathways for the formation of PC, suggesting that PC might play an important role during the life cycle of the *Rhizobiaceae*.

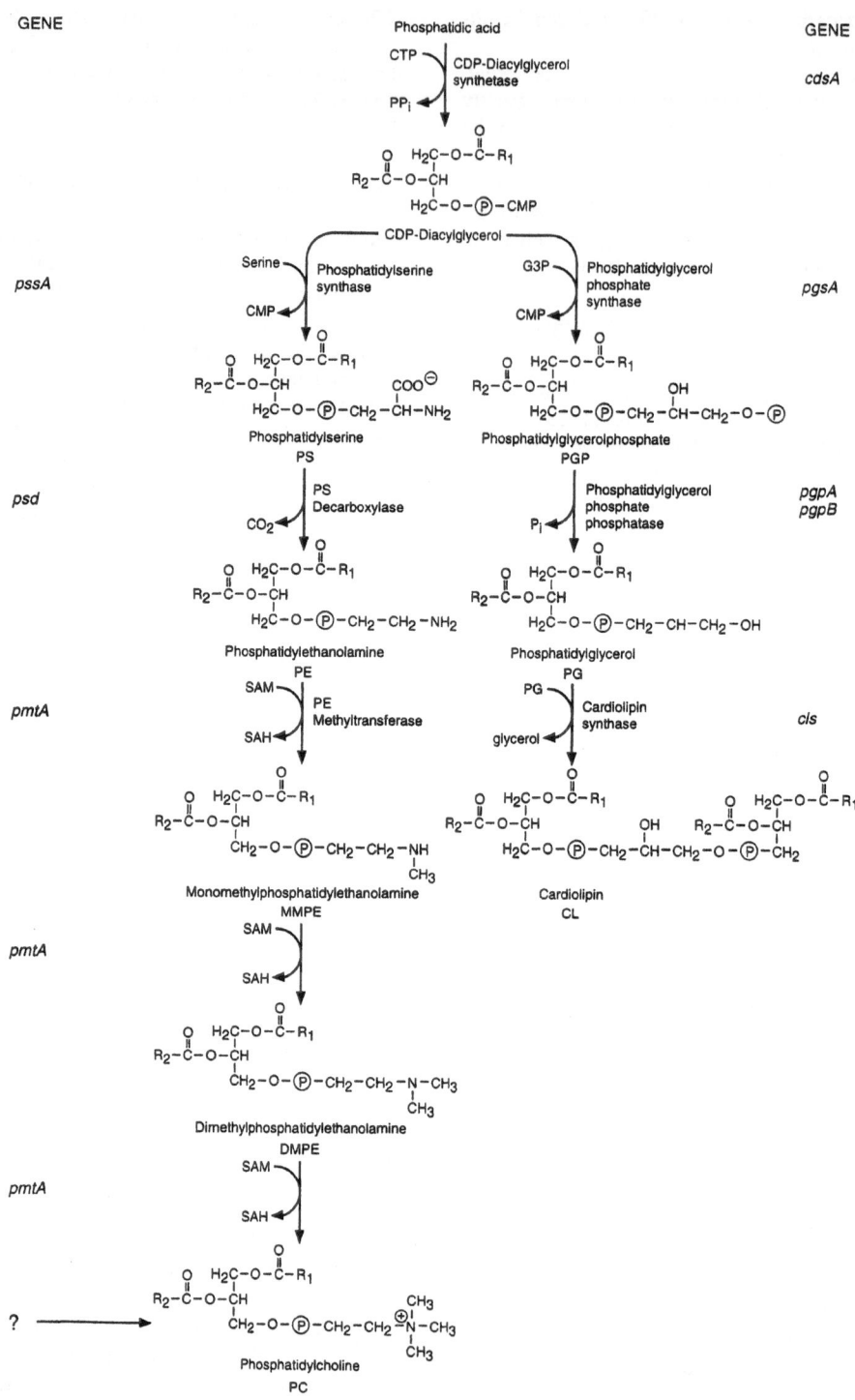

Figure 9. Phospholipid biosynthesis.

VIII. Phospholipid Turnover

Phospholipids can function as intermediates in the synthesis of other molecules. Two examples will be discussed.

VIII.A. DIACYLGLYCEROL CYCLE

Periplasmic glucans, like the membrane-derived oligosaccharides (MDO) in *E. coli* or the cyclic glucans of the *Rhizobiaceae*, can be substituted with anionic or with zwitterionic residues. In some cases these polar residues are biosynthetically derived from the head groups of phospholipids. Cyclic glucans substituted with *sn*-1-phosphoglycerol residues from *Agrobacterium* (Miller *et al.*, 1987) or *S. meliloti* obtain their phosphoglycerol substituent from phosphatidylglycerol (Geiger *et al.*, 1991b; Miller *et al.*, 1988) and may only occur when cyclic glucans are transported across the inner membrane (Breedveld *et al.*, 1994). During transfer of the polar head group to cyclic glucans diacylglycerol (diglyceride, DG) is formed as a second product and has to be phosphorylated to phosphatidic acid in order to reenter phospholipid biosynthesis. The structural gene for diacylglycerol kinase from *S. meliloti* has been cloned and sequenced (Miller *et al.*, 1992).

The β-1,3;1,6-linked cyclic glucans in *Bradyrhizobium japonicum* can be substituted with phosphocholine residues (Rolin *et al.*, 1992). It is likely that this phosphocholine head group is derived from phosphatidylcholine.

Periplasmic glucans are described in more detail in chapter 5.

VIII.B. 2-ACYLGLYCEROLPHOSPHO-ETHANOLAMINE CYCLE

The maturation of the major outer membrane lipoprotein of *E. coli* is a complicated process: The primary gene product is a prolipoprotein that contains a 20-residue N-terminal signal peptide. A glycerol residue is then transferred from the polar head group of PG to Cys21 of the lipoprotein to form a thioether. Then 2 fatty acids are esterified to the glycerylcysteine. The signal peptide is removed by a specific endopeptidase (signal peptidase II) that cleaves the bond between Gly20 and diacylglycerylcysteine 21 to form apolipoprotein. The last step in lipoprotein maturation is the transfer of an additional fatty acid to the new N-terminus.

This fatty acid is derived from the 1-position of PE. The other resulting product, 2-acylglycerolphosphoethanolamine is reacylated from acyl-ACP by a specific acyl transferase (Aas) (Cooper *et al.*, 1989; Hsu *et al.*, 1989; Jackowski *et al.*, 1994).

IX. Regulation of Phospholipid Biosynthesis

At low potassium phosphate concentrations of approximately 0.1 mM in the culture medium, phosphate limits the growth of most Gram-negative bacteria. Under those conditions nonphosphorous-containing lipids are formed instead of phospholipids in *Pseudomonas* and *Rhodobacter*. In *Rhodobacter sphaeroides* such phosphorous-free lipids are glycolipids (glucosylgalactosyl diacylglycerol and monohexosyl diacylglycerol), a sulfolipid (sulfoquinovosyl diacylglycerol), a betaine lipid (1,2-di-*O*-acyl-[4'-(*N*, *N*, *N*-trimethyl)-homoserine]glycerol), and an ornithine lipid (Benning *et al.*, 1995).

So far the molecular mechanisms controlling the switch from phospholipid biosynthesis to the biosynthesis of alternative, phosphorous-free lipids are not known.

Interestingly, the locus *ndvF*, essential for normal nodule development by *S. meliloti* (Charles *et al.*, 1991), was found to contain four genes, *phoCDET*, which encode an ABC-type transport system for the uptake of inorganic phosphate (Pi) (Bardin *et al.*, 1996). *ndvF* mutants, unable to take up Pi, contain low concentrations of cytoplasmic Pi and studies on their lipid composition will help to understand how the biosynthesis of alternative, phosphorous-free lipids is regulated in rhizobia.

X. Alternative Membrane Lipids

X.A. GLYCOLIPIDS

Glycosyl diacylglycerols are widespread membrane glycolipids in plants, animals, and Gram-positive bacteria but they are rarely found in Gram-negative bacteria. A diglycosyl diacylglycerol (BF-7) was reported to show various responses, like thick and short roots, root hair deformation, and cortical cell division, on white clover, the host plant of *R. leguminosarum* bv. *trifolii* (Orgambide *et al.*, 1994). BF-7 was isolated from *R. leguminosarum* bv. *trifolii* ANU843 and its covalent structure was

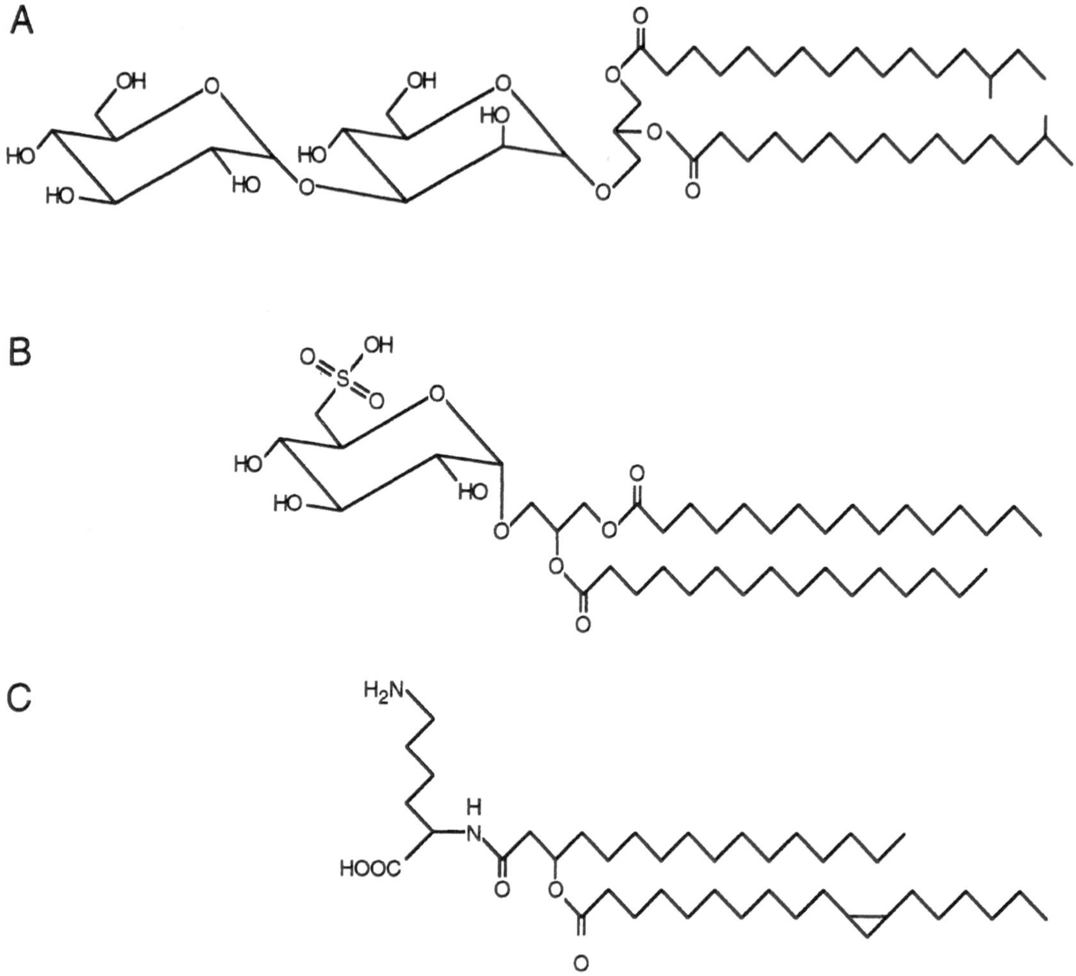

Figure 10. Membrane lipids without phosphorous.

A. Rhizobial Glycolipid. Diglycosyl diacylglycerol (BF-7) isolated from *R. leguminosarum* bv. *trifolii* ANU843 (Orgambide *et al.*, 1994) B. Rhizobial sulfolipid. Structure of the major component in the family of the sulfoquinovosyl diacylglycerols in *S. meliloti* 2011. The most prominent fatty acyl species are hexadecanoic acid, octadecenoic acid, and a methylene octadecenoic acid (Cedergren and Hollingsworth, 1994). C. Agrobacterial lysine-containing lipid (Tahara *et al.*, 1976).

reported to be 1,2-diacyl-3-*O*-(α-D-glucopyranosyl-(1-3)-*O*-α-D-mannopyranosyl)-glycerol (see Figure 10A). The formation of BF-7 seems to depend on the functional activation of the common *nod* genes. However, there is no direct biochemical proof that BF-7 is made directly through the action of *nod* gene products. From the proposed biochemical functions of NodC, NodB, and NodA it seems unlikely that either one of them can be involved in the biosynthesis of BF-7. Alternatively, the synthesis of

BF-7 might be a secondary effect starting only after synthesis and accumulation of *nod* gene-dependent lipo-chitin oligosaccharides has begun to alter membrane composition and therefore causing secondary, regulatory biosynthesis in order to counterbalance the effect of lipo-chitin oligosaccharides on membranes. In this connection it would be interesting to compare the time course of the syntheses of lipo-chitin oligosaccharide and BF-7.

X.B. SULFOLIPIDS

Sulfolipids are integral membrane parts of photosynthetic bacteria and of plant chloroplast membranes and were long thought to be associated with the photosynthetic process. Recently, sulfolipid-deficient mutants of the photosynthetic bacterium *Rhodobacter sphaeroides* were isolated (Benning *et al.*, 1993; Benning and Somerville, 1992a; Benning and Somerville, 1992b). Such mutants showed intact photosynthesis and the only detectable phenotype was a reduced growth rate under phosphate-limiting conditions (Benning *et al.*, 1993). Phosphate limitation of the wild type caused a significant reduction in the amount of all phospholipids and an increased amount of sulfolipid. By contrast, the sulfolipid-deficient mutant had reduced levels of PC and PE but maintained a normal level of PG. Therefore the sulfolipid 6-sulfo-α-D-quinovosyl-diacylglycerol may function as a surrogate for phospholipids, particularly phosphatidylglycerol, under phosphate-limiting conditions (Benning *et al.*, 1993). At least 4 structural genes (*sqdA, sqdB, sqdC,* and *sqdD*) are involved in sulfolipid biosynthesis in *R. sphaeroides* (Benning and Somerville, 1992a; Benning and Somerville, 1992b; Rossak *et al.*, 1995).

Labeling studies with [35S]-sulfate showed that sulfolipids were present in *Rhizobium meliloti, R. leguminosarum,* and *Rhizobium* NGR234, whereas *Agrobacterium* and *E. coli* seemed to be sulfolipid-free (Cedergren and Hollingsworth, 1994). One of the two sulfolipid classes from *R. meliloti* was isolated and its structure was determined to be sulfoquinovosyl diacylglycerol (Cedergren and Hollingsworth, 1994) (Figure 10 B). So far the role of sulfolipids in rhizobia is unclear and it is obvious that rhizobial mutants deficient in sulfolipids are needed to define their role.

X.C. AMINO ACID-CONTAINING LIPIDS

Ornithine lipids, in which two acyl substituents are linked to ornithine, are regularly found in Gram-negative bacteria. Similar lipids are produced by *Agrobacterium tumefaciens* but in this instance lysine replaces ornithine. In this lipid, 3-OH-palmitate is amide-linked to the α-amino group of lysine and its hydroxy group is esterified mainly by lactobacillic acid (Figure 10C), with lesser proportions of palmitate and vaccenate (Tahara *et al.*, 1976).

X.D. HOPANOID LIPIDS

Hopanoids are a class of pentacyclic triterpenoid lipids occurring predominantly in bacteria. They are thought to comprise a class of membrane stabilizers important to the proper functioning of the bacterial cell membrane. Hopanoids are found in a wide range of Gram-negative and Gram-positive bacteria. They have been described in the nitrogen-fixing *Azotobacter* and *Beijerinckia* bacteria and in symbiotic nitrogen-fixing *Frankia* and *Bradyrhizobium* (Kannenberg *et al.*, 1995) bacteria. Bradyrhizobia contain the triterpenoids hopene, diplopterol, the methylated derivatives thereof, elongated forms of hopanoids, as well as squalene, the biosynthetic precursor in hopanoid biosynthesis (Figure 11). Recently, the gene encoding a key enzyme of hopanoid biosynthesis, the squalene-hopene cyclase (Shc) has been cloned from

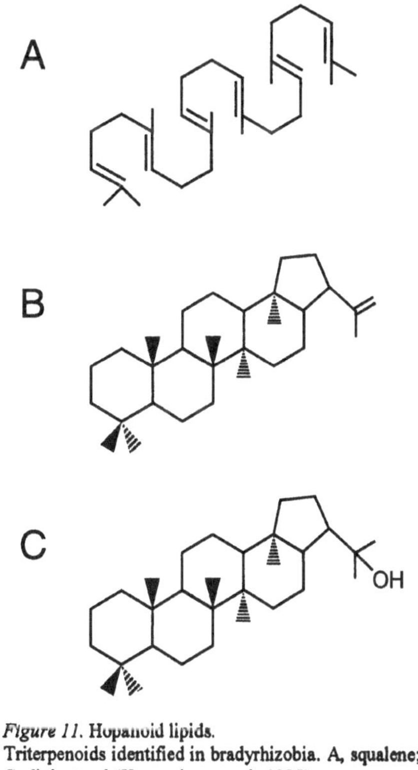

Figure 11. Hopanoid lipids.
Triterpenoids identified in bradyrhizobia. A, squalene; B, hopene; C, diplopterol (Kannenberg *et al.*, 1995).

Bradyrhizobium japonicum. Upon expression of the *shc* gene in *E. coli*, the recombinant enzyme is able to catalyze the cyclization of squalene to hopanoid derivatives *in vitro* (Kannenberg *et al.*, 1996; Perzl *et al.*, 1996). In some bradyrhizobia hopanoids can make up to 50 % (w/w) of the total lipid fraction. Surprisingly however, no triterpenoids have been detected in representatives of the genus *Rhizobium* (Kannenberg *et al.*, 1995).

So far the role of hopanoids in nitrogen-fixing organisms is unclear. In *Frankia* nitrogen fixation occurs within specialized multicellular structures termed vesicles. A vesicle is surrounded by a multilamellate, lipid-containing envelope that apparently functions as a barrier to oxygen diffusion. The envelope consists primarily of two hopanoid lipids, bacteriohopanetetrol and bacteriohopanetetrol phenylacetate monoester, the latter of which is vesicle-specific (Berry *et al.*, 1993). Also in bradyrhizobia hopanoids might play a role in the adaption to microaerobic conditions. They might be involved in other mechanisms of cell adaption, such as the resistance of bradyrhizobia to acidic growth conditions, a considerably higher resistance than that of *Rhizobium* bacteria. This assumption is supported indirectly by the finding that the amount of hopanoid lipids in the thermoacidophilic bacterium *Alicyclobacillus* (formerly *Bacillus*) *acidocaldarius* increases in response to decreasing pH values in the culture medium.

Hopanoid-deficient mutants had been isolated from *Zymomonas mobilis* (Tahara *et al.*, 1988) and show reduced tolerance to high glucose concentrations in comparison to the wild type. Clearly, to define the exact role of hopanoids in bradyrhizobia the isolation of hopanoid-deficient mutants and a thorough characterization of their respective phenotypes is needed.

X.E. CAROTENOIDS

Carotenoids are not common in rhizobia, but have been characterized for a *Bradyrhizobium* species (*Lupinus*) and an atypical species isolated from *Lotononis bainesii*. Interestingly, the major carotenoids in the latter species are C30 carotenoid acids esterified with one or two acylated glucose residues (Kleinig and Broughton, 1982). These have also been found in *Pseudomonas rhodos*. The main pigments in the *Bradyrhizobium* species are bicyclic C40 carotenoids and can be described as hydroxy and mono-oxo derivatives of the β-carotene nostoxanthin (2,3,2',3'-di-*trans*-tetrahydroxy-β,β'-carotene, 44%; and 2,3,2',3'-di-*trans*-tetrahydroxy-β,β'-carotene-4-one, 48%) (Kleinig *et al.*, 1977). The corresponding dione is the most abundant carotenoid (52%) in a mutant of this organism (Beyer *et al.*, 1979) (Figure 12).

XI. Amphiphilic Fatty Acid Derivatives

In contrast to membrane lipids, amphiphilic fatty acid derivatives, like lipo-chitin oligosaccharides or *N*-acyl homoserine lactones, can be detected to some extent in the watery phase. The concentrations of

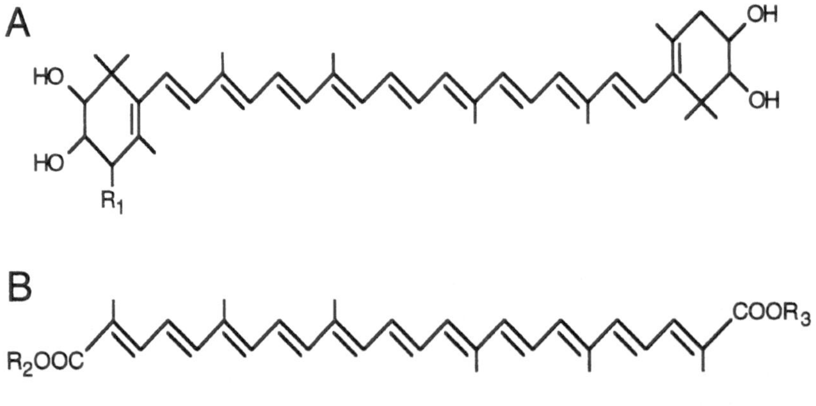

Figure 12. Carotenoids in *Rhizobiaceae*.

A. Main pigments of *Bradyrhizobium lupini*. R1 = O or R1 = 2H (Kleinig *et al.*, 1977). B. Main pigments of *Rhizobium* strain isolated from *Lotononis bainesii* (Kleinig and Broughton, 1982). R2 = glucose (acylated with a fatty acid), R3 = H or glucose (acylated with a fatty acid).

these amphiphilic molecules are sufficient to have them act as signalling compounds. The biosynthesis of lipo-chitin oligosaccharides is described elsewhere (Chapter 20).

XI.A. BIOSYNTHESIS OF *N*-ACYL HOMOSERINE LACTONE SIGNALS

Many bacteria, including several pathogens of plants and humans, use a pheromone called an autoinducer to regulate gene expression in a cell density-dependent manner. The *Agrobacterium* autoinducer (AAI) of *A. tumefaciens* was shown to be *N*-(β-ketooctanoyl)-L-homoserine lactone (Zhang *et al.*, 1993). AAI stimulates the conjugational transfer of the tumor-inducing (Ti) plasmid (Piper *et al.*, 1993). AAI is synthesized by the TraI protein (Hwang *et al.*, 1994), which is encoded by the Ti plasmid. Purified TraI uses β-ketooctanoyl-ACP and *S*-adenosylmethionine to make the β-ketooctanoyl and the homoserine lactone moieties of AAI, respectively (Moré *et al.*, 1996).

Surprisingly, the structural identification of bacteriocin *small* from *Rhizobium leguminosarum* revealed a novel *N*-(β-D-hydroxy-7-*cis*-tetradece-noyl)-L-homoserine lactone (Schripsema *et al.*, 1996). This novel autoinducer, together with RhiR, seems to be required for the activation of both the rhizosphere-expressed *rhiABC* operon and a growth-inhibiting function encoded by the symbiosis plasmid pRL1JI (Gray *et al.*, 1996). In *S. meliloti* a different autoinducer must be synthesized (Gray *et al.*, 1996).

N-acyl homserine lactones are described in more detail in chapter 10.

XII. Perspectives

In the *Rhizobiaceae* a variety of saccharidic components (lipo-chitin oligosaccharides, cyclic glucans, exopolysaccharides, lipopolysaccharides, and capsular polysaccharides) are important for the interaction with the plant host. Although a broad spectrum of unusual membrane lipid components is found in the *Rhizobiaceae* so far in no case a function can be attributed to any of those compounds. All members of the *Rhizobiaceae* interact with plants giving rise to nitrogen-fixing root nodules or tumorigenic proliferations thereby displaying a very complex phenotype. It can be expected that even minor changes in membrane

properties of the bacterium would affect the interaction with the plant host. In order to define the role of a certain lipid component, mutants defective in its synthesis need to be isolated and phenotypically characterized, especially with regard to their ability to interact with plants.

XIII. References

Arondel, V., Benning, C. and Somerville, C.R. (1993) J. Biol. Chem. 268, 16002-16008.

Bardin, S., Dan, S., Osteras, M. and Finan, T.M. (1996) J. Bacteriol. 178, 4540-4547.

Bell, R. M. (1974) J. Bacteriol. 117, 1065-1076.

Benning, C., Beatty, J. T., Prince, R.C. and Somerville, C.R. (1993) Proc. Natl. Acad. Sci. USA 90, 1561-1565.

Benning, C., Huang, Z.-H. and Gage, D.A. (1995) Arch. Biochem. Biophys. 317, 103-111.

Benning, C. and Somerville, C.R. (1992a) J. Bacteriol. 174, 2352-2360.

Benning, C. and Somerville, C.R. (1992b) J. Bacteriol. 174, 6479-6487.

Bergler, H., Fuchsbichler, S., Högenauer, G. and Turnowsky, F. (1996) Eur. J. Biochem. 242, 689-694.

Berry, A.M., Harriott, O.T., Moreau, R. A., Osman, S.F., Benson, D.R. and Jones, A.D. (1993) Proc. Natl. Acad. Sci. USA 90, 6091-6094.

Beyer, P., Kleinig, H., Englert, G., Meister, W. and Noack, K. (1979) Helv. Chim. Acta 62, 2551-2557.

Bhat, U.R., Carlson, R.W., Busch, M. and Mayer, H. (1991a) Int. J. Syst. Bacteriol. 41, 213-217.

Bhat, U.R., Mayer, H., Yokota, A., Hollingsworth, R.I. and Carlson, R.W. (1991b) J. Bacteriol. 173, 2155-2159.

Bibb, M.J., Biro, S., Motamedi, H., Collins, J.F. and Hutchinson, C.R. (1989) EMBO J. 8, 2727-2736.

Bloch, K. (1971) β-hydroxydecanoyl thioester dehydrase, in P.D. Boyer (ed.) The Enzymes, New York: Academic Press. pp. 441-464.

Bogdanov, M. and Dowhan, W. (1995) J. Biol. Chem. 270, 732-739.

Breedveld, M.W., Yoo, J.S., Reinhold, V.N. and Miller, K.J. (1994) J. Bacteriol. 176, 1047-1051.

Bremer, J. and Greenberg, D.M. (1961) Biochim. Biophys. Acta 46, 205-216.

Brozek, K.A., Carlson, R.W. and Raetz, C.H.R. (1996) J. Biol. Chem. 271, 32126-32136.

Bunn, C.R. and Elkan, G.H. (1971) Can. J. Microbiol. 17, 291-295.

Bunn, C.R., McNeill, J.J. and Elkan, G.H. (1970) J. Bacteriol. 102, 24-29.

Cai, H., Erhardt, P., Troppmair, J., Diaz-Meco, M.T., Sithanandam, G., Rapp, U.R., Moscat, J. and Cooper, G.M. (1993) Mol. Cell. Biol. 13, 7645-7651.

Cane, D.E. (1994) Science 263, 338-340.

Cedergren, R.A. and Hollingsworth, R.I. (1994) J. Lipid Res. 35, 1452-1461.

Charles, T.C., Newcomb, W. and Finan, T.M. (1991) J. Bacteriol. 173, 3981-3992.

Cloutier, J., Laberge, S. and Antoun, H. (1997) Mol. Plant-Microbe Interact. 10, 401-406.

Collins, M.D. and Jones, D. (1981) Microbiol. Rev. 45, 316-354.

Colonna-Romano, S., Arnold, W., Schlüter, A., Boistard, P., Pühler, A. and Priefer, U.B. (1990) Mol. Gen. Genet. 223, 138-147.

Cooper, C.L., Boyce, S.G. and Lueking, D.R. (1987) Biochemistry 26, 2740-2746.

Cooper, C.L., Hsu, L., Jackowski, S. and Rock, C.O. (1989) J. Biol. Chem. 264, 7384-7389.

Cronan, J.E. and Rock, C.O. (1996) Biosynthesis of membrane lipids, in F.C. Neidhardt (ed.) Escherichia coli and Salmonella, Washington, D.C.: ASM Press. pp. 612-636.

Crosby, J., Sherman, D.H., Bibb, M.J., Revill, W.P., Hopwood, D. A. and Simpson, T.J. (1995) Biochim. Biophys. Acta 1251, 32-42.

Das, P. K., Basu, M. and Chatterjee, G.C. (1979) J. Gen. Appl. Microbiol. 25, 1-9.

Debellé, F., Plazanet, C., Roche, P., Pujol, C., Savagnac, A., Rosenberg, C., Prome, J.-C. and Denarie, J. (1996) Mol. Microbiol. 22, 303-314.

Debellé, F. and Sharma, S.B. (1986) Nucleic Acids Res. 14, 7453-7471.

Debellé, F., Yang, G.P., Ferro, M., Truchet, G., Prome, J.-C. and Denarie, J. (1997) Rhizobium nodulation factors in perspective, in A. Legocki, H. Bothe, A. Pühler (eds.) Biological fixation of nitrogen for ecology and sustainable agriculture, Berlin, Heidelberg, New York, Springer. pp 15-23.

de Hertogh, A.A., Mayeux, P.A. and Evans, H.J. (1964) J. Biol. Chem. 239, 2446-2453.

de Moll, E. (1996) Biosynthesis of biotin and lipoic acid, in F. C. Neidhardt (ed.) Escherichia coli and Salmonella, Washington, D. C.: ASM Press. pp. 704-709.

Demont, N., Ardourel, M., Maillet, F., Prome, D., Rerro, M., Prome, J.-C. and Denarie, J. (1994) EMBO J. 13, 2139-2149.

Demont, N., Debellé, F., Aurelle, H., Dénarié, J. and Promé, J.-C. (1993) J. Biol. Chem. 268, 20134-20142.

de Rudder, K.E.E., Thomas-Oates, J.E. and Geiger, O. (1997) J. Bacteriol. 179, 6921-6928.

Dryden, S.C. and Dowhan, W. (1996) J. Bacteriol. 178, 1030-1038.

English, C.K., Wear, D.J. and Margileth, A.M. (1988) JAMA 259, 1347-1352.

Epple, G., van der Drift, K.G.M., Thomas-Oates, J.E. and Geiger, O. (1997) submitted .

Faizova, G.K., Borodulina, Y.S. and Samsonova, S.P. (1971) Mikrobiologiya 40, 471-474.

Firmin, J.L., Wilson, K.E., Carlson, R.W., Davies, A.E. and Downie, J.A. (1993) Mol. Microbiol. 10, 351-360.

Froelich, J.E., Poorman, R., Reardon, E., Barnum, S.R. and Jaworski, J.G. (1990) Eur. J. Biochem. 193, 817-825.

Fuqua, C., Winans, S.C. and Greenberg, E.P. (1996) Annu. Rev. Microbiol. 50, 727-751.

Geiger, O., Glushka, J., Lugtenberg, B.J.J., Spaink, H.P. and Thomas-Oates, J.E. (1997) Mol. Plant-Microbe Interact. in press.

Geiger, O., Ritsema, T., van Brussel, A.A.N., Tak, T., Wijfjes, A.H.M., Bloemberg, G.V., Spaink, H.P. and Lugtenberg, B.J.J. (1994a) Plant Soil 161, 81-89.

Geiger, O., Spaink, H.P. and Kennedy, E.P. (1991a) J. Bacteriol. 173, 2872-2878.

Geiger, O., Thomas-Oates, J.E., Glushka, J., Spaink, H.P. and Lugtenberg, B.J.J. (1994b) J. Biol. Chem. 269, 11090-11097.

Geiger, O., Weissborn, A.C. and Kennedy, E.P. (1991b) J. Bacteriol. 173, 3021-3024.

Gerson, T. and Patel, J.J. (1975) Appl. Microbiol. 30, 193-198.

Gerson, T., Patel, J.J. and Nixon, L.N. (1975) Lipids 10, 134-139.

Ghose, R., Geiger, O. and Prestegard, J.H. (1996) FEBS Lett. 388, 66-72.

Goldfine, H. and Ellis, M.E. (1964) J. Bacteriol. 87, 8-15.

Goodridge, A.G. (1991) Fatty synthesis in eucaryotes, in D.E. Vance and J. Vance (ed.) Biochemistry of Lipids, Lipoproteins and Membranes, Amsterdam, London, New York, Tokyo: Elsevier. pp. 111-139.

Gouill, C.L., Desmarais, D. and Dery, C.V. (1993) Mol. Gen. Genet. 240, 146-150.

Gray, K.M., Pearson, J.P., Downie, J.A., Boboye, B.E.A. and Greenberg, E.P. (1996) J. Bacteriol. 178, 372-376.

Grogan, D.W. and Cronan, J.E. (1984) J. Bacteriol. 158, 286-295.

Grogan, D.W. and Cronan, J.E. (1986) J. Bacteriol. 166, 872-877.

Hale, R.S., Jordan, K.N. and Leadlay, P.F. (1987) FEBS Lett. 224, 133-136.

Heacock, P.N. and Dowhan, W. (1987) J. Biol. Chem. 262, 13044-13049.

Heath, R.J. and Rock, C.O. (1996) J. Biol. Chem. 271, 27795-27801.

Hildebrand, J.G. and Law, J.H. (1964) Biochemistry 3, 1304-1308.

Hiraoka, S., Matsuzaki, H. and Shibuya, I. (1993) FEBS Lett. 336, 221-224.

Høj, P.B. and Svendsen, I. (1983) Carlsberg Res. Commun. 48, 285-305.

Hollingsworth, R.I. and Carlson, R.W. (1989) J. Biol. Chem. 264, 9300-9303.

Hopwood, D. A. and Sherman, D. H. (1990) Annu. Rev. Genet. 24, 37-66.

Horvath, B., Kondorosi, E., John, M., Schmidt, J., Török, I., Györgypal, Z., Barabas, I., Wieneke, U., Schell, J. and Kondorosi, A. (1986) Cell 46, 335-343.

Hsu, L., Jackowski, S. and Rock, C. O. (1989) J. Bacteriol. 171, 1203-1205.

Hubac, C., Guerrier, D., Ferran, J., Tremolieres, A. and Kondorosi, A. (1992) J. Gen. Microbiol. 138, 1973-1983.

Hughes, P.E. and Tove, S.B. (1982) J. Bacteriol. 151, 1397-1402.

Hwang, I., Li, P.-L., Zhang, L., Piper, K.R., Cook, D.M., Tate, M.E. and Farrand, S.K. (1994) Proc. Natl. Acad. Sci. USA 91, 4639-4643.

Issartel, J-P., Koronakis, V. and Hughes, C. (1991) Nature (London) 351, 759-761.

Jackowski, S., Cronan, J.E. and Rock, C.O. (1991) Lipid metabolism in procaryotes, in D. E. Vance and J. Vance (ed.) Biochemistry of lipids, lipoproteins and membranes, Amsterdam, London, New York Tokyo: Elsevier. pp. 45-85.

Jackowski, S., Jackson, P.D. and Rock, C.O. (1994) J. Biol. Chem. 269, 2921-2928.

Jarvis, B.D.W. and Tighe, S.W. (1994) Plant Soil 161, 31-41.

Kaneshiro, T. and Law, J.H. (1964) J. Biol. Chem. 239, 1705-1713.

Kaneshiro, T. and Marr, A.G. (1962) J. Lipid Res. 3, 184-189.

Kannenberg, E.L., Perzl, M. and Härtner, T. (1995) FEMS Microbiol. Lett. 127, 255-262.

Kannenberg, E.L., Perzl, M., Müller, P., Härtner, T. and Poralla, K. (1996) Plant Soil 186, 107-112.

Kennedy, E.P. and Weiss, S.B. (1956) J. Biol. Chem. 222, 193-214.

Khosla, C., Ebert-Khosla, S. and Hopwood, D.A. (1992) Mol. Microbiol. 6, 3237-3249.

Khosla, C., McDaniel, R., Ebert-Khosla, S., Torres, R., Sherman, D.H., Bibb, M.J. and Hopwood, D.A. (1993) J. Bacteriol. 175, 2197-2204.

Kiss, E., Reuhs, B.L., Kim, J.S., Kereszt, A., Petrovics, G., Putnoky, P., Dusha, I., Carlson, R.W. and Kondorosi, A. (1997) J. Bacteriol. 179, 2132-2140.

Kleinig, H. and Broughton, W.J. (1982) Arch. Microbiol. 133, 164.

Kleinig, H., Heumann, W., Meister, W. and Englert, G. (1977) Helv. Chim. Acta 60, 254-258.

Kusters, R., Breukink, E., Gallusser, A., Kuhn, A. and de Kruijff, B. (1994) J. Biol. Chem. 269, 1560-1563.

Kusters, R., Dowhan, W. and de Kruijff, B. (1991) J. Biol. Chem. 266, 8659-8662.

Ladd, S.L., Sommer, S.A., LaBerge, S. and Toscano, W. (1993) Clin. Neuropharmacol. 16, 540-549.

Lambalot, R.H. and Walsh, C.T. (1995) J. Biol. Chem. 270, 24658-24661.

Law, J.H., Zalkin, H. and Kaneshiro, T. (1963) Biochim. Biophys. Acta 70, 143-151.

Lerouge, P., Roche, P., Faucher, C., Maillet, F., Truchet, G., Promé, J-C. and Dénarié, J. (1990) Nature (London) 344, 781-784.

Lowe, R.H. and Evans, H.J. (1962) Soil Sci. 94, 351-356.

Lugtenberg, B.J.J. and Peters, R. (1976) Biochim. Biophys. Acta 441, 38-47.

Maagd, R.A. d. and Lugtenberg, B.J.J (1986) J. Bacteriol. 167, 1083-1085.

Mac Kenzie, S.L., Lapp, M.S. and Child, J.J. (1979) Can. J. Microbiol. 25, 68-74.

Manasse, R.J. and Corpe, W.A. (1967) Can. J. Microbiol. 13, 1591-1603.

Mileykovskaya, E.I. and Dowhan, W. (1993) J. Biol. Chem. 268, 24824-24831.

Miller, K.J., Gore, R.S. and Benesi, A.J. (1988) J. Bacteriol. 170, 4569-4575.

Miller, K.J., McKinstry, M.W., Hunt, W.P. and Nixon, B.T. (1992) Mol. Plant-Microbe Interact. 5, 363-371.

Miller, K.J., Reinhold, V.N., Weissborn, A.C. and Kennedy, E.P. (1987) Biochim. Biophys. Acta 901, 112-118.

Miller, K.J., Shon, B.C., Gore, R.S. and Hunt, W.P. (1990) Curr. Microbiol. 21, 205-210.

Miller, R.W. and Tremblay, P.A. (1983) Can. J. Biochem. Cell Biol. 61, 1334-1340.

Morbidoni, H.R., de Mendoza, D. and Cronan, J.E. (1996) J. Bacteriol. 178, 4794-4800.

Moré, M.I., Finger, L.D., Stryker, J.L., Fuqua, C., Eberhard, A. and Winans, S.C. (1996) Science 272, 1655-1658.

Oliver, J.D. and Colwell, R.R. (1973a) Int. J. Syst. Bacteriol. 23, 442-458.

Oliver, J.D. and Colwell, R.R. (1973b) J. Bacteriol. 114, 897-908.

Orgambide, G.G., Philip-Hollingsworth, S., Hollingsworth, R.I. and Dazzo, F.B. (1994) J. Bacteriol. 176, 4338-4347.

Perzl, M., Müller, P., Poralla, K. and Kannenberg, E.L. (1996) Microbiology 143, 1235-1242.

Petrovics, G., Putnoky, P., Reuhs, B., Kim, J., Thorp, T.A., Noel, K.D., Carlson, R.W. and Kondorosi, A. (1993) Mol. Microbiol. 8, 1083-1094.

Piper, K.R., Beck von Bodman, S. and Farrand, S.K. (1993) Nature (London) 362, 448-450.

Platt, M.W., Miller, J., Lane, W.S. and Kennedy, E.P. (1990) J. Bacteriol. 172, 5440-5444.

Pluschke, G., Hirota, Y. and Overath, P. (1978) J. Biol. Chem. 253, 5048-5055.

Possmeyer, F. (1989) Metabolism of phosphatidylcholine in lung, in D.E. Vance (ed.) Phosphatidylcholine metabolism, Boca Raton, Florida: CRC Press. pp. 205-224.

Post-Beittenmiller, M.A., Schmid, K.M. and Ohlrogge, J.B. (1989) Plant Cell 1, 889-899.

Raetz, C.R.H. (1975) Proc. Natl. Acad. Sci. USA 72, 2274-2278.

Randle, C.L., Albro, P.W. and Dittmer, J.C. (1969) Biochim. Biophys. Acta 187, 214-220.

Rawlings, M. and Cronan, J.E. (1992) J. Biol. Chem. 267, 5751-5754.

Reuhs, B. L. (1996) Acidic capsular polysaccharides (K antigens) of Rhizobium, in G. Stacey, B. Mullin and P.M. Gresshoff (ed.) Biology of Plant-Microbe Interactions, St. Paul: International Society for Molecular Plant-Microbe Interactions. pp. 331-336.

Reuhs, B.L., Carlson, R.W. and Kim, J.S. (1993) J. Bacteriol. 175, 3570-3580.

Revill, W.P., Bibb, M.J. and Hopwood, D.A. (1996) J. Bacteriol. 178, 5660-5667.

Revill, W.P. and Leadlay, P.F. (1991) J. Bacteriol. 173, 4379-4385.

Ritsema, T., Geiger, O., van Dillewijn, P., Lugtenberg, B.J.J. and Spaink, H.P. (1994) J. Bacteriol. 176, 7740-7743.

Ritsema, T., Wijfjes, A.H.M., Lugtenberg, B.J.J. and Spaink, H.P. (1996) Mol. Gen. Genet. 251, 44-51.

Ritsema, T., Gehring, A.M., Stuitje, A.R., Dandal, I., van der Drift, K.M.G.M., Lambalot, R.H., Walsh, C.T., Thomas-Oates, J.E., Lugtenberg, B.J.J. and Spaink, H.P. (1997) Mol. Gen. Genet. in press.

Roche, P., Maillet, F., Plazanet, C., Debelle, F., Ferro, M., Truchet, G., Prome, J-C. and Denarie, J. (1996) Proc. Natl. Acad. Sci. USA 93, 15305-15310.

Rock, C.O. and Cronan, J.E. (1985) Lipid metabolism in procaryotes, in D.E. Vance and J.E. Vance (ed.) Biochemistry of lipids and membranes, Menlo Park: The Benjamin/Cummings Publishing Company. pp. 73-115.

Rock, C.O. and Cronan, J.E. (1979) J. Biol. Chem. 254, 9778-9785.

Rolin, D.B., Pfeffer, P.E., Osman, S.F., Szwergold, B.S., Kappler, F. and Benesi, A.J. (1992) Biochim. Biophys. Acta 1116, 215-225.

Ronson, C.W. and Primrose, S.B. (1979) J. Gen. Microbiol. 112, 77-88.

Rossak, M., Tiedje, C., Heinz, E. and Benning, C. (1995) J. Biol. Chem. 270, 25792-25797.

Schofield, P.R. and Watson, J.M. (1986) Nucl. Acids Res. 14, 2891-2903.

Schripsema, J., de Rudder, K.E.E., van Vliet, T.B., Lankhorst, P.P., de Vroom, E., Kijne, J.W. and van Brussel, A.A.N. (1996) J. Bacteriol. 178, 366-371.

Schultze, M., Quiclet-Sire, B., Kondorosi, E., Virelizier, H., Glushka, J.N., Endre, G., Gero, S.D. and Kondorosi, A. (1992) Proc. Natl. Acad. Sci. USA 89, 192-196.

Shearman, C. A., Rossen, L., Johnston, A.W.B. and Downie, J.A. (1986) EMBO J. 5, 647-652.

Shen, Z. and Byers, D.M. (1996) J. Bacteriol. 178, 571-573.

Sherman, D.H., Malpartida, F., Bibb, M.J., Kieser, H.M., Bibb, M.J. and Hopwood, D.A. (1989) EMBO J. 8, 2717-2725.

Sherr, S. I. and Law, J. H. (1965) J. Biol. Chem. 240, 3760-3765.

Shi, W., Bogdanov, M., Dowhan, W. and Zusman, D.R. (1993) J. Bacteriol. 175, 7711-7714.

Silbert, D.F. (1967) Proc. Natl. Acad. Sci. USA 58, 1579-1586.

Simoni, R.D., Criddle, R.S. and Stumpf, P.K. (1967) J. Biol. Chem. 242, 573-581.

Slabas, A. R. and Fawcett, T. (1992) Plant Mol. Biol. 19, 169-191.

Souciet, G. and Weil, J-H. (1992) 81, 215-225.

Spaink, H.P., Bloemberg, G.V., van Brussel, A.A.N., Lugtenberg, B.J.J., van der Drift, K.M.G.M., Haverkamp, J. and Thomas-Oates, J.E. (1995) Mol. Plant-Microbe Interact. 8, 155-164.

Spaink, H.P., Sheeley, D.M., van Brussel, A.A.N., Glushka, J., York, W.S., Tak, T., Geiger, O., Kennedy, E.P., Reinhold, V.N. and Lugtenberg, B.J.J. (1991) Nature (London) 354, 125-130.

Streit, W.R., Joseph, C.M. and Phillips, D.A. (1996) Mol. Plant-Microbe Interact. 9, 330-338.

Tahara, Y., Ogawa, Y., Sakakibara, T. and Yamada, Y. (1986) Agric. Biol. Chem. 50, 257-259.

Tahara, Y., Ogawa, Y., Sakakibara, T. and Yamada, Y. (1987) Agric. Biol. Chem. 51, 1425-1430.

Tahara, Y., Yamada, Y. and Kondo, K. (1976) Agr. Biol. Chem. 40, 1449-1450.

Tahara, Y., Yamashita, T., Kondo, M. and Yamada, Y. (1988) Agric. Biol. Chem. 52, 3189-3190.

Tahara, Y., Yamashita, T., Sogabe, A. and Ogawa, Y. (1994) J. Gen. Appl. Microbiol. 40, 389-396.

Therisod, H. and Kennedy, E.P. (1987) Proc. Natl. Acad. Sci. USA 84, 8235-8238.

Therisod, H., Weissborn, A.C. and Kennedy, E.P. (1986) Proc. Natl. Acad. Sci. USA 83, 7236-7240.

Thompson, E.A., Kaufman, A.E., Johnston, N.C. and Goldfine, H. (1983) Lipids 18, 602-606.

van der Groot, F.G., Didat, N., Pattus, F., Dowhan, W. and Letellier, L. (1993) Eur. J. Biochem. 213, 217-221.

Vance, D.E. and Ridgway, N.D. (1988) Prog. Lipid Res. 27, 61-79.

Wang, A.Y. and Cronan, J.E. (1994) Mol. Microbiol. 11, 1009-1017.

Yamada, Y., Takinami-Nakamura, H., Tahara, Y., Oyaizu, H. and Komagata, K. (1982) J. Gen. Appl. Microbiol. 28, 7-12.

Yamashita, S., Oshima, A., Nikawa, J. and Hosaka, K. (1982) Eur. J. Biochem. 128, 589-595.

Zhang, L., Murphy, P.J., Kerr, A. and Tate, M.E. (1993) Nature (London) 362, 446-448.

Chapter 5

Cell-surface β-glucans

Michaël W. Breedveld and Karen J. Miller

I. Introduction

The associations between rhizobia and agrobacteria and their host plants have been the subject of intensive study for many years, and the cell-surface carbohydrates of these bacteria are now recognized to provide functions during the infection process. In this chapter, the structural features of the cyclic β-glucans of the *Rhizobiaceae* will be summarized. The identification and characterization of the genetic loci and enzyme systems involved in cyclic β-glucan biosynthesis will also be reviewed. Finally, the functions of these unique molecules in the free-living bacteria as well as during plant infection will be

considered. The reader is also referred to an earlier review on the cyclic β-glucans (Breedveld and Miller 1994).

II. Structural Characteristics of the Cyclic β-Glucans

Cyclic β-glucans, originally described as "crown-gall polysaccharides", were reported for the first time in 1942 in *Agrobacterium tumefaciens* cultures (McIntire *et al.*, 1942). Since then, cyclic β-glucans have been found in cultures of all *Agrobacterium*, *Rhizobium* and *Bradyrhizobium* species thus far examined. In *Agrobacterium* and *Rhizobium* species, these molecules are linked solely by β-(1,2)-glycosidic bonds. The linkage has been established by methylation analysis, periodate oxidation and Smith degradation, while the β-anomeric configuration at the C-1 carbon atoms was confirmed by ^{1}H and ^{13}C NMR spectroscopy (Amemura *et al.*, 1983; Dell *et al.*, 1983; Hisamatsu *et al.*, 1983) (for a review, see Breedveld and Miller 1994). The macrocyclic, unbranched form was originally proposed because of the absence of reducing and non-reducing terminal residues (York *et al.*, 1980; Zevenhuizen and Scholten-Koerselman 1979), and this was subsequently confirmed by ^{13}C-NMR spectroscopy and fast-atom bombardment mass spectrometry (Dell *et al.*, 1983).

The cyclic β-(1,2)-glucans of *R. leguminosarum* and *A. tumefaciens* strains consist of a mixture of rings with degrees of polymerization (DPs) ranging from 17 to 25, while much larger cyclic β-(1,2)-glucans (up to DP 40) have been detected within cultures of *S. meliloti* (formerly known as *R. meliloti*) (Breedveld and Miller 1994; Hisamatsu 1992). The size distribution of the cyclic β-(1,2)-glucans appears to result from competing elongation and cyclization reactions (Williamson *et al.*, 1992).

Interestingly, cyclic β-(1,2)-glucans with DP below 17 have never been found. Consistent with this, molecular modelling studies have predicted that β-(1,2)-glucan rings with DP lower than 17 are not energetically favored (Palleschi and Crescenzi 1985; York 1995; York *et al.*, 1993). These studies have also attempted to model the size of the cavity of the cyclic β-(1,2)-glucans. In one modelling study, the cavity was predicted to be relatively hydrophobic, approximately 14 angstroms in diameter, and to have the ability to form inclusion complexes with hydrophobic guest molecules (Palleschi and Crescenzi 1985) (Figure 1). In a more recent modeling study (Andre *et al.*, 1995), a non-symmetric molecule with a small hydrophilic cavity (3.7 Angstroms) was proposed. Andre *et al.*, (Andre *et al.*, 1995) also predicted that the cyclic (1,2)-β-glucans should not readily form inclusion complexes with guest molecules. While there have been reports of inclusion complex formation with a variety of hydrophobic guest molecules (Koizumi *et al.*, 1984; Morris *et al.*, 1991; Okada *et al.*, 1985), other researchers have failed to detect such complexes (Andre *et al.*, 1995).

Species of *Bradyrhizobium* synthesize cyclic β-glucans containing both β-(1,3) and β-(1,6) glycosidic linkages (Miller *et al.*, 1990). These molecules contain 10 to 13 glucose residues and appear to be branched in structure (Rolin *et al.*, 1992). The structure of the bradyrhizobial cyclic β-(1,6)-β-(1,3)-glucan of DP 13 has been proposed to consist of a backbone of 12 glucose residues containing triplets of β-(1,3) linked glucose residues separated by triplets of β-(1,6)-linked glucose residues (Rolin *et al.*, 1992). In this model, one residue is present as a branch on the C-6 of a β-(1,3)-linked glucose residue (Figure 1).

Although early studies with the cyclic β-(1,2)-glucans suggested that these molecules were unsubstituted and strictly neutral in character, it is now known that the cyclic β-(1,2)-glucans may become charged through the addition of anionic substituents such as sn-1-phosphoglycerol, succinic acid and methylmalonic acid (Breedveld and Miller 1994). The predominant substituent on the cyclic β-(1,2)-glucans of *S. meliloti* and *A. tumefaciens* strains is sn-1-phosphoglycerol which is derived from the head group of phosphatidylglycerol and is linked to glucose through a phosphodiester linkage at position C-6 (Miller *et al.*, 1988). The levels of substituents on the cyclic β-(1,2)-glucans vary greatly among different species of *Agrobacterium* and *Rhizobium*. Generally, a large fraction of the glucans during logarithmic growth contain between 1 to 4 sn-1-phosphoglycerol substituents per glucan backbone, while much higher relative levels of neutral glucans are present in stationary phase cultures (Breedveld and Miller 1994). Furthermore, neutral forms of the cyclic β-(1,2)-glucans are the biosynthetic precursors of the anionic forms (Breedveld and Miller 1995; Geiger *et al.*, 1991), and anionic substituents are added to the cyclic β-(1,2)-glucan backbone within the periplasmic compartment (Breedveld and Miller 1995).

Despite the fact that the cyclic β-(1,2)-glucans of *S. meliloti* and *A. tumefaciens* become highly modified with anionic substituents, some rhizobia (e.g. several strains of *R. leguminosarum*) appear to lack the capacity to add substituents to their cyclic β-(1,2)-glucans. In *Bradyrhizobium* species, the cyclic β-(1,6)- β-(1,3)-glucans are uncharged in character yet contain the zwitterionic substituent phosphocholine, which appears to be linked through the C-6 of a (1,3)-linked glucose residue (Rolin *et al.*, 1992).

We note that cyclic β-glucans are synthesized by other bacterial genera including *Azospirillum*, *Brucella*, and *Xanthomonas* (Table 1). It is intriguing that these bacteria have the capacity to infect

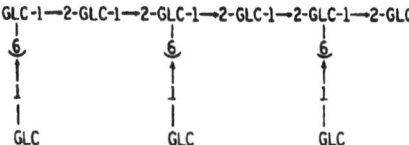

Figure 1. Representative structures of a cyclic β-(1,2)-glucan with DP-18 (upper left), a cyclic β-(1,6)-β-(1,3)-glucan (DP=12) with one branched β-(1,6)-linked glucose residue (upper right), and a MDO-molecule with 9 glucose residues (bottom). Non-sugar substituents are not shown.

eukaryotic hosts, strongly suggesting that the cyclic nature of these molecules may be important during host infection.

III. Localization of the Cyclic β-Glucans

Initial studies of the cyclic β-(1,2)-glucans of *Rhizobium* and *Agrobacterium* species focused on the extracellular medium of these cultures. Consequently, the presence of cell-associated cyclic ß-glucans was overlooked. It has now been established that cyclic β-(1,2)-glucans are cell-associated with levels ranging from 5-20% of the total cellular dry weight, depending on the species and on cultural conditions. Therefore, the cyclic β-glucans are major cellular constituents. For example, cellular concentrations of the cyclic β-(1,2)-glucans of *R. leguminosarum* have been reported to range between 50 to 100 mg/g cellular dry weight with highest levels accumulated during logarithmic growth (Zevenhuizen 1981). The cellular concentrations of the cyclic β-(1,2)-glucans of *S. meliloti* are slightly higher (*i.e.* 200 mg/g cellular dry weight) and appear to remain constant throughout the logarithmic and stationary phases of growth (Zevenhuizen 1981; Zevenhuizen and Van Neerven 1983). The cellular concentrations of the cyclic β-(1,6)- β-(1,3)-glucans within *B. japonicum* cultures (eg. 50 mg/g cellular dry weight) are slightly lower (Miller and Gore 1992).

Further examinations with respect to their localization have revealed that the cyclic β-glucans are predominantly localized within the periplasmic compartment. Assuming the relative periplasmic volume of these rhizobia is similar to that measured for *E.coli* and *S. typhimurium* (Stock *et al.*, 1977), it can be estimated that cyclic glucans reach concentrations between 10-100 mM within the

Gene/ Locus	Organism	Size of predicted product (kDa)	Function
ndvA	S. meliloti	67	export
chvA	A. tumefaciens	65	export
ndvB	S. meliloti	319	backbone synthesis
	R. fredii	?	backbone synthesis
	R. leguminosarum	>319*	backbone synthesis
	B. japonicum	91*	backbone synthesis
ndvC	B. japonicum	62	synthesis β-(1,6) linkages
chvB	A. tumefaciens	319*	backbone synthesis
cgm	S. meliloti	?	addition of phosphoglycerol substituents

Table 1. Chromosomal loci associated with cyclic glucan biosynthesis
*gene sequence not yet known; size of the protein estimated by protein gel electrophoresis

periplasmic compartment (Breedveld and Miller 1994; Breedveld *et al.*, 1992). This estimated concentration range is very close to the measured solubility limit of these molecules of 250 g/l (Koizumi *et al.*, 1984; Okada *et al.*, 1985) (which corresponds to a concentration range of 60 - 90 mM for glucan molecules containing 17 to 25 glucose residues).

The presence of cyclic β-glucans in the extracellular medium varies greatly among different species and is influenced strongly by growth stage and cultural conditions. High levels of extracellular cyclic β-glucans are generally detected within stationary phase cultures (Breedveld and Miller 1994; Geiger *et al.*, 1991; Zevenhuizen 1981; Zevenhuizen and Van Neerven 1983). The secretion of cyclic β-(1,2)-glucans has also been shown to be strongly enhanced (eg. grams per liter) in certain growth media and at elevated temperatures (Breedveld *et al.*, 1993; Breedveld *et al.*, 1990; Breedveld *et al.*, 1992).

IV. Biosynthesis of the Cyclic β-Glucans

IV.A. GENETIC LOCI INVOLVED IN CYCLIC β-GLUCAN BIOSYNTHESIS

Agrobacterium chvA and *chvB* mutants and *S. meliloti ndvA* and *ndvB* mutants are impaired for the biosynthesis of cyclic β-(1,2)-glucans (Cangelosi *et al.*, 1989; Geremia *et al.*, 1987; Stanfield *et al.*, 1988). Originally, Douglas and coworkers identified the *chvA* and *chvB* (CHromosomal Virulence) genes on the chromosome of *Agrobacterium tumefaciens* as

loci required for virulence and for attachment of the bacterium to plant cells (Douglas *et al.*, 1985). Subsequently, Dylan and coworkers (Dylan *et al.*, 1986) demonstrated that *S. meliloti* contained two genetic loci on the chromosome (*ndvA* and *ndvB*; Nodule DeVelopment) that were required for nodulation, and that were functionally and structurally homologous to *chvA* and *chvB*. Indeed, mutants of *A. tumefaciens* at either locus can be fully complemented with *S. meliloti* cosmid clones containing *ndvA* and *ndvB* (Dylan *et al.*, 1986). Thus, these studies identified genetic loci involved in cyclic β-(1,2)-glucan biosynthesis and also provided evidence that the cyclic β-(1,2)-glucans function during plant infection. Further analysis of these loci has revealed that the *chvA(ndvA)* and *chvB(ndvB)* genes are adjacent to each other and are transcribed in a convergent fashion (Douglas *et al.*, 1985; Dylan *et al.*, 1986). Furthermore, loci homologous to *chvA(ndvA)* and/or *chvB(ndvB)* have been found in *R. leguminosarum* (Dylan *et al.*, 1986) and *Azospirillum brasilense* (Raina *et al.*, 1995). Although various studies have reported that *Bradyrhizobium* species do not contain sequences homologous to *chvA(ndvA)* or *chvB(ndvB)* (Bhagwat and Keister 1992; Bhagwat and Keister 1995; Bhagwat *et al.*, 1993; Dylan *et al.*, 1986), Thomas and coworkers have shown that mesquite-infecting bradyrhizobia contain sequences homologous to *ndvB* if hybridization stringency conditions are reduced (Thomas *et al.*, 1994). Recently, 2 genetic loci in *B. japonicum* involved in cyclic β-(1,3)-β-(1,6)-glucan biosynthesis have been identified and described as *ndvB*-like and *ndvC*. Mutants at these loci are impaired for both cyclic β-glucan biosynthesis and soybean nodulation (Bhagwat *et al.*, 1996; Dunlap *et al.*, 1996; Pfeffer *et al.*, 1996). Finally, Breedveld *et al.*, (Breedveld *et al.*, 1995) have recently identified the *cgm* (Cyclic Glucan Modification) locus in *S. meliloti*. A *S. meliloti* mutant at this locus is not able to add sn-1-phosphoglycerol substituents to the cyclic β-(1,2)-glucan backbone. However, in contrast to the mutants described above, the *cgm* mutant is able to effectively nodulate its host plant (alfalfa) (Breedveld *et al.*, 1995).

IV.B. NDVB/CHVB PROTEINS OF *RHIZOBIUM* AND *AGROBACTERIUM*: MEMBRANE-ASSOCIATED GLYCOSYLTRANSFERASES

The *ndvB* gene in *S. meliloti* and the *chvB* gene in *A. tumefaciens* have each been shown to encode a high molecular weight cytoplasmic membrane protein that is involved in the biosynthesis of the cyclic β-(1,2)-glucans from UDP-glucose (Geremia *et al.*, 1987; Ielpi *et al.*, 1990; Zorreguieta *et al.*, 1988; Zorreguieta and Ugalde 1986). Based on its sequence, the *ndvB* gene of *S. meliloti* is predicted to encode a 319 kDa protein. Castro *et al* (Castro *et al.*, 1995) have provided evidence that the NdvB protein of *R. leguminosarum* is slightly larger than the NdvB protein of *S. meliloti*.

During the biosynthesis of the cyclic β-(1,2)-glucans, the NdvB and ChvB proteins have each been shown to form a covalent intermediate with the glucan backbone. This intermediate has been detected in in vitro analyses of cyclic β-(1,2)-glucan biosynthesis from UDP-glucose using membrane preparations derived from *Rhizobium* and *Agrobacterium* (Geremia *et al.*, 1987; Zorreguieta *et al.*, 1988; Zorreguieta and Ugalde 1986; Zorreguieta *et al.*, 1985). The structures of the neutral cyclic β-(1,2)-glucans produced in vitro are indistinguishable from those produced in vivo (Amemura 1984; Breedveld *et al.*, 1992; Williamson *et al.*, 1992). In an attempt to further purify the cyclic β-(1,2)-glucan synthetase activity from membrane preparations of *A. radiobacter*, Kinoshita and coworkers (Kinoshita *et al.*, 1991) identified 2 activities on the basis of binding behavior on an anion exchange column. One synthetase (referred to as synthetase I) was further characterized, was found to have an apparent Mr of 350,000, and showed the same catalytic activity as crude membrane preparations. It is possible that this activity corresponds to the purified ChvB protein from this bacterium.

The results from the above studies suggest that the NdvB and ChvB proteins are involved in several stages of cyclic β-(1,2)-glucan biosynthesis (initiation, elongation, and cyclization). Further evidence for this is suggested from the following: i) UDP-glucose serves as the only substrate, and no oligosaccharides (*e.g.* sophorose) or other primers are needed for the synthesis of the glucan ring (Amemura 1984; Breedveld *et al.*, 1992; Williamson *et al.*, 1992; Zorreguieta *et al.*, 1985; Zorreguieta *et al.*, 1985). (ii) Only cyclic forms of the β-(1,2)-glucans are detected after release from these proteins (Amemura 1984; Breedveld *et al.*, 1992; Williamson *et al.*, 1992; Zorreguieta *et al.*, 1985). (iii) Although lipid-linked intermediates are involved in the biosynthesis of many other bacterial polysaccharides, such intermediates have not been detected during the biosynthesis of the cyclic β-glucans (Breedveld and Miller 1994; Castro *et al.*, 1995). A recent study provides strong evidence that the *ndvB*/*chvB*-encoded glucosyltransferase is responsible for all stages of neutral cyclic β-(1,2)-glucan backbone biosynthesis. In this study, Castro *et al.*, (Castro *et al.*, 1996) demonstrated that solubilized *A. tumefaciens* inner membrane protein, separated on native protein gels, was able to form the 319 kDa protein-linked oligosaccharide intermediate in situ. Cyclic β-(1,2)-glucan was also formed upon incubation of the gel portion containing the 319 kDa protein intermediate with UDP-[^{14}C]-glucose, demonstrating that all three enzymatic activities required for neutral cyclic β-(1,2)-glucan formation (initiation, elongation, and cyclization) are associated with this protein.

It has been proposed that the NdvB/ChvB proteins have domains involved in functions other than cyclic β-(1,2)-glucan biosynthesis, since mutagenesis studies have indicated that up to 40% of the NdvB/ChvB protein (from the carboxyl terminus) is not required for β-glucan biosynthesis (Bhagwat *et al.*, 1992; Ielpi *et al.*, 1990; Zorreguieta *et al.*, 1988). Although lower quantities of glucans are synthesized by "downstream" mutants missing their carboxyl domains, structural analysis of these glucans has revealed them to be a mixture of cyclic glucans with size distribution and phosphoglycerol substituent profile similar to those found in wild-type cells (Breedveld *et al.*, 1994).

It has been reported that a strain of the diazotroph *Azospirillum brasilense* has a genetic locus (*cvi*, Chromosomal VIrulence) in which the NH$_2$-terminus of the predicted protein has significant sequence homology with the NH$_2$-terminus of the NdvB protein of *S. meliloti* (Raina *et al.*, 1995). Furthermore, the cloned *A. brasilense* locus can complement an *A. tumefaciens chvB* mutant with respect to tumor formation in leaf disks of tobacco plants (Raina *et al.*, 1995). In an earlier study, a glucosyltransferase activity within inner membranes of *A. brasilense* was characterized and found to function during the biosynthesis of β-(1,6)-β-(1,3)-glucans from UDP-glucose (Inon de Iannino and Ugalde 1993). Bohin and coworkers have provided evidence that the β-(1,6)-β-(1,3)-glucans of *A. brasilense* are cyclic in character (Bohin, 1996, personal communication). Curiously, cyclic β-(1,2)-glucans have not been detected in *A. brasilense* cultures. These results therefore suggest that the NH$_2$-termini of proteins involved in the synthesis of different cyclic β-glucans are very similar in structure.

As discussed above, it is very likely that the NdvB and ChvB proteins of *Rhizobium* and *Agrobacterium* species, respectively, mediate all stages of backbone biosynthesis beginning with UDP-glucose as substrate. A number of studies have examined cyclic β-(1,2)-glucan biosynthesis in vitro using membrane preparations derived from *R. leguminosarum*, *S. meliloti*, *A. tumefaciens*, and *A. radiobacter* (Amemura 1984; Breedveld *et al.*, 1992; Williamson *et al.*, 1992; Zorreguieta *et al.*, 1985; Zorreguieta and Ugalde 1986). In all cases, activation by Mn^{2+} or Mg^{2+} has been observed. Furthermore, the structures of the cyclic β-(1,2)-glucans produced in vitro are indistinguishable from those produced in vivo. As far as we are aware, there have been no published in vitro studies that have examined the transfer of substituents to the cyclic β-(1,2)-glucans.

In vitro cyclic β-glucan biosynthesis has also been examined using membrane preparations derived from *B. japonicum* (Cohen and Miller 1991). Again, Mg^{2+} or Mn^{2+} have been shown to stimulate this activity. It is noted, however, that the products synthesized in vitro have a much higher level of β-(1,3)-glycosidic linkages when compared to those synthesized in vivo (Miller *et al.*, 1990; Rolin *et al.*, 1992). It has been proposed that the in vitro products contain a cyclic β-(1,3) backbone containing 11 glucose residues with variable numbers of β-(1,6)-glucose branch points (Inon de Iannino and Ugalde 1993). These branch points are believed to serve as primers for the formation of β-(1,3)-linked linear branches. Because the structure of the bradyrhizobial glucans is more complex than that of

the cyclic β-(1,2)-glucans, it would seem necessary that proteins in addition to NdvB are involved in their biosynthesis.

IV.C. NDVA/CHVA PROTEINS OF *RHIZOBIUM* AND *AGROBACTERIUM* ARE INVOLVED IN CYCLIC β-(1,2)-GLUCAN TRANSPORT

The *ndvA* and *chvA* genes have been shown to encode proteins of 67 and 65 kDa, respectively (Cangelosi *et al.*, 1989; Inon De Iannino and Ugalde 1989; Stanfield *et al.*, 1988). These proteins share 76% identity at the amino acid level and both also share homology with ABC (ATP Binding Cassette) transporters (Palmen *et al.*, 1994). This family of proteins is associated with a variety of distinct biological processes in both prokaryotes and eukaryotes, most of them involving active transport of (hydrophilic) molecules across the cytoplasmic membrane. All of these proteins share a conserved domain of about 200 amino acid residues, including an ATP-binding site. Based on this homology, it has been proposed that the NdvA and ChvA proteins are involved in the transport of the cyclic β-(1,2)-glucans to the periplasm and extracellular medium (Cangelosi *et al.*, 1989; Stanfield *et al.*, 1988). Consistent with this proposal, it has been shown that the excretion of cyclic β-(1,2)-glucans to the extracellular medium is greatly impaired in *A. tumefaciens chvA* mutants (Inon De Iannino and Ugalde 1989). Furthermore, *ndvA* and *chvA* mutants do not produce periplasmic glucans but, instead, accumulate cyclic glucans in the cytoplasm (Breedveld *et al.*, 1994; Inon De Iannino and Ugalde 1989). Interestingly, Stanfield *et al.*, (Stanfield *et al.*, 1988) reported that the oligosaccharide repeating unit of the exopolysaccharide (EPS) of *S. meliloti* could not be detected in culture supernatants of *ndvA* mutants. Whether this is a transport-related phenomenon remains to be determined, since the mutants also produce more of the high molecular mass EPS. Of further interest is the discovery by Becker *et al.*, (Becker *et al.*, 1995) of the *exsA* gene from *S. meliloti* which is highly homologous to *ndvA*. Based on this homology and the results of Stanfield *et al.*, (Stanfield *et al.*, 1988), Becker and coworkers suggest that the ExsA and NdvA proteins both participate in EPS export (Becker *et al.*, 1995). Structural characterization of the cell-associated cyclic β-(1,2)-glucans of an *ndvA* mutant of *S. meliloti* 102F34 revealed a mixture of cyclic glucans with a ring size distribution indistinguishable from wild-type cells (Breedveld *et al.*, 1994). Thus, *ndvA* (*chvA*) is apparently not required for cyclization of the β-(1,2)-glucan backbone. However, anionic cyclic β-(1,2)-glucan biosynthesis is dramatically reduced in *chvA* and *ndvA* mutants (Breedveld *et al.*, 1994; Inon De Iannino and Ugalde 1989). The inability of these mutants to synthesize anionic cyclic β-(1,2)-glucans is apparently an indirect consequence of the failure of these cells to transport cyclic β-(1,2)-glucans to the periplasm, since it has been shown that the sn-1-phosphoglycerol and succinyl substituents are added to the cyclic β-(1,2)-glucan backbone in this compartment (Breedveld *et al.*, 1995; Breedveld and Miller 1995).

Little is known about the energetics of transport of cyclic glucans across the cytoplasmic membrane. Because NdvA and ChvA fall within the family of ABC transporters, it has been predicted that hydrolysis of ATP is required for transport (Palmen *et al.*, 1994). Partial trypsin digestion studies have indicated that ChvA and ChvB may form complexes within the inner membrane of *A. tumefaciens* (Inon De Iannino and Ugalde 1989). Whether or not the same protein machinery mediates the transport of the cyclic β-(1,2)-glucans to the periplasm and to the extracellular medium remains to be determined. Studies with *R. leguminosarum* suggest that export from the periplasm to the extracellular medium is not an active process, but results from an increased permeability of the outer membrane (Breedveld *et al.*, 1992).

IV.D. *NDV* GENES IN *BRADYRHIZOBIUM JAPONICUM*

In 1995, Bhagwat and coworkers (Bhagwat and Keister 1995) identified a *ndvB*-like locus in *B. japonicum* that is involved in the biosynthesis of cyclic β-(1,6)-β-(1,3)-glucans. A mutation at this locus (created by site-directed mutagenesis) resulted in the loss of cyclic glucan biosynthesis by *B. japonicum*. This mutant was also unable to form effective nodules on soybean (Bhagwat and Keister 1995), indicating that cyclic glucans may be required during this symbiosis. It is further noted that the cloned *ndvB*-like locus of *B. japonicum* can complement the symbiotic behavior of a *S. meliloti* *ndvB* mutant (Bhagwat *et al.*, 1993). Whether or not this *ndvB*-like locus shares any sequence homology with the *ndvB* gene of *S. meliloti* is unknown,

however, it is noted that Bhagwat and coworkers were unable to detect hybridization between the *ndvB* gene of *S. meliloti* and genomic DNA derived from *B. japonicum* (Bhagwat and Keister 1992). It is possible that the *ndvB*-like locus of *B. japonicum* encodes the 90 kDa inner membrane protein identified by de Iannino and Ugalde (Inon de Iannino and Ugalde 1993) which apparently becomes covalently linked to the bradyrhizobial glucan during biosynthesis. This finding by de Iannino and Ugalde gives strength to the hypothesis that the presence of a protein-linked intermediate is a general requirement for the synthesis of cyclic β-glucans.

Recently, Bhagwat and coworkers have identified a second *ndv* locus from *B. japonicum* involved in cyclic β-(1,3)-β-(1,6) glucan biosynthesis (Bhagwat *et al.*, 1996). This locus, termed *ndvC*, is apparently involved in the synthesis of β-(1,6) linkages. *B. japonicum* mutants at this locus synthesize glucans containing predominantly β-(1,3) linkages. Interestingly, these mutants are impaired for soybean nodulation yet are only slightly sensitive to hypoosmotic growth conditions. Sequence analysis revealed that *ndvC* is predicted to encode a membrane protein of 62 kDa.

IV.E. THE *CGM* LOCUS OF *S. MELILOTI*

Breedveld *et al.*, (Breedveld *et al.*, 1995) utilized a thin layer chromatographic screening procedure to identify *S. meliloti* mutants impaired for cyclic β-(1,2)-glucan biosynthesis. A screen of 2150 randomly generated Tn5 mutants revealed one selectively blocked in its ability to transfer phosphoglycerol substituents to the cyclic β-(1,2)-glucan backbone. The locus containing the Tn5 insertion has been designated *cgm* for "Cyclic Glucan Modification".

Despite the fact that the *cgm* mutant synthesizes cyclic β-(1,2)-glucans devoid of phosphoglycerol substituents, it apparently compensates for this defect by adding higher levels of succinyl substituents to the cyclic β-(1,2)-glucan backbone. Indeed, the overall negative charge on the glucans synthesized by the *cgm* mutant is similar to that found in wild-type cells. This result suggests that the enzymes which mediate the transfer of succinyl substituents to the cyclic β-(1,2)-glucan backbone compete with those which mediate the transfer of sn-1-phosphoglycerol substituents. This is also consistent with experiments, which show both modification reactions occur within the periplasm (Breedveld et al., 1995; Breedveld and Miller 1995).

V. Roles for Cyclic β-Glucans During Hypoosmotic Adaptation

V.A. THE CYCLIC β-GLUCANS SHARE PROPERTIES WITH THE MEMBRANE-DERIVED OLIGOSACCHARIDES OF E.COLI

Periplasmic glucans were first discovered in E.coli where they are referred to as "Membrane-derived oligosaccharides" (MDO). The MDO contain a β-(1,2)-linked backbone, β-(1,6)-linked branches, and range in size from 6 to 12 glucose residues (Kennedy 1996) (Figure 1). Furthermore, these glucans are highly modified with sn-1-phospho-glycerol, succinate, and phospho-ethanolamine substituents (for a review on MDO, see Kennedy 1987; Kennedy 1996). It is interesting that the linear, branched MDO of E.coli and the cyclic β-(1,2)-glucans of S. meliloti and A. tumefaciens contain sn-1-phosphoglycerol substituents derived from phosphatidylglycerol (Miller et al., 1988; Miller et al., 1987). Furthermore, biosynthesis of the MDO by E.coli and the cyclic β-glucans by most Rhizobiaceae is greatest during growth in low osmolarity media (Breedveld and Miller 1994), indicating that the accumulation of these molecules within the periplasm provides a mechanism for the cell to cope with hypoosmotic environments (Breedveld and Miller 1994; Miller et al., 1986). However, it should be noted that in several strains of R. leguminosarum and one strain of S. meliloti the level of cell-associated cyclic β-glucans is independent of medium osmolarity (see below) (Breedveld et al., 1992; Soto et al., 1992).

The osmotic regulation of periplasmic glucan biosynthesis was first observed for the MDO by Kennedy (Kennedy 1982) who proposed that these anionic glucans and their counter-ions constitute the major osmotically-active solutes within the periplasmic compartment when cells are cultured in media of low osmotic strength. Similarly, it can be concluded that the cyclic β-glucans of the Rhizobiaceae are major osmotically-active solutes within the periplasmic compartments of these bacteria. Because these glucans accumulate to high levels during growth at low osmolarity, it may be predicted that their accumulation (i) may provide a mechanism for the cell to regulate the relative volume of the periplasmic compartment, (ii) should lead to a reduction in turgor pressure across the cytoplasmic membrane, and (iii) may lead to the development of a Donnan potential across the outer membrane. Anionic forms of the periplasmic glucans would be expected to be the most effective form of periplasmic solute because their counterions should also contribute to periplasmic osmolarity. Furthermore, only anionic periplasmic glucans would increase the ionic strength of the periplasm and have the capacity to contribute to the establishment of a Donnan potential across the outer membrane.

V.B. MUTANTS DEFECTIVE FOR PERIPLASMIC GLUCAN BIOSYNTHE-SIS ARE IMPAIRED FOR GROWTH IN HYPOOSMOTIC MEDIA

ndv and chv mutants of S. meliloti and A. tumefaciens have been shown to be specifically impaired for growth in hypoosmotic media (Cangelosi et al., 1990; Dylan et al., 1990). When solutes are added to the growth medium, growth is restored to wild-type levels (Cangelosi et al., 1990; Dylan et al., 1990). These mutants also show alterations in a variety of cell-surface properties when grown in media of low osmolarity. These properties include: (i) a loss of motility with reduced numbers of flagella; (ii) a greater resistance to certain bacteriophages; (iii) an increased sensitivity to certain antibiotics; (iv) an increased production of extracellular polysaccharides; and (v) modified cell-surface protein composition (Cangelosi et al., 1990; Dylan et al., 1990; Dylan et al., 1986; Geremia et al., 1987; Nagpal et al., 1992; Soto et al., 1992). Similar alterations in cell surface properties have been observed with MDO mutants of E.coli (Fiedler and Rotering 1988; Geiger et al., 1992; Holtje et al., 1988) although it is noted that mdoA mutants of E.coli apparently do not show impaired growth at low osmolarity (Kennedy 1982). Of interest is the finding that the cell-surface alterations observed in ndv and chv mutants can be partially suppressed when cells are grown at elevated osmolarity (Cangelosi et al., 1990; Dylan et al., 1990). It is also noted that mutants of B. japonicum at the ndvB-like locus are also motility impaired and impaired for growth in hypoosmotic media (Bhagwat et al., 1996).

Additional evidence for an important role for the cyclic β-(1,2)-glucans during hypoosmotic adaptation comes from the fact that after repeated attempts to isolate second site "osmorevertants" of *ndv* mutants of *S. meliloti*, revertants with only partially restored osmotolerance have been identified (Dylan *et al.*, 1990; Quandt *et al.*, 1992). Further evidence is derived from studies with *exoC* mutants of *S. meliloti*, which are also impaired for growth in hypoosmotic media (Dickstein *et al.*, 1988). The *exoC* gene encodes for phosphoglucomutase, an enzyme involved in the biosynthesis of UDP-glucose, resulting in defects in the biosynthesis of EPS, LPS, and cyclic β-(1,2)-glucans (Uttaro *et al.*, 1990). Since no other polysaccharide-defective mutants of *S. meliloti* have been reported to be selectively impaired for growth at low osmolarity, the results from these studies are also consistent with a role for cyclic β-(1,2)-glucans during hypoosmotic adaptation.

Interestingly, the *cgm* mutant of *S. meliloti*, which lacks the ability to transfer sn-1-phosphoglycerol substituents to the cyclic glucans, is not impaired for growth in hypoosmotic media. However, the cyclic glucans of this mutant contain higher levels of succinyl substituents which results in an overall anionic charge on these molecules similar to that found in wild-type cells (Breedveld *et al.*, 1995). Mutants of *B. japonicum* at the *ndvC* locus also display wild-type growth rates in hypoosmotic media. However, this probably results from the fact that the *ndvC* mutants produce wild-type levels of periplasmic glucans (although the structure of these glucans is altered) (Bhagwat *et al.*, 1996).

V.C. OSMOTIC REGULATION OF CYCLIC β-GLUCAN BIOSYNTHESIS

The biosynthesis of cyclic β-glucans is osmotically regulated in a wide variety of *Rhizobium*, *Agrobacterium*, and *Bradyrhizobium* strains, strongly suggesting a general role for periplasmic cyclic β-glucans in the hypoosmotic adaptation of the *Rhizobiaceae*. However, this conclusion is complicated by observations that the level of cell-associated cyclic β-glucans in several strains of *R. leguminosarum* is independent of growth medium osmolarity (Breedveld *et al.*, 1991). Furthermore, these strains of *R. leguminosarum* excrete large amounts of cyclic β-(1,2)-glucans during growth at elevated osmolarity, indicating that the integrity of the outer membrane of these bacteria is modified

under this growth condition (Breedveld *et al.*, 1992). It is important to note that these *R. leguminosarum* strains synthesize only neutral, unsubstituted cyclic β-(1,2)-glucans (Breedveld *et al.*, 1991; Zevenhuizen *et al.*, 1990). Thus, it is possible that it is only the anionic forms of the periplasmic cyclic β-(1,2)-glucans which are critical for hypoosmotic adaptation. It must be noted, however, that the cyclic β-glucans of *Bradyrhizobium* species (whose synthesis is osmotically regulated) are neutral in character yet are modified with the zwitterionic substituent phosphocholine (Miller and Gore 1992; Rolin *et al.*, 1992; Tully *et al.*, 1990).

Osmotic regulation of periplasmic glucan biosynthesis by *S. meliloti* appears to occur at a post-translational level because of the following observations: (i) The expression of *ndvA* and *ndvB* is not stimulated when cells are transferred from high to low osmolarity media (Dylan *et al.*, 1990); (ii) in vitro cyclic β-(1,2)-glucan synthetase activity is similar within membrane preparations isolated from cells grown at low and high osmolarity (Zorreguieta *et al.*, 1990); (iii) NdvB/ChvB levels are similar within cells grown in low and high osmolarity media (Zorreguieta *et al.*, 1990). Therefore, it appears that the activities of the enzyme systems involved in cyclic glucan biosynthesis are inhibited within cells grown in high osmolarity media. Consistent with this possibility, it has been shown that cyclic glucan synthetase activity is inhibited in vitro by high ionic strength (Zorreguieta *et al.*, 1990). Curiously, non-ionic solutes do not inhibit cyclic glucan synthetase activity (Zorreguieta *et al.*, 1990). Similar results have been reported with *R. leguminosarum* membrane preparations (Breedveld *et al.*, 1992) and for MDO biosynthesis in vitro (Rumley *et al.*, 1992), however, Lacroix *et al.*, (Lacroix *et al.*, 1991) have shown that osmotic regulation of MDO biosynthesis also occurs at the level of transcription.

Breedveld and Miller (Breedveld and Miller 1995) examined the effects of growth medium osmolarity on the transfer of sn-1-phosphoglycerol substituents to the cyclic β-(1,2)-glucans of *S. meliloti*. This study revealed that the putative sn-1-phosphoglycerol transferase is present but its activity is inhibited when cells are grown at elevated osmolarity.

In addition to osmotic regulation, the biosynthesis of cyclic β-glucans has also been shown to be regulated by end-product inhibition (Breedveld *et al.*, 1992). Similar results have been reported for MDO biosynthesis (Rumley *et al.*, 1992). The concentrations of cyclic β-(1,2)-glucans which are inhibitory fall within a range similar to that found within the periplasm (eg. 15 mM), indicating that this inhibitory effect has physiological significance. It may be speculated that end-product inhibition occurs via a periplasmic domain in the NdvB protein. Under conditions in which the outer membrane becomes leaky (*e.g.* when *R. leguminosarum* cells are grown at elevated temperature or at high osmolarity (Breedveld *et al.*, 1992)), a constant loss of cyclic β-(1,2)-glucans from cells prevents end-product inhibition and results in the accumulation of glucans within the extracellular medium.

VI. Cyclic β-Glucans and Bacterium-Plant Interactions

VI.A. BEHAVIOR OF CYCLIC β-(1,2)-GLUCAN DEFICIENT MUTANTS DURING INFECTION OF THE HOST PLANT

It has been established that cyclic β-glucans play a role during infection of the host plant. *A. tumefaciens chvA* and *chvB* mutants, affected in the biosynthesis of cyclic β-(1,2)-glucans, do not attach to plant cells and are avirulent (Douglas *et al.*, 1985; Puvanaserajah *et al.*, 1985). Similarly, *S. meliloti* and *R. fredii ndvA* and *ndvB* mutants form ineffective white pseudonodules on alfalfa and soybean, respectively. Bacteroids cannot be isolated from these pseudonodules, which contain a small number of infection threads that abort at an early stage (Bhagwat *et al.*, 1992; Dylan *et al.*, 1990). Interestingly, *ndvB*-like mutants of *B. japonicum* form ineffective nodules on the soybean from which bacteroids can be isolated (Bhagwat and Keister 1995), suggesting that the cyclic β-glucans may play roles during different stages of nodulation in different symbioses.

Additional studies with *S. meliloti ndv* mutants have revealed that the cyclic β-(1,2)-glucans are important, but not essential for legume nodulation. This is concluded from the existence of pseudorevertants of *ndv* mutants which regain the capacity to generate nitrogen-fixing nodules yet remain defective for cyclic β-(1,2)-glucan biosynthesis. These pseudorevertants also remain osmosensitive (Dylan *et al.*, 1990). It is noted, however, that the total number of infection threads elicited by these pseudorevertants is substantially

lower than that elicited by wild-type cells (Dylan *et al.*, 1990). Furthermore, these pseudorevertants remain impaired in their ability to attach to plant cells (Dylan *et al.*, 1990).

VI.B. INFECTION AND OSMOREGULATION

It can be expected that osmoadaptive responses are important during nodulation because *Rhizobium* is likely to encounter a range of osmotic environments within the rhizosphere, infection thread, and symbiosome. Similarly, it is clear that *Agrobacterium* must cope with different osmotic challenges in the soil and during infection of the plant through wounds and proliferation in the intercellular spaces. Therefore, the accumulation of periplasmic cyclic β-glucans may allow these bacteria to more readily adapt to changing osmotic conditions during plant infection. Consistent with a continual need for periplasmic glucans during nodulation, high levels of cyclic β-glucans are detected within bacteroids isolated from mature root nodules (Gore and Miller 1993).

It may be that the failure of *ndv* and *chv* mutants to infect plants is an indirect consequence of the lack of cyclic β-(1,2)-glucans. For example, it may be the inability of these bacteria to properly osmoregulate that is primarily responsible for their failure to infect plants. Quandt *et al.*, (Quandt *et al.*, 1992) have provided evidence that the functions of the cyclic β-(1,2)-glucans during root nodulation are linked to their osmoadaptive functions. Specifically, these researchers isolated "osmorevertants" of *S. meliloti ndvB* mutants and later showed these mutants had regained the capacity to infect alfalfa nodules (although this infection did not restore nitrogen fixation)(Quandt *et al.*, 1992).

As stated above, in most rhizobia, the synthesis of cyclic β-glucans is highest during growth in hypoosmotic media. Upon a shift to high osmotic strength media, the synthesis of these molecules decreases dramatically. It is, therefore, somewhat surprising that cyclic β-glucans are detected within bacteroids isolated from mature root nodules (Gore and Miller 1993) because it is likely that bacteroids experience a moderately elevated osmotic environment within the root nodule (Botsford and Lewis 1990). In fact, levels of cyclic β-(1,6)-β-(1,3)-glucans within *B. japonicum* bacteroids isolated from mature soybean nodules are similar to levels present within aerobic, free-living cultures grown at low osmolarity (Gore and Miller 1992). How is it possible that cyclic glucan synthesis continues throughout the entire infection process when the bacteria or bacteroids are presumed to encounter osmotic environments, which do not favor cyclic glucan accumulation in free culture? It could be that rhizobia encounter or experience an environment of lower osmolarity within the plant compartment than expected. Alternatively, the biosynthesis of cyclic β-glucans may be regulated differently within the plant host. Evidence for this possibility is provided by a study of the peanut rhizobia (Ghittoni and Bueno 1995). Cyclic glucan biosynthesis by these bacteria was stimulated during growth at high osmolarity if glycine betaine was provided in the growth medium. Whether such a stimulation might explain the continued synthesis of cyclic glucans during infection remains to be investigated, however, it is noted that bacteroids are able to accumulate various osmoprotectants such as glycine betaine in response to high osmotic strength (Fougere and Le Rudulier 1990), and this compound is synthesized by many plants (Wyn Jones and Storey 1981).

VI.C. CYCLIC β-GLUCANS, INFECTION, AND THE CELL ENVELOPE

Another consequence of the failure of cells to synthesize cyclic β-glucans is an alteration of the properties of the cell envelope. The pleiotropic cell-surface character of the *ndv/chv* mutants (see above) suggests that the accumulation of cyclic β-glucans within the periplasmic compartment greatly influences the overall structure of the cell-envelope. Furthermore, studies with "downstream" *ndvB* mutants indicate that the concentration of the cyclic β-(1,2)-glucans within the periplasm has an important influence on cell-envelope structure. The cell surface and symbiotic properties of the "downstream" *ndvB* mutants are intermediate in character when compared to wild-type cells and "upstream" *ndvB* mutants (Breedveld *et al.*, 1994; Ielpi *et al.*, 1990). As discussed above, "downstream" *ndvB* mutants accumulate lower levels of periplasmic cyclic β-(1,2)-glucans than wild type cells. Based on the above, it is possible that the cell envelope alterations of *ndv* and *chv* mutants are responsible for the failure of these bacteria to properly osmoregulate or infect plants. The strongest evidence for an indirect role for the cyclic β-(1,2)-

glucans during plant infection has been provided by studies showing that *A. tumefaciens chvB* mutants fail to synthesize active rhicadhesin (Smit *et al.*, 1992; Swart *et al.*, 1994b; Swart *et al.*, 1993). Rhicadhesin is a calcium-binding, surface protein that has been implicated in the attachment of *Agrobacterium* (and other members of the *Rhizobiaceae*) to plant cells (Smit *et al.*, 1992). *A. tumefaciens chvB* mutants synthesize active rhicadhesin only when cells are grown in media of elevated osmotic strength and in the presence of millimolar concentrations of calcium. Under these same growth conditions, virulence can be restored (Swart *et al.*, 1994a; Swart *et al.*, 1994b; Swart *et al.*, 1993). Furthermore, the addition of purified cyclic β-(1,2)-glucans to *A. tumefaciens chvA* (O'Connell and Handelsman 1989) or *chvB* (Cangelosi *et al.*, 1989) cultures does not restore virulence, however, the addition of rhicadhesin partially restores attachment ability and virulence. These results suggest that it is rhicadhesin, not the cyclic β-(1,2)-glucans, that is essential for virulence and attachment. It is, however, possible that the cyclic β-(1,2)-glucans also mediate the attachment of *Rhizobium* and *Agrobacterium* to specific sites on the plant cell surface. While there have been no attempts to characterize such binding sites, studies do suggest that the cyclic β-(1,2)-glucans have the capacity to bind to pea lectins (Planque and Kijne 1976).

VI.D. COULD THE CYCLIC β-GLUCANS ACT AS SIGNAL MOLECULES DURING PLANT INFECTION?

The overall ring size distributions of the cyclic β-(1,2)-glucans have been shown to be very similar among diverse species of *Rhizobium* and *Agrobacterium*. Thus, it is unlikely that the cyclic β-(1,2)-glucan backbones confer specificity during legume nodulation. In fact, although different biovars of *R. leguminosarum* have distinct host specificities, there are no notable differences among the structures of the cyclic β-(1,2)-glucans synthesized by these various biovars (Breedveld and Miller 1994). Furthermore, both *R. fredii* and *B. japonicum* are symbionts of the soybean, yet *R. fredii* synthesizes cyclic β-(1,2)-glucans and *B. japonicum* synthesizes cyclic β-(1,6)-β-(1,3)-glucans (Bhagwat and Keister 1992; Miller *et al.*, 1992). Therefore, it may be the cyclic character, not the arrangement of glycosidic linkages, which represents the critical structural feature of these molecules. For similar reasons, the identity of the substituents on the cyclic β-glucans may not play a crucial role in determining host specificity, since *R. fredii* produces sn-1-phosphoglycerol-substituted, anionic cyclic glucans, while *B. japonicum* synthesizes phosphocholine-substituted, neutral cyclic glucans (Bhagwat and Keister 1992; Rolin *et al.*, 1992). Furthermore, a *S. meliloti cgm* mutant, unable to add sn-1-phosphoglycerol substituents to the cyclic glucans, is fully infective on its host plant (Breedveld *et al.*, 1995). It is noted, however, that this mutant produces anionic glucans, which contain elevated levels of succinyl substituents (Breedveld *et al.*, 1995). Therefore, it is possible that the overall anionic character is important for the functions of these molecules during the *S. meliloti*-alfalfa symbiosis. Further indications for the importance of anionic substituents comes from the fact that in phosphorus-limited cultures of *S. meliloti*, anionic glucans are produced, although no phosphoglycerol-substituents can be detected. Also under these conditions, the glucans are highly modified with succinyl substituents (Breedveld *et al.*, 1995).

It is possible that cyclic β-glucans act as signalling molecules during plant infection. For example, it has been shown for both the *S. meliloti*-alfalfa (Bhagwat *et al.*, 1992; Dylan *et al.*, 1990) and *R. leguminosarum* bv *trifolii*-clover (Abe *et al.*, 1982) symbioses that the addition of purified cyclic β-(1,2)-glucan preparations enhances both nodule number and the kinetics of nodule formation. It is noted, however, that attempts to complement the symbiotic phenotype of *ndv* mutants of *S. meliloti* by the addition of exogenous cyclic β-(1,2)-glucans have been unsuccessful (Dylan *et al.*, 1990). This failure suggests that the periplasmic location of the cyclic β-(1,2)-glucans is important during nodulation.

Additional evidence for signalling functions is derived from studies of the cyclic β-(1,6)-β-(1,3)-glucans of *B. japonicum*. These glucans share structural features with glucan fragments derived from the mycelial walls of fungal pathogens of the soybean (Darvill and Albersheim 1984). These fungal wall glucan fragments are potent elicitors of phytoalexin production by the soybean. For this reason, Miller *et al.*, (Miller *et al.*, 1994) examined the bradyrhizobial glucans for possible elicitor activity. Their study revealed that the bradyrhizobial glucans do, in fact, elicit isoflavonoid production within soybean cotyledons. Ebel and coworkers (Ebel *et al.*, 1995) performed a similar study and showed that bradyhizobial glucans compete with fungal glucan fragments for binding to a putative elicitor receptor of the soybean. These latter researchers have proposed that the bradyrhizobial glucans may act as suppressors of inducible plant defense responses. While this is an intriguing possibility, it is noted that *R. fredii* also has the capacity to nodulate soybean yet does not synthesize cyclic β-(1,6)-β-(1,3)-glucans. Instead, this bacterium synthesizes cyclic β-(1,2)-glucans (Bhagwat *et al.*, 1992).

The recent study by Bhagwat and coworkers (Bhagwat *et al.*, 1996) provides further evidence that the structure of the cyclic β-(1,6)-β-(1,3)-glucans of *B. japonicum* is important during soybean nodulation. *B. japonicum* mutants at the *ndvC* locus synthesize wild-type levels of periplasmic β-glucans, however, the glucans contain predominantly β-(1,3)-linked glucose and very. little β-(1,6)-linked glucose. These mutants induce only pseudonodules on soybean roots (Bhagwat *et al.*, 1996).

VI.E. WHY ARE THE PERIPLASMIC GLUCANS OF THE *RHIZOBIACEAE* CYCLIC?

There may be several possible advantages to synthesizing a cyclic molecule. For example, it is possible that the cyclic backbone may be more resistant to enzymatic degradation than a linear molecule. Such increased resistance could be important in the rhizosphere environment. A second possible advantage is that cyclic glucans could be used as a reserve material selectively by the *Rhizobiaceae*, and may not be available to other competing bacteria in the rhizosphere. For example, such a situation occurs with cyclodextrin-producing bacteria (e.g. *Klebsiella oxytosa*). These bacteria synthesize a cyclodextranase which allows them to utilize cyclodextrins as a reserve material where other competing microorganisms cannot (Feederle *et al.*, 1996) (The cyclodextrins, synthesized from starch by cyclodextrin glycosyltransferases, are cyclic glucans containing 6 to 8 glucose residues linked by α-(1,4)-glycosidic bonds (Szejtli 1990).). Although such a strategy seems possible for the *Rhizobiaceae*, there have been no reports of rhizobial or bradyrhizobial β-glucanases capable of hydrolyzing the cyclic β-glucans, and rhizobial cultures are apparently unable to utilize cyclic β-glucans as a source of carbon, phosphorus or

Genus	Structure/Linkage	DP	Literature
Acetobacter xylinum	linear β-(1,2)	6-42	Amemura *et al.*, 1985b; Semino and Dankert 1993
Agrobacterium radiobacter	cyclic β-(1,2)	17-25	Amemura 1984; Kinoshita *et al.*, 1991; Zevenhuizen *et al.*, 1990
Agrobacterium tumefaciens	cyclic β-(1,2)	17-25	Breedveld and Miller 1994
Alcaligenes	cyclic β-(1,2)		Harada 1983
Azospirillum brasilense[1]	cyclic β-(1,6)-β-(1,3)	10-13	Bohin 1996
Bradyrhizobium japonicum	cyclic β-(1,6)-β-(1,3)		Bhagwat and Keister 1992; Inon de Iannino and Ugalde 1993; Miller *et al.*, 1990; Rolin *et al.*, 1992
Brucella abortus[2]	cyclic β-(1,2)	17-24	Bundle *et al.*, 1988
B. melitensis	cyclic β-(1,2)	17-24	Bundle *et al.*, 1987
Burkholderia solanacearum	cyclic β-(1,2) 1 α-(1,6)-linkage	13	Talaga *et al.*, 1996
Erwinia herbivora	linear β-(1,2)-β-(1,6)		Smith *et al.*, 1995
Escherichia coli	linear β-(1,2)-β-(1,6)	6-12	Kennedy 1987
Klebsiella pneumoniae	linear β-(1,2)-β-(1,6)		Amemura and Cabrera-Crespo 1986
Pseudomonas syringae	linear β-(1,2)-β-(1,6)		Talaga *et al.*, 1994
Rhizobium (peanut)	"β-glucan"	20	Ghittoni and Bueno 1995
Rhizobium(tropical strain)[3]	linear β-(1,2)	6-19	Amemura *et al.*, 1985a
Rhizobium fredii	cyclic β-(1,2)	17-25	Bhagwat *et al.*, 1992
R. leguminosarum	cyclic β-(1,2)	17-25	Breedveld and Miller 1994
S. meliloti	cyclic β-(1,2)	17-40	Breedveld and Miller 1994
Xanthomonas spp[4]	cyclic β-(1,2) 1 α-(1,6)-linkage	16	Talaga *et al.*, 1996
Xanthomonas	linear β-(1,2)	8-20	Amemura and Cabrera-Crespo 1986

Table 2. Occurrence of low molecular mass cellular (periplasmic) β-glucans in Gram-negative bacteria. The Periplasmic location has not been confirmed in some of these bacteria.

[1] Although the glucans have been proposed to be linear (Altabe *et al.*, 1994), a recent study by Bohin and co-workers showed that β-glucans of *A. brasilense* are cyclic (unpublished results).

[2] Analysis of the 16S rRNA and lipid A composition of *B. abortus* has revealed a close phylogenetic relationship with *A. tumefaciens* and *S. meliloti* (Moreno *et al.*, 1990). Furthermore, an *ndv*B-like gene from *B. abortus* can complement an *ndv*B mutant of *S. meliloti* (Inon De Iannino *et al.*, 1996).

[3] The only reported exception to the rule that periplasmic β-glucans of members of the family *Rhizobiaceae* are cyclic.

[4] *X. oryzae, X. phaseoli, X. campestris*

energy (Breedveld *et al.*, 1993; Pfeffer *et al.*, 1994; Zevenhuizen 1981).

A third possible advantage is that a cyclic structure may permit these glucans to form inclusion complexes with (hydrophobic) guest molecules (Koizumi *et al.*, 1984; Okada *et al.*, 1985). Indeed, there is some evidence that such inclusion complexes may form. Specifically, it has been shown that the solubility of naringenin (a legume-derived flavonoid) is increased in the presence of cyclic β-(1,2)-glucan preparations (Morris *et al.*, 1991). Because flavonoids are inducers of the nodulation genes of *Rhizobium* species (see elsewhere in this book), the formation of inclusion complexes between flavonoids and cyclic β-(1,2)-glucans could potentially influence the effectiveness of nodulation. This could perhaps provide an explanation for the enhancement of nodulation observed when cyclic β-(1,2)-glucan preparations are added to rhizobium-legume nodulation systems (Abe *et al.,* 1982; Bhagwat *et al.,* 1992; Dylan *et al.,* 1990). Interestingly, it has also been shown that γ-cyclodextrin preparations (approximately 70 μM) enhance nodule formation when added to the *R. leguminosarum* bv *trifolii*-clover nodulation assay system (Abe *et al.,* 1982). Further studies aimed at examining the formation of cyclic β-glucan inclusion complexes are needed, as there is some debate as to whether or not such complexes can actually form (Andre *et al.,* 1995).

VII. Concluding Remarks and Future Perspectives

All *Rhizobium, Bradyrhizobium*, and *Agrobacterium* strains synthesize periplasmic, cyclic β-glucans. These cyclic β-glucans are major cell envelope constituents that have been shown to provide functions for the free-living forms of these bacteria during hypoosmotic adaptation as well as during the process of plant infection. The greatest insight concerning the possible roles for the cyclic β-glucans has been gained from *ndv* and *chv* mutants of *Rhizobium* and *Agrobacterium*.

Mutants with mutations at the *ndv* and *chv* loci are impaired for growth in hypoosmotic media and are defective in their ability to infect plants. Whether or not the cyclic β-glucans act directly or indirectly during osmotic adaptation and host infection remains to be clarified; however, accumulation of cyclic β-glucans within the periplasmic compartment strongly influences the overall cell envelope architecture of rhizobia. Studies with *ndv/chv* mutants have also revealed the involvement of two proteins (NdvA/ChvA and NdvB/ChvB) during cyclic β-(1,2)-glucan backbone biosynthesis and export. More recent studies have provided insight concerning the enzymes responsible for the addition of substituents to the cyclic β-(1,2)-glucan backbone. Both phosphoglycerol and succinyl substituents are added in the periplasmic compartment and a genetic locus, *cgm*, has been identified which is involved in phosphoglycerol addition to the cyclic β-(1,2)-glucans of *S. meliloti*. Whether or not additional proteins are involved in cyclic β-(1,2)-glucan biosynthesis, modification, and export is unknown. By comparison, much less is known concerning the synthesis of the cyclic β-(1,6)-β-(1,3)-glucans of *Bradyrhizobium* species, however,

Bhagwat and coworkers (Bhagwat *et al.,* 1996) have recently identified two genetic loci involved in the synthesis of these molecules. Clearly, future efforts should be made to identify additional genetic loci associated with cyclic β-glucan biosynthesis in *Rhizobium, Agrobacterium* and *Bradyrhizobium* species.

It is noteworthy that periplasmic glucans have now been found in several Gram-negative genera (Table 2), and that almost all of these organisms are known to infect eukaryotes. This suggests that periplasmic glucans provide important functions during host infection. Interestingly, the genetic locus which is apparently required for the synthesis of periplasmic β-glucans in *Pseudomonas syringae*, is also required for this bacterium to incite disease symptoms on its host plant (Loubens *et al.,* 1993). Additionally, the synthesis of periplasmic glucans by *P. syringae* is osmoregulated (Talaga *et al.,* 1994). Further research is needed to establish whether the presence of periplasmic β-glucans is a general requirement for Gram-negative bacteria to osmoregulate properly, and/or to establish successful symbiotic or pathogenic relationships. These studies should also provide insight concerning structure-function relationships for these glucans (*e.g.* cyclic vs. linear, branched vs. unbranched, substituents vs. no substituents).

Finally, the continuation of periplasmic β-glucan synthesis by rhizobial bacteroids throughout nodule development is intriguing, since the osmotic environment within the nodule is likely to be potentially high enough to inhibit glucan biosynthesis in free-living rhizobia. Therefore, it is possible that cyclic β-glucan biosynthesis may be regulated differently in the bacteroid vs the free-living bacterium. Studies aimed at examining periplasmic β-glucan biosynthesis within bacteroids may provide important insight concerning the physiological rationale for the continued synthesis and accumulation of these molecules throughout legume nodulation.

VIII. Acknowledgements

Research in Karen Miller's laboratory has been supported by grant MCB-9505706 from the National Science Foundation.

IX. References

Abe, M., Amemura, A. and Higashi, S. (1982) Plant Soil 64, 315-324.

Altabe, S.G., Inon-de-Iannino, N., de-Mendoza, D. and Ugalde, R.A. (1994) J. Bacteriol. 176, 4890-4898.

Amemura, A. (1984) Agric. Biol. Chem. 48, 1809-1817.

Amemura, A. and Cabrera-Crespo, J. (1986) J. Gen. Microbiol. 132, 2443-2452.

Amemura, A., Footrakul, P., Koizumi, K., Utamura, T. and Taguchi, H. (1985a) J. Ferment. Technol. 63, 115-120.

Amemura, A., Hashimoto, T., Koizumi, K. and Utamura, T. (1985b) J. Gen. Microbiol. 131, 301-307.

Amemura, A., Hisamatsu, M., Mitani, H. and Garada, T. (1983) Carbohydr. Res. 114, 277-285.

Andre, I., Mazeau, K., Taravel, F.R. and Tvaroska, I. (1995) Int. J. Biol. Macromol. 17, 189-198.

Becker, A., Kuster, H., Niehaus, K. and Puhler, A. (1995) Molecular and General Genetics 249, 487-497.

Bhagwat, A.A., Gross, K.C., Tully, R.E. and Keister, D.L. (1996) J. Bacteriol. 178, 4635-4642.

Bhagwat, A.A. and Keister, D.L. (1992) Can. J. Microbiol. 38, 510-514.

Bhagwat, A.A. and Keister, D.L. (1995) Mol. Plant-Microbe Interact. 8, 366-370.

Bhagwat, A.A., Tully, R.E. and Keister, D.L. (1992) Mol. Microbiol. 6, 2159-2165.

Bhagwat, A.A., Tully, R.E. and Keister, D.L. (1993) Fems Microbiol. Lett. 114, 139-144.

Bohin, J.P. (1996). personal communication.

Botsford, J.L. and Lewis, T.A. (1990) Appl. Environ. Microbiol. 56, 488-494.

Breedveld, M.W., Benesi, A.J., Marco, M.L. and Miller, K.J. (1995) Appl. Environ. Microbiol. 61, 1045-1053.

Breedveld, M.W., Dijkema, C., Zevenhuizen, L.P.T.M. and Zehnder, A.J.B. (1993) Gen. Microbiol. 139, 3157-3163.

Breedveld, M.W., Hadley, J.A. and Miller, K.J. (1995) J. Bacteriol. 177, 6346-6351.

Breedveld, M.W. and Miller, K.J. (1994) Microbiol. Rev. 58, 145-161.

Breedveld, M.W. and Miller, K.J. (1995) Microbiology 141, 583-588.

Breedveld, M.W., Yoo, J.S., Reinhold, V.N. and Miller, K.J. (1994) J. Bacteriol. 176, 1047-1051.

Breedveld, M.W., Zevenhuizen, L.P.T.M., Canter-Cremers, H.C.J. and Zehnder, A.J.B. (1993) Antonie Van Leeuwenhoek 64, 1-8.

Breedveld, M.W., Zevenhuizen, L.P.T.M. and Zehnder, A.J.B. (1990) Appl. Environ. Microbiol. 56, 2080-2086.

Breedveld, M.W., Zevenhuizen, L.P.T.M. and Zehnder, A.J.B. (1991) Arch. Microbiol. 156, 501-506.

Breedveld, M.W., Zevenhuizen, L.P.T.M. and Zehnder, A.J.B. (1992) J. Bacteriol. 174, 6336-6342.

Bundle, D.R., Cherwonogrodzky, J.W. and Perry, M.B. (1987) FEBS Lett. 216, 261-264.

Bundle, D.R., Cherwonogrodzky, J.W. and Perry, M.B. (1988) Infect. Immun .56, 1101-1106.

Cangelosi, G.A., Martinetti, G., Leigh, J.A., Lee, C.C., Theines, C. and Nester, E.W. (1989) J. Bacteriol. 171, 1609-1615.

Cangelosi, G.A., Martinetti, G. and Nester, E.W. (1990) J. Bacteriol. 172, 2172-2174.

Castro, O.A., Zorreguieta, A., Ielmini, V., Vega, G. and Ielpi, L. (1996) J. Bacteriol. 178, 6043-6048.

Castro, O.A., Zorreguieta, A., Semino, C. and Ielpi, L. (1995) Arch. Microbiol. 163, 454-462.

Cohen, J.L. and Miller, K.J. (1991) J. Bacteriol. 173, 4271-4276.

Darvill, A.G. and Albersheim, P. (1984) Annu. Rev. Plant Physiol. 35, 243-275.

Dell, A., York, W.S., McNeill, M., Darvill, A.G. and Albersheim, P. (1983) Carbohydr. Res. 117, 185-200.

Dickstein, R., Bisseling, T., Reinhold, V.N. and Ausubel, F.M. (1988) Genes Dev. 2, 677-687.

Douglas, C.J., Staneloni, R.J., Rubin, R.A. and Nester, E.W. (1985) J. Bacteriol. 161, 850-860.

Dunlap, J., Minami, E., Bhagwat, A.A., Keister, D.L. and Stacey, G. (1996) Mol. Plant-Microbe Interact. 9, 546-555.

Dylan, T., Helinski, D.R. and Ditta, G.S. (1990) J. Bacteriol. 172, 1400-1408.

Dylan, T., Ielpi, L., Stanfield, S., Kashyap, L., Douglas, C., Yanofsky, M., Nester, E., Helinski, D.R. and Ditta, G. (1986) Proc. Natl. Acad. Sci. USA 83, 4403-4407.

Dylan, T., Nagpal, P., Helinski, D.R. and Ditta, G.S. (1990) J. Bacteriol. 172, 1409-1417.

Ebel, J., Bhagwat, A.A., Cosio, E.G., Feger, M., Kissel, U., Mithoefer, A. and Waldmueller, T. (1995) Can. J. Bot. 73, S506-S510.

Feederle, R., Pajatsch, M., Kremmer, E. and Bock, A. (1996) Arch. Microbiol. 165, 206-212.

Fiedler, W. and Rotering, H. (1988) J. Biol. Chem. 263, 14684-14689.

Fougere, F. and Le Rudulier, D. (1990) J. Gen. Microbiol. 136, 157-163.

Geiger, O., Russo, F.D., Silhavy, T.J. and Kennedy, E.P. (1992) J. Bacteriol. 174, 1410-1413.

Geiger, O., Weissborn, A.C. and Kennedy, E.P. (1991) J. Bacteriol. 173, 3021-3024.

Geremia, R.A., Cavaignac, S., Zorreguieta, A., Toro, N., Olivares, J. and Ugalde, R.A. (1987) J. Bacteriol. 169, 880-884.

Ghittoni, N.E. and Bueno, M.A. (1995) Can. J. Microbiol. 41, 1021-1030.

Gore, R.S. and Miller, K.J. (1992) J. Bacteriol. 174, 7838-7840.

Gore, R.S. and Miller, K.J. (1993) Plant Physiology 102, 191-194.

Harada, T. (1983) Biochem. Soc. Symp. 48, 97-116.

Hisamatsu, M. (1992) Carbohydr. Res. 231, 137-146.

Hisamatsu, M., Amemura, A., Koizumi, K., Utamura, T. and Okada, Y. (1983) Carbohydr. Res. 121, 31-40.

Holtje, J.V., Fiedler, W., Rotering, H., Walderich, B. and van-Duin, J. (1988) J. Biol. Chem. 263, 3539-3541.

Ielpi, L., Dylan, T., Ditta, G.S., Helinski, D.R. and Stanfield, S.W. (1990) J. Biol. Chem. 265, 2843-2851.

Inon De Iannino, N., Briones, G., Tolmasky, M.E. and Ugalde, R.A. (1996) ASM Annual Meeting, H-137.

Inon De Iannino, N. and Ugalde, R.A. (1989) J. Bacteriol. 171, 2842-2849.

Inon de Iannino, N. and Ugalde, R.A. (1993) Arch. Microbiol. 159, 30-38.

Kennedy, E.P. (1982) Proc. Natl. Acad. Sci. USA 79, 1092-1095.

Kennedy, E.P. (1987) Membrane-derived oligosaccharides. In, F. C. Neidhardt , J. L. Ingraham , K. B. Low , B. Magasanik , M. Schaechter and H. E. Umbarger (Ed) *Escherichia coli* and *Salmonella typhimurium*, cellular and molecular biology, Vol 1 American Society for Microbiology, Washington, D.C.

Kennedy, E.P. (1996) Membrane-derived oligosaccharides (periplasmic beta-D-glucans) of *Escherichia coli*. In, F. C. Neidhardt , R. Curtis , J. L. Ingraham , C. C. Lin , K. B. Low , B. Magasanik , W. S. Reznikoff , M. Riley , M. Schaechter and H. E. Umbarger (Ed) *Escherichia coli* and *Salmonella typhimurium*, cellular and molecular biology, Vol 1 American Society for Microbiology, Washington, D.C.

Kinoshita, S., Nakata, M., Amaemura, A. and Taguchi, H. (1991) J. Ferment. Bioeng. 72, 416-421.

Koizumi, K., Okada, Y., Horiyama, S., Utamura, T., Higashiura, T. and Ikeda, M. (1984) J. Inclusion Phenom. 2, 891-899.

Lacroix, J.M., Loubens, I., Tempete, M., Menichi, B. and Bohin, J.P. (1991) Mol. Microbiol. 5, 1745-1753.

Loubens, I., Debarbieux, L., Bohin, A., Lacroix, J.M. and Bohin, J.P. (1993) Mol. Microbiol. 10, 329-340.

McIntire, F.C., Peterson, W.H. and Riker, A.J. (1942) J. Biol. Chem. 143, 491-496.

Miller, K.J. and Gore, R.S. (1992) Curr. Microbiol. 24, 101-104.

Miller, K.J., Gore, R.S. and Benesi, A.J. (1988) J. Bacteriol. 170, 4569-4575.

Miller, K.J., Gore, R.S., Johnson, R., Benesi, A.J. and Reinhold, V.N. (1990) J. Bacteriol. 172, 136-142.

Miller, K.J., Hadley, J.A. and Gustine, D.L. (1994) Plant Physiology 104, 917-923.

Miller, K.J., Kennedy, E.P. and Reinhold, V.N. (1986) Science 231, 48-51.

Miller, K.J., Reinhold, V.N., Weissborn, A.C. and Kennedy, E.P. (1987) Biochim. Biophys. Acta 901, 112-118.

Moreno, E., Stackebrandt, E., Dorsch, M., Wolters, J., Busch, M. and Mayer, H. (1990) J. Bacteriol. 172, 3569-3576.

Morris, V.J., Brownsey, G.J., Chilvers, G.R., Harris, J.E., Gunning, A.P. and Stevens, B.H.J. (1991) Food Hydrocoll. 5, 185-188.

Nagpal, P., Khanuja, S.P. and Stanfield, S.W. (1992) Mol. Microbiol. 6, 479-488.

O'Connell, K.P. and Handelsman, J. (1989) Mol. Plant-Microbe Interact. 2, 11-16.

Okada, Y., Horiyama, S. and Koizumi, K. (1985) Yakugaku Zasshi 106, 240-247.

Palleschi, A. and Crescenzi, V. (1985) Gazz. Chim. Ital. 115, 243-245

Palmen, R., Driessen, A.J.M. and Hellingwerf, K.J. (1994) Biochim. Biophys. Acta 1183, 417-451.

Pfeffer, P.E., Becard, G., Rolin, D.B., Uknalis, J., Cooke, P. and Tu, S. (1994) Appl. Environ. Microbiol. 60, 2137-46.

Pfeffer, P.E., Osman, S.F., Hotchkiss, A., Bhagwat, A.A., Keister, D.L. and Valentine, K.M. (1996) Carbohydr. Res.296,23-37 .

Planque, K. and Kijne, J.W. (1976) FEBS Lett. 73, 64-66.

Puvanaserajah, V., Schell, F.M., Stacey, G., Douglas, C.J. and Nester, E.W. (1985) J. Bacteriol. 164, 102-106.

Quandt, J., Hillemann, J., Niehaus, K., Arnold, W. and Puhler, A. (1992) Mol. Plant-Microbe Interact. 5, 420-427.

Raina, S., Raina, R., Venkatesh, T.V. and Das, H.K. (1995) Mol. Plant- Microbe Interact. 8, 322-326.

Rolin, D.B., Pfeffer, P.E., Osman, S.F., Szwergold, S., Kappler, F. and Benesi, A.J. (1992) Biochim. Biophys. Acta 1116, 215-225.

Rumley, M.K., Therisod, H., Weissborn, A.C. and Kennedy, E.P. (1992) J. Biol. Chem. 267, 11806-11810.

Semino, C.E. and Dankert, M.A. (1993) J. Gen. Microbiol. 139, 2745-2756.

Smit, G., Swart, S., Lugtenberg, B.J. and Kijne, J.W. (1992) Mol. Microbiol. 6, 2897-2903.

Smith, A.R.W., Rastall, R.A., Blake, P. and Hignett, R.C. (1995) Microbios. 83, 27-39.

Soto, M.J., Lepek, V., Lopez, L.I.M., Olivares, J. and Toro, N. (1992) Mol. Plant-Microbe Interact. 5, 288-293.

Stanfield, S.W., Ielpi, L., O'Brochta, D., Helinski, D.R. and Ditta, G.S. (1988) J. Bacteriol. 170, 3523-3530.

Stock, J.B., Rauch, B. and Roseman, S. (1977) J. Biol. Chem. 252, 7850-7861.

Swart, S., Logman, T.J.J,. Lugtenberg, B.J.J., Smit, G. and Kijne, J.W. (1994a) Arch. Microbiol. 161, 310-315.

Swart, S., Lugtenberg, B.J.J., Smit, G. and Kijne, J.W. (1994b) J. Bacteriol. 176, 3816-3819.

Swart, S., Smit, G., Lugtenberg, B.J.J. and Kijne, J.W. (1993) Mol. Microbiol. 10, 597-605.

Szejtli, J. (1990) Carbohydr. Polym. 12, 375-392.

Talaga, P., Fournet, B. and Bohin, J.P. (1994) J. Bacteriol. 176, 6538-6544.

Talaga, P., Stahl, B., Wieruszeski, J.M., Hillenkamp, F., Tsuyumu, S., Lippens, G. and Bohin, J.P. (1996) J. Bacteriol. 178, 2263-2271.

Thomas, P.M., Golly, K.F., Zyskind, J.W. and Virginia, R.A. (1994) Appl. Environ. Microbiol. 60, 1146-1153.

Tully, R.E., Keister, D.L. and Gross, K.C. (1990) Appl. Environ. Microbiol. 56, 1518-1522.

Uttaro, A.D., Cangelosi, G.A., Geremia, R.A., Nester, E.W. and Ugalde, R.A. (1990) J. Bacteriol. 172, 1640-1646.

Williamson, G., Damani, K., Devenney, P., Faulds, C.B., Morris, V.J. and Stevens, B.J. (1992) J. Bacteriol. 174, 7941-7947.

Wyn Jones, R.G. and Storey, R. (1981) Betaines. In, L. G. Paleg and D. Aspinall (Ed) The Physiology and Biochemistry of Drought Resistance in Plants (pp 171-204). Acad. Press, New York.

York, W.S. (1995) Carbohydr. Res. 278, 205-225.

York, W.S., McNeil, M., Darvill, A.G. and Albersheim, P. (1980) J. Bacteriol. 142, 243-248.

York, W.S., Thomsen, J.U. and Meyer, B. (1993) Carbohydr. Res. 248, 55-80.

Zevenhuizen, L.P.T.M. (1981) Antonie Van Leeuwenhoek 47, 481-497.

Zevenhuizen, L.P.T.M. and Scholten-Koerselman, H.J. (1979) Antonie Van Leeuwenhoek 45, 165-175.

Zevenhuizen, L.P.T.M., van-Veldhuizen, A. and Fokkens, R.H. (1990) Antonie Van Leeuwenhoek 57, 173-178.

Zevenhuizen, L.P.T.M. and Van Neerven, A.R.W. (1983) Carbohydr. Res. 118, 127-134.

Zorreguieta, A., Cavaignac, S., Geremia, R.A. and Ugalde, R.A. (1990) J. Bacteriol. 172, 4701-4704.

Zorreguieta, A., Geremia, R.A., Cavaignac, S., Cangelosi, G.A., Nester, E.W. and Ugalde, R.A. (1988) Mol. Plant-Microbe Interact. 1, 121-127.

Zorreguieta, A., Tolmasky, M.E. and Staneloni, R.J. (1985) Arch. Biochem. Biophys. 238, 368-372.

Zorreguieta, A. and Ugalde, R.A. (1986) J. Bacteriol. 167, 947-51.

Zorreguieta, A., Ugalde, R.A. and Leloir, L.F. (1985) Biochem. Biophys. Res. Commun. 126, 352-357.

Chapter 6

Production of Exopolysaccharides

Anke Becker and Alfred Pühler

I. Introduction

A broad variety of bacteria including the *Rhizobiaceae* are able to secrete polysaccharides. Sugar polymers that form an adherent cohesive layer on the cell surface are designated capsular polysacharides (CPS), whereas the term exopolysaccharide (EPS) is used for polysaccharides with little or no cell association. Due to the variation of monosaccharide sequences, condensation linkages and non-carbohydrate decorations, an infinite array of structures can be provided by this class of macromolecules. Different rheological properties depend on the structure and the molecular weight of EPS. These properties and the location of EPS,

forming the outer layer of the cell surface, contribute to the cell protection against environmental influences, attachment to surfaces, nutrient gathering and to antigenicity (Costerton *et al.*, 1987, Sutherland 1988, Whitfield 1988, Beveridge and Graham 1991). The structural diversity of oligosaccharides derived from EPS enables them to function additionally as informational molecules in cell-cell-communications. Finally, many symbiotic bacteria of the *Rhizobiaceae* use oligosaccharides as signal molecules in the interaction with their host plant.

The objective of this review is to give an overview on the molecular mechanisms of rhizobial EPS biosynthesis and EPS function. We intend to assemble data from genetic and biochemical approaches and mainly focus on *Sinorhizobium meliloti* (formerly *Rhizobium meliloti*) as a well characterized system providing data on gene function, regulation of gene expression, EPS biosynthetic pathways and EPS functions. The production and function of EPS in other rhizobia is discussed in a comparative way.

II. Principles of Exopolysaccharide Structure and Biosynthesis

Despite the wide range of different EPS structures common principles reflecting the mode of biosynthesis can be identified.

II.A. EXOPOLYSACCHARIDE STRUCTURE

Bacterial EPS are either homopolymers composed of a single sugar or heteropolysaccharides containing different sugars. The carbohydrate components found in rhizobial EPS are mainly common monosaccharides like D-glucose, D-galactose, D-mannose, D-glucuronic acid and D-galacturonic acid. Non-carbohydrate substituents often identified are O-acetyl groups as well as ketal-linked pyruvate and succinyl half ester groups. Less abundant are hydroxybutanyl, propionyl and glyceryl groups. In addition to uronic acids, negatively charged decorations are responsible for the acidic character of most rhizobial EPS. The regular structure of heteropolysaccharides results from their composition of repeating units. Apart from linear EPS polymers, molecules carrying side-chains were also identified.

II.B. EXOPOLYSACCHARIDE BIOSYNTHESIS

The biosynthesis of EPS involves a series of sequential steps. Nucleotide diphospho-sugars represent activated sugars serving as precursors for the formation of polysaccharides (Shibaev 1986). Different modes of biosynthesis were reported. Processive sequential transfer of sugars to a growing polysaccharide chain attached to an acceptor was described for many homopolymers. This mode of polymerization results in a polysaccharide chain growing at the non-reducing terminus (Figure 1). Group II CPS composed of sialic acid of *E. coli* K1

and K5 (Rohr and Troy 1980, Finke *et al.*, 1991) and the mannan O-polysaccharide of *E. coli* O9 (Kido *et al.*, 1995) are examples for this mechanism of polymerization. Undecaprenol phosphate was identified as the acceptor for the polymerization of most homopolymers, although in some cases, *e.g.* for the *E. coli* K5 CPS, no evidence was obtained for the existence of an undecaprenol-linked intermediate (Finke *et al.*, 1991). Since proteins were identified associated to the reducing terminus of the *E. coli* K1 CPS (Weisgerber and Troy 1990) and attached to intermediates of β-glucan biosynthesis of several bacteria (Breedveld and Miller 1994), proteins were also discussed as potential acceptors for polysaccharide polymerization.

The second mode of biosynthesis involves the formation of undecaprenol-linked intermediates, which are subsequently polymerized in a blockwise manner resulting in the repeating unit structure of heteropolysaccharides. The transfer of the growing lipid-linked polymer to the non-reducing terminus of a monomeric lipid-linked precursor results in a polymer chain growing at the reducing terminus (Figure 1). The first example of this biosynthetic mechanism extensively studied today was the production of xanthan by *Xanthomonas campestris* (Capage *et al.*, 1987, Coplin and Cook 1990, Ielpi *et al.*, 1993).

In contrast to the participation of lipid-linked intermediates examples were reported for an acceptor-independent mode of polymerization. Cellulose biosynthesis by *Acetobacter xylinum* is performed by cellulose synthase with the cellulose remaining attached to the enzyme during polymerization (Valla *et al.*, 1989, Wong *et al.*, 1990). The cellulose synthase is probably located in the inner membrane. Membrane bound proteins involved in polymerization and export of alginate were also identified in *Pseudomonas aeruginosa* (May *et al.*, 1991). Lipid-linked intermediates were also not identified in the biosynthesis of alginate (Sutherland 1982).

The involvement of lipid carriers or membrane-bound proteins in EPS biosynthesis gave evidence for the synthesis of polysaccharides from cytoplasmic nucleotide diphosho-precursors at the inner membrane. Polymerization occurs at the inner or at the outer face of the cytoplasmic membrane. An example for the polymerization at the inner side is the assembly of the mannose polymer in *E. coli* O9 strains (Kido *et al.*, 1995). In contrast, the synthesis of lipid-linked intermediates of the *Salmonella*

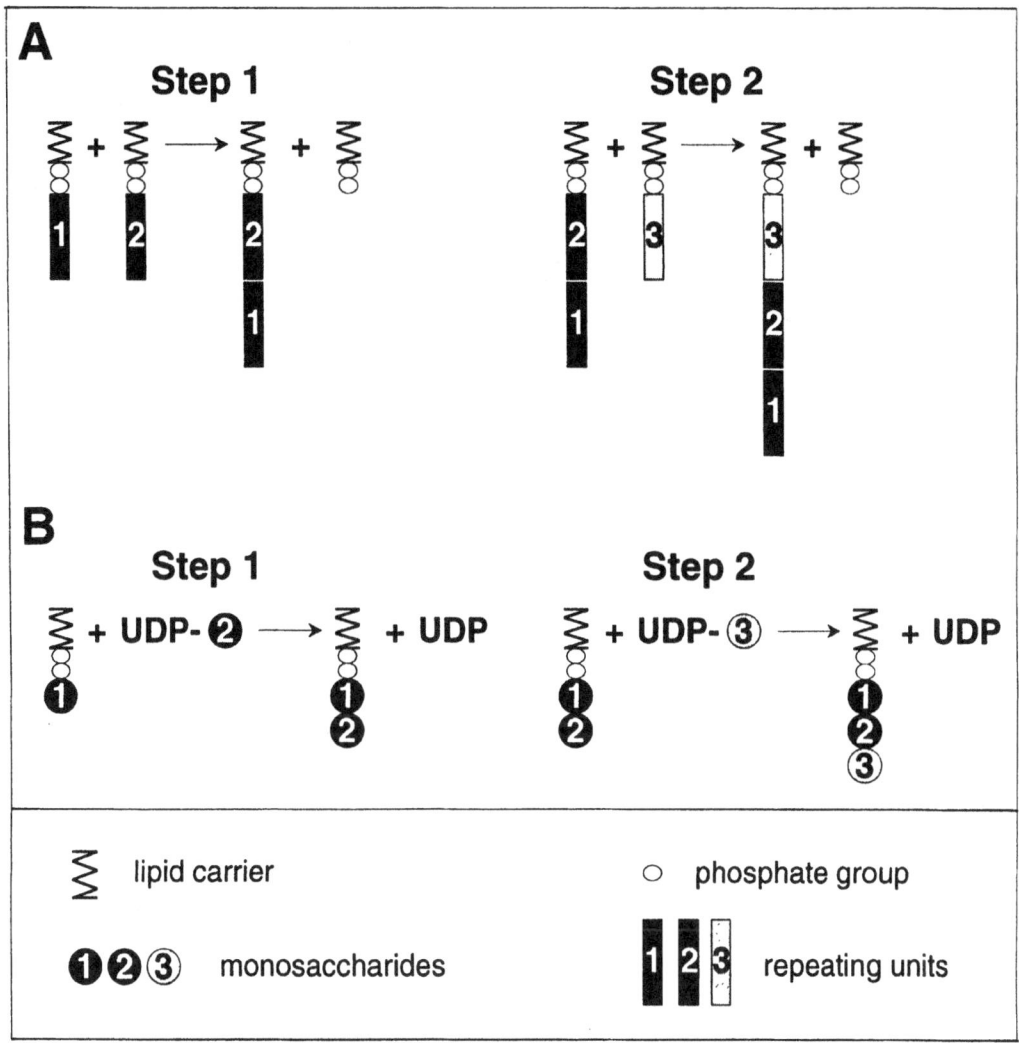

Figure 1. Mechanisms of polysaccharide polymerization. A. Blockwise polymerization by transfer of the growing lipid-linked polymer to the non-reducing terminus of a lipid-linked repeating unit results in a polymer chain growing at the reducing end. B. The processive sequential transfer of monosaccharides to the non-reducing terminus of a polysaccharide chain results in the growth at the non-reducing end.

enterica serogroups B and E O antigens takes place at the inner face whereas the blockwise polymerization is performed at the periplasmic face (McGrath and Osborn 1991).

Modifications of polysaccharides were reported at the level of lipid-linked intermediates prior to polymerization (Capage *et al.*, 1987, Vanderslice *et al.*, 1989) or as postpolymerization substitutions (May and Chakrabarty 1994).

The export of polysaccharides to the cell surface is poorly understood. ATP binding cassette (ABC) transporters seem to be involved in the transport of homopolymers polymerized in a processive manner across the inner membrane (Whitfield and Valvano 1993, Whitfield 1995). Although outer membrane proteins involved in group II CPS and alginate production were identified (May *et al.*, 1991, Frosch *et al.*, 1992) the translocation of polysaccharides accross the outer membrane is hardly characterized. Implication of membrane-adhesion sites (Bayer junctions) (Bayer 1979; 1991) in translocation was suggested (Kröncke *et al.*, 1990a;b), although experimental results were contradictory (Whitfield and Valvano 1993).

II.C. EXAMPLES OF STRUCTURES OF RHIZOBIAL EXOPOLYSACCHARIDES

One of the most extensively studied rhizobial EPS is the succinoglycan produced by several *Sinorhizobium meliloti*, *Agrobacterium* and *Alcaligenes* strains (Hisamatsu *et al.*, 1980; Zevenhuizen 1990). The structure of succinoglycan synthesized by the *S. meliloti* SU47 derivative strain Rm1021 (Aman *et al.*, 1981; Reinhold *et al.*, 1994) and *S. meliloti* YE-2 (Matulová *et al.*, 1994) was studied in detail (Figure 2). Succinoglycan is composed of octasaccharide repeating units containing one galactose and seven glucose residues joined by β-1→3, β-1→4 and β-1→6 glycosidic linkages. The repeating unit can be decorated by acetyl, pyruvyl and succinyl groups. Apart from succinoglycan, often referred to as EPS I, these *S. meliloti* strains have the ability to produce a second EPS, termed galactoglucan or EPS II (Her *et al.*, 1990; Zevenhuizen 1990) (Figure 2). EPS II contains glucose and galactose residues in the ratio of one to one joined by α-1→3 and β-1→3 glycosidic bonds. Decorations by acetyl- and pyruvyl groups were reported (Her *et al.*, 1990). Galactoglucan was also isolated from *Agrobacterium radiobacter* culture supernatants (Zevenhuizen 1989).

The EPS produced by *Rhizobium* sp. NGR234 is structurally related to succinoglycan (Djordjevic *et al.*, 1986) (Figure 2). With respect to the five sugars of the reducing terminal part of the repeating unit the structure of succinoglycan and the *Rhizobium* sp. NGR234 EPS is identical. Acetyl- and pyruvyl groups were also identified in the EPS of *Rhizobium* sp. NGR234. Many *R. leguminosarum* strains of different biovars nodulating different plant hosts have a similar basic EPS structure (Figure 2), although the pattern of decoration may vary (McNeil *et al.*, 1986; Hollingworth *et al.*, 1988; Canter Cremers *et al.*, 1990; 1991a;b). Several strains of *Rhizobium etli* produce EPS which differ from the basic structure shown in Figure 2 (McNeil *et al.*, 1986; Gil-Serrano *et al.*, 1990).

Bradyrhizobium japonicum strains produce EPS containing glucose, mannose, galactose, galacturonic acid and 4-O-methylglucose whereas *Bradyrhizobium elkanii* (formerly type II *B. japonicum*) produces EPS composed of rhamnose and 4-O-methylglucuronic acid (Dudman 1978; Mort and

Bauer 1980; 1982; Puvanesarajah *et al.*, 1987; Minamisawa 1989) (Figure 2). Several strains of *B. japonicum* and *B. elkanii* produce large amounts of polysaccharide in *Glycine max* nodules (Newcomb 1981; Streeter *et al.*, 1992; 1994; An *et al.*, 1995). These polysaccharides that accumulate in the symbiosome are called NPS (nodule polysaccharides). *B. elkanii* EPS and NPS have the same structure (An *et al.*, 1995), whereas *B. japonicum* NPS differ structurally from the EPS produced in cultures (Mort and Bauer 1982; Streeter *et al.*, 1992; 1994).

Several rhizobial strains produce EPS species differing in molecular weight. Two major fractions, high molecular weight (HMW) polymers and low molecular weight (LMW) oligosaccharides, were identified in the culture supernatants of *S. meliloti* (Amemura *et al.*, 1983; Zevenhuizen and van Neerven 1983; Leigh and Lee 1988; González *et al.*, 1996), *Rhizobium* sp. NGR234 (Djordjevic *et al.*, 1987), *R. leguminosarum* bv. *trifolii* (Djordjevic *et al.*, 1987), *Rhizobium* sp. GRH2 (Lopez-Lara *et al.*, 1993) and *B. japonicum* strains (Kosch *et al.*, 1995).

III. The Genetics of Rhizobial Exopolysaccharide Biosynthesis

Early studies on bacterial EPS production were confined to the purification and structural investigation of EPS. Genetic approaches have broadened the scope of this research. Techniques of molecular genetics facilitated the analysis of genes involved in EPS biosynthesis and regulation. The isolation of transposon-tagged mutants set the basis for gene identification and biochemical approaches providing data on EPS biosynthetic pathways.

III.A. ORGANIZATION OF RHIZOBIAL GENE CLUSTERS DIRECTING EXOPOLYSACCHARIDE BIOSYNTHESIS

In many cases genes directing the biosynthesis of bacterial EPS are situated in large gene clusters, although examples were reported, where genes involved in EPS biosynthesis were organized as single loci dispersed in the genome. Genes involved in rhizobial EPS production, were mapped on the chromosome and on large plasmids. The EPS biosynthesis gene clusters usually contain genes

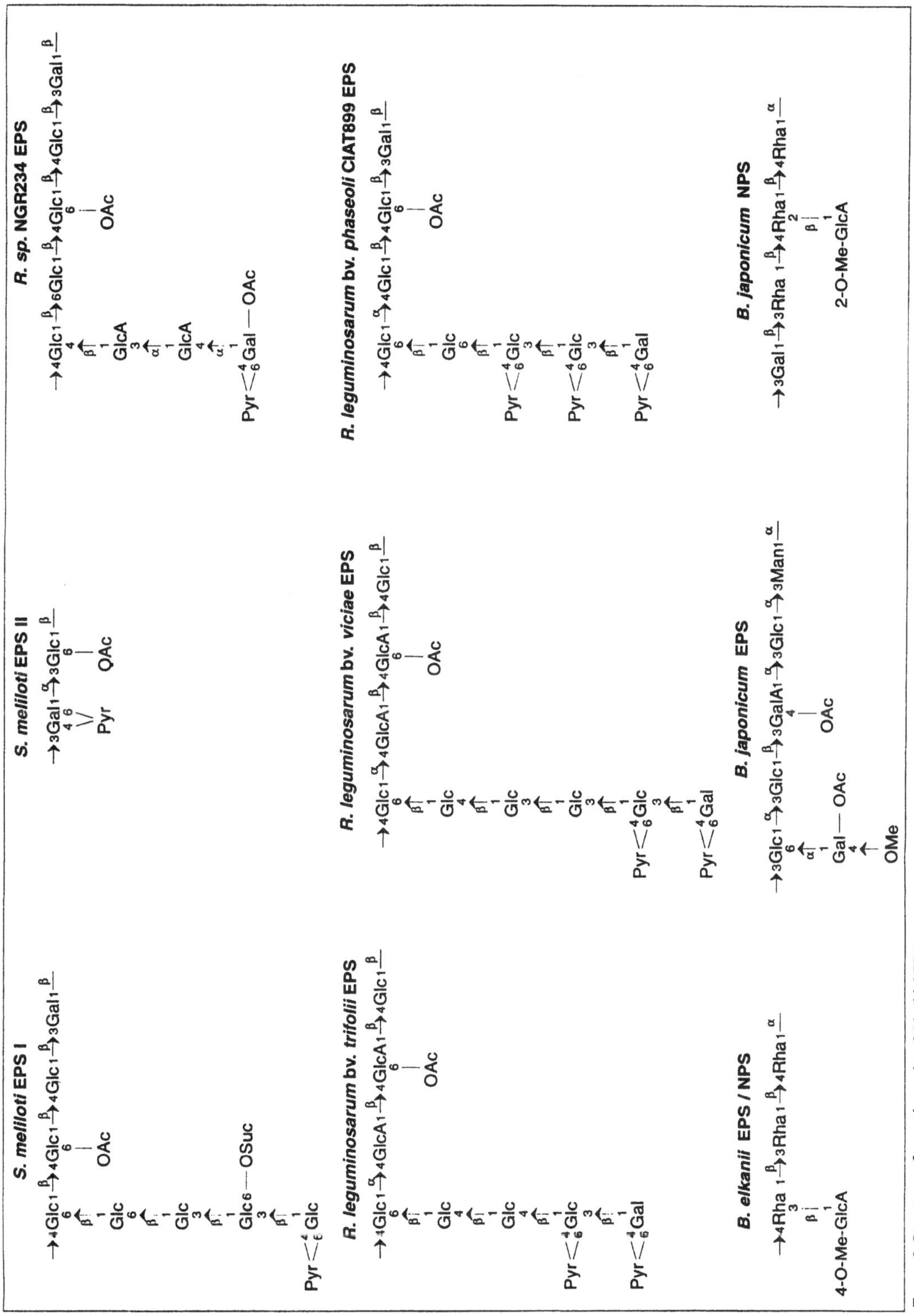

Figure 2 Structures of repeating units of rhizobial EPS.
Abbreviations: Ac, acetate; Gal, galactose; Glc, glucose; GalA, galacturonic acid; GlcA, glucuronic acid; Man, mannose; Me, methyl; NPS, nodule polysaccharide; Pyr, pyruvate; Rha, rhamnose; Suc, succinate.

encoding enzymes directing the assembly of the repeating unit and its decoration as well as the polymerization and export. In some cases, enzymes involved in the biosynthesis of nucleotide diphospho-sugar precursors are also encoded by genes of these clusters. In cases where these genes encode housekeeping functions they often represent a second copy of the housekeeping gene. Genes involved in the regulation of EPS production are generally not included in biosynthesis gene clusters.

The EPS I biosynthesis of *S. meliloti* represents one of the best studied examples for bacterial EPS production and therefore serves as a model system to discuss the genetics and the biosynthetic pathway of a complex rhizobial heteropolysaccharide. Where information on the genetics and biosynthesis of EPS produced by other *Rhizobiaceae* are available, these data are discussed in comparison to the EPS I biosynthesis of *S. meliloti*.

III.A.1. The exo/exs-Gene Cluster of S. meliloti SU47 Directing the Biosynthesis of EPS I

S. meliloti SU47 mutants deficient in EPS I production were initially isolated by transposon mutagenesis (Leigh *et al.*, 1985; Finan *et al.*, 1985; Leigh *et al.*, 1987). Loci affected by transposon insertions were mapped on the chromosome and on the second megaplasmid (Finan *et al.*, 1986; Hynes *et al.*, 1986; Long *et al.*, 1988a;b; Müller *et al.*, 1988; Charles and Finan 1990). A large gene cluster involved in EPS I biosynthesis comprising at least 27 kb was identified on megaplasmid 2 of the *S. meliloti* SU47 derivative strains Rm1021 and Rm2011. Analysis of the nucleotide sequence revealed the presence of 21 genes termed *exo* or *exs* genes (Figure 3) (Buendia *et al.*, 1991; Reed *et al.*, 1991a; Becker *et al.*, 1993a;b;c; Glucksmann *et al.*, 1993a;b; Müller *et al.*, 1993; Becker *et al.*, 1995b). The *exsH* gene identified by York and Walker (1997) is located 7.4 kb downstream of *exsB*. In addition, five loci affecting EPS I biosynthesis were identified on the chromosome. These include *exoR* (Doherty *et al.*, 1988; Reed *et al.*, 1991b), *exoS* (Doherty *et al.*, 1988) and *mucR* (Zhan *et al.*, 1989; Keller *et al.*, 1995) which are implicated in regulation, as well as *exoC* (Finan *et al.*, 1986) and *exoD* (Reed and Walker 1991a;b).

Homologies of the deduced *exo* and *exs* gene products to proteins of known function suggested that the *exo* and *exs* genes encode proteins required for the synthesis of precursors and the repeating unit, decoration, polymerization, export and degradation of EPS I as well as for the regulation of EPS I production (Buendia *et al.*, 1991; Reed *et al.*, 1991a; Becker *et al.*, 1993a;b;c; Glucksmann *et al.*, 1993a;b; Müller *et al.*, 1993; Becker *et al.*, 1995a;b; York and Walker 1997) (Table 1).

The *exo/exs*-gene cluster is organized in several operons (Buendia *et al.*, 1991; Becker *et al.*, 1993a;b;c; Müller *et al.*, 1993; Becker *et al.*, 1995b) (Figure 3). In total, at least eighteen promoters were identified by plasmid integration mutagenesis of the *S. meliloti* Rm2011 wild type strain or mutants carrying *exo-lacZ* and *exs-lacZ* transcriptional fusions. The transcription directed by several promoters results in the formation of overlapping transcripts. This is particularly obvious in the *exoHKLAMONP*, *exoWV* and *exoYFQ* region.

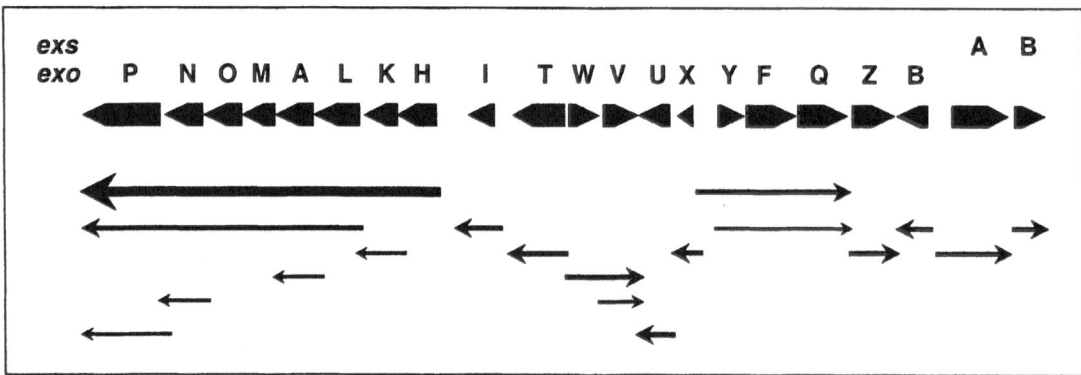

Figure 3. Gene and operon structure of the *S. meliloti* SU47 *exo/exs* gene cluster. The extent of operons identified is indicated by the length of the arrows, whereas the strength of promoters identified is indicated by their width.

Protein	Protein location[a]	Function deduced from experimental evidence or homology	Gene location[b]	References
ExoC	CY	phosphoglucomutase	CHR	Uttaro et al. 1990
ExoB	CY	UDP-glucose-4-epimerase	MP 2	Buendia et al. 1991
ExoN	CY	UDP-glucose pyrophosphorylase	MP 2	Becker et al. 1993b, Glucksmann et al. 1993b
ExoY	CMa	galactosyl transferase	MP 2	Müller et al. 1993, Reuber & Walker 1993a
ExoA	CMa	glucosyl transferase	MP 2	Becker et al. 1993b, Glucksmann et al. 1993a, Reuber & Walker 1993a
ExoL	CY	glucosyl transferase	MP 2	Becker et al. 1993b, Glucksmann et al. 1993a, Reuber & Walker 1993a
ExoM	CY	glucosyl transferase	MP 2	Becker et al. 1993b, Glucksmann et al. 1993a, Reuber & Walker 1993a
ExoO	CY	glucosyl transferase	MP 2	Becker et al. 1993b, Glucksmann et al. 1993a, Reuber & Walker 1993a
ExoU	CMa	glucosyl transferase	MP 2	Becker et al. 1993c, Glucksmann et al. 1993a, Reuber & Walker 1993a
ExoW	CY	glucosyl transferase	MP 2	Becker et al. 1993c, Glucksmann et al. 1993a, Reuber & Walker 1993a
ExoZ	CMi	acetyl transferase	MP 2	Buendia et al. 1991, Reuber & Walker 1993b
ExoH	CMi	succinyl transferase	MP 2	Becker et al. 1993a, Leigh et al. 1987
ExoV	CY	pyruvyl transferase	MP 2	Becker et al. 1993c, Glucksmann et al. 1993b, Reuber & Walker 1993a
ExoF	PP	required for transfer of galactose	MP 2	Müller et al. 1993, Reuber & Walker 1993a
ExoQ	CMi	export and polymerization	MP 2	Müller et al. 1993, Reuber & Walker 1993a
ExoT	CMi	export and polymerization	MP 2	Becker et al. 1993c, Glucksmann et al. 1993b, Reuber & Walker 1993a
ExoP	CMi	export and polymerization	MP 2	Becker et al. 1993b, 1995a, Glucksmann et al. 1993b, Reuber & Walker 1993a
ExsA	CMi	export	MP 2	Becker et al. 1995b
ExoK	PP or EC	endo-β-1,3-1,4-glycanase	MP 2	Becker et al. 1993a
ExsH	EC	endoglycanase	MP 2	York and Walker 1997
ExoX	CMa	negative regulator	MP 2	Zhan & Leigh 1990, Reed et al. 1991a, Müller et al. 1993
ExoI	PP or EC	unknown	MP 2	Becker et al. 1993c
ExsB	CY	negative regulator	MP 2	Becker et al. 1995b
ExoR	CY	negative regulator	CHR	Doherty et al. 1988, Reed et al. 1991b, Ozga et al. 1994
ExoS	CMa	negative regulator	CHR	Doherty et al. 1988, Ozga et al. 1994, Østerås et al. 1995
MucR	CY	positive regulator	CHR	Zhan et al. 1989, Keller et al. 1995
ExoD	CY	unknown	CHR	Reed and Walker 1991a,b

Table 1. Properties of genes and gene products involved in EPS I biosynthesis of S. meliloti SU47.

[a] Subcellular location of the proteins as predicted from the hydrophobicity of the amino acid sequence or from gene fusion experiments; CY, cytoplasm; CM, cytoplasmic membrane (i, integral membrane protein; a, attached to or anchored in the membrane); PP, periplasm; EC, extracellular;
[b] CHR, chromosome; MP 2, megaplasmid 2.

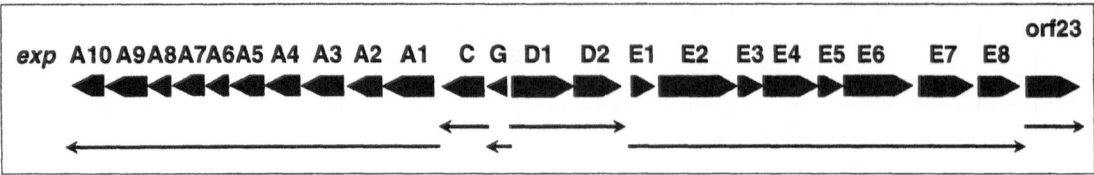

Figure 4. Gene and operon structure of the *S. meliloti* SU47 *exp* gene cluster. The extent of operons identified is indicated by the length of the arrows.

III.A.2. The exp Gene Cluster of S. meliloti SU47 Directing the Biosynthesis of EPS II

The ability of *S. meliloti* SU47 to produce galactoglucan (EPS II) was first observed in the presence of a mutation in either of the chromosomally located genes *expR* (Glazebrook and Walker 1989) or *mucR* (Zhan *et al.*, 1989; Keller *et al.*, 1995). The *exp* gene cluster directing the biosynthesis of EPS II was identified on megaplasmid 2 (Glazebrook and Walker 1989). It is separated from the *exo/exs* gene cluster by approximately 200 kb (Charles and Finan 1991). A characterization of the *exp* gene cluster by transposon mutagenesis revealed that it comprises at least 23 kb and contains at least six complementation groups required for EPS II biosynthesis (Glazebrook and Walker 1989). Nucleotide sequence analysis of a 32 kb DNA fragment carrying the complete *exp* gene cluster resulted in the identification of 22 *exp* genes (Becker *et al.*, 1997) (Figure 4). Homologies of the deduced *exp* gene products to proteins of known function suggested that the *exp* genes encode proteins involved in nucleotide diphospho-sugar biosynthesis, sugar polymerization, export and regulation of *exp* gene expression (Table 2).

Protein	Location[1]	Putative function deduced from homologies
ExpG	CY	transcriptional regulator
ExpA7	CY	glucose-1-phosphate thymidylyltransferase
ExpA8	CY	dTDP-4-dehydrorhamnose 3,5-epimerase
ExpA9	CY	dTDP-glucose 4,6-dehydratase
ExpA10	CY	dTDP-4-dehydrorhamnose reductase
ExpA2	CY	glycosyl transferase
ExpE2	CY	glycosyl transferase
ExpA3	CY	glycosyl transferase
ExpC	CMa	glycosyl transferase
ExpE4	CY	glycosyl transferase
ExpE7	CY	glycosyl transferase
ExpD1	CMi	ABC transporter protein of a proteinase export complex
ExpD2	CM and OM	MFP protein of a proteinase export complex
ExpE1	OM or EC	secreted Ca^{2+} binding protein
ExpE3	CY	methyltransferase
ORF23	PP	periplasmic binding protein
ExpA1	CY	unknown
ExpA4	CY	unknown
ExpA5	CY	unknown
ExpA6	PP	unknown
ExpE5	CMa	unknown
ExpE6	CMi	unknown
ExpE8	CY	unknown

Table 2. Putative functions of the *exp* gene products of *S. meliloti* SU47 (Becker *et al.*, 1997).

---

[1] Subcellular location of the proteins as predicted from the hydrophobicity of the amino acid sequences; CY, cytoplasm; CM, cytoplasmic membrane (i, integral membrane protein; a, attached to or anchored in the membrane); PP, periplasm; OM, outer membrane; EC, extracellular.

III.A.3. Genes with Homologies to
S. meliloti SU47 exo Genes are Present in
Several Other Rhizobial Species

Several genes involved in EPS biosynthesis of rhizobia displayed homologies to *exo* genes of *S. meliloti* SU47. In *Rhizobium* sp. NGR234 a 15 kb *exo* gene cluster comprising four complementation groups with striking similarities to the *S. meliloti* SU47 *exo* gene cluster was identified on the chromosome (Chen *et al.*, 1988; Gray *et al.*, 1990; Zhan *et al.*, 1990). An additional complementation group, termed *exoG*, which did not map to the *exo* gene cluster was identified in *Rhizobium* sp. NGR234 (Chen *et al.*, 1988). *S. meliloti* SU47 mutants in *exoA*, *exoB*, *exoY*, *exoL*, *exoM*, *exoN* and *exoP* could be complemented by the *exo* gene cluster of *Rhizobium* sp. NGR234 (Zhan *et al.*, 1990). In contrast, *exoC*, *exoD*, *exoK* and *exoH* mutants were not complemented by this *exo* gene cluster. Since *exoC* and *exoD* of *S. meliloti* SU47 are not linked to the *exo* gene cluster the failure to complement mutants in these genes indicate that the corresponding genes of *Rhizobium* sp. NGR234 might also be unlinked (Zhan *et al.*, 1990). The *exoH* gene product involved in succinylation of EPS I (Leigh *et al.*, 1987) is likely to be specific for *S. meliloti* SU47, since EPS produced by *Rhizobium* sp. NGR234 does not contain succinyl groups (Djordjevic *et al.*, 1986). Additionally, mutants in the *exoB* complementation group of *Rhizobium* sp. NGR234 were not complemented by the *S. meliloti* SU47 *exo* gene cluster implying that this complementation group is specific for *Rhizobium* sp. NGR234. Hybridization experiments revealed that the *exoD* complementation group of *Rhizobium* sp. NGR234 corresponds to *exoL*, *exoA* and *exoM* of *S. meliloti* SU47 (Zhan *et al.*, 1990). Moreover, the *Rhizobium* sp. NGR234 *exoY* and *exoC* complementation groups were assigned to *exoY* and *exoB* of *S. meliloti*, SU47, respectively. DNA sequences of the *exoX-exoY* region of both species are almost identical (Gray *et al.*, 1990) indicating that large parts of both *exo* gene clusters are closely related. This is also consistent to the similar structures of the acidic EPS produced by *Rhizobium* sp. NGR234 and EPS I synthesized by *S. meliloti* (Aman *et al.*, 1981; Djordjevic *et al.*, 1986; Reinhold *et al.*, 1994). Interestingly, a gene similar to *exoX* was also identified on the symbiotic replicon pNGR234a in *Rhizobium* sp. NGR234 (Freiberg *et al.*, 1997).

In *R. leguminosarum* bv. *viciae* and *R. leguminosarum* bv. *phaseoli* genes homologous to the *S. meliloti* SU47 *exoY* were identified (Borthakur *et al.*, 1986; 1988; Latchford *et al.*, 1991; Ivashina *et al.*, 1994). Furthermore, the *psiA* gene encoding a protein similar to ExoX of *S. meliloti* SU47 was found in *R. leguminosarum* bv. *phaseoli* (Borthakur *et al.*, 1985; 1988; Latchford *et al.*, 1991; Mimmack *et al.*, 1994a;b). Ivashina *et al.*, (1995; 1996) identified a 5.5 kb chromosomal fragment involved in EPS biosynthesis of *R. leguminosarum* bv. *viciae*. This fragment is not linked to the *pssA* or *psiA* gene. Six genes designated *pssC*, *pssD*, *pssE*, *pssF*, *pssG* and *pssH* were identified on this fragment (Ivashina *et al.*, 1996). The predicted gene products of *pssC*, *pssD*, *pssE* and *pssF* resemble ExoM, ExoO and ExoW of *S. meliloti* SU47 (Ivashina *et al.*, 1996). This indicates that the proteins encoded by these *pss* genes also function as glycosyl transferases forming β-glycosidic linkages. Furthermore, the predicted gene product of *pssH* displays similarities to glycosyl transferases of *Streptomyces* C5 and *Erwinia herbicola* (Ivashina *et al.*, 1996). In addition, van Workum *et al.*, (1997) identified four genes (*pssA*, *pssC*, *pssD* and *pssE*) in *R. leguminosarum* bv. *trifolii* involved in EPS biosynthesis. The *pssA* gene product is almost identical to the Pss4 protein of *R. leguminosarum* bv. *viciae* and homologous to ExoY of *S. meliloti*. *pssD* and *pssE* form an operon closely linked to *pssC*. These three genes that are not closely linked to *pssA* encode proteins with similarities to several glycosyl transferases (van Workum *et al.*, 1997). A genetic locus functionally homologous to *exoB* of *S. meliloti* SU47 was also identified in *R. leguminosarum* bv. *viciae* (Canter Cremers *et al.*, 1990) and *R. leguminosarum* bv. *trifolii* (Sanchez-Andujar *et al.*, 1997). In addition, a genetic locus which was proposed to encode a galactosyl transferase was identified in *R. leguminosarum* bv. *viciae* (Breedveld *et al.*, 1993).

Król *et al.*, (1994; 1996) isolated Tn5 mutants in an *exo* gene cluster located on a 300 kb non-symbiotic plasmid of *R. leguminosarum* bv. *trifolii*. Two genes encoding proteins homologous to two components of ABC transporters involved in signal peptide-independent secretion of proteins were found in this region (Król *et al.*, 1994; 1996). Interestingly, similar proteins involved in EPS II biosynthesis of *S. meliloti* SU47 are encoded by the *expD1* and *expD2* genes (Becker *et al.*, 1997). In *B. japonicum* genes encoding proteins homologous to *exoB* and *exoP* of *S. meliloti* SU47 were identified (Parniske *et al.*, 1993; Kosch *et al.*, 1995).

III.B. REGULATION OF EXOPOLYSACCHARIDE PRODUCTION

Adaptation of EPS production to various environmental conditions requires a complex regulatory network involving cross talk between different regulatory components. Regulation of EPS production can occur on different levels of gene expression. Regulatory mechanisms affecting transcription or translation of EPS biosynthesis genes as well as the activity of proteins involved in EPS biosynthesis are described.

III.B.1. Regulation of EPS Production in S. meliloti

Limitation of non-carbon nutrients, e.g. nitrogen, phosphate and sulfur, was found to stimulate EPS I production of *S. meliloti* SU47 (Leigh *et al.*, 1985; Doherty *et al.*, 1988), whereas phosphate limitation stimulated EPS II production (Zhan *et al.*, 1991). Studies on the regulation of EPS biosynthesis suggested that different osmotic conditions modify EPS production in *S. meliloti* SU47 (Breedveld *et al.*, 1990b). Low osmotic pressure resulted in the predominant production of low molecular weight EPS I (Breedveld *et al.*, 1990b). The production of high molecular weight EPS I was stimulated by an increased osmotic pressure. *S. meliloti* YE produced high molecular weight EPS II and low molecular weight EPS I in the presence of low concentrations of NaCl (Zevenhuizen and Faleschini 1991). An increase of the NaCl concentration of the medium resulted in an enhanced production of EPS I at the expense of EPS II and low molecular weight EPS I (Zevenhuizen and Faleschini 1991).

Two *S. meliloti* SU47 mutations affecting the chromosomally located genes *exoR* and *exoS* were isolated which caused an increased production of EPS I (Doherty *et al.*, 1988) and affected the expression of *exo* genes (Reed *et al.*, 1991b; Leigh *et al.*, 1993; Ozga *et al.*, 1994). Limitation of ammonia usually stimulates production of EPS I. Doherty *et al.*, (1988) reported that an *exoR* mutant produced increased amounts of EPS I regardless of the presence of ammonia in the medium, whereas the absence of ammonia still stimulated EPS I production of an *exoS* mutant. This is contrary to results of Ozga *et al.*, (1994) who reported that *exoR* was not required for the regulation of EPS I

biosynthesis by ammonia. Analysis of *exo* mRNA levels demonstrated that ExoR influences the transcriptional level of several *exo* genes (Reed *et al.*, 1991b). The increased expression of *exoP*-, *exoA*-, *exoF*- and *exoQ*-*phoA* translational fusions in an *exoR* and *exoS* mutant background is consistent with this finding (Reuber *et al.*, 1991). No effect of these mutations on the expression of *exoT*- and *exoB*-*phoA* translational fusions was found (Reuber *et al.*, 1991). In addition, the expression of an *exoY*-*lacZ* transcriptional fusion was stimulated in an *exoR* mutant background (Ozga *et al.*, 1994). EPS I overproduction by the *exoR* mutant was suppressed by mutations that mapped to the *exoS* locus (Leigh *et al.*, 1993; Ozga *et al.*, 1994). These *exoS** suppressor mutations lowered the expression of an *exoY*-*lacZ* transcriptional fusion in the *exoR* mutant background (Ozga *et al.*, 1994).

Whereas ExoR displayed no homologies to known regulators (Reed *et al.*, 1991b), ExoS was found to be identical to the ChvG protein (Østerås *et al.*, 1995; G.C. Walker, personal communication). The ChvG protein is homologous to sensor proteins of two-component regulatory systems (Østerås *et al.*, 1995). Therfore, ExoS might be involved in sensing environmental factors that influence EPS I production of *S. meliloti* SU47. The second DNA binding component ChvI of this two-component system might also be involved in the regulation of EPS I production.

The ratio of *exoX* to *exoY* affects the level of EPS I production of *S. meliloti* SU47 (Zhan and Leigh 1990; Reed *et al.*, 1991a; Müller *et al.*, 1993). Both genes are located in the *exo/exs* gene cluster on megaplasmid 2. Mutants in *exoX* overproduced EPS I, whereas the presence of multiple copies of *exoX* inhibited the production of EPS I. The presence of *exoX* and *exoY* on a multi-copy plasmid overcame the inhibitory effect of multiple copies of *exoX*. Since *exo* gene expression was unaffected by the gene dosage of *exoX* and *exoY* it was suggested that ExoX functions as a posttranscriptional inhibitor of ExoY which encodes a galactosyl transferase catalyzing the initial step of EPS I biosynthesis (Reed *et al.*, 1991a; Müller *et al.*, 1993; Reuber and Walker 1993a). Both proteins are probably associated with the inner membrane (Müller *et al.*, 1993) suggesting that protein-protein interaction might occur. Moreover, the *exsB* gene also located in the *exo/exs* gene cluster might encode a negative regulator affecting a posttranscriptional level of *exo* gene expression (Becker *et al.*, 1995b). Overexpression of *syrM* in *S. meliloti* SU47, a gene involved in regulation of *nod* gene expression, caused the enhanced production of EPS, although the regulatory mechanism is unknown (Mulligan and Long 1989). Since *exoA* and *exoB* were required for the enhanced synthesis of EPS and *exoA* is specific for the production of EPS I it is likely that EPS I biosynthesis was stimulated (Mulligan and Long 1989). The *syrA* gene, which maps closely to *syrM* was also required for the enhanced production of EPS (Mulligan and Long 1989). For more details on *syrM* refer to chapter 19.

Reuber *et al.,* (1991) reported that *exoA-*, *exoP-* and *exoF-phoA* translational fusions were mainly expressed in the infection zone of alfalfa nodules suggesting that *exo* gene expression is downregulated in the symbiotic zone.

Usually, *S. meliloti* SU47 almost exclusively produces EPS I. To date the only condition identified to cause EPS II biosynthesis is the limitation of phosphate (Zhan *et al.,* 1991). Phosphate limitation stimulated the β-galactosidase activity mediated by an *exp-lacZ* transcriptional fusion and therefore affected the transcription of at least one *exp* gene (Zhan *et al.,* 1991). Two *S. meliloti* SU47 mutants were isolated which constitutively produced EPS II. The *mucR* gene affected in one of these mutants is located on the chromosome. Mutations in *mucR* resulted in the constitutive biosynthesis of EPS II and in a reduction of EPS I production (Zhan *et al.,* 1989; Keller *et al.,* 1995; Bertram-Drogatz *et al.,* 1997b). In addition, only low molecular weight EPS I oligosaccharides were identified in the culture supernatants of *mucR* mutants (Keller *et al.,* 1995; Bertram-Drogatz *et al.,* 1997b). The MucR protein (Keller *et al.,* 1995) is highly homologous to *Agrobacterium* Ros proteins which constitute repressors of the *virC* and *virD* operons as well as activators of EPS production (Cooley *et al.,* 1991; d'Souza-Ault *et al.,* 1993; Brightwell *et al.,* 1995). Moreover, the *mucR* and *ros* genes are negatively autoregulated (Cooley *et al.,* 1991; Keller *et al.,* 1995). A potential zinc finger motif of the C_2H_2 type was identified in the MucR and Ros amino acid sequences. D'Souza-Ault *et al.,* (1993) demonstrated that the *A. tumefaciens* Ros protein can specifically bind to the *virC/virD* promoter region. The *ros*, *virC* and *virD* genes are preceded by an inverted repeat called Ros box that has been discussed to mediate the specific binding of Ros (d'Souza-Ault *et al.,* 1993).

The expression of *exp-lacZ* transcriptional fusions located in either of the five *exp* complementation groups was induced in the *S. meliloti* SU47 *mucR* mutant background (Keller *et al.,* 1995; Becker *et al.,* 1997). In contrast, the transcription of most *exo-lacZ* fusions was not affected by the *mucR* mutation (Keller *et al.,* 1995). Only the β-galactosidase activities of strains carrying *exoY-*, *exoF-*, and *exoK-lacZ* transcriptional fusions were slightly decreased in a *mucR* mutant background in comparison to the wild type background (Leigh *et al.,* 1993; Keller *et al.,* 1995). In contrast, the expression of *exoH-lacZ* and *exoX-lacZ* transcriptional fusions in the *mucR* mutant background was slightly enhanced (Bertram-Drogatz *et al.,* 1997b). Positive regulation of the transcription directed by the promoter situated upstream of *exoY* influences the expression of the *exoYFQ* operon. Since the ratio of *exoX* to *exoY* affects the level of EPS I production and *exoY* is involved in the initial step of EPS I biosynthesis, activation of *exoY* transcription by *mucR* might be one of the mechanisms regulating EPS I biosynthesis.

It was demonstrated that MucR can specifically bind to the *mucR*, *exoY* and *exoH* promoter regions (Bertram-Drogatz *et al.,* 1997a;b). In addition, inverted repeats homologous to the Ros box identified upstream of genes regulated by Ros in *A. tumefaciens* were found upstream of the *mucR* and *exoY* coding regions of *S. meliloti* (Bertram-Drogatz *et al.,* 1997a;b). Therefore, MucR itself exerts a negative autoregulation as well as a weak activation of *exoY* and a weak repression of *exoH* transcription. Interestingly, the ORF2 protein, which is 50% similar to MucR, was identified in *S. meliloti* (Barnett and Long 1997). No function could be assigned to the orf2 locus that maps close to *syrB*.

In addition to *mucR* mutants, the *S. meliloti* SU47 *expR* mutant isolated by Glazebrook and Walker (1989) constitutively produces EPS II due to the induction of *exp* gene expression. The spontaneous mutation responsible for the phenotype of the *expR* mutant was mapped to the chromosome, but the corresponding gene has not been cloned yet (Glazebrook and Walker 1989). In contrast to *mucR* mutants exclusively producing high molecular weight EPS II, the *expR* mutant produced a high and a low molecular weight EPS II fraction (González *et al.,* 1996).

Finally, a transcriptional regulator might be encoded by the *expG* gene which is located in the *exp* gene cluster on megaplasmid 2 (Becker *et al.,* 1997). The *expG* gene product displayed weak homologies to proteins of the MarR family of transcriptional

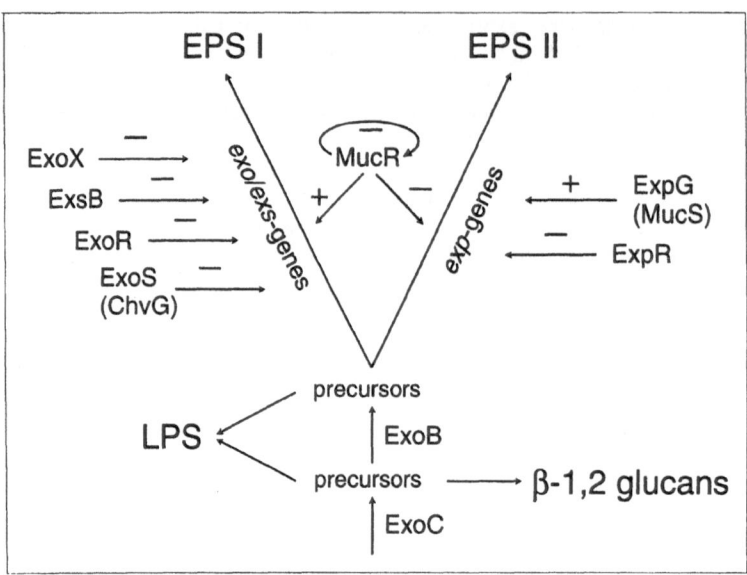

Figure 5. Schematic representation of the regulation of EPS biosynthesis in *S. meliloti* SU47. This figure is based on a model proposed by Leigh *et al.*, (1993) and was modified according to results reported by Glazebrook and Walker (1989), Becker *et al.*, (1995b, 1997) and Astete and Leigh (1996).

regulators, which bind DNA through a helix-turn-helix motif (Becker *et al.*, 1997). Furthermore, in *S. meliloti* SU47 the *muc*S gene, which is identical to expG, was identified by Astete and Leigh (1996). This gene was found to be required for the induction of EPS II production by low phosphate concentrations and by extra copies of the *exp* gene cluster (Astete and Leigh 1996). The authors also reported that *muc*S (*expG*) is required for the activation of at least the expression of one gene of the *expE* complementation group by phosphate starvation. MucS (ExpG) might therefore function as an activator of *exp* gene expression. In Figure 5 a summary of the regulation of EPS biosynthesis in *S. meliloti* SU47 is shown.

III.B.2. Regulation of EPS Production in Rhizobium sp. NGR234 and R. leguminosarum

Genes homologous to *S. meliloti* SU47 genes involved in the regulation of EPS production were identified in *Rhizobium* sp. NGR234, *R. leguminosarum* bv. *viciae* and *R. leguminosarum* bv. *phaseoli*. The *exoX* and *exoY* genes of *S. meliloti* SU47 and *Rhizobium* sp. NGR234 are highly homologous (Gray *et al.*, 1990). As in *S. meliloti* SU47, the ratio of *exoX* to *exoY* also affected the

level of EPS production in *Rhizobium* sp. NGR234 (Gray *et al.*, 1990). A high copy number of the *Rhizobium* sp. NGR234 *exoX* gene abolished EPS production in the homologous strain as well as in *S. meliloti* SU47 and *R. fredii* (Gray *et al.*, 1990). Since transcriptional regulation of *exoX* by *exoY* or *exoY* by *exoX* could not be demonstrated, a post-transcriptional mechanism was proposed (Gray and Rolfe 1990, Gray and Rolfe 1992).

The *pss4* (*pss2, pssA*) gene of *R. leguminosarum* bv. *viciae* and the *pssA* gene of *R. leguminosarum* bv. *phaseoli* were found to encode proteins homologous to the *S. meliloti* SU47 ExoY protein (Borthakur *et al.*, 1986; 1988; Latchford *et al.*, 1991; Ivashina *et al.*, 1994). As ExoY, the PssA protein was found to be associated to the inner membrane (Latchford *et al.*, 1991). In addition, *psiA* encoding a protein similar to ExoX of *S. meliloti* SU47 and *Rhizobium* sp. NGR234 was identified on the symbiotic plasmid pRP2JI in *R. leguminosarum* bv. *phaseoli* (Borthakur *et al.*, 1985; 1988; Latchford *et al.*, 1991; Mimmack *et al.*, 1994a;b). The hydropathic profiles of ExoX and PsiA are similar. Both proteins contain a hydrophobic N-terminal and a hydrophilic C-terminal half (Gray *et al.*, 1990). In addition, association of PsiA to the cytoplasmic membrane was demonstrated (Latchford *et al.*, 1991; Mimmack *et al.*, 1994b). Multiple copies of *psiA* in

R. leguminosarum bv. *phaseoli* abolished EPS production (Borthakur *et al.*, 1985; Latchford *et al.*, 1991). The *psrA* gene, which is also located on the symbiotic plasmid pRP2JI was isolated, since multiple copies of this gene suppressed the failure to produce EPS caused by multiple copies of *psiA* (Borthakur and Johnston 1987; Mimmack *et al.*, 1994a). It was demonstrated that PsrA represses *psiA* transcription and resembles transcriptional regulators which contain a helix-turn-helix-DNA binding domain (Borthakur and Johnston 1987; Mimmack *et al.*, 1994a). Using *psiA*- and *pssA-gusA* fusions Latchford *et al.*, (1991) showed that *psiA* is expressed in the symbiotic zone of *Phaseolus* nodules, whereas *pssA* is not. This indicates that *psiA* might be involved in downregulation of *pss* gene expression in the nodule.

The *mucR* gene of S. meliloti SU47 involved in regulation of EPS production was identified in *Rhizobium* sp. NGR234, *R. fredii*, *R. leguminosarum* bv. *phaseoli* and *A. tumefaciens* by hybridization experiments (Keller *et al.*, 1995). The *ros* genes highly homologous to *mucR* were sequenced from *A. tumefaciens* and *A. radiobacter* (Cooley *et al.*, 1991; d'Souza-Ault *et al.*, 1993; Brightwell *et al.*, 1995). Recently, Bittinger *et al.*, (1997) identified the *rosR* gene in *Rhizobium etli*. This gene is highly homologous to *mucR* and affects nodulation competitiveness and cell surface hydrophobicity of *R. etli*. Moreover, two genes showing similarities to *mucR* were found on the symbiotic replicon pNGR234a of *Rhizobium* sp. NGR234 (Freiberg *et al.*, 1997). Tiburtius *et al.*, (1996) reported that alkaline phosphatase acivities mediated by *exoY-phoA* translational fusions decreased in the background of *ros* mutations in *A. radiobacter*. This indicates that the expression of *exoY* involved in an early step of EPS biosynthesis is activated by the *ros* gene product in *A. radiobacter*. In addition, Tiburtius *et al.*, (1996) identified an imperfect inverted repeat upstream of *exoY* which is similar to the Ros box preceding the *ros* gene and the *virC* and *virD* genes in *A. tumefaciens* (D'Souza-Ault *et al.*, 1993). Similar sequences were also identified upstream of *exoY* and *mucR* in S. meliloti (Bertram-Drogatz *et al.*, 1997; Bertram-Drogatz, personal communication). A specific binding activity of the Ros protein to the *virC/virD* promoter region of *A. tumefaciens* and of the MucR protein to the promoter regions of *mucR*, *exoY* and *exoH* of S. meliloti, all containing sequences similar to the Ros box, was demonstrated (D'Souza-Ault *et al.*, 1993; Bertram-Drogatz *et al.*, 1997a;b). This indicates that the Ros proteins and the

MucR protein exert the regulation directly by binding to sites located upstream of the target genes for regulation.

Furthermore, a gene homologous to the S. meliloti *exoR* gene was identified in *R. leguminosarum* bv. *viciae* (Reeve *et al.*, 1997).

III.C. EXAMPLES FOR PATHWAYS OF EXOPOLYSACCHARIDE BIOSYNTHESIS

Few biosynthetic pathways of rhizobial EPS were analyzed in detail. The combination of genetic and biochemical approaches resulted in detailed models for the production of EPS I in S. meliloti SU47. The biosynthesis of this EPS as an example for the production of a heteropolysaccharide will be discussed.

III.C.1. The Biosynthetic Pathway of EPS I in S. meliloti SU47

The biosynthetic pathway of EPS I resembles that of other heteropolysaccharides composed of repeating units (Figure 6). Tolmasky *et al.*, (1980; 1982) demonstrated that the repeating unit is assembled on a lipid carrier resulting in a polyprenyl diphosphate octasaccharide intermediate. UDP-galactose and UDP-glucose were identified as the cytoplasmic precursors for the formation of the lipid-linked repeating unit (Buendia *et al.*, 1991; Becker *et al.*, 1993b; Glucksmann *et al.*, 1993b; Reuber and Walker 1993a). ExoC, ExoB and ExoN are involved in the biosynthesis of the precursors. The phosphoglucomutase encoded by *exoC* converts glucose-6-phosphate to glucose-1-phosphate (Uttaro *et al.*, 1990). This enzymatic activity is essential for the biosynthesis of EPS I, EPS II, LPS and β-(1-2)glucans (Zhan *et al.*, 1989; Uttaro *et al.*, 1990). ExoB displays UDP-glucose-4-epimerase activity (Buendia *et al.*, 1991). The conversion of UDP-glucose to UDP-galactose by ExoB is essential for EPS and LPS production. In contrast to *galE* mutants of *E. coli* (Fukasawa and Nikaido 1961), S. meliloti SU47 *exoB* mutants were able to grow on medium containing galactose as the sole carbon source, since S. meliloti can use the De Ley-Douderoff way for the conversion of galactose (Arias *et al.*, 1986). UDP-galactose can only be formed by epimerization of UDP-glucose, since galactokinase activity is absent in S. meliloti. Therefore, *exoB* mutants were unable to produce

EPS I, EPS II and LPS, whereas the biosynthesis of β-(1-2)glucans was not affected (Zhan *et al.*, 1989; Canter Cremers *et al.*, 1990; Buendia *et al.*, 1991). ExoN displays UDP-glucose pyrophosphorylase activity (Becker, unpublished) catalyzing the synthesis of UDP-glucose from glucose-1-phosphate and UTP. Mutations in *exoN* resulted in a reduced production of EPS I. The decrease of EPS I production might be due to at least one further gene encoding this enzyme, although a gene homologous to *exoN* was not identified by Southern hybridization (Glucksmann *et al.*, 1993b). A dTDP-glucose pyrophosphorylase might substitute for ExoN, since this enzyme can also display some UDP-glucose pyrophosphorylase activity (Mello and Glase 1965).

Functions were assigned to most of the *exo* genes by the analysis of lipid-linked biosynthetic intermediates that accumulated in *exo* mutants (Reuber and Walker 1993a). The sequential activities of eight glycosyl transferases and three decorating enzymes result in the formation of the lipid-linked octasaccharide repeating unit. ExoY and ExoF were identified to be involved in the initial transfer of the galactosyl residue to the lipid carrier. ExoA, ExoL,

ExoM, ExoO, ExoU and ExoW are involved in the addition of the subsequent sugars. The glycosyl transferase responsible for the addition of the terminal sugar of the repeating unit remains to be identified. Analysis of EPS I produced by *exoH* and *exoZ* mutants revealed that ExoH is involved in succinylation (Leigh *et al.*, 1987), whereas ExoZ is implicated in acetylation (Reuber and Walker 1993b). The *exoM* mutant blocked in the addition of the fourth sugar accumulated the non-acetylated and acetylated lipid-linked trisaccharide (Reuber and Walker 1993a). In addition, mutants in *exoP*, *exoQ* or *exoT* involved in polymerization or export of EPS I accumulated lipid-linked octasaccharide subunits that were acetylated, succinylated and pyruvylated to various extent. This indicates that decoration might occur prior to polymerization of the repeating units. Non-acetylated and non-succinylated EPS I can still be polymerized and exported. Since *exoV* mutants blocked in pyruvylation did not secrete EPS I polymer pyruvylation seems to be essential for polymerization and export (Becker *et al.*, 1993c; Glucksmann *et al.*, 1993b; Reuber and Walker 1993a).

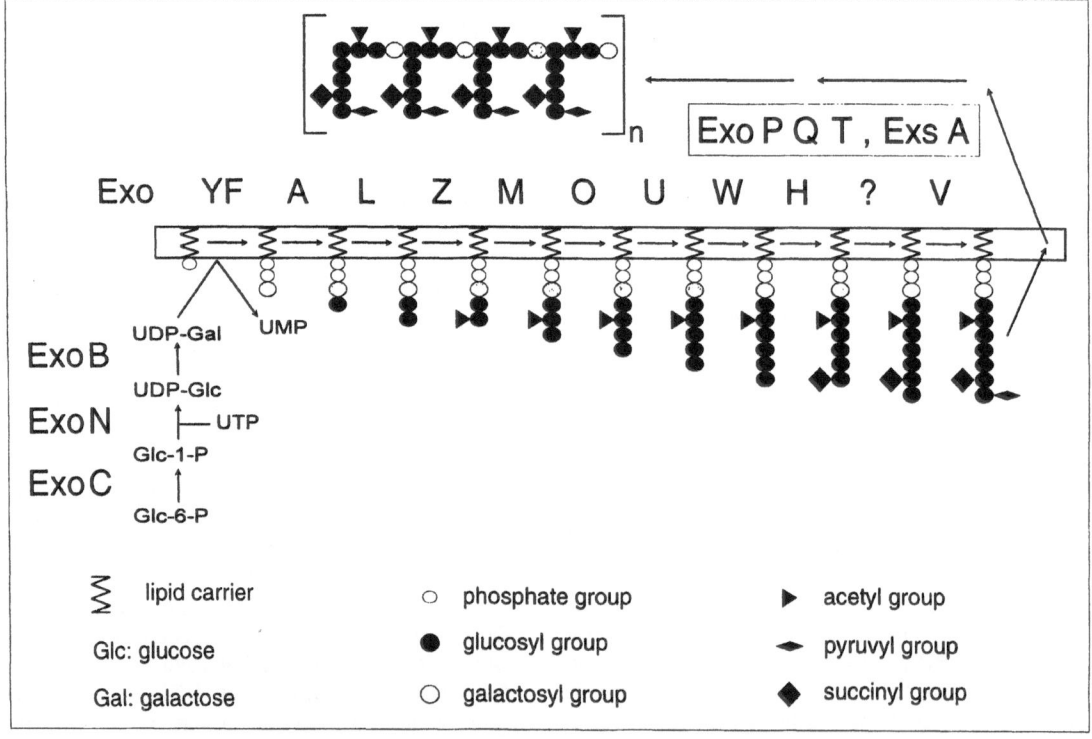

Figure 6. Schematic model of the biosynthetic pathway of *S. meliloti* SU47 EPS I (data from: Leigh *et al.*, 1987, Uttaro *et al.*, 1990, Buendia *et al.*, 1991, Reuber and Walker 1993a,b; Becker *et al.*, 1995a,b).

Recently, Gonzáles *et al.*, (1996b) reported evidence for the ExoQ protein to be involved in the biosynthesis of High Molecular Weight EPS I, whereas ExoT was suggested to be involved in the synthesis of EPS I octasaccharide trimers and tetramers. Since octasaccharide monomers were exclusively detected in culture supernatants of *exoP* mutants, it was concluded that ExoP is essential for the biosynthesis of High Molecular Weight and Low Molecular Weight EPS I polymerization products (Gonzáles *et al.*, 1996b). The membrane topology of the ExoP protein was analyzed by translational fusions of *lacZ* and *phoA* reporter genes to the *exoP* gene. Based on this analysis the ExoP protein consists of an N-terminal domain mainly located in the periplasm and a cytoplasmic C-terminal domain (Becker *et al.*, 1995a). The N-terminal domain of ExoP resembles CLD proteins involved in the determination of the O-antigen chain length of several enterobacteria. A short consensus motif common to CLD proteins and ExoP homologues was found in the amino acid sequence of the N-terminal domain (Becker *et al.*, 1995a; Stingele *et al.*, 1996), but the functional significance of this motif remains to be investigated. The C-terminal domain of ExoP displays homologies to ATPases and contains a putative nucleotide binding site. This C-terminal domain is not present in the CLD proteins, but was found in homologous proteins involved in exopolysaccharide polymerization and export, e.g. EpsB of *Pseudomonas solanacearum* (Huang and Schell 1995), the Wzc protein of *Escherichia coli* K12 (Stevenson *et al.*, 1996) or AmsA of *Erwinia amylovora* (Bugert and Geider 1995). The N- and C-terminal domains of the proteins homologous to the *S. meliloti* SU47 ExoP protein can be encoded by one gene or by two consecutive genes as is the case for CpsC and CpsD of *Streptococcus thermophilus* Sfi6 (Stingele *et al.*, 1996). *S. meliloti* SU47 mutants carrying truncated *exoP** genes exclusively encoding the N-terminal domain produced reduced amounts of EPS I (Becker *et al.*, 1995a). This reduction could be suppressed by a mutation in the negative regulatory gene *exoR*. The ratio of high molecular weight to low molecular weight EPS I was significantly decreased in *exoP** mutants, whereas LMW EPS I was exclusively detected in *exoP*/exoR* double mutants (Becker *et al.*, 1995a). A similar phenotype was reported for *B. japonicum* mutants affected in the C-terminal domain of ExoP (P. Müller, Marburg, Germany, personal communication; Kosch *et al.*, 1995). It is tempting to speculate that ExoP proteins might be implicated in processes determining the ratio of high molecular weight to low molecular weight EPS.

The two proteins ExoK and ExsH are able to specifically degrade EPS I (York and Walker 1997). Both proteins were found to contribute to the production of Low Molecular Weight EPS I.

III.C.2. The Biosynthetic Pathway of EPS II in S. meliloti SU47

The biosynthetic pathway of EPS II has not yet been elucidated, but based on homologies of predicted *exp* gene products to proteins of known function some suggestions concerning their function and role in the biosynthesis of EPS II can be made. Four Exp proteins (ExpA7, ExpA8, ExpA9 and ExpA10) might be involved in the formation of dTDP-rhamnose (Becker *et al.*, 1997). The first intermediate of the dTDP-rhamnose biosynthetic pathway is dTDP-glucose, which might serve as the precursor for glucose in EPS II. In contrast, UDP-glucose was identified as the precursor for glucose in EPS I (Reuber and Walker 1993a), whereas UDP-galactose is needed for EPS I and EPS II production (Zhan *et al.*, 1989; Buendia *et al.*, 1991). The use of different NDP-glucose precursors in the biosynthesis of EPS I and EPS II might be important for the differential regulation of EPS production in *S. meliloti* SU47. Since Exp proteins probably involved in later steps of dTDP-rhamnose synthesis were found to be essential for EPS II production, it was proposed that dTDP-rhamnose is implicated in EPS II biosynthesis (Becker *et al.*, 1997), although this sugar was not detected in EPS II isolated from culture supernatants (Her *et al.*, 1990). Therefore, EPS II might contain only traces of rhamnose or this sugar might be involved in EPS II biosynthesis, but is not contained in the final polymer. Rhamnose might be required for the synthesis of a primer to start a new EPS II chain or as the donor of a sugar that might be added to the terminus of the EPS II polymer to stop chain extension.

Six potential glycosyl transferases are encoded by *exp* genes. Two of these (ExpA2 and ExpE2) might form β-glycosidic linkages, whereas the four other potential transferases (ExpA3, ExpC, ExpE4 and ExpE7) might form α-glycosidic bonds (Becker *et al.*, 1997). Since EPS II contains β-1→3 and α-1→3 linkages, at least one glycosyl transferase forming α-glycosidic linkages and one glycosyl transferase forming β-glycosidic bonds is expected to be involved in the formation of EPS II.

The ExpE1 protein displayed homologies to *Rhizobium* NodO proteins (Becker *et al.*, 1997). These proteins contain a C-terminal secretion signal and are secreted into the medium. They were shown to have Ca^{2+} binding activity as well as pore-forming activity in lipid bilayers (Economou *et al.*, 1990; Sutton *et al.*, 1994) indicating that ExpE1 might be implicated in EPS II export. As NodO, ExpE1 also contained a putative C-terminal secretion signal (Becker *et al.*, 1997). Homology data suggest that ExpD1 and ExpD2 are components of a transporter complex involved in signal peptide-independent secretion of a protein that contains a C-terminal secretion signal (Becker *et al.*, 1997). *S. meliloti* mutants in *expD1* and *expD2* were blocked in EPS II production indicating that these two gene products are essential for EPS II biosynthesis (Glazebrook and Walker 1989; Becker *et al.*, 1997).

IV. Functions of Rhizobial Exopolysaccharides in the Free-Living State and and Their Role in the Root Nodule Symbiosis

The rheological properties and the location of EPS, forming the outer layer of the cell surface, contribute to cell protection against environmental influences, attachment to surfaces, nutrient gathering and antigenicity. Furthermore, the structural diversity enables EPS to function as informational molecules in cell to cell communications. EPS play a specific role in the establishment of the symbiosis between rhizobia and legumes, particularly in the invasion of root nodules.

IV.A. ROLE OF EXOPOLYSACCHARIDES IN OSMOPROTECTION

Alterations of osmotic pressure induce modifications of EPS production in *S. meliloti* (Breedveld *et al.*, 1990b; Zevenhuizen and Faleschini 1991). A difference in the relative production of high molecular weight and low molecular weight EPS I depending on the environmental osmolarity was reported for *S. meliloti* SU47 but not for *A. tumefaciens* (Breedveld *et al.*, 1990b). In contrast to *A. tumefaciens* and *R. leguminosarum* strains of a number of biovars that combine low periplasmic concentrations of β-(1,2)-glucans with capsule formation in the stationary phase, *S. meliloti* does not form capsules and maintains constant high periplasmic β-(1,2)-glucan concentrations during different growth phases (Zevenhuizen 1981; 1984; Breedveld *et al.*, 1990b). It was suggested that deposition of capsule material and high periplasmic concentrations of β-(1,2)-glucans can be a mechanism for different members of the Rhizobiaceae to maintain cell integrity (Breedveld *et al.*, 1990b). At critical conditions of osmolarity, *R. leguminosarum* bv. *viciae* and *trifolii* strains secrete β-(1,2)-glucans into the culture medium (Breedveld *et al.*, 1990a). This was not observed for *S. meliloti* SU47 (Breedveld *et al.*, 1990b). In contrast, *S. meliloti* SU47 *exo* mutants deficient in EPS I production secrete considerable amounts of β-(1,2)-glucans (Leigh and Lee 1988). This might indicate that EPS I fulfills a function in osmoprotection and that the function of β-(1,2)-glucan secretion by EPS I deficient mutants is to maintain cell integrity.

IV.B. ROLE OF EXOPOLYSACCHARIDES IN ROOT NODULE SYMBIOSIS

The ability to establish an effective symbiosis is severely affected in mutants of many rhizobia deficient in EPS production. This is particularly the case for the formation of indeterminate nodules, although there are few examples for EPS mutants inducing effective indeterminate nodules, as was reported for EPS mutants of astragali rhizobia on *Astragalus sinicus* (Chen *et al.*, 1996) and for some EPS mutants of *Rhizobium* sp. strain TAL1145 on tree legumes (Parveen *et al.*, 1997). EPS deficient mutants of *R. leguminosarum* bv. *viciae* and *trifolii* as well as *S. meliloti*, *Rhizobium* sp. NGR234 and *Rhizobium* sp. strain TAL1145 do not nodulate their host plants, which form indeterminate nodules or induce nodules devoid of bacteria due to the formation of abortive infection threads (Chakravorty *et al.*, 1982; Borthakur *et al.*, 1986; 1988; Chen *et al.*, 1985; Finan *et al.*, 1985; Leigh *et al.*, 1985; Djordjevic *et al.*, 1987; Diebold and Noel 1989; Bialek *et al.*, 1995; Skropuska *et al.*, 1995; Parveen *et al.*, 1996; Rolfe *et al.*, 1996). EPS mutants of broad host range rhizobia able to nodulate plants forming indeterminate and determinate nodules were described which displayed different symbiotic phenotypes on these different hosts. Callus-like structures were induced by EPS deficient mutants of *Rhizobium* sp. NGR234 on *Leucaena leucocephala* plants which form indeterminate nodules, whereas non-nitrogen fixing nodules were induced on

Macroptilium atropurpureum forming determinate nodules (Djordjevic *et al.*, 1987). Furthermore, EPS mutants of *M. loti* induced nitrogen-fixing determinate nodules on *Lotus pendiculatus* and ineffective callus-like structures on *Leucaena leucocephala* (Hotter and Scott 1991). Therefore, it was proposed that EPS is not essential for nodulation of plants forming determinate nodules (Borthakur *et al.*, 1986; Diebold and Noel 1989; Ko and Gayda 1990; Hotter and Scott 1991). A correlation of the different symbiotic phenotypes of EPS mutants with the differences in infection thread formation in determinate and indeterminate nodules was suggested (Kijne 1992). Although EPS seems to be not essential for the formation of determinate nitrogen-fixing nodules on *Glycine max*, early stages of the development of nodules infected with *B. japonicum* EPS mutants were disturbed. Delayed nodulation and reduced competitiveness were reported (Bhagwat *et al.*, 1991; Kosch *et al.*, 1994; Parniske *et al.*, 1993; 1994; Eggleston *et al.*, 1996).

In summary, the deficiencies of EPS mutants in the establishment of the symbiosis with plants forming indeterminate nodules seem to be more severe than with a host forming determinate nodules. The role of EPS in the symbiotic interaction of *S. meliloti* SU47 and alfalfa was extensively analyzed. It will be discussed in more detail and will be compared to related symbiotic interactions.

IV.B.1. Symbiotic Phenotypes of S. meliloti SU47 Mutants Deficient in EPS Production

Nodules induced by *exo* mutants of *S. meliloti* SU47 on *Medicago sativa* were devoid of bacteria and failed to fix nitrogen (Finan *et al.*, 1985; Leigh *et al.*, 1985). These nodules are termed pseudonodules. Delayed root hair curling and infection threads that aborted within peripheral cells of the developing nodule were described (Yang *et al.*, 1992). Nodules developed more frequently on secondary roots than on the primary root and lacked a discrete apical meristem (Yang *et al.*, 1992). In contrast, several centers of mitotic activity were frequently present. Pseudonodules display brown necrotic areas on their surfaces indicating plant defense responses (Niehaus *et al.*, 1993). Cortical cell walls of pseudonodules were abnormally thick and incrusted with autofluorescent material. In comparison to wild-type nodules an increase of phenolic compounds incorporated into the cell walls of pseudonodules was found (Niehaus *et al.*, 1993). An increase of peroxidase activity and increased levels of

phenylalanine ammonia-lyase (PAL), 3-O-methyltransferase (COMT) and isoflavone reductase (IFR) transcripts in pseudonodules indicate a typical plant defense response (Niehaus *et al.*, 1993; 1994). Niehaus *et al.*, (1993) suggested that plant defense responses induced by EPS I deficient mutants of *S. meliloti* are responsible for their impairment in infection. Van Workum *et al.*, (1995) reported that reduction of ethylene production by *Vicia sativa* ssp. *nigra* plants enabled infection by *R. leguminosarum* bv. *viciae* mutants deficient in EPS production. They suggested that ethylene induced during infection influences root cell growth and therefore inhibits nodule development and infection. Delayed nodulation might result in accumulation of ethylene that probably causes the defects in nodulation and infection. In addition, van Workum *et al.*, (1995) proposed that plant defense responses induced by EPS I deficient mutants of *S. meliloti* (Niehaus *et al.*, 1993) might be a side effect of ethylene production and not primarily responsible for the defective infection of these mutants. EPS deficient mutants of *R. leguminosarum* bv. *viciae* impaired in root nodule formation on *V. sativa* ssp. *nigra* (van Workum *et al.*, 1995) were delayed in root hair curling and infection thread formation on *Trifolium repens* and *Trifolium subterraneum* (Rolfe *et al.*, 1996). Only empty nodules were induced on *T. subterraneum*. If ethylene production of these *Trifolium* plants is induced by these mutants, its main effect is probably on nodule development and not on nodule infection.

Delayed infections of lobes of pseudonodules by EPS I deficient *S. meliloti* SU47 mutants were reported (Niehaus *et al.*, 1993). Six weeks post inoculation approximately 25 % of the plants carried few infected nodules. Infection thread like structures were observed in cortical cells of these nodules (Niehaus *et al.*, 1993). In contrast to infection threads of wild type nodules these structures were thicker and surrounded by a rigid wall. Niehaus *et al.*, (1993) proposed that EPS I deficient mutants can overcome the defense response of alfalfa plants after prolonged incubation. A delayed infection was also reported for *exoH* mutants of *S. meliloti* SU47 producing non-succinylated EPS I, although these nodules already appeared after three to five weeks (Leigh *et al.*, 1987; Leigh and Lee 1988). This implies that *exoH* mutants might overcome the plant defense earlier than *S. meliloti* SU47 mutants which fail to produce EPS I. Since the *exoH* transposon mutants analyzed almost completely lacked the low molecular weight EPS I fraction (Leigh *et al.*, 1987; Leigh and Lee 1988), it is not clear if the lack of succinylation or EPS I oligosaccharides is responsible for the deficiency in infection.

Overproduction of EPS I can also have severe effects on nodule development. Whereas *exoX* mutants of *S. meliloti* SU47 overproducing EPS I are able to infect alfalfa nodules (Zhan and Leigh 1990; Reed *et al.*, 1991a; Müller *et al.*, 1993), the corresponding *R. leguminosarum* bv. *phaseoli* mutants in *psiA* induced non-nitrogen fixing nodules on *Phaseolus* beans (Borthakur *et al.*, 1985). Negative regulation of EPS I production occuring at an earlier level of *exo* gene expression might be responsible for the ability of *S. meliloti* SU47 *exoX* mutants to induce effective nodules. In contrast to *exoS* mutants of *S. meliloti* SU47 that predominantly induced nitrogen-fixing nodules on alfalfa although some variation in the fixation phenotype was reported, *exoR* mutants that are extreme overproducers of EPS I were unable to invade nodules successfully (Doherty *et al.*, 1988; Ozga *et al.*, 1994). Suppressor mutations in the *exoS* locus allowed *exoR* mutants to invade nodules due to reduction of EPS I overproduction (Ozga *et al.*, 1994). EPS I production might be deleterious in the bacteroid state.

IV.B.2. Role of EPS I and EPS II in the Invasion of Alfalfa Nodules by S. meliloti SU47

Coinoculation experiments showed that *nod* gene mutants unable to induce nodules can help EPS deficient mutants in nodule invasion. *L. leucocephala* plants were coinoculated with Sym plasmid cured mutants of *Rhizobium* sp. NGR234 deficient in nodule induction and *exo* mutants that failed to produce EPS (Chen and Rolfe 1987; Djordjevic *et al.*, 1987). Nitrogen fixing nodules were formed that contained both mutants. Similar results were obtained from coinoculation experiments of *M. sativa* with *S. meliloti* SU47 *exo* and *nod* mutants (Klein *et al.*, 1988; Pühler *et al.*, 1988; Kapp *et al.*, 1990). The observation that nitrogen-fixing nodules always contained both mutants suggested that both coinoculation partners have to be present during the whole infection process.

In some symbiotic systems addition of purified EPS restored the ability of EPS mutants to infect nodules. Djordjevic *et al.*, (1987) reported that the exogeneous supplementation with purified EPS or EPS derived oligosaccharides restored the ability of *Rhizobium* sp. NGR234 EPS deficient mutants to induce nitrogen fixing nodules on *L. leucocephala*.

Purified EPS or oligosaccharides also promoted invasion of *Trifolium repens* nodules by EPS mutants of *R. leguminosarum* bv. *trifolii* (Djordjevic *et al.*, 1987). Heterologous EPS did not promote nodule invasion by EPS mutants indicating that EPS fulfills a specific function (Djordjevic *et al.*, 1987). The symbiotic deficiency of *S. meliloti* SU47 EPS I mutants was overcome by supplementation with low molecular weight EPS I (Urzainqui and Walker 1992; Battisti *et al.*, 1992). High molecular weight EPS I did not promote nodule invasion of *S. meliloti* SU47 *exo* mutants (Battisti *et al.*, 1992). The most charged tetramer of the EPS I octasaccharide repeating unit was identified as the active compound in the low molecular weight fraction (Battisti *et al.*, 1992). Since induction of nitrogen-fixing nodules by *exo* mutants exogenously supplemented with low molecular weight EPS I was delayed in comparison to the formation of fixing nodules by the wild type, it was suggested that EPS I is needed within the nodule during nodule invasion or that High Molecular Weight EPS I might also contribute to the invasion process (Battisti *et al.*, 1992).

S. meliloti SU47 double mutants blocked in EPS I biosynthesis which produced EPS II due to the *expR* mutation induced nitrogen fixing nodules on *M. sativa*, but not on *Medicago cerulea, Medicago truncatula, Melilotus alba* and fenugreek (Glazebrook and Walker 1989). All these plants form effective nodules after inoculation with the *S. meliloti* SU47 wild type strain. It was therefore concluded that EPS II can substitute for EPS I in invasion of *M. sativa* nodule. This observation was in contradiction to the symbiotic phenotype of the *S. meliloti* SU47 *exoY/mucR* double mutant which was also blocked in EPS I production and synthesized EPS II (Keller *et al.*, 1995). This mutant induced pseudonodules on *M. sativa* (Keller *et al.*, 1995). González *et al.*, (1996) reported that the *expR* mutation caused the production of a high molecular weight and a low molecular weight EPS II fraction, whereas the *mucR* mutant exclusively produced the high molecular weight EPS II form. The ability of *S. meliloti* SU47 mutants deficient in EPS I and EPS II biosynthesis could be restored by the exogenous addition of low molecular weight EPS II, but not by supplementation with high molecular weight EPS II. A purified subfraction containing oligosaccharides of 15 to 20 EPS II repeating units promoted nodule invasion (González *et al.*, 1996).

A role of Low Molecular Weight EPS I in the suppression of plant defence responses was proposed

(Niehaus *et al.*, 1993; 1994). To test this hypothesis elicitor responsive cell cultures of the host plant *M. sativa* and as a control of the non host plants *Nicotiana tabacum* and *Lycopersicum esculentum* were established and several purified substances were analyzed for possible elicitor or suppressor functions (Niehaus *et al.*, 1997). Apart from other defense related reactions, all cell cultures reacted to the addition of small amounts of the non specific yeast-elicitors with a strong transient alkalinization of their culture medium. In alfalfa cell cultures the elicitor induced alkalinization could be suppressed by the simultaneous application of Low Molecular Weight EPS I. Neither High Molecular Weight EPS I, High Molecular Weight EPS II nor the heterologous EPS xanthan from *Xanthomonas campestris* provoked a reduction of the elicitor response. None of the carbohydrate preparations were able to suppress the elicitor-induced alkalinization in the cell cultures of the non host plants tobacco and tomato. These data provide evidence for a specific function of Low Molecular Weight EPS I as a suppressor of plant defense-related processes.

V. Conclusions and Perspectives

On the one hand, EPS are ubiquitous molecules produced by many bacteria including the *Rhizobiaceae*. They are most likely important for the survival of bacteria in many aspects, since these molecules may contribute to protection of the cells, attachment to surfaces and nutrient gathering (Costerton *et al.*, 1987; Sutherland 1988; Whitfield 1988; Beveridge and Graham 1991). On the other hand, extracellular oligosaccharides derived from EPS biosynthetic pathways play specific roles in the establishment of the symbiotic interaction between rhizobia and the host plant.

The ubiquitous character of EPS suggests that these molecules were primarily evolved for functions in the free-living state of bacteria and were later recruited for functions in the bacteria-plant interaction. Complex regulatory mechanisms controlling EPS biosynthesis may be necessary to suit the requirements of diverse EPS functions.

Biochemical and genetic approaches resulted in detailed models for the biosynthetic pathways of EPS in some systems. This knowledge will certainly facilitate the analysis of EPS biosynthesis in other systems. In contrast, the regulation of EPS production is poorly understood. Many different regulators were identified in rhizobia. However, the environmental factors influencing EPS production and the mechanisms of regulation need still to be investigated.

Understanding how rhizobia use the various structural properties of EPS in the free-living and the symbiotic state may give new insights in the complex interaction between bacteria and their environment.

VI. References

Aman, P., McNeil, M., Franzen, L.-E., Darvill, A.G. and Albersheim, P. (1981) Carbohydr. Res. 95, 263-282.

Amemura, A., Harada, T., Abe, M. and Higashi, S. (1983) Carbohydr. Res. 115, 165-174.

An, J., Carlson, R.W., Glushka, J. and Streeter, J.G. (1995) Carbohydr. Res. 269, 303-317.

Arias, A. and Cervenansky, C. (1986) J. Bacteriol. 167, 1092-1094.

Astete, S.G. and Leigh, J.A. (1996) Mol. Plant-Microbe Interact. 9, 395-400.

Barnett, M.J. and Long, S.R. (1997) Mol. Plant-Microbe Interact. 10, 550-559.

Battisti, L., Lara, J.C. and Leigh, J.A. (1992) Proc. Natl. Acad. Sci. USA 89, 5625-5629.

Bayer, M.E. (1979) in M. Inouye (ed.), Bacterial outer membranes: biogenesis and functions, John Wiley & Sons, Inc., New York, USA, 167-202.

Bayer, M.E. (1991) J. Struct. Biol. 107, 268-280.

Becker, A., Kleickmann, A., Arnold, W. and Pühler, A. (1993a) Mol. Gen. Genet. 238, 145-154.

Becker. A., Kleickmann, A., Keller, M., Arnold, W. and Pühler, A. (1993b) Mol. Gen. Genet. 241, 367-379.

Becker, A., Kleickmann, A., Küster, H., Keller, M., Arnold, W. and Pühler, A. (1993c) Mol. Plant-Microbe Interact. 6, 735-744.

Becker, A., Niehaus, K. and Pühler, A. (1995a) Mol. Microbiol. 16, 191-203.

Becker, A., Küster, H., Niehaus, K. and Pühler, A. (1995b) Mol. Gen. Genet. 249, 487-497.

Becker, A., Rüberg, S., Küster, H., Roxlau, A.A., Keller, M., Ivashina, T., Cheng, H.-P., Walker, G.C. and Pühler, A. (1997) J. Bacteriol.179, 1375-1384.

Bertram-Drogatz, P.A., Rüberg, S., Becker, A. and Pühler, A. (1997a) Mol. Gen. Genet. 254, 529-5388.

Bertram-Drogatz, P.A., Quester, I., Becker, A. and Pühler, A. (1997b) Mol. Gen. Genet., submitted.

Beveridge, T.J. and Graham, L.L. (1991) Microbiol. Rev. 55, 684-705.

Bhagwat, A.A., Tully, R.E. and Kleister D.L. (1991) Appl. Environ. Microbiol. 57, 3496-3501.

Bialek, U., Skorupska, A., Yang, W.-C., Bisseling, T. and van Lammeren, A.A.M. (1995) Planta 197, 184-192.

Bittinger, M.A., Milner, J.L., Savilla, B.J. and Handelsman, J. (1997) Mol. Plant-Microbe Interact. 10, 180-186.

Borthakur, D., Downie, J.A., Johnston, A.W.B. and Lamb, J.W. (1985) Mol. Gen. Genet. 200, 278-282.

Borthakur, D., Barber, C.E., Lamb, J.W., Daniels, M.J., Downie, J.A. and Johnston, A.W.B. (1986) Mol. Gen. Genet. 203, 320-323.

Borthakur, D. and Johnston, A.W.B. (1987) Mol. Gen. Genet. 207, 149-154.

Borthakur, D., Barker, R.F., Latchford, J.W., Rossen, L. and Johnston, A.W.B. (1988) Mol. Gen. Genet. 213, 155-162.

Breedveld, M.W., Zevenhuizen, L.P.T.M. and Zehnder, A.J.B. (1990a) Appl. Environ. Microbiol. 56, 2080-2086.

Breedveld, M.W., Zevenhuizen, L.P.T.M. and Zehnder, A.J.B. (1990b) J. Gen. Microbiol. 136, 2511-2519.

Breedveld, M.W., Canter Cremers, H.C.J., Batley, M., Posthumus, M.A., Zevenhuizen, L.P.T.M., Wijffelman, C.A. and Zehnder, A.J.B. (1993) J. Bacteriol. 175, 750-757.

Breedveld, M.W. and Miller, K.J. (1994) Microbiol. Rev. 58, 145-161.

Brightwell, G., Hussain, H., Tiburtius, A., Yeoman, K.H. and Johnston A.W.B. (1995) Mol. Plant-Microbe Interact. 8, 747-754.

Buendia, A.M., Enenkel, B., Köplin, R., Niehaus, K., Arnold, W. and Pühler, A. (1991) Mol. Microbiol. 5, 1519-1530.

Bugert, P. and Geider, K. (1995) Mol. Microbiol. 15, 917-933.

Canter Cremers, H.C.J., Batley, M., Redmond, J.W., Eydems, L., Breedveld, M.W., Zevenhuizen, L.P.T.M., Pees, E., Wijffelman, C.A. and Lugtenberg B.J.J. (1990) J. Biol. Chem. 265, 21122-21127.

Canter Cremers, H.C.J., Batley, M., Redmond, J.W., Wijfjes, A.H.M., Lugtenberg, B.J.J., Wijffelman C.A. (1991a) J. Biol. Chem. 266, 9556-9564.

Canter Cremers, H.C.J., Stevens, K., Lugtenberg, B.J.J., Wijffelman, C.A., Batley, M., Redmond, J.W., Breedveld, M.W. and Zevenhuizen, L.P.T.M. (1991b) Carbohydr. Res. 218, 185-200.

Capage, M.A., Doherty, D.H., Betlach, M.R. and Vanderslice, R.W. (1987). International patent WO87/05938.

Chakravorty, A.K., Zurkowski, W., Shine, J. and Rolfe, B.G. (1982) J. Mol. Appl. Genet. 1, 585-596.

Charles, T.C. and Finan, T.M. (1991) Genetics 127, 5-20.

Chen, H., Batley, M., Redmond, J.W. and Rolfe, B.G. (1985) J. Plant Physiol. 120, 331-349.

Chen, H. and Rolfe, B.G. (1987) J. Plant Physiol. 127, 307-322.

Chen, H., Gray, J.X., Najudu, M., Djordjevic, M.A., Batley, M., Redmond, J.W. and Rolfe, B.G. (1988) Mol. Gen. Genet. 212, 310-316.

Chen, H., Long, B.-G. and Song, H.-Y. (1996) Plant and Soil 179, 217-221.

Cooley, M.B., d'Souza, M.R. and Kado, C.I. (1991) J. Bacteriol. 173, 2608-2616.

Coplin, D.L. and Cook, D. (1990) Mol. Plant-Microbe Interact. 3, 271-279.

Costerton, J.W., Cheng, K.J., Geesey, G.G., Ladd, T.I., Nickel, J.C., Dasgupta, M. and Marrie, T.J. (1987) Ann. Rev. Microbiol. 41, 435-464.

Diebold, R. and Noel, K.D. (1989) J. Bacteriol. 171, 4821-4830.

Djordjevic, S.P., Rolfe, B.G., Batley, M. and Redmond, J.W. (1986) Carbohydr. Res. 148, 87-99.

Djordjevic, S.P., Chen, H., Batley, M., Redmond, J.W. and Rolfe, B.G. (1987) J. Bacteriol. 169, 53-60.

Doherty, D., Leigh, J., Glazebrook, J. and Walker, G.C. (1988) J. Bacteriol. 170, 4249-4256.

d'Souza-Ault, M.R., Cooley, M.B. and Kado, C.I. (1993) J. Bacteriol. 175, 3486-3490.

Dudman, W.F. (1978) Carbohydr. Res. 66, 9-23.

Eggleston, G., Huber, M.C., Liang, R., Karr, A.L. and Emerich, D.W. (1996) Mol. Plant-Microbe Interact. 9, 419-423.

Finan, T.M., Hirsch, A.M., Leigh, J.A., Johansen, E., Kuldau, G.A., Deegan, S., Walker, G.C. and Signer, E.R. (1985) Cell 40, 869-877.

Finan, T.M., Kunkel, B., de Vos, G.F. and Signer, E.R. (1986) J. Bacteriol. 167, 66-72.

Finke, A., Bronner, D., Nikolaev, A.J., Jann, B. and Jann, K. (1991) J. Bacteriol. 173, 4088-4094.

Freiberg, C., Fellay, R., Bairoch, A., Broughton, W.J., Rosenthal, A. and Perret, X. (1997) Nature 387, 394-401.

Frosch, M., Müller, D., Bousset, K. and Müller, A. (1992) Infection and Immunity 60, 798-803.

Fukasawa, T. and Nikaido, H. (1961) Biochim. Biophys. Acta 48, 470-483.

Glazebrook, J. and Walker, G.C. (1989) Cell 56, 661-672.

Glucksmann, M.A., Reuber, T.L. and Walker, G.C. (1993a) J. Bacteriol. 175, 7033-7044.

Glucksmann, M.A., Reuber, T.L. and Walker, G.C. (1993b) J. Bacteriol. 175, 7045-7055.

Gil-Serrano, A., del Junco, A.S., Tejero-Mateo, P., Megias, M. and Caviedes, M.A. (1990) Carbohydr. Res. 204, 103-107.

González, J.E., Reuhs, B.L. and Walker, G.C. (1996a) Proc. Natl. Acad. Sci. USA 93, 8636-8641.

González, J.E., York, G.M. and Walker, G.C. (1996b) Gene 179, 141-146.

Gray, J.X., Djordjevic, M.A. and Rolfe, B.G. (1990) J. Bacteriol. 172, 193-203.

Gray, J.X. and Rolfe, B.G. (1990) Mol. Microbiol. 4, 1425-1431.

Gray, J.X. and Rolfe, B.G. (1992) Arch. Microbiol. 157, 521-528.

Her, G.-R., Glazebrook, J., Walker, G.C., Reinhold, V.N. (1990) Carbohydr. Res.198, 305-312.

Hisamatsu, M., Abe, J., Amemura, A. and Harada, T. (1980) Agric. Biol. Chem. 44,1049-1055.

Hollingworth, R.I., Dazzo, F.B., Hallenga, K., Musselman, B. (1988) Carbohydr. Res. 172, 97-111.

Hotter, G.S. and Scott, D.B. (1991) J. Bacteriol. 173, 851-859.

Huang, J, and Schell, M. (1995) Mol. Microbiol. 16, 977-989.

Hynes, M.F., Simon, R., Müller, P., Niehaus, K., Labes, M. and Pühler, A. (1986) Mol. Gen. Genet. 202, 356-362.

Ielpi, L., Couso, O. and Dankert, M.A. (1993) J. Bacteriol. 175, 2490-2500.

Ivashina, T.V., Khmelnitsky, M.I., Shlyapnikov, M.G., Kanapin, A.A. and Ksenzenko, V.N. (1994) Gene 150, 111-116.

Ivashina, T.V., Sadykhov, M.R., Chatuev, B.M., Dmitriev, V.V., Suzina, N.E., Duda, V.I. and Ksenzenko, V.N. (1995) in I.A. Tikhonovich, N.A. Provorov, V.I. Romanov, and W.E. Newton (eds.), Nitrogen fixation: fundamentals and aplications, Kluwer Academic Publishers, Dordrecht, Netherlands, 404.

Ivashina, T., Sadykov, M., Senchenkova, S., Shashkov, A., Shibaev, V. and Ksenzenko, V. (1996) in A. Wójtowicz, J. Stepkowska, and A. Szlagowska (eds.), 2nd European Nitrogen Fixation Conference & NATO Advanced Research Workshop, Biological fixation of nitrogen for ecology and sustainable agriculture, Scientific Publishers OWN, Poznan, Poland, 127.

Kapp, D., Niehaus, K., Quandt, J., Müller, P. and Pühler, A. (1990) Plant Cell 2, 139-151.

Keller, M., Roxlau, A., Weng, W.M., Schmidt, M., Quandt, J., Niehaus, K., Jording, D., Arnold, W. and Pühler, A. (1995) Mol. Plant-Microbe Interact. 8, 267-277.

Kido, N., Torgov, V.I., Sugiyama T., Uchiya, K., Sugihara, H., Komatsu, T., Kato, N. and Jann, K. (1995) J. Bacteriol. 177, 2178-2187.

Kijne, J.W. (1992) in G.S. Stacey, R.H. Burris and J. H.J. Evans (eds.), Biological nitrogen fixation, Chapman and Hall, New York, USA, 349-398.

Klein, S., Hirsch, A.M., Smith, C.A. and Signer, R. (1988) Mol. Plant-Microbe Interact. 1, 94-100.

Ko, Y.H. and Gayda, R. (1990) J. Gen. Microbiol.136, 105-113.

Kosch, K., Jacobi, A., Parniske, M., Werner, D. and Müller, P. (1994) Z. Naturforsch. 49c, 727-736.

Kosch, K., Batinic, T., Niehaus, K., Werner, D. and Müller, P. (1995) in I.A. Tikhonovich, N.A. Provorov, V.I. Romanov, and W.E. Newton (eds.), Nitrogen fixation: fundamentals and applications, Kluwer Academic Publishers, Dordrecht, Netherlands, 329.

Król, J., Wojcieszuk, A. and Skorupska, A. (1994) in G.B. Kiss, and G. Endre (eds.), Proceedings of the 1st European Nitrogen Fixation Conference, Officina Press, Szeged, Hungary, 319.

Król, J., Janczarek, M. and Skorupska, A. (1996) in A. Wójtowicz, J. Stepkowska, and A. Szlagowska (eds.), 2nd European Nitrogen Fixation Conference & NATO Advanced Research Workshop, Biological fixation of nitrogen for ecology and sustainable agriculture, Scientific Publishers OWN, Poznan, Poland, 126.

Kröncke, K.D., Boulnois, G., Roberts, I., Bitter-Suermann, D., Golecki, J.R., Jann, B., Jann, K. (1990a) J. Bacteriol. 172, 1085-1091.

Kröncke, K.D., Golecki, J.R. and Jann, K. (1990b) J. Bacteriol. 172, 3469-3472.

Latchford, J.W., Borthakur, D. and Johnston, A.W.B. (1991) Mol. Microbiol. 5, 2107-2114.

Leigh, J.A., Signe, E.R. and Walker, G.C. (1985) Proc. Natl. Acad. Sci. USA 82, 6231-6235.

Leigh, J.A., Reed, J.W., Hanks, J.F., Hirsch, A.M. and Walker, G.C. (1987) Cell 51, 579-587.

Leigh, J.A. and Lee, C.C. (1988) J. Bacteriol. 170, 3327-3332.

Leigh, J.A., Battisti, L., Lee, C.C., Ozga, D.A., Zhan, H. and Astete, S. (1993) in E.W. Nester, and D.P.S. Verma (eds.), Advances in molecular genetics of plant-microbe interactions, Kluwer Academic Publishers, Dordrecht, Netherlands, 175-181.

Long, S., Reed, J.W., Himawan, J. and Walker, G.C. (1988a) J. Bacteriol. 170, 4239-4248.

Long, S., McCune, S. and Walker, G.C. (1988b) J. Bacteriol. 170, 4257-4265.

Lopez-Lara, I.M., Orgambide, G., Dazzo, F.B., Olivares, J. and Toro N. (1993) J. Bacteriol. 175, 2826-2832.

Matulová, M., Toffanin, R., Navarini, L., Gilli, R., Paoletti, S. and Cesáro, A. (1994) Carbohydr. Res. 265, 167-179.

May, T.B., Shinabarger, D., Maharaj, R., Kato, J., Chu, L., DeVault, J.D., Roychoudhury, S., Zielinski, N., Berry, A., Rothmel, R.K., Mistra, T.K. and Chakrabarty A.M. (1991) Clin. Microbiol. Rev. 4, 191-206.

May, T.B. and Chakrabarty, A.M. (1994) Trends Microbiol. 2, 151-156.

McGrath, B.C. and Osborn, M.J. (1991) J. Bacteriol. 173, 649-654.

McNeil, M., Darvill, J., Darvill, A.G., Albersheim, P., van Veen, R., Hooykaas, P., Schilperoort, R. and Dell, A. (1986) Carbohydr. Res. 146, 307-326.

Mello, A. and Glase, L. (1965) J. Biol. Chem. 240, 398-405.

Mimmack, M.L., Hong, G.F. and Johnston, A.W.B. (1994a) Microbiology 140, 455-461.

Mimmack, M.L., Borthakur, D., Jones, M.A., Downie, J.A. and Johnston, A.W.B. (1994b) Microbiology 140, 1223-1229.

Minamisawa, K. (1989) Plant Cell Physiol. 30, 877-884.

Mort, A.J. and Bauer, W.D. (1980) Plant Physiol. 66, 158-163.

Mort, A.J. and Bauer, W.D. (1982) J. Biol. Chem. 257, 1870-1875.

Müller, P., Hynes, M., Kapp, D., Niehaus, K. and Pühler, A. (1988) Mol. Gen. Genet. 211, 17-26.

Müller, P., Keller, M., Weng, W.M., Quandt, J., Arnold, W. and Pühler, A. (1993) Mol. Plant-Microbe Interact. 6, 55-65.

Mulligan, J.T. and Long, S.R. (1988) Genetics 122, 7-18.

Newcomb, W. (1981) in K.L. Giles, and A.G. Atherly (eds.), Biology of the Rhizobiaceae. Academic Press, New York, USA.

Niehaus, K., Kapp, D. and Pühler, A. (1993) Planta 190, 415-425.

Niehaus, K., Kapp, D., Lorenzen, J., Meyer-Gattermann, P., Sieben, S. and Pühler, A. (1994) Acta Horticulturae 381, 258-264.

Niehaus, K., Baier, R., Kohring,B., Flaschl, E. and Pühler. A. (1997) in A. Legocki, H. Bothe, and A. Pühler (eds.), Biological fixation of nitrogen for ecology and sustainable agriculture, Springer Verlag, Heidelberg, Germany, 110-114.

Østerås, M., Stanley, J. and Finan, T.M. (1995) J. Bacteriol. 177, 5485-5494.

Ozga, D.A., Lara, J.C. and Leigh, J.A. (1994) Mol. Plant-Microbe Interact. 7, 758-765.

Parniske, M., Kosch, K., Werner, D. and Müller, P. (1993) Mol. Plant-Microbe Interact. 6, 99-106.

Parniske, M., Schmidt, P.E., Kosch, K. and Müller, P. (1994) Mol. Plant-Microbe Interact. 7, 631-638.

Parveen, N., Webb, D.T. and Borthakur, D. (1996) Mol. Plant-Microbe Interact. 9, 364-372.

Parveen, N., Webb, D.T. and Borthakur, D. (1997) Microbiology 143, 1959-1967.

Pühler, A., Enenkel, B., Hillemann, A., Kapp, D., Keller, M., Müller, P., Niehaus, K., Priefer, B., Quandt, J. and Schmidt, C. (1988) in H. Bothe, F. de Bruijn, W.E. Newton (eds.), Nitrogen fixation: hundred years after, Gustav Fischer, New York, USA, 423-430.

Puvanesarajah, V., Schell, F.M., Gerhold, D. and Stacey, G. (1987) J. Bacteriol. 169, 137-141.

Reed, J.W. and Walker, G.C. (1991a) Genes & Development 5, 2274-2287.

Reed, J.W. and Walker, G.C. (1991b) J. Bacteriol. 173, 664-677.

Reed, J.W., Capage, M. and Walker, G.C. (1991a) J. Bacteriol. 173, 3776-3788.

Reed, J.W., Gazebrook, J. and Walker, G.C. (1991b) J. Bacteriol. 173, 3789-3794.

Reeve, W.G., Dilworth, M.J., Tiwari, R.P., Glenn, A.R. (1997) Microbiology 143, 1951-1958.

Reinhold, B.B., Chan, S.Y., Reuber, T.L., Marra, A., Walker, G.C. and Reinhold, V.N. (1994) J. Bacteriol. 176, 1997-2002.

Reuber, T.L., Long, S. and Walker, G.C. (1991) J. Bacteriol. 173, 426-434.

Reuber, T.L. and Walker, G.C. (1993a) Cell 74, 269-280.

Reuber, T.L. and Walker, G.C. (1993b) J. Bacteriol. 175, 3653-3655.

Rohr, T.E. and Troy, F.A. (1980) J. Biol. Chem. 255, 2332-2342.

Rolfe, B.G., Carlson, R.W., Ridge, R.W., Dazzo, F.B., Mateos, P.F. and Pankhurst, C.E. (1996) Aust. J. Plant Physiol. 23, 285-303.

Sanchez-Andujar, B., Philip-Hollingsworth, S., Dazzo, F.B. and Palomares, A.J. (1997) Mol. Gen. Genet. 255, 131-140.

Shibaev, V.N. (1986) Adv. Carboh. Chem. 44, 277-339.

Skorupska, A, Bialek, U, Urbanik-Sypniewska, T. and van Lammeren, A (1995) J. Plant Physiol. 147, 93-100.

Stevenson, G. andrianopoulos K., Hobbs, M. and Reeves, P.R. (1996) J. Bacteriol. 178, 4885-4893.

Stingele, F., Neesler, J.-R. and Mollet, B. (1996) J. Bacteriol. 178, 1680-1690.

Streeter, J.G., Salminen, S.O., Whitmoyer, R.E. and Carlson, R.W. (1992) Appl. Environ. Microbiol. 58, 607-613.

Streeter, J.G., Salminen, S.O., Beuerlein, J.E. and Schmidt, W.H. (1994) Appl. Environ. Microbiol. 60, 2939-2943.

Sutherland, I.W. (1982) Adv. Microbial Physiol. 23, 79-150.

Sutherland, I.W. (1988) Int. Rev. Cytol. 113, 187-231.

Sutton, J.M., Lea, E.J.A. and Downie, J.A. (1994) Proc. Natl. Acad. Sci. USA 91, 9990-9994.

Tiburtius, A, de Luca, N.G., Hussain, H. and Johnston, A.W.B. (1996) Microbiology 142, 2621-2629.

Tolmasky, M.E., Staneloni, R.J., Ugalde, R.A. and Leloir, L.F. (1980) Arch. Biochem. Biophys. 203, 358-364.

Tolmasky, M.E., Staneloni, R.J. and Leloir, L.F. (1982) J. Biol. Chem. 257, 6751-6757.

Urzainqui, A, Walker, G.C. (1992) J. Bacteriol. 174, 3403-3406.

Uttaro, A.D., Cangelosi, G.A., Geremia, R.A., Nester, E.W. and Ugalde, R.A. (1990) J. Bacteriol. 172, 1640-1646.

Valla, S., Coucheron, D.H., Fjaervik, E., Kjosbakken, J., Weinhouse, H., Ross, P., Amikam, D. and Benziman, M. (1989) Mol. Gen. Genet. 217, 26-30.

Vanderslice, R.W., Doherty, D.H., Capage, M.A., Betlach, M.R., Hassler, R.A., Henderson, N.M., Ryan-Graniero, J. and Tecklenburg, M. (1989) in V. Crescenzi, I.C.M. Dea, S. Paoletti, S.S. Stivala, and I.W. Sutherland (eds.), Biomedical and biotechnological advances in industrial polysaccharides, Gordon and Breach, New York, USA, 145-156.

van Workum, W.A.T., van Brussel, A.A.N., Tak, T., Wijffelman, C.A. and Kijne, J.W. (1995) Mol. Plant-Microbe Interact. 8, 278-285.

van Workum, W.A.T., Canter Cremers, H.C.J., Wijfjes, A.H.M., van der Kolk, C., Wijffelman C.A. and Kijne, J.W. (1997) Mol. Plant-Microbe Interact. 10, 290-301.

Weisgerber, C. and Troy, F.A. (1990) J. Biol. Chem. 265, 1578-1587.

Whitfield, C. (1988) Can. J. Microbiol. 34, 415-420.

Whitfield, C. and Valvano, M.A. (1993) Adv. Microbial Physiol. 35, 135-246.

Whitfield, C. (1995) Trends Microbiol. 3, 178-185.

Wong, H.C., Fear, A.L., Calhoon, R.D., Eichinger, G.H., Mayer, R., Amikam, D., Benziman, M., Gelfand, G.H., Meade, J.H., Emerick, A.W., Bruner, R., Ben-BAssat, A. and Tal, R. (1990) Proc. Natl. Acad. Sci. USA 87, 8130-8134.

Yang, C., Signer, E.R. and Hirsch, A.M. (1992) Plant Physiol. 98, 143-151.

York, G.M. and Walker, G.C. (1997) Mol. Microbiol. 25, 117-134

Zevenhuizen, L.P.T.M. (1981) Antonie van Leeuwenhoek 47, 481-497.

Zevenhuizen, L.P.T.M. (1984) Appl. Microbiol. Biotech. 20, 393-399.

Zevenhuizen, L.P.T.M. and van Neerven, A.R.W. (1983) Carbohydr. Res. 124, 166-171.

Zevenhuizen, L.P.T.M. (1989) in V. Cresenzi, I.C.M. Dea, S. Paoletti, S.S. Stivala and I.W. Sutherland (eds.), Biomedical and biotechnological advances in industrial polysaccharides, Gordon and Breach, New York, USA, 301-311.

Zevenhuizen, L.P.T.M. (1990) in E.A. Dawes (ed.), Novel biodegradable microbial polymers, Kluwer Academic Publishers, Dordrecht, Netherlands, 387-402.

Zevenhuizen, L.P.T.M. and Faleschini, P. (1991) Carbohydr. Res. 209, 203-209.

Zhan, H., Levery, S.B., Lee, C.C. and Leigh, J.A. (1989) Proc. Natl. Acad. Sci. USA 86, 3055-3059.

Zhan, H., Gray, J.X., Levery, S.B., Rolfe, B.G. and Leigh, J.A. (1990) J. Bacteriol. 172, 5245-5253.

Zhan, H. and Leigh, J.A. (1990) J. Bacteriol. 172, 5254-5259.

Zhan, H., Lee, C.C. and Leigh, J.A. (1991) J. Bacteriol. 173, 7391-7394.

Chapter 7

Lipopolysaccharides and K-Antigens: Their Structures, Biosynthesis, and Functions

Elmar L. Kannenberg, Bradley L. Reuhs, L. Scott Forsberg and Russell W. Carlson

I. Introduction

The bacterial surface is the first line of defense against antimicrobial molecules and stress caused by changes in the environment surrounding the bacterium. In the case of plant- and animal-microbe interactions, many bacterial cell surface molecules are important virulence determinants. Thus, in order to understand the molecular basis for bacterial-plant interactions, it is important to characterize the molecular architecture of the bacterial cell surface,

and how the bacterium modifies this architecture in response to its different environments, including its *in planta* environment.

The members of the family *Rhizobiaceae* are Gram-negative bacteria often found in association with plants (the family includes the plant nodule-forming symbionts *Rhizobium*, *Azorhizobium*, *Bradyrhizobium*, *Mesorhizobium*, and *Sinorhizobium*, which will be collectively referred to as rhizobia; and *Phyllobacterium*, as well as *Agrobacterium*, the gall-forming plant pathogen (Jordan, 1984, Young *et al.*,

1996). The outer surface of these bacteria consists of a complex array of different molecules, which include lipopolysaccharides (LPSs), capsular polysaccharides (CPSs), cyclic glucans (see chapter 5), extracellular polysaccharides (EPSs) (see chapter 6), porins, fimbri, and flagella. Rhizobia are also known to produce additional molecules, including lipo chitin oligosaccharides (LCOs) in response to the plant (see chapters 16, 19, 20 and 21), and glycosylated diacylglycerol molecules (Cedergren *et al.*, 1996). Virtually nothing has been reported concerning the cell surface molecules of *Phyllobacterium*. In the case of *Agrobacterium*, some information on the surface molecules has been included in this chapter, but the information available is rather limited. The focus of this chapter is, therefore, predominantly on the rhizobial LPSs and CPSs.

The LPSs are complex glycolipid molecules, which constitute the outer leaflet of the Gram-negative bacterial outer membrane. Conceptually, LPSs can be divided into three structural regions, the O-chain polysaccharide, core oligosaccharide, and lipid regions, Figure 1. The O-chain polysaccharide is a polymerized repeating oligosaccharide, which is attached to the core oligosaccharide region of the LPS. The core oligosaccharide is, in turn, attached to a glucosamine (GlcN) residue that is part of the lipid region of the LPS. A 3-deoxy-D-*manno*-2-octulo-sonosyl (Kdo) residue is involved in this

oligosaccharide-lipid linkage and is acid labile due to its ketosidic glycoside bond. The lipid, known as lipid-A, commonly consists of a disaccharide sugar backbone involving glucosamine residues, which are both O- and N-fatty, acylated. The lipid-A region of the LPS serves to anchor the entire molecule in the bacterial outer membrane.

Two types of rhizobial CPSs have been reported. One type of CPS consists of a neutral polysaccharide which has been purified from strains of *R. leguminosarum* (Breedveld *et al.*, 1993; Zevenhuizen *et al.*, 1983). Its role, if any, in symbiotic infection is unknown. The second type CPS is representative of a class of acidic polysaccharides with varying structures, all of which are comprised of a polymerized oligosaccharide. A common feature of most of these acidic CPSs from rhizobia is the presence of Kdo or a variant of Kdo (Reuhs *et al.*, 1993). This feature makes these CPSs analogous to the group II K-antigens reported in *Escherichia coli* (Jann *et al.*, 1990), and therefore they will be referred to as K-antigens. These K-antigens have been found in all rhizobia examined thus far, as well as in some other plant pathogenic bacteria (Reuhs *et al.*, 1994b). While K-antigens and LPSs are expected to play important roles in symbiosis, the structure-function relationships of these molecules are only partly understood and are the focus of this chapter.

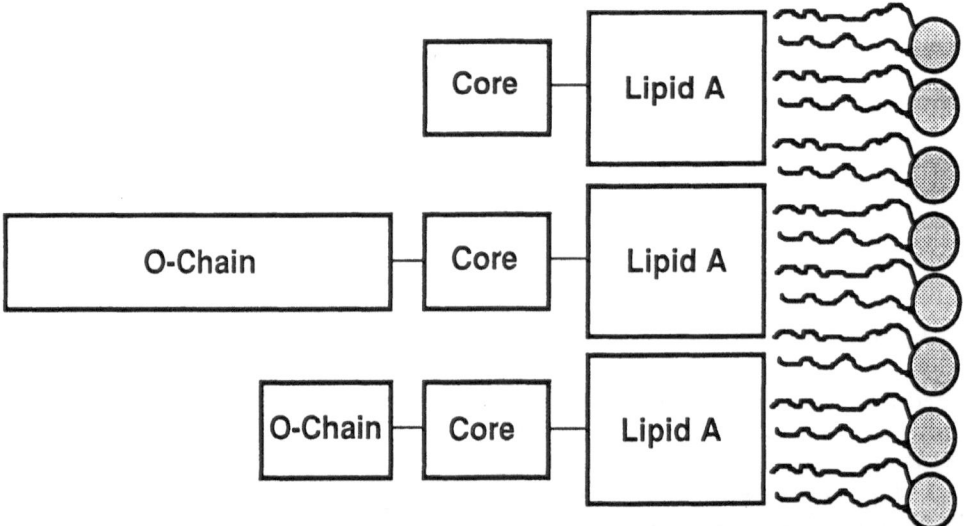

Figure 1. A schematic diagram showing the three structural regions present in LPS. The surface of a Gram-negative bacterium can have complete LPSs which contain all three structural regions (the "smooth" LPS, or LPS I), LPS which lacks the O-chain polysaccharide (the "rough" LPS, or LPS II), and LPS which contains truncated versions of the O-chain polysaccharide (referred to as "semi-rough" or "semi-smooth" LPS, or LPS III, IV, V, etc.)

II. Lipopolysaccharides

Early work suggested that LPSs might play crucial roles in the interaction between members of the *Rhizobiaceae* and plants. There was precedence for their functional importance in the infection of animal host cells by enteric bacterial pathogens, and for the interaction of bacteria with bacteriophages. In addition, LPSs constitute one of the major components of the outer membrane and, therefore, were likely candidates for playing a determining role in such cell-cell recognition processes. The LPSs from *Agrobacterium* were reported to be involved in the attachment of this pathogen to host cells (Whatley *et al.*, 1977; Whatley *et al.*, 1970). Early reports also suggested that there was a specific attachment of the symbiont *Rhizobium* to the legume host root which was thought to be mediated by host plant lectins and the rhizobial LPSs (Wolpert *et al.*, 1976; Kato *et al.*, 1981; Kato *et al.*, 1980; Kato *et al.*, 1979). The subject of attachment of rhizobia and agrobacteria to plant cells is discussed in Chapter 11. Subsequent reports regarding rhizobia-legume interactions (discussed further below) showed that the LPS is important in later infection events.

While LPSs were shown to be crucial for symbiotic infection (Stacey *et al.*, 1991; Priefer 1989; de Maagd *et al.*, 1989; Perotto *et al.*, 1994; Cava *et al.*, 1989; Carlson *et al.*, 1987a; Noel *et al.*, 1986), the function of the LPS during this process is not known. Certain symbiotically defective mutants completely lacked the O-chain polysaccharide of their LPSs. The complete loss of this portion of the LPS is a rather large change in the rhizobial cell surface. Thus, it was not known whether the symbiotic defect was a direct or indirect result of this alteration in the rhizobial cell wall. However, subsequent work, using monoclonal antibodies specific to the LPS of bacteria or bacteroids, showed that changes in LPS structure occur during symbiotic infection, and that these structural changes likely reside in the O-chain polysaccharide portion of the LPS (Kannenberg *et al.*, 1994a; Kannenberg *et al.*, 1994b; Kannenberg *et al.*, 1992; Sindhu *et al.*, 1990; Wood *et al.*, 1989; Noel *et al.*, 1996b; Tao *et al.*, 1992; Noel 1992). These results suggested that the LPS O-chain may have a more direct role during the symbiotic infection process.

The changes in LPS structure, and the potential functions of LPS in symbiosis are discussed in section II.F. The following section describes the purification, structures, and some biosynthetic aspects of LPSs from *Rhizobiaceae* bacteria.

II.A. THE ISOLATION OF LIPOPOLYSACCHARIDES

II.A.1. Extraction Procedures.

The LPSs from *Rhizobiaceae* bacteria, as with enteric bacterial LPSs, are complex molecules and a single LPS preparation contains a heterogeneous mixture of molecules. This mixture usually contains molecules which contain various lengths of the O-chain polysaccharide, core oligosaccharide, and lipid-A, and also molecules which lack the O-chain polysaccharide, see Figure 1. Those with O-chain polysaccharide are often referred to as "smooth" LPS, or LPS I, and those without O-chain as "rough" LPS, or LPS II. Occasionally LPS molecules with truncated O-chains are also present, particularly in some mutants (Carlson *et al.*, 1995). These molecules are known as "semi-rough" or "semi-smooth" LPSs, or LPS III, IV, V, etc., depending on the extent of truncated O-chain that is attached.

Crude rhizobial or agrobacterial LPS preparations can be obtained by a variety of the usual procedures developed originally for the purification of enteric bacterial LPSs. These procedures include cold trichloroacetic acid (TCA) extraction, petroleum ether/chloroform/phenol (Galanos *et al.*, 1969), and the hot phenol-water procedure (Westphal *et al.*, 1965). To date almost all rhizobial LPSs have been extracted using the hot-phenol water procedure. At elevated temperatures, a homogeneous phase of water and phenol is used for extraction which separates into two phases after cooling. The more hydrophobic compounds are found in the phenol phase and the more hydrophilic compounds in the aqueous phase.

Phenol-water extraction of *R. leguminosarum* strains (this includes *R. etli*) results in LPS in both the phenol and water layers with the majority being found in the water layer. The LPS found in the phenol layer contains a higher proportion of longer fatty acyl chains than that found in the water layer. Extraction of several LPS mutants, which contain LPS with truncated O-chains, results in LPS in both the water and phenol layers, but with a larger proportion found in the phenol layer when compared to the parent LPS. The LPSs from *S. meliloti*, *S. fredii*, and *A. tumefaciens* are also found in both the phenol and water layers, but primarily in the water layer, after phenol-water extraction (Russa *et* al., 1996; Reuhs et al., 1994a; Reuhs et al., 1993; Weibgen et al., 1993; Russa et al., 1991; Carlson et al., 1985).

The LPSs from B. japonicum and B. elkanii behave quite differently from those of Rhizobium species during phenol-water extraction in that the majority of LPS is found in the phenol layer (Carrion et al., 1990). In fact, except for B. japonicum USDA110, extraction of strains representing a number of serogroups resulted in LPS being present exclusively in the phenol layer (Carrion et al., 1990). Analysis of at least one of these LPSs revealed that the O-chain polysaccharide was extensively O-acetylated and that both the O-chain and core oligosaccharide were devoid of charged glycosyl residues except for Kdo (Carlson et al., 1992; Carrion et al., 1990). Thus, it is likely that B. japonicum LPSs are generally quite hydrophobic and, therefore, are extracted preferentially into the phenol layer.

While hot phenol-water extraction of rhizobia has proved effective in obtaining the initial crude LPS preparation, there are a number of disadvantages to this procedure. As we have described above, certain forms of LPS may be preferentially extracted into the water layer and, therefore, some forms of LPS may be overlooked entirely. In the case of some species, e.g. B. japonicum or B. elkanii, the LPSs are extracted exclusively into the phenol layer instead of the water layer making their subsequent purification more difficult. For certain mutants, e.g. those containing LPSs with truncated O-chains, phenol-water extraction has proven to be very inefficient resulting in poor yields of LPS. Also, often large batches of bacteria are required in order to obtain sufficient amounts of LPS for structural or biological studies. Hot-phenol water extraction of bacteria from 100 to 400 L cultures can be a tedious and somewhat hazardous process due to the amounts of phenol required.

Thus, new extraction protocols have been developed. Generally, it has been found that large amounts of all forms of rhizobial and agrobacterial LPSs can be extracted simply by stirring the cell pellet in a solution of EDTA-triethanol amine (TEAoL) (pH 7), or EDTA-triethylamine (TEA) (pH 7) for 30 to 60 minutes at room temperature. The EDTA-TEA extraction procedure is improved if it is made 5% phenol and the extraction is done at 60°C. This procedure is simple, rapid, and allows one to extract large batches of bacteria (Ridley and Carlson, manuscript in preparation). It can also be used on a micro-scale level for screening changes or differences in LPS structures (Valverde et al., 1997).

II.A.2. Purification of LPS

Once the crude LPS is obtained by one of the above extraction procedures, it must be further purified. Several methods have been used to purify LPSs and the method of choice is somewhat dependent on the bacterial species.

Gel filtration chromatography using Sepharose 4B and an EDTA-TEA buffer has been used to purify a number of rhizobial LPSs; particularly those from various R. leguminosarum strains (Carlson et al., 1992; Carrion et al., 1990). This method allows the separation of LPS from contaminating extracellular polysaccharide (EPS), as well as from contaminating low molecular weight carbohydrates, e.g. the cyclic β-2-linked glucans.

Sepharose 4B gel filtration has also been used to purify the LPSs from S. meliloti and S. fredii strains. However, in these two instances the crude LPS preparations were found to be contaminated by large amounts of the Kdo-rich K-antigens (Reuhs et al., 1993) which were not completely separated from the LPS during this procedure (Reuhs et al., 1993). One method by which these LPSs could be further separated from the K-antigens was by gel-filtration chromatography using Sephadex G-150 in the presence of a buffer solution containing 1% deoxycholate (DOC) (Reuhs et al., 1993). The presence of DOC de-aggregates the LPS and, thereby, allows it to be separated from the relatively larger K-antigen. An interesting feature of this procedure is that it also separates the various molecular sizes of LPS; e.g. LPS I from LPS II (Reuhs et al., 1994a).

Preparative gel electrophoresis is another method of LPS purification (Kim et al., 1996b). This method allows one to (a) separate LPS from K-antigen or other types of CPSs from various bacterial species, (b) separate CPSs from lipid-bound CPSs, and (c) separate discreet size classes of CPSs, K-antigens, and LPSs. This method can also be applied to the isolation of discreet size classes of EPS (Kim et al., 1996b). As much as 100 mg of crude LPS extract can be applied to the preparative gel electrophoresis apparatus. Initially the sample is electrophoresed in the absence of DOC. Without DOC the LPS remains as a very large molecular weight aggregate and does not enter the separating gel. However, the acidic K-antigen (or CPSs) migrates through, and elutes from, the gel. Once the K-antigen has eluted from the gel, DOC is added to the electrophoresis buffer. The presence of DOC de-aggregates the LPS which then migrates through the gel and is separated into its different size classes.

Possibly the most convenient and efficient method for obtaining *Rhizobium* LPS in a highly purified form is the use of affinity chromatography on a column containing polymyxin-agarose; polymyxin is an acylated cyclic peptide antibiotic that strongly binds to the lipid-A portion of an LPS (Morrison *et al.*, 1976; David *et al.*, 1992). Crude LPS extracts from any of the above extraction procedures can be applied to the column. Following sample application, the polymyxin column is eluted sequentially with solutions of increasing ionic and chaotropic strength in order to elute the non LPS components bound through non-specific weak interactions. When all of the non-LPS components have been eluted from the column, it is washed with 1% DOC, or a strong chaotropic agent such as 4 M guanidine, in order to elute the LPS. Polyacrylamide gel electrophoresis (PAGE) analysis of the eluted LPS in the presence of DOC shows a complete absence of non-LPS contaminants. The purified LPS resulting from polymyxin-affinity chromatography is, frequently, of higher purity than that obtained using standard gel filtration methods. All LPSs from *Rhizobiaceae* examined to date bind to polymyxin-agarose, including LPSs from *R. etli*, *R. leguminosarum*, *S. meliloti*, *S. fredii*, *Rhizobium* sp., *Agrobacterium*, etc. Such polymyxin-purified preparations consist of a mixture of all the various size classes of LPSs produced by the bacterium. These variously sized LPSs can be separated from one another as described above.

The combination of the simplified extraction procedure using EDTA-TEAoL or EDTA-TEA with polymyxin-affinity chromatography allows the purification of LPS in two steps. This procedure has been used to purify both large (Ridley and Carlson, unpublished) and small batches (Valverde *et al.*, 1997) of LPS.

II.B. THE STRUCTURES OF LIPOPOLYSACCHARIDES

In general, the structural elucidation of LPSs proceeds along the lines developed for enteric LPSs, or, indeed, for all glycoconjugates. Structural characterization requires determination of (a) glycosyl and fatty acid composition, (b) glycosyl linkages, (c) glycosyl anomeric and stereochemical configuration, and (d) the sequence of glycosyl residues and location of fatty acyl substituent groups. These analyses require the use of combined gas liquid chromatography (GLC)-mass spectrometry

(MS), nuclear magnetic resonance spectroscopy (NMR), and MS techniques such as fast atom bombardment (FAB)-MS, electrospray ionization (ESI)-MS, and matrix assisted laser desorption time of flight (MALDITOF)-MS.

Lipopolysaccharides, due to their overall complexity and high degree of structural heterogeneity, present a variety of problems during structural analysis. The analysis is perhaps even more difficult with *Rhizobium* LPSs which lack certain of the common structural formats found within the enteric LPSs, and which appear to contain a somewhat wider variety of less common and/or labile sugars in "non-standard" locations compared to the typical enteric LPSs. Nevertheless, all LPSs from members of the *Rhizobiaceae* examined to date appear to share the same overall structural motif as those from the enteric bacteria, that is, they can be divided, both structurally and functionally, into three regions, the O-chain polysaccharide, the core region oligosaccharide, and the lipid-A moiety (Figure 1).

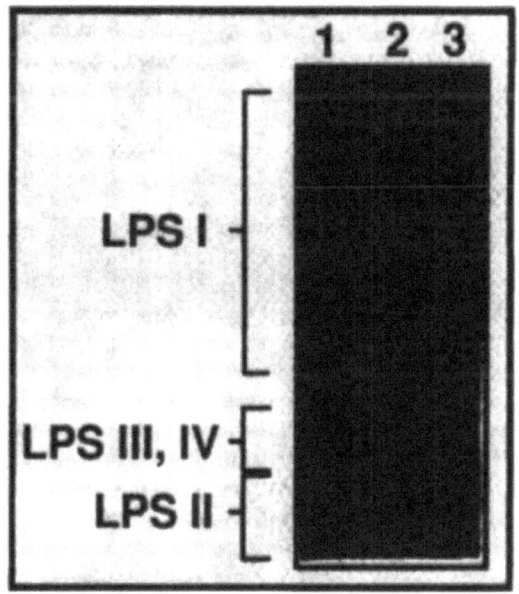

Figure 2. A PAGE analysis of the LPSs from *R. leguminosarum* bv. viciae 128C53 (well 1), *R. etli* CE3 (well 2), and *R. leguminosarum* bv. trifolii ANU843 (well 3). The various LPS species (*i.e.* LPS I, II, III, IV, and V) have the general structural features as defined in the legend to Figure 1

II.B.1. The LPSs From Rhizobium leguminosarum and R. etli.

Perhaps the best characterized rhizobial LPSs are those from *R. etli* and *R. leguminosarum* strains. The O-chains of *R. leguminosarum* strains are highly variable from strain to strain and are comprised of deoxy and methylated deoxyglycosyl residues, as well as uronic acid residues, and, in some cases, heptosyl residues (Wang *et al.*, 1994; Carlson *et al.*, 1978; Carlson *et al.*, 1987a; Carlson 1984; Carlson *et al.*, 1987b). These O-chains consist of a polymerized repeating oligosaccharide. Figure 2 shows examples of PAGE banding patterns for various LPSs from *R. leguminosarum* and *R. etli* strains. For some strains the number of repeats in the O-chain varies greatly and PAGE analysis results in a "ladder" pattern (Carlson *et al.*, 1983; Carlson 1984). For other strains the number of repeats is restricted to a narrow range and PAGE analysis results in a only few major bands (Carlson *et al.*, 1987a; Carlson *et al.*, 1987b; Carlson 1984). An interesting feature of *R leguminosarum* O-chains isolated after mild acid hydrolysis of the LPS is that they have a Kdo residue at their reducing ends (Carlson *et al.*, 1987a; Carlson 1984), a feature that is not true of enteric LPSs. The O-chain structures of one strain each of *R. leguminosarum* bv. *trifolii*, *R. leguminosarum* bv. *viciae*, and *R. etli* are shown in Table 1. In the case of *R. etli* CE3 (formerly *R. leguminosarum* bv. *phaseoli* CE3), the O-chain repeating unit consists of glucuronic acid (GlcA), fucose (Fuc), and 3-O-methylrhamnose (3MeRha) (Carlson *et al.*, 1987a). Even though GlcA is present, O-chain isolated after mild acid hydrolysis of the LPS does not bind to a DEAE column at pH 8.0 indicating that it is a largely neutral molecule. Thus, the GlcA carboxyl group must be derivatized in some, as yet, unknown manner which renders it neutral, *e.g.* GlcA may actually be a glucuronamide residue. The examination of other structures is important since the presence of an O-chain is essential for proper nodule development, and since structural changes in the O-chain occur during symbiotic infection (see section II.F).

The complete structure of the lipid-A-core region of *R. etli* CE3 LPS is shown in Figure 3. The glycosyl components consist of Kdo, glucosamine, 2-aminogluconate (GlcNonate), mannose (Man), galactose (Gal), and galacturonic acid (GalA). Two core oligosaccharides released by mild acid hydrolysis of the labile Kdo glycoside bonds were structurally elucidated; these being the α-D-GalA-

1→5[α-D-GalA-1→4]-Kdo trisaccharide, and a α-D-Gal-1→6[α-D-GalA-1→4]-α-D-Man-1→5-Kdo tetra-saccharide (Bhat *et al.*, 1991a; Carlson *et al.*, 1990). Identical oligosaccharides have also been reported in *R. leguminosarum* bv. *trifolii* and *R. leguminosarum* bv. *viciae* LPSs (Hollingsworth *et al.*, 1994; Hollingsworth *et al.*, 1989b; Hollingsworth *et al.*, 1990; Carlson *et al.*, 1988; Zhang *et al.*, 1992). It should also be noted that 3-deoxyheptulosaric acid has also been reported in the LPS core region of some *R. leguminosarum* bv. *trifolii* strains (Russa *et al.*, 1991).

The lipid-A from CE3 was isolated after mild acid hydrolysis and its structure determined (Bhat *et al.*, 1994). This lipid-A has a very unusual structure compared to that from many other bacteria. It is devoid of phosphate, it does not have acyloxyacyl substituents, it contains only hydroxy fatty acids including the very long chain 27-hydroxyoctacosanoic acid (27-OHC28:0) (Hollingsworth *et al.*, 1989a), a GalA residue is at the 4'- position instead of phosphate, and it has 2-aminogluconate in place of GlcN-1-phosphate. In addition, unlike enteric lipid-A in which the *N*-acyl substituent is exclusively β-hydroxymyristate, this lipid-A contains GlcN which is heterogeneously *N*-acylated with either β-hydroxymyristate, β-hydroxypalmitate, or β-hydroxystearate (Bhat *et al.*, 1991c; Bhat *et al.*, 1994).

The arrangement of the core oligosaccharides and the point of attachment of the O-chain in the LPS was established by the analysis of LPSs from mutants which either completely lacked the O-chain or contained severely truncated forms of the O-chain (Carlson *et al.*, 1995, Forsberg and Carlson, 1997). This structural arrangement is as depicted in Figure 3. It should be noted that an LPS preparation exists as a heterogeneous mixture of molecules at various stages of biosynthetic completion. Thus, LPS molecules, in a single LPS preparation, which lack the O-chain can have different core structures due to various missing hexose or hexuronosyl residues (Forsberg and Carlson, 1997). However, a complete LPS molecule which contains the O-chain polysaccharide has the core structure indicated in Figure 3. Chemical analysis of the *R. etli* LPS has shown that the O-chain polysaccharide is attached to the galactosyl residue of the core region (see Figure 3), i.e. to the "tetrasaccharide" core element. Thus, any defect in this region of the core should lead to a "rough" LPSs. This, in fact, is the case for *R. etli* (Carlson *et al.*, 1995), and also appears to be

Strain	Structure	Reference
R. leg. bv. *trifolii* 4S	→3)- α-L-Rha-(1→3)- α-L-Rha-(1→4)- β-D-GlcNAc-(1→3)- α-L-Rha-(1→ 2) ↑ α-D-ManNAc-(1	Wang *et al.*, 1994
R. leg. bv. *viciae* 128C53	→3)- α-L-Rha-(1→3)- α-L-Fuc-(1→3)- α-L-Fuc-(1→ 2) ↑ α-D-Man-(1	Chen 1987
R. etli CE3	→4)- β-D-GlcA*-(1→4)- α-L-Fuc-(1→ 3) ↑ α-3-O-MeRha-1)	Forsberg *et al.*, in preparation
R. tropici CIAT899	→4)- β-D-Glc-(1→3)- α-D-2-O-Ac-6-deoxy-Tal-(1→3)- α-L-Fuc-(1→	Gil-Serrano *et al.*, 1995

Table 1: Rhizobial O-chain polysaccharide structures
Except of the O-chain polysaccharide of *R. leg.* bv. *viciae* 128C53, all of the O-chains contain O-acetyl groups. The precise locations of these O-acetyl substituents, except for strain CIAT899, are not known. Abbreviations: Glc, glucose; Man, mannose, ManNAc, *N*-acetylmannosamine; Rha, rhamnose; Tal, talose; GlcA*, glucuronic acid derivatized in an unknown manner that renders it neutral (see text); Fuc, fucose; Me, methyl; Ac, acetyl

the case for *R. leguminosarum* bv. *viciae* (Hollingsworth *et al.*, 1994; Zhang *et al.*, 1992). That the galactosyl residue is the site of O-chain attachment in *R. leguminosarum* LPS is also supported by the finding that an *exoB* mutant, defective in UDP-galactose epimerase, has an LPS which lacks the O-chain polysaccharide (Van Workum *et al.*, 1995). There is also one report which shows that the LPS from *R. leguminosarum* bv. *trifolii* 4S which lacks the tetrasaccharide portion of the core region also lacks the O-chain polysaccharide (Cedergren *et al.*, 1996). That report also states that, in this strain, the O-chain polysaccharide is attached to a diacylglycerol moiety rather than to the LPS core region.

II.B.2. Lipopolysaccharide Structures From Other Rhizobiaceae Bacteria.
Very little work has been done regarding the detailed analysis of the LPS from *Agrobacterium*. However, composition work has shown that *A. tumefaciens* biovars 1 and 2 contain O-chains that are enriched in rhamnose (Weibgen *et al.*, 1993). PAGE analysis has indicated that biovar 3 has a "rough" type of LPS that lacks an O-chain polysaccharide (Weibgen *et al.*, 1993)
In addition to the structures mentioned above, the structure of the O-chain from *R. tropici* CIAT899

has been published (Gil-Serrano *et al.*, 1995) and is shown in Table 1. As with the *R. etli* and *R. leguminosarum* O-chains, this structure has a repeating oligosaccharide unit that is rich in deoxyhexoses. In fact, the compositions of a number of rhizobial O-chains show that they are all enriched in deoxyhexoses and methylated deoxyhexoses and can also contain 6-deoxy amino sugars and *N*-methyl 6-deoxy amino sugars (Weibgen *et al.*, 1993; Zevenhuizen *et al.*, 1980; Russa *et al.*, 1995; Glowacka *et al.*, 1986; Carrion *et al.*, 1990; Carlson 1982; Carlson *et al.*, 1978; Carlson 1984; Gil-Serrano *et al.*, 1995). In addition, NMR analysis has shown that the O-chains can be highly acetylated (Bhat *et al.*, 1992; Carrion *et al.*, 1990; Gil-Serrano *et al.*, 1995). Therefore, rhizobial LPS O-chains can be, for a polysaccharide, quite hydrophobic. The O-chains from *B. japonicum* and *B. elkanii* LPSs, which also contain deoxy sugars, are extensively acetylated and, as mentioned above, are sufficiently hydrophobic to be extracted into the phenol layer during hot phenol-water extraction (Bhat *et al.*, 1992; Carrion *et al.*, 1990).
The O-chains for many *Rhizobium* and *Bradyrhizobium* species are the dominant antigenic determinants. When antibodies are raised against these bacteria, most are directed against the O-chain polysaccharide (Carrion *et al.*, 1990; Carlson *et al.*,

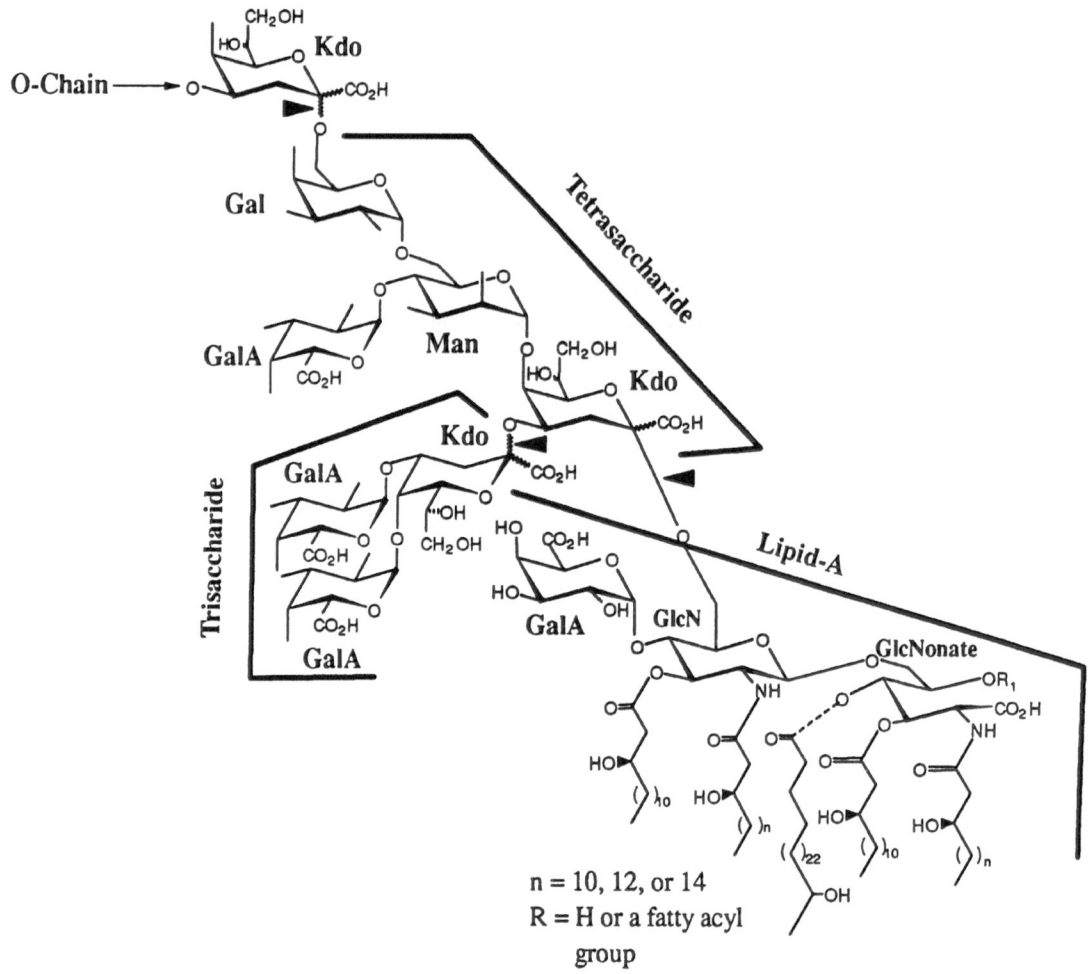

Figure 3. The structure of the core oligosaccharide and lipid-A region *R. etli* CE3. Current data indicate that this structure is common among strains of *R. leguminosarum* and *R. etli* (discussed further in the text) with the possible exception of some variation in the fatty acylation pattern (Bhat *et al.,* 1991c). The "wavy" bonds of the Kdo linkages indicate that the anomeric configurations (*i.e.* α or β) of these bonds are not known; however, based on other LPS structures these Kdo residues are likely to be α-linked. Acid labile bonds are indicated with an arrow.

1987a; Carlson *et al.,* 1987b). Usually, there is only a slight reaction with the "rough" LPS which lacks the O-chain (Lucas *et al.,* 1996).

The O-chains of *S. fredii, S. meliloti* and *R.* sp. NGR234 appear to be quite different from those of *Rhizobium* or *Bradyrhizobium.* Composition analysis indicates that they can be homopolymers, *e.g.* a glucan (Reuhs *et al.,* 1994a). In addition, in these species, the core oligosaccharide is the dominant antigenic region of the LPS (Reuhs *et al.,* 1995; Petrovics *et al.,* 1993; Reuhs *et al.,* 1993), *i.e.* antiserum to these bacterial species reacts strongly with the "rough" LPS. These results indicate that the O-chains of these species are not very antigenic, or that the ratio of "smooth" to "rough" LPSs is much lower than that observed for *R. leguminosarum,*

R. etli, and *Bradyrhizobium* species. Thus, the abundance of rough LPS on the cell surface of *S. meliloti*, *S. fredii*, and *R.* sp. NGR234 makes it the dominant LPS antigen in these species. *Rhizobium galegae* is another species in which the rough LPS is the abundant form of LPS (Lipsanen *et al.*, 1989). The relatedness of this species to other rhizobia is not yet fully established.

The core oligosaccharides from a few other rhizobial LPSs have also been characterized. In the case of *B. elkanii* 61A101c, the core region consists of two oligosaccharides; a α-4-O-MeMan-(1→5)-Kdo disaccharide, and a α-Man-(1→4)-α-Glc-(1→4)-Kdo trisaccharide. These oligosaccharides were released by mild acid hydrolysis of LPS isolated from a mutant that lacks the O-chain polysaccharide (Carlson *et al.*, 1992). Mild acid hydrolysis of the parent LPS releases only the trisaccharide and the O-chain, which contains 4-O-MeMan. Therefore, it is likely that the O-chain polysaccharide is linked to the 4-O-MeMan residue of the core region. Unlike the LPSs from *R. leguminosarum*, the *B. elkanii* LPS core region does not contain any acidic sugars except Kdo.

The smooth and rough LPSs from a strain *S. fredii* have been separated from one another, and analysis of the "rough" LPS showed that the core region consisted of Kdo, Glc, Gal, GalA, and GlcA (Reuhs *et al.*, 1994a). The linkage positions and sequence of these glycosyl residues have not been determined. The high performance anion exchange chromatography (HPAEC) profiles of core oligosaccharides from rough and smooth LPSs are identical to one another (Reuhs *et al.*, 1994a) indicating that these two LPS forms have identical core structures. Furthermore, HPAEC analyses indicate that the core regions from different strains of *S. fredii* all have a common set of oligosaccharides, and a second set of oligosaccharides which vary among the different strains (discussed further below, Figure 4B).

Sinorhizobium meliloti and *Rhizobium* sp. NGR234 LPSs have core regions that are closely related to these of *S. fredii* LPSs (this relationship is discussed further in the next section). It is also reported that the LPS from *S. meliloti* has a core region which contains the unusual sugar, 3-deoxy-2-heptulosaric acid (Russa *et al.*, 1996; Russa *et al.*, 1991); a sugar that is normally part of a plant cell wall component called rhamnogalacturononan II (Stevenson *et al.*, 1988). This unusual sugar has also been reported to be a core component of *Agrobacterium* LPSs (Weibgen *et al.*, 1993). It has also been reported that *S. meliloti* LPSs are sulfated (Cedergren *et al.*, 1995).

This result may explain why there are several sets of sulfation genes in *S. meliloti*, in addition to those (*nodP*, *nodQ*, and *nodH*) responsible for the sulfation of the lipo chitin oligosaccharide Nod factors (Schwedock *et al.*, 1992).

The lipid-As of rhizobial LPSs vary in structure among the different rhizobial species. The glycosyl compositions of the various lipid-A regions are summarized in Table 2. In addition to the *R. leguminosarum* lipid-A whose structure is described above, *S. meliloti* lipid-A is reported (Urbanik-Sypniewska *et al.*, 1989) to consist of an acylated bis-phosphorylated glucosamine disaccharide similar to that found in numerous enteric bacteria. *Bradyrhizobium japonicum*, *B. elkanii*, *Bradyrhizobium* sp. (*Lupinus*), and *M. loti* lipid-As all contain 2,3-diaminoglucose (Russa *et al.*, 1995; Mayer *et al.*, 1989b). The one *B. elkanii* lipid-A examined also contains mannose (Carlson, unpublished), and the *M. loti* lipid-A is phosphorylated (Russa *et al.*, 1995). Detailed glycosyl composition analysis of isolated *Agrobacterium* lipid-A has not been reported, however analysis of the intact LPSs indicates that the lipid-A contains glucosamine (Weibgen *et al.*, 1993). The types of fatty acids found in various rhizobial lipid-A molecules can be quite variable among the different species (Russa *et al.*, 1985; Bhat *et al.*, 1991c; Russa *et al.*, 1995; Mayer *et al.*, 1989b), however usually the major fatty acids include various β-hydroxy fatty acids, 27-hydroxyoctacosanoic acid, and smaller amounts of various saturated and mono-unsaturated fatty acids. Some lipid-As also contain small amounts of oxo-fatty acids (Russa *et al.*, 1995).

II.C. LIPOPOLYSACCHARIDE STRUCTURAL VARIATION AND PHYLOGENETIC RELATEDNESS AMONG THE *RHIZOBIACEAE* BACTERIA

Structural details of the lipid-A and core regions of LPSs have been used to recognize phylogenetic relationships of gram-negative bacteria (Neumann *et al.*, 1995; Mayer *et al.*, 1990; Bhat *et al.*, 1991b; Mayer *et al.*, 1989a; Mayer *et al.*, 1984). The relationships determined by 16S or 5S rRNA analyses have been successively correlated with lipid-A structures. Therefore it is of interest to compare LPS structures with the known phylogenetic relationships among the *Rhizobiaceae* bacteria.

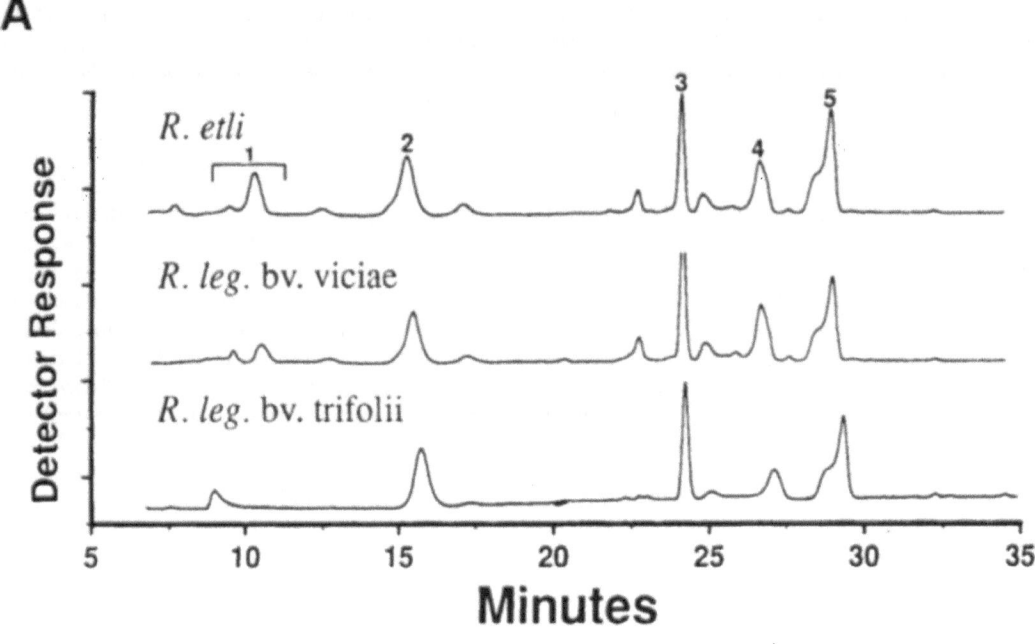

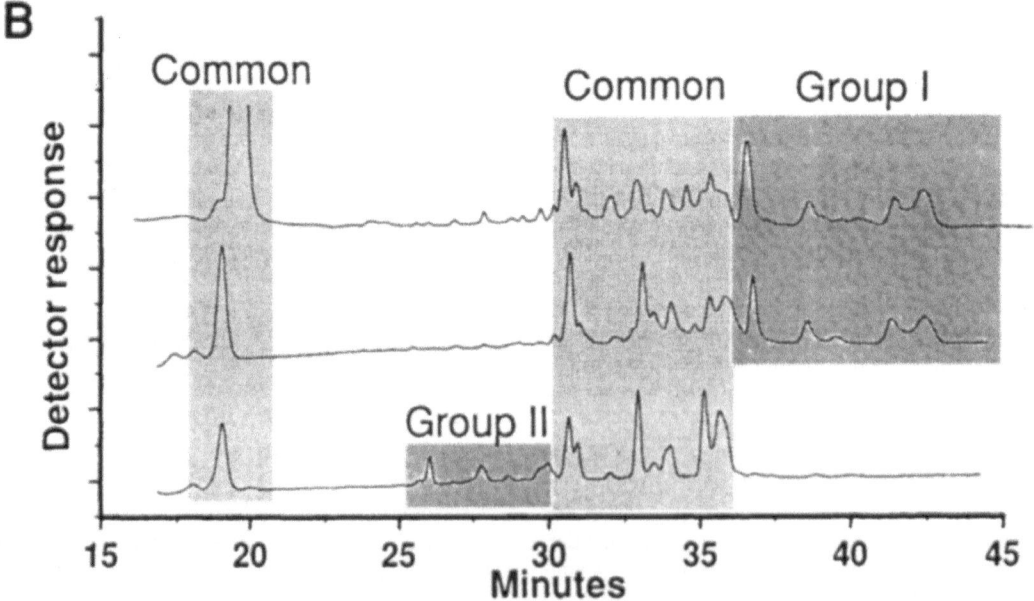

Figure 4. (A). The high performance anion exchange chromatography (HPAEC) profiles of the core oligosaccharides released by mild acid hydrolysis of the LPSs from *R. etli* CE3 (top), *R. leguminosarum* bv. *viciae* 3841 (middle), and *R. leguminosarum* bv. *trifolii* ANU843 (bottom). The peaks are as follows (Carlson *et al.,* 1995): 1 = monomeric Kdo (including various anhydro versions), 2 = monomeric GalA, 3 = the tetrasaccharide, 4 = the tetrasaccharide with an anhydro Kdo residue, 5 = the trisaccharide. (B). The HPAEC profiles of the core oligosaccharides released by mild acid hydrolysis of the LPSs from *S. fredii* USDA205 (top), *Rhizobium* sp. NGR234 (middle), and *S. meliloti* Rm1021 (bottom). These LPSs contain oligosaccharides that are present in all species with slight quantitative variation (the "common" oligosaccharides), oligosaccharides that are unique to "group 1" LPSs, and others that are unique to "group 2" LPSs.

One of the earliest novelties concerning rhizobial lipid-A was the discovery that *R. leguminosarum* biovar trifolii lipid-A contained, in addition to other fatty acyl residues, the very long chain fatty acid, 27-OHC28:0 (Hollingsworth *et al.*, 1989a), which was subsequently found in the lipid-A from all members of the *Rhizobiaceae* examined with the one exception (a result that should be re-examined) being *A. caulinodans* (Bhat *et al.*, 1991c). Since it had been determined from 16S rRNA homology studies (Woese 1987) that the *Rhizobiaceae* family belongs to the alpha sub-class of *Proteobacteria*, the lipid-As from a number of other species in this group which include phototrophic, nitrifying, nodulating, and intracellular bacteria, including both plant and animal pathogens, were examined. Other lipid-As which contained 27-OHC28:0 were those from *Rhodopseudomonas viridis, R. palustris, Nitrobacter winogradskyi, N. hamburgensis, Oligotropha carboxydovorans* (formerly *Pseudomonas carboxydovorans*), *Brucella abortus, Afipia felis* (formerly known as a bacterium associated with cat-scratch disease), *A. tumefaciens, A. radiobacter, A. rhizogenes,* and *Thiobacillus* spp. (Bhat *et al.*, 1991b). While not all of the alpha sub-class species contain lipid-A with 27-OHC28:0, when this fatty acid was present the species was found to belong to this phylogenetic group (Bhat *et al.*, 1991b).

Examination of the glycosyl residues found in the various rhizobial lipid-As revealed variations in glycosyl components which differed in accordance with known phylogenetic relationships, Table 2. Based on these lipid-A compositions, the *Rhizobiaceae* species examined could be divided into five clusters: I, *S. meliloti* and *S. fredii*; II, *R. leguminosarum* bv. *trifolii, R. leguminosarum* bv. *viciae, R. leguminosarum* bv. *phaseoli*, and *R. etli*; III, *B. japonicum* and *B.* sp. (*Lupinus*); IV, *B. elkanii*; and V, *M. loti*. Thus, the various designated clusters based on lipid-A compositions were similar to those based on other phylogenetic studies (Young *et al.*, 1996; Martínez-Romero *et al.*, 1996; Yanagi *et al.*, 1993; Jarvis *et al.*, 1992). The one exception is the relatively recent re-classification of the *R leguminosarum* bv. *phaseoli* strain used in this study (CE3) as being a member of a new species, *R. etli* (Segovia *et al.*, 1993). In the case of *R. leguminosarum* bv. *trifolii*, bv. *viciae*, and *R. etli* CE3, not only do their lipid-As have the same glycosyl compositions, but they also all appear to have the structure described in the previous section and shown in Figure 3. Not enough information is

available on *Agrobacterium* lipid-A to include this species in Table 2.

The relationship between LPS structure and phylogeny is further verified when the comparison of the core oligosaccharide regions is included. As with the lipid-A composition data, the same groupings can be deduced from a comparison of the various LPS core oligosaccharides. A comparison of the various core regions can be made by HPAEC analysis of the LPS mild acid hydrolysate. The HPAEC profile normally shows only the core oligosaccharides since the O-chain polysaccharide elutes in the void volume, or, if acidic, at much later retention times than the core oligosaccharides. Figure 4A compares the HPAEC core profiles of *R. etli* CE3, *R. leguminosarum* bv. *viciae*, and *R. leguminosarum* bv. *trifolii* ANU843, showing that they are essentially identical. For these *R. leguminosarum* profiles the structure of each HPAEC peak has been determined (Carlson *et al.*, 1995). Thus far, most *R. leguminosarum* and *R. etli* strains examined contain an LPS core region that consists of the tetra- and trisaccharides (Bhat *et al.*, 1991a; Hollingsworth *et al.*, 1994; Zhang *et al.*, 1992; Hollingsworth *et al.*, 1990; Carlson *et al.*, 1988) described in the previous section and, most likely, structurally arranged as shown in Figure 3. One reported exception was the LPS from *R. leguminosarum* bv. *trifolii* 4S which is reported to lack the tetrasaccharide core component as well as the O-chain polysaccharide (Cedergren *et al.*, 1996). Even though the structures of other rhizobial LPS core oligosaccharides have not been determined, such HPAEC "fingerprinting" can be used to compare the core regions of various LPSs. Figure 4B compares the HPAEC profiles of strains *S. fredii* USDA205, *S. meliloti* Rm1021, and *R.* sp. NGR234. The HPAEC profiles are very different from those for *R. leguminosarum* LPSs, but are similar to one another in that there are a number of oligosaccharides that are common to all three species, and other oligosaccharides which vary. The presence of oligosaccharides that are common among all three species supports previous work (Young *et al.*, 1996; Martinez-Romero 1994; Elkan 1992) which shows this close phylogenetic relatedness. The structural relatedness of the LPS core regions among various rhizobia can also be examined by polyacrylamide gel electrophoresis and immunoblotting. Using this technique, a series of monoclonal antibodies (MAbs) were found that were specific to the core region of *R. leguminosarum* bv. *viciae* LPS. All of these core MAbs were able to

Group	Species	GlcN	DAG	Man	GalA	GlcN-onate	Phosphate	Reference
I	*S. meliloti*	+	-	-	-	-	+	Urbanik-Sypniewska *et al.*, 1989
	S. fredii	+	-	-	-	-	n.d.	Bhat *et al.*, 1991c
II	*R. leg.* bv. *trifolii*	+	-	-	+	+	-	Bhat *et al.*, 1991c
	R. leg. bv. *viciae*	+	-	-	+	+	-	Bhat *et al.*, 1991c
	R. etli	+	-	-	+	+	-	Bhat *et al.*, 1994; Bhat *et al.*, 1991c
III	*B. japonicum*	-	+	-	-	-	-	Mayer *et al.*, 1989b
	B. sp. (*Lupinis*)	-	+	-	-	-	-	Mayer *et al.*, 1989b
IV	*B. elkanii*	-	+	+	-	-	n.d.	Carlson *et al.*, unpub.
V	*M. loti*	-	+	-	-	-	+	Russa *et al.*, 1995

Table 2. Glycosyl components of the lipid A from bacteria belonging to the *Rhizobiaceae*.
GlcN, glucosamine; DAG, 2,3-diaminoglucose; Man, mannose; GalA, galacturonic acid; GlcN-onate, 2-aminogluconate; n.d. not determined

bind to LPSs from every strain of *R. leguminosarum* tested, as well as to the LPS from *R. etli* CE3 (Kannenberg *et al.*, 1996; Lucas *et al.*, 1996). These data confirm the structural identity of the core region among *R. etli* and *R. leguminosarum* strains. In the case of *S. fredii*, *S. meliloti*, and *R.* sp. NGR234, polyclonal antiserum against any one strain reacts strongly against the "rough" form of its LPS (LPS II). Using polyclonal antiserum prepared against *S fredii* USDA205 or against *S. meliloti* Rm41, the strains of these species could be divided into four groups based on the reactivity of their "rough" LPSs with either USDA205 or Rm41 antiserum, see Table 3. These results would predict that future 16S rRNA studies using a wide range of *S. fredii*, and *S. meliloti* will show that strains of these species can be divided into at least four different, but closely related, phylogenetic groups. Further work is in progress to determine if the HPAEC "fingerprinting" of the LPSs from other species, *e.g. B. japonicum*, *B. elkanii*, *R. tropici*, *M. loti*, *A. tumefaciens*, etc., also reflects the known phylogenetic relationships.

II.D. THE BIOSYNTHESIS OF LIPID-A IN *R. ETLI* (AND *R. LEGUMINOSARUM*)

Biosynthesis of the lipid-A portion of LPS is crucial for viability of *E. coli* (Raetz 1993; Sirisena *et al.*,

1992; Raetz 1990). It is also known that this portion of the LPS from enteric bacteria is responsible for its toxicity, which occurs as a result of an over stimulation of the host's immune system; *e.g.* causing the production of lethal amounts of tumor necrosis factor and other cytokines (Takada *et al.*, 1992; Rietschel *et al.*, 1994; Rietschel *et al.*, 1993; Rietschel *et al.*, 1992; Rietschel *et al.*, 1990). Structural features of the lipid-A that are essential for this toxicity include the presence of a glucosamine disaccharide backbone, phosphate groups, and certain fatty acyl residues.

Due to the unique structure of the *R. leguminosarum* lipid-A and to the requirement of lipid-A for the viability of the Gram-negative bacterial cell, it was of interest to examine its biosynthetic pathway. In *E. coli*, the details of the lipid-A biosynthetic pathway have been worked out by Raetz and co-workers (Clementz *et al.*, 1997; Clementz *et al.*, 1996; Raetz 1993; Raetz 1990). Using the various *E. coli* lipid-A precursors, it was found that *R. leguminosarum* contained the same enzyme activities as those found in *E. coli* which synthesize Kdo_2lipid-IV_A (a precursor to *E. coli* lipid-A) from UDP-GlcNAc (Price *et al.*, 1994), see Figure 5. In *E. coli*, the steps leading to the synthesis of Kdo_2lipid-IV_A are crucial for cell viability (Raetz 1993; Sirisena *et al.*, 1992; Raetz 1990), and the presence of the 4'-phosphate is required (Brozek *et al.*, 1989; Belunis *et al.*, 1992) for the transfer of the

Group	Species	Strain	Anti-USDA205	Anti-Rm41
I	*S. fredii*	USDA205, USDA208, USDA191	++	--
	Rhizobium sp.	NGR234		
II	*S. fredii*	USDA257, USDA201, USDA192, USDA196, USDA197, HH103	--	++
	S. meliloti	NRG133, NRG247, NRG286, AK631, Rm1021, NRG185		
III	*S. meliloti*	NRG23, NRG53	+/-	+/-
IV	*S. fredii*	HH303	--	--

Table 3. The LPS II serogroups of *R. fredii, R. meliloti,* and *R.* sp. NGR234 using polyclonal antiserum generated against *S. fredii* USDA205 or *S. meliloti* Rm41. --, negative reaction; ++, positive reaction; +/-, weakly positive reaction.

two Kdo residues to lipid-IV_A from CMP-Kdo catalyzed by Kdo transferase (KdtA). After the synthesis of Kdo_2lipid-IV_A further processing occurs, *i.e.* the addition of the acyloxyacyl fatty acids, to form the mature *E. coli* lipid-A. These results showed that *R. leguminosarum* likely makes a very close structural analog of Kdo_2lipid-IV_A, indicating that the biosynthetic steps leading to Kdo_2lipid-IV_A are crucial for the cell viability of a very wide range of Gram-negative bacteria.

The above results also suggested that *R. leguminosarum* possesses unique enzymes which convert this Kdo_2lipid-IV_A precursor into the mature rhizobial lipid-A structure, Figure 5. Therefore, these results predict that *R. leguminosarum* should not possess acylating enzymes for the formation of acyloxyacyl substituents, it should have 4'- and 1-phosphatases, an oxidation system capable of converting the reducing-end glucosamine to 2-aminogluconate, a transferase which transfers a galacturonosyl residue to the 4'-position, and a unique acyl transferase system for the incorporation of 27-OHC28:0. Recently, 4'-phosphatase activity was reported (Price *et al.,* 1995) in *R. leguminosarum.* This enzyme was found in *R. etli* CE3, and in several strains of *R. leguminosarum,* but not in *E. coli.* This phosphatase activity prefers the presence of Kdo in that Kdo_2lipid-IV_A is an efficient substrate but not lipid-IV_A, and it is membrane bound. Another report describes the presence of the 1-phosphatase activity in *R. leguminosarum* (Brozek *et al.,* 1996b). Also *R. leguminosarum* cell extracts, as predicted, had no detectable acylating activity for the formation of acyloxyacyl groups. In addition, recently a unique ACP from *R. leguminosarum* has been isolated that is required for the transfer of 27-OHC28:0 to Kdo_2lipid-IV_A (Brozek *et al.,* 1996a). The sequence of this unique ACP revealed

that it is encoded by open reading frame (*orf**), an *lps* gene earlier identified and partially characterized (Selbitschka *et al.,* 1991). While some enzyme activities involved in the biosynthesis of *R. leguminosarum* lipid-A have been deduced, no other information concerning the genes encoding these enzymes has been published. Further work is in progress regarding the biosynthesis of *R. leguminosarum* lipid-A.

II.E. THE GENETICS OF *RHIZOBIUM* LPS SYNTHESIS

The genetics of LPS biosynthesis has been examined to a certain degree in *R. leguminosarum* and *R. etli.* A number of *lps* regions (*e.g.* α-, β-, and λ-regions) have been identified in strain *R. etli* CE3 (Noel, 1992; Cava *et al.,* 1990; Cava *et al.,* 1989). One region, the β-region, is on a plasmid (not the symbiotic plasmid) (Garcia-de los Santos and Brom, 1997; Brom *et al.,* 1992; Cava *et al.,* 1990; Cava *et al.,* 1989), while the other regions are located on the chromosome. The α-region encodes proteins involved in both core oligosaccharide and O-chain polysaccharide synthesis (Carlson *et al.,* 1995; Tao *et al.,* 1992), and consists of nine different complementation groups (A to I) on a 17 kb region of DNA (Tao *et al.,* 1992). Mutations in complementation groups A through F affect O-chain synthesis in that these mutants most frequently contain complete core regions but truncated O-chains, while mutations in G through I affect core oligosaccharide synthesis. Both the β- and λ-regions are required for core oligosaccharide synthesis (Brom *et al.,* 1992; Cava *et al.,* 1990; Cava *et al.,* 1989). Mutations in either of these two regions result

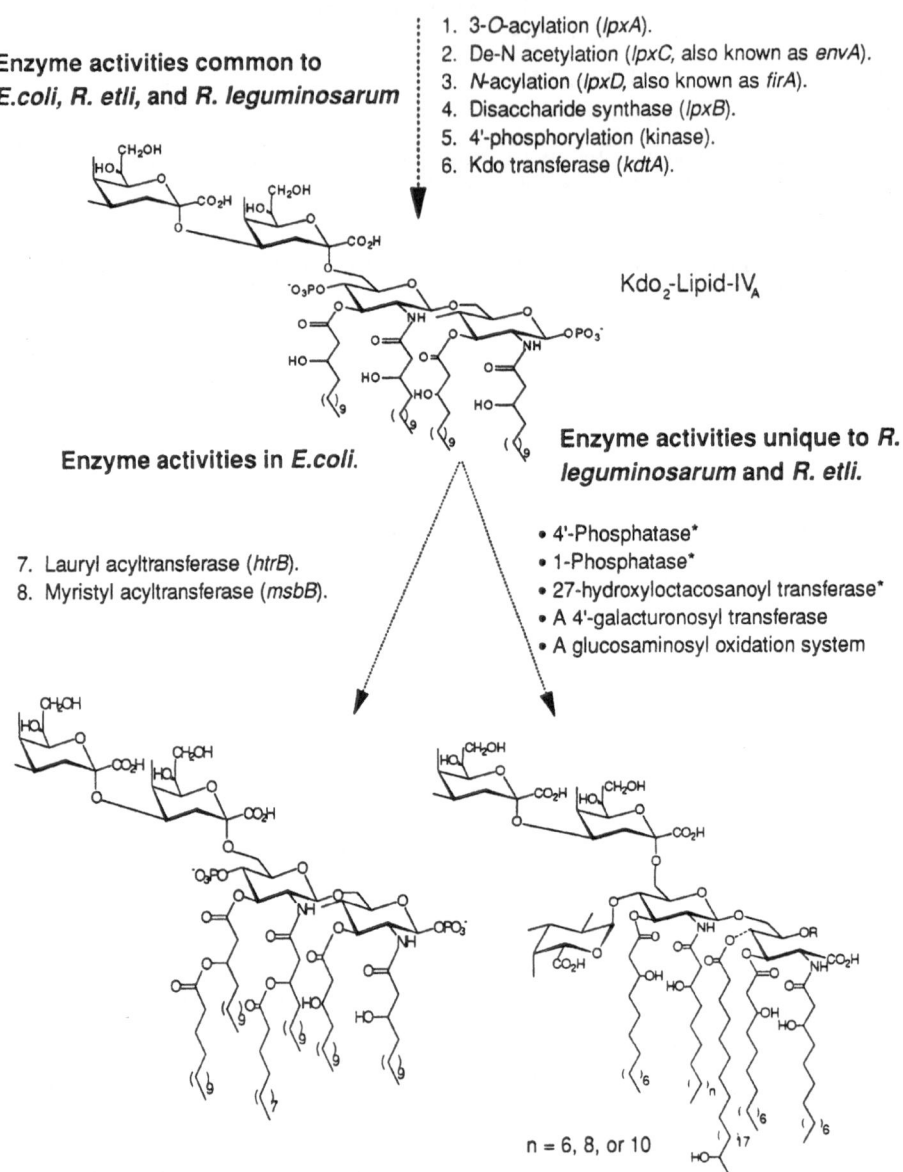

UDP-GlcNAc

Enzyme activities common to
E.coli, R. etli,* and *R. leguminosarum

1. 3-*O*-acylation (*lpxA*).
2. De-N acetylation (*lpxC,* also known as *envA*).
3. *N*-acylation (*lpxD,* also known as *firA*).
4. Disaccharide synthase (*lpxB*).
5. 4'-phosphorylation (kinase).
6. Kdo transferase (*kdtA*).

Kdo$_2$-Lipid-IV$_A$

Enzyme activities in *E.coli*.

7. Lauryl acyltransferase (*htrB*).
8. Myristyl acyltransferase (*msbB*).

Enzyme activities unique to *R.* leguminosarum and *R. etli*.

- 4'-Phosphatase*
- 1-Phosphatase*
- 27-hydroxyloctacosanoyl transferase*
- A 4'-galacturonosyl transferase
- A glucosaminosyl oxidation system

n = 6, 8, or 10

Figure 5. A diagram showing the enzyme activities that are common to the biosynthesis of both *E. coli, R. etli,* and *R. leguminosarum* lipid-As, and the point of divergence in their biosynthesis. The enzymes that are common to both species are those which are required for cell viability in *E. coli*. The unique processing enzymes present in *R. etli* and *R. leguminosarum* are hypothesized not to be required for cell viability. Those marked with "*" have been detected and partially characterized (see text). The genes encoding the various enzymes are given in the parentheses.

in LPSs that completely lack the O-chain polysaccharide (Brom *et al.*, 1992; Bhat *et al.*, 1991a; Cava *et al.*, 1990; Cava *et al.*, 1989). In enteric bacteria, *e.g. E. coli* or *Salmonella*, the genes for core synthesis are clustered on a region of the chromosome known as *rfa*, and those for O-chain polysaccharide synthesis on another chromosomal region called *rfb* (Schnaitman *et al.*, 1993). Thus, as stated by Cava (Cava *et al.*, 1990), these results show that the arrangement of the *lps* genes in *R. etli* CE3 is quite different from that in *Salmonella* or *E. coli*.

Chromosomal regions responsible for the presence of O-chain polysaccharides have also been identified in other strains of *R. leguminosarum* bv. *viciae* (Kannenberg *et al.*, 1992; de Maagd *et al.*, 1989b; Priefer 1989), and *R. leguminosarum* bv. *trifolii* (Brink *et al.*, 1990). An *R. leguminosarum* bv. *trifolii* mutant of ANU843 was reported to be complemented by the α-region from *R. etli* CE3 resulting in an LPS which contained the CE3 O-chain and not the ANU843 O-chain (Brink *et al.*, 1990). This result supports the above structural data showing that the core regions for *R. etli*, *R. leguminosarum* bv. *viciae*, and *R. leguminosarum* bv. *trifolii*, *i.e.* the acceptors for O-chain transfer, must have identical structures. In addition to the α-, β-, and λ- regions, another region has been identified that is required for "smooth" LPS synthesis; *i.e.* mutations in this region result in mutants that have only "rough" LPS (Allaway *et al.*, 1996; Poole *et al.*, 1994). This *lps* region is clustered with genes involved in dicarboxylic acid transport (*dct* genes) (Allaway *et al.*, 1996; Poole *et al.*, 1994). One such mutant appears to lack the terminal Kdo residue that is attached to O-6 of the core galactosyl residue (Allaway *et al.*, 1996).

It is apparent from the above studies that the genetics of *R. leguminosarum* LPS synthesis is an area that still needs considerable work. Of particular interest are those genes, which convert the $Kdo_2lipid-IV_A$ precursor into the unique *R. leguminosarum* lipid-A. These genes are probably not required for cell viability, and, therefore, it should be possible to obtain mutants in which the LPS contains lipid-A that is phosphorylated, that does not contain the 27-OHC28:0 fatty acid (*e.g.* an *orf** mutant), or does not contain the galacturonosyl residue at the 4'-position. Given the importance of the bacterial membrane in symbiosis, the effects of such mutations on the core region, on the presence or absence of an O-chain, and on infection of the host legume would be of great interest.

II.F. RHIZOBIAL LPS STRUCTURAL VARIATION DURING SYMBIOTIC NODULE DEVELOPMENT

Historically, the *Rhizobium* LPS was first addressed chemically (Carlson 1984; Carlson 1982; Carlson *et al.*, 1978) and biochemically (Dazzo 1979; Kamberger 1979). The chemical approach, to date, has lead to considerable insight into *Rhizobium* LPS structure (discussed above). The biological functions are less clear and still provide the focus for our current efforts in LPS research. The combination of genetic, immunochemical, cytological, and immunocytological approaches has shaped current understanding regarding the functions of LPSs in symbiosis.

II.F.1. The Rhizobium LPS and its Structural Adaptations in Plant Nodules and in Free-Living Culture.

Detailed biochemical investigations of LPS expression in free-living rhizobia and in nodule bacteria have been employed to investigate whether or not the LPS undergoes structural modifications during symbiosis. Because the plant challenges the bacteria with a series of very different microenvironments, surface adaptations are to be expected. Also, investigations into the LPSs of animal pathogens have suggested that LPSs respond to changes in bacterial environments and can exhibit structural modifications (see below and, for a recent review, Proctor *et al.*, 1995). LPS structural modifications and adaptations in rhizobia are, therefore, areas that are currently being investigated in some detail.

Pertinent to this question has been the development of sets of monoclonal antibodies (MAbs) with specificity for the O-chain moiety of the LPS (Wright 1990; Tao *et al.*, 1992; Sindhu *et al.*, 1990; Johansen *et al.*, 1984; Olsen *et al.*, 1994; Kannenberg *et al.*, 1994c Kannenberg *et al.*, 1992; de Maagd *et al.*, 1989a; Brewin *et al.*, 1986; Bradley *et al.*, 1988), or for LPS core structures (Lucas *et al.*, 1996). These MAbs have served as molecular probes to investigate subtle changes in LPS structures that occur during symbiosis, or when bacteria are subjected to culture conditions thought to mimic those within the nodule.

LPS Epitope Expression on Free-Living and Nodule Rhizobia. Biochemical investigations with monoclonal antibodies (Mabs) have revealed

changes in LPS epitope expression during symbiosis. Thus far, these LPS structural changes have been located in the O-chain portion of the LPS. Differential expression of epitopes has been observed in rhizobia from free-living cultures that had been subjected to different growth conditions and in nodule-derived bacteria. These observations have led to our current understanding that different types of LPSs can be expressed in rhizobia and that this expression depends on the environment experienced by the bacteria (Kannenberg et al., 1994c; Sindhu et al., 1990; Kannenberg et al., 1989; VandenBosch et al., 1989; Kannenberg et al., 1992; Tao et al., 1992; Olsen et al., 1994; Wood et al., 1989; Bradley et al., 1988; Brewin et al., 1986). To date, LPS epitope expression has been investigated most thoroughly in strains of R. leguminosarum and R. etli. LPS epitopes are expressed either constitutively or their expression is regulated. Among regulated epitopes, expression, or lack of expression, has been observed as a consequence of environmental changes. From these investigations, LPS epitopes have been grouped into different classes. One class of regulated epitopes is predominantly expressed on the LPSs from free-living bacteria, while another class is normally only found on the LPS of nodule bacteria (Sindhu et al., 1990; Kannenberg et al., 1994c Kannenberg et al., 1992; VandenBosch et al., 1989; Tao et al., 1992).

To reveal the effect of different symbiotic hosts on LPS expression, investigators have used pairs of near isogenic strains of R. leguminosarum differing only in their symbiotic plasmids (Sindhu et al., 1990), one contains a plasmid permitting nodulation of pea, and the second contains a plasmid permitting nodulation of bean. While LPSs from these isogenic strains show no differences in size and antigenicity under free-living conditions, LPSs isolated from bacteria within pea or bean nodules differ in molecular size and in LPS-I antibody reactivity from each other and from the LPSs of free-living bacteria (Sindhu et al., 1990). These host-related modifications have been observed in two pairs of serologically independent R. leguminosarum strains, demonstrating that they are not peculiar to individual strains (Sindhu et al., 1990). Hence, rhizobia (which in free-living cultures express similar or identical LPSs) in different symbiotic hosts, such as pea and bean, have modifications in LPS expression that are somewhat host-dependent. It is also possible that these LPS modifications are not so much host-dependent but dependent on whether the host forms indeterminate (pea) or determinate (bean) nodules. In any case, it

seems that LPS structures are finely tuned to the needs of different host nodule environments and are governed by nodule environmental cues that are currently being investigated in some detail.

LPS Structural Adaptation During Pea Nodule Development. In pea nodules induced by strain R. leguminosarum 3841, VandenBosch et al., (VandenBosch et al., 1989) showed first that an epitope on the LPS O-chain (recognized by monoclonal antibody MAC 203) is predominantly expressed in bacteroids after their release into the plant cytoplasm inside the symbiosomes but well before expression of the nitrogenase, see Figure 6C, F. By employing a panel of antibodies with specificity for LPS O-chain epitopes, a more complete picture of epitope expression in pea nodules emerged (Kannenberg et al., 1994c). Constitutively expressed epitopes, as expected, were expressed throughout all invaded parts of the nodule, Figure 6B. Regulated LPS epitopes displayed two general patterns of expression: epitopes that follow an approximately radial pattern of expression, Figure 6C, or lack of expression, Figure 6D, and those that follow a linear axis of symmetry along the axis of nodule development, Figure 6E. For comparison, the expression of nitrogenase, which, too, is expressed along the linear axis of nodule development, is depicted in Figure 6F. Furthermore, in pea nodule development, serologically independent R. leguminosarum strains give essentially the same patterns of epitope expression (Kannenberg et al., 1994c), showing that LPS epitope expression is not a random phenomenon of an individual R. leguminosarum strain.

The observed changes in LPS epitope expression in pea nodules would suggest that most are taking place after release of the bacteria into the symbiosome and during bacterial multiplication (VandenBosch et al., 1989). This result implicitly suggests that the LPSs expressed by bacteria within the infection threads largely resemble those of free-living bacteria. Variations in epitope expression have also been observed during transition to mature bacteroids as well as at later stages of bacteroid development (Kannenberg et al., 1994c). The abrupt transition in LPS epitope expression (Goosen-de Roo et al., 1991; Kannenberg et al., 1994c VandenBosch et al., 1989) casts some doubt on whether the observed modifications are manufactured by *de novo* biosynthesis of LPS molecules. Goosen-de Roo et al. (1991), who investigated LPS-I epitope expression

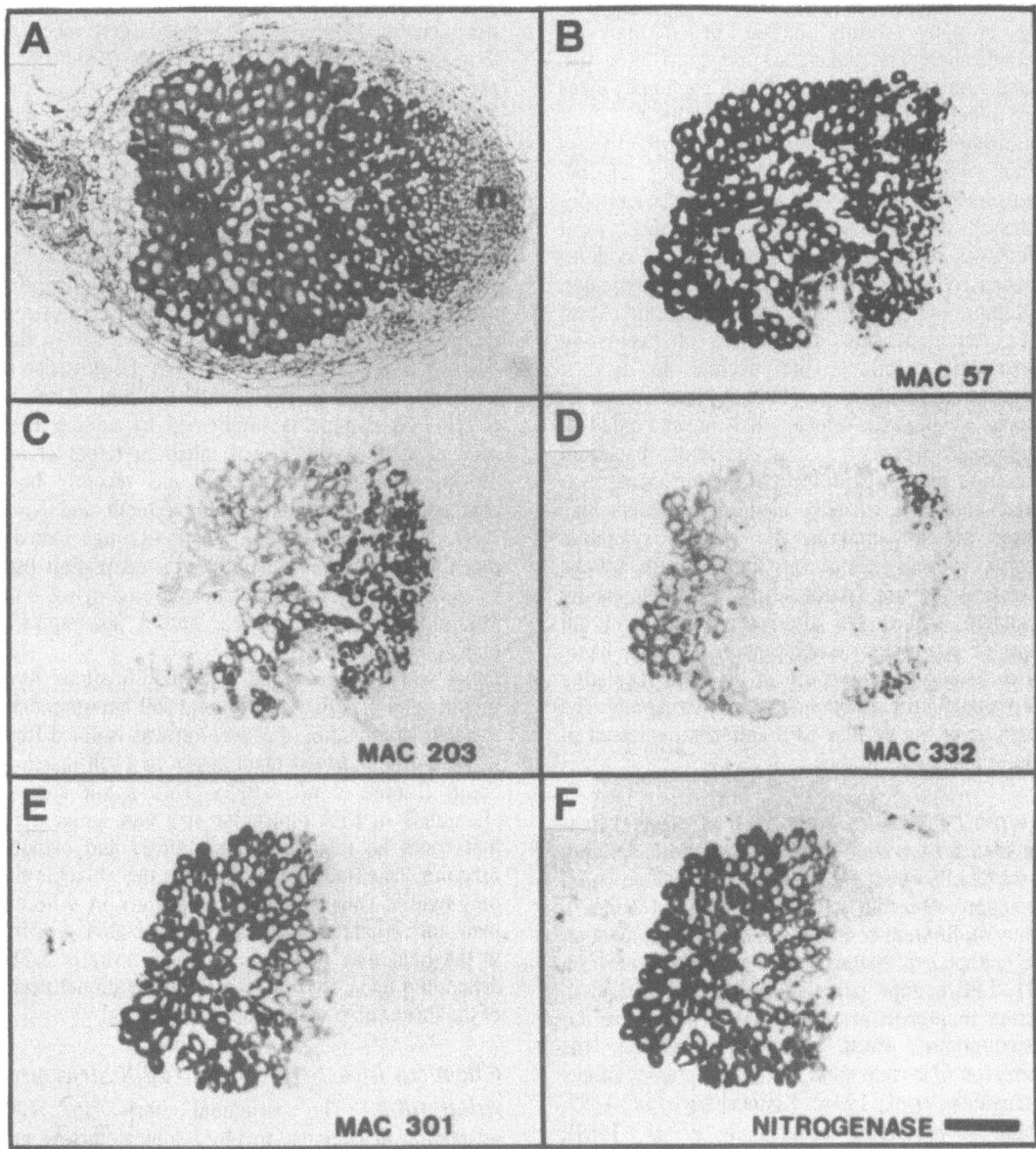

Figure 6. Longitudinal section of pea root nodules induced by *Rhizobium leguminosarum* bv. *viciae* three weeks after infection. (A) Toluidine blue-stained nodule section containing a wild-type strain 3841: m, uninfected apical meristem, followed towards the root (r) by the infection thread containing the invasion zone and the central tissue, which contains the nitrogen-fixing bacteroids; the central tissue is surrounded by an outer sheath of uninfected parenchyma. (B-H) Adjacent longitudinal sections after treatment with rat monoclonal antibodies (the antibodies are named in the picture) to reveal *in situ* expression of LPS-I epitopes followed by immunogold staining and silver enhancement. Section (F) was stained with rabbit antisera specific for nitrogenase. (Note: The sections were not counter stained with toluidine blue. Hence, only bacterial staining can be seen; plant tissues [root and nodule base, nodule cortex, and uninfected apical meristem] are not seen.) The staining pattern for a constitutively expressed epitopes is shown in (B); staining patterns of regulated epitopes are shown in (C, D, E), leading in (C) to expression and in (D) to suppression of the epitopes. The epitopes in (C, D) show in their expression pattern approximately radial symmetry; sections (E) shows the epitopes that seem to relate to developmental stages (for comparison see the expression of the nitrogenase in section [F]) and are expressed in an axial pattern along the axis of development in the pea nodule. Bar = 0.5 mm (Photographs courtesy of S. Perotto, reproduced with permission from Ref. Kannenberg *et al.,* 1994c)

of *R. leguminosarum* in *Vicia sativa* nodules, observed that epitopes present on rhizobia disappear very abruptly shortly before or during cell internalization. The authors suspected that an active degradation or modification of LPS-I epitopes takes place rather than repression of their synthesis.

The mechanism by which these LPS epitope changes are regulated is unclear. Kannenberg *et al.*, (Kannenberg *et al.*, 1994c) noted that the epitopes expressed in a radial fashion could reflect local differences in physiological conditions of nodules experienced by *endo*symbiotic bacteria. In particular, differences in oxygen supply to the bacteroids seem to be important since the pattern of expression approximately follows the decline in oxygen concentration (Witty *et al.*, 1986). At least one of the epitopes expressed along the linear axis of development seems to correlate with bacteroid maturation and the expression of the nitrogenase. (It is not known if rhizobia unable to express this epitope are impaired in the normal symbiotic infection process.) Since variations in LPS epitope expression are not confined in a developmentally dependent manner but also occur throughout all stages of pea nodule development, it seems likely that a number of factors are involved in regulating and producing the different structural variations. The structures of the various LPS epitopes expressed in nodules are not yet known.

In vitro LPS epitope expression. Several lines of evidence have shed some light on the underlying causes of LPS epitope expression and LPS structural adaptation. Rhizobia have been cultured under a variety of different conditions to study their effect on LPS epitope expression. In *R. leguminosarum* strain 3841, LPS epitope expression has been studied in relation to growth at acidic pH or under low O_2 concentrations. Both conditions induced the expression of certain epitopes and suppressed others (Kannenberg *et al.*, 1994c; Kannenberg *et al.*, 1992; Sindhu *et al.*, 1990; Kannenberg *et al.*, 1989), indicating that physiological factors are involved in LPS epitope expression. In another investigation, LPSs in bacteroids from strain *R. etli* CE3 reacted weakly with two monoclonal antibodies which normally react strongly with the LPS from laboratory cultured bacteria. The same reactivity was observed in LPSs isolated from bacteria cultured under a range of regimes: acidic pH, high temperature, low phosphate, or low oxygen concentrations (Tao *et al.*, 1992). In strain *R. leguminosarum* 3841,

investigators found that bacteria cultured in complex medium at reduced oxygen concentrations and at a near-neutral pH express LPSs that largely resemble those of the corresponding bacteroids (Kannenberg *et al.*, 1992).

The LPS epitopes which correlated with developmental stages could not be significantly expressed under laboratory culture conditions using defined culture medium (Kannenberg *et al.*, 1994c). This finding led to speculation that a plant-derived compound may be involved in the expression of this type of epitope. Recently it has been shown that plant factors play a role in LPS epitope expression. Reuhs *et al.*, (Reuhs *et al.*, 1994a) have shown that *S. fredii* LPS changes in response to the presence of host root extract in the growth medium. Also, an *R. etli* LPS epitope is suppressed by adding bean seed exudate to the growth medium (Noel *et al.*, 1996b). The active compound has recently been characterized as an anthocyanin (Duelli and Noel, 1997; Noel *et al.*, 1996a). These findings indicate that LPS expression and molecular adaptation may be generated through mechanisms employing both physiological conditions and plant-specific molecular signals.

Other workers examining LPS modifications have investigated the effect of different soil environments. A halotolerant strain of *S. meliloti* was isolated from nodules of a *Melilotus* plant grown in a salt marsh in Spain (Lloret *et al.*, 1995). This strain showed alterations in LPS molecular size and antigenicity that could be related to ionic stress and osmotic pressure. This finding may indicate that rhizobia not only exhibit considerable heterogeneity in different plant micro-habitats, but also that rhizobia growing in the field may display a complex array of LPSs depending upon, and serving the adaptational needs of the local micro-environments in the soil.

Chemical Investigations into LPS Structural Adaptation. The structural basis for LPS adaptation in response to physiological factors and during symbiosis is not yet adequately characterized. Bhat *et al.*, (Bhat *et al.*, 1992) have shown that the O-chain of the LPS from *R. etli* strain CE3 (a derivative of CFN42) grown at pH 4.8 has an altered methylation pattern. These changes correlate with the weak binding of two monoclonal antibodies to the LPS from bacteria cultured at low pH (4.8) compared with the strong binding when bacteria are cultured at neutral pH. This switch in the methylation pattern of LPS was also observed in the

LPS from nodule bacteria (Bhat *et al.*, 1992). The LPSs from free-living and pea nodule bacteria of strain *R. leguminosarum* 3841 have also been isolated by hot phenol-water extraction (see above) and partially characterized. The LPSs from free-living bacteria extracted predominantly into the water phase, while those from nodule bacteria were more evenly distributed between the two phases. These findings suggest that LPSs expressed in pea nodules are considerably more hydrophobic in character than those obtained from free-living bacteria. Partial chemical characterization indicated that this increase in hydrophobicity in nodule bacterial LPS may be due to changes in the carbohydrate part of the molecule, and to a higher proportion of long-chain fatty acids (Kannenberg *et al.*, 1994b).

II.F.2.Rhizobium LPS Mutants in Nodule Development

Symbiotic phenotypes of mutants with defects in their LPSs have been widely assessed to evaluate possible LPS functions in nodule development. As pointed out by Noel (Noel 1992), the majority of *Rhizobium* LPS mutants do indeed form nodules. Many instances have been reported in which nodule development is impaired by LPS mutants (López-Lara *et al.*, 1995; Niehaus *et al.*, 1994; Chapman *et al.*, 1990; Stacey *et al.*, 1991; Kannenberg *et al.*, 1992; de Maagd *et al.*, 1989b; Priefer 1989; Noel *et al.*, 1986), often leading to a Fix⁻, or strongly reduced nitrogen-fixing ability. Mutants of this kind, which can have incomplete nodule development, have also been designated as "nodule development mutants" or Ndv⁻.

The most severe phenotypes have been reported for LPS mutants, which are symbionts of hosts that form determinant nodules. For example, a mutant of *B. elkanii* strain 62A101c lacking the O-chain moiety of the LPS has, depending on the soybean variety, three different phenotypes (Stacey *et al.*, 1991). On the roots of a number of soybean varieties this mutant did not induce any nodules. On others, it caused white empty bumps or callus-like structures with severe abnormalities (*i.e.* there is no evidence for bacterial invasion or infection threads, and these structures have central vascular tissue and not the lateral vascular tissue of a normal nodule) and localized cell death of epidermal cells. A non-nodulating LPS-mutant of *B. japonicum* USDA110 has also been described (van de Wiel *et al.*, 1990). This mutant lacks the O-chain of the LPS

but has also additional poorly characterized modifications in surface composition (van de Wiel *et al.*, 1990). A second non-nodulating *B. elkanii* mutant (strain 61A76) apparently lacks only certain sugars of the O-chain (Maier *et al.*, 1978).

Rhizobium etli mutants lacking the LPS O-chain polysaccharide are also symbiotically defective in that they cause severe aberrations, inducing small white nodules on *Phaseolus* beans (another host forming determinate nodules) which are devoid of leghemoglobin (Noel *et al.*, 1986). A nodule meristem is formed, and development seems normal for the first few days but becomes aberrant later, showing central, rather than the normal peripheral, vascular bundles. With these mutants, infection is aborted in the root hair cell in strangely bloated infection threads with the bacteria embedded in a matrix structure not seen in the normal infection thread (Noel *et al.*, 1986). From the protein content of nodules induced by LPS-impaired rhizobia on beans, Noel *et al.*, (Noel *et al.*, 1986) concluded that these nodules resemble mature roots rather than normal nodules even though nodule development seems normal for the first few days. The molecular basis for the subsequent deterioration of the mutant-induced nodules is not known. Finally, no LPS-impaired mutant rhizobia have been isolated that carry infection beyond the above-described abnormalities in determinate nodules, making it difficult to assess LPS functions for later stages of cell invasion and development in this type of nodule. On hosts forming indeterminate nodules (*e.g.*, pea, vetch, and clover), *R. leguminosarum* bv. *viciae* LPS mutants completely lacking the LPS O-chain develop nodules with abnormal infection threads and only sporadically release of bacteria into nodule cells (Perotto *et al.*, 1994; de Maagd *et al.*, 1989b; Priefer 1989). The bacteroids senesce early and fix little or no nitrogen. An LPS mutant of *R. leguminosarum* bv. *trifolii* strain ANU843 lacking the O-chain moiety of the LPS induced impaired nodules on clover with similar defects (Chapman *et al.*, 1990). Detailed cytological analysis in pea nodules revealed that the walls of infection threads show secondary modifications and possibly callose deposition, accumulation of an intercellular matrix largely composed of a plant-derived glycoprotein, and sporadic cell death. These modifications have been interpreted as mild plant host defense reactions (Perotto *et al.*, 1994). Furthermore, LPS-defective mutants released into pea plant cells differentiated into abnormally swollen and elongated bacteroids, Figure 7. Apparently, rhizobial cell division is impaired, and the synchronized growth and division of the bacterial and peribacteroid membranes (PBM), which occurs during development of indeterminate nodules in pea, is disrupted (Perotto et al., 1994).

The symbiotic phenotypes of LPS mutants from S. meliloti are somewhat confusing. Clover et al., (Clover et al., 1989) concluded that the LPS plays no role in nodule formation on alfalfa. This conclusion was based on the observation that LPS mutants of S. meliloti strain SU47 form normal nodules under most circumstances. However, these mutants have not been well characterized chemically. Analysis of the extracted LPS by PAGE showed differences in mobility from that of the wild-type strain but these differences have not been further characterized. All mutants apparently had both LPS-I and LPS-II. Lagares et al., (Lagares et al., 1992) studied a mutant derivative of strain S. meliloti Rm2011. The LPS of the S. meliloti Rm2011 mutant, although not fully characterized chemically, seems also to contain an LPS O-chain, albeit a modified one (Lagares et al., 1992). The mutant demonstrated reduced and delayed abilities to nodulate primary alfalfa (Medicago sativa) roots and was less competitive in a nodulation assay than the wild-type strain. In contrast, the same mutant formed aberrant nodules on M. truncatula (Niehaus et al., 1994). The mutant cells were released from the infection threads but failed to multiply within the plant's cytoplasm, which, instead, showed numerous vesicles. Also, fresh nodule sections showed brown necrotic cells within the central nodule. Overall, this is a phenotype with some similarity to the one described for LPS mutants of R. leguminosarum. Sinorhizobium meliloti mutants, as well as S. fredii mutants, which completely lack the O-chain of the LPS remain to be isolated and assessed for their symbiotic phenotypes. Finally, evidence is growing that the surface molecules of Sinorhizobium strains display some contrasting properties to those of other rhizobia, which may further complicate the picture about structure function-relationships in the different rhizobia (for a more detailed discussion of this aspect, see section II.B.2.).

Three other classes of LPS mutants complete the current picture of what is known about LPS structures necessary for symbiotic functioning. One class comprises LPS mutants with severely truncated O-chains but still retaining a certain amount of O-chain. In nodulation assays with bean and pea, these mutants had the same phenotypes as those which completely lack the O-chain (Perotto et al., 1994; Kannenberg et al., 1992; Cava et al., 1989; de

Maagd et al., 1989b). This class of mutants shows that a single, or even a very few, O-chain repeating unit is not sufficient to restore full symbiotic competence. A second class of mutants has been described that produces an LPS-I structure but at approximately one third the normal level (Cava et al., 1989). This mutant is as deficient in infection ability as mutants that entirely lack an LPS-I, a result that has been put forward to argue against a signal function for LPSs since such a function would probably be satisfied by even low concentrations of LPS-I. However, this conclusion may be premature for two reasons. The structure of LPS expressed under laboratory conditions may differ from that expressed in the nodule. Furthermore, this particular mutant lacks quinovosamine, a sugar normally found in the LPS O-chain (Tao et al., 1992, L.S. Forsberg, personal communication). Hence, this mutation produces pleiotropic effects on the LPS, altering both the structure and the amount of LPS-I expressed. A third class of mutants has been described that expresses distinctly different or modified O-chain polysaccharides than does the parent strain (Kannenberg et al., 1992; Brink et al., 1990). This class of mutants was as fully effective in symbiosis as was the parent, suggesting that the specific structure of the O-chain is not essential for LPS function. However, it is possible that the LPS O-chain serves as a rather non-specific scaffold for more specific structural decorations; e.g. acetyl or methyl groups, or a branching sugar, that are important in symbiotic infection. This line of thinking would indicate that there might be a class of enzymes that modify certain O-chain sugars independently of the specific LPS O-chain structure.

Drawing firm conclusions from the phenotypes of LPS mutants with regard to the LPS structure/function relationship is complicated for several reasons. First, mutants with major defects in their LPS structures could be pleiotropic with regard to the entire bacterial surface. Complete removal of the O-chain polysaccharide most certainly exposes other outer membrane components to the plant cell; e.g. the LPS core region, proteins, and other carbohydrates such as the K-antigens. The interaction of these other molecules with the plant cell surface may be the cause of the deleterious effects on symbiosis. These mutants may also contain a weakened and possibly leaky membrane allowing exchange of molecules that are deleterious to symbiosis. Also, the bacterium may compensate for the dramatic alteration in LPS by modifying and/or producing more of other surface components.

In fact, this has been reported for one *R. leguminosarum* LPS mutant in which there was a dramatic increase in the amount of cyclic β-2-linked glucan and a ten-fold reduction in the amount of a neutral capsular polysaccharide (Breedveld *et al.*, 1993). In spite of these complexities, all of the above results, have shaped our current understanding of LPS structural necessities. For proper function in symbiosis, complete LPS molecules are needed, *i.e.*, they must be composed of a sizable O-chain, core oligosaccharide, and lipid A and, possibly, in sufficient amounts (Noel 1992; Kannenberg *et al.*, 1992).

II.F.3. Possible Roles of LPS in Nodule Development

Throughout nodule development rhizobia must modulate plant development, all the while avoiding the elicitation of plant defenses and maintaining cell vitality. No clear-cut functions in any of these generalized aspects of symbiotic development have so far been assigned to LPSs. In this section, the possible roles of *Rhizobium* LPSs in the following stages of nodule development are discussed: infection initiation, root tissue invasion, bacterial release, bacteroid development, and bacteroid senescence.

Infection Initiation. Initiation of infection thread formation requires live rhizobia for LCO synthesis, but also their physical presence at the initiation site (Van Spronsen *et al.*, 1994; Kijne 1990; Brewin 1991). It is not fully understood why this is the case, but close proximity of the plant and bacterial surface molecules may be a requirement for infection thread formation and may also explain why attachment of rhizobia to root hairs is apparently required for infection initiation (Kijne *et al.*, 1992; Kijne 1990).

The fact that most LPS mutants are able to invade plant tissue to some degree would suggest that the LPS is not involved in initiating infection. However, Dazzo *et al.*, (1991) found that the LPS of *R. trifolii* 0403 binds rapidly to root hair tips of clover and apparently promotes infection tread formation. This promotion effect was not observed with LPSs from heterologous rhizobia. While this observation suggests a host-symbiont specific enhancing effect of the LPS on infection thread formation during nodule initiation, the mechanism of this influence is not clear. Two other lines of indirect evidence may corroborate the proposition of LPS involvement in attachment. First, the failure of *Bradyrhizobium* LPS

mutants to nodulate supports the involvement of LPS in infection thread formation on a determinant nodule-forming host (soybean). Second, the *S. meliloti* LPS mutant described by Lagares *et al.*, (Lagares *et al.*, 1992) leads to a reduced number of infection events on its indeterminate host, alfalfa, which may be a contributory factor in the reduced competitiveness of that strain.

Root Tissue Invasion. Generally, in hosts forming determinate nodules the tissue invasion process by LPS mutants seems more severely affected than in hosts forming indeterminate nodules, and observations of necrosis and cell death indicate that host defenses are elicited in these nodules (Stacey *et al.*, 1991; Noel *et al.*, 1986). The reasons for the observed differences in the phenotypes of LPS mutants between determinate and indeterminate nodule-forming hosts (section II.F.2.) are unclear, but may reflect differences in host plant microenvironments, modifications in the barrier properties of the mutant cell walls, and, possibly, more refined LPS plant interactions.

Brewin (Brewin 1991) has argued that the difference in phenotypes may reflect their different modes of infection. In the case of indeterminate nodule-forming hosts, LPS-dependent modifications preferentially affect cell internalization and bacteroid development leading to an interruption in symbiosis at a relatively late stage. It is also possible that in the case of determinate nodule-forming hosts, the close proximity of the plant and mutant bacterial cell surfaces in the infection threads may have a detrimental effect, or the volume of the remaining lumen within the infection thread be so small that relatively high concentrations of root-exuded antimicrobial compounds, such as phytoalexins, develop. Phenotypes of LPS-mutant rhizobia could then be explained by increased sensitivity towards toxic compounds in these mutants, resulting in the observed abortion of infection threads. In an attempt to address directly the sensitivity of LPS-mutant rhizobia towards antimicrobial agents, Eisenshenk *et al.*, (Eisenschenk *et al.*, 1994) showed that at least some mutants of *R. etli* CE3 lacking the LPS O-chain are considerably more sensitive than the parent strain to root extract compounds. However, the relevance of this observation is not yet clear since another LPS mutant (also inducing impaired nodule development) did not show the increased sensitivity to root extract. The phenotypes induced by LPS-deficient rhizobia on hosts which form determinate nodules show some similarity to those described for EPS-deficient rhizobia on hosts forming indeterminate nodules (*i.e.* no nodules at all or nodules almost or completely devoid of bacteria (Noel 1992)). It has been shown that small amounts of EPS fractions can rescue the EPS-deficient mutants (Battisti *et al.*, 1992; Djordjevic *et al.*, 1987; Gonzales *et al.*, 1996). These findings indicate that EPSs may be involved in a signal function. However, EPS-deficient mutants can still effectively nodulate determinate nodule-forming hosts, indicating that either EPS is not needed or that its function is substituted by some other surface compound. Based on the plant phenotypes, it is tempting to speculate if LPS could functionally play an EPS-like role in the nodulation of determinate nodule-forming hosts. However, to verify this two important control experiments have yet to be performed. It has not been reported if LPS, or the LPS O-chain, can rescue the defective nodulation of determinate hosts by the LPS-deficient mutants; it is also not clear if in the nodulation of determinate hosts EPS is functionally substituted by another recently discovered surface polysaccharide, a K-antigen (see section III), as it is the case in the nodulation of alfalfa by EPS-deficient mutants of some strains of *S. meliloti* (Reuhs *et al.*, 1995; Petrovics *et al.*, 1993).

In the spreading infection threads of indeterminate nodules, rhizobia are surrounded by a plant cell wall that forms a dense layer between the bacterial cell surface and the cytoplasmic plant cell membrane. Ions, such as boron and calcium, have been implicated in nodule development (Bolaños *et al.*, 1994; Munns 1970; Sethi *et al.*, 1981) and are components of the plant cell wall. O'Neill *et al.*, (O'Neill *et al.*, 1996) showed recently that the pectic polysaccharides of plant cell walls may be covalently cross-linked by boron, and the importance of calcium as counter-ions for uronides and other negatively charged molecules has been discussed in chapter 22 and elsewhere (Brewin 1991). Interestingly, Bolaños *et al.*, (Bolaños *et al.*, 1996) have reported infection thread distortions in the development of boron-deficient pea nodules: enlarged and abnormally shaped infection threads that frequently burst and release bacteria into damaged host cells, an over-production of plant matrix material in which the rhizobial cells were embedded, and impaired bacterial release. The authors point out that these symptoms somewhat resemble the weak defense reaction observed with LPS mutants on pea (Perotto *et al.*, 1994). Boron forms ester bonds with hydroxyl groups, which are particularly stable with certain diols or polyols (Loomis *et al.*, 1992). Hence, hydroxyl groups on the exposed LPS core region in LPS-mutant rhizobia or on some other surface component expressed or exposed as a result of a certain mutation could sequester boron away from the plant cell wall leading to very low concentrations and, thus, causing detrimental effects on plant cell wall stability. Perhaps this could also explain why different classes of LPS-mutant rhizobia (O-chain-deficient mutants, mutants with truncated LPS, and those with reduced amounts of LPS-I) all seem to lead to the same or very similar impairments in their respective hosts. Another possibility (and the possibilities are not mutually exclusive) is that a released surface compound in LPS-mutant rhizobia has an effect on plant cell wall-cytoplasmic membrane connections (Mohnen *et al.*, 1993; Penel *et al.*, 1996), leading to impaired infection thread development.

Bacterial Release. Bacterial release takes place when the tips of the infection threads meet a newly formed meristematic host cell. At this point the tip of the infection thread is no longer sheeted by a cell wall, and the rhizobia are engulfed by the "naked" cytoplasmic membrane to form symbiosomes (Roth *et al.*, 1989; Brewin 1991). The mechanism seems to resemble endocytosis. However, mechanistically it is not clear if endocytosis or phagocytosis is taking place and if it is receptor- or integrin-mediated (Verma *et al.*, 1996). In either case, close contact of plant cytoplasmic membrane with the rhizobial surface may be critical; however, the molecules involved in that step are unknown. Also, the bacterial cell wall as a whole is presumably undergoing major changes at this point. Rhizobia, which in the infection threads are likely to be encapsulated by other surface carbohydrates (*e.g.* EPS or CPS), seem to lose their capsule prior to or during cell internalization (Latchford *et al.*, 1991; Roth *et al.*, 1991; Brewin 1991). There is little understanding of what triggers these changes (Kannenberg *et al.*, 1994a).

The drastically reduced numbers of invaded cells formed by LPS mutant rhizobia on pea or vetch, indicate that LPSs may be involved in cell internalization (Perotto *et al.*, 1994; de Maagd *et al.*, 1989b; Priefer 1989). The mechanisms of LPS involvement are not clear. However, a model in which LPSs play a role in attachment to the cytoplasmic membrane has some precedence in animal systems where the involvement of LPSs in cell adherence and invasion is under discussion (for a recent review see Jacques 1996).

Bacteroid Development. Inside the symbiosome compartment, the bacteria first multiply then develop into pleiomorphic, nitrogen-fixing mature bacteroids. In newly released symbiosomes, the expanding plant membrane divides with the dividing bacteria and develops into the functional PBM. In pea symbiosomes, this process happens in a synchronized manner, leading to bacteroids individually surrounded by the PBM. In bean symbiosomes, this synchrony breaks down, leading eventually to several bacteroids per PBM envelope (Brewin 1991). In pea symbiosomes, the PBM is apparently not covered by plant cell wall material (Brewin 1991). The fact that bacterial exopolysaccharide production also seems suppressed in symbiosomes (Latchford *et al.*, 1991; Roth *et al.*, 1991) would allow the bacterial cell surface to come in close contact with the PBM. Physical contact between the PBM and the bacteroids has been observed in symbiosomes (Roth *et al.*, 1991; Robertson *et al.*, 1985), but the function of this contact and the principle molecules involved are not understood.

The fate of bacteroids derived from rhizobia with impaired LPS structures has only been investigated in any detail in pea nodules (see section II.F.2.). There, the observed impairments in bacteroid developments have been taken as evidence that a complete LPS-I is essential for normal bacteroid development (Perotto *et al.*, 1994; Kannenberg *et al.*, 1992). During the stage of intracellular multiplication, the loss of synchronous division between the multiplying rhizobia and the PBM is puzzling, for details, see section II.F.2. and Figure 7. It could indicate that adhesion between the bacterial membrane and the PBM is involved in maintaining the synchrony and that this process is disrupted in bacteroids derived from mutant rhizobia with modified LPS structures. Some evidence for adhesion between the bacterial surface and the PBM has been observed by analysis of isolated membranes from nodule homogenates (Bradley *et al.*, 1986), but the molecular basis for this interaction is not understood (Brewin 1991). Therefore, the question of whether LPS is involved directly or indirectly in such an adhesion process is still unresolved.

Some clues about what can be expected in the case of direct LPS involvement in adhesion processes can be derived from the biochemical analysis of LPS

structural adaptation during bacteroid development. In peas, rhizobial bacteroids apparently have reduced amounts of LPS expressed (van Brussel *et al.*, 1977), with little or only trace amounts of LPS-II (Sindhu *et al.*, 1990). The LPS-I displays modified antigenicity, and molecular size (Kannenberg *et al.*, 1992; Sindhu *et al.*, 1990; Brewin *et al.*, 1986). Direct chemical and immunochemical characterization has shown that these LPS-I forms are more hydrophobic (Kannenberg *et al.*, 1996; Kannenberg *et al.*, 1994b). These findings may indicate that molecular size and hydrophobicity are important factors in adhesion processes, possibly facilitating the interaction between the PBM and the bacterial membrane. In accordance with these assumptions, the majority of the modifications in LPS epitopes seems to take place during rhizobial multiplication and after release into the symbiosomes, suggesting *de novo* biosynthesis of the more hydrophobic LPS forms during that stage of symbiosis (Kannenberg *et al.*, 1994c; Goosen-de Roo *et al.*, 1991; VandenBosch *et al.*, 1989). The fate of the pre-existing LPS forms during this process is not clear. It has been suggested that the rhizobial surface is sloughed off (Bal *et al.*, 1980). However, this mode of action is questionable and needs further substantiation. This gradual replacement of pre-existing LPS by modified LPS forms arising from *de novo* synthesis has also been questioned (Goosen-de Roo *et al.*, 1991) since it has been observed that LPS epitope changes occur rapidly suggesting that the pre-existing LPS was modified and not gradually replaced by newly synthesized molecules (Kannenberg *et al.*, 1994c; Goosen-de Roo *et al.*, 1991).

Understanding the functions of LPS in bacterial developments has been further complicated through the discovery that some LPS epitopes are expressed at defined developmental stages. Such changes in epitope expression suggest that, in addition to the overall increase in LPS hydrophobicity, there are specific LPS structural requirements. For example, in pea nodules the LPS epitope recognized by the monoclonal antibody MAC301 is expressed in close proximity to nitrogenase expression, Figure 6. However, this developmental stage-specific LPS epitope expression is apparently independent of nitrogenase expression since the MAC301 epitope is also expressed in nodules induced by rhizobial *nifH* mutants unable to fix nitrogen, (Kannenberg *et al.*, 1994c). Additionally, this developmental stage-specific expression pattern cannot be easily explained as being a function of the nodule oxygen status, and therefore,

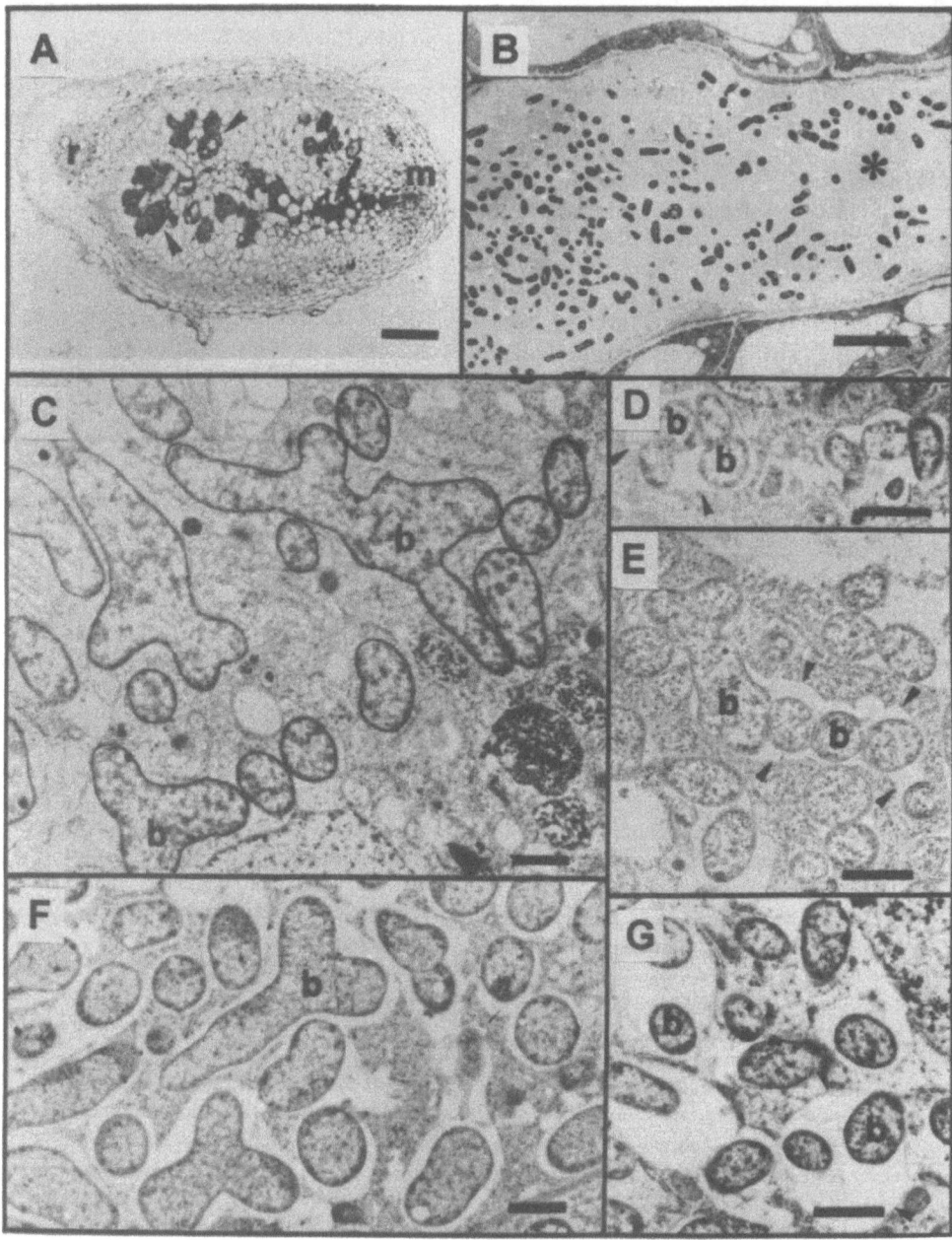

Figure 7. A pea nodule occupied by a mutant derivative of *R. leguminosarum* 3841 with a defective LPS four weeks after infection. (A) Toluidine blue-stained nodule section showing that a few infected cells contain bacteroids (arrowheads), large numbers of uninvaded cells, and a large invasion structure (arrow; scale bar = 0.5 mm). (B) Electron microscopic micrographs showing the large invasion structures in greater detail, revealing an abundant, amorphous matrix (asterisk) that embeds numerous bacteria (scale bars = 0.5 μm). The electron micrograph (C-G) show the development of bacteroids of the LPS mutant derivative and for comparison that of wild-type strain *R. leguminosarum* 3841. Mature bacteroids from the LPS-defective strain (C) that are enlarged and highly branched are compared with those from the wild-type strain (F); the early stages of LPS-defective bacteroids (D and E) showing higher numbers of bacteroids enclosed within the peribacteroid membrane (the position of this membrane is indicated by arrowheads); for comparison, the early stage of wild-type bacteroid development (G) showing no more than two bacteroids per symbiosome; Scale bars = 100 μm. (Photographs courtesy of S. Perotto, reproduced with permission from Ref. Perotto *et al.*, 1994)

additional factors (possibly plant-derived such as the anthocyanin compounds) could be involved in this type of LPS modification.

One of the general functions of the LPS is believed to be its role in maintaining the stability of Gram-negative bacterial outer membranes (Vaara 1992). In fact, the premature senescence of LPS-mutant bacteroids in pea nodules has been explained as a stability problem due to modifications in the outer membrane (Perotto *et al.*, 1994) and increased sensitivity toward the physiological stresses encountered in the symbiosomes. This interpretation finds some backing from a recent investigation into a pea mutant with impaired bacteroid development (Borisov *et al.*, 1992). In nodules induced by wild-type rhizobia, symbiosomes contained several bacteroids per envelope, a feature similar to bacteroid development of LPS-defective mutant rhizobia in a normal pea (Perotto *et al.*, 1994). While the bacteroids in nodules of this pea mutant failed to differentiate into nitrogen-fixing bacteroids, they did not degenerate quickly and in four-week-old nodules did not show the symptoms of premature senescence observed for LPS-mutant bacteroids in nodules of a normal pea. This observation could be taken as an indication that the wild-type rhizobial cell can withstand mutant pea symbiosome conditions, and that the premature senescence of LPS-mutant bacteroids is related to their cell wall modifications which may result in increased sensitivity to the *in planta* physiological conditions or antimicrobial compounds.

The differences in the modifications in LPS structure observed in bacteroids from near isogenic strains of *R. leguminosarum* (differing only in their symbiotic plasmids and exhibit radically different LPSs in pea versus bean nodules) presumably reflect differences in the nodule physiology between peas and beans. It is also possible that plant-derived components could be involved in regulating these putative host-specific changes. It is not known if there is a shift towards more hydrophobic LPSs in bean bacteroids. The differences observed in LPS expression in bean and pea bacteroids add to the growing list of developmental differences between determinate and indeterminate nodules that affect tissue and cell invasion by rhizobia and bacteroid development in symbiosomes (Roth *et al.*, 1991; Brewin 1991; Kijne 1990; Sutton *et al.*, 1981). Besides that, it seems likely that many LPS structural adaptations, especially those that are physiologically controlled, are not necessarily restricted to certain nodule environments (*i.e.*, they are not strictly nodule-specific) but may also have *ex*

planta functions in the soil. This more generalized role for LPSs in rhizobial growth may also help to explain why different physiological conditions can lead to similar variations in LPS epitope expression (Tao *et al.*, 1992; Kannenberg *et al.*, 1994c; Sindhu *et al.*, 1990; Kannenberg *et al.*, 1989).

Nodule Senescence. There is little understanding of what actually governs bacteroid senescence. It is generally assumed that the symbiosome constitutes a stressful environment for the occupying bacteroids (Roth *et al.*, 1991; Brewin 1991). It has been suggested that the symbiosomes may constitute a lytic compartment (Werner 1992) with pH values balanced by the metabolic and nitrogen-fixing activities of the bacteroids (Brewin 1991; Kannenberg *et al.*, 1989). In this model, acidification of the symbiosomes by plant action would tend to activate the lytic enzymes, eventually leading to bacteroid senescence. As acidification has also been shown to influence LPS epitope expression, it is noteworthy that certain LPS epitopes that had been suppressed in the younger parts of the nodule are expressed again in the older parts of the nodule (Kannenberg *et al.*, 1994c). It is not clear if the expression of these epitopes in bacteroids is directly related (*i.e.* the cause or consequence) to their senescence.

II.G. LPS STRUCTURAL ADAPTATION IN OTHER CELL-CELL INTERACTION MODELS

The study of pathogenic bacteria in animals has shown that LPSs seem to share certain features of structural adaptation and function with plant-associated bacteria. Functionally, in animal-bacterial interactions the LPSs also seem to have an important role for infection of, and survival in, the host, and, therefore, are important virulence factors (for recent general reviews, see Moran *et al.*, 1996; Goldberg *et al.*, 1996; Jacques 1996; Proctor *et al.*, 1995; Schletter *et al.*, 1995; Rietschel *et al.*, 1994; Raetz 1993). For example, when bacterial pathogens are examined immediately after isolation from infected tissue, they are found to possess LPSs with O-chain polysaccharides, and variants or mutants which lack the O-chain are avirulent (Whitfield *et al.*, 1993; Whitfield 1995; Schnaitman *et al.*, 1993; Lynn *et al.*, 1992; Foster *et al.*, 1995). These results indicate that complete LPSs (containing the O-chain, core, and lipid-A) are critical in animal-bacterial interactions. Animal pathogens are challenged by complement-mediated lysis and/or by phagocytosis. Recently, it was found that *S. enteritidis* isolated from chick spleens have LPSs with large O-chain:core ratios, indicating a high number of O-chain repeats in the LPS molecules (Rahman *et al.*, 1997). In general, bacteria that display LPSs with bulky O-chains are more resistant to complement-mediated killing and/or phagocytosis. The bulky LPS molecules are thought to act by forcing the complement membrane attack complex to form too far away from the bacterial membrane to be effective (Valvano 1992). This shows that certain LPS sizes and concentrations can be critical in cell-cell interactions.

Recently, in *Neisseria* it has been shown that modification (*e.g.* sialylation) of the LPS occurs and is thought to minimize the defense response by mimicking the host's cell surface glycoconjugates (Preston *et al.*, 1996; Van Putten 1993), also showing that specificity can be crucial in cell-cell interactions. However, the role of LPSs in animal pathogenicity is also far from being established and is still a matter very much under discussion. In particular the molecular mechanisms and functions of LPS action in the bacterial cell and in cell-cell interactions are still largely unknown. It is especially interesting that LPSs have recently been implicated in attachment to cell structures and to extracellular matrix components (Jacques 1996). These findings could develop into an exiting new avenue to elucidate the mechanisms of cell entrance by bacteria.

The above findings also demonstrate the striking flexibility in LPS molecular structure. It can be tailored in molecular size, quantity, and composition to the specific needs of the bacteria (survival and cell-cell interactions) in diverse microhabitats. Additionally, advanced genetic analysis of animal pathogens (*e.g.*, recently genes have been described that are involved in O-chain tailoring; for a review see Whitfield 1995) will allow investigators to see if such genes are conserved in rhizobia.

Investigations into the LPSs of plant pathogens, however, are largely in the descriptive stage (for a recent review on polysaccharides in plant pathogens, see Denny 1995). Denny (1995) suspects that the primary function of LPSs in pathogenic interactions may lie in their barrier properties against toxic plant compounds and in survival of the bacterial cells. Recent findings with LPS mutants of *Xanthomonas campestris* with a modified LPS core region seem to corroborate this suspicion (Dow *et al.*, 1995). These mutants showed altered behavior in a range of host and non-host plants resulting in a drop in the number of recoverable bacteria within the first 24 hours after inoculation.

Other reports argue for a more direct involvement of LPSs in plant virulence. Newmann *et al.*, (Newman *et al.*, 1995) have shown that isolated LPSs of *X. campestris* can induce defense-related gene expression in *Brassica campestris*, and it is known that inoculation of tobacco plants with purified LPSs from certain bacteria can prevent the hypersensitive response to incompatible bacteria and reduce symptoms in compatible interactions (Mazzucchi 1983). LPSs have also been implicated in the induction of systemic resistance against plant pathogens (Leeman *et al.*, 1995; Sequeira 1983). However, at the present time there is little understanding of underlying molecular mechanisms. Also, very little is known about whether LPSs are structurally modified in response to environmental stresses experienced by bacteria within plant habitats during pathogenic interactions.

III. K-Antigens.

The acidic capsular antigens of *E. coli* were named K-antigens (from "Kapselantigene") since they proved to be bacterial cell surface antigens that were distinct from the LPSs (i.e. O-antigens), and EPSs. K-Antigens, are tightly associated with the bacterial cells, do not impart a mucoid colony morphology, and are common components of many well studied bacteria (Whitfield *et al.*, 1993), forming an acidic polysaccharide sheath over the surface of the bacterial cell (see reviews, Jann *et al.*, 1991; Jann *et al.*, 1990). These polysaccharides are believed to protect the bacteria from host defense responses and are known to be essential for pathogenic virulence. The K-antigens of *E. coli* are divided into two subclasses, group I and group II, based on several criteria (Jann *et al.*, 1990): Group I K-antigens are characterized by a temperature insensitive expression, the lack of a phosphatidic acid anchor, and the predominance of large polysaccharide repeating units, containing uronic acids and substituents such as pyruvate as the acidic components. In this sense, the group I K-antigens are analogous to the EPS of rhizobia. It has been suggested that the *E. coli* group I K-antigens may be anchored to the OM via lipid A (Jann *et al.*, 1990), but this has not been convincingly established. The group I K-antigens have not been as thoroughly studied as the group II polysaccharides which are discussed further below.

III.A. GROUP II K-ANTIGENS.

The group II K-antigens include those that contain Kdo, sialic acid, amino acids, or phosphate as the acidic components (Jann *et al.*, 1991; Jann *et al.*, 1990). In contrast to the group I polysaccharides, the group II polysaccharides comprise relatively small repeating units of only one, two or three sugars, and have a much higher charge density (Jann *et al.*, 1991; Jann *et al.*, 1990). Phosphatidic acid is thought to be a membrane anchor for the group II K-antigens since this lipid component has been found linked to the reducing end of the K-antigen via Kdo (Jann *et al.*, 1991; Jann *et al.*, 1990). In *E. coli*, the expression of group II K-antigen capsules is temperature regulated, with capsular expression occurring at 20°C or higher (Orskov *et al.*, 1984).

Several reports have contributed to the understanding of the genetics of group II K-antigen expression in *E. coli* (Finke *et al.*, 1991; Boulnois *et al.*, 1993; Petit *et al.*, 1995; Boulnois *et al.*, 1987; Bronner *et al.*, 1993; Pazzani *et al.*, 1993). The genes are arranged in three adjacent regions, termed the *kps* gene cluster. Region 1 (~7 kb) encodes for six proteins, designated KpsF, KpsE, KpsD, KpsU, KpsC, and KpsS (Pazzani *et al.*, 1993). Functions have been assigned for some of the proteins based on the predicted amino acid sequences and the biochemical defects of the respective mutants. KpsE and KpsD are postulated to function in the export of the PS through the cell envelope, following translocation from the cytosol (Silver *et al.*, 1987; Wunder *et al.*, 1994; Rosenow *et al.*, 1995a). This step is thought to be accompanied by lipid-anchor substitution at the reducing end since mutations in these genes result in the accumulation of lipid-linked polysaccharide in the periplasm (Pazzani *et al.*, 1993; Wunder *et al.*, 1994; Rosenow *et al.*, 1995a). The predicted amino acid sequence of KpsE indicates that it is integrated into the cytoplasmic membrane (CM) (Rosenow *et al.*, 1995a), and that of KpsD shows a signal sequence, suggesting a periplasmic locale (Wunder *et al.*, 1994). KpsU is likely a Kdo synthetase specific for K-antigen biosynthesis since it shows sequence similarity to CMP-Kdo synthetase, the enzyme responsible for the synthesis of Kdo required for LPS biosynthesis (Rosenow *et al.*, 1995b; Finke *et al.*, 1989; Finke *et al.*, 1990). The presence of this second gene for Kdo synthesis indicates that K-antigens and LPSs do not share a common biosynthetic pathway. KpsC and KpsS are thought to function in the transport of the K-antigen to the cell surface (Bron-

ner *et al.*, 1993). Both lack the N-terminal signal sequences and hydrophobic domains typical of membrane proteins, indicating that they function in the cytosol. Both genes appear to affect the translocation across the CM and also the chain length of the polysaccharide (Bronner *et al.*, 1993).

Region 2 is strain specific and contains genes encoding proteins for the synthesis of various monosaccharides, glycosyl transferases, and repeating unit polymerases required for the synthesis of the strain specific K-antigen (Pazzani *et al.*, 1991; Boulnois *et al.*, 1990). Region 2 is flanked by regions 1 and 3, which are conserved in those strains with a *kps* (group II K-antigen) gene cluster (Pazzani *et al.*, 1991; Boulnois *et al.*, 1990).

Region 3 (~1.6 kb) consists of a single operon containing two genes (*kpsM* and *kpsT*) (Saier, Jr. 1994; Pavelka *et al.*, 1994; Pavelka Jr *et al.*, 1991). The predicted amino acid sequences of KpsT and KpsM were similar to ATP-dependent transport proteins (ABC cassette) (Saier, Jr. 1994; Pavelka *et al.*, 1994; Pavelka Jr *et al.*, 1991), and, therefore, may function in the transport of the nascent polysaccharide across the CM. Mutations in these genes result in the localization of low molecular weight polysaccharide in the cytosol.

III.B. THE STRUCTURES OF THE K-ANTIGENS FROM RHIZOBIA AND OTHER PLANT ASSOCIATED BACTERIA.

The rhizobial K-antigens are distinguished from other surface polysaccharides by several criteria (Kim *et al.*, 1996b; Petrovics *et al.*, 1993; Reuhs *et al.*, 1993): First, they are distinct from rhizobial EPS in that they are not excreted into the culture medium, they adhere tightly to the cell surface, they are antigenic, and have different structures (Reuhs *et al.*, 1994b). Second, there is no evidence thus far that the rhizobial K-antigen polysaccharide is attached to a lipid anchor such as phosphatidic acid or lipid-A as is the case for other K-antigens and LPSs. Third, rhizobial K-antigens are structurally distinct from LPS, containing a higher proportion of Kdo or a Kdo variant which yields a positive response to the thiobarbituric acid (TBA) assay, and having a much higher charge density with negative charges present on each repeating unit, sometimes on each sugar. Several past reports (Brzoska *et al.*, 1991; Williams *et al.*, 1990a; Williams *et al.*, 1990b; Putnoky *et al.*, 1990; Carlson *et al.*, 1985) mistakenly identified crude preparations of K-antigens as LPSs based on the presence of Kdo (TBA positive material) since Kdo was assumed to be an LPS-marker component. This led to inaccurate conclusions regarding the identity of genes affected in certain mutants, and the role(s) of LPSs in symbiosis.

Rhizobial K-antigens were first found in the water layer of hot phenol/water extracts during the purification of LPSs from *S. fredii* strains and were initially identified as being derived from LPS (Carlson *et al.*, 1985). Subsequently it was found these K-antigens were distinct molecules, which could be separated from the LPSs by gel-filtration chromatography in the presence of DOC (Reuhs *et al.*, 1993). Thus far, the structures of three K-antigens from strains of *S. fredii* have been determined (Forsberg *et al.*, 1997; Reuhs *et al.*, 1993), and several others from *S. fredii*, *S. meliloti*, and *Rhizobium* sp. NGR234 are under further investigation. These structures are summarized in Table 4. They vary in structure from strain to strain within a single species, and also from species to species. However, within this variability there appears to be a conserved structural motif of a disaccharide repeat involving a variable glycosyl residue bonded to a 1-carboxy-2-keto-3-deoxy sugar (Kdx). In *S. fredii* strains, this latter sugar is Kdo, while in *S. meliloti* and in *Rhizobium* sp. NGR234 this sugar is some type of nonulosonic acid; in one case *N*-acetylneuraminic acid. One exception is the K-antigen from *S. fredii* HH303 in which no Kdo was detected. It should be noted that strain HH303 also has different host specificity than other *S. fredii* strains in that it is able to nodulate some American varieties of soybean.

The study of the K-antigen polysaccharides has been facilitated by the use of PAGE with or without DOC, and by using an Alcian Blue/silver staining procedure (Corzo *et al.*, 1991; Reuhs *et al.*, 1993). The K-antigens and LPSs are easily distinguished from one another since the former silver-stain only when fixed in the presence of Alcian Blue, and the latter do not migrate into the gel in the absence of DOC (Kim *et al.*, 1996b; Reuhs *et al.*, 1993). The K-antigens consistently yield an Alcian Blue-specific ladder pattern or smear. The ladder pattern is due to the heterogeneity in the degree of polymerization, and the smear is probably due to additional microheterogeneity in the various substituent groups (when present). Using this protocol, a systematic evaluation of many plant-associated Gram-negative bacteria was undertaken to determine if the presence of these K-antigen polysaccharides may be a common feature of these bacteria. This study, included examples of many well studied plant pathogens, in addition to rhizobia; namely *A. tumefaciens*, *Erwinia amylovora*, *E. carotovora*, *E. stewartii*, *Pseudomonas solanacearum*, *P. syringae*, and *X. campestris*, (Reuhs *et al.*, 1994b; Reuhs *et al.*, 1997). The results showed that cell-associated, acidic polysaccharides are, in fact, common components of these bacterial cell extracts. Composition analysis of these polysaccharides from species of *Rhizobium* other than *Sinorhizobium* showed that they did not have the conserved structural motif given in Table 3 (Reuhs *et al.*, 1994b, Reuhs *et al.*, unpublished), i.e. they do not have the Kdx sugar. However, analysis of the K-antigen from *A. tumefaciens* indicated that it did have a Kdx sugar (Reuhs *et al.*, 1997). It is apparent that an acidic capsular polysaccharide is a general feature of a broad range of plant symbiont and pathogenic bacteria.

III.C. K-ANTIGEN EXPRESSION IN RHIZOBIA.

Genetic investigations have shown that capsule expression in *S. meliloti* AK631 involves several gene regions, comprising a minimum of fifteen genes, but most likely many more. One plasmid-borne gene region carries the *rkpZ* gene whose expression lowers the size range of the exported K-antigen (Reuhs *et al.*, 1995). It is the low molecular weight form of the K-antigen that is required for infection by the EPS mutants (discussed below). The activity of RkpZ is non-specific with regard to the K-antigen structure since introduction of *rkpZ* from *S. meliloti* AK631 into *S. fredii* USDA257 resulted in a reduction in the size range of the USDA257 K-antigen even though it is structurally distinct from that of *S. meliloti* AK631 (Reuhs *et al.*, 1995).

A second rhizobial K-antigen locus is the chromosomal *fix-23* gene region (Petrovics *et al.*, 1993; Putnoky *et al.*, 1990). One complementation unit of *fix-23*, which comprises six ORFs (termed *rkpABCDEF*), appears to encode a lipid carrier involved in the biosynthesis and export of the capsule (Petrovics *et al.*, 1993) since both the gene arrangement in the cistron, and the individual ORF sequences were similar to those for the fatty acid biosynthesis genes in rats.

Three other complementation units harbor four genes (*rkpGHIJ*) which encode products that may modify

Strain	Structure	Reference
S. fredii USDA205	[→3)- α-D-Gal*p*-(1→5)- β-D-Kdo*p*-(2→]$_n$	Reuhs *et al.*, 1993
	[→2-O-MeMan→Kdo→]$_n$	Reuhs *et al.*, 1993
S. fredii USDA257	[→3)- α-D-Man*p*-(1→5)- β-D-Kdo*p*-(2→]$_n$	Forsberg *et al.*, 1997
	[→3)- α-D-2-OMeMan*p*-(1→5)- β-D-Kdo*p*-(2→]$_n$	Forsberg *et al.*, 1997
S. fredii USDA201	[→Gal→Kdo→2-OMeHex→Kdo→]$_n$	Reuhs *et al.*,1997
S. fredii USDA208	[→Gal→Kdo→]$_n$	Reuhs *et al.*, 1997
S. meliloti NRG185	[→GlcNAc→Kdo→]$_n$	Reuhs *et al.*, 1997
S. meliloti NRG247	[→Glc→NeuNAc→]$_n$	Reuhs *et al.*, 1997
S. meliloti AK631	[→ (R$_1$)Hex→(R$_2$)Kdn5NAc7NAc→]$_n$	Reuhs *et al.*, unpublished
Rhizobium sp. NGR234	[→Glc→Kdn5NAc7NAc→]$_n$	Reuhs *et al.*, 1997
S. fredii HH303	[Rha, GalA]$_n$	Reuhs *et al.*, 1997
Consensus Structure	[→ (R$_1$)Sug→(R$_2$)Kdx→]$_n$	

Table 4: Structures of the rhizobial K-antigens
Abbreviations: Gal, galactose; Kdo, 3-deoxy-D-*manno*-2-octulosonic acid; Man, mannose; Hex, undetermined hexose; Me, a methyl group; Kdn5NAc7NAc, 2-keto-3,5,7,9-tetradeoxy-5,7-diaminononulosonic acid; NeuNAc, N-acetylneuraminic acid; R$_1$and R$_2$, acetyl or β-hydroxybutyryl substituents; Sug, any sugar; Kdx, a 1-carboxy-2-keto-3-deoxy sugar. The only exception to the consensus structure is that from strain HH303 which is discussed further in the text.

the lipid carrier or the polymerization process. Two mutants in *rkpGHIJ* retained sensitivity to a phage which binds K-antigen. These mutants were shown by ELISA and immunoblot assays to partially export K-antigen (Kiss *et al.*, 1997). In contrast, most mutations in the *fix-23* region result in the intracellular accumulation of incompletely polymerized K polysaccharide (Kiss *et al.*, 1997). The fact that a *rkpABCDEF*-like cluster of genes has never been reported in *E. coli* indicates that the mechanism for K-antigen expression in soil bacteria is fundamentally different from that in *E. coli*. In spite of these possible differences in K-antigen expression, RkpZ shows significant similarity to KpsC (Pazzani *et al.*, 1993) of *E. coli*, a protein which also affects the chain length of the K-antigen. Additionally, the predicted amino acid sequence of RkpJ shows significant homology to KpsS of *E. coli*, another protein involved in polymerization and export of the group II K-antigens (Pazzani *et al.*, 1994; Bronner *et al.*, 1994).

The data gathered thus far from studies of *S. meliloti* and *S. fredii* suggest a model (Reuhs *et al.*, 1995) for K-antigen expression in these rhizobia: (1) The disaccharide repeating units may be produced by gene products from as yet uncharacterized gene regions, and an initial polymerization of the repeating units may take place on a lipid carrier yielding polysaccharide subunits of ~5000-7000

daltons (8-15 repeating units). (2) Exportable polysaccharides may then be assembled on a capsule-specific lipid carrier, determined by the *fix23* region, resulting in high molecular weight polymers formed from the above subunits. (3) The polymerization process would be modified by RkpZ, which promotes the export of the smaller polysaccharides. Interestingly, in enteric bacteria both K-antigen and LPS O-chain sizes are reported to be regulated by a chain length determining gene (*cld*) (Dodgson *et al.*, 1996; Batchelor *et al.*, 1992; Franco *et al.*, 1996; Stevenson *et al.*, 1995), and it may be that RkpZ functions by interacting, in some manner, with a rhizobial "*cld*-like" gene product. (4) Upon termination of the polymerization process, the K-antigens would be exported to the cell surface, and during this transport the lipid carrier may be removed and recycled. In the case of *E. coli* K-antigens, KpsD is a possible candidate for removal of the lipid during transport to the surface since this protein has homology to PgpB, a phosphatidylglycerophosphatase (Icho 1988a; Icho 1988b). It is quite likely that a similar protein could be involved in the export of the rhizobial K-antigen.

A gene region that controls host range in *S. fredii*-soybean interactions has also been shown to affect the expression of the K-antigen (Kim *et al.*, 1996a; Balatti *et al.*, 1990). Mutations in the *nolWXBTUV* genes of *S. fredii* USDA257 result in extending the host range of this strain to include cultivars of soybeans which are not normally infected by this strain. These extended host-range mutants are significantly altered in their surface chemistry, with clear changes in both the LPS and K-antigens (Reuhs, unpublished). The exact mechanism by which the *nol* gene products produce these changes is unclear at this time. However, the functions of the *nol* gene products appear to be secondary to the more general aspects of K-antigen expression since the *nolWXBTUV* operon is specific to *S. fredii* and has not been reported in *S. meliloti*. Interestingly, several of the genes in the *nolWXBTUV* region are homologous to the *hrp* genes of *Xanthomonas* and *Pseudomonas* (Meinhardt *et al.*, 1993; Huguet *et al.*, 1997).

III.D. THE BIOLOGY OF *RHIZOBIUM* K-ANTIGENS.

Strain *S. meliloti* AK631 is an *exoB⁻* mutant of Rm41 that produces neither EPS I nor EPS II yet is Fix⁺ *Medicago sativa* (Putnoky *et al.*, 1990; Putnoky *et al.*, 1988) indicating that another surface component can functionally substitute for EPS I or EPS II. A mutation in either the *rkpZ* gene or the *rkpABCDEFGHIJ* gene region of strain AK631 results in a Fix⁻ phenotype (Williams *et al.*, 1990a; Williams *et al.*, 1990b; Putnoky *et al.*, 1988). For *R. meliloti* strains carrying *rkpZ*, it was established that the K-antigen, in the absence of EPS, could promote the infection of alfalfa.

A recent study addressed the effects of purified K-antigens on the expression of certain plant genes involved in the host defense response. Whole cells of *R. meliloti* Rm41, mutant AK631, and three capsule mutants of AK631 (*rkpA-*, *rkpH-*, and *rkpJ-*), as well as purified K-antigen from strain AK631, were employed in this study (Becquart-de Kozak *et al.*, 1997). The infusion of the whole cells of AK631 or the purified K-antigen into the leaves of alfalfa seedlings resulted in a significant accumulation of chalcone synthase (CHS) mRNA, an enzyme in the phenyl propanoid biosynthetic pathway. This suggested a signal-based response of the host-plant to this bacterial product. In contrast, there was no response to whole cells of the *rkp* mutants, which lack the K-antigen capsule and cannot infect the host plant, or to purified EPS (Becquart-de Kozak *et al.*, 1997), indicating that the induction of mRNA to CHS was due (at least in part) to the K-antigen. Importantly, the kinetics of this CHS induction was

different than that observed from the infusion of plant pathogens (Becquart-de Kozak *et al.*, 1997), indicating that the response to *S. meliloti* AK631 cells is different from the typical defense response to a potential pathogen. Additionally, live cells of *S. meliloti* AK631 elicited a two phase induction of the CHS mRNA, whereas dead cells or the purified K-antigen elicited only the first phase. This indicated that some component not normally expressed was responsible for the second phase of induction. It is possible that the host-plant, in response to the K-antigen, produces a signal molecule affecting a further modification of the bacterium resulting in the production of the second-phase elicitor. In fact, the expression of K-antigens has been shown to be affected by host plant-derived compounds: The addition of apigenin, a *nod*-inducing flavonoid, to the growth media of *S. fredii* USDA205 was shown to increase the minor to major K-antigen ratio (Reuhs *et al.*, 1994a). Also, host-root extract was shown to greatly increase the expression of the K-antigen by *S. meliloti* AK631 (Reuhs *et al.*, 1995). These effects may be important in the infection process.

The rhizobial K-antigen expression is also affected by abiotic factors, including temperature and pH. One study was initiated to ascertain if K-antigen expression would be shut down at temperatures below 20°C, which occurs with group II K-antigens of *E. coli* (Jann *et al.*, 1990; Orskov *et al.*, 1984). Based on the very different environment that rhizobia occupy compared with that of *E. coli*, it was predicted that this would not be the case. The rhizobial K-antigens were, in fact, expressed at very low temperatures (7-18° C), and changes in size range were also observed (Geller and Reuhs, unpublished). This second observation may be significant since the size range of the K-antigens and EPS II is important in the infection process (Gonzales *et al.*, 1996; Reuhs *et al.*, 1995).

IV. Conclusions

It is clear that the outer membrane and the surrounding capsule from rhizobia are composed in part of a complex array of LPS and K-antigen molecules. These classes of molecules are functionally crucial for establishing successful symbiotic associations with plant hosts. However, to date no clear-cut functions have been assigned. To relate structure to function has been complicated by the structural complexity of the molecules involved, and by the recently discovered structural adaptations of these molecules in response to various bacterial growth environments. On the other hand, a recent report by Freiberg et al., (Freiberg et al., 1997) may prove very helpful in revealing structure-function relationships. That report describes the sequence of the entire symbiotic plasmid from *Rhizobium* sp. NGR234 and shows that, in addition to genes required for nodulation and nitrogen fixation, this plasmid contains a number of genes with similarities to polysaccharide synthesis genes. While the function of the products encoded by these genes are not known, it may be that they are involved in some of the structural adaptations that occur to the various surface polysaccharides during symbiosis.

The study of rhizobial K-antigens is an area of emerging interest. Although much less is known about the K-antigens than about LPS , some insights have arisen. The K-antigens are common products of rhizobia, and probably many other plant associated bacteria. Studies of *S. fredii* and *S. meliloti* have shown that these capsular polysaccharides are surface antigens which are structurally variable and strain-specific with a conserved basic structural motif within a species. Capsule expression involves at least fifteen genes in several distinct regions. Its expression is modulated by host-derived compounds and certain environmental conditions. The K-antigens have been implicated in the occurrence of infection related defense responses towards rhizobia in certain host plants.

Structural analyses of LPSs have shown that there is a correlation between the structures of the lipid-A-core regions and the taxonomic classification of the various species of *Rhizobiaceae*. Structural elucidation has also revealed that these LPSs can be very different in structure from those of enteric bacteria, yet the synthesis of the lipid-A portion, which is crucial for cell viability of Gram-negative bacteria, has many steps in common with those of an enteric bacterium such as *E. coli*. It is only after the synthesis of the minimal lipid-A precursor required for cell viability that structural modifications take place leading to the unique *R. leguminosarum* lipid-A.

Analysis of the LPS from various LPS mutants has shown that (a) effective symbiosis of either determinate or indeterminate nodule-forming hosts require the O-chain portion, and (b) mutants with small amounts of LPS carrying the O-chain, or with truncated O-chain are not effective symbionts. Analysis of indeterminate nodules formed by

R. leguminosarum LPS mutants on pea or vetch has revealed that symbiosis is disrupted during, and/or after release of the bacteria into the host cell. However LPS mutants of determinate nodule-forming hosts are defective at the earlier stage of infection thread development. The defective phenotypes may be due to an increased sensitivity of a weakened bacterium to antimicrobial compounds produced by the host plant. Chemical and immunocytological studies show that the LPSs, during indeterminate nodule formation, undergo an increase in hydrophobicity as well as a change in certain O-chain structural features during bacterial release and bacteroid differentiation. While the exact functions of the LPSs are not known, it is possible that they are important in specific interactions between bacterial and host cell surfaces (possibly membrane-membrane interactions). In determinate nodule-forming hosts, this interaction would first occur during infection thread formation and development. In indeterminate hosts this type of interaction would primarily occur during release into the host cortical cell and during the synchronous division of bacteria and the PBM during bacteroid formation. The idea that LPSs play a role(s) in these putative cell-cell surface (possibly membrane) interactions is consistent with the timing of the different phenotypes observed by LPS mutants in determinate and indeterminate nodules. The exact nature of the molecular interaction between the LPS of the bacterial envelope and the infection thread membrane or PBM remains to be determined.

Symbiotic infection is also accompanied by structural changes in the LPS O-chain polysaccharide as detected using MAbs. However their functions with regard to symbiotic development are not known. The LPS epitope changes, as well as the increase in hydrophobicity, can also be produced by growing the bacteria at low O_2, or low pH, indicating that some of the LPS structural modifications may reflect a general method by which these bacteria adapt to physiological changes in their micro-environment, whether that environment be in the soil or inside the host plant. Other epitope changes were found to be correlated with different stages of nodule development. These changes may be dictated by host-specific, or nodule-type specific (*i.e.* determinate vs. indeterminate) molecules such as the anthocyanins reported by Noel et al., (Noel et al., 1996a). In the case of pea nodules, the O-chain epitope changes occur during internalization of the bacterium within the host root cell and during bacteroid development. It is precisely these stages of symbiosis that are disrupted during nodulation by LPS mutants that are deficient in O-chain polysaccharide.

Determining the roles that LPSs and K-antigens play in symbiotic infection and nodule development is an exciting area of research. Future work delineating the functions of LPSs and K-antigens in this process will have important implications, not only for understanding the nitrogen-fixing symbiotic process, but for understanding the overall functions of these molecules as virulence factors in the infection of eucaryotic cells by Gram-negative bacteria.

V. Acknowledgements

This work has been supported by grants from the NIH (GM895832, RWC), the NSF (IBN9305022, RWC and BLR), from the Center for Plant and Microbial Complex Carbohydrate Research funded by the DOE (DE-FG09-87ER13810, to the CCRC), and from the Deutsche Forschungsgemeinschaft (PO 117/16-1) and the University of Tübingen (to ELK). The authors also thank M. O'Neill and A. Perlick for helpful discussion, and R. Nuri and A. Dunn for secretarial help. In addition, many individuals in the laboratory of RWC have made contributions to the research described in this chapter. These include: U. R. Bhat, T-B. Chen, D. Geller, B. Jeyaretnam, J. Kim, B. Krishnaiah, N. Price, and B. Ridley.

VI. References

Allaway, D., Jeyaretnam, B., Carlson, R.W. and Poole, P.S. (1996) J. Bacteriol. 178, 6403-6406.

Bal, A.K., Shantharam, S. and Verma, D.P.S. (1980) Can. J. Microbiol. 26, 1096-1103.

Balatti, P.A. and Pueppke, S.G. (1990) Plant Physiol. 94, 1276-1281.

Batchelor, R.A., Alifano, P., Biffali, E., Hull, S.I. and Hull, R.A. (1992) J. Bacteriol. 174, 5228-5236.

Battisti, L., Lara, J.C. and Leigh, J.A. (1992) Proc. Natl. Acad. Sci. USA. 89, 5625-5629.

Becquart-de Kozak, I., Reuhs, B.L., Buffard, D., Breda, C., Kim, J.S., Esnault, R. and Kondorosi, A. (1997) Mol. Plant Microbe Interact. 10, 114-123.

Belunis, C.J. and Raetz, C.R.H. (1992) J. Biol. Chem. 267, 9988-9997.

Bhat, U.R., Bhagyalakshmi, S.K. and Carlson, R.W. (1991a) Carbohydr. Res. 220, 219-227.

Bhat, U.R., Carlson, R.W., Busch, M. and Mayer, H. (1991b) Int. J. Syst. Bacteriol. 41, 213-217.

Bhat, U.R., Mayer, H., Yokota, A., Hollingsworth, R.I. and Carlson, R.W. (1991c) J. Bacteriol. 173, 2155-2159.

Bhat, U.R. and Carlson, R.W. (1992) J. Bacteriol. 174, 2230-2235.

Bhat, U.R., Forsberg, L.S. and Carlson, R.W. (1994) J. Biol. Chem. 269, 14402-14410.

Bolaños, L., Esteban, E., De Lorenzo, C., Fernández-Pascual, M., De Felipe, M.R., Gárate, A. and Bonilla, I. (1994) Plant Physiol. 104, 85-90.

Bolaños, L., Brewin, N.J. and Bonilla, I. (1996) Plant Physiol. 110, 1249-1256.

Borisov, A.Y., Morzhina, E.V., Kulikova, O.A., Tchetkova, S.A., Lebsky, V.K. and Tikhonovich, I.A. (1992) Symbiosis. 14, 297-313.

Boulnois, G.J., Roberts, I.S., Hodge, R., Hardy, K.R., Jann, K.B. and Timmis, K.N. (1987) Mol. Gen. Genet. 208, 242-246.

Boulnois, G.J. and Roberts, I.S. (1990) Curr. Top. Microbiol. Immunol. 150, 1-18.

Boulnois, G.J. and Jann, K. (1993) Mol. Microbiol. 3, 1819-1823.

Bradley, D.J., Butcher, G.W., Galfre, G., Wood, E.A. and Brewin, N.J. (1986) J. Cell Sci. 85, 47-61.

Bradley, D.J., Wood, E.A., Larkins, A.P., Galfre, G., Butcher, G.W. and Brewin, N.J. (1988) Planta. 173, 149-160.

Breedveld, M.W., Canter Cremers, H.C.J., Batley, M., Posthumus, M.A., Zevenhuizen, L.P.T.M., Wijffelman, C.A. and Zehnder, A.J.B. (1993) J. Bacteriol. 175, 750-757.

Brewin, N.J., Wood, E.A., Larkins, A.P., Galfre, G. and Butcher, G.W. (1986) J. Gen. Microbiol. 132, 1959-1968.

Brewin, N.J. (1991) Annu. Rev. Cell Biol. 7, 191-226.

Brink, B.A., Miller, J., Carlson, R.W. and Noel, K.D. (1990) J. Bacteriol. 172, 548-555.

Brom, S., García de los Santos, A., Stepkowsky, T., Flores, M., Dávila, G., Romero, D. and Palacios, R. (1992) J. Bacteriol. 174, 5183-5189.

Bronner, D., Sieberth, V., Pazzani, C., Roberts, I.S., Boulnois, G.J., Jann, B. and Jann, K. (1993) J. Bacteriol. 175, 5984-5992.

Bronner, D., Sieberth, V., Pazzani, C., Roberts, I.S., Boulnois, G.J., Jann, B. and Jann, K. (1994) J. Bacteriol. 175, 5984-5992.

Brozek, K.A., Hosaka, K., Robertson, A.D. and Raetz, C.R.H. (1989) J. Biol. Chem. 264(12), 6956-6966.

Brozek, K.A., Carlson, R.W. and Raetz, C.R.H. (1996a) J. Biol. Chem. 271, 32126-32136.

Brozek, K.A., Kadrmas, J.L. and Raetz, C.R.H. (1996b) J. Biol. Chem. 271, 32112-32118.

Brzoska, P.M. and Signer, E.R. (1991) J. Bacteriol. 173, 3235-3237.

Carlson, R.W., Sanders, R.E., Napoli, C. and Albersheim, P. (1978) Plant Physiol. 62, 912-917.

Carlson, R.W. (1982) Surface chemistry. Pages 199-234 in, Nitrogen Fixation, Vol. 2, Rhizobium. W.J. Broughton. ed. Clarendon Press, Oxford.

Carlson, R.W. and Lee, R.P. (1983) Plant Physiol. 71, 223-228.

Carlson, R.W. (1984) J. Bacteriol. 158, 1012-1017.

Carlson, R.W. and Yadav, M. (1985) Appl. Environ. Microbiol. 50, 1219-1224.

Carlson, R.W., Kalembasa, S., Turowski, D., Pachori, P. and Noel, K.D. (1987a) J. Bacteriol. 169, 4923-4928.

Carlson, R.W., Shatters, R., Duh, J.-L., Turnbull, E., Hanley, B., Rolfe, B.G. and Djordjevic, M.A. (1987b) Plant Physiol. 84, 421-427.

Carlson, R.W., Hollingsworth, R.L. and Dazzo, F.B. (1988) Carbohydr. Res. 176, 127-135.

Carlson, R.W., Garcia, F., Noel, K.D. and Hollingsworth, R.I. (1990) Carbohydr. Res. 195, 101-110.

Carlson, R.W. and Krishnaiah, B.S. (1992) Carbohydr. Res. 231, 205-219.

Carlson, R.W., Reuhs, B., Chen, T.-B., Bhat, U.R. and Noel, K.D. (1995) J. Biol. Chem. 270, 11783-11788.

Carrion, M., Bhat, U.R., Reuhs, B. and Carlson, R.W. (1990) J. Bacteriol. 172, 1725-1731.

Cava, J.R., Elias, P.M., Turowski, D.A. and Noel, K.D. (1989) J. Bacteriol. 171, 8-15.

Cava, J.R., Tao, H. and Noel, K.D. (1990) MGG. 221, 125-128.

Cedergren, R.A. and Hollingsworth, R.I. (1994). J. Lipid Res. 35, 1452-1461.

Cedergren, R.A., Lee, J., Ross, K.L. and Hollingsworth, R.I. (1995) Biochemistry. 34, 4467-4477.

Cedergren, R.A., Wang, Y. and Hollingsworth, R.I. (1996) J. Bacteriol. 178, 5529-5532.

Chapman, K., Pankratz, S., Philip-Hollingsworth, S., Dazzo, F. and Wright, S. (1990) Mutations that alter O-antigen structure of R. trifolii ANU843. Page 254 Nitrogen Fixation, Achievements and Objectives. P.M. Gresshoff, L.E. Roth, G. Stacey, and W.E. Newton. eds. Chapman and Hall, New York.

Chen, T.-B. (1987) The structure of the O-antigen from lipopolysaccharide of Rhizobium leguminosarum 128C53 and its nod⁻fix⁻ mutant. (Abstract)

Clementz, T., Bednarski, J. and Raetz, C.R.H. (1996) J. Biol. Chem. 271, 12095-12102.

Clementz, T., Zhou, Z. and Raetz, C.R.H. (1997) J. Biol. Chem. 272, 10353-10360.

Clover, R.H., Kieber, J. and Signer, E.R. (1989) J. Bacteriol. 171, 3961-3967.

Corzo, J., Pérez-Galdona, R., León-Barrios, M. and Gutiérrez-Navarro, A.M. (1991) Electrophoresis. 12, 439-441.

David, S.A., Balasubramanian, K.A., Mathan, V.I. and Balaram, P. (1992) Biochim. Biophys. Acta Lipids Lipid Metab. 1165, 147-152.

Dazzo, F.B. (1979) ASM Nws. 45, 238-240.

Dazzo, F.B., Truchet, G.L., Hollingsworth, R.I., Hrabak, E.M., Pankratz, H.S., Philip-Hollingsworth, S., Salzwedel, J.L., Chapman, K., Appenzeller, L., Squartini, A., Gerhold, D. and Orgambide, G. (1991) J.Bacteriol. 173, 5371-5384.

de Maagd, R.A., de Rijk, R., Mulders, I.H.M. and Lugtenberg, B.J.J. (1989a) J. Bacteriol. 171, 1136-1142.

de Maagd, R.A., Rao, A.S., Mulders, I.H.M., Roo, L.G., van Loosdrecht, M.C.M., Wijffelman, C.A. and Lugtenberg, B.J.J. (1989b) J. Bacteriol. 171, 1143-1150.

Denny, T.P. (1995) Annu. Rev. Phytopathol. 33, 173-197.

Djordjevic, S.P., Chen, H., Batley, M., Redmond, J.W. and Rolfe, B.G. (1987) J. Bacteriol. 169, 53-60.

Dodgson, C., Amor, P. and Whitfield, C. (1996) J. Bacteriol. 178, 1895-1902.

Dow, J.M., Osbourn, A.E., Wilson, T.J.G. and Daniels, M.J. (1995) Mol. Plant Microbe Interact. 8, 768-777.

Duelli, D. M. and Noel, K.D. (1997) Mol. Plant-Microbe Interact. 10, 903-910.

Eisenschenk, L., Diebold, R., Perez-Lesher, J., Peterson, A.C., Peters, N.K. and Noel, K.D. (1994) Appl. Environ. Microbiol. 60, 3315-3322.

Elkan, G.H. (1992) Can. J. Microbiol. 38, 446-450.

Finke, A., Roberts, I., Boulnois, G., Pzzani, C. and Jann, K. (1989) J. Bacteriol. 171, 3074-3079.

Finke, A., Jann, B. and Jann, K. (1990) FEMS Microbiol. Lett. 69, 129-134.

Finke, A., Bronner, D., Nikolaev, A.V., Jann, B. and Jann, K. (1991) J. Bacteriol. 173, 4088-4094.

Forsberg, L.S. and Carlson, R.W. (1997) J. Biol. Chem. in press.

Forsberg, L.S. and Reuhs, B.L. (1997) J. Bacteriol. 179, 5366-5371.

Foster, J.W. and Spector, M.P. (1995) How Salmonella survive against the odds. 145.(Abstract)

Franco, A.V., Liu, D. and Reeves, P.R. (1996) J. Bacteriol. 178, 1903-1907.

Freiberg, C., Fellay, R., Bairoch, A., Broughton, W.J., Rosenthal, A. and Perret, X. (1997) Nature. 387, 394-401.

Garcia-de los Santos, A. and Brom, S. (1997) Mol. Plant-Microbe Interact. 10, 891-902.

Galanos, C., Luderitz, O. and Westphal, O. (1969) Eur. J. Biochem. 9, 245.(Abstract)

Gil-Serrano, A.M., González-Jiménez, I., Mateo, P.T., Bernabé, M., Jiménez-Barbero, J., Megías, M. and Romero-Vázquez, M.J. (1995) Carbohydr. Res. 275, 285-294.

Glowacka, M., Russa, R. and Lorkiewicz, Z. (1986) Acta Microbiol. Pol. 35, 199-205.

Goldberg, J.B. and Pier, G.B. (1996) Trends in Microbiol. 4, 490-494.

Gonzales, J.E., Reuhs, B.L. and Walker, G.C. (1996) Proc. Natl. Acad. Sci. USA. 93, 8636-8641.

Goosen-de Roo, L., de Maagd, R.A. and Lugtenberg, B.J.J. (1991) J. Bacteriol. 173, 3177-3183.

Hollingsworth, R.I. and Carlson, R.W. (1989a) J. Biol. Chem. 264, 9300-9303.

Hollingsworth, R.I., Carlson, R.W., Garcia, F. and Gage, D.A. (1989b) J. Biol. Chem. 264, 9294-9299.

Hollingsworth, R.I., Carlson, R.W., Garcia, F. and Gage, D.A. (1990) J. Biol. Chem. 265, p.12752

Hollingsworth, R.I., Zhang, Y. and Priefer, U.B. (1994) Carbohydr. Res. 264, 271-280.

Huguet, E. and Bonas, U. (1997) MPMI. 10, 488-498.

Icho, T. (1988a) J. Bacteriol. 170, 5117-5124.

Icho, T. (1988b) J. Bacteriol. 170, 5110-5116.

Jacques, M. (1996) Trends in Microbiol. 4, 408-410.

Jann, B. and Jann, K. (1990) Curr. Top. Microbiol. Immunol. 150, 19-42.

Jann, K. and Jann, B. (1991) Biochem. Soc. Trans. 19, 623-628.

Jarvis, B.D.W., Downer, H.L. and Young, J.P.W. (1992) J. Syst. Bacteriol. 42, 93-96.

Johansen, E., Finan, T.M., Gefter, M.L. and Signer, E.R. (1984) J. Bacteriol. 160, 454-457.

Jordan, C.D. (1984) Family III, Rhizobiaceae. Pages 235-256 in, Bergey's Manual of Systematic Bacteriology, vol. 1. N.R. Krieg, and J.G. Holt, eds. Williams & Wilkins, Baltimore/London.

Kamberger, W. (1979) FEMS Microbiol. Lett. 6, 361-365.

Kannenberg, E., Forsberg, L.S. and Carlson, R.W. (1996) Plant and Soil. 186, 161-166.

Kannenberg, E.L. and Brewin, N.J. (1989) J. Bacteriol. 171, 4543-4548.

Kannenberg, E.L., Rathbun, E.A. and Brewin, N.J. (1992) Mol. Microbiol. 6(17), 2477-2487.

Kannenberg, E.L. and Brewin, N.J. (1994a) Trends in Microbiol. 2, 277-283.

Kannenberg, E.L., Brewin, N.J. and Carlson, R.W. (1994b) Biochemical Separation of Lipopolysaccharides Expressed in Rhizobium Nodule Bacteria. Page 320. Proceedings of the 1st European Nitrogen Fixation Conference. G.B. Kiss and G. Endre. eds. Officina Press, Szeged, Hungary.

Kannenberg, E.L., Perotto, S., Bianciotto, V., Rathburn, E.A. and Brewin, N.J. (1994c) J. Bacteriol. 176, 2021-2032.

Kato, G., Maruyama, Y. and Nakamura, M. (1979) Agric. Biol. Chem. 43, 1085-1092.

Kato, G., Maruyama, Y. and Nakamura, M. (1980) Agric. Biol. Chem. 44, 2843-2855.

Kato, G., Maruyama, Y. and Nakamura, M. (1981) Plant Cell Physiol. 22, 759-771.

Kijne, J. (1990) The Rhizobium infection process. Biological nitrogen fixation. G. Stacey, R.H. Burris, and H.J. Evans. eds. Chapman and Hall, New York.

Kijne, J.W., Lugtenberg, B.J.J. and Smit, G. (1992) Attachment, lectin and initiation of infection in (Brady)rhizobium-legume interactions. Pages 281-294 in, Molecular signals in plant-microbe communications. D.P.S. Verma. ed. CRC Press, Boca Raton, Ann Arbor, London.

Kim, J.S. and Reuhs, B.L. (1996a) Mol. Plant-Microbe Interact. 8th Int. MPMI Congr.(Abstract)

Kim, J.S., Reuhs, B.L., Rahman, M.M., Ridley, B. and Carlson, R.W. (1996b) Glycobiology. 5, 433-437.

Kiss, E., Reuhs, B.L., Kim, J.S., Kereszt, A., Petrovics, G., Putnoky, P., Dusha, I., Carlson, R.W. and Kondorosi, A. (1997) J. Bacteriol. 179, 2132-2140.

Lagares, A., Caetano-Anollés, G., Niehaus, K., Lorenzen, J., Ljunggren, H.D., Pühler, A. and Favelukes, G. (1992) J. Bacteriol. 174, 5941-5952.

Latchford, J.W., Borthakur, D. and Johnston, A.W.B. (1991) Mol. Microbiol. 5, 2107-2114.

Leeman, A., Van Pelt, J.A., Den Ouden, F.M., Heinsbroek, M., Bakker, P.A.H.M. and Schippers, B. (1995) Am. Phytopathol. Soc. 85, 1021-1027.

Lipsanen, P. and Lindstrom, K. (1989) FEMS Microbiol. Lett. 58, 323-328.

Lloret, J., Bolaños, L., Lucas, M.M., Peart, J.M., Brewin, N.J., Bonilla, I. and Rivilla, R. (1995) Appl. Environ. Microbiol. 61, 3701-3704.

Loomis, W.D. and Durst, R.W. (1992) Biofactors. 3, 229-239.

López-Lara, I.M., Orgambide, G., Dazzo, F.B., Olivares, J. and Toro, N. (1995) Microbiology. 141, 573-581.

Lucas, M.M., Peart, J.L., Brewin, N.J. and Kannenberg, E.L. (1996) J. Bacteriol. 178, 2727-2733.

Lynn, W.A. and Golenbock, D.T. (1992) Immunol. Today. 13, 271-276.

Maier, R.J. and Brill, W.J. (1978) J. Bacteriol. 133, 1295-1299.

Martinez-Romero, E. (1994) Plant and Soil. 161, 11-20.

Martínez-Romero, E. and Caballero-Mellado, J. (1996) Crit. Rev. Plant Sci. 15, 113-140.

Mayer, H. and Weckesser, J. (1984) Unusual lipid A's, structures, taxanomical relevance and potential value for endotoxin research. Pages 221-247 in, Handbook of Endotoxins, Vol. 1, Chemistry of Endotoxins. E.T. Reitschel. ed. Elsevier, Amsterdam.

Mayer, H., Bhat, U.R., Masoud, H., Radziejewska-Lebrecht, J., Widemann, C. and Krauss, J.H. (1989a) Pure Appl. Chem. 61, 1271-1282.

Mayer, H., Krauss, J.H., Urbanik-Sypniewska, T., Puvanesarajah, V., Stacey, G. and Auling, G. (1989b) Arch. Microbiol. 151, 111-116.

Mayer, H., Moreno, E., Stackebrandt, E., Dorsch, M., Wolters, J. and Busch, M. (1990) J. Bacteriol. 172(7), 3569-3576.

Mazzucchi, U. (1983) Recognition of Bacteria by Plants. Pages 299-324 in, Biochemical Plant Pathology. J.A. Callow. ed. John Wiley & Sons Ltd.

Meinhardt, L.W., Krishnan, H.B., Balatti, P.A. and Pueppke, S.G. (1993) Mol. Microbiol. 9, 17-29.

Mohnen, D. and Hahn, M.G. (1993) Cell Biology. 4, 93-102.

Moran, A.P., Appelmelk, B.J. and Aspinall, G.O. (1996) J. Endotoxin Res. 3, 521-531.

Morrison, D.C. and Jacobs, D.M. (1976) Immunochemistry. 13, 813-818.

Munns, D.N. (1970) Plant Soil. 32, 90-102.

Neumann, U., Mayer, H., Schiltz, E., Benz, R. and Weckesser, J. (1995) Microbiology. 141, 2013-2017.

Newman, M.A., Daniels, M.J. and Dow, J.M. (1995) Mol. Plant-Microbe Interact. 8, 778-780.

Niehaus, K., Lorenzen, J., Lagares, A. and Pühler, A. (1994) A Rhizobium meliloti lipopolysaccharide mutant effective on Medicago sativa (alfalfa) but ineffective on medicago truncatula. 7th ISMPMI. (Abstract)

Noel, K.D., VandenBosch, K.A. and Kulpaca, B. (1986) J. Bacteriol. 168, 1392-1401.

Noel, K.D. (1992) Rhizobial polysaccharides required in symbioses with legumes. Pages 341-357 in, Molecular signals in plant-microbe communications. D.P.S. Verma. ed. CRC Press, Boca Raton, Ann Arbor, London.

Noel, K.D., Duelli, D.M. and Neumann, V.J. (1996a) Molecular Plant-Microbe Interactions, St. Paul.

Noel, K.D., Duelli, D.M., Tao, H. and Brewin, N.J. (1996b) MPMI. 9, 180-186.

O'Neill, M.A., Warrenfeltz, D., Kates, K., Pellerin, P., Doco, T., Darvill, A.G. and Albersheim, P. (1996) J. Biol. Chem. 271, 22923-22930.

Olsen, P., Wright, S., Collins, M. and Rice, W. (1994) Appl. Environ. Microbiol. 60, 654-661.

Orskov, F., Sharma, V. and Orskov, I. (1984) J. Gen. Microbiol. 130, 2681-2684.

Pavelka Jr, M.S., Wright, L.F. and Silver, R.P. (1991) J. Bacteriol. 173, 4603-4610.

Pavelka, M.S., Hayes, S.F. and Silver, R.P. (1994) J. Biol. Chem. 269, 20149-20158.

Pazzani, C., Roberts, I. and Boulnois, G. (1991) Biochem. Soc. Trans. 19, 628-630.

Pazzani, C., Rosenow, C., Boulnois, G.J., Bronner, D., Jann, K. and Roberts, I.S. (1993) J. Bacteriol. 175, 5978-5983.

Pazzani, C., Rosenow, C., Boulnois, G.J., Bronner, D., Jann, K. and Roberts, I.S. (1994) J. Bacteriol. 175, 5978-5983.

Penel, C. and Greppin, H. (1996) Plant Physiol. Biochem. 34, 479-488.

Perotto, S., Brewin, N.J. and Kannenberg, E.L. (1994) Mol. Plant Microbe Interact. 7, 99-112.

Petit, C., Rigg, G.P., Pazzani, C., Smith, A., Sieberth, V., Stevens, M., Boulnois, G., Jann, K. and Roberts, I.S. (1995) Mol. Microbiol. 17, 611-620.

Petrovics, G., Putnoky, P., Reuhs, B., Kim, J., Thorp, T.A., Noel, D., Carlson, R.W. and Kondorosi, A. (1993) Mol. Microbiol. 8, 1083-1094.

Poole, P.S., Schofield, N.A., Reid, C.J., Drew, E.M. and Walshaw, D.L. (1994) J. Gen. Microbiol. 140, 2797-2809.

Preston, A., Mandrell, R.E., Gibson, B.W. and Apicella, M.A. (1996) Crit. Rev. Microbiol. 22, 139-180.

Price, N.P.J., Kelly, T.M., Raetz, C.R.H. and Carlson, R.W. (1994) J. Bacteriol. 176, 4646-4655.

Price, N.P.J., Jeyaretnam, B., Carlson, R.W., Kadrmas, J.L., Raetz, C.R.H. and Brozek, K.A. (1995) Proc. Natl. Acad. Sci. USA. 92, 7352-7356.

Priefer, U.B. (1989) J. Bacteriol. 171, 6161-6168.

Proctor, R.A., Denlinger, L.C. and Bertics, P.J. (1995) Lipopolysaccharide and Bacterial Virulence. Pages 173-194 in, Virulence Mechanisms of Bacterial Pathogens. J.A. Roth, C.A. Bolin, K.A. Brogden, F.C. Minion, and M.J. Wannemuehler. eds. American Society for Microbiology, Washington, D.C.

Putnoky, P., Grosskopf, E., Ha, D.T.C., Kiss, G.B. and Kondorosi, A. (1988) J. Cell Biol. 106, 597-607.

Putnoky, P., Petrovics, G., Kereszt, A., Grosskopf, E., Ha, D.T.C., Bánfalvi, Z. and Kondorosi, A. (1990) J. Bacteriol. 172, 5450-5458.

Raetz, C.R.H. (1990) Annu. Rev. Biochem. 59, 129-170.

Raetz, C.R.H. (1993) J. Bacteriol. 175, 5745-5753.

Rahman, M.M., Guard-Petter, J. and Carlson, R.W. (1997) J. Bacteriol. 179, 2126-2131.

Reuhs, B.L., Carlson, R.W. and Kim, J.S. (1993) J. Bacteriol. 175, 3570-3580.

Reuhs, B.L., Geller, D.G, Kim, J.S., Fox, J.F., Kolli, V.S.K. and Pueppke, S.G. (1997) Appl. Environ, Microbiol., submitted.

Reuhs, B.L., Kim, J.S., Badgett, A. and Carlson, R.W. (1994a) Mol. Plant-Microbe Interact. 7, 240-247.

Reuhs, B.L., Kim, J.S., Geller, D.A., Carlson, R.W., Williams, M.N.V. and Pueppke, S.G. (1994b) Mol. Plant-Microbe Interact. Edinburgh, 71.(Abstract)

Reuhs, B.L., Williams, M.N.V., Kim, J.S., Carlson, R.W. and Côté, F. (1995) J. Bacteriol. 177, 4289-4296.

Reuhs, B.L., Kim, J.S. and Matthysse, A.G. (1997) J. Bacteriol. 179, 5372-5379.

Rietschel, E.T., Brade, L. et al., (1990) Adv. Exp. Med. Biol. 256, 81-100.

Rietschel, E.T. and Brade, H. (1992a) Sci. Am. 267, 54-61.

Rietschel, E.T., Kirikae, T., Schade, F.U., Ulmer, A.J., Holst, O., Brade, H., Schmidt, G., Mamat, U., Grimmecke, H.-D., Kusumoto, S. and Zähringer, U. (1993) Immunobiology. 187, 169-190.

Rietschel, E.T., Kirikae, T., Schade, F.U., Mamat, U., Schmidt, G., Loppnow, H., Ulmer, A.J., Zähringer, U., Seydel, U., Di Padova, F., Schreier, M. and Brade, H. (1994) FASEB J. 8, 217-225.

Rietschel, E.T., Brade, L., Lindner, B. and Zahringer, U. (1992b) Biochemistry of lipopolysaccharides. Pages 3-41 in, Bacterial endotoxic lipopolysaccharides, Volume I, Molecular biochemistry and cellular biology. D.C. Morrison and J.L. Ryan. eds. CRC Press, Boca Raton, Ann Arbor, London, Tokyo.

Robertson, J.G., Wells, B., Brewin, N.J., Wood, E., Knight, C.D. and Downie, J.A. (1985) The Legume-Rhizobium Symbiosis, A cell surface interaction. Pages 317-331 in, The Cell Surface in Plant Growth and Development. K. Roberts, A.W.B. Johnston, C.W. Lloyd, P. Shaw, and H.W. Woolhouse. eds.

Rosenow, C., Esumeh, F., Roberts, I.S. and Jann, K. (1995a) J. Bacteriol. 177, 1137-1143.

Rosenow, C., Roberts, I.S. and Jann, K. (1995b) FEMS Microbiol. Lett. 125, 159-164.

Roth, L.E. and Stacey, G. (1989) European Journal of Cell Biology. 49, 13-23.

Roth, L.E. and Stacey, G. (1991) Rhizobium-Legume Symbiosis. Pages 255-301 in, Microbial Cell-Cell Interactions. M. Dworkin. ed. American Society of Microbiology, Washington, D.C.

Russa, R., Luderitz, O. and Rietschel, E.T. (1985) Arch. Microbiol. 141, 284-289.

Russa, R., Urbanik-Sypniewska, T., Choma, A. and Mayer, H. (1991) FEMS Microbiol. Lett. 84, 337-344.

Russa, R., Urbanik-Sypniewska, T., Lindström, K. and Mayer, H. (1995) Arch. Microbiol. 163, 345-351.

Russa, R., Bruneteau, M., Shashkov, A.S., Urbanik-Sypniewska, T. and Mayer, H. (1996) Arch. Microbiol. 165, 26-33.

Saier, M.H., Jr. (1994) Microbio. Rev. 58, 71-93.

Schletter, J., Holger, H., Ulmer, A.J. and Rietschel, E.T. (1995) Arch. Microbiol. 164, 383-389.

Schnaitman, C.A. and Klena, J.D. (1993) Microbiol. Rev. 57, 655-682.

Schwedock, J.S. and Long, S.R. (1992) Genetics. 132, 899-909.

Segovia, L., Young, J.P.W. and Martínez-Romero, E. (1993) Int. J. Syst. Bacteriol. 43, 374-377.

Selbitschka, W., Arnold, W., Priefer, U.B., Rottschäfer, T., Schmidt, M., Simon, R. and Pühler, A. (1991) Mol. Gen. Genet. 229, 86-95.

Sequeira, L. (1983) Annu. Rev. Microbiol. 37, 51-79.

Sethi, R.S. and Reporter, M. (1981) Protoplasma. 105, 321-325.

Silver, R.P., Aaronson, W. and Vann, W.F. (1987) J. Bacteriol. 169, 5489-5495.

Sindhu, S.S., Brewin, N.J. and Kannenberg, E.L. (1990) J. Bacteriol. 172, 1804-1813.

Sirisena, D.M., Brozek, K.A., MacLachlan, P.R., Sanderson, K.E. and Raetz, C.R.H. (1992) J. Biol. Chem. 267, 18874-18884.

Stacey, G., So, J.-S., Roth, L.E., Bhagya Lakshmi, S.K. and Carlson, R.W. (1991) Mol. Plant Microbe Interact. 4, 332-340.

Stevenson, G., Kessler, A. and Reeves, P.R. (1995) FEMS Microbiol. Lett. 125, 23-30.

Stevenson, T.T., Darvill, A.G. and Albersheim, P. (1988) Carbohydr. Res. 179, 269-288.

Sutton, W.D., Pankhurst, C.E. and Craig, A.S. (1981) The Rhizobium Bacteroid State. Pages 149-177 in, International Review of Cytology. G.H. Bourne, J.F. Danielli, and K.W. Jeon. eds. Academic Press, New York.

Takada, H. and Kotani, S. (1992) Structure-function relationships of lipid A. Pages 107-134 in, Bacterial endotoxic lipopolysaccharides, Volume I, Molecular biochemistry and cellular biology. D.C. Morrison and J.L. Ryan. eds. CRC Press, Boca Raton, Ann Arbor, London, Tokyo.

Tao, H., Brewin, N.J. and Noel, K.D. (1992) J. Bacteriol. 174, 2222-2229.

Urbanik-Sypniewska, T., Seydel, U., Greck, M., Weckesser, J. and Mayer, H. (1989) Arch. Microbiol. 152, 527-532.

Vaara, M. (1992) Microbiol. Rev. 56, 395-411.

Valvano, M.A. (1992) Can. J. Microbiol. 38, 711-719.

Valverde, C., Hozbor, D.F. and Lagares, A. (1997) BioTechniques. 22, 230-232.

van Brussel, A.A.N., Planque, K. and Quispel, A. (1977) J. Gen. Microbiol. 101, 51-56.

van de Wiel, C., Norris, J.H., Bochenek, B., Dickstein, R., Bisseling, T. and Hirsch, A.M. (1990) Plant Cell. 2, 1009-1017.

Van Putten, J.P.M. (1993) EMBO J. 12, 4043-4051.

Van Spronsen, P.C., Bakhuizen, R., van Brussel, A.A.N. and Kijne, J.W. (1994) Eur. J. Cell Biol. 64, 88-94.

Van Workum, W.A.T., van Brussel, A.A.N., Tak, T., Wijffelman, C.A. and Kijne, J.W. (1995) Mol. Plant-Microbe Interact. 8, 278-285.

VandenBosch, K.A., Brewin, N.J. and Kannenberg, E.L. (1989) J. Bacteriol. 171, 4537-4542.

Verma, D.P.S. and Hong, Z. (1996) Trends in Microbiol. 4, 364-368.

Wang, Y. and Hollingsworth, R.I. (1994) Carbohydr. Res. 260, 305-317.

Weibgen, U., Russa, R., Yokota, A. and Mayer, H. (1993) Syst. Appl. Microbiol. 16, 177-182.

Werner, D. (1992) Physiology of Nitrogen-Fixing Legume Nodules, Compartments and Functions. Pages 399-431 in, Biological Nitrogen Fixation. G. Stacey, R.H. Burris, and H.J. Evans. eds. Chapman & Hall, New York.

Westphal, O. and Jann, K. (1965) Meth.Carbohydr.Chem. 5, 83-91.

Whatley, M.H., Bodwin, J.S., Lippincott, B.B. and Lippincott, L.A. (1970) Infect. Immun. 13, 1080-1083.

Whatley, M.H. and Spiess, L.D. (1977) Plant Physiol. 60, 765-766.

Whitfield, C. and Valvano, M.A. (1993) Adv. Microb. Physiol. 35, 136-246.

Whitfield, C. (1995) Trends in Microbiol. 3, 178-185.

Williams, M.N.V., Hollingsworth, R.I., Brzoska, P.M. and Signer, E.R. (1990a) J. Bacteriol. 172, 6596-6598.

Williams, M.N.V., Hollingsworth, R.I., Klein, S. and Signer, E.R. (1990b) J. Bacteriol. 172, 2622-2632.

Witty, J.F., Minchin, F.R., Skot, L. and Sheehy, J.E. (1986) Nitrogen Fixation and Oxygen in Legume Root Nodules. Pages 275-314 in, Oxford Surveys of Plant Molecular & Cell Biology. B.J. Miflin. ed. Oxford University Press, New York.

Woese, C.R. (1987) Microbiol. Rev. 51, 221-271.

Wolpert, J.S. and Albersheim, P. (1976) Biochem. Biophys. Res. Commun. 70, 729.(Abstract)

Wood, E.A., Butcher, G.W., Brewin, N.J. and Kannenberg, E.L. (1989) J. Bacteriol. 171, 4549-4555.

Wright, S.F. (1990) Appl. Environ. Microbiol. 56, 2262-2264.

Wunder, D.E., Aaronson, W., Hayes, S.F., Bliss, J.M. and Silver, R.P. (1994) J. Bacteriol. 176, 4025-4033.

Yanagi, M. and Yamasato, K. (1993) FEMS Microbiol. Lett. 107, 115-120.

Young, J.P.W. and Haukka, K.E. (1996) New Phytol. 133, 87-94.

Zevenhuizen, L.P.T.M., Posthumus, M.A. and Scholten-Koerselman, H.J. (1980) Arch. Microbiol. 125, 1-8.

Zevenhuizen, L.P.T.M. and van Neerven, A.R.W. (1983) Carbohydr. Res. 124, 166-171.

Zhang, Y., Hollingsworth, R.I. and Priefer, U.B. (1992) Carbohydr. Res. 231, 261-271.

Chapter 8

Soil Biology of the *Rhizobiaceae*

Michael J. Sadowsky and Peter H. Graham

I. Introduction

The family *Rhizobiaceae* is comprised of six genera of plant-associating bacteria, *Rhizobium*, *Bradyrhizobium*, *Sinorhizobium*, *Azorhizobium*, *Agrobacterium*, and *Phyllobacterium*. A seventh genus, *Mesorhizobium*, has been proposed recently (Jarvis *et al.*, 1997; Young and Haukka, 1996). In Bergey's manual (Kersters and DeLey, 1984), the first four genera are distinguished from *Agrobacterium* and *Phyllobacterium* by their ability to form root- or stem-nodule symbioses with members of the plant family Leguminosae in which atmospheric nitrogen may be fixed into ammonia, and used by the host. On the other hand, many members of the genus *Agrobacterium* are described in Bergey's manual as phytopathogenic, and are capable of producing hypertrophies, tumors, and abnormal root growth on diverse plant species.

Recent findings have tended to blur the distinction between species. Thus, strains of *Agrobacterium tumefaciens* transformed with plasmids from *Rhizobium* will form N_2-fixing nodules with *Phaseolus vulgaris* (Martinez et al., 1987). Young and Haukka (1996) grouped *A. rhizogenes* with *R. tropici*, and *R. galegae* with *A. vitis*. It is also possible that *A. tumefaciens* will be renamed. The latter organism has been intensively studied for its ability to cause crown gall disease of many dicotyledonous plants (for reviews see Hooykaas and Schilperoort, 1992; Kado, 1991; Winans, 1992), and in the transformation of several different plant species (Christou, 1997). While much information is available concerning the genetics and physiology of tumor induction by *A. tumefaciens* (Winans, 1992), there is little information concerning the soil biology of this important bacterium. While the survival of *A. tumefaciens* in soil and symptomless plant tissue has been shown to cause occasional outbreaks of plant disease (Bouzar *et al.*, 1987; Burr *et al.*, 1995; Schroth *et al.*, 1971), non-tumorigenic agrobacteria are frequently isolated from soils (Burr *et al.*, 1987). Recently, the polymerase chain reaction (PCR) technique has been used to detect and enumerate *A. tumefaciens* in soil (Haas *et al.*, 1995; Picard et al., 1992) and specific primers have been designed to detect Ti and Ri plasmids from phytopathogenic *Agrobacterium* strains (Sawada *et al.*, 1995).

Members of the genera *Rhizobium, Bradyrhizobium, Mesorhizobium, Sinorhizobium,* and *Azorhizobium,* are unusual among soil organisms in that they have an extensive soil phase as free-living heterotrophs, and form specific symbiotic interaction with leguminous plants. This dual mode of existence presents unique problems and opportunities. When soils are cropped to a particular legume for the first time, few infective rhizobia are likely to be present, and inoculation will generally be needed for adequate nodulation and nitrogen (N_2) fixation (Date, 1991; Diatloff, 1977; Graham, 1985). Significant yield increases following inoculation are then common (Abel and Erdman, 1964; Date, 1991; Graham, 1985; Roughley *et al.*, 1993). In contrast, and irrespective of whether inoculation has been practiced, soils in the areas where a particular crop has been planted for a number of years usually contain abundant rhizobia, and even uninoculated plants will be well nodulated. However, strains used in inoculation studies will usually give rise to only a small percentage of the nodules formed (Date, 1991; Ellis *et al.*, 1984; Ham, 1978). Yield increases following inoculation under such conditions are uncommon (Berg *et al.*, 1988; Date, 1991; Ham, 1978). Consequently, since many of the soil rhizobia are less than fully effective in N_2-fixing ability (Gibson *et al.*, 1975; Guar and Lowther, 1980; Ham, 1978; Singleton and Tavares, 1986), the legume involved will rarely achieve its full yield potential. Despite intensive investigation in recent years, we are only now beginning to understand some of the factors influencing the survival and the persistence of rhizobia in the soil, and their competitiveness for nodulation sites on the host. This review focuses on the biotic and abiotic factors influencing the survival of the root/stem nodule bacteria in soil, and on the traits in host and microsymbiont, which are important in competition for nodulation. Throughout this review we will collectively refer to members of the genera *Bradyrhizobium, Rhizobium, Sinorhizobium, Mesorhizobium* and *Azorhizobium* as "rhizobia".

II. Soil Populations of the Rhizobia and Their Diversity

Rhizobia comprise a relatively small proportion of the total soil bacteria. It has been estimated that *Rhizobium* and *Bradyrhizobium* represent between 0.1 to 8.0% of the total bacterial population in bulk or rhizosphere soil and 0.01 to 0.14% of its biomass (Bottomley, 1992; Brockwell *et al.*, 1995; Schortemeyer *et al.*, 1997). Discrepancies between population densities obtained using most probable number (MPN), plate count, and immunofluorescence analyses suggest that not all the rhizobia are infective. Soberon-Chavez and Najera (1989) reported the isolation of non-nodulating *R. leguminosarum* strains, and non-symbiotic rhizobia have been isolated from the rhizosphere of *Lotus corniculatus* (Sullivan *et al.*, 1996). Moreover, Jarvis and coworkers (1989) demonstrated that some non-rhizobial soil bacteria are capable of nodulating legumes, provided they contain *Rhizobium* nodulation genes. It has been postulated for several years that plasmid transfer between soil bacteria accounts for genetic diversity among soil populations of *Rhizobium* and it has been demonstrated that symbiotic plasmids can transfer nodulation ability to non-nodulating rhizobia (Kinkle and Schmidt, 1990). Recently, Sullivan and coworkers (1995) demonstrated that nodulating strains of *Rhizobium loti* arise through the acquisition of chromosomally located symbiotic genes.

Rhizobia may be applied to agricultural soils as commercial inoculants; introduced as aerial or seed-borne contaminants; or may initially have been present as the microsymbiont of an indigenous legume. When seeds are inoculated, the number of rhizobia applied per hectare will vary with seed size, planting density and method of inoculation. High quality inoculants applied to seed as recommended by their manufacturer, achieve inoculation rates of 2.5×10^3 to 1×10^6 rhizobia seed^{-1} (FAO, 1991; Somasegaran and Hoben, 1994; Vincent, 1970). This corresponds to application rates of about 8×10^{10} rhizobia ha^{-1} for bean and soybean, while for subterranean clover the number of rhizobia introduced is approximately 6×10^{10} ha^{-1} (Brockwell and Bottomley, 1995). Rates for granular soil-applied inoculants or for cover inoculation in irrigation water are even higher (5×10^{11} to 2.6×10^{15} rhizobia ha^{-1}) (Ciafardini and Lombardo, 1991; Gault *et al.*, 1994; Smith, 1992; Smith *et al.*, 1981; Smith and del Rio Escurra, 1982). It is not surprising, given these numbers, that the inoculant strain(s) often dominates in nodulation in the first-year of a newly-introduced crop (Brockwell *et al.*, 1982; Gibson *et al.*, 1976; Singleton and Tavares, 1986), or that a rapid buildup of rhizobia in the soil occurs once nodules senesce and large numbers of viable rhizobia are released (McDermott *et al.*, 1987; Sutton, 1983). In many cases, this ensures domination by the inoculant strain(s) even 5-15 years after initial inoculation (Brunel et al, 1988; Diatloff, 1977; Lindstrom *et al.*, 1990).

Not all introduced legumes do receive inoculation, nor are all inoculants correctly applied. In such situations, seed, soil or aerial contamination will usually lead to some initial nodule formation, and over a period of 4-5 years, to a gradual buildup in the soil rhizobial population. The number of soil rhizobia that result will vary with a number of factors, as discussed in subsequent sections, but can range from 10^2 to 10^7 cells g^{-1} soil (Brockwell and Bottomley, 1995; Carter *et al.*, 1995; Ellis *et al.*, 1984; Hirsch, 1996; Roughley *et al.*, 1995; Singleton and Tavares, 1986; Slattery and Coventry, 1993; Strain *et al.*, 1994; Triplett *et al.*, 1993). Most field or ecological studies designed to count or measure the diversity of these organisms involve a serial dilution of the soil, trapping the rhizobia by nodule formation on sterile seedlings, and the isolation of the rhizobia directly from nodules (Somasegaran and Hoben, 1994). Extraction procedures can under-estimate *Rhizobium* numbers in the soil, and mask their diversity (Dye *et al.*, 1995). Moreover, the host used in trapping the rhizobia is likely to have a marked impact on the recovery of particular groups of rhizobia (Bottomley *et al.*, 1994; Bromfield *et al.*, 1995; Brunel *et al.*, 1996; Keatinge *et al.*, 1995; Kumar Rao *et al.*, 1981; van Berkum *et al.*, 1995). Selective culture media exist for some species of rhizobia (Gault and Schwinghamer, 1993; Tong and Sadowsky, 1993), and may be used in direct isolation of these organisms from soil. The direct purification and probing of soil DNA is increasingly more common place (Streit *et al.*, 1993). To date, the influence of these latter methods on estimations of apparent population size and diversity has not been established. The use of genetically marked strains also facilitates population studies with particular organisms (Hirsch, 1996; Wilson, 1995; Wilson *et al.*, 1995).

Interest in the diversity of soil rhizobia has increased in recent years with the development of improved techniques for strain characterization and grouping, and the recommendation for polyphasic taxonomy in *Rhizobium* (Graham *et al.*, 1991). Methods used have included: evaluation of plasmid profiles (van Berkum *et al.*, 1995); serology including the ELISA technique (Leung *et al.*, 1994; Madrzak *et al.*, 1995; Sadowsky *et al.*, 1987a, van Berkum *et al.*, 1995); restriction fragment length polymorphism (RFLP) coupled to the use of a) DNA probes (Anyango *et al.*, 1995; Bromfield *et al.*, 1995; Demezas *et al.*, 1995; Hirsch *et al.*, 1993; Lindstrom *et al.*, 1990; Madrzak *et al.*, 1995; Nour *et al.*, 1995; Sadowsky, *et al.*, 1990; Sullivan *et al.*, 1996; Thomas *et al.*, 1994), b) REP-, or ERIC- or other PCR methodology (Brunel *et al.*, 1996; de Bruijn, 1993; Judd *et al.*, 1993; Labes *et al.*, 1996; Madrzak *et al.*, 1995; Richardson *et al.*, 1995; Sadowsky and Hur, 1996), and c) RAPDs (Dye *et al.*, 1995; Paffetti *et al.*, 1996; Rome *et al.*, 1996); multilocus enzyme electrophoresis (MEE) (Bottomley *et al.*, 1994; Demezas *et al.*, 1991; Demezas *et al.*, 1995; Dupuy *et al.*, 1994; Eardly *et al.*, 1990; Gordon *et al.*, 1995; Hagen and Hamrick, 1996; Leung *et al.*, 1994; Piñero *et al.*, 1988; Souza *et al.*,1994; Strain *et al.*, 1994; Strain *et al.*, 1995), and SDS-PAGE analysis of cell proteins or lipopolysaccharides (Dupuy *et al.*, 1994; Roberts *et al.*, 1980; Sadowsky *et al.*, 1987b).

Population diversity among rhizobia associated with a particular legume is likely to be greatest in the center of origin for that host (Lie *et al.*, 1987). In the case of *Phaseolus vulgaris*, Piñero *et al.*, (1988) recorded a mean genetic distance per enzyme locus of 0.691 for 51 isolates of *Rhizobium etli* from the Mesoamerican center of origin. This value is higher than that recorded for any other species of bacteria. Similarly, Souza *et al.*, (1994) isolated 372 bean strains, and were able to group them into 7 clusters comprising a total of 95 electrophoretic types (ET). The rhizobia from wild populations of *Phaseolus vulgaris* and *P. coccineus* grouped by location and host, whereas rhizobia isolated from cultivated beans were very heterogeneous. Surprisingly, soils in this region contained an abundance of non-infective rhizobia, Segovia *et al.*, (1991) found 40 non-infective rhizobia for each nodulating isolate recovered. The non-nodulating isolates could be divided into four clusters with 85 ETs. These clustered with nod$^+$ *R. etli* in enzyme electrophoresis studies, and gave RFLP hybridization patterns equal to those of *R. etli* when probed with a rRNA operon probe. In parallel studies done with populations of *Sinorhizobium meliloti* from Asia and the Middle East, Eardly *et al.*, (1990) demonstrated two evolu-tionary lineages within what until then had been considered a single species. This finding has been confirmed in other studies (Brunel *et al.*, 1996; Gordon *et al.*, 1995; Paffetti *et al.*, 1996) and has implications for host relationships and for pH tolerance in soil.

Quite diverse rhizobial populations can also develop in association with species that are not native to a particular region. Leung *et al.*, (1994) identified 53 ET among 200 isolates of *R. leguminosarum* in Oregon, and clustered them into 7 groups, and Amarger *et al.*, (1994) found beans in France to be nodulated by two different species of rhizobia, and identified 40 different plasmid profiles among the 287 isolates tested. The species of legume cultivated can also markedly affect the nature of the soil population. While most of the studies mentioned above have reported a clonal population structure, a number suggest the possibility of recombination between particular groups of soil rhizobia (Demezas *et al.*, 1995; Sullivan *et al.*, 1995) and both plasmid transfer between strains (Jarvis *et al.*, 1989; Kinkle and Schmidt, 1990; Thomas *et al.*, 1994; Young and Wexler, 1988) and chromosomal symbiotic gene transfer (Sullivan *et al.*, 1995) have been noted.

III. Rhizobial Persistence in Soil and Rhizosphere

There are many examples in the literature where strains introduced as inoculants into essentially rhizobia-free soil, are not encountered in the nodules of their host 2-3 years later. Brockwell *et al.*, (1982) evaluated the first- and second-year nodulation of 19 strains of *R. leguminosarum* bv. *trifolii* in a soil with less than one *Rhizobium* g^{-1}. All of the strains produced more than 80% of the nodules formed in the first year, but in the second year nodule occupancy by the inoculant strains ranged from 19 to 94%. The remainder of the nodules was presumably occupied by indigenous and contaminating rhizobia, indicating that even under very favorable conditions, inoculant strains can fail to nodulate their target hosts. Similar results have been reported for greenleaf *Desmodium* in Queensland (Date, 1991) and for *Phaseolus vulgaris* by Vlassak *et al.*, (1996). It should be noted, however, that the inability of a strain in soil to compete for, or to be involved in nodule formation, does not necessarily mean its displacement from the soil. Bromfield *et al.*, (1995) compared the populations of *S. meliloti* recovered from nodules and directly from soil, and found the

frequency with which particular genotypes were recovered from each was significantly different.

It is easy to envision a differential effect of biotic or abiotic stresses on particular rhizobial strains. This is evident in the second year mortality problem detailed by Marshall *et al.*, (1963) and Chatel and Parker (1973); in the variable persistence of *S. meliloti* in the acid soils of South-Western Australia (Howieson and Ewing, 1986), and in studies done by Carter *et al.*, (1995) with *Vicia faba*. However, individual strain responses to stress do not explain all such differences in strain persistence. That specific inoculant rhizobia can persist in soil is evident in the earlier-cited papers of Diatloff (1997), Brunel *et al.*, (1988) and Lindstrom *et al.*, (1990). Similar results have been reported by Sanginga *et al.*, (1994). Chatel *et al.*, (1968) used the term "saprophytic competence" to describe the ability of strains to persist beyond the first year. Unfortunately, few studies have sought to determine what traits are important to this ability. Bushby (1990) noted that in the bradyrhizobia, surface electrophoretic charge was correlated with soil pyrophosphate extractable iron and with soil pH, and hypothesized that the surface charge characteristics of these organisms are matched to the soils in which they occur. Several publications indicate that population levels of introduced *B. japonicum* strains decline to a predictable equilibrium value in different soils (Corman *et al.*, 1987; Crozat *et al.*, 1982 and 1987). The ability of rhizobia to degrade aromatic compounds is discussed below. In the only study where this has been related to persistence, there was no correlation between catabolic capability and persistence (Rynne *et al.*, 1994). This is obviously an area where considerable additional research effort is needed.

IV. Influence of the Legume Host on Rhizobial Populations and Diversity in Soil

The legume host can dramatically influence the prevalence and type of rhizobia in soils (Bottomley, 1992). This may be due to: a) the relatively non-specific enhancement of rhizobia because of their ability to metabolize a substance present in root exudate; b) the result of multiplication and release of rhizobia from the nodule; or c) the ability of the host legume to select particular groups of rhizobia from a mixed population. While both *Rhizobium* and *B. japonicum* strains can exist in soils for a relatively long time in the absence of a host plant (Bottomley, 1992; Brunel *et al.*, 1988; Kucey and Hynes 1989; Sanginga *et al.*, 1994; Slattery and Coventry, 1993; Weaver *et al.*, 1972), specific biovars of *Rhizobium* have been shown to be influenced by the host. For example, Kucey and Hynes (1989) showed that under pea, population densities of *R. leguminosarum* bv. *viceae* were about three orders of magnitude greater than those of *R. leguminosarum* bv. *phaseoli* in the same field. However, in a neighboring field under bean, the trend was reversed. Similar results have been reported for *R. leguminosarum* bvs. *viceae* and *trifolii* strains under vetch and clover (Bottomley, 1992). There is conflicting data on whether the rhizosphere of a legume selectively stimulates the growth of specific populations of rhizobia (Bottomley, 1992). However, Woomer *et al.*, (1988) provided quantitative data to support the contention that numbers of rhizobia were greatest in the presence of their respective legume host. Moreover, van Egeraat (1975) showed that *R. leguminosarum* bv. *viciae* benefits from the presence of homoserine in the root exudate of pea. The need for the presence of the appropriate legume host in achieving sustainable levels in the soil appears to vary with species of organism. Hirsch (1996) noted the survival of *R. leguminosarum* in Rothamsted soil at levels of 10^4 to 10^6 cells g^{-1} in the absence of their host. However, Hirsch (1996) and Triplett (1993) found that *S. meliloti* soil populations are highly dependent on the presence of the host. With *Bradyrhizobium*, Roughley *et al.*, (1995) reported the survival of *B. japonicum* in soil during three crops of paddy rice and one of triticale, a total of 5.5 years between crops of soybean, and Triplett *et al.*, (1993) reported no significant decline in soybean rhizobia for up to 4 years after the last crop. It has been suggested that such ability to survive over long periods of time in soil might be helped to some degree by the protective action of a nodule-specific polysaccharide (Streeter *et al.*, 1995) though these substances appear unlikely to persist for long periods of time in soil.

IV.A. HOST PREFERENCE IN NODULATION

Various degrees of host preference in nodulation are evident in many of the papers considered in the previous section. The existence of two genomic species among strains nodulating *Medicago* has already been mentioned. Brunel *et al.*, (1996) isolated strains of *S. meliloti* from soil using *Medicago polymorpha*, *M.*

truncatula, *M. rigidula*, *M. orbicularis* and *M. minima* as trap hosts, and characterized them using PCR-RFLP. Two divergent groups (R1 and R2) were noted, with R1 strains eliciting only rudimentary nodules on *M. polymorpha*, but nodulating normally with the other five species. In contrast, R2 strains were highly effective in symbiosis with *M. polymorpha*, but not *M. rigidula*. Bromfield *et al.*, (1995) found similar differences in *Sinorhizobium* genotypes associated with *M. sativa* and *M. alba*, but also noted major differences in the frequency with which particular insertion sequence (IS) genotypes were recovered from soil and nodules. Similar specificities have been found with isolates of *Bradyrhizobium*. Bottomley *et al.*, (1994) identified 17 ETs among 95 strains isolated from *Lupinus* and *Ornithopus* species. Of these, 73% fell into two closely related ETs, and dominated in the nodulation of white lupin, serradella and siratro. Strains of a third ET were dominant in nodules of blue lupin. Lange and Parker (1960a, b) earlier had noted differences in the response of lupin species with bradyrhizobia indigenous to southwestern Australia. Similarly Spoerke *et al.*, (1996) found 15 different ETs among the bradyrhizobia associated with biotypes C and S of *Amphicarpa bracteata*, with only one ET associated with both plant biotypes. Three of the ETs produced, on average, less than 1 nodule plant^{-1} with biotype S hosts. In earlier studies with this host, Parker (1995) contrasted nodulation and nitrogen fixation with eight different plant populations obtained from New York, Illinois, and Virginia, and showed that hosts inoculated with a local strain of bradyrhizobia yielded an average of 39% more biomass per plant. Specificity in the symbiotic association between *Lotus corniculatus* and *R. loti* from natural populations has also been reported (Lieven-Antoniou and Whittam, 1997).

Identification of host and inoculant-quality strain combinations in which host preference leads to high nodule occupancy by the inoculant strain in soils containing abundant indigenous rhizobia is yet to be exploited at a commercial level. There are, however, several promising examples. Hardarson and Gareth Jones (1979a; 1979b) and Jones and Hardarson (1979) found significant differences among three varieties of *Trifolium repens* in their preference for particular strains of clover rhizobia, and showed differences in response with temperature and pH. Much of their data appeared to result from the effects of host, *Rhizobium*, and environment on time to appearance of first nodules. Similarly, the bean cultivar RAB39 nodulated equally well with both *R. etli*

UMR1632 and *R. tropici* UMR1899 when either was applied alone, but strongly selected for the latter strain when a mixture of these strains was applied (Montealegre *et al.*, 1995; Montealegre and Graham, 1996). This preference was little affected by pH or temperature, and was still evident when inoculation with UMR1899 was delayed by as much as 8 hours relative to that with UMR1632. Preference in nodulation has also been reported for *Medicago* (Hardarson *et al.*, 1981 and 1982; Materon, 1994) and *Cicer* (Chandra and Parker, 1985). It is an interesting contrast that symbiotic promiscuity with indigenous strains of *Bradyrhizobium* (Nangju, 1980; Pulver *et al.*, 1982) is actively sought for tropical soybean cultivars (Sanginga *et al.*, 1996).

IV.B. HOST RESTRICTION OF NODULATION

In addition to their ability to select specific strains of rhizobia, there are several examples where the host plant restricts nodulation by specific strains or serogroups of rhizobia. The ability to control which strains nodulates, subsequently influences the numbers of that strain released into the soil during nodule senescence. Preempting nodulation by ineffective or inefficient strains of rhizobia has been proposed as a means to control competition for nodulation (Cregan and Keyser, 1986; Fobert *et al.*, 1991).

One of the earliest examples of host-controlled restriction of nodulation was found with European *Rhizobium leguminosarum* bv. *viceae* strains and *Pisum sativum* cv. Afghanistan (Lie, 1978). While most European strains of *R. leguminosarum* bv. *viceae* fail to nodulate cv. Afghanistan, strain TOM nodulates quite efficiently. Subsequently it has been shown that many Middle Eastern strains of *Rhizobium leguminosarum* bv. *viceae* have the ability to nodulate this host genotype (Ma and Iyer 1990; Winarno and Lie 1979). A single recessive host-gene, *sym-2*, found in cultivar Afghanistan has been shown to condition restriction of nodulation by European strains (Holl 1975; Lie 1984). Moreover, the cultivar specificity determinant in strain TOM, *nodX*, which specifically interacts with *sym-2* has been cloned and sequenced (Davis *et al.*, 1988). Host-controlled restriction of nodulation has also been reported in the *R. leguminosarumn* bv. *trifolii*-clover symbiosis (Gibson, 1968; Gibson and Brockwell 1968). While strain TA1 nodulates and forms effective nodules on the majority of clover cultivars, it fails to nodulate cv. Woogenellup. A single recessive gene in this

cultivar, *rwt1*, was shown to condition restriction of nodulation by strain TA1 (Lewis-Henderson and Djordjevic, 1991). Lewis-Henderson and Djordjevic (1991) have shown that two genes in *R. leguminosarumn* bv. *trifolii* strain TA1, *nodM* and *csn*-1, specifically interact with host alleles to control nodulation specificity. More recently, Montealegre and Kipe-Nolt (1994) reported that *Phaseolus vulgaris* genotypes, G21117 and G10002 restrict nodulation by several *R. etli* strains.

Similar strains by cultivar (or genotype) interactions have also been reported in the *B. japonicum*-soybean symbiosis. Host-controlled restriction of nodulation by *B. japonicum* strains can occur at the strain or serogroup level. For example, the single recessive host gene *rj1* and the two host genes *rj5* and *rj6* condition restriction of nodulation by all bradyrhizobia. On the other hand, the single dominant soybean genes, *rj2*, *rj3*, and *rj4*, condition restriction of nodulation by *B. japonicum* strains in serogroups 122 and C1, 33, and 61 and 123, respectively (Devine *et al.*, 1990; Sadowsky *et al.*, 1992; Vest *et al.*, 1973). Recently, Lohrke and coworkers (1996) reported that a single recessive gene in *Glycine max* PI 417566 conditions restriction of nodulation by *B. japonicum* strain USDA 110 and that a single negatively-acting *B. japonicum* gene, *noeD*, controls nodulation specificity with this soybean genotype (Lohrke *et al.*, 1995). Host-controlled restriction of nodulation by USDA 110 has also been reported to be conditioned by a single gene in soybean cv. Peking (Ferrey *et al.*, 1994), although the inheritance of the host allele was not determined. Soybean-controlled nodulation restriction is not limited to *B. japonicum* strains. *Sinorhizobium fredii* strain USDA 257 nodulates primitive lines of soybean, but is unable to nodulate improved varieties such as cv. McCall (Heron *et al.*, 1989). The *S. fredii nolC* (Krishnan and Puepke, 1991) and *nolBTUVW* (Meinhart *et al.*, 1993) genes have been shown to prevent nodulation of cv. McCall and a single dominant host gene has been shown to control the symbiotic interaction (Trese, 1995).

V. Biotic and Abiotic Effects on Soil Rhizobia

V.A. SOIL ACIDITY

Soil acidity constrains agricultural production on as much as 25% of the earth's croplands (Munns,

1986), a proportion that is likely to increase as population pressures in third world countries result in the use of ever more marginal lands for agriculture (Vance and Graham, 1995). For the nodulated legume, pH effects include reduced survival and growth of rhizobia in soil or on the seed, reduced attachment and root hair infection, and poor plant growth. This may be due to the result of hydrogen ion concentration *per se,* to aluminum and manganese toxicity, or to the induced deficiency of essential nutrients including molybdenum, calcium, phosphorus and cobalt. This topic has been extensively reviewed in recent years (Flis *et al.,* 1993; Glenn and Dilworth, 1994; Graham, 1992; Munns, 1986). The reader is also referred to related studies on the adaptive acid tolerance response in *Salmonella* and other bacteria (Foster, 1993; Hall *et al.,* 1995). This Chapter will emphasize only selected aspects of soil acidity.

Acid pH has a marked influence on specific rhizobia, with *S. meliloti* being particularly acid sensitive. Many *Rhizobium* strains are unable to grow in culture media of pH 5.0 or less, and only *R. tropici, R. loti* and some *Bradyrhizobium* sp. are able to tolerate pH 4.5 (Brockwell *et al.,* 1991; Date and Halliday, 1979; Graham *et al.,* 1982, 1994; Keyser and Munns, 1979a,b). These differences are generally reflected in the numbers of particular rhizobia found in acid soils. Brockwell *et al.,* (1991) found an average of 8.9 x 10^4 *S. meliloti* g^{-1} in soils of pH 7.0 or more, but only 37 g^{-1} in soils of pH less than 6.0. Two years after inoculation into soils of pH 4.8 to 5.7, Carter *et al.,* (1995) found only three of eight strains of *R. leguminosarum* bv. *viciae* that were able to maintain soil populations in excess of 100 rhizobia g^{-1}. The case of *Rhizobium etli* is perhaps even more striking. *Phaseolus vulgaris* is native to the highland regions of Mesoamerica and Andean South America, and in both areas beans have been domesticated for more than 7,000 years (Gepts and Debouck, 1991; Kaplan, 1965). In these somewhat arid regions, with soils that are generally neutral to alkaline in pH, beans associate overwhelmingly with *R. etli* (Piñero *et al.,* 1988; Souza *et al.,* 1994). In relatively recent times, Mesoamerican beans were introduced via the Caribbean into Brazil, and Andean bean types into Europe and Africa (Gepts, 1988). Survival of the acid-sensitive *Rhizobium etli* in the acid soils of Brazil and Africa must have been limited, because in some of these areas this organism has been largely replaced by *R. tropici* as the microsymbiont for beans. Anyango *et al.,* (1995) found two soils from

Kenya to each contain 1.47 x 10^4 rhizobia g^{-1}; whereas only one of 40 isolates from the Naivasha soil (pH 6.8) belonged to *R. tropici,* 30 of 35 isolates from the Daka-ini soil (pH 4.5) were from this species. Similarly in Brazil, 111 of 155 isolates from beans grown in Sao Paulo were able to grow in medium of pH 4.5, while 116 of 140 isolates tested nodulated *Leucaena leucocephala* (Vargas and Denardin, 1992). The origins of *R. tropici* are unclear, although a number of *Macroptilium* spp. are native to this region (Fevereiro, 1987) and this organism has been recovered from native hosts such as *Lonchocarpus* and *Gliricidia* (Hungria *et al.,* 1993).

The basis for strain differences in pH tolerance remain uncertain, but include differences in cytoplasmic pH maintenance (Graham *et al.,* 1994; O'Hara *et al.,* 1989), accumulation of compatible solutes (Aarons and Graham, 1991; Graham *et al.,* 1994), membrane permeability (Chen *et al.,* 1993; Graham *et al.,* 1994), calcium metabolism (Howieson *et al.,* 1992; Reeve *et al.,* 1993; Tiwari *et al.,* 1992), and proton extrusion (Chen *et al.,* 1993). Acid-shock proteins and pH dependent structural epitopes have also been identified in the rhizobia (Aarons and Graham, 1991; Bhat and Carlson, 1992), and Tiwari *et al.,* (1996) have demonstrated the existence of a two component signal transduction system in *S. meliloti* that is pH responsive.

Strains tolerant to acid soil conditions will often, but not always, nodulate better than acid-sensitive strains at low soil pH (Graham *et al.,* 1982; Howieson and Ewing, 1986). Thus, while *R. etli* inoculant strains are generally more competitive in nodule formation for *P. vulgaris* than are strains of *R. tropici* (Martinez and Rosenbleuth, 1991; Chaverra and Graham, 1992), the latter organisms will generally dominate under acid soil conditions (Frey and Blum, 1994; Streit *et al.,* 1992; Vargas and Graham, 1988). Similarly Buendia-Claveria *et al.,* (1994) found *S. fredii* generally more competitive than *B. japonicum* in more neutral Spanish soils. *B. japonicum* USDA 110 and 3-15-b3 totally outcompeted *S. fredii* HH103 in the nodulation of soybean at pH 4.9. Differences in host plant tolerance can parallel those found with the rhizobia. The most striking examples are with *Medicago polymorpha* and *M. murex* in Australia (Ewing *et al.,* 1989; Howieson *et al.,* 1991) and with *G. max* in the cerrado soils of Brazil (Spchar, 1994). The selection of *Medicago* cultivars and *Sinorhizobium* strains for acid-pH tolerance has permitted the extension of annual pasture legumes into a large area of southwestern Australia and has resulted in herbage and seed yield increases of 51% and 31%, respectively. Production statistics are highly correlated with rhizobial persistence and nodulation in an acid soil (Howieson *et al.*, 1991). Calcium is needed at higher concentration for both acid-tolerant and -sensitive strains of *S. meliloti* (O'Hara *et al.*, 1989; Howieson *et al.*, 1992; Reeves *et al.*, 1993; Tiwari *et al.*, 1994), but no explanation for this requirement is known.

V.B. TEMPERATURE

Soil temperature influences the growth and survival of rhizobia in soil, its interaction with the legume host, and competition for nodulation. While many soil microbes have the capacity to survive at relatively low soil temperatures, their growth rates and metabolic activities often decline below optimal temperatures (Atlas and Bartha, 1994). Rhizobia from temperate regions are tolerant to 4°C, but little growth generally occurs (Trinick, 1982). However, rhizobia isolated from the Canadian high arctic have been reported to grow well at 5°C (Prévost *et al.*, 1987). High soil temperatures lead to the death of many soil microbes, including rhizobia. Strain differences among cowpea rhizobia were, however, evident in studies done with isolates from Nigeria (Eaglesham and Ayanaba, 1984). It should be noted that as with much other microbial growth and survival parameters, the effect of temperature on microbes is interactive; moisture and pH affect microbial survival at different temperatures. Since biological systems respond to changes of temperature in predictable ways, it is not surprising that relatively high and low temperatures negatively impact rhizobia and the nodulation process. Relatively high root temperature has been shown to inhibit bacterial infection, N_2 fixation ability and the general growth of legume plants (Arayankoon *et al.*, 1990; Frings, 1976; Hungria and Franco, 1993; Kishinevsky and Weaver 1992; Michiels *et al.*, 1994; Munevar and Wollum, 1982; Rainbird *et al.*, 1983). Since both the macro- and micro-symbionts are differentially affected by temperature stress, it is to be expected that temperature effects on nodulation and nitrogen fixation show strong strain by cultivar interactions (Arayankoon *et al.*, 1990; La Favre and Eaglesham, 1986; Munevar and Wollum, 1982). Zhang *et al.*, (1995) and Lynch and Smith (1994) have reported that cool root-zone temperatures limit nodulation and nitrogen fixation in the soybean-

B. japonicum symbiosis. Similarly, Montanez *et al.*, (1995) and Rice *et al.*, (1995) reported that nodulation and nitrogen fixation in the *B. japonicum*/soybean and alfalfa/*S. meliloti* symbioses were negatively affected by low temperatures. However, Rice and coworkers reported that *S. meliloti* has the ability to nodulate and fix nitrogen with alfalfa between 10 and 12°C. On the otherhand, rhizobia have been isolated from *Oxytropis* and *Astragalus* in the high Canadian arctic (Prevost *et al.*, 1987). These rhizobia also have the ability to nodulate sainfoin (*Onobrychis*) and have been shown to effectively fix nitrogen at 5°C. Likewise, Scandinavian *R. leguminosarum* bv. *trifolii* strains nodulated better and fixed more nitrogen at 10°C, than did those strains isolated from more temperate regions (Ek-Jander and Fahraeus 1971). Surprising, few autoecological studies in non-sterile natural soils have been done to assess the growth and persistence of rhizobia at different temperatures. However, Kennedy and Wollum (1988) reported that populations of *B. japonicum* strains decreased in non-sterile soils that were exposed to elevated temperature and Woomer *et al.*, (1988) reported that total rhizobial populations along an elevation transect in Hawaii were significantly correlated with soil temperature.

V.C. WATER AVAILABILITY

Soil water can influence the growth of soil microorganisms, plants, and their interactions through the processes of diffusion, mass flow, and concentration of nutrients (Paul and Clark, 1989). While much is known concerning the plant's response to water and water stress, effects of soil water on the deposition of plant exudates in the rhizosphere (Drew, 1990) and the subsequent growth of rhizosphere microorganisms have been little studied. However, decreased soil water can be postulated to directly affect the growth of rhizosphere microorganisms by decreasing water activity below critical growth tolerance limits and indirectly by altering the types and quantities of root exudates released into the rhizosphere and by altering plant growth and consequently root architecture.

One possible reason for poor nodulation of legume species in arid soils is the decrease in population levels of rhizobia in soil during the dry season. Many studies have reported differences in drought tolerance among rhizobia (mostly laboratory strains), however such studies remain empirical in the absence of any knowledge of the mechanisms involved in drought tolerance. Fuhrmann *et al.*, (1986) showed that strains of clover rhizobia can exhibit distinctive desiccation tolerance in sterile soil systems. The same authors have also reported that *Rhizobium leguminosarum* bv. *trifolii* strain C16 showed excellent survival over a 21-day period, even at a water potential of -500 MPa. The commercial strain TA1 produced the lowest observed populations after 42 days of incubation. Attempts to select desiccation resistant strains of *S. meliloti* by Mary *et al.*, (1985) were not conclusive, perhaps due to the short, 14 day, incubation period. Some studies have suggested that bradyrhizobia are more resistant to water stress than rhizobia (Bushby and Marshall, 1977; Chatel and Parker, 1973; Mary *et al.*, 1994; Woomer *et al.*, 1988). However, other studies have produced contradictory results (Mahler and Wollum, 1981; Pena-Cabriales and Alexander, 1979; van Rensberg and Strijdom, 1980). It should be noted, however, that these studies are often difficult to interpret since the number of rhizobia and bradyrhizobia examined is often limited and the differences shown may only reflect a large degree of intrageneric strain diversity.

V.D. NUTRITIONAL FACTORS

Brockwell *et al.*, (1995) reviewed the nutritional factors influencing the ecology of rhizobia in soil. Their discussion emphasizes the diverse nutritional capability of these organisms, especially in relation to their metabolism of aromatic compounds in soil. Both *Rhizobium* and *Bradyrhizobium* strains possess pathways for the degradation of protocatechuate (Chen *et al.*, 1984; Parke and Ornston, 1986), and the catechol degradative pathway has been shown in *R. leguminosarum* (Chen *et al.*, 1984: Parke and Ornston, 1984). Furthermore, Rao and Cooper (1995) suggest that soybean rhizobia can degrade daidzein and genistein to yield aromatic compounds. The degradation of poly-chlorinated biphenyls (Damaj and Ahmad, 1996) and mimosine (Soedarjo *et al.*, 1994) by rhizobia further extends the metabolic diversity of these organisms, and must influence their ability to survive in soil. Aromatic compounds could be of particular importance to the growth and persistence of "deep" rhizobia associated with *Acacia albida* (Dupuy *et al.*, 1994) and *Prosopis glandulosa* (Jenkins *et al.*, 1987; Thomas *et al.*, 1994). Surprisingly, there have been few studies on the metabolism of aromatic compounds by rhizobia in soil and rhizosphere, though Rynne *et al.*, (1994) have shown that mutants of *R. leguminosarum* which have lost the ability to degrade aromatic compounds retain their competitiveness.

Rhizopine synthesis within the legume nodule is a trait associated with particular symbiotic plasmid types (Wexler *et al.*, 1996), found in 10-14 % of *S. meliloti* and *R. leguminosarum* bv. *viciae* isolates (Murphy *et al.*, 1995; Rossbach *et al.*, 1995). In these strains, synthesis of rhizopine is invariably associated with its catabolism by free-living rhizobia (Murphy *et al.*, 1995). When rhizopine-producing and non- producing strains are evaluated for competitiveness, the one degrading rhizopine is the more competitive. The manipulation of rhizopine synthesis and catabolism genes in rhizobia is but one aspect of rhizosphere engineering so that "plants resist soilborne pathogens more effectively, be better hosts to beneficial microorganisms, remediate toxic waste, or attract communities of soil organisms that enhance plant health" (O'Connell *et al.*, 1996).

Hopanoid lipids constitute up to 40% of the total lipids in *Bradyrhizobium*, but not *Rhizobium* strains (Kannenberg *et al.*, 1995). Their role in the bacterial membrane, in the stress response of bradyrhizobia, and their metabolism are of considerable interest.

While the studies mentioned above emphasize naturally occurring compounds, Fouilleux *et al.*, (1996) obtained better *Bradyrhizobium* survival in soil, and increased early nodulation and overall N_2 fixation using granular inoculants amended with glycerol and sodium glutamate.

Amendment of soil with sludge can reduce both the numbers of rhizobia in the soil and their diversity (Giller *et al.*, 1989; Hirsch *et al.*, 1993). Giller *et al.*, (1989) reported only a single ineffective strain of *R. leguminosarum* bv. *trifolii* persisting in soils amended with sludge over a 20 year period. Giller *et al.*, (1993) found better survival of *S. meliloti* in the same soil, while Kinkle *et al.*, (1987) found numbers of soybean rhizobia greatest in soil amended with 112 tons h^{-1} sludge, with no evidence of serogroup change.

VI. Competition in Soil and Rhizosphere

Competition among rhizobia for the nodulation of legumes has been studied for many years and has recently been reviewed by Triplett and Sadowsky

(1992), Streeter (1994), and Toro (1996). Competitive success has been operationally defined as the ability of one or more rhizobial strains to occupy nodules of a suitable legume host when challenged with simple mixtures or indigenous soil populations of rhizobia. When competing strains are mixed uniformly through the soil, competitive success can also be defined mathematically (Armager and Lobreau, 1982; Beattie et al., 1989; Ireland and Vincent, 1968). A majority of competition studies, however, have been performed in artificial plant growth media and results from these studies are unlikely to mimic conditions in soil (Bottomley, 1992). Moreover, most studies on competition for nodulation have used genetically dissimilar and serologically distinct strains. These studies, however, do not reflect ecologically significant interactions among microbial populations. In nature, competition for nodulation among rhizobia is more important among genetically related populations.

The inability of a particular inoculant strain to occupy a substantial portion of legume nodules under field conditions has been referred to as the "competition problem". Numerous abiotic and biotic factors are known to influence the competitiveness of specific rhizobial inoculants. These abiotic factors include: plant and microbial nutrient limitations and requirements; soil moisture, type, location, pH, and temperature, soil texture and organic matter content (Bottomley, 1992; Dowling and Broughton, 1986; Triplett and Sadowsky, 1993; Turco and Sadowsky, 1995). Biotic factors effecting competition for nodulation include: bacteriocin and antibiotic production (Triplett et al., 1988; Triplett and Sadowsky, 1992), numbers and types of indigenous microorganisms, selective predation by protozoa (Danso et al., 1975), speed of nodulation coupled to the autoregulatory response (McDermott and Graham, 1990), exopolysaccharide production (Zdor and Pueppke, 1991; Araujo and Handelsman, 1990; Bhagwat and Keister, 1991), inoculant placement (McDermott and Graham, 1989), and motility (Caetano-Annoles et al., 1988, Liu et al., 1989, Mellor et al., 1987).

While microbial numbers are not the absolute determinant of nodulation success in soil (Bottomley, 1992; Moawad et al., 1984), inoculation will rarely be beneficial in soils which contain more than 10^2 to 10^3 rhizobia g^{-1} (Evans et al., 1996; Singleton and Tavares, 1986). Thies et al., (1991) and Saginga et al., (1996) suggest reduction in the percentage of nodules formed by the inoculant strain

when the indigenous strains occur at levels of only 10 rhizobia g^{-1}.

When inoculants are applied to soils that already contain indigenous rhizobia, the inoculant strain(s) is at a severe disadvantage. Frequently, introduced strains are outnumbered by as much as 250 to 1 (Brockwell et al., 1995), are not evenly distributed throughout the soil (Bushby, 1993; McDermott and Graham, 1989), and are often not well adapted to general soil conditions. As a result, inoculant rhizobia may only form 5-10% of the nodules produced in the first year, with the frequency with which they are found in nodules declining even further in subsequent crops. This has usually been perceived as a problem of competition between the inoculant rhizobia and those that are indigenous to the soil and has led to an intensive research effort emphasizing: (1) the selection of strains for competitive ability (Berg et al., 1988; Caldwell, 1969; Johnson and Means, 1964; Klubek et al., 1988; Smith and Wollum, 1989), (2) the use of larger than normal inoculation rates (Kapusta and Rouwenhoest, 1973; Weaver and Frederick, 1974a;b), (3) development of improved inoculant carriers and delivery systems (Boonkerd et al., 1978; Kremer and Peterson, 1983; Stoddard, 1976; Wilson, 1975), (4) detailed ecological studies of the indigenous strains, and especially of those belonging to B. japonicum serogroup 123 (Ellis et al., 1984; Moawad et al., 1984; Sadowsky et al., 1987b; 1990; Viteri and Schmidt, 1987; 1989), (5) studies on the molecular basis for competitive interaction, (6) repeated inoculation (Dunigan et al., 1984; Howieson, 1995; McLoughlin et al., 1990), and (7) host preference and restriction in nodulation (Cregan et al., 1986; Ferrey et al., 1994; Sadowsky et al., 1987b; Sadowsky et al., 1990; Sadowsky and Cregan, 1992). Strains of microorganisms which are stable members of the microbial community (members of the "local" microbial flora) appear to have a selective, competitive advantage in occupying available niches in that local environment (Atlas and Bartha 1993). However, primary or early preemptive colonization, followed by prolonged periods of stable maintenance in a soil population most likely leads to the establishment of the "indigenous" state (Atlas and Bartha 1993; Turco and Sadowsky 1995).

In contrast to the approaches adopted above, the studies of Kamicker and Brill (1989), McDermott and Graham (1989) and Wadisirisuk et al., (1989) suggest that inoculant rhizobia are competitive for nodulation sites in the crown region of the plant, near where they are applied, but that they are relatively immobile, and are diluted out as the root system elongates. McDermott and Graham (1989) found inoculant strains to occupy 33-39% of nodules in the crown region of the plant, but not to give rise to any lateral root nodule more than 8 cm from the inoculated seed coat. A number of other studies have also documented decline in inoculant strain representation in the rhizosphere with increasing distance from the point of inoculation (Salema and Parker, 1982). This suggests that we should distinguish in our research between competitiveness *per se*, and overcoming the problem of inoculant mobility and decreased lateral root nodulation. One area where this might have immediate impact on experimental approach would be in the evaluation of motility as an aid to competitive performance. Numerous studies have contrasted the infective performance of strains differing in motility (Mellor *et al.*, 1987; Catlow *et al.*, 1990a;b; Broek and van der Leyden, 1995) or chemotaxis (Broek and van der Leyden, 1995; Caetano Anolles *et al.*, 1988a;b; Kape *et al.*, 1991). While such traits could be of great significance for specific microsites on the root, inoculant strains are unlikely to have the ability to migrate over the large distances needed for the nodulation of lateral roots. Graham and McDermott (1990) have suggested that rhizobia in soil and rhizosphere are not even in direct competition, but rather are fighting the clock to complete a sequence of infection events before root hairs mature (Bhuvaneswari *et al.*, 1981) or regulatory responses are triggered (Heron and Pueppke, 1987; Pierce and Bauer, 1983). Cells failing to complete these events within the time frame available will form few nodules, and appear non-competitive. If this hypothesis were correct, speed of infection (estimated from the nodulation of root-tip marked plants as described by Bhuvaneswari *et al.*, 1981) should be closely correlated with competitive ability. Such a relationship has been reported in several studies (Chaverra and Graham, 1992; Lupwayi *et al.*, 1996; McDermott and Graham, 1990; Stephens and Cooper, 1988), but interestingly was not observed when one of the strains used was preferred by the legume host (Montealegre and Graham, 1996).

VI.A. GENETICS OF COMPETITION FOR NODULATION

Recent advances in molecular biology and genetics have elucidated some of the genes involved in the competition for nodulation process. The reader is cautioned, however, to remember that almost all bacterial genes that affect microbial growth and saprophytic competence in soil will most likely result in a competition defective phenotype. Consequently, there have been no clear examples of the isolation of "competition genes", *per se*, merely genes that affect competition. Nevertheless, there have been several reports of the identification of genes that affect competition for nodulation. How these genes are involved in nodulation competitiveness is not known, although in many instances these genes have been shown to affect the efficiency of nodulation, cell surface characteristics and the synthesis of exopolysaccharides, motility, antibiotic production, and symbiotic effectiveness.

VI.A.1. Cell Surface Characteristics
Various alterations in competition for nodulation are seen when genes controlling cell surface characteristics are disrupted. For example, nonmucoid mutants of *R. fredii* strain USDA 208 were shown to be more competitive than the wild-type for nodulation of *Glycine max* cv. Peking (Zdor and Pueppke, 1991). In contrast, several *Tn5* mutants of *R. leguminosarum* bv. *phaseoli*, *R. tropici*, and *B. japonicum* that were deficient in exopolysaccharide (EPS) synthesis were less competitive than their respective wild-type strains (Araujo and Handelsman, 1990; Bhagwat *et al.*, 1991). The wild-type region corresponding to the *Tn5* insertion in *B. japonicum* that could complement the mutant for both competitiveness and EPS phenotypes was isolated (Bhagwat and Keister, 1991), but has not yet been thoroughly characterized at the genetic level. A *R. meliloti* lipopolysaccharide mutant has also been shown to have a delay in nodule formation, produce fewer nodules, and have reduced competitiveness relative to the wild-type strain (Lagares *et al.*, 1992).

VI.A.2. Motility
While motility and chemotaxis are not essential for nodulation, they have been shown to be important for competition for nodulation (Ames and Bergman, 1981; Caetano-Anolles *et al.*, 1988b), at least for organisms occurring in the same root region. In several instances, decreased motility appears to cause a deficiency in competitiveness. Mellor *et al.*, (1987) isolated nonmotile mutants of *R. leguminosarum* bv. *trifolii*, and these mutants were less competitive for nodulation than wild-type. Liu *et al.*, (1989) studied a nonmotile *Tn7*-induced

mutant of *B. japonicum*. Although this mutant demonstrated decreased competitiveness, this may have been simply due to its decreased growth rate. Ames and Bergman (1981) isolated ethyl methanesulfonate-induced nonmotile mutants of *R. meliloti*, which were both flagellated and nonflagellated. These mutants were deficient in competitiveness, although growth and nodule formation rates were unchanged.

VI.A.3. Antibiosis
Genes encoding antibiosis factors can also confer increased competitiveness, and can have the same effect when transferred to another strain (Triplett, 1988). Triplett and Barta (1987) reported that the trifolitoxin gene of the highly competitive *R. leguminosarum* bv. *trifolii* strain T24 affects competition for nodulation between trifolitoxin-sensitive and -resistant strains. When this gene was stably integrated into the genome of the less competitive strain TA1, it conferred increased nodulation competitiveness compared to the wild-type strain (Triplett, 1990; Triplett et al., 1987; Triplett et al., 1989). More recently, trifolitoxin production in *Rhizobium etli* has been shown to increase nodulation competitiveness in soil (Roberto *et al.*, 1997).

VI.A.4. Nodulation Efficiency
The ability of a microsymbiont to efficiently nodulate it's legume host also has a direct affect on competition for nodulation. Sanjuan and Olivares (1989) identified a DNA region in *R. meliloti* which is involved in nodulation efficiency and competition for nodulation on alfalfa. These authors termed this DNA region *nfe*, for nodule formation efficiency. Mutations in the *nfe* locus cause a severe delay in nodule formation and reduce the competitiveness of mutant strains. The expression of *nfe* genes is dependent on the NifA-RpoN regulatory system (Sanjuan and Olivares, 1989; Soto *et al.*, 1993). A *nfe*-like gene, *nfeC*, which also effects competitiveness has been identified in *B. japonicum* (Chun and Stacey, 1994) and mutations in the *B. japonicum nodVW* have been reported to cause a delay in nodulation (Göttfert *et al.*, 1990, Göttfert, 1993). In *R. fredii* strain USDA 201, a mutation in the *nolJ* gene also causes a delay in nodule formation, reduced efficiency of nodulation, an a decrease in competitiveness (Boundy-Mills *et al.*, 1994).

VI.A.5. Symbiotic Effectiveness
Mutations in several nodulation (*nod*) genes can affect infection rates, nodulation competitiveness, and nodule development (Hahn and Hennecke, 1988; Sargent et al., 1987). For example, a mutant of *R. meliloti* which is defective in lipooligosaccharide production exhibits altered competitiveness for nodulation of alfalfa (Lagares *et al.*, 1992). There are, however, conflicting reports about whether mutations in *nif* or *fix* genes affect competitiveness. While in some studies plants appear to select for effective nitrogen-fixing microsymbionts (George and Robert, 1991), in others no such discrimination exists (Franco and Vincent, 1976). Perhaps the best study to illustrate this point is by Hahn and Studer (1986). These workers isolated a *Tn5* mutant of *B. japonicum* that was Fix⁻; however, this strain still equally competed equally with the wild-type strain for nodulation of soybean.

VI.A.6. Uncharacterized Factors
Several studies have identified genetic alterations which cause changes in competitiveness, but the functions of the genetic elements have not been identified. McLoughlin *et al.*, (1987) compared 600 random *Tn5* mutants of *R. fredii*, and found four that exhibited reduced competitiveness. Unfortunately, they were unable to identify the physiological functions of the genes involved. In a contrasting study, Beattie and Handelsman (1993) used a cosmid library from the highly competitive strain *R. leguminosarum* bv. *phaseoli* strain KIM5. When transferred into the less competitive strain CE3, nine isolates demonstrated increased competitiveness. However, the increase in competitiveness was due to alterations elsewhere in the genome rather than the presence of the new cosmid. Recently, Mavingui *et al.*, (1997) used random DNA amplification (RDA) on a *R. tropici* strain and demonstrated that recombinants had increased competitiveness for nodule formation.

VI.B. INVOLVEMENT OF PLASMIDS IN COMPETITIVENESS

Several studies have localized genes involved in competition for nodulation on the symbiotic plasmid of various fast-growing *Rhizobium* species (Brewin *et àl.*, 1983; Brom *et al.*, 1992; Bromfield *et al.*, 1985; Michiels *et al.*, 1995; Milner *et al.*, 1992; Soto *et al.*, 1993). In the *Pisum sativum* cv. Afghanistan-*Rhizobium leguminosarum* bv. *viceae* symbiosis, Lie *et al.*, (1978;1987) found that a non-nodulating strain thoroughly blocked the nodulation of an otherwise compatible strain, indicating that the competitive ability of a given strain may not be related to the ability to form nodules. A subsequent study with this symbiotic system showed that the critical period for the competitive blocking effect occurred very early, up to 24 hr, after inoculation (Winarno and Lie, 1979). The genes responsible for the competitive blocking phenotype were shown to be located on the symbiotic plasmid (Dowling and Broughton, 1987; Dowling *et al.*, 1989). In another study, strains of *R. leguminosarum* bv. *phaseoli* which had been cured or deleted of each resident plasmid were analyzed to determine the contribution of each plasmid to nodulation, competition, and growth-related functions (Brom *et al.*, 1992). These workers found that disruption of four of the six resident plasmids significantly diminished competitiveness. Thus additional plasmids, other than pSym, may be necessary for increased competitiveness. This is supported by a study of Mathis *et al.*, (1985), in which a *R. fredii* derivative which had been cured of its *Sym* plasmid was still able to nodulate and fix nitrogen. This derivative, however, exhibited delayed nodulation and was less competitive than the wild-type strain.

Transfer of Sym plasmids from one strain to another has also been shown to alter competitiveness. Dejong *et al.*, (1981;1982) showed that strains of *R. leguminosarum* bv. *trifolii* containing a recombinant *Sym* plasmid exhibited increased symbiotic effectiveness which, can indirectly affect the competitiveness of a strain. In another study, Brewin *et al.*, (1983) transferred four different Sym plasmids into *R. leguminosarum* strains which lacked *Sym* plasmid-linked determinants. In all four cases, although growth in the rhizosphere was unaffected, the nodulation rate was influenced. In one combination, a cryptic plasmid from a *R. meliloti* strain placed in a *R. leguminosarum* bv. *phaseoli* recipient improved competitiveness. Taken together, these results indicate the importance of plasmids in determining the competitiveness of several fast-growing species of *Rhizobium*.

VII. References

Aarons, S.R. and Graham, P.H. (1991) Plant Soil 134, 145-151.

Abel, G.H. and Erdman, L.W. (1964) Agron. J. 56, 423-424.

Amarger, N., Bours, M., Revoy, F., Allard, M.R. and Laguerre, G. (1994) Plant Soil 161, 147-156.

Amarger, N. and Lobreau, J.P. (1982) Appl. Environ. Microbiol. 44, 583-588.

Ames, P., and Bergman, K. (1981) J. Bacteriol. 148, 728-729.

Anyango, B., Wilson, K.J., Beynon, J.L. and Giller, K.E. (1995) Appl. Environ. Microbiol. 61, 4016-4021.

Araujo, R.S., and Handelsman, J. (1990) in P.M. Greshoff, L.E. Roth, G. Stacey, and W.E. Newton (eds.), Nitrogen Fixation: Achievements and Objectives, Chapman and Hall, New York.

Arayankoon, T., Schomberg, H.H., and Weaver, R.W. (1990) Plant Soil 126,209-213.

Atlas, R.M., and Bartha, R. (1993) Microbial ecology: fundamentals and applications. The Benjamin/Cummings Publishing Company, Redwood City, California.

Beattie, G.A., Clayton, M.K. and Handelsman, J. (1989) Appl. Environ. Microbiol. 55, 2755-2761.

Beattie G.A., and Handelsman, J. (1993) J. Gen. Microbiol. 139, 529-538.

Berg, R.K., Loynachan, T.E., Zablotowicz, R.M. and Liebermann, M.T. (1988) Agron. J. 80, 876-891.

Bhagwat, A.A., and Keister, D.L. (1991) Appl. Environ. Microbiol. 57, 3496-3501.

Bhat, U.M. and Carlson, R.W. (1992) J. Bacteriol. 174, 2230-2235.

Bhuvaneswari, T.V., Bhagwhat, A.A. and Bauer, W.D. (1981) Plant Physiol. 68, 1144-1149.

Boonkerd, N., Weber, D.F., and Bezdicek, D.F. (1978) Agron. J. 70, 547-549.

Bordeleau, L., and Prévost, D. (1994) Plant Soil 161, 115-125.

Bottomley, P.J. (1992) G. Stacey *et al.*, (Eds.) Chapman and Hall, New York. pp 293-348.

Bottomley, P.J., Cheng, H.H. and Strain, S.R. (1994) Appl. Environ. Microbiol. 60, 1754-1761.

Boundy-Mills, K.L., Kosslak, R.M., Tully, R.E., Pueppke, S.G., Lohrke, S.M., and Sadowsky, M.J. (1994) Mol. Plant-Microbe Interact. 7, 305-308.

Brewin, N.J., Wood, E.A., and Young, J.P.W. (1983) J. Gen. Microbiol. 129, 2973-2977.

Brockwell, J. and Bottomley, P.J. (1995) Soil Biol. Biochem. 27, 683-697.

Brockwell, J., Bottomley, P.J. and Thies, J.E. (1995) Plant Soil 174, 143-180.

Brockwell, J. and Dudman, W.F. (1968) Aust. J. Agric. Res. 19, 749-757.

Brockwell, J., Gault, R.R., Zorin, M. and Roberts, M.J. (1982). Aust. J. Agric. Res. 33, 803-815.

Brockwell, J., Pilka, A. and Holliday, R.A. (1991) Aust. J. Exp. Agric. 31, 211-219.

Broek, A.V. and van der Leyden, J. (1995) Mol. Plant. Microbe Interact. 8, 800-810.

Brom, S., Garcia de los Santos, A., Stepkowsky, T., Flores, M., Davila, G., Romero, D., and Palacios, R. (1992) J. Bacteriol. 174, 5183-5189.

Bromfield, E.S.P., Barran, L.R. and Wheatcroft, R. (1995) Mol. Ecol. 4, 183-188.

Bromfield, E.S.P., Lewis, D.M., and Barran, L.R. (1985) J. Bacteriol. 164, 410-413.

Brown, A.D. (1990) Microbial water stress physiology, John Wiley and Sons, West Sussex, England, pp. 313.

Brunel, B., Cleyet-Marel, J.-C., Normand, P., and Bardin, R. (1988) Appl. Environ. Microbiol. 54, 2636-2242.

Brunel, B., Rome, S., Ziani, R. and Cleyet-Marel, J.C. (1996). FEMS Microbial Ecol. 19, 71-82.

Buendia-Claveria, A.M., Rodrigues Navarro, D.N., Santamaria Linaza, C., Ruiz Sainz, J.E. and Tempranovera, F. (1994) Syst. Appl. Bacteriol. 17, 155-160.

Burr, T.J., Katz, B.H., and Bishop, A.L. (1987) Plant Disease 71, 617-620.

Burr, T.J., Reid, C.L., Yoshimura, M., Momol E.A., and Bazzi, C. (1995) Plant Disease 79, 677-682.

Bushby, H.V.A. (1990) Soil Biol. Biochem. 22, 1-9.

Bushby, H.V.A. (1993) Soil Biol. Biochem. 25, 597-605.

Bushby, H.V.A., and Marshall, K.C. (1977) J. Gen. Microbiol. 99, 19-27.

Caetano-Anolles, G., Crist-Estes, D.K. and Bauer, W.D. (1988a) J. Bacteriol. 170, 3164-3169.

Caetano-Anolles, G., Wall, L.G., Micheli, A.T. de, Macchi, E.M., Bauer, W.D. and Favelukes, G. (1988b) Plant Physiol. 86, 1228-1235.

Caldwell, B.E. (1969) Agron. J. 61, 813-815.

Carter, J.M., Tieman, J.S. and Gibson, A.H.(1995) Soil Biol. Biochem. 27, 617-623.

Catlow, H.Y. , Glenn, A.R. and Dilworth, M.J. (1990a) Soil. Biol. Biochem. 22, 331-336.

Catlow, H.Y., Glenn, A.R. and Dilworth, M.J. (1990b) Soil Biol. Biochem. 22, 573-575.

Chandra, R. and Parker, R.P. (1985) Trop. Agric. 62, 90-94.

Chatel, D.L., Greenwood, R.M. and Parker, C.A. (1968) Proc. 9th Int. Con. Soil Science, Adelaide 2, 65-73.

Chatel, D.L., and Parker, C.A. (1973) Soil Biol. Biochem. 5, 415-423.

Chaverra, M.H. and Graham, P.H. (1992) Crop Sci. 32, 1432-1436.

Chen, Y.P., Glenn, A.R. and Dilworth, M.J. (1984) FEMS Microbiol. Letts. 21, 201-205.

Chen, H., Richardson, A.E. and Rolfe, B.G. (1993) Appl. Environ. Microbiol. 59, 1798-1804.

Chun, J.Y. and Stacey, G. (1994) Mol. Plant. Microbe Inter. 7, 248-255.

Christou, P. (1997) Field Crop Res. 53, 83-97.

Ciarfardini, G. and Lombardo, G.M. (1991) Agron. J. 83, 622-625.

Corman, A., Crozat, Y., and Cleyet-Marel, J.-C. (1987) Biol. Fertil. Soils 4, 79-84.

Cregan, P.B. and Keyser, H.H. (1986) Crop Sci. 26, 911-916.

Crozat, Y., Cleyet-Marel, J.-C., and Corman, A. (1987) Biol. Fertil. Soils 4, 85-90.

Crozat, Y., Cleyet-Marel, J.-C., Giraud, J.J., and Obaton, M. (1982) Soil Biol. Biochem. 14, 401-405.

Damaj, M. and Ahmad, D. (1996) Biochem. Biophys. Res. Comm. 218, 908-915.

Danso, S.K.A., Keya, S.O., and Alexander, M. (1975) J. Microbiol. 21, 884-895.

Date, R.A. (1991) Soil Biol. Biochem. 23, 533-541.

Date, R.A. and Halliday, J. (1979) Nature (Lond) 277, 62-64.

Davis E.O., Evans, I.J. and Johnston, A.W.B. (1988) Mol. Gen. Genet. 212, 531-535.

de Bruijn, F.J. (1992). Appl. Environ. Microbiol. 58, 2180-2187.

Dejong, T.M., Brewin, N.J. and Phillips, D.A. (1981) J. Gen. Microbiol. 124-127.

Dejong, T.M., Brewin, N.J., Johnston, A.W.B. and Phillips, D.A. (1982) J. Gen. Microbiol. 128, :1829-1838.

Demezas, D.H., Reardon, T.B., Strain, S.R., Watson, J.M. and Gibson, A.H. (1995) Mol. Ecol. 4, 209-22.

Demezas, D.H., Reardon, T.B., Watson, J.M. and Gibson, A.H. (1991) Appl. Environ. Microbiol. 57, 3489-3495.

Devine, T., Kuykendall, L.D., and O'Neill, J.J. (1990) Theor. Appl. Genet. 80, 33-37.

Diatloff, A. (1977) Soil Biol. Biochem. 9, 85-88.

Dowling, D.N., and Broughton, W.J. (1986). Annu. Rev. Microbiol. 40, 131-157.

Dowling, D.N., Samrey, U., Stanley, J. and Broughton, W.J. (1987) J. Bacteriol. 169, 1345-1348.

Dowling, D.N., Stanley, J., and Broughton, W.J. (1989) Mol. Gen. Genet. 216, 170-174.

Drew, M.C. (1990) Root function, development, growth and mineral nutrition, in J. Lynch (ed.), The Rhizosphere, John Wiley and Sons, West Sussex, England, pp. 35-58.

Dunigan, E.P., Bollich, P.K., Hutchinson, R.L., Hicks, P.M., Zaunbrecher, F.C., Scott S.G., and Mowers, R.P. (1984) Agron. J. 76, 463-466.

Dupuy, N., Willems, A., Pot, B., Dewettinck, D., Vandenbruaene, I., Maestrojuan, G., Dreyfus, B., Kersters, K., Collins, M.D. and Gillis, M. (1994). Int. J. Syst. Bacteriol. 44, 461-473.

Dye, M., Skot, L., Mytton, L.R., Harrison, S.P., Dooley, J.J. and Cresswell, A. (1995) Can. J. Microbiol. 41, 336-344.

Eaglesham, A.R.J. and Ayanaba, A. (1984) Current developments in biological nitrogen fixation, N.S. Subba Rao (Ed.) Edward Arnold, London, pp 1-35.

Eardly, B.D., Materon, L.A., Smith, N.H., Johnson, D.A., Rumbaugh, M.D. and Selander, R.K. (1990) Appl. Environ. Microbiol. 56, 187-194.

Ek-Jander, J., and Fahraeus, G. (1971) Plant Soil, Spec. Vol., 129-137.

Ellis, W.R., Ham, G.E. and Schmidt, E.L. (1984) Agron. J. 76, 573-576.

Evans, J., Gregory, A., Dobrowolski, N., Morris, S.G., O'Connor, G.E. and Wallace, C. (1996) Soil Biol. Biochem. 28, 247-255.

Ewing, M.A., Pannell, D.J. and Morrison, D.A. (1989). Pastures and profit in the Australian ley farming system, Proc. XVI Intern. Grassl. Conf. (Nice) pp 1349-1350.

Ferrey, M.L., Graham, P.H., and Russelle, M.P. (1994). Can. J. Microbiol. 40, 456-460.

Fevereiro, V.P.B. (1987) *Macroptilium* (Benth). Urban do Brasil. (Leguminosae-Faboideae-Phaseoleae-Phaseolinae), Arqu, Jardin Rio de Janiero 28, 109-180.

Flis, S.E., Glenn, A.R. and Dilworth, M.J. (1993) Soil Biol. Biochem. 25, 403-417.

Fobert, P.R., Roy, N., Nash, J.H., and Iyer, V.N. (1991) Appl. Environ. Microbiol. 57, 1590-1594.

Food and Agriculture Organization of the United Nations (FAO) (1991). Expert consultation on legume inoculant production and quality control, FAO, Rome. pp 145.

Foster, J.W. (1993) J. Bacteriol. 175, 1981-1987.

Fouilleux, G., Revellin, C., Hartmann, A. and Catroux, G. (1996) FEMS Microbial Ecol. 20, 173-183.

Franco, A.A. and Vincent, J.M. (1976) Plant Soil 45, 27-48.

Frey, S.D. and Blum, L.K. (1994) Plant Soil 163, 157-164.

Fuhrmann, F., Davey, C.B., and Wollum, A.G. (1986) Soil Sci. Soc. Am. J. 50, 639-644.

Gault, R.R., Bernardi, A.L., Thompson, J.A., Andrews, J.A., Banks, L.W., Hebb, D.M. and Brockwell, J. (1994) Aust. J. Exp. Agric. 34, 401-409.

Gault, R.R. and Schwinghamer, E.A. (1993). Soil Biol. Biochem. 25, 1161-1166.

George, M.L.C. and Robert, F. (1991) Planta 180, 303-311.

Gepts P. (1988) A middle American and an Andean common bean gene pool, in P. Gepts (ed.) Genetic resources of *Phaseolus* beans, Kluwer, Dordrecht. pp 375-390.

Gepts, P. and Debouck, D. 1991. Origin, domestication, and evolution of the common bean *(Phaseolus vulgaris* L.), in A. van Schoonhoven and O. Voysest (eds.) Common beans, Research for crop improvement. CIAT, Cali, Colombia. pp 7-53.

Gibson, A.H. (1968) Aust. J. Agric. Res. 19, 907-918.

Gibson, A.H. and Brockwell, J. (1968) Aust. J. Agric. Res. 19, 891-905.

Gibson, A.H., Curnow, B.C., Bergersen, F.J., Brockwell, J. and Robinson, A.C. (1975) Soil Biol. Biochem. 7, 95-102.

Gibson, A.H., Date, R.A., Ireland, J.A. and Brockwell, J. (1976) Soil Biol. Biochem. 8, 395-401.

Giller, K.E., McGrath, S.P. and Hirsch, P.R. (1989) Soil Biol. Biochem. 21, 841-848.

Giller, K.E., Nussbaum, R., Chaudri, A.M. and McGrath, S.P. (1993) Soil Biol. Biochem. 25, 273-278.

Glenn, A.R. and Dilworth, M.J. (1994) FEMS Microbiol. Letts. 123, 1-10.

Göttfert, M., Grob, P., and Hennecke, P. (1990) Proc. Natl. Acad Sci. (USA) 87, 2680-2684.

Göttfert, M. (1993) FEMS Microbiol. Rev. 104, 39-64.

Gordon, D.M., Wexler, M., Reardon, T.B. and Murphy, P.J. (1995) Soil Biol. Biochem. 27, 491-499.

Graham, P.H. (1985) Problems of soybean inoculation in the tropics, Proc. III World Soybean Conf. (Ames). R. Shibles (Ed.) pp 951- 959.

Graham, P.H. (1992) Can. J. Microbiol. 38, 475-484.

Graham, P.H., Draeger, K.J., Ferrey, M.L., Conroy, M.J., Hammer, B.E., Martinez, E., Aarons, S.R. and Quinto, C. (1994) Can. J. Microbiol. 40, 198-207.

Graham, P.H. and McDermott, T.R. (1990) The physiology, biochemistry, nutrition, and bioengineering of soybeans, implications for future management. FAR, St Louis, pp 1-18.

Graham, P.H., Sadowsky, M.J., Keyser, H.H., Barnet, Y.M., Bradley, R.S., Cooper, J.E., de Ley, D.J., Jarvis, B.D.W., Roslycky, E.B., Strijdom, B.W. and Young, J.P.W. (1991) Int. J. Syst. Bacteriol. 41, 582-587.

Graham, P.H., Viteri, S.E., Mackie, F., Vargas, A.A.T. and Palacios, A. (1982) Field Crops Res. 5, 121-128.

Guar, Y.D. and Lowther, W.L. (1980) J. Agric. Res. 23, 529-532

Hagen, M.J. and Hamrick, J.L. (1996) Mol. Ecol. 5, 177-186.

Hahn, M., and Hennecke, H. (1988) Appl. Environ. Microbiol. 54, 55-61.

Hahn, M. and Studer, D. (1986) FEMS Microbiol. Lett. 33, 143-148.

Hall, H.K., Karem, K.L. and Foster, J.W. (1995) Adv. Microbial. Physiol. 37, 229-272.

Ham, G.E. (1978) Advances in legume science, R.J. Summerfield and A.H. Bunting (eds.) Royal Botanic Gardens, Kew pp 289-296.

Hardarson, G,. Heichel, G.H., Barnes, D.K. and Vance, C.P. (1982) Crop Sci. 22, 55-58.

Hardarson, G., Heichel, G.H., Vance, C.P. and Barnes, D.K. (1981) Crop Sci. 21, 562-567.

Hardarson, G. and Gareth Jones, D. (1979a) Ann. Appl. Biol. 92, 229-236.

Hardarson, G. and Gareth Jones, D. (1979b) Ann. Appl. Biol. 92, 329-333.

Haas, J.H., Moore, L.W., Ream, W., and Manulis, S. (1995) Appl. Environ. Microbiol 61, 2879-2884.

Heron D.S., Ersek, T., Krishan, H.B. and Pueppke, S.G. (1989) Plant-Microbe Interact. 2, 4-10.

Heron, D.S., and Pueppke, S.G. (1987) Plant Physiol. 84, 1391-1396.

Hirsch, P.R. (1996) New Phytol. 133, 159-171.

Hirsch, P.R., Jones, M.J., McGrath, S.P. and Giller, K.E. (1993) Soil Biol. Biochem. 25, 1485-1490.

Holl, F.B. (1975) Euphytica 24, 767-770.

Howieson, J.G. (1995) Soil Biol. Biochem. 27, 603-610.

Howieson, J.G. and Ewing, M.A. (1986) Aust. J. Agric. Res. 37, 55-64.

Howieson, J.G., Ewing, M.A., Thorn, C.W. and Revell, C.K. (1991) R.J. Wright *et al.*, (eds.) Plant-soil interactions at low pH. Kluwer Academic Publishers, Netherlands. pp 589-595.

Howieson, J.G., Robson, A.D. and Abbott, L.K. (1992) Aust. J. Agric. Res. 43, 765-772.

Hooykaas, P.J.J., and Schilperoort, R.A. (1992) Plant Mol. Biol. 19, 15-38.

Hungria, M., and Franco, A.A. (1993) Plant Soil 149, 95-102.

Hungria, M., Franco, A.A. and Sprent, J.I. (1993) Plant Soil 149, 103-109.

Ireland, J.A. and Vincent, J.M. (1968) Trans. 9th. Int. Conf. Soil Sci, Adelaide 2, 85-93.

Jarvis, B.W.D., Ward, L.J.H., and Slade, E.A. (1989) Appl. Environ. Microbiol. 55, 1426-1434.

Jenkins, M.B., Virginia, R.A. and Jarrell, W.M. (1987) Appl. Environ. Microbiol. 53, 36-40.

Jarvis, B.D.W., van Berkum, P., Chen, W.X., Nour, S.M., Fernandez, M.P., Cleyet-Mavel, J.C. and Gilles, M. (1997) Intern. J. Syst. Bacteriol. 47, 895-898.

Johnson, H.W., and Means, U.M. (1964) Agron. J. 56, 60-62.

Jones, D.G. and Hardarson, G. (1979) Ann. Appl. Biol. 92, 221-228.

Judd, A.K., Schneider, M., Sadowsky, M.J. and de Bruijn, F.J. (1993) Appl. Environ. Microbiol. 59, 1702-1708.

Kamicker, B.J., and Brill, W.J. (1987) Appl. Environ. Microbiol. 53, 1737-1742.

Kannenberg, E.L., Perzl, M. and Hartner, T. (1995) FEMS Microbiol. Letts 127, 255-262.

Kape, R., Parnische, M. and Werner, D. (1991) Appl. Environ. Microbiol. 57, 316-319.

Kaplan, L. (1965) Econ. Bot. 19, 358-368.

Kapusta, D., and Rouwenhoest, D.L. (1973 Agron. J. 65, 916-919.

Kado, C.I. (1991). Crit. Rev. Plant Sci. 10, 1-32.

Keatinge, J.D.H., Beck, D.P., Materon, L.A., Yurtsever, N., Karuc, K. and Altuntas, S. (1995 Exper. Agric. 31, 501-507.

Kennedy, A.C., and Wollum, A.G. (1988) Soil Biol. Biochem. 20, 933-937.

Kersters, K., and DeLey J. (1984) Genus III, *Agrobacterium*, in N.R. Krieg and J.G. Hol (eds.), Bergey's Manual of Sytematic Bacteriology, Williams & Wilkins, Baltimore, pp. 244-254./

Keyser, H.H. and Munns, D.N. (1979a) Soil Sci. Soc. Amer. J. 43, 500-503.

Keyser H.H. and Munns, D.N. (1979b) Soil Sci. Soc. Amer. J. 43, 519-523.

Kinkle, B.K., Angle, J.S. and Keyser, H.H. (1987) Appl. Environ. Microbiol. 53, 315-319.

Kinkle, BK., and Schmidt, E.L. (1991) Appl. Environ. Microbiol. 57, 3264-3269.

Kipe-Nolt, J.A., Montealegre, M., and Thome, J. (1992) New Phytol. 120, 489-494.

Kishinevsky, B.D., Sen, D., and Weaver, R.W. (1992) Plant Soil 143, 275-282.

Klubek, B.P., Hendrickson, L.L., Zablotowicz, R.M., Skwara, J.E., Varsa, E.C., Smith, S., Isleib, T.G., Maya, J., Valdes, M., Dazzo, F.B., Todd, R.L., and Walgenback, D.D. (1988) Soil Sci. Amer. J. 52, 662-666.

Kremer, R.J., and Peterson, H.L. (1983) Agron. J. 75, 139-143.

Krishnan, H.B., and Pueppke, S.G. (1991) Mol. Microbiol. 5, 737-745.

Kucey, R.M.N., and Hynes, M.F. (1989) Can. J. Microbiol. 35, 661-667.

Kumar Rao, J.V.D.K., Dart, P.J., and Usha Khan, M. (1982) P.H. Graham and S.C. Harris (Eds.) BNF technology for tropical agriculture, CIAT, Cali Colombia, pp 291-295.

Labes, G., Ulrich, A. and Lentsch, P (1996) Appl. Environ. Microbiol. 62, 1717-1722.

La Favre, A.K., and Eaglesham, A.R.J. (1986) Can. J. Microbiol. 32, 22-27.

Lagares, A., Caetano-Annoles, G., Niehaus, K., Lorenzen, J., Lunggren, H.D., Puhler, A., and Favelukes, G. (1992) J. Bacteriol. 174, 5841-5952.

Lange, R.T. and Parker, C.A. (1960a) Plant Soil 13, 137-146.

Lange, R.T. and Parker, C.A. (1960b) Nodulation patterns of legumes, Nature (Lond.) 186, 178-179.

Leung, K., Strain, S.R., de Bruijn, F.J. and Bottomley, P.J. (1994) Appl. Environ. Microbiol. 60, 416-426.

Lewis-Henderson, W.R., and Djordjevic, M.A. (1991). J. Bacteriol. 173, 2791-2799.

Lie, T.A. (1978) Ann. Appl. Biol. 88, 462-465.

Lie, T.A. (1984) Plant Soil 82, 415-425.

Lie, T.A. , Goktan, D., Pijnenborg, J. and Anlarsal, E. (1987) Plant Soil 100, 171-181.

Lieven-Antoniou, C.A. and Whittam, T.S. (1997) Mol. Ecol. 6, 629-639.

Lindstrom, K., Lipsanen, P. and Kaijalanen, S. (1990) Appl. Environ. Microbiol. 56, 444-450.

Liu, R., Tran, V.M., and Schmidt, E.L. (1989) Appl. Environ. Microbiol. 55, 1895-1900.

Lohrke, S.M., Orf, J.H., and Sadowsky, M.J. (1995) Characterization of a negatively-acting genotype-specific nodulation gene from *Bradyrhizobium japonicum* USDA 110, Abstracts of the 15th North American Symbiotic Nitrogen Fixation Conference, Raleigh, North Carolina.

Lohrke, S.M., Orf, J.H., and Sadowsky, M.J. (1996) Crop Sci. 36, 1271-1276.

Lupwayi, N.Z., Stephens, P.M. and Noonan, M.J. (1996) Symbiosis 21, 233-248.

Lynch, D.H., and Smith, D.L. (1994) Physiologia Plantarum 90, 105-113.

Ma, S.-W., and Iyer, V.N. (1990) Appl. Environ. Microbiol. 56, 2206-2212.

Madrzak, C.J., Golinska, B., Kroliczak, J., Pudelko, K., Lazewska, D., Lampa, B. and Sadowsky, M.J. (1995) Appl. Environ. Microbiol. 61, 1194-1200.

Mahler, R.L., and Wollum, A.G. (1981) Soil Sci. Soc. Am. J. 45, 761-766.

Mavingui, P., Flores, M., Romero, D., Martínez-Romero, E. and Palacios, R. (1997) Nature Biotechnol. 15, 564-569.

Marshall, K.C., Mulcahy, M.J. and Chowhury, M.S. (1963) J. Aust. Inst. Agric. Sci. 29, 160-164.

Martinez, E., Palcios, R., and Sanchez, F. (1987) J. Bacteriol. 169, 2828-2834.

Martinez-Romero, E. and Rosenblueth, M. (1990) Appl. Environ. Microbiol. 56, 2384-2388.

Mary, P., Dupuy, N., Dolhembiremon, C., Delfives, C., and Tailliez, R. (1994) Soil Biol. Biochem. 26, 1125-1132.

Mary, P., Ochin, D., and Tailliez, R. (1985) Appl. Environ. Microbiol. 50, 207-211.

Materon, L.A. (1994) Appl. Soil. Ecology 1, 255-260.

Mathis, J.N., Barbour, W.M. and Elkan, G.H. (1985) Appl. Environ. Microbiol, 49, 1385-1388.

McDermott, T.R. and Graham, P.H. (1989) Appl. Environ. Microbiol. 55, 2493-2498.

McDermott, T.R. and Graham, P.H. (1990) Appl. Environ. Microbiol. 56, 3035-3039.

McDermott, T.R., Graham, P.H. and Brandwein, D.H. (1987) Arch. Microbiol. 148, 100-106.

McLoughlin, T.J., Alt, S.G., and Merlo, P.A. (1990) Can. J. Microbiol. 36, 794-800.

McLoughlin, T.J., Merlo, A.O., Satola, S.W. and Johansen, E. (1987) J. Bacteriol. 169, 410-413.

Meinhardt, L.W., Krishnan, H.B., Balatti, P.A., and Pueppke, S.G. (1993) Mol Microbiol. 9, 17-27.

Mellor, H.Y., Glenn, A.R., and Dilworth, M.J. (1987) Arch. Microbiol. 148, 34-39.

Michiels, J., Verreth, C., and Van Derleyden, J. (1994) Appl. Environ. Microbiol. 60, 1206-1212.

Milner, J.L., Araujo, R.S., and Handelsman, J. (1992) Mol. Microbiol. 6, 3137-3147.

Moawad, H.A., Ellis, W.R., and Schmidt, E.L. (1984) Appl. Environ. Microbiol. 47, 607-612.

Montanez, A., Danso, S.K.A., and Hardarson, G. (1995) Appl. Soil Ecol. 2, 165-174.

Montealegre, C. and Graham, P.H. (1996) Can. J. Microbiol. In press.

Montealegre, C., Graham, P.H. and Kipe-Nolt, J.A. (1995) Can. J. Microbiol. 41, 992-998.

Montealegre, C. and Kipe-Nolt, J.A. (1994) Arch. Microbiol. 162, 352-356.

Munevar, F., and Wollum, A.G. (1982) Agron. J. 74, 138-142.

Munns, D.N. (1986) Adv. Plant Nutr. 2, 63-91.

Murphy, P.J., Wexler, W., Grzemski, W., Rao, J.P. and Gordon, D. (1995) Soil Biol. Biochem. 27, 525-529.

Nangju, D. (1980) Agron. J. 72, 403-406.

Nour, S.M., Cleyet-Marel, J.C., Normand, P. and Fernandez, M.P. (1995) Int. J. Syst. Bacteriol. 45, 640-648.

O'Connell, K.P., Goodman, R.M. and Handelsman, J. (1996) Trends Biotech. 14, 883-88.

O'Hara, G.W., Goss, T.J., Dilworth, M.J. and Glenn, A.R. (1989) Appl. Environ. Microbiol. 55, 1870-1876.

Paffetti, D., Scotti, C., Gnocchi, S., Fancelli, S. and Bazzicalupo, M. (1996) Appl. Environ. Microbiol. 62, 227-2285.

Parke, D. and Ornston, L.N. (1984) J. Gen. Microbiol. 130, 1743-1750.

Parke, D. and Ornston, L.N. (1986) J. Bacteriol. 165, 288-292.

Parker, M.A. (1995) Ecology 76, 1525-1535.

Paul, E.A., and Clark, F.E. (1989) Soil Microbiology and Biochemistry. Academic Press, San Diego, CA, pp 271.

Pena-Cabrieles, J.J., and Alexander, M. (1979) Soil Sci. Soc. Am. J. 43, 962-966.

Picard, C., Ponsonnet, C., Paget, E., Nesme, X., and Simonet, P. (1992) Appl. Environ. Microbiol. 58, 2717-2722.

Pierce, M. and Bauer, W.D. (1983) Plant Physiol. 73, 286-290.

Piñero, D., Martinez, E. and Selander, R.K. (1988) Appl. Environ. Microbiol. 54, 2825-2832.

Prévost, D., Bordeleau, L.M., Caudry-Reznick, S., Schulman, H.M., and Antoun, H. (1987) Plant Soil 98, 313-324.

Pulver, E.L., Brockman, F. and Wien, H.C. (1982) Crop Sci. 22, 1065-1070.

Rainbird, R.M., Atkins, C.A., and Pate, J.S. (1983) Plant Physiol. 73, 392-394.

Rao, J.R. and Cooper, J.E. (1995) Mol. Plant. Microbe Interact. 8, 855-862.

Reeve, W.G., Tiwari, R.P., Dilworth, M.J. and Glenn, A.R. (1993) Soil Biol. Biochem. 25, 581-586.

Rice, W.A., Olsen, P.E., and Collins, M.M. (1995) Plant Soil 170, 351-358.

Richardson, A.E., Viccars, L.A., Watson, J.M. and Gibson, A.H. (1995) Soil Biol. Biochem. 27, 515-524.

Roberto, E.A., Scupham, A.J., and Triplett, E.W. (1997) Molec. Plant-Microbe Int. 10, 228-233.

Roberts, G.P., Leps, W.T., Silver, L.E. and Brill, W.J. (1980) Appl. Environ. Microbiol. 39, 414-422.

Rome, S., Brunel, B., Normand, P., Fernandez, M. and Cleyet-Marel, J.P. (1996) Arch. Microbiol. 165, 285-288.

Rossbach, S., Rasul, G., Schneider, M., Eardly, B. and de Bruijn, F.J. (1995) Mol. Plant Microbe Interact. 8, 549-559.

Roughley, R.J., Gault, R.R., Gemell, L.G., Andrews, J.A., Brockwell, J., Thompson, J.A. (1995) Plant Soil 176, 1-14.

Roughley, R.J., Gemell, L.G., Thompson, J.A. and Brockwell, J. (1993) Soil. Biol. Biochem. 25, 1453-1458.

Rynne, F.G., Glenn, A.R., and Dilworth, M.J. (1994) Soil Biol. Biochem. 26, 703-710.

Sadowsky, M.J., Bohlool, B.B., and Keyser, H.H. (1987a) Appl. Environ. Microbiol. 53, 1785-1789.

Sadowsky, M.J., and Cregan, P.B. (1992) Appl. Environ. Microbiol. 58, 720-723.

Sadowsky, M.J., Cregan, P.B., and Keyser, H.H. (1990) Appl. Environ. Microbiol. 56, 1468-1474.

Sadowsky, M.J., and Hur, H.-G. (1996) J.R. Lupski, G. Weinstock, and F.J. de Bruijn (eds.), Bacterial Genomes: Structure and Analysis, Chapman and Hall, In Press.

Sadowsky, M.J., Olson, E.R., Foster, V.E., Kosslak, R.M., and Verma, D.P.S. (1988) J. Bacteriol. 170, 171-178.

Sadowsky, M.J., Tully R.E., Cregan, P.B., and Keyser, H.H. (1987b) Appl. Environ. Microbiol. 53, 2624-2630.

Salema, M.P., Parker, C.A., Kidby, D.K. and Chatel, D.L. (1982) BNF technology in tropical agriculture, P.H. Graham and S.C. Harris, (Eds.) CIAT, Cali, Colombia, pp 213-217.

Sanginga, N., Abaidoo, R., Dashiell, K., Carsky, R.J. and Okogun, A. (1996) Appl. Soil. Ecol. 3, 215-224.

Sanginga, N., Danso, S.K.A., Mulongoy, K., and Ojeifo, A.A. (1994) Plant Soil 159, 199-204.

Sanjuan, J., and Olivares, J. (1989) J. Bacteriol. 171, 4154-4161.

Sanjuan, J., and Olivares, J. (1991) Arch. Microbiol. 155, 543-548.

Sargent, L., Huang, S.Z., Rolfe, B.G., and Djordjevic, M.A. (1987) Appl. Environ. Microbiol. 53, 1611-1619.

Sawada, H., Ieki, H., and Matsuda, I. (1995) Appl. Environ. Microbiol. 61, 828-831.

Schortemeyer, M., Hartwig, U.A., Hendrey, G.R., and Sadowsky, M.J. (1996) Soil Biol. Biochem. 28, 1717-1724.

Schroth, M.N., Weinbold, A.R., McCain, A.H., Hildebtrand, D.C., and Ross, N. (1971) Hilgardia 40, 537-552.

Segovia, L., Piñero, D., Palacios, R. and Martinez-Romero, E. (1991) Appl. Environ. Microbiol. 57, 426-433.

Singleton, P.W. and Tavares, J.W. (1986) Appl. Environ. Microbiol. 51, 1013-1018.

Slattery, J.F., and Coventry, D.R. (1993) Soil Biol. Biochem. 25, 1725-1730.

Smith, G.B. and Wollum, A.M. (1989) Appl. Environ. Microbiol. 55, 1957-1962.

Smith, R.S. (1992) Can. J. Microbiol. 38, 485-492.

Smith, R.S. and del Rio Escurra, G.A. (1982) J. Agric. Univ. P.R. 66, 241-249.

Smith, R.S., Ellis, M.A. and Smith, R.E. (1981) Agron. J. 73, 505-508.

Soberon-Chavez, G., and Najera, R. (1989) Can. J. Microbiol. 35, 464-468.

Soedarjo, M., Hemscheidt, T.K. and Borthakur, D. (1994) Appl. Environ. Microbiol. 60, 4268-4272.

Somasegaran, P. and Hoben, H.J. (1994) Handbook for rhizobia. Springer Verlag, New York, pp 450.

Soto, M.J., Zorano, A., Mercadoblanco, J., Lepek, V., Olivares, J., and Toro, N. (1993) J. Mol. Biol. 229, 570-576.

Souza, V., Eguiarte, L., Avila, G., Cappello, R., Gallardo, C., Montoya, J. and Piñero, D. (1994) Appl. Environ. Microbiol. 60, 1260-1268.

Spehar, C.R. (1994) Field Crops Res. 41, 141-146.

Spoerke, J.M., Wilkinson, H.H. and Parker, M.A. (1996) Evolution 50, 146-154.

Stephens, P.M. and Cooper, J.E. (1988) Soil Biol. Biochem. 20, 465-470.

Stoddard, C.D. (1976) Farm. Chem. 139, 50-52.

Strain, S.R., Leung, K., Whittam, T.S., de Bruijn, F.J. and Bottomley, P.J. (1994) Appl. Environ. Microbiol. 60, 2772-2778.

Strain, S.R., Whittam, T.S. and Bottomley, P.J. (1995) Mol. Ecol. 4, 105-114.

Streeter, J.G. (1994) Can. J. Microbiol. 40, 513-522.

Streeter, J.G., Peters, N.K., Salminen, S.O., Pladys, D. and Zhaohua, P. (1995) Plant Physiol. 107, 857-864.

Streit, W., Bjourson, A.J., Cooper, J.E. and Werner, D. (1993) FEMS Microbial Ecol. 13, 59-68.

Streit, W., Kosch, K. and Werner, D. (1992) Biol. Fertil. Soils. 14, 140-144.

Sullivan, J.T., Eardly, B.D., Van Berkum, P., and Ronson, C.W. (1996) Appl. Environ. Microbiol. 62, 2818-2825.

Sullivan, J.T., Patrick, H.N., Lowther, W.L., Scott, D.B., and Rosson, C.W. (1995) Proc. Natl. Acad. Sci. (USA) 92, 8985-8989.

Sutton, W.D. (1983) W.J. Broughton (Ed.) Nitrogen fixation Volume 3. Legumes. Oxford university Press, New York. pp 144-212.

Thies, J.E., Singleton, P.W., and Bohlool, B.B. (1991) Appl. Environ. Microbiol. 57, 19-28.

Thomas, P.M., Golly, K.F., Zyskind, J.W. and Virginia, R.A. (1994) Appl. Environ. Microbiol. 60, 1146-1153.

Tiwari, R.P., Reeve, W.G. and Glenn, A.R. (1992) FEMS Microbiol. Letts. 100, 107-112.

Tiwari, R.P., Reeve, W.G., Dilworth, M.J. and Glenn, A.R. (1996) Microbiology 142, 601-610.

Tong, Z. and Sadowsky, M.J. (1993) Appl. Environ. Microbiol. 60, 581-586.

Toro, N. (1996) World J. Microbiol. Biotech. 12, 157-162.

Trese, A.T. (1995) Euphytica 81, 279-282.

Trinick, M.J. (1982) Biology, in W.J. Broughton (ed.), Nitrogen Fixation, Vol. 2: *Rhizobium*, Clarence Press, Oxford, England, pp. 76-146.

Triplett, E.W. (1988) Proc. Natl. Acad. Sci. (USA) 85, 3810-3814.

Triplett, E. W. (1990) Appl. Environ. Microbiol. 56, 98-103.

Triplett, E.W., Albrecht, K.A. and Oplinger, E.S. (1993) Soil Biol. Biochem. 25, 781-784.

Triplett, E. W. and Barta, T. M. (1987) Plant Physiol. 85, 335-342.

Triplett, E.W. and Sadowsky, M.J. (1992) Ann. Rev. Microbiol. 46, 399-428.

Triplett, E. W., Schink, M. J., and Noeldner, K. L. (1989) Mol. Plant-Microbe Interact. 2, 202-208.

Turco, R.F., and Sadowsky, M.J. (1995) Bioremediation: Science and Applications, Soil Science Special Publication No. 43, Soil Science Society of America, Madison, Wisconsin.

van Berkum, P., Beyene, D., Vera, F.T. and Keyser, H.H. (1995) Appl. Environ. Microbiol. 61, 2649-2653.

van Egaraat, A.W.S.M. (1975) Plant Soil 42, 381-386.

van Rensberg, H.J., and Strijdom, B.W. (1980) Soil Biol. Biochem. 12, 353-356.

Vance, C.P.and Graham, P.H. (1995) I. Tikonovich et al., (eds.) Nitrogen fixation, fundamentals and applications. Kluwer Academic Press, Dordrecht. pp 77-86.

Vargas, A.A.T. and Denardin, N.D. (1992) R. Bras. Ci. Solo 16, 337-342.

Vargas, A.A.T. and Graham, P.H (1988) Field Crops Res. 19, 91-101.

Vest G., Weber, D.F., and Sloger, C. (1973) B. E. Caldwell (ed.), Soybeans: Improvement, Production, and Uses. Agronomy 16, 353-390. Am. Soc. of Agron., Madison, Wis.

Vincent, J.M. (1970) A manual for the practical study of root-nodule bacteria. IBP Handbook No15, Blackwell Scientific Publications, Oxford, pp 164.

Viteri, S.E., and Schmidt, E.L. (1987) Appl. Environ. Microbiol. 53, 1872-1875.

Viteri, S.E., and Schmidt, E.L. (1989) Soil Biol. Biochem. 21, 461-463.

Vlassak, K. van der Leyden, J. and Franco, A. (1996) Biol. Fert. Soils 21, 61-68.

Wadisirisuk, P., Danso, S.K.A. Hardarson, G. and Bowen, G.D. (1989) Appl. Environ. Microbiol. 55, 1711-1716.

Weaver, R.W. and Frederick, L.R. (1974a) Agron. J. 66, 233-236.

Weaver, R.W. and Frederick, L.R. (1974b) Agron. J. 66, 233-236.

Weaver, R.W., Frederick, L.R., and Dumenil, L.C. (1972) Soil Sci. 114, 137-141.

Wexler, M., Gordon, D.M. and Murphy, P.J. (1996) Microbiology 142, 1059-1066.

Wilson, D.O. (1975) Agron. J. 67, 76-78.

Wilson, K.J. (1995) Soil Biol. Biochem. 27, 501-514.

Wilson, K.J., Sessitsch, A., Corbo, J.C., Giller, K.E., Akkermans, A.D.L. and Jefferson, R.A. (1995) Microbiology 141, 1691-1705.

Winans, S.C. (1992) Microbiol. Rev. 56, 12-31.

Winarno, R. and Lie, T.A. (1979) Plant and Soil 51, 135-142.

Woomer, P., Singleton, P.W., and Bohlool, B.B. (1988) Appl. Environ. Microbiol. 54, 1112-1116.

Young, J.P.W., and Haukka, K.E. (1996) New Phytol. 133, 87-94.

Young, J.P.W. and Wexler, M. (1988) J. Gen. Microbiol. 134, 2731-2739.

Zdor, R.E., and Pueppke, S.G. (1991) Can. J. Microbiol. 37, 52-58.

Zhang, F., Lynch, D.H., and Smith, D.L. (1995) Environ. Experimen. Bot. 35, 279-285.

Chapter 9

Opines and Opine-Like Molecules Involved in Plant-*Rhizobiaceae* Interactions

Yves Dessaux, Annik Petit, Stephen K. Farrand and Peter J. Murphy

I. Introduction

The first reports on opines date back to the mid-fifties, when Morel (1956) and Lioret (1956) independently presented their results on, respectively, the metabolism of arginine and the identification of unusual amino acids in *Agrobacterium*-induced crown gall tumors, at a meeting of the French Society for Plant Physiology. These tumor compounds were later purified and identified (Figure 1) as lysopine (Biemann *et al.*, 1960), octopine (Ménagé and Morel, 1964), octopinic acid (Ménagé and Morel, 1965), nopaline (Goldmann *et al.*, 1969) and collectively termed opines (reviews: Tempé and Schell, 1977; Tempé and Goldmann, 1982). The significance of the metabolic perturbation undergone by the crown gall cells remained unclear for several years mostly because the specificity of opines as markers of these tissues was long debated (reviewed by Tempé and Goldmann, 1982). Three observations, however, were milestones in the understanding of the role of opines in the interaction:

1) *Agrobacterium* can degrade opines (Lejeune and Jubier, 1968); 2) the nature of opines synthesized in a tumor depends on the inciting *Agrobacterium tumefaciens* strains, not on the plant as it had been initially proposed (Goldmann *et al.*, 1968 ; Petit *et al.*, 1970); and 3) the correlation between opine degradation by agrobacteria and opine synthesis in plant cells is strict: that is, a given *A. tumefaciens* strain can only degrade the opines synthesized by the tumors it induced (Petit *et al.*, 1970; reviewed by Tempé and Petit, 1983). Interestingly, these features were immediately interpreted as an indication of a possible gene transfer by Petit and Tourneur (1972) - a hypothesis that was first suggested by Braun (1947) and Klein (1954) (reviewed in Braun, 1982 and in Tempé and Petit, 1983) - whereas it was understood only years later that the production of opines by the crown gall cells provided the pathogen with a selective growth advantage (Schell *et al.*, 1979; Tempé *et al.*, 1979). This understanding resulted also from the discovery of the Ti plasmids (Zaenen *et al.*, 1974; Van Larebeke

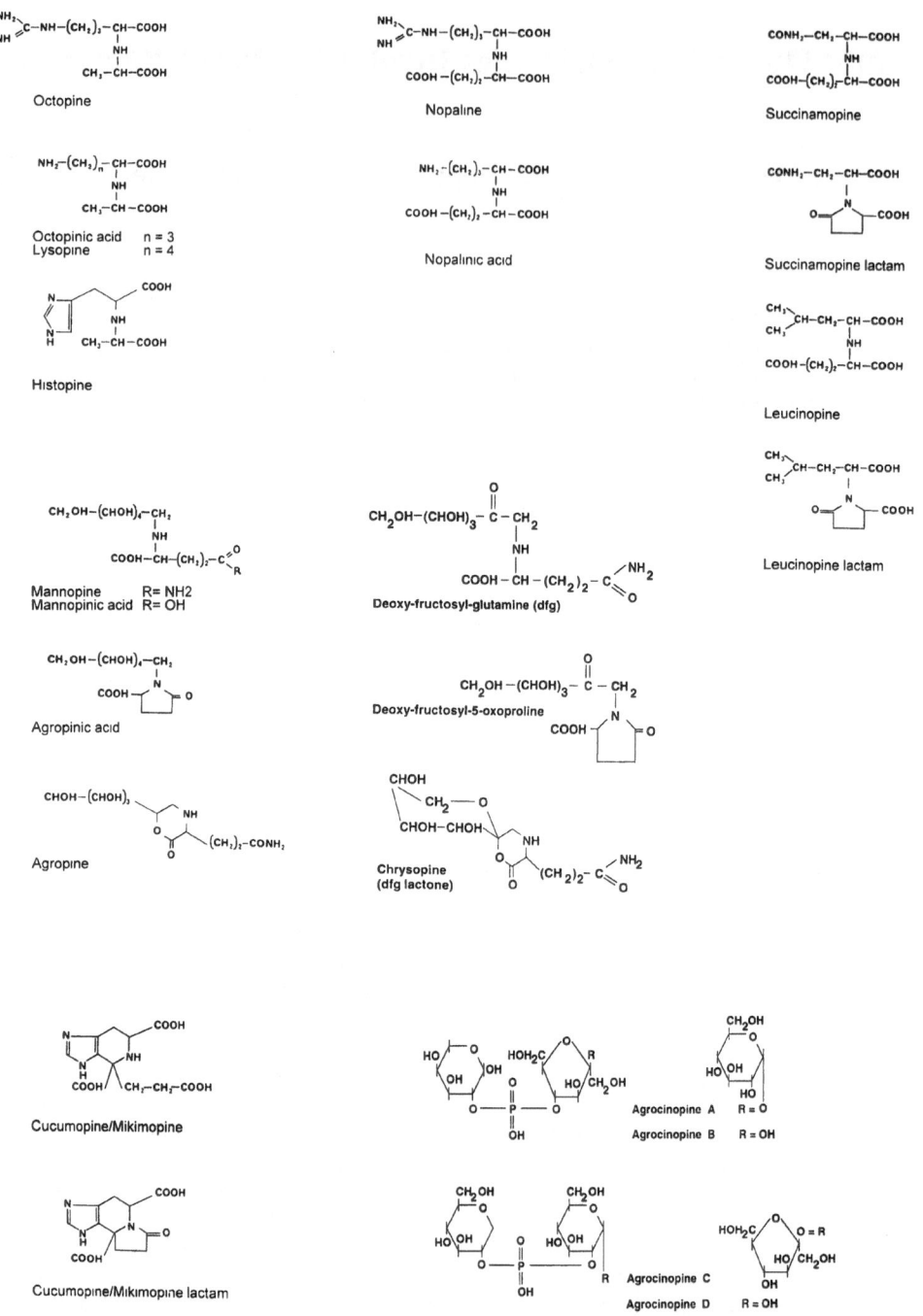

Figure 1. Structures of *Agrobacterium*-induced opines.
For some of these opines (such as succinamopine) D,L and L,L isomers are produced and degraded by agrobacteria (Chilton *et al.*, 1984a, 1984b). For clarity, isomers are not presented except for cucumopine and mikimopine.

et al., 1974; Watson *et al.*, 1975), the elucidation of the tumorigenesis mechanism (Chilton *et al.*, 1977; De Beuckeleer *et al.*, 1978; Thomashow *et al.*, 1980; Willmitzer *et al.*, 1980, see also other chapters in this book), the localization of the genes involved in opine synthesis and degradation on the Ti plasmids (Bomhoff *et al.*, 1976; Montoya *et al.*, 1977) and the demonstration of the opine-induced, conjugal activity of these plasmids (Kerr *et al.*, 1977; Petit *et al.*, 1978a). All these data were combined and the fundamental role of opines in the *Agrobacterium*-plant interaction was rationalized by the originators of "the opine concept" (Tempé *et al.*, 1979) and "the genetic colonization theory" (Schell *et al.*, 1979) which define opines as follows. Opines are small-size molecules, the presence of which in the crown gall tumor is triggered by the pathogen to support its multiplication and to promote the dissemination of its virulence determinants (previous reviews on this topics: Dessaux *et al.*, 1992, 1993; Gelvin, 1992).

In our opinion, the value of the work on opines first comes from its biological significance since data collected on these molecules allowed scientists to give a comprehensive description of the *Agrobacterium*-plant interaction. The work on opines also has brought valuable information on more subtle aspects of this interaction, such as biocontrol of the disease and evolution of Ti and Ri plasmids. Remarkably, the interest of this work also extends beyond the *Agrobacterium*-plant interaction since opine-like molecules (termed rhizopines) have also been found in nodules incited on legume plants by various *Rhizobium* strains (see below and a previous review by Murphy and Saint, 1992). As described below, the presence of rhizopines in nodules results from a different mechanism since these compounds are synthesized by the bacteroids and not by transformed plant cells. However, as opines do, rhizopines serve as selective growth substrates, although, as discussed below, this may not be their principal role in plant-microbe interactions. The purpose of this chapter is to review all the above-mentioned aspects of the work on opines, in both the *Agrobacterium*-plant and the *Rhizobium*-plant interactions.

II. The Development of the Opine Concept

Formulated for a limited number of pathogenic *A. tumefaciens* strains, the validity of the opine concept was immediately challenged by the occurrence of " null-type " tumors which appeared at this time to contain no known opines. Using healthy plant tissues as controls, it was shown that null-type tumor extracts contained one or more compounds (Firmin and Fenwick, 1977; Tate *et al.*, 1982) which were specifically degraded by the inciting null-type *Agrobacterium* strains (Guyon *et al.*, 1980). These compounds were identified as the mannityl opines (mannopine, agropine, mannopinic and agropinic acids; Figure 1) (Firmin and Fenwick, 1977; Tempé *et al.*, 1980; Tate *et al.*, 1982). A similar protocol (based on growth analysis of various *Agrobacterium* strains on healthy tissue or tumor extracts) was used a few years later to identify the opines of the succinamopine family in other null-type tumors (Chilton *et al.*, 1984a, 1984b). These results certainly corroborated the opine concept. However, they only dealt with the *A. tumefaciens*-plant interactions and a need for a significant extension of the opine concept therefore still existed at this time.

The major extension of the opine concept came in part from the work on the *Agrobacterium rhizogenes*-plant interaction. For the pathologist, *A. rhizogenes* and *A. tumefaciens* differ by the types of disease they cause (Kersters and De Ley, 1984). *A. tumefaciens* induces tumorous growth of the plant cells (Smith and Townsend, 1907), whereas *A. rhizogenes* incites root formation at the inoculation site (Riker *et al.*, 1930). This latter type of interaction, however, was chosen for further studies on the validity of the opine concept because it exhibited more analogies to, than differences with, the *A. tumefaciens*-plant interaction. The pertinence of this choice was confirmed as opines were clearly detected in hairy roots culture tissues using simple opine detection and opine degradation assays (Petit *et al.*, 1983; Isogai *et al.*, 1988; Davioud *et al.*, 1988). Combined with the above results, the fact that opines were synthesized by hairy roots in a strain-specific manner provided the first indication that the *A. rhizogenes*-plant interaction is another natural instance of gene transfer (Birot *et al.*, 1987; Tepfer, 1989). Physiological, genetic and molecular data met here to show the involvement of a plasmid (called Ri, root-inducing) as the main determinant of pathogenicity in *A. rhizogenes* (White and Nester, 1980), and the occurrence of a transfer of DNA from the bacteria to the hairy root cell during the infection process (Chilton *et al.*, 1982; White *et al.*, 1982; Spano *et al.*, 1982). Further studies demonstrated that genes involved in opine biosynthesis are located

Classification of Ti and Ri Plasmids		
Type of Plasmids	Relevant Opine Marker	Opine(s) inducing the transfer of the Ti or Ri Plasmid
Ti Plasmids		
Octopine	Octopine, octopinic acid, lysopine, histopine, agropine, mannopine, agropinic and mannopinic acids	Octopine, octopinic acid, lysopine
Nopaline	Nopaline, nopalinic acid, agrocinopines A + B	Agrocinopines A + B
Agropine	Agropine, mannopine, agropinic and mannopinic acids, agrocinopines C + D leucinopine, leucinopine lactam, L,L succinamopine	Agrocinopines C + D
Succinamopine	D,L succinamopine, succinamopine lactam, succinopine, nopaline[a]	Unknown
Lippia	Agrocinopines C + D[a]	Unknown
Chrysopine/ succinamopine	Chrysopine, deoxy-fructosyl-5-oxoproline (dfop), L,L succinamopine, L,L leucinopine	Unknown
Chrysopine/ nopaline	Chrysopine, deoxy-fructosyl-5-oxoproline (dfop), nopaline	Unknown
Grapevine I	Octopine[b], cucumopine	Unknown
Grapevine II	Nopaline	Unknown
Grapevine III	Vitopine, rideopine	Unknown
Ri Plasmids		
Agropine	Agropine, mannopine, agropinic and mannopinic acids, agrocinopines A + B[a]	Unknown
Mannopine	Mannopine, agropinic and mannopinic acids, agrocinopines C + D[a]	Unknown
Cucumopine	Cucumopine, cucumopine lactam	Unknown
Mikimopine	Mikimopine, mikimopine lactam	Unknown

Table 1. The opine-based classification of Ti and Ri plasmids
[a] These compounds are probably degraded by the bacteria harboring these plasmids though their presence in tumors or hairy roots remains unclear. [b] Presence of octopinic acid and lysopine in the tumors was not investigated.

on the Ri and Ti plasmid T-DNA (Barker *et al.*, 1983; Gielen *et al.*, 1984; Slightom *et al.*, 1986). Interestingly, the genetic determinants involved in opine degradation are most often borne on the Ri plasmids. However, and in contrast with the *A. tumefaciens* case, some of these determinants are also located on another replicon which can cointegrate with the Ri plasmid in some strains of *A. rhizogenes* (Petit *et al.*, 1983; Jouanin, 1984; Hufmann *et al.*, 1984; Jouanin *et al.*, 1986). Also in contrast with the *Agrobacterium* case, none of the opines present in the hairy root extract induce the conjugal transfer of the Ri plasmid, which appears to occur constitutively (A. Petit, unpublished). Overall, these results confirmed that opines are key chemical mediators of the *Agrobacterium*-plant interaction. Synthesized at the expense of the pool of metabolites of the transformed plant cells, opines provide pathogenic agrobacteria with the nitrogen, carbon and energy sources necessary for their propagation, and hence the dissemination of their Ti or Ri plasmids.

At present, over 20 opine molecules are known. As indicated above, their occurrence in crown gall and hairy root hyperplasia depends on the Ti or Ri plasmid hosted by the pathogenic bacteria (reviews: Dessaux *et al.*, 1992, 1993). This feature provided the basis for a rudimentary but useful classification of these plasmids, as shown in Table 1. However, this classification may not be the most pertinent one, as suggested by numerous studies on the physical and functional relationships of the Ti plasmids of *A. vitis* (the agent of crown gall tumors on *Vitis vinifera*). These studies indicate that Ti plasmids have probably evolved as "mosaic" structures, composed of conserved regions (the T-DNA, the virulence genes) and variable regions such as those determining opine degradation (Paulus *et al.*, 1991a, 1991b; Otten *et al.*, 1992, 1993; Van Nuenen *et al.*, 1993; Otten and De Ruffray, 1994). A classification based on this only criteria may therefore artificially increase the "distance" between two plasmids which, otherwise, could be highly related. The mosaic structure probably also applies to the *A. tumefaciens* Ti plasmids which are composed in part of related regions (T-DNA, *vir* genes, etc., see Figure 2).

III. Opine-like Molecules in the *Rhizobium*-Legume Interaction

In the early 80's Tempé and Petit (1983), whilst investigating the generality of the opine concept in *Agrobacterium*, extended their study on opine-like compounds to *Rhizobium*. The rationale behind this was that plant pathogenic and symbiotic interactions are similar and rhizobia and agrobacteria are both members of the family *Rhizobiaceae*. Especially, since *Rhizobium meliloti* and *Agrobacterium* are very similar, initial studies were undertaken with this species. From a study of 20 *S. meliloti* strains one, L5-30, was identified which produced an opine-like compound when associated with the plant. This compound was identified by inoculating alfalfa plants with the rhizobia, harvesting the nodules produced, macerating these and removing non-specific carbon and nitrogen sources by incubation with agrobacteria. The original strain was then tested for growth on minimal media supplemented with this extract. Growth indicated the presence of an opine-like compound thus linking the bacterial induction of the compound in the plant with its catabolism by free-living bacteria. This compound was later identified as 3-*O*-methyl-*scyllo*-inosamine (3-*O*-MSI; Figure 3) and termed a rhizopine (Murphy *et al.*, 1987). Rhizopines were therefore discovered by analogy with *Agrobacterium* opines. Rhizopines are now defined as nodule-specific compounds produced in bacteroids and catabolized by the nodule-inducing rhizobia (Murphy and Saint 1992; Murphy *et al.*,

1995). Their function is to affect intra-species competition for nodulation (Gordon *et al.*, 1996).

Implicit in the opine concept is the fact that opine genes are transferred to the plant via the T-DNA and are expressed under plant promoters (see other chapters in this book). Molecular studies (as discussed below) indicate this is not the case with the rhizopine system. However, the analogy to opines is considerable as, although rhizopine genes are not incorporated into the plant or controlled by plant promoters, they are expressed in the bacteroid (the site of nitrogen fixation), influence symbiosis and are controlled by a symbiotic promoter.

In rhizobial-plant microbe interactions a number of plant-derived compounds such as calystegins, betaines, homoserine and mimosine produced by host and non-host plants can act as carbon and nitrogen substrates for rhizobia (Van Egeraat, 1975; Tepfer *et al.*, 1988a, 1988b; Boivin *et al.*, 1990a; 1990b; and Soedarjo *et al.*, 1994) and can collectively be termed nutritional mediators (see below). None of the above are similar to opines or rhizopine as they are not induced by the catabolizing strain. Catabolism of rhizopine has been shown to be highly specific to those strains producing it, indeed more specific than opines, which might be catabolized by a broader range of microorganisms than just the tumor-inducing strain (Beauchamp *et al.*, 1990). To date, rhizopines have only been detected in rhizobia and not in other plant-associated bacterial species (Rossbach *et al.*, 1995).

Generally, *Agrobacterium tumefaciens* strains fall into one of the seven classes described in Table 1 according to the opines degraded by the bacteria and being present in the tumors they induce. This is not the case with rhizopines. In an extensive study, covering in excess of 330 strains, 25 rhizopine strains were found amongst the species *S. meliloti* and *R. leguminosarum* bv. *viciae* but not in the species *R. etli*, *R. tropici* or *R. leguminosarum* bv. *trifolii* and bv. *phaseoli* (Wexler *et al.*, 1995). Of these 25 rhizopine strains found, all but one induced the production of 3-*O*-MSI with the remaining strain producing a closely related compound: *scyllo*-inosamine (SI; Figure 3) (Wexler *et al.*, 1995). The inability to detect rhizopine strains in certain species may simply reflect that they are present in very low frequency. Alternatively, plant precursors may not be available to produce the rhizopine(s), or there may be no advantage conferred by rhizopines within these species.

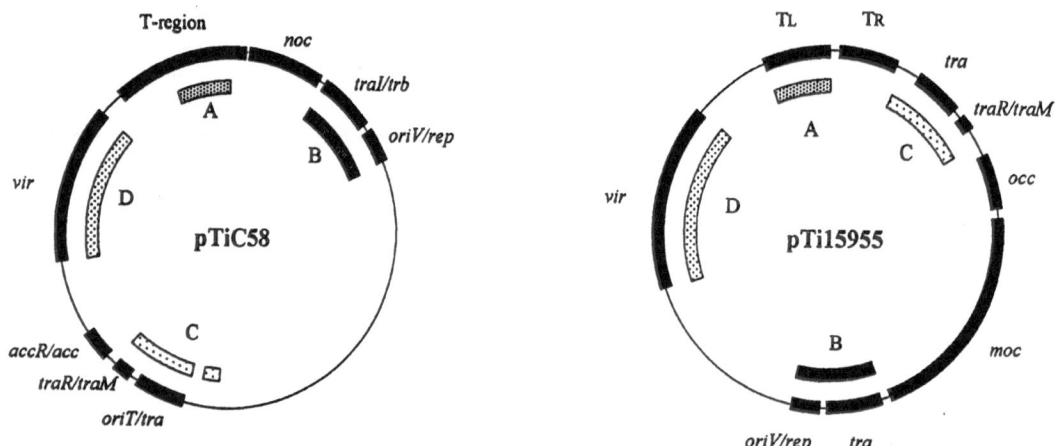

Figure 2. The functional organization of Ti plasmids
The functional organization of the nopaline-type pTiC58 is shown on the left panel while that of the octopine-type pTi15955 is shown on the right panel. For pTiC58 legends are: *noc*, nopaline catabolism; *traI/trb*, transfer region (including the *traI* gene involved in autoinducer production, see the chapter by Farrand, this book); *oriV/rep*, origin of replication of pTiC58; *oriT/tra*, origin of transfer of pTiC58; *traR/traM*, region involved in the regulation of the transfer of the pTiC58; *accR/acc*, agrocinopine catabolism (including the *accR* gene coregulating the transfer of the pTi transfer and the degradation of agrocinopines); and, *vir*, virulence region (see other chapters in this book). For pTi15955 legends are as for pTiC58, except: TL and TR, left and right parts of the T-DNA respectively; *occ*, octopine catabolism; and *moc*, mannityl opine catabolism. Regions A, B, C and D of pTi15955 exhibit strong sequence similarity with the corresponding A, B, C, D regions of pTiC58.

Within the rhizopine producing species, approximately 11% of *S. meliloti* strains and 12% of *R. leguminosarum* bv. *viciae* have the rhizopine phenotype. As stated above, rhizopines play a role in intra-species competition. Therefore it is rational that not all strains within a species produce them, since if they did none would have an advantage. As well, other strains may produce, as yet unidentified, compounds which can be defined as rhizopines. Nevertheless, it is still a quandary as to why only approximately 10% of these strains produce rhizopine. To address this question, the genetic relationship amongst rhizopine-producing strains was studied. Rhizopine strains from both *S. meliloti* and *R. leguminosarum* bv. *viciae* were analyzed by

multi locus enzyme electrophoresis (MLEE), using 14 loci. The rhizopine phenotype was not associated with chromosome-type but did show some relationship to Sym plasmid-type. In general, the phenotype was widely dispersed amongst genetic backgrounds (Wexler *et al.*, 1996). As well, the geographic distribution of rhizopine strains was widespread, suggesting that this is perhaps the result of their dissemination with commercial *Medicago* and *Pisum* species, rather than that of independent rhizopine gene evolution (Wexler *et al.*, 1995).
Initial studies of rhizopine genes were performed on *S. meliloti* strains L5-30 and Rm220-3, which were isolated from widely separated geographical regions. In both these strains, the genes involved in rhizopine

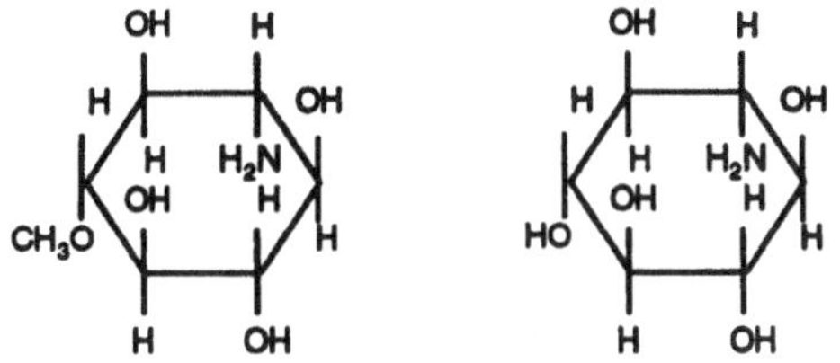

region) in common with the *nifH* gene (encoding a nitrogenase subunit, Hageman and Burris, 1978) and is also flanked by a *nifA/ntrA* regulated promoter (as is the *nifH* gene). Since ORF1 is not translated in *Rhizobium* and a frameshift mutation does not prevent rhizopine production (Murphy *et al.*, 1993; Rao *et al.*, 1995), it is considered that its major role is linking the *mos* operon to a *nifA*-regulated promoter. It therefore appears that the *mos* operon has sequestered a copy of a *nifH* gene, including its promoter, ensuring that rhizopine biosynthesis is symbiotically regulated.

The function of MosA was determined by a combination of chemical and genetic studies. Since the *mosA* gene is completely absent in Rm220-3 (Rao *et al.*, 1995) and this strain produces *scyllo*-inosamine (SI) rather than 3-*O*-MSI (Saint *et al.*, 1993), it was concluded that MosA is involved in the methylation of the rhizopine. This was confirmed by the removal of *mosA* from the L5-30 *mos* locus resulting in the production of SI rather than 3-*O*-MSI. Surprisingly, MosA does not have homology to methylases; therefore, its mode of action in affecting methylation remains unclear. MosB is likely to have a regulatory role since the central domain has extensive homology with a diverse range of proteins involved in either antibiotic or outer cell wall biosynthesis and is common to regulatory proteins DegT and DnrJ having a putative helix-turn-helix motif. A regulatory role for MosB could be to redirect pre-existing biochemical pathways diverting them to rhizopine biosynthesis. However, as a considerable portion of the MosB protein lies outside this region it is possible that it may also have an enzymatic role (Murphy *et al.*, 1993). Since MosC has 12 membrane spanning regions and immuno gold labelling studies indicate it is associated with bacteroid membranes, it likely plays a transport role either bringing a precursor of rhizopine production into the bacteroid or transporting the final product out. Although rhizopine is produced in bacteroids, it is rapidly removed from the site of manufacture since it can not be detected in isolated bacteroids (Grzemski and Murphy, unpublished data).

Overall, the *mos* locus is a mosaic structure having homology to many genes in symbiotic nitrogen fixation (for example *fixU*, *fixA*, *nifH* see Murphy *et al.*, 1993). Many of these represent small regions of homology and the functional significance would seem obscure. Such a structure is reminiscent of a hot spot for rearrangement occurring within the locus. As well, close to the *mos* locus, there are

3-O-methyl-scyllo-inosamine   scyllo-inosamine
(3-O-MSI)   (SI)

Figure 3. Structures of *Rhizobium*-induced rhizopines.

biosynthesis (*mos* genes) and those involved in its catabolism (*moc* genes), are closely linked and on a large bacterial *nod-nif* symbiotic plasmid essential for symbiotic nitrogen fixation (Murphy *et al.*, 1987; Saint *et al.*, 1993). In addition, of 25 other rhizopine strains isolated, all have both synthesis and catabolic genes that are Sym plasmid located (Wexler *et al.*, 1995). The co-occurrence of rhizopine catabolic and synthesis genes and their Sym plasmid location suggest that these genes have coevolved as part of a bi-component functional unit involved in symbiosis. Importantly, the *mos* locus is regulated by symbiotic nitrogen fixation *nifA/ntrA* regulatory genes (Murphy *et al.*, 1988; Saint *et al.*, 1993). This ensures that the locus is coordinately regulated with the nitrogen fixation process (Gussin *et al.*, 1986) and implies that it is controlled by low oxygen levels (Ditta *et al.*, 1987), emphasizing the symbiotic nature of these genes.

The structures of the rhizopine biosynthesis *mos* loci from *S. meliloti* strains L5-30 and Rm220-3 have been determined (Figure 4). Both consist of an operon regulated by a *nifA/ntrA* type promoter. The L5-30 *mos* locus has four ORFs, termed ORF1 and *mosABC* (Murphy *et al.*, 1993). The *mosA* gene is absent from strain Rm220-3 (resulting in the production of SI rather than 3-*O*-MSI) but the remaining sequence shares 98% homology with that of L5-30 (Rao *et al.*, 1995). ORF1 has regions (including the leader and 5'-region of the coding symbiotic promoters which could potentially transcribe small open reading frames. It is tempting to speculate that these are examples of natural promoter fusions representing false starts for genes to come under the control of symbiotic promoters.

Similar false starts have been observed in the vicinity of nitrogen fixation genes in other *Rhizobium* strains (Better *et al.*, 1983).
The biosynthetic pathway of rhizopines is unknown. Based on structural similarities, part of the

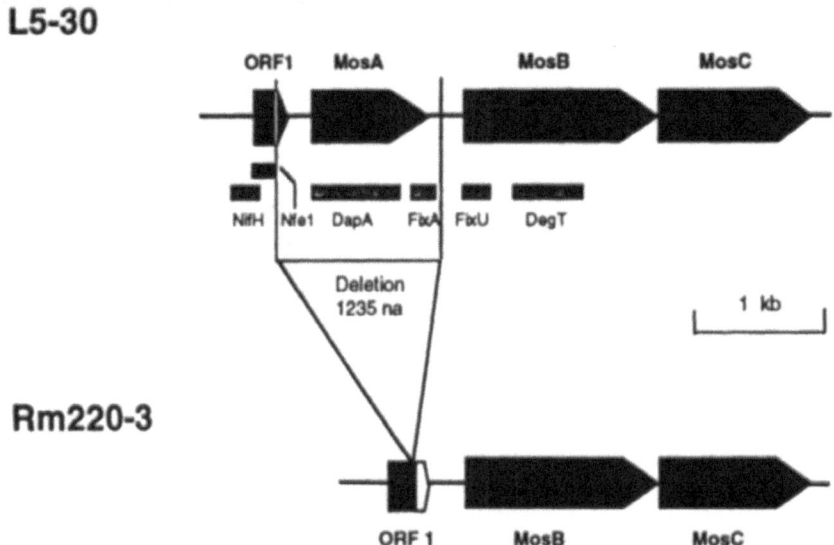

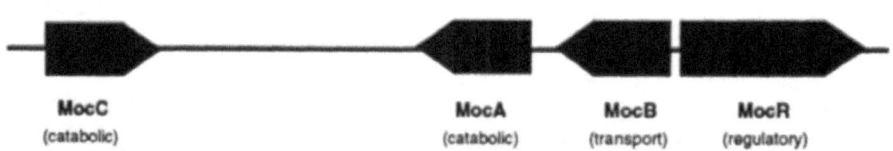

Figure 4. Rhizopine-related genes in *Rhizobium meliloti*
A. Rhizopine biosynthesis genes (*mos*). The black arrows indicate homologous ORFs between strains L5-30 and Rm220-3. The hatched bars indicate homologous regions to known genes. After Rao *et al.*, (1995). B. Rhizopine catabolic genes (*moc*) in strain L5-30. Likely functions are indicated below the genes. After Rossbach *et al.*, (1994).

biosynthetic pathway may resemble that involved in the early steps of streptidine biosynthesis in *Streptomyces* spp. (Walker, 1995). Regardless of the pathway, since rhizopines are produced in bacteroids, they almost certainly are synthesized from plant-derived precursors. However, they are not produced from, or reducing equivalents are not supplied by, the C_4 carboxylic acids (which drive nitrogen fixation in *S. meliloti*, Dilworth and Glenn, 1984) since a *dctA* mutant affecting the transport of these compounds into the nodule does not affect rhizopine production (Grzemski and Murphy, unpublished data). Taken together with the above results, the Mos functions might nevertheless represent an efficient way of parasitising plant products by employing transport processes and regulating biochemical pathways likely already present in the plant and nodule.

Rhizopine catabolic (*moc*) genes enabling rhizobia to grow on rhizopines as a sole carbon source are expressed in free-living bacteria with no evidence that they are also expressed in the bacteroid, as is the case with other nutritional mediators such as trigonelline (Boivin *et al.*, 1990a; 1990b). These genes were originally isolated from *S. meliloti* L5-30 as a 15 kb fragment closely linked to the biosynthetic *mos* genes (Murphy *et al.*, 1987). More recently they have been delimited to at least 4 genes (Figure 4) termed *mocABRC* (Rossbach *et al.*, 1994a). Based on sequence homologies, *mocA,B* and *R* genes likely encode for inositol degradation, transport and regulatory functions, respectively. The *mocC* gene shows no homologies to sequences in the DNA databases. MocR shows homology to the GntR class of bacterial regulatory proteins. Once again, as with the *mos* genes, this suite of genes is likely interacting with genes already present in the bacteria to complete rhizopine catabolism. Rhizopine catabolic genes are highly conserved both within and between species. There is only 0.5% divergence in *R. leguminosarum* bv. *viciae* (12 strains) and 1.6 % divergence within *S. meliloti* (13 strains), as shown by RFLP analysis (Wexler *et al.*, 1996).

Since rhizopines were found by analogy with *Agrobacterium* opines and they act as carbon sources, it was initially hypothesized that they may function in a similar manner to opines. That is, rhizopines would serve as selective growth substrates thereby increasing the proportion of the rhizopine-catabolizing strains in the rhizosphere and as a consequence increasing the percentage of these bacteria occupying the nodule. Early experiments using mutants with *Tn5* inactivated rhizopine genes

in competition studies on plants grown in agar indicated that rhizopines were indeed affecting competition for nodulation (Putnoky, Murphy and Kondorosi, unpublished results; Murphy and Saint 1992; Murphy *et al.*, 1995). In a more recent and extensive study carried out over a 12-month period and conducted in soil, these findings were extended. The wild-type, rhizopine-producing *S. meliloti* strain L5-30 and mutant L5-30 strains (generated with *Tn5*), either deficient for rhizopine synthesis or catabolism, were inoculated onto alfalfa host plants in competition experiments. These experiments demonstrated that no apparent advantage resulted from the ability to synthesize a rhizopine. However, the ability to catabolize rhizopine provided a clear advantage when in competition with a strain without this ability. These results suggest that in competition with a catabolism deficient mutant, the ability to catabolize rhizopine results in enhanced rates of nodulation. Since the competitive advantage occurred within the first few weeks and did not increase with time, the experiments were not consistent with the hypothesis that the sole role of rhizopines is to act as a proprietary growth substrate for the free-living population of the producing strain. Although rhizopine acts as a proprietary growth substrate *ex planta* (and in this sense is similar to opines), it seems that another mechanism may be operating *in planta*.

Rhizopine genes appear particularly relevant to the evolution of the *Rhizobium*-legume symbiosis. Where the genes have been studied in detail, evidence suggests that the same promoter (from a *nifH* gene) has been sequestered to both supply the plant with nitrogen as well as to produce a symbiotic product - a rhizopine - which provides the bacteria with an advantage in competition for nodulation. Thus both the bacteria and the plant can obtain a specific advantage from the symbiotic interaction, a point which is implied in the symbiotic relationship but never completely resolved. Although the initial isolation of rhizopines leant heavily upon similarities with opines, it would now seem that they are important plant-microbe products but likely function in a different manner to opines.

IV. Recent Advances in Opine Research in the *Agrobacterium*-Plant Interaction

Within the last years, exciting results have been obtained by the scientists "on the opine trail" (Tempé and Petit, 1983). One aspect of these studies, dealing with the fine regulation of the conjugation of the Ti plasmids, is reviewed in this book by S.K. Farrand (Chapter 10). Here, we will essentially focus on recent advances on three other opine-related themes, *i.e.* (i) opines as chemoattractants; (ii) agrocinopines and agrocins; and (iii) evolution of opine-related functions, the study of which will eventually lead to a more comprehensive understanding of the role of opines in the *Agrobacterium*-plant interaction.

IV.A. CHEMOTAXIS TO OPINES

As with most motile gram-negative bacteria, members of the genus *Agrobacterium* are chemoattracted by a variety of substrates (Loake *et al.*, 1988; recently reviewed in Vande Broek and Vanderleyden, 1995). Of particular interest is the observation by Hawes and co-workers that *Agrobacterium* strains are strongly attracted to exudates produced by root border cap cells (Hawes *et al.*, 1988; Hawes and Smith, 1989). Moreover, although there is some dispute as to the requirement for *virA* and *virG*, work from two groups indicates that phenolic compounds with *vir* gene inducer activity are chemoattractants to agrobacteria (Parke *et al.*, 1987; Ashby *et al.*, 1988; Shaw *et al.*, 1988). All of this suggests that chemotaxis plays a role in attracting agrobacteria to their host plants and perhaps to susceptible wound sites. This notion is strengthened by the observation that mutants of *A. tumefaciens* blocked in general chemotactic functions, while tumorigenic in standard plant inoculation assays, are avirulent when incorporated into soil in which the host has been planted (Hawes and Smith, 1989).

Recently, Kim and Farrand (submitted) have shown that *Agrobacterium* strains can be chemoattracted by opines. Chemotaxis is dependent upon plasmids that encode functions for the catabolism of the opine in question. Strains with nopaline/agrocinopine-type Ti plasmids are attracted to nopaline and agrocinopines A+B but not to octopine, while strains with octopine/mannityl opine-type Ti plasmids are attracted to octopine but not to mannopine, nopaline or agrocinopines A+B. Not surprisingly, genetic and molecular analysis localized determinants required for chemoattraction to the opine catabolic loci of the Ti plasmids. Moreover, analysis of mutants and subclones suggests that only the gene encoding the periplasmic binding protein associated with the transport system of the locus is required. Mutations

affecting other components of the transport systems, or the genes encoding the opine catabolic enzymes, had no effect on chemotaxis. This is consistent with observations in other, more well studied systems, in which only the periplasmic binding proteins associated with ABC-type transport systems are required for chemotaxis to the cognate substrates (Higgins, 1992).

Intriguingly, some Ti plasmid gene products show high amino acid sequence similarities to methyl-accepting chemotaxis proteins (Mcp) of other bacteria. These latter proteins are the membrane-associated primary sensory transducers of the chemotaxis response system (Stock *et al.*, 1987; reviews Lukat and Stock, 1993; Stock and Mowbray, 1995). Moreover, these *mcp*-like genes are components of operons associated with opine catabolism functions. For example, an operon located between *tra* and *acc* of pTiC58 encodes an *mcp*-like gene (Hwang and Farrand, unpublished results). Expression of this gene, as well as that of the rest of the operon, are regulated by agrocinopines A+B through AccR (see below), suggesting that the *mcp*-like gene product may serve to sense these sugar-phosphate opines. However, insertion mutations in this gene had no effect on chemotaxis towards agrocinopines A+B: clearly this *mcp*-like gene is dispensable for chemotaxis to opines, at least in the bacterial strains tested (Hwang and Farrand, unpublished results). This suggests that these bacteria encode MCP receptors, most likely from genes located on the bacterial chromosome, that recognize the periplasmic protein complex that binds opine(s). However, it is conceivable that the MCP proteins of some agrobacteria, or other hosts able to propagate Ti plasmid replicons, may not recognize these signals, and thus require the MCP proteins encoded by the Ti plasmid. This interesting hypothesis remains to be tested.

Within the opine concept, one could envision several roles for opine-mediated chemotaxis. First, it could provide a mechanism whereby opine-catabolizing agrobacteria that wander, or are swept away from the vicinity of the tumor, can make their way back to the nutrient-rich environment. Second, it may serve to attract *A. tumefaciens* strains with somewhat different Ti plasmids. These, then, may recombine with the Ti plasmids resident in the tumor inducing strain to produce new Ti plasmid types. Finally, it could provide a mechanism to attract non-tumorigenic, but opine-catabolizing *A. radiobacter* strains to tumors. This, of course, is of value to such agrobacteria since they themselves cannot induce tumors and therefore are reduced to parasitizing the neoplasias generated by virulent strains. In *A. radiobacter* strain K84, this concept is taken even one step further (see below). This bacterium displays chemotaxis to tumors producing nopaline and agrocinopines A+B, substrates it can catabolize by virtue of its large opine-catabolism plasmid (Clare *et al.*, 1990; Hayman and Farrand, 1990). Moreover, once in the tumor, it can gain a competitive advantage by the production of agrocin 84, an antibiotic active against certain virulent agrobacteria (Roberts *et al.*, 1977; Murphy and Roberts, 1979; Kerr, 1980; see also below). That the antibiotic is taken up and activated by the agrocinopine catabolism system located on the Ti plasmid of the target *A. tumefaciens* strains (Hayman and Farrand, 1988; 1990) only adds the final flourish to this elegant parasitic system.

IV.B. THE AGROCINOPINE OPINES

IV.B.1. Discovery and Chemistry

The discovery of the agrocinopine opines charts an interesting course in the application of the opine concept. It had been known for some time that the anti-agrobacterial antibiotic agrocin 84, produced by *A. radiobacter* strain K84 is highly specific. This agent inhibits the growth of only certain isolates of *A. tumefaciens* and *A. rhizogenes*, and sensitivity is associated with the type of Ti or Ri plasmid present in the strain (for review, see Farrand, 1991). Ellis and Murphy (1981) reasoned that the specificity of agrocin 84 might be due to its ability to mimic some unknown opine produced by the neoplasias induced by these susceptible strains. If true, then the presence of this unknown opine should be demonstrated by its ability to induce higher levels of sensitivity to the agrocin antibiotic. This proved to be the case: an unknown compound was specifically found in tumors induced only by the agrocin 84-sensitive strains of *A. tumefaciens* that induced super-sensitivity of these strains to the antibiotic (Ellis and Murphy, 1981). This compound could be taken up and catabolized only by agrobacteria harboring Ti plasmids that 1) induced the tumors from which the compound was isolated, and 2) conferred susceptibility to agrocin 84. This established the opine nature of the agent, and firmly linked it to agrocin 84 sensitivity. The activity was associated with two closely migrating compounds, and because of the relationship with agrocin 84, the new opines were named agrocinopines A and B (Figure 1).

Chemical analysis showed these new opines to be unlike any of the previously-characterized tumor metabolites. Indeed, agrocinopines A and B proved to be sugar-phosphodiesters. Agrocinopine A contains arabinose in linkage from C-2 to the C-4 of the fructose moiety of sucrose, while agrocinopine B differs only in the absence of the glucose moiety of the sucrose residue (Figure 1; Ryder *et al.*, 1984). It is not clear whether agrocinopine B forms from agrocinopine A, or whether the two opines are synthesized in parallel. This opine family is associated with classical nopaline-type Ti plasmids and with certain opine catabolism plasmids present in *A. rhizogenes* and *A. radiobacter* isolates (Hayman and Farrand, 1990). In their studies, Ellis and Murphy (1981) identified a second set of two agrocinopine-type opines, called agrocinopines C and D. Chemically, agrocinopine C is a phosphodiester of glucose and sucrose, with the linkage being between C-2 of the two glucose units. Agrocinopine D is related, but lacks the fructose moiety of the sucrose subunit (M.E. Tate, personal communication). Moreover, the glucose moiety exists in an equilibrium between the furanoside and the pyranoside forms. Agrocinopines C+D are found in tumors induced by strains harboring the classical agropine-type Ti plasmid, pTiBo542 (Ellis and Murphy, 1981), or the mega-Ti plasmids of the *Lippia*-type agrobacteria (Unger *et al.*, 1985; Ryder and Farrand, unpublished results).

Isolates of *Agrobacterium* spp. that are innately susceptible to agrocin 84 can catabolize agrocinopines A+B, and, surprisingly, agrocinopine D, but not agrocinopine C (Ellis and Murphy, 1981). Interestingly, although it cannot be catabolized, agrocinopine C protects agrocinopine A+B-type strains from inhibition by agrocin 84. Strains that can catabolize agrocinopines C+D are not innately susceptible to agrocin 84. However, if these strains are precultured in the presence of either of these two opines, they express sensitivity to the antibiotic.

Several lines of evidence suggest that susceptibility to agrocin 84 is associated with the plasmid-encoded gene set responsible for the uptake and catabolism of the agrocinopine-type opines. First, only that agrobacteria able to catabolize the agrocinopine opines are susceptible to agrocin 84. Second, mutations abolishing susceptibility to agrocin 84 invariably result in the inability to take up and catabolize the opines. Third, mutations resulting in constitutive high-level expression of the opine catabolism locus make the cells super-sensitive to agrocin 84. Fourth, presence of agrocinopines inhibits the uptake of radiolabelled agrocin 84. In this regard Murphy and Roberts (1979) showed that agrocin 84 is taken up *via* a high-affinity, energy-dependent transport system. Moreover, they showed that the system contains a periplasmic protein that specifically binds the antibiotic, and that the presence of this binding protein is dependent upon a Ti plasmid that confers susceptibility to the antibiotic and catabolism of the opine.

IV.B.2. The acc Operon and its Regulation

More recently, the region of pTiC58 encoding catabolism of agrocinopines A+B has been cloned and sequenced (Hayman and Farrand, 1988; Hayman *et al.*, 1993; Kim and Farrand, in preparation). The gene system, located between coordinates 125 and 134 kb of the Ti plasmid, is composed of eight open reading frames, *accR*, and *accA* through *accG* (Figure 5). These eight genes are entirely sufficient to confer the ability to take up and utilize the opine as sole carbon source (Hayman *et al.*, 1993). The first gene, *accR* encodes a repressor protein that regulates expression of the *acc* operon (Beck von Bodman, *et al.*, 1992). Consistent with this, mutations in this gene result in super-sensitivity to agrocin 84 and in a greatly enhanced rate of uptake of both the opine and the antibiotic (Hayman and Farrand, 1988; Beck von Bodman, *et al.*, 1992). The next five genes, *accABCDE*, encode an ABC-type, ATP-driven, periplasmic binding protein-dependent transport system (Kim and Farrand, in preparation). Furthermore, AccA, which at the amino acid sequence level is related to periplasmic substrate binding proteins of other ABC-type transporters, is located in the periplasm (Hayman, *et al.*, 1993). This protein most likely is identical to the agrocin 84 binding protein first described by Murphy and Roberts (1979). Overall, the five components of this transport system are most closely related to those of the dipeptide and oligopeptide transport systems of *Escherichia coli* and *Salmonella typhimurium* (Kim and Farrand, in preparation).

The last two genes, *accF* and *accG* probably encode the enzymes involved in catabolizing the opines. AccF shows sequence similarities to UgpQ, a glycerol phosphodiester phosphodiesterase of *E. coli*, suggesting that this protein may be a phosphodiesterase that cleaves the opines into their respective sugar subunits. Consistent with this assignment, mutations in *accF* abolish the ability of the bacteria to grow with agrocinopines A+B as sole source of carbon (Kim and Farrand, in preparation). The mutation also abolishes susceptibility to agrocin 84, a

finding which indicates that this antibiotic must be processed before it becomes toxic to the cells. Interestingly, *accF* mutants still take up agrocin 84, confirming that this gene is probably not required for transport of the antibiotic or the opine. *accG*, the last gene of the locus, could encode a product that is related at the amino acid sequence level to myo-inositol monophosphatases of eukaryotic origin (Kim and Farrand, in preparation). The gene is required for the utilization of agrocinopines A+B as sole carbon source, but is not necessary for transport of the opine or for susceptibility to agrocin 84.

Presumably, the activity encoded by *accG* could be involved in dephosphorylating the sugar phosphate produced from agrocinopines A+B by the action of AccF. But it is not clear whether the substrate is arabinose-2-phosphate, or sucrose (fructose-4)-phosphate. Intriguingly, Swords *et al.*, (1996) recently reported that the protein product of *accF* is related to the putative product of *avrBS2*, an avirulence gene from *Xanthomonas campestris*, pv. *vesicatoria*. However, *X. campestris* pv. *vesicatoria* is resistant to agrocin 84, even when preincubated with agrocinopines A+B (K. Swords, personal communication). How the protein product of this gene or the metabolic product of its enzymatic activity induces the R-gene-dependent HR in resistant plants is unknown.

The *acc* operon of pTiC58 is expressed at a low but measurable basal level, thus accounting for susceptibility to agrocin 84 even in the absence of the opine. The observation that cells become super-sensitive to the antibiotic in the presence of agrocinopines suggested that expression of the *acc* operon is inducible by the opines. Using *lacZ* fusions, Hayman and Farrand (1988) showed this to be the case. As described above, primary regulation of the *acc* gene is mediated by AccR (Hayman and Farrand, 1988; Kim and Farrand, in preparation), a transcriptional repressor encoded by the first gene of the *acc* locus. This repressor is related to classic repressors of other sugar catabolism systems including FucR, DeoR and LacR (Beck von Bodman *et al.*, 1992).

IV.B.3 Other Agrocinopine Systems

Very little is known concerning the gene systems involved in uptake and catabolism of agrocinopines C+D encoded by the Ti plasmids of Bo542 and the *A. tumefaciens* isolates from *Lippia* (Unger *et al.*, 1985). As described above, the locus from pTiBo542 does confer susceptibility to agrocin 84, but only after induction with the agrocinopines C or D. There

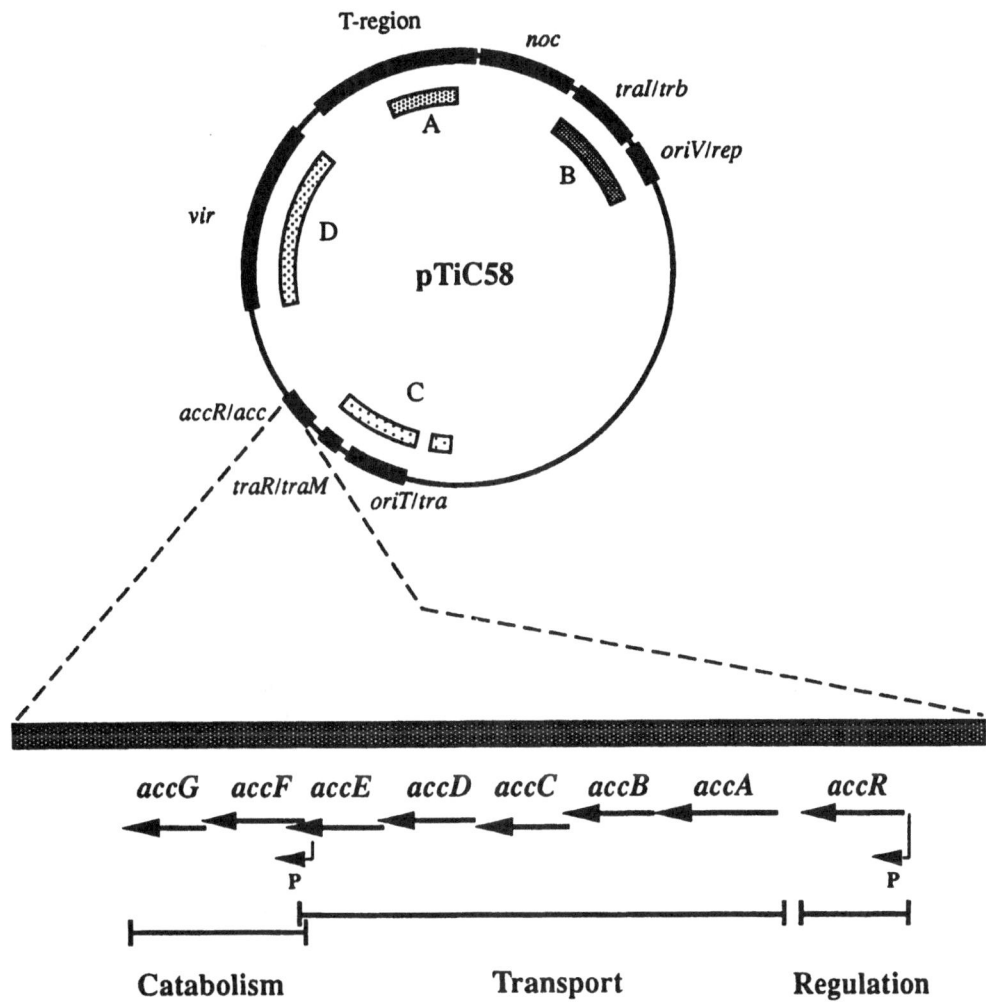

Figure 5. The agrocinopine catabolic (*acc*) operon
The structural organization of the pTiC58 is as shown in Figure 2. Legends are as in Figure 2. The *acc* region has been magnified to show the regulatory gene *accR*, the genes involved in agrocinopine uptake and those involved in degradation of this opine.

is no detectable homology between the *acc* (agrocinopine A+B catabolism) genes of pTiC58 and those of pTiBo542 as assessed by Southern analysis (Hayman and Farrand, 1990) suggesting that the genes involved in utilization of the two agrocinopine-type opine families are quite different. Surprisingly, the *Lippia*-type Ti plasmids harboring the agrocinopine C+D catabolic loci and the pTiC58 harboring the agrocinopine A+B catabolic genes behave similarly. Both are susceptible to agrocin 84 even in the absence of the opines, but are induced to

supersensitivity when agrocinopines C+D are provided in the assay medium (Unger *et al.*, 1985, Ryder and Farrand, unpublished results).
Hayman and Farrand (1990) divided the plasmid-encoded loci specifying utilization of agrocinopines A and B into three types (Table 2). Type I systems, as exemplified by *acc* of pTiC58, confer uptake and utilization of agrocinopines A+B as well as susceptibility to agrocin 84 even in the absence of the opines. Such systems were detected on nopaline-type Ti plasmids from diverse sources, as well as on a no-

Class	Representative Plasmid	Trait		
		Agrocinopine Transport	Response to Agrocin 84	Similarity with *acc* of pTiC58
I	pTiC58	+	S[b]	Prototype
	pTiT37	+	S	+++[d]
	pTiK27	+	S	+++
	pTiJ73	N.D.[a]	S	+++
	pAtK299	N.D.	S	++
II	pArA4a	N.D.	S	+/-
III	pAtK84b	+	R	- to +/-
N.D.	pTiBo542	-	S[c]	-

Table 2. Classes and characteristics of regions specifying catabolism of agrocinopines A+B and/or sensitivity to agrocin 84 on *Agrobacterium* plasmids

[a] N.D., not determined

[b] S, susceptible to agrocin 84 without prior induction with agrocinopines A and B; R, resistant to agrocin 84 even when induced with agrocinopines A and B

[c] pTiBo542 confers susceptibility to agrocin 84 only when the host strain is induced with agrocinopines C and D

[d] +++, strong similarity (no RFLP observed); ++, strong similarity (RFLP observed); +/- weak similarity; -, no similarity

paline-catabolic plasmid present in a non-tumorigenic *A. radiobacter* isolate, K299. These loci are all closely related (as judged by Southern analysis) although that from pAtK299 showed weaker hybridization and a different fragment pattern when probed with *acc* from pTiC58. Type II system, which is encoded by the opine-catabolic plasmid pArA4a found in the *A. rhizogenes* strain A4, confers a phenotype similar to that of the Type I loci, but shows only barely detectable homology with the *acc* probe from pTiC58. The Type III system is quite interesting. This *acc* determinant encodes uptake and catabolism of agrocinopines A+B, but does not confer susceptibility to agrocin 84 under any tested conditions. This locus also does not confer uptake of the antibiotic, even when the bacteria were tested following preculture with the inducer opines. Significantly, this locus is associated with the opine-catabolic plasmid of *A. radiobacter* strain K84, which itself produces agrocin 84. This result is somewhat surprising. Plasmids pAtK84b and pTiC58 show overall sequence similarities, and are strongly related in many of the determinants that they share in common (Clare *et al.*, 1990). These include the *noc* loci encoding catabolism of nopaline, the *tra* and *trb* regions encoding conjugal transfer functions, and the *oriV/rep* regions responsible for plasmid replication, partitioning and stability. Moreover, both plasmids utilize agrocinopines A+B as the conjugal opines (see chapter 10). These observations suggest that pTiC58 and pAtK84b share a recent common origin (Clare,

et al., 1990). However, Southern analysis indicates that the *acc* locus of pAtK84b is only weakly, at best, related to that of pTiC58 (Clare *et al.*, 1990; Hayman and Farrand, 1990). Coupled with the differences in phenotypes associated with the two *acc* determinants, one can conclude that either the two gene sets are of different origins, or they have diverged from each other at a rate much higher than that of other determinants common to the two plasmids.

IV.B.4. Agrocinopines, Agrocin 84 and Competition for Colonizing Crown Gall Tumors

The characteristics of *Agrobacterium* strain K84 lead to some interesting speculations. As described above, this bacterium exhibits two intriguing and related phenotypes. First, it is, itself, non-tumorigenic, but is capable of catabolizing two opine families, nopaline and agrocinopines A+B. It expresses these traits due to the presence of a plasmid, pAtK84b, that is closely related to pTiC58, a classical nopaline-, agrocinopine-type Ti plasmid. Second, it produces a highly active antiagrobacterial antibiotic called agrocin 84. This antibiotic specifically targets pathogenic agrobacteria harboring Ti or Ri plasmids encoding catabolism of agrocinopine-type opines. One can envision that strain K84 functions as a parasite on tumors induced by tumorigenic agrobacteria. That it produces an antibiotic specific to the pathogens already colonizing tumors, the opines of which it is equipped to utilize, provides a degree of elegance to the system. Strain K84 has developed a strategy to displace its primary competitors, and this strategy is directed against a target gene system guaranteed to be induced in the competitor strain that it wishes to displace (Farrand, 1991). This, then, represents an interesting, fratricidal twist to the Opine Concept.

IV.C. EVOLUTION OF OPINE RELATED FUNCTIONS

A key feature of the Ti plasmids is to bear two sets of genes, one (the virulence genes, the opine catabolic functions, etc.) being expressed in a prokaryotic background (the bacteria) and the other (the T-DNA genes) in an eukaryotic background (the plant) (see several chapters in this book and Figure 2 for the functional organization of Ti plasmids). How prokaryotic and eukaryotic functions have evolved in the bacteria, and what are their origins remain in part unexplained. Amongst the established data, the sequencing of the T-DNA borders and that of the virulence region revealed their relations to genes or sequences involved in plasmid conjugal transfer (Buchanan-Wollaston et al, 1987; Cook and Farrand, 1992; Pansegrau et al, 1993; Shirasu and Kado, 1993; Farrand et al., 1996; reviews: Citovsky et al., 1992; Kado, 1994; Less and Lanka, 1994; see also several chapters in this book), suggesting a bacterial ancestry for these two regions. Similar work on the T-DNA genes demonstrated that several of them share homology to genes (or sequence) detected in other plant pathogenic bacteria (Powell and Morris, 1986; Yamada et al., 1985, 1986; reviews: Morris, 1986; Gaudin et al, 1994). In contrast, the detection of an intron-like sequence in one of the loci involved in root induction of the T-DNA of A. rhizogenes indicates that this sequence could be of eukaryotic origin (Magrelli et al., 1994). The interpretation of recent and earlier findings on opine structure and metabolism may provide more comprehensive answers to the above evolutionary questions essentially because opine synthesis is controlled by eukaryotic-type genes while opine degradation is determined by prokaryotic genes.

IV.C.1. The Mannityl Opine- and the Amadori Opine-Type Plasmids: a Soil Origin?

Amongst know opines, the mannityl opines (agropine, mannopine, mannopinic and agropinic

acid; Figure 1) have received considerable attention, probably because of their broad occurrence in crown gall and hairy root hyperplasias (review: Dessaux et al, 1992) and also because crown gall tumor cells synthesize large amounts of these compounds (up to 7% of their dry weight; Tate et al., 1982). The functions involved in mannityl opine degradation have been extensively studied on the 195 kb octopine-type Ti plasmid of strain 15955 , where they occupy ca. 40 kb of DNA (Figure 2) (Dessaux et al., 1987; Dessaux et al., 1988; Farrand et al., 1990). Mannityl opine degradation is regulated, as originally deduced from the selection of regulatory mutants on chemically-synthesized mannityl opine analogues (Chilton and Chilton, 1984). The regulation is complex: cross-induction and multiple regulation occur as shown for mannopinic and agropinic acid catabolic functions. Similarly, mannopine and agropine induce the degradation of the four mannityl opines (Dessaux et al., 1988; Hong et al., 1993). Furthermore, mannopine degradation is catabolite-repressed (for instance by succinate) unless the opine is the sole available source of nitrogen (Nautiyal et al., 1992; Hong et al., 1993).

The catabolic pathway of these opines have been investigated. In the mannopinic acid-induced, mannopinic acid degradative pathway, this opine is degraded to mannose (and probably glutamate), the sugar moiety being isomerized to fructose (Dessaux et al., 1988). A second catabolic pathway for mannopinic acid exists: it probably involves the enzyme(s) responsible for mannopine degradation (Dessaux et al., 1988; Hong et al., 1993). This opine is degraded to deoxy-fructosyl-glutamine (dfg) (Kim et al., 1996), a molecule which might be in turn converted to mannose or fructose (or derivatives) and glutamine. Interestingly, mannopine is also converted to agropine by bacteria and cell free-extracts (and vice-versa; Dessaux et al., 1986a), a metabolic pathway identical to the agropine biosynthetic pathway occurring in crown gall tumors (Ellis et al., 1984; Salomon et al., 1984). The lactonization of mannopine to agropine was therefore first attributed to the activity of the expression of ags, a T-DNA gene responsible for agropine synthesis in tumors. This hypothesis was ruled out by the analysis of an insertion mutant harboring a disrupted ags gene, which retains the capacity to lactonize mannopine to agropine (Hong et al., 1997) Further genetic studies demonstrated that the agropine-mannopine lactonase, also termed mannopine cyclase, was the only enzymatic activity required to degrade agropine provided that the bacterial strain can transport agropine and degrade mannopine (Hong and Farrand, 1994). These data strongly suggest that agropine degradation proceeds via mannopine and dfg. Contrasting with the amount of data on the degradation of agropine, mannopine and mannopinic acid, the catabolic pathway for agropinic acid remains unclear, though recent findings suggests that it might also involve dfg as an intermediate (Vaudequin-Dransart, unpublished data).

A 21 kb region involved in mannopine and agropine degradation has been cloned (Dessaux *et al.*, 1987), mutagenized (Hong *et al.*, 1993) and partly sequenced (Kim and Farrand, 1996; Kim *et al.*, 1996). Results appear in Figure 6. The upper region determines mannopine and agropine transport systems, while the lower region encodes the enzymic activities responsible for agropine and mannopine degradation. Two additional open reading frames, *mocB* and *mocA*, were identified in this region (it should be noted that the *moc* genes in this system are not the same as in the rhizopine system in *Rhizobium*). They are homologous to two genes encoding 6-phosphohexose dehydogenase and dehydratase, respectively. Their role in mannopine or dfg degradation remains unclear. However, a gene (*mocE*) determining a kinase activity possibly involved in sugar phosphorylation is also present in this mannopine/agropine catabolic region. The *mocR* gene is probably one of the regulators of the expression of the other catabolic genes for it resembles a putative repressor for agropine degradation sequenced from a Ri plasmid (Kim and Farrand, 1996). Strikingly, genes involved in mannopine and agropine synthesis are closely related to those involved in their degradation *e.g.* the locus encoding bacterial conversion of mannopine to agropine (termed *mocF or agcA*) shares ca. 70% similarity at the nucleic acid level with the open reading frame of *ags*. These results suggest that both the gene expressed in the eukaryotic background (*ags*) and that expressed in the prokaryotic background (*agcA*) have evolved from a common ancestor (Hong *et al.*, 1997). Similar studies revealed that genes *mocC* and *mocD* responsible for conversion of mannopine to dfg (*mocC*) and degradation of dfg (*mocD*) share noticeable similarity to the *mas*1 and *mas*2 T-DNA genes involved in mannopine and dfg synthesis, respectively. Surprisingly, the 3' end of the *mas*2 gene is homologous to part of the kinase gene, *mocE* (Kim and Farrand, 1996; Kim *et al.*, 1996). Here also, the eukaryotic and prokaryotic genes may therefore have evolved from a common ancestor.

We are now left with the question: what was the ancestor? The recent identification of a new class of opines, the Amadori opines (Chilton *et al.*, 1995), specific for tumors naturally induced on *Ficus* and *Chrysanthemum* by a new class of *Agrobacterium* strains (Bush and Pueppke, 1991; Kovács and Pueppke, 1994; Bouzar *et al.*, 1995; Vaudequin-Dransart *et al.*, 1995), shed new light on this question. One of the three opines defining this class is dfg (Figure 1). The two others are chrysopine, the dfg lactone which resembles agropine, and deoxy-fructosyl-oxo-proline (dfop), the lactam derivative of dfg which resembles agropinic acid (Chilton *et al.*, 1995; Figure 1). Interestingly, aside from octopine discovered in the muscles of various *Octopus* sp. (see below), dfg is the only opine naturally and abundantly occurring outside crown gall tumors, as it appears spontaneously in most (if not all) decaying plant material (Anet 1957; Anet and Reynolds, 1957; Heyns and Paulsen, 1959). Such materials are often in contact with the soil, the natural habitat of *Agrobacterium*. This feature may have provided " the selective pressure for selection and conservation of dfg catabolic genes in *Agrobacterium*, even before Ti plasmids existed " (Vaudequin-Dransart *et al.*, 1995). This hypothesis is strongly supported by the observation that most assayed pathogenic and non-pathogenic *Agrobacterium* strains degrade dfg, a result which implies that the relevant catabolic functions are not located on the Ti plasmids but elsewhere in the *Agrobacterium* genome. Recent findings indeed suggest that these determinants are located on the cryptic plasmid of *Agrobacterium* (Vaudequin-Dransart *et al.*, in preparation). The opine dfg might therefore be an ancestral opine from which some other opine molecules have evolved. The above results which identified dfg as an intermediate in the synthesis and degradation of both the mannityl opines (Ellis *et al.*, 1984; Salomon *et al.*, 1984) and the Amadori opines (Chilton *et al.*, 1995; Vaudequin-Dransart *et al.*, unpublished) indeed support this proposal.

The comprehensive model accounting for the evolution of the Ti plasmids is still speculative (Dessaux *et al.*, 1992; Vaudequin-Dransart *et al.*, 1995). It predicts that the original dfg catabolic plasmids acquired the functions allowing the bacteria to convert this widely occurring molecule to mannopine and agropine, supposedly two substrates less accessible to competing soil microorganisms. At the same time or later, genes responsible for conversion of dfg to mannopine and agropine could have been duplicated and integrated to a "proto T-DNA". Interestingly, the evolution of agrocinopine catabolic and anabolic

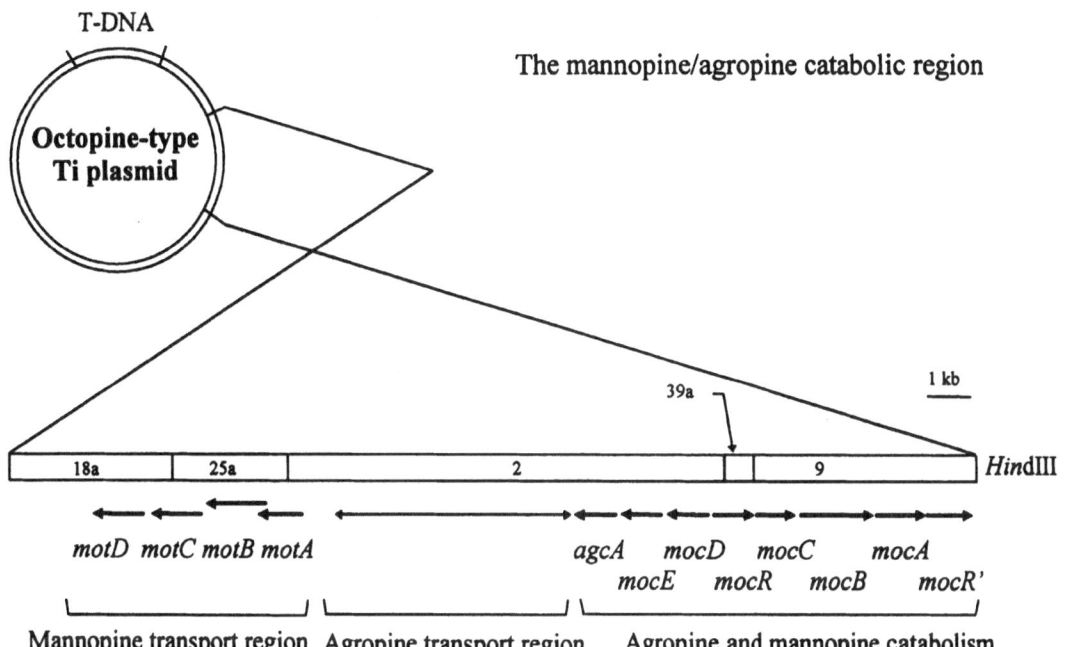

Figure 6. The mannopine/agropine catabolic region of the octopine-type Ti plasmid
These results were taken from Hong *et al.,* (1993), Kim and Farrand (1996) and Kim *et al.,* (1996). The *Hind*III restriction map from the octopine-type Ti plasmid is from De Vos *et al.,* (1981). Gene legends are: *mot*, mannopine transport; *agcA* (or *mocF*), mannopine cyclase

functions also involves the duplication strategy first identified for the mannityl opines. Thus, the product of the above mentioned *accF* gene involved in agrocinopine A and B catabolism, is related at the amino acid sequence level to the product of *acs*, the agrocinopine synthase gene encoded on the T-region of the nopaline/agrocinopine-type Ti plasmids (Kim and Farrand, in preparation). Whatever the validity of the current evolutionary model, the studies on opine catabolic and biosynthetic functions provided an elegant way to address the question of the "making" of the Ti and Ri plasmids. Interestingly, data collected on this topic strongly support the description of Ti and Ri plasmids as being essentially catabolic replicons.

IV.C.2. The Octopine- and Nopaline-Type Ti Plasmid: a Marine Origin?

As stated above, octopine and nopaline were amongst the first discovered opines. Consequently,

the first studies on the biological role of opines dealt with these molecules. Octopine appeared to be the inducer of the transfer of the octopine-type Ti plasmid (whereas nopaline is not the inducer of the transfer of the nopaline-type Ti plasmid) and a valuable nutrient for *Agrobacterium* (Kerr *et al.,* 1977; Klapwijk *et al.,* 1978; Petit *et al.,* 1978a; Petit *et al.,* 1978b; Tempé *et al.,* 1978; Ellis *et al.,* 1982a, 1982b). Both degradation of octopine and Ti plasmid transfer appeared to be inducible and co-regulated by octopine and related opines (lysopine, octopinic acid), a result confirmed by the analysis of regulatory mutants selected on various non-inducing opine analogues such as nor-octopine or histopine (Klapwijk *et al.,* 1976; Montoya *et al.,* 1977; Petit and Tempé 1978; Petit *et al.,* 1978a; Tempé *et al.,* 1978; Klapwijk and Schilperoort, 1979). Three classes of mutants were found: some were constitutive for both functions whereas the two other groups were constitutive either for opine degradation or conjugal transfer (Klapwijk *et al.*, 1978; Tempé *et al.*, 1978). The catabolic pathways for octopine and lysopine were investigated. Octopine and lysopine are cleaved back to their constituents: pyruvate and arginine or lysine, respectively (Jubier 1972; Bomhoff, 1976). Arginine is degraded to ornithine, and this latter molecule is directly converted to proline, the catabolism of which is achieved in a two-step pathway, *via* delta-1-pyrroline-5-carboxylate (P5C) to glutamate (Dessaux *et al.*, 1986b; Farrand and Dessaux, 1986; Sans *et al.*, 1987; Cho *et al.*, 1996). The conversion of ornithine to proline is achieved by an intriguing enzyme termed ornithine cyclodeaminase (OCDase). This activity is remarkable because *Agrobacterium* was the first gram-negative bacteria in which it was detected (Dessaux *et al.*, 1986b; Sans *et al.*, 1987). OCDase also shows strong homology to the *Rhizobium meliloti* Nfe protein involved in nodulation efficiency (Soto *et al.*, 1994). More surprisingly, the major consitutent of the eye lens of various Australian marsupials, the μ-cristallin, is related to this unusual *Agrobacterium* enzyme. Since OCDase activity yields proline, a molecule generally regarded as an osmoprotectant, μ-cristallin could exhibit a similar biological function to maintain the transparency of this essential eye component under hydric stress conditions (Kim *et al.*, 1992).

As shown in Figure 7, the octopine catabolic determinants are clustered in two divergently transcribed regulons (reviewed in Schröder *et al.*, 1990). The first one contains the regulatory gene *occR* (Von Linting *et al.*, 1991; Habeeb *et al.*, 1991), a member of the *lys*R gene family, which also regulates the conjugal transfer of the plasmid (Klapwijk *et al.*, 1978; Tempé *et al.*, 1978). The *occR* negatively regulates its own transcription (whether octopine is present or not) and positively regulates the transcription of the catabolic genes clustered in the second operon in the presence of octopine (Wang *et al.*, 1992; Von Lintig *et al.*, 1994). These genes determine an octopine transport system (*occQ, occM, occJ/T, occP*) (Valdivia *et al.*, 1991; Zanker *et al.*, 1992). OccJ/T might be the octopine-binding periplasmic protein while the three others appear to be membrane-associated transport proteins These *occ* genes show similarity to other genes encoding various components of basic amino acid transport systems, such as HisP (histidine permease), HisM and HisQ (two membrane proteins), HisJ and ArgT (histidine and arginine periplasmic binding proteins) (Valdivia *et al.*, 1991; Zanker *et al.*, 1992). Interestingly, octopine and

nopaline are arginine derivatives (see above) and histopine is a histidine derivative (Kemp, 1977). Downstream of the genes determining the transport system, are *ooxB* and *ooxA*, which encode the two subunits of the octopine oxidase that cleaves octopine to arginine (Zanker *et al.*, 1992). The last gene of this operon is *ocd* (Schrindler *et al.*, 1989); it encodes the OCDase discussed above. Genes encoding arginase, and P5C dehydrogenase are not located on octopine-type Ti plasmids (Dessaux *et al.*, 1986b). The nopaline degradative region is closely related to the octopine catabolic region, but organized differently (Schröder *et al.*, 1990; Zanker *et al.*, 1992; Cf. Figure 7). The genes responsible for nopaline transport (*nocP, nocJ/T, nocQ* and *nocM*) are also divergently transcribed from a promotor that also controls the transcription of the regulatory gene *nocR* (Von Lintig *et al.*, 1994). Interestingly, the genes encoding the catabolic enzymes are located downstream to those encoding nopaline uptake and transcribed from another opine-inducible promotor (Zanker *et al.*, 1992; Von Lintig *et al.*, 1994). In addition to *noxB* and *noxA* (nopaline oxidase), and in between the *noxA* and *ocd* (OCDase) genes, are two genes specific for the nopaline catabolic region: *arc*, encoding an arginase and *40k*, an open reading frame of unknown function encoding a 40 kDa protein (Sans *et al.*, 1987, 1988; Schrell *et al.*, 1989). With these two exceptions, genes encoding nopaline uptake and degradation are closely related to those encoding octopine degradation, suggesting a common lineage (Schindler *et al.*, 1989; Zanker *et al.*, 1992; Zanker *et al.*, 1994). In most octopine and nopaline-type strains (a major exception being strain C58), another copy of the *ocd* gene is present in the bacterial genome, which allows growth on arginine as sole carbon source (Dessaux *et al.*, 1986b; Cho *et al.*, 1996). In bacteria lacking this extra copy, such as C58, mutants constitutive for nopaline or octopine degradation can therefore be selected on arginine. Similarly, octopine is a noninducing substrate for the nopaline uptake and degradation systems (Klapwijk *et al.*, 1977; Petit and Tempé, 1975; Petit and Tempé, 1978; Petit *et al.*, 1978b).

While genes involved in dfg, mannopine and agropine synthesis are related to their catabolic counterparts, this situation is not observed for octopine and nopaline degradation. Nopaline and octopine are synthesized in crown gall tumors by the enzymes octopine synthase and nopaline synthase, respectively, which show no functional homologies to the corresponding catabolic oxidases. At the gene level, the *ocs* and *nos* genes determining the

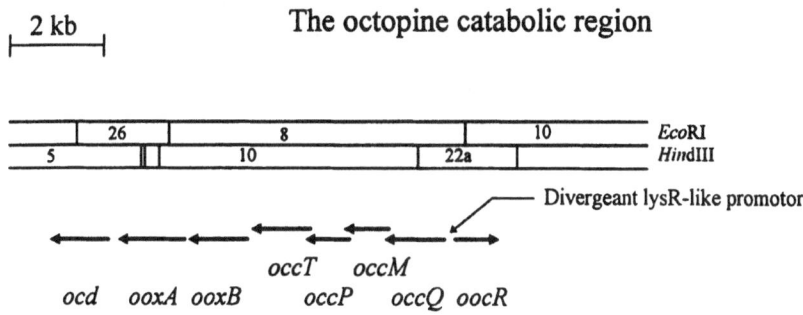

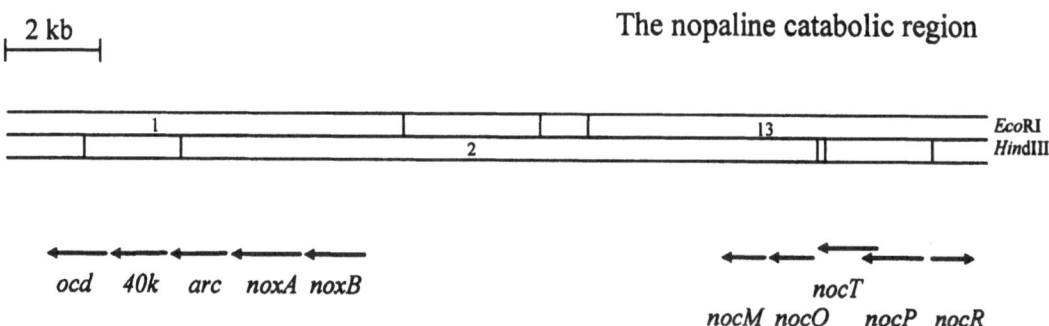

Figure 7. The octopine (*occ*) and nopaline (*noc*) catabolic regions of the octopine-type Ti plasmid
These results were essentially from Zanker *et al.,* (1992). The restriction map from the octopine-type Ti plasmid is from De Vos *et al.,* (1981) while that from the nopaline type Ti plasmid is from Zanker *et al.,* (1992). This latter is slightly different from the original map of Depicker al. (1980). Gene legends are: *ocd*, ornithine cyclodeaminase, *oox*, octopine oxidase, *occ*, octopine catabolism; *40k*, ORF encoding a 40kDa protein of unknown function; *arc*, arginine catabolism (arginase); *nox*, nopaline oxidase; and *noc*, nopaline catabolism.

synthesis of octopine and nopaline in tumors exhibit clear similarity but appear to be unrelated to the *oox* and *nox* catabolic genes (Zanker *et al.,* 1994). The origin of both the catabolic and anabolic genes still remains unclear. Though opine catabolic genes have been found in other *Rhizobiaceae* (Bergeron *et al.,* 1993), several studies suggest a possible marine origin for octopine and nopaline metabolic functions. First, octopine was originally found in *Octopus* spp. as an arginine derivative produced by anaerobic muscle work (Thoai and Robin, 1959). This compound was later identified in other marine animals, such as various cephalopods and bivalves (review: Tempé, 1983). In some of these animals, fast swimming requires high ATP concentrations provided by the anaerobic glycolysis and by the transfer of phosphate from phosphoarginine to ADP, two pathways leading to an accumulation of NADH

and arginine (Grieshaber and Gäde, 1976). The biological role of octopine synthase (also termed octopine dehydrogenase) is to combine arginine to pyruvate, a reaction which requires the oxidation of NADH, and eventually lowers the concentrations of both NADH and arginine in muscle cells. This reaction insures that glycolysis and ATP regeneration from phosphoarginine are not inhibited and therefore enables the animal to maintain high muscular activity over long periods of time (Zammit and Newsholme, 1976). In terms of evolution, a remarkable result was the recent identification of marine species of *Agrobacterium* isolated from the northeastern Atlantic Ocean bottom sediments (Ruger and Hofle, 1992). The identification of these microorganisms as agrobacteria seems well-established as it relies on phenotypic and molecular data (DNA/DNA homology, DNA composition, analysis of 5S rRNA and tRNA genes). Though the opine catabolic properties of these isolates were not investigated, it is very tempting to speculate that some of them might degrade octopine, a hypothesis consistent with the presence of octopine in the environment (review by Tempé, 1983) of these marine *Agrobacterium* species and with the isolation of opine-degrading marine bacteria from mollusks (Dion, 1986).

V. Opines as Key Compounds in the *Agrobacterium*-Plant Interaction

From the data presented above, the role of opines in the *Agrobacterium*-plant interaction is clear: as stated by the opine concept, they serve to feed the pathogen and to favor its dissemination (see also chapter 10 on bacterial conjugation). Indeed, within the last five years, ecological studies dealing with opine production by the plants and opine utilization by the bacteria have provided some experimental support to the theory. A first valuable result came from the comparison of the growth of two near isogenic *Pseudomonas putida* strains, one being a naturally-occurring mannopine utilizer and the second one an insertion mutant of the previous strain unable to degrade this opine. In the rhizosphere of transgenic plants producing mannopine, the wild-type mannopine utilizer outgrew the nonutilizing mutant (Lam *et al.*, 1991; Lam and Gaffney, 1993). Similar experiments involved gnotobiotic devices in which near-isogenic, untransformed and transformed, opine-producing *Lotus corniculatus* plants were grown under hydroponic conditions. The co-inoculation of opine-utilizing and opine-nonutilizing agrobacteria in these model rhizosphere systems allowed Guyon *et al.*, (1993) to show that the production of the opines mannopine, mannopinic acid and agropinic acid by the transgenic plants specifically and significantly promoted the growth of the opine-degrading agrobacteria. Finally, based on a similar experimental design, Savka and Farrand (1997) observed that an agropine-utilizing *Pseudomonas fluorescens* strain outgrew its nonutilizing counterpart in a gnotobiotic model rhizosphere of agropine-producing tobacco plants. Overall, these experiments strengthened the opine concept, not only because they corroborate the assumption that opine production by the transformed plant cells provides a selective advantage to the associated opine-utilizing bacteria, but essentially

because they constitute direct demonstrations of a theory which long remained only predictive.

However, some results dealing with the biological role of opines remain unexplained. For instance, some of these compounds enhance the expression of the virulence genes of *Agrobacterium* in the presence of the phenolic inducer acetosyringone. This phenomenon is mediated by the products of the *virA* and *virG* genes that sense the presence of the phenolic inducers (VirA) and induce the expression (VirG) of the other virulence genes (Veluthambi *et al.*, 1989). This result is in agreement with the demonstration that some opine uptake occurs in Ti plasmid-free strains (Krishnan *et al.*, 1991), and with the demonstration that the cryptic plasmid of *Agrobacterium* is involved in dfg degradation and may favor the utilization of octopine as sole nitrogen source (Vaudequin-Dransart *et al.*, in preparation). Interestingly, experiments performed under octopine limitation in chemostats have shown that these bacteria degrade octopine more efficiently as a nitrogen source than as a carbon source (Bell, 1990).

Recent studies on opines have brought some additional refinements to the opine concept and more precise ideas on the *Agrobacterium*-plant interaction. The data presented in this review were chosen to enlighten these refinements. For instance, recent advances in opine-related chemotactism confirmed that opines are signal molecules intervening at early and late stages of the interaction. Furthermore, the examples of agrocinopines and agrocin K84 illustrate that opines are also involved in bacteria-bacteria interactions in the tumor environment. Finally, the analysis of the opine catabolic and biosynthetic functions ascertains that opines are key molecules in the *Agrobacterium*-plant interaction not only as explained by the opine concept, but also because they probably played a major role in the evolution of the Ti and Ri plasmids of *Agrobacterium*.

VI. Where Does the "Opine Trail" Lead?

In their paper entitled La piste des opines (The opine trail), Petit and Tempé (1983) concluded with the question "Où va la piste des opines" which translates to where does the opine trail lead? At this time, these authors stressed the urgent need for ecological studies to prove the validity of the opine concept. Some 15 years later, it appears that they have been listened to. From the results briefly described above, ecology appears as one of the

" territories " towards which the opine trail has led. This assertion will certainly receive more experimental support in the near future since numerous assays involving transgenic plants producing opines and associated bacteria are being performed. Interestingly, most of these investigations now aim at evaluating the possibility to engineer plant-bacterium associations by the establishment of an opine-based, trophic link between the two organisms. Since the validity of the opine concept seems ascertained, it is reasonable to speculate that the production of opines by the plants will specifically favor the growth of opine-degrading bacteria in the phytosphere, hence that of any opine-degrading, man-introduced, bacterial inoculant (Dessaux et al., 1987; Guyon et al., 1993; Farrand et al., 1994; Wilson and Lindow 1994; Rossbach et al., 1994b; Wilson et al., 1995; review: O'Connell et al., 1996). If these bacteria are potentially beneficial to the plant (as would be a disease control agent), the engineered association may then be turned into an " artificial symbiosis " (Dessaux et al., 1992) where the micro-organism protects the plant that feeds it.

Though there still is a long way from now to " artificial symbiosis ", two noticeable results have already been reported. The first relates to the colonization of the leaf surface. Using transgenic tobacco plants producing opines, Wilson et al., (1995) were able to alter the epiphytic colonization ability of a strain of Pseudomonas syringae by conferring opine degradation upon this bacterium. The second experiment was performed at the rhizosphere level. Operating in near-field conditions, Oger et al., (1997) demonstrated that the growth of mannopine-utilizing bacteria was drastically enhanced in the rhizosphere of mannopine-producing Lotus corniculatus plants, an observation that was extended to nopaline- and cucumopine-producing L. corniculatus plants, and nopaline- and cucumopine-degrading microorganisms. All these results confirm that forcing a plant to produce opines may constitute an elegant way to specifically favor the growth of opine-degrading bacterial species. This strategy has received different names such as " artificial symbiosis " (Dessaux et al., 1992), altered " nutritional resource partitioning " (Wilson and Lindow, 1994), or " biased rhizosphere " (Rossbach et al., 1994b). In this respect, there is a common agreement that investigation on the rhizosphere, a " largely unexplored frontier for genetic engineering " (O'Connell et al., 1996) will undoubtedly benefit from earlier studies on opines and opine production by transformed plant cells.

Petit and Tempé (1983) also called for additional studies on the role of opines in the Rhizobium-legume interactions. Here also, scientists adhered to this proposal. To be convinced of this, one only need to consider the amount of data reviewed here on the occurrence of rhizopines in nodules resulting from the Rhizobium-legume interaction. Interestingly, while research on rhizopine developed, other data on various compounds produced by plants (or plant organs) and specifically degraded by bacteria belonging to the genus Rhizobium have been gathered. Examples of these molecules are the calystegins (Tepfer et al., 1988a). These polyhydroxy-alkaloids exhibit glycosidase inhibitory properties (Molyneux et al., 1993) and are detected in large quantities in the root system and in exudates of a very limited number of plant species such as Calystegium sepium, Convolvulus arvensis and Atropa belladona (Tepfer et al., 1988 b). Interestingly, Rhizobium meliloti strain Rm41 seems to be one of the only bacterial soil isolates able to degrade these compounds, a feature which may account for the detection of this strain in the rhizosphere of Calystegium plants, though these are non-host-plants for symbiotic nitrogen fixation (Tepfer et al., 1988a, 1988b; Boivin et al., 1990b). In a similar fashion, mimosine is a toxic amino acid synthesized in large amounts by the roots, foliage and seeds of plants belonging to the legume genera Mimosa and Leucaena. The Leucaena-nodulating Rhizobium strains utilize mimosine as the only carbon and nitrogen source whereas other Rhizobium strains do not (Soedarjo et al., 1994). Strikingly, the biological significance of mimosine utilization resembles that of rhizopine utilization (though these latter compounds are synthesized by the nodule bacteroids) i.e. both contribute to the selection of the appropriate population(s) of nodulating rhizobacteria. A similar observation has been made with A. vitis strains which appear to degrade tartrate, a major organic acid produced by the host plant Vitis vinifera (Crouzet and Otten, 1995).

Finally, a recurring question remains; have opine-like molecules been detected beyond the Rhizobiaceae-plant interactions? Limited and essentially negative data have been obtained so far. However, very recent studies on the fasciation disease caused by the gram-positive bacteria Rhodococcus fascians, has shed some new light on this question. Using the differential growth assay which was originally developed for identifying opines in the Agrobacterium-induced tumors, Goethals and co-workers (personal communication) found that fasciation galls contain specific compounds which are degraded by the inciting *Rhodococcus* pathogen in a strain-dependant fashion. The structure of these molecules and whether they can be regarded as rhodopines is still under investigation. Beyond plants, the capability of the bacteria *Bacteroides thataiotaomicron* to degrade chondroin sulfate, heparin, and other mucopolysaccharides might play a role in the colonization of the extracellular matrix of the mammalian gastroinstestinal tract (Hwa and Salyers, 1992; Cheng *et al.*, 1992). Similarly, the enteric bacteria *Salmonella typhimurium* exhibit a well-adapted sialidase activity which efficiently cleaves sialoglycoconjugates which are highly abundant molecules of the mammalian gut (Vimr, 1994). It is therefore tempting to speculate that the bacterial capability to specifically catabolize host metabolites (the key feature of the opine concept) probably extends beyond the plant-bacterium interactions to other two-organism relationships (pathogenic, symbiotic or saprophytic) where it may control, at least in part, the mutual recognition and adaptation of bacteria and their hosts.

VII. Acknowledgments

The authors warmly thank Koen Goethals for communication of unpublished material on *Rhodococcus fasciens*, Pascale Leroux for her help in gathering references, and Nathalie Jansion (ISV-CNRS) for excellent artwork.

VIII. References

Anet, E.F.L.J. (1957). Aust. J. Chem. 10, 193-197.
Anet, E.F.L.J. and Reynolds, T.M. (1957). Aust. J. Chem. 10, 182-192.
Ashby, A.M., Watson, M.D., Loake, G.J. and Shaw, C.H. (1988). J. Bacteriol.170, 4181-4187.
Barker, R.F., Idler, K.B., Thompson, D.V. and Kemp, J.D. (1983). Plant Mol. Biol. 2, 335-350.
Beauchamp, C.J., Chilton, W.S., Dion, P. and Antoun, H. (1990). Appl. Environ. Microbiol. 56, 150-155.
Beck von Bodman, S., Hayman, G.T. and Farrand, S.K. (1992). Proc. Natl. Acad. Sci. USA 89, 643-647.
Bell, C.R. (1990). Appl. Environ. Microbiol. 56, 1775-1781.
Bergeron, J., Beaulieu, C., Levesque, R., Kondorosi, A. and Dion, P. (1993). Can. J. Microbiol. 39, 1041-1050.
Better, M., Lewis, B., Corbin, D., Ditta, G. and Helinski, D.R. (1983). Cell 35, 479-485.
Biemann, K., Lioret, C., Asselineau, J., Lederer, E. and Polonski, J. (1960). Bull. Soc. Chi. Biol. 42, 979-991.

Birot, A.-M., Bouchez, D., Casse-Delbart, F., Durand-Tardiff, M., Jouanin, L., Pautot, V., Robaglia, C., Tepfer, D., Tepfer, M., Tourneur, J. and Vilaine, F. (1987). Plant Physiol. Biochem. 25, 323-355.
Boivin, C., Camut, S., Malpica, C.A., Truchet, G. and Rosenberg, C. (1990a). The Plant Cell 2, 1157-1170.
Boivin, C., Malpica, C., Rosenberg, C., Goldmann, A., Fleury, V., Maille, M., Message, B., Pamboukdjian, N. and Tepfer, D. (1990b). Symbiosis 9, 147-154.
Bomhoff, G. (1976). Studies on crown-gall - a plant tumor. Ph.D. Thesis, University of Leiden, The Netherlands.
Bomhoff, G.H., Klapwijk, P.M., Kester, H.C., Schilperoort, R.A., Hernalsteens, J-P. and Schell, J. (1976) Mol. Gen. Genet., 145, 188-181.
Bouzar, H., Chilton, W.S., Nesme, X., Dessaux, Y., Vaudequin, V., Petit, A., Jones, J.B. and Hodge, N.C. (1995). Appl. Environ. Microbiol. 61, 65-73.
Braun, A.C. (1947). Am. J. Bot. 34, 234-240.
Braun, A.C. (1982). In, Kahl G and Schell J (eds) Molecular biology of plant tumors, pp 155-210. Academic Press, San Diego.
Buchanan-Wollaston, V., Passiatore, J.E. and Cannon, F. (1987). Nature 328, 172-175.
Bush, A. and Pueppke, S.G. (1991). Appl. Environ. Micorbiol. 57, 2468-2472.
Cheng, Q., Hwa, V. and Salyers, A.A. (1992). J. Bacteriol. 174, 7185-7193.
Chilton, M.D., Drummond, M.H., Merlo, D.J., Sciaky, D., Montoya, A.L., Gordon, M.P., Nester, E.W. (1977) Cell 11, 263-271.
Chilton, M.D., Tepfer, D.A., Petit, A., David, C., Casse-Delbart, F. and Tempé, J. (1982). Nature 295, 432-434.
Chilton, W.S. and Chilton, M.D. (1984). J. Bacteriol. 158, 650-658
Chilton, W.S., Rinehart, Jr. K.L. and Chilton, M.D. (1984a). Biochemistry 23, 3290-3297.
Chilton, W.S., Tempé, J., Matzke, M. and Chilton, M.D. (1984b). J. Bacteriol. 157, 357-362.
Chilton, W.S., Stomp, A.M., Beringue, V., Bouzar, H., Vaudequin-Dransart, V., Petit, A. and Dessaux, Y. (1995). Phytochemistry 40, 619-628.
Cho, K., Fuqua, C., Martin, B.S. and Winans, S.C. (1996). J. Bacteriol. 178, 1872-1880.
Citovsky, V., Gail McLean, B., Greene, E., Howard, E., Kuldau, G., Thorstenson, Y., Zupan, J. and Zambryski, P. (1992). In, Verma DPS (ed) Molecular signals in plant-microbe communications, pp. 169-199. CRC Press, Roca Raton.
Clare, B.G., Kerr, A. and Jones, D.A. (1990). Plasmid 23, 126-137.
Cook, D.M. and Farrand, S.K. (1992). J. Bacteriol. 174, 6238-6246.
Crouzet, P. and Otten, L. (1995). J. Bacteriol. 177, 6518-6526.
Davioud, E., Petit, A., Tate, M.E., Ryder, M.H. and Tempé, J. (1988). Phytochemistry 27, 2429-2433.
De Beuckeleer, M., De Block, M., De Greve, H., Depicker, A., De Vos, R., De Vos, G., De Wilde, M., Dhaese, P., Dobbelaere, M.R., Engler, G., Genetello, C., Hernalsteens, J-P., Holsters, M., Jacobs, A., Schell, J., Seurinck, J., Silva, B., Van Haute, E., Van Montagu, M., Van Vliet, F., Villaroel, R. and Zaenen, I. (1978). In, Ridé M (ed) Proc. 4th Int. Conf. on Plant Pathogenic Bacteria, pp 115-126. INRA, Angers.
Depicker, A., De Wilde, M., De Vos, G., Van Vliet, F., De Brock, M., Villaroel, R., Van Montagu, M. and Schell, J. (1980). Plasmid 3, 192-211.
Dessaux, Y., Guyon, P., Farrand, S.K., Petit, A. and Tempé, J. (1986a) J. Gen. Microbiol. 132, 2549-2559.

Dessaux, Y., Petit, A., Tempé, J., Demarez, M., Legrain, C. and Wiame, J.M. (1986b). J. Bacteriol. 166, 44-50.

Dessaux, Y., Tempé, J. and Farrand, S.K. (1987). Molec. Gen. Genet. 208, 301-308.

Dessaux, Y., Guyon, P., Petit, A., Tempé, J., Demarez, M., Legrain, C., Tate, M.E. and Farrand, S.K. (1988). J. Bacteriol. 170, 2939-2946.

Dessaux, Y., Petit, A. and Tempé, J. (1992). In, Verma DPS (ed) Molecular signals in plant-microbe communications, pp. 109-136. CRC Press, Roca Raton.

Dessaux, Y., Petit, A. and Tempé, J. (1993). Phytochemistry 34, 31-38.

De Vos, G., De Beuckeleer, M., Van Montagu, M. and Schell, J. (1981). Plasmid 6, 249-253.

Dilworth, M.J. and Glenn, A. (1984). Trends Biochem. Sci. 9, 519-523.

Dion, P. (1986). Can. J. Microbiol. 32, 959-963.

Ditta, G., Virts, E., Palomares, A.J. and Kim, C.H. (1987). J. Bacteriol. 169, 3217-3223.

Drummond, M.H., Gordon, M.P., Nester, E.W. and Chilton, M.D. (1977). Nature 269, 535-536.

Ellis, J.G. and Murphy, P.J., (1981). Molec. Gen. Genet. 181, 36-43

Ellis, J.G., Murphy, P.J. and Kerr, A. (1982a). Molec. Gen. Genet. 186, 275-281.

Ellis, J.G., Kerr, A., Petit, A. and Tempé, J. (1982b). Molec. Gen. Genet. 186, 269-274.

Ellis, J.G., Ryder, M.H. and Tate, M.E. (1984). Molec. Gen. Genet. 195, 466-473.

Farrand, S.K. (1991). In, Pimentel D (ed) Pest management handbook, Vol. 2, pp. 311-329. CRC Press, Boca Raton.

Farrand, S.K. and Dessaux, Y. (1986). J. Bacteriol. 167, 732-734.

Farrand, S.K., Hwang, I. and Cook, D.M. (1996). J. Bacteriol. 178, 4233-4247.

Farrand, S.K., Tempé, J. and Dessaux, Y. (1990). Molec. Plant-Microbe Interact. 3, 259-267.

Farrand, S.K., Wilson, M., Lindow, S.E. and Savka, M.A. (1994). In, Ryder, M.H., Stephens, P.M. and Bowen, G.D. (eds) Improving plant productivity with rhizosphere bacteria, CSIRO, Glen Osmond.

Firmin, J.L. and Fenwick, R.G. (1978). Nature 276, 842-844.

Gaudin, V., Vrain, T. and Jouanin, L. (1994). Plant Physiol. Biochem. 32, 11-29.

Gielen, J., De Beuckeleer, M., Seurinck, J., Deboeck, F., De Greve, H., Lemmers, M., Van Montagu, M. and Schell, J. (1984). EMBO J. 3, 835-846.

Gelvin, S.B. (1992). In Verma DPS (ed) Molecular signals in plant-microbe communications, pp. 137-167. CRC Press, Roca Raton.

Goldmann, A., Tempé, J. and Morel, G. (1968). C.R. Soc. Biol. 162, 630-631.

Goldmann, A., Thomas, D.W. and Morel, G. (1969). C.R. Acad. Sci. (Ser. D) 268, 852-854.

Gordon, D.M., Ryder, M.H., Heinrich, K. and Murphy, P.J. (1996). Appl. Environ. Microbiol. 62, 3991-3996.

Grieshaber, M. and Gäde, G. (1976). J. Comp. Physiol. 108, 225-232.

Gurley, W.B., Kemp, J.D., Albert, M.J., Sutton, D.W. and Callis, J. (1979). Proc. Natl. Acad. Sci. USA 77, 2828-2832.

Gussin, G.N., Ronson, C.W. and Ausubel, F.M. (1986). Ann. Rev. Genet. 20, 567-591.

Guyon, P., Chilton, M.D., Petit, A. and Tempé, J. (1980). Proc. Natl. Acad. Sci. USA 77, 2693-2697.

Guyon, P., Petit, A., Tempé, J. and Dessaux, Y. (1993). Molec. Plant-Microbe Interact. 6, 92-98.

Habeeb, L.F., Wang, L. and Winans, S.C. (1991). Molec. Plant-Microbe Interact. 4, 379-385.

Hageman, R.V. and Burris, R.H. (1978). Proc. Natl. Acad. Sci. 75, 2699-2702.

Hawes, M.C., Smith, L.Y. and Howarth, A.J. (1988). Mol. Plant-Microbe Interact. 1, 182-186.

Hawes, M.C. and Smith, L.Y. (1989). J. Bacteriol. 171, 5668-5671.

Hayman, G.T. and Farrand, S.K. (1988). J. Bacteriol. 170, 1759-1767.

Hayman, G.T. and Farrand, S.K. (1990). Molec. Gen. Genet. 223, 465-473.

Hayman, G.T., Beck von Bodman, S., Kim, H., Jiang, P. and Farrand, S.K. (1993). J. Bacteriol. 175, 5575-5584.

Higgins, C.F. (1992). Ann. Rev. Cell. Biol. 8, 67-113.

Heyns, K. and Paulsen, H. (1959). Liebigs Annalen der Chimie 622, 160-174.

Hong, S.B., Dessaux, Y., Chilton, W.S. and Farrand, S.K. (1993). J. Bacteriol. 175, 401-410.

Hong, S.B. and Farrand, S.K. (1994). J. Bacteriol. 176, 3576-3583.

Hong, S.B. I. Hwang, Y. Dessaux, P. Guyon, K.S. Kim, Farrand, S.K. (1997). J. Bacteriol. in press.

Huffmann, G.A., White, F.F., Gordon, M.P. and Nester, E.W. (1984). J. Bacteriol. 157, 269-276.

Hwa, V. and Salyers, A.A. (1992). Appl. Environ. Microbiol. 58, 869-876.

Isogai, A., Fukuchi, N., Hayashi, M., Kamada, H., Harada, H. and Suzuki, A. (1988). Agric. Biol. Chem. 52, 3235-3237.

Jouanin, L. (1984). Plasmid 12, 91-102.

Jouanin, L., Tourneur, J., Tourneur, C. and Casse-Delbart, F. (1986). Plasmid 16, 124-134.

Jubier, M.F. (1972). FEBS Lett. 28, 129-132.

Kado, C.I. (1994). Molec. Microbiol. 12, 17-22.

Kemp, J.D. (1977). Biochem. Biophys. Res. Commun. 74, 862-868.

Kerr, A. (1980). Plant Dis. 64, 25-30.

Kerr, A., Manigault, P. and Tempé, J.(1977). Nature 265, 560-561.

Kesters, K. and De Ley, J. (1984) In, Krieg, N.G. and Holt, J.G. (eds) Bergey's Manual of systematic bacteriology, Vol. 1, pp. 244-254. Williams and Wilkins, Baltimore.

Kim, K.S. and Farrand, S.K. (1996). J. Bacteriol. 178, 3275-3284.

Kim, K.S., Chilton, W.S. and Farrand, S.K. (1996). J. Bacteriol. 178, 3285-3292.

Kim, R.Y., Gasser, R. and Wistow, G.J. (1992). Proc. Natl. Acad. Sci. USA 89, 9292-9296.

Klapwijk, P.M. and Schilperoort, R.A. (1979). J. Bacteriol. 139, 424-431.

Klapwijk, P.M., Hooykaas, P.J.J., Kesters, H.C.M., Schilperoort, R.A. and Rorsch, A. (1976). J. Gen. Microbiol. 96, 155-163.

Klapwijk, et al., (1977). J. Gen. Microbiol. 102,1-11.

Klapwijk, P.M., Scheulderman, T. and Schilperoort, R.A. (1978). J. Bacteriol. 136, 775-785.

Klein, R.M. (1954). Brookhaven Symp. Biol, abnormal and pathological plant growth, pp 97-114.

Kovacs, L.G. and Pueppke, S.G. (1994). Molec. Gen. Genet. 242, 327-336.

Krishnan, M., Burgner, J.W., Chilton, W.S. and Gelvin, S.B. (1991). J. Bacteriol. 173, 903-905.

Lam, S.T. and Gafney, T. (1993). In, Chet I (ed) Biotechnology in plant disease control, pp. 291-320. Wiley-Liss, New York.

Lam, S.T., Torkewitz, N.R., Nautiyal, C.S. and Dion, P. (1991). Phytopathology 81, 1163-1164.

Lioret, C. (1956). Bull. Soc. Fr. Physiol. Veg. 2, 76.

Lejeune, B. and Jubier, M.F. (1968). C. R. Acad. Sci. (Ser. D) 264, 1803-1805.

Less, M. and Lanka, E. (1994). Cell 77, 321324.

Loake, G.J., Ashby, A.M. and Shaw, C.M. (1988). J. Gen. Microbiol. 134,1427-1432.

Lukat, G.S. and Stock, J.B. (1993). J. Cell. Biochem. 51, 41-46.

Magrelli, A., Langenkemper, K., Dehio, C., Schell, J. and Spena, A. (1994) Science 266, 1986-1988.

Ménagé, A. and Morel, G. (1964). C. R. Acad. Sci. 259, 4795-4796.

Ménagé, A. and Morel, G. (1965). C.R. Soc. Biol. 159, 561-562.

Molyneux, R.J., Pan, Y.T., Goldmann, A., Tepfer, D.A. and Elbein, A.D. (1993). Arch. Biochem. Biophys. 304, 81-88.

Montoya, A., Chilton, M.D., Gordon, M.P., Sciaky, D., Nester, E.W. (1977). J. Bacteriol. 129, 101-107.

Morel, G. (1956). Bull. Soc. Fr. Physiol. Veg. 2, 75-76.

Morris, R.O. (1986). Annu. Rev. Plant Physiol. 37, 509-538.

Murphy, P.J. and Roberts, W.P. (1979). J. Gen. Microbiol. 114, 207-213.

Murphy, P.J. and Saint, C.P. (1992) In, Verma DPS (ed) Molecular Signals in Plant-Microbe Communication, pp. 377-390. CRC Press, Boca Raton.

Murphy, P.J., Heycke, N., Banfalvi, Z., Tate, M.E., de Bruijn, F., Kondorosi, A., Tempé, J. and Schell, J. (1987). Proc. Natl. Acad. Sci. USA 84, 493-497.

Murphy, P.J., Heycke, N., Trenz, S.P., Ratet, P., de Bruijn, F.J. and Schell, J. (1988). Proc. Natl. Acad. Sci. USA 85, 9133-9137.

Murphy, P.J., Trenz, S.P., Grzemski, W., de Bruijn, F.J. and Schell, J. (1993). J. Bacteriol. 175, 5193-5204.

Murphy, P.J., Wexler, M., Grzemski, W., Rao, J.P. and Gordon, D.M. (1995). Soil Biology and Biochemistry 27, 525-529.

Nautiyal, C.S., Dion, P. and Chilton, W.S. (1992). J. Bacteriol. 174, 2215-2221.

O'Connell, K.P., Goodman, R.M. and Handelsman, J. (1996). Trends in Biotechnol. 14, 83-88.

Oger, P., Petit, A. and Dessaux, Y. (1997). Nature/Biotechnology 15, 369-372.

Otten, L., Canaday, J., Gérard, J.C., Fournier, P., Crouzet, P. and Paulus, F. (1992). Mol. Plant Microbe Interact. 5, 279-287.

Otten, L., Gérard, J.C. and De Ruffray, P. (1993). Plasmid 29, 154-159.

Otten, L. and De Ruffray, P. (1994). Mol. Gen. Genet. 245, 493-505.

Pansegrau, W., Schoumacher, F., Hohn, B. and Lanka, E. (1993). Proc. Natl. Acad. Sci USA 90, 11538-11542.

Parke, D., Ornston, L.N. and Nester, E.W. (1987). J. Bacteriol. 169, 5336-5338.

Paulus, F., Canaday, J. and Otten, L. (1991a). Mol Plant Microbe Interact. 4, 190-197.

Paulus, F., Canaday, J., Vincent, F., Bonnard, G., Kares, C. and Otten, L. (1991b) Plant Mol. Biol. 16, 601-614.

Petit, A., Delhaye, S., Tempé, J. and Morel, G. (1970). Physiol. Vég. 8, 205-213.

Petit, A. and Tourneur, J. (1972). C.R. Acad. Sci. (Ser D.) 275, 137-139.

Petit, A. and Tempé, J. (1975). C.R. Acad. Sci. 281, 467-469.

Petit, A. and Tempé, J. (1978). Molec. Gen. Genet. 167, 147-155.

Petit, A., Tempé, J., Kerr, A., Holsters, M., Van Montagu, M. and Schell, J. (1978a). Nature 271, 570-571.

Petit, A., Dessaux, Y. and Tempé, J. (1978b). In, Ridé M (ed) Proc. 4th Int. Conf. on plant pathogenic bacteria, pp. 143-152. INRA, Angers.

Petit, A., David, C., Dahl, G.A., Ellis, J.G., Guyon, P., Casse-Delbart, F. and Tempé, J. (1983). Mol. Gen. Genet. 190, 204-214.

Powell, G.K. and Morris, R.O. (1986). Nucleic Acid Res. 14, 2555-2565.

Rao, J.P., Grzemski, W. and Murphy, P.J. (1995). Microbiology 141, 1683-1690.

Riker, A., Banfield, W., Wright, W., Keitt, G. and Sagen, K. (1930). J. Agric. Res. 41, 507-540.

Roberts, W.P., Tate, M.E. and Kerr, A. (1977). Nature 265, 379-380.

Rossbach, S., Kulpa, D.A., Rossbach, U. and de Bruijn, F.J. (1994a). Mol. Gen. Genet. 245, 11-24.

Rossbach, S., McSpadden, B., Kulpa, D., Rasul, G., Ganoff, M. and De Bruijn, F.J. (1994b). Molec. Ecol. 3, 610-611.

Rossbach, S., Rasul, G., Schneider, M., Eardly, B. and de Bruijn, F.J. (1995). Molec. Plant Microbe Interact. 8, 549-559.

Ruger, H.J. and Holfe, M.G. (1992). Int. J. Sys. Bacteriol. 42, 133-143

Ryder, M.H., Tate, M.E. and Jones, G.P. (1984). J. Biol. Chem. 259, 9704-9710.

Saint, C.P., Wexler, M., Murphy, P.J., Tempé, J., Tate, M.E. and Murphy, P.J. (1993). J. Bacteriol. 175, 5205-5215.

Salomon, F., Deblaere, R., Leemans, J., Hernalsteens, J.P., Van Montagu, M. and Schell, J. (1984). EMBO J. 3, 141-146.

Sans, N., Schröder, G. and Schröder, J. (1987). Eur. J. Biochem. 167, 81-87.

Sans, N., Schindler, U. and Schröder, J. (1988). Eur. J. Biochem. 173, 123-130.

Savka, M.A. and Farrand, S.K. (1997). Nature/Biotechnology 15, 363-368.

Schell, J., Van Montagu, M., De Beuckeller, M., De Block, M., Depicker, A., De Wilde, M., Engler, G., Genetello, C., Hernalsteens, J.P., Holsters, M., Seurinck, J., Silva, B., Van Vliet, F. and Villaroel, R. (1979). Proc. R. Soc. London (Ser B) 204, 251-156.

Schindler, U., Sans, N. and Schröder, J. (1989). J. Bacteriol. 171, 847-854.

Schrell, A., Alt-Mörbe, J., Lanz, T. and Schröder, J. (1989). Eur. J. Biochem. 184, 635-641.

Schröder, J., Von Lintig, J. and Zanker, H. (1990). In, Henneke, H. and Verma, D.P.S. (eds) Advances in molecular gentics of the plant-microbe interactions, Vol. 1, pp. 28-31. Kluwer Acad. Publishers, Doordrecht.

Shaw, C.H., Ashby, A.M., Brown, A., Royal, C., Loake, G.J. and Shaw, C.H. (1988). Mol. Microbiol. 2, 413-417.

Shirazu, K. and Kado, C.I. (1993). Nucleic Acid Res. 21, 353-354.

Slightom, J.L., Durand-Tardiff, M., Jouanin, L. and Tepfer, D. (1986). J. Biol. Chem. 261, 108-121.

Smith, E. and Townsend, C. (1907). Science 24, 671-673.

Soedarjo, M., Hemscheidt, T.K. and Borthakur, D. (1994). Appl. Environ. Microbiol. 60, 4268-4272.

Soto, M.J., Zorzano, A., García-Rodriguez, F.M., Mercado-Blanco, J., López-Lara, I.M., Olivares, J. and Toro, N. (1994). Molec. Plant-Microbe Interaction 7, 703-707.

Spano, L, Pomponi, M., Costantino, P., Van Slogteren, G.M.S. and Tempé, J. (1982). Plant Mol. Biol. 1, 291-300.

Stock, A.M. and Mowbray, S.L. (1995). Curr. Opin. Struct. Biol. 5, 744-751.

Stock, A., Schaeffer, E., Koshland, Jr. D.E. and Stock, J. (1987). J. Biol. Chem. 262, 8011-8014.

Swords, K.M., Dahlbeck, D., Kearney, B., Roy, M. and Staskawicz, B.J. (1996). J. Bacteriol. 178, 4661-4669.

Tate, M.E., Ellis, J.G., Kerr, A., Tempé, J., Murray, K.E. and Shaw, K.J. (1982). Carbohydr. Res. 104, 105-120.

Tempé, J. (1983). In, Weinstein B (ed) Chemistry and biochemistry of amino acids, peptides and proteins, Vol 7, pp113-203. Marcel Dekker Inc., New York and Basel.

Tempé, J. and Schell, J. (1977). In, Legocky AB (ed) Translation of natural and synthetic polynucleotides, pp 415-420. University of Agriculture, Poznan.

Tempé, J. and Goldamnn, A. (1982). In, Kahl G and Schell J (eds) Molecular biology of plant tumors, pp 427-449. Academic Press, San Diego.

Tempé, J. and Petit, A. (1983). In, Pühler A (ed.) Molecular genetics of the bacteria-plant interaction, pp. 14-32. Springer, Berlin.

Tempé, J., Estrade, C. and Petit, A. (1978). In, Ridé M (ed) Proc. 4th Int. Conf on plant pathogenic bacteria, pp. 153-160. INRA, Angers.

Tempé, J., Guyon, P., Tepfer, D. and Petit, A. (1979). In, Timmis KN, Puhler A (eds) Plasmids of medical, commercial and environmental importance, pp. 353-363. Elsevier/North Holland Biomedical Press, Amsterdam.

Tempé, J., Guyon, P., Petit, A., Ellis, J.G., Tate, M.E. and Kerr, A. (1980). C.R. Acad. Sci. (Ser. D) 290, 1173-1176.

Tepfer, D. (1989). In, Kosuge T, Nester E (eds) Plant-Microbe Interactions, pp 294-342. McGraw Hill, New York.

Tepfer, D., Goldmann, A., Pamboukdjian, N., Maille, M., Lepingle, A., Chevalier, D., Dénarié, J. and Rosenberg, C. (1988a). J. Bacteriol. 170, 1153-1161.

Tepfer, D., Goldman, A., Fleury, V., Maille, M., Message, B., Pamboukdjian, N., Boivin, C., Dénarié, J., Rosenberg, C., Lallemand, J.Y., Descoins, C., Charpin, I. and Amarger, N. (1988b). In, Palacios R and Verma DPS (eds) Molecular genetics of plant-microbe interactions, pp. 139-144.

Thoai, N.V. and Robin, Y. (1959). Biochim. Biophys. Acta 35, 446-453.

Thomashow, M.F., Nutter, R., Montoya, A.L., Gordon, M.P. and Nester, E.W. (1980). Cell, 19, 729-739.

Unger, L., Ziegler, S.F., Huffman, G.A., Knauf, V.C., Peet, R., Moore, L.W., Gordon, M.P. and Nester, E.W. (1985). J. Bacteriol. 164, 723-730.

Valdivia, R.H., Wang, L. and Winanas, S.C. (1991). J. Bacteriol. 173, 6398-6405.

Vande Broek, A. and Vanderleyden, J. (1995). Molec. Plant Microbe Interactions 8, 800-810.

Van Egeraat, A.W.S.M. (1975). Plant and Soil 42, 381-386.

Van Larebeke, N., Engler, G., Holsters, M., Van den Elsacker, S., Zaenen, I., Schilperoort, R.A. and Schell, J. (1974). Nature 252, 169-170.

Van Nuenen, M., De Ruffray, P. and Otten, L. (1993) Mol. Gen. Genet. 240, 49-57.

Vaudequin-Dransart, V., Petit, A., Poncet, C., Ponsonnet, C., Nesme, X., Jones, J.B., Bouzar, H., Chilton, W.S. and Dessaux, Y. (1995). Molec Plant-Microbe Interact. 8, 311-321.

Veluthambi, K., Krishnan, M., Gould, J.H., Smith, R.H. and Gelvin, S.B. (1989). J. Bacteriol. 171, 3696-3703.

Vimr, E.R. (1994). Trends in Microbiol. 2, 271-277.

Von Lintig, J., Zanker, H. and Schröder, J. (1991). Molec. Plant-Microbe Interact. 4, 370-378.

Von Lintig, J., Kreusch, D. and Schröder, J. (1994). J. Bacteriol. 176, 495-503.

Walker, J.B. (1995). J. Bacteriol. 177, 818-822.

Wang, L., Helmann, J.D. and Winans, S.C. (1992). Cell 69, 659-667.

Watson, B., Currier, T.C., Gordon, M.P., Chilton, M.D. and Nester, E.W. (1975). J. Bacteriol. 123, 255-264.

Wexler, M., Gordon, D.M. and Murphy, P.J. (1995). Soil Biol. Biochem. 27, 531-537.

Wexler, M., Gordon, D.M. and Murphy, P.J. (1996). Microbiology 142, 1059-1066.

White, F.F. and Nester, E.W. (1980). J. Bacteriol. 141, 1134-1141.

White, F.F., Ghidossi, G., Gordon, M.P. and Nester, E.W. (1982). Proc. Natl. Acad. Sci. USA, 79, 3193-3197.

Willmitzer, L., De Beuckeleer, M., Lemmers, M., Van Montagu, M. and Schell, J. (1980). Nature 287, 359-361.

Wilson, M. and Lindow, S.E. (1994). Appl. Environ. Microbiol. 60, 4468-4477.

Wilson, M., Savka, M.A., Hwang, I., Farrand, S.K. and Lindow, S.E. (1995). Appl. Environ. Microbiol. 61, 2151-2158.

Yamada, T., Lee, P.D. and Kosuge, T. (1986). Proc. Natl. Acad. Sci. USA 83, 8263-8267.

Yamada, T., Palm, C.J., Brooks, B. and Kosuge, T. (1985). Proc. Natl. Acad. Sci. USA 82, 6522-6526.

Zaenen, I., Van Larebeke, N., Teuchy, H., Van Montagu, M. and Schell, J. (1974). J. Mol. Biol. 86, 109-127.

Zammit, V.A. and Newsholme, E.A. (1976). Biochem. J. 160, 447-462.

Zanker, H., Von Lintig, J. and Schröder, J. (1992). J. Bacteriol. 174, 841-849.

Zanker, H., Lurz, G., Langridge, U., Langridge, P., Kreusch, D. and Schröder, J. (1994). J. Bacteriol. 176, 4511-4517.

Chapter 10

Conjugal Plasmids and Their Transfer

Stephen K. Farrand

I. Introduction

Virtually all members of the family *Rhizobiaceae* that have been examined contain one or more plasmids. Most of these elements are large and some are known to play key roles in the biology of these bacteria, especially with respect to the association between these microorganisms and their plant hosts. Structure-function studies indicate that, in a given species, a particular plasmid specific to that isolate may contain regions that are conserved among plasmids present in other isolates of the same species, in isolates of other species of that genus, or even in isolates of other genera of the family. This suggests several important concepts that tie together the various species and genera of the family *Rhizobiaceae*. First, that these plasmids often share substantial regions of similar sequence suggests that the individual elements have arisen by recombination with other plasmids. However, these plasmids often exhibit equally large regions of non-similarity indicating that they maintain diversity. Thus, some *sym* plasmids share substantial sequence relatedness with some Ti plasmids. But these two plasmid types also contain large regions of unrelated sequence. This, of course is what defines one set as *sym* plasmids and the other as Ti plasmids. However, that they share regions of relatedness suggests that some core portion of these plasmids may be common and indeed essential for the maintenance of these elements as well as for the recombination events that lead to their diversity. This suggests, in turn, that these plasmids are capable of transferring between bacteria of the same family and perhaps bacteria of more distantly related taxa. Recent analyses indicate that the replication regions of many *Rhizobium* plasmids are related to those of several Ti and Ri plasmids (Turner and Young, 1995; Turner, *et al.*, 1996) and that these plasmids may all fall into a superfamily of extrachromosomal elements sharing a common core replicator region. More intriguingly, as will be discussed below, this conserved replicator region is intimately linked to a bank of conserved genes associated with conjugal transfer in at least two *Rhizobium* plasmids (Turner, *et al.*, 1996; Freiberg, *et al.*, 1997), as well as in several Ti plasmids (Alt-Mörbe, *et al.*, 1997; Li and Farrand, unpublished). Thus, a common replication machinery, linked with a conjugal transfer system, may form the core structure of many of the important plasmids present in the various members of the family *Rhizobiaceae*. Furthermore, available sequence data suggests that these large plasmids take on and evolve their characters by recombination with other plasmids or components of the bacterial genome. Thus, the conjugal transfer systems may be of particular importance to the evolution, as well as the dissemination of these plasmids. The nature of the conjugal transfer systems of these plasmids and the significance of conjugation with respect to the variability found within the plasmids of the members of this family forms the nucleus of this review.

II. Occurrence of Transmissible Elements in the Family *Rhizobiaceae*

Genetic elements transmissible by mating have been reported in virtually all members of the family *Rhizobiaceae* examined to date. The bulk of the screening studies have focused on members of the genus *Rhizobium*. During the 1970's and 1980's a number of reports appeared in the literature describing conjugal transmission of plasmids from *Rhizobium* donors to *Rhizobium* and *Agrobacterium* recipients. However, with only a few exceptions, the nature of the conjugal systems of these plasmids has not been further investigated. Within the genus *Agrobacterium* transmissible plasmids have been reported in *A. tumefaciens*, *A. radiobacter*, *A. rhizogenes* and *A. vitis*. Most of these studies have focused on the virulence or opine-catabolism plasmids of these isolates. Thus, Ti, Ri and At plasmids in many isolates of these strains have been reported to be transmissible. However, in many of these cases results should be judged with caution. Often the plasmids of interest co-exist with other large plasmids, especially in primary isolates from the field. But even commonly-used laboratory strains of *Rhizobium* and *Agrobacterium* spp. usually harbor more than one plasmid. Moreover, in many cases, these plasmids can and do form cointegrates. This is especially well illustrated by the agropine- and mannopine-type *A. rhizogenes* strains such as A4 and 8196 in which the Ri plasmid readily cointegrates with a large co-resident plasmid (Petit, *et al.*, 1983). Thus, transfer of a given plasmid may be due to self-conjugation, but it also might result from cointegration with another self-conjugal plasmid resident within the donor, or to non-cointegrative, *trans*-active mobilization by some other conjugal element present in the strain. Further genetic analysis of progeny in which only the plasmid of interest is present is necessary in order to differentiate between these possibilities. Failure to detect transfer also must be interpreted with caution. Lack of strongly selectable markers can make conjugation at low frequencies difficult to detect. In addition, the expression of the conjugal transfer systems may be regulated. Thus, unless matings are carried out under precisely the proper conditions, conjugal transfer of any given element may not be detectable. This later point is wonderfully illustrated by the Ti plasmids of *A. tumefaciens* in which conjugal transfer is detectable only when the proper signal molecules are made available to the donor.

II.A. TRANSMISSIBLE PLASMIDS OF *RHIZOBIUM*

The plasmids present in the members of the genus *Rhizobium* often are subdivided into those carrying genes essential to the processes of nodulation and nitrogen fixation, the *sym* plasmids (Hooykaas, *et al.*, 1981) and those which confer ancillary functions, or are genetically cryptic, the non-*sym* plasmids (Mercado-Blanco and Toro, 1996). Members of both classes are known to be transmissible. However, most rhizobial isolates contain several to many plasmids and, as noted above, the transmission of a given element may occur by self conjugation, or by mobilization by one of several mechanisms. Thus, in some instances the transferable plasmids were shown to be non-self conjugal; transmission being mediated by mobilization via other conjugal elements present in the donor.

To further complicate the matter, most *Rhizobium* plasmids lack easily selectable markers. In many of these studies, this problem was addressed by isolating a larger number of derivatives of the strain of interest, each harboring an independent insertion of a transposon such as Tn5. In some subset of these isolates, according to the rationale, the transposon should have inserted into one of the plasmids present in the strain; thus marking that plasmid with a selectable trait. In some of the studies pooled populations of transposon-containing cells then were used as donors in matings with appropriate recipients, followed by selection for the transfer of resistance to kanamycin. Alternatively, in other studies a representative number of independent progeny recovered from the transposition tagging experiments were independently mated with a recipient, again selecting for the transfer of the transposon marker. Using this strategy, several *sym* plasmids appear certainly to be self conjugal. These include pRL1JI (Johnston, *et al.*, 1978) and its Tn5-marked derivative, pJB5JI, as well as pRL5JI (Brewin, *et al.*, 1980), both from *R. leguminosarum* bv. *viciae*, and pSym5 from *R. leguminosarum* bv. *trifolii* (Hooykaas *et al.*, 1981). The *sym* plasmid pRP2JI of *R. leguminosarum* bv. *phaseoli* also was reported to be self-transmissible (Lamb, *et al.*, 1982), but this was not confirmed by out-matings from donors known to lack other conjugal systems. Among the non-*sym* plasmids, several have been reported to be self-conjugal. Thus, following marking with Tn5, Johnston, *et al.* (1982) showed that two plasmids from *R. leguminosarum* 1062 (a derivative of strain 300), pRL8JI and pIJ1001 most likely

are self conjugal. Neither of these plasmids codes for functions essential to nodulation or nitrogen fixation. Interestingly, both of these plasmids, which normally transfer at low frequency, can be mobilized at high frequency by the self-conjugal *sym* plasmid, pRL1JI (Johnston, *et al.*, 1982). Similarly, following tagging with a Tn5 derivative, Mercado-Blanco and Olivares (1993) showed that a ca. 170 kb cryptic plasmid of *Sinorhizobium meliloti* GR4 is self conjugal. Interestingly, this cryptic plasmid can mobilize transfer of another non-conjugal 210 kb cryptic plasmid also present in strain GR4.

A completely different strategy depends upon allowing the plant to select for transconjugants. In this approach a Sym⁺ donor is mated with a Sym⁻ recipient and the mixture is plated on medium selective only for the recipient. The colonies that arise on the selection plates are collected as a pool and inoculated to the legume host correct for the *sym* plasmid present in the donor. Any nodules that form should have been induced by recipients that inherited the *sym* plasmid from the donor. This can be verified by examining markers associated with the rhizobia recovered from the nodules. Thus, for example, Rao, *et al.* (1994) reported that four isolates of *R. leguminosarum* bv. *trifolii* transferred *nod* and *nif* host range characters to a derivative of *R. leguminosarum* bv. *trifolii* cured of its endogenous *sym* plasmid. This transmission occurred on plates and also in soil microcosms. However, the plasmids in the strains recovered from the nodules were not characterized at the molecular level.

The *sym* plasmids of *S. meliloti* present a more difficult subject. These plasmids generally are very large, usually three to ten times the size of the *sym* plasmids present in other rhizobia. These plasmids generally are not thought to be self-conjugal, although they can be mobilized by other self-conjugal elements present in the cell (Kondorosi, *et al.*, 1982). Introducing an *oriT* site from a self-conjugal plasmid such as RP4, also allows these very large plasmids to be mobilized by the appropriate conjugal helper plasmid (Pretorius-Güth, *et al.*, 1990).

Several *sym* plasmids have been described as being non-self conjugal but mobilizable by other plasmids. Thus, pSym9, the ca 400 kb *sym* plasmid of *R. leguminosarum* bv. *phaseoli* RCC3622 is non-conjugal but can be mobilized by the R plasmid RL180 (Hooykaas, *et al.*, 1985). Similarly, pSym1, the 225 kb *sym* plasmid of *R. leguminosarum* RCC1001 is not capable of self-transfer, but can be mobilized by RL180 (Hooykaas, *et al.*, 1982). And as noted above, the smaller cryptic plasmid in *S. meliloti*

GR4 apparently can mobilize transfer of the larger cryptic plasmid present in this strain.

The plethora of plasmids in donor strains, as well as the lack of suitable plasmid-free *Rhizobium* recipients, has hindered the study of the conjugal systems of *Rhizobium* plasmids. Thus, beyond the fact that they exist, until recently, virtually nothing was known concerning the conjugation systems associated with these elements.

II.B. TRANSMISSIBLE PLASMIDS OF *AGROBACTERIUM*

The Ti plasmids of *A. tumefaciens* are self-conjugal elements. However, this was not at all clear when studies involving *in vitro* matings were first initiated in the early 1970's. In fact, the first evidence that the Ti plasmids were transmissible came from *in situ* matings and these experiments were conducted before the existence of the Ti plasmids was even proved (Kerr, 1969; 1971). It wasn't until the correct conditions for activating their transfer systems were identified that the Ti plasmids were finally shown to be conjugal elements (Kerr, *et al.*, 1977; Genetello, *et al.*, 1977). Since then, the classic octopine/mannityl opine- and nopaline/agrocinopine A+B-type Ti plasmids as well as the agropine-type Ti plasmid pTiBo542 have been shown to be self-conjugal. Similarly, the newly identified chrysopine-type Ti plasmids such as pTiChry5 apparently are self-conjugal (Kovács and Pueppke, 1994; Vaudequin-Dransart, *et al.*, 1996), as are the limited host range mega-Ti plasmids present in the so-called Lippia strains of *A. tumefaciens* (Unger, *et al.*, 1985). Several Ti plasmids from the biovar 3 grapevine strains, now called *A. vitis* (Szegedi, *et al.*, 1992), also have been shown to be self-conjugal. In addition to the more well-known Ti plasmids of *Agrobacterium*, other elements, such as opine catabolic plasmids (Ellis, *et al.*, 1982b), pAgK84 which codes for production of agrocin 84 (Farrand, *et al.*, 1985) and several Tr plasmids, which confer tartrate utilization (Szegedi, *et al.*, 1992), have been reported to be transmissible. The question is more complex in *A. rhizogenes*. Most of the mannopine- and agropine-type *A. rhizogenes* strains harbor three large plasmids. In these, the two smaller plasmids cointegrate to form the largest element. It is clear from the work of Petit, *et al.* (1983) that the mannopine-type Ri plasmid pRi8196 is transmissible and probably self-conjugal. Whether the Ar plasmid in this strain is conjugal currently is unknown. It also is clear from this study that the Ri

and Ar plasmids in two agropine-type strains, A4 and 15834 are transmissible. However, whether one or both of these elements is self-conjugal has yet to be determined. Interestingly, the capacity to utilize the four mannityl opines present in hairy roots induced by these two strains is split between the two plasmids; the Ri plasmid confers growth on agropine but not mannopine and mannopinic acid, while the Ar plasmid confers growth on mannopine and mannopinic acid, but not agropine (Petit, *et al.*, 1983). Most *Agrobacterium* strains also harbor a very large plasmid, somewhere in the order of 350-450 kb in size. Until recently, there was no evidence that the large plasmid in any one isolate is the same as, or even significantly related to, the large plasmid in any other isolate. However, recent studies indicate that the large plasmids in at least some strains confer catabolism of Amadori compounds that are produced spontaneously from decaying vegetation (Kim and Farrand, 1996; Vaudequin-Dransart, *et al.*, 1995). This suggests that at least some of these plasmids may share gene systems in common. One such plasmid, pAtC58 apparently is self-conjugal (van Montagu and Schell, 1979). However, nothing is known about the nature of the conjugal transfer system associated with this plasmid.

We are not aware of any studies concerning the nature of the transfer systems of the Tr, Ar or Ri plasmids. However, some information is available concerning the conjugal transfer system of pAgK84. Insertion and deletion analysis delimited the transfer locus of pAgK84 to a ca 3 kb region (Farrand, *et al.*, 1985; Farrand, *et al.*, 1992). This is too small to code for a conventional conjugal transfer system, suggesting that transfer of pAgK84 occurs by mobilization using the conjugal functions of some other system in the *Agrobacterium* donor. Alternatively, pAgK84 may code for a novel conjugation apparatus requiring less coding capacity than conventional systems. The conjugal nature of this plasmid will be considered in more detail in Section VI.

Little is known about the transfer systems of plasmids present in members of the genus *Rhizobium*. As described above, some, but not all of the *sym* plasmids certainly are self conjugal, but the characteristics of the conjugation systems of these plasmids and the conditions in which the plasmids transfer have not yet been explored in detail. However, there is a considerable amount of information concerning conjugal transfer of plasmids present in members of the genus *Agrobacterium*. From the elegantly simple experiments demonstrating the conjugal nature of the

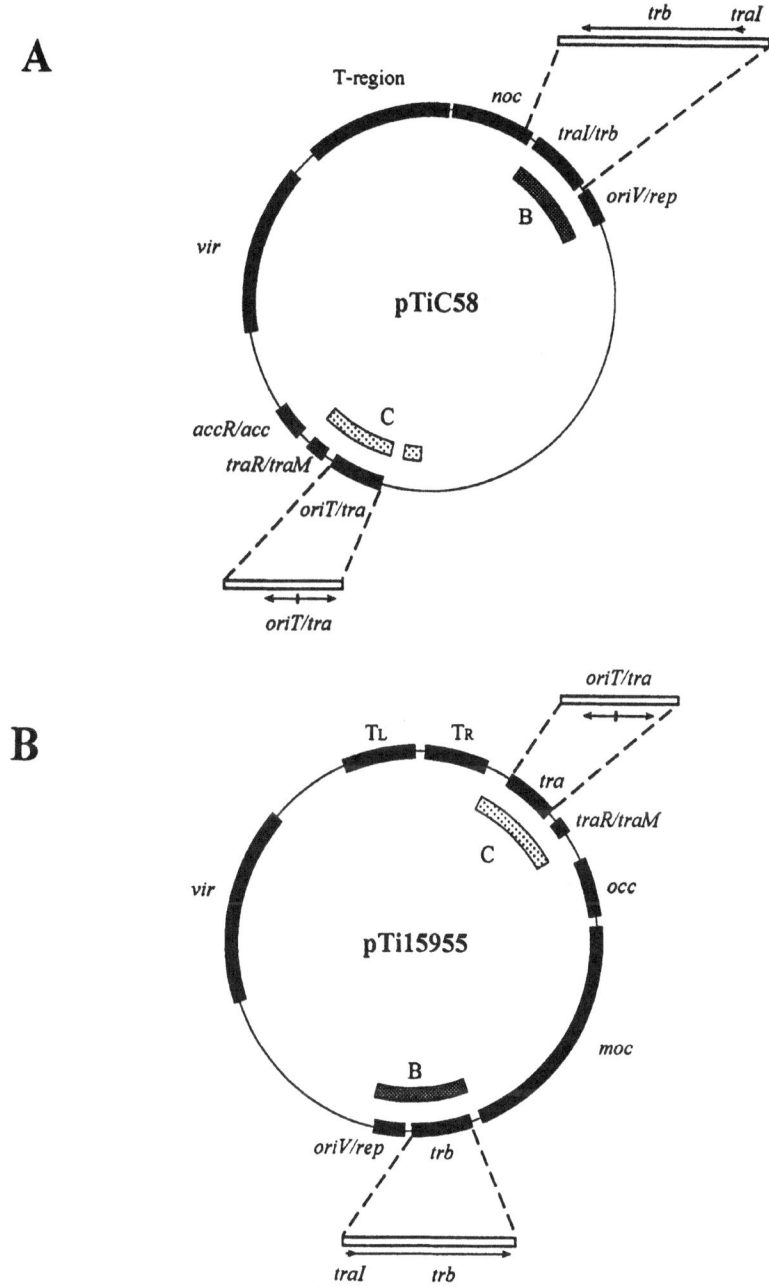

Figure 1. Physico-genetic maps of the two classical Ti plasmid types. A. pTiC58, a nopaline/agrocinopine-type Ti plasmid. B. pTi15955, an octopine/mannityl opine-type Ti plasmid. pTiA6, pTiAch5, pTiB6 and pTiR10 are virtually indistinguishable from pTi15955 at the RFLP level. Regions of known function are indicated by the thick black lines and are: T-region, transferred DNA; T_L and T_R, the left and right T-regions respectively; *noc*, nopaline catabolism; *occ*, octopine catabolism; *acc* agrocinopine A+B catabolism; *moc*, mannityl opine catabolism; *oriV/rep* the replicator region; *vir*, the virulence regulon; *traR/traM* transcriptional activator of the *tra* regulon and its antiactivator, respectively; *traI/trb*, the *trb* operon of the *tra* regulon; and *oriT/tra*, the two divergent *tra* operons flanking the *oriT* region. The basic structure of the *traI/trb* and *oriT/tra* regions are shown as blow-ups. The stippled arcs labeled B and C denote two of the four regions of close relatedness between the two Ti plasmids as determined by the heteroduplex analysis of Engler, *et al.* (1981).

Ti plasmids conducted more than 20 years ago by Kerr's group to the complex studies being carried out today concerning the structure and regulation of the Ti plasmid transfer system at the molecular level, these elements are becoming the paradigm for understanding the mechanisms and implications of conjugal gene transfer among bacteria.

III. Conjugal Transfer Systems of *Agrobacterium* Ti Plasmids

The Ti plasmids are notable in that they encode two functionally discrete conjugal transfer systems. One has evolved to translocate a segment of the Ti plasmid from the bacterium to plant cells. The second has retained its role in mediating the transfer of the entire plasmid from a bacterial donor to a bacterial recipient. The composition of each of these conjugation systems now is known at the DNA sequence level. The two systems are completely independent; *vir*-mediate transfer of T-DNA to plants does not require *tra* functions (Rogowsky, *et al.*, 1990) and the *tra*-mediated conjugal transfer of these plasmids to bacterial hosts does not require any functions encoded by *vir* (Cook, *et al.*, 1997). Moreover, studies on phylogenetic relatedness indicate that the two systems derived from different ancestors, although some reassortment of genetic determinants between the two primordial systems may have occurred.

III.A. STRUCTURE OF THE Ti PLASMID TRANSFER SYSTEM

Conjugal transfer functions of bacterial plasmids are conceptually divided into those associated with DNA metabolism, the Dtr functions, and those associated with the transport of the plasmid from donor to recipient, the Mpf functions. The Dtr system is comprised of the *cis*-acting *oriT* site at which conjugal transfer initiates, as well as the genes coding for the proteins responsible for the DNA processing reactions. The proteins interact with the DNA molecule at *oriT* to form the relaxosome, a protein-DNA complex at which the enzymology of strand nicking and processing occurs. The Mpf system is a complex, largely uncharacterized structure that most likely spans the inner and outer membranes of the donor bacterium. This structure forms the gated channel through which the DNA intermediate is transferred, as well as the structures that are involved in recognizing and forming the attachment with the

recipient. Mpf functions also include those involved in the actual fusion that presumably is required to form the transfer pore. It is not known if these functions of recognition, attachment, fusion and translocation are combined in one structure, or are divided among two or more structures that work independently, or which interact with each other.

The genetic structure of the conjugal transfer systems of the nopaline/agrocinopine A+B- and octopine/mannityl opine-type Ti plasmids now has been characterized at the gene level. The genes of the conjugal transfer systems of these two Ti plasmid types are very closely related and clearly they have derived from a common origin. Moreover, Dtr and Mpf functions of the two Ti plasmids are interchangeable (Cook and Farrand, 1992; Cook, *et al.*, 1997), which speaks to a conservation of function between the two Ti plasmid types.

Mutational analyses suggested that the conjugal transfer functions of both Ti plasmid types are located in two gene clusters (Holsters, *et al.*, 1980; DeGreve, *et al.*, 1981; Beck von Bodman, *et al.*, 1989). This proved to be the case. One set of genes, encoding the Dtr functions, is linked to the locus coding for catabolism of the conjugal opine; octopine for pTiR10 and agrocinopines A+B for pTiC58 (Figure 1). This gene set corresponds to homology region C of Engler, *et al.* (1981). The second set, representing the Mpf functions, is closely linked to the replication regions of the two Ti plasmids and corresponds to homology region B.

III.A.1. The tra Operons

The genes coding for the DNA metabolism functions of both Ti plasmid types are organized as two operons divergently transcribed from an intergenic region that contains the *oriT* (Figure 2A). On pTiC58, one operon, expressed in the clockwise direction, codes for three genes, *traA*, *traF* and *traB*. The second, anticlockwise-reading operon also codes for three genes, *traC*, *traD* and *traG* (Farrand, *et al.*, 1996).

The minimum *oriT* sequence is contained within a 65 bp region located between the two *tra* operons (Cook and Farrand, 1992). This sequence contains a 12 bp domain that is almost identical to the *nic* domain of the broad host range IncQ plasmid RSF1010. The region also contains an inverted repeat, which is a common feature of *oriT* sites. However, the IR is not related to that present adjacent to the *nic* site of RSF1010. While this 65 bp domain represents the minimal unit, conjugal transfer at wild-type frequencies requires additional *cis*-active inverted

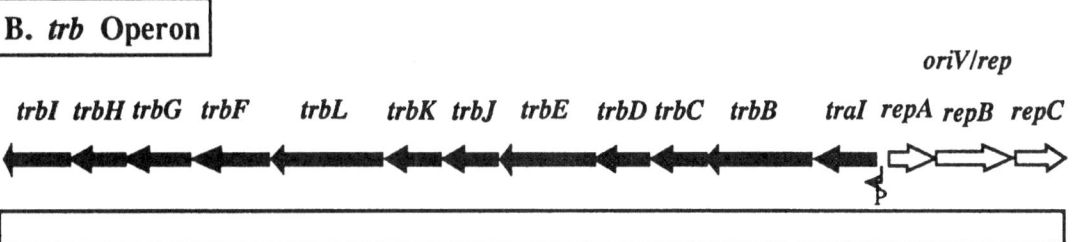

Figure 2. Genetic structures of the components of the Ti plasmid *tra* regulon. A. The two divergently transcribed *tra* operons. The *oriT* with its *nic* site, as well as the divergent promoter (p) regulated by TraR is located in the interval between the *traCDG* and the *traAFB* operons. B. The *trb* operon. On both Ti plasmid types this operon is closely linked to and divergently transcribed from the three genes of the replication region. The first gene of the operon, *traI*, codes for production of *Agrobacterium* autoinducer. The remaining *trb* genes are required for the Mpf functions.

repeat elements located both to the left and the right of the core sequence (Cook and Farrand, 1992). This dependence upon sequence elements able to impart secondary structure to the DNA molecule is common within regions with *oriT* activity. Interestingly, while the 12 bp domain of pTiC58 is completely conserved in the *oriT* region of pTiR10, the nucleotide sequences of the flanking regions, which contain the IR structures, differ significantly between the two Ti plasmids (Alt-Mörbe, *et al.*, 1996). Yet the 65 bp minimum active *oriT* domain of pTiC58 allows mobilization by the *trans*-acting conjugal transfer functions of the octopine/mannityl opine-type Ti plasmid (Cook and Farrand, 1992).

Genetic analysis and considerations based on sequence suggest that *traA*, *traF*, *traC*, *traD* and *traG* are required for conjugation. On the other hand, *traB*, although necessary for maximum transfer efficiency, is not essential (Farrand, *et al.*, 1996). The function of several of the encoded products, while yet to be shown

biochemically, can be inferred from their predicted amino acid sequences. TraA, a 123 kD protein, contains a domain that is conserved in site-specific single stranded endonucleases from other conjugal plasmids (Farrand, *et al.*, 1996; Alt-Mörbe, *et al.*, 1996). These enzymes are responsible for catalyzing the nick at their respective *oriT* sites from which conjugal transfer initiates. TraA also contains a domain related to the helicase domain of TraI, the *oriT*-specific single-stranded endonuclease from plasmid F. Presumably this activity is involved in unwinding the replicating DNA molecule during the strand transfer process. These similarities suggest that TraA serves as the site-specific single strand endonuclease responsible for introducing the nick within the *oriT* region of the Ti plasmid. Such an assignment awaits confirmation by biochemical analyses. TraC and TraD, both of which are small proteins, are not related to any proteins in the data bases. The Dtr regions of transfer systems from other conjugal plasmids code for similarly small proteins that function in relaxosome assembly and stability (Pansegrau, *et al.*, 1990; Scherzinger, *et al.*, 1992). TraC and TraD may play a similar role in the relaxosome of the Ti plasmid. TraF is related to a protein of the same name encoded by the *tra* system of the IncP1α plasmid RP4 (Farrand, *et al.*, 1997). In RP4, TraF is not part of the Dtr system, but is required for a functional Mpf (Waters, *et al.*, 1992). The significance of this protein, as well as its role in assembly of the Mpf, will be discussed later in this section. TraB is not significantly related to any protein sequence in the data bases and, as described above, it is not essential for conjugal transfer of pTiC58.

There is a duplication of a contiguous DNA sequence beginning in the 3′ coding region of *traA*, extending through *traF* and terminating in the 5′ portion of the coding region of *traB* that is located between *virE2* and *virF* in the *vir* region of the octopine/mannityl opine-type Ti plasmids (Farrand, *et al.*, 1996). The conservation at the nucleotide sequence level is quite high, especially for the *traF* gene. However, mutational analysis suggests that this region of the *vir* regulon is not involved in tumorigenesis (Melchers, *et al.*, 1990). Furthermore, the duplication apparently is not present in the *vir* region of pTiC58. Although it is conceivable that the interval has some function, it is more likely that the duplicated region is non-functional and represents sequences retained from some past recombinational or transpositional event.

TraG is of considerable interest. Virtually all conjugal transfer systems examined to date contain homologs of this protein. These include TraG of RP4, TrwB of R388, TraD of F and VirD4 of the Ti and Ri plasmid *vir* systems (Lessl, *et al.*, 1992). Where examined, these proteins are essential to their respective DNA transfer systems. In RP4, while TraG is required for conjugation, it is not, *per se*, a component of the relaxosome or of the mating pore. However, TraG of RP4 and TraG and VirD4 of the Ti and Ri plasmids contain putative membrane spanning domains. Consistent with this, VirD4 from pRiA4 fractionates to the membrane compartment (Okamoto, *et al.*, 1991). Lessl, *et al.* (1993) have suggested that TraG of RP4 functions to interface the relaxosome with the mating bridge assembly during active export of the DNA transfer intermediate to the recipient cell. Genetic analysis indicates that TraG of pTiC58 is essential for conjugation (Farrand, *et al.*, 1996) and given its relatedness, may serve a function similar to that of TraG in the RP4 conjugal transfer system.

Analysis using *lacZ* fusion transposons suggests that the two *tra* operons of pTiC58 each contain downstream genes that are expressed as part of these transcriptional units. However, the insertions forming these fusions have no effect on conjugal transfer of the full sized Ti plasmid (Farrand, *et al.*, 1996). Sequence analysis of these regions in pTiR10 identified several potential open reading frames; one downstream of *traG* and four downstream of *traB* (Alt-Mörbe, *et al.*, 1996). However, these putative genes and their encoded products are not related to sequences present in the data bases. This gene organization is reminiscent of that of the *tra* region of RP4; several of the distal genes of the leader and primase operons are not required for conjugal transfer of the R plasmid under any of the conditions tested (Pansegrau, *et al.*, 1994).

III.A.2. The trb Operon

Genes presumably responsible for forming the mating bridge assembly are organized as a single cluster that is tightly linked to the *oriV/rep* region of both pTiR10 and pTiC58 (Alt-Mörbe, *et al*, 1996; Li and Farrand, unpublished). This cluster, called *trb*, could code for 12 genes (Figure 2B). The first gene, *traI* codes for the *Agrobacterium* autoinducer synthase and will be discussed in Section IV. The remaining 11 genes, called *trbB* through *trbL*, are homologs of genes of the same names located in the *trb* operon of RP4 (Alt-Mörbe, *et al.*, 1996). That they code for products that are associated with the mating bridges derives from work with the *trb* region of RP4. These genes are required for conjugal transfer of the R plasmid as well as for surface-associated functions such as entry exclusion (Haase, *et al.*, 1996) and sensitivity to infection by RP4-specific bacteriophages (Grahn, *et al.*, 1997). Furthermore, like their RP4 counterparts, many of the putative translational products of the Ti plasmid *trb* genes contain N-terminal secretion signals and several contain putative membrane spanning domains. Although the 11 *trb* genes of the Ti plasmids code for products that are closely related to their counterparts from RP4, the gene orders are somewhat different. *trbJKL* which comprise the last three essential genes of the RP4 *trb* operon are located between *trbE* and *trbF* of the Ti plasmid *trb* locus.

Each of a series of six Tn5 insertion mutations mapping to the *trb* region of pTiC58 abolishes conjugal transfer (Beck von Bodman, *et al.* 1989). In more recent experiments, Li and Farrand (unpublished results) have constructed non-polar insertion mutations in all 11 *trb* ORFs. Insertions in all but *trbK* have profound negative effects on conjugal transfer of the Ti plasmid. This is consistent with similar studies using RP4. With the exception of *trbK* and the distal genes *trbM*, *trbN*, *trbO* and *trbP*, for which there are no homologs in the Ti plasmid system, all of the *trb* genes are essential for conjugal transfer of the R plasmid (Lessl, *et al.*, 1993; Haase, *et al.*, 1995).

The *trb* regions of both Ti plasmids may be expressed as a single unit from a promoter located just upstream of *traI*. However, complementation analysis of the Tn5 insertion mutations described above suggests that the *trb* region of pTiC58 may be organized as at least two transcriptional units (Beck von Bodman, *et al.*, 1989). More recent analysis is consistent with this interpretation, but the location of the second promoter and its regulatory characteristics, remain to be determined (Li and Farrand, unpublished results). Similar analyses have not been reported for the *trb* region of octopine/mannityl opine-type Ti plasmids.

The genetic and sequence data strongly suggest that the *trb* regions of pTiC58 and pTiR10 code for the conjugal mating bridge. However, to date no biochemical studies have been reported concerning the nature of the proteins coded for by this region, or the location of these proteins in the soluble, or inner or outer membrane compartments of the donor cell. There also is the question of sex pili. While there is uniform agreement that RP4 codes for a pilus, the physical manifestation of this structure has proved difficult to characterize at the biochemical level (Haase, *et al.*, 1995; E. Lanka, personal communication). Based on sequence considerations, it is believed that *trbC* codes for the pilus structural protein of RP4. Furthermore, TraF, which apparently has protease activity, can cleave TrbC protein (Haase and Lanka, 1997). Such post-translational processing is common in the maturation of pilus structural proteins and the assembly of conjugal pili (Silverman, 1997). Given that the *tra*/*trb* gene sets contain homologs of *traF* and *trbC*, it is reasonable to expect that the Ti plasmids produce a sex pilus. However, attempts to detect pili associated with the conjugal transfer system of pTiC58 by electron microscopy have not been successful (E. Lanka, personal communication).

Finally, the location of *trb* on the map of the octopine/mannityl opine-type Ti plasmids accounts for a peculiarity exhibited by one of the more well-studied strains of *A. tumefaciens*. Strain A348, which has been worked on extensively by the Nester group, is a laboratory construct in which pTiA6NC, taken from a non-clumping variant of stain A6, was introduced into A136 (Knauf and Nester, 1982). A136, in turn, is a rifampicin- and nalidixic acid-resistant mutant of

NT1, a Ti plasmid-cured derivative of the nopaline/agrocinopine-type strain C58 (Watson, *et al.*, 1975). Thus, A348 is of the C58 chromosomal lineage. The Ti plasmid in this strain is completely non-conjugal, even under optimum conditions of induction by octopine (see Section IV). Sciaky, *et al.* (1977) showed that pTiA6NC contains a *ca.* 12 kb deletion mapping to the 5 o'clock region of the plasmid. In their sequence analysis of the *trb* region of the octopine/mannityl opine-type Ti plasmids, Alt-Mörbe, *et al.* (1996) showed that this deletion removes most of *trbI*, the last gene of the *trb* operon, along with a portion of the region required for catabolism of mannopinic acid. This deletion most likely accounts for the transfer-deficient character of this plasmid. Interestingly, other stock cultures of strain A6 contain conjugation-proficient Ti plasmids (Zhang and Kerr, 1991), suggesting that the deletion occurred after the dissemination of this strain to various laboratories around the world.

III.B. FUNCTION AND PHYLOGENY

As described above, genetic and sequence analysis suggest that genes required for conjugal transfer of the Ti plasmid are located in two discrete clusters, *tra* and *trb*. However, it is conceivable that additional functions required for conjugation are located at other sites on these plasmids. In fact, two reports in the literature indicate that conjugal transfer also requires functions provided by the *vir* regulon. Insertion mutations in several *vir* genes, in addition to abolishing tumorigenicity, drastically decreased conjugal transfer frequencies of the mutant Ti plasmids (Gelvin and Habeck, 1990; Steck and Kado, 1990). Furthermore, normal transfer frequencies were restored when the mutations in *vir* were complemented *in trans* with the wild-type alleles.

More recent studies using a binary *tra* system indicate that *vir* does not contribute to conjugal transfer (Cook, *et al.*, 1997). In this work a system was constructed in which *tra*/*oriT* is contained on one plasmid while *trb* is contained on a second. The conjugal functions encoded by the two independent replicons catalyze transfer of that replicon containing the Ti plasmid *oriT* site. Transfer of the *oriT* plasmid occurred at frequencies indistinguishable from that of the intact Ti plasmid in the absence of any components of the *vir* regulon. Cook, *et al.* (1997) suggested that the influence of the mutations in *vir* on conjugal transfer was due to dominant-negative effects of the altered *vir* gene products on the conjugal transfer system. This explanation is consistent with the observation that the influence of *vir* mutations on conjugation is allele-specific; some insertion mutations in a given *vir* gene affect conjugation frequencies while others in the same gene do not.

Analyses using the binary system also confirmed the non-essential nature of the genes mapping downstream of *traG* and of *traB* in the two *tra* operons. Moreover, using two different *tra/oriT* constructs, Cook, *et al.* (1997) confirmed the observation that *traB*, while required for maximum transfer frequencies, is not essential for conjugation.

Comparative sequence analysis of the *tra* systems from nopaline/agrocinopine A+B- and octopine/ mannityl opine-type Ti plasmids indicates that the two are closely related (Farrand, *et al.*, 1996; Alt-Mörbe, *et al.*, 1996; Li and Farrand, unpublished). This is consistent with the observation that a plasmid harboring the *oriT* region from pTiC58 can be mobilized by the *tra* system of pTi15955 and that the *trb* system of pTi15955 can function with the *tra/oriT* components of pTiC58 (Cook and Farrand, 1992; Cook, *et al.*, 1997). Clearly the *tra* and *trb* systems of the two Ti plasmids are functionally interchangeable. However, while they are related, they are not identical. In fact, there is remarkable divergence between the Tra systems of the two Ti plasmid types, especially in the protein products of some of the *trb* genes. While most give pairwise relatedness values ranging from 80-90%, others, such as TrbD, TrbH and especially TrbK, are significantly less conserved (Li and Farrand, unpublished results).

Comparative sequence analysis also sheds light on the origins of the Ti plasmid conjugal transfer system. Clearly this system is chimeric (Table 1). The *oriT* shares a common lineage with that of the broad host range, mobilizable IncQ R plasmid RSF1010 (Cook and Farrand, 1992). Consistent with this, a domain of TraA, the putative *oriT* nicking enzyme, is related to MobA, the *oriT* nicking enzyme of RSF1010 (Farrand, *et al.*, 1996; Alt-Mörbe, *et al.*, 1996). This is a widely disseminated *oriT*-nickase system. Related *oriT* regions are found in diverse plasmids from both Gram-negative and Gram-positive bacteria including pTF1, pSC101, pG01 and pIP501 (Climo, *et al.*, 1996). However, as described above, TraA of the Ti plasmid contains a second domain showing sequence relatedness to the helicase domain of TraI, the *oriT* nicking enzyme from plasmid F (Farrand, *et al.*, 1996; Alt-Mörbe, *et al.*, 1996). This domain is not present in the MobA proteins of RSF1010 or the otherwise unrelated plasmid pTF1. It is conceivable

that *traA* arose by a fusion of a *mobA*-like gene with a gene encoding a helicase, or a helicase domain such as *traI* of F.

Although the primary Dtr functions of the Ti plasmid *tra* system resemble those of RSF1010, the Mpf functions are coded for by genes that share a common lineage with those of the RP4 conjugal transfer system. Thus the *trb* genes and also *traF* of the Ti plasmid show strong relatedness to those of RP4, although the gene order is somewhat altered. In addition, *traG*, the gene that most likely confers specificity between the relaxosome and the mating bridge, is related to that of RP4. Interestingly, this gene, in the form of *virD4*, also is conserved in the Ti plasmid *vir* system. Despite these similarities, the Mpf functions of the Ti plasmid apparently have diverged significantly from those of RP4. The Ti plasmid will not mobilize transfer of a plasmid carrying the *oriT* of RP4 (Cook and Farrand, 1992). Nor will RP4 mobilize a recombinant plasmid containing the *tra/oriT* region of pTiC58. In addition, the Ti plasmid does not confer susceptibility to PRR1 or PRD1, two bacteriophages that are dependent upon the RP4 Mpf system for attachment and infection (Li and Farrand, unpublished results; see Section V).

Taken together, the *tra* system of the nopaline/ agrocinopine A+B- and octopine/mannityl opine-type Ti plasmids contains components related to the transfer systems of RSF1010 and RP4. This chimerism is reminiscent of that observed for the *vir* system. The Dtr functions of *vir*, ie, T-region borders and the nicking enzyme, VirD2, are of the same lineage as those of RP4 (Pansegrau and Lanka, 1991), while the Mpf system, ie *virB* shares a common ancestry with that of the IncN plasmid pKM101 (Pohlman, *et al.*, 1994). Interestingly, in both *tra* and *vir*, the protein believed responsible for bridging the relaxosome complex with the mating pore assembly shares ancestry with that of the IncP1 transfer system. Thus, both TraG of the Ti plasmid *tra* system and VirD4 of the T-DNA transfer system are most closely related to TraG of RP4 (Lessl, *et al.*, 1992; Farrand, *et al.*, 1996).

This chimerism manifests itself in some interesting specificity characteristics. For example, RSF1010 contains its own Dtr system in the form of an *oriT* and the MobA, MobB and MobC components of the relaxosome (Scholtz, *et al.*, 1989; Scherzinger, *et al.*, 1992). However, the IncQ plasmid lacks an Mpf system and does not code for a TraG-like function. Thus, this plasmid is non-self conjugal, but it can be mobilized by RP4. The IncP1 plasmid provides the Mpf functions. But more importantly, TraG of RP4

Ti plasmid	Homologous system	Known or proposed function
OriT	RSF1010/R1162 pSC101 pTF1 pG01 pIP501	Site for initiation of transfer
DNA metabolism functions TraA: N-terminal domain TraA: C-terminal domain	 MobA of RSF1010 and R1162 C-terminal domain of TraI of F	 *oriT*-specific endonuclease Helicase
Mating bridge Trb proteins TrbK TraF	 Trb proteins of RP4 TrbK of RP4 TraF of RP4	 Mating bridge/pilus Entry exclusion Protease, prepilin protein processing
Relaxosome/mating bridge TraG	 TraG of RP4 VirD4 of Ti and Ri plasmid *vir*	 Interface relaxosome to mating bridge

Table 1. Phylogenetic relationships of Ti plasmid *tra* system components

recognizes the relaxosome of RSF1010 and can interface the nucleoprotein complex to the RP4 mating pore. Interestingly, the Ti plasmid *vir* system also can mobilize transfer of RSF1010 to both plant and bacterial recipients (Buchanan-Wollaston, *et al.*, 1989;Beijersbergen, *et al.*, 1992). This implies that VirD4 also recognizes the RSF1010 relaxosome and can interface the plasmid transfer intermediate to the VirB mating pore. Surprisingly, although the *oriT* and putative nickase of the Ti plasmid *tra* system share a common ancestry with those of RSF1010, the Ti plasmid cannot mobilize transfer of the IncQ R plasmid (Cook and Farrand, 1992). This, in spite of the fact that the Mpf system of the Ti plasmid is closely related to that of RP4 which, as noted above, can mobilize RSF1010. This set of paradoxical observations was clarified by the recent report that the inability of the Ti plasmid *tra* system to mobilize RSF1010 is due to the specificity conferred by TraG. Thus, when the native TraG of pTiC58 was replaced by TraG from RP4, the Ti plasmid gained the capacity to mobilize transfer of RSF1010 (Hamilton, *et al.*, 1996). Presumably, TraG of the Ti plasmid no longer recognizes the relaxosome of RSF1010. But, these results also indicate that TraG of RP4 does recognize the mating bridge of the Ti plasmid *tra* system. Moreover, *traG* of RP4 complements a Ti plasmid *traG* mutation indicating that the RP4 protein can recognize the Ti plasmid relaxosome (Hamilton, *et al.*, 1996). Thus, while *traG* of the Ti plasmid *tra* system

evidently has diverged to a point of high specificity, the RP4 homolog has retained a broader degree of relaxosome recognition.

III.C. TRANSFER SYSTEMS OF OTHER LARGE PLASMIDS

Until recently, no detailed information was available concerning the *tra* systems of other large plasmids found in members of the family *Rhizobiaceae*. However, the entire sequence of the *sym* plasmid from *Rhizobium* sp. NGR234 has now been reported (Freiberg, *et al*, 1997). This 536 kb plasmid contains a set of *tra* genes that are remarkably similar in organization and derived amino acid sequences of their gene products to those of the nopaline/ agrocinopine A+B- and octopine/mannityl opine-type Ti plasmids. The core *oriT* region is identical to that of pTiC58 and the key inverted repeat adjacent to the core region is almost perfectly conserved between the two plasmid types. Moreover, the *trb* operon contains a homolog of the *traI* gene and the operon is directly linked to the divergently-transcribed plasmid replicator region in a manner similar to that observed in the two Ti plasmids. Consistent with this, Freiberg, *et al.* (1997) have reported that pNGR234a is self-conjugal. However, within the *trb* operon of pNGR234a *trbE* is divided into two open reading frames. Either the function conferred by this gene can be supplied by two proteins, or there is an error in the DNA sequence of this region of the *sym* plasmid. Regardless, clearly, pNGR234a and the two Ti plasmid types share a recent common origin at least with respect to the core functions of replication and conjugation. Interestingly, while the *trb/oriV-rep* and the *tra/oriT* regions are closely linked on pNGR234a, these two groups of genes are separated by almost 100 kb on the two Ti plasmid types.

The *tra* and *trb* loci most likely also are present in pAtK84b, a 173 kb plasmid conferring catabolism of nopaline and agrocinopines A and B. This plasmid, which is present in the biovar 2 non-tumorigenic *A. radiobacter* isolate K84, is self-conjugal and, like the Ti plasmids, conjugation is inducible by a particular set of opines (see Section VI). Hybridization analysis indicates that much of pAtK84b is closely related to segments of pTiC58, including the *tra/oriT* and *trb/oriV-rep* regions (Clare, *et al.*, 1990). However, pAtK84b lacks detectable relatedness with the *vir* and T-regions of pTiC58. The overall relatedness between the two elements prompted Clare, *et al.* (1990) to suggest that pAtK84b arose from a pTiC58-like Ti plasmid following deletions that removed the *vir* and T-region. If such is the case, it is perhaps not surprising that this plasmid should code for Ti plasmid-like *tra* functions.

Similarly, the *trb/oriV-rep* and *tra/oriT* regions of pTiB₆806, an octopine/mannityl opine-type Ti plasmid, hybridized strongly with other *Agrobacterium* plasmids including pTiT37, pTiK27, pTi223, pTiIIBV7 and the opine catabolic plasmid, pAtAg19b (Drummond and Chilton, 1978). pTiT37 and pTiK27 both are known to be self-conjugal (Ellis, *et al.*, 1982b). In another set of studies, Otten's group identified strong relatedness between regions corresponding to the *trb/oriV-rep* and *tra/oriT* regions of both pTiC58 and pTiAch5 and the wide host range, octopine/cucumopine- and nopaline-type Ti plasmids present in *Agrobacterium vitis* (Otten, *et al.*, 1993; Otten and De Ruffray, 1994). In fact, a large sector of these two *A. vitis* Ti plasmids which contains regions of homology to both *tra* loci is co-linear between the two and also with the region of pTiC58 mapping clockwise from *trb/oriV-rep* to *tra/oriT*. This region, common to the three otherwise distinct Ti plasmids, must have derived from a common ancestor.

The issue is less clear with other Ti plasmids. While there does appear to be some relatedness between the *trb/oriV-rep* region of pTiB₆806 and the agropine-type Ti plasmid pTiBo542, the hybridization signals are weak (Drummond and Chilton, 1978). Moreover, pTiBo542 exhibits aberrant incompatibility properties

with pTiAch5 suggesting that its replicator region, if at one time related to that of pTiAch5, has diverged from that of the IncRh1-type Ti plasmids (Komari, *et al.*, 1986). However, pTiBo542 is self-conjugal and conjugation is induced by an opine, which is a hallmark of Ti plasmid conjugation systems (Ellis, *et al.*, 1982a; Ellis, *et al.*, 1982b; see Section IV).

Virtually nothing is known concerning the conjugal transfer systems of Ri plasmids. As noted in Section II, several groups have shown that the classical agropine- and mannopine-type Ri plasmids can be transferred from donors to a suitable recipient (Moore, *et al.*, 1979; Petit, *et al.*, 1983). However, in the most well characterized agropine- and mannopine-type isolates, the Ri plasmid generally co-exists and readily forms cointegrates with another large plasmid also present in the bacterium (White and Nester, 1980; Petit, *et al.*, 1983). The additional plasmids in the agropine-type isolates code for catabolism of several of the opines found in the hairy roots induced by these strains while that in the mannopine-type strain 8196 has no known function (Petit, *et al.*, 1983). Analysis of progeny from matings suggests that the mannopine-type Ri plasmid may be self-conjugal. However, similar analyses of matings involving agropine-type donors are more difficult to interpret. In fact, the co-inheritance data of Petit, *et al.* (1983) is most consistent with a model in which the opine-catabolic plasmid in these strains mobilizes the transfer of the Ri plasmid. Consistent with this, Southern analysis indicates only weak relatedness between the *trb/oriV-rep* region of pTiB₆806 and the agropine-type Ri plasmid pHRI (Jouanin, 1984). But no relatedness was reported between the Ti plasmid *tra/oriT* region and this Ri plasmid. Further work will be required to determine the conjugal properties of the plasmids present in these types of *A. rhizogenes* isolates.

With the exception of the sequence information concerning pNGR234a described above, no details are available concerning the *tra* systems of plasmids from *Rhizobium* species. However, Prakash and Schilperoort (1982), in their examination of regions of relatedness between Ti plasmids and several *Rhizobium sym* plasmids, reported that pSym9 from *R. leguminosarum* bv. *phaseoli* and pSym5, from *R. leguminosarum* bv. *trifolii* contain regions that hybridize strongly with the *tra/oriT* and *trb/oriV-rep* regions of pTiAch5. pSym5 is self-transmissible (Hooykaas *et al.*, 1981), suggesting that, like pNGR234a, this *sym* plasmid contains a functional Ti plasmid-like replication-conjugal transfer core region.

IV. Regulation of Ti Plasmid Conjugation

The conjugal transfer system of Ti plasmids is interesting from a second perspective. We have known for many years that conjugation of several Ti plasmid types is strongly regulated and that Ti plasmid transfer responds to environmental cues produced by the tumors initiated by the *Agrobacterium* strain. Subsets of opines produced by these tumors serve to induce the conjugal transfer apparatus. This level of gene regulation is mediated by the transcriptional regulator cognate to the particular opine catabolism system encoded by the Ti plasmid. However, while the opines control expression of their catabolic systems directly, they regulate conjugation in an indirect manner. Expression of the *tra* genes is regulated directly by a transcriptional activator and a small, diffusible second messenger molecule produced by the donor population. This system is one of a growing family of regulatory signalling systems that utilize the accumulation of *N*-acyl-homoserine lactones (acyl-HSL's), produced by the bacteria themselves, to induce specific sets of response genes. These signalling systems function to sense the density of the bacterial population. In *Agrobacterium*, this quorum-sensing system serves to initiate expression of the *tra* genes only when the donor population has reached some threshold level. Thus, the phenomenon of conjugal transfer is regulated by two sets of conditions; the proximity of a source of specific nutrients, that is the opines, and the density of the donor population. Regulation of conjugation by the opines and by the quorum-sensing system is linked by the simple expediency of placing the expression of the *tra* transcriptional activator under the control of the opine-responsive gene regulator. Intriguingly, while the strategy for accomplishing this is the same in nopaline/agrocinopine A+B- and octopine/mannityl opine-type Ti plasmids, the genetic organization responsible for this cascade of control mechanisms is completely different between the two Ti plasmid types. Thus, the two systems arose independently, suggesting that it is important to the biology of *Agrobacterium* or the Ti plasmids themselves, that conjugation fall under the umbrella of the opine regulon. This, in turn, emphasizes the importance of opines to the biology of *Agrobacterium* and speaks to the elegance of the interaction that has evolved between *Agrobacterium* and its host plants.

IV.A. MOLECULAR MECHANISMS

IV.A.1. Regulation by Opines

Our recognition that certain opines are required to induce conjugation traces back to the observation of Kerr (1969; 1971) that virulence could be transferred from a pathogen to a non-pathogenic *Agrobacterium* recipient, but only when the two were grown together

Plasmid type	Associated opines	Representative plasmids	Conjugal opine
Nopaline	Nopaline, Agrocinopines A+B	pTiC58 pTiK27 pTiT37	Agrocinopines A+B Agrocinopines A+B Constititutive[1]
Octopine	Octopine, Mannityl opines	pTi15955 pTiA6 pTiR10 pTiAch5 pTiB6	Octopine Octopine Octopine Octopine Octopine
Agropine	Mannityl opines, Succinamopine, Agrocinopines C+D	pTiBo542	Agrocinopines C+D
Opine-catabolic	Nopaline, Agrocinopines A+B	pAtK84b pAtK112	Agrocinopines A+B and Nopaline

Table 2. Conjugal Opines of *Agrobacterium* Plasmids

[1] Transfer of pTiT37 at a low frequency does not require induction by opines, but pregrowth of the donor with agrocinopines A or B results in a considerable increase in transfer frequency (Ellis, *et al.*, 1983b).

on crown gall tumors. Attempts to transfer virulence on standard laboratory medium generally were unsuccessful. In fact, this work preceded the discovery of the Ti plasmids as physical entities. The actual demonstration of the conjugal nature of the Ti plasmids and that transfer is induced by tumor-specific signals, resulted from the elegant experiments of Kerr *et al.*(1977) and of Genetello *et al.* (1977) showing that transfer of the Ti plasmid was dependent upon pre-growth of the donor on medium containing the appropriate opine. There then followed a number of genetic studies exploring this phenomenon (reviewed in Farrand, 1993), the conclusions of which can be summarized as follows.

1. For a given Ti plasmid, generally only one opine type would serve as the conjugal inducer. Thus, conjugal transfer of the nopaline/agrocinopine A+B-type Ti plasmid pTiC58 is inducible by agrocinopines A and B, but not by nopaline, or by any other opine family tested (Ellis, *et al.*, 1982b). On the other hand, conjugal transfer of octopine/mannityl opine-type Ti plasmids, such as pTi15955, pTiR10 and pTiB6, is inducible by octopine (or other members of this family such as lysopine) but not by the mannityl opines or by any other opines tested.

2. The conjugal opine also induced expression of the gene set required for uptake and catabolism of that opine. Thus it was proposed early on that conjugation and catabolism of the conjugal opine are co-regulated by a single transcriptional effector (Petit, *et al.*, 1978; Klapwijk, *et al.*, 1978). This was entirely consistent with several studies showing that mutants isolated as being constitutive for one trait often exhibited constitutive expression of the other (Petit and Tempé, 1978; Petit, *et al.*, 1978; Klapwijk and Schilperoort, 1979).

To date, only four of the eight identified opine families are known to induce conjugal transfer. These include the octopine and nopaline families of arginyl opines and the two phosphodiester opine classes, agrocinopines A and B and agrocinopines C and D. Furthermore, although most Ti plasmids code for the production and utilization of two or more opine families, usually only one serves as the conjugal signal. Given this, the known Ti plasmids can be grouped into one of several classes depending upon the specific opine that induces conjugal transfer (Table 2). There are exceptions, however. pAtK84b, the Ti plasmid-like opine catabolic element present in *A. radiobacter* strain K84 can be induced for conjugal transfer by nopaline as well as by agrocinopines A and B (Ellis, *et al.*, 1982b). Similarly, the uncharacterized

plasmid responsible for catabolism of opines by *A. radiobacter* strain K112 can be induced for conjugation by these same two sets of opines.

Over the last several years the nature of this coordinate control by opines has been established at the molecular level for the nopaline/agrocinopine- and the octopine/mannityl opine-type Ti plasmids. Detailed descriptions of the relevant opine catabolic systems are presented in Chapter 9. Thus, we present only a summary of the characteristics of these two gene systems.

Early reports indicated that nopaline induced conjugation of the classical nopaline/agrocinopine-type Ti plasmids such as pTiC58 and pTiK27 (Kerr, *et al.*, 1977; Petit, *et al.*, 1978). This proved to be incorrect. Rather, either of the two sugar phosphodiester opines, agrocinopines A or B (see Figure 3A), serve as the conjugal opines for this class of Ti plasmids (Ellis, *et al.*, 1982a; Ellis, *et al.*, 1982b). Moreover, these two opines can be utilized as sole source of carbon and energy by agrobacteria harboring the appropriate Ti plasmid. The *acc* operon, which is responsible for the transport and catabolism of these two opines is located at about 8 o'clock on the map of pTiC58 and is closely linked to the *tra/oriT* gene cluster (Figure 1). The locus, which has been completely sequenced, codes for eight genes, all expressed in the same transcriptional orientation (Figure 4A). The product of the first of these, *accR* is a transcriptional repressor that responds to agrocinopines A and B (Beck von Bodman, *et al.*, 1992). The next five genes code for an ABC-type periplasmic protein-dependent transport system, while the last two code for functions required for the utilization of the two sugar phosphate opines as sole source of carbon (Kim and Farrand, 1997). In the wild-type Ti plasmid *acc* is expressed at a relatively high basal level and is inducible some five- to ten-fold by addition of agrocinopines A and B to the culture medium (Hayman and Farrand, 1988; Kim and Farrand, unpublished results). Recent results indicate that purified AccR repressor protein binds to *cis* elements located in the 5' untranslated region upstream of *accR* (Kim and Farrand, unpublished results). Moreover, addition of agrocinopines abolishes this binding suggesting that the unmodified opines are the true inducers of the system. Deletion or insertion mutations in *accR* lead to constitutive expression of *acc* as well as high level constitutive conjugation of the Ti plasmid. Thus, AccR is a repressor and apparently is the master regulator responsible for the transcriptional co-regulation of

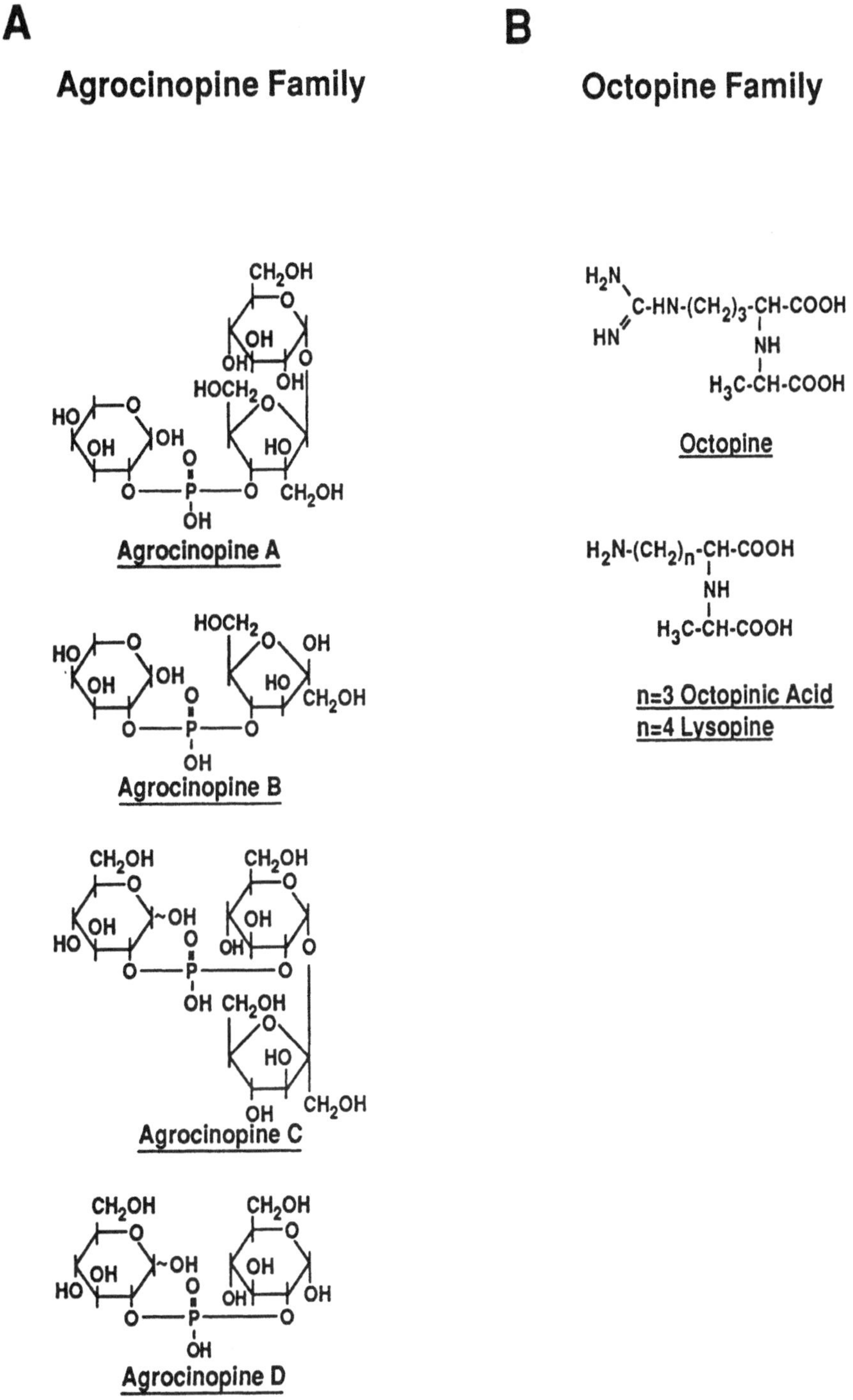

A

Agrocinopine Family

B

Octopine Family

Agrocinopine A

Agrocinopine B

Agrocinopine C

Agrocinopine D

Octopine

n=3 Octopinic Acid
n=4 Lysopine

Figure 3. Chemical structures of the conjugal opines. A. The agrocinopine family. Agrocinopines A and B induce transfer of nopaline/agrocinopine-type Ti plasmids, while agrocinopines C and D induce conjugation of the agropine/agrocinopine C+D-type Ti plasmid pTiBo542. B. The octopine family. Octopine and several of its homologs induce conjugation of the octopine/mannityl opine-type Ti plasmids.

conjugation and the uptake and catabolism of the conjugal opines.

The mechanism by which opines regulate conjugation is somewhat different in the classical octopine/mannityl opine-type Ti plasmids such as pTiA6, pTi15955, pTiAch5, pTiR10 and pTiB6. With these Ti plasmids, octopine, or members of the octopine family including octopinic acid and lysopine serve as the conjugal opines (Figure 3B). Uptake and catabolism of these arginyl opines is dependent upon the *occ* regulon located at about 2 o'clock on the Ti plasmid map (Figure 1). Like the *acc* operon of pTiC58, the *occ* operon is closely linked to the *tra/oriT* locus on these octopine/mannityl opine-type Ti plasmids. Early genetic analysis of mutants that were constitutive for catabolism of octopine and also for conjugal transfer, suggested that the two traits were co-regulated by repression (Petit and Tempé, 1978; Kapwijk *et al.*, 1978; Klapwijk and Schilperoort, 1979). More recent studies showed this to be incorrect. The *occ* regulon, which has been completely sequenced (Valdivia, *et al.*, 1991; Zanker, *et al.*, 1992; Zanker, *et al.*, 1994), is organized as two divergently transcribed gene sets (Figure 4B). The first is monocistronic and codes for *occR*, the translational product of which is a LysR-like transcriptional activator. The product of this gene, OccR, regulates expression of *occ* in response to octopine (Habeeb, *et al.*, 1991; Von Lintig, *et al.*, 1991). The second encodes the seven structural genes of the pathway. The first four of these code for an ABC-type transport system, the next two for the subunits of octopine oxidase and the last for ornithine cyclodeaminase (Figure 4B). This set of seven genes is expressed as a single transcriptional unit from a promoter located between *occR* and *occQ* (Wang, *et al.*, 1992; Cho, *et al.*, 1997). Transcriptional activation from the *occQ* promoter requires OccR and octopine (Cho and Winans, 1993). Thus, null mutations in *occR* result in the inability to catabolize octopine, as well as a Tra⁻ phenotype, even when the donor cells are grown in the presence of the conjugal opine. However, as with other LysR-like activators, OccR also has repressor activity; in the absence of octopine, OccR actively represses expression of its own gene, *occR* as well as the *occQ* operon (Cho and Winans, 1993). Transfer-constitutive mutants of octopine-type Ti plasmids can be isolated and such mutants invariably are constitutive for expression of octopine catabolic functions (Petit and Tempé, 1978; Kapwijk, *et al.*, 1978; Klapwijk and Schilperoort, 1979; our unpublished results). While these most likely represent altered-function alleles of OccR, the nature

of such mutations at the molecular level has not been determined. However, Cho and Winans (1993) have isolated an allele of *occR*, called *occRE23G*, that activates expression of *occ* and also the *tra* regulon in the absence of the opine ligand. When *in trans* to a wild-type Ti plasmid the *occRE23G* allele is dominant and activates conjugation even in the absence of octopine (Fuqua and Winans, 1996a).

IV.A.2. Regulation by Autoinduction

There is no doubt that opines are required as key signals for the induction of conjugal transfer of at least some Ti plasmids. It also is quite clear that control of conjugation by the conjugal opine is mediated by the same transcriptional regulator that is involved in controlling expression of the opine catabolic gene system. However, while both AccR and OccR directly regulate expression of *acc* and *occ* respectively, these two transcriptional regulators control transcription of the *tra* and *trb* operons in an indirect manner. This was first indicated by the observation that a recombinant clone containing a *lacZ* fusion to *traG* from pTiC58 did not express even though it lacked *accR* (Piper, *et al.*, 1993). Similarly, a *lacZ* fusion to a *tra* gene in a recombinant clone from pTiR10 failed to activate even when *occR* was provided *in trans* (Fuqua and Winans, 1994). However, in both systems the *tra::lacZ* reporter could be activated *in trans* by a recombinant clone containing a small fragment mapping near the *tra/oriT* region, suggesting that expression of the *tra* operon is regulated by activation. This proved to be the case; sequence analysis of these clones identified a single gene in both, called *traR* (Piper, *et al.*, 1993; Fuqua and Winans, 1994). The translational product of this gene is related to LuxR, the transcriptional activator responsible for the induction of expression of the *lux* operon in the chemiluminescent marine bacterium *Vibrio fischeri* (see Dunlap and Greenberg, 1991 for a review of the *V. fischeri lux* system). Mutations in *traR* in either Ti plasmid abolish conjugal transfer, as well as the ability to induce expression of *lacZ* reporter fusions to *tra* and *trb* genes (Piper, *et al.*, 1993; Fuqua and Winans, 1994). However such mutations have no effect on expression of *acc* or *occ*. Thus, with both Ti plasmid types, TraR is required for induction of expression of the *tra* regulon, but plays no detectable role in the regulation of expression of the genes involved in uptake and catabolism of the conjugal opines.

LuxR and other members of the LuxR family all require small molecules as co-activators to induce

A. The *acc* Operon

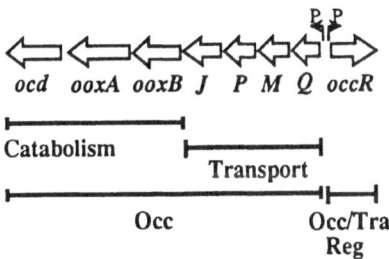

B. The *occ* Operon

Figure 4. Genetic structures of the operons coding for catabolism of the conjugal opines. A. The *acc* operon from pTiC58 which confers uptake and catabolism of agrocinopines A and B. B. The *occ* operon from pTiR10 which codes for transport and catabolism of octopine and its relatives. This operon also has been characterized from the closely related plasmid pTiB6S3 (Zanker, *et al.*, 199) While essentially identical to *occJ*, Zanker, *et al.*, have named the fourth gene *occT*. Both code for the periplasmic binding protein associated with the octopine transport system. Genes and their functions, where known, are described in the text.

gene expression. These signal molecules, which are produced by the bacteria themselves, are *N*-acyl-homoserine lactones. VAI (*Vibrio* autoinducer), the co-activator associated with the *V. fischeri lux* system, is *N*-(3-oxo-hexanoyl)-L-homoserine lactone (Eberhard, *et al.*, 1981). Because of their lipophilic nature, these molecules can diffuse out of and back into the bacteria in which they are produced (Kaplan and Greenberg, 1985). As such, then, they accumulate

in the environment in which the bacteria are growing. The acyl-HSL signals, in conjunction with their cognate transcriptional activators, comprise the key elements of a gene regulatory strategy called autoinduction, or more recently, quorum-sensing (Fuqua, *et al.*, 1994). The system is designed to regulate the expression of target genes in response to the population density of the bacteria. Thus, since the signal molecules accumulate as a function of cell growth, the bacteria can sense their relative population density by monitoring the level of the acyl-HSL in the environment. According to the model, when the amount of the acyl-HSL signal reaches a threshold level it interacts with LuxR, or its relevant homolog such as TraR. This interaction is believed to convert the R protein into a functional transcriptional activator, although the mechanism of this interaction and of the activation of the transcriptional regulator is not known. This gene regulatory paradigm has been the subject of innumerable reviews over the last several years (see, for example Fuqua, *et al.*, 1996; Swift, *et al.*, 1996; Fuqua, *et al.*, 1994).

Work from Kerr's laboratory first suggested that a diffusible signal molecule plays a role in the regulation of Ti plasmid conjugal transfer (Zhang and Kerr, 1991). In their analysis of conjugal transfer, they observed that certain Ti plasmids, which they called Tra-efficient or Tra[e], could be induced to transfer at high levels by addition of the correct conjugal opine. Others, while showing some induction by the conjugal opine, did not transfer at nearly the same frequencies as did the Tra[e] plasmids. These they called Tra-inefficient, or Tra[ie] Ti plasmids. In one of those excellent examples of observational science, Zhang and Kerr (1991) noted that when Tra[e] and Tra[ie] donors were cultivated close to each other on solid opine-induction medium, the Tra[ie] donors now transferred their Ti plasmids at high frequencies. They demonstrated that this phenomenon was due to the production by the Tra[e] donor of a diffusible factor that stimulated the Tra[ie] strains. Moreover, they showed that production of the diffusible agent, which they called Conjugation Factor, or CF, required the presence of, but could not substitute for the conjugal opine. CF has been purified and its structure determined by mass and infrared spectrometry (Zhang, *et al.*, 1993). The signal molecule, now called *Agrobacterium* autoinducer or AAI, is *N*-(3-oxo-octanoyl)-L-homoserine lactone, making it a close analog of the *Vibrio fischeri* autoinducer (Figure 5). Despite the similarity in structure of the signal molecules, there is considerable specificity to the acyl-HSL signal system. While full activation of a

A

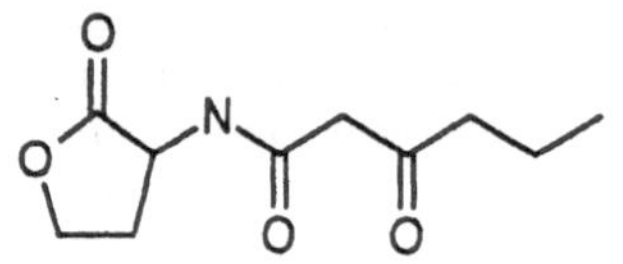

B

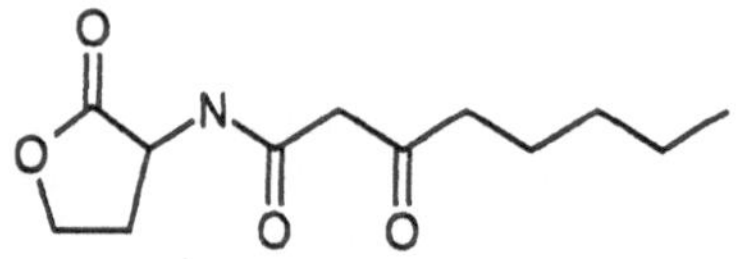

Figure 5. Chemical structures of *N*-acyl-homoserine lactone autoinducers. A. VAI, the primary autoinducer of the bioluminescence system of *Vibrio fischeri*. This signal molecule is *N*-(3-oxo-hexanoyl)-L-homoserine lactone. B. AAI, the autoinducer of the *A. tumefaciens* Ti plasmid conjugal transfer system. This compound is *N*-(3-oxo-octanoyl)-L-homoserine lactone.

tra::lacZ reporter fusion by TraR requires only nM amounts of AAI, it requires about 1000 times more VAI (Zhang, *et al.*, 1993).

Production of AAI by *Agrobacterium* donors is dependent upon the Ti plasmid and requires the *traI* gene. *traI* is a homolog of *luxI*, the gene required for synthesis of VAI by *V. fischeri*. Interestingly, as discussed in Section III, *traI* is the first gene of the *trb* operon in both pTiC58 and pTiR10 (Hwang, *et al.*, 1994; Fuqua and Winans, 1994. See Figure 2B). Thus, organization of the *traI/trb* operon is similar to that of *lux* in which *luxI* is the first gene of the structural operon encoding the bioluminescence functions. Expression of the *trb* operon is dependent upon activation by TraR, making the induced production of AAI dependent upon activation by autoinduction. This again resembles the bioluminescence system of *V. fischeri* in which the activation of expression of *luxI* and therefore the induced production of VAI by the bacterium is dependent upon LuxR and the VAI that has accumulated as a result of basal level expression of the *lux* operon. TraI is necessary and sufficient for the synthesis of AAI. Although homoserine lactone was predicted to be an intermediate (Huisman and Kolter, 1994), using an *in vitro* system, Moré, *et al.*, (1996) conclusively showed that purified TraI catalyzes the synthesis of AAI from *S*-adenosyl-methionine and 3-oxo-octanoyl-ACP. This is consistent with the pathway for synthesis of VAI

first proposed by Eberhard, *et al.* (1991), and now has been confirmed by Schaefer, *et al.* (1996)

Several lines of evidence indicate that *trb* and the two *tra* operons are activated for transcription directly by TraR. First, as noted above, small reporter clones containing only the *tra* or *trb* promoter region and a *lacZ* fusion to an appropriate *tra* or *trb* gene express strongly, but only when provided with TraR and AAI (Piper, *et al.*, 1993; Hwang, *et al.*, 1994; Fuqua and Winans, 1994). Omitting AAI or TraR completely abolishes expression by these reporters. Second, each of these promoter regions contains a signature 18 bp almost-perfect inverted repeat structure. Called the *tra* box, this IR is very similar in sequence to a 20 bp IR, called the *lux* box, located in the promoter region of the *lux* structural operon (Hwang, *et al.*, 1994; Fuqua and Winans, 1994; Farrand, *et al.* 1996). Deletions in the *lux* box abolish the ability of LuxR and VAI to activate expression of downstream genes (Baldwin, *et al.*, 1989). Mutations within the *tra* box have a similar deleterious effect on the TraR/AAI-activation of *tra* gene reporters (Fuqua and Winans, 1996b; Cook, 1996). Studies with an active form of LuxR indicate that this activator, in conjunction with RNA polymerase, binds specifically to the *lux* box present in the *lux* promoter region (Stevens, *et al.*, 1994; Stevens and Greenberg, 1997).

IV.A.3. The Regulatory Hierarchy

Induction of conjugation requires both the conjugal opine-mediated regulator and the quorum-sensing system dependent upon the accumulation of AAI. In both Ti plasmid types the link between these two regulatory systems is accomplished by placing *traR* under the control of the opine regulon. In the nopaline-type Ti plasmid *traR* is the fourth gene of a five-gene set divergently transcribed from the *accR-acc* promoter region (Piper and Farrand, unpublished, see Figure 6A). Several lines of evidence indicate that these genes constitute an operon that is regulated by AccR in response to the agrocinopine opines (Piper and Farrand, unpublished results). First, analysis of a *lacZ* fusion to *traR* shows that expression of this gene is repressed by AccR and is derepressed when AccR is absent, or when agrocinopines are added to the growth medium. Second, Tn5 insertions located in the genes upstream of *traR* abolish induction of transfer by agrocinopines. These mutations can be complemented by a fragment that encodes *traR* but does not overlap the sites of the insertions. This suggests that the Tra⁻ phenotype associated with these insertions results from their

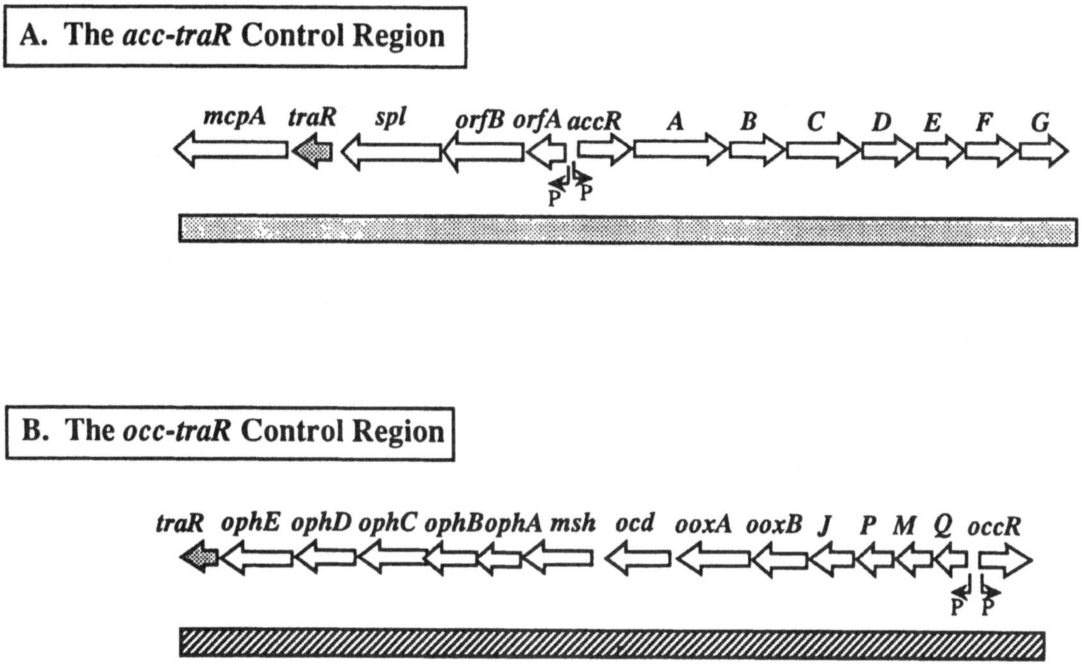

A. The *acc-traR* Control Region

B. The *occ-traR* Control Region

Figure 6. Structure relationships of the genes of the *traR* and conjugal opine operons. A. The gene arrangements from the opine catabolism-conjugation control region of the nopaline/agrocinopine-type Ti plasmid, pTiC58. The operon containing *traR* is transcribed divergently from that which codes for the catabolism of the conjugal opines, agrocinopines A and B. Transcription in both directions is regulated from the promoter/operator complex (p) located between *orfA* and *accR*. This promoter is regulated by repression by AccR in response to the agrocinopines. B. The gene arrangements from the analogous region of the octopine/mannityl opine-type Ti plasmid pTiR10. *traR* is the distal gene of a 14-gene operon beginning at *occQ*. Expression of this operon from a promoter/operator complex (p) between *occQ* and *occR* is regulated by the transcriptional activator, OccR, in response to octopine. For both Ti plasmis, genes and their functions, where known, are described in the text.

polar effect on the expression of the downstream *traR* gene. Third, RT-PCR and primer extension analyses indicate that the 6.8 kb region is expressed as a single RNA transcript from a promoter located just upstream of *orfA*, the first gene of the five-gene unit. Finally, promoter resection analysis indicates that AccR-regulated expression of *traR* initiates from the region located 5′ to *orfA*. Thus, the ca 245 bp segment located between *accR* and *orfA* constitutes a divergent promoter/operator complex that allows AccR to regulate transcription in both directions. Consistent with this, as described in Section IV.A. above, gel retardation assays indicate that this region contains

cis-active elements to which AccR binds (Kim and Farrand, unpublished results).

The nature of the genes with which *traR* is associated is intriguing. While the translational product of *orfA* is not related to any protein sequences in the data bases, *orfB* could code for a product that is related to that of a gene, *yjbB* of *Escherichia coli*. This gene, located at coordinates 4,225,310 to 4,226,941 of the *E. coli* genome (Blattner, *et al.*, 1997), has no known function. The third gene, called *spl*, could code for a protein that is closely related to the enzyme sucrose phosphorylase present in several bacteria (Piper and Farrand, unpublished results). The last gene, located just downstream of *accR* could code for a protein closely related to McpA, a methyl-accepting chemotaxis protein from *Caulobacter crescentus*. Clearly these genes do not define a set of conjugation functions. More likely they are, or were at one time involved in the sensing and catabolism of an unknown substrate. The putative sucrose phosphorylase is particularly intriguing considering that the agrocinopine opines are phosphodiesters of arabinose and sucrose. As detailed in Chapter 9, sequence analysis suggests that AccF codes for a phosphodiesterase that could split the agrocinopines into two products, one of which would be a phosphorylated sugar (see Kim and Farrand, 1997). The product of the *spl* gene might be involved in the further metabolism of this intermediate. However, the *traR* operon is not required for the uptake or catabolism of agrocinopines, or for the use of these opines as sole carbon and energy source (Hayman, *et al.*, 1993; Kim and Farrand, 1997). Nor is *mcpA* required for chemotaxis to agrocinopines (Kim and Farrand, 1998). Thus, the function of the genes contained in the *traR* operon remains to be determined.

The gene arrangement that places *traR* under control of *occR* in the octopine/mannityl opine-type Ti plasmids is quite different, although it accomplishes the same result. In this system, *traR* is the last gene of an operon composed of 14 genes, the first seven of which comprise *occ* (Fuqua and Winans, 1996a; Figure 6B). The first of those following *occ*, called *msh*, could code for a protein related to MetH of *E. coli* while the next five, which directly precede *traR* could code for proteins related to those of the oligopeptide permease family of ABC-type substrate transporters. However, uptake studies failed to identify a substrate for this putative transport system (Fuqua and Winans, 1996a). Genetic analysis indicated that, except for *traR*, none of these genes are required for conjugation. Strongly polar insertion mutations within this region abolish octopine-inducible conjugation and also expression of the distal *traR* gene (Fuqua and Winans, 1996a). These upstream mutations all are complementable by providing *traR in trans*. Thus, like the *traR* operon of pTiC58, *traR* of pTiR10 is contained within an operon the expression of which is controlled by the regulator, in this case OccR, responsive to the conjugal opine. Consistent with this, expression of *traR* is controlled directly by OccR and promoter analysis indicates that this control is exerted at the *occQ* promoter which lies at the very 5′ end of the 14 kb gene set (Fuqua and Winans, 1996a).

Thus, in both the octopine- and nopaline-type Ti plasmids *tra* and *trb* gene expression requires the transcriptional activator, TraR and its cognate quorum-sensing co-inducer, AAI. Yet in both cases, conjugation also requires the opine signal produced by the crown gall tumor. The two gene regulatory systems are linked by the straightforward strategy of placing the expression of *traR* under the control of the opine regulon. Remarkably, while both Ti plasmids employ this tactic to assure the hierarchal relationship of the two regulatory circuits, the actual gene arrangements involving *traR* that are responsible for implementing the system are quite different. This indicates that the event responsible for placing *traR* under control of the agrocinopine regulon occurred independent of that which placed the same gene under the control of the octopine regulon. This suggests, in turn, that it is important to the Ti plasmid that conjugation be regulated by the appropriate opine.

The arrangements and positionings of the *tra* and *trb* gene clusters on the two Ti plasmid types also are interesting. As noted above, the *traI-trb* operon is tightly linked to the *oriV/rep* region on both pTiC58 and pTiR10. But also, the conjugal opine catabolism-*traR* gene cluster of both Ti plasmids is closely linked to the *traI/oriT* gene complex (Figure 7). Remarkably, while the *traI/oriT-traR* interval is very strongly conserved between the two Ti plasmids, the operons in which *traR* is located and the operons comprising the structural and regulatory genes for the transport and catabolism of the conjugal opines are entirely unrelated between the two Ti plasmid types. Thus, the interval from *traR* through the *traCDG* operon is, with exception of a few gaps, almost perfectly conserved in the two Ti plasmids, exactly as predicted by the electron microscopic heteroduplex analysis of Engler, *et al.* (1981).

IV.B. A MODULATING FUNCTION

The TraR-mediated gene regulatory system contains an additional regulatory element. This function, coded for by *traM*, serves to modulate the ability of TraR to activate transcription of the *tra* regulon. *traM* is present in nopaline-type and in octopine-type Ti plasmids and in both the gene is located between *traR* and the *traAFB* operon (Hwang, *et al.*, 1995; Fuqua, *et al.*, 1995; see Figure 7). In both Ti plasmid types mutations in *traM* result in an opine-independent transfer-constitutive phenotype. Yet these mutations have no effect on the regulation of

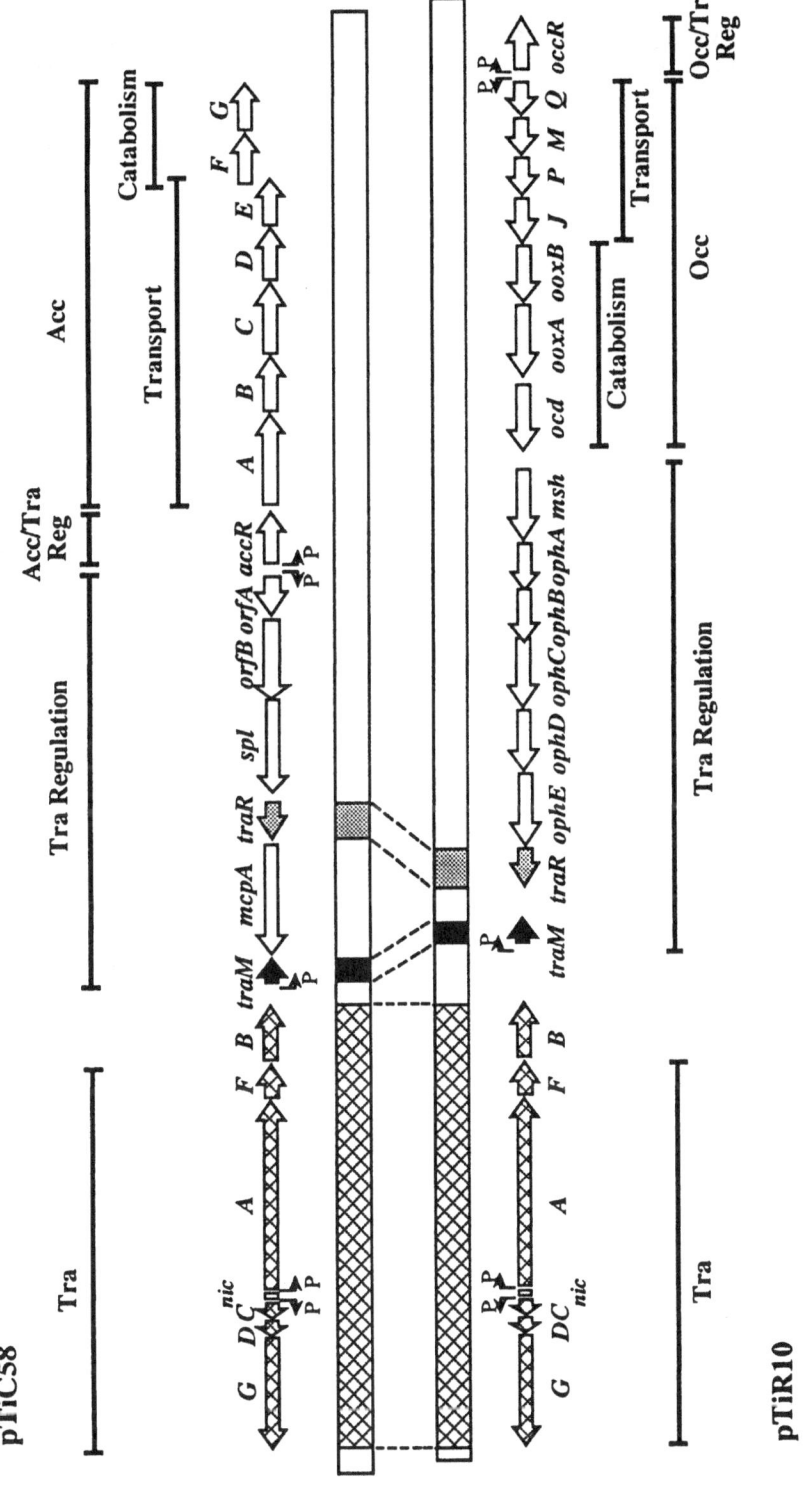

Figure 7. The juxtaposition of the *tra* and *tra*-regulatory operons in the two Ti plasmid types. The regions from pTiC58 and pTiR10 are aligned with respect to the *tra/oriT* locus. In both cases, the *tra* regulatory region, composed of the *traR* operon and the operons for catabolism of the conjugal opines, is immediately adjacent to the *tra/oriT* region. Genes phylogenetically conserved between the two Ti plasmid are depicted by shadings. The hatched shading denotes the *tra* genes, the stippled shading denotes *traR* and the black fill denotes *traM*. The remaining genes are phylogenetically unrelated between the two Ti plasmids. Segments of the two Ti plasmid *tra* regions connected by the dotted lines represent homology region C as determined by the electron microscopic heteroduplex analysis of Engler, *et al.* (1981).

expression of *acc* or *occ* indicating that TraM does not influence the function of AccR or OccR, the primary opine regulator proteins. Consistent with this, Ti plasmids that are transfer-constitutive due to mutations in *accR* or *occR* become hyperconjugal upon introduction of a mutation in *traM* (Hwang, *et al.*, 1995; Fuqua, *et al.*, 1995). Yet TraM does not directly regulate transcription of the *tra* regulon, or even *traR* itself (Hwang, *et al.*, 1995; Fuqua, *et al.*, 1995).

The effect of TraM is dependent upon its level of expression. Thus, *traM* exerts its inhibitory effect when expressed at levels higher than that of *traR*. On the other hand, when *traR* is overexpressed with respect to *traM*, no inhibition occurs (Hwang, *et al.*, 1995; Fuqua, *et al.*, 1995). These observations suggest that TraM inhibits TraR function, perhaps by directly interacting with the transcriptional activator. This is consistent with the fact that the inhibitory effect of TraM cannot be overcome by addition of excess AAI (Zhang and Kerr, 1991; Hwang, *et al.*, 1995) unless the balance between the two is only slightly in favor of TraM (Fuqua, *et al.*, 1995). Recent analysis using the yeast two-hybrid system indicates that TraM and TraR can interact (Hwang, Smyth and Farrand, unpublished results). Moreover, mutations in *traM* that abolish the suppressive effect on conjugation also abolish the interaction of TraM with TraR as assayed in the yeast tester system.

There are no known homologs of TraM in any of the quorum-sensing systems examined to date. In fact, with the exception of the product of a *traM*-like gene on pNGR234a, TraM is not related at the amino acid sequence level to any protein sequence in the data bases. Furthermore, TraM does not affect LuxR-mediated activation of the *lux* operon or LasR-mediated activation of *lasB* (Hwang and Farrand, unpublished results). This suggests that the antiactivator is unique to the *traR* gene regulatory system.

IV.C. BIOLOGY OF THE REGULATORY SYSTEM

Quorum-sensing systems based on a LuxR-like transcriptional regulator and a cognate acyl-homoserine lactone signal are believed to provide bacteria with a method to sense their population levels. Characteristically, genes regulated by quorum-sensing express only after the bacteria have reached a critical population size. Moreover, the addition of exogenous autoinducer to the culture at low cell densities can short-circuit the system resulting in premature gene induction. Such is the case with the Ti plasmid *tra* system. Following opine induction, there is a substantial lag before donors become active for transfer (Fuqua and Winans, 1994; Piper and Farrand, unpublished results). As measured using *lacZ* reporter fusions, this lag represents a delay in the activation of transcription of the *tra* regulon. As predicted by the model, the delay in gene expression and also in conjugation can be shortened substantially by adding authentic AAI and the conjugal opine to the culture at the same time (Fuqua and Winans, 1994; Piper and Farrand, unpublished results). Consistent with this, donors that are transfer-constitutive because of mutations in the opine regulatory protein produce induced levels of AAI. Such mutants show kinetics of induction similar to those observed with wild-type donors in which the conjugal opine and AAI are added at the same time.

These results are consistent with a model, presented in Figure 8, in which conjugal transfer of the Ti plasmid is dependent upon a suitable donor population size. But this parameter is not sufficient. That conjugation is regulated by opines indicates that the availability of these nutritional resources is important to the system. Thus, in the absence of opines, expression of the opine-catabolic genes is repressed. In addition, because the *traR* operon also is controlled by the primary opine regulator, the transcriptional activator required for expression of the *tra* regulon also is not produced. In plasmids such as pTiC58, this control is by repression; in plasmids such as pTiR10, regulation occurs by activation. But in both, the result is the same; in the absence of the conjugal opine, *traR* is not expressed at levels sufficient to activate expression of the *tra* regulon. Furthermore, because high level expression of the *tra/trb* operon requires activated TraR, very little AAI is produced by the bacteria in the absence of the conjugal opine.

In environments in which the conjugal opine is available, this signal is taken up by the bacteria where it is free to interact with its transcriptional effector. The interaction results in the derepression, in the case of *acc*, or the activation, in the case of *occ*, of the respective opine catabolic operons, as well as the *traR* operons. The TraR protein, produced as a result of this induction by the opines, remains inactive, however, until sufficient amounts of AAI have accumulated as a function of cell growth. When this finally occurs, activated TraR activates expression of the *tra* regulon and the Ti plasmid becomes conjugal.

What, then, is the role of the antiactivator, TraM? In both Ti plasmid types mutations in *traM* lead to a

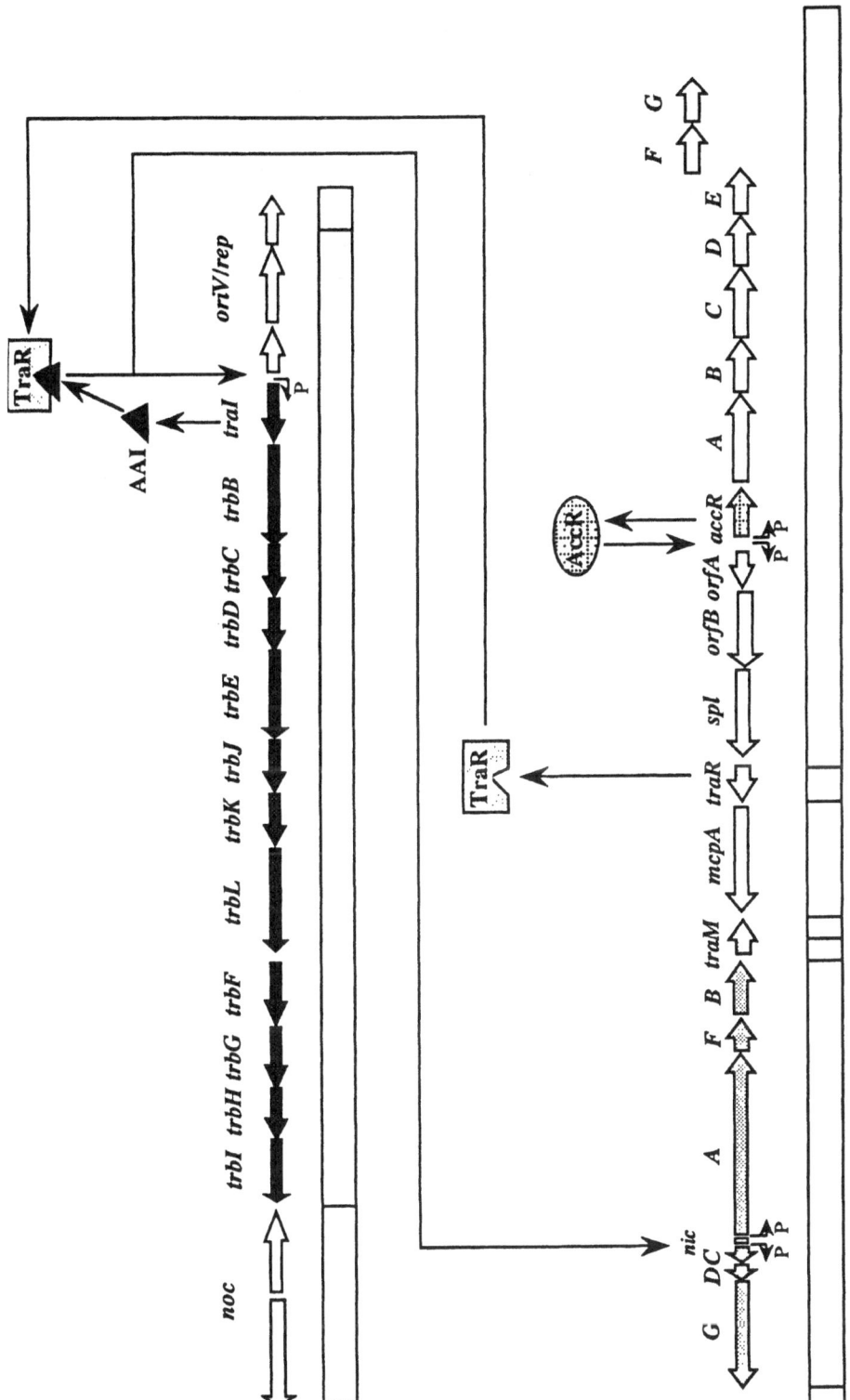

Figure 8. The opine/quorum-sensing regulatory hierarchy that controls expression of the *tra* regulon. The model presents the relationship between the two regulatory circuits using those of pTiC58 as the example. In the absence of the conjugal opine, AccR represses expression of *acc* and also the *traR* operon. The *traAFB*, *traCDG* and *traI/trb* operons, all of which require TraR, are not expressed at levels required for conjugation under these conditions. However, the *traI/trb* operon apparently is expressed at a very low level, sufficient to produce small amounts of AAI. When the conjugal opines are made available by the crown gall tumors, these signals interact with AccR apparently destabilizing the interaction between this repressor and the *traR/acc* operator/promoter complex. This results in divergent transcription of the two operons leading to the production of TraR. When the AAI accumulates to its threshold level, it interacts with TraR and the activator-ligand complex can then activate transcription of the three *tra* operons. TraM, which functions as an antiactivator, is excluded from the model for simplicity. See text for additional details.

transfer-constitutive phenotype that is independent of the opines. This suggests that even under non-inducing conditions the operons in which *traR* is located express at a level adequate to produce sufficient amounts of TraR to activate the *tra* regulon. Presumably, in the absence of the conjugal opines, TraM is present in excess over TraR. Under these conditions, 1 interaction of TraM with TraR prevents the activator from binding AAI. Alternatively, this interaction may prevent the TraR-AAI complex from activating transcription at the *tra* and *trb* promoters. According to this model, upon induction by the conjugal opines, TraR is expressed at levels that overcome the available TraM. Thus, free activator is available to bind AAI, or once bound with AAI, to activate transcription. No matter the mechanism, the model predicts that TraM serves to prevent premature conjugation that otherwise would occur due to basal level expression of *traR* in the absence of the conjugal opine.

The conservation of both regulatory systems among several plasmid types also suggests that it is important to conjugation that at least two sets of conditions be present. The opine signal most likely serves to inform the Ti plasmid that nutritional conditions are favorable to its bacterial host. The purpose of the quorum-sensing system is less obvious. Perhaps by coordinately inducing the conjugal transfer system in the entire donor population, the Ti plasmid insures that any recipient happening into the locale will be mated. Central to the biology of *Agrobacterium* this implies that the Ti plasmid recognizes the crown gall tumor as a habitat in which horizontal transfer is beneficial to its well-being.

IV.D. TRAR ASSOCIATED WITH OTHER SYSTEMS

Homologs of *traR* are associated with other opine systems. Oger, *et al.* (1998) recently reported on an allele of *traR* linked to a set of four genes required for the transport of mannopine from the octopine-mannityl opine-type Ti plasmid pTi15955. Zhu and Winans (1998) also have examined this gene in the closely-related Ti plasmid, pTiR10. The gene, called *trlR*, is transcriptionally co-expressed with the four upstream transporter genes and this expression is induced by mannopine. *trlR*, then, represents a second opine-associated *traR*-like gene present in this class of Ti plasmids. Interestingly, *trlR* actually is a frameshift allele of *traR*. Deletion of a T:A base pair at gene coordinate 542 results in a translational product that is

identical to TraR associated with the octopine regulon except for an altered carboxy-terminal domain. Based on an analysis of LuxR, this region is believed to comprise the DNA binding domain of TraR. Thus, not surprisingly, growth in the presence of mannopine, while inducing expression of *trlR*, does not induce conjugal transfer of the Ti plasmid. On the contrary, expression of *trlR* inhibits octopine-mediated induction of conjugation of pTi15955 and pTiR10 (Oger, *et al.*, 1998; Zhu and Winans, 1998). This suppressive effect results from the inhibition of expression of the *tra* and *trb* operons. However, TrlR does not inhibit expression of the octopine-regulated *traR* gene of this Ti plasmid. Given that LuxR is believed to function as a homomultimer, these observations led both groups to propose that *trlR* is a dominant negative allele of *traR*. Presumably, when both genes are expressed, TrlR and TraR can, with some probability, interact with each other to form inactive heteromultimers.

The discovery of a second *traR*-like gene associated with the octopine-mannityl opine-type Ti plasmids prompted Oger, *et al.* (1998) to examine other Ti, Ri, and opine-catabolic plasmids for sequences homologous to this regulatory gene. Based on Southern hybridization analyses, several other *Agrobacterium* plasmids, including the nopaline/agrocinopine-type Ti plasmid pTiC58, contain more than one set of sequences related to *traR*. Other Ti plasmids, most notable pTiBo542 and its close relative, pTiChry9 contain only a single fragment that hybridizes with the *traR* probe. The probe also hybridized to opine-catabolic plasmids isolated from two unrelated *A. radiobacter* strains. This suggests that regulation of conjugation by opines via the *traR*/AAI-dependent quorum-sensing system may not be restricted to Ti plasmids. While none of these hybridizing fragments have been sequenced to date, the results suggest that the conjugal Ti-like plasmids of *Agrobacterium* contain several copies of *traR* and that these may be associated with other opine systems. Interestingly, the *traR* probe did not hybridize with fragments from pRiA4, the Ri plasmid of the agropine/agrocinopine-type *A. rhizogenes* strain A4. This is consistent with hybridization studies indicating that pRiA4 does not contain sequences detectably related to the *tra* or *trb* regions of the Ti plasmid conjugal transfer system (see Section III).

The presence of more than one *traR* gene on pAtK84b is particularly intriguing. This plasmid, isolated from *A. radiobacter* K84, codes for catabolism of nopaline and agrocinopines A and B. Moreover, although closely related to pTiC58, pAtK84b lacks sequences detectably related to *vir* and T-DNA (Clare, *et al.*, 1990). Regions of this plasmid are strongly related to the *tra* and *trb* operons of pTiC58. Furthermore, like pTiC58, conjugation is inducible by agrocinopines A and B (Ellis, *et al.*, 1983b). However, unlike pTiC58, conjugation of pAtK84b also is inducible by growth with nopaline. It is conceivable that the two fragments of pAtK84b that hybridize with the probe represent *traR* genes associated with the agrocinopine and nopaline catabolic gene systems.

How *traR* becomes associated with opine-catabolic units is unknown. Three iterations of this gene arrangement have been characterized at the nucleotide sequence level (Fuqua and Winans, 1996; Oger, *et al.*, 1998; Piper and Farrand, unpublished results). In all three cases, while there is usually 100-200 base pairs of DNA between *traR* and the upstream and downstream genes, there is no evidence for the involvement of IS sequences or other transpositional elements. Moreover, there is little if any conservation in the intergenic sequences 5′ and 3′ to the *traR* genes in any of the *traR* operons.

traR is not confined to plasmids of the genus *Agrobacterium*. As described in Section III, sequence analysis of pNGR234a showed that this conjugal *sym* plasmid contains a *tra* regulon remarkably similar to that of the Ti plasmids. pNGR234a also contains homologs of *traR* and *traM* (Freiberg, *et al.*, 1997). There is no proof to date that the *traR* homolog is involved in regulating conjugal transfer of this plasmid. However, it is likely to be the case. There is a homolog of *traI* associated with the *trb* operon of the *sym* plasmid and strain NGR234 produces an *N*-acyl-homoserine lactone similar, if not identical in structure to *Agrobacterium* autoinducer (Freiberg, *et al.*, 1997). Furthermore, as with the Ti plasmids, the *traI/trb* operon is closely linked to the *oriV/rep* determinants. However, rather than being associated with a regulated operon as it is in the Ti plasmids, sequence analysis suggests that *traR* of pNGR234a is monocistronic. This is consistent with the observation that this plasmid is constitutive for conjugal transfer. It remains to be determined if conjugation of pNGR234a is dependent upon donor cell density in a quorum-sensing manner as it is with the Ti plasmids.

Similarly, the *sym* plasmid of *R. etli* strain CFN42 contains sequences homologous to *traI* linked to the *oriV/rep* region (our analysis of sequence in data base acquisition U80928 from Ramírez-Romero, *et al.*, 1997). This strain also produces several compounds with autoinducer activity (Rosemeyer, *et al.*, 1998; Cha and Farrand, unpublished results). Recall that *traI* is responsible for production of AAI and that, on the

Ti plasmid, the gene is the first element of the *trb* operon. Moreover, the *traI/trb* operon is closely linked to the *oriV/rep* region of the Ti plasmid, just as it is in pNGR234a and in the *R. etli sym* plasmid. These anecdotal observations suggest that the association of the *traR/traI* quorum-sensing gene regulatory system with the genes coding for the plasmid replication machinery may be a feature common among the large conjugal plasmids present in the family *Rhizobiaceae*. The implications of these gene linkages will be considered in more detail in Section VII.

V. Characteristics of Ti Plasmid Conjugal Transfer

In most mechanistic respects, the Ti plasmid *tra* system probably is not significantly different from those of other plasmids such as F and RP4. However, some differences exist in transfer characteristics among the various conjugal plasmids and how the Ti plasmid system compares with these is only poorly documented. Significant progress has been made over the past several years concerning several of these characteristics of Ti plasmid conjugation.

V.A. ENVIRONMENTAL CONDITIONS

Aside from the requirement for signal molecules to induce conjugal transfer, several environmental conditions influence transfer of the Ti plasmids. Conjugal transfer of the octopine/mannityl opine-type Ti plasmid pTiB6 long has been known to be temperature-sensitive. This element transfers at high frequency at 27°C, but the frequencies drop dramatically when matings are conducted at 33°C (Tempé, *et al.*, 1977). However, this character most likely is particular to pTiB6. Several other Ti plasmids, including the almost-indistinguishable plasmid pTiA6, transfer at high frequency at temperatures up to 35°C (L. Zhang and A. Kerr, personal communication). This relatively broad temperature range contrasts sharply with the profile exhibited by the *vir* transfer system. Recent results, using mobilization of an RSF1010 derivative, showed that *vir* mediated DNA transfer is maximal at 22°C and drops off to almost undetectable levels at temperatures above 29°C (Fullner and Nester, 1996). That the two transfer systems differ so remarkably in their temperature-activity profiles suggests that it is unlikely that *vir* is required for maximum functioning of the *tra*-mediated transfer system.

Conjugal transfer of pTiA6 occurs over a pH range of 5.9 to 7.7, but declines sharply at values below pH 5.9 (L. Zhang and A. Kerr, personal communication). Whether this reflects the functioning of the conjugal transfer apparatus, or of a pH dependence of the transcriptional regulatory circuitry is not known. Again, however, this pH-activity profile contrasts with the absolute requirement of the *vir* transfer system for conditions below pH 5.75. Thus, the *vir* transfer system operates at maximal efficiency within a pH range that is strongly inhibitory to the Ti plasmid conjugal transfer system. This difference in pH-activity profiles suggests that the two systems operate under rather different environmental conditions. This, in turn, implies that the environment encountered by the bacterium at the site of infection within the plant differs significantly from the environment provided to the bacterium by the crown gall tumor that eventually forms at that site.

Transfer of octopine/mannityl opine-type Ti plasmids is inducible on minimal medium containing the appropriate conjugal opine. However, transfer remains repressed on rich medium, even when the conjugal opine is provided in excess (Hooykaas, *et al.*, 1979). The basis for this inhibition by rich medium is not understood. However, the sulfur-containing amino acids methionine and cysteine are strongly inhibitory to conjugation (Hooykaas, *et al.*, 1979). Considering that *S*-adenosylmethionine is a precursor for the synthesis of the quorum-sensing signal molecule, *N*-(3-oxo-octanoyl)-L-homoserine lactone (Moré, *et al.*, 1997; see Section IV), it is tempting to speculate that these two amino acids may somehow interfere with the synthesis of this essential signal molecule. However, this does not appear to be the case (S. Winans, personal communication). The mechanism by which these sulfur-containing amino acids inhibit conjugal transfer remains to be determined.

V.B. MATING SUBSTRATE

The conjugal transfer systems of some plasmids function more efficiently when matings occur on solid surfaces, while others function with equal efficiency in liquid or on solid surfaces. Virtually all studies concerning plasmid transfer from members of the *Rhizobiaceae* have employed matings on agar surfaces or on micropore membranes. To our knowledge there have been no studies reporting the relative efficiency

of plasmid conjugation when matings were carried out in liquid versus on solid surfaces. Recent work by Piper and Farrand (unpublished results) clearly demonstrates that pTiC58 transfers only when the matings occur on solid surfaces. The plasmid did not transfer at a detectable frequency when the matings were conducted in liquid medium. This makes sense; the normal habitat for *Agrobacterium* spp. and probably all other members of the family *Rhizobiaceae* contains very little free water. Rather, the bacterium most likely spends most of its life associated with solid surfaces such as soil particles and root and tumor surfaces.

Fuqua and Winans (1994) reported that induction of conjugal transfer also requires that donors be grown on solid surfaces. Based on a comparative analysis of the transfer properties of wild-type and transfer-constitutive mutants, these workers concluded that induction of the *tra* gene system of the octopine/mannityl opine-type Ti plasmid pTiR10 requires a factor or condition specific to solid surfaces. However, a recent study by Oger, *et al.* (1998) found no difference in gene activation properties, or in conjugal transfer frequencies between solid-grown and liquid grown donors harboring either pTi15955 or pTiC58.

V.C. ENTRY EXCLUSION

Some conjugal plasmids, when present in a recipient, prevent that bacterium from inheriting a second conjugal plasmid from a donor. This property, called surface exclusion, or entry exclusion (Eex), is highly specific. The Eex system of a given plasmid usually functions to exclude only certain other, closely related plasmids. In a study done some years ago, Hooykaas *et al.* (1980) did not detect any entry exclusion activity associated with octopine- or nopaline-type Ti plasmids. However, as described in Section III, these Ti plasmids both contain within their *trb* regions a homolog of the RP4 *trbK* gene. In this IncP1 plasmid, *trbK* is both necessary and sufficient for the entry exclusion phenotype (Haase, *et al.*, 1996). Alt-Mörbe, *et al.* (1996) reported that the cloned *trbK* gene of pTiR10 did not confer entry exclusion on *Agrobacterium* recipients when mated with a Ti plasmid donor. Mature, processed TrbK of RP4 is acylated at the N-terminal cysteine residue. The putative TrbK protein of pTiR10 lacks this cysteine prompting Alt-Mörbe, *et al.* (1996) to propose that absence of an acylation site accounts for the lack of entry exclusion associated with this protein.

Recently we have found that pTiC58 exerts entry exclusion against itself and also, to a lesser extent, against pTi15955 (Nelson, Bratis and Farrand, unpublished results). On the other hand, consistent with the reports of Hooykaas *et al.* (1980) and Alt-Mörbe, *et al.* (1996) pTi15955 does not exert detectable entry exclusion against itself or against pTiC58. The putative TrbK protein of pTiC58, like that of pTiR10 lacks the N-terminal cysteine residue suggesting that entry exclusion is not strictly dependent upon acylation. This is consistent with the observation by Haase, *et al.* (1996) that mutating the N-terminal cysteine of RP4 TrbK diminishes but does not abolish entry exclusion. They concluded that acylation of TrbK is not essential for its role in surface exclusion. These workers also showed that TrbK of RP4 contains an N-terminal secretion signal and that the protein is associated with the outer surface of the cytoplasmic membrane. They propose that the signal sequence is responsible for translocation to the periplasm while the lipid moiety functions to anchor the protein to the outer surface of the inner membrane. Sequence analysis suggests that the TrbK proteins of pTiR10 and pTiC58 also contain N-terminal secretion signals, but neither of these proteins has been localized in the bacterial cell.

The *trbK* genes are not strongly conserved between the octopine and nopaline-type Ti plasmids. This might explain why pTiC58 expresses entry exclusion while pTi15955 apparently does not; the *trbK* gene of the latter may have accumulated changes that abolished function. There is even less conservation between the *trbK* genes of the Ti plasmids and that of RP4. Thus it is not surprising that the R plasmid and the Ti plasmids fail to exert entry exclusion against each other (Nelson, Bratis and Farrand, unpublished results). The *trb* region of pNGR234a also contains a *trbK* homolog (Freiberg, *et al.*, 1997). The sequence of the translational product of this gene is quite divergent from that of the two Ti plasmids and also from that of RP4. Whether this plasmid exerts entry exclusion against any other plasmid has not been reported.

The entry exclusion capacity of pTiC58, while significant, is several orders of magnitude less efficient as compared to that of RP4. This coupled with the observation that the *trbK* genes of pTiC58, pTiR10 and pNGR234a are highly divergent suggests that the entry exclusion character of these plasmids is not essential to the biology of the system.

Interestingly, entry exclusion exerted by pTiC58 is dependent upon induction of the conjugal transfer system (Nelson, Bratis and Farrand, unpublished results). Thus, cells harboring the uninduced Ti

plasmid do not exclude entry of pTiC58 or pTiR10. However, a derivative of pTiC58 that is mutationally constitutive for conjugal transfer strongly exerts surface exclusion, preventing entry of the Ti plasmid from the donor. This suggests that in nature, under conditions in which the *tra* system is induced entry exclusion may prevent donors harboring the same Ti plasmid from mating with each other. However, cells harboring a Ti plasmid can accept different, otherwise excluded Ti plasmids, if the *tra* system of the Ti plasmid in the recipient is repressed. Such conditions might exist in the vicinity of tumors that do not produce the conjugal opine active for the Ti plasmid present in the recipient.

V.D. BACTERIOPHAGE SENSITIVITY

Certain bacteriophages infect and propagate only in hosts harboring conjugal plasmids. Bacteriophages PRD1 and PRR1, for example, plaque only on bacteria that harbor IncP, IncN, or IncW plasmids. Often, these bacteriophages recognize some component of the plasmid-encoded Mpf apparatus as substrate for attachment or infection. A recent study by Grahn, *et al.* (1997) suggested that TrbC, TrbE and TrbL of RP4 are particularly important for infection by PRD1. Considering the close relatedness between the *trb* genes of the Ti plasmid and those of the IncP1 plasmids, it would seem likely that one or both of these phages should infect *Agrobacterium* hosts. Surprisingly, this is not the case. Neither of these bacteriophages causes productive infections in *A. tumefaciens* strains harboring a transfer-constitutive derivative of pTiC58 (Li and Farrand, unpublished results). Moreover, neither bacteriophage even attaches at a detectable level to the *A. tumefaciens* host. On the other hand, *A. tumefaciens* strains harboring RP4 plaque both bacteriophages with efficiencies comparable to *E. coli* strains harboring the IncP1 R plasmid (Li and Farrand, unpublished results). Thus, the failure of these bacteriophages to plaque on *A. tumefaciens* strains harboring a Ti plasmid is not due to some inherent defect in bacteriophage replication or maturation. Rather it appears that the epitopes of the RP4 mating bridge complex required for bacteriophage adsorption are not sufficiently conserved in the Ti plasmid system.

V.E. CONJUGAL HOST RANGE

We now know that most conjugal transfer systems are extremely promiscuous. What limits acquisition is not transmission but rather replication. In this regard, the replicative range of the known Ti plasmids is relatively narrow, being restricted to members of the family *Rhizobiaceae*. The Ti plasmids can be conjugated to and stably propagated in several species of the genus *Rhizobium* (Hooykaas, *et al.*, 1977). Moreover, *Rhizobium* transconjugants generally express Ti plasmid-encoded traits. On the other hand, matings with recipients outside of the family *Rhizobiaceae* do not yield stable transconjugants. However, if assays are employed that do not rely on replication, it is quite clear that the conjugal host range of these plasmids is substantially broader than was originally thought. Some years ago Sprinzel and Geider (1988) showed, using a zygotic induction assay, that pTiB6 is transmissible to *E. coli* recipients. Two additional lines of evidence support these results. First, the full size Ti plasmid will mobilize transfer of a broad host range recombinant plasmid containing the Ti plasmid *oriT* region from *Agrobacterium* donors to *E. coli* recipients (Cook and Farrand, 1992). Moreover, such transfer requires only the Ti plasmid *tra/oriT* and *trb* functions; the binary *tra* system developed by Cook, *et al.* (1997) transfers the *tra/oriT* component to *E. coli* and *Pseudomonas fluorescens* recipients at frequencies comparable to those obtained with *Agrobacterium* recipients. Second, a Ti plasmid marked with Tn5 will transfer the transposon from an *Agrobacterium* donor to *E. coli* and *P. fluorescens* recipients (Hwang and Farrand, unpubl. results). In essence, the Ti plasmid acts as a suicide transposon donor, with kanamycin-resistant progeny arising as a result of random transposition of Tn5 into the genome of the recipient. Matings with *P. fluorescens* recipients generate transconjugants from such transposition events at a predictably low frequency. However, microcolonies resistant to kanamycin arise from such matings at a frequency several orders of magnitude higher. Most of these microcolonies do not grow and cannot be subcultured on medium containing the antibiotic. However, a few give rise to large kanamycin-resistant sectors which can be recultured (Hwang and Farrand, unpubl. results). This suggests that the Ti plasmid is capable of limited, abortive replication in *P. fluorescens*. The large colonies that occasionally arise from the micro-colonies presumably represent events in which Tn5 has transposed from the Ti plasmid into the genome of one or more cells at some point after the development of the microcolony.

VI. Conjugal Transfer Systems of Other Plasmids

VI.A. CONJUGAL TRANSFER OF pAGK84

As noted earlier, pAgK84, the plasmid responsible for production of agrocin 84 by the crown gall biocontrol strain *A. radiobacter* K84, apparently is selftransmissible. However, genetic analysis indicates that no more than 3 kb of this plasmid is responsible for conjugation (Farrand, *et al.*, 1985; Farrand, *et al.*, 1992). Among the known conjugation systems of plasmids from Gram-negative bacteria all are several times this size. The genes encoding the Mpf functions alone generally occupy 8-12 kb of DNA. This brings into question the conclusion by Farrand, *et al.* (1985) that pAgK84 is entirely self-conjugal. However, if the plasmid requires a mobilizing element, it is not dependent on the conjugation system of pAtC58, pTiC58, or pAtK84b. The agrocin plasmid can be conjugated from C58-type donors lacking pAtC58 and pTiC58 and from derivatives of K84 that lack pAtK84b (Farrand, *et al.*, 1985).

Alternatively, pAgK84 may harbor some novel type of gene transfer system. This is unlikely. Our recent sequence analysis has identified at least two genes of the *tra* region as being homologs of the *traI/virD2* and *traG/virD4* gene families (Peñalver and Farrand, unpubl. results). The former most likely codes for the site-specific single stranded endonuclease that nicks at the *oriT* of the plasmid. The latter probably codes for the Mpf-associated protein responsible for interfacing the relaxosome with the mating bridge. Although much work remains to be done, these results suggest that pAgK84 contains a conventional conjugation system. However, the plasmid may lack the genetic determinants for a mating bridge and, if so, must rely on that provided by some other transmissible replicon present in the agrobacterial cell. Alternatively, pAgK84 may also code for its own Mpf functions. How then to explain the observation that only those Tn5 insertions mapping to the 3 kb *tra* region influence conjugal transfer (Farrand, *et al.*, 1985)? It is conceivable that conjugation functions coded for by some otherwise unknown transmissible element in the *Agrobacterium* genome can complement mutations in some, if not all of the *trans*-acting Dtr and Mpf genes of pAgK84.

There is some practical interest in understanding the conjugation system of pAgK84. *Agrobacterium radiobacter* strain K84 is used commercially throughout the world to control crown gall disease on agronomically-important plants (Farrand, 1991 recent review). Production of the anti-agrobacterial antibiotic agrocin 84 by strain K84 is believed to be a major component of the control process. pAgK84 codes not only for production of agrocin 84, but also for immunity to this antibiotic. Transmission of pAgK84 to resident pathogenic agrobacteria in the soil could lead to a breakdown in the efficacy of use of this biocontrol agent. In fact, transfer of pAgK84 to pathogens has been detected under field conditions and these once-susceptible pathogens no longer are sensitive to agrocin 84 (Vicedo, *et al.*, 1993; Stockwell, *et al.*, 1996). Furthermore, there have been reports of the failure of strain K84 to control crown gall in areas in which it once was effective (L. W. Moore, pers. comm.). Given the potential of *in situ* plasmid transfer to compromise the biocontrol system, Jones, *et al.* (1988) developed a derivative of strain K84, called K1026, in which the known *tra* region of pAgK84 was deleted. This strain controls crown gall as well as does strain K84. And, as expected, it fails to transfer the agrocin plasmid to suitable recipient strains during *in vitro* matings. More importantly, studies suggest that strain K1026 does not transfer the modified agrocin plasmid under field conditions (Vicedo, *et al.*, 1993).

VI.B. OTHER LARGE PLASMIDS IN *AGROBACTERIUM* SPP.

Many isolates of *A. tumefaciens* harbor epigenetic elements substantially larger than the Ti plasmid. Most of these plasmids confer no known phenotype and are considered cryptic. Recent studies on the pathway for catabolism of the mannityl opines indicates that such plasmids in some isolates may code for uptake and catabolism of one member of a set of Amadori-type compounds containing deoxyfructose and glutamine called the chrysopine family of opines (Kim and Farrand, 1996; Y. Dessaux and V. Vaudequin-Dransart, personal communication; see Chapter 9). The classical nopaline/agrocinopine-type *A. tumefaciens* strain C58 harbors such a plasmid called pAtC58. A derivative of this *ca.* 450 kb plasmid marked with Tn1 was reported to be self-conjugal (van Montagu and Schell, 1979). In addition, Cook and Farrand (1992) reported that pAtC58 can mobilize transfer of RSF1010 at a low frequency. Beyond this, nothing is known about the conjugal characteristics of this plasmid. Nor is it known if pAtC58 is identical to, or related in any part to the cryptic plasmids found in other *Agrobacterium* isolates.

VI.C. OPINE-CATABOLIC PLASMIDS

Many field isolates of *Agrobacterium* spp. are not tumorigenic but can catabolize opines. Where examined, these strains harbor large plasmids that encode the opine utilization traits. As noted in Sections III and IV, one such plasmid, pAtK84b, is self-conjugal. Furthermore, transfer is induced by two sets of opines. Southern hybridizations indicate that this plasmid shares sequences in common with pTiC58 corresponding to the *tra/oriT* and *trb/rep-oriV* regions (Clare, *et al.*, 1990). These regions most likely account for the conjugal properties of this plasmid. Furthermore, pAtK84b contains two non-contiguous DNA fragments that hybridize with *traR* (Oger, *et al.*, 1998). Given the fact that conjugation of pAtK84b is inducible by nopaline and also by agrocinopines A and B, it is most probable that this plasmid contains two opine-regulated *traR* genes. The locations of these genes and their relationship, both physical and functional, with the opine catabolic determinants of this plasmid remain to be determined.

As described in Section III, many agropine-type isolates of *A. rhizogenes* harbor, in addition to the Ri plasmid, another large plasmid. In strains like A4 and 15834 this second plasmid codes for catabolism of a subset of the opines produced by the hairy roots induced by these strains (Petit, *et al.*, 1983). Thus, these elements are, in fact, opine catabolic plasmids. As noted previously, these plasmids most probably are self-conjugal (Petit, *et al.*, 1983) and there is reason to suspect that they can mobilize transfer of the Ri plasmid co-resident in the cell (see Section III). pArA4a, the opine catabolic plasmid from *A. rhizogenes* strain A4 confers catabolism of agrocinopines A and B (Petit, *et al.*, 1983; Hayman and Farrand, 1988). These sugar phosphodiesters serve as the conjugal opines of the nopaline/agrocinopine-type Ti plasmids as well as of pAtK84b described above. Whether these opines induce conjugal transfer of pArA4a has yet to be determined.

VI.D. TARTRATE PLASMIDS

Many isolates of *Agrobacterium vitis* utilize tartrate as sole source of carbon. In some cases the genes required for uptake and catabolism of this substrate are associated with the Ti plasmid. In other cases, these traits map to another large plasmid co-resident in the bacterium. Among the latter class, several of these tartrate plasmids are self-conjugal (Szegedi, *et al.*, 1992). Nothing is known about the *tra* genes on these plasmids. However, several of these plasmids conjugate at much higher frequencies when matings are performed on crown galls as compared to matings conducted on agar medium (Szegedi, et al., 1992). This is reminiscent of the requirement for opines exhibited by the Ti plasmids and suggests that the tartrate plasmids may contain a similar conjugal transfer system.

VI.E. OTHER TRANSMISSIBLE ELEMENTS

There is reason to believe that the *Agrobacterium* genome harbors one or more additional elements that are either conjugal or transmissible by mating. Cook and Farrand (1992) in their studies of the *oriT* region of pTiC58, observed that a derivative of *A. tumefaciens* C58 lacking pTiC58 and pAtC58 mobilized the transfer of pDSK519, a derivative of RSF1010, at a low but detectable frequency. This strain does not contain any detectable plasmids. Although the frequency of transfer was very low, the phenomenon was reproducible. This suggests that strain C58 harbors a conjugal element that is independent of pTiC58 and pAtC58, but which can mobilize transfer of the R plasmid.

The existence of such an element gains support from another set of observations. As noted above, pTiA6NC, which is completely deficient in conjugation, contains a deletion that extends into *trbI* (Alt-Mörbe, et al., 1996). However, when *traR* is expressed in this strain from a high copy number recombinant plasmid, such donors transfer pTiA6NC, albeit at a frequency several orders of magnitude lower than that of the related plasmid, pTiR10 (Alt-Mörbe, et al., 1997). The *trbI* homolog from RP4 is essential for conjugal transfer of the R plasmid (Haase, et al., 1995) and it is likely that the product of this gene plays a similar essential role in conjugal transfer of the Ti plasmid. How, then, does one account for the low-frequency conjugal transfer of pTiA6NC? It is conceivable that the defect in TrbI of this Ti plasmid is compensated for by an otherwise-cryptic conjugal element in this strain. The genetic and physical properties of this transmissible element remain to be determined.

In another set of studies, Cho, et al. (1997) demonstrated that the octopine/mannityl opine-type *A. tumefaciens* strain R10 could transfer *arcAB* and *putA*, markers associated with catabolism of arginine and proline, to a recipient strain. Such transfer was dependent upon the Ti plasmid and was inducible by octopine. These genes are not located on the Ti plasmid, but are associated with a ca 240 kb linear DNA element, called pAtR10b, in strain R10. Hybridization analysis suggested that this element also is present in derivatives of strain C58. Dessaux, et al. (1989) reported similar results showing that pTiR10 could transfer a set of markers usually associated with chromosomes to recipients. Interestingly, this latter study suggested that only certain of such markers could be transferred in this fashion. Among these are traits associated with catabolism of arginine and most likely represent the *arcAB* and *putA* genes studied by Cho, et al. (1997). The evidence is consistent with a model in which this linear element is a component of the *Agrobacterium* genome and can be mobilized for transfer by the Ti plasmid. pAtR10b apparently is not related to the 2.1 Mb linear chromosome that comprises a portion of the *Agrobacterium* chromosomal complement (Allardet-Servent, et al., 1993).

VII. Biological Significance and Implications

The phenomenon of conjugal transfer is significant to the biology of the *Rhizobiaceae* at levels ranging from the structure and evolution of the plasmids themselves through the roles played by these elements in the biology of the bacteria, to their function in establishing and maintaining diversity in the genetic structure of the family.

VII.A. PLASMID SURVIVAL

Conceptually, a plasmid is faced with two problems concerning its survival. The first is survival within the entire population of its resident host. The second, more subtle concern is the survival of the plasmid in the face of competition from bacteria more suitable to the habitat than its current host of residence. Conjugal transfer is one strategy by which a plasmid can protect itself from both threats. In the first case, conjugal transfer insures stability through horizontal transmission within the population. In the second, it insures survival by horizontal dissemination to other organisms. This latter property is especially important. Conjugal transfer to other rhizobia and agrobacteria in soils gives these plasmids the opportunity to test other chromosomal backgrounds as hosts that might be better adapted to the local environment. This is an intriguing possibility. There is ample evidence in the literature that Ti plasmids as well as *sym* and non-*sym* plasmids transfer in natural environments (Kerr, 1969, 1971; Kinkle and Schmidt, 1991; Pretorius-Güth, *et al.*, 1990; Rao, *et al.*, 1994; Schofield, *et al.*, 1987). It also is the case that soils can contain a high population of non-pathogenic *Agrobacterium* strains that lack Ti plasmids. These easily could constitute a reservoir of recipients. But what about rhizobia? Although we think of them as agrobacteria, based on 16S rRNA sequence, members of the biovar 2, non-pathogenic *A. radiobacter* group (now called *A. rhizogenes*) are virtually indistinguishable from many species of fast-growing rhizobia (Willems and Collins, 1993; Sawada, *et al.*, 1993). It is intriguing to consider the possibility that these organisms may serve as a diversity reservoir for *sym* plasmids as well as for Ti plasmids. By this scenario, if such a bacterium inherits a Ti plasmid, it becomes *A. tumefaciens*; if it inherits a *sym* plasmid, it becomes a species of *Rhizobium*.

VII.B. PLASMID EVOLUTION

Conjugal transfer systems insure plasmid maintenance within a population of cells. They also contribute to the horizontal distribution of plasmids to bacterial hosts of other chromosomal backgrounds. These two roles are important to the plasmids as well as to the bacteria in which they propagate. However, conjugal transfer systems may play a central and critical role in the evolution of the structure and the coding diversity of the plasmids themselves. In this regard, the gene structure of the core components of the Ti plasmids may be especially relevant. As described in several previous sections, at least one component of the *tra* regulon is intimately linked to the replication region of the classical IncRh1 Ti plasmids and also with that of at least one *sym* plasmid. In each of these, the *trb* operon, including *traI*, is divergently oriented from the *rep* genes and the two are separated by only about 400-800 bp (see, for example, Figure 2). In pNGR234a, the second component of the *tra* regulon, the *traI/oriT* cluster, also is closely linked to *rep*, while in the Ti plasmids, this locus is some distance away. Nor is this organization unique to pNGR234a among the large plasmids of *Rhizobium* species. The strong relatedness between the Ti plasmid *trb/rep* region and several *sym* plasmids from *R. leguminosarum* strains (Prakash and Schilperoort, 1982) indicates that this gene arrangement may be common among *Rhizobium* plasmids. These *sym* plasmids also contain regions that hybridize strongly to probes encompassing the *traI/oriT* component of the Ti plasmid conjugal

transfer system. Remarkably, the three genes of the *rep* locus are highly conserved between the Ti plasmids and pNGR234a (Freiberg, *et al.*, 1997). There is reason to suspect that the *rep/oriV* of these plasmids are members of a superfamily of replication systems common to the large plasmids present in members of the family. Thus Turner, *et al.* (1995), using PCR analysis, have identified a replication motif centering on conservation of the *repC* gene present in a variety *sym* and non-*sym* plasmids. Their sequence analysis of these *repC* genes indicates that, while they all are related, they show branches of divergence, indicating that the plasmid replicons may be separating from one another. Although only limited sequence is available, one such replication region, that of pRL8JI, contains at least a portion of an open reading frame that could code for a protein closely related to TrbB of the Ti plasmid *trb* operon (our analysis of sequences in data base accession X89447; see Turner and Young, 1995). Like the arrangement in the Ti plasmids, this gene is oriented divergently from the *rep* locus. Unlike the Ti plasmid, there apparently is no homolog of *traI* located in this region of pRL8JI. A picture emerges from these observations in which the Ti plasmids and at least some of the *sym* and non-*sym* plasmids represent different but related lineages of the same basic plasmid replicon. One class is distinguished from another only by the nature of the genes associated with the basic replicon. Thus, Ti plasmids contain *vir*, T-regions and opine-catabolic loci while the *sym* plasmids contain genes for nodulation and nitrogen fixation. But these plasmids, as large as they are, also contain a bewildering array of other genes and gene systems, the functions of which are not understood. It is likely that these genes are required for some facet of the biology of the bacterial host in which the plasmid resides. Intriguingly, some such functions may share parallels of biological significance. Thus, the opine systems of the Ti plasmids are conceptually mirrored by the *scyllo*-inosamine system encoded on the *sym* plasmid of *S. meliloti* L-530 (Murphy, *et al.*, 1987; Murphy, *et al.*, 1988; Murphy, *et al.*, 1993; see Chapter 9).

Why, then, the association of the *tra* regulon with this conserved replicon? Given the conservation of the gene sets, their internal organization and in some cases, linkage with each other, it seems clear that the *tra* regulon is part of the core replicon structure of these plasmids. Largely from the elegant work of Otten's group on the structures of the Ti plasmids from *A. vitis*, we now understand that the Ti plasmids evolve by recombining large blocks of DNA (Otten, *et al*, 1992). Thus, diversity among these plasmids comes not from mutation, but from acquisition and reassortment of existing gene sets. With the sequence of pNGR234a now available and with the considerable amount of information available concerning relatedness between various plasmids of *Rhizobium*, it is most likely that the *sym* and non-*sym* plasmids evolve by the same mechanisms. Recombination bringing in new traits requires gene transfer. Thus, the conjugal transfer systems of these plasmids may well be indispensable for their evolution.

In this respect, one can envision that Ti (or *sym*) plasmids in the recipient may serve as the substrate for recombination with another Ti (or *sym*) plasmid entering from a donor. Indeed, work by Hooykaas *et al.* (1980) demonstrated that cointegrates do form when a Ti plasmid of one type enters a recipient harboring a different Ti plasmid. The product of recombination could be better suited for a particular plant or a particular environment than either of the two parental plasmids. The chimeric structural relatedness of the nopaline/agrocinopine- and the octopine/mannityl opine-type Ti plasmids is a predictive outcome of this hypothesis. The fact that the nopaline- and the octopine/cucumopine-type Ti plasmids of *A. vitis* contain a large segment that is essentially co-linear with a region of pTiC58 (Otten and De Ruffray, 1994)) lends additional support to this scenario. Such recombinations most certainly occur among the plasmids of *Rhizobium* spp. The *sym* plasmid pBR1AN, isolated from a mating between *R. leguminosarum* bv. *viciae* and *R. leguminosarum* bv. *trifolii* is a derivative of the pea *sym* plasmid pJB5JI that has acquired *nif* and *nod* determinants from the clover *sym* plasmid (Djordjevic, *et al.* 1983). Strains harboring pBR1AN nodulate both peas and clover. Thus, an unforced recombination event has generated a chimeric *sym* plasmid that confers a broader host range than either of its two parents. Alternatively, opine-catabolic (or non-*sym*) plasmids may serve as similar recombinational partners, thus generating novel Ti or *sym* plasmids. It is even conceivable that *sym* plasmids might acquire traits from Ti plasmids by such recombination events. In support of this, the *sym* megaplasmid of *S. meliloti* A3 contains a region that is closely related at the DNA sequence level to the *noc* operon of the nopaline/agrocinopine A+B-type Ti plasmids (Au and Dion, 1994). This isolate of *S. meliloti* can utilize nopaline as sole carbon source.

VII.C. GENOME EVOLUTION

Although considerably more speculative, it is possible that the conjugal systems of these plasmids play an important role in the evolution of the chromosomal gene compliment in members of the family *Rhizobiaceae*. There is abundant literature concerning the fluid nature of the *Rhizobium* genome. The genomes of both *Agrobacterium* and *Rhizobium* contain a number of active insertion sequences and transposons. Some of these elements have transposed to the indigenous plasmids in these strains. One can easily imagine that these conjugal plasmids could be vehicles for moving chromosomal genes from one bacterium to another. The observations by Dessaux, *et al.* (1989) and more recently by Cho, *et al.* (1997) concerning mobilization of select markers by Ti plasmid transfer serves to emphasize this role. However, we do not have enough information, either primary or comparative, concerning the chromosomal genome of these bacteria to draw firm conclusions about the role of plasmid transfer in their evolution.

In summary, given these speculations, we propose that the conserved linkage of the *tra* regulon with the *rep* region reflects the need for these plasmids to conjugate in order to assure their survival and to promote their diversity. But it also is possible that the conserved core of *tra* and *rep* functions is essential to the mechanisms by which the chromosomal components of the genomes of these bacteria develop and maintain their diversity.

VIII. Acknowledgements

Work in the author's laboratory was supported by grants R01 GM 52465 from the NIH, and AG93-37301-8943 and AG95-37312-1639 from the USDA. I thank Pei-Li Li, Audra Smyth, Ramon Peñalver, Philippe Oger, Ingyu Hwang, Zhaoquin Luo, Kevin Piper, Stephen Winans, Clay Fuqua, Andrew Binns, Erich Lanka, Anath Das and Walt Ream for helpful discussions during the preparation of this review.

IX. References

Allardet-Servent, A., Michaux-Charachon, S., Jumas-Bilak, E., Karayan, L. and Ramuz, M. (1993) J. Bacteriol. 175, 7869-7874.

Alt-Mörbe, J., Stryker, J.L., Fuqua, C., Li, P-L., Farrand, S.K. and Winans, S.C. (1996) J. Bacteriol. 178, 4248-4257.

Au, S., Bergeron, J. and Dion, P. (1994) Abstr. 7th Internatl. Symp. Mol. Plant-Microbe Interact. Abstr. 1, p 15.

Baldwin, T.O., Devine, J.H., Heckel, R.C., Lin, J-W. and Shadel, G.S. (1989) J. Biolum. Chemilum. 4, 326-341.

Beck von Bodman, S., Hayman, G.T. and Farrand, S.K. (1992) Proc. Natl. Acad. Sci. (USA) 89, 643-647.

Beck von Bodman, S., McCutchan, J.E. and Farrand, S.K. (1989) J. Bacteriol. 171, 5281-5289.

Beijersbergen, A., den Dulk-Ras, A., Schilperoort, R.A. and Hooykaas, P.J.J. (1992) Science 256, 1324-1327.

Blattner, F.R., Plunkett III, G., Bloch, C.A., Perna, N.T., Burland, V., Riley, M., Collado-Vides, J., Glasner, J.D., Rode, C.K., Mayhew, G.F., Gregor, J., Davis, N.W., Kirkpatrick, H.A., Goeden, M.A., Rose, D.J., Mau, B. and Shao, Y. (1997) Science 277, 1453-1462.

Brewin, N.J., Beringer, J.E. and Johnston, A.W.B. (1980) J. Gen. Microbiol. 120, 413-420.

Buchanan-Wollaston, V., Passiatore, J.E. and Cannon, F. (1987) Nature (London) 328, 172-175.

Cho, K. and Winans, S.C. (1993) J. Bacteriol. 175, 7715-7719.

Cho, K., Fuqua, C. and Winans, S.C. (1997) J. Bacteriol. 179, 1-8.

Clare, B.G., Kerr, A. and Jones, D.A. (1990) Plasmid 23, 126-137.

Climo, M.W., Sharma, V.K. and Archer, G.L. (1996) J. Bacteriol. 178, 4975-4983.

Cook, D.M. (1996) Ph.D. thesis, University of Illinois, Urbana, Illinois USA

Cook, D.M. and Farrand, S.K. (1992) J. Bacteriol. 174, 6238-6246.

Cook, D.M., Li, P-L., Ruchaud, F., Padden, S. and Farrand, S.K. (1997) J. Bacteriol. 179, 1291-1297.

De Greve, H., Decraemer, H., Seurinck, J., van Montagu, M. and Schell, J. (1981) Plasmid 6, 235-248.

Dessaux, Y., Petit, A., Ellis, J.G., Legrain, C., Demarez, M., Wiame, J-M., Popoff, M. and Tempé, J. (1989) J. Bacteriol. 171, 6363-6366.

Djordjevic, M.A., Zurkowski, W., Shine, J. and Rolfe, B.G. (1983) J. Bacteriol. 156, 1035-1045.

Drummond, M.H. and Chilton, M-D. (1978) J. Bacteriol. 136, 1178-1183.

Dunlap, P.V. and Greenberg, E.P. (1991) In, Dworkin, M (ed.), Microbial cell-cell interactions, pp 219-253. American Society for Microbiology, Washington DC.

Eberhard, A., Burlingame, A.L., Eberhard, C., Kenyon, G.L., Nealson, K.H. and Oppenheimer, N.J. (1981) Biochemistry 20, 2444-2449.

Eberhard, A., Longin T., Widrig, C.A. and Stranick, S.J. (1991) Arch. Microbiol. 155, 294-297.

Engler, G., Depicker, A., Maenhaut, R., Villarroel, R., van Montagu, M. and Schell, J. (1981) J. Mol. Biol. 152, 183-208.

Ellis, J.G., Murphy, P.J. and Kerr, A. (1982a) Mol. Genet. Genet. 186, 275-281.

Ellis, J.G., Kerr, A., Petit, A. and Tempé, J. (1982b) Mol. Gen. Genet. 186, 269-274.

Farrand, S.K. (1991) In, Pimentel, D (ed.) CRC handbook of pest management in agriculture, 2nd ed., Vol II, pp311-329, CRC Press, Boca Raton, FL.

Farrand, S.K. (1993) In, Clewell, DB (ed) Bacterial conjugation, pp. 255-291, Plenum Press, New York, NY.

Farrand, S.K., Slota, J.E., Shim, J-S. and Kerr, A. (1985) Plasmid 13, 106-117.

Farrand, S.K., Wang, C-L., Hong, S-B., O'Morchoe, S.B. and Slota, J.E. (1992) Plasmid 28, 201-212.

Farrand, S.K., Hwang, I. and Cook, D.M. (1996) J. Bacteriol. 178, 4233-4247.

Freiberg, C., Fellay, R., Bairoch, A., Broughton, W.J., Rosenthal, A. and Perret, X. (1997) Nature (London) 387, 394-401.

Fullner, K.J. and Nester, E.W. (1996) J. Bacteriol. 178, 1498-1504.

Fuqua, C. and Winans, S.C. (1994) J. Bacteriol. 176, 2796-2806.

Fuqua, C., Burbea, M. and Winans, S.C. (1995) J. Bacteriol. 177, 1367-1373.

Fuqua, C. and Winans, S.C. (1996a) Mol. Microbiol. 20, 1199-1210.

Fuqua, C. and Winans, S.C. (1996b) J. Bacteriol. 178, 435-440.

Fuqua, C., Winans, S.C. and Greenberg, E.P. (1994) J. Bacteriol. 176, 269-275.

Fuqua, C., Winans, S.C. and Greenberg, E.P. (1996) Annu. Rev. Microbiol. 50, 727-751.

Gelvin, S.B. and Habeck, L.L. (1990) J. Bacteriol. 172, 1600-1608.

Genetello, C., van Larebeke, N., Holsters, M., De Picker, A., van Montagu, M. and Schell, J. (1977) Nature (London) 265, 561-563.

Grahn, A.M., Haase, J., Lanka, E. and Bamford, D.H. (1997) J. Bacteriol. 179, 4733-4740.

Haase, J., Lurz, R., Grahn, A.M., Bamford, D.H. and Lanka, E. (1995) J. Bacteriol. 177, 4779-4791.

Haase, J., Kalkum, M. and Lanka, E. (1996) J. Bacteriol. 178, 6720-6729.

Haase, J. and Lanka, E. (1997) J. Bacteriol. 179, 5728-5735.

Habeeb, L., Wang, L. and Winans, S.C. (1991) Mol. Plant-Microbe Interact. 4, 379-385.

Hamilton, C.M., Cook, D.M., Lanka, E. and Farrand, S.K. (1996) Abstr. 96th Gen. Mtg. Amer. Soc. Microbiol. Abstr H77, p. 496.

Hayman, G.T. and Farrand, S.K. (1988) J. Bacteriol. 170, 1759-1767.

Hayman, G.T., Beck von Bodman, S., Kim, H., Jiang, P. and Farrand, S.K. (1993) J. Bacteriol. 175, 5575-5584.

Holsters, M., Silva, B., van Vliet, F., Genetello, C., de Block, M., Dhaese, P., Depicker, A., Inzé, D., Engler, G., Villarroel, R., van Montagu, M. and Schell, J. (1980) Plasmid 2, 212-230.

Hooykaas, P.J.J., Klapwijk, P.M., Nuti, M.P., Schilperoort, R.A. and Rörsch, A. (1977) J. Gen. Microbiol. 98, 477-484.

Hooykaas, P.J.J., Roobol, C. and Schilperoort, R.A. (1979) J. Gen. Microbiol. 110, 99-109.

Hooykaas, P.J.J., den Dulk-Ras, H., Ooms, G. and Schilperoort, R.A. (1980) J.Bacteriol. 143, 1295-1306.

Hooykaas, P.J.J., van Brussel A.A.N., den Dulk-Ras, H., van Slogteren, G.M.S. and Schilperoort, R.A. (1981) Nature (London) 291, 351-353.

Hooykaas, P.J.J., Snijdewint, F.G.M. and Schilperoort, R.A. (1982) Plasmid 8, 73-82.

Hooykaas, P.J.J., Den Dulk-Ras, H., Regensburg-Tuïnk, A.J.G., Van Brussel, A.A.N. and Schilperoort, R.A. (1985) Plasmid 14, 47-52.

Huisman, G.W. and Kolter, R. (1994) Science 265, 537-539.

Hwang, I., Cook, D.M. and Farrand, S.K. (1995) J. Bacteriol. 177, 449-458.

Hwang, I., Li, P-L., Zhang, L., Piper, K.R., Cook, D.M., Tate, M.E. and Farrand S.K. (1994) Proc. Natl. Acad. Sci. (USA) 91, 4639-4643.

Johnston, A.W.B., Hombrecher, G., Brewin, N.J. and Cooper, M.C. (1982) J. Gen . Microbiol. 128, 85-93.

Johnston, A.W.B., Beynon, J.L., Buchanan-Wollaston, A.V., Setchell, S.M., Hirsch, P.R. and Beringer, J.E. (1978) Nature (London) 276, 634-636.

Jones, D.A., Ryder, M.H., Clare, B.G., Farrand, S.K. and Kerr, A. (1988) Mol. Gen. Genet. 212, 207-214.

Jouanin, L. (1984) Plasmid 12, 91-102.

Kaplan, H.B. and Greenberg, E.P. (1985) J. Bacteriol. 163, 1210-1214.

Kerr, A. (1969) Nature (London) 223, 1175-1176.

Kerr, A. (1971) Physiol. Plant Pathol. 1, 241-246.

Kerr, A., Manigault, P. and Tempé, J. (1977) Nature (London) 265, 560-561.

Kondorosi, A., Kondorosi, E., Pankhurst, C.E., Broughton, W.J. and Banfalvi, Z. (1982) Mol. Gen. Genet. 188, 433-439.

Kim, H. and Farrand, S.K. (1997) J. Bacteriol. 179, 7559-7572.

Kim, H. and Farrand, S.K. (1998) Mol. Plant-Microbe Interact., 11, 131-143.

Kim, K.S. and Farrand, S.K. (1996) J. Bacteriol. 178, 3275-3284.

Kinkle, B.K. and Schmidt, E.L. (1991) Appl. Environ. Microbiol. 57, 3264-3269.

Klapwijk, P.M. and Schilperoort, R.A. (1979) J. Bacteriol. 139, 424-431.

Klapwijk, P.M., Scheulderman, T. and Schilperoort, R.A. (1978) J. Bacteriol. 136, 775-785.

Knauf, V.C. and Nester, E.W. (1982) Plasmid 8, 45-54.

Komari, T., Halperin, W. and Nester, E.W. (1986) J. Bacteriol. 166, 88-94.

Kovács, L.G. and Pueppke, S.G. (1994) Mol. Gen. Genet. 242, 327-336.

Lamb, J.W., Hombrecher, G. and Johnston, A.W.B. (1982) Mol. Gen. Genet. 186, 449-452.

Lessl, M., Pansegrau, W. and Lanka, E. (1992) Nucleic Acids Res. 20, 6099-6100.

Lessl, M., Balzer, D., Weyrauch, K. and Lanka, E. (1993) J. Bacteriol. 175, 6415-6425.

Melchers, L.S., Maroney, M.J., den Dulk-Ras, A., Thomspon, D.V., van Vuuren, H.A.J., Schilperoort, R.A. and Hooykaas, P.J.J. (1990) Plant Mol. Biol. 14, 249-259.

Mercado-Blanco, J. and Olivares, J. (1993) Arch. Microbiol. 160, 477-485.

Mercado-Blanco, J. and Toro, N. (1996) Mol. Plant-Microbe Interact. 9, 535-545.

Moore, L.W., Warren, G. and Strobel, G. (1979) Plasmid 2, 617-626.

Moré, M., Finger, L.D., Stryker, J.L., Fuqua, C., Eberhard, A. and Winans, S.C. (1996) Science 272, 1655-1658.

Murphy, P.J., Heycke, N., Banfalvi, Z., Tate, M.E., de Bruijn, F., Kondorosi, A., Tempé, J. and Schell, J. (1987) Proc. Natl. Acad. Sci. (USA) 84, 493-497.

Murphy, P.J., Heycke, N., Trenz, S.P., Ratet, P., de Bruijn, F., Kondorosi, A., Tempé, J. and Schell, J. (1988) Proc. Natl. Acad. Sci. (USA) 85, 9133-9137.

Murphy, P.J., Trenz, S.P., Grzemski, W., de Bruijn, F.J. and Schell, J. (1993) J. Bacteriol. 175, 5193-5204.

Oger, P., Kim, K.S., Sackett, R.L., Piper, K.R. and Farrand, S.K. (1998) Mol. Microbiol. 27, 277-288.

Okamoto, S., Toyoda-Yamamoto, A., Ito, K., Takebe, I. and Machida, Y. (1991) Mol. Gen. Genet. 228, 24-32.

Otten, L. and De Ruffray, P. (1994) Mol. Gen. Genet. 245, 493-505.

Otten, L., Canaday, J., Gérard, J-C., Fournier, P., Crouzet, P. and Paulus, F. (1992) Mol. Plant-Microbe Interact. 5, 279-287.

Otten, L., Gérard, J-C. and De Ruffray, P. (1993) Plasmid 29, 154-159.

Pansegrau, W. and Lanka, E. (1991) Nucleic Acids Res. 19, 345.

Pansegrau, W., Lanka, E., Barth, P.T., Figurski, D.H., Guiney, D.G., Haas, D., Helinski, D.R., Schwab, H., Stanisch, V.A. and Thomas, C.M. (1994) J. Mol. Biol. 239, 623-663.

Pansegrau, W., Balzer, D., Kruft, V., Lurz, R. and Lanka, E. (1990) Proc. Natl. Acad. Sci. (USA) 87, 6555-6559.

Petit, A. and Tempé, J. (1978) Mol. Gen. Genet. 167, 147-155.

Petit, A., David, C., Dahl, G.A., Ellis, J.G., Guyon, P., Casse-Delbart, E. and Tempé, J. (1983) Mol. Gen. Genet. 190, 204-214.

Piper, K.R., Beck von Bodman, S. and Farrand, S.K. (1993) Nature (London) 362, 448-450.

Pohlman, R.F., Genetti, H.D. and Winans, S.C. (1994) Mol. Microbiol. 14, 655-668.

Prakash, R.K. and Schilperoort, R.A. (1982) J. Bacteriol. 149, 1129-1134.

Pretorius-Güth, I-M., Pühler, A. and Simon, R. (1990) Appl. Environ. Microbiol. 56, 2354-2359.

Ramírez-Romero, M.A., Bustos, P., Girard, O.R., Cevallos, M.A. and Dávila, G. (1997) Microbiology 143, 2825-2831.

Rao, J.R., Fenton, M. and Jarvis, B.D.W. (1994) Soil Bio. Biochem. 26, 339-351.

Rogowsky, P.M., Powell, B.S., Shirasu, K., Lin, T-S., Morel, P., Zyprian, E.M., Steck, T.R. and Kado, C.I. (1990) Plasmid 23, 85-106.

Rosemeyer, V., Michiels, J., Verreth, C. and Vanderleyden, J. (1998) J. Bacteriol. 180, 815-821.

Sawada, H., Ieki, H., Oyaizu, H. and Matsumoto, S. (1993) Internatl. J. Syst. Bacteriol. 43, 694-702.

Schaefer, A.L., Hanzelka, B.L., Cronan, Jr. J.E. and Greenberg, E.P. (1996) Proc. Natl. Acad. Sci. (USA) 93, 9505-9509.

Scherzinger, E., Lurz, R., Otto, S. and Dobrinski, B. (1992) Nucleic Acids Res. 20, 41-48.

Schofield, P.R., Gibson, A.H., Dudman, W.F. and Watson, J.M. (1987) Appl. Environ. Microbiol. 53, 2942-2947.

Sciaky, D., Montoya, A.L. and Chilton, M-D. (1977) Plasmid 1, 238-253.

Scholz, P., Haring, V., Wittmann-Liebold, B., Ashman, K., Bagdasarian, M. and Scherzinger, E. (1989) Gene 75, 271-288.

Silverman, P.M. (1997) Mol. Microbiol. 23, 423-429.

Sprinzel, M. and Geider, K. (1988) J. Gen. Microbiol. 134, 413-424.

Steck, T.R. and Kado, C.I. (1990) J. Bacteriol. 172, 2191-2193.

Stevens, A.M. and Greenberg, E.P. (1997) J. Bacteriol. 179, 557-562.

Stevens, AM, Dolan, KM and Greenberg, EP. (1994) Proc. Natl. Acad. Sci. (USA) 91, 12619-12623.

Stockwell, V.O., Moore, L.W. and Loper, J.E. (1993) Appl. Environ. Microbiol. 59, 2112-2120.

Swift, S., Throup, J.P., Williams, P., Salmond, G.P.C. and Stewart, G.S.A.B. (1996) Trends in Biol. Sci. 21, 214-219.

Szegedi, E., Otten, L. and Czako, M. (1992) Mol. Plant-Microbe Interact. 5, 435-438.

Tempé, J., Petit, A., Holsters, M., van Montagu, M. and Schell, J. (1977) Proc. Natl. Acad. Sci. (USA) 74, 2848-2849.

Turner, S.L. and Young, J.P.W. (1995) FEMS Microbiol. Lett. 133, 53-58.

Turner, S.L., Rigottier-Gois, L., Power, R.S., Amarger, N. and Young, J.P.W. (1996) Microbiology 142, 1705-1713.

Unger, L., Ziegler, S.F., Huffman, G.A., Knauf, V.C., Peet, R., Moore, L.W., Gordon, M.P. and Nester, E.W. (1985) J. Bacteriol. 164, 723-730.

Valdivia, R.H., Wang, L. and Winans, S.C. (1991) J. Bacteriol. 173, 6398-6405.

Van Montagu, M. and Schell, J. (1979) In, Timmis, K.N. and Pühler, A. (eds) Plasmids of medical, environmental and commerical importance, pp71-95, Elsevier/North Holland Biomecical Press, Amsterdam.

Vaudequin-Dransart, V., Petit, A., Poncet, C., Ponsonnet, C., Nesme, X., Jones, J.B., Bouzar, H., Chilton, W.S. and Dessaux, Y. (1995) Mol. Plant-Microbe Interact. 8, 311-321.

Vicedo, B., Peñalver, R., Asins, M. and López, M.M. (1993) Appl. Environ. Microbiol. 59, 309-315.

Von Lintig, J., Zanker, H. and Schröder, J. (1991) Mol. Plant-Microbe Interact. 4, 370-378.

Wang, L., Helmann, J.D. and Winans, S.C. (1992) Cell 69, 659-667.

Waters, VL, Stack, B, Pansegrau, W, Lanka, E and Guiney, DG. (1992) J. Bacteriol. 174, 6666-6673.

Watson, B., Currier, T.C., Gordon, M.P., Chilton, M-D. and Nester, E.W. (1975) J. Bacteriol. 123, 255-264.

White, F.F. and Nester, E.W. (1980) J. Bacteriol. 144, 710-720.

Willems, A. and Collins, M.D. (1993) Internatl. J. Syst. Bacteriol. 43, 305-313.

Zanker, H., Lurz, G., Langridge, U., Langridge, P., Kreusch, D. and Schröder, J. (1994) J. Bacteriol. 176, 4511-4517.

Zanker, H., Von Lintig, J. and Schröder, J. (1992) J. Bacteriol. 174, 841-849.

Zhang, L. and Kerr, A. (1991) J. Bacteriol. 173, 1867-1872.

Zhang, L., Murphy, P.J., Kerr, A. and Tate, M.E. (1993) Nature (London) 362, 446-448.

Zhu, J. and Winans, S.C. (1998) Mol. Microbiol. 27, 289-297.

Chapter 11

Attachment of *Rhizobiaceae* to Plant Cells

Ann G. Matthysse and Jan W. Kijne

I. Introduction

Rhizobiaceae are a family of free-living soil bacteria. Thanks to special properties, these bacteria can escape from poor soil conditions by spending a part of their lifetime in or on a plant. During this stage of life, many of them are attached to the surface of plant cells. The varied structure of plant cell walls offers several possibilities for physicochemical interactions, ranging from hydrophobic interactions to ligand-receptor binding. Among the *Rhizobiaceae*, *Agrobacterium* and (*Sino/Meso/Brady/Azo*) *Rhizobium* (the latter group referred to as rhizobia) have been studied in some detail with regard to their attachment properties. These genera will be discussed in this chapter.

Plant roots attract agrobacteria and rhizobia. The bacteria are chemotactic to substances released from plants such as sucrose and amino acids (Gaworzewska and Carlile, 1982; Hawes and Smith, 1989). Following chemotaxis the bacteria can colonize plant roots forming microcolonies or biofilms or entering wound sites. These wounds may be made by foreign agents (including scientists) or may be the result of tissue damage occurring during normal plant growth (for example, the cracks produced at the site of emergence of lateral roots).

I.A. *AGROBACTERIUM*-PLANT INTERACTIONS

Once *A. tumefaciens* have entered a wound, they bind loosely to the surface of the plant cells (Matthysse, 1986). This loose binding is followed by bacterial cellulose synthesis which results in tight irreversible binding of the bacteria to the plant cells (Matthysse, 1983). Phenolic substances (acetosyringone and related compounds) combined with the acid pH and sugars commonly found in plant wounds cause the induction of the bacterial *vir* genes, which are carried on a plasmid (pTi) (Binns and Thomashow, 1988). The *vir* genes are involved in the formation of a pilus encoded by the *virB* operon (Fullner *et al.*, 1996; Kado, 1994; Lessi and Lanka, 1994; Pohlman *et al.*, 1994; Sanders *et al.*, 1991), and the transfer of T-DNA sequences from the bacterial pTi to the plant cell mediated by *virB, C, D,* and *E* gene products (and possibly the products of other genes) (Klee *et al.,* 1983). The T-DNA is integrated into plant chromosomes (Chilton *et al.,* 1977). The expression of the T-DNA in the plant cell leads to the production of increased levels of auxin and cytokinin resulting in uncontrolled growth of the cell to form a crown gall tumor (Binns and Thomashow, 1988; Akiyoski *et al.,* 1982). Other genes, which are expressed in the plant cell, are included in the T-DNA, most notably the genes for the enzymes for opine synthesis by the plant cell (Tempe *et al.,* 1984). The bacterium possesses on its Ti plasmid the genes required for the ability to use these opines as carbon and nitrogen sources to support bacterial growth. It is the induction of opine production by the plant cell and the use of opines as substrates for growth by the bacteria which thought to be the explanation for the benefit to the bacterium of tumor formation (see also Chapter 12).

The initial bacterial attachment to plant cells discussed in this chapter appears to be required for a later secondary attachment (presumably mediated by the pilus) and for the transfer of T-DNA from the bacterium to the plant cell. Most agrobacteria appear to be able to carry out this DNA transfer to a wide range of host plants including dicots, monocots, and gymnosperms. The bacteria also show binding to wound sites on these plants (Matthysse, 1986).

In addition, initial attachment to the surface of plant cells may play a role in colonization of intact roots by agrobacteria. Whether attachment to wound sites leading to DNA transfer and attachment to intact roots leading to colonization involve similar or even identical mechanisms is unknown. One group of agrobacteria, *A. vitis* that is primarily a pathogen of grapes, can invade the root and colonize the xylem (Burr *et al.,* 1995). The mechanism by which these bacteria move from the soil to the xylem is unclear (see also Chapter 10).

I.B. *RHIZOBIUM*-PLANT INTERACTIONS

Rhizobia are also able to enter wounds on roots and bind to plant cell surfaces. But rhizobia can take this process one step further. These bacteria are able to invade intact plant cells, namely growing root hairs and root cortical cells. Certain rhizobia can enter roots between epidermal cells. Rhizobial root infection is host-plant-specific and is restricted to

roots of leguminous plants (*Fabaceae*) and of *Parasponia,* a woody member of the elm family. In response to flavonoid compounds produced by the plant root, rhizobia produce specific signal molecules, lipo-chitin oligosaccharides (LCOs) (Lerouge *et al.,* 1990; Spaink *et al.,* 1991). The ability to produce LCOs is encoded by *nod* (nodulation) genes, present either on a plasmid (pSym) or on the chromosome. Root infection requires production of LCOs. These signal molecules activate host root cells. Epidermal and outer root cortical cells either allow intercellular penetration or form a tip-growing tube-like structure, the infection thread, through which the bacteria enter a root hair and the underlying cortical cells (Van Brussel *et al.,* 1992). Infection thread formation requires the presence of living bacteria. Evidence is accumulating that, in addition to LCOs, rhizobial surface polysaccharides are essential factors in the root infection process (Leigh and Walker, 1994). It is assumed that infection thread formation starts at the site of rhizobial attachment to a root hair tip and is initiated by local production of LCOs by immobilized rhizobia. This attachment process resembles that of agrobacteria in that loose binding is followed by anchoring with bacterial cellulose fibrils (Smit *et al.,* 1987). However, the following multiplication of the rhizobia either in the root hair curl and in the infection thread or between the root cortical cells requires the bacteria to be free and unattached. Thus, whereas agrobacteria show their most significant interaction with plant cells in a state of attachment, attachment of rhizobia during root infection is a transient phenomenon (see also Chapter 22).

Root infection by rhizobia ultimately results in the formation of root nodules. The bacteria are taken up by nodule cells, and develop into nitrogen-fixing bacteroids (Chapter 23). Since few rhizobia infect a host plant root and many more leave a senescent nodule, a root nodule can be considered as a host-plant-specific enrichment culture of rhizobia.

Rhizobia are also able to colonize the plant root surface. As with agrobacteria, this type of attachment has not yet been studied as a separate topic.

I.C. COMPARISON OF AGROBACTERIA AND RHIZOBIA

Most probably, rhizobia do not conjugate with plant cells. However, transfer of a Ti plasmid into certain rhizobia confers upon these bacteria the ability to behave like agrobacteria and to induce plant tumor formation (Hooykaas *et al.*, 1977). This suggests that rhizobial attachment to the surface of plant cells in wound sites is similar to that of agrobacteria. In general, rhizobia produce smaller tumors than do agrobacteria, if at all, and apparently conjugative behavior with plant cells requires more than most rhizobia can offer. Still, the attachment mechanism may be similar. *Vice versa*, agrobacteria can attach to legume root hairs. Transfer of nodulation genes to agrobacteria confers upon these bacteria the ability to produce LCOs and to induce nodule formation (Hooykaas *et al.*, 1981). However, in many legumes infection thread formation is poor and obviously requires specific rhizobial properties.

These observations have suggested to some researchers that agrobacteria and rhizobia use the same two-step mechanism for attachment to plant cells (Dazzo and Truchet, 1983, Smit *et al.*, 1992), whereas additional interactions such as conjugation (agrobacteria) and infection thread formation (rhizobia) are controlled by genus-specific genes. It should be noted that such specific genes may influence the attachment conditions. For example, rhizobia can produce LCOs when present in the rhizosphere. Within a few minutes, these LCOs can influence growth of root hairs, the target cells for rhizobial attachment (Heidstra *et al.*, 1994; see also Chapter 22). A change in root hair growth most probably includes a change in the surface of the root hair tip. Thus, by secreting LCOs, rhizobia may influence the surface to which they are going to attach. This possibility will play a role in attachment assays which allow rhizobia to produce LCOs and allow root hairs to grow.

However, observations of the broad host range of most agrobacteria and the ability of these bacteria to attach to root surfaces and wound sites of plants ranging from tobacco, carrot and *Arabidopsis thaliana* to moss when compared with the narrow host range of most rhizobia (generally a limited group of legumes) have suggested to other researchers that mechanisms of the first step of the initial attachment may differ between these related bacteria (Matthysse, 1996). In addition, agrobacteria can induce tumors on many parts of the plant including roots, tubers, stems, and leaves while rhizobia-plant interactions are generally limited to roots. The second step of the initial attachment involving the synthesis of cellulose fibrils is commonly agreed to be similar in both groups of bacteria.

In the following part of this chapter, we will discuss whether certain similarities and differences in interactions of Rhizobiaceae with plant cells can be based on attachment characteristics. Crucial in this discussion is an evaluation of methods in use for the study of bacterial attachment to plant cells and knowledge about the growth conditions for agrobacteria and rhizobia in plant wounds and in the rhizosphere, respectively.

II. Methods

Study of bacterial attachment requires the use of an attachment assay. A good assay meets two conditions: (1) use of circumstances which resemble *in vivo* circumstances, and (2) the possibility of direct quantification. These prerequisites can be incompatible and a critical evaluation of an attachment assay is an essential part of a critical evaluation of attachment results.

II.A. AGROBACTERIA

The early work on the attachment of *A. tumefaciens* to plant cells relied on an indirect assay using the inhibition of tumor formation on bean leaves by various substances or bacterial strains as a measure of their ability to interfere with bacterial adhesion (Lippincott and Lippincott, 1969). While this assay has the advantage that the only attachment measured is that required for virulence, it has the considerable disadvantage that anything which interferes with tumor formation is assumed to act via the inhibition of bacterial attachment. Thus, it is unclear whether the inhibition of tumor formation by lipopolysaccharide (LPS) from the bacteria or by methylated pectin from the plant represents inhibition of attachment by these substances or is due to interference with some other aspect of tumor formation.

In the late 1970s, this indirect assay was replaced by assays measuring the attachment of bacteria to tissue culture cells (Ohyama *et al.*, 1979; Matthysse *et al.*, 1978). The use of these cells has the advantage that bacterial adhesion can be measured directly either microscopically, by using radioactively labeled bacteria, or by counts of viable bacterial cells. The major disadvantage of the use of tissue culture cells is that they may not represent the cell types or microenvironments, which the bacteria normally encounter during tumor formation.

Microscopic assays are difficult to quantitate and necessitate the use of large numbers of bacteria. However, they allow the determination of the spatial localization of the adherent bacteria. Assays using radioactive bacteria also require large numbers of bacteria. They can be quantitated relatively easily, provided some correction is made for quenching. However, bacteria that are trapped in the tissue, but not actually bound to it, are counted as bound in these assays. The same holds true for bacteria that are bound to other bacteria rather than to the plant cell surface. Viable cell counts of free and bound bacteria have the advantage that relatively low numbers of bacteria can be used in the assay. The determination of free bacteria is relatively straight forward, but it is difficult to be certain that the release of bound bacteria is complete. In this assay, as in most other quantitative assays, trapped bacteria and bacteria bound to other bacteria will be counted as bound. Furthermore, all of the nonmicroscopic assays of bacterial adhesion suffer from the difficulty that the tissue must be washed and collected. The method of washing may remove loosely bound bacteria. In addition, methods of separating free bacteria from tissue with bound bacteria usually fractionate large bacterial aggregates with the plant cells. These bacterial aggregates even if unattached to plant cells are then counted as bound bacteria (Matthysse, 1995).

Other assays for attachment of *A. tumefaciens* have used tissue slices, which have the advantage that the slices eventually form tumors. However, the problem of trapped bacteria becomes very large with tissue slices, which generally have uneven surfaces where bacteria can become caught. These uneven surfaces also hamper microscopy of tissue slices. Root cap cells and the surface of roots have also been used to examine the attachment of the bacteria with the light microscope (Hawes and Brigham, 1992). However, the surface of these cells differs in many respects from that of cells in wound sites, and the adhesion observed may have more to do with bacterial colonization of the root than with pathogenesis and tumor formation (Matthysse, 1995).

II.B. RHIZOBIA

In comparison with the study of agrobacterial attachment, study of rhizobial attachment has an advantage in that the root surface is the normal site of action. The primary target cells for root infection in many rhizobia-legume associations, young emerging root hairs, are easily visible. For measurement of attachment, which is (supposedly) required for nodulation, the obvious choice of many authors is a microscopic counting method. Non-microscopic methods such as determination of radiolabeled bacteria or plate counting of detached bacteria will be more useful in studying rhizobial root colonization in general. It should be noted that the most frequent infection of roots of many legumes occurs in the zone where root hairs are not yet present at the time of inoculation (Bhuvaneswari *et al.*, 1981). However, a rhizobial attachment assay based on selective use of epidermal cells that are about to form a root hair has not been described.

An essential factor in direct counting methods using root hairs is the physiological state of the root hairs. Young root hairs are protuberances of plant root epidermal cells, which grow by apical growth (tip growth). Depending on the conditions, the root hairs may or may not continue tip growth during the assay. Disturbance of normal growth may lead to changes in the surface of the root hair tip. These changes can be induced, for example, by special properties of the incubation medium or by rhizobial signal molecules. A useful but rather laborious microscopic counting method, which favors normal interactions between rhizobia and root hairs, makes use of so called Fåhraeus slides, in which seedling roots are incubated between a cover slip and a slide in a medium that allows root infection and nodulation (Fåhraeus, 1957, Dazzo *et al.*, 1976).

Cultured soybean, rice and asparagus cells have been used for assaying binding of *Rhizobium* and *Bradyrhizobium* cells (Ho *et al.*, 1988; Terouchi and Syono, 1990). Advantages and disadvantages of this method have already been mentioned for agrobacterial attachment.

As pointed out, microscopic assays necessitate the use of large numbers of bacteria in order to obtain reliable results (one million cells per ml, or more) (for example, Smit *et al.*, 1986). However, root nodules can be formed after inoculation of a root with hundred cells per ml, or even less. Dilute inocula better represent the amount of rhizobia naturally encountered in many soils. Favelukes and colleagues developed an interesting hybrid assay in which low numbers of bacteria are allowed to adhere to the surface of a seedling and, after washing, are quantitated by culturing the roots embedded in nutrient agar and counting and localizing the resulting microcolonies along the root surface after a few days (Caetano-Anolles and Favelukes, 1986a). Disadvantages of this method are (i) measurement of adherence to noninfectable plant cells, (ii) ignorance of adsorbed but nondividing cells, and (iii) given the period between adsorption and counting, interference by factors which may influence colony formation rather than attachment. Nevertheless, the method has been useful in discovering host-symbiont specific phenomena during early interactions between rhizobia and host roots (Caetano-Anolles and Favelukes, 1986b).

III. Genetics of Bacterial Adhesion

Adhesins can be identified by a genetic approach involving gain or loss of function. For *Rhizobiaceae*, gain-of-function experiments have not yet been reported. For loss of function, there are theoretically at least four classes of mutations which can result in the failure of the bacteria to bind to plant cells: (1) steps required before bacterial adhesion can start may be missing, (2) the bacterial binding site may be missing or altered, (3) the entire bacterial surface may be altered so that the binding site is missing, incorrectly exposed or secreted into the medium, or (4) the bacterial surface may be covered with material, such as extracellular polysaccharides (EPS), which masks the binding site. Nonattaching bacterial mutants of each of these phenotypes have been identified for biotype 1 of *A. tumefaciens*. Currently a few mutants from classes 2 and 3 have been found for rhizobia.

III.A. AGROBACTERIA

All agrobacterial genes presently known to be involved in adhesion are chromosomal. A group of nonattaching mutants, *att*, appears to include both mutants blocked in steps prior to adhesion and mutants which lack the bacterial adhesin (Reuhs *et al.*, 1997). Att mutations are found in a large region of the bacterial chromosome (more than 30 kb). Some of them, AttA1 through AttH, can be complemented by the addition of conditioned medium to the binding assay (Matthysse *et al.*, 1996; Matthysse, 1994). Conditioned medium is produced by incubating wild type agrobacteria and plant cells together for several hours followed by filter sterilization of the medium. Neither partner is able to produce effective conditioned medium by itself. Sequential incubation of each partner separately produced effective conditioned medium only if the plant cells were added before the bacteria and not vice versa. These results suggest that the plant cells produce a signal to which the bacteria respond prior to attachment. AttA1-H mutants may be defective in sensing the plant signal or in responding to it. Interestingly, the *attA1-H* genes show homology to ATP-binding cassette (ABC) transport system genes (Matthysse *et al.*, 1996). Att mutants AttJ and AttR can not be complemented by conditioned medium and may lack the bacterial binding site (Matthysse *et al.*, 1996; Reuhs *et al.*, 1997). Significantly, each Att mutant is avirulent on all host plants tested.

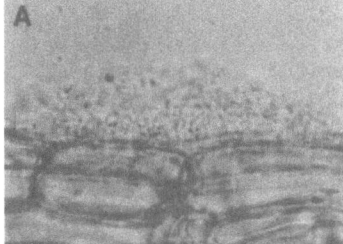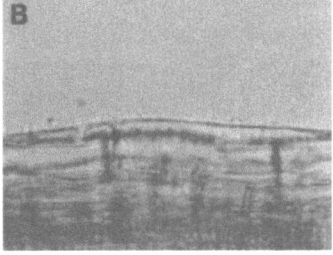

Figure 1. Binding of *A. tumefaciens* to roots of *A. thaliana* as observed in the light microscope using Nomarski optics and living tissue. A. Binding of bacteria to roots in 0.4% sucrose 1 mM CaCl₂. Note the numerous bacteria adhering to the root epidermis. Similar binding was seen on the cut end of the root and on the root hairs. B. Binding in the presence of the acetylated polysaccharide purified from wild type bacteria. Note the almost complete inhibition of binding. A similar inhibition of binding to cut ends and root hairs was seen.

The *attR* gene shows homology to transacetylases (Reuhs *et al.*, 1997). When surface polysaccharides of wild type and AttR mutant bacteria were compared, the AttR mutants were found to lack an acetylated capsular polysaccharide present in the parent strain. This polysaccharide was low molecular weight (between 3 and 5 kDa). It contained glucose, glucosamine, and an unidentified sugar related to 2-keto-3-deoxyoctulosonic acid (KDO). It is probably a type of K antigen (Chapter 7). Purified preparations of the polysaccharide from the wild type bacteria inhibited bacterial binding to carrot cells and *A. thaliana* roots (Figure 1).

Three of the other known nonattaching agrobacterial mutants appear to involve nonspecific alterations in the bacterial surface. ChvA and ChvB (for Chromosomal Virulence) mutants are unable to synthesize β-1,2-glucans which are found in the periplasmic space and aid the bacterium in its response to low external osmotic pressure (Douglas *et al.*, 1982; Puvanesarajah *et al.*, 1985). Mutations in these genes cause avirulence on some, but not all, host plants. The mutations are pleiotropic: in addition to lack of β-1,2-glucan molecules, the mutants show reduced motility and appear to overproduce acidic EPS. Similar mutants of *Rhizobium* are called NdvA and NdvB (for Nodule Development).

The effect of a *chvB* mutation on nodulation has been tested after transfer to *A. tumefaciens* of the genes encoding the enzymes involved in synthesis of LCOs. Such transgenic β-1,2-glucan-deficient agrobacteria are hardly able to attach to pea root

hairs and are unable to infect legume roots (Swart *et al.*, 1993).

PscA (ExoC) mutants of Agrobacterium are unable to make glucose-1-phosphate (Zorreguita *et al.*, 1988). Not surprisingly, these mutants show many alterations in surface polysaccharides and have a very low growth rate on Luria agar (Cangelosi *et al.*, 1987; Thomashow *et al.*, 1987).

Attachment of each agrobacterial mutant has been assayed by testing binding to tissue culture cells. The fact that the mutants are avirulent on most host plants suggests that this assay measures something essential in the bacterial plant interaction. However, there may be additional genes required for bacterial attachment to wound sites, which could be missed in a tissue culture assay. In addition, with the exception of *PscA*, which is a strain A6 mutant, all of these mutants are in the C58 chromosomal background. Thus, genes required for bacterial attachment, which differ between strains, have not been examined.

As already mentioned, initial attachment of *A. tumefaciens* and many rhizobia to plant cells is a two-step process. The genes discussed above are involved in the first step, a loose binding of the bacteria to the host surface. The second step in bacterial attachment may involve cellulose fibrils from the bacteria, produced either before or after attachment. Under appropriate conditions, cellulose fibrils may also contribute to direct binding to the plant cell surface. Cellulose binds very strongly to itself. Thus, bacterial cellulose synthesis results in tight binding of the bacteria to the plant cell wall. In addition, cellulose is sticky and free bacteria become

Figure 2. Binding of *A. tumefaciens* to carrot suspension culture cells as seen in A. the light microscope using living tissue and in B. the scanning electron microscope. Note the presence of large aggregates of bacteria held together by cellulose fibrils.

entrapped in the cellulose fibrils made by the attached bacteria. This results in the formation of large aggregates of bacteria on the plant surface (Matthysse, 1983) (Figure 2). Agrobacterial genes required for cellulose synthesis, *celABCDE*, have been cloned and sequenced (Matthysse *et al.*, 1995). Mutants with defective *cel* genes bind loosely to tissue culture cells and can be removed by vortexing. The mutants fail to form aggregates on the plant cell surface. Cel mutants also bind loosely to wound sites on leaves of *Bryophyllum daigremontiana* and can be removed by gentle water washing (Matthysse, 1983). These mutants are virulent, but are much reduced in virulence (10 to 1000-fold depending on the particular mutation) when compared to the wild type parent (Minnemeyer *et al.*, 1991).

Little is known about the genes required for the attachment of other biotypes of agrobacteria to host cells. Nonattaching mutants of *A. rhizogenes* have been isolated but the mutant phenotypes have not been characterized (Robertson Crews *et al.*, 1990). Nonattaching mutants of biotype 3 Agrobacteria have not been reported.

III.B. RHIZOBIA

To our knowledge, rhizobial genes directly involved in attachment to plant cells have not been characterized, with the exception of *ndvA* and *ndvB*. These genes are the rhizobial equivalents of *chvA* and *chvB*, and absence of the corresponding gene products results in pleiotropic cell surface changes, including a deficiency in production of β-1,2-glucan. *S. meliloti ndv* mutants are impaired in root infection and nodulation (Dylan *et al.*, 1986).

A well-documented case of rhizobial attachment mutants concerns *B. japonicum* cells, unable to produce the Bj38 adhesin (Ho *et al.*, 1990). This rhizobial surface lectin recognizes galactose-containing epitopes on the surface of cultured soybean cells, and may be associated with fimbriae. Bj38-deficient mutants show reduced nodulation of soybean (Ho *et al.*, 1993). To our knowledge, the Bj38 gene has not been cloned.

Like the situation with *Agrobacterium*, the second step in attachment of many rhizobial strains involves bacterial cellulose fibrils. These fibrils contribute to the formation of large aggregates of bacteria on the root hair surface (Figure 3B), but may also be involved in direct binding to these plant cells. Cellulose-deficient *R. leguminosarum* cells are still able to attach to the root hair surface, and do not

show significant nodulation defects under laboratory conditions (Smit *et al.*, 1987).

Several authors have obtained indications that rhizobial *nod* genes, involved in synthesis of LCOs, play a role in host-plant-specific attachment (Dazzo *et al.*, 1988). These indications are based on a comparison of attachment kinetics of homologous wild type strains with those of certain *nod* mutants and heterologous strains. However, these results do not necessarily characterize LCOs as adhesins. After inoculation, LCOs quickly induce changes in host root hairs. If such changes stimulate attachment, homologous rhizobia will show enhanced binding to the root hair surface in comparison with *nod* mutants and heterologous strains.

IV. Bacterial Adhesins

The nature of the adhesin(s) involved in the initial binding of the bacteria to the plant cell surface has been examined using a variety of techniques. Most assays for bacterial adhesins are fundamentally inhibition assays. For example, Lippincott and Lippincott (Lippincott and Lippincott, 1969) used inhibition of tumor formation by various agrobacterial preparations to identify the fraction containing the putative adhesin. Inhibition of bacterial binding to tissue culture cells and *A. thaliana* roots has been tested by the group of Matthysse, while the group of Kijne tested inhibition of agrobacterial and rhizobial binding to pea root hairs. Inhibition assays have identified various candidates for the bacterial adhesin, such as lipopolysaccharide (LPS), other surface polysaccharides including K antigens, and proteins. Recently these assays have been applied to the testing of mutant bacteria.

IV.A. AGROBACTERIA

Inhibition of tumor formation on bean leaves suggested that the agrobacterial adhesin was contained in LPS preparations from wild type bacteria (Whately *et al.*, 1976). This activity was missing in extracts of bacteria that were avirulent due to the presence of the plasmid pSa (New *et al.*, 1983). However, binding to tissue culture cells appeared to be unaffected by LPS addition (Gurlitz *et al.*, 1987). In both cases crude preparations of LPS were used. These may have been contaminated to varying extents with other surface polysaccharides particularly K antigens. Binding to tissue culture cells was also unaffected by treatment of the bacteria with an inhibitor of LPS biosynthesis which resulted in bacteria which had lost all smooth LPS and retained only rough LPS (Goldman *et al.*, 1992).

Binding of *A. tumefaciens* to pea root hair tips was inhibited (but not completely prevented) by an agrobacterial extract containing a 14 kD calcium-binding surface protein, rhicadhesin (Smit *et al.*, 1989). Binding could be inhibited by a rhicadhesin preparation from *Rhizobium* as well. These results suggest a common mechanism for attachment to root hairs by both types of Rhizobiaceae (Figure 3). Rhicadhesin activity has also been found in other members of this family, but not from bacteria other than Rhizobiaceae. It can not be excluded that purified rhicadhesin preparations contained low amounts of surface polysaccharides. However, the activity could be abolished by treatment with proteases or with heat. Rhicadhesin activity has not been tested with other types of plant cells, such as carrot suspension cells.

Restoration of the binding and virulence of nonattaching mutants has also been used as an assay to identify the bacterial adhesin. This type of assay is problematic in that a substance, which is supposed to reside in the membrane or cell wall, is unlikely to be able to sit in its proper location when added exogenously and thus may not restore binding even if it is an adhesin. An exogenously added adhesin is only likely to test positive in this assay if it is able to bridge the two surfaces when it is not anchored in either of them. Furthermore, an added substance may mask inhibitory compounds, for example by lowering electrostatic repulsion. In addition, attachment of mutants that are blocked at a step prior to bacterial attachment may be restored in a variety of ways by exogenous substances, which do not necessarily include the adhesin. For example, a group of bacterial chromosomal genes (*attA1A2BCDEFGH*) with homology to ABC transport systems is required for agrobacterial attachment (see above). Binding of these mutants to tissue culture cells was restored by low molecular weight substances present in conditioned medium. Virulence of an AttD mutant on carrot discs was also restored by conditioned medium. It is believed that the active substance in conditioned medium is involved in signaling between the bacteria and the host plant (Matthysse, 1994; Matthysse *et al.*, 1996).

Restoration of binding of ChvB mutants to roots has been studied in detail. Although ChvB mutants lack the enzyme for the production of β-1,2-glucan (Castro *et al.*, 1996), this oligosaccharide was unable to restore binding or virulence of the mutants (it also did not inhibit binding of wild type bacteria). However, binding and virulence were restored by the addition of a rhicadhesin preparation (Swart *et al.*, 1993). Significantly, rhicadhesin was present in nonattaching ChvB mutants but it was not longer active in the inhibition of binding of wild-type bacteria to root hairs. This suggests that ChvB mutants lack a component, which acts together with rhicadhesin. Most probably, this component is not β-1,2-glucan, since addition of β-1,2-glucan did not enhance attachment, either in the absence or in the

 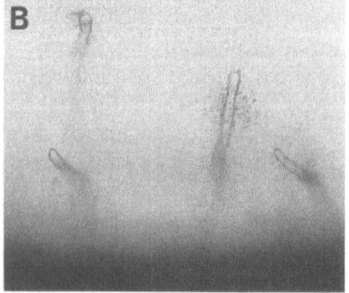

Figure 3. Binding of bacteria to root hairs of alfalfa as seen with living tissue in the light microscope with Nomarski optics. A. Binding of *A. tumefaciens*. Note that the root hairs are straight and that bacteria can be seen bound all along their length. B. Binding of *S. meliloti*. The root hairs are curled due to LCO production by the bacteria and can not be photographed in one plane. Bacterial aggregates at the root hair tips are shown.

presence of rhicadhesin. Moreover, rhicadhesin activity and attachment ability could be restored by growth under high-osmotic-strength conditions, whereas these conditions did not restore β-1,2-glucan production. The presence of calcium in the high-osmotic-strength medium appeared to be required for the production of active rhicadhesin. Calcium probably plays a role in anchoring, stabilization and/or activation of rhicadhesin (Smit *et al.*, 1991). Since high osmotic strength of the growth medium influences the ratio and, probably, the conformation of surface polysaccharides of Rhizobiaceae (Breedveld *et al.*, 1990), rhicadhesin may be activated in combination with a capsular factor.

Taken together, the results suggest a role for rhicadhesin in agrobacterial attachment to pea root hairs, whereas an acetylated K antigen appears to be involved in the adhesion of *A. tumefaciens* to carrot suspension culture cells and *A. thaliana* roots.

IV.B. RHIZOBIA

Several rhizobial surface components have been proposed to function as an adhesin: extracellular polysaccharides (EPS), capsular polysaccharides (CPS), lipopolysaccharides (LPS), and surface proteins such as rhicadhesin and Bj38.

EPS, CPS and LPS can be involved in binding, for example, by gelling interactions with plant polysaccharides (Tako and Nakamura, 1986) or by acting as ligands for plant adhesins (Dazzo and Truchet, 1983). Loss-of-function experiments may not identify one of these saccharides as an adhesin, since remaining saccharides may take over its function. Current evidence indicates that rhizobial mutants deficient in EPS production or in synthesis of the O-antigenic part of LPS are still able to attach to root hair tips (Van Workum, 1997; Smit *et al.*, 1987). CPS or K antigen mutants have not been studied with regard to adhesion.

Rhizobial surface polysaccharides have been proposed to function as adhesins by serving as saccharide ligands for plant sugar-binding proteins (lectins) present at the root hair surface. A clover lectin, trifoliin A, is present at the outer periphery of clover root hairs, and is secreted into the rhizosphere (Sherwood *et al.*, 1984a). Depending on the culture conditions, possible ligands for binding by trifoliin A are LPS, CPS and EPS oligosaccharide fragments, coming from the clover symbiont *R. leguminosarum* bv *trifolii* (*R.l. trifolii*) (Hrabak *et al.*, 1981;

Sherwood *et al.*, 1984b; Abe *et al.*, 1984). Interestingly, an *R.l. trifolii* mutant lacking its Sym plasmid was unable to bind trifoliin A, and showed significantly diminished attachment to clover root hairs (Dazzo *et al.*, 1988).

In contrast, a Sym plasmid-deficient *R.l. viciae* strain is perfectly able to attach to pea root hairs, its accumulation at root hair tips being facilitated by pea lectin. For one *R.l. viciae* strain, a glycan associated with LPS appeared to be an appropriate pea lectin ligand (Planqué and Kijne, 1977). The discrepancy of the data obtained with the two rhizobial biovars could be explained by the different attachment conditions used. The group of Dazzo made use of Fåhraeus slides in which growing clover root hairs were challenged with rhizobia at pH 6. The pea results obtained by the group of Kijne are based on assays at pH 7.5, with use of phosphate buffer, under conditions in which growth of root hairs may have been impaired.

Rhicadhesin has been proposed to represent an essential adhesin for rhizobia. Unfortunately, rhicadhesin-deficient mutants have not yet been found. Evidence for a role of rhicadhesin in rhizobial attachment has been obtained from experiments testing inhibition as well as restoration of attachment. The results have been summarized in a model, in which rhicadhesin is calcium-bound to the rhizobial surface (Smit *et al.*, 1991). With *R.l. viciae*, this binding is acid-sensitive. At pH 5, rhicadhesin is released into the medium probably due to a defect in binding of calcium to the rhizobial surface (Swart, 1994). Under acidic conditions, surface saccharides rather than rhicadhesin may play a predominant role in attachment to root hair tips.

In contrast to agrobacterial chvB mutants, ndvB mutants of *S. meliloti* show (diminished) attachment to alfalfa roots, initiation of infection thread formation, and induction of the formation of pseudonodules. Furthermore, a cell surface preparation of a *S. meliloti* NdvB mutant contained rhicadhesin activity (Swart, unpublished results). These results may be due to the relatively high osmotolerance of *S. meliloti* (Breedveld *et al.*, 1990). Many rhizobia produce cellulose fibrils. As discussed above for *Agrobacterium*, such fibrils may function as adhesins. Addition of isolated cellulose fibrils or commercial cellulose preparations inhibited attachment of *R.l. viciae* to pea root hairs when added to the bacteria just before incubation (Smit *et al.*, 1987). This could be caused by aggregation of the bacteria, with a concomitant decrease in the number of bacteria free to attach. Cellulose-deficient mutants or cellulase-treated wild-type cells were still able to attach, but were unable to form aggregates on root hair tips. Apparently, rhizobial cellulose fibrils play a more important role in attachment to other rhizobia than in binding to the root hair surface. Indeed, fibril-overproducing mutants strongly flocculate, in contrast to fibril-deficient mutants (Smit *et al.*, 1987).

V. Plant Adhesins

Assays for plant adhesins (like those for bacterial adhesins) have relied heavily on the inhibition of bacterial virulence or attachment. Assays involving the inhibition of tumor formation on bean leaves suggested that the active substance was a pectin derivative, possibly methylated pectin (Lippincott *et al.*, 1977). Pectin fractions of tomato inhibited bacterial binding to tomato tissue culture cells (Neff *et al.*, 1987). However, commercial pectin did not inhibit attachment of bacteria to carrot cells (Gurlitz *et al.*, 1987). More recent work has shown that short chain pectins can act as elicitors of phytoalexin production (which may inhibit tumor formation) (Hahn *et al.*, 1981). Thus, the interpretation of these results may not be as straight forward as originally thought. Although the active tomato cell wall fractions contain pectin, the activity was sensitive to proteases. Treatment of carrot, but not tomato, tissue culture cells with proteinases followed by addition of excess proteinase inhibitor prevented bacterial attachment (Gurlitz *et al.*, 1987). In an attempt to identify a hapten of either the bacterial or plant adhesin, many sugars and amino acids have been tested for inhibition of attachment. All have been found to be without effect. The effect on bacterial attachment of the addition of exogenous polysaccharides and proteins has been used to identify a plant adhesin. Extensin and other available plant cell wall proteins have no effect on bacterial adhesion.

Plants have recently been shown to possess a surface protein related to human vitronectin (Zhu *et al.*, 1991; Pressner, 1991). This plant protein has not been purified, but human vitronectin was found to inhibit attachment of *A. tumefaciens* to carrot cells (Wagner and Matthysse, 1992). Antibodies to vitronectin also inhibited attachment. Wild-type *A. tumefaciens* biotype 1 bound radioactive human vitronectin. The various nonattaching mutants described above failed to bind vitronectin. Extraction of carrot cells with dilute Triton solutions was previously shown to remove the adhesin (Gurlitz *et al.*, 1987). After these treatments, the plant cells were not killed and recovered the ability to bind bacteria after 3 to 6 hours. The material removed with Triton was analyzed by PAGE and found to contain several polypeptides. When a Triton extract of carrot cells was examined using Western blots with anti-vitronectin antibodies, two bands were identified which reacted with the antibodies (Wagner and Matthysse, 1992). These proteins are currently the best candidates for the carrot adhesin. However, the evidence that a plant vitronectin-like protein serves as an adhesin is not conclusive. The gene for this protein has not yet been cloned and the actual relationship of the protein to human vitronectin is unknown.

Human vitronectin contains a sequence (amino acids RGD) which is involved in the binding of the protein to integrins found in the plasma membrane. Thus one would not expect this sequence to be exposed to the medium. Therefore it was not surprising that the addition of a hexapeptide containing RGD to the medium did not effect the attachment of *A. tumefaciens* to carrot cells or to *A. thaliana* roots. It is not known if the plant protein contains an RGD sequence. It is possible that the inhibition of attachment by human vitronectin or antivitronectin antibodies represents a steric blockage by vitronectin and antivitronectin antibodies of an actual adhesion site located close to a vitronectin-like protein on the plant cell surface.

An examination of ecotypes of *A. thaliana* which were unable to take up T-DNA from *A. tumefaciens* has revealed that some ecotypes of this plant are defective in their ability to bind *A. tumefaciens*. It is believed that it is the failure of the bacteria to bind to these plants that is responsible for the failure of the bacteria to transfer T-DNA to, and form tumors on, them (Nam *et al.*, 1997). The nature of the defect in these plants, which is responsible for the failure to bind agrobacteria, is unknown.

Vitronectin may also play a role in rhizobial attachment. The (human) protein was found to inactivate rhicadhesin in an attachment-inhibition assay with *R. leguminosarum* bv *viciae* and pea root hairs (Swart *et al.*, 1994). A hexapeptide containing an RGD sequence could inhibit attachment of both *Rhizobium* and *Agrobacterium* to pea root hairs, and suppressed inhibition of attachment by rhicadhesin. A control peptide, containing an RGE sequence, did not inhibit attachment and did not suppress rhicadhesin activity.

Rhicadhesin receptor[1]	A-D-A-D-A-L-Q-D-L-C-V-A-D-Y-A-S . V-I-L-V-N-G-F-A-S-K-P
Germin[2]	T-D-P-D-P-L-Q-D-F-C-V-A-D-L-D-G-K-A-V-S-V-N-G-H-T-C-K-P
Pseudogermin[2]	A-D-P-D-P-L-Q-D-F-?-V-A-D-L-?-D-N-A-V-?-V

Figure 4. Homology of the N-terminal sequences of a putative binding receptor for *Rhizobium leguminosarum* rhicadhesin, germin and pseudogermin (from Swart *et al.*, 1994, Lane *et al.*, 1992, respectively), ([1], Swart *et al.*, 1994; [2], Lane *et al.*, 1992).

Rhicadhesin activity could also be suppressed by a 32 kD glycoprotein, isolated from pea root cell walls (Swart *et al.*, 1994). This protein did not cross-react with polyclonal antibodies to human vitronectin. The putative adhesin could be extracted from pea root cell walls by $CaCl_2$, and bound to hydroxyapatite. At the time of purification and N-terminal sequencing, no homology with known proteins was found, but a recent search revealed a possible membership of the germin family (Figure 4). Germin, an oxalate oxidase associated with the extracellular matrix in plants (Lane, 1994), has been proposed to play a role in plant cell expansion, and may well be active in growing root hairs. Since germin also associates with hemicellulosic polysaccharides, a germin homologue may well interact with rhizobial surface polysaccharides. However, it should be noted that in each case of inhibition of rhizobial attachment by putative plant adhesins or peptides, inhibition was only partial and a considerable number of rhizobia were shown to attach to the root hair surface.

Lectins are well-characterized plant adhesins. According to a recent proposal, lectins are proteins with at least one noncatalytic binding site for sugar residues (Peumans and Van Damme, 1995). Many observations suggest that a certain family of lectins, the legume Asn-Ca lectins (Kijne *et al.*, 1997), is involved in binding of rhizobia to legume root hair tips. Presence of legume lectins on the surface of root hairs has been demonstrated for various legumes (for a recent example, see Diaz *et al.*, 1996). For the pea lectin PSL, its pattern of location on the pea root surface, its secretion into the rhizosphere, its contribution to accumulation of infective pea rhizobia on root hair tips, plus the fact that pea rhizobia produce several ligands for pea lectin, i.e., EPS, LPS and a capsular glycan, are all consistent with the hypothesis of Hamblin and Kent (1973) that legume lectins bind symbiotic rhizobia to the roots. However, neither mutants of *R.l. viciae* which are defective in binding of pea lectin nor pea mutants unable to produce PSL are currently available.

The results obtained for PSL are consistent with the earlier work of Dazzo and co-workers showing that trifoliin A, a clover lectin, is involved in attachment of *R.l. trifolii* to clover root hairs (Dazzo and Truchet, 1983). This protein, however, has not been fully characterized and the corresponding gene awaits cloning. Likewise, the role of other lectins such as the peanut root lectin PRA II (Kalsi *et al.*, 1995) in rhizobial attachment merit further study.

VI. Environmental Factors

Results of an attachment assay are largely determined by environmental factors. For example, the attachment mechanism of *R.l. viciae* to pea root hairs is significantly influenced by the growth conditions of the bacteria. Carbon-limited rhizobia, grown in tryptone-yeast extract (TY) medium, attach well to pea root hair tips at pH 7.5, but fail to do so at pH values below 6 (Figure 5, Swart, 1994). However, manganese-limited rhizobia, grown in yeast extract-mannitol (YM) medium, are still able to attach at pH 6 (although attachment is optimal at pH 7.5). Usually, nodulation assays are performed with plants grown under slightly acidic conditions. Furthermore, the pH of the extracellular matrix of growing plant cells, such as root hairs, is believed to be moderately low. TY-grown rhizobia in contrast to YM-grown rhizobia are poorly infectious, if at all (Kijne *et al.*, 1988). This may suggest that (i) attachment is an essential part of the infection process, and (ii) study of the attachment process of TY-grown rhizobia is less relevant for the study of nodulation (but may be very relevant for colonization).

Various reports have highlighted the role of calcium in rhizobial attachment either by showing its requirement in the incubation medium for specific attachment (Caetano-Anolles *et al.*, 1989; Lodeiro *et al.*, 1995) or improved attachment of rhizobia grown under calcium-limiting conditions (Wisniewski and Delmotte, 1996). Calcium (and magnesium) may play an important role in anchoring of adhesins to the bacterial surface (Swart, 1994) or in regulating the surface charge of plant cells and bacteria. These phenomena are pH-dependent and accentuate the importance of physicochemical interactions in adhesion.

Growth-inhibited *R.l. viciae* cells have better attachment ability and are more virulent than exponentially growing cells (Kijne *et al.*, 1988). The influence of growth conditions can be accurately studied with the use of a chemostat. Enhanced attachment could be induced not only by manganese limitation, but also by nitrogen limitation, iron limitation, and oxygen limitation. Various authors have reported that the attachment ability of rhizobia is dependent on the growth phase of the bacteria. In most cases, however, the nutritional properties of the media used were not defined. Unfortunately, the

primary limiting factors for growth of rhizobia in the soil and in the rhizosphere are still unknown and remain to be determined. Furthermore, as discussed above, better attachment may result from improved production of stimulatory LCOs.

Environmental factors probably play a less significant role in the adhesion of agrobacteria to the host plant surface. There appears to be a requirement for live bacteria for binding to plant cells. All methods of killing the bacteria which have been tested, including high temperature, ultraviolet light, antibiotics and glutaraldehyde, produce bacteria which are unable to bind to plant cells. Thus, (micro)environments which affect bacterial viability are also likely to decrease bacterial adhesion. This requirement for living bacteria may be due to the

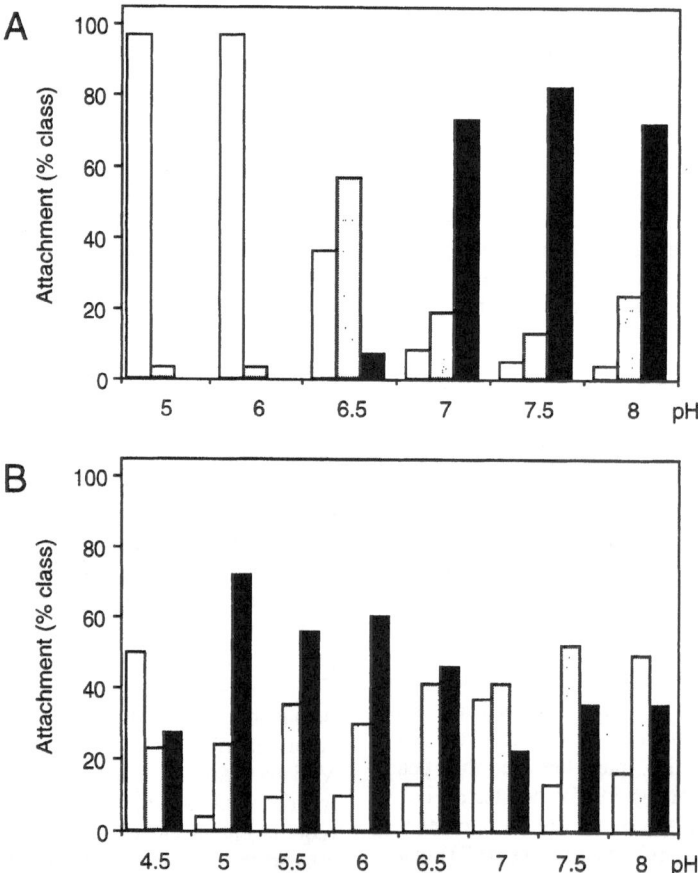

Figure 5. pH dependency of attachment of *Rhizobium leguminosarum* bv. *viciae* strain 248 (A) and *Agrobacterium tumefaciens* strain LBA1010 (B) to pea root hair tips. Attachment was quantified by randomly screening developing root hairs, and was distinguished into three classes: class 1 (open bars), root hairs without attached bacteria; class 2 (dotted bars), root hairs with few bacteria, attached directly to the root hair; class 3 (black bars), root hairs with many attached bacteria, forming an aggregate. The percentage of each class was calculated (from Swart, 1994).

necessity for some exchange of signals between the bacteria and the plant before adhesion occurs.

The binding of biotype 1 agrobacteria to carrot culture suspension cells was relatively unaffected by alterations in the incubation medium including changes in pH and presence or absence of divalent cations or chelating agents (Gurlitz *et al.*, 1987). Attachment to tissue culture cells or *A. thaliana* roots was also unaffected by the medium in which the bacteria were grown.

Growth on TY medium, a medium which favors bacterial conjugation, is favorable for agrobacterial attachment and tumor formation as well. In contrast to the situation for rhizobia, rhicadhesin is not released from the agrobacterial surface at low pH (Swart, 1994). TY-grown agrobacteria bound well to root hair tips at pH values between 5 and 6 (Figure 5). They were able to nodulate when provided with a Sym plasmid. This observation accentuates that TY-grown agrobacteria have properties other than those of TY-grown rhizobia, and that agrobacteria and rhizobia are not simply interchangeable by exchange of a Ti plasmid or nodulation genes.

Binding of biotype 1 agrobacteria did not require calcium or magnesium. Chelating agents such as EDTA or EGTA did not inhibit attachment. Neither high ionic strength (up to 0.25 M NaCl) nor low ionic strength had an effect on attachment. The kinetics of attachment were the same in 4% sucrose as in MS medium. High osmotic pressure (0.4 M mannitol) did not affect bacterial attachment. Variation in pH between 5.5 and 7.5 was without effect. Higher and lower pH values inhibited bacterial growth and attachment (Gurlitz *et al.*, 1987; Sykes and Matthysse, 1988). Temperatures between 20° and 37° C also did not affect attachment (Matthysse, 1988). Temperatures above and below this range inhibited both bacterial growth and attachment.

A. tumefaciens biotypes 2 and 3 are more sensitive to environmental factors. High ionic strength (more than 0.09 M) inhibited the attachment of all biotype 2 strains and some biotype 3 strains. However, the growth of these strains was also inhibited by high ionic strength. All biotypes of agrobacteria were capable of binding to carrot cells in 4% sucrose. There seemed to be no requirement for any specific substances in the medium except possibly those which were released by the plant (Sykes and Matthysse, 1988).

VII. Conclusion

Obviously, major factors in the attachment of agrobacteria and rhizobia to plant cells remain to be identified: the nature of bacterial adhesins, the gene(s) which encode their synthesis, the nature of plant adhesins, essential steps which precede adhesion, and the relationship between adhesion involved in virulence and adhesion leading to root colonization.

From the results described above, it may be concluded that Rhizobiaceae produce various molecules able to play a role in attachment to various plant surfaces. Environmental conditions partly determine which of these factors play a key role. Adhesins can be directly involved in attachment by cross-bridging cell surfaces. Adhesins may also indirectly be involved by lowering electrostatic repulsion or by enabling hydrophobic interactions.

The available data allow us to speculate that Rhizobiaceae have a basic attachment mechanism in which rhicadhesin is involved. With *Rhizobium* this mechanism is acid-sensitive in contrast to the situation with *Agrobacterium*. Interactions with plant cells leading to DNA transfer (agrobacteria) or infection thread formation (rhizobia) may require an additional attachment mechanism, in which bacterial surface polysaccharides are important factors. This mechanism is inducible by plant factors in agrobacteria, and by nutritional limitation in rhizobia. Surface polysaccharides may function as adhesins by binding to plant proteins such as vitronectin, germin homologues, or lectins. Calcium and magnesium ions may modify the charge of surface PS, thereby influencing their interactions with plant and bacterial proteins. Isolation of nonpleiotropic rhicadhesin-deficient mutants will be necessary to reveal the relationship between the two putative attachment mechanisms. Differences in the attachment mechanisms of agrobacteria and rhizobia, such as a different acid-sensitivity, may be related to the fact that agrobacteria attach and conjugate, whereas rhizobia attach but then infect the plant tissue.

VIII. References

Abe, M., Sherwood, J.E., Hollingsworth, R.I. and Dazzo, F.B. (1984) J. Bacteriol. 160, 517-520.

Akiyoski, D.E., Morris, R.O., Hinz, R., Mishke, B.S., Kosuge, T., Garfinkel, D.J., Gordon, M.P. and Nester, E.W. (1982) Proc Natl Acad Sci USA 80, 407-411.

Bhuvaneswari, T.V., Bhagwat, A.A. and Bauer, W.D. (1981) Plant Physiol. 68,1144-1149.

Binns, A.N. and Thomashow, M.F. (1988) Ann. Rev. Microbiol. 42, 575-606.

Breedveld, M.W., Zevenhuizen, L.P.T.M. and Zehnder, A.J.B. (1990) J. Gen. Microbiol. 136, 2511-2519.

Burr, T.J., Reid, C.L., Yoshimura, M., Momol, E.A. and Bazzi, C. (1995) Plant Dis. 79, 677-682.

Caetano-Anolles, G. and Favelukes, G. (1986a) Appl. Environ. Microbiol. 52,371-376.

Caetano-Anolles, G. and Favelukes, G. (1986b) Appl. Environ. Microbiol. 52,377-382.

Caetano-Anolles, G., Lagares, A. and Favelukes, G. (1989) Plant Soil. 117, 67-74.

Cangelosi, G.A., Hung, L., Puvanesarajah, V., Stacey, G., Ozaga, D.A., Leigh, J.A. and Nester, E.W. (1987) J. Bacteriol. 159, 2086-2091.

Castro, O.A., Zorreguieta, A., Ielmini, V., Vega, G. and Ielpi, L. (1996) J. Bacteriol. 178,6043-6048.

Chilton, M.D., Drummond, M.H., Merlo, D.J., Sciaky, D., Montoya, A.L., Gordon, M.P. and Nester, E.W. (1977) Cell 11, 263-271.

Dazzo, F.B., Napoli, C.A. and Hubbell, D.H. (1976) Appl. Environ. Microbiol. 32, 166-171.

Dazzo, F.B. and Truchet, G. (1983) J. Membrane Biol. 73, 1-16.

Dazzo, F., Hollingsworth, R., Philip-Hollingsworth, S., Robeles, M., Olen, T., Salzwedel, J., Djordjevic, M. and Rolfe, B. (1988) In, Bothe, H., De Bruijn, F.J. and Newton, W.E. (eds) Nitrogen Fixation, Hundred Years After, pp 431435. Gustav Fisher Verlag, Stuttgart, New York.

Diaz, C.L., Logman, T.J.J., Stam, H.C. and Kijne, J.W. (1996) Plant Physiol. 109, 1167-1177.

Douglas, C.J., Halperin, W. and Nester, E.W. (1982) J. Bacteriol. 152, 1265-1275.

Dylan, T., Ielpi, L., Stanfield, S., Kashyap, L., Douglas, C., Yanowsky, M., Nester, E., Helinski, D.R. and Ditta, G. (1986) Proc. Natl. Acad. Sci. USA 83, 4403-4407.

Fåhraeus, G. (1957) J. Gen. Microbiol. 16, 374-381.

Fullner, K.J., Lara, J.C. and Nester, E.W. (1996) Science 273, 1107-1109.

Gaworzewska, E.T. and Carlile, M.J. (1982) J. Gen. Microbiol. 128,1179-1188.

Goldman, R.C., Capobianco, J.O., Doran, C.C. and Matthysse, A.G. (1992) J. Gen. Microbiol. 138, 1527-1533.

Gurlitz, R.H.G., Lamb, P.W. and Matthysse, A.G. (1987) Plant Physiol. 83, 564-568.

Hahn, M.G., Darvill, A.G. and Albersheim, P. (1981) Plant Physiol. 68, 1161-1169.

Hamblin, J. and Kent, S.P. (1973) Nature New Biol. 245, 28-29.

Hawes, M.C. and Brigham, L.A. (1992) Advances Plant Pathol. 8, 119-147.

Hawes, M.C. and Smith, L.Y. (1989) J. Bacteriol. 171, 5668-5671.

Heidstra, R., Geurts, R., Franssen, H., Spaink, H.P., Van Kammen, A. and Bisseling, T. (1994) Plant Physiol. 105, 787-797.

Ho, S-C., Schindler, M. and Wang, J.L. (1990) J. Cell Biol. 111, 1639-1643.

Ho, S-C., Wang, J.L., Schindler, M. and Loh, J. (1993) In, Palacios, P., Mora, J. and Newton, W.E. (eds) New Horizons in Nitrogen Fixation, Kluwer Acad. Publ., Dordrecht, p. 237

Ho, S-C., Ye, W., Schindler, M. and Wang, J.L. (1988) J. Bacteriol. 68, 3882-3890.

Hooykaas, P.J.J., Klapwijk, P.M., Nuti, M.P., Schilperoort, R.A. and Rörsch, A. (1988) J. Gen. Microbiol. 98, 477-484.

Hooykaas, P.J.J., Van Brussel, A.A.N., Den Dulk-Ras, H., Van Slogteren, G.M.S. and Schilperoort, R.A. (1981) Nature 291, 351-353.

Hrabak, E.M., Urbano, M.R. and Dazzo, F.B. (1981) J. Bacteriol. 148, 697-711.

Kado, C.I. (1994) Mol. Microbiol. 12, 17-22.

Kalsi, G., Babu, C.R. and Das, R.H. (1995) Glycoconjugate J. 12, 45-50.

Kijne, J.W., Bauchrowitz, M.A. and Diaz, C.L. (1997) Plant Physiol. 115, 869-873.

Kijne, J.W., Smit, G., Diaz, C.L. and Lugtenberg, B.J.J. (1988) J. Bacteriol. 170, 2994-3000.

Klee, H.J., White, F.F., Iyer, V.N., Gordon, M.P. and Nester, E.W. (1983) J. Bacteriol. 153, 878-883.

Lane, B.G. (1994) FASEB J. 8, 294-301.

Lane, B.G., Cuming, A.C., Frégeau, Carpita, N.C., Hurkman, W.J., Bernier, F., Dratewka-Kos, E. and Kennedy, T.D. (1992) Eur. J. Biochem. 209, 961-969.

Leigh, J.A. and Walker, G.C. (1994) Trends in Genetics 10, 63-67.

Lerouge, P., Roche, P., Faucher, C., Maillet, F., Truchet, G., Promé, J.C. and Dénarié, J. (1990) Nature 344, 781-784.

Lessi, M. and Lanka, E. (1994) Cell 77, 321-324.

Lippincott, B.B. and Lippincott, J.A. (1969) J. Bacteriol. 97, 620-628.

Lippincott, B.B., Whately, M.H. and Lippincott, J.A. (1977) Plant Physiol. 59, 388-390.

Lodeiro, A.R., Lagares, A., Martinez, E.N. and Favelukes, G. (1995) Appl. Environ. Microbiol. 61, 1571-1579.

Matthysse, A.G. (1983) J. Bacteriol. 154, 906-915.

Matthysse, A.G. (1986) CRC Crit. Rev. Microbiol. 13, 281-307.

Matthysse, A.G. (1988) In, Palacios, R. and Verma, D.P.S. (Eds) Molecular Genetics of Plant-Microbe Interactions. APS Press, Minneapolis, pp 63-68.

Matthysse, A.G. (1994) Protoplasma 183, 131-136.

Matthysse, A.G. (1995) In, Methods in Enzymology. Academic Press, San Diego, CA, pp 189-206.

Matthysse, A.G. (1996) In, Fletcher, M. and Savage, D. (Eds) Molecular and Ecological Diversity of Bacterial Adhesion. John Wiley & Sons, Inc. New York, NY, pp 129-153.

Matthysse, A.G., Lightfoot, R. and White, S. (1995) J. Bacteriol. 177, 1069-1075.

Matthysse, A.G., Wyman, P.M. and Holmes, K.V. (1978) Infect. Immun. 22, 516-522.

Matthysse, A.G., Yarnall, H.A. and Young, N. (1996) J. Bacteriol. 178, 5302-5308.

Minnemeyer, S.L., Lightfoot, R. and Matthysse, A.G. (1991) J.Bacteriol. 173, 7723-7724.

Nam, J., Matthysse, A.G. and Gelvin, S. (1997) Plant Cell 9, 317-333.

Neff, N.T., Binns, A.N. and Brandt, C. (1987) Plant Physiol. 83, 525-528.

New, P.B., Scott, J.J., Ireland, C.R., Farrand, S.K., Lippincott, B.B. and Lippincott, J.A. (1983) J. Gen. Microbiol. 129, 3657-3660.

Ohyama, K., Pelcher, L.E., Schaefer, A. and Fowkes, L.C. (1979) Plant Physiol. 63, 382-387.

Peumans, W.J. and Van Damme, E.J.M. (1995) Plant Physiol. 109, 347-352.

Planqué, K. and Kijne, J.W. FEBS Lett. 73, 64-66.

Pohlman, R.F., Genetti, H.D. and Winans, S.C. (1994) Mol. Microbiol. 14, 655-668.

Pressner, K.T. (1991) Ann. Rev. Cell. Biol. 7, 275-310.

Puvanesarajah, V., Schell, F.M., Stacey, G., Douglas, C.J. and Nester, E.W. (1985) J. Bacteriol. 164, 102-106.

Reuhs, B.L., Kim, J.S. and Matthysse, A.G. (1997) J. Bacteriol. 179, 5372-5379.

Robertson Crews, J.L., Colby, S. and Matthysse, A.G. (1990) J. Bacteriol. 172, 6182-6188.

Sanders, L., Wang, C., Walling, L. and Lord, E. (1991) Plant Cell 3, 629-635.

Sherwood, J.E., Truchet, G. and Dazzo, F.B. (1984a) Planta 162, 540-547.

Sherwood, J.E., Vasse, J.M., Dazzo, F.B. and Truchet, G. (1984b) J. Bacteriol. 159, 145-152.

Smit, G., Kijne, J.W. and Lugtenberg, B.J.J. (1986) J. Bacteriol. 168, 821-827.

Smit, G., Kijne, J.W. and Lugtenberg, B.J.J. (1987) J. Bacteriol. 169, 4294-4301.

Smit, G., Logman, T.J.J., Boerrigter, M.E.T.I., Kijne, J.W. and Lugtenberg, B.J.J. (1989) J. Bacteriol. 171,4054-4062.

Smit, G., Tubbing, D.M.J., Kijne, J.W. and Lugtenberg, B.J.J. (1991) Arch. Microbiol. 155, 278-283.

Smit, G., Swart, S., Lugtenberg, B.J.J. and Kijne, J.W. (1992) Mol. Microbiol. 6, 2897-2903.

Spaink, H.P., Sheeley, D.M., Van Brussel, A.A.N., Glushka, J., York, W.S., Tak, T., Geiger, O., Kennedy, E.P., Reinhold, V.N. and Lugtenberg, B.J.J. (1991) Nature 354, 125-131.

Swart, S. (1994) Rhicadhesin-mediated attachment of Rhizobiaceae. PhD Thesis, Leiden University, Leiden, The Netherlands.

Swart, S., Logman, T.J.J., Smit, G., Lugtenberg, B.J.J. and Kijne, J.W. (1994) Plant Mol. Biol. 24, 171-183.

Swart, S., Smit, G., Lugtenberg, B.J.J. and Kijne, J.W. (1993) Mol. Microbiol. 10, 597-605.

Sykes, L. and Matthysse, A.G. (1988) Phytopathology 78, 1322-1326.

Tako, M. and Nakamura, S. (1986) FEBS Lett. 204, 33-36.

Tempe, J., Petit, A. and Farrand, S.K. (1984). In, Verma, D.P.S. and Hohn, T. (Eds) Genes involved in microbe-plant interactions. Springer Verlag, New York, pp 271-286.

Terouchi, N. and Syono, K. (1990) Plant Cell Physiol. 31, 119-127.

Thomashow, M.F., Karlinsky, J.E., Marks, J.R. and Hurlburt, R.E. (1987) J. Bacteriol. 169, 3209-3216.

Van Brussel, A.A.N., Bakhuizen, R., Van Spronsen, P.C., Spaink, H.P., Tak, T., Lugtenberg, B.J.J. and Kijne, J.W. (1992) Science 257, 70-72.

Van Workum, W.A.T. (1994) Biosynthesis and function of exopolysaccharides from *Rhizobium leguminosarum*. PhD Thesis, Leiden University, Leiden, The Netherlands.

Wagner, V.T. and Matthysse, A.G. (1992) J. Bacteriol. 174, 5999-6003.

Whately, M.H., Bodwin, J.S., Lippincott, B.B. and Lippincott, J.A. (1976) Inf Immun 13, 1080-1083.

Wisniewski, J-P. and Delmotte, F.M. (1996) Can. J. Microbiol. 42,234-242.

Zhu, J., Shi, J., Sing, U. and Carpita, N.C. (1991) Plant Physiol. 96s, 10.

Zorreguita, A., Geremia, R.A., Cavaignac, S., Cangelosi, G.A., Nester, E.W. and Ugalde, R.A. (1988) Mol. Plant-Microbe Interact. 1, 121-127.

Chapter 12

The *Agrobacterium* Oncogenes

Andrew N. Binns and Paolo Costantino

I. Crown Galls and Hairy Roots as Neoplastic Disease

I.A. HISTORY

Virulent strains of *Agrobacterium tumefaciens* and *Agrobacterium rhizogenes* induce non-self limiting - neoplastic - growths on susceptible plants, generally in dicotyledonous species from the Angiosperms. In most cases, *A. tumefaciens* induces unorganized "crown gall" tumors (so named because the growths were often observed at the crown of the plant), though certain isolates can induce teratomatous tumors that exhibit a choatic array of plant structures (Figure 1). *A. rhizogenes* infection results in the continuous proliferation of "hairy roots" from the infection site (Figure 2). The demonstration that *A. tumefaciens* (then called *Bacterium tumefaciens*) is the causal agent of crown gall tumors was first presented by Smith and Townsend (1907) who showed that this bacterium could fulfill Koch's postulates. Later, Riker *et al.* (1930) showed *A. rhizogenes* was, similarly, the causal agent of the hairy root disease.

In the period of time just after these discoveries (1920's and 30's) most of the leading students of crown gall were convinced that the observed growth promotion at the infection site required the continuous presence of the bacteria (see review by Braun, 1982 for early history). During the 1940's, however, Armin Braun carried out several series of experiments that fundamentally altered the way scientists approached the crown gall tumor problem.

Figure 1. Crown gall teratoma induced on stem of *Kalenchōe daigremontiana* by strain T37 of *A. tumefaciens.* (photographed by Professor Armin C. Braun)

First, in experiments utilizing "secondary tumors", that is, tumors that arose distal to the infection site, White and Braun (1942) showed that the tumors did not contain observable bacteria, despite extensive efforts to find them and such tumors were capable of continuous, non-self limiting and bacteria-free growth when grafted onto healthy host plants. The authors concluded that the induction of crown gall tumors represented an authentic transformation to the neoplastic state. Second, Braun and Laskaris (1942)

carried out studies on plants infected by an attenuated strain (A66), originally isolated in Riker's lab (Hendrickson *et al.*, 1934). These studies showed that when tested on decapitated plants the attenuated strain was almost avirulent, but if auxin were provided, via lanolin paste, on the decapitation site above the wound, then large tumors developed. The investigators proposed that the tumorigenesis had two phases - an inception phase and a development phase. They proposed that the attenuated strain had a

Figure 2. Hairy roots induced on stem of *Kalenchöe daigremontiana* by *A. rhizogenes*.

deficiency in the development phase. It is now known that this strain has an insertion mutation in one of the *Agrobacterium* oncogenes (Binns *et al.*, 1982). Third, by delaying inoculation until various times after wounding, Riker (1926), and later, Braun and Mandle (1948), found that the bacteria functioned to induce tumors only within the first few days after wounding. Braun and colleagues then exploited the observation of Riker (1926) that *A. tumefaciens* can not induce tumors at temperatures >30°C even though host and bacteria can both grow at this temperature. Plants were exposed to high temperatures at various times after inoculation with the virulent bacteria and shown to require ~36 hours of permissive temperatures in order to exhibit tumor induction (Braun and Mandle, 1948). Based on these experiments Braun proposed that some factor emanating from the bacteria, the "Tumor Inducing Principle" (TIP), must be transmitted to the plant cells during an approximate 36 hour period of time between 24 and 96 hours after infection at a fresh wound site. Finally, based on studies in which he demonstrated a reversal of the tumorous state in certain cases, Braun (1951; 1959) proposed that the

TIP may be a self-replicating entity, one that replicated in the plant independent of its chromosomes. Although this last concept proved to be incorrect, Braun's amazing experiments during the 40's and 50's provided the experimental foundation for the study of *Agrobacterium* mediated transformation.

During the time that Braun's studies provided the biological underpinnings for the study of the crown gall tumor problem, the investigations of Morel and colleagues concerning the biochemistry of such tumors provided the key insight that led to the identification of the elusive TIP. These investigators noted that crown gall tumors produced unusual amino acid-sugar conjugates - later termed opines - that were not normally found in the plant (for review of early work see Tempé and Goldmann, 1982). Moreover, they showed that the type of opine produced by a tumor was specified by the bacterial strain, not the plant (Goldmann *et al.*, 1968; Petit *et al.*, 1970). For example, strain B6 induces octopine producing tumors while strain T37 induces tumors that contained nopaline, regardless of the species of plant investigated. These studies strongly suggested the TIP was genetic information that was transferred from bacterium to plant, and this information varied from bacterium to bacterium, at least in terms of specifying the opine production. The issue of opines and the reasons for their importance to the bacterium are discussed in other chapters in this book.

I.B. Ti AND Ri PLASMIDS

I.B.1. Ti Plasmids and T-DNA

After numerous attempts to demonstrate the nature of the TIP, the crucial observation that led to an understanding of the molecular basis of the crown gall tumor transformation was the transfer of a virulence factor between Agrobacteria. In these experiments, originally carried out by Kerr in the late 60's, the capacity for virulence as well as the specificity of the tumor's opine, was transferred from virulent to avirulent bacteria when both were inoculated into tumors originally induced by the virulent strain (Kerr, 1969). The second major clue was that strains C58 and Ach5 could be cured of their capacity to induce tumors by heat treatment (Hamilton and Fall, 1971). Because genetic transfer between bacteria is often mediated by conjugative plasmids, and because heat treatment was known to cure bacteria of plasmids in some cases, a search for plasmids responsible for virulence was initiated. This search resulted in the discovery that virulent strains carried large (>200 kbp) plasmids and that the transfer of such plasmids to avirulent strains rendered them virulent (Zaenen et al., 1974; van Larebeke et al., 1975; Watson et al., 1975).

The experiments described above demonstrated that the TIP described by Braun was encoded by the so-called tumor inducing (Ti) plasmids. Immediately, a search was initiated to determine whether the TIP was, in fact, the Ti plasmid itself and whether this plasmid was transferred into plant cells during the tumor inception phase. After two years of intensive investigation Chilton et al (1977) used solution hybridization methodology to demonstrate that the entire Ti plasmid was not transferred to plant cells, but that at least a small portion of it was and this was named the transfer DNA (T-DNA). Shortly thereafter, Southern blot analysis more precisely defined the extent and exact boundaries of the T-DNA in plants (Lemmers et al., 1980; Thomashow et al., 1980). One intriguing observation was that two separate pieces of T-DNA, termed T_L-DNA and T_R-DNA were found in the octopine type Ti plasmid, whereas one contiguous stretch of T-DNA was found in the nopaline type T-DNA. During this period of time, several labs independently discovered that the T-DNA from strains of Agrobacterium that resulted in tumors with different opines had what were described as "conserved regions" and "non-conserved" regions, and, in octopine type Ti plasmids, the conserved region falls within the T-DNA only (Chilton et al. 1978; Depicker et al. 1978). We now know that genes within the conserved regions encode proteins required for neoplastic growth (see below), and are shared by all Ti plasmids so far examined, whereas genes in the non conserved regions encode enzymes involved in opine synthesis (see Chapter 9).

I.B.2. Ri Plasmids and T-DNA

Attempts to unravel the mechanism of root induction by A. rhizogenes began much later than the work on crown gall tumors and took great advantage from the investigations of the latter. Although early experiments seemed to suggest that hairy roots could be elicited by cell-free filtrates of A. rhizogenes (Hopkins and Durbin, 1971), more recent work on this system was in fact based on the assumption that the molecular events underlying hairy root induction would be similar to those responsible for crown gall tumors. Albinger and Beiderbeck (1977) were able to transfer root-inducing capability from A. rhizogenes to A. tumefaciens by means of in planta "Kerr-type" conjugal transfers (Kerr, 1969). This fact suggested an extrachromosomal (plasmid) location of the virulence determinants of A. rhizogenes. The involvement of a plasmid, originally denominated Hr, in hairy root induction was subsequently confirmed by Moore et al. (1979) and White and Nester (1980). In the meantime, Ti plasmids and T-DNA had been characterized, and the search for root-inducing plasmids was vigorously pursued. Several A. rhizogenes strains were analyzed for plasmid content and all virulent strains were indeed found, by polyacrylamide gel electrophoresis and electron microscopy, to harbor plasmids comparable in size (98 to 190 MDa) to the Ti plasmids (Costantino et al., 1981). In the same year, the opine mannopine was detected in hairy roots (Tepfer and Tempé, 1981), a strong indication that the molecular mechanism of induction of these neoplastic roots is indeed based upon T-DNA transfer. A year later, M.D. Chilton, in the laboratory of J. Tempé, provided conclusive southern-blot evidence that hairy roots contain T-DNA, originating from the virulence plasmid of A. rhizogenes (Chilton et al.,

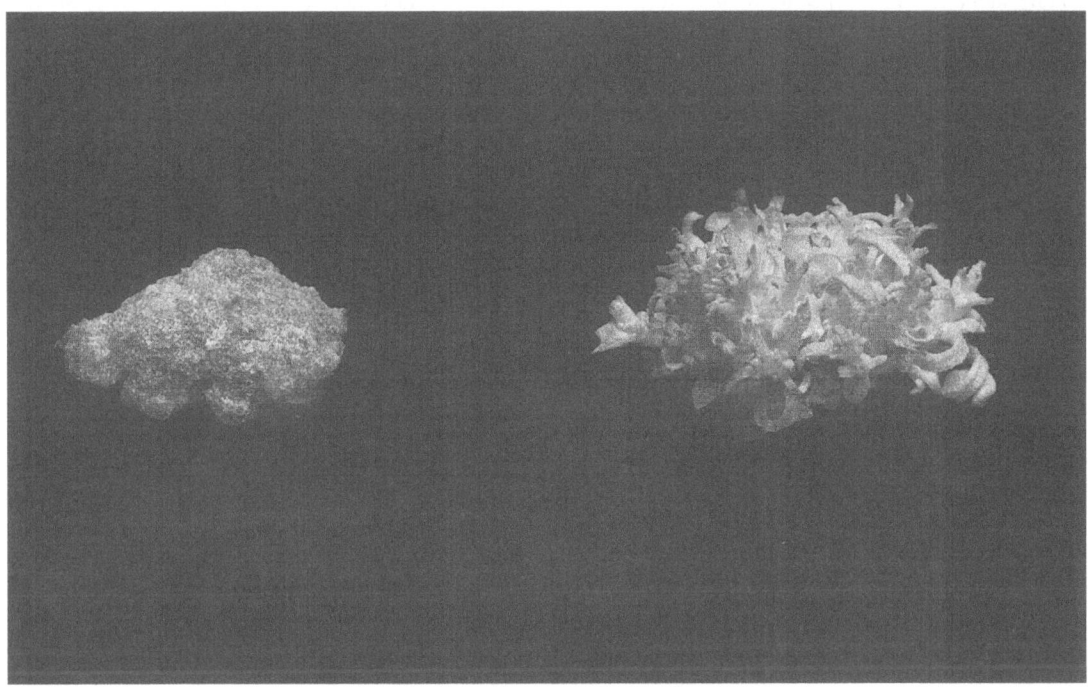

Figure 3. Cloned tumor lines of tobacco transformed by *A. tumefaciens* strains carrying either wild type (left) or "shooty mutant" (right) octopine type Ti plasmid, grown on hormone-free plant tissue culture medium.

1982). These plasmids were then denominated Ri (*root inducing*). Later in 1982, the presence of T-DNA in hairy roots was confirmed by a number of other laboratories (Spanò *et al.*, 1982; White *et al.*, 1982; Willmitzer *et al.*, 1982).

Subsequent characterization of the Ri T-DNAs of strains that induce mannopine (pRi8196) or cucumopine (pRi2659) producing roots demonstrated a single continuous T-DNA segment, some 15 (core T-DNA) to 30 (extended T-DNA) kilobases long (Byrne *et al.*, 1983; David and Tempé, 1987). In contrast, the transferred-DNA of agropine type Ri plasmids (pRi1855, pA4 and p15834, all virtually identical) consists of two stretches of DNA, denominated T_L and T_R, separated by a long sequence which is not integrated in the plant genome (Huffman *et al.*, 1984; De Paolis *et al.* 1985). The root inducing genes are located on the T_L-DNA, while the T_R-DNA encodes agropine and auxin

biosynthesis (see below). These latter functions are absent from both mannopine and cucumopine T-DNAs. The agropine T_L-DNA shares with mannopine and cucumopine T-DNA two strongly conserved regions separated by a much less conserved stretch of DNA (Filetici *et al.*, 1987; Brevet *et al.*, 1988).

II. T-DNA Oncogenes

II.A. ONCOGENES ON THE Ti PLASMID

II.A.1. Discovery and Genetics

Following the discovery of the conserved and non-conserved regions of the T-DNA of the Ti plasmids, genetic analysis. identified a series of mutations that affected tumor morphology and phenotype (Garfinkel *et al.*, 1981; Ooms *et al.*, 1981; Leemans *et al.*, 1982) (Figure 3). For example, transformation of tobacco by strains carrying these mutations resulted in either "shooty" or "rooty" tumors, and these phenotypes were maintained when the tumors were cultured (Ooms *et al.*, 1981; Binns *et al.*, 1982; Leemans *et al.*, 1982). Two separate genes were identified that, when mutated, resulted in the shooty phenotype (*tumor morphology shooty, or tms*), whereas mutations in only one gene resulted in the rooty phenotype (*tumor morphology rooty, or tmr*). Strains carrying mutations in both the "shooty" and "rooty" loci were avirulent but could be shown to transfer DNA into plant cells as indicated by opine production at the wound site (Hille *et al.*, 1983; Ream *et al.*, 1983). Based on the known effects of auxin and cytokinin on morphogenesis and cell division in tobacco tissue cultures (Skoog and Miller, 1957), and the fact that wild type crown gall tumors required neither of these hormones for growth in culture (Braun, 1958), these results suggested that the genes in question were responsible for plant hormone production or enhanced plant hormone perception. In support of this hypothesis "shooty" transformants, cultured under conditions that did not allow shoot formation, were shown to be auxin requiring (Binns *et al.*, 1982). The observation that after prolonged culture on hormone free medium wild type tumors usually consisted of mainly non-transformed cells, while shooty tumors had a much greater percentage of transformed cells (Binns *et al.*, 1982; van Slogteren *et al* 1983) suggested that plant hormone production by the wild type transformants

could "cross-feed" non-transformed cells and this capacity was drastically reduced in the mutants. Similarly, tobacco cells transformed by the *tmr*-nopaline type Ti plasmid were cytokinin requiring in culture (Binns, 1983). Finally, analysis of hormone levels in both primary tumors and cultured tumors demonstrated that wild type transformants had high levels of both auxin and cytokinin, whereas auxin levels were reduced in the shooty tumors and cytokinin levels were reduced in the rooty tumors (Akiyoshi *et al.*, 1983; van Onckelen *et al.*, 1984). These latter studies conclusively demonstrated that plant hormone accumulation was specified by the T-DNA and, at the very least, partially responsible for the tumorous growth and phenotypes of the transformed cells.

II.A.2. Biochemistry of Ti Plasmid Oncogenic Proteins

Because most of the early genetics exploited transposon mutagenesis, the identification and cloning of the genes responsible for auxin and cytokinin biosynthesis was rapidly accomplished (Figure 4). These genes contain no introns and are expressed, at low abundance, as polyadenylated mRNAs in transformed plant cells (Willmitzer *et al.*, 1982b; Willmitzer *et al.*, 1983). More difficult was the identification of the biochemical role played by the proteins encoded by the different genes. Ultimately, the two genes proposed to be responsible for auxin production (originally referred to as *tms1/gene 1* and *tms2/gene 2*) were shown to encode

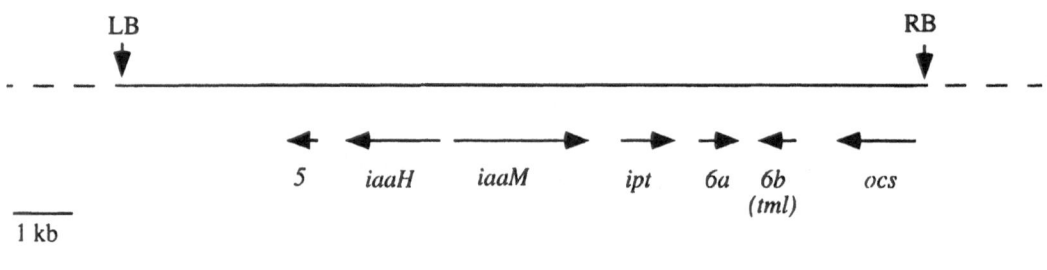

Octopine pTiA6 T$_L$-DNA

Figure 4. Map of the T$_L$-DNA of octopine type plasmid pTiA6 showing approximate location and orientation of several oncogenes as well as octopine synthase (*ocs*) and the left and right borders repeats (LB, RB) of the T-DNA.

a two step biosynthetic pathway. The *iaaM* (*tms1*) gene product, tryptophan monooxygenase, converts tryptophan to indole-3-acetamide (Thomashow *et al.*, 1986; van Onckelen *et al.*, 1986) which is then converted to indole-3-acetic acid through the activity of the *iaaH* (*tms2*) gene product, indole-3-acetamide hydrolase (Inzé *et al.*, 1984; Schröder *et al.*, 1984; Thomashow *et al.*, 1984). This observation was surprising because this pathway is not known to occur in higher plants but has been identified in pathogenic strains of *Pseudomonas* that induce hyperplasia on many plants (Yamada *et al.*, 1985). The gene (*tmr/gene 4*) responsible for cytokinin production was shown to encode dimethylallyl-pyrophosphate (DMAPP) transferase (or isopentenyl transferase) and is now known as *ipt* (Akiyoshi *et al.*, 1984; Barry *et al.*, 1984; Buchmann *et al.*, 1985). The enzyme converts DMAPP and AMP to isopentenyl-AMP, which, in turn, can be rapidly converted to a broad array of cytokinins by host encoded enzymes (for review see Binns, 1994). Homologous genes encoding *ipt*-like enzymes have also been found in numerous pathogenic bacteria (Crespi *et al.*, 1992; Lichter *et al.*, 1995). Apparently, both auxin and cytokinin biosynthetic genes were somehow 'captured' by *Agrobacterium* and modified so as to allow expression in eukaryotic cells.

The elucidation of the biochemical activities of the auxin and cytokinin biosynthetic loci ended over forty years of investigation designed to elucidate the identity and activity of the oncogenic elements within *Agrobacterium*. Strains carrying only these three genes on the T-DNA could induce wild type tumors on most plants (Inzé *et al.*, 1984). Yet the conserved DNA of the Ti plasmid continues to retain many of its secrets. For example, one gene in the octopine type Ti plasmid has been identified that, when mutated, yields larger tumors than wild type (Ream *et al.*, 1983). The biochemical activity of this gene, referred to as *tumor morphology large* (*tml*), or *gene6b*, has not been clarified. Several reports suggest that it is involved in hormone responsiveness, and, in the cases, a different form of hormone independence (Hooykaas *et al* 1988; Spanier *et al.*, 1989; Tinland *et al.*, 1989; Wabiko and Minemura, 1996). Similarly, mutations in a T-DNA gene known only as "*gene 5*" exhibit a phenotype only when other T-DNA genes are mutant: increased shootiness in *tms* mutants (Leemans *et al.*, 1982). *Gene5* was shown to encode an indole lactate synthase (Körber *et al.*, 1991), but the physiological relevance of indole-3-lactic acid is disputed. Detailed tests (Sprunck *et al.*, 1995) of its proposed role as an auxin antagonist (Körber *et al.* 1991) were negative, leaving the function of this indole derivative within the tumor unclear.

II.A.3. Effects of Oncogenes in Transgenic Plants

Following the identification and functional characterization of the auxin and cytokinin biosynthetic genes, several labs have engineered plants to contain various versions of them. Most of these analyses utilized some type of controllable promoter - for example the heat shock, tissue specific, hormone inducible wound inducible, and/or senescence specific promoters (Smigocki and Owens, 1988; Medford *et al.*, 1989; Schmülling *et al.*, 1989a; Estruch *et al.*, 1991c; Smart *et al.*, 1991; Smigocki, 1991; Li *et al.*, 1992; Gan and Amasino, 1995) - to characterize the response of the plant to unusual hormone conditions. In general, the results of these studies have been in agreement with exogenous application of hormones to either the whole plant or parts of it. For example, expression of *ipt* has resulted in release of apical dominance, delay of senescence and flowering abnormalities. Similarly, plants overexpressing *iaaM*, and hence producing large quantities of indole-3-acetamide, show effects predicted by application of exogenous auxin (*e.g.* adventitious root formation, strict apical dominance, thickened stem) (Klee *et al.*, 1987). These phenotypes appear to result from auxin produced through the non-specific conversion of indole-3-acetamide to IAA by host amidohydrolases. While the *iaaM* and *ipt* genes have, in general, caused the expected phenotype when in transgenic plants, the situation is not nearly so clear in transgenic plants containing either *gene 5* or *gene6b* (*tml*). Recent studies in which *N. tabacum* leaf explants were transformed with *gene 6b* alone suggest that it is involved in organogenesis (Wabiko and Minemura, 1996). The plants regenerated from these transformants had defects indicative of excess cytokinin-like activity, but the cytokinin levels were as in wild type. While this suggests that *gene6b* may be involved in a down stream step in cytokinin signaling, different results have been obtained with the same gene obtained from different Ti plasmids (Spanier *et al.*, 1989; Tinland *et al.*, 1989). These studies indicated *gene6b* interferes with or reduces cytokinin activity. Further work is clearly necessary to understand the role(s) played by *gene5* and *6b* in tumorigenesis and, conceivably, plant development.

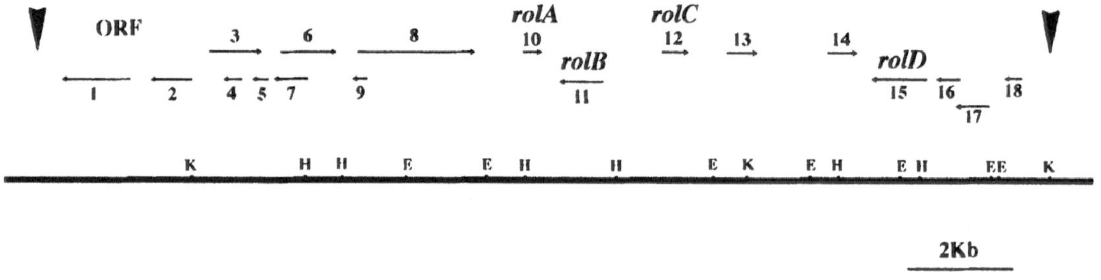

Agropine pRi1855 TL-DNA map

Figure 5. Map of the T_L-DNA of agropine type plasmid pRi1855 showing approximate location and orientation of the *rol* genes. Arrows indicate border repeats of the T-DNA.

II.B. ONCOGENES ON THE RI T-DNA

II.B.1.Discovery and Genetics of rol and other Ri Genes

As mentioned above, the T_R-DNA of agropine Ri plasmids harbors auxin biosynthetic (*aux*) loci, revealed by their very high DNA sequence homology with the equivalent genes (*iaaM and iaaH)* in the Ti plasmids (Huffman *et al.,* 1984; De Paolis *et al.,* 1985). The presence of *aux* genes caused an initial misinterpretation of the hairy root phenomenon. The Ri plasmid T-DNA was originally considered to be similar to *tmr* mutants of the Ti plasmid since, as discussed above, these mutants result in auxin over-production in infected cells and root formation. However, mannopine and cucumopine T-DNAs lack *aux* genes but induce hairy roots indistinguishable, on most plant hosts, from those due to the agropine strains. On some hosts, however, the mannopine and cucumopine strains exhibit significantly reduced virulence. Functional equivalence of Ri *aux* and Ti *iaa*-genes was demonstrated in complementation assays by Offringa *et al.,* (1986). The same authors showed also that the T_R-DNA could induce formation of tumors on some plant species, such as decapitated pea.

The relative roles of the *aux* genes and of the genes on the T_L-DNA were sorted out by means of a complex series of infections utilizing different *A. rhizogenes* strains that contained different complements of Ri genes of the agropine type Ri plasmid. The *aux* genes were shown to play a rather ancillary role in hairy root induction, merely providing auxin which, whenever the endogenous auxin level of the host is limiting, triggers root differentiation in cells whose response to the hormone is somehow "amplified" by the presence of T_L-DNA genes (Cardarelli *et al.,* 1987a). Hairy root induction is thus not the result of a gross auxin unbalance and the genetic determinants of Ri-induced neoplastic roots are different from those of the crown gall tumors. True hairy root inducing genes reside on the T_L-DNA which does not share sequence homology with the Ti T-DNA (Risuleo *et al.,* 1982; Huffman *et al.,* 1984).

An extensive transposon-tagging genetic analysis of the agropine-type T-DNA was reported in 1985 by White *et al.*(1985). These authors assayed the virulence properties of bacteria harboring Tn5-mutagenized T-DNAs and could localize four loci in the T_L-DNA that affect induction and growth of hairy roots. The loci were denominated *rol* (root loci) *A, B, C* and *D* (Figure 5). However, this genetic analysis did not yield readily interpretable information on the Ri T-DNA genes as did the similar experiments in the case of the Ti T-DNA mutant analysis described above. Rather, either no roots (when *rolB* was inactivated) or modified roots (*rolA, C* and *D*) were produced, observations that could not be ascribed to the well described auxin-cytokinin relationship that was so useful in interpreting many of the Ti plasmid mutants. One year later, Slightom *et al.* (1986) published the complete 21 kbp sequence of the T_L-DNA of the agropine-type T_L-DNA, which revealed the presence of at least 18 open-reading frames. Of these, ORF10, 11, 12 and 15 were found to coincide with, respectively, *rolA, B, C* and *D*.

Genetic analysis of hairy root induction showed that a segment of Ri T-DNA encompassing only the *rolA, B* and *C* genes is almost as effective in inducing neoplastic roots as the whole of the Ri T-DNA (Capone *et al.*, 1989a). Even more intriguing was the finding that *rolB*, alone among the Ri T-DNA genes, is individually capable of inducing roots on all different plants tested. Inoculation of different organs of different plants with recombinant *Agrobacterium* strains carrying only *rolB* resulted in neoplastic root growth on carrot discs (Capone *et al.*, 1989a; Capone *et al.*, 1989b) tobacco stems (Cardarelli *et al.*, 1987b), tobacco leaves (Spena and Schell, 1987) and *Kalanchöe* leaves (Spena and Schell, 1987). Furthermore, roots induced by *rolB* infections show the characteristic plagiotropic and branched growth pattern *in vitro* of hairy roots and are all transformed (as opposed to most roots induced by *aux* genes or exogenously supplied auxin) pointing to a cell-autonomous effect of this oncogene (Capone *et al.*, 1989b). In contrast, these studies showed that *rolA* and *rolC* are capable of inducing limited root growth on tobacco leaves, the latter only when driven by the strong CaMV 35S promoter. No roots could be induced by infection of tobacco stems with *rolD* (Mauro *et al.*, 1996). These results are in agreement with the above reported transposon-mutagenesis of White *et al.* (1985), where *rolB* was shown to be the only T-DNA locus that, if inactivated, would totally suppress root induction by *A. rhizogenes*. In contrast, as mentioned, inactivation of *rolA, C* (and of *rolD*) resulted only in an attenuation of root induction. Thus, each of the *rol* genes affects plant development and morphogenesis in its own characteristic way and induces neoplastic growth of plant cells. Each is, in other words, a true plant oncogene.

II.B.2. Effects of Ri T-DNA Genes on Transgenic Plants

A major difference between hairy roots and crown gall tumors is that fertile plants containing full-length T-DNA can be regenerated from the former. As hairy roots are of clonal origin (David and Tempé, 1987; Capone *et al.*, 1989a) , plants obtained from single roots are derived from a single T-DNA-transformed cell. It was shown that the Ri T-DNA is transmitted through subsequent generations and segregates in a Mendelian fashion (Costantino *et al.*, 1984; Tepfer, 1984). However, plants containing the Ri T-DNA develop abnormally and show a series of characteristic morphological alterations denominated, as a whole, the "hairy root syndrome" (Tepfer, 1984). The most consistent developmental anomalies in various plant species are wrinkled leaves, altered internode length, non-geotropic roots, altered flower morphology, and reduced seed production.

In order to identify functions on the Ri T-DNA, a number of Ri ORFs and all *rol* genes were cloned in suitable plant vectors. Recombinant *Agrobacterium* strains containing Ri ORFs and *rol* genes individually and as combinations were utilized by various laboratories for infections on different plant hosts and for the production of transgenic plants. As described above, the "hairy root syndrome" could be almost completely accounted for and dissected in the individual and synergistic effects of just the *rol* genes *A, B,* and *C*. Transgenic plants containing *rolA* have wrinkled leaves and reduced internodal distance (Sinkar *et al.*, 1988) while *rolB* induces flower heterostyly and abundant adventitious rooting (Cardarelli *et al.*, 1987b; Schmülling *et al.*, 1988) and *rolC* causes reduced apical dominance, altered leaf morphology and reduced seed production (Schmülling *et al.*, 1988). Very recently, *rolD* has been shown to be responsible for a dramatic acceleration in the vegetative to reproductive phase transition in transgenic plants. Tobacco plants harboring this oncogene set flowers in one third the time required by non-transformed plants and at a much earlier developmental stage (Mauro *et al.*, 1996). Transgenic plants carrying another Ri gene, ORF13 were shown to produce a number of graft-transmissible morphological anomalies such as small and very wrinkled leaves, short flowers and poorly developed root system (Hansen *et al.*, 1993).

The *rolB* oncogene, with its strong root-inducing capability, attracted most of the attention in the hope that its characterization would help in elucidating key molecular events in root differentiation. More recently, though, a broader role of *rolB* in plant morphogenesis has been unveiled. By means of cultured thin cell layer (TCL) explants it was shown that this oncogene, in addition to stimulating, as expected, the *in vitro* rooting program, strongly enhances the differentiation of flowers and of shoots in the respective differentiative media (Altamura *et al.*, 1994). In explants from transgenic plants carrying *rolB* the neoformed meristems are much more numerous than in the controls and these new meristems will give rise to primordia of different organs, depending on the prevailing hormonal conditions. Thus, *rolB* does not just specifically promote root differentiation but acts, more in general, as a meristem-inducing gene. Root induction appears to be only one of the possible outcomes of the interaction of the *rolB* gene product with the differentiation programs of plant cells. This makes sense given that roots seem to be the most frequent adventitious organ in a plant and are elicited by a variety of endogenous and exogenous stimuli (Andersen, 1986; Jarvis, 1986).

II.B.3. Biochemistry of Ri Gene Functions

Work intended to elucidate the biochemical functions of the Ri oncogenes has been partly based on the effects on development and morphology of transgenic plants. For the reasons described above, most of the attention has thus been focused on the *rol* genes *A, B* and *C*. More recently, the effects and biochemistry of ORF13 and *rolD* have also been investigated.

The morphological anomalies of the *rolB* transgenic plants described above, the capability of these plants and explants to give rise to adventitious roots, and the growth pattern of the latter - addition of auxin to a culture of non-transformed roots produces the branching and lack of gravity response typical of *rolB* roots (Capone *et al.*, 1989b) - led to the suggestion that plant cells expressing this oncogene are perturbed in the activity of the hormone auxin. An elegant and simple explanation of the effects of *rolB* seemed to be provided by the enzymatic activity (a β-glucosidase) which had been at one point ascribed to the RolB protein. Substrates for this activity were shown to be indoxyl-glucosides, structurally similar to indole-3-acetic acid (IAA) glucosides (Estruch *et al.*, 1991d). This β-glucosidase activity suggested that RolB could raise the level of auxin by releasing free, active IAA from its inactive glucose conjugates (Estruch *et al.*,

1991d). However, IAA-glucosides were later shown not to be a substrate for RolB and the level of IAA was found to be substantially the same in *rolB*-transformed and normal plant tissues (Nilsson *et al.*, 1993a).

Alternatively, the altered auxin activity of *rolB*-transformed cells could be due to an altered perception of the hormone stimuli. In other words, rather than to a higher concentration of auxin, *rolB*-cells might owe their properties to their higher sensitivity to this hormone. In line with this second hypothesis, *rolB*-transformed plant cells were shown to be far more sensitive to auxin than normal ones. A typical short-term effect of auxin on protoplasts, the induction of a transmembrane electric potential, requires an auxin concentration three to five orders of magnitude lower in *rolB* protoplasts than in non-transformed ones (Maurel *et al.*, 1991). In addition, the same authors showed that the increased sensitivity to auxin of *rolB*-protoplasts is not due to increased levels of the hormone (Delbarre *et al.*, 1994). Also in favor of a perturbed perception/transduction of auxin stimuli seems to be the tyrosine phosphatase activity of RolB recently demonstrated by means of overexpression of this protein in *Escherichia coli* (Filippini *et al.*, 1996). In view of the key role played by kinases and phosphatases in hormonal response in animals, this activity of RolB, if confirmed on the purified protein, would represent an invaluable tool to shed light on the signal transduction pathway of auxin - a still very elusive domain in spite of decades of work on this most intensively studied plant hormone.

Of the other *rol* oncogenes, the biochemical function of *rolC* is possibly the best defined. The morphological traits of plants transformed by *rolC* are reminiscent of the overall effects of the plant hormones cytokinins. Accordingly, the gene product of *rolC* was shown to be a cytoplasmic protein with cytokinin-β-glucosidase activity, capable of releasing cytokinins from N-glucoside conjugates (Estruch *et al.*, 1991a; Estruch *et al.*, 1991b). However, subsequent measurements of hormone levels in *rolC* transgenic plants surprisingly showed reduced, rather than increased, levels of cytokinins and an unexpected increase in a third class of plant hormones, gibberellins (Nilsson *et al.*, 1993b). A reduction of the gibberellic acid content in plants expressing *rolA* has been suggested to be causally involved in the altered phenotype induced by this oncogene (Dehio *et al.*, 1993). Finally, investigations using transgenic plants carrying ORF13 suggest that this gene may encode an as yet unidentified cytokinin-like factor (Hansen *et al.*, 1993). Work on this oncogene has however progressed very little in recent years.

To summarize, despite considerable evidence that correlates the effects of the Ri oncogenes with plant hormone effects and despite the efforts of several laboratories through the years, the biochemical functions of none of them has been as yet firmly assigned and the molecular basis of hairy root induction remains quite elusive.

III. Interaction of *Agrobacterium* Oncogenes and Plant Oncogenes

III.A. ROLE OF THE HOST IN DEFINING PHENOTYPE

III.A.1. Wound Response

The response of plants to the *Agrobacterium* onco-genes is defined not only by the biochemical activi-ties of the gene products but also by the capacity of the plant to respond to oncogene activities. One parameter that is crucial in defining the phenotype of transformed cells is the wound response. Some of the earliest studies of *Agrobacterium* demonstrated that wounding was necessary in order to get crown gall formation (see Kahl, 1982 for review). In particular, Braun demonstrated that a "window of competence" for crown gall formation existed, showing that a wound site had the potential to be transformed from 36-96 hours after wounding (Braun 1982). We now know that this response, which includes phenolic production, apoplastic acidification and polysaccha-ride synthesis, is required for the expression of the virulence (*vir*) genes of the Ti and Ri plasmids (see review in this volume). Beyond induction of the virulence genes, however, the early cell divisions at and around the wound site appear to be critical in order for transformation to occur. This conclusion is based on the observations that the timing of maximum transformation frequency observed in both wounded tissues of intact plants and isolated cells correlates with the initial round of cell division observed in these cases (for review see Binns and Thomashow, 1988). Several authors have suggested that integration of the T-DNA into plant nuclear DNA may be dependent on cell cycle events, in particular, DNA synthesis (*e.g.* Braun, 1978). This remains to be experimentally demonstrated.

III.A.2. Complementation of T-DNA Onco-gene Mutations by Host Genes.

Early investigations of crown gall tumorigenesis demonstrated that diverse plant species would vary in their response to a particular strain of *A. tumefaciens* (see Braun, 1982 for review). The response of various hosts and even different tissues from the same host to strains of *A. tumefaciens* carrying mutations in their oncogenes has provided evidence for "plant oncogenes". The concept is that certain hosts (or host tissues) infected with mutant bacterial strains activate physiological systems that overcome the absence of a T-DNA oncogene, resulting in tumors indistinguishable from those induced by wild type strains. For example, *Agrobacterium* strains carrying mutations in the *iaaH* and/or *iaaM* genes induce wild-type appearing (non-shoot forming) tumors in *Nicotiana rustica*, *N. debneyi* or *N. glutinosa* (Ooms *et al.*, 1981; Binns *et al.*, 1982). When such tumors of *N. glutinosa* were put into tissue culture, they exhibited an auxin independent growth phenotype in the absence of shoot formation. This demonstrated that the observed "complementation" was not due to a high level of auxin in the plant, but, rather, the activation of a host encoded auxin autonomy system (Binns *et al.*, 1982). Intriguingly, this type of complementation was also observed in several other *Nicotiana* species (Binns *et al.*, 1982; Binns, 1984) that, when crossed with *N. langsdorfii*, yield tumor prone hybrids (for review see Smith, 1988). These results suggest that this group of *Nicotiana* has an endogenous oncogene (or oncogenes) that can be activated by stimuli provided either by certain genes of the T-DNA or by coexistence with *N. langsdorffii* genes. One hypothesis that could explain the observed complementation of *iaaH* or *iaaM* mutants is that, in the presence of some hormone imbalance - e.g. the presence of an active *ipt* gene - genetic systems are activated that yield auxin independent growth. This hypothesis was tested by transforming *N. glutinosa* with a T-DNA that contained only the *ipt* gene and nopaline synthase. The resultant tumors were wild type in appearance on the plant and auxin and cytokinin independent in culture (Binns *et al.*, 1987). The most striking finding in those studies was that feeding exogenous cytokinin to non-transformed tissues did not induce the auxin autonomous phenotype. A similar case was observed when *N. tabacum* leaf explants were transformed by an octopine type *Agrobacterium* strain mutated in the *ipt* gene but wild type for the *iaaH* and *iaaM* loci:

the resultant transformants were cytokinin independent, even though leaf explants provided with exogenous auxin were cytokinin requiring (Black et al., 1994).

The observation that some cells can respond differently to a hormone they are producing than if they are receiving that hormone from outside the cell makes sense from a developmental perspective. For example, one well known response to cytokinins is the initiation of tracheary element formation (e.g. Fukada and Komamine, 1980). Assuming that cytokinins are produced in meristems, it is clear that the tracheary element response to cytokinins would be detrimental to meristem function. The signaling systems that tell a cell it is producing a hormone rather than receiving it are unknown, though variation in the metabolism of the hormone may play an important role. Motyka et al. (1996) recently demonstrated that induction of the ipt gene in transgenic tobacco tissue cultures (utilizing the "tet regulatable" system - see Gatz et al., 1992) induces a dramatic increase in both the endogenous cytokinins and cytokinin oxidase - an enzyme that cleaves the isopentenyl side chain from molecules such as isopentenyladenosine. While this study indicates that endogenous production of cytokinin can affect the metabolic activities within the cells, it did not seek to determine whether there was an observable difference if the hormone had been applied exogenously rather than through induction of the ipt gene.

III.A.3. Cellular rol Genes

Most amazingly, at least some plants contain true homologues of the rol oncogenes. Early Southern blots, intended to demonstrate the presence of Ri plasmid sequences in tobacco hairy roots DNA, reproducibly showed hybridization of the T-DNA probe also with DNA from non-transformed control tissue [Costantino, unpublished results]. Sequences homologous to a Ri T-DNA probe, were also detected in the genome of Daucus carota (Spanò et al, 1982). Later, sequences homologous to the core region of the T-DNA were clearly shown to be present in the genome of Nicotiana glauca (White et al., 1983) and several other members of the genus Nicotiana, including tobacco (Furner et al., 1986). The homology of the "cellular T-DNA" (cT-DNA) of N. glauca, as it is denominated, with the bacterial T-DNA encompasses the region of the rol genes B, C and the cT-DNA is present as two arms of an imperfect repeat, both approximately colinear with

the Ri T-DNA (Furner et al., 1986). The nucleotide and amino acid sequences of the Ng (Nicotiana glauca) rol genes are, respectively 83% and 75% homologous with their bacterial counterparts and "the integrity of these open-reading frames might be taken as evidence for conservation, and thus function, in the genus Nicotiana" (Furner et al., 1986). However, no transcription of the cT-DNA could be detected by means of Northern blots. Furner et al. (1986) regarded the cT-DNA as the vestige of an ancestral introduction of T-DNA from A. rhizogenes, which then became part of the genome of the genus Nicotiana.

A few years later, however, Ichikawa et al. (1990) showed that the NgrolB and C genes are transcriptionally active in tumors from Nicotiana glauca x Nicotiana langsdorffii tumorigenic hybrids. These authors put forward the hypothesis that the spontaneous formation of these long-studied genetic tumors in these hybrids might be due to the endogenous N.glauca rol genes - absent from the genome of the other parental plant - which are turned on due to exposure to the genetic background of N. langsdorffii. The same group in Japan later cloned the N. glauca homologues of Ri ORF13 and ORF14, thus expanding the patrimony of cellular Ri oncogenes, and showed that also these genes are silent in the plant but expressed in the genetic tumors. Finally, Meyer et al. (1995) demonstrated that the endogenous version of rolC is transcriptionally active in tobacco plants and its expression is tissue specific, up-regulated by cytokinin and down-regulated by auxin. These authors offer the intriguing suggestion that a "gene of bacterial origin introduced during evolution can have a function in a modern plant".

III.B. REGULATION OF AGROBACTE-RIUM ONCOGENES IN PLANTS

III.B.1 Ti Oncogenes

Beyond understanding the biochemical activities of the Ti oncogenes, it is apparent that their activities can be regulated within the plant cell. This concept was first indicated in studies by Braun and Wood (1976), who characterized the shoots that could be regenerated from shoot forming crown gall teratomas of tobacco induced by strain T37. While these shoots could not form roots, they could be grown to an adult, reproductive stage by grafting them onto non-transformed hosts. The intriguing observation was that while the tumorous state (uncontrolled proliferation) was suppressed in these shoots, it could be reactivated if tissues from them were placed in culture (Braun and Wood, 1976). In particular, tissues normally expected to have relatively high levels endogenous auxin (e.g. young leaves) grew spontaneously on hormone free medium whereas tissues with low levels of auxin (e.g. pith) required brief exposure to exogenous auxin before the tumorous state - that is capacity for growth on hormone free medium - was reinitiated. While the molecular basis of the suppression of uncontrolled growth is not known, Black *et al.* (1986) demonstrated that the *iaaH* gene is not active in differentiated tissue of plants carrying the T37 auxin biosynthetic genes, but becomes activated during culture. In contrast, the genes encoding opine synthesis are active regardless of the differentiated state of the plant cells (Wood *et al.*, 1978). Even in cases where the genes encoding auxin biosynthesis are active, the actual levels of both the enzyme and its product can vary enormously during the growth cycle of a cultured tumor line (Kuleck *et al.*, 1991). Although the transcriptional regulation of the *iaaH* and *iaaM* genes from the Ti plasmid have not been carefully studied, the homologous genes from the Ri plasmid are apparently most active in dividing cells (Gaudin and Jouanin, 1995).

In several cases, the expression characteristics of the T-DNA oncogenes have been studied by creating oncogene promoter:reporter gene fusions in transgenic plants and monitoring reporter activity. For example, expression of an *ipt* promoter:GUS fusion in transgenic potato demonstrated 10-100x higher GUS activity in stems and roots compared to leaves and tubers (Dymock *et al.*, 1991). The expression of this construct from leaf tissues could, however, be activated to high levels by culturing the tissues *in vitro*, indicating that the *ipt* promoter is, indeed, responsive to specific environmental conditions. This contrasts with results from a similar construct tested in tobacco that yielded a more constitutive response (Neuteboom *et al* 1993). Analysis of transgenic tobacco plants carrying a *gene 5* promoter:*aphII* fusion demonstrated that this gene was expressed in stems and petioles, but not in leaf tissues and that expression from leaf protoplasts was dependent on the presence of both auxin and cytokinin (Körber *et al.*, 1991). In the case of *gene6b* promoter:GUS fusions in tobacco, expression was greatest in roots, and low in leaves, cotyledons, hypocotyls and stems (Bagyan *et al.*, 1995). This study also demonstrated that wounding induced expression of *gene6b*, particularly from leaf tissues,

and auxin further stimulates the expression from leaf explants.

To summarize, the activity of the Ti plasmid T-DNA oncogenes can be regulated in plant cells via poorly defined molecular mechanisms. Given the potential toxicity of high concentrations of plant hormones, this capacity for regulation appears to be adapted to insure the survival and continuous proliferation of the transformed cells.

III.B.2. Ri Oncogenes

Among the Ri T-DNA genes, the transcriptional regulation of *rolB* is the most thoroughly characterized. Most of the analysis has been carried out on transgenic plants expressing gene fusions between deletions of the 5' upstream non-coding region of *rolB* and the GUS reporter gene. This analysis unveiled the fine regulation that *rolB* is subject to in the host plant and the complex architecture of the promoter of this gene. Under the control of about 1200 bp of its 5' non coding region, *rolB* is developmentally regulated. In early stages of embryo development the gene is inactive and is then synchronously activated in all cells at the end of the globular stage; *rolB* is then expressed in all cells of the embryo up to almost complete maturation (Chichiriccò *et al.*, 1992). The sudden and synchronous activation of *rolB* in the embryo cells occurs at the stage of development when embryo cells acquire the capability to respond to auxin (Lo Schiavo *et al.*, 1991). In mature embryos, *rolB* is mainly expressed in the radical pole and, to a lesser extent, in the central cylinder (Altamura *et al.*, 1991; Capone *et al.*, 1991). In adult plants, *rolB* expression is tissue-specific, localized mainly in the initial cells of the root, shoot and flower meristems and, to a lesser extent in parenchymatic cells of the vascular system (Altamura *et al.*, 1991; Capone *et al.*, 1991). The *rolB* promoter is also activated by auxin. In leaf protoplasts freshly prepared from transgenic plants *rolB* is expressed at a very low level. Following 24h incubation with auxin, a 15-20 fold induction is observed, with the earliest detectable increase after 5-6 hrs. (Maurel *et al.*, 1990; Capone *et al.*, 1991). Several domains, denominated A to E, have been identified within the *rolB* promoter region. Different combinations of these domains direct gene expression in different populations of cells of the plant meristems. Of particular interest is domain B, whose deletion suppresses expression in all meristematic cells and also suppresses responsiveness of the promoter to auxin (Capone *et al.*, 1994).

The fact that the same regulatory domain is necessary for gene expression in meristems and for auxin responsiveness and the fact that *rolB* is activated during embryogenesis at the precise time when embryo cells acquire the capability to respond to auxin may suggest a link between hormonal control and tissue-specificity. Cell-specific expression and developmental control of *rolB* might thus be ultimately due to differences in hormone concentration and/or sensitivity between different cells and different stages of development. It should be pointed out that the *rolB* promoter is very likely a mosaic of plant regulatory sequences and is controlled by endogenous plant regulatory proteins. Thus, the same plant regulatory proteins that control *rolB* must also control the expression of endogenous plant genes. Access to these genes, possibly involved in plant morphogenesis is an intriguing perspective offered by the work on *rolB* regulation. Indeed, a plant regulatory factor binding to domain B has been recently identified. This protein, NtBBF1, binds to DNA trough a single zinc finger of a novel type (de Paolis *et al.*, 1996). Interestingly, the same zinc finger domain is shared by a whole new wide family of plant transcription factors, present in phylogenetically very distant plants such as maize, tobacco, Arabidopsis and Cucurbis sp.

The transcriptional regulation of the other *rol* genes has been much less thoroughly investigated. Histochemical analysis of transgenic plants expressing GUS constructs with the 5' regulatory regions of *rolA* (Schmülling *et al.*, 1989b), *rolC* (Schmülling *et al.*, 1989b; Sugaya *et al.*, 1989) and *rolD* (Trovato *et al*, 1997), revealed that each of these oncogenes is regulated in the host plant in its own characteristic way. In particular, expression of *rolD* is under developmental control and correlates with the elongation/expansion and maturation phases of the plant tissues. It would be interesting to correlate patterns of expression with functions of the *rol* oncogenes to decipher their complex interplay aimed at (and evolved for) the control of plant cell growth and differentiation.

IV. Concluding Remarks

The oncogenes of *Agrobacterium tumefaciens* and *Agrobacterium rhizogenes* fall into two major categories. The first category includes genes that encode enzymes involved in plant hormone biosynthesis. Expression of these genes leads to higher than usual levels of endogenous auxin or cytokinin, resulting in growth effects related to these substances. The second, and less well characterized, category are genes whose protein products appear to be involved in either the metabolism of plant hormones or the response of plant cells to those hormones. The *rol* genes of the Ri plasmid are particularly intriguing examples of the second category and the fact that at least some plants appear to have incorporated these genes into their active genome makes them even more unusual. Phenotypes of plant cells transformed by these oncogenes depend, however, not only on their biochemical activities but also the physiological and biochemical activities of host "oncogenes", that is host genes that affect cell proliferation and/or differentiation. In the future, research on the *Agrobacterium* oncogenes will likely focus on how plants may be manipulated so as to achieve altered levels of endogenous hormones, thereby yielding desired phenotypes and how plants can be manipulated with these oncogenes to acquire tissue or organ specific alterations in hormone perception or response. Finally, the *Agrobacterium* oncogenes (in particular the *rol* genes) will likely offer intriguing insights in the study of plant regulatory and structural genes relevant to morphogenetic and developmental processes.

V. References

Akiyoshi, D.E., Klee, H., Amasino, R.M., Nester, E.W. and Gordon, M.P. (1984) Proc. Natl. Acad. Sci. USA 81, 5994-5998.

Akiyoshi, D.E., Morris, R.O., Hinz, R., Mischke, B.S., Kosuge, T., Garfinkel, D.J., Gordon, M.P. and Nester, E.W. (1983) Proc. Natl. Acad. Sci. USA 80, 407-411.

Albinger, G. and Beiderbeck, R. (1977) Phytopath. Z 90, 306-310.

Altamura, M.M., Archilletti, T., Capone, I. and Costantino, P. (1991) New Phytol. 118, 69-78.

Altamura, M.M., Capitani, F., Gazza, L., Capone, I. and Costantino, P. (1994) New Phytol. 126, 283-293.

Andersen, A.S. (1986). in M. B. Jackson, (ed) New Root Formation in Plants and Cuttings, Martinus Nijhoff Publishers, Dordrecht, pp. 223-253.

Bagyan, I.L., Revenkova, E.V., Posmogova, G.E., Kraev, A.S. and Skryabin, K.G. (1995) Plant Mol. Biol. 29, 1299-1304.

Barry, G.F., Rogers, S.G., Fraley, R.T. and Brand, L. (1984) Proc. Natl. Acad. Sci. USA 81, 4776-4780.

Binns, A.N. (1983) Planta 158, 272-279.

Binns, A.N. (1984) Oxford Surveys of Plant Mol. and Cell. Biol. 1, 133-160.

Binns, A.N. (1994) Annu. Rev. Plant Physiol. Plant Mol. Biol. 45, 173-196.

Binns, A.N., Labriola, J. and Black, R.C. (1987) Planta 171, 539-548.

Binns, A.N., Sciaky, D. and Wood, H.N. (1982) Cell 31, 605-612.

Binns, A.N. and Thomashow, M.F. (1988) Ann. Rev. Microbiol. 42, 575-606.

Black, R.C., Binns, A.N., Chang, C.-F. and Lynn, D.G. (1994) Plant Physiol. 105, 989-998.

Black, R.C., Kuleck, G.A. and Binns, A.N. (1986) Plant Physiol. 80, 145-151.

Braun, A.C. (1951) Cancer Research 11, 839-844.

Braun, A.C. (1958) Proc. Natl. Acad. Sci. USA 44, 344-349.

Braun, A.C. (1959) Proc. Natl. Acad. Sci. USA 45, 932-938.

Braun, A.C. (1978) Biochim. Biophys. Acta 516, 167-191.

Braun, A.C. (1982). in G. Kahl and J. Schell, (eds) Molecular Biology of Plant Tumors, Academic Press, New York, pp. 155-210.

Braun, A.C. and Laskaris (1942) Proc. Natl. Acad. Sci. USA 28, 468-477.

Braun, A.C. and Mandle, R.J. (1948) Growth 12, 255-269.

Braun, A.C. and Wood, H.N. (1976) Proc. Natl. Acad. Sci. USA 73, 496-500.

Brevet, J. and Tempé, J. (1988) Plasmid 19,75-83.

Buchmann, I., Marner, F.-J., Schröder, G., Waffenschmidt, S. and Schröder, J. (1985) EMBO J. 4, 853-859.

Byrne, M.C., Koplow, J., David, C., Tempe, J. and Chilton, M.-D. (1983) J. Mol. Appl. Genet. 2, 201-209.

Capone, I., Cardarelli, M., Mariotti, D., Pomponi, M., De Paolis, A. and Costantino, P. (1991) Plant Mol. Biol. 16, 426-443.

Capone, I., Cardarelli, M., Trovato, M. and Costantino, P. (1989a) Mol. Gen. Genet. 216, 239-244.

Capone, I., Frugis, G., Costantino, P. and Cardarelli, M. (1994) Plant Mol. Biol. 25, 681-691.

Capone, I., Spanò, L., Cardarelli, M., Bellincampi, D., Petit, A. and Costantino, P. (1989b) Plant Mol. Biol. 13, 43-52.

Cardarelli, M., Mariotti, D., Pomponi, M., Spanò, L., Capone, I. and Costantino, P. (1987b) Mol. Gen. Genet. 209, 475- 480.

Cardarelli, M., Spanò, L., Mariotti, D., Mauro, M.L. and Costantino, P. (1987a) Mol. Gen. Genet. 208, 457-463.

Chichiriccò, G., Costantino, P. and Spanò, L. (1992) Plant Cell Physiol. 33, 827-832.

Chilton, M.-D., Drummond, M.H., Merlo, D.J., Sciaky, D., Montoya, A.L., Gordon, M.P. and Nester, E.W. (1977) Cell 11, 263-271.

Chilton, M.-D., Tepfer, D.A., Petit, A., Casse-Delbart, F. and Tempé, J. (1982) Nature 295, 432-434.

Costantino, P., Mauro, M.L., Micheli, G., Risuleo, G., Hooykaas, P.J.J. and Schilperoort, R.A. (1981) Plasmid 5, 170-182.

Costantino, P., Spanò, L., Pomponi, M., Benvenuto, E. and Ancora, G. (1984) J. Mol. Appl. Genet. 2, 465-470.

Crespi, M., Messens, E., Caplan, A.B., Van Montagu, M. and Desomer, J. (1992) EMBO J. 11, 795-804.

David, C. and Tempé, J. (1987) Plant Mol. Biol. 9, 585-592.

de Paolis, A., Mauro, M.L., Pomponi, M., Cardarelli, M., Spanò, L. and Costantino, P. (1985) Plasmid 13, 1-7.

de Paolis, A., Sabatini, S., De Pascalis, L., Costantino, P. and Capone, I.A. (1996) Plant J. 10, 215-223.

Dehio, C., Grossmann, K., Schell, J. and Schmülling, T. (1993) Plant Mol. Biol. 23, 1199-1210.

Delbarre, A., Muller, P., Imhoff, V., Barbier-Brygoo, H., Maurel, C., Leblanc, N., Perrot-Rechenmann, C. and Guern, J. (1994) Plant Physiol. 105, 563-569.

Dymock, D., Risiott, R., de Pater, S., Lancaster, J., Tillson, P. and Ooms, G. (1991) Plant Mol. Biol. 17, 711-725.

Estruch, J.J., Chriqui, D., Grossmann, K., Schell, J. and Spena, A. (1991a) EMBO J. 10, 2889-2895.

Estruch, J.J., Parets-Soler, A., Schmülling, T. and Spena, A. (1991b) Plant Mol. Biol. 17, 547-550.

Estruch, J.J., Prinsen, E., Onckelen, H.V., Schell, J. and Spena, A. (1991c) Science 254, 1364-1367.

Estruch, J.J., Shell, J. and Spena, A. (1991d) EMBO J. 10, 3125-3128.

Filetici, P., Spanò, L. and Costantino, P. (1987) Plant Mol. Biol. 9, 19-26.

Filippini, F., Rossi, V., Marin, O., Trovato, M., Downey, P.M., Costantino, P., Lo Schiavo, F. and Terzi, M. (1996) Nature 379, 499-500.

Fukada, H. and Komamine, A. (1980) Plant Physiol. 65, 57-66.

Furner, I.J., Huffman, G.A., Amasino, R.M., Garfinkel, D.J., Gordon, M.P. and Nester, E.W. (1986) Nature 319, 422-427.

Gan, S. and Amasino, R.M. (1995) Science 255, 1986-1988.

Garfinkel, D.J., Simpson, R.B., Ream, L.W., White, F.F., Gordon, M.P. and Nester, E.W. (1981) Cell 27, 143-153.

Gatz, C., Frohberg, C. and Wendenburg, R. (1992) Plant J. 2, 397-404.

Gaudin, V. and Jouanin, L. (1995) Plant Mol. Biol. 28, 123-136.

Goldmann, A., Tempé, J. and Morel, G. (1968) C R Seances Soc. Biol. Ses. Fil. 162, 630-631.

Hamilton, R.C. and Fall, M.Z. (1971) Experientia 27, 229-230.

Hansen, G., Vaubert, D., Héron, J.N., Clerot, D., Tempé, J. and Brevet, J. (1993) Plant J. 4, 581-585.

Hendrickson, A.A., Baldwin, I.L. and Riker, A.J. (1934) J. Bacteriol. 28, 597-618.

Hille, J., Wullems, G.J. and Schilperoort, R.A. (1983) Plant Mol. Biol. 2, 155-164.

Hooykaas, P. J. J., Dulk-Ras, H. and Schilperoort, R. A. (1988) Plant Mol. Biol. 11, 791-794.

Hopkins, D.L. and Durbin, R.D. (1971) Can. J. Microbiol. 17, 1409-1412.

Huffman, G.A., White, F.F., Gordon, M.P. and Nester, E.W. (1984) J. Bacteriol. 157, 269-276.

Ichikawa, T., Ozeki, Y. and Syono, K. (1990) Mol. Gen. Genet. 220, 177-180.

Inzé, D., Follin, A., Van Lijsebettens, M., Simoens, C., Genetello, C., Van Montagu, M. and Schell, J. (1984) Mol. Gen. Genet. 194, 265-274.

Jarvis, B.C. (1986) in M. B. Jackson, (eds) New Root Formation in Plants and Cuttings, Martinus Nijhoff Publishers, Dordrecht, pp. 191-222.

Kahl, G. (1982) in J. Schell and G. Kahl, (eds) Molecular Biology of Plant Tumors, Academic Press, New York, pp., 211-267,

Kerr, A. (1969) Nature 223, 1175-1176.

Klee, H.J., Horsch, R.B., Hinchee, M.A., Hein, M.B. and Hoffmann, N.L. (1987) Genes and Devel. 1, 86-96.

Körber, H., Strizhov, N., Staiger, D., Feldwisch, J., Olsson, O., Sandberg, G., Palme, K., Schell, J. and Koncz, C. (1991) EMBO J. 10, 3983-3991.

Kuleck, G.A., Binns, A.N. and Black, R.C. (1991) Plant Physiol. 95, 1004-1011.

Leemans, J., Deblaere, R., Willmitzer, L., DeGreve, H., Hernalsteens, J.P., Van Montagu, M. and Schell, J. (1982) EMBO J. 1, 147-152.

Lemmers, M., De Beuckeleer, M., Holsters, M., Zambryski, P., Depicker, A., Hernalsteens, J.P., Van Montagu, M. and Schell, J. (1980) Jour Mol. Biol. 144, 353-376.

Li, Y., Hagen, G. and Guilfoyle, T.J. (1992) Devel. Biol. 153, 386-395.

Lichter, A., Barash, I., Valinsky, L. and Manulis, S. (1995) J. Bacteriol. 177, 4457-4465.

Lo Schiavo, F., Filippini, F., Cozzani, F., Vallone, D. and Terzi, M. (1991) Plant Physiol. 97, 1303-1308.

Maurel, C., Barbier-Brygoo, H., Spena, A., Tempé, J. and Guern, J. (1991) Plant Physiol. 97, 212-216.

Maurel, C., Brevet, J., Barbier-Brygoo, H., Guern, J. and Tempé, J. (1990) Mol. Gen. Genet. 223, 58-64.

Mauro, M.L., Trovato, M., De Paolis, A., Gallelli, A., Costantino, P. and Altamura, M.M. (1996) Dev. Biol. 180, 693-700.

Medford, J.I., Horgan, R., El-Sawi, Z. and Klee, H.J. (1989) Plant Cell 1, 403-413.

Meyer, A.D., Ichikawa, T. and Meins, F. (1995) Mol. Gen. Genet. 249, 265-273.

Moore, L., Guylyn, W. and Strobel, G. (1979) Plasmid 2, 617-626.

Motyka, V., Faiss, M., Strnad, M., Kaminek, M. and Schmülling, T. (1996) Plant Physiol. 112, 1035-1043.

Neuteboom, S. T. C., Hulleman, E., Schilperoort, R. A. and Hoge, H. C. (1993) Plant Mol. Biol. 22, 933-929.

Nilsson, O., Crozier, A., Schmülling, T., Sandberg, G. and Olsson, O. (1993a) Plant J. 3, 681-689.

Nilsson, O., Moritz, T., Imbault, N., Sandberg, G. and Olsson, O. (1993b) Plant Physiol. 102, 363-371.

Offringa, I.A., Melchers, L.S., Regensburg-Tuink, A.J.G., Costantino, P., Schilperoort, R. and Hooykaas, P.J.J. (1986) Proc. Natl. Acad. Sci. USA 83, 6935-6939.

Ooms, G., Hooykaas, P.J.J., Noleman, G. and Schilperoort, R.A. (1981) Gene 14, 33-50.

Petit, A., Delhaye, S., Tempé, J. and Morel, G. (1970) Physiol. Veg. 8, 205-213.

Ream, L.W., Gordon, M.P. and Nester, E.W. (1983) Proc. Natl. Acad. Sci. USA 80, 1660-1664.

Riker, A.J. (1926) J. Agr. Res. 32, 83-96.

Riker, A.J. (1930) Studies on infectious hairy roots of nursery apple trees. J. Agric. Res. 41, 438-446.

Risuleo, G., Battistoni, P. and Costantino, P. (1982) Plasmid 7, 45-51.

Schmülling, T., Beinsberger, S., De Greef, J., Schell, J., Van Onckelen, H. and Spena, A. (1989a) FEBS Lett. 249, 401-406.

Schmülling, T., Schell, J. and Spena, A. (1989b) Plant Cell 1, 665-670.

Schmülling, T., Schell, J. and Spena, A. (1988) EMBO J. 7, 2621-2629.

Schröder, G., Waffenschmidt, S., Weiler, E. and Schröder, J. (1984) Eur. J. Biochem. 138, 387-391.

Sinkar V. P., Pythoud F., White F. F., and Nester, E. W. (1988) Genes and Devel. 2, 688-697.

Skoog, F. and Miller, C.O. (1957) Symp. Soc. Exp. Biol. 11, 118-131.

Slightom, J.L., Durand-Tardif, M., Jouanin, L. and Tepfer, D. (1986) J. Biol. Chem. 261, 108-121.

Smart, C., M., Scofield, S.R., Bevan, M.W. and Dyer, T.A. (1991) Plant Cell 3, 647-656.

Smigocki, A.C. (1991) Plant Mol. Biol. 16, 105-115.

Smigocki, A.C. and Owens, L.D. (1988) Proc. Natl. Acad. Sci. USA 85, 5131-5135.

Smith, E.F. and Townsend, C.O. (1907) Science 25, 671-673.

Smith, H.H. (1988) J. Hered. 79, 277-283.

Spanier, K., Schell, J. and Schreier, P.H. (1989) Mol. Gen. Genet. 219, 209-216.

Spanò, L., Pomponi, M., Costantino, P., Van Slogteren, G.M.S. and Tempé, J. (1982) Plant Mol. Biol. 1, 291-300.

Spena, A.S., T.Koncz, C. and Schell, J. (1987) EMBO J. 6, 3891-3899.

Sugaya, S., Hayakawa, K., Handa, T. and Uchimiya, H. (1989) Plant Cell Physiol. 30, 649-653.

Tempé, J. and Goldmann, A. (1982). in, G. Kahl and J. Schell, (eds) Molecular Biology of Plant Tumors, Academic Press, New York, pp. 428-450.

Tepfer, D. (1984) Cell 37, 959-967.

Tepfer, D.A. and Tempé, J. (1981) CR Hebd. Séances Acad. Sci. Ser. III, 153-156.

Thomashow, L.S., Reeves, S. and Thomshow, M.F. (1984) Proc. Natl. Acad. Sci. USA 81, 5071-5075.

Thomashow, M.F., Hugly, S., Buchholz, W.G., Reeves, S. and Thomashow, L.S. (1986) Science 231, 616-618.

Thomashow, M.F., Nutter, R., Montoya, A.L., Gordon, M.P. and Nester, E.W. (1980) Cell 19, 729-739.

Tinland, B., Huss, B., Paulus, F., Bonnard, G. and Otten, L. (1989) Mol. Gen. Genet. 219, 217-224.

Trovato, M., Mauro, M.L., Costantino, P. and Altamura, M. (1987) Protoplasma, in press.

van Larebeke, N., Gentello, C., Schell, J., Schilperoort, R.A., Hermans, A.K., Heranlsteens, J.P. and Van Montagu, M. (1975) Nature 255, 742-743.

van Onckelen, H., Prinsen, E., Inzé, D., Rudelsheim, P., Van Lijsebettens, M., Follin, A., Schell, J., Van Montagu, M. and De Greef, J. (1986) FEBS Lett. 198, 357-360.

van Onckelen, H., Rudelsheim, P., Inzé, D., Rudelsheim, P., Van Lijsebettens, M., Van Montagu, M. and Schell, J. (1984) Plant and Cell Physiol. .5, 1017-1025.

Wabiko, H. and Minemura, M. (1996) Plant Physiol 112, 939-951.

Watson, B., Currier, T.C., Gordon, M.P., Chilton, M.-D. and Nester, E.W. (1975) J. Bacteriol. 123, 255-264.

White, F.F., Garfinkel, D.J., Huffman, G., Gordon, M.P. and Nester, E.W. (1983) Nature 301, 348-350.

White, F.F., Ghidossi, G., Gordon, M.P. and Nester, E.W. (1982) Proc. Natl. Acad. Sci. USA 79, 3193-3197.

White, F.F. and Nester, E.W. (1980) J. Bacteriol. 141, 1134-41.

White, F.F., Taylor, B.H., Huffman, G.A., Gordon, M.P. and Nester, E.W. (1985) J. Bacteriol. 164, 33-44.

White, P. and Braun, A.C. (1942) Cancer Research 2, 597-617.

Willmitzer, L., Dhaese, P., Schreider, P.H., Schmalenbach, W., Van Montagu, M. and Schell, J. (1983) Cell 32, 1045-1056.

Willmitzer, L., Sanchez-Serrano, J., Buschfeld, E. and Schell, J. (1982a) Mol. Gen. Genetics 186, 16-22.

Willmitzer, L., Simons, G. and Schell, J. (1982b) EMBO J. 1, 139-146.

Wood, H.N., Binns, A.N. and Braun, A.C. (1978) Differentiation 11, 175-180.

Yamada, T., Palm, C.J., Brooks, B. and Kosuge, T. (1985) Proc. Natl. Acad. Sci. USA 82, 6522-6526.

Zaenen, I., Van Larabeke, N., Teuchy, H., Van Montagu, M. and Schell, J. (1974) J. Mol. Biol. 86, 109-127.

Chapter 13

Organization and Regulation of Expression of the *Agrobacterium* Virulence Genes

Tonny M. Johnson and Anath Das

I. Introduction

The gram-negative soil bacterium *Agrobacterium tumefaciens* causes the neoplastic crown gall tumor disease on most dicotyledonous plants upon infection at a wound site (Smith and Townsend, 1907). Tumor formation results from the stable transfer of a segment of bacterial DNA into the plant nuclear genome. The transferred (T-)DNA is a segment of a large Ti-(tumor inducing) plasmid and encodes, among others, genes that catalyze the biosynthesis of the plant hormones auxin and cytokinin. Constitutive expression of these genes in a transformed plant leads to an uncontrolled cell growth, a cancerous phenotype (reviewed in Das, 1997; Hooykaas and Beijersbergen, 1994; Zupan and Zambryski, 1995). The related bacterium *Agrobacterium Rhizogenes* donates its Ri-(root inducing) plasmid DNA causing the hairy root disease (Moore *et al.*, 1979; White and Nester, 1980).

The characteristic features of a crown gall tumor are the hormone independent phenotype for growth in culture and the ability to synthesize novel low molecular mass compounds known as opines (Braun, 1958; Menage and Morel, 1964; Guyon *et al.*, 1980; Petit and Tempe, 1985). The opine biosynthetic genes are encoded within the T-DNA while genes for opine catabolism are located in another region of the Ti-plasmid. The opines can be used as carbon, nitrogen and energy source for the growth of the bacterium. Different strains of agrobacteria produce different opines, *e.g.*, octopine, nopaline, leucinopine and succinamopine, which are used in the naming of the bacterial strains.

II. Functions Essential for DNA Transfer

The T-DNA does not encode genes essential for its own transfer. The only essential element in the T-DNA is a cis-acting site, a 24 bp direct repeat sequence known as the border sequence (Yadav *et al.*, 1982; Zambryski *et al.*, 1982; Holsters *et al.*, 1983). This sequence found at the two ends of the transferred DNA defines the T-DNA. The deletion of the right border sequence of the nopaline Ti-plasmid renders the strain avirulent; the virulence of the mutant can be restored by a synthetic 24 bp DNA (Wang *et al.*, 1984; Peralta and Ream, 1985). The octopine Ti-plasmid contains a second conserved sequence, the overdrive, which is essential for efficient DNA transfer. The overdrive is a 24 bp sequence situated on the right of the border sequence and functions as an enhancer. It stimulates DNA transfer when placed in either orientation and at a variable distance from the border sequence (Peralta *et al.*, 1986; van Haaren *et al.*, 1987). The ability to transfer a piece of DNA placed *in cis* with a border sequence to plant cells by the *Agrobacterium*-mediated gene transfer has revolutionized genetic manipulation of plants.

Genes essential for DNA transfer are encoded in the virulence (*vir*) region of the Ti-plasmid (Garfinkel and Nester, 1980; Ooms *et al.*, 1980). The *vir* region of the octopine Ti-plasmid pTiA6 is about 44 kb in size and is composed of eight complementation groups, *virA-virH* (Stachel and Nester, 1986; Kanemoto *et al.*, 1989; Melchers *et al*, 1990). While mutations in *virA*, *virB*, *virD* and *virG* lead to an avirulent phenotype, mutations in the other loci lead to an attenuated phenotype or to the restriction of host range (Stachel and Nester, 1986). The nopaline Ti plasmid *vir* region lacks *virF* and *virH* but contains another accessory gene *tzs*. The organization of the octopine and nopaline virulence genes are shown in Figure 1.

The DNA sequence of the *vir* genes of the octopine Ti plasmids pTiA6 and the nopaline Ti plasmid pTiC58 has been determined. This region can encode 22-27 proteins seventeen of which are essential for virulence. Two proteins, VirA and VirG, are the regulatory proteins that control expression of all the *vir* genes (Stachel and Zambryski, 1986). VirA and VirG are homologous to the proteins of the bacterial two-component regulatory system (Winans *et al.*, 1986; Melchers *et al.*, 1986; Leroux *et al.*, 1987).

The *virB* operon contains eleven open reading frames (Thompson *et al.*, 1988; Ward *et al.*, 1988; Kuldau *et al.*, 1990; Rogowski *et al.*, 1990). Analysis of the primary sequence of the proteins indicates that at least nine are membrane proteins because they contain one or more membrane spanning domain, a signal sequence that can transport a protein into the periplasm or both. Biochemical studies demonstrated that all eleven proteins are associated with the bacterial membranes (Thorenson *et al.*, 1993; Ward *et al.*, 1990; Berger and Christie, 1993; Shirasu and Kado, 1993). Since *virB* is not required for the formation of the T-strand DNA and the encoded polypeptides are membrane or membrane-associated proteins, it is postulated that the VirB proteins participate in the formation of a pore structure that allows the exit of the T-DNA from the bacterium. In addition to the VirB proteins, VirD4 is required for DNA transfer (Beijersbergen and Hooykaas, 1992; Yoshioka *et al.*, 1994).

The *virD* operon can encode five proteins (Jayaswal *et al.*, 1987). Three, VirD1, VirD2 and VirD4, are essential for DNA transfer (Stachel and Nester, 1986; Vogel and Das, 1992, Koukolikova-Nicola *et al.*, 1993). VirD1 and VirD2 together catalyze nicking at the T-DNA borders initiating the processing of the T-DNA for its transfer to the plant cell (Yanofsky *et al.*, 1986; Jayaswal *et al.*, 1987). VirD1 contains a potential helix-turn-helix motif that maps within an essential domain (Porter *et al.*, 1987; Vogel and Das, 1994). While the two proteins must bind to the border sequence for nicking to occur, no specific binding to this sequence has been observed *in vitro*. VirD4, an integral membrane protein, is essential for the transfer of the T-DNA from bacteria to plant cell (Yoshioka *et al.*, 1994; Okamoto *et al.*, 1991).

The *virC* operon can encode two proteins. Mutations in *virC* affect the host range of the strain (Yanofsky *et al.*, 1985). *virC* mutants have an attenuated phenotype on *K. daigremontiana* but are fully virulent on *N. glauca*. The *virE* operon encodes two

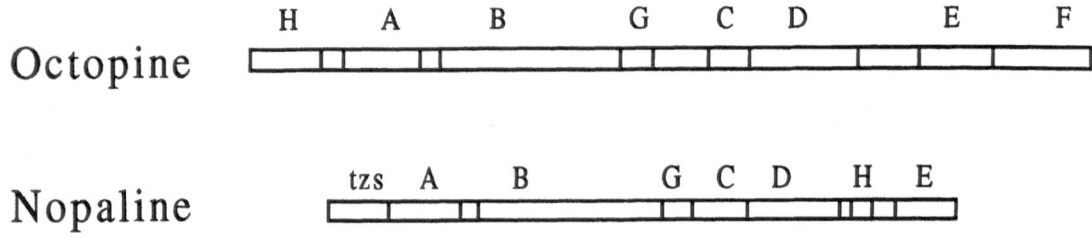

Figure 1. A physical map of the virulence regions of the octopine and nopaline Ti-plasmids. All genes are transcribed left to right except for the octopine Ti-plasmid *virC* and the nopaline Ti-plasmid *virC*, *virH* and *tzs* genes that are transcribed right to left.

proteins, VirE1 and VirE2 (Winans *et al.*, 1987). VirE1, a 7 kDa polypeptide, is essential for DNA transfer (McBride and Knauf, 1988). It is required for the transfer of VirE2 to plant cells (Sundberg *et al.*, 1996). The larger protein, VirE2, is a single-stranded DNA binding protein that cooperatively binds ssDNA in a sequence-independent manner (Citovsky *et al.*, 1989; Sen *et al.*, 1989). It is postulated that VirE2 protects the T-strand DNA against cellular nucleases (Das, 1988).

The octopine Ti-plasmid contains two additional *vir* loci, *virF* and *virH*. Mutations in *virF* result in an attenuated phenotype. The lack of *virF* in the nopaline strain makes it unable to efficiently form tumors on *N. glauca* (Melchers *et al.*, 1990). *virF* encodes one protein that has similarity with VirE2 in some respects (see next paragraph). *virH* (previously called *pinF*) is not an essential locus and a mutation in *virH* has no effect on tumor formation (Stachel and Nester, 1986). This locus was initially identified as a plant-inducible locus and was subsequently found to be required for tumor formation only when a low amount of bacteria is used for infection (Stachel and Nester, 1986; Kanemoto *et al.*, 1989). The *virH* region contains four open reading frames, two of which are homologous to cytochrome P-450 enzymes. Due to the lack of mutations in the other ORFs the role of the first and the fourth ORF in tumorigenesis is not known. Mutations in either ORF2 or ORF3 gave an attenuated phenotype indicating the polypeptide represented by ORF3 or both ORFs are essential of VirH function.

An unusual property of *virE* and *virF* is the ability of a T-DNA minus strain to complement a mutation in these genes by coinoculation of two *Agrobacterium* strains on a wound site (Otten *et al.*, 1984; 1985). For complementation all essential *vir* genes as well as the chromosomal virulence genes must be present in the complementing strain (Otten *et al.*, 1984; Christie *et al.*, 1989). It was postulated that *virE* and *virF* encode diffusible products that can be shared by different bacteria on the wound site (Otten *et al.*, 1985). Subsequent studies with plants that express VirE2 and VirF showed that the transgenic plants are susceptible to infection by the respective *vir* mutant indicating that these proteins function in the plant cell (Citovsky *et al.*, 1992; Regensburg-Tüink and Hooykaas, 1993). Regensburg-Tüink and Hooykaas (1993) concluded that the T-DNA transport pore can transport Vir proteins to the plant cell and extracellular complementation involves transport of VirE2/VirF and T-DNA from the two strains. Studies

of Binns *et al.*, (1995) and Sundberg *et al.*, (1996) support this conclusion.

III. Regulation of *vir* Gene Expression

The use of the transposon tn*3::lacZ* for the analysis of the *vir* region is largely responsible for our current understanding of the virulence region (Stachel and Nester, 1986). That study defined the boundary of individual locus, identified the direction of transcription, led to the discovery of *virG* and defined the regulation of expression of the *vir* genes. Under vegetative growth conditions only *virA* and *virG* are expressed (Stachel and Zambryski, 1986). Transcription of all the *vir* genes including *virA* and *virG* is induced when bacteria and plant cells are mixed together (Stachel and Zambryski, 1986; Winans *et al.*, 1988). A small molecule(s) present in the plant cell culture medium functions as an inducer of *vir* gene expression (Stachel et. al., 1986). Two inducers, 3',5' dimethoxy-4'-hydroxy acetophenone (commonly known as acetosyringone) and α-hydroxy acetosyringone, were first identified from tobacco culture cells by Stachel *et al.*, (1985). In conjunction with an inducer the two constitutively expressed genes, *virA* and *virG*, are essential for the transcription activation of the *vir* genes (Stachel *et al.*, 1986). Induction of *vir* gene expression requires a low pH of the growth medium (pH 5.3 - 6) and a low ($\leq$30°C) incubation temperature.

Transcription of *virG* is controlled by other factors, *e.g.*, phosphate limitation, acid pH and other stress stimuli, in a *virA-virG* independent manner (Winans, 1990; Aoyama *et al.*, 1991; Mantis and Winans, 1992). It is postulated that the level of VirG is limiting in vegetatively grown bacteria. The ability of the *virG* promoters to respond to the other environmental conditions increases the level of the protein such that it can now positively autoregulate expression of itself and of the other *vir* genes in response to plant signal molecules.

Two *vir* genes, *virC* and *virD*, are subjected to negative regulation as well. The divergent *virC* and *virD* genes share a common regulatory region. A chromosomal gene, *ros*, acts as a repressor to keep these genes turned off under vegetative growth conditions (Close *et al.*, 1985; 1987). *ros* is not required for the tumor-forming ability of the bacterium. The encoded product, Ros, is a 15.5 kDa protein that binds to an inverted repeat sequence located upstream of the *virCD* promoters (Cooley *et al.*, 1991).

III.A. CIS-ACTING SEQUENCES

The existence of a common regulatory mechanism that controls the expression of all the *vir* genes suggested that a conserved element probably functions as a signal for transcription activation. DNA sequence analysis of the promoter-regulatory regions of the pTiA6 *vir* genes led to the identification of two hexamers that are present in most of these genes (Das *et al.*, 1986). It was postulated that these sequences may play a role in the regulation of *vir* gene expression. A reexamination of these sequences revealed that one hexamer, dGCAATT, is a subset of a larger 12 bp element, dTNCAATTGAAAPy, that is present in one or more copies in all the *vir* genes (Winans *et al.*, 1987). Similar sequences are found upstream of the nopaline Ti-plasmid *vir* genes as well (Close *et al.*, 1988). The conserved sequence was named the 'vir box' and was postulated to serve as an essential cis-acting element for *vir* gene expression. The Ri-plasmid *vir* genes contain one or more blocks of a six base pair sequence, dTG(A/T)AAPy, that is similar to the 3' half of the 12 mer *vir* box sequence of the pTiA6 and pTiC58 *vir* genes (Aoyama *et al.*, 1989). This sequence was postulated to play a role in the regulation of the pRi *vir* genes.

Analysis of the pTiA6 *virB* promoter-regulatory region led to the identification of sequences essential for its expression (Das and Pazour, 1989). That study demonstrated that 68 nucleotides upstream of the transcription initiation site are sufficient for the plant inducible expression of *virB*. Within this region lie two *vir* box sequences that are 11 bp apart. Deletion into the distal *vir* box led to the complete loss of *virB* expression. An insertion or deletion in the spacer between the two *vir* boxes is deleterious indicating that the relative position of the two *vir* boxes and/or that of the promoter and the distal *vir* box is important for *virB* expression. A further analysis of these sequences indicated that the *vir* box is a tetradecameric sequence with the consensus dPuPyTDCAATTGHAAPy (D = A, G or T; H = A, C or T) (Pazour and Das, 1990a). The central six residues of the sequence, dCAATTG, is the most conserved. Mutation of the two T residues to CC led to a 80 percent reduction in the inducible expression of *virB* (Das and Pazour, 1989). The *vir* box is a palindrome and the presence of only the right half of the *vir* box is not sufficient for *virB* expression. The right half of the pTiA6 *virB* vir box (dTTGAAAT) is homologous to the pRi *vir* box. Therefore, the half

site found in the pRiA4 *vir* genes is not sufficient for the pTiA6 *virB* expression.

The number of *vir* boxes in the different *vir* genes is highly variable (Winans *et al.*, 1987; Steck *et al.*, 1988; Das and Pazour, 1989). While the pTiA6 *virA* and *virE* contain one *vir* box, the *virB*, *virC*, *virD* and *virG* contain two to five *vir* boxes. No correlation between the number of *vir* boxes and the level of induction of a gene was observed. The highly inducible *virE* requires one *vir* box while the low inducible *virG* requires two *vir* boxes (Pazour and Das, 1990; Winans, 1990). A thorough analysis of the role of the various *vir* boxes in *vir* gene expression was undertaken by Pazour and Das (1990). Of the five *vir* boxes the divergent *virCD* genes share, only one is essential for *virD* expression. For *virC* expression, one box is essential while a second one is required for a high level of expression. Of the two *virB* vir boxes one is essential and the presence of the other augments gene expression. The conservation of the *vir* box in the *vir* gene regulatory regions and their essential role in *vir* gene expression suggest that this sequence serves as the recognition site for the regulatory protein VirG.

There appears to be a difference in the nature of the *vir* box sequence essential for the expression of the Ti- and Ri-plasmid virulence genes. While the pTiA6 and pTiC58 *vir* boxes are 14 residues in length the proposed *vir* box for the pRiA4 *vir* genes is 6 residues in length (Pazour and Das, 1990; Aoyama *et al.*, 1989). The 6 residue pRiA4 *vir* box, that is the 3' half of the pTi *vir* box, is not sufficient for pTiA6 *virB* expression (Das and Pazour, 1989). Because the VirG proteins of the three plasmids are virtually identical (85-94% identity; 92-98% similarity) it is difficult to conceive that the binding sites of the proteins will vary considerably. An analysis of the regulatory region of the pRi *vir* genes shows that all inducible pRi *vir* genes contain a sequence homologous to the pTi *vir* box (Table 1). In contrast to the pTi *vir* genes only one such element is found in each of the pRi *vir* genes. We propose that these sequences function as the *vir* box for the pRiA4 *vir* genes.

III.A.1. Location of the Cis-Acting Sequences

All functional *vir* boxes in both pTiA6 and pRiA4 *vir* genes lie 42-76 nucleotides upstream of the transcription initiation sites (Table 2). The only exception is pTiA6 *virG* vir box III that is centered at -24.5. The essential *vir* boxes are found

The *vir* box sequence of the Ri plasmid *vir* genes	
vir Box	DNA Sequences in the *vir* gene nontranscribed region
virAII	G T T T C A T T T G A A A C A A A C -66
virB	T T T T C G C T T C A A A T G A A A T C G A A -52
virC	T A A T T G T T A C A T T T G C A A C T A T T -42
virD	A T T G T T G C A A A T G T A A C A A T T -62
virE	C A C G A A T T G C A G T T G A A A C A C G A -42
virG1	A C A A A A T T A C A T T T G T A G C A A A -52
Consensus	Pu T T D C A A/ T/ T G N A A C A/ A A/
	T A C T
pTiA6 Consensus	Pu Py T D C A A T T G A/ A A Py
	T

Table 1. DNA sequence and the nomenclature of the pRi *vir* gene *vir* boxes are from Aoyama *et al.*, (1990). The numbers on the right indicate the location of the rightmost residue relative to the transcription initiation site (+1).

predominantly at three positions, ~ -52.5, -62.5 and -72.5, that are one helical turn away from one another suggesting that in all *vir* genes the VirG protein binds to one face of the template DNA and it is the same face that RNA polymerase binds.

How does an activator sitting at -52.5 or at -72.5 makes similar contacts with RNA polymerase to activate transcription? While the physical proximity of the protein at -52.5 to the promoter can allow necessary interactions with the polymerase, the protein at -72.5 will be two additional helical turns away. It is conceivable that two different mechanisms are used to contact RNA polymerase in the two classes of genes. This situation is analogous to that with the *E. coli* c-AMP binding protein (CRP) which activates transcription from many *E. coli* promoters (reviewed in Kolb *et al.*, 1993). These promoters fall into two groups where the CRP binding site is centered either at -41.5 or at -61.5. In the latter group of

promoters DNA bending is used as a mechanism to bring CRP and RNA polymerase close to each other. A similar mechanism may be used for the transcription activation of a subset of *vir* promoters.

III.B. TRANS-ACTING FACTORS

III.B.1. Environmental Signals
Expression of the virulence genes is dependent on plant cell growth medium (Stachel *et al.*, 1985). Natural metabolites of plant cells function as inducers for *vir* gene expression. The first two inducers identified were acetosyringone (AS), and α-hydroxy acetosyringone from tobacco culture cells (Stachel *et al.*, 1985). An analysis of several chemical derivatives of AS indicated the importance of the phenolic hydroxyl group at position 4 and the two methoxy groups at positions 3 and 5 (Figure 2).

The position of the *vir* box sequence of the pTi and pRi *vir* genes		
Gene	Residue *vir* box centered at	
	Ti *vir* gene	Ri *vir* gene
virA	-73.5	-76.5
virB Box 1	-61.5	-62.5
Box 2	-42.5	--
virC Box C2D4	-53.5	-52.5
Box C4D2	-42.5	--
virD	-75.5	-72.5
virE	-61.5	-52.5
virG Box 1	-57.5	-62.5
Box 3	-24.5	--

Table 2. The requirement of the pTi *vir* gene *vir* boxes has been established (Das and Pazour, 1989; Pazour and Das, 1990; Winans, 1990). The pRi *vir* gene *vir* boxes are from Table 1.

A comprehensive study on the essential structural features of the *vir* inducing compounds was undertaken by Melchers *et al.,* (1989b). That study demonstrated that the basic unit is an *ortho*-methoxy phenol. A second methoxy group at R5 position significantly increases the potency of the inducer. The R1 position on the other hand can be occupied by many groups including -H, -CHO, COOH, -COCH₃ and CH=CH·CO₂H. The presence of a -C=C- at the R1 position increases the potency of the compound. Most recently it was shown that an unsubstituted phenol can induce *vir* gene expression in *Agrobacterium* KU12, a wild-type strain isolated from S. Korea, suggesting that a phenol is the basic molecule for *vir* gene induction (Y-W. Lee *et al.,* 1995). Several naturally occurring phenol derivatives, viz., methyl syringate from grapevine, dehydroconiferyl from tobacco, ethyl ferulate from wheat and flavanoids from petunia, have been identified as *vir* gene inducers (Spencer *et al.,* 1990; Usami *et al.,* 1988; Messens *et al.,* 1990; Zerback *et*

al., 1989). A phenyl propanoid glucoside of Douglas fir, coniferin, was identified as a strong inducer of *vir* gene expression (Morris and Morris, 1990). This is the only known inducer that lacks a free phenolic -OH group. However, a survey of several *Agrobacterium* strains showed that strains that have a high β-glucosidase activity are more tumorigenic on Douglas fir. It is postulated the cellular glucosidase converts coniferin to coniferyl alcohol which is the true inducer of *vir* gene induction.

The presence of certain sugars greatly increases the sensitivity of *Agrobacterium* towards the phenolic inducers (Cangelosi *et al.,* 1990; Shimoda *et al.,* 1990). A group of aldolases, viz., L-arabinose, D-xylose, D-lyxose, D-glucose, D-mannose, D-idose, D-galactose, D-fucose and the non-metabolized sugars 2-deoxy-D-glucose and 6-deoxy-D-glucose, most of which have an identical C-3 stereochemical structure, are active as enhancers of acetosyringone-inducible *vir* gene expression. The sugars act through a chromosomal protein, ChvE

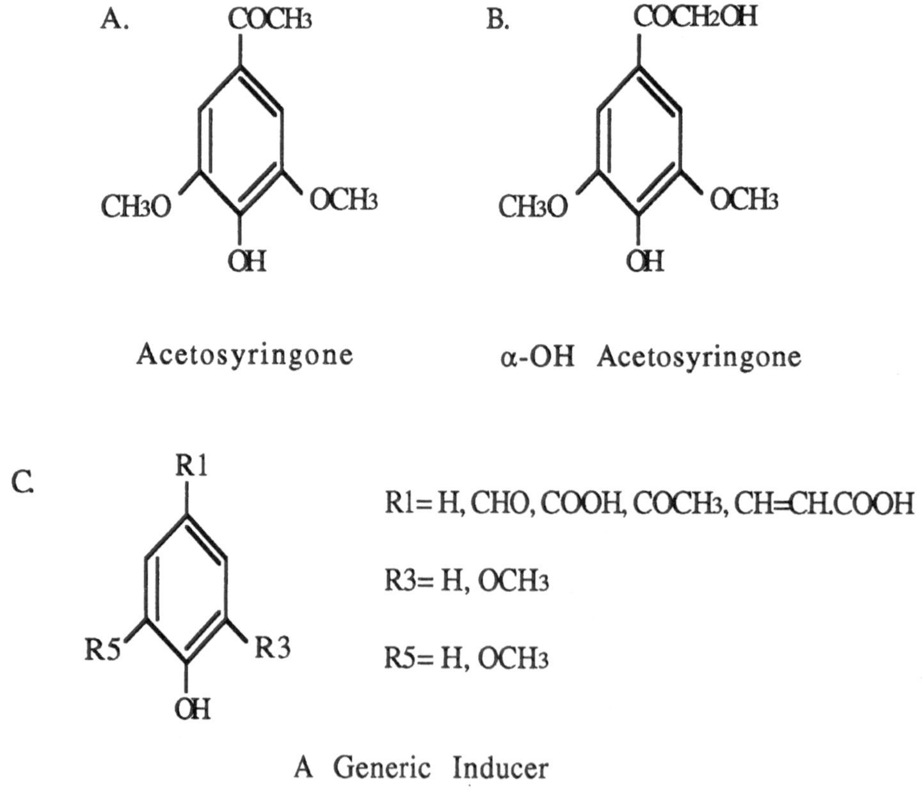

A. Acetosyringone

B. α-OH Acetosyringone

C. A Generic Inducer

R1= H, CHO, COOH, COCH₃, CH=CH.COOH

R3= H, OCH₃

R5= H, OCH₃

Figure 2. Inducers of the *Agrobacterium vir* genes. Tobacco produces acetosyringone and α-hydroxy acetosyringone.

(Cangelosi *et al.*, 1990). ChvE is a glucose binding protein that interacts with the sensor VirA. Other environmental factors that affect *vir* gene expression are pH and temperature. Stachel *et al.*, (1985) observed that a low pH (pH 5.3-6) is an absolute requirement for *vir* gene induction. Induction is also temperature sensitive with an optimal temperature of 28-30°C. The sensor for all the environmental factors is the sensor protein VirA which senses the phenolic inducer, is largely responsible for the low pH requirement and is inactive at high temperature rendering the entire process temperature sensitive (Melchers *et al.*, 1989a; Turk *et al.*, 1991; Jin *et al.*, 1993a).

III.B.2. Protein Factors

VirA. VirA and VirG are the two regulators of *vir* gene expression (Stachel and Zambryski, 1986). The Ti and Ri plasmid *virA* gene products are 829-833 amino acid residues in length. VirA contains two hydrophobic regions (at positions 18-39 and 260-278 of the pTiA6 VirA) that function as transmembrane domains (Melchers *et al.*, 1989). Most of the protein, residues 279-end, lies in the cytoplasm. The cytoplasmic region consists of two conserved domains: a kinase domain and a receiver domain. The kinase domain, residues 419-691 in the pTiA6 VirA, is conserved in the family of the sensor proteins (Leroux *et al.*, 1987; Melchers *et al*, 1987). The receiver domain is homologous to the N-terminal domain of VirG and is conserved in the VirG family of proteins (Stock *et al*, 1989). This domain is found only in a subset of the sensor proteins.

VirA chimeras were used for the functional analysis of the N-terminal domains of VirA (Melchers *et al.*, 1989a; Turk *et al.*, 1994). Only the first transmembrane domain could be exchanged with a similar domain of the *E. coli* Tar receptor indicating that the first transmembrane domain is essential only for the proper topology of the protein (Melchers *et al.*, 1989a). The mapping of a constitutive mutant to this domain in a later study indicated that this domain plays an additional essential role in VirA function (Pazour *et al.*, 1991). The periplasmic domain does not function in inducer recognition because the deletion of most of the domain (residue 63-240 of the pTiA6 VirA) has no effect on *vir* gene induction. The deletion mutant, however, is insensitive to sugars suggesting that the periplasmic domain is essential for the sugar responsiveness of *vir* gene expression. This domain is believed to

interact with the sugar binding protein, ChvE. Mutational studies showed that single amino acid changes and short deletions within the periplasmic domain abolish the ability of VirA to respond to sugars. Three single mutants, *virAE210V*, *virAG211D* and *virAE255L*, and a double mutant, *virAT44ST45R*, do not respond to sugars and are constitutive for sugar-mediated augmentation of *vir* gene expression (Machida *et al.*, 1993; Turk *et al.*, 1993; Banta *et al.*, 1994). A loss of response to sugar was also observed when either of the two transmembrane domains was exchanged with a similar domain of the *E. coli* Tar protein (Turk *et al.*, 1994). The latter observation supports the conclusion of Pazour *et al.*, (1991) that both transmembrane domains play an important role in VirA function.

Mutational analysis suggests that the N-terminal segment of the cytoplasmic domain is essential for acetosyringone-induced *vir* gene expression (Chang and Winans, 1992). Although direct evidence for its role in inducer sensing is yet lacking, recent studies support this conclusion (Turk *et al.*, 1994; Doty *et al.*, 1996). An important unanswered question is how acetosyringone interacts with VirA. The failure to observe direct binding of an inhibitor of *vir* gene expression, α-bromo-5-iodoacetovanillone, to VirA and its ability to bind specifically to two chromosomal proteins, p10 and p21, led K. Lee *et al.*, (1992) to suggest that other proteins mediate signal transduction from the inducer to VirA. Recent genetic studies of Lee *et al.*, (1995), however, strongly suggest that VirA is the sensor of the inducer. The latter study analyzed several Ti-plasmids that respond to different inducers. By exchanging Ti-plasmids and chromosomal background it was shown that *virA*, and not the bacterial chromosome, is the determinant for the *vir* gene inducer.

The central segment of the cytoplasmic domain encodes the kinase domain, the domain found conserved in all sensor proteins (reviewed in Stock *et al.*, 1989; Parkinson and Kofoid, 1992). This region contains a conserved histidine, His 474 of pTiA6 VirA, that is the site of phosphorylation. In response to the inducer VirA phosphorylates itself at the histidine residue and transfers the phosphate to aspartic acid 52 of VirG. Inducer-independent autophosphorylation of a VirA derivative containing the cytoplasmic domain has been demonstrated *in vitro* (Jin *et al.*, 1990b; Huang *et al.*, 1990). The mapping of three acetosyringone independent mutants, *virAA469V*, *virAG471R* and *virAG471E*, near histidine 474 suggests that the local structure of the region plays an important role in *vir* gene expression (Ankenbauer *et al.*, 1991; Pazour *et al.*, 1991). Two other constitutive mutants, *virAL658F* and *virAG665D*, were mapped to the C-terminus of this domain (Pazour *et al.*, 1991; McLean *et al.*, 1994). These mutants lie within or near a glycine rich sequence that is postulated to function in nucleotide binding (Pazour *et al.*, 1991). The codominant phenotype of several constitutive mutants led Pazour *et al.*, (1991) to propose that VirA functions as an oligomer. In support of this hypothesis Pan *et al.*, (1993) demonstrated that VirA exists as a dimer and the inducer has no effect of dimer formation.

An unusual feature of VirA structure is the presence of a domain homologous to the N-terminal domain of the receiver in the extreme C-terminus (Stock *et al.*, 1989). A similar feature is found in some sensor proteins, viz., *E. coli* ArcB, *E. coli* RcsE, *B. pertussis* BvgS, *B. thetaiotaomicron* RteA, *P. syringae* LemA, *X. campestris* RffC and *M. xanthus* AsgA (reviewed in Parkinson and Kofoid, 1992). The mapping of a constitutive mutant to this domain and its structural similarity with the eukaryotic protein kinases led Pazour *et al.*, (1991) to postulate that this domain is reminiscent of the 'autoinhibitory' domain of the protein kinases (Soderling, 1990) and is a negative regulator of VirA function. Subsequent studies demonstrated that deletion of this domain renders VirA constitutive (Chang and Winans, 1992; Gubba *et al.*, 1995). Recent studies of Chang *et al.*, (1996) suggest that overproduction of this domain *in trans* restores a low basal activity of a C-terminal deletion mutant of VirA. That study also showed that removal of this domain allows non-stimulatory phenolics such as 4-hydroxy acetophenone to stimulate *vir* gene expression suggesting that another function of this domain is to restrict the range of phenolics that can activate *vir* gene transcription.

VirG. The second regulatory protein, VirG, is the activator that positively controls expression of all the *vir* genes. DNA sequence analysis and genetic studies indicated that the VirG proteins of the Ti- and Ri-plasmids are 241 residues in length and share 85-94 percent sequence identity among one another (Winans *et al.*, 1986; Melchers *et al.*, 1986; Powell *et al.*, 1987; Aoyama *et al.*, 1989; Pazour and Das, 1990b). VirG is composed of two domains: a N-terminal receiver domain and a C-terminal output domain. In response to the plant phenolics VirA becomes phosphorylated and transfers the phosphate

to an aspartic acid in the N-terminal domain of VirG. *In vitro* studies of Jin *et al.*, (1990a) demonstrated that VirA phosphorylates the aspartic acid at position 52. An alteration of aspartic acid 52 to a nonphosphorylatable homolog asparagine or glutamic acid abolishes VirG function (Roitsch *et al.*, 1990; Pazour *et al.*, 1992).

In addition to the aspartic acid 52, two other aspartic acid at positions 8 and 9 and a lysine at position 102 are conserved in the VirG family of regulatory proteins (Stock *et al.*, 1989). Site-specific mutagenesis showed that the aspartic acid at position 8 and an additional one at position 98 are essential for VirG function (Roitsch *et al.*, 1990). The C-terminal domain of VirG, the output domain, encodes the DNA binding function (Powell and Kado, 1990). This domain is homologous to a domain of the *E. coli* OmpR, PhoB and SfrA (Winans *et al.*, 1986; Melchers *et al.*, 1986). The analyses of fusion proteins, the comparison of primary sequences of several DNA-binding proteins and the characterization of VirG mutants suggest that residues 160-210 constitute the DNA binding domain (Powell and Kado, 1990; Suzuki, 1993; Mallik and Das, unpub. results). The requirement of a *vir* box in *vir* gene expression and its presence in the nontranscribed regions of the inducible genes led to the suggestion that the *vir* box is the target for VirG binding. Using protein purified from overproducing *E. coli* strains specific binding of VirG to the *vir* box sequences of pTiA6 and pRiA4 *vir* promoters was observed (Jin *et al.*, 1990c; Pazour and Das, 1990b; Tamamoto *et al.*, 1990). In the case of the *virCD* and *virG* promoters, promoters that contain multiple *vir* boxes, VirG bound specifically only to those boxes that are essential for gene expression *in vivo* (Pazour and Das, 1990a; b).

Mutagenesis of *virG* has been used extensively to study the functional domains. Three mutations that allow inducer independent constitutive expression of a *vir* gene have been identified (Pazour *et al.*, 1992; Gubba *et al.*, 1995; Han *et al.*, 1992; Jin *et al.*, 1993b; Scheeren-Groot *et al.*, 1994). These mutants map to the receiver domain and have a substitution of asparagine 54 to aspartic acid (*virGN54D*), isoleucine 77 to valine (*virGI77V*) and isoleucine 106 to leucine (*virGI106L*). One, *virGN54D*, is not responsive to *virA* and acetosyringone while the other two are activated by VirA (Pazour *et al.*, 1992; Gubba *et al.*, 1995). Two mutants, *virGN54D* and *virGI106L*, require aspartic acid 52 for function. The third one, *virGI77V*, does not require aspartic acid 52 and in one study was isolated as a second site mutation that restores the function of a nonfunctional mutant virGD52E (Gubba et al., 1995). This mutant is supersensitive to acetosyringone and is induced by a low concentration of the inducer (Scheeren-Groot et al., 1994). Two other mutants with a similar supersensitive phenotype, virGM13T and virGH15R, have also been isolated (Han et al., 1992). The mapping of all the constitutive mutants to the receiver domain and the repeated failures to isolate one that maps to the C-terminal domain suggest that a single amino acid substitution in the output domain is probably not sufficient for VirG activation. Activation of VirG must involve a conformational change in the C-terminal domain. This change is induced by phosphorylation of the N-terminal receiver domain or by a mutation in the N-terminal domain but not by a single amino acid substitution in the C-terminal domain.

A large number of mutants defective in vir gene induction were identified by random mutagenesis of virG (Scheeren-Groot et al., 1994; Mallik and Das, unpublished results). These mutants mapped throughout the coding region of the protein. Several mutants in the N-terminal domain did not affect phosphorylation although the effect of the mutations on the kinetics of the reaction is not known. A number of mutants mapping to the C-terminal domain were defective in transcription activation but proficient in both phosphorylation and DNA binding suggesting that these mutations affect interaction between VirG and RNA polymerase (or the transcription machinery). Three mutants in the C-terminal domain, virGL156P, VirGR194K and virGE208G are defective in DNA binding in vitro (Scheeren-Groot et al., 1994). Two of these virGL156P and VirGE208G are also defective in phosphorylation. Unless these two mutant proteins failed to refold properly during purification (the purification scheme involved denaturation and renaturation of the proteins), the existence of these mutants would suggest that the C-terminal domain affects the structure of the N-terminal domains as well.

Transcription activation by VirG requires phosphorylation. In vitro VirA phosphorylates VirG at aspartic acid 52 (Jin et al., 1990a). How phosphorylation activates VirG remains to be elucidated. Studies on VirG homologues, OmpR, PhoB and NtrC, suggest possible mechanisms for vir gene transcription. Transcription requires the formation of the template DNA-RNA polymerase closed complex followed by its isomerization to the open complex. In vitro RNA polymerase does not

efficiently bind to a vir promoter. The formation of DNA-RNA polymerase complex is promoted by the constitutive mutant protein VirGN54D but not by VirG indicating that the formation of the closed promoter complex is promoted by activated VirG (Mallik and Das, unpublished results). A similar mechanism is used by the E. coli activators OmpR and PhoB (Tsung et al., 1990; Makino et al., 1993). The other VirG homolog NtrC promotes transcription by promoting the conversion of the closed promoter complex to an open complex (Ninfa et al., 1987; Popham et al., 1989).

Phosphorylation of several VirG homologs affects DNA binding. The affinity of the protein for the cognate binding site is increased ~10 fold upon phosphorylation (Aiba et al., 1989; Roggiani and Dubuau, 1993; Nakashima et al., 1993). A similar mechanism is likely to be applicable to VirG as well. A fusion protein of the constitutive mutant VirGN54D has a higher affinity for the vir box than a similar fusion with the wild type protein (Han and Winans, 1994). Phosphorylation of NtrC on the other hand does not affect the affinity for the DNA but promotes cooperativity of binding (Weiss et al., 1992). A third effect of phosphorylation is oligomerization of the protein (Fiedler and Weiss, 1995). The receiver domain of the regulators appears to be the oligomerization domain. It is postulated that the output domain inhibits the dimerization domain in the absence of the inducer. Upon phosphorylation the receiver domain dimerizes activating a new function(s). Phosphorylation of NtrC and PhoB induces oligomerization. Oligomerization of NtrC activates its ATPase activity but not DNA binding (Mettke et al., 1995). Phosphorylation of OmpR induces conformation change that results in an altered sensitivity to proteases (Kenney et al., 1995).

IV. Conclusions

The essential features of vir gene regulation are summarized in Figure 3.

Plant cells release phenolics that act as signal molecules for vir gene induction. Phenol and substituted phenols are the inducing agents. These molecules are natural components of the plant cell. The Ti-pasmid encoded virA and virG genes are essential for vir gene induction. Studies to date suggest that VirA is the sensor of the plant signal molecules. Interaction of VirA with the inducer

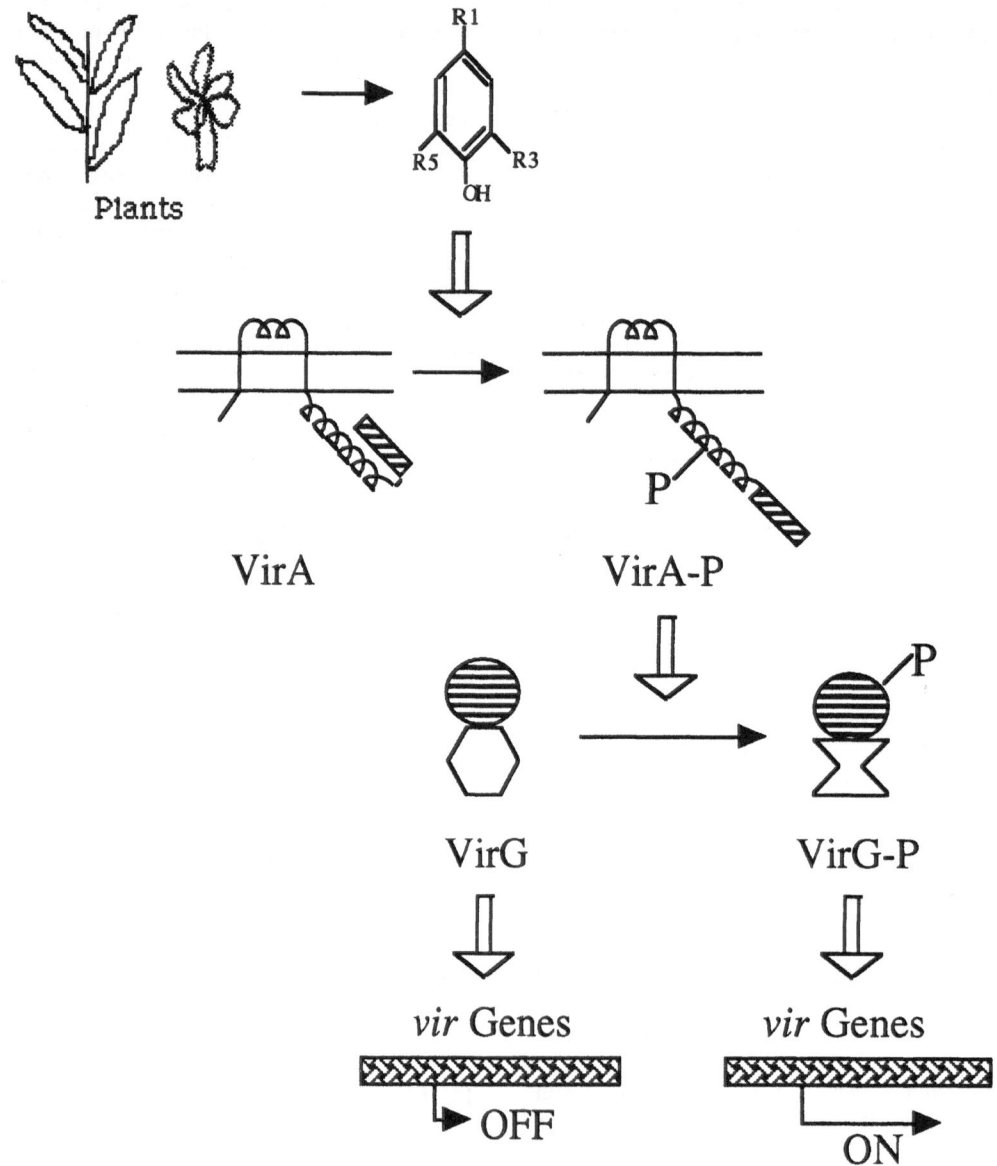

Figure 3. A model for the induction of *vir* gene expression. See text for details.

activates its protein kinase activity leading to autophosphorylation at a conserved histidine at position 474. The activation of the kinase function may be a consequence of a structural alteration in the C-terminal domain of the protein. The affinity of VirA for the inducer is enhanced by ChvE, a protein encoded by the bacterial chromosome. Phosphorylated VirA phosphorylates VirG at a conserved aspartic acid residue at position 52 leading to a conformational change in the C-terminal DNA binding domain of the protein. Phospho-VirG activates transcription of all *vir* genes. Transcription activation is probably a result of increased affinity of phospho-VirG for its binding site as well as that of a productive interaction with RNA polymerase.

V. Acknowledgements

We thank our colleagues for sharing unpublished results prior to publication. Work in the authors' laboratory was supported by a grant from the US National Institutes of Health and by an American Cancer Society Faculty Research Award.

VI. References

Aiba, H., Nakasai, F., Mizushima, S. and Mizuno, T. (1989) J. Biochem. 106, 5-7.

Ankenbauer, R., Best, E., Palanca, C. and Nester, E. (1991) Mol. Plant-Micro. Inter. 4, 400-406.

Aoyama, T., Hirayama, T., Tamamoto, S. and Oka, A. (1989) Gene 78, 173-178.

Aoyama, T., Takanami, M., Makino, K. and Oka, A. (1991) Mol. Gen. Genet. 227, 385-390.

Aoyama, T., Takanami, M. and Oka, A. (1989) Nucl. Acids Res. 17, 8711-8725.

Banta, L., Joerger, R., Howitz, V., Campbell, A. and Binns, A. (1994) J. Bacteriol. 176, 3242-3249.

Beijersbergen, A., Dulk-Ras, A. D., Schilperoort, R. A. and Hooykaas, P. J. (1992) Science 256, 1324-1327.

Berger, B. R. and Christie, P. J. (1993) J. Bacteriol. 175, 1723-1734.

Binns, A., Beaupre, C. and Dale, E. (1995) J. Bacteriol. 177, 4890-4899.

Braun, A. C. (1958) Proc. Natl. Acad. Sci. USA 44, 344-349.

Cangelosi, G., Ankenbauer, R. and Nester, E. (1990) Proc. Natl. Acad. Sci. USA 87, 6708-6712.

Chang, C.-H. and Winans, S. (1992) J. Bacteriol. 174, 7033-7039.

Chang, C. H., Zhu, J. and Winans, S. C. (1996) J. Bacteriol. 178, 4710-4716.

Christie, P. J., Ward, J. E., Winans, S. and Nester, E. W. (1988) J. Bacteriol. 170, 2659-2667.

Citovsky, V., Wong, M. L. and Zambryski, P. (1989) Proc. Natl. Acad. Sci. USA 86, 1193-1197.

Citovsky, V., Zupan, J., Warnick, D. and Zambryski, P. (1992) Science 256, 1802-1805.

Close, T., Rogowsky, R., Kado, C., Winans, S., Yanofsky, M. and Nester, E. (1987) J. Bacteriol. 169, 5113-5118.

Close, T. J., Tait, R. C. and Kado, C. I. (1985) J. Bacteriol. 164, 774-781.

Cooley, M. B., D'Souza, M. R. and Kado, C. I. (1991) J. Bacteriol. 173, 2608-2616.

Das, A. (1988) Proc. Natl. Acad. Sci. USA 85, 2909-2913.

Das, A. (1997) Subcellular Biochemistry: Plant Microbe Interactions, Plenum Publishing Corp, London, pp. (in press).

Das, A. and Pazour, G. (1989) Nucl. Acids Res. 17, 4541-4550.

Das, A., Stachel, S., Ebert, P., Allenza, P., Montoya, A. and Nester, E. (1986) Nucl. Acids Res. 14, 1355-1364.

Doty, S. L., Yu, M. C., Lundin, J. I., Heath, J. D. and Nester, E. W. (1996) J. Bacteriol. 178, 961-970.

Fernandez, D., Spudich, G., Zhou, X.-R. and Christie, P. (1996) J. Bacteriol. 178, 3168-3176.

Fiedler, U. and Weiss, V. (1995) EMBO J. 14, 3696-3705.

Garfinkel, D. and Nester, E. (1980) J. Bacteriol. 144, 732-743.

Gubba, S., Xie, Y.-H. and Das, A. (1995) Mol. Plant-Micro. Inter. 8, 788-791.

Guyon, P., Chilton, M.-D., Petet, A. and Tempe, J. (1980) Proc. Natl. Acad. Sci. USA 65, 2693-2697.

Han, D. and Winans, S. (1994) Mol. Microbiol. 12, 23-30.

Han, D. C., Chen, C., Chen, Y. and Winans, S. C. (1992) J. Bacteriol. 174, 7040-7043.

Holsters, M., Villarroel, R., Geilen, J., Seurinck, J., DeGreve, H., Van Montagu, M. and Schell, J. (1983) Mol. Gen. Genet. 190, 35-41.

Hooykaas, P. J. and Beijersbergen, A. G. (1994) Annu. Rev. Phytopathol. 32, 157-179.

Huang, M., Cangelosi, G., Halperin, W. and Nester, E. (1990) J. Bacteriol. 172, 1814-1822.

Huang, Y., Morel, P., Powell, B. and Kado, C. (1990) J. Bacteriol. 172, 1142-1144.

Jayaswal, R., Veluthambi, K., Gelvin, S. and Slightom, J. (1987) J. Bacteriol. 169, 5035-5045.

Jin, S., Prusti, R., Roitsch, T., Ankenbauer, R. and Nester, E. (1990a) J. Bacteriol. 172, 4945-4950.

Jin, S., Roitsch, T., Ankenbauer, R., Gordon, M. and Nester, E. (1990b) J. Bacteriol. 172, 525-530.

Jin, S., Roitsch, T., Christie, P. and Nester, E. (1990c) J. Bacteriol. 172, 531-537.

Jin, S., Song, Y., Deng, W., Gordon, M. and Nester, E. (1993a) J. Bacteriol. 175, 6830-6835.

Jin, S. G., Song, Y. N., Pan, S. Q. and Nester, E. W. (1993b) Mol. Microbiol. 7, 555-562.

Kanemoto, R. H., Powell, A. T., Akiyoshi, D. E., Regier, D. A., Kerstetter, R. A., Nester, E. W., Hawes, M. C. and Gordon, M. P. (1989) J. Bacteriol. 171, 2506-2512.

Kenney, L., Bauer, M. and Silhavy, T. (1995) Proc. Natl. Acad. Sci. USA 92, 8866-8870.

Kolb, A., Busby, S., Buc, H., Garges, S. and Adhya, S. (1993) Annu. Rev. Biochem. 1993, 749-795.

Koukolikova-Nicola, Z., Raineri, D., Stephens, K., Ramos, C., Tinland, B., Nester, E. W. and Hohn, B. (1993) J. Bacteriol. 175, 723-731.

Kuldau, G. A., De Vos, G., Owen, J., McCaffrey, G. and Zambryski, P. (1990) Mol. Gen. Genet. 221, 256-266.

Lee, K., Dudley, M., Hess, K., Lynn, D., Joerger, R. and Binns, A. (1992) Proc. Natl. Acad. Sci. USA 89, 8666-8670.

Lee, Y.-W., Jin, S., Sim, W.-S. and Nester, E. W. (1995) Proc. Natl. Acad. Sci. USA 92, 12245-12249.

Leroux, B., Yanofsky, M., Winans, S., Ward, J., Ziegler, S. and Nester, E. (1987) EMBO J. 6, 849-856.

Machida, Y., Shimoda, N., Yamamoto-Toyoda, A., akahashi, Y., Nishihama, R., Aoki, S., Matsuoka, K., Nakamura, K., Yoshioka, Y., Ohba, T. and Obata, T. (1993) Advances in Molecular Genetics of Plant-Microbe Interactions, Kluwer Academic Publishers, Utrecht, pp. 85-96.

Makino, K., Amemura, M., Kim, S.-K., Nakata, A. and Shinagawa, H. (1993) Genes & Development 7, 149-160.

Mantis, N. and Winans, S. (1992) J. Bacteriol. 174, 1189-1196.

McBride, K. E. and Knauf, V. C. (1988) J. Bacteriol. 170, 1430-1437.

McLean, B. G., Greene, E. A. and Zambryski, P. C. (1994) J. Biol. Chem. 269, 2645-51.

Melchers, L., D., T., Idler, K., Schilperoort, R. and Hooykaas, P. (1986) Nucl. Acids Res. 14, 9933-9940.

Melchers, L., Maroney, M. J., Dulk-Ras, A., Thompson, D. S., van Vuuren, H., Schilperoort, R. A. and Hooykaas, P. (1990) Plant Mol. Biol. 14, 249-259.

Melchers, L. S., Regensburg-Tuink, T., Bourret, R., Sedee, J., Schilperoort, R. and Hooykaas, P. (1989a) EMBO J. 8, 1919-1925.

Melchers, L., Regensburg-Tuink, A., Schilperoort, R. and Hooykaas, P. (1989b) Mol. Microbiol. 3, 969-977.

Melchers, L. S., Thompson, D., Idler, K., Neuteboom, S., de Maagd, R., Schilperoort, R. and Hooykaas, P. (1987) Plant Mol. Biol. 9, 635-645.

Menage, A. and Morel, G. (1964) C. R. Acad. Sci. 259, 4795-4796.

Messens, E., Dekeyser, R. and Stachel, S. E. (1990) Proc. Natl. Acad. Sci. USA 87, 4368-4372.

Mettke, I., Fiedler, U. and Weiss, V. (1995) J. Bacteriol. 177, 5056-5061.

Moore, L., Warren, G. and Strobel, G. (1979) Plasmid 2, 617-626.

Morris, J. and Morris, R. O. (1990) Proc. Natl. Acad. Sci. USA 87, 3614-3618.

Nakashima, K., Sugiura, A., Kanamaru, K. and Mizuno, T. (1993) Mol. Microbiol. 7, 109-116.

Ninfa, A. J., Reitzer, L. J. and Magasanik, B. (1987) Cell 50, 1039-1046.

Okamoto, S., Toyoda-Yamamoto, A., Ito, K., Takebe, I. and Machida, Y. (1991) Mol. Gen. Genet. 228, 24-32.

Ooms, G., Kalpwijk, P. M., Poulis, J. A. and Schilperoort, R. A. (1980) J. Bacteriol. 144, 82-91.

Otten, L., DeGreve, H., Leemans, J., Hain, R., Hooykaas, P. and Schell, J. (1984) Mol. Gen. Genet. 175, 159-163.

Otten, L., Piotrowiak, G., Hooykaas, P., Dubois, M., Szegedi, E. and Schell, J. (1985) Mol. Gen. Genet. 199, 189-193.

Pan, S., Charles, T., Jin, S., Wu, Z.-L. and Nester, E. (1993) Proc. Natl. Acad. Sci. USA 90, 9939-9943.

Parkinson, J. and Kofoid, E. (1992) Annu. Rev. Genet. 26, 71-112.

Pazour, G. J. and Das, A. (1990a) Nucl. Acids Res. 18, 6909-6913.

Pazour, G. J. and Das, A. (1990b) J. Bacteriol. 172, 1241-1249.

Pazour, G. J., Ta, C. N. and Das, A. (1991) Proc. Natl. Acad. Sci. USA 88, 6941-6945.

Pazour, G. J., Ta, C. N. and Das, A. (1992) J. Bacteriol. 174, 4169-4174.

Peralta, E. G., Hellmiss, R. and Ream, W. (1986) EMBO J. 5, 1137-1142.

Peralta, E. G. and Ream, W. (1985) Proc. Natl. Acad. Sci. USA 82, 5112-5116.

Petit, A. and Tempe, J. (1985) in (van Vloten-Doting, L., Groot, G. and Hall, T., eds.), Molecular Form and Function of the Plant Genome, Plenum, New York, pp. 625-636.

Popham, D., Szeto, D., Keener, J. and Kustu, S. (1989) Science 243, 629-635.

Porter, S. G., Yanofsky, M. F. and Nester, E. W. (1987) Nucl. Acids Res. 15, 7503-7517.

Powell, B. and Kado, C. (1990) Mol. Microbiol. 4, 1-8.

Powell, B., Powell, G., Morris, R., Rogowsky, P. and Kado, C. (1987) Mol. Microbiol. 1, 309-316.

Regensburg-Tuink, A. and Hooykaas, P. J. (1993) Nature 363, 69-71.

Roggiani, M. and Dubnau, D. (1993) J. Bacteriol. 175, 3181-3187.

Rogowsky, P. M., Powell, B., Shirasu, K., Lin, T., Morel, P., Zyprian, E., Steck, T. and Kado, C. (1990) Plasmid 23, 85-106.

Roitsch, T., Wang, H., Jin, S. and Nester, E. (1990) J. Bacteriol. 172, 6054-6060.

Scheeren-Groot, E., Rodenburg, K., Dulk-Ras, A., Turk, S. and Hooykaas, P. (1994) J. Bacteriol. 176, 6418-6426.

Sen, P., Pazour, G. J., Anderson, D. and Das, A. (1989) J. Bacteriol. 171, 2573-2580.

Shimoda, N., Toyoda-Yamamoto, A., Nagamine, J., Usami, S., Katayama, M., Sakagami, Y. and Machida, Y. (1990) Proc. Natl. Acad. Sci. USA 87, 6684-6688.

Shirasu, K. and Kado, C. (1993) FEMS Micro. Letter 111, 287-294.

Smith, E. F. and Townsend, C. O. (1907) A plant tumor of bacterial origin, Science 25, 671-673.

Soderling, T. (1990) J. Biol. Chem. 265, 1823-1826.

Spencer, P., Tanaka, A. and Towers, G. (1990) Phytochemistry 29, 3785-3788.

Stachel, S., Messens, E., Van Montagu, M. and Zambryski, P. (1985) Nature 318, 624-629.

Stachel, S. and Nester, E. (1986) EMBO J. 5, 1445-1454.

Stachel, S., Timmerman, B. and Zambryski, P. (1986) Nature 322, 706-712.

Stachel, S. and Zambryski, P. (1986) Cell 46, 325-333.

Stachel, S. E., Nester, E. W. and Zambryski, P. C. (1986) Proc. Natl. Acad. Sci. USA 83, 379-383.

Steck, T., Morel, P. and Kado, C. (1988) Nucl. Acids Res. 16, 8736.

Stock, J., Ninfa, A. and Stock, A. (1989) Microbiol. Rev. 53, 450-490.

Sundberg, C., Meek, L., Carroll, K., Das, A. and Ream, W. (1996) J. Bacteriol. 178, 1207-11212.

Suzuki, M. (1993) EMBO J. 12, 3221-3226.

Tamamoto, S., Aoyama, T., Takanami, M. and Oka, A. (1990) J. Mol. Biol. 215, 537-547.

Thompson, D. V., Melchers, L. S., Idler, K. B., Schilperoort, R. A. and Hooykaas, P. J. (1988) Nucl. Acids Res. 16, 4621-4636.

Thorstenson, Y. R., Kuldau, G. A. and Zambryski, P. C. (1993) J. Bacteriol. 175, 5233-5241.

Tsung, K., Brissette, R. and Inouye, M. (1990) Proc. Natl. Acad. Sci. USA 87, 5940-5944.

Turk, S., Melchers, L., den Dulk-Ras, H., Regenburg-Tuink, A. and Hooykaas, P. (1991) Plant Mol. Biol. 16, 1051-1059.

Turk, S., Nester, E. and Hooykaas, P. (1993) Mol. Microbiol. 7, 719-724.

Turk, S., van Lange, R., Regenburg-Tuink, T. and Hooykaas, P. (1994) Plant Mol. Biol. 25, 899-907.

Usami, S., Okamoto, S., Takebe, I. and Machida, Y. (1988) Proc. Natl. Acad. Sci. USA 85, 3748-3752.

van Haaren, M. J. J., Sedee, N., Schilperoort, R. and Hooykaas, P. (1987) Nucl. Acids Res. 15, 8983-8997.

Vogel, A. M. and Das, A. (1992) J. Bacteriol. 174, 303-308.

Vogel, A. and Das, A. (1994) Mol. Microbiol. 12, 811-817.

Wang, K., Herrerra-Estrella, L., Van Montagu, M. and Zambryski, P. (1984) Cell 38, 455-462.

Ward, J. E., Akiyoshi, D. E., Regier, D., Datta, A., Gordon, M. P. and Nester, E. (1988) J. Biol. Chem. 263, 5804-5814.

Ward, J. E., Akiyoshi, D. E., Regier, D., Datta, A. and Gordon, M. P. (1990) J. Biol. Chem. 265, 4768.

Ward, J. E., Dale, E. M., Nester, E. W. and Binns, A. N. (1990) J. Bacteriol. 172, 5200-5210.

Weiss, V., Claverie-Martin, F. and Magasanik, B. (1992) Proc. Natl. Acad. Sci. USA 89, 5088-5092.

White, F. and Nester, E. (1980) J. Bacteriol. 144, 710-720.

Winans, S. (1990) J. Bacteriol. 172, 2433-2438.

Winans, S., Allenza, P., Stachel, S., McBride, K. and Nester, E. (1987) Nucl. Acids Res. 15, 825-836.

Winans, S., Ebert, P., Stachel, S., Gordon, M. and Nester, E. (1986) Proc. Natl. Acad. Sci. USA 83, 8278-8282.

Winans, S., Jin, S., Komari, T., Johnson, K. and Nester, E. (1987) in (von Wettstein, D. a. N. H. C., eds.), Plant Molecular Biology, Plenum Press, New York, pp. 573-582.

Winans, S., Kerstetter, R. and Nester, E. (1988) J. Bacteriol. 170, 4047-4054.

Yadav, N. S., J., V., Bennett, D. R., Barnes, W. M. and Chilton, M. D. (1982) Proc. Natl. Acad. Sci. USA 79, 6322-6326.

Yanofsky, M. F., Lowe, B., Montoya, A., Rubin, R., Krul, W., Gordon, M. and Nester, E. W. (1985) Mol. Gen. Genet. 201, 237-246.

Yanofsky, M. F. and Nester, E. W. (1986) J. Bacteriol. 168, 237-243.

Yoshioka, Y., Takahashi, Y., Matsumoto, S., Kojima, S., Matsuoka, K., Nakamura, K., Ohshima, K., Okada, N. and Machida, Y. (1994) in (Kado, C. and Crosa, J., eds.), Molecular Mechanisms of Bacterial Virulence, Kluwer Academic Publishers, Dordrecht, The Netherlands, pp. 231-248.

Yusibov, V., Steck, T., Gupta, V. and Gelvin, S. (1994) Proc. Natl. Acad. Sci. USA 91, 2994-2998.

Zambryski, P., Depicker, A., Kruger, K. and Goodman, H. (1982) J. Mol. Appl. Genet. 1, 361-370.

Zerback, R., Dressler, R. and Hess, D. (1989) Plant Science 62, 83-91.

Zupan, J. and Zambryski, P. (1995) Plant Physiol. 107, 1041-1047.

Chapter 14

Function of the Ti-Plasmid Vir Proteins: T-Complex Formation and Transfer to the Plant Cell

Fernando de la Cruz and Erich Lanka

I. Introduction

T-DNA transfer from *Agrobacterium tumefaciens* to plant cells is apparently the unique process of trans-kingdom genetic exchange that occurs naturally with detectable efficiency. This transfer mechanism seems to be exceptionally broad host-range, since *Agrobacterium spp.* can also transfer the T-DNA to yeast (Bundock *et al.*, 1995; Piers *et al.*, 1996). In spite of its uniqueness, the mechanism used by *Agrobacterium* to infect plant cells is very similar to the universal mechanism for bacterial genetic exchange: conjugation. It would be rewarding to understand one day why a specific bacterium, and only this bacterium, possesses such a powerful adaptation mechanism of survival as an interkingdom transfer mechanism, which it exercises very efficiently, while the rest of bacteria, which would have doubtlessly also benefited from it, has been incapable of acquiring this evolutionary tool.

This is particularly intriguing since bacterial plasmids are in fact able to transfer DNA to yeast by conjugation at least under laboratory conditions (Heinemann *et al.*, 1989). A better understanding of the molecular details of the underlying mechanism might be able to shed light on the causes of this extreme example of ecological adaptation. Additionally, gene transfer to plants is of obvious biotechnological interest.

In spite of the theoretical and practical importance of a better understanding of *Agrobacterium* T-DNA transfer, we know relatively little of the process. The initial steps of T-DNA processing have been analyzed in some detail, but the steps by which the T-strands are activated for transport and actually transported to the plant cell are still largely a mystery. The aims of this work are to describe what we know, and to identify some questions for which an answer would be particularly welcome. Several recent reviews have addressed aspects of the mechanism of T-DNA transfer from Ti plasmids (Baron *et al.*, 1996; Hooykaas *et al.*, 1994; Kado, 1994; Zambryski, 1992); others stressed the analogies between T-DNA transfer and bacterial conjugation (Firth *et al.*, 1996; Lanka *et al.*, 1995; Lessl *et al.*, 1994; Pansegrau *et al.*, 1996; Winans *et al.*, 1996).

T-DNA transfer can be divided into several functional steps: (*a*) contact formation between *Agrobacterium* and plant cells, (*b*) activation of expression of the operons involved in the synthesis of Vir proteins, (*c*) processing of the T-DNA by formation of a single stranded VirD2*T-DNA*VirE2 complex, (*d*) T-strand activation for transport, (*e*) transport of the T-DNA complex to the plant cytoplasm via the VirB DNA-transport structure, (*f*) transport to the plant cell nucleus, and (*g*) chromosomal integration. As discussed in section II, we believe that several of these steps [*c*, *d* and *e*] are common to bacterial conjugation and we will review this in particular detail. Others seem to be unique to this specialized DNA transfer system [*a*, *b*, *f* and *g*] and are discussed in other chapters of this book. We are assuming contact formation between *Agrobacterium* and the plant cell has taken place (as discussed in chapter 11), and transcriptional activation of the *vir* operons led to the synthesis of the VirA, B, C, D, E, F, G, H and J proteins (as discussed in chapter 13). The next step (step *c*) is the processing of the T-DNA to form a T-strand in the *Agrobacterium* cell. This step will be further discussed in section III, T-strand activation (step *d*) in section IV, and transport of the DNA (step *e*) in section V. Most significantly, the same transport complex used for T-DNA transport is also used for export of proteins VirE2 and VirF, as discussed in section VI. Finally, all data are interpreted in a T-DNA transport model discussed in section VII. Steps *f* and *g* will in turn be discussed in chapter 15.

II. T-DNA Transfer is a Specialized Form of Bacterial Conjugation

A segment of DNA from the Ti plasmids of *Agrobacterium tumefaciens*, called transferred DNA or T-DNA, can be efficiently transmitted from bacteria to plant cells. It arises from the DNA contained between two specific sequences called right and left borders (RB and LB) in the Ti-plasmid. In molecular detail, the segment containing the T-DNA is cleaved at both RB and LB, a single strand (the T-strand) is displaced and transported to the plant cell by a Ti-encoded multiprotein transport apparatus. Once inside the plant cell the T-DNA, still as ssDNA, is further transported to the nucleus probably due to nuclear localization signals present both in VirD2 and VirE2, and integrated by illegitimate recombination more or less randomly in one of the host chromosomes, with or without being previously replicated to produce a dsDNA molecule. Several functional and structural characteristics of this process indicate strongly that transfer of the T-DNA to plant cells closely resembles bacterial conjugation.

II.A. FUNCTIONAL SIMILARITY

Unfortunately many steps in T-DNA transfer are scarcely known, as is the case for conjugation. Nevertheless, for those steps or reactions for which more is known, the similarities between both processes are striking:

(a) *T-DNA Processing*. The first step in T-strand production is the cleavage of the T-DNA of the Ti-plasmids at RB and LB. This process is catalyzed by VirD1 and VirD2. When the process was reproduced *in vitro*, VirD1/VirD2 catalyzed the same reactions as the relaxases of conjugative plasmids. VirD2 alone, as occurs with relaxases, catalyzed cleavage and strand transfer using model oligonucleotides and plasmids containing the *nic* site (Pansegrau *et al.*, 1993b). The process required Mg^{2+} ions, no energy source, and the reactions were reversible. As a consequence of the cleavage reaction, a specific VirD2 tyrosyl residue becomes covalently linked to the 5' end of the cleaved phosphodiester bond. The phosphotyrosyl bond is hydrolyzed/joined during the strand transfer reaction (Pansegrau *et al.*, 1993b). VirD1, as occurs with relaxase-accessory proteins, was required, together with VirD2, for cleavage of supercoiled DNA containing *nic* (Scheiffele *et al.*, 1995). In the latter reaction, the 3'-end of the cleavage site was held tightly by the complex, so that it could not be extended by DNA polymerases. Thus, the biochemistry of the initial processing reactions in T-DNA transfer and in conjugation is essentially equivalent.

Transfer of the T-DNA requires, in *cis*, only RB. The process is, however, more efficient in the presence of an adjacent sequence called *overdrive* (van Haaren *et al.*, 1987; Peralta *et al.*, 1985). The combination of these two sequences can be considered analogous to a conjugative *oriT*. Deletion of the RB abolishes tumorigenesis, while removal of the LB does not detectably affect virulence (Joos *et al.*, 1983). Thus T-DNA transfer seems to occur unidirectionally. Reversal of the orientation of the RB leads to transfer of very long T-strands (>170 kb) encompassing the entire Ti plasmid (Miranda *et al.*, 1992). These properties are very similar to those of conjugative DNA transfer.

(b) *Plasmid Mobilization*. *vir*-genes can mediate mobilization of small plasmids like RSF1010 both to plant cells or to other bacteria (Beijersbergen *et al.*, 1992; Buchanan-Wollaston *et al.*, 1987). The VirB complex and the VirD4 protein are required for mobilization. Other Vir proteins, such as VirD1/VirD2, or VirE2, are not required. These are the same requirements as for RSF1010 mobilization by conjugative plasmids such as RP4: the mating-pair-formation components as well as a TraG-like protein are required, while the relaxase and accessory proteins of the conjugative helper plasmids are dispensable (Cabezón *et al.*, 1994; Cabezón *et al.*, 1997; Lessl *et al.*, 1993). Thus, the presumed T-DNA transport complex can act upon conjugative relaxosomes as well as upon T-strand intermediates, underlining their functional equivalence. Interestingly, RSF1010 inhibits the VirB-mediated transfer of both T-DNA (partially) and VirE2 (completely) to plant cells (Binns *et al.*, 1995). This result was interpreted as a competition for transport sites.

(c) *Sex Pilus*. Production of a conjugative pilus is the most conspicuous morphological feature in conjugating bacteria. When other structural and functional data suggested the close analogy between T-DNA transfer and conjugation, the involvement of a pilus in T-DNA transfer was a logical proposal in the field. It was this conviction that led the search for a Vir-dependent pilus, and discovery of its existence strongly reinforced the view of conjugation and T-DNA transfer as functionally related processes. In *Agrobacterium*, a pilus-like apparatus is produced in the presence of *virB* gene expression together with VirD4. There is a correlation between pilus production and the ability to transfer RSF1010, since the same components are required for both processes (Fullner *et al.*, 1996b). Besides, for *Agrobacterium* to produce Vir-pili, bacteria had to be grown at 19-22°C. At 28°C, the usual growth temperature for

Agrobacterium, there were no pili. Interestingly, RSF1010 mobilization by Vir, both to bacteria and to plants, was also temperature-sensitive, underscoring the relationship between both events (Fullner *et al.*, 1996a). The VirB-related conjugative mating-pair-formation proteins, although not the TraG-like proteins, are in turn required for pilus formation by conjugative plasmids (Pansegrau *et al.*, 1996), (see below).

II.B. SEQUENCE SIMILARITY

In general, it appears that the *virD* and *virB* operons are required for transfer steps held in common with the conjugation process, while other operons (*virA*, *virG*, *virF*, *virH*, *virJ*) are involved in regulation or other unique aspects of T-DNA transfer. The operons *virC* and *virE* have unique features but also probably share similar features with components of conjugative systems (section VI). In the common operons there are extensive sequence similarities:

(a) *Vir Proteins*. There is substantial sequence similarity between the *vir*-gene products and the *tra* gene products of a variety of plasmids (including the Ti-plasmid-encoded conjugative system active in DNA transfer between *Agrobacterium* (chapter 10), and the conjugative system of *Rhizobium* plasmid pNGR234a, responsible for symbiosis with legumes (Freiberg *et al.*, 1997)). The similarity extends from the operon comprising most of the transport proteins, *virB* (see section IV), to those implicated in DNA-processing (like *virC*, *virD* and *virE*). These similarities have been reviewed in previous articles (Kado, 1994; Lessl *et al.*, 1994). Besides, there is also extensive sequence similarity between VirB proteins and the products of two gene clusters implicated in bacterial virulence: the Ptl proteins, responsible for export of the *Bordetella pertussis* toxin (Weiss *et al.*, 1993), and the Cag proteins, playing an as yet undefined role in a pathogenicity island of *Helicobacter pylori* (Censini *et al.*, 1996; Tomb *et al.*, 1997). These two systems will be discussed further in section VII.

(b) *Gene Organization*. Figure 1 represents the *virB* operon aligned to the mating-pair-formation operons of a series of sequenced conjugative plasmids. As reported previously, the *vir* and *tra* operons match in their gene organization (Kado, 1994; Lessl *et al.*, 1994). Figure 1 shows strict conservation of the gene order between *virB* and the

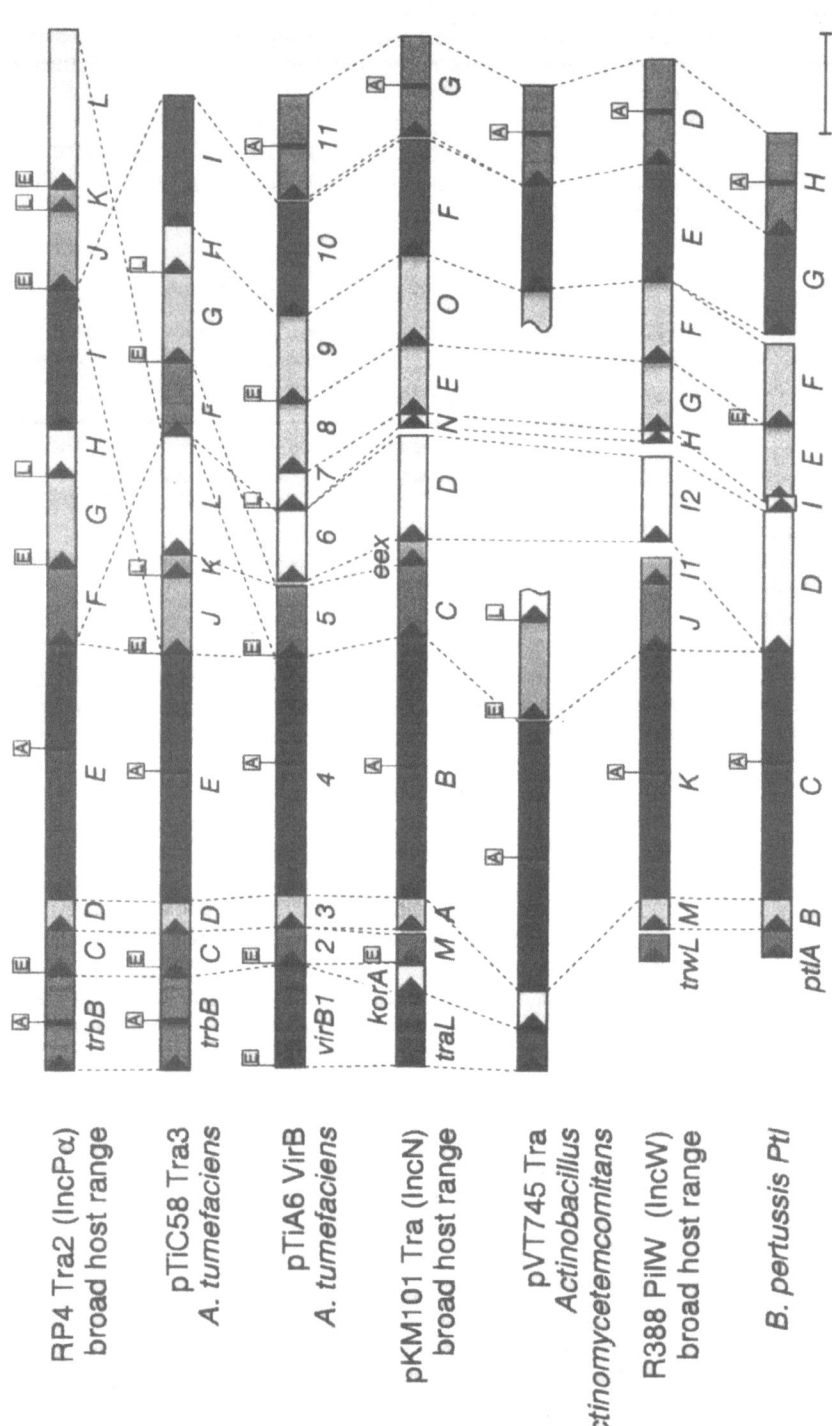

Figure 1. Conserved gene organization among bacterial protein export systems related to DNA transport. Genes encoding similar products are connected by broken lines or are shown in the same color. Tags marked with E represent signal cleavage sites for *E. coli* signal peptidase Lep. Tags marked with L indicate lipoprotein signatures at the N terminus of the respective protein. Conserved nucleotide binding motifs of type A are marked by tags with A. In the Ptl operon genes D and I overlap slightly, indicated by a staggered arrangement of the bars. References/GenBank accession numbers: IncPα RP4 Tra2 ((Lessl *et al.*, 1992a)/ M93696); pTiC58 Tra3 ((Alt-Mörbe *et al.*, 1996)/U43674 and U43675); pTiA6 VirB ((Ward *et al.*, 1988)/J03216); pKM101 ((Pohlman *et al.*, 1994)/U09868); pVT745 (D. Galli and D. Leblanc, personal communication); R388 Pil_w ((Bolland, 1992; Rivas *et al.*, 1997)/X81123); *B. pertussis* Ptl ((Weiss *et al.*, 1993)/L10720). The conjugative transfer region of the large *Rhizobium* plasmid pNGR234 ((Freiberg *et al.*, 1997)/U00090) is similar in gene organization to the Tra region of the *Agrobacterium* Ti plasmids, and it is not included in the Figure for simplicity.

Transport system	cagI-region	cagII-region			
Helicobacter pylori	CagE	Orf10	Orf11	Orf13	Orf15
RP4	TrbE	TraG	TrbB	TrbI	?
R388	TrwK	TrwB	TrwD	TrwE	TrwF
Ti/*vir*	VirB4	VirD4	VirB11	VirB10	VirB9

Table 1. Sequence similarities of the *cag*-region of *Helicobacter pylori* with DNA-transport systems

tra operon of pKM101 (IncN). In the case of R388 (IncW) only the *virB1* homologue is missing, although it could be located in an upstream, as yet unsequenced region (Bolland, 1992). It should be mentioned that *virB1* is not essential for T-DNA transport (section IV). Additional divergence is found between *virB* and the closely related RP4-Tra2 and Ti-Tra3 operons. In these cases, *trbB*, the *virB11* homologues, are located upstream of the rest of the genes, while *trbL*, the *virB6* homologue of RP4 is located downstream in the operon. VirB8 and VirB9 homologues are missing. Other presumed gene shuffling events are shown in Figure 1. A much weaker similarity has been described among five VirB proteins and five transfer gene products of F, TraA (VirB2), TraB (VirB10), TraC (VirB4), TraE (VirB5), and TraL (VirB3) (Kado, 1994). These phylogenetic relationships indicate that the transport systems discussed here are widespread among different organisms.

The *virD*, *virE*, and *virC* operons also contain homologues in conjugative systems, as reviewed (Lessl *et al.*, 1994; Pansegrau *et al.*, 1996). In particular, the *virD* operons can be aligned to the operons involved in plasmid conjugative DNA processing, which contain the relaxase and accessory proteins. Table 2 displays the names of homologous proteins in different systems, for reference.

(c) *Classification*. Conjugative systems have been classified in four families, according to sequence conservation between the relaxases, as well as between their *nic* sites (Lanka *et al.*, 1995). Within a given family there is extensive amino acid conservation between the relaxases, and nucleotide conservation between the *nic* sites. Among families, there is only conservation of a series of motifs, which are the signatures of the relaxase proteins, as shown in Figure 2. According to this classification, VirD2 and T-borders fall nicely into the P-family of conjugative transfer systems. There is extensive

similarity among the P-type relaxases, as shown previously (Pansegrau *et al.*, 1991; Pansegrau *et al.*, 1994), as well as conservation of a P-type *nic* site (Figure 3). As can be seen in Figure 2, conjugative relaxases, including VirD2, constitute a broad protein superfamily, in which proteins from almost all conjugative or mobilizable systems are represented, from plasmids in gram-positive bacteria to conjugative transposons (with the notorious absence of ColE1). The most conspicuous motifs in the superfamily are motifs I and III. Motif I contains the tyrosyl residue that forms the covalent bond with the DNA. Motif III, which contains a series of three conserved histidines, is thought to be directly involved in the reaction mechanism (Pansegrau *et al.*, 1996). In addition, Figure 3 shows that the T-borders and *nic* sites, do show loose conservation of a consensus sequence. Although the similarity shown is not compelling, it adds to the notion that most bacterial conjugative systems are related, and the conserved features among them are also present in the T-DNA transfer system of the Ti plasmids.

III. Production of T-Strands

T-DNA processing can be explained following the phage ϕX174 replication analogy (Kornberg *et al.*, 1992). ϕX174 replication occurs by displacement of the (+)-strand by the complex formed of RFI DNA, the phage encoded gpA protein and the *E.coli* Rep helicase. Displacement occurs immediately after the gpA-catalyzed cleavage reaction at the double stranded origin (*dso*) by ATP-driven unwinding in the presence or absence of DNA polymerase. T-DNA processing occurs similarly by displacement of the T-strand after VirD1/D2 cleavage at the T-borders. The *dso* is equivalent to the T-borders or the *nic* sites of conjugative systems (Pansegrau *et al.*, 1996). A diagram showing the processing and transport of the T-strand is shown in Figure 4.

nic region Family [a]	Plasmid (Inc group)	Relaxase	Cleavage Accessory Protein [b]	Additional oriT-binding Proteins [c]	Coupling Protein [d]
P	Ti plasmid	VirD2	VirD1	-	VirD4
P	RP4 (IncP)	TraI	TraJ	TraK	TraG
P	R6K (IncX)	TaxC	TaxA	DDP3	TaxB
P	R64 (IncI1)	NikB	NikA	n.d.	n.d.
F	F (IncF1)	TraI	TraY	TraM	TraD
F	R388 (IncW)	TrwC	TrwA	-	TrwB
F	CloDF13	MobC	MobA	-	MobB
Q	RSF1010 (IncQ)	MobA	-	MobC	

Table 2. DNA processing proteins in various systems
[a] *nic* region families correspond to the classification of Lanka and Wilkins (1995) and Pansegrau and Lanka (1996). Most references to sites and relevant proteins can be found in these reviews. More recently characterized *oriT*-processing systems include R388 (Llosa *et al.*, 1995; Moncalián *et al.*, 1997), R6K (Avila *et al.*, 1996; Núñez *et al.*, 1997) and CloDF13 (Núñez B, Avila P, de la Cruz F (in preparation)).
[b] Protein absolutely required for site cleavage (P-family), or just enhancing cleavage (F-family). TraM of the IncFII plasmid R1 has also been shown to enhance relaxation (E. Zechner, unpublished).
[c] Protein essential for functional activity of the transfer system, but not for *nic* cleavage.
[d] Protein that presumably couples the relaxosome to the transport complex. References to the specific proteins can be found in (Cabezón *et al.*, 1997

T-DNA processing is catalyzed by the VirD1/D2 proteins. Ti plasmid mutants in either *virD1* or *virD2* genes are incapable of T-border cleavage (Stachel *et al.*, 1987; Veluthambi *et al.*, 1988). Furthermore, the reaction was reproduced in vitro using purified VirD1 and VirD2 proteins (Scheiffele *et al.*, 1995). Protein VirD2 catalyzed cleavage of model oligonucleotides containing T-border sequences (Pansegrau *et al.*, 1993b). The reaction was reversible and required only Mg^{2+}. Addition of a second oligonucleotide led to a strand-transfer reaction, the result of which was the formation of a hybrid oligonucleotide containing the 3'-half of one, and the 5'-half of the second oligonucleotide. Similar reactions were shown for gpA of ϕX174. The same reactions, occurring with ongoing rolling-circle replication on a supercoiled DNA template, would result in a ssDNA circle and the re-generation of the starting molecule. A schematic summary is given in Figure 4.

When T-DNA processing is induced by acetosyringone, normally secreted by plants when wounded, *vir*-induction results in the expression of Vir proteins and in the production of T-strands as single-stranded DNA in the *Agrobacterium* cell in a process preceding and independent of DNA transfer to the plant (Stachel *et al.*, 1986; Stachel *et al.*, 1987). The possibility that the ssDNA molecules found are artifactual intermediates, that is, a side-product of a normal reaction, is unlikely since T-strands are produced in relatively high amounts of about one molecule per bacterial cell (Stachel *et al.*, 1986). Thus, apparently,

synthesis of the Vir proteins directly results in T-DNA processing to the stage of free T-strands in the *Agrobacterium* cell. On the other hand, in the in vitro VirD1/VirD2 catalyzed border cleavage reaction on supercoiled DNA, denaturing agents are required to detect cleavage, exactly as observed in the case of conjugative transfer origins.

In order to reconcile this apparent contradiction, there is an observation that can be very significant. One interesting peculiarity of the VirD2 mediated cleavage reaction was that the specificity of the enzyme for its corresponding oligonucleotide sequence was clearly more relaxed than that of conjugative relaxases. Conjugative relaxases are very specific for the sequence of their *nic* sites (Llosa *et al.*, 1996; Pansegrau *et al.*, 1993a), while VirD2 could use the RP4-specific oligonucleotide as well as the T-border oligonucleotide (Pansegrau *et al.*, 1993b). This result was interpreted as a specialization of this enzyme to produce random integration of the T-DNA in the plant chromosomal DNA. However, recent results cast doubt on the idea that chromosomal integration is catalyzed by VirD2. It appears that integration is determined by the host. In plants, illegitimate recombination is the predominant form, while in yeast, it can be general or illegitimate recombination, depending on the existence of homology between the T-DNA and the yeast chromosomes (Bundock *et al.*, 1995; Bundock *et al.*, 1996). In the case of VirD2 being somehow involved in the process of target selection, it seems

Plasmid	Gene		I		II		III	Size	Sequence Name
R751	TraI	(16-128)	DFAGLAN Y IT	41	DKTYHLIV.SFRAGE	22	HQRISAV H NDT....DNL H I H IAINKIHPTR	(747)	TRI5_ECOLI
RP4	TraI	(16-128)	DFAELVK Y IT	41	DKTYHLIV.SFRAGE	22	HQRVSAV H HDT....DNL H I H IAINKIHPTR	(732)	TRI6_ECOLI
R64	NikB	(17-177)	SFEDLVS Y VS	90	DPVFHYIL.SWQSHE	22	HQYVSAV H TDT....DNL H V H VAVNRVHPET	(899)	B38529
pTF-FC2	MobA	(20-132)	RVSRLTG Y IR	42	DTINHYVL.SWREGE	22	HQAIYGL H ADT....DNL H L H LAINRVHPET	(465)	MOBA_THIFE (1)
ps194	Rlx	(12-113)	SASRAIN Y AE	31	GVQAHTVIQSFKPGE	20	HQVAVYT H TDK....DHY H N H IIINSVDLET	(320)	RLX1_STAAU
pC223	Rlx	(12-113)	STSRAIN Y AE	31	GNEGHVVIQSFKPNE	20	HQVAVYT H NDT....DHY H N H IVINSIDLET	(330)	RLX3_STAAU
pC221	Rlx	(12-112)	SASRAIN Y AE	31	GIQAHTVIQSFKPGE	20	HQVAVYT H TDK....DHY H N H IVINSVDLET	(315)	RLX2_STAAU
pTiC58	VirD2	(23-145)	QIINQLE Y LS	48	ELTTHIIV.SFPAGT	25	YNYLTAF H IDR....DHP H L H VVVNRELLG	(447)	VID2_AGRT5
pRiA4b	VirD2	(23-145)	QIINQLE Y LS	48	ELTTHIIV.SFPAGT	25	YNYLTAF H IDR....DHP H L H VVVNRELLG	(424)	VID2_AGRT6
pTiA6NC	VirD2	(23-145)	QIINQLE Y LS	48	DLTTHIIV.SFPAGT	25	YNYLTAY H VDR....DHP H L H VVVNRELLG	(436)	VID2_AGRRA
R6K	TaxC	(65-184)	GIKNSID Y MS	46	KITQNIVF.SPPVSA	21	NRFVLGY H EDK....KEHP H V H VVFRIKDTDG	(385)	X95535 (332..1489)
Tn4399	MocB	(10-121)	SFSGCVC Y VL	37	VGHTSLNF.SPEDGE	24	TQYIIVA H IDK....EHP H C H IVFNRVNDNDG	(318)	B48487
pLV22a	MbpB	(10-129)	HGVAALE Y DL	37	NNCLRFEV.SPSIEE	24	HQYIIVAR H SGTESKKEQA H L H ILANRVSLSG	(264)	U25716 (415..1209)
pMV158	Pre	(38-145)	PSRSHLN Y EL	31	VLCDEWII.TSDKDF	26	NIAYASV H LDE....STP H M H MGVVPFENGK	(494)	PRE_STRAG
pLB4	PreA	(38-144)	VSRSHLN Y DL	30	VLVNEWII.TSDKDF	26	NIRYAVV H MDE....KTP H M H MGIVPFDDDK	(361)	PREA_LACPL
pLAB1000	PreA	(38-144)	VSHSHLN Y DL	30	VLVNEWII.TSDKDF	26	NIRYAVV H MDE....KTP H M H MGIVPFDDDK	(361)	PREA_LACHI
pUB110	Pre2	(38-144)	HERTREN Y DL	30	VLVNELIV.TSDRDF	26	NIAYATV H NDE....QTP H M H LGVVPMRDGK	(420)	PRE2_STAAU
pTB913	Pre	(38-144)	KERSHEN Y DL	30	VLVNELIV.TSDRHF	26	NIAYATV H VDE....KTP H M H LGVVPMRDGK	(415)	PRE_BACSP
pGI2	PreA	(39-143)	YSKSEQN Y DL	30	VVLSEFVV.TASPDY	26	NTLYAMV H MDE....ATP H M H IGVMPITEDN	(445)	PREA_BACTU
pE194	PreI	(39-145)	HEETYKN Y DL	30	IRHVDGLV.TSDKDF	26	NMLYATV H LDE....RVP H M H FGFVPLTEDG	(403)	PRE1_STAAU
Tn4451	TnpZ	(41-164)	VIEERIP Y NV	32	TLFNELVI.DVNTMY	28	NVISAVM H ADE<13>YHY H L H AMVLPVVEKE	(421)	U15027 (5028..6293)
R46	TraHI	(11-172)	NVTSVVG Y YS	59	RLGYDLTF.SAPKGV	48	LVVATFR H ETSRALDPDL H T H AFVMNMTQRE	(1078)	U43676 (8..3244)
R100	TraI	(9-169)	SAGSAGN Y YT	58	RPGYDLTF.SAPKSV	48	LVMALFN H DTSRDQDPQL H T H VVVANVTQHN	(1756)	TRI2_ECOLI
F	TraI	(9-169)	SAGSAGN Y YT	58	RPGYDLTF.SAPKSV	48	LVMALFN H DTSRDQEPQL H T H AVVANVTQHN	(1756)	TRI1_ECOLI
R388	TrwC	(11-173)	DIGRAAS Y YE	60	RIGLDLTF.SAPKSV	48	LVIGKFR H ETSRERDPQL H T H AVILNMTKRS	(966)	S43878
pSC101	Mob	(15-135)	SASPHAD Y IA	44	CTYREIEI.ALPREL	20	HAYQFAI H NPK<6> EQP H A H IMFSERINDG	(371)	RLX1_SALTY
RSF1010	MobA	(18-132)	SARAKAD Y IQ	42	RLFKEVEF.ALPVEL	21	LPYTLAI H AGGG...ENP H C H LMISERINDG	(709)	MBA2_ECOLI
pTF1	MobL	(18-167)	SATGAAA Y RA	52	VLVREIEI.SLPTEL	21	VAADVAL H APR<26>GNW H A H ILLSACHVQP	(378)	MOBL_THIFE
pIP501	ORFI	(12-139)	SLIAMAS Y RS	52	QLCREVNV.ALPIEL	21	MIADVAI H RDD....ENNP H A H IMLTMREVDS	(654)	L39769 (1415..3379)
pGO1	Nes	(18-143)	SATAKSA Y NS	54	QVAREIII.GLPNEF	20	MIVDLNI H KINE...ENP H A H LLCTLRGLDK	(665)	U50629 (214..2211)
pTiC58	TraA	(18-154)	SVVLSAA Y QH	65	QLARDLTI.ALPLEL	21	MVADWVY H DNP....GNP H I H LMTTLRPLTE	(1101)	U40389 (3239..6544)
pTiR10	TraA	(18-154)	SVVLSAA Y RH	65	QLARDLTI.ALPLEL	21	MVADWVY H ENP....GNP H I H LMTTLRPLSD	(1100)	U43674 (7977..11279)
pNGR234a	TraA	(18-154)	SAVLSAA Y RH	65	QLAKDVTI.ALPTEL	21	MVADWVY H DAP....GNP H V H LMTTLRPLTA	(1102)	TRAA_RHISN

Figure 2. The VirD2 relaxase belongs to the Mob protein superfamily of RCR replication proteins (Ilyina *et al.*, 1992; Pansegrau *et al.*,1991). There are four families shown in different blocks that correspond to the four *nic* site families shown in Fig.3. There is a fifth family (the ColE1 family) which does not show conservation of the motifs shown. The figure shows the three amino acid sequence motifs conserved in the Mob superfamily, and highlights the four completely conserved amino acids: the tyrosine that forms the covalent bond with the 5'-end of *nic*, and a set of three histidines, possibly involved in coordination of the Mg^{2+} ion. The last column refers to the database names for the proteins (in SwissProt or PIR), when available, or for the DNAs (in GenBank), including in these cases the coordinates used for translation.

clear that the illegitimate recombination events which happen during integration do not occur by the same biochemical pathway used for T-strand cleavage and strand-transfer (Tinland *et al.*, 1995).

Alternatively, we propose that the lack of specificity of VirD2 cleaving reaction can lead to a relaxed affinity for the 3'-end of *nic*, and can be interpreted as a specialization of the processing reaction, with two molecular results. First, piloting to the recipient requires that VirD2 frees its clamp on the 3' end of the T-strand in the donor, because single stranded linear DNA was found in the donor cells. This can be achieved if VirD2 affinity to the 3'-end is constitutively relaxed, compared to TraI of RP4. Second, as seen in Figure 4, the relaxed grip on the 3'-ends by both the first (RB) and the second (LB) VirD2 molecules, allows the first VirD2 molecule to leave, after the unwinding reaction, with the T-strand and a free 3'-end. If this were the case, significant differences will be found in the mechanisms of T-DNA transfer and conjugation in this step.

Another point worth mentioning is that only the T-DNA sequence between RB and LB was transferred, compared to bacterial conjugation where the whole plasmid is normally transferred. But this is only an apparent difference. It has been shown that plasmids containing two directly repeated *oriT*s can transfer the segment of DNA between the two copies (Erickson *et al.*, 1993). In turn, Ti-derived binary vectors are often transferred entirely to plants (van der Graaff *et al.*, 1996). Besides Ti-plasmids with a reversed RB can transfer most of the Ti plasmid DNA (Miranda *et al.*, 1992). There is some discussion about the form in which the T-DNA is transported to the plant, since linear dsDNA corresponding to the T-DNA was also observed in *Agrobacterium* cells, but accumulating evidence suggests transport of ssDNA as the more likely (Tinland *et al.*, 1995).

IV. T-Strand Activation

IV.A. T-DNA TRANSFER AND BACTERIAL CONJUGATION MAY USE DIFFERENT STRATEGIES TO CONTROL T-DNA PROCESSING

As mentioned before, the Vir proteins can process the T-DNA up to free T-strands upon activation. On the other hand, in vitro assembled conjugative relaxosomes are unable to carry out the unwinding reaction, since the cleaved 3'-end of *nic* is tightly bound to the relaxase, and not accessible to the DNA polymerase (Pansegrau *et al.*, 1990; Scherzinger *et al.*, 1992; Willetts *et al.*, 1984). No reagent is known today which allows conversion of relaxosomes into a replication machinery. This "relaxosomal clamp" can be explained as due to the high affinity of the 3'-end of *nic* for the relaxase, as shown in Pansegrau *et al.*, (1996).

Conjugative plasmids can be naturally repressed for DNA transfer, like R100 or R6K, or derepressed, like F, RP4, or R388. Derepressed plasmids have their conjugative machinery constitutively available, while repressed plasmids either conjugate at low frequency due to occasional escape from repression, or are transiently derepressed by a first mating event, resulting in epidemic spread through recipients. Available evidence suggests that, before donor cells engage in conjugation, relaxosomes are already preassembled, irrespective of the repression state of the *tra* genes. This was shown for RP4 (Pansegrau *et al.*, 1993a), plasmid F (Sherman *et al.*, 1994) and R6K (Avila *et al.*, 1996) among others, since negatively superhelical DNA could be isolated in the form of relaxosomes, in the absence of recipient cells (Fürst *et al.*, 1989).

As a particularly interesting case, plasmid R6K contains approximately 20 copies per cell, and still most of the molecules are present in the form of relaxosomes, in spite of being naturally repressed. R6K*drd1*, a conjugatively derepressed mutant of R6K, transfers at a frequency 1000-fold higher than R6K. Nevertheless cells carrying either plasmid contain most of the plasmid molecules as preassembled relaxosomes (Avila *et al.*, 1996). Thus, the number of relaxosomes is unrelated to the conjugation frequency and it can be assumed that relaxosome formation precedes initiation of conjugation. Therefore, triggering of conjugative DNA processing, that should occur immediately after contact with the recipient is established, should act at a later stage as a relaxosome activation process, or further downstream. Therefore, it is generally assumed that a mating signal triggers conjugation by releasing the "relaxosomal clamp", and thus allowing the unwinding of the T-DNA and subsequent transport of the T-strand.

The case of Ti plasmid T-DNA transfer could probably shed light on this problem. The starting situation might be different, in that *vir* gene expression is subject to repression, and T-borders are apparently free of relaxosomes. This is inferred from

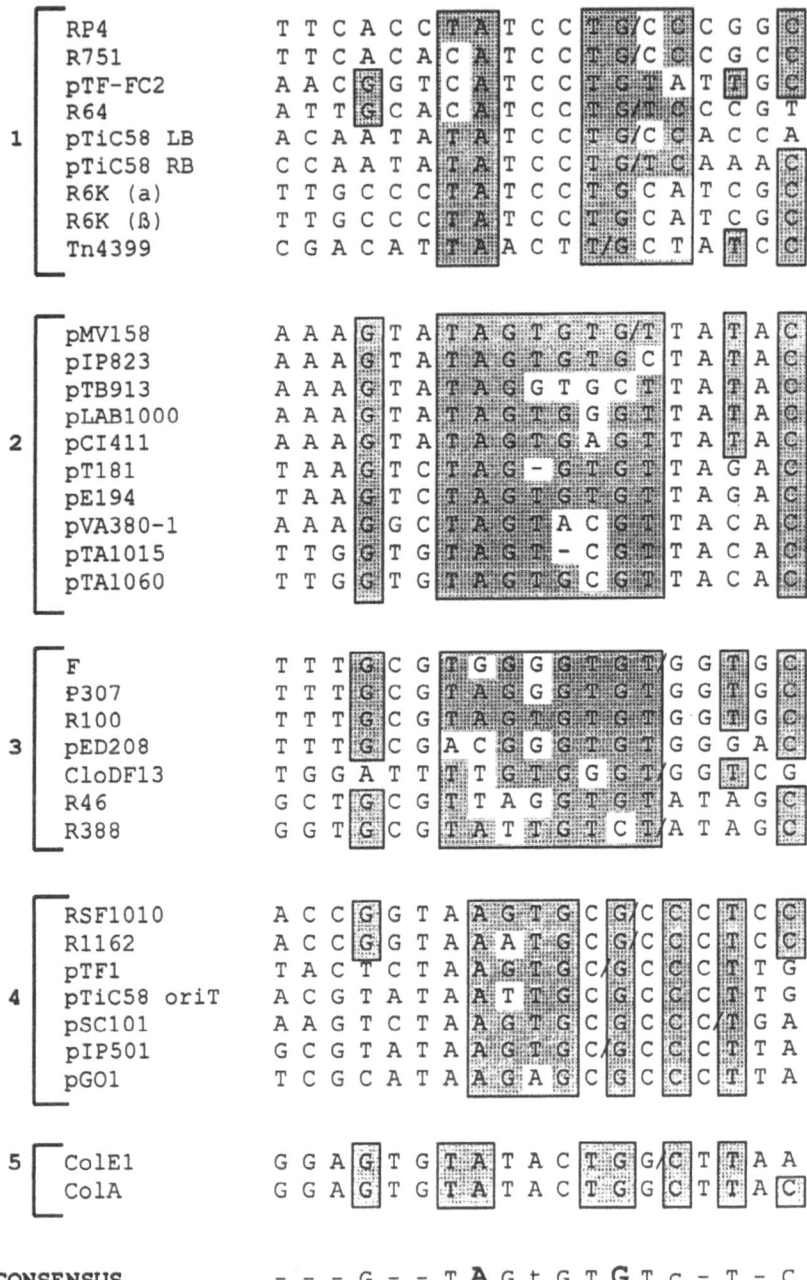

Figure 3. The Ti plasmid T-borders belong to the RP4 class of origins of conjugative transfer. The figure shows the sequences recognized by five classes of proteins from the Mob superfamily, with residues identical to a proposed consensus highlighted. Instead of marking conservation within the families, we emphasize conservation of a consensus sequence through the families. Although this alignment can be considered rather controversial, it may indicate that all relaxases recognize specific variants of an ancestor core sequence. References are given in the footnote of Table 2. Recently characterized *oriT* sequences are: Tn399 (Murphy *et al.*, 1995), Group 2 (Guzman *et al.*, 1997), and pG01 (Chimo *et al.*, 1996).

the fact that induction is required to see cleavage at border sequences or single T-strands (Stachel *et al.*, 1986; Stachel *et al.*, 1987). Once induced, the Vir proteins are able to process the T-DNA up to the stage of free T-strands thus bypassing the "relaxosomal clamp". On the other hand, the Vir system is able to promote conjugation (and transfer to plants) of RSF1010 so that, whatever the differences are in the DNA processing reactions, the RSF1010 relaxosome produced by the RSF1010 Mob proteins can be acted upon by the Vir transport system. This is a very important fact that suggests that the corresponding complexes should not be basically different, if they are to be recognized by the same "trigger". Furthermore, T-strands are not secreted into the medium by the T-transfer system of *Agrobacterium* (cited in Stachel *et al.*, (1986) as data not shown). Thus, even a pre-existing T-strand should have to be activated for transport. Thus we believe it should be assumed that the "trigger" for DNA transfer acts after synthesis of the T-strands. How can it be imagined that the same "trigger" acts on RSF1010 and VirD2/VirD1 relaxosomes? The trigger must be a rather general substance or signal, which can be passed on from a mating-pair-formation apparatus to several types of relaxosomes. The trigger thus would most reasonably act upon a vehicle that interacts both with the relaxosome and with the Mpf system. This concept puts VirD4 in center stage.

IV.B. VIRD4 AS MOLECULAR MOTOR AND A PLAUSIBLE TRANSMITTER OF THE MATING SIGNAL

VirD4 is essential for T-DNA transfer to plant cells (Koukolíková-Nicola *et al.*, 1993). According to immuno-localization studies, VirD4 was localized to the cytoplasmic membrane. However, fractionation analysis showed it to be present both in the inner membrane fraction and in an intermediate fraction, which suggests a possible binding also to the outer membrane (Okamoto *et al.*, 1991). The same authors constructed a series of alkaline-phosphatase fusions to show that only the region around amino acids 30-41 resulted in PhoA+ fusions, and so this domain of the protein faces the periplasm. This fact, together with the predicted α-helical trans-membrane segments in VirD4 suggests that VirD4 is a bitopic membrane protein, traversing the inner membrane twice, with the N- and C- termini facing the cytoplasm.

VirD4 belongs to a class of proteins called TraG-like proteins. These large (50-70 kDa) proteins are defined by sequence and functional similarities (Balzer *et al.*, 1994; Lessl *et al.*, 1992; Ziegelin *et al.*, 1991). TraG-like proteins of conjugative plasmids are in all cases essential components of their transfer apparatuses. TraG-like proteins are associated with the cytoplasmic membrane, as is VirD4 (Okamoto *et al.*, 1991; Panicker *et al.*, 1992; Haase *et al.*, in preparation). Their sequences contain more or less well conserved type A and type B nucleotide binding motifs, transmembranal and amphiphilic helical segments. In the case of RP4 TraG the importance of the potential NTP-binding motifs for function has been verified by site-directed mutagenesis (Balzer *et al.*, 1994). Replacement of certain residues in the motifs inactivates TraG function, confirming the significance of these domains. In only one case, F TraD, rather preliminary evidence was presented for DNA binding (Panicker *et al.*, 1992).

Old experiments demonstrated that TraD, the TraG-homologue of plasmid F, acts in a late step in conjugation, since mutations in *traD* resulted in a partial defect in replacement strand synthesis in a conjugative donor cell (Kingsman *et al.*, 1978). It was not required for pilus formation nor for relaxosome processing, so it was directly implicated in DNA transport. Thus, TraD was considered a key element in DNA transport, possibly the actual molecular motor of the transmission. Although no direct evidence for this role has been reported, indirect evidence is consistent with it. It has been concluded from genetic and biochemical data that TraG and their analogues interact with the relaxosome, maybe via protein-protein interaction with the relaxase (Balzer *et al.*, 1994; Cabezón *et al.*, 1994; Cabezón *et al.*, 1997; Lessl *et al.*, 1993; Pansegrau *et al.*, 1996; Waters *et al.*, 1992). However, TraG-like proteins are not required for nicking or T-DNA production (Stachel *et al.*, 1987) In general TraG-like proteins are not needed for pilus production either but they are needed for the mobilization of small non-self-transmissible plasmids by conjugative helper plasmids. The exception is the VirD4 protein, the presence of which is required for both transfer of RSF1010 to plant cells and in addition it was required for the synthesis or assembly of the Vir pilus (Fullner *et al.*, 1996). Sequence analysis has not yet provided an obvious explanation for this observation. However it could well be that VirD4 includes an activity which is normally exerted by the pilus assembly machinery as a separate function, but in the case of the Vir system it may have been fused to VirD4 (Haase *et al.*, 1997).

TraG-like proteins were only found among DNA transfer systems but not in systems thought to be involved in protein export only, like the VirB-related Ptl system of *Bordetella pertussis* responsible for the *pertussis* toxin export (Weiss *et al.*, 1993). Thus it is tempting to speculate that TraG-like proteins are part of a machinery to convert (or adapt) a protein transport system into a DNA and protein transfer system. The finding that *Helicobacter pylori* specifies a chromosomally encoded TraG analogue might shed some light on this problem. It will be interesting to find out whether the protein encoded by the virulence region *cag* belongs to a conjugative DNA transfer system or is a remnant of such a system (Censini *et al.*, 1996).

V. The VirB-Transport Structure

V.A. WHAT IS THE FUNCTION OF THE VIRB COMPLEX IN *AGROBACTERIUM* T-DNA TRANSFER?

In this section there are still more questions than answers. An important question concerns the molecular differences between so-called liquid and surface maters. IncN, P or W plasmids, which showed closer similarity to VirB (Figure 1), are surface maters. Cells containing these plasmids conjugate much better on the surface of semi-solid media than in broth. This characteristic is correlated to the production of thick, rigid pili. On the other hand, data on the function of pili in conjugation were mainly obtained with the F-system. Since F-containing cells produce thinner, flexible pili, and conjugate much better in liquid medium than on semi-solid surfaces, the observations made with F may not be valid for plasmid transfer occurring with higher frequencies on semi-solid surfaces. This possibility is clearly manifested in plasmid R64. This large IncI1 plasmid mates in liquid as well as in solid surfaces. However, it produces two types of pili: a thin pilus, which is required only for liquid mating, and a thick, rigid pilus, which is required both for solid and liquid mating (Kim *et al.*, 1997). Thus, the evidence seems to indicate that liquid mating is a superimposed evolutionary acquisition by some plasmid systems. As happens in R64, the complete sets of genes that are required for liquid mating are

probably irrelevant to the DNA transport mechanism.

When it became clear that the VirB operon was structurally and functionally analogous to mating-pair-formation systems of conjugative plasmids assembling rigid pili, the hunt for a pilus-like structure seemed an obvious research goal (Kado, 1994; Winans *et al.*, 1996). Not surprisingly, it was found, and its appearance was shown to depend on each of the VirB proteins (Fullner *et al.*, 1996). The presence of VirD4 was also required for the synthesis of Vir-dependent pili. Thus, the pilus found by Fullner *et al.* (Fullner *et al.*, 1996) is likely to function as a "conjugative" pilus. But, important and clarifying as this finding certainly was, it did not help much in solving the question of the mechanism of T-DNA transport. The reason is that the debate about the role of the conjugative pilus in bacterial conjugation is far from being settled (see, for example, recent discussions about the topic in Dreiseikelmann, 1994; Firth *et al.*, 1996; Haase *et al.*, 1995; Silverman, 1997). Firstly, DNA could not be found trapped in conjugative pili when they were isolated (Frost *et al.*, 1991). Secondly, pili retract before transfer of DNA starts, so typically mating cells have their outer membranes apposed (Dürrenberger *et al.*, 1991). Thus it is reasonable to ask the question whether the apparatus for pilus assembly and erection is a constituent of the DNA transfer pore. Third, there is concern about the lack of complete correlation between the presence of the RP4 Trb-dependent extracellular filamentous structures and the transfer ability (Haase *et al.*, 1995).

These results, all casting doubt on the direct involvement of pili in DNA transport were only counteracted by a single report in which conjugative transfer of plasmid F was claimed to proceed through an extended pilus (Harrington *et al.*, 1990). However, presentation of these results in a plasmid meeting (Schloss Ringberg, Germany, 1991) provoked a lot of arguments. The work was not pursued further, thus there is some suspicion about the real meaning of the Harrington and Rogerson experiment. Available data, in summary, amounts to the statement that VirB proteins are required for pilus synthesis. The involvement of VirB proteins in T-DNA transfer directly remains unclear since there are no experiments tackling this question. Three possible roles can be ascribed to the VirB-dependent pilus, the same that were proposed for conjugative pili. Firstly, bacterial conjugation could resemble filamentous phage infection, with the transferring-

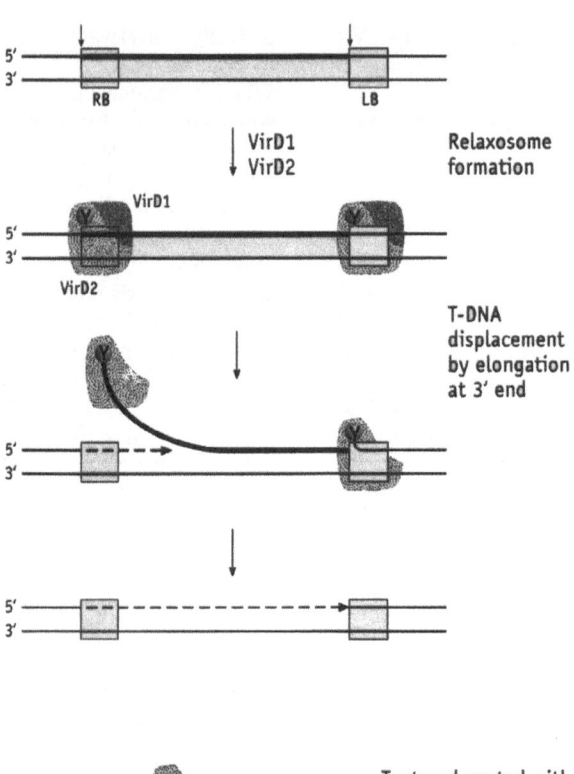

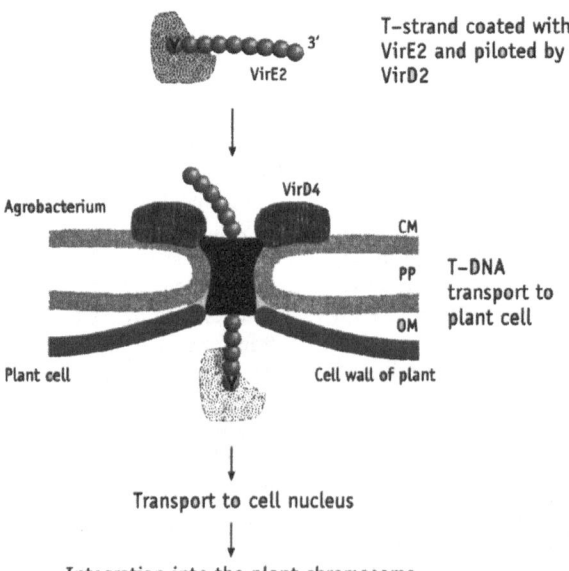

Figure 4. Scheme of T-DNA strand processing. 1) Relaxosome formation by VirD1/D2 is followed by reversible cleavage at both RB and LB; 2) Release of the 3' grip at RB, and strand displacement replication using the 3' end as primer; the displaced strand is progressively covered by VirE2; 3) Second cleavage at LB, and liberation of the T-strand bound to VirD2; 4) The nucleoprotein complex (T-complex) formed by the T-DNA coated by VirE2 and piloted by VirD2 is led to the bacterial transport site by VirD4, and transported to the plant cell by the T-DNA transport system.

strand passing through the lumen of a pilus-like "bridge" (Harrington *et al.*, 1990). As a variation on this theme, DNA transfer could occur, after retraction of the pili, through the pilus assembly machinery; i.e. the basal plate of the pilus. Secondly, the pilus could merely act as a "grappling hook", bringing conjugating cells together so that conjugative junctions (Dürrenberger *et al.*, 1991) can be formed. Thirdly, the pilus could participate in the transmission of the mating signal to the relaxosome, thus triggering conjugation. These three roles, bridging, grappling or sensing, do not need to be absolute alternatives; according to our view, they could all be performed by the VirB and analogous multiprotein machineries of conjugative plasmids.

V.B. COMPONENTS OF THE VIRB "COMPLEX"

During the last few years, much of the work carried out on and around the VirB proteins was designed to generate materials (mutants, proteins, antibodies, *etc.*) that should clear the way to find out functions for the VirB proteins. For example, Berger and Christie (Berger *et al.*, 1993) constructed in frame deletions within each of the eleven *virB* genes, to analyze their individual contribution to the transfer process. VirB2 to VirB11 were found to be essential components of the transport apparatus, while the mutation in VirB1 reduced virulence about 100-fold indicating either that a host-encoded component could replace it, or that it functions only as an optimizing factor. Several additional reports suggest the existence of physical interactions among a group of the VirB proteins, indicating that the proteins may form a complex (as detailed below). Also, deletion of some of the *virB* genes led to reduction of the steady state levels of other VirB proteins, further suggesting that physical interactions are involved in the assembly or stabilization of neighboring proteins (Berger *et al.*, 1994; Fernandez *et al.*, 1996a; Jones *et al.*, 1994). Finally, the VirB proteins were localized to the different compartments of the bacterial cell by a combination of techniques such as proteinase K susceptibility, Western blotting of isolated membrane fractions, or phosphatase activity of translational fusions (Table 3). In conclusion, all data taken together are suggestive of the existence of a membrane-spanning complex that is composed of VirB proteins. In addition, some characteristics of the individual proteins, and of their most conspicuous interactions are emerging at a growing

rate. The intriguing question that remains to be solved is whether the components are held together tightly enough so that the complex can be isolated as such. A survey of the literature allowed us to speculate, and in some cases, draw conclusions about the contributions of individual proteins in the still hypothetical VirB complex.

VirB1 It is synthesized as a 28 kDa-protein containing an N-terminal signal sequence but no obvious helical trans-membrane segments. The signal peptide is removed by the chromosomally-encoded Sec machinery to produce a 27 kDa protein. According to localization studies, this results in an inner membrane-bound form of the protein. Further processing after amino acid residue 179 of VirB1 results in a secreted soluble form of the protein, which could be purified (Baron *et al.*, 1997a). The N-terminal part resembles a lysozyme-like protein (β-1,4-glycosidase), as judged by structural similarity to chicken egg white lysozyme (Bayer *et al.*, 1995; Mushegian *et al.*, 1996). Putative catalytic amino acids were mutated and resulted in a loss of virulence (Mushegian *et al.*, 1996). A similar effect resulted from an in-frame deletion of the gene (Berger *et al.*, 1994). Thus it can be assumed that these point mutations knock out the activity of the protein. It is almost obligatory to speculate that the N-terminal domain of VirB1 would locally disintegrate some envelope polysaccharides to help in the passage of the T-DNA or otherwise allowing the assembly of the mating pore. The processed C-terminal part probably is likely to function on the cell surface, and may interact with the plant cell. It may also be significant that the cell-bound VirB1 is interacting with the VirB7/B9 complex (Baron *et al.*, 1997a).

VirB2 The 12.3 kDa-protein predicted from the DNA sequence is converted to a 7.2 kDa polypeptide by cellular signal peptidases. Processing occurs both in *Agrobacterium* and in *E.coli* in the absence of other *vir*-encoded proteins. The processed protein associates predominantly with the inner membrane (Jones *et al.*, 1996). Both the processed form and the precursor VirB2 proteins are extremely hydrophobic according to their amino acid composition. In all these respects, VirB2 is analogous to TraA, the major structural component of the F-pilus. Therefore, it apparently is a good candidate for being a pilin (Jones *et al.*, 1996; Shirasu *et al.*, 1993). In fact, when Vir-dependent pili were reported for

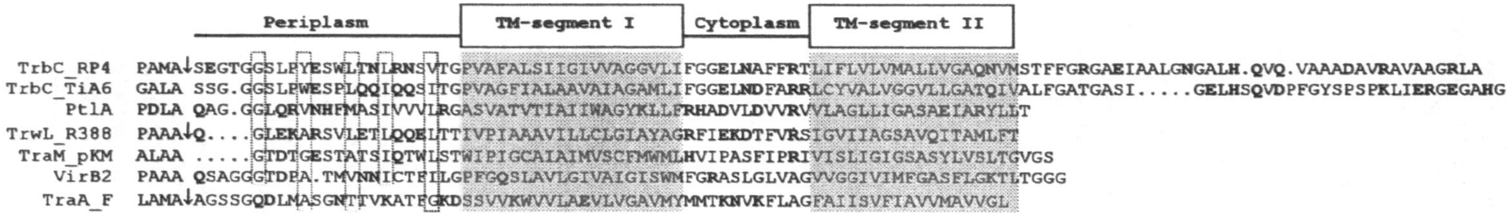

	Periplasm	TM-segment I	Cytoplasm	TM-segment II

```
TrbC_RP4    PANA↓SEGTGGSLRPYESWLITNLRNSVTGPVAFALSIIGIVVAGGVLIFGGELNAFFRTLLFLVLVNALLVGAQNVMSTFFGRGAEIAALGNGALH.QVQ.VAAADAVRAVAAGRLA
TrbC_TiA6   GALA SSG.GGSLPWESPLQQIQQSITGPVAGFIALAAVAIAGAMLIFGGELNDFARRLCYALVGGVLLGATQIVALFGATGASI....GELHSQVDPFGYSPSPKLLERGEGAHG
  PtlA      PDLA QAG.GGIQRVNHFMASIVVLRGASVATVTIAIIWAGYKLLFRHADVLDVVRVLAGLLIGASAEIARYLLT
TrwL_R388   PAAA↓Q....GLEKARSVLETLQQELITTVPIAAAVILLCLGIAYAGRFIEKDTFVRSLGVIIAGSAVQITAMLFT
TraM_pKM    ALAA ....GIDTIGESTATSIQTWLSTWIPIGCALAIMVSCFMWMLHVIPASFIPRIVISLIGIGSASYLVSLTGVGS
  VirB2     PAAA QSAGGGIDEA.TMVNNICTEILGFFGQSLAVIGTVAIGISWMFGRASLGLVAGPVGGTVIMFGASFLGKTLTGGG
  TraA_F    LANA↓AGSSGGQDLNASGNTIVKATEGKDSSVKWVVLAEVLVGAVMYMMTKWVKFLAGFAIISVFIAVVMAVVGL
```

Figure 5. Structural organization of the N-terminally processed forms of proposed conjugative pilins. The proteins were aligned manually according to experimentally determined cleavage sites (vertical arrows), R388 TrwL (Bolland, 1992); F TraA (Paiva *et al*., 1996); RP4 TrbC (Lessl *et al*., 1992; Haase *et al*., 1997). Elements of secondary structure for F-pilin were taken from (Paiva *et al*., 1992). A potential periplasmic amphiphilic α-helix is represented by the residues forming the non-polar side (dotted boxes). The two transmembrane α-helical segments are shown in a shaded background. Highly polar and charged residues are shown in bold. TraA of F, TrwL of R388 and TrbC of RP4 are subject to N-terminal processing by *E. coli* signal peptidase Lep (Haase *et al*., 1997). Additional C-terminal processing of TrbC by an chromosomally-encoded peptidase and TraF results in a mature protein of approx. 7 kDa which matches the size of the other proteins listed. TraF functions as a RP4 specific protease. A TraF analog among the Ti plasmid was found by sequence comparison. It is known as Orf 2 and maps adjacent to the *virF* locus (Farrand *et al*., 1996).

Agrobacterium, VirB2 was considered by the authors as one of the most plausible candidates for the major structural protein (Fullner *et al.*, 1996). The pilin of plasmid F and F-like plasmids plays a central role in conjugation so it has been the subject of considerable study (Firth *et al.*, 1996; Silverman, 1997). If VirB2 is in fact related to F-pilin, and functions in contacting the plant cell, a lot of information may be gained from comparison of both systems. The structure, synthesis and topology of the F-pilus subunit, as well as the assembly of the F-pilus filament was recently reviewed (Firth *et al.*, 1996; Silverman, 1997). Using the structural information gathered by analysis of the F-pilus as a basis (Paiva *et al.*, 1996), we have aligned the pilin sequences as shown in Figure 5. As can be seen, VirB2 follows the putative structural organization of the proposed conjugative pilins. Their amino acid sequences, and secondary structure predictions, are consistent with the mature forms being almost exclusively α-helical, with two potential large hydrophobic α-helixes crossing the bacterial membrane. The N-terminus of the mature protein faces the periplasm, while there is a short and conserved loop in the cytoplasm. It may be significant that the RP4 and Ti plasmid conjugative pilins contain a C-terminal extension, which is not present in the rest of the pilins. This part of the protein is cleaved off in a TraF catalyzed maturation process (Haase *et al.*, 1997).

VirB3 When co-synthesized with the rest of the VirB proteins, VirB3 localized predominantly to the outer membrane. However, its synthesis or stability was found to depend on co-expression of VirB4. VirB3 has no typical signal sequence, and contains one predicted trans-membrane α-helix. Thus in theory it should localize to the inner membrane. Indeed this was found in the absence, but not in the presence of VirB4. Thus, it is thought that VirB4 stabilizes VirB3 and translocates it to the outer membrane (Jones *et al.*, 1994).

VirB4 It is the largest of the VirB proteins (84 kDa.). Although devoid of trans-membrane segments, it consistently localized to the inner membrane indicating that it either interacts with other integral membrane proteins, or it interacts directly with the membrane by other type of structures such as β-sheets. A study using translational fusions and protease sensitivity, allowed a refinement of the topology of VirB4: the data indicated that VirB4 is an integral inner membrane

protein with two short periplasmic exposed segments (Dang *et al.*, 1997). The protein, purified from inclusion bodies, exerted ATPase activity upon refolding (Shirasu *et al.*, 1994). The functional importance of the NTP-binding domain has been verified in several other reports (Berger *et al.*, 1993; Fullner *et al.*, 1994). These authors also showed that mutations in K-439, the invariant lysine in the Walker motif A, which determines the configuration the ATP binding pocket, resulted in a dominant negative phenotype. This result was taken to imply that VirB4 works within a DNA-transport complex (Fullner *et al.*, 1994) although no direct evidence of such a complex is available.

VirB5 According to its amino acid sequence, VirB5 contains a signal sequence and one trans-membrane segment. It localized to the cytoplasmic membrane, but also to the cytoplasm (Thorstenson *et al.*, 1993). However, VirB5 shares sequence similarity to TraC of plasmid pKM101 (Pohlman *et al.*, 1994). Since it has been reported that *traC* mutants can be complemented extracellularly (Winans *et al.*, 1985), it is possible that VirB5 can either assemble from the extracellular side or, alternatively, it can act directly on the recipient cell.

VirB6 Together with VirB2, it is the most hydrophobic of the VirB proteins, showing six potential trans-membrane segments, but no signal sequence. Accordingly, it localized to the cytoplasmic membrane (Table 3).

VirB7 It is the smallest protein (4.5 kDa) in the VirB operon, but possibly the most thoroughly analyzed. It is an outer-membrane-associated lipoprotein exposed at the periplasmic surface (Fernandez *et al.*, 1996b). VirB7 interacts with and stabilizes VirB4, VirB9, VirB10 and VirB11 (Fernandez *et al.*, 1996a). The interaction with VirB9 is particularly obvious, since they form an isolable disulfide-linked protein complex (Anderson *et al.*, 1996; Baron *et al.*, 1997b; Spudich *et al.*, 19976). VirB7 could be labeled with (^{3}H)-palmitic acid in *A. tumefaciens* verifying its lipoprotein nature (Baron *et al.*, 1997b). The fatty acid residues could be responsible for its membrane association.

VirB8 It localizes to the cytoplasmic membrane as predicted by the lack of a signal sequence and the presence of one trans-membrane segment. This was verified according to immunogold electron

Protein	Amino acids[1]	Signal sequence[2]	TM-segments[3]	Enzymatic activity[4]	Cellular localization[5]	References
VirB1	239	type II (exptal)	no	lysozyme?	IM (Thorstenson et al., 1993) PhoA+	
VirB2	121	type I (exptal)	2	pilin?	IM+OM (Shirasu et al., 1993) PhoA+	Fullner et al., 1996b; Jones et al., 1996; Shirasu et al., 1993
VirB3	108	no	1 (long)	-	OM (Shirasu et al., 1993) PhoA-	Jones et al., 1994
VirB4	789	no	no	ATPase stabilization of VirB3	IM (Thorstenson et al., 1993) PhoA-/PhoA+ K-/K+	Dang et al., 1997; Fullner et al., 1994; Shirasu et al., 1994
VirB5	220	type I	no?	-	PP PhoA+ K+	
VirB6	295	no	5-6	-	IM PhoA+	
VirB7	55	type II	no	stabilization of VirB4/9/10/11 complex B7/B9	OM facing PP PhoA+ K+	Anderson et al., 1996; Baron et al., 1997b; Fernandez et al., 1996a; Fernandez et al., 1996b; Spudich et al., 1996
VirB8	237	type I?	1	-	IM+OM (Thorstenson et al., 1993) PhoA- K+	Thorstenson et al., 1994
VirB9	293	type I	no	complex B7/B9 complex B9/B10/B11?	IM+OM (Shirasu et al., 1993, Thorstenson et al., 1993) PhoA- K+	see VirB7
VirB10	377	no	1	complex B9/B10/B11?	IM+OM (Thorstenson et al., 1993) PhoA+ K+	Beaupré et al., 1997; Ward et al., 1990
VirB11	343	no	no	ATPase? complex B9/B10/B11?	IM (Thorstenson et al., 1993) PhoA- K-	Christie et al., 1989; Rashkova et al., 1997; Stephens et al., 1995

Table 3. Experimental and/or computer predicted features of VirB proteins.
[1]length of the unprocessed form.
[2]predicted or experimentally determined cleavage by type I or type II (lipoprotein) signal peptidase.
[3]according to the algorithm of Klein et al. (Klein et al., 1985), for α-helical transmembrane segments.
[4]The VirB7/B9 complex was reported in (Anderson et al., 1996; Baron et al., 1997b; Spudich et al., 1996). The possible VirB9/B10/B11 complex was reported in (Finberg et al., 1995).
[5]Western blotting of isolated cellular fractions carried out by Shirasu and Kado (Shirasu et al., 1993), or Thorstenson et al. (Thorstenson et al., 1993): CP = cytoplasm; IM = inner membrane; PP = periplasm; OM = outer membrane. TnPhoA fusions carried out by Beijersbergen et al. (Beijersbergen et al., 1994): PhoA+ = active fusions were found; PhoA- = no active fusions were found. Proteinase K susceptibility studies with bacterial spheroplasts (complemented also with fractionation studies) were from Fernandez et al. (Fernandez et al., 1996b)

microscopy and the use of Tn*phoA* fusions (Table 3). Previously, cell fractionation localized VirB8 to both cytoplasmic and outer membranes (together with VirB9 and VirB10) (Thorstenson *et al.*, 1994).

VirB9 has a signal sequence but no trans-membrane helices. It interacts strongly with VirB7 as explained above. VirB9, when stabilized by VirB7, is responsible for the formation of high molecular weight complexes, probably oligomers, of another VirB protein, VirB10 (Beaupré *et al.*, 1997).

VirB10 The 48 kDa-protein VirB10 shows one trans-membrane helix but no signal sequence. It localized as an integral IM protein of 48 kDa (instead of the predicted 40.7 kDa) which was found to form high molecular weight aggregates (Ward *et al.*, 1990), in a process which is dependent on VirB9. VirB10 has a theoretical pI of 5.4, which could explain its low mobility in SDS-PAGE.

VirB11 was initially reported to possess ATPase and autophosphorylation activity (Christie *et al.*, 1989), but a later analysis refuted this finding. Thus no ATP binding or hydrolysis activity can be definitively ascribed to the protein (Stephens *et al.*, 1995). The latter authors used the soluble fraction of a N-terminal fusion of VirB11 to the maltose binding protein to determine enzymatic activities *in vitro*. Although the fusion protein could complement a *virB11* mutant *in vivo*, it neither showed ATPase nor autophosphorylation activity when assayed *in vitro*. However, when the invariant Lys in the potential Walker motif A was mutagenized, the mutant proteins were inactive *in vivo*, and genetically recessive to the wild type. Thus, the putative ATP-binding site seemed to be essential for functional activity. This apparent contradiction might be explained by the findings of Rashkova *et al.* (1997). These authors carried out a detailed study showing that about 70% of VirB11 is associated with the cytoplasmic face of the IM, while the remaining 30% remained in the soluble fraction. The same result was obtained irrespective of the presence of other VirB proteins, which is surprising considering the effects of *virB7* mutants on VirB11 stability. The membrane-bound fraction was only partially solubilized with urea. Besides, mutants in the potential Walker motif A partitioned almost exclusively to the IM, indicating that the membrane association domain must belong to a different portion of the protein. These and other results

suggest a complex, perhaps dynamic interaction of VirB11 with the bacterial membrane, with NTP binding/hydrolysis playing a central role. If this were the case, the soluble protein fraction chosen by Stephens *et al* (Stephens *et al.*, 1995) may be devoid of nucleoside triphosphatase binding or hydrolyzing activity, but not the particular fraction previously used by Christie *et al.*, (1989). Both N- and C-terminal halves of the protein could interact with the membrane independently (Rashkova *et al.*, 1997). The C-terminal half showed a dominant negative phenotype in complementation assays. This property was also common to a point mutation located near, but outside the Walker motif A. These results often occur in homo- or hetero-oligomeric proteins. In this respect, it is significant that the VirB11-homolog of plasmid RP4, TrbB, is known to form homohexamers in the presence of ATP and Mg^{2+} ions (Pansegrau *et al.*, 1994; Lanka, unpublished results). Finally, TrwD, the VirB11-homolog of plasmid R388, shows a number of interesting differences and similarities with VirB11: The purified protein shows ATPase activity *in vitro*. Also, a mutant in the ATP-binding motif was trans-dominant, suggesting that ATP binding and oligomerization are related processes. TrwD fractionated to the OM as well as to the soluble fraction, both in the presence and in the absence of other Trw proteins. It could be solubilized from the cell membrane fraction by 0.5 M NaCl, suggesting that it is a peripheral membrane protein (Rivas *et al.*, 1997).

V.C. SIMILARITIES OF VIRB PROTEINS TO PILUS-ERECTING PROTEINS OF CONJUGATIVE TRANSFER SYSTEMS

As shown in Figure 1, and widely reviewed, the VirB operon shares close similarities to conjugative transfer systems specialized in surface mating, and assembling rigid pili. However, no biochemical data is available for any of the constituent proteins of these systems, with the exception of the VirB11-homologues discussed before. Unfortunately, the transfer system which has been more thoroughly examined in its *Mpf* proteins is plasmid F, which is less related to VirB than the surface maters pKM101, RP4 or R388. Only VirB proteins B2, B3, and B4 show very weak similarity to TraA, TraL, and TraC respectively (Shirasu *et al.*, 1993; Shirasu *et al.*, 1994). By virtue of this weak similarity, it would seem that VirB2-B3-B4 form the core (less variable part) of the pilus assembly/erecting structure. The lack of homology between the F pilus assembly proteins and the remaining VirB proteins is intriguing, particularly when compared to the obvious similarity of the DNA processing functions cited in Table 2. In particular, it is very surprising that F does not contain a VirB11 homologue, taking into account that this protein is the most widely spread of the DNA-transport proteins outside the conjugative systems (Hobbs *et al.*, 1993; Rivas *et al.*, 1997). Since VirB11-like proteins are present in type II, III and IV protein transport systems, but not in F, it is possible that the F-pilus assembly mechanism is only very distantly related to the group shown in Figure 1.

VI. The VirE and VirF Operons

The *VirE* operon consists of two genes, *virE1* and *virE2*. In octopine Ti plasmids there is a single *virF* gene, which is not present in nopaline Ti plasmids, and results in a broader host range of plant infection by octopine plasmids. The *virE* and *virF* operons are important to this chapter since *Agrobacterium* strains defective in either *virE2* or *virF* can be complemented for tumorigenesis by co-infection with a helper strain containing a complete set of *vir* genes, but containing no T-region (reviewed in (Hooykaas *et al.*, 1994)). Furthermore, both VirE2 and VirF are transported to the plant cells, and can even be provided by transgenic plants, indicating that they act at the plant side and are not obligatorily implicated in T-DNA transport through the bacterial membranes (Citovsky *et al.*, 1992; Regenburg-Tuïnk *et al.*, 1993). Available genetic evidence is taken to imply that transport of VirF and of VirE2 from an *Agrobacterium* cell to a plant cell can occur in the absence of T-DNA transfer and depends entirely on the VirB transport system. However, transport of the proteins has not yet been shown directly. VirE2 (60.5 kDa), a protein that binds to ssDNA tightly and cooperatively, but without sequence specificity, can coat the total length of the T-strand (Christie *et al.*, 1988). The 20 kb T-strand produced by the nopaline plasmid would bind approximately 600 molecules of VirE2 yielding a complex of about 50×10^3 kDa. The C-terminal half of VirE2 contains the DNA binding domain, the N-terminal half apparently is involved in the cooperativity of binding to DNA (Chapter 15; Dombek *et al.*, 1997). VirE2 contains two bipartite nuclear localization sequences, both of which are required for nuclear import of VirE2 in tobacco protoplasts (Citovsky *et al.*, 1992). The coupling of ssDNA binding and nuclear localization activities suggested that VirE2 alone could mediate nuclear localization of ssDNA (Zupan *et al.*, 1996).

IncP-plasmids encode a VirE2 analogous protein, TraC1, a primase and DNA binding protein, which was shown to be transmitted from the donor to the recipient cell during conjugation (Rees *et al.*, 1990). In addition to its DNA binding property, TraC1 exerts primase activity to initiate conversion of the ssDNA to dsDNA. A similar activity has not been found with VirE2 in vitro (Lanka, unpublished). The DNA binding property of TraC1, however, may play a role in conjugation that is comparable to VirE2, i.e. coating the DNA during transport. "Packaging" of DNA may provide protection against nuclease degradation and/or facilitate the transport of DNA as a nucleoprotein complex. The intriguing question that remains to be solved is whether TraC1 transport to the recipient is dependent on the *virB* homologues, the *trb* genes.

VII. Analogies to other Protein Transport Systems: T-Complex Transfer and Conjugation Use a Type IV Protein Export Pathway

DNA-independent VirE2 and VirF transport by the VirB system provided a very welcome link to the initially puzzling finding of a protein export system that was closely related to the *Agrobacterium tumefaciens* Vir system: the Ptl system of *Bordetella pertussis* (Figure 1). However, an obvious structural relationship between the pertussis toxin, a protein complex of five different subunits, and the two Vir proteins has not been found. Furthermore, neither protein bears resemblance to any of the pilin proteins. These results suggest the transport apparatus of the T-complex is capable of recognizing a wide variety of target proteins. The existence of the Ptl system promoted speculation about the chemical nature of the molecules, which are actually recognized, and transported, by the Ti-virulence/conjugative export mechanism. Regensburg-Tuïnk and Hooykaas (Regenburg-Tuïnk *et al.*, 1993) were the first to propose that T-DNA transfer could be considered in fact as nucleoprotein transport and thus is related to other protein export mechanisms. The opposite premise proposes that a DNA transport system was recruited for the export of pathogenic macromolecules (Winans *et al.*, 1996). Citovski and Zambryski (1993) reviewed the variety of nucleic acid transport systems in nature, and could draw a set of general rules involved in such transfer. They realized that, in most cases, transport systems use a single-stranded nucleic acid that is invariably associated with specific proteins. Attention is drawn to a potential *Helicobacter pylori* secretion system for the export of virulence determinants (Table 1). A pathogenicity island was found in the *Helicobacter pylori* chromosome, which shows obvious homologues to virB4, B9, B10 and B11, as well as to VirD4 (Censini *et al.*, 1996; Covacci *et al.*, 1997; Tomb *et al.*, 1997). The homology to VirD4 has provoked exciting new theoretical considerations. In our protein export pathway scheme, TraG-like proteins were considered to be the unique hallmark of type IV secretion systems, since they were found in all conjugative plasmids, including F (Lessl *et al.*, 1992a; see Nuñez *et al.*, 1996 for the most recent occurrence in plasmid R6K). Conversely, export systems unrelated to DNA transfer never showed a TraG-homologue. This was consistent to the proposed role of these proteins as an interface between the relaxosome and the transport apparatus (section IV.B). Now the finding of ORF10 in *cagII* of *Helicobacter pylori* defies our proposal. It is very exciting to learn that pathogenicity islands are probably mobile, being transferred from one bacteria to another by as yet unknown mechanisms, apparently only active within infected organisms (Mel *et al.*, 1996). If we are correct, *cagII* could be a remnant of a conjugative-like DNA transport system, since it contains a TraG-homologue. We will not be surprised if a TraG-homologue is found in turn in *Bordetella pertussis*. And following this line of reasoning, searching for TraG-homologues could unveil a plethora of bacterial DNA transfer mechanisms, capable of transferring genetic material not only to other bacteria, but also to plant cells and maybe other eukaryotic cells.

The homology of VirB11 to the corresponding proteins of type II export systems, suggested that the use of VirB-like proteins in pure protein export mechanisms could be the rule, more than the exception. It has been found recently that type III secretion systems deliver bacterial proteins into eukaryotic cells (Lee, 1997). Besides, type III secretion systems can be called "contact-dependent" secretion systems, since contact to the eukaryotic cell is required to activate protein export (Galán, 1996). Furthermore, the secretion activity is accompanied by the extracellular accumulation of filaments (Parsot *et al.*, 1995), which is characteristic of and conceptually analogous to the sex pilus in conjugative mechanisms. All these phenomena direct us to the inescapable conclusion that the general protein secretion pathway (type II secretion (Pugsley, 1993)) has suffered a stepwise specialization to become either a protein injection system (type III secretion), or a DNA injection system, that is, a conjugative system (type IV secretion). First, the exported substrate could be a filamentous protein, so that extracellular appendages could be formed facilitating, for instance, adhesion to target cells (type II-like pilin producing systems). Second, the exported substrate, such as a toxin, could be injected into the target cell by the extracellular appendages stabbing into them, or otherwise bringing the recipient membrane into intimate contact by retraction (type III). Finally, one of these systems "evolved" to use a nucleoprotein-complex as an export substrate, so that the nucleic acid could be transferred to another cell. The conjugative pathway, that was called type IV protein export pathway (Salmond, 1994), seems to be unique and reasonably different from other protein transport processes (Salmond *et al.*, 1993). A rather similar proposal has recently been made (Christie, 1997) following the original idea of Regensburg-Tuïnk and Hooykaas (1993).

The proposed type IV protein export pathway will not only include Ti-plasmid T-DNA transmission and DNA transfer by bacterial conjugation, but could also include other transport systems like M13 phage extrusion and import of DNA by transformation (competence) in Gram-positive bacteria. We would like to promote the continued consideration of the common aspects of these processes in nucleoprotein and protein transport, which eventually might lead to a general scheme for describing DNA transport between cells.

VIII. Acknowledgements

We are indebted to Ellen Zechner and Matxalen Llosa for critical reading and correction of the manuscript. This work was supported by a grant (PB95-1269) from the D.G.I.C.Y.T. to F.C. and Sonderforschungsbereich grant B344/A8 of the Deutsche Forschungsgemeischaft to E.L.

IX. References

Alt-Mörbe, J., Stryker, J.L., Fuqua, C., Li, P.L., Farrand, S.K. and Winans, S.C. (1996) J. Bacteriol. 178, 4248-4257.

Anderson, L.B., Hertzel, A.V. and Das, A. (1996) Proc. Natl. Acad. Sci. USA 93, 8889-8894.

Avila, P., Núñez, B. and de la Cruz, F. (1996) J. Mol. Biol. 261, 135-143.

Balzer, D., Pansegrau, W. and Lanka, E. (1994) J. Bacteriol. 176, 4285-4295.

Baron, C. and Zambryski, P.C. (1996) Curr. Biol. 6, 1567-1569.

Baron, C., Llosa, M., Zhou, S. and Zambryski, P.C. (1997a) J. Bacteriol. 179, 1203-1210.

Baron, C., Thorstenson,, Y.R. and Zambryski, P.C. (1997b) J. Bacteriol. 179, 1211-1218.

Bayer, M., Eferl, R., Zellnig, G., Teferle, K., Dijstra, A., Koraimann, G. and Högenauer, G. (1995) J. Bacteriol. 177, 4279-4288.

Beaupré, C.E., Bohne, J., Dole, E.M. and Binns, A.N. (1997) J. Bacteriol. 179, 78-89.

Beijersbergen, A., Dulk-Ras, A.D., Schilperoort, R.A. and Hooykaas, P.J.J. (1992) Science 256, 1324-1327.

Beijersbergen, A., Smith, S.J. and Hooykaas, P.J.J. (1994) Plasmid 32, 212-218.

Berger, B.R. and Christie, P.J. (1993) J. Bacteriol. 175, 1723-1724.

Berger, B.R. and Christie, P.J. (1994) J. Bacteriol. 176, 3646-3660.

Binns, A.N., Beaupre, C.E. and Dale, E.M. (1995) J. Bacteriol. 177, 4890-4899.

Bolland, S. (1992) Genes involved in plasmid R388 conjugative pilus production. Ph.D. thesis, University of Cantabria, Spain.

Buchanan-Wollaston, V., Passiatore, J.E. and Cannon, F. (1987) Nature 328, 172-175.

Bundock, P., den Dulk-Ras, A., Beijersbergen, A. and Hooykaas, P.J.J. (1995) EMBO J. 14, 3206-3214.

Bundock, P. and Hooykaas, P.J.J. (1996) Proc. Natl. Acad. Sci. USA 93, 15272-15275.

Cabezón, E., Lanka, E. and de la Cruz, F. (1994) J. Bacteriol. 176, 4455-4458.

Cabezón, E., Sastre, J.I. and de la Cruz, F. (1997) Mol. Gen. Genet. 254, 400-406.

Censini, S., Lange, C. Xiang, Z., Crabtree, J.E., Ghiara, P., Brodovsky, M., Rappuoli, R. and Covacci, A. (1996) Proc. Natl. Acad. Sci. USA 93, 14648-14653.

Christie, P.J. (1997) J. Bacteriol. 179, 3085-3094.

Christie, P.J., Ward, J.E., Winans, S.C. and Nester, E.W. (1988) J. Bacteriol. 170, 2659-2667.

Christie, P.J., Ward, J.E., Gordon, M.P. and Nester, E.W. (1989) Proc. Natl. Acad. Sci. USA 86, 9677-9681.

Citovsky, V. and Zambryski, P. (1993) Annu. Rev. Microbiol. 47, 167-197.

Citovsky, V., Zupan, J., Warnick, D. and Zambryski, P. (1992) Science 256, 1802-1805.

Chimo, M.W., Sharma, V.K. and Archer, G.L. (1996) J. Bacteriol. 178, 4975-4983.

Covacci, A., Falkow, S., Berg, D.E. and Rappuoli, R. (1997) Trends Microbiol, 5, 205-208.

Dang, T.A.T. and Christie, P.J. (1997) J. Bacteriol. 179, 453- 462.

Dombek, P. and Ream, W. (1997) J. Bacteriol. 179, 1165-1173.

Dreiseikelmann, B. (1994) Microbiol. Revs. 58,293-316.

Dürrenberger, M.B., Villiger, W. and Bächi, T. (1991) J. Struct. Biol. 107, 146-156.

Erickson, M.J. and Meyer, R.J. (1993) Mol. Microbiol. 7, 289-298.

Farrand, S.K., Hwang, I. and Cook, D.M. (1996) J. Bacteriol. 178, 4223-4247.

Fernandez, D., Spudich, G.M., Zhou, X. and Christie, P.J. (1996a) J. Bacteriol. 178, 3168-3176.

Fernandez, D., Dang, T.A.T., Spudich, G.M., Zhou, X., Berger, B.R. and Christie, P.J. (1996b) J. Bacteriol. 178, 3156-3167.

Finberg, K.E., Muth, T.R., Young, S.P., Maken, J.B., Heitritter, S.M., Binns, A.N. and Banta, L.M. (1995) J. Bacteriol. 177, 4881-4889.

Firth, N., Ippen-Ihler, K. and Skurray, R.A. (1996) in Neidhardt FC "Escherichia coli and Salmonella" (2nd edition). ASM Press, Washington, DC.

Freiberg, C., Fellay, R., Bairoch, A., Broughton, W.J., Rosenthal, A. and Perret, X. (1997) Nature 387, 384-401.

Frost, L.S. and Bazett-Jones, D.P. (1991) J. Bacteriol. 173,7728-7731.

Fullner, K.J., Stephens, K.M. and Nester, E.W. (1994) Mol. Gen. Genet. 245, 704-715.

Fullner, K.J. and Nester, E.W. (1996a) J. Bacteriol. 178,1498-1504.

Fullner, K.J., Lara, J.C. and Nester, E.W. (1996b) Science 273, 1107-1109.

Fürste, J.P., Pansegrau, W., Ziegelin, G., Kröger, M. and Lanka, E. (1989) Proc. Acad. Sci. USA 86, 1771-1775.

Galán, J.E. (1996) Mol. Microbiol. 20, 263-271.

van der Graaff, E., den Dulk-Ras, A. and Hooykaas, P.J.J. (1996) Plant Molec. Biol. 31, 677-681.

Guzman, L.M. and Espinosa, M. (1997) J. Mol. Biol. 266, 688-702.

Haase, J., Lurz, R., Grahn, M., Bamford, D.H. and Lanka, E. (1995) J. Bacteriol. 177,4779-4791.

Haase, J. and Lanka, E. (1997) J. Bacteriol. 179,, 5728-5735.

van Haaren, M.J.J., Sedee, N.J.A., Schilperoort, R.A. and Hooykaas, P.J.J. (1987) Plant Mol. Biol. 8, 95-104.

Harrington, L.C. and Rogerson, A.C. (1990) J. Bacteriol. 172. 7263-7264.

Heinemann, J.A. and Sprague, Jr. G.F. (1989) Nature 340, 205-209.

Hobbs, M. and Mattick, J.S. (1993) Mol. Microbiol. 10, 233-243.

Hooykaas, P.J.J. and Beijersbergen, A. (1994) Annu. Rev. Phytopathol. 32, 157-179.

Ilyina, T.V. and Koonin, E.V. (1992) Nucleic Acids Res, 13, 3279-3285.

Jones, A.L., Shirasu, K. and Kado, C.I. (1994) J. Bacteriol. 176, 5255-5261.

Jones, A.L., Lai, E., Shirasu, K. and Kado, C.I. (1996) J. Bacteriol. 178, 5706-5711.

Joos, H., Inze, D., Caplan, A., Sormann, M., van Montagu, M. and Schell, J. (1983) Cell 32, 1057-1067.

Kado, C.I. (1994) Mol. Microbiol. 12, 17-22.

Kim, S-R. and Komano, T. (1997) J. Bacteriol. 179, 3594-3603.

Kingsman, A. and Willetts, N. (1978) J. Mol. Biol. 122, 287-300.

Klein, P., Kanehisa, M. and DeLisi, C. (1985) Biochim. Biophys. Acta 815, 468-476.

Kornberg, A. and Baker, T. (1992) DNA replication. WH Freeman & Co., NY (2nd edition).

Koukolíková-Nicola, Z., Raineri, D., Stephens, K., Ramos, C., Tinland, B., Nester, E.W. and Hohn, B. (1993) J. Bacteriol. 175, 723-731.

Lanka, E. and Wilkins, B.M. (1995) Annu. Rev. Biochem. 64, 141-169.

Lee, C.A. (1997) Trends Microbiol. 5, 148-156.

Lessl, M., Pansegrau, W. and Lanka, E. (1992a) Nucleic Acids Res. 20, 6099-6100.

Lessl, M., Balzer, D., Pansegrau, W. and Lanka, E. (1992b) J. Biol. Chem. 267, 20471-20480.

Lessl, M., Balzer, D., Weyrauch, K. and Lanka, E. (1993) J. Bacteriol. 175, 6415-6425.

Lessl, M. and Lanka, E. (1994) Cell 77, 321-324.

Llosa, M., Grandoso, G. and de la Cruz, F. (1995) J. Mol. Biol. 246, 54-62.

Llosa, M., Grandoso, G., Hernando, M.A. and de la Cruz, F. (1996) J. Mol. Biol. 264, 56-67.

Mel, S.F. and Mekalanos, J.J. (1996) Cell 87, 795-798.

Miranda, A., Janssen, G., Hodges, L., Peralta, E.G. and Ream, W. (1992) J. Bacteriol. 174, 2288-2297.

Moncalián, G., Grandoso, G., Llosa, M. and de la Cruz, F. (1997) J. Mol. Biol. 270, 188-200.

Murphy, C.G. and Malamy, M.H. (1995) J. Bacteriol. 177, 3158-3165.

Mushegian, A.R., Fullner, K.J., Koonin, E.V. and Nester, E.W. (1996) Proc. Natl. Acad. Sci. USA 93, 7321-7326.

Nuñez, B., Avila, P. and de la Cruz, F. (1997) Mol. Microbiol. 24, 1157-1168.

Okamoto, S., Toyoda-Yamamoto, A., Ito, K., Takebe, I. and Machida, Y. (1991) Mol. Gen. Genet. 228, 24-32.

Paiva, W.D., Grossman, T. and Silverman, P.M. (1992) J. Biol. Chem. 267, 26191-26197.

Paiva, W.D. and Silverman, P.M. (1996) Mol. Microbiol. 19, 1277-1286.

Panicker, M.M. and Minkley, Jr. E.G. (1992) J. Biol. Chem. 267, 12761-12766.

Pansegrau, W., Balzer, D., Kruft, V., Lurz, R. and Lanka, E. (1990) Proc. Natl. Acad. Sci. USA 87, 6555-6559.

Pansegrau, W. and Lanka, E. (1991) Nucleic Acids Res. 19, 3455.

Pansegrau, W., Schröder, W. and Lanka, E. (1993a) Proc. Natl. Acad. Sci. USA 90, 2925-2929.

Pansegrau, W., Schoumacher, F., Hohn, B. and Lanka, E. (1993b) Proc. Natl. Acad. Sci. USA 90, 11538-11542.

Pansegrau, W., Schröder, W. and Lanka, E. (1994) J. Biol. Chem. 269, 2782-2789.

Pansegrau, W. and Lanka, E. (1996) Prog. Nucleic Acid Res. and Mol. Biol. 54, 197-251.

Parsot, C., Ménard, R., Gounon, P. and Sansonetti, P.J. (1995) Mol. Microbiol. 16, 291-300.

Peralta, E.G. and Ream, L.W. (1985) In Szalay, A.A. and Legocki, R.P. (ed.), Advances in the molecular genetics of the bacteria-plant interaction. Cornell University Publishers, Ithaca, N.Y.

Piers, K.L., Heath, J.D., Liang, X., Stephens, K.M. and Nester, E.W. (1996) Proc. Natl. Acad. Sci. USA 93, 1613-1618.

Pohlman, R.F., Genetti, H.D. and Winans, S.C. (1994) Mol. Microbiol. 14, 655-668.

Pugsley, A.P. (1993) Microbiol. Rev. 57, 50-108.

Rashkova, S., Spudich, G.M. and Christie, P.J. (1997) J. Bacteriol. 179, 583-591.

Rees, C.E.D. and Wilkins, B.M. (1990) Mol. Microbiol. 4, 1199-1205.

Regensburg-Tuïnk, A.J.G. and Hooykaas, P.J.J. (1993) Nature 363, 69-71.

Rivas, S., Bolland, S., Cabezón, E., Goñi, F.M. and de la Cruz, F. (1997) J. Biol. Chem. 272, 25583-25590.

Salmond, G.P. and Reeves, P.J. (1993) Trends Biochem. Sci. 18, 7-12.

Salmond, G.P. (1994) Annu. Rev. Phytopathol. 32, 181-200.

Scheiffele, P., Pansegrau, W. and Lanka, E. (1995) J. Biol. Chem. 270, 1269-1276.

Scherzinger, E., Lurz, R., Otto, S. and Dobrinski, B. (1992) Nucleic Acids Res. 20, 41-48.

Sherman, J.A. and Matson, S.W., (1994) J. Biol. Chem. 269, 26220-26226.

Shirasu, K. and Kado, C.I. (1993) FEMS Microbiol. Lett. 111, 287-294.

Shirasu, K., Koukolikova-Nicola, Z., Hohn, B. and Kado, C.I. (1994) Mol. Microbiol. 11, 581-588.

Silverman, P.M. (1997) Mol. Microbiol. 23, 423-429.

Spudich, G.M., Fernandez, D., Zhou, X. and Christie, P.J. (1996) Proc. Natl. Acad. Sci. USA 93, 7512-7517.

Stachel, S.E., Timmerman, B. and Zambryski, P. (1986) Nature 322, 706-712.

Stachel, S.E., Timmerman, B. and Zambryski, P. (1987) EMBO J. 6, 857-863.

Stephens, K.M., Roush, C. and Nester, E.W. (1995) J. Bacteriol. 177, 27-36.

Thorstenson, Y.R., Kuldau, G.A. and Zambryski, P.C. (1993) J. Bacteriol. 175, 5233-5241.

Thorstenson, Y.R. and Zambryski, P.C. (1994) J. Bacteriol. 176, 1711-1717.

Tinland, B., Schoumacher, F., Gloeckler, V., Bravo-Angel, A.M. and Hohn, B. (1995) EMBO J. 14, 3585-3595.

Tomb, J.F., White, O., Kerlavage, A.R., Clayton, R.A., Sutton, G.G., Fleischmann, R.D., Ketchum, K.A., Klenk, H.P., Gill, S., Dougherty, B.A., Nelson, K., Quackenbush, J., Zhou, L., Kirkness, E.F., Peterson, S., Loftus, B., Richardson, D., Dodson, R., Khalak, H.G., Glodek, A., Mckenney, K., Fitzgerald, L.M., Lee, N., Adams, M.D., Hickey, E.D., Berg, D.E., Gocayne, J.D., Utterback, T.R., Peterson, J.D., Kelly, J.M., Cotton, M.D., Weidman, J.M., Tujii, C., Bowman, C., Watthey, L., Wallin, E., Haynes, W.S., Borodovsky, M., Karp, P.D., Smith, H.O., Fraser, C.M. and Venter, J.C. (1997) Nature 388, 539-547.

Veluthambi, K., Ream, W. and Gelvin, S. (1988) J. Bacteriol. 170, 1523-1532.

Ward, J.E., Akiyoshi, D.E., Regier, D., Datta, A., Gordon, M.P. and Nester, E.W. (1988) J. Biol. Chem. 263, 5804-5814.

Ward, J.E., Dale, E.M., Nester, E.W. and Binns, A.N. (1990) J. Bacteriol. 172, 5200-5210.

Waters, V.L., Strack, B., Pansegrau, W., Lanka, E. and Guiney, D.G. (1992) J. Bacteriol. 174, 6666-6673.

Weiss, A.A., Johnson, F.D. and Burns, D.L. (1993) Proc. Natl. Acad. Sci. USA 90, 2970-2974.

Willetts, N. and Wilkins, B.M. (1984) Microbial. Rev. 48, 24-41.

Winans, S.C. and Walker, G.C. (1985) J. Bacteriol. 161, 402-410.

Winans, S.C., Burns, D.L. and Christie, P.J. (1996) Trends Microbiol. 4, 64-68.

Zambryski, P.C. (1992) Annu. Rev. Plant Physiol. Plant Mol. Biol. 43, 465-490.

Ziegelin, G., Pansegrau, W., Strack, B., Balzer, D., Kröger, M., Kruft, V. and Lanka, E. (1991) DNA Seq. 1, 303-327.

Zupan, J.R., Citovsky, V. and Zambryski, P.C. (1996) Proc. Natl. Acad. Sci. USA 93, 2392-2397.

Chapter 15

Role of Virulence Proteins of *Agrobacterium* in the Plant

Luca Rossi, Bruno Tinland and Barbara Hohn

I. Introduction

A. tumefaciens is a plant pathogen which is able to transfer a defined piece of its own DNA, the transferred DNA or T-DNA to the plant cell of many dicotyledonous plants. The T-DNA is then integrated into the plant genome of the host plants where the genes that the T-DNA is carrying are expressed (see Chapter 12). This transferred DNA is delimited by two 25 bp long direct repeats called the right and the left border and is situated on a large plasmid called Ti-plasmid. As soon as the bacterium finds a host for transformation it induces the expression of a series of genes called virulence genes located on the Ti-plasmid adjacent to the T-DNA. The "bottom" strand of the two border sequences is cleaved by the site-specific endonuclease, coded for by the VirD1 and VirD2 proteins. This ss-DNA molecule has then to be detached from the Ti-plasmid to be transferred out of the bacteria to the plant cell. This step of the processing reaction may not be independent of T-DNA export from the bacterium. The VirD2 protein remains covalently attached to the 5' end of this ss-T-DNA for the rest of its extrachromosomal life in the plant. Due to its position on the T-DNA, the VirD2 protein was suspected to contain signals which are piloting the T-DNA through the various steps of the transfer. It was suggested that VirD2 directs the complex out of the *A. tumefaciens* cell by recognising the pore through which the complex passes, that it guides the complex to the plant cell cytoplasm, that it targets the complex to the nucleus of the plant cell, that it protects the 5' end of the T-DNA from nucleases and that it even directs integration of the T-DNA into the chromosomal DNA. Another virulence protein which is believed to accompany the T-DNA on its way to the plant cell nucleus is the ss-DNA binding VirE2 protein. This protein is believed to coat the ss-T-DNA molecule and contribute to its efficient transfer to the plant cell nucleus.

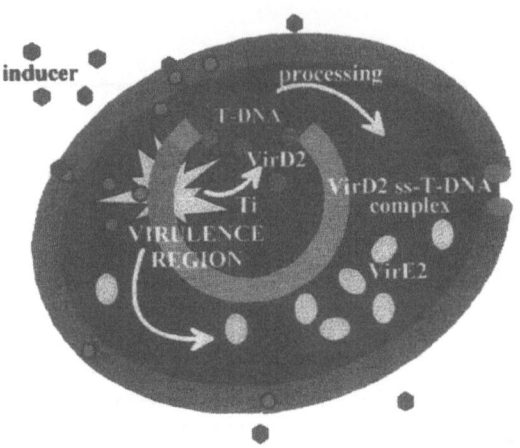

Figure 1a. T-DNA processing

A. tumefaciens is transferring only few molecules of T-DNA to every single plant cell it attacks. Nevertheless it can efficiently transform it. In this chapter we will try to understand why *A. tumefaciens* is so efficient in doing this. The main reason for its efficiency is most probably related to the fact that the T-DNA does not travel to the plant cell as a naked DNA molecule but instead as a protein-DNA complex.

II. T-DNA Processing

Wounded plant cells begin to produce lignin in the process of wound healing. The lignin precursors are phenolic compounds such as acetosyringone, that are chemoattractive for *A. tumefaciens* and induce the expression of the *vir* genes, which then mediate T-DNA transfer (Figure 1a). The constitutively expressed *virA* gene encodes a sensor protein which acts directly or indirectly as a receptor for acetosyringone (for details see chapter 12), for low external pH (between pH 5.0 and 5.8), and for several sugars, in particular monomers of major cell wall polysaccharides. These elements all contribute to the induction of the autophosphorylation of the VirA protein. The phosphate group of the activated VirA protein is then transferred to the constitutively expressed VirG protein. The phosphorylated VirG protein acts as a transcriptional activator of the *vir* operons probably by binding to the vir-box which is present in the 5'-non-coding region of all *vir* operons. The products of the *vir* genes liberate the

T-DNA from the Ti-plasmid (i.e. processing) and transport it to the plant cell nucleus.

The first two genes of the *virD* operon, *virD1* and *virD2*, are responsible for the processing reaction. The genes of the *virC* operon, and in particular VirC2, are believed to help during this reaction. VirD2 catalyses site-specific cleavage of single-stranded DNA containing the border sequence, *in vitro* (Pansegrau *et al.*, 1993) but requires VirD1 for the site-specific cleavage of the double-stranded border sequence (Scheiffele *et al.* 1995). Both *in vivo* and *in vitro*, the VirD2 protein generates single-stranded nicks between the third and fourth base of the border repeat in the bottom strand of the T-DNA and thereby attaches tightly to the 5' end of the nicked DNA (Ward and Barnes, 1988; Young and Nester, 1988; Dürrenberger *et al.*, 1989) probably via the tyrosine 29 residue (Vogel and Das, 1992; Pansegrau *et al.*, 1993): upon displacement, most probably resynthesis of the bottom strand, the ss-T-DNA-VirD2 complex is formed.

The VirE2 protein is the most abundant protein produced in *A. tumefaciens* after induction of the virulence genes by plant phenolic compounds (Stachel, 1986a). The VirE2 protein binds unspecifically and in an highly cooperative manner to single-stranded DNA *in vitro* (Gietl, 1987; Sen, 1989; Christie, 1988; Citovsky, 1988; Das, 1988; Citovsky, 1989). However, it is not binding to double-stranded DNA or to RNA (Das, 1988). An *in vitro* assembled ss-DNA-VirE2 complex has been estimated to have a thickness of 2 nm and a length of 0.18 nm per nucleotide, as visualised by electronmicroscopy upon shadowing with platinum particles (Citovsky *et al.*, 1989). This represents an increase in length of the ss-DNA of about 55%. A 20 kb T-strand would accommodate more than 600 VirE2 molecules and have a molecular weight of 50.10^6 daltons. Interestingly visualisation of the ss-DNA complex by scanning electron microscopy (STEM) revealed that VirE2 packages ss-DNA into a rigid and hollow 12,8 nm wide cylindrical filaments with a telephone cord like structure (Lartey and Citovsky, 1997). Stretching of this coil structure results in a 4.3 nm wide filament which is comparable to the 2,0 nm rod described above (Lartey and Citovsky, 1997).

Due to its single-stranded DNA binding properties VirE2 was believed to coat the T-DNA-VirD2 complex in *A. tumefaciens* and form thereby the T-complex. This hypothesis is now questioned for two reasons: first, VirE2 overexpressed in transgenic plants can complement an *A. tumefaciens* strain mutated in *virE2* to full virulence (Citovsky *et al.*, 1992); second, a mutation or complete deletion of the *virE2* gene in *A. tumefaciens* does not affect the processing reaction (Stachel, 1986b, Rossi *et al.*,

1996). VirE2 is thus not required for the formation of the transferred complex, it is required for efficient transfer of the T-DNA (Stachel, 1986b; Christie, 1988; Gardner, 1986; Grimsley, 1989; Rossi, 1996) and it is necessary only in the plant cell (Citovsky *et al.*, 1992). These findings do not exclude a cooperative binding of VirE2 to the T-strand prior to transfer, but the necessity for it is not documented. Direct evidence for T-strand-VirE2 interaction in the bacterium comes from experiments employing immunoprecipitation of T-strands by VirE2 antibodies (Christie *et al.*, 1988). However it cannot be excluded that this complex assembles during the isolation procedure. Similarly, the AcvB protein, an octopine VirJ homologue present in strain A208, seems to be associated with the T-strand in the bacteria (Wirawan *et al.*, 1996). An *acvB* Tn5 insertion was shown to be avirulent (Wirawan *et al.*, 1996, Yoshioka *et al.*, 1996) suggesting that this protein, possibly in association with the T-strand, is involved in some essential step of the transport of the T-DNA.

The VirF protein, another virulence protein specific

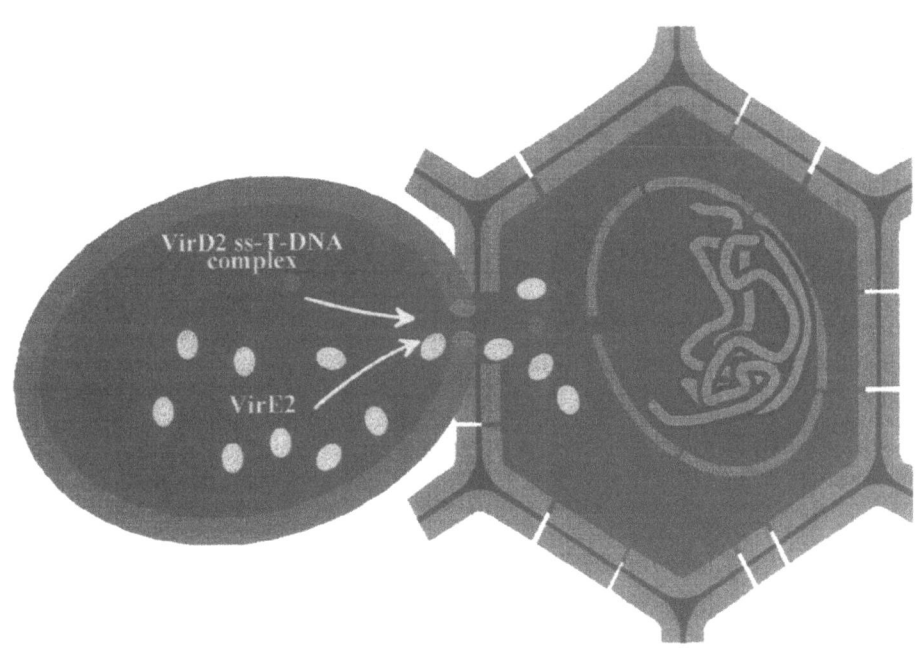

Figure 1b. T-DNA transfer to the plant cell

for the octopine strain, was shown to be important inside the plant cell for T-DNA transfer to *N. glauca*. Nopaline strains or octopine strains mutated in *virF* have a weakened virulence on *tomato* (Hooykaas *et al.*, 1984) and *N. glauca* (Melchers *et al.*, 1990). Full virulence can be restored by culturing these strains with *N. glauca* plants expressing the VirF protein. VirF is thus most probably transferred to the plant cell during cocultivation of octopine type *A. tumefaciens* strains with *N. glauca* and contributes to the T-DNA transfer (Regensburg-Tuïnk and Hooykaas, 1993).

III. The ss-T-DNA-VirD2-VirE2 Complex (T-Complex).

III.A. TRANSPORT OF THE SS-T-DNA-VIRD2 COMPLEX AND VIRE2 TO THE PLANT CELL (FIGURE 1B)

Plant transformation mediated by *A. tumefaciens* requires the attachment of the bacteria to a wounded site of the plant. This step occurs most probably during *vir* gene induction and is mediated by a series of chromosomal genes (described in Chapter 11). Attachment is a requirement for T-DNA transfer to plant cells, although even bacteria artificially introduced into plant cells can transfer their DNA, thereby bypassing attachment (Escudero *et al.*, 1995). Upon attachment to the plant cell or introduction of bacteria in the plant cell *A. tumefaciens* exports the ss-T-DNA-VirD2 complex and VirE2 to the plant cell. The form of T-DNA found in *A. tumefaciens* as a result of the processing reaction and found after entry into the nucleus of the plant cell is the single-stranded form (Tinland *et al.*, 1994; Yusibov *et al.*, 1994). We therefore believe that this form is travelling through the different barriers that it needs to pass to reach the plant cell nucleus.

The first barrier is represented by the *A. tumefaciens* membrane. The export of the ss-T-DNA-VirD2 complex from the bacterium is probably mainly mediated by the *virD4* and *virB* gene products. 9 of the 11 ORFs contained in the *virB* operon encode proteins that contain hydrophobic membrane spanning regions. The majority of VirB proteins therefore may associate with bacterial cell membranes. The VirB4 (Berger and Christie, 1993), VirB8 (Dale *et al.*, 1993), VirB9 and VirB10 (Ward

et al. 1990) and VirB11(Christie *et al.* 1989) proteins as well as the membrane spanning VirD4 protein (Lin and Kado, 1993) have been shown to be essential for tumorigenicity. These proteins are thus probably forming the structure necessary for transfer of the ss-T-DNA-VirD2 complex to the plant cell. The VirB4 and VirB11 proteins possess ATP binding sites and ATPase activity, thereby likely providing the energy necessary for transport of the complex across the bacterial membrane (Christie, 1997).

Since the main function(s) of VirE2 reside in the plant cell, VirE2 has to be exported out of *A. tumefaciens*. The export of VirE2 also requires the function of the *virB* gene products. VirE2 is thus believed to follow the same route which may be used by the ss-T-DNA-VirD2 complex to be exported from *A. tumefaciens*. An interesting experiment was performed for the first time in 1984 by Otten *et al.*, They showed that VirE2 protein expressed in a T-DNA free *A. tumefaciens* strain can complement to full efficiency the T-DNA transfer from a strain mutated in *virE2* (Otten *et al.*, 1984). VirE2 is thus exported in absence of the T-DNA and the T-DNA in absence of VirE2. This process is called extracellular complementation. Further experiments showed that the strain donating the VirE2 protein, called helper strain, as well as the T-DNA donor strain, has to contain functional VirB proteins and a functional VirD4 protein as well as attachment functions (Christie *et al.*, 1988). In contrast to the T-DNA donor strain the helper strain does not need to carry the VirD2 protein (Fullner *et al.*, 1994). We recently asked the question whether an *A. tumefaciens* strain carrying non-functional VirB proteins is deficient in plant transformation solely because it cannot export the VirE2 protein. We thereby tested if T-DNA transfer from a VirB mutated strain could be complemented by the VirE2 protein produced by an helper strain devoid of a T-DNA. We found no complementation and concluded that export of the T-DNA *per se* requires the *virB* gene products. This indicates that VirE2 and the T-DNA-VirD2 complex independently require the *virB* gene products to reach the plant cell (Rossi *et al.*, unpublished).

Recently it was found that export of VirE2, but not of the T-DNA-VirD2 complex, is dependent on the function of the VirE1 protein (Sundberg *et al.*, 1996). A strain deleted in *virE1* accumulates wild type levels of VirE2 protein that can not be exported to the plant cell and its plant transformation efficiency is thereby drastically reduced.

Nevertheless uncoated T-DNA molecules can efficiently be exported from this strain indicating that VirE2 does not interfere with the export of the T-DNA to the plant cell. Since VirE2 is absolutely required for virulence one has to postulate that VirE2 activity is exerted only in the plant cell. This hypothesis is strengthened by two experimental approaches: VirE2 transgenic plants can complement an *A. tumefaciens* strain mutated in *virE2* to full virulence (Citovsky *et al.*, 1992) (see above); and recent results from export competition experiments in which RSF1010 plasmid derivatives were introduced into *A. tumefaciens*. This plasmid belongs to the broad host range IncQ incompatibility group and it can be transferred by conjugation between bacteria. The transfer reaction requires the MobA protein, which cleaves the origin of transfer of the plasmid and then attaches covalently to the DNA. *A. tumefaciens* is able to mediate the transfer of RSF1010 to plant cells although this plasmid lacks the border sequences necessary for VirD2 mediated T-DNA transfer. Several Vir proteins encoded in the Ti-plasmid are required for this transfer (Buchanan-Wollaston *et al.*, 1987). A multicopy RSF1010 plasmid derivative has been shown to compete for T-DNA export to plant cells, most probably by sequestering some limiting factors for the transfer. Both transfer intermediates, the T-DNA-VirD2 and the RSF1010-MobA complex, need the VirE2 protein for efficient transfer to the plant cell nucleus. Overexpression of VirE2 in the bacteria containing both transfer intermediates did not restore the capacity to transfer the T-DNA to the plant cell nucleus and has little effect on RSF1010 transfer. In contrast RSF1010 mediated transformation was increased by extracellular complementation using a VirE2 containing strain (Binns *et al.*, 1995). These results suggest that the transferred molecules are uncoated DNA-VirD2 and DNA-MobA complexes and that the VirE2 protein associates with the ss-DNA only in the plant cell.

III.B. ASSOCIATION OF VIRE2 PROTEIN WITH THE SS-T-DNA-VIRD2 COMPLEX IN THE PLANT CELL

The properties of the VirE2 proteins, as found *in vitro*, suggest that VirE2 covers the ss-T-DNA-VirD2 complex at some step during the transfer process. Support for this hypothesis was provided only recently. An *A. tumefaciens* strain deleted in the *virE2* gene was constructed, plants were transformed using this strain (with very low efficiency) and the pattern of integration of these inserts was compared to the one obtained studying the inserts of plant transformed with the wild type strain. This analysis revealed that mainly severely truncated forms of the T-DNA are found integrated into the plant genome after T-DNA transfer mediated by the VirE2 minus strain. In contrast only very short deletions are found in plants transformed by the wild type strain (see below). In bacteria the number and length of T-stands formed after induction in the wild type strain and in the VirE2 minus strain is similar. VirE2 is thus required in the plant cell to protect the ss-T-DNA molecules from being degraded by plant nucleases (Rossi *et al.*, 1996). The most straightforward explanation for this function is that VirE2 is physically protecting the ss-T-DNA against plant nucleases by coating the entire length of the DNA molecules. These results are consistent with the finding that *in vitro* assembled ss-DNA-VirE2 complexes are resistant to digestion by nucleases (Sen *et al.*, 1988, Citovsky *et al.*, 1989). These results are also consistent with the finding that a reduced amount of T-DNA molecules can be found in the cytoplasm of protoplasts cocultivated with a *virE2* mutant as compared to the wild type (Yusibov *et al.*, 1994). Finally, these results are consistent with the requirement of extra copies of VirE2 for transfer of binary bacterial artificial chromosomes to plants (Hamilton *et al.*, 1996).

As described earlier in this chapter VirE2 and the T-DNA-VirD2 complex can travel independently to the plant cell thereby requiring mostly the same, but partially different elements (i.e. the VirE1 protein). Although it cannot be ruled out that in the wild type situation the T-complex, complete with VirE2 proteins, travels to the plant cell, it is unlikely, in our view, that different mechanisms are used to transfer the ss-T-DNA-VirD2 complex and the complete T-complex. This then leads to the conclusion that VirE2 protein molecules in the bacteria do not "see" the ss-T-DNA-VirD2 complex because otherwise they immediately would "grab" it. This in turn most likely means that the ss-T-DNA-VirD2 complex immediately upon production is secreted into plants. It would be very interesting to find out whether this complex associates with VirE2 at the plant cell plasma membrane, in the plant cell cytoplasm, on the outside periphery of the plant cell nucleus, or only in the plant cell nucleus. One may speculate that VirE2 meets the ss-DNA-VirD2 complex at the entry point of the plant cell thereby facilitating entry of the

DNA. Alternatively the partners may associate only inside the plant cell nucleus.

III.C. COMPARISON OF T-DNA TRANSFER TO BACTERIAL CONJUGATION

Conjugation is the process used by bacteria to exchange genetic material between them. The mechanism of T-DNA transfer has strong similarities to the bacterial conjugative transfer system encoded by the broad host range plasmid RP4 (IncP; Lessl and Lanka, 1994) (Winans et al., 1996). Both transfer systems have an origin of transfer, the *oriT* for conjugational transfer and the right border for T-DNA transfer, and a site-specific endonuclease which is responsible for initiation of DNA transmission. The origin of transfer sequences are very similar and the enzymes processing them contain highly conserved motifs in the N-terminal part of the protein. In both cases the DNA is transferred as ss-DNA protein complex in which the respective site-specific endonuclease (TraI for RP4 and VirD2 for Ti-plasmid) attaches to the 5' end of the ss-DNA molecule. Furthermore the proteins thought to establish the channel used for the transfer of protein-DNA complex the recipient cell are similar in sequence and gene organisation (Tra2 and TraG of RP4 and VirB and VirD4 of Ti-plasmid, respectively). However the structure of the transfer channels used for conjugation of RP4 or T-DNA

transfer has not been identified. In analogy with the F conjugation system pilus like structures may be expected. The presence/absence of pili in various mutants of *Agrobacterium tumefaciens* was found recently to correlate with the capacity to effect T-DNA transfer between different *A. tumefaciens* cells. This suggests that pili are required for T-DNA transfer to plants (Fullner et al., 1996) (for details see Chapter 14).

In the two systems the organisation of the recipient cell is considerably different. In bacterial conjugation the recipient is again a bacterial cell whereas for T-DNA transfer it is a plant cell. In bacterial conjugation the transfer origin is also the transfer terminus and the entry point of the plasmid in the recipient cell is already the target compartment. There the plasmid recircularises to be maintained extrachromosomally or, if it contains sequence homology to the recipient chromosome, it integrates. In addition it retains the ability to move to other bacteria. In contrast, start and stop of T-DNA transfer are located at two border sequences (although one border on a small plasmid is sufficient) and the entry point into the plant cell is not yet the target compartment of the plant cell. The T-DNA has to be targeted to the nucleus and since it cannot be maintained extracellularly it has to integrate into the plant genome for survival. Furthermore the T-DNA does not encode the genes required for its movement and is therefore irreversibly integrated into the genome. Finally in

Property	T-DNA transfer	Bacterial conjugation
I Similarities		
origin of transfer	+ (borders)	+ (oriT)
transfer protein, nicking, covalent attachment	+	+
activity of transfer protein *in vitro*	+	+
single stranded intermediate	+	+(where analysed)
transfer apparatus	+	+
II Differences		
establishment of transferred DNA in the recipient	integration by illegitimate recombination	site specific integration.
nuclear uptake	+	-
gene activity of transferred DNA inside donor organism	-	+
recipient becomes a donor after transfer	-	+

Table 1. Similarities and differences between T-DNA transfer and bacterial conjugation.

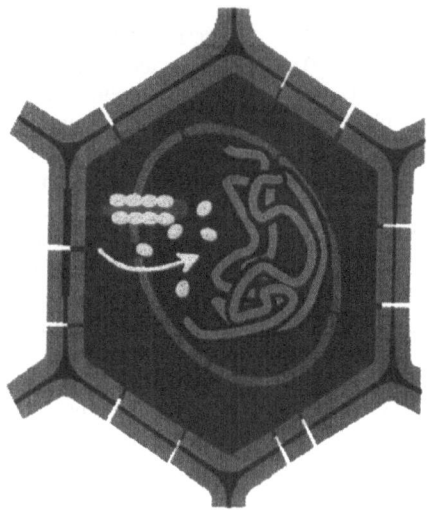

Figure 1c. Nuclear import of the T-complex

bacterial conjugation the transferred genes are expressed in the donor and in the recipient cell whereas in T-DNA transfer the genes encoded in the T-DNA have eukaryotic regulatory elements and are expressed only in the recipient cell (Table 1).

IV. Nuclear Import of the T-Complex

IV.A. TRANSPORT OF THE T-COMPLEX TO THE PLANT CELL NUCLEUS (FIGURE 1C)

After entry in the plant cell by whatever route the T-complex moves to the nucleus. Nuclear import of proteins requires the presence of specific signal sequences called nuclear localisation signals (NLS) (for review see Garcia Bustos *et al.*, 1991). To be defined as an NLS, a sequence must be necessary to direct a nuclear protein into the nucleus and sufficient to redirect a cytoplasmic protein into the nucleus. Although there is no strict NLS consensus most signals are rich in basic amino acids. This import process is mediated by a well-controlled active transport mechanism and can be generally separated in two distinct steps. The first step is an NLS dependent binding of the nuclear import substrate to the cytoplasmic side of the nuclear pore complex (NPC). This step is mediated by a family of 50-60-kDa soluble NLS receptors which act as carriers for transport into the nucleus (for review see Adam, 1995). The interaction of the NLS receptor-NLS containing protein complex to the nuclear pore complex seems to be mediated by the p97 protein. The second step is an energy dependent translocation through the nuclear pore (for review see Görlich and Mattaj, 1996). This step involves the small GTPase Ran and a small protein NTF2. According to one model Ran hydrolyses GTP at the periphery of the NPC where the NLS containing protein is bound. This allows progression of the transport towards the central channel where interaction with the NPC is mediated by NTF2.

Since the T-DNA arriving in the plant cell is associated with the virulence proteins VirD2 and most probably VirE2, these proteins were considered prime candidates for providing permission for nuclear entry to the T-DNA. Indeed the VirD2 and the VirE2 proteins could be shown to contain short peptides resembling NLSs. These sequences, when fused to a reporter protein that is normally localised in the cytoplasm, rendered it nuclear (Herrera-Estrella *et al.*, 1990; Tinland *et al.*, 1992; Howard *et al.*; 1992; Citovsky *et al.*, 1992). The significance of these sequences for ss-T-DNA transfer was further analysed for VirD2. It was shown that targeting to the plant cell nucleus was dependent on the C-terminal NLS of VirD2, whereby the quantitative aspect of the contribution of VirD2 to targeting was dependent on the plant analysed and the assay used (Shurvington *et al.*, 1992; Rossi *et al.*, 1993, Narasimhulu *et al.*, 1996, Yoshikoka *et al.*, 1996). These results suggest that the C-terminal NLS of VirD2 is piloting the ss-T-DNA to the plant cell nucleus. On the other hand the contribution of the NLSs of VirE2 protein to T-DNA transfer could not be directly tested because mutations in the NLS domain were found to affect the ss-DNA binding properties of VirE2 (Citovsky *et al.*, 1992). The functions carried by VirE2 were found to be independent of the functions carried by the NLSs of VirD2 (Rossi *et al.*, 1996). Thus VirE2 does not carry the same function as the NLS of VirD2. This result indicates that VirE2 is not piloting the complex to the nucleus in an NLS dependent manner but instead the NLS of VirD2 is carrying this function. This does not exclude that VirE2 can contribute to the targeting the T-complex to the nucleus by facilitating the second step of the nuclear import process, namely the translocation of the complex through the nuclear pore. We consider it unlikely that VirE2 contributes to this second step in an NLS dependent manner (whereby this function would have to be different from the function carried by the NLS of VirD2), since the second step of protein import into the nucleus is in general NLS independent. In addition, the NLS domains of VirE2 overlap with the DNA binding domain (Citovsky *et al.*, 1992) and are probably masked in the interior of the VirE2 ss-DNA complex. Instead we believe that VirE2 protein could contribute to the second step of the nuclear import of the T-complex by elongating the complex in order to facilitate its translocation (Citovsky *et al.*, 1989). Consistent with this hypothesis is the finding that fluorescently labelled ss-DNA complexed with VirE2 accumulates in the plant cell nucleus when microinjected into the cytoplasm of stamen hair cells of the flowering plant *Tradescantia virginiana* (Zupan *et al.*, 1996). The fluorescently labelled DNA microinjected alone

remained cytoplasmic whereas deletion of the N-terminal NLS sequence of VirE2 which reduces nuclear localisation of the protein and retains partial ss-DNA binding activity, retards nuclear import of the complex considerably. This mutated protein binds ss-DNA, but not cooperatively and is found both in the nucleus and in the cytoplasm. The structure of the ss-DNA-mutant VirE2 complex may thereby be altered which would hinder the passive import of the complex into the nucleus. Alternatively, as the authors of this work speculate, the reduced targeting strength of the NLS of this VirE2 mutant may decrease the efficiency of the active import of the ss-DNA-VirE2 complex into the nucleus. The study of a *virE2* mutant deleted in the NLS sequences but retaining the ss-DNA binding activity as the wild type protein would help to differentiate between these two hypotheses.

The nuclear transport process plays an essential role in eukaryotes and the constituents of the import apparatus are probably of common evolutionary origin in different kingdoms (Lake and Rivera, 1994). The structure of the nuclear pore, the energy requirement for import and the receptors for the imported proteins seem to be conserved in plants, vertebrates and yeast. Differences are also found: the carbohydrate modifications of the nuclear pore complex proteins are different in yeast and plants. Furthermore some import inhibitors such as Wheat germ agglutinin or N-ethylmaleimide inhibit nuclear import in vertebrates but not in plants (for review see Hicks and Raikhel, 1995). The bipartite NLS of VirD2 has been shown to function in plants and yeast (Tinland *et al.*, 1992). On the other hand purified and fluorescently labelled VirE2 protein, when injected into the cytoplasm of *Drosophila* embryos and *Xenopus* oocytes was found to be localised in the cytoplasm of these cells (Gurlanick *et al.*, 1996). This may not be so surprising since also in plants the NLSs of VirE2 are functioning suboptimally (Citovsky *et al.*, 1994). Modification of

KRp**R**drhdgelgg**RKR**a**R**	VirD2 C-terminal
Kl**R**ped**R**yiqte**K**yg**R**	VirE2
Kt**R**ygsdtei**K**l**K**s**K**	VirE2
KRtygsdtei**K**l**K**s**K**	modified VirE2 (Gurlanick *et al.*, 1996)
KRpaatkkagqa**KKKK**	nucleoplasmin NLS

Table 2. NLSs of VirD2 and VirE2

the NLS of VirE2 to a motif which fits the bipartite motif consensus sequence perfectly rendered the protein nuclear in both *Xenopus* and *Drosophila* (Table 2). It will be important to analyse this mutant also in plants, as protein and in the context of T-DNA transfer.

IV.B. COMPARISON OF THE NUCLEAR IMPORT OF THE T-COMPLEX TO THE IMPORT OF VIRUSES

Transport of nucleic acids to the nucleus has been analysed for *A. tumefaciens* T-DNA transfer and for viral DNA or RNA transfer. One example of nuclear import of mammalian viral DNA is represented by the transport of the viral HIV-1 preintegration complex to the nucleus (for review see Goldfarb, 1995). After infection of cells by HIV-1, nascent viral DNA in the form of a high molecular weight nucleoprotein preintegration complex is transported to the nucleus and integrated into the host genome. This complex is composed of viral nucleic acid, the *gag* integrase and matrix proteins, and of the *pol* reverse transcriptase protein (Bukrinsky *et al.*, 1993a). The HIV-1 *gag* matrix proteins have been shown to contain NLSs which contribute to the karyophilic properties of the viral preintegration complex (Bukrinsky *et al.*, 1993b). The integrase protein contained in this complex was described to be necessary for the integration of retroviral DNA into the host DNA (Farnet *et al.*, 1991). The preintegration complex of HIV-1 thus contains proteins which contribute to nuclear entry and integration of the viral DNA.

The only recognised group of plant viruses that possesses ss-DNA genomes is represented by the geminiviruses. This virus replicates inside the nucleus of the plant cell and has thereby to travel in and out of the nucleus at some points of its life cycle. One of the proteins encoded by the virus, the BR1 protein, has been shown to localise to the nucleus of the plant cell and to bind single-stranded DNA (Pascal *et al.*, 1994). About 30 minutes after microinjection of fluorescently labelled DNA into spongy mesophyll cells of tobacco, the DNA concentrated in the vicinity of the nucleus. Subsequent microinjection of the BR1 protein resulted in the release of the DNA from the nucleus (Noueiry *et al.*, 1994). On the basis of these experiments the authors of this work proposed that while entry of viral DNA into the nucleus does not

require viral proteins, export of the DNA out of the nucleus requires the functions of the BR1 protein.

Another plant virus that encodes a protein which localises into the nucleus is the tobacco etch potyvirus. The replication of the RNA virus takes place in the cytoplasm. Interestingly the NLS containing N1a protein encoded by this virus localises primarily to the nucleus but a fraction of the protein is attached covalently to the 5' terminus of the genomic cytoplasmatically localised RNA. N1a has been postulated to be required for initiation of RNA synthesis or for processing of RNA intermediates, but its role in the nucleus has not yet been elucidated. It is interesting to note that the NLSs of N1a overlap or are very close to the tyrosine residue that mediates covalent attachment of the N1a molecule to the 5' end of the RNA. This interaction is most probably responsible for the suppression of the nuclear transport of the complex (Carrington *et al.*, 1991). This situation is reminiscent of the one of the N-terminal NLS of VirD2 which was shown to be functional in targeting a reporter protein to the nucleus (Tinland *et al.*, 1992) but was shown to be dispensable for targeting of the ss-DNA-VirD2 complex to the nucleus (Rossi *et al.*, 1993). Also in this case the NLS sequence is very near the tyrosine that mediates covalent attachment of VirD2 to the ss-T-DNA.

While the tobacco etch potyviral RNA remains cytoplasmatic the viral RNA of influenza virus is transported to the nucleus (O'Neil *et al.*, 1995). The nucleoprotein (NP) of Influenza Virus is coating its viral genomic RNA. The naked RNA localises in the cytoplasm whereas in presence of NP the RNA is able to dock to and enter the nucleus. These results suggest that uptake of influenza RNA segments is mediated by the NLS of NP.

In the examples described here it is interesting to note that generally proteins are responsible for the movement in or out of the nucleus of viral DNA or RNA. Furthermore it is clear that the localisation of the free proteins does not always reflect the localisation of the protein when complexed with their respective nucleic acid molecule. In 79% of nuclear DNA or RNA binding proteins with proven NLS sequence, the NLS domains are found to overlap or to be immediately adjacent to the DNA binding domains of the proteins (LaCasse *et al.*, 1995). This is consistent with a role of many of these proteins only when inside the nucleus of the cell, for instance as transcription factors, but not with a role of targeting of nucleic acids into the nucleus. For this purpose these two domains have to be separated since they have to function at the same time.

It is believed that nuclear export and import of proteins could have a similar general mechanism although different transport signals are involved (Gerace, 1995). The export of a specific premessenger RNP particle through the NPC to the cytoplasm has been directly visualised by Mehlin *et al.*, 1995. During translocation, the particle becomes extended and passes through the nuclear pore with the 5' end of the transcript in the lead (Mehlin et al, 1992). It is most probably the 5' end with its cap structure that is important for the recognition of the particle at the NPC and which is thereby positioning

V. Integration of the T-DNA

V.A. T-DNA INTEGRATION IS MEDIATED MAINLY BY PLANT ENZYMES

Once in the plant nucleus, the T-DNA becomes integrated in the plant genome. This integration follows a mode of illegitimate recombination and occurs randomly in the plant cell genome (Chyi *et al.*, 1986; Wallroth *et al.*, 1986; Gheysen *et al.*, 1991; Mayerhofer *et al.*, 1991; for review see Tinland and Hohn, 1995; Tinland 1996).

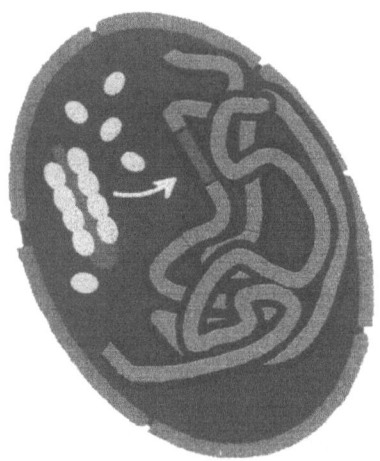

Figure 1d. Integration of T-DNA

the particle before translocation. By analogy to these examples the virulence proteins composing the T-complex could mediate nuclear uptake of the T-DNA. The mechanism of T-complex import into the plant nucleus with the VirD2 bound 5' terminus leading the passage may resemble that of RNP transport described above although passage through the nuclear membrane occurs in the opposite orientation.

Furthermore the T-DNA has been described to integrate preferentially in transcriptionally active sequences (Koncz *et al.*, 1989; Herman *et al.*, 1990). Apart from T-DNA, no other elements are known to invade the plant cell and to integrate into the nuclear genome. Autonomous plant transposons move within a cell and for this movement they need an enzyme which is encoded in the transposed DNA. Furthermore, excision and reintegration of these

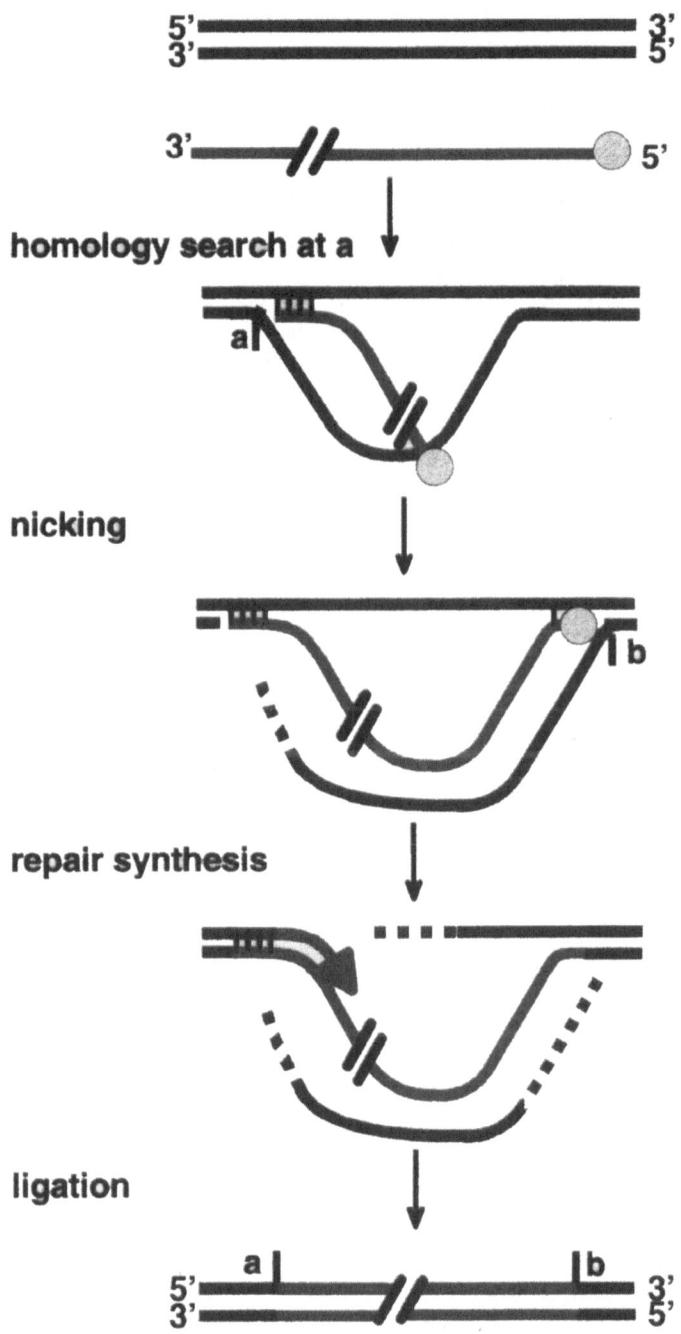

Figure 2. Integration of T-DNA (single strand model)

elements require special sequences at the termini of the DNA molecules. In contrast, the T-DNA does not encode functions necessary for its movement and does not carry special sequences at the end of the molecule which allow its integration into the plant genome (the border sequences are only required inside the bacterium). Moreover, once integrated into the genome it cannot excise anymore. Some analogy to T-DNA integration can be found in animal systems. As mentioned above, retroviruses such as HIV-1 integrate their viral DNA into the genome with the help of viral integrase. Retroviral DNA integration resembles transposable elements in that both integrate site-specifically on the side of the viral but not the host DNA.

As described above the T-DNA travels to the plant cell as a ss-T-DNA-VirD2 complex and is most probably reaching the plant cell nucleus as a complex coated by VirE2 proteins. An important question therefore is whether these two virulence proteins carry an integrase function, similar to retroviruses, which would mediate integration of the T-DNA into the plant genome. To address this question a mutant of the VirD2 protein was constructed which resulted in defective integration of the T-DNA into the plant genome (Tinland et al., 1995). This mutation, although inefficient in T-DNA transfer, did not affect the efficiency with which the T-DNA was integrating. These results suggest that VirD2 is not an integrase. Absence of the VirE2 protein did not affect the efficiency of T-DNA integration either (Rossi et al., 1996). This function is most probably taken over by plant DNA damage repair enzymes. In agreement with this hypothesis are recent results which showed that Arabidopsis mutants hypersensitive to UV or γ radiations have a reduced capacity to integrate T-DNA (Sonti et al., 1996). The process of T-DNA integration is thus fundamentally different from retrovirus integration.

Using the sequences of the flanking regions at the insertion sites of T-DNA, sequences of the plant genome prior to T-DNA integration were isolated. Analysis of such preinsertion sequences revealed that the process of T-DNA integration leads to deletions of about 10 to 70 bp of the plant genome. In this aspect the illegitimate mode of recombination of the T-DNA with the plant DNA resembles the excision repair mechanism described for yeast (Prakash et al., 1993) and humans (Hoeijmakers et al., 1994). In excision repair the 3' and 5' incisions are thought to be coordinated and such excisions result in deletion of short DNA fragments. A similar mechanism may

be used for T-DNA integration (Tinland and Hohn 1995; Tinland 1996).

V.B. THE VIRD2 AND VIRE2 PROTEINS ARE REQUIRED FOR PRECISE INTEGRATION OF COMPLETE T-DNA UNITS.

The T-DNA is processed at the RB and LB in A. tumefaciens by the VirD2 protein. This cleavage creates a ss-DNA fragment of defined length and sequence. By cloning and sequencing the region of the genome where the T-DNA integrated it is possible to compare the integrated sequence with the original one. Analysis of sequences of many transformants led to the definition of a typical pattern of T-DNA integration. Independent of the plant used as a recipient for T-DNA (Arabidopsis or tobacco) it was generally found that the 5' end of the T-DNA up to the nucleotide to which the VirD2 protein is covalently linked is frequently integrating into the plant genome. In contrast, at the left T-DNA/plant DNA junction usually more nucleotides are lost from the T-DNA (up to 50) than on the right junction. This suggests that VirD2 might have roles in protecting the integrity of the T-DNA at the 5' end and in integration of the 5' end of the T-DNA. Indeed, the VirD2 protein was shown recently to be involved in the ligation of the 5' end of the T-DNA into the plant cell DNA (Tinland et al., 1995). As mentioned before the inactivation of this function does not affect the efficiency of integration of the T-DNA indicating that ligation of the right T-DNA terminus to plant T-DNA is not dependent on VirD2. However, a T-DNA terminus associated with a VirD2 protein in its wild type configuration directs ligation mediated by plant enzymes to the 5' most nucleotide. Since this does not seem to be the efficiency limiting step of integration it was suggested that the 3' end of the ss-T-DNA realises the first synapsis with the plant DNA by finding some homology to the plant DNA and then the very 5' end of the T-DNA, protected by VirD2, ligates to a 3'OH of nicked plant DNA (Tinland et al., 1995, Tinland 1996) (Figure 2).

As mentioned above in absence of VirE2 protein only severely truncated forms of the T-DNA are found integrated into the plant genome whereby the efficiency of integration is not dependent on the presence of VirE2. The functions of VirE2 therefore are protection of the ss-T-DNA and possibly nuclear

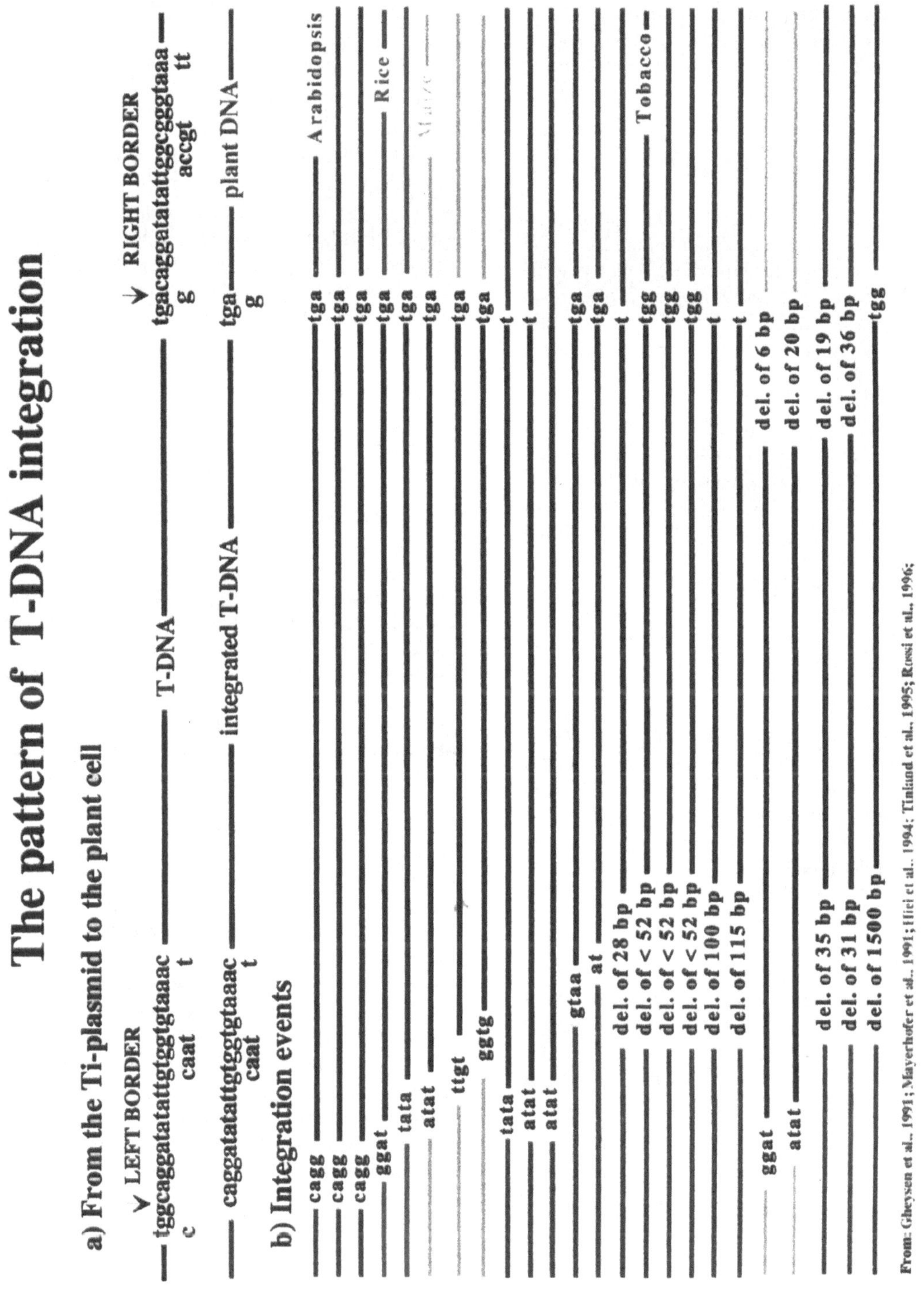

The pattern of T-DNA integration

a) From the Ti-plasmid to the plant cell

LEFT BORDER

RIGHT BORDER

tggcaggatatattggtgtaaac   T-DNA   tgacaggatatattggcggggtaaa
c   caat   t   g   accgt   tt

caggatatattggtgtaaac   integrated T-DNA   tga   plant DNA
caat   t   g

b) Integration events

cagg		tga	Arabidopsis
cagg		tga	
cagg		tga	
ggat		tga	Rice
tata		tga	
atat		tga	Maize
ttgt		tga	
ggtg		tga	
tata		t	
atat		t	
atat		t	
gtaa		tga	
at		tga	
del. of 28 bp		t	
del. of < 52 bp		tgg	Tobacco
del. of < 52 bp		tgg	
del. of < 52 bp		tgg	
del. of 100 bp		t	
del. of 115 bp			
ggat	del. of 6 bp		
atat	del. of 20 bp		
del. of 35 bp	del. of 19 bp		
del. of 31 bp	del. of 36 bp		
del. of 1500 bp		tgg	

From: Gheysen et al. 1991; Mayerhofer et al. 1991; Hiei et al. 1994; Tinland et al. 1995; Rossi et al. 1996;

Figure 3. The Pattern of T-DNA integration

entry (see above). The efficiency of plant transformation by *A. tumefaciens* therefore relies on the coordinated action of VirE2 and VirD2 in protecting the entire length of the T-DNA and in efficiently targeting of the T-DNA to the nucleus of the plant cell.

VI. Horizontal Gene Transfer

Horizontal gene transfer refers to the mobilisation and stabilisation of genetic information from one organism to another. This type of exchange is well-studied in prokaryotes but examples for genomic flux can also be found between prokaryotes and eukaryotes. Examples of such genetic transfer are found between *E. coli* and the yeast *Saccharomyces cerevisiae* (Heinemann and Sprague, 1989; Nishikawa *et al.*, 1992), *A. tumefaciens* and plants and *A. tumefaciens* and yeast (Bundock *et al.*, 1995, Piers *et al.*, 1996). Interestingly the mechanism of transfer of conjugative plasmids from *E.coli* to yeast is very similar to the one involved in bacterial conjugation. This transfer is not restricted to a specific bacterial transfer system since two plasmids belonging to different incompatibility groups could be mobilised to yeast cells (Heinemann and Sprague, 1989, Nishikawa *et al.*, 1992).

In contrast to the genetic exchange between *E.coli* and yeast, *A. tumefaciens* evolved a specific mechanism for the transfer of a defined piece of its DNA to the plant cells. This mechanism resembles in some aspects to bacterial conjugation system and has probably evolved from such a system (see section III.C.). Consistent with this hypothesis is the finding that the products of the virulence region of *A. tumefaciens* normally required for T-DNA transfer can direct the conjugative transfer of an IncQ plasmid between agrobacteria (Beijersbergen *et al.*, 1992). However one difference between DNA transfer to agrobacteria and to plants is the role of the VirE2 protein. VirE2 is not required for the transfer of the IncQ plasmid between bacteria indicating either that another ss-DNA binding is taking over this function in the donor or recipient cell, or that VirE2 carries functions with are specifically needed only in the plant cell.

T-DNA transfer to *S. cerevisiae* was found to be dependent on the virulence genes encoded by the Ti-plasmid (Bundock *et al.*, 1995; Piers *et al.*, 1996). The requirements for this process of the virulence proteins are very similar to those to plants.

Only the functions necessary for attachment of *A. tumefaciens* to the plant cell do not seem to be required for yeast transformation (Piers *et al.*, 1996). T-DNA transfer to yeast also depends on the VirB proteins and the VirD4 protein, confirming the idea that these proteins are most probably forming the transfer apparatus between *A. tumefaciens* and the recipient cell. Interestingly the C-terminal part of VirD2 carrying the NLS sequence is also essential for yeast transformation (Bundock *et al.*, 1995). This is consistent with the findings mentioned above that the NLSs of VirD2 are active also in yeast cells (Tinland *et al.*, 1992). In contrast a transposon insertion in the VirE2 coding sequence reduced the efficiency of T-DNA mediated yeast transformation only by a factor of 10. This indicates a diminished role of VirE2 in yeast T-DNA transfer as compared to transfer to plants, possibly pointing to differences in the recipient cells with respect to nucleases and/or nuclear import requirements.

Integration into the yeast genome was found to occur via homologous recombination, the predominant mode of integration of any foreign DNA into yeast DNA. Only when the T-DNA lacks homology with yeast genome, it integrates at random positions via illegitimate recombination (Bundock and Hooykaas, 1996) In contrast, in plants T-DNA integrates by illegitimate recombination and only at very low frequencies by homologous recombination (referenced and discussed in Puchta and Hohn, 1996). This confirms the hypothesis described earlier (see section V.A.) that plant enzymes and not *A. tumefaciens* encoded proteins are responsible for the integration of the T-DNA into the host genome.

The natural hosts of *A. tumefaciens* are a wide range of dicotyledonous plants. Nevertheless T-DNA transfer to monocots could be detected using agroinfection (Grimsley *et al.*, 1987). Recently also transgenic rice (Hiei *et al.*, 1994) and maize plants (Ishida *et al.*, 1996) were obtained using *A. tumefaciens*. Interestingly the pattern of T-DNA integration found in these plants is very similar to the one found in dicotyledonous plants (Figure 3) indicating that the same group of plant enzymes is responsible for the integration of the T-DNA in monocotyledonous and dicotyledonous plants.

In summary the transfer machinery of *A. tumefaciens* (i.e. VirB and VirD4) can be used to deliver DNA between *A. tumefaciens* cells and to deliver DNA to yeast cells and plant cells. (Figure 4).

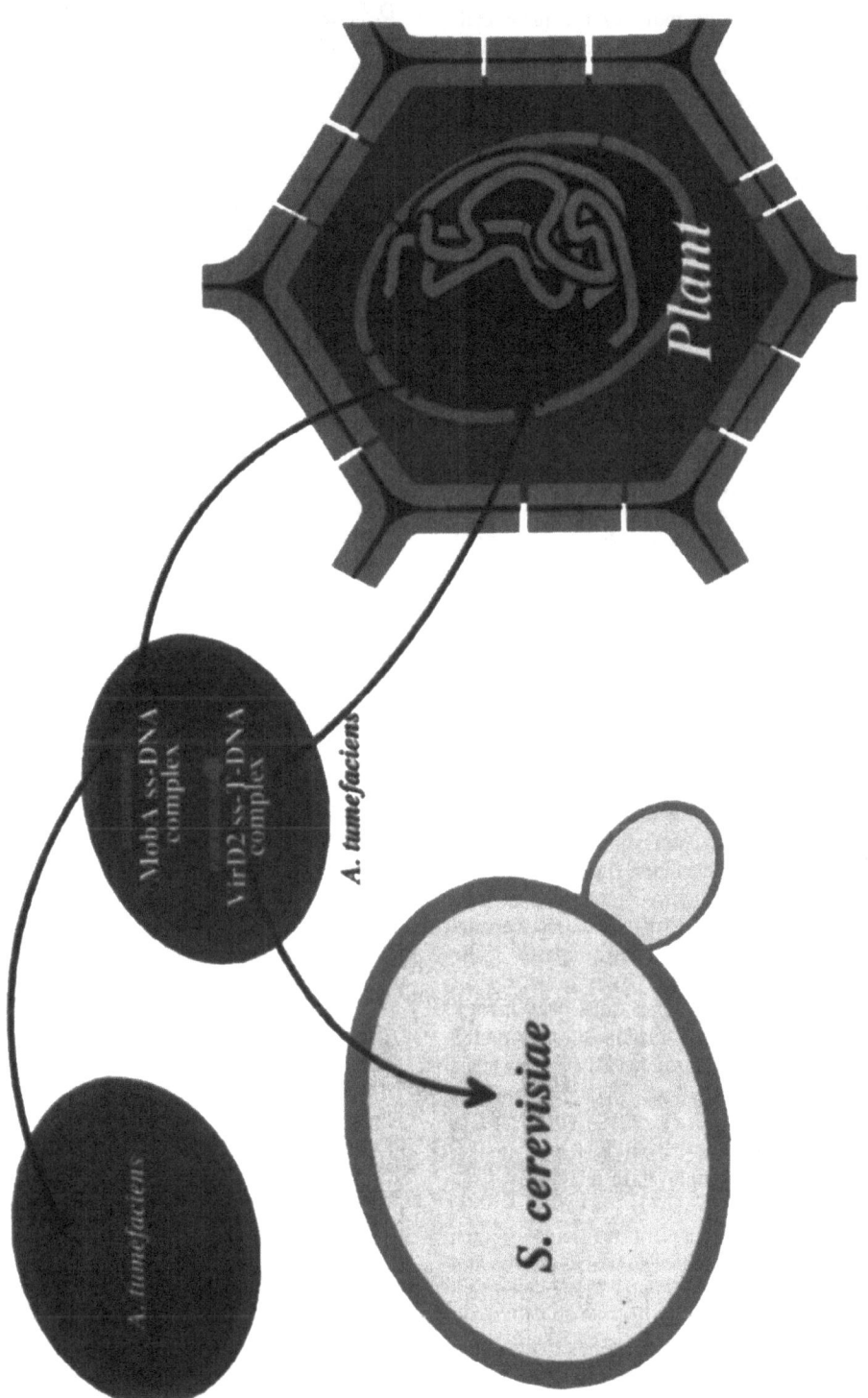

Figure 4. Horizontal gene transfer from *Agrobacterium*, mediated by the virulence transport system

VII. Perspectives

Several virulence proteins move to the plant cell, separately and/or together with T-DNA, and there perform functions essential for nuclear entry and integration of T-DNA, indirectly or directly. The proteins described in this review are mainly VirD2 and VirE2, but we do not have information on any other that may be essential for *Agrobacterium's* normal life and for T-DNA transfer, acting (also) in the plant. Likewise, we do not know if and how virulence proteins guide the T-DNA out of the bacterial transport apparatus, into the plant cell. Another question that needs to be addressed is how the cell nucleus is reached: are T-DNA complexes guided along the cytoskeleton to the nuclear pores; are virulence proteins such as VirE2 and VirD2 interacting with cytoskeletal structures, such as has been documented for tobacco mosaic virus movement protein (Heinlein *et al.*, 1995)? Also relating to activities preceding nuclear targeting, how is the transfer of VirE2 protein molecules, if indeed it occurs independently of T-DNA/VirD2 movement, into the plant cell accomplished? Where and when do the VirE2 molecules get to meet the T-DNA-VirD2 complex? It should be noted at this stage, that so far none of the virulence proteins implicated in movement to the plant cell has ever physically be localised there. It therefore remains a challenge for the future to actually find them in the cell infected by *Agrobacterium*.

How does the T-complex interact with components of the nuclear import machinery? Which importins are involved? Is VirD2 indeed the leading head localising the pore, entering the translocation machinery? Are the NLSs of VirE2 exposed to the surface of the complex, or (as we propose) is the function of this protein, apart from protection of T-DNA, to shape the complex such that neither charges nor coiling of the flexible single-stranded filament obviates translocation to the nucleus? Electronmicroscopical studies, *in vitro* nuclear import assays (Merkle *et al.*, 1996; Hicks *et al.*, 1996) and the yeast two hybrid system, amongst others, will have to be employed to get closer to the answers.

Analysis of T-DNA integration is a fascinating area of research with great potential for basic science and applied technology. The T-complex is delivered into the nucleus as a virus like particle in which VirD2 protects the 5' terminus and VirE2 binds to the entire length of the single-stranded T-DNA. Whereas the phosphotyrosine bond of VirD2 is replaced by a phosphodiester bond between the T-DNA and plant DNA, VirE2 must be displaced before or at the stage of integration. It is not known if T-DNA integrates as single- or double-strand, although the frequent inverted repeats found in many transformants (de Neve *et al.*, 1997) require second-strand synthesis before ligation and integration. Certainly these T-DNA molecules, if not all, have to loose their VirE2 coat before ligation and integration.

It will require integrated approaches to analyse the enzymology of T-DNA integration. Use of the yeast-two-hybrid system may yield proteins interacting with protein components of the T-DNA complex; *in vitro* analysis of T-DNA integration, with the use of model plant-DNAs, simulated T-complexes, and plant extracts, will be useful; complementary approaches include analysis, at the molecular level, of plant ecotypes or radiation sensitive mutants impaired specifically in T-DNA integration (Sonti *et al.* 1995, Nam *et al.*, 1997).

A question which may or may not relate to virulence proteins in the plant cell, is: How does the T-DNA "select" its integration locus? Many of the target regions are (potentially) transcriptionally active, but the sites selected were preselected for expression of the complete marker gene located on the T-DNA (Koncz *et al.*, 1989). Could one find silent T-DNAs? There is ample evidence for silenced T-DNAs, frequently in conjunction with repeats within a T-DNA locus, as well as between different transgenes or a transgene/endogenous gene-pair (reviewed in Matzke & Matzke 1995); but does T-DNA also integrate in silent loci? Generally plant DNA is not found as a chemically pure molecule, but as chromatin; can chromatin structure and T-DNA insertion preferences be correlated? Different, but related questions concern the cell cycle: is chromatin a particular stage in the cell cycle, such as S, especially receptive for T-DNA (Villemont *et al.*, 1997); is availability of replication or repair (or transcription) complexes a precondition for efficient integration?

Finally, there are special requests from plant engineers in the laboratory and in the fields, to work out regimes for clean integration of single copy T-DNA units. Precise integration will be a requirement for legal acceptance of transgenic crop, and a single integrated T-DNA unit seems to yield the most dependable expression levels. It remains to be seen if, amongst other possibilities, manipulation of virulence genes, including VirD2 and VirE2, can improve the situation. It also remains to be explored if gene targeting, an efficient version of which is in high demand, can be made routine by changing activities or regulation of virulence proteins.

VIII. Acknowledgements

We acknowledge the critical help of our enthusiastic collaborators. We thank many colleagues who have contributed unpublished information; this includes Yasunori Machida, Walt Ream, Anath Das, Stanton Gelvin and Paul J J Hooykaas.

IX. References

Adam, S. A. (1995) Trends in Cell. Biol. 5, 189-191.

Beijersbergen, A., Den Dulk-Ras, A., Schilpoort, R.A. and Hooykaas, P.J.J. (1992). Science, 256, 1324-1327.

Berger, B.R. and Christie, P.J. (1993). J. Bacteriol. 175, 1723-1734.

Binns, A., Beaupré, C. and Dale, E.M. (1995). J. Bacteriol. 177, 4890-4899.

Bukrinsky, Sharova, N., Mc Donald, T., Pushkaraskaya, T., Tarpley, W.G. and Stevenson, M. (1993a). Proc. Nat. Acad. Sci. USA 90: 6125-6129.

Buchanan-Wollaston, V., Passiatore, J.E. and Cannon, F. (1987) Nature 328, 172-175.

Bundock, P., den Dulk-Ras, A., Beijersbergen, A. and Hooykaas, P.J.J. (1995). EMBO J. 14, 3206-3214.

Bundock, P., and Hooykaas, P.J.J. (1996) Proc. Natl. Acad. Sci. USA 93, 15272-15275.

Bukrinsky, M., Haggerty, S., Dempsey, M., Sharova, N., Adzhubel, A., Spitz, L., Lewis, P., Goldfarb, D., Emermen, M. and Stevenson, M. (1993b) Nature 365, 666-668.

Carrington, J.C., Freed, D.D. and Leinicke, A.J. (1991) The Plant Cell 3, 953-962.

Christie, P.J., Ward, J.E., Winans, S.C. and Nester, E.W. (1988) J. Bacteriol. 170, 2659-2667.

Christie, P.J., Ward, Jr. J.E., Gordon, M.P. and Nester, E.W. (1989) Proc. Natl. Acad. Sci. USA 86, 9677-9681.

Christie, P.J. (1997). J. Bacteriol. 179, 3085-3094.

Chyi, Y.-S., Jorgensen, R.A., Goldstein, D., Tanksley, S.D. and Loaiza-Figueroa, F.L. (1986) Mol. Gen. Genet. 204, 64-69.

Citovsky, V., de Vos, G. and Zambryski, P. (1988) Science 240, 501-504.

Citovsky, V., Wong, M.L. and Zambryski, P. (1989) Proc. Natl. Acad. Sci. USA 86, 1193-1197.

Citovsky, V., Zupan, J., Warnick, D. and Zambryski, P. (1992) Science 256, 1802-1805.

Citovsky, V., Warnick, D. and Zambryski, P. (1994) Proc. Natl. Acad. Sci. USA, 91, 3210-3214.

Dale, E.M., Binns, A.N. and Ward, J.E. (1993). J. Bacteriol. 175, 887-891.

Das, A. (1988) Proc. Natl. Acad. Sci. USA 85, 2909-2913.

De Neve, M., De Buck, S., Jacobs, A., Van Montagu, M., Depicker, A. (1997) Plant J. 11, 15-29.

Dürrenberger, F., Crameri, A., Hohn, B. and Koukolíková-Nicola, Z. (1989) Proc. Natl. Acad. Sci. USA 86, 9154-9158.

Escudero, J., Neuhaus, G. and Hohn, B. (1995) Proc. Natl. Acad. Sci. USA 92, 230-234.

Farnet, C. and Haseltine, W. (1991) J. Virol. 65, 1910-1915.

Fullner, J.K., Stephens, K.M. and Nester, E.W. (1994) Mol. Gen. Genet. 245, 704-715.

Fullner, J.K., Lara, J.C. and Nester, E.W. (1996) Science 273, 1007-1009.

Garcia Bustos, J., Heitman, J. and Hall, M.N. (1991) Biochim. Biophys. Acta Rev. 1071, 83-101.

Gardner, R.C. and Knauf, V.C. (1986) Science 231, 725-727.

Gerace, L. (1995) Cell 82, 341-344.

Gheysen, G., Villarroel, R. and van Montagu, M. (1991) Genes & Dev., 5, 287-297.

Christie, P.J. (1997). J. Bacteriol. 179, 3085-3094.

Gietl, C., Koukolíková-Nicola, Z. and Hohn, B. (1987) Proc. Natl. Acad. Sci. USA 84, 9006-9010.

Goldfarb, D.S. (1995) Curr. Biol. 5, 570-573.

Görlich, D., Mattaj, I.W. (1996) Science 271, 1513-1518.

Grimsley, N., Hohn, T., Davies, J.W., Hohn, B. (1987) Nature, 325, 177-179.

Gurlanick, B., Thomsen, G. and Citovsky, V. (1996) The Plant Cell, 8, 363-373.

Hamilton, C.M., Frary, A., Lewis, C. and Tanksley, S.D. (1996) Proc. Natl. Acad. Sci. USA 93, 9975-9979.

Heineman, J.A. and Sprague, Jr. G.F. (1989) Nature, 340, 205-209.

Heinlein, M., Epel, B.L., Padgett, H.S., Beachy, R.N. (1995) Science 270, 1983-1985.

Herman, L., Jacobs, A., Van Montagu, M. and Depicker, A. (1990) Mol. Gen. Genet. 224, 248-256.

Herrera-Estrella, A., Van Montagu, M. and Wang, K (1990) Proc. Natl. Acad. Sci. USA, 87, 9534-9537.

Hicks, G. and Raikhel, N. (1995) Annu. Rev. Cell Dev. Biol. 11, 155-188.

Hicks, G.R., Smith, H.M., Lobreaux, S. and Raikhel, N.V. (1996) Plant Cell 8, 1337-1352.

Hiei, Y., Ohta, S., Komari, T. and Kumashiro, T. (1994). Plant J. 6, 271-282.

Hoeijmakers, J.H.J. and Bootsma, D. (1994) Nature, 371, 654-655.

Hooykaas, P.J.J., Den Dulk-Ras, H.M. and Schilperoort, R.A. (1984) Plasmid 11, 195-205.

Howard, E.A., Zupan, J.R., Citovsky, V. and Zambryski, P.C. (1992) Cell 68, 109-118.

Ishida, Y., Saito, H., Ohta, S., Hiei, Y., Komari, T. and Kumashiro, T. (1996) Nature Biotechnol. 14, 745-750.

Koncz, C., Martini, N., Mayerhofer, R., Koncz-Kalman, Z., Körber, H., Redei, G.P. and Schell, J. (1989) Proc. Natl. Acad. Sci. USA 86, 8467-8471.

Koukolíková-Nicola, Z., Raineri, D., Stephens, K., Ramos, C., Tinland, B., Nester, E.W. and Hohn, B. (1993) J. Bacteriol. 175, 723-731.

Lake, and Rivera, (1994) Proc. natl. Acad. Sci. USA 91, 2880-2881.

LaCasse, E.C. and Lefebvre, Y.A. (1995) Nucl. Acids Res. 23, 1647-1656,

Lartey, R. and Citovsky, V. (1997) In Gen. Eng. Principles and Methods 19, 201-214.

Lessl, M. and Lanka, E. (1994) Cell 77, 321-324.

Lin, T-S. and Kado, C.I. (1993) Mol. Microbiol. 9, 803-812.

Matzke, M.A. and Matzke, A.J.M. (1995) Plant Physiol. 107, 6679-6685.

Mayerhofer, R., Koncz-Kalman, Z., Nawrath, C., Bakkeren, G., Crameri, A., Angelis, A., Redei, G.P., Schell, J., Hohn, B. and Koncz, C. (1991) EMBO J. 10, 697-704.

Mehlin, H., Daneholt, B. and Skoglund, U. (1992) Cell 69, 605-613.

Mehlin, H., Daneholt, B. and Skoglund, U. (1995) J. Cell. Biol. 129, 1205-1216.

Melchers, L.S., Maroney, M.J., den Dulk-Ras, A., Thompson, D.V., van Vuuren, H.A.J., Scilperoort, R.A. and Hooykaas, P.J.J. (1990) Plant Mol. Biol., 14, 249-259.

Merkle, T., Leclerc, D., Marshallsay, C., Nagy, F. (1996) Plant J. 10, 1177-1186.

Nam, J., Matthysse, A.G., Gelvin, S.B. (1997) Plant Cell 9, 317-333.

Narasimhulu, S.B., Deng, X., Sarria, R. and Gelvin, S.B. (1996) The Plant Cell 8, 873-886.

Nishikawa, M., Suzuki, K. and Yoshida, K. (1992) Curr. Genet. 21, 101-108.

Noueiry, A.O., Lucas, W.J. and Gilbertson, R.L. (1994) Cell, 76, 925-932.

O'Neill, R., Jaskunas, R., Blobel, G., Palese, P. and Moroianu, J. (1995) J. Biol. Chem. 270, 22701-22704.

Otten, L., De Greve, H., Leemans, J., Hain, R., Hooykaas, P.J.J. and Schell, J. (1984) Mol. Gen. Genet. 195, 159-163.

Pansegrau, W., Schoumacher, F., Hohn, B. and Lanka, E. (1993) Proc. Natl. Acad. Sci. USA, 90, 11538-11542.

Pascal, E., Sanderfoot, A.A., Ward, B.M., Medville, R., Turgeon, R. and Lazarowitz , S.G. (1994) The Plant Cell, 6, 995-1006.

Piers, K.L., Heath, J.D., Liang, X., Stephens, K.M. and Nester, E.W. (1996) Proc. Natl. Acad. Sci. USA, 93, 1613-1618.

Prakash, S., Sung, P. and Prakash, L. (1993) Annu. Rev. Genet., 27, 33-70.

Puchta, H. and Hohn, B. (1996) Trends in Plant Sciences 1, 340-348.

Regensburg-Tuïnk, A.J.G. and Hooykaas, P.J.J. (1993) Nature, 363, 69-71.

Rossi, L., Hohn, B. and Tinland, B. (1993) Mol. Gen. Genet. 239, 345-353.

Rossi, L., Hohn, B. and Tinland, B. (1996) Proc. Natl. Acad. Sci. USA 93, 126-130.

Scheiffele, P.R., Pansegrau, W. and Lanka, E. (1995) J. Biol. Chem. 270, 1269-1276.

Sen, P., Pazour, G.J., Anderson, D. and Das, A. (1989). J. Bacteriol. 171, 2573-2580.

Shurvinton, C.E., Hodges, L. and Ream, W. (1992) Proc. Natl. Acad. Sci. USA 89, 11837-11841.

Sonti, R., Chiurazzi, M., Wong, D., Davies, C., Harlow, G. and Mount, D. (1995) Proc. Natl. Acad. Sci. USA 92, 11786-11790.

Stachel, S.E., Nester, E.W. and Zambryski, P.C. (1986a) Proc. Natl. Acad. Sci. USA 83, 379-383.

Stachel, S.E., Timmerman, B. and Zambryski, P. (1986b) Nature 322, 706-712.

Sundberg, C., Meek, l., Carroll, K., Das, A. and Ream, W. (1996). J. Bacteriol. 178, 1207-1212.

Sweet, D.J. and Gerace, L. (1995) Trends in Cell Biol. 5, 444-447.

Tinland, B., Koukolikova-Nicola, Z., Hall, M.N. and Hohn, B. (1992) Proc. Natl. Acad. Sci. USA, 89, 7442-7446.

Tinland, B., Hohn, B. and Puchta, H. (1994) Proc. Natl. Acad. Sci. USA, 91, 8000-8004.

Tinland, B. and Hohn, B. (1995). In Genetic Engineering, Principles and Methods. Setlow JK (ed.) Plenum New York 17, 209-229.

Tinland, B., Schoumacher, F., Gloeckler, V., Bravo Angel, A.M. and Hohn, B. (1995) EMBO J. 13, 5764-5771.

Tinland, B. (1996) Trends in Plant Sciences 1, 178-184.

Villemont, E., Dubois, F., Sangwan, D.R., Vasseur, G., Bourgeois, Y. and Sangwan-Norreel, (1997) Planta., 201, 160-172.

Vogel, A.M. and Das, A. (1992) J. Bacteriol., 174, 303-312.

Wallroth, M., Gerats, A.G.M., Rogers, S.G., Fraley, R.T. and Horsch, R.B. (1986) Mol. Gen. Genet. 202, 6-15.

Ward, E.R. and Barnes, W.M. (1988) Science, 242, 927-930.

Ward, J.E., Dale, E.M., Christie, P.J., Nester, E.W. and Binns, A.N. (1990) J. Bacteriol., 172, 5187-5199.

Winans, S.C., Burns, D.L. and Christie, P.J. (1996) Trends Microbiol., 4, 64-68.

Wirawan, I.G.P. and Kojima, M. (1996) Biosci. Biotech. Biochem. 60, 44-49.

Yoshioka, Y., Takahashi, Y., Matsuoka, K., Nakamura, K., Koizumi, J., Kojima, M. and machida, Y. (1996) Plant and Cell Physiol. 37, 782-789.

Young, C. and Nester, E.W. (1988) J. Bacteriol. 170, 3367-3374.

Yusibov, V.M., Steck, T.R., Gupta, V. and Gelvin, S.B. (1994). Proc. Natl. Acad. Sci. USA, 91, 2994-2998.

Zupan, J., Citovsky, V. and Zambrysky, P. (1996) Proc. Natl. Acad. Sci. USA 93, 2392-2397.

Chapter 16

Determinants of Host Specificity of *Agrobacterium* and their Function

Wanyin Deng and Eugene W. Nester

I. Introduction

The host range of *Agrobacterium* spp. is extremely broad in contrast to other plant bacterial pathogens or the symbiotic *Rhizobium* spp. It can induce tumor formation on a wide range of dicotyledonous plants, as well as on a number of gymnosperms and a few monocots, following artificial inoculation (De Cleene 1985; De Cleene and De Ley 1976; Smith and Hood 1995). However, the host range of *Agrobacterium* spp. is much more restricted in nature, since the trademark crown gall disease is limited to grapevines, stone and pome fruit trees, nut trees, and a few ornamentals, on which most economic damage is done (Kerr 1992). The host range of *Agrobacterium* was traditionally assayed by tumor formation, but some plants, including a

number of monocotyledons, do not respond with tumor formation although they permit *Agrobacterium*-mediated T-DNA transfer. Jarchow *et al.*, (1991) proposed that these plants be termed "silent hosts". In this section, we will not make the distinction between natural host range and artificial host range except when necessary. Any plant which can be infected by *Agrobacterium* spp. and allows T-DNA transfer and integration, naturally or artificially, will be viewed as a host.

The ability of *Agrobacterium* spp. to infect so many different plants is surprising if one considers the complexity of the T-DNA-mediated tumorigenesis process (see other Chapters on *Agrobacterium*). The fact that artificially induced *Agrobacterium* can transfer T-DNA into yeast (Bundock *et al.*, 1995; Piers *et al.*, 1996) and that *A. tumefaciens* has a very wide host range in general (De Cleene and De Ley 1976) suggests that *Agrobacterium* exploits very fundamental properties and functions of eukaryotic cells during the T-DNA transfer process. It also implies that the host factors involved in the transformation process are highly conserved among plants and that *Agrobacterium* spp. might have yet unidentified ways of evading the plant defense system. However, differences in host range do exist among individual *Agrobacterium* isolates, and even wide-host-range (WHR) strains cannot induce tumors on all plants. Certain plants, which include most of the economically important cereals and legumes, are recalcitrant to tumor formation.

Many reasons can account for these limitations in the host range of *Agrobacterium* spp. (also see Binns and Thomashow 1988, and Chapter 12). First, T-DNA transfer might not occur. In theory, this can result from poor bacterial growth and/or survival in the plant wound environment, the lack of *vir* gene inducers or receptors for bacterial attachment in the plants, or defective *vir* genes which result in inefficient T-DNA processing or transfer. Further, the T-DNA might not induce tumors because of its failure to integrate in the plant, the lack of a complete set of the T-DNA genes, or poor or over-expression of the T-DNA genes. Also, different plants might respond differently to T-DNA gene expression and the ensuing changes in phytohormone levels and effects, since plant sensitivity to alterations in hormone levels do differ. *Agrobacterium* spp. exhibit species- and sometimes cultivar- specific host specificity. This specificity might be due to differences in the biochemistry, physiology as well as the anatomy of the hosts. Certain plants may not be sensitive to the phytohormone alterations brought about by the expression of the T-DNA oncogene complement, and therefore fail to form tumors following *Agrobacterium* infection, while others may be overly sensitive, and respond by a necrotic reaction. In addition, plants in different developmental stages may also respond differently to the infection of *Agrobacterium*. Thus, the type of Ti (Ri) plasmid, the chromosomal makeup of the bacteria and its interaction with the residing Ti (Ri) plasmids, and biochemical and physiological as well as anatomical variations within and among plant hosts can all contribute to the observed differences in the host range of *Agrobacterium* spp. While the importance of many of the bacterial determinants have been delineated as discussed below, the plant factors involved remain poorly understood.

II. Ti Plasmids and Host Specificity

Although the genus *Agrobacterium* as a group and some *Agrobacterium* strains in particular have a remarkably wide host range (De Cleene and De Ley 1976), host specificity for individual strains or isolates is common. Some *A. vitis* strains isolated from grapevine tumors have been shown to have a limited host range, unable to induce tumors on certain plants (Panagopoulos and Psallidas 1973). Anderson and Moore (1979) showed that most, if not all, *Agrobacterium* isolates exhibited host specificity. Earlier studies had shown that the host range of the transconjugants of different Ti plasmids and various chromosomal backgrounds seems to track with the Ti plasmid (Loper and Kado 1979; Thomashow *et al.,* 1980; Knauf *et al.,* 1982), suggesting that the Ti plasmid is the primary determinant of host specificity. Further studies by various laboratories have demonstrated that a number of components of the Ti plasmid, which usually affect either the efficiency of *vir* gene induction, T-DNA processing and transfer, or the nature of the T-DNA make-up, play a role in determining the host specificity of *Agrobacterium*. Since different plants have varying susceptibility to crown gall disease, as a general rule we will view any *Agrobacterium* gene as a potential host range determinant if mutations in the gene attenuate (not abolish) bacterial virulence on any plant host.

II.A. *VIR* GENES AND HOST SPECIFICITY

II.A.1. *virA and vir Gene Induction*
One of the early events in tumor formation induced by *Agrobacterium* is the recognition of a wounded plant by the bacterium. As discussed in Chapter 13, the VirA/G two-component system controls the sensing and perception of plant signal molecules and the expression of *vir* genes. The VirA protein is located in the inner membrane, and acts as a sensor for specific phenolic compounds and other environmental conditions. Genetic studies have indicated that the VirA protein senses phenolic compounds directly (Lee *et al.,* 1995), similar to the NodD protein of *Rhizobium* which senses flavonoids and regulates the expression of nodulation genes (Horvath *et al.,* 1987; Spaink *et al.,* 1987; Chapter 19). In addition, the VirA proteins from different *Agrobacterium* strains seem to prefer different phenolic compounds (Lee *et al.,* 1995). Since the variety,

amount, and ratio of phenolic compounds excreted by different plant species upon being wounded presumably vary, it is conceivable that the *virA* gene of a particular *Agrobacterium* strain can dictate what plants it can infect, just like the *nodD* gene in *Rhizobium* (reviewed by Long 1989).

Indeed, the *virA* gene has been suggested to be a host range determinant for some *Agrobacterium* strains. As discussed earlier, some *A. vitis* strains have a limited host range (LHR). These strains are specifically associated with grapevines, and are tumorigenic on only one or a few plant species (Panagopoulos *et al.*, 1978; Knauf *et al.*, 1982). One such strain, Ag162, has been studied extensively. The host range of this strain can be extended to new plants by introducing the *virA* gene from a WHR strain (Yanofsky *et al.*, 1985a). It has been shown that its *vir* genes cannot be induced by acetosyringone, a potent natural inducer for WHR strains, at physiologically relevant concentrations due to its *virA* gene (Leroux *et al.*, 1987). Even very high concentrations of acetosyringone only induce the expression of LHR *vir* genes to a limited extent (Spencer *et al.*, 1990; Turk *et al.*, 1991). At least part of the low inducibility of the LHR *virA* by acetosyringone can be attributed to the *virA* promoter (Turk *et al.*, 1993). Another phenolic compound, syringic acid methyl ester, isolated from grapevine wound exudates has been shown to induce *vir* gene expression in LHR strains, but again only at very high, probably physiologically irrelevant concentrations (Spencer *et al.*, 1990). Since the LHR strains do induce tumors on some plants, and their *vir* genes must be induced somehow, it is possible that another yet unidentified inducer is present in the host plants of these LHR strains. Alternatively, the level of *vir* gene induction required to form a crown gall tumor might be very low for these hosts.

The VirA protein is also responsible for sensing monosaccharides via a chromosomally encoded periplasmic galactose-binding protein (ChvE) (Cangelosi *et al.*, 1990; Shimodo *et al.*, 1990). The sugars first bind to ChvE, which in turn interacts, with VirA to synergistically induce *vir* gene expression in the presence of phenolic compounds (see Chapter 13). Mutants of *virA* which fail to interact with ChvE and therefore have lost the sugar effect show a decreased host range (Shimodo *et al.*, 1993; Doty *et al.*, 1996). They can infect certain plants but not others, having the same phenotype as a *chvE* mutant (Doty *et al.*, 1996; Huang *et al.*, 1990). See sections III.A.2. and III.B.1. in this chapter for additional information on the relationship between ChvE or proper ChvE-VirA

interactions and host specificity. In addition, different environmental conditions (pH and temperature) can differentially affect *vir* gene induction in different *Agrobacterium* strains by affecting their VirA function, suggesting a role of VirA in host range determination under those environmental conditions (Jin *et al.*, 1993a; Turk *et al.*, 1991).

The importance of the *virA* gene in host range determination is also revealed by another type of study, agroinfection of monocots, especially maize. "Agroinfection" is a technique in which *Agrobacterium* is used as a vector to deliver viral genomes into plants by the T-DNA transfer process. The transfer into the plant cell nucleus is detected by the development of viral disease symptoms on the plants (Grimsley *et al.*, 1986). It has been found that the ability to agroinfect maize is Ti/Ri plasmid specific, and similar results have been obtained on other members of the *Gramineae*, such as rice and wheat (Grimsley *et al.*, 1987; Boulton *et al.*, 1989; Jarchow *et al.*, 1991; Dasgupta *et al.*, 1991; Woolston *et al.*, 1988). In these studies, many octopine-type strains of *A. tumefaciens* are unable, or only weakly able, to agroinfect the monocots, while nopaline-type strains of *A. tumefaciens* and agropine- and mannopine-type strains of *A. rhizogenes* are highly agroinfectious. The major difference in agroinfecting ability, as reported by Raineri *et al.*, (1993), is the *virA* gene, which encodes the plant-signal receptor. It is suggested that the octopine-type VirA activity is uninduced or specifically inhibited by maize extracts (Raineri *et al.*, 1993). This is supported by the observation that octopine-type, acetosyringone-independent signaling *virA* mutants, which express *vir* genes in the absence of inducing plant signals (Ankenbauer *et al.*, 1991), can agroinfect at high levels (Raineri *et al.*, 1993). This interpretation is also consistent with the report that *Agrobacterium*-mediated gene transfer efficiency to maize by an octopine-type strain was greatly improved upon introduction of *virGN54D*, which codes for an Asn-54 to Asp amino acid change in the VirG protein causing constitutive *vir* gene expression independent of VirA and stimulating plant signals (Hansen *et al.*, 1994; Pazour *et al.*, 1992; Jin *et al.*, 1993b).

II.A.2. virE and virC

Mutations in both *virC* and *virE* loci result in an alteration of host range. Both *virC* and *virE* mutants show an attenuated or avirulent phenotype on some plants, while they remain fully virulent on others (Klee *et al.*, 1983; Hille *et al.*, 1984; Hooykaas *et al.*,

1984; Hirooka *et al.*, 1987; Gardner and Knauf 1986; Grimsley *et al.*, 1989). It should be emphasized here that all attenuated *Agrobacterium* strains are likely to carry mutations which can alter their host range since different plants have different susceptibilities to crown gall. It has been shown that the *virE2* gene of the *virE* operon encodes a single stranded DNA binding protein (see Chapter 15). There is evidence that VirE2 might have multiple functions in helping T-DNA transfer and integration inside the plant cell, since VirE2 possesses nuclear localization signals and transgenic plants for the *virE2* gene can serve as a host for *Agrobacterium virE* mutants (Gardner and Knauf 1986; Citovsky *et al.*, 1992). On the other hand, one of the *virC* operon-encoded proteins, VirC1, functions within the bacteria by binding to a T-DNA transfer enhancer (the *overdrive* sequence) and promoting T-DNA processing and T-strand formation (Toro *et al.*, 1989a, 1989b; Horsch *et al.*, 1986; De Vos and Zambryski 1989). The studies on *virE* and *virC* suggest that both loci are involved in increasing *Agrobacterium* infection efficiency. It is possible that some plants need a higher efficiency of T-DNA transfer from the bacteria in order to form visible tumors. This could mean that either more plant cells at the wound site are transformed or each transformed plant cell receives more copies of the T-DNA from a number of bacteria.

In addition to a T-DNA gene, *virC*, along with *virA*, from a WHR Ti plasmid must also be introduced into an LHR strain in order to restore a WHR phenotype (Yanofsky *et al.*, 1985a). Furthermore, the *virC* locus in WHR strains is involved in eliciting a necrotic response on certain *Vitis* cultivars, and *virC* mutants fail to induce the response apparently due to a lower transformation frequency (Yanofsky *et al.*, 1985a; Deng *et al.*, 1995b). Agroinfection studies on *Zea mays* or *Brassica* using *virC* and *virE* mutants support the host range determining role of these genes (Grimsley *et al.*, 1989). Agroinfection of maize with maize streak virus using strains of *A. tumefaciens* carrying mutations in the nopaline-type pTiC58 virulence region showed an almost absolute dependence on the product of the *virC* operon. In contrast, agroinfection of *Brassica rapa* with cauliflower mosaic virus was less dependent on the *virC* gene products. Mutants in *virE* were attenuated, not abolished, in agroinfecting both plants (Grimsley *et al.*, 1989). Although further studies are needed, these results suggest that *virC* might play a more important role in monocot transformation compared to dicot transformation.

II.A.3. Ti-Specific vir Genes: virF, pinF (virH), tzs, and virJ

The *vir* region is one of the most conserved regions in Ti plasmids. The linear orders of *vir* genes are the same for all Ti plasmids studied so far (see Chapter 13). However, differences do exist even in this highly conserved region. For example, three operons, *virF*, *virH*, and *virJ* are believed to be unique to the *vir* region of octopine Ti plasmids, while nopaline Ti plasmids contain a specific *tzs* gene (Kalogeraki and Winans 1995; Pan et al., 1995; for reviews see: Binns and Thomashow 1988; Hooykaas and Beijersbergen 1994). Although these *vir* genes are different from *virA*, *virC*, and *virE* genes described above in that they are Ti-specific, they are under the control of the VirA/VirG two-component regulatory system and their expression is induced by plant signal molecules just like other *vir* genes (see Chapter 13). They can be viewed as auxiliary virulence genes since they are present only in certain Ti plasmids. Because of their nonessential or accessory role in *Agrobacterium*-mediated tumorigenesis, they are likely to be involved in determining the host range of the strain.

virF. *A. tumefaciens* strains with nopaline Ti plasmids are only weakly virulent on *Nicotiana glauca* in contrast to strains with octopine or leucinopine Ti plasmids (Melchers *et al.*, 1990). Melchers *et al.*, (1990) demonstrated that this difference between octopine and nopaline strains in virulence is due to a gene in the *vir* region, and not their T-DNA genes. This gene, termed *virF*, is present in octopine, but not nopaline Ti plasmids. It is necessary for efficient tumorigenesis on *N. glauca* since *virF* mutants of octopine strains have weakened virulence on the plant (Otten *et al.*, 1984; Melchers *et al.*, 1990).

The *virF* gene is of particular interest in that *virF* mutants can be complemented for tumorigenesis by coinfection with a helper strain carrying an intact *vir* region but lacking the T-DNA (Otten *et al.*, 1984; Melchers *et al.*, 1990). The helper strains must express not only the *virF* gene but also the *vir* operons *virA*, *virB*, *virD*, and *virG* (Melchers *et al.*, 1990). Interestingly, similar results were obtained for *virE* mutants (Otten *et al.*, 1984; Christie *et al.*, 1988). Also similar to *virE* (Citovsky *et al.*, 1992), transgenic *N. glauca* plants expressing the *virF* gene can complement *virF* mutants of octopine strains for tumorigenesis, and are converted into hosts for nopaline strains (Regensburg-Tuïnk and Hooykaas

1993), strongly suggesting that the VirF protein plays a role within the plant cell. It appears that both VirE and VirF proteins are transported from the bacteria into plant cells via the same channel used by the T-DNA complex. It will be interesting to locate the common signals present in these proteins recognized by the transport pore. It has also been reported in agroinfection studies that *virF* can partially inhibit T-DNA transfer to maize by octopine strains (Jarchow *et al.*, 1991), but the mechanism is not clear. Although the function of the VirF protein is still unknown, the available evidence points to the possibility that VirF plays some role in helping T-DNA transfer inside the host plant cell.

virH (pinF). Mutants in the *virH* locus of octopine strains showed an attenuated virulence on a variety of dicotyledonous plants, and this attenuation became more pronounced with decreasing numbers of bacterial cells in the inoculum (Kanemoto *et al.*, 1989). Two of the polypeptides encoded by the *virH* operon are similar to each other and to known protein sequences for cytochrome P-450 enzymes, a well characterized family of proteins which catalyze the NADH-dependent addition of oxygen to a number of different substrates (Kanemoto *et al.*, 1989). The observation that a large number of *virH*-negative cells are needed to induce tumors suggests a decrease in the transformation frequency of plant cells by these mutants. It is possible that *virH*-encoded proteins may be involved in the detoxification of toxic agents present in the plant wound sap and thereby help the bacteria survive and grow *in planta*. It was originally suggested that *virH* is an octopine Ti plasmid-specific operon since the corresponding location of *virH* in nopaline Ti plasmids is occupied by the *tzs* gene (see next section) (Hooykaas and Beijersbergen 1994; Jarchow *et al.*, 1991). However, by using a *virH*-specific probe, we have discovered that *virH*-homologous sequences are present in both octopine and nopaline Ti plasmids as well as succinamopine-type Ti plasmids (Hood et al 1986; Kovacs and Pueppke 1994) and *A. rhizogenes* strains, but are absent from the *A. vitis* strains examined (Deng and Nester, unpublished data). This raises the possibility that some of the seemingly Ti plasmid-specific *vir* genes are actually present on more Ti plasmids but at different locations. The presence of *virH* in so many different Ti/Ri plasmids suggests that it plays an important role in *Agrobacterium* pathogenesis.

tzs. The *tzs* gene encodes a cytokinin prenyl transferase responsible for *trans*-zeatin synthesis (Akiyoshi *et al.*, 1985, 1987; Beaty *et al.*, 1986). Its expression is induced by plant signal molecules via VirA/VirG (John and Amasino 1988; Powell *et al.*, 1988). The *tzs* gene has been found in the *vir* region of nopaline Ti plasmids and in agropine and mannopine Ri plasmids, but not in octopine or succinamopine Ti plasmids (Akiyoshi *et al.*, 1987; Beaty *et al.*, 1986). Its role in plant cell transformation has yet to be demonstrated, although the importance of cytokinins in *Agrobacterium*-mediated transformation and the absence of the *tzs* gene from certain Ti plasmids seem to suggest that *tzs* ought to play some role in host range determination. It has been shown that root induction on flax cotyledon explants by an *A. rhizogenes* agropine strain is markedly increased by co-inoculation with a disarmed *A. tumefaciens* strain containing a plasmid carrying the *tzs* gene (Zhan *et al.*, 1990). This supports the idea that the *tzs* gene, while not essential for transformation, may promote transformation, possibly by stimulating plant cell division at wound sites and conditioning the plant cells for more efficient transformation. Whether the *tzs* gene present in the Ri plasmids of some *A. rhizogenes* strains plays a similar role remains to be studied.

virJ. The *virJ* gene lies between *virA* and *virB*, and its expression is regulated by plant signals under the control of VirA/VirG. This gene is Ti plasmid-specific since it is present in octopine but not nopaline Ti plasmids (Kalogeraki and Winans 1995; Pan et al., 1995). It first appeared to be a gene nonessential for virulence since mutations in *virJ* had no effect on virulence (Melchers *et al.*, 1987; Stachel and Nester 1986). No data yet support a role of *virJ* in host range determination. However, this gene is particularly interesting because its gene product and a chromosome-encoded protein AcvB apparently have similar functions (Kalogeraki and Winans 1995; Pan et al., 1995; Wirawan *et al.*, 1993). The *acvB* gene is present in the chromosome of both octopine and nopaline strains, and this explains why *virJ* mutations in octopine strains have no effect on virulence. The importance of the VirJ/AcvB function in virulence has been shown by the observations that *acvB* mutants of nopaline strains are avirulent (Wirawan *et al.*, 1993) and that *virJ* can complement the non-tumorigenic phenotype of the *acvB* mutants (Kalogeraki and Winans 1995; Pan et al., 1995). This suggests that the two genes are interchangeable in terms of their function in tumorigenesis and that the function is redundant in octopine strains. Unlike *virJ*, the expression of *acvB* is not regulated by VirA/VirG and plant signal molecules. VirJ has a putative signal peptide and is found primarily in the periplasmic space (Pan *et al.*, 1995). Although its exact function is still unknown, there is a report that the AcvB protein has a single stranded DNA binding activity (Kang *et al.*, 1994). However, this observation has not been confirmed (Mushegian, Pan, and Nester, unpublished data). The strain lacking both *acvB* and *virJ* failed to transfer T-DNA into plant cells, although T-strands were made. This suggests that VirJ or AcvB protein is required for the transfer of T-DNA from *A. tumefaciens* to plant cells (Kang *et al.*, 1994; Pan *et al.*, 1995). It is plausible that VirJ may extend the host range of octopine strains to plants where more efficient T-DNA transfer is needed for tumor induction. VirJ can do so during infection by ensuring sufficient quantities of the AcvB function for efficient tumorigenesis when the amount of AcvB protein is limiting.

The seeming dispensability of some *vir* genes for virulence, such as *virF*, *virH*, *tzs*, *virJ*, *virD3*, and *virD5* (for a review see: Hooykaas and Beijersbergen 1994), raises an interesting question. Although these genes are present in many Ti plasmids and are controlled by the same regulatory mechanism as the other *vir* genes, mutations in these genes do not affect the virulence of the respective strains or only lead to a minor attenuation of tumorigenicity on some plants. Are at least some of these genes dispensable because there are two copies of the same gene, one on the Ti plasmid and the other on the chromosome, a situation similar to *virJ*? It appears that an *acvB* gene duplication occurred and somehow one of the copies was recruited into the Ti plasmid to become *virJ*. The presence of sequences resembling insertion sequences surrounding the *virJ* gene suggests that the movement of the duplicated chromosomal *acvB* gene to the Ti plasmid might result from an insertion sequence-mediated transposition (Hooykaas and Beijersbergen 1994; Pan et al., 1995). The VirJ protein is homologous only to the C-terminal half of AcvB. Pan *et al.*, (1995) hypothesized that during evolution the coding region of the N-terminal half of AcvB was removed and the *vir* regulatory sequences added to the promoter region. The resulting VirJ was presumably more specialized for a role in tumorigenesis. A similar event might have happened for *virH*, since *Agrobacterium* should have chromosome-encoded P-

450 enzymes and since the remnants of insertion sequences have also been found in the upstream region of the *virH* gene (Deng *et al.*, 1995a; Kanemoto *et al.*, 1989). Available evidence suggests that transposable elements have played an important role in the evolution of Ti plasmids (Otten *et al.*, 1992; Deng *et al.*, 1995a). Gene duplications have been found in *Rhizobium* spp. and other bacteria (reviewed by Kalogeraki and Winans 1995). Gene transfer from one genome to another in a single cell is not uncommon. In fact, gene transfer from organellar (chloroplast or mitochondrion) genomes to the nuclear genome has been demonstrated in eukaryotes (Nugent and Palmer 1991).

The functions of these seemingly dispensable *vir* genes await further studies. Since different plants likely need different efficiencies of T-DNA transformation for tumor formation, the host range determining role of these genes can only be decided by testing a wide variety of plants, but this remains to be done. Furthermore, these *vir* genes have only been assayed under artificial conditions. In nature, where the environmental conditions and the population of bacteria available for infection are more subtle and delicate, these genes may be crucial for successful tumorigenesis by *Agrobacterium*. The study of more strains of *Agrobacterium* may reveal even more subtle differences in the constitution and organization of the virulence regions of octopine, nopaline, and other types of Ti plasmids. It is possible that different *Agrobacterium* strains have acquired new features in the *vir* (and/or T-DNA) regions of their Ti plasmids during evolution in order to adapt to certain plant hosts or certain environmental conditions.

II.B. T-DNA GENES AND HOST SPECIFICITY

II.B.1. T-DNA Genes

For wild-type *Agrobacterium* strains, whose host range is judged by the end product of tumor or hairy root induction and not by T-DNA transfer and integration, the importance of T-DNA genes in host range determination is obvious. T-DNA gene expression and the resulting overproduction and/or activity of plant hormones are the ultimate cause for neoplastic plant cell growth. As in *Rhizobium* species where the type of *nod* factors a strain produces determines the species of legume plants it can infect (see Chapters on *Rhizobium*, this book), the type of the T-DNA make-up will determine whether an

Agrobacterium strain can induce tumors or hairy roots on a plant. It is possible that some *Agrobacterium* strains acquired their host specificity pattern through T-DNA mutations and/or rearrangements that led to optimal growth induction on a particular host plant. In general, T-DNA genes can be divided into three categories based on their function: (1) biosynthesis of plant growth regulators (auxin/cytokinin); (2) modulation of plant growth hormone activities; and (3) synthesis and secretion of opines (for a detailed description of the T-DNA genes see Chapter 12). Of these three kinds of T-DNA genes, the first two have been implicated in host range determination.

The *tms* (*iaaM* and *iaaH*) and *tmr* (*ipt*) genes are responsible for the biosynthesis of two major plant growth regulators, auxin and cytokinin. The *tmr* gene is involved in the biosynthesis of cytokinin, and the *tms* genes encode functions for auxin production. These genes are present in the T-DNA region of most virulent, tumorigenic *Agrobacterium* strains, and these strains usually have a wide host range, inciting tumors on a large number of plant species. Some *A. vitis* strains have a defective or weak *tmr* gene and an LHR phenotype. The addition of a *tmr* oncogene from a WHR Ti plasmid to the T-DNA region of the LHR strains (*A. vitis* Ag63 or Ag162) expands their host range (Buchholz and Thomashow 1984; Hoekema et al., 1984; Yanofsky et al., 1985a, 1985b). However, the resulting host range is substantially less than that conferred by the WHR Ti plasmid. At least two more genes, *virA* and *virC*, from the *vir* region of the WHR strain must also be introduced into the LHR strain in order to restore a wide host range (Yanofsky et al., 1985a; also see the preceding section II.A. of this chapter).

Another example of T-DNA genes determining host range relates to a biotype 2, LHR *A. tumefaciens* strain, AB2/73. This strain induces tumors on *Lippia canescens* but few other plants (Unger et al., 1985). This strain apparently has a set of weak or nonfunctional auxin biosynthesis-related genes, since the introduction of the *tms* genes from a WHR strain to its T-DNA region expands its host range considerably (Raineri, Nguyen, Cao, Gordon, Nester, and Unger, unpublished data).

The *tms* or *iaa* genes are also very important for the host range of *A. rhizogenes* (White et al., 1985). This indicates that different plant species are sensitive to different levels and ratios of auxins and cytokinins. Some of the T-DNA genes, namely genes 5 and 6b of the Ti plasmid and the *rol* genes of Ri plasmids, modulate the activities of auxin and/or cytokinin and

therefore the effect of *tms* and *tmr* genes (also see Chapter 12). These genes seem to be good candidates for contributing to host specificity, but their role in this respect has not been demonstrated except for gene 6b. It has been shown that gene 6b by itself provokes small tumor formation on plants, but this tumorigenic effect is only seen on particular plant species such as *Nicotiana glauca* and *Kalanchoe tubiflora*, strongly suggesting a host-range-determining role for gene 6b (Hooykaas et al., 1988).

The host range of an *Agrobacterium* strain reflects the genetic make-up of its T-DNA to a certain degree. The number of functional T-DNA genes varies greatly among the *A. vitis* isolates, mostly due to mutations and deletions caused by the transposition and rearrangements mediated by insertion sequences (reviewed by Otten et al., 1992). Although all of these strains are closely associated with grapevines, some of them can infect other plant species in artificial inoculations and show a wide host range (Anderson and Moore 1979; Knauf et al., 1982; Loper and Kado 1979; Thomashow et al., 1980). For a number of LHR isolates, they invariably carry some defect in their T-DNA make-up (Buchholz and Thomashow 1984; Hoekema et al., 1984; Yanofsky et al., 1985a). These strains provide an excellent system for studying the role of T-DNA genes in host specificity. The correlation between the host range of the strains and the number and make-up of their T-DNA genes needs further studies.

II.B.2. T-DNA Border Sequences

T-DNA borders define the length of the T-DNA region, and dictate what T-DNA genes are transferred into plant cells. Paulus et al., (1991) reported that the right border of the T_A-DNA region of *A. vitis* LHR strain Ag57 was deleted during evolution. This strain can still induce tumors on *Nicotiana rustica* and *Vitis vinifera*, presumably using a pseudoborder sequence. However, this strain is non-tumorigenic on *N. tabacum*. It becomes tumorigenic on *N. tabacum* only if the T_A-DNA right border sequence is restored, indicating the right TA border is required for tumor induction on some hosts, but not on others by the LHR strain Ag57 (Paulus et al., 1991). This also indicates that certain plants are more sensitive to inefficient T-DNA transfer than others.

The T-DNA processing enhancer, the so-called *overdrive*, can also contribute to host specificity (Peralta and Ream 1985; Peralta et al., 1986; Culianez-Macia and Hepburn 1988). Its effect on host range is most likely mediated by *virC*, since the

VirC protein has been shown to bind to the sequence, suggesting a role of VirC in promoting T-DNA processing (Toro *et al.*, 1989b; also see section II.A.2. on *virC* and *virE*, this Chapter). Whether the *overdrive* sequence is required for tumor induction on a plant will again reflect the sensitivity of the plant to the efficiency of the T-DNA transfer.

II.B.3. The Necrotic Response

Some of the *A. vitis* LHR isolates are known to have defective and less than a full complement of T-DNA genes, and they cannot be induced by the most commonly used plant phenolic inducers (see sections II.A.1. and II.B.1. on "*virA* and *vir* gene induction" and "T-DNA genes", this Chapter). Yet, these strains are highly tumorigenic on many varieties of grapevines, indicating that the *Vitis* species are very sensitive hosts for tumorigenesis. It is therefore surprising that some of the highly virulent WHR strains of *A. tumefaciens* cannot induce tumor formation on grapevines. Instead, as first reported by Yanofsky *et al.*, (1985a), these WHR strains induce a necrotic response defined as plant cell death at the site of *Agrobacterium* inoculation. Although both the *vir* region and the T-DNA are required, the T-DNA is directly responsible for inducing the response (Pu and Goodman 1992; Yanofsky *et al.*, 1985a). Mutations in the *virC* locus prevent the necrotic response and allow the induction of tumors by the strains, implying that high T-DNA transfer efficiency is important (Yanofsky *et al.*, 1985a). Deng *et al.*, (1995b) studied the necrogenesis induced by the supervirulent *A. tumefaciens* strain A281 (Hood *et al.*, 1986; Jin *et al.*, 1987) on grapevines, and reported that two T-DNA genes, *tms1* and 6b, are responsible for inducing the necrotic response. Since the two genes are involved in the synthesis and activity modulation of auxin, it is proposed that the necrotic response induced by *A. tumefaciens* on grapevines results from auxin toxicity (Deng *et al.*, 1995b). This hypothesis is consistent with the stem polarity effect on grapevine necrogenesis and the unusual T-DNA structure of strain A281 (Deng *et al.*, 1995b). These studies suggest that high T-DNA transfer efficiency and a full set of the T-DNA genes in an *Agrobacterium* strain do not always ensure successful tumorigenesis on a plant. The adaptation of a strain on a particular host relies on the fine-tuning and balancing of its *vir* and T-DNA gene make-up when it interacts with the plant during evolution. It is possible that the necrotic response can also be used to explain the apparent resistance of plants, other than grapevines, to tumor formation induced by some *Agrobacterium* strains.

III. *Agrobacterium* Chromosomal Factors and Host Specificity

III.A. CHROMOSOMAL GENES AFFECTING HOST SPECIFICITY

Although the Ti plasmid is the primary determinant of *Agrobacterium* host range, the chromosomal background also plays an important role. This became clear since some Ti plasmids confer different host specificity when present in different *Agrobacterium* chromosomal backgrounds (Hamada and Farrand 1980; Knauf *et al.*, 1982). Besides the *vir* genes on the Ti plasmids, several chromosomal genes have been shown to be important for *Agrobacterium*-mediated tumor induction (for a review see: Hooykaas and Beijersbergen 1994). These chromosomal genes can be divided into two groups based on their proposed functions in the tumorigenesis process. One group codes for proteins, which interact with the host plant. Their functions include factors which allow attachment to the plant cell and growth and survival in the plant wound environment. The other group is involved in *vir* gene induction or T-DNA transfer, and these genes seem to promote conditions, which allow the Ti plasmids to function optimally for tumor induction. As discussed below, the role of some of these chromosomal genes in host specificity has been clearly demonstrated. Although these genes were identified because of their involvement in *Agrobacterium* virulence, most of them play dual roles, one in the physiology of the bacteria and the other in tumor induction. This is reflected in the pleiotropic phenotype of their mutants. Therefore additional chromosomal genes involved in *Agrobacterium* virulence and interaction with plant cells are likely to be identified especially if mutants other than null mutants are isolated.

III.A.1. Chromosomal Genes Concerned with Interacting with the Host Plant

One of the early steps in tumor induction is the site-specific attachment of the bacteria to the host cell (Lippincott and Lippincott 1969). Many of the first chromosomal genes which were identified in *Agrobacterium* and shown to be important for tumor induction are involved in the attachment of agrobacteria to plant cells. These are *chvA* and *chvB* (Douglas *et al.*, 1982, 1985), *exoC* (*pscA*) (Cangelosi *et al.*, 1987; Thomashow *et al.*, 1987), and *att* (Matthysse 1987). Mutations in these genes result in avirulence or severely attenuated virulence on many plant species. It appears that the cellulose fibrils on the cell surface of *A. tumefaciens* also help the bacteria attach tightly to the plant host cell, since *cel*-negative mutants have a reduced virulence (Matthysse 1983; Minnemeyer *et al.*, 1991).

Among these genes, the *att* loci seem to be specifically involved in attachment, but the function of their gene products remains to be elucidated. The genes *chvB* and *chvA* code for products involved in the synthesis and export of cyclic β-1,2-glucan, respectively (Puwanesarajah *et al.*, 1985; Zorreguieta *et al.*, 1988), and *exoC* encodes a function essential for the production of extracellular polysaccharides (Cangelosi *et al.*, 1987; Uttaro *et al.*, 1990). How *chvA/B* and *exoC* genes are involved in attachment is still not clear, but it is likely linked to the changes in the cell envelope present in their mutants. The *chvB* mutants exhibit a pleiotropic phenotype, including attachment deficiency, non-motility, and avirulence among others (Douglas *et al.*, 1982; Swart *et al.*, 1994a). It has been reported that attachment ability and virulence of *chvB* mutants can be restored by growing the mutants in the presence of high calcium in a high osmotic medium (Swart *et al.*, 1994b). Part of the reason for the restoration of virulence is the production of the active protein termed rhicadhesin, a bacterial Ca^{2+}-binding protein that mediates the first step of direct attachment of bacteria to plant cells. Unfortunately no mutants have been isolated which lack rhicadhesin, so it is difficult to evaluate the significance of this molecule in attachment. It has been suggested that cyclic β-1,2-glucan is not required for attachment and virulence, but only indirectly influences these properties through an osmoregulatory effect (O'Connell and Handelsman 1989; Swart *et al.*, 1994b). The role of the *chvB* gene in host range determination is suggested by the observation that although avirulent on most plants, *chvB* mutants can incite tumors on a few plant species, including *Solanum tuberosum* (Hooykaas and Schilperoort 1986). The plant receptors to which *Agrobacterium* attaches have been studied by only one laboratory. Wagner and Matthysse (1992) reported that human vitronectin and anti-vitronectin antibodies both inhibited the binding of *A. tumefaciens* to carrot cells, suggesting that

A. tumefaciens might utilize a vitronectin-like protein as the receptor for its attachment to plant cells. Interestingly, *chvB*, *exoC*, and *att* mutants all showed reduced ability to bind vitronectin (Wagner and Matthysse 1992). Genes involved in host cell attachment are likely candidates for host specificity determination. Although these studies did not establish the direct role of *chvA/B*, *exoC*, and *att* genes in host range determination, they did show the importance of bacterial attachment in virulence. Because of the absolute requirement of bacterial attachment for virulence, it is not surprising that null mutants of these genes are avirulent on almost all the plants tested. Their role in host specificity may be better evaluated if subtler changes such as point mutations in these genes can be obtained.

Agrobacterium chromosomal genes affecting bacterial growth and survival in the plant wound environment have also been identified. Mutants of a chromosomally encoded two-component sensory transduction system, ChvG/ChvI, were inhibited by plant wound sap in growth, and were highly sensitive to acidic pH (Charles and Nester 1993; Mantis and Winans 1993). As will be discussed later, these genes are also involved in *vir* gene induction. The *chvG/I* mutants were either avirulent or highly attenuated in virulence on a number of plant species. Although it is not clear whether the two genes are involved in host specificity, the identification of the regulated genes and the environmental signal(s) recognized by this two-component system is likely to reveal some host range determinants. However, one should be cautious in drawing any conclusions regarding the role of these genes in tumorigenesis because of the pleiotropic effects of their mutations.

III.A.2. Chromosomal Genes Affecting vir Gene Expression or T-DNA Transfer

A number of chromosomal genes that affect *vir* gene expression have been identified. As discussed earlier in this chapter, the level of *vir* gene expression and the efficiency of T-DNA processing and transfer play an important role in host specificity. It is therefore possible that many of the chromosomal genes affecting *vir* gene expression or T-DNA transfer are also likely to play some role in host range determination.

chvE The *chvE* gene is probably the best studied *chv* gene. Its function in *vir* gene expression and role in host specificity have been clearly established. The *chvE* gene encodes a periplasmic galactose-binding protein (Cangelosi et al., 1990; Huang et al., 1990), and is required for the synergistic sugar effect in VirA-mediated vir gene activation by acetosyringone and chemotaxis towards a variety of sugars (Cangelosi et al., 1990; Shimodo et al., 1990). The role of chvE in host specificity is evident in that mutations in chvE result in a strain which is avirulent on some, but not all plants (Huang et al., 1990). In addition, virA mutants which no longer display the sugar effect and probably no longer interact with ChvE have the same host range as a chvE mutant (Doty et al., 1996). The chvE gene provides a good example of a chromosomal gene of Agrobacterium which plays dual roles, one in the physiology of Agrobacterium, and the other in the signal transduction pathway required for plant cell transformation.

cbg Genes. Some A. tumefaciens strains virulent on gymnosperms carry specific genes in their chromosome encoding functions which can convert coniferin, normally not an inducer for vir genes, into coniferyl alcohol, a potent inducer (Castle et al., 1992; Morris and Morris 1990). It has been shown that Agrobacterium strains with high β-glucosidase activity respond to coniferin and infect Douglas fir seedlings, whereas most strains with low β-glucosidase fail to respond to coniferin and are therefore avirulent on this plant (Castle et al., 1992). Two of the β-glucosidase genes (cbg) were cloned, and their products hydrolyze coniferin when expressed in E. coli (Castle et al., 1992). However, how much of the host range variation on Douglas fir can be accounted for by the cbg genes remains to be determined since mutants of these genes have not been isolated (Morris, personal communication).

miaA and ros. The miaA gene encodes a tRNA::isopentenyltransferase. When mutated, it results in a 2- to 10-fold reduction in the induced expression of vir genes by acetosyringone (Gray et al., 1992). The decreased vir gene expression in the mutant is probably due to a reduced translation efficiency. Although this effect may not be specific for vir gene expression and miaA mutants are virulent on most plants, a slight reduction in the virulence of these mutants on red potato plants were observed, implying that the miaA gene might have some host range determining role (Gray et al., 1992). Another chromosomal gene, ros, encodes a repressor of the promoter region of the virC and virD operons, but not other vir operons (Cooley et al., 1991; D'Souza-

Ault et al., 1993). Mutations in ros cause the constitutive expression of the two operons and other phenotypes (Close et al., 1987). Since the virC operon has been implicated in assisting T-DNA processing and in host range determination and the virD operon is important for T-DNA processing and transfer (see previous sections), the ros gene may therefore modulate the level of VirC and VirD proteins for efficient infection of certain hosts. However, its host range effect has not been demonstrated.

ivr, chvG/chvI, chvD, and acvB. Of these chromosomal genes, some are absolutely required for virulence, while others seem to be host range determinants, although the mechanism remains to be elucidated. Metts et al., (1991) reported on three chromosomal loci (ivr) that are necessary for vir gene expression and virulence. Mutants of two of the genes were avirulent on all plants tested, but the mutants of the third gene showed full virulence on some plants and were avirulent on others (Metts et al., 1991). This host range defect might be due to the reduction of vir gene expression in the mutant, but this remains to be proven because of the pleiotropic phenotype of these mutants.

The chvG/chvI genes are not only involved in bacterial survival and growth in the plant wound environment as discussed earlier, but also implicated in the control of vir gene expression since chvG and chvI mutants fail to induce their vir genes (Charles and Nester 1993; Mantis and Winans 1993). However, whether and how they affect the host range is not clear, and the pleiotropic phenotype of their mutants makes it difficult to study. It is possible that some of the genes regulated by chvG/chvI might be host range determinants in response to certain environmental stimuli.

The chvD gene encodes a periplasmic ATPase (Winans et al., 1988). Mutations in chvD attenuated virG induction by acidic pH and phosphate starvation. This gene might be involved in host specificity since chvD mutants showed reduced virulence on some plants (Winans et al., 1988).

The acvB gene encodes a function redundant in octopine strains since the virJ gene has the same function in T-DNA transfer (Kalogeraki and Winans 1995; Pan et al., 1995; Wirawan et al., 1993), but this gene is essential for virulence in nopaline strains which do not possess the virJ gene. It is plausible that the acvB gene can affect the host range of octopine strains when the function shared by VirJ and AcvB is a limiting factor.

III.B. HOST SPECIFICITY AND THE INTERACTIONS BETWEEN CHROMOSOMAL AND Ti PLASMID GENES

Based on the discussion in the previous sections, it is evident that neither the chromosome nor the Ti plasmid contains all of the genetic information required for T-DNA transfer and tumor induction. A chromosomal background can sometimes modify or condition the virulence properties of a Ti plasmid. Numerous studies have suggested that the interaction between a Ti plasmid and a chromosomal background are important for *Agrobacterium* virulence (Hood *et al.*, 1986, 1987; Kovacs and Pueppke 1993; Paulus *et al.*, 1989a, 1989b). This is also reflected in the fact that Ti plasmids can be conjugatively transferred between bacteria (see Chapter 10). In a number of cases, transconjugants of one Ti plasmid in different chromosomal backgrounds exhibit different virulence properties and host range. The importance of the interactions between Ti plasmids and different types of chromosomal background is exemplified by the "supervirulence" phenomenon and the close association of certain Ti plasmids with certain chromosomes in grapevine strains.

III.B.1. The "Supervirulence" Phenomenon

A. tumefaciens A281 is a hybrid strain containing the Ti plasmid from the succinamopine strain Bo542 in the chromosomal background of the nopaline-type strain C58 (Sciaky *et al.*, 1978). Strain A281 incites larger and earlier-appearing tumors on a wide range of plants when compared with its parental strains Bo542 and C58. It is also tumorigenic on alfalfa and soybean, a host range trait not exhibited by the Ti plasmid donor strain Bo542 (Hood *et al.*, 1986, 1987; Kovacs and Pueppke 1993). This supervirulent phenotype appears to be due to a *vir* region containing *virB* and *virG* genes in the Ti plasmid pTiBo542 (Hood *et al.*, 1986; Jin *et al.*, 1987). And indeed, non-supervirulent *Agrobacterium* strains carrying these genes have been successfully used to transform some recalcitrant plant species (Hiei *et al.*, 1994; Ishida *et al.*, 1996; Komari 1990). However, the contribution of the chromosome and the interaction between the chromosome and the Ti plasmid cannot be underestimated since the strain carrying the same Ti plasmid in its native chromosomal background (Bo542) is not nearly as

virulent as A281 (Hood *et al.*, 1987). This implies that the chromosomal background influences the extent to which a Ti plasmid can realize its tumorigenic potential.

The phenomenon that a chromosomal background can potentiate the tumorigenic ability of a Ti plasmid is also found in the studies on a naturally occurring *A. tumefaciens* strain Chry5, which is indistinguishable from A281 in virulence (Kovacs and Pueppke 1993). It has been found that the Chry5 chromosome enhances the virulence on soybean of all the three Ti plasmids tested (pTiAch5, pTiT37, and pTiBo542) compared to their wild-type parental strains, although pTiBo542 and pTiChry5 have the highest tumorigenic potentials (Kovacs and Pueppke 1993, 1994). The two plasmids pTiBo542 and pTiChry5 are highly similar, and share one opine marker, L,L-succinamopine (Kovacs and Pueppke 1994). They may be phylogenetically related. Interestingly, their parental strains Bo542 and Chry5 were originally isolated from taxonomically related plant hosts (Miller 1975; Sciaky *et al.*, 1978). These studies suggest that the virulence-conditioning effects of different chromosomal backgrounds might be due to positively or negatively acting factors. Whether these factors are encoded by some of the chromosomal virulence genes discussed earlier remains to be determined.

One chromosomal gene, *chvE*, was shown to have different effects on the *vir* gene induction by acetosyringone in different Ti plasmids. The hybrid strain A348, which has the Ti plasmid of the octopine strain A6 in the chromosomal background of the nopaline strain C58, can be induced by low levels of acetosyringone in the presence of sugars which synergistically activate *vir* gene expression via ChvE (Cangelosi *et al.*, 1990; Shimodo *et al.*, 1990). However, the sugar effect is not needed to induce the *vir* genes in A348 if high levels of acetosyringone are present. Doty *et al.*, (1996) demonstrated that C58 VirA required the ChvE protein in order to respond to even high levels of acetosyringone while *vir* gene induction under these conditions was unaffected in the *chvE* mutant of A348. This suggests that the ChvE protein encoded by the C58 chromosome interacts with pTiC58-encoded VirA differently from pTiA6-encoded VirA. Another transconjugant CB100, which harbors the Ti plasmid from the biotype II *A. tumefaciens* strain D10B/87 in the chromosomal background of strain C58, was shown to be defective in *vir* gene induction while its parental strain D10B/87 was induced normally (Bélanger, Loubens, Nester, and Dion, unpublished data). The defect in *vir* gene induction in strain D10B/87 was due, at least in part, to the incompatible interaction between D10B/87 Ti plasmid-encoded VirA and C58 chromosome-encoded ChvE (Bélanger, Loubens, Nester, and Dion, unpublished data). These studies suggest that the *vir* gene induction and virulence properties of a given *Agrobacterium* strain are the end result of coevolution of a chromosome and a Ti plasmid, and the optimization of interactions between them.

The importance of the chromosomal background for virulence is also suggested by the fact that nopaline-type Ti plasmids which do not have the *virJ* gene can only confer virulence in a chromosomal background with a functional *acvB* gene, which codes the same function as *virJ* (Kalogeraki and Winans 1995; Pan et al., 1995; Wirawan *et al.*, 1993). Taken together, this type of study holds the potential to identify *A. tumefaciens* chromosome-Ti plasmid combinations that are especially virulent on certain plant species. Such strains would be of considerable practical value in the genetic engineering of recalcitrant plants.

III.B.2. The Grapevine Strains

Agrobacterium strains can be divided into three biotypes based on their chromosomal markers (reviewed by Kerr 1992). The biotype III strains, now named *A. vitis*, have been found to be specifically associated with grapevines in nature, although some of the strains can induce tumors on other plant species when artificially inoculated (Burr *et al.*, 1987; Knauf *et al.*, 1982; Panagopoulos *et al.*, 1978; Perry and Kado 1982). Although the full picture for this host specificity is still obscure, it is apparent that several factors contribute, at least in part, to the close association between *A. vitis* and grapevines.

One of these factors is an *A. vitis*-specific polygalacturonase gene. In addition to crown gall, *A. vitis* also causes a root decay that is specific to grape roots (Burr *et al.*, 1988). This root decay is caused by a polygalacturonase encoded by the *A. vitis* chromosome (Rodriguez-Palenzuela *et al.*, 1991). This enzyme plays an important role in host specificity since it is not present in other biotypes of *Agrobacterium* and its mutant not only fails to induce root necrosis but also is substantially less tumorigenic on grapevines (Rodriguez-Palenzuela *et al.*, 1991).

The ability of most *A. vitis* isolates to use tartrate, an abundant compound in grapevines, as a sole carbon source may also contribute to their host specificity. It has been shown that unlike most *A. tumefaciens* biotype I strains, both biotype II and *A. vitis* strains utilize L-tartrate as a sole carbon source, but biotype II strains prefer glucose to tartrate whereas *A. vitis* uses tartrate even in the presence of glucose (Kerr and Panagopoulos 1977; Szegedi 1985). This trait is plasmid encoded (Szegedi *et al.*, 1992; Otten *et al.*, 1995), and may be another important host range factor for *A. vitis*.

The Ti plasmids carried by *A. vitis* strains can be divided into at least three groups, octopine/cucumopine-type, nopaline-type, and vitopine-type (for reviews see: Paulus *et al.*, 1989b; Otten *et al.*, 1992). Considerable differences in their T-DNA structure caused by insertion sequences have been observed. By following the distribution of these insertion sequences, Paulus *et al.*, (1989a and 1989b) noticed that the octopine/cucumopine-type Ti plasmids seem to be associated with particular chromosomal backgrounds of *A. vitis*. Why such diverse groups of Ti plasmids, with few exceptions, are specifically associated with the biotype III chromosomal background of *A. vitis* is intriguing. Further studies in this area will inevitably extend our appreciation of the interactions between Ti plasmids and their chromosomal backgrounds, and help us design better ways to control crown gall disease on grapevines.

IV. Plant Factors and *Agrobacterium* Host Specificity

The ultimate outcome resulting from the transfer of the T-DNA into a plant cell is determined by the plant cell's response to the changes in hormone levels and effects caused by the T-DNA gene expression. However, host factors involved in *Agrobacterium*-mediated tumorigenesis are still the most poorly understood area in this bacterium-plant interaction. Plant factors apparently play an important role in the host range determination of *Agrobacterium*, since variations in susceptibility to *Agrobacterium* infection exist among closely related plant species or even different cultivars. This has been studied extensively in grapevines, peas, and soybean (Hawes *et al.*, 1989; Hobbs *et al.*, 1989; Lowe and Krul 1991; Owens and Cress 1985; Szegedi *et al.*, 1989). Resistance to tumor formation induced by *A. vitis* in grapevine *Vitis amurensis* is a dominant trait (Szegedi and Kozma 1984). This species-specific and sometimes cultivar-specific host susceptibility implies that host range determinants encoded by *Agrobacterium* Ti plasmid and chromosome might interact with, or are regulated by, certain highly conserved plant factors. Unfortunately, very little is known about the nature and genetic basis of this species- and cultivar-specific host susceptibility.

The ability of a plant to synthesize and secrete *vir* gene-inducing compounds should play a role in *Agrobacterium* host range determination. It has been reported that some hard-to-transform plants do not secrete *vir* gene-inducing molecules that are secreted by easily-transformed dicot plants (Usami *et al.*, 1987), but this does not explain why some seemingly non-transformable plants also synthesize strong *vir* gene inducers (Messens *et al.*, 1990; Usami *et al.*, 1988). The host specificity might also be influenced by the metabolites of a plant. Sahi *et al.*, (1990) reported that a corn metabolite (DIMBOA) strongly inhibited both the growth of *A. tumefaciens* and the induction of its *vir* genes. This may explain the difficulty of stably transforming maize using *Agrobacterium*.

A plant receptor for *Agrobacterium* attachment might be involved in the interaction. The receptor seems to be a protein since extraction of carrot cells with dilute detergent as well as treatment of the cells with trypsin or other proteases has been shown to remove the bacterial binding sites (Gurlitz *et al.*, 1987). Vitronectin-like proteins have been implicated in attachment of *Agrobacterium* to plant cells (Wagner and Matthysse 1992). Swart *et al.*, (1994) reported the isolation of a putative glycoprotein receptor for *A. tumefaciens* rhicadhesin from pea root cell walls. However, definitive evidence is still lacking for a *bona fide* plant cell receptor to which *Agrobacterium* cells bind. Fullner *et al.*, (1996) have recently reported that in addition to T-DNA processing and transfer, the *vir* genes code for the assembly of pili in *Agrobacterium*. There is evidence from a comparison of nucleotide and protein sequences that the VirB2 protein is the pilin subunit (Kado 1994; Shirasu and Kado 1993). It is possible that VirB2 can serve as an adhesin for *Agrobacterium* to tightly bind to plant cells. It will be interesting to see whether plant cells possess a receptor for the VirB2 protein and whether variations in the putative receptor can determine the susceptibility of a plant to *Agrobacterium* infection.

Host factors have also been found to affect transformation after T-DNA is transferred into plant cells. It has been reported that certain *Arabidopsis thaliana* irradiation-hypersensitive mutants are deficient in T-DNA integration (Sonti *et al.*, 1995). How much the plant factors discussed above can account for the species- and cultivar-specific host specificity is not clear. Identification of more plant determinants in their susceptibility to *Agrobacterium* will in no doubt help us better understand the host specificity of *Agrobacterium*.

V. Transformation of Monocots and Recalcitrant Plants by *Agrobacterium*: Successes and Challenges

An extensive study of the host range of crown gall caused by *A. tumefaciens* showed that the majority of the gymnosperms and the dicotyledonous angiosperms were sensitive to the disease, while only a few monocots were susceptible (De Cleene 1985; De Cleene and De Ley 1976). Therefore it was originally thought that there is a taxonomic boundary for the host range of *Agrobacterium*. Differences in biochemistry and physiology as well as anatomy between dicots and monocots have been suggested to explain this barrier (for reviews see: Binns and Thomashow 1988; Smith and Hood 1995). However, crown gall tumor formation depends on the age, cell type (stems vs. leaves), and physiological state of the plants as well as the infection procedure even in some sensitive dicotyledonous plants. Monocotyledons might not be natural hosts for *Agrobacterium*-mediated crown gall formation, but this does not rule out the possibility that *Agrobacterium* is able to transfer its T-DNA into a monocotyledonous plant. The advantages of *Agrobacterium*-mediated transformation over other transformation methods such as protoplast transformation and biolistic methods (Ishida *et al.*, 1996; Potrykus 1990; also see Chapter 17 on "the use of *Agrobacterium* for plant genetic engineering") and the economic importance of some of the monocotyledons, namely the family *Gramineae*, have encouraged researchers to attempt to transform a number of monocotyledons using *Agrobacterium* as the vector. These attempts have identified several critical parameters in *Agrobacterium*-mediated transformation of monocotyledons.

The problematic steps that can limit the efficiency of monocotyledon transformation do not seem to be in the attachment of bacteria or the availability of *vir* gene-inducing compounds in at least some monocots. Although *Agrobacterium* may attach to the cells of some monocots poorly, it was shown convincingly that *A. tumefaciens* can attach to many monocots, including maize and wheat, in the same manner as they attach to dicots (Douglas *et al.*, 1985; Draper *et al.*, 1983; Lippincott and Lippincott 1978; Graves *et al.*, 1988; and references therein). Lack of *vir* gene inducers cannot be used to explain why some monocots are difficult to transform either since inducing compounds are clearly present in a number of monocots, although some monocots, such as maize, can also produce growth and *vir* gene induction inhibitors (Messens *et al.*, 1990; Sahi *et al.*, 1990; Usami *et al.*, 1988). Some monocots, such as rice (Xu *et al.*, 1996), may synthesize *vir* gene inducing compounds, but only do so in a specific developmental stage or a specific tissue.

Although most monocots are resistant to crown gall disease, *Agrobacterium* can induce tumors on monocots in the families *Liliaceae* and *Amaryllidaceae* (De Cleene 1985; De Cleene and De Ley 1976; Hernalsteens *et al.*, 1984; Hooykaas-Van Slogteren *et al.*, 1984). Unlike most other monocotyledons, monocots in these two families are easily amenable to plant tissue culture and produce a wound response similar to many dicotyledonous plants (Hernalsteens *et al.*, 1984; Bytebier *et al.*, 1987; and references therein). As a general rule, the amenability of a plant to tissue culture correlates well with its susceptibility to tumor induction by *Agrobacterium* since in both cases plant growth regulators are the key players. It is therefore not surprising that another monocot *Commelina communis*, which produces a wound response similar to dicotyledonous plants and is amenable to plant tissue culture, can be easily induced to form tumors by *A. tumefaciens* (Deng and Shao 1989). This implies that for many monocotyledons only certain cells in the plant are susceptible to *Agrobacterium* infection, or that T-DNA transfer and integration do occur, but tumor formation does not result because of the insensitivity of these plants to hormone alterations. Differences in the basic anatomy and location of meristematic cell types between dicots and monocots and the tendency of monocotyledonous cells in general to lose their ability to dedifferentiate at a very early stage in development can all be reasons why most monocotyledons do not form tumors in response to *Agrobacterium* infection (Graves *et al.*, 1988; Smith and Hood 1995).

Indeed, by using improved inoculation and transformation methods, wild-type *A. tumefaciens* was shown to infect a number of important members of the *Gramineae*, including maize, rice, and barley and wheat (Deng *et al.*, 1990; Graves and Goldman 1988; Raineri *et al.*, 1990). Inoculation of *Agrobacterium* into meristematic regions or easily-dedifferentiated tissues was the key to success. By using the agroinfection technique, Grimsley *et al.*, (1988) also showed that the shoot apical meristematic region of maize seedlings is the tissue susceptible to *A. tumefaciens* infection. The same agroinfection approach was used to show that *A. tumefaciens* can deliver T-DNA into many, if not all, cereals and grasses, including rice, wheat, oat, and onion (Dasgupta *et al.*, 1991; Dommisse *et al.*, 1990; Donson *et al.*, 1988; Woolston *et al.*, 1988). Similarly, maize shoot tips were found to be susceptible to T-DNA transformation (Gould *et al.*, 1991; Shen *et al.*, 1993). The importance of using meristematic or embryogenic tissue was also stressed in *Agrobacterium*-mediated transformation of rice (Chan *et al.*, 1993; Raineri *et al.*, 1990). In addition, the use of phenolic compounds or wound exudate from a dicot plant during the cocultivation of *Agrobacterium* and plant tissues was also shown to be critical for the transformation of some monocotyledons, implying *vir* gene-inducing molecules in these plants could be the limiting factor (Chan *et al.*, 1993; Schafer *et al.*, 1987). Although a growing number of monocots have been reported to be susceptible to *Agrobacterium* infection (for a comprehensive review see: Smith and Hood 1995), these experiments were criticized by some investigators who felt that they lacked unequivocal molecular and genetic data to support the claims (Potrykus 1990). It was not until recently that such unequivocal evidence was provided for *Agrobacterium*-mediated transformation of two of the most important cereal crops, rice and maize (Hiei *et al.*, 1994; Ishida *et al.*, 1996).

A research group from Japan Tobacco Inc. has reported that high efficiency transformation of some cultivars or varieties of rice and maize mediated by *A. tumefaciens* can be achieved (Hiei *et al.*, 1994; Ishida *et al.*, 1996). Hundreds of fertile transgenic rice and maize plants were obtained with an efficiency similar to that of transformation in dicotyledons, and clear Mendelian inheritance of the transgenes was confirmed by genetic and molecular analysis. This success was attributed to a number of factors, which had been previously implicated in *Agrobacterium*-mediated transformation experiments on other monocots and dicots. These included the choice of plant material used for transformation, the conditions of tissue culture, bacterial strains, and appropriate vectors and selectable markers. Among these variables, the infection of actively dividing, most often embryogenic plant tissues, the supplementation of *vir* gene-inducing compounds, and the use of a supervirulent vector appear to be the most important factors (Hiei *et al.*, 1994; Hood *et al.*, 1986; Ishida *et al.*, 1996; Jin *et al.*, 1987; Komari 1990). However, some rice and maize cultivars or varieties were clearly more difficult to transform than others (Hiei *et al.*, 1994; Ishida *et al.*, 1996). By using a differently modified vector, Hansen *et al.*, (1994) also showed that the transformation frequency of maize could be greatly improved. These studies clearly demonstrate that *Agrobacterium* uses the same mechanism to transfer T-DNA into both dicotyledons and monocotyledons. It is encouraging to note that a few other recalcitrant and important crop plants, such as pea and cassava, have been successfully transformed using *Agrobacterium* as the vector (Li *et al.*, 1996; Schroeder *et al.*, 1993; Shade *et al.*, 1994). Recently, by using embryogenic suspension cultures, an efficient *Agrobacterium*-mediated transformation procedure has been established for an important forage crop, the monocot ryegrasses (*Lolium perenne* L. and *Lolium multiflorum* Lam.) (Wang, Hong, Schubert, Zhang, and Posselt, personal communication). Some of the parameters important for *Agrobacterium*-mediated transformation of rice and maize were reiterated in these studies. Not coincidentally, these successes were preceded by the establishment of a highly reproducible and efficient tissue culture procedure for these plants. The challenge remaining for the transformation of these important crops is how to overcome the cultivar barrier.

VI. Concluding Remarks

During the past twenty years, considerable progress has been made in elucidating the mechanism by which *Agrobacterium* induces crown gall disease. Advances have also been made in understanding the host specificity of *Agrobacterium*. We now understand that the Ti plasmid and a number of chromosomal genes of *Agrobacterium* as well as some plant factors are important host range determinants. We now also appreciate the fact that the ability of an *Agrobacterium* strain to infect a plant can be modified by constructing better vectors and improving infection procedures and tissue culture techniques. This ever-expanding knowledge

about the biology and host specificity of *Agrobacterium* has already led to the transformation of a number of important crops, including rice and maize, which were previously thought to be resistant to *Agrobacterium*. We are now at the threshold where we are able to use *Agrobacterium* to transform most, if not all, important crops at will.

To make that happen, however, we need to know more about the plant factors involved in determining the host specificity of *Agrobacterium*, which are now still very poorly understood. The elucidation of the plant factors participating in the T-DNA transfer and integration process should allow us to improve the efficiency of *Agrobacterium*-mediated transformation. In addition, we need to learn more about the biology of plant tissue culture and regeneration upon which an efficient procedure for the production of transgenic plants must depend. Without this knowledge, the genotype or cultivar barrier observed in the transformation of rice and maize will be hard to cross. However, if the recent success of *Agrobacterium*-mediated transformation of maize, rice, and other crops has told us anything, that is that there seems to be no limit to what plants *Agrobacterium* can transform if we keep trying!

VII. References

Akiyoshi, D.E., Regier, D.A. and Gordon, M.P. (1987) J. Bacteriol. 169, 4242-4248.

Akiyoshi, D.E., Regier, D.A., Jen, G. and Gordon, M.P. (1985) Nucleic Acids Res. 13, 2773-2788.

Anderson, A. and Moore, L. (1979) Phytopathology 69, 320-323.

Ankenbauer, R.G., Best, E.A., Palanca, C.A. and Nester, E.W. (1991) Mol. Plant-Microbe Interact. 4, 400-406.

Beaty, J.S., Powell, G.K., Lica, L., Regier, D.A., MacDonald, E.M.S., Hommes, N.G. and Morris, R.O. (1986) Mol. Gen. Genet. 203, 274-280.

Binns, A.N. and Thomashow, M.F. (1988) Annu. Rev. Microbiol. 42, 575-606.

Boulton, M.I., Buchholz, W.G., Marks, M.S., Markham, P.G. and Davies, J.W. (1989) Plant Mol. Biol. 12, 31-40.

Buchholz, W.G. and Thomashow, M.F. (1984) J. Bacteriol. 160, 327-332.

Bundock, P., den Dulk-Ras, A., Beijersbergen, A. and Hooykaas, P.J.J. (1995) EMBO J. 14, 3206-3214.

Burr, T.J., Bishop, A.L., Katz, B.H., Blanchard, L.M. and Bazzi, C. (1988) Phytopathology 77, 1424-1427.

Burr, T.J., Katz, B.H. and Bishop, A.L. (1987) Plant Dis. 71, 617-620.

Bytebier, B., Deboeck, F., Greve, H.D., Van Montagu, M. and Hernalsteens, J.-P. (1987) Proc. Natl. Acad. Sci. USA. 84, 5345-5349.

Cangelosi, G.A., Ankenbauer, R.G. and Nester, E.W. (1990) Proc. Natl. Acad. Sci. USA. 87, 6708-6712.

Cangelosi, G.A., Hung, L., Puvanesarajah, V., Stacey, G., Ozga, D.A., Leigh, J.A. and Nester, E.W. (1987) J. Bacteriol. 169, 2086-2091.

Castle, L.A., Smith, K.D. and Morris, R.O. (1992) J. Bacteriol. 174,1478-1486.

Chan, M.T., Chang, H.H., Ho, S.L., Tong, W.F. and Yu, S.M. (1993) Plant Mol. Biol. 22, 491-506.

Charles, T. and Nester, E.W. (1993) J. Bacteriol. 175, 6614-6625.

Christie, P.J., Ward, J.E., Winans, S.C. and Nester, E.W. (1988) J. Bacteriol. 170, 2659-2667.

Citovsky, V., Zupan, J., Warnick, D. and Zambryski, P. (1992) Science 256, 1802-1805.

Close, T.J., Rogowsky, P.M., Kado, C.I., Winans, S.C., Yanofsky, M.F. and Nester, E.W. (1987) J. Bacteriol. 169, 5113-5118.

Cooley, M.B., D'Souza, M.R. and Kado, C.I. (1991) J. Bacteriol. 173, 2608-2616.

Culianez-Macia, F.A. and Hepburn, A.G. (1988) Plant Mol. Biol. 11, 389-399.

Dasgupta, I., Hull, R., Eastop, S., Poggi-Pollini, C., Blakebrough, M., Boulton, M.I. and Davies, J.W. (1991) J. Gen. Virol. 72 (Pt 6), 1215-1221.

De Cleene, M. (1985) Phytopath. Z. 113, 81-89.

De Cleene, M. and De Ley, J. (1976) Bot. Rev. 42, 389-466.

Deng, W., Gordon, M.P. and Nester, E.W. (1995a) J. Bacteriol. 177, 2554-2559.

Deng, W., Lin, X. and Shao, Q. (1990) Science in China (Series B) (English Edition) 33, 27-33.

Deng, W., Pu, X.-A., Goodman, R.N., Gordon, M.P. and Nester, E.W. (1995b) Mol. Plant-Microbe Interact. 8, 538-548.

Deng, W. and Shao, Q. (1989) Chinese Science Bulletin (English Edition) 34, 1815-1819.

De Vos, G.D. and Zambryski, P. (1989) Mol. Plant-Microbe Interact. 2, 43-52.

Dommisse, E.M., Leuing, D.W.M., Shaw, M.L. and Conner, A.J. (1990) Plant Sci. 69, 249-257.

Donson, J., Gunn, H.V., Woolston, C.J., Pinner, M.S., Boulton, M.I., Mullineaux, P.M. and Davies, J.W. (1988) Virology 162, 248-250.

Doty, S.L., Yu, M.C., Lundin, J.I., Heath, J.D. and Nester, E.W. (1996) J. Bacteriol. 178, 961-970.

Douglas, C., Halperin, W., Gordon, M.P. and Nester, E.W. (1985) J. Bacteriol. 161, 764-766.

Douglas, C., Halperin, W. and Nester, E.W. (1982) J. Bacteriol. 152, 1265-1275.

Draper, J., Mackenzie, I.A., Davey, M.R. and Freeman, J.P. (1983) Plant Sci. Lett. 29, 227-236.

D'Souza-Ault, M., Cooley, M. and Kado, C.I. (1993) J. Bacteriol. 175, 3486-3490.

Fullner, K.J., Lara, J.C. and Nester, E.W. (1996) Science 273, 1107-1109.

Gardner, R.C. and Knauf, V.C. (1986) Science 231, 725-727.

Gould, J., Devey, M., Hasegawa, O., Ulian, E.C., Peterson, G. and Smith, R.H. (1991) Plant Physiol. 95, 426-434.

Graves, A.E., Goldman, S.L., Banks, S.W. and Graves, A.C.F. (1988) J. Bacteriol. 170, 2395-2400.

Graves, A.E. and Goldman, S.L. (1986) Plant Mol. Biol. 7, 43-50.

Gray, J., Wang, J. and Gelvin, S. (1992) J. Bacteriol. 174, 1086-1098.

Grimsley, N., Hohn, B., Ramos, C., Kado, C. and Rogowsky, P. (1989) Mol. Gen. Genet. 217, 309-316.

Grimsley, N., Hohn, T., Davis, J.W. and Hohn, B. (1987) Nature 325, 177-179.

Grimsley, N., Hohn, B., Hohn, T. and Walden, R. (1986) Proc. Natl. Acad. Sci. USA. 83, 3282-3286.

Grimsley, N., Ramos, C., Hein, T. and Hohn, B. (1988) Bio/technology 6, 185-189.

Gurlitz, R.H.G., Lamb, P.W. and Matthysse, A.G. (1987) Plant Physiol. 83, 564-568.

Hamada, S.E. and Farrand, S.K. (1980) J. Bacteriol. 144, 732-743.

Hansen, G., Das, A. and Chilton, M.-D. (1994) Proc. Natl. Acad. Sci. USA. 91, 7603-7607.

Hawes, M.C., Robbs, S.L. and Pueppke, S.G. (1989) Plant Physiol. 90, 180-184.

Hernalsteens, J.-P., Thia-Toong, L., Schell, J. and Van Montagu, M. (1984) EMBO J. 3, 3039-3041.

Hiei, Y., Ohta, S., Komari, T. and Kumashiro, T. (1994) Plant J. 6, 271-282.

Hille, J., van Kan, J. and Schilperoort, R. (1984) J. Bacteriol. 158, 754-756.

Hirooka, T., Rogowsky, P.M. and Kado, C.I. (1987) J. Bacteriol. 169, 1529-1536.

Hobbs, S.L.A., Jackson, J.A. and Mahon, J.D. (1989) Plant Cell Rep. 8, 274-277.

Hoekema, A., de Pater, B.S., Fellinger, A.J., Hooykaas, P.J.J. and Schilperoort, R.A. (1984) EMBO J. 3, 3043-3047.

Hood, E.E., Helmer, G.L., Fraley, R.T. and Chilton, M.-D. (1986) J. Bacteriol. 168, 1291-1301.

Hood, E.E., Fraley, R.T. and Chilton, M.-D. (1987) Plant Physiol. 83, 529-534.

Hooykaas, P.J.J. and Beijersbergen, A.G.M. (1994) Annu. Rev. Phytopathol. 32, 157-179.

Hooykaas, P.J.J., den Dulk-Ras, H. and Schilperoort, R.A. (1988) Plant Mol. Biol. 11, 791-794.

Hooykaas, P.J.J., Hofker, M., den Dulk-Ras, H. and Schilperoort, R.A. (1984) Plasmid 11,   195-205.

Hooykaas, P.J.J. and Schilperoort, R.A. (1986) In, B. Lugtenberg (ed.), Recognition in microbe-plant symbiotic and pathogenic interactions. Springer-Verlag, Berlin. pp. 189-202.

Hooykaas-Van Slogteren, G.M.S., Hooykaas, P.J.J. and Schilperoort, R.A. (1984) Nature 311, 763-764.

Horsch, R.B., Klee, H.J., Stachel, S., Winans, S.C., Nester, E.W., Rogers, S.G. and Fraley, R.T. (1986) Proc. Natl. Acad. Sci. USA. 83, 2571-2575.

Horvath, B., Bachem, C.W.B., Schell, J. and Kondorosi, A. (1987) EMBO J. 6, 841-848.

Huang, M.-L.W., Cangelosi, G.A., Halperin, W. and Nester, E.W. (1990) J. Bacteriol. 172, 1814-1822.

Ishida, Y., Saito, H., Ohta, S., Hiei, Y., Komari, T. and Kumashiro, T. (1996) Nature Biotechnology 14, 745-750.

Jarchow, E., Grimsley, N.H. and Hohn, B. (1991) Proc. Natl. Acad. Sci. USA. 88, 10426-10430.

Jin, S., Komari, T., Gordon, M.P. and Nester, E.W. (1987) J. Bacteriol. 169, 4417-4425.

Jin, S., Song, Y.-N., Deng, W., Gordon, M.P. and Nester, E.W. (1993a) J. Bacteriol. 175, 6830-6835.

Jin, S., Song, Y.-N., Pan, S.Q. and Nester, E.W. (1993b) Mol. Microbiol. 7, 555-562.

John, M.C. and Amasino, R.M. (1988) J. Bacteriol. 170, 790-795.

Kado, C.I. (1994) Mol. Microbiol. 12,17-22.

Kalogeraki, V.S. and Winans, S.C. (1995) J. Bacteriol. 177,892-897.

Kanemoto, R.H., Powell, A.T., Akiyoshi, D.E., Regier, D.A., Kerstetter, R.A., Nester, E.W., Hawes, M.C. and Gordon, M.P. (1989) J. Bacteriol. 171,2506-2512.

Kang, H.W., Wirawan, I.G.P. and Kojima, M. (1994) Biosci. Biotech. Biochem. 58, 2024-2032.

Kerr, A. (1992) In, Balows, K. et al., (ed.), The prokaryotes, 2nd edition, Vol. III. Springer-Verlag, Berlin. pp. 2214-2235.

Kerr, A. and Panagopoulos, C.G. (1977) Phytopathol. Z. 90, 172-179.

Klee, H.J., White, F.F., Iyer, V.N., Gordon, M.P. and Nester, E.W. (1983) J. Bacteriol. 153, 878-883.

Knauf, V.C., Panagopoulos, C.G. and Nester, E.W. (1982) Phytopathology 72, 1545-1549.

Komari, T. (1990) Plant Cell Rep. 9, 303-306.

Kovacs, L.G. and Pueppke, S.G. (1993) Mol. Plant-Microbe Interact. 6,601-608.

Kovacs, L.G. and Pueppke, S.G. (1994) Mol. Gen. Genet. 242, 327-336.

Lee, Y.-W., Jin, S., Sim, W.-S. and Nester, E.W. (1995) Proc. Natl. Acad. Sci. USA. 92, 12245-12249.

Leroux, B., Yanofsky, M.F., Winans, S.C., Ward, J.E., Ziegler, S.F. and Nester, E.W. (1987) EMBO J. 6, 849-856.

Li, H.-Q., Sautter, C., Potrykus, I. and Puonti-Kaerlas, J. (1996) Nature Bio/technology 14, 736-740.

Lippincott, B.B. and Lippincott, J.A. (1969) J. Bacteriol. 97, 820-628.

Long, S.R. (1989) Cell 56,203-214.

Loper, J.E. and Kado, C.I. (1979) J. Bacteriol. 139, 591-596.

Lowe, B.A. and Krul, W.R. (1991) Plant Physiol. 96, 121-129.

Mantis, N. and Winans, S.C. (1993) J. Bacteriol. 175, 6626-6636.

Matthysse, A.G. (1983) J. Bacteriol. 154, 906-915.

Matthysse, A.G. (1987) J. Bacteriol. 169, 313-323.

Melchers, L.S., Maroney, M.J., den Dulk-Ras, A., Thompson, D.V., van Vuuren, H.A.J., Schilperoort, R.A. and Hooykaas, P.J.J. (1990) Plant Mol. Biol. 14, 149-159.

Melchers, L.S., Thompson, D.V., Idler, K.B., Neuteboom, S.T.C., de Maagd, R.A., Schilperoort, R.A. and Hooykaas, P.J.J. (1987) Plant Mol. Biol. 9, 635-645.

Messens, E., Dekeyser, R. and Stachel, S.E. (1990) Proc. Natl. Acad. Sci. USA. 87, 4368-4372.

Metts, J., West, J., Doares, S. and Matthysse, A. (1991) J. Bacteriol. 173, 1080-1087.

Miller, H.N. (1975) Phytopathology 65, 805-811.

Minnemeyer, S.L., Lightfoot, R. and Matthysse, A.G. (1991) J. Bacteriol. 173,7723-7724.

Morris, J.W. and Morris, R.O. (1990) Proc. Natl. Acad. Sci. USA. 87, 3614-3618.

Nugent, J.M. and Palmer, J.D. (1991) Cell 66,473-481.

O'Connell, K.P. and Handelsman, J. (1989) Mol. Plant-Microbe Interact. 2, 11-16.

Otten, L., De Greve, H., Leemans, J., Hain, R., Hooykaas, P. and Schell, J. (1984) Mol. Gen.Genet. 195, 159-163.

Otten, L., Canaday, J., Gerard, J.-C., Fournier, P., Crouzet, P. and Paulus, F. (1992) Mol. Plant-Microbe Interact. 5, 279-287.

Otten, L., Crouzet, P., Salomone, J.-Y., de Ruffray, P. and Szegedi, E. (1995) Mol. Plant-Microbe Interact. 8, 138-146.

Owens, L.D. and Cress, D.E. (1985) Plant Physiol. 77, 87-94.

Pan, S.Q., Jin, S.G., Boulton, M.I., Hawes, M., Gordon, M.P. and Nester, E.W. (1995) Mol. Microbiol. 17,259-269.

Panagopoulos, C.G. and Psallidas, P.G. (1973) J. Appl. Bacteriol. 36, 233-240.

Panagopoulos, C.G., Psallidas, P.G. and Alivazatos, A.S. (1978) In, Proc. 4th Int. Conf. Plant Pathol. Bacteria, Angers, France. pp.221-228.

Paulus, F., Ride, M. and Otten, L. (1989a) Mol. Gen. Genet. 219,145-152.

Paulus, F., Huss, B., Tinland, B., Ride, M., Szegedi, E., Tempe, J., Petit, A. and Otten, L. (1989b) Mol. Plant-Microbe Interact. 2,64-74.

Paulus, F., Huss, B., Tinland, B., Herrmann, A., Canaday, J. and Otten, L. (1991) Mol. Plant- Microbe Interact. 4, 163-172.

Pazour, G.J., Ta, C.V. and Das, A. (1992) J. Bacteriol. 174, 4169-4174.

Peralta, E.G., Hellmiss, R. and Ream, W. (1986) EMBO J. 5, 1137-1142.

Peralta, E.G. and Ream, W. (1986) Proc. Natl. Acad. Sci. USA. 82, 5112-5116.

Perry, K.L. and Kado, C.I. (1982) J. Bacteriol. 151, 343-350.

Piers, K., Heath, J.D., Liang, X., Stephens, K.M. and Nester, E.W. (1996) Proc. Natl. Acad. Sci. USA. 93, 1613-1618.

Potrykus, I. (1990) Bio/technology 8, 535-542.

Powell, G.K., Hommes, N.G., Kuo, J., Castle, L.A. and Morris, R.O. (1988) Mol. Plant-Microbe Interact. 1, 235-242.

Pu, X.-A. and Goodman, R.N. (1992) Physiol. Mol. Plant Pathol. 41, 241-254.

Puvanesarajah, V., Schell, F.M., Stacey, G., Douglas, C.J. and Nester, E.W. (1985) J. Bacteriol. 164, 102-106.

Raineri, D.M., Bottino, P., Gordon, M.P. and Nester, E.W. (1990) Bio/technology 8, 33-38.

Raineri, D.M., Boulton, M.I., Davies, J.W. and Nester, E.W. (1993) Proc. Natl. Acad. Sci. USA. 90, 3549-3553.

Regensburg-Tuïnk, A.J.G. and Hooykaas, P.J.J. (1993) Nature 363, 69-71.

Rodriguez-Palenzuela, P., Burr, T.J. and Collmer, A. (1991) J. Bacteriol. 173, 6547-6552.

Sahi, S.V., Chilton, M.-D. and Chilton, W.S. (1990) Proc. Natl. Acad. Sci. USA. 87, 3879-3883.

Schafer, W., Gorz, A. and Kahl, G. (1987) Nature 327, 529-532.

Schroeder, H.E., Schotz, A.H., Wardley-Richardson, T., Spencer, D. and Higgins, T.J.V. (1993) Plant Physiol. 101, 751-757.

Sciaky, D., Montoya, A. and Chilton, M.-D. (1978) Plasmid 1, 238-253.

Shade, R.E., Schroeder, H.E., Pueyo, J.J., Tabe, L.M., Murdock, L.L., Higgins, T.J.V. and Chrispeels, M.J. (1994) Bio/technology 12, 793-796.

Shen, W.-H., Escudero, J., Schlappi, M., Ramos, C., Hohn, B. and Koukolikova-Nicola, Z. (1993) Proc. Natl. Acad. Sci. USA. 90, 1488-1492.

Shimoda, N., Toyoda-Yamamoto, A., Aoki, S. and Machida, Y. (1993) J. Biol. Chem. 268, 26552-26558.

Shimoda, N., Toyoda-Yamamoto, A., Nagamine, J., Usami, S., Katayama, M., Sakagami, Y. and Machida, Y. (1990) Proc. Natl. Acad. Sci. USA. 87, 6684-6688.

Shirasu, K. and Kado, C.I. (1993) FEMS Microbiol. Lett. 111,287-294.

Smith, R.H. and Hood, E.E. (1995) Crop Sci. 35,301-309.

Sonti, R.V., Chiurazzi, M., Wong, D., Davies, C.S., Harlow, G.R., Mount, D.W. and Signer, E.R. (1995) Proc. Natl. Acad. Sci. USA. 92, 11786-11790.

Spaink, H.P., Wijffelman, C.A., Pees, E., Okker, R.J.H. and Lugtenberg, B.J.J. (1987) Nature 328, 337-340.

Spencer, P.A., Tanaka, A. and Towers, G.H.N. (1990) Phytochemistry 29, 3785-3788.

Stachel, S.E. and Nester, E.W. (1986) EMBO J. 5, 1445-1454.

Swart, S., Logman, T.J.J., Lugtenberg, B.J.J., Smit, G. and Kijne, J.W. (1994a) Arch.Microbiol. 161, 310-315.

Swart, S., Lugtenberg, B.J.J., Smit, G. and Kijne, J.W. (1994b) J. Bacteriol. 176, 3816-3819.

Swart, S., Logman, T.J.J., Smit, G., Lugtenberg, B.J.J. and Kijne, J.W. (1994c) Plant Mol. Biol. 24, 171-183.

Szegedi, E. (1985) Acta Phytopathol. Acad. Sci. Hung. 20, 17-22.

Szegedi, E. and Kozma, P. (1984) Vitis 23, 121-126.

Szegedi, E., Korbuly, J. and Otten., L. (1989) Physiol. Mol. Plant Pathol. 35, 35-43.

Szegedi, E., Otten, L. and Czako, M. (1992) Mol. Plant-Microbe Interact. 5, 435-438.

Thomashow, M.F., Karlinsey, J.E., Marks, J.R. and Hurlbert, R.E. (1987) J. Bacteriol. 169, 3209-3216.

Thomashow, M.F., Panagopoulos, C.G., Gordon, M.P. and Nester, E.W. (1980) Nature 283, 794-796.

Toro, N., Datta, A., Yanofsky, M. and Nester, E.W. (1989a) Proc. Natl. Acad. Sci. USA. 85, 8558-8562.

Toro, N., Datta, A., Carmi, O.A., Young, C., Prusti, R.K. and Nester, E.W. (1989b) J. Bacteriol. 171, 6845-6849.

Turk, S., Nester, E. and Hooykaas, P. (1993) Mol. Microbiol. 7, 719-724.

Turk, S.C.H.J., Melchers, L.S., den Dulk-Ras, H., Regensburg-Tuink, A.J.G. and Hooykaas, P.J.J. (1991) Plant Mol. Biol. 16, 1051-1059.

Unger, L., Ziegler, S.F., Huffman, G.A., Knauf, V.C., Peet, R., Moore, L.W., Gordon, M.P. and Nester, E.W. (1985) J. Bacteriol. 164, 723-730.

Usami, S., Morikawa, S., Takebe, I. and Machida, Y. (1987) Mol. Gen. Genet. 209, 221-226.

Usami, S., Okamoto, S., Takebe, I. and Machida, Y. (1988) Proc. Natl. Acad. Sci. USA. 85, 3748-3752.

Uttaro, A.D., Cangelosi, G.A., Geremia, R.A., Nester, E.W. and Ugalde, R.A. (1990) J. Bacteriol. 172, 1640-1646.

Wagner, V.T. and Matthysse, A.G. (1992) J. Bacteriol. 174, 5999-6003.

White, F.F., Taylor, B.H., Huffman, G.A., Gordon, M.P. and Nester, E.W. (1985) J. Bacteriol. 164, 33-44.

Winans, S.C., Kerstetter, R.A. and Nester, E.W. (1988) J. Bacteriol. 170, 4047-4054.

Wirawan, I.G.P., Kang, H.W. and Kojima, M. (1993) J. Bacteriol. 175, 3208-3212.

Woolston, C.J., Barker, R., Gunn, H., Boulton, M.I. and Mullineaux, P.M. (1988) Plant Mol. Biol. 11, 35-43.

Xu, D., Li, B., Liu, Y., Huang, Z. and Gu, L. (1996) Science in China (Series C) 39, 8-16.

Yanofsky, M., Lowe, B., Montoya, A, Rubin, R., Krul, W., Gordon, M. and Nester, E. (1985a) Mol. Gen. Genet. 201, 237-246.

Yanofsky, M., Montoya, A., Knauf, V., Lowe, B., Gordon, M. and Nester, E. (1985b) J. Bacteriol. 163, 341-348.

Zhan, X., Jones, D.A. and Kerr, A. (1990) Plant Mol. Biol. 14, 785-792.

Zorreguieta, A., Geremia, R.A., Cavaignac, S., Cangelosi, G.A., Nester, E.W. and Ugalde, R.A. (1988) Mol. Plant-Microbe Interact. 1, 121-127.

Chapter 17

The Use of *Agrobacterium* for Plant Genetic Engineering

Kathleen D'Halluin and Johan Botterman

I. Introduction

Agrobacterium tumefaciens and *A. rhizogenes* are soil pathogens that induce crown gall tumors and hairy roots, respectively, on plants. During the infection a specific segment of the tumor-inducing (Ti) or root-inducing (Ri) plasmid, the T-DNA, is transferred from the bacterium to the plant cell, followed by its integration in the plant genome and subsequent expression. The T-DNA contains 1) genes encoding enzymes involved in the synthesis of phytohormones (*onc* genes) that cause proliferation of crown gall and hairy root cells and 2) genes which code for the production of opines, that serve as specific nutrients for the infecting bacteria. The T-DNA region on the Ti-plasmid is flanked by directly repeated 25-bp border sequences that are required in cis for its precise excision and transfer to the plant cell (Zambryski, 1992). Excision and transfer functions of the T-DNA are mediated by *vir* genes, located on the Ti-plasmid outside the border sequences. This DNA transfer system based on *Agrobacterium* T-DNA transfer has been exploited to develop host-vector systems for the introduction of gene constructs in plant species.

II. Vector Design (Figure 1)

Since the bacterial genes on the T-DNA are not necessary for its transfer, these genes can be deleted and replaced by a gene of interest (Zambryski *et al.*, 1983). Such "disarmed" vectors still contain T-DNA border sequences and the DNA comprised between the border repeats, is transferred to the plant genome. Co-integration vectors have been constructed in which a gene of interest, cloned on an *E.coli* plasmid is inserted into the T-DNA by homologous recombination (Deblaere *et al.*, 1985). Binary vector systems, in which the *Agrobacterium* strain carries two plasmids, are also used. One plasmid has 1) a selectable marker for propagation in the bacterial host and 2) cloning sites between T-DNA borders and 3) origins of replication that allow replication in both *E. coli* and *Agrobacterium* host strains. Virulence functions required for T-DNA transfer are complemented in trans by virulence genes, present on a Ti-plasmid derivative (Hoekema *et al.*, 1983).

Modified binary vectors, carrying extra virulence genes from the supervirulent strain A281, (Hiei *et al.*, 1994; Komari *et al.*, 1990 and Saito *et al.*, 1992) or carrying two separate T-DNAs, have been developed (Komari *et al.*, 1996). In the latter a fragment from the virulence region of pTiBo542 (Komari *et al.*, 1986) was used as spacer fragment between the two T-DNAs and seems to be useful for the production of marker free plants. High co-transformation frequencies favoring unlinked loci were obtained with these types of vectors (Komari *et al.*, 1996).

Most constructs contain less than 25 kb between the T-DNA borders. Recently, a binary bacterial artificial chromosome (BIBAC) vector was

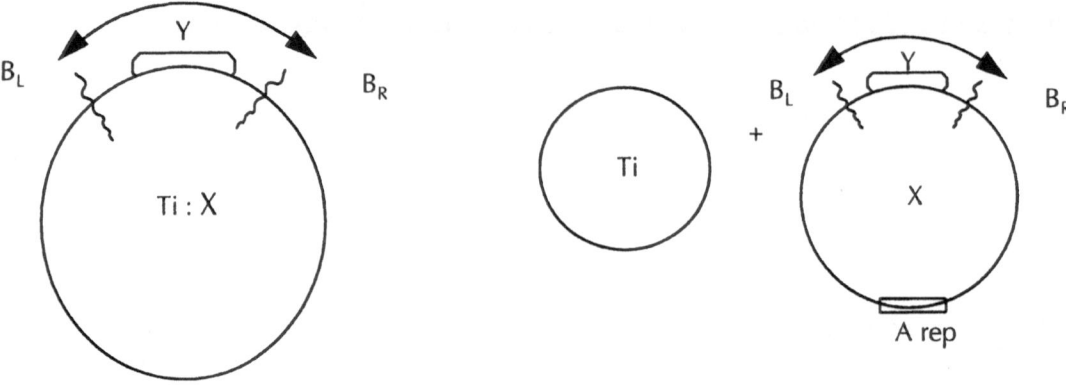

Ti : X : co-integration vector

    y: gene of interest
    B_L: left border
    B_R: right border

Ti + X: binary vector system
A rep: Agro replicon

Figure 1. Vector Design

described which enables to transfer at least 150 kb of foreign DNA into the nuclear genome of tobacco (Hamilton *et al.*, 1996).

III. Method

Agrobacterium has routinely and widely been used as vector to transfer foreign genes into dicotyledonous plants. The method of T-DNA transfer is relatively simple, protoplasts are not longer required and DNA can be introduced into plant tissues. Most methods are a modification of the leaf disc transformation system of Horsch *et al.*, (1985). Sectioned tissues such as leaf, root, stem, hypocotyl and cotyledon segments, embryogenic or organogenic callus and even whole organs, such as zygotic or somatic embryos and seeds can be used as starting material for co-cultivation with *Agrobacterium*. *Agrobacterium* transformation in *planta* has also been demonstrated for *Arabidopsis* (Bechtold *et al.*, 1993, Feldmann *et al.*, 1994). Any gene of

interest can be inserted within the T-DNA and transformed into a wide variety of dicotyledonous plants. Nearly all major dicotyledonous crops have been transformed by *Agrobacterium*. Many attempts have been made to produce stably transgenic cereals, but until recently without much success. Recently, Hiei *et al.*, (1994) demonstrated that *Agrobacterium*-mediated transformation of rice has become an applicable strategy. *Agrobacterium*-mediated transformation of rice scutellum-derived callus yielded transformants at a high frequency (10-30%). An improved binary vector, a so-called superbinary vector, comprising a fragment from the virulence region of pTiBo542 containing *virG* and the 3'end of *virB* was used. Success in stable transformation of maize was also recently reported by using the same type of vector (Ishida *et al.*, 1996). It may be expected that by further unraveling the process of *Agrobacterium*-mediated transfer of DNA in dicotyledonous plants, and by further development of more appropriate vector and tissue culture systems, *Agrobacterium* will be routinely applicable as well for cereal transformation in the future.

IV. *Agrobacterium*-Mediated Transformation Versus Direct Gene Transfer Methods

As it was believed for a long time that monocotyledonous plants were not susceptible for *Agrobacterium*, alternative transformation systems based on the direct delivery of DNA into cells and tissues have been developed. Although successful transformation methods based on direct DNA delivery methods are nowadays available, the availability of an *Agrobacterium* based transformation procedure for agronomically important crops such as cereals would offer major advantages (Chilton, 1993).

The main advantages are 1) DNA fragments as comprised between the T-DNA border repeats are transferred, 2) relatively large segments of DNA comprising linked gene constructs can be transferred, 3) the integration patterns are relatively simple with a high frequency of single copy integration events which appear to be less prone to methylation and gene silencing and 4) the integrated DNA is not rearranged relative to the designed gene constructs in the bacterial plasmid. It has been well documented that the integrated T-DNA is genetically stable and segregates in a Mendelian fashion.

This is in contrast with transformation events from direct gene transfer methods where transformants with multi-copy insertions are mostly obtained. Non-functional pieces of the gene and vector parts as well are also often integrated in a random fashion in the genome. Unusual patterns of segregation or no inheritance at all are often observed (Walters *et al.*, 1990; Register *et al.*, 1994). These observations significantly complicate the evaluation of transgenic lines in downstream screening programs. In addition, under the current regulatory requirements for product development with crop species carrying transgenic traits, the complex integration patterns do create additional hurdles to get regulatory approval. As only plants with simple integration patterns and without undesired DNA sequences are recommended for product development, many transformants obtained by direct DNA delivery methods have to be discarded. More tedious screening programs and a significantly larger pool of transformants have to be produced by direct DNA delivery methods compared with *Agrobacterium*-mediated transformation to select a line with an adequate insertion event.

V. Gene Tagging

Large numbers of independent transformants are required for insertional mutagenesis of the genome. The high *Agrobacterium* transformation frequency in *Arabidopsis* has been used to create large numbers of T-DNA-tagged *Arabidopsis* lines (Koncz *et al.*, 1992; Feldmann *et al.*, 1994). The integration of T-DNA into the plant genome can be used to identify plant promoters by creating fusions of reporter genes to plant regulatory sequences and to identify plant promoters as T-DNA integrates preferentially into actively transcribed regions of the plant genome (Koncz *et al.*, 1989; Kertbundit *et al.*, 1991, Mandal *et al.*, 1993). Promoter sequences that direct specific patterns of expression can be tagged through random T-DNA-mediated insertion of a promoter-less *gus* gene (Lindsey *et al.*, 1993). For example Goddijn *et al.*, (1993) demonstrated the feasibility of tagging of promoters responding to nematode feeding by *Agrobacterium*-mediated transformation of *Arabidopsis* with promoter trap vectors, comprising a promoterless *gus* gene near its border.

The tag can also be engineered to contain transcriptional enhancers which causes a deregulation of the expression of the flanking genes after insertion of the tag. This leads to the production of dominant mutations. This approach has been used to isolate genes implicated in mediating auxin signal transduction (Walden *et al.*, 1994) and modifying polyamine biosynthesis (Fritze *et al.*, 1995).

VI. Gene Targeting

There exists considerable variation in expression levels of introduced genes between independent transformants. This variability seems to be the result from the different sites of integration of the introduced gene in the recipient genome (position effect) (Jones *et al.*, 1985). This causes high quantitative and qualitative variation in expression of the transgene. A possible approach to overcome the variation in transgene expression is the targeting of the gene construct into specific chromosomal positions of the genome. Gene targeting could result in a more reproducible quantitative and qualitative expression of a particular transgene. Unfortunately, gene targeting by homologous recombination seems not efficient in plants (Lee *et al.*, 1990; Halfter *et al.*, 1992; Offringa *et al.*, 1993, Morton *et al.*, 1995). Possible improvements to increase partly the frequency of homologous recombination in plants may include the use of longer stretches of homology, the application of more efficient negative selectable marker genes, or the induction of double strand breaks (DSBs) at the target locus. Recently Puchta *et al.*, (1996) demonstrated an increase in the frequency of homologous recombination at a specific locus by the induction of a DSB via an endonuclease. The natural *TGA3* locus of Arabidopsis was recently targeted by using a total homology of 7kb through *Agrobacterium*-mediated transformation with a T-DNA vector carrying a selectable marker gene flanked by two genomic fragments derived from the locus of interest (Miao *et al.*, 1995). Out of 2580 transformed calli, only one targeted callus line could be isolated from which no plants could be regenerated. In conclusion, gene targeting seems not yet an applicable technique for plants at this moment.

The demonstration by Odell *et al.*, (1990, 1994) that plants respond to the bacteriophage P1 derived loxP-Cre site specific recombination system could open a possibility to use this system in *Agrobacterium*-mediated transformation to direct site-specific integration and regulation of gene expression. Site specific recombination systems may also be useful for removing of selectable marker genes from the plant genome and for clustering transgenes. Moreover the site-specific placement of recombination sites within a genome could lead to the large-scale restructuring of chromosomes (Ow *et al.*, 1995).

VII. Application of *Agrobacterium*-Mediated Transformation for Engineering New Traits in Plant Species

Plant genetic engineering, using *Agrobacterium* mediated transformation can help the plant breeder by providing lines with specific modifications of certain genes or engineered with new genes that can result in improved resistance to herbicide, disease and environmental stress, improved hybrid breeding, improved plant architecture and photosynthetic efficiency, improved plant-based remediation techniques, improved industrial characteristics such as quality of seeds and fruits, higher qualities of specific oils, starch and proteins, the production of peptides with various applications to healthcare, the production of novel specialty chemicals etc.

In order to prevent tedious screening of transgenic lines to identify a line with an adequate insertion event of the transgene, *Agrobacterium*-mediated transformation is preferred. A brief overview of some of the major gene constructs used to engineer plants based on *Agrobacterium*-mediated transformation is summarized.

Resistance has been engineered to different non-selective herbicides such as phosphinothricin (De Block *et al.*, 1987), glyphosate (Shah *et al.*, 1986), sulphonylureas and imidazolinones (Haughn *et al.*, 1988). The engineered herbicide resistance can be used not only to control weeds, but also to screen the progeny of transgenic lines in plant breeding. The screening of agronomically useful phenotypes in the breeding program can be facilitated by physically linking the herbicide resistance gene to other genes conferring agronomically useful traits that can otherwhise not be immediately selected as for example the coupling of a genetic male sterility gene to a herbicide resistance gene such as the *bar* gene (Mariani *et al.*, 1990).

As plant breeders and farmers have little defense against virus diseases different strategies have been exploited to engineer virus resistance either by introducing in plants genes encoding viral capsid or coat proteins (Abel *et al.*, 1986; Loesch-Fries *et al.*, (1988) or cDNA copies of satellite RNAs from a number of RNA viruses (Harrison *et al.*, (1987), Gerlach *et al.*, (1987) or by the expression of antisense RNA (Cuozzo *et al.*, 1988).

Insect control as target for transformation has become very important. The use of fewer or less toxic agrochemicals has become increasingly important in insect control as a result of concerns about health risks and environmental pollution. Success in engineering insect resistance in transgenic plants has been achieved by the introduction of genes by *Agrobacterium*-mediated transformation of genes that encode for insecticidal crystal proteins (ICPs) (Vaeck *et al.*, 1987; Fischhoff *et al.*, 1987) or by the introduction of a proteinase inhibitor gene (Hilder *et al.*, 1987).

Approaches to engineer resistance to fungal diseases also receive much attention. Transgenic tobacco and canola seedlings expressing a bean chitinase gene under control of the constitutive CaMV35S promoter showed an increased ability to survive in soil infested with the fungal pathogen *Rhizoctonia* and a delayed development of disease symptoms (Broglie *et al.*, 1991). Another possible approach to engineer fungus-resistant plants by expressing the bacterial ribonuclease, *barnase*, under control of a promoter which is exclusively responsive to pathogen attack was reported by Strittmatter *et al.*, (1995).

Environmental stress is one of the major limiting factors in plant productivity. Genetic manipulation of the expression of enzymes involved in scavenging reactive oxygen intermediates has led to new insights in the role of these enzymes in oxidative stress protection and in new approaches to improve stress tolerance of plants by genetic engineering (Allen, 1995).

Hybrid breeding has been limited to those plants in which male sterility lines have been identified that could be restored to fertility or to those in which, mechanical removal of the anthers has been economically feasible. Hybrid breeding can be facilitated tremendously and broadened to a wide range of plants by the possibility of engineering nuclear male sterility. The introduction of a cytotoxic ribonuclease gene, *barnase*, under control of a tapetum-specific promotor produced male-sterile plants because of tapetal cell destruction (Mariani *et al.*, 1990). The fertility of these genetically engineered male-sterile plants could be restored by the introduction of a chimeric RNase inhibitor gene, *barstar*, under control of the same anther-specific promoter pTA29 (Mariani *et al.*, 1992).

Transgenic manipulation of plant architecture offers opportunities to improve productivity. As the major determinant of plant architecture in the field is the response of the plant to environmental light signals, genetic engineering of the phytochrome (PHY) gene family opens possibilities for the manipulation of yield. When plants are grown at high densities, the shade avoidance syndrome is involved, manifested by a strong elongation growth, occurring at the expense of leaf growth, storage organ production and reproductive development. In field grown tobacco plants with overexpression of PHYA, shade avoidance was suppressed, leading to enhanced allocation of assimilates to the leaves and an increase in harvest index (Robson *et al.*, 1996).

The *rol* genes of *Agrobacterium rhizogenes* can be used to interfere with plant development such as formation of adventitious roots and release of axillary buds from apical dominance, regulation of flower induction, flower development and fertility. Major drawbacks however is that plants transformed with the rol genes show many pleiotropic side effects. Pleiotropic effects may be overcome by *rol* genes that are driven by organ- or tissue specific promoters. An application of *rol* genes for crop improvement was demonstrated in roses consisting of an untransformed scion grafted on a rootstock transformed with a *rol* gene. An accelerated release of axillary buds of the untransformed scion was observed, but no other side effects of the *rol* genes were seen (van der Salm, 1996).

Pollution of our environment with heavy metals poses a significant threat to human health. As the metals cannot be degraded chemically, there are no safe and inexpensive approaches for reducing the toxic metal pollution in our ecosystem. By using *Agrobacterium* mediated transformation, a mercuric ion reductase gene was introduced into *Arabidopsis* and transgenic plants were produced that convert toxic ionic mercury to a less toxic vapor form. This opens perspectives that genetically engineered plants can one day be used as approach to the phytoremediation of metal ion pollution (Rugh *et al.*, 1996).

Improvements to industrial characteristics of agricultural products by genetic engineering have been accomplished. Examples are the extended shelf-life for tomato fruits by introducing a gene that decreases the level of endogenous ethylene (Hamilton *et al.*, 1990; Oeller *et al.*, (1991), or that reduce the activity of the cell-wall-degrading enzyme polygalacturonase (Gray *et al.*, 1994), the increased levels of unsaturated fatty acids in oilseed by modulating the expression of stearoyl-ACP desaturase (Kridl *et al.*, 1991), the increased level of starch by the expression of a mutant *E. coli* ADP glucose (Stark *et al.*, 1991), the higher nutritional value by introducing for example a chimeric gene encoding a Brazil nut methionine rich seed protein (Altenbach *et al.*, 1989), the potential for the production of novel biopolymers in plants, such as polyhydroxybutyrate, a biodegradable thermoplastic in transgenic Arabidopsis plants (Poirier *et al.*, 1992), the exploration of transgenic plants as a new source of products with various applications to healthcare such as vaccines (Haq et al., 1995), recombinant antibodies (Fiedler and Conrad, 1995) etc.

During the last decade *Agrobacterium* mediated transfer of gene constructs in plant species has been very valuable for the engineering of new traits in crop species and for the development of tools to unravel gene structures and functions. The further identification, mapping and isolation of genes which control morphological and biochemical aspects of plant development together with further attempts to increase the frequency of homologous recombination between the transgene and native genomic sequences, will lead to the more "precise" engineering of specific traits leading to plants with improved agronomic and/or industrial characteristics.

VIII. References

Abel, P., Nelson, R., De, B., Hoffman, N., Rogers, S, Fraley, R. and Beachy, R. (1986) Science 232, 738-743.

Allen, R. (1995) Plant Physiol. 107, 1049-1054.

Altenbach, S., Pearson, K., Meeker, G., Staraci, L. and Sun, S. (1989) Plant Molecular Biology 13, 513-522.

Bechtold, N., Ellis. J. and Pelletier, G. (1993) C.R. Acad. Sci. 316, 1194-1199.

Bensen, R., Johal, G., Crane, V., Tossberg, J., Schnable, P., Meely, R. and Briggs, S. (1995) The Plant Cell, 7, 75-84.

Broglie, K., Chet, I., Holliday, M., Cressman, R., Biddle, P., Knowlton, S. Mauvais, C. and Broglie, R. (1991) Science 254, 1194-1197.

Chilton, M.D. (1993) Proc. Natl. Acad. Sci. USA 90, 3119-3120.

Cuozzo, M., O'Connell, K., Kaniewski, W., Fang, R., Chua, N. and Tumer, N. (1988) Bio/Technol. 6, 549-557.

Deblaere, R., Bytebier, B., De Greve, H., Deboeck, F., Van Montagu, M., Schell, J. and Leemans, J. (1985) Nucleic Acids Res. 13, 4777-4788.

De Block, M., Botterman, J., Vandewiele, M., Dockx, J., Thoens, C., Gosselé, V., Movva, N., Thompson, C., Van Montagu, M. and Leemans, J. (1987) EMBO J. 6, 2513-2518.

Feldmann, K., Malmberg, R.L. and Dean, C. (1994) in Arabidopsis (Meyerowitz, E.M. and Somerville, C.R., eds), pp. 137-172, Cold Spring Harbor Laboratory Press.

Fiedler, U. and Conrad, U. (1995) Bio/Technology 13, 1090-1093.

Fischhoff, D., Bowdish, K., Perlak, F., Manone, P., McCormick, S., Niedermeyer, J., Fean, D., Kusano-Kretzmer, K., Mayer, E., Rochester, D., Rogers, S. and Fraley, T. (1987) Bio/Technology 5, 807-813.

Fritze, K., Czaja, I. and Walden, R. (1995) Plant J. 7, 261-272.

Gerlach, W., Llewellyn, D. and Haseloff, J. (1987) Nature 328, 802-805.

Goddijn, O., Lindsey, K., van der Lee, F., Klap, J. and Sijmons, P. (1993) Plant J. 4, 863-873.

Gray, J., Picton, S., Giovannoni, J. and Grierson, D. (1994) Plant Cell Environ. 17, 557-571.

Halfter, U., Morris, p. and Willmitzer, L. (1992) Mol. Gen. Genet. 231, 186-193.

Hamilton, A., Lycett, G. and Grierson, D. (1990) Nature 346, 284-287.

Hamilton, C., Frary, A., Lewis, C. and Tanksley, S. (1996) Proc. Natl. Acad. Sci. USA 93, 9975-9979.

Harrison, B., Mayo, M. and Baulcombe, D. (1987) Nature 328, 799-802.

Haughn, G., Smith, J., Mazur, B. and Somerville, C. (1988) Mol. Gen. Genet. 211, 266-271.

Hiei, Y., Ohta, S., Komari, T. and Kumashiro, T. (1994) Plant J. 6, 271-282.

Hilder, V., Gatehouse, A., Sheerman, S., Barker, R. and Boulter, D. (1987) Nature 330, 160-163.

Hoekema, A., Hirsch, P., Hooykaas, P.J.J. and Schilperoort, R.A. (1983) Nature 303, 179-180.

Horsch, R., Fry, J., Hoffman, N., Eichholtz, D., Rogers, S. and Fraley, R. (1985) Science 227, 1229-1231.

Ishida, Y., Saito, H., Ohta, S., Hiei, Y., Komari, T. and Kumashiro, T. (1996) Nature Biotechnology 14, 745-750.

Jones, J., Duinsmuir, P. and Bedbrook, J. (1985) EMBO J. 4, 2411-2418.

Kertbundit, S., De Greve, F., Deboeck, M., Van Montagu, M. and Hernalsteens, J.P. (1991) Proc. Natl. Acad. Sci. USA 88, 5212-5216.

Komari, T. (1990) Plant Cell Rep. 9, 303-306.

Komari, T., Halperin, W. and Nester, E. (1986) J. Bacteriol. 166, 88-94.

Komari, T., Hiei, Y., Saito, Y., Murai, N. and Kumashiro, T. (1996) Plant J. 10, 165-174.

Koncz, C., Martini, N., Mayerhofer, R., Koncz-Kalman, Z., Körber, H., Redei, G. and Schell, J. (1989) Proc. Natl. Acad. Sci. USA 86, 8467-8471.

Koncz, C., Németh, K., Redei, G. and Schell, J. (1992) Plant Mol. Biol. 20, 963-976.

Kridl, J., Knutzon, D., Johnson, W., Thompson, G., Radke, S., Turner, J. and Knauf, V. (1991) Molecular Biology of Plant Growth and Development, Tucson, 723.

Lee, K., Lund, P., Lowe, K. and Dunsmuir, P. (1990) Plant Cell 2, 415-425.

Lindsey, K., Wei, W., Clarke, M., McArdle, H., Rooke, L. and Topping, J. (1993) Transgenic Res. 2, 33-47.

Loesch-Fries, L., Merlo, D., Zinnen, T., Burhop, L., Hill, K., Krahn, K., Jarvia, N., Nelson, S. and Halk, E. (1988) EMBO J. 6, 1845-1851.

Mandal, A., Lang, V, Orczyk, W. and Palva, E.T. (1993) Theor. Appl. Genet. 86, 621-628.

Mariani, C., De Beuckeleer, M., Truettner, J., Leemans, J. and Goldberg, R. (1990) Nature 347, 737-741.

Mariani, C., Gosselé, V., De Beuckeleer, M., De Block, M., Goldberg, R., De Greef, W. and Leemans, J. (1992) Nature 357, 384-387.

Miao, Z.H. and Lam, E. (1995) Plant J. 7, 359-365.

Morton, R. and Hooykaas, P. (1995) Mol. Breed. 1, 123-132.

Odell, J., Caimi, P., Sauer, B. and Russell, S. (1990) Mol. Gen. Genet. 223, 369-378.

Odell, J., Hoopes, J. and Vermerris, W. (1994) Plant Physiol. 106, 447-458.

Oeller, P., Min-Wong, L., Taylor, L., Pike, D. and Theologis, A. (1991) Science 254, 437-439.

Offringa, R., Franke-van Dijk, M., De Groot, M., van den Elzen, P. and Hooykaas, P. (1993) Proc. Natl. Acad. Sci. USA 90, 7346-7350.

Ow, D. and Medberry, S. (1995) Critical Reviews in Plant Sciences 14, 239-261.

Poirier, Y., Dennis, D., Klomparens, K. and Somerville C. (1992) Science 256, 520-523.

Puchta, H., Dujon, B. and Hohn, B. (1996) Proc. Natl. Acad. Sci. USA 93, 5055-5060.

Register, J., Peterson, D., Bell, P., Bullock, W., Evans, I., Frame, B., Greenland, A., Higgs, N., Jepson, I., Jiao, S., Lewnau, C., Sillick, J. and Wilson, H. (1994) Plant Mol. Biol. 25, 951-961.

Robson, P., McCormac, A., Irvine, A. and Smith, H. (1996) Nature Biotechnology 14, 995-998.

Rugh, C., Wilde, H., Stack, N., Thompson, D., Summers, A. and Meagher, R. (1996) Proc. Natl. Acad. Sci. USA 93, 3182-3187.

Saito, Y., Komari, T., Masuta, C., Hayashi, Y., Kumashiro, T. and Takanami, Y. (1992) Theor. Appl. Genet. 83, 679-683.

Shah, D., Horsch, R., Klee, H., Kishore, G., Winter, I., Tumer, N., Hironaka, C., Sanders, P., Gasser, C., Aykent, S., Siegel, N., Rogers, S. and Fraley, R. (1986) Science 233, 476-481.

Stark, D., Barry, G., Muskopf, Y. , Timmerman, K. and Kishore, G. (1991) Molecular Biology of Plant Growth and Development, Tucson, 714.

Strittmatter, G., Janssens, J., Opsomer, C. and Botterman, J. (1995) Bio/Technology 13, 1085-1089.

Vaeck, M., Reynaerts, A., Höfte, H., Jansens, S., De Beuckeleer, M., Dean, C., Zabeau, M., Van Montagu, M. and Leemans, J. (1987) Nature 328, 33-37.

van der Salm, T., Hänisch ten Cate, C. and Dons, H. (1996) Plant Molecular Biology Reporter 14, 207- 228.

Walden, R., Fritze, K., Hayashi, H., Miklashevichs, E., Harling, H. and Schell, J. (1994) Plant Mol. Biol. 26, 1521-1528.

Walters, D., Vetsch, C., Potts, D., Lundquist, R. (1992) Plant. Mol. Biol. 18, 189-200.

Zambryski, P., Joos, H., Genetello, C., Leemans, J., Van Montagu, M. and Schell, J. (1983) EMBO J. 4, 277- 284.

Zambryski, P. (1992) Annu. Rev. Plant Physiol. 43, 465-490.

Chapter 18

Diversity of Root Nodulation and Rhizobial Infection Processes

Az-Eddine Hadri, Herman P. Spaink, Ton Bisseling and Nicholas J. Brewin

I. Rhizobial Host Specificity

The occurrence of nodules on the roots of plants of the family of Leguminosae has attracted the interest of scientist since the Middle Ages. This is illustrated by the description of root nodules by Dalechamps in his work Historia Generalis Plantarum, published in two volumes in 1586 and 1587 (Figure 1). Up to know there is still only one exception known to the general rule that rhizobia establish a nitrogen-fixing symbiosis exclusively with legumes: several species of *Parasponia*, a genus belonging to the family Ulmaceae (Trinick, 1979; Becking, 1992), are known to interact with bacteria from the genera *Rhizobium* and *Bradyrhizobium*. Despite extensive studies of plant-microbe interactions over several decades, it is still not known what distinctive features of legumes enable them to be colonised by rhizobia, nor what unique attributes of *Parasponia* allow this plant to respond to *Rhizobium*-derived nodulation signals as a legume mimic (Marvel *et al.*, 1987).

Unlike their host plants, rhizobia are in general a very diverse group of gram negative bacteria including *Rhizobium, Bradyrhizobium, Azorhizobium, Sinorhizobium and Mesorhizobium* (see Chapter 1) (Young and Haukka, 1996). Depending on the interacting rhizobial strain, different groups of host plants can be defined among the Leguminosae family where certain rhizobia are able to nodulate a specific subset of leguminous plants. Based on this criterion, the term "cross-inoculation group" was assigned defining classes of rhizobial strains that are able to nodulate a particular group of host plants. However, many species of rhizobia have not been extensively tested for their host range. Though not very strict, two major groups of host ranges can be distinguished, the narrow host range and the broad host range (Table 1 and 2 respectively). In addition, some bacteria are adapted to particular cultivars of host species, some examples are given in Table 2. A particular example is the interaction between Afghan pea and *R. l.* bv. *viciae* strain TOM which has been proposed to be an example of a specific interaction based on a gene-for-gene relationship. As illustrated by the examples of Table 1, there is no obvious correlation between bacterial speciation and the host range of the species. An example is the plant species

Bacterial species	Host plants
Rhizobium leguminosarum	
biovar *viciae* [1]	*Pisum, Viciae, Lathyrus,* Lens
biovar *trifolii* [2]	*Trifolium*
biovar *phaseoli* [3]	*Phaseolus*
Sinorhizobium meliloti	*Medicago, Melilotus, Trigonella*
Rhizobium etli [4]	*Phaseolus*
Mesorhizobium loti [5]	*Lotus, Anthyllis, Lupinus*
Rhizobium galegae	*Galega*
Sinorhizobium fredii [6]	*Glycine, Vigna*
Rhizobium tropici [7]	*Leucaena, Phaseolus, Medicago, Macroptileum*
Bradyrhizobium japonicum	*Glycine, Macroptilium, Vigna*
Bradyrhizobium elkanii	*Glycine, Macroptilium, Vigna*
Azorhizobium caulinodans	*Sesbania rostrata*
Rhizobium sp.	
BR816	*Leucaena, Phaseolus, Parasponia*
NGR234, MPIK1030	Broad host range and also the non-legume *Parasponia*
Sinorhizobium fredii USDA 257	Broad host range

Table 1: Example of rhizobial host range
Formerly used names of bacterial species: [1] *R. leguminosarum*; [2] *R. trifolii*; [3] *Rhizobium meliloti*; [4] *Rhizobium phaseoli* or *R. leguminosarum* biovar *phaseoli* type I strains; [5] *Rhizobium loti*; [6] *Rhizobium fredii*; [7] *R. leguminosarum* biovar *phaseoli* type II strains.

Phaseolus that can be nodulated by various bacterial species including narrow and broad host rhizobia. A major determinant of host specificity is the rhizobial signal molecules named Nod factors (Chapter 21) that each specific strain makes and which is well correlated with the host group it can interact with. The determination and comparison of Nod factors in different rhizobial species point out that there is indeed a correlation between rhizobial host range and Nod factor structure (Dénarié *et al.*, 1996).

II. Evolutionary Diversity of the Nodulation Process

Other groups of plants establish root nodule symbioses with a nitrogen-fixing gram-positive actinorhizal bacterium of the genus *Frankia* (Pawlowski and Bisseling, 1996). The known hosts of *Frankia* include over 200 species in 24 genera distributed among eight families, all of them woody dicotyledonous plants (Baker and Mullin, 1992). Whereas legume nodules all have peripheral vasculature, the actinorhizal nodules take the form of modified lateral roots (with central vasculature). However, other features of tissue and cell invasion show similarities between actinorhizal and rhizobial

symbioses. *Frankia* infections, like those of *Rhizobium*, can either involve root-hair deformation or intercellular penetration (Miller and Baker, 1986; Callaham and Torrey, 1977); invading bacterial hyphae are always swathed in a zoogloeal matrix of plant-produced polysaccharides and glycoproteins (Liu and Berry, 1991); and there is enhanced expression of plant haemoglobin genes (Jacobsen-Lyon *et al.*, 1995; Appleby, 1992). It has recently been recognised that the actinorhizal plants and the Leguminosae constitute an evolutionary clade, the rosid clade I (Soltis *et al.*, 1995), within which there is apparently a predisposition towards the development of a root nodule symbiosis. From within this clade there seem to have been several independent evolutionary origins for both the actinorhizal and *Rhizobium* symbioses (Doyle, 1994; Swensen, 1996). Since *Parasponia* and the rest of the Ulmaceae fall within this clade, it is possible that *Parasponia* has adapted some of the components that normally operate in an actinorhizal symbiosis to function in the context of a symbiosis with *Rhizobium* (Lancelle and Torrey, 1984). Thus, the modified lateral root morphology and the induction of haemoglobin are characteristics that might have been expected if *Parasponium* had interacted with *Frankia* rather than with *Rhizobium*. In an evolutionary context, it is important to remember that the Leguminosae (Fabaceae) are a

	Host	Bacterium strain	Reference
Cultivar host range	Afghanistan pea	*R.l.* bv. *viciae* TOM	Lie, 1978; Kozik *et al.*, 1995
	Woogenellup clover[1]	*R.l.* bv. *trifolii* TA1	Lewis-Henderson and Djordjevic, 1991
	Rj4 soybean	USDA 61 [2]	Vest and Caldwell, 1972; Sadowsky an Cregan, 1992
Narrow host range	*Trifolium* spp.	*R.l.* bv. *trifolii*	
Broad host range	> 50 genera	NGR234	Lewin *et al.*, 1987
	> 26 genera	*S. fredii*	Pueppke and Broughton, 1996

Table 2: Examples of three types of host range
[1] *T. Subterranean* cv Woogenellup; [2] *Bradyrhizobium japonicum* strain USDA 61 and strains belonging to the USDA 123 serogroup are not able to nodulate this host while other strains do.

very large and diverse family of dicotyledonous plants comprising some 20,000 species (Allen and Allen, 1981). Two sub-families are predominantly woody trees and shrubs, the Caesalpinioideae with about 180 genera and 2800 species and the Mimosoideae with about 65 genera and 2,900 species. The third major sub-family, the Papilionoideae, with over 500 genera and an estimated 14,500 species, both woody and herbaceous, includes most of the grain and forage legumes of importance to agriculture. As discussed by Quispel (Chapter 25), it is probable that the first distinct ancestors of the legume group arose in the late Cretaceous period, c. 70 million years ago, and the fossil evidence indicates that the three sub-families diverged some 50 million years ago (Sprent and Raven, 1992). Current members of all three sub-families establish a nitrogen-fixing symbiosis with *Rhizobium*, *Bradyrhizobium* or *Azorhizobium* spp., suggesting that the capacity for symbiotic interaction with *Rhizobium* developed very early in the evolution of the family. However, only seven of the genera from the Caesalpiniodiae have been found to nodulate (J. Sprent, personal communication), whereas in the other two subfamilies the frequency of nodulation is over 90% (de Faria *et al.*, 1989). Current opinion suggests that the symbiosis may have evolved independently on several occasions (Young and Johnston, 1989; Soltis *et al.*, 1995; Doyle, 1994).

III. Comparative Anatomy and Structure of the Root Nodule

III.A. RHIZOBIAL SYMBIOSIS

Bearing in mind the evolutionary diversity of the rhizobia-legume symbiosis and the small number of species that have been examined in any detail, it is clearly dangerous to make generalisations about structural changes that are the basis of the establishment of the symbiotic state. Some indication of this diversity is given in Table 3 which documents the principal mechanism of tissue and cell invasion for a small number of well-studied legumes (both herbaceous and woody plants) and also for the non-legume *Parasponia*. Each host listed in Table 3 has been selected because it exhibits some distinctive feature in relation to the mechanism of tissue and cell invasion by *Rhizobium*, although it should be emphasised that more than one invasion route can sometimes occur simultaneously within a single host plant. Because there are many reported instances of the same strain of rhizobium nodulating different hosts by different routes, it is assumed that the characteristics of infection and nodule morphogenesis are almost completely determined by the host plant (Kijne, 1992).

In the following, we will describe root nodule formation and nodule tissue organization in *Rhizobium*-legume symbiosis based on well-studied legume plants (*e.g.* pea, alfalfa, soybean and clover). In the second part, we will highlight the differences at the level of tissue and cell invasion among legume plants and the non-legume *Parasponia*. In the last part, we will deal with actinorhizal symbiosis and a similar comparison with the rhizobial nodule will be drawn.

III.A.1. Root Nodule Formation and Infection in Rhizobia-Legume Symbiosis

The shapes and development of nodules can be very diverse as exemplified in Figure 2. Two major types can be distinguished: indeterminate nodules, which contain a persistent meristem which results in cylindrical-shaped structures (*e.g.* Figure 2C), and determinate, which do not form a persistent

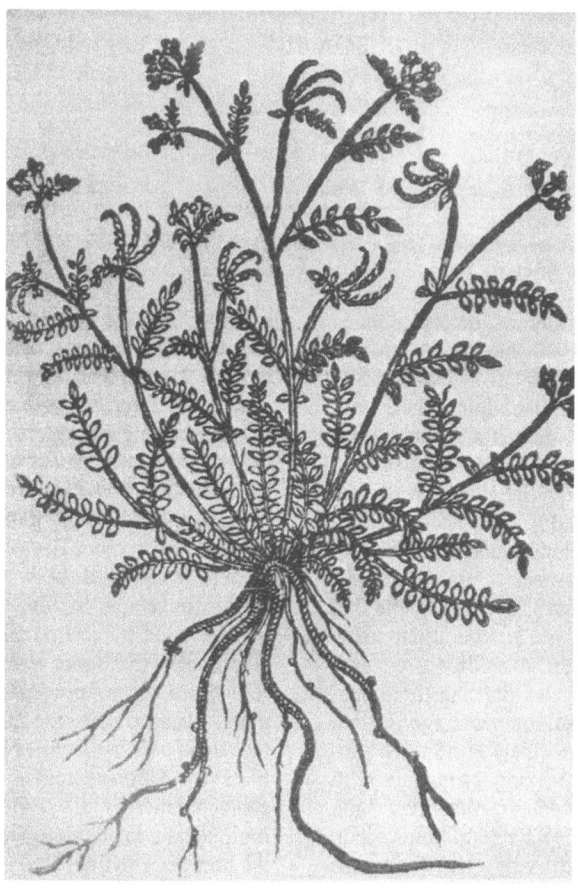

Figure 1. Reproduction of an original wood cut by Dalechamps (1586) describing the habitus of *Ornithopodium tuberosum* Dalech.

meristem, and therefore remain spherical-shaped (*e.g.* Figure 2D). Furthermore, nodules can also be found localized on the stem of various tropical legumes such as *Sesbania rostrata* (Figure 2B). Despite these large differences, a common characteristic for all root and stem nodulation processes is its localisation opposite protoxylem poles (Figure 3).

In general during root nodule formation, two processes, infection and nodule organogenesis, occur simultaneously. In order to infect the root, rhizobia first attach to the root hair, and then induce root hair deformation and curling followed by infection thread formation (Kijne *et al.*, 1992, Turgeon and Bauer

1985). Concurrent with infection, root cortical cells dedifferentiate and start dividing. This results in the formation of a nodule primordium from which the nodule will develop. Already in early stages, the nodule primordia can be distinguished from lateral root primordia, which arise in the pericycle (Figure 4A).

The association of *R. leguminosarum* bv. *viciae* with the garden pea (*Pisum sativum*) (the first listed herbaceous legume in Table 3 and shown in Figure 2A), is one of the most studied symbioses and represents a developmental sequence considered typical of many legume-*Rhizobium* combinations

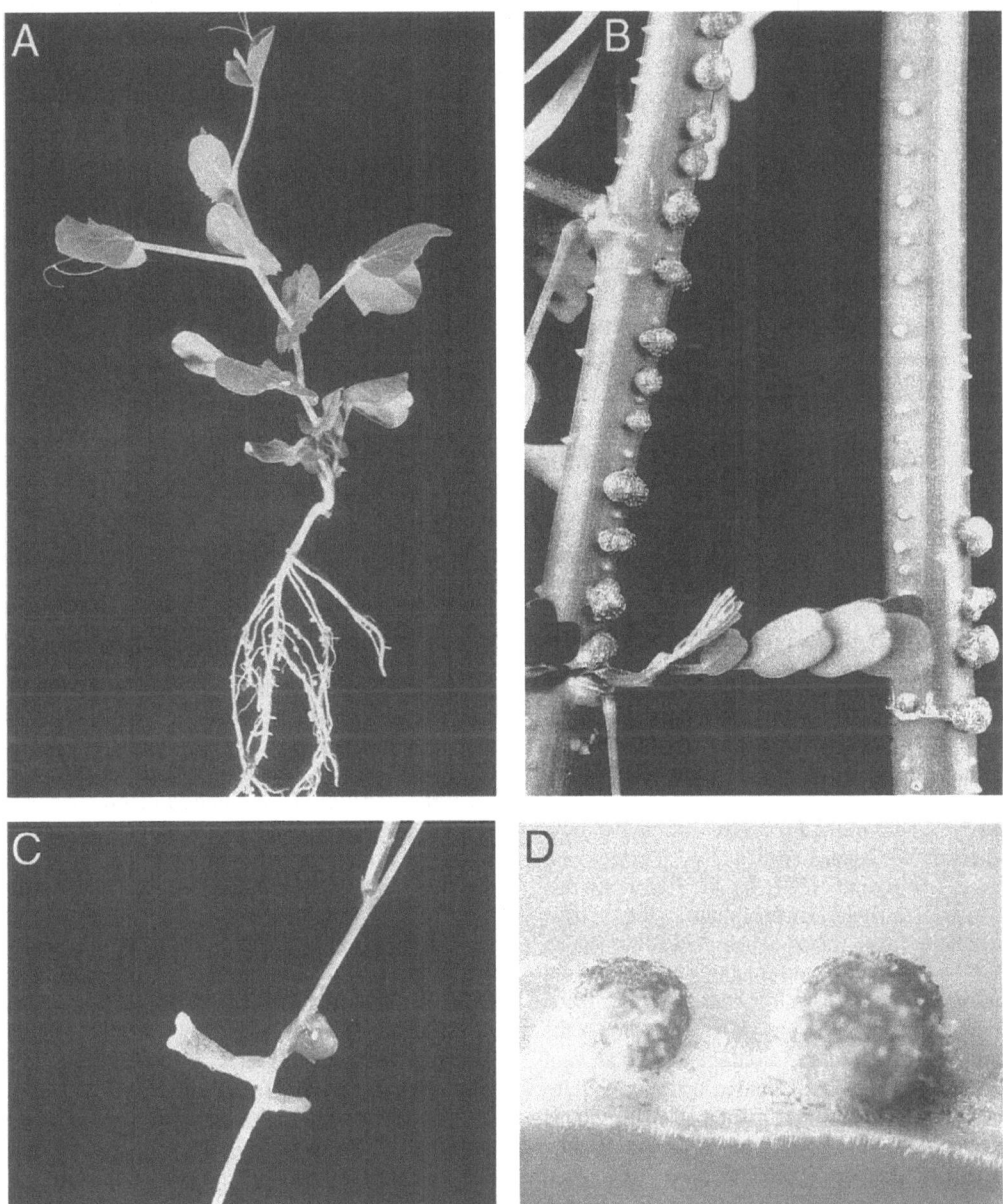

Figure 2. Examples of different types of nodules induced by rhizobia. (A) Indeterminate nodules induced on the root of a pea plant. (B) Determinate nodules induced on the stem of *Sesbania rostrata*. Nodules are induced on dormant root primordia of the stem and therefore the process is comparable to root nodulation. (C) A close-up of indeterminate root nodules of *Vicia hirsuta*. (D) A close-up of determinate root nodules induced in *Lotus preslii*. The picture in panel B is courtesy of Dr. M.Holsters.

	1	2	3	4	5	6	7	8	9	10	11	12	13	References
Host Legume														
Pisum sativum	+	+	-	+	+	-	-	+	-	-	+	+	-	Kijne, 1992
Glycine max	+	+	-	+	+	-	+	+	-	-	+	+	-	Turgeon and Bauer, 1985
Arachis hypogeae	+	-	+	+	-	+	+	+	-	+	+	+	-	Chandler, 1978
Stylosanthes spp	-	-	+	+	-	+	+	+	-	+	+	+	-	Chandler, 1982
Neptunia natans	-	-	+	+	-/+	+	+	+	-	+	+	+	-	Subba-Rao et al., 1995
Andira spp	-	-	+	+	-	+	-	+	-	-	+	-	+	de Faria et al., 1987
Non-legume														
Parasponia spp.	+	-	+	+	+	-	-	-	+	+	+	-	+	Lancelle and Torrey, 1984

Table 3. Tissue and cell invasion by rhizobia in a variety of host plants.

Root Epidermis
1. Root hair curling or deformation
2. Root hair infection thread
3. Crack entry through epidermis

Root Cortex
4. Cortical cell proliferation
5. Cell-to-cell spread by infection threads
6. Colonisation of intercellular spaces
7. Colonisation of mitotic cells

Nodule Development
8. Vascular tissue peripheral
9. Vascular tissue central
10. Nodules at sites of lateral root emergence

Symbiotic Nitrogen Fixation
11. Haemoglobin formation
12. N_2-fixing bacteroids in symbiosomes
13. N_2-fixing bacteria in fixation threads

(*e.g.* those involving *Medicago*, *Trifolium* and *Vicia* spp.). The invasion route is summarised in Figure 5, and has been extensively described in the literature (Newcomb, 1976; Kijne, 1992; Brewin, 1991). The infection thread filled with proliferating rhizobia grows from the root hair towards the nodule primordium localized in the inner cortex (Bakhuizen, 1988) (Figure 4B). There, the rhizobia are released into the cytoplasm of the host cells, surrounded by a plant-derived membrane called the peribacteroid membrane (Newcomb, 1981; Kijne, 1992). At this stage the rhizobia differentiate into their endosymbiotic form, the so-called bacteroids, which together with the surrounding peribacteroid membrane are called symbiosomes (Kijne, 1992, Roth and Stacey, 1989b). After infection, the primordia develop into dinitrogen fixing root nodules. In the case of the garden pea, the nodules are of an indeterminate type, which is characterised by an uninfected apical meristem. During nodule development, rhizobia occupy a series of specialized ecological niches as the physical interaction with host cells becomes progressively more intimate (Figure 5 a-e).

The above-described mode of tissue and cell invasion is typical of plants that develop nodules of the indeterminate type. A significant variation on this infection mechanism is seen with determinate nodulating plants such as soybean (*Glycine max*), common bean (*Phaseolus vulgaris*) and *Vigna spp.* (Turgeon and Bauer, 1985). In all of these cases, inoculation with *Rhizobium* induces root hairs to curl shortly after emergence, thereby trapping rhizobia at the cell junction, from which the infection thread initiates at the root hair base. Another difference is that the first *Rhizobium*-induced cell divisions occur in the outer cortex (the hypodermis) (Figure 4C) and these dividing cells are soon penetrated by infection threads, which release bacteria (as symbiosomes) into the cytoplasmic space. In such cases, colonisation of the central tissue of the nodule is achieved by division of rhizobia and symbiosomal units within the cytoplasm of the meristematic cells themselves. These meristems subsequently abort, leading to the development of spherical nodules of the determinate type (Brewin *et al.*, 1992).

III.A.2. Tissue Organization in Determinate and Indeterminate Nodules

Determinate and indeterminate nodules have a similar tissue organization (Figure 6 a-b), a central tissue where bacteria are hosted, surrounded by several peripheral tissues (for review see Rolfe and Gresshoff, 1988; Brewin, 1991; Franssen *et al.*, 1992). The peripheral tissues include the nodule cortex, the endodermis and the nodule parenchyma

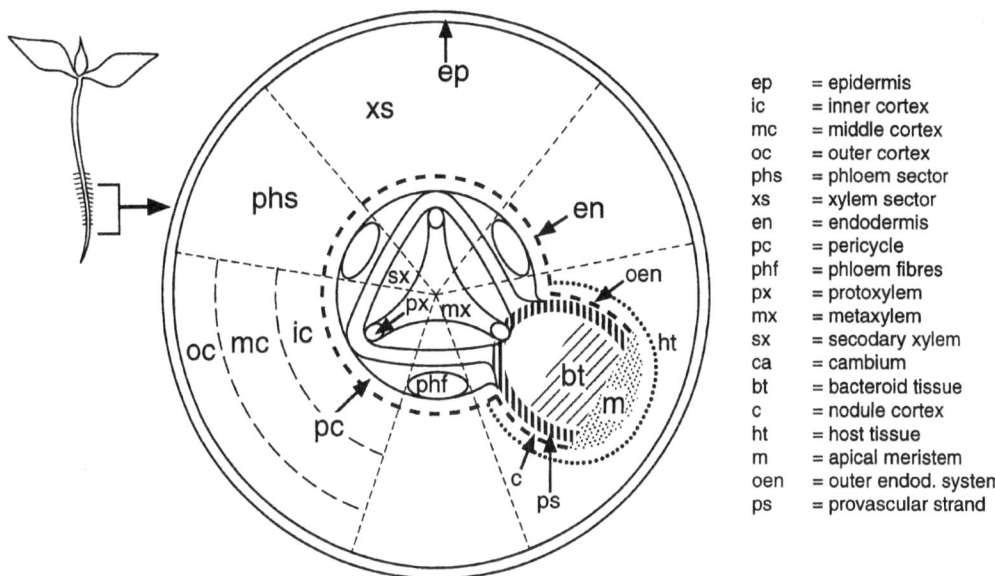

Figure 3. Localisation of the process of root nodulation. Nodules are formed in the region of the root where new root hairs are formed. A pea root nodule is given as an example of the radial localisation of the nodule forming process. Common to all root and stem nodulation processes is that the nodules are formed in front of the xylem poles, therby making the connection to the vascular system possible.

(Van de Wiel *et al.*, 1990). The latter tissue harbours the nodule vascular bundles. The central tissue is composed of two cell types, namely infected and uninfected cells. The infected cell type is fully packed with bacteria where nitrogen fixation takes place. A few cell layers of uninfected cells -named boundary layers- separate the central tissue from the nodule parenchyma (Gresshoff and Delves, 1986; Franssen, *et al.*, 1992). Meristems of indeterminate nodules are uninfected and continuously differentiate into the different nodule tissues. This results in nodule tissues composed of cells at successive stages of development. Therefore the central tissues can be divided into adjacent zones representing successive stages of development. The ultrastructural changes in both plant cells and rhizobia, starch accumulation and the onset of nitrogen fixation were used to classify different zones (Vasse *et al.*, 1990). The meristem at the apex is designated as zone I. The meristem is formed of small non-vacuolate cells that are not infected by rhizobia. The prefixation zone II immediately follows this zone. In the distal part of this zone II, infection threads penetrate the post-

meristematic cells, and bacteria are released into plant cytoplasm and differentiation of both symbionts begins. In the proximal part of the prefixation zone II, plant cells elongate and symbiosomes proliferate. In the nitrogen fixation zone III, the plant cells have reached their maximal size. The interzone II-III is located in between the nitrogen fixation zone III and the prefixation zone II. This zone is characterized by the start of starch accumulation in the infected cells, the presence of rhizobia with a specific morphology (Vasse *et al.*, 1990) and the induction of *nif* genes (Yang *et al.*, 1991). In older nodules, a senescent zone IV is present. The zonation proposed by Vasse *et al.* (1990) is applicable to other indeterminate nodules (Franssen *et al.*, 1992). During the different steps of nodule development, nodule-specific or nodule-enhanced plant genes are expressed; these are the so-called nodulin genes (Van Kammen, 1984). These genes have been divided in two groups, namely those expressed at early stages of development (*ENOD*) and the nodulin genes expressed around the onset of nitrogen fixation (*NOD*) (Nap and Bisseling, 1990).

III.A.3. Other Modes of Tissue and Cell Invasion in Rhizobia-Legume Symbiosis

Arachis hypogaea differs from the herbaceous legumes described above in that, although root hair deformation was observed when roots were inoculated with rhizobia (Chandler, 1978), any colonisation of root hair cells was apparently abortive. Instead, infection through the epidermis involves intercellular penetration ("crack entry"), *i.e.* entry at the point of emergence of lateral roots. Cell divisions are induced in the cortex and colonisation of these tissues is initially by invasion of intercellular spaces and ultimately by penetration of cell walls. Thus, there are no cell-to-cell infection threads, but other events in *Arachis* are similar to those in *Glycine. Arachis* nodules are an example of the 'aeschynomenoid' type, characterised by their compact structure and the complete lack of uninfected host cells in the central nodule tissue (Sprent and Raven, 1992).

Early infection in *Stylosanthes* has been shown to differ slightly from that observed with *Arachis* (Chandler *et al.,* 1982). No root hair curling associated with rhizobia inoculation was observed, although some root hair branching occurred and root hairs were most frequent around the site of lateral roots. Rhizobia entered intercellular spaces at the point of emergence of lateral roots, penetrating into the cortex by causing cell wall alterations in cortical cells, entering them and eliciting progressive cell collapse. Finally, the cell wall was penetrated by irregular intrusions and bacteria were released as bacteroids into dividing cells derived from the inner cortex. These infected host cells eventually became the central tissue of the nodule. This form of intercellular penetration apparently involves collapse and compression of root cortical cells before the mutualism can be effectively established. In other characteristics, root nodules of *Stylosanthes* differed in only minor detail from *Arachis. Mimosa scabrella* offers another interesting variation in that penetration of the epidermis was apparently by dissolution of the middle lamella of the radial cell walls (de Faria *et al.,* 1988). Interestingly, sub-epidermal root hair cells are occasionally formed upon inoculation.

Neptunia natans is an aquatic legume without root hairs (Subba-Rao *et al.,* 1995). Initial penetration was observed at the point of emergence of lateral roots and cytological studies revealed an intercellular mode of colonisation in the root cortex, followed by an intracellular route of dissemination within developing nodule cells. Other aquatic legumes, *e.g.*

Aeschynomene fluminensis and *Sesbania rostrata* in association with *Azorhizobium spp,* develop nodules on their stems or, more precisely, at the base of emerging adventitious roots derived from stems (Figure 2B). The invasion strategy is basically similar to that described here for *Neptunia.* Depending on environmental conditions, *Aeschynomene fluminensis* and *Sesbania rostrata* can induce nodules on either stems or roots (Loureiro *et al.,* 1995; Ndoye *et al.,* 1994), and these nodules can exhibit different infection pathways.

Andira inermis is an example of a woody legume which shows a very interesting primitive feature, namely that the nitrogen-fixing rhizobia are not released into the host cytoplasm as symbiosomal units but are retained within long and persistent infection threads that occupy a substantial volume of the cytoplasmic space within host cells. Interestingly, this structural feature is also seen in the non-legume *Parasponia.* In a survey of root nodule structure in genera of tree legumes, de Faria *et al.* (1987) reported persistent infection threads in 12 genera studied, including all the representatives from the sub-family Caesalpinioideae, some of the Papilionoideae (including *Andira*), but none of the representatives from the more advanced subfamily Mimosoideae. From their structural studies of *Andira,* these authors concluded that bacteria spread via intercellular spaces unconfined by threads. No infection threads were observed to cross cell boundaries. Rhizobia entered developing nodule cells by intrusions through the cell wall and thereafter were confined by host-produced cell wall material and membranes, forming persistent infection threads, within which nitrogen fixation takes place.

III.A.4. Generalisations

In view of the great variety of infection mechanisms operating in different host-rhizobia combinations, it is only possible to describe the essential features of the rhizobia-legume symbiosis in terms of a few very broad generalisations. 1. Colonisation of host tissue by rhizobia involves growth of bacteria within the intercellular space between plant cells: in many cases, plant cell wall growth may be modified to accommodate the invading bacteria. 2. Rhizobia induce cortical cell divisions leading to the development of a mass of new tissue (generally regarded as a new plant organ), in which the central tissue is colonised by bacteria and the outer tissue carries the vasculature. 3. Following tissue and cell

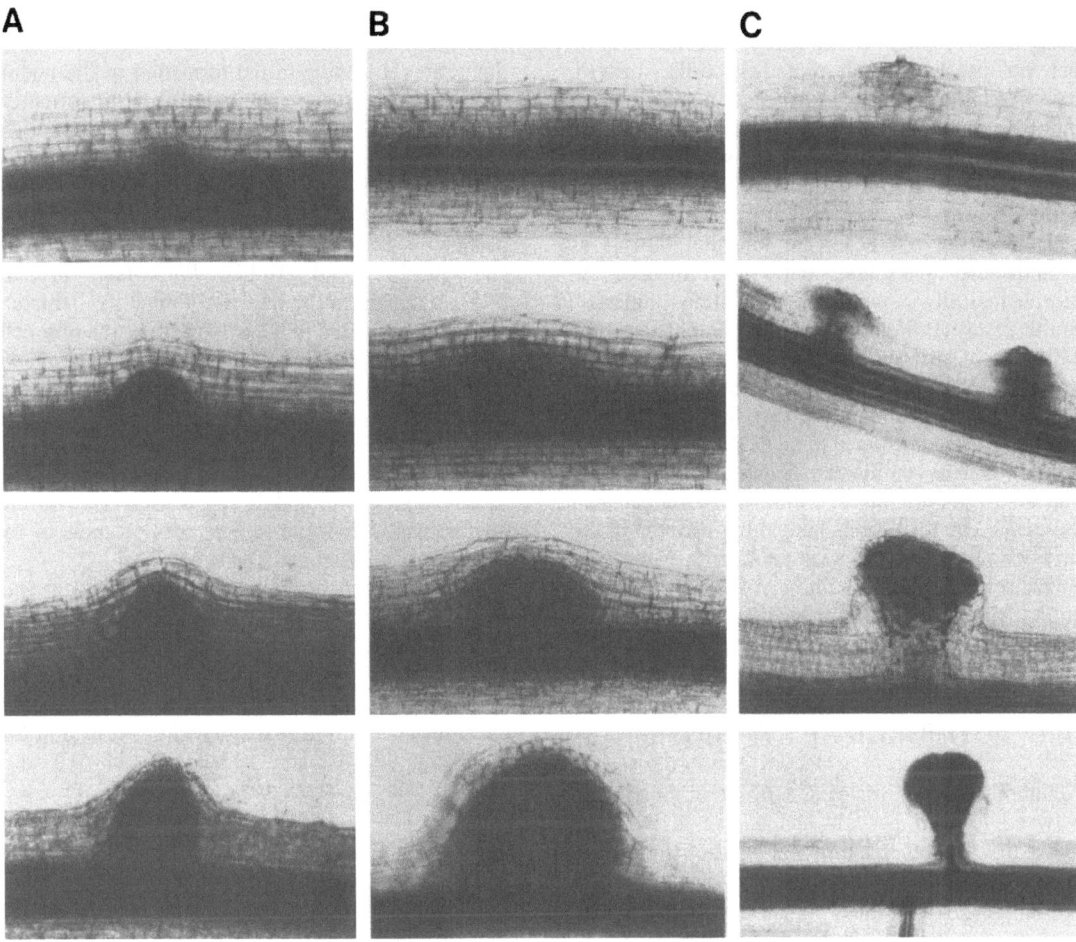

Figure 4. Various stages in the development of lateral roots (A), inner cortical root primordia (B), and outer cortical root primordia (C). Shown are roots of *Vicia hirsuta* (A and B) or *Phaseolus vulgaris* (C) bleached by hypochlorite treatment and stained with methylene blue. Reproduced from Spaink (1996) with kind permision of CRC Press, Inc.

colonisation, the nitrogen-fixing state is (more-or-less) an endosymbiotic relationship: the capacity for bacterial nitrogen-fixation is induced in a low oxygen environment in a space that can best be described as intracellular but extracytoplasmic. 4. The intracellular morphology of host cells harbouring nitrogen-fixing rhizobia is highly variable between different host legumes, but in all cases the bacteria are surrounded (individually or collectively) by a plant-derived membrane, termed the symbiosomal (or peribacteroid) membrane, which appears to control the metabolic interactions between the endophytic rhizobia and the host plant

cytoplasm. 5. Nodules lack a root cap and are not geotropic.

III.A.5. Rhizobium-Non Legume Symbiosis

The remarkable association of *Rhizobium* with the non-legume *Parasponia andersonii* was first reported by Trinick (1979), who pointed out that the nodules possessed central vascular strands with infected cells occupying a swollen cortical zone in a crescent around the central bundle. This type of structure is more reminiscent of modified lateral roots than legume nodules. *Parasponia* nodules therefore resemble the coralloid-type nodules of actinorhizal plants. During the early stages of interaction between *Rhizobium* and the non-legume host, there is some evidence of root hair modification but no penetration of root hair cells. Instead, infection is via intercellular penetration at the base of induced multi-cellular root hairs (Lancelle and Torrey, 1984) and infection threads were observed to pass from cell to cell within the proliferating root cortex. However, Trinick (1979) noted that rhizobial cells were not released from infection threads as symbiosomal packages, but were retained in morphologically specialised "fixation threads". Smith *et al.* (1986) subsequently demonstrated the existence of striking differences between the thick cell walls of invasive infection threads that traverse nodule cortical cells and the very thin cell walls of threads that are presumed to be the locus of nitrogen fixation activity within the nodule. Thus, the host appears to programme two different developmental pathways, one for threads harbouring rhizobia in the invasive phase and the other for threads harbouring rhizobia in the phase of nitrogen fixation. Whether and to what extent "fixation threads" are structurally and functionally distinct from symbiosomes is a matter for conjecture and debate (Becking, 1992).

III.B. ACTINORHIZAL SYMBIOSIS

Actinorhizal symbiosis is the result of interaction between *Frankia* bacteria and a diverse group of plants that belongs to 8 different families (for review see Benson and Silvester, 1993). The result of this symbiosis is the formation of actinorhizal nodules. *Frankia* is a filamentous actinomycetous bacterium (reviewed in Simonet *et al.*, 1990; Baker and Mullin, 1992) which normally grows in hyphal form and additionally is able to form specialized morphological structures, vesicles, where N_2 fixation occurs.

Frankia can infect the root system in two ways depending on the host plant. The first mode of infection takes place *via* the root hair. This applies to *Alnus*, *Casuarina*, *Comptonia* and *Myrica* (Torrey, 1976; Berry *et al.*, 1986; Callaham *et al.*, 1979). The second mode of infection occurs through intercellular penetration, which is reported for *Elaeagnus* and *Ceanothus* (Miller and Baker, 1985; Liu and Berry, 1991). *Frankia* bacteria that invade root hairs form tube-like structures called encapsulations, which resemble the infection threads in legumes (Berry and Sunell, 1990). *Frankia*, like *Rhizobium*, induces mitotic activity in the root cortical cells, proximal to the infected root hair. This

results in the formation of the so-called prenodule (Burgess and Peterson, 1987), the function of which is unclear. In addition, cell divisions are induced in the pericycle leading to the formation of the nodule primordium, a process that is similar to the formation of a lateral root primordium. The nodule primordium develops into a root nodule lobe where specialized cells become fully packed with the microsymbiont that is surrounded by a membrane derived from the plasma membrane of the host (for review, see Berry and Sunell, 1990). Unlike *Rhizobium*, *Frankia* bacteria remain in the infection thread-like structure and are not released in the cytoplasm of the host cell. Moreover, although *Frankia*, like *Rhizobium*, induces cell divisions in the nodule cortex, only the cell divisions induced in the root pericycle lead to the formation of the final nodule primordium. The primordia of actinorhizal root nodules, as well as *Rhizobium*-legume nodule primordia, are formed predominantly opposite to a protoxylem pole of the root stele (Callaham and Torrey, 1977).

In the case where intercellular penetration of the root and colonization of the intercellular spaces takes place (Miller and Baker, 1985; Liu and Berry, 1991), *Frankia* hyphae enter the root in between epidermal cells and proceed strictly intercellularly. This type of infection does not lead to prenodule formation but following intercellular colonization of the root cortex, cell divisions are induced in the root pericycle resulting in the formation of the nodule primordium.

III.B.1. Tissue Organization in Actinorhizal Symbiosis

Actinorhizal nodules are perennial structures that consist of multiple lobes (Bond, 1974). A striking difference between actinorizal nodules and legume nodules is reflected in their tissue organization. An actinorhizal nodule lobe represents a modified lateral root with a central vascular cylinder. In contrast to the lateral root, actinorhizal nodule lobes have a superficial periderm, lack a root cap and contain both infected and noninfected cortical cells. The growth pattern of these lobes is indeterminate due to the presence of an apical meristem that differentiates continuously in a proximal direction (Berry and Sunell, 1990). It has been shown that actinorhizal nodule lobes can be divided into zones representing successive stages of development (Ribeiro *et al.*, 1995) (Figure 6c). Zone 1, the meristematic zone, consists of small dividing cells that do not contain bacteria. Zone 2, the infection zone, corresponding to

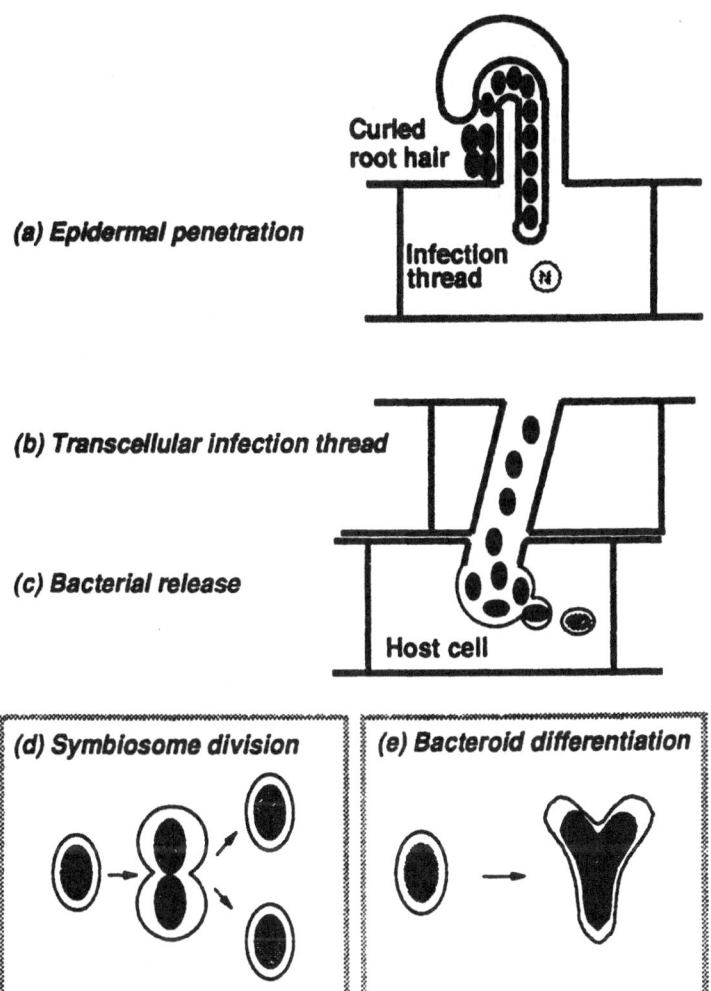

Figure 5. Stages in host cell invasion *Pisum sativum* by *Rhizobium*. (a) Epidermal penetration. Infective rhizobia cause characteristic curling and deformation on emerging and growing root hairs. Bacteria invade the plant through a newly formed tunnel, the infection thread, created by the inversion of apical growth at the tip of the root hair cell. Inward growth of the infection thread is coordinated by the plant cytoskeleton in association with the nucleus (N). (Alternative forms of epidermal penetration include "crack entry" at the point of emergence of lateral roots, *e.g. Arachis*, or dissolution of radial cell walls of the epidermis, *e.g. Mimosa*.) (b) By continuing growth of transcellular infection threads, bacterial invasion spreads from cell to cell through the root cortex and into the invasion zone in the post-meristematic zone of the incipient nodule. [Alternative forms of tissue invasion include colonisation of intercellular spaces, *e.g. Arachis*, or bacterial release (as symbiosomes) into the cytoplasm of host cells that divide mitotically to generate the central infected tissue, *e.g. Phaseolus*.] (c) Endocytosis. Individual bacteria are released through the plasmalemma from an unwalled infection droplet, which arises as an extrusion from the infection thread. Rhizobia engulfed by plasma membrane are termed bacteroids. They occupy a discrete cytoplasmic compartment, termed the symbiosome, which is bounded by a plant-derived symbiosomal (or peribacteroid) membrane. (Alternatively, bacteria may be released from the unwalled tip of an invasion structure which penetrates a host cell but does not exit from the other face (the infection peg), *e.g. Phaseolus, Glycine max* . In woody legumes, *e.g. Andira*, and in *Parasponia*, membrane-enclosed symbiosomes do not exist, but nitrogen-fixing bacteria develop within tubular intrusions termed 'fixation threads'.) (d) Symbiosome division. As the enclosed bacteria divide inside the host cell, the bounding symbiosomal membrane divides concomitantly. (Alternatively, as in determinate nodules, *e.g. Phaseolus*, this synchrony breaks down in the later stages of development and bacteroids are eventually packaged in clusters of 6-12.) (e) Bacteroid differentiation. When bacteroids have ceased division, they develop the capacity for biological nitrogen fixation. This is accompanied by substantial enlargement and morphological differentiation of bacteroids in host cells of indeterminate nodules. (Alternatively, there is no morphological differentiation for the nitrogen-fixing bacteroids present in the multiply-occupied symbiosomes of determinate nodules, nor for those in the fixation threads of woody legumes and *Parasponia*.) Reproduced from Kannenberg and Brewin (1994) by kind permission of Elsevier Publishers.

the prefixation zone in legume nodules, contains enlarging cortical cells. Some of these cells are infected and become larger than uninfected cells while being gradually filled with hyphae (Schwintzer *et al.*, 1982; Lalonde, 1979). Upon full packing of the infected cells with hyphae, swellings are formed on the termini of hyphea or on short branches; these structures represent the provesicles (Fontaine *et al.*, 1984). In zone 3, the fixation zone, provesicles differentiate into vesicles while the synthesis of nitrogenase, the enzyme responsible for the reduction of atmospheric nitrogen into ammonia, is induced (Huss-Daneil and Bergman, 1990). In zone 4, the senescence zone, cortical cells become senescent and the microsymbiont as well as the host cytoplasm is degraded.

In contrast to the *Rhizobium*-legume system, the molecular mechanisms that trigger actinorhizal nodule formation are not known. However, since root hair deformation in actinorhizal plants can occur without direct contact with *Frankia* (Burggraaf *et al.*, 1983; Prin and Rougier, 1987), it is possible that a molecule is secreted by the bacterium that triggers root hair deformation, as Nod factors do in the *Rhizobium*-legume interaction.

IV. Plant Regulation of Nodulation

Many plant mutants have been described which are disturbed in their nodulation and/or infection characteristics (for a review see Caetano-Anolles and Gresshoff, 1991). For instance a large set of non-nodulating mutants of pea, soybean have been characterized. Interestingly, some of these mutants are also disturbed in their interaction with mycorrhiza (Gianinazi-Pearson, 1996). This suggests common characteristics of mycorrhizal interaction with the rhizobial symbiosis.

In both actinorhizal and legume plants nodule

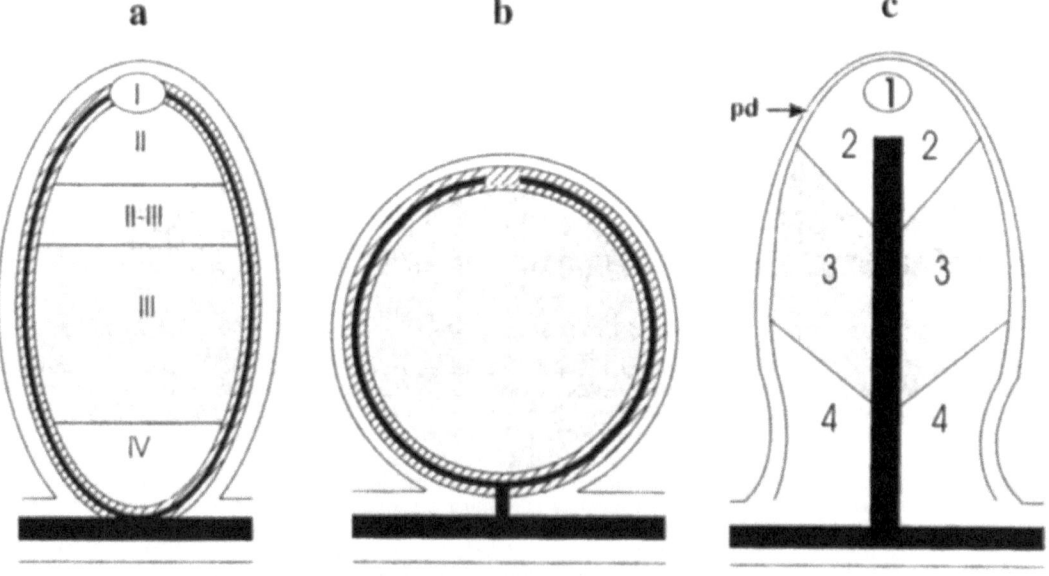

Figure 6. Schematic representation of nodule structure in indeterminate, determinate, and actinorhizal nodule.
(a) Indeterminate legume nodule. The central tissue can be divided into 5 adjacent zones (Vasse *et al.*, 1990). The meristem at the apex is designated as zone I which is immediately followed by the prefixation zone II. In this zone, cells become infected. In the interzone II-III, bacterial nitrogen fixation begins (Yang *et al.*, 1991) and proceeds throughout the nitrogen fixation zone III. Zone IV represents the senescence zone where bacteria are degraded. The nodule parenchyma forms the oxygen diffusion barrier. (b) Determinate legume nodule. The cells in the central tissue have a similar developmental stage. (c) Actinorhizal nodule lobe from *Alnus glutinosa*. The tissue organization in actinorhizal nodule is similar to that of indeterminate legume nodule (Ribeiro *et al.*, 1995). Zone 1 represents the meristem. Zone 2 harbours cells that become infected by and gradually filled with *Frankia* hyphae (Schwintzer *et al.*, 1982). In zone 3, nitrogen fixation occurs. In zone 4, the senescence zone, plant cytoplasm and bacteria are degraded.

number is strictly autoregulated (Caetano-Anolles and Gresshoff, 1991). The regulation of the number of nodules occurs at least at two levels. During initial nodulation of legumes most infections and primordia are eventually arrested so that a restricted number of nodules are established in the susceptible zone. After the first successful infections and primordia have been formed, nodulation does not occur in the newly developing part of the root, a phenomenon known as autoregulation of nodule number.

Ethylene is a key player at the initial level of regulation because mutants insensitive to this hormone form many more nodules (Varma Penmetsa and Cook, 1997) and compounds that block ethylene perception or synthesis can also increase nodule number. However, these mutants have a normal autoregulation mechanism and as a result these plants become densely nodulated in a small region of the root system. Ethylene not only controls nodule number but also provides positional information in the root. In wild type roots, primordia are almost exclusively formed opposite protoxylem poles. However, when plants are treated with blockers of ethylene perception/synthesis or the ethylene insensitive mutant plants, they form nodules opposite the phloem poles (Heidstra et al., 1997). These observations are in good agreement with the fact that ethylene appears to be synthesised in regions of the root pericycle opposite the phloem poles.

Autoregulation mutants form large amounts of nodules over the entire root system and such mutants have been useful to study the mechanisms controlling autoregulation. This process appears to involve several signalling events. Probably there is a compound released from the primordia that is sensed in a systemic manner by other parts of the plant. Grafting experiments between wild type and supernodulation mutants revealed that also the shoot plays an important role in autoregulation. Most likely a leaf signal is formed in response to the nodule-derived signal and the interaction between shoot and root tissues subsequently results in an inhibition of new nodulation. The signalling between the shoot and root is rather complex and probably involves inhibitory and stimulatory compounds (Gresshoff et al., 1988, 1989; Francisco and Harper, 1995). The identification of several Lotus and Medicago supernodulation mutants will provide the means to characterise the genes involved and to study the molecular mechanisms controlling autoregulation. Determinants of host specificity have not been extensively studied. In pea the mannose binding lectin PSL has been implied to be a major

determinant of host specificity (Diaz et al., 1995, 1996). Introduction of this pea lectin gene psl into clover extended the host range of R. l. bv. viciae, however the function of lectin is not known. During the development of the legume root nodule, lectins are present in at least three distinct locations, and putative functions have been described based on this distribution at the surface of root hairs: lectins may promote the aggregation of rhizobia in the zone of the root that becomes infected; in the nodule primordium they may stimulate mitotic activity by lowering the threshold of response to Rhizobium-derived nodulation factors; and in the central tissue of mature nitrogen fixing nodules they may function as part of a transient nitrogen reserve (Brewin and Kardailsky, 1997).

V. Acknowledgements

We are grateful to Janet Sprent for useful discussions on comparative legume anatomy.

VI. References

Allen, O.N. and Allen, E.K. (1981) The Leguminosae, Macmillan Publishing Company, London, pp. 1-831.

Appleby, C.A. (1992) Sci. Prog. 76, 365-398.

Baker, D.D. and Mullin, B.C. (1992) In G. Stacey, R.H. Burris and H.J. Evans (eds.) Biological Nitrogen Fixation, Chapman & Hall, New York, pp. 259-292.

Bakhuizen, R., (1988) The Plant Cytoskeleton in the Rhizobium-Legume Symbiosis, PhD Thesis, Leiden University, Leiden, The Netherlands.

Becking, J.H. (1992) In G. Stacey, R.H. Burris and H.J. Evans (eds.) Biological Nitrogen Fixation, Chapman & Hall, New York, pp. 497-559.

Bensen, D.R. and Silvester, W.B. (1993) Microbiol. Rev. 57, 293-319.

Berry, A.M. and Sunell, L.A. (1990) In C.R. Schwintzer and J.D. Tjepkema (eds.) The Biology of Frankia and Actinorhizal Plants, Academic Press, New York, pp. 61-81.

Berry, A.M., McIntyre, L. and McCully, M.E. (1986) Can. J. Bot. 64, 292-305.

Bond, G. (1974) In A. Quispel (ed.) The Biology of Nitrogen Fixation, North-Holland Publishing Compagny, Amsterdam, pp. 342-378.

Brewin, N.J. (1991) Annu. Rev. Cell Biol. 7, 191-226.

Brewin, N.J. and Kardailsky, J.V. (1997) Trend in Plant Sci. 2, 92-98.

Brewin, N.J., Downie, J.A. and Young, J.P.W. (1992) In J. Lederberg (ed.) Encyclopedia of Microbiology, Volume 3, Academic Press Inc., San Diego, pp. 239-248.

Burgess, D. and Peterson, R.L. (1987) Can. J. Bot. 65, 1647-1657.

Burggraaf, A. J.P., van der Linden, J. and Tak, T. (1983) Plant Soil 74, 175-188.

Caetano-Anolles, G. and Gresshoff, P.M. (1991) Annu. Rev. Microbiol. 45, 345-382.

Callaham, D. and Torrey, J.G. (1977) Can. J. Bot. 55, 2306-2318.

Callaham, D., Newcomb, W., Torrey, J.G. and Peterson, R.L. (1979) Bot. Gaz. 140 (Suppl.), S1-S9.

Chandler, M.R. (1978) J. Exp. Bot. 29, 749-755.

Chandler, M.R., Date, R.A. and Roughley, R.J. (1982) J. Exp. Bot. 33, 47-57.

de Faria, S.M., Hay, G.T. and Sprent, J.I. (1988) J. Gen. Microbiol. 134, 2291-2296.

de Faria, S.M., Lewis, G.P., Sprent, J.I. and Sutherland, J.M. (1989) New Phytol. 111, 607-619.

de Faria, S.M., McInroy, S.G. and Sprent, J.I. (1987) Can. J. Bot. 65, 553-558.

Dénarié, J., Debellé, F., Promé, J.C. (1996) Annu. Rev. Biochem. 65, 503-535.

Diaz, C.L., Spaink, H.P. and Kijne, J.W. (1996) In G. Stacey, B. Mullin and P.M. Gresshoff (eds.) Biology of Plant-Microbe Interactions, ISMPMI, St. Paul, pp. 399-402.

Diaz, C.L., Spaink, H.P., Wijffelman, C.A. and Kijne, J.W. (1995) Mol. Plant-Microbe Interact., 348-356.

Doyle, J.J. (1994) Annu. Rev. Ecol. Syst. 25, 325-349.

Fontaine, M.F., Lancelle, S.A., Torrey, J.G. (1984) J. Bacteriol. 160, 921-927.

Francisco, P.J. Jr. and Harper, J.E. (1995) Plant Science 107, 167-176.

Franssen, H.J., Vijn, I. Yang. W.-C. and Bisseling, T. (1992) Plant Mol. Biol. 19, 89-107.

Gianinazzi-Pearson, V. (1996) Plant Cell 8, 1871-1883.

Gresshoff, P.M. and Delves, A.C. (1986) In A.D. Blonstein and P.J. Kings (eds.) Plant Gene Research III, Springer Verlag, Wien, pp. 159-206.

Gresshoff, P.M., Krotzkey, A.J., Mathews, A., Day, D.A., Schuller, K.A. Olsson, J.E., Delves, A.C. and Carrol, B.J.(1988) J. Plant Physiol. 132, 417-423.

Gresshoff, P.M., Olsson, J.E., Li, Z.Z. and Caetano-Anolles, G. (1989) Current Topics Plant Biochem. Plant Physiol. 8, 125-139.

Heidstra, R., Yang ,W.-C., Yalcin, Y., Peck, S., Emons, A., Van Kammen, A. and Bisseling, T. (1997) Development 124, 1781-1787.

Huss-Daneil, K. and Bergman, B. (1990) New Phytol. 116, 443-455.

Jacobsen-Lyon, K., Østergaard Jensen, E., Jørgensen, J.E., Marcker, K.A., Peacock, W.J. and Dennis, E.S. (1995) Plant Cell 7, 213-223.

Kijne, J.W. (1992) In G. Stacey, R.H. Burris and H.J. Evans (eds.) Biological Nitrogen Fixation, Chapman & Hall, NY, London, pp. 349-398.

Kozik, A., Heidstra, R., Horvath, B., Kulikova, O., Tikhonovich, I., Noel Ellis, T.H., Van Kammen, A., Lie, T.A. and Bissling, T. (1995) Plant Science 108, 41-49.

Lalonde, M. (1979) Bot. Gaz. 140 (Suppl.), S35-S43.

Lancelle, S.A. and Torrey, J.G. (1984) Can. J. Bot. 63, 25-35.

Lewin, A., Rosenberg, C., Meyer, Z.A., Wong, C.H., Nelson, L., Monen, J.F., Stanley, J., Dowling, D.N., Dénarié, J. and Broughton, W.J. (1987) Plant Molecular Biology 8, 447-459.

Lewis-Henderson, W.R. and Djordjevic, M.A. (1991) J. Bacteriol. 173, 2791-2799.

Lie, T.A. (1978), Annals of Applied Biology 88, 445-487.

Liu, Q. and Berry, A.M. (1991) Protoplasma 163, 82-92.

Loureiro, M.F., James, E.K., Sprent, J.I. and Franco, A.A. (1995) New Phytol. 130, 531-544.

Marvel, D.J., Torrey, J.G. and Ausubel, F.M. (1987) Proc. Natl. Acad. Sci. USA 84, 1319-1323.

Miller, I.M. and Baker D.D. (1985) Protoplasma 128, 107-119.

Nap, J.-P. and Bisseling, T. (1990) Science 250, 948-954.

Ndoye, I., de Billy, F., Vasse, J., Dreyfus, B. and Truchet, G. (1994) J. Bacteriol. 176, 1060-1068.

Newcomb, W. (1981) In K.L. Giles and A.G. Atherly (eds.) Biology of the Rhizobiaceae, Academic Press, New York, pp. 247-297.

Newcomb, W.E. (1976) Can. J. Bot. 54, 2163-2186.

Pawlowski, K. and Bisseling, T. (1996) Plant Cell 8, 1899-1913.

Penmetsa, V.R. and Cook, D.R. (1997) Science 275, 527-530.

Prin, Y. and Rougier, M. (1987) Plant Physiol. (life Sci Adv) 6, 99-108.

Pueppke, S.G. and Broughton, W.J. (1996) In Li, Lie, Chen, Zhou (eds.) Diversity and Taxonomy of Rhizobia, China Agricultural Scientech Press, Beijing, pp. 126-127.

Ribeiro, A., Akkerman, A.D.L., Van Kammen, A., Bisseling, T. and Pawlowski, K. (1995) Plant Cell 7, 785-794.

Rolfe, B.G. and Gresshoff, P.M. (1988) Annu. Rev. Plant Physiol. Plant Molec. Biol. 39, 297-319.

Roth, L.E. and Stacey, G. (1989a) Eur. J. Cell Biol. 49, 24-32.

Roth, L.E. and Stacey, G. (1989b) Eur. J. Cell Biol. 49, 13-23.

Sadowsky, M.J. and Cregan, P.B. (1992) Applied Environmental Microbiology 58, 720-723.

Schwintzer, C.R., Berry, A.M. and Disney, L.D. (1982) Can. J. Bot. 60, 746-757.

Simonet, P., Normand, P., Hirsch, A.M. and Ackermans, A.D.L. (1990) In P.M. Gresshoff (ed.) The Molecular Biology of Symbiotic Nitrogen Fixation, CRC Press, Bocoraton, FL, pp. 77-109.

Smith, C.A., Skvirsky, R.C. and Hirsch, A.M. (1986) Can. J. Bot. 64, 1474-1483.

Soltis, D.E., Soltis, P.S., Morgan, D.R., Swensen, S.M., Mullin, B.C., Dowd, J.M. and Martin, P.G. (1995) Proc. Natl. Acad. Sci. USA 92, 2647-2651.

Spaink, H.P. (1996) Crit. Rev. Plant Sciences 15: 559-582.

Sprent, J.I. and Raven, J.A. (1992) In G. Stacey, R.H. Burris and H.J. Evans (eds.) Biological Nitrogen Fixation, Chapman and Hall, New York, pp. 461-496.

Subba-Rao, N.S., Mateos, P.F., Baker, D., Pankratz, H.S., Palma, J., Dazzo, F.B. and Sprent, J.I. (1995) Planta 196, 311-320.

Swensen, S. (1996) Am. J. Bot. 83, 1503-1512.

Torrey, J.G. (1976) Am. J. Bot. 63, 335-344.

Trinick, M.J. (1979) Can. J. Microbiol. 25, 565-578.

Turgeon, B.G. and Bauer, W.D. (1985) Planta 163, 328-349.

Van de Wiel, C., Scheres, B., Franssen, H., Van Lierop, M.J., Van Lammeren, A. and Bisseling, T. (1990) EMBO J. 9, 1-7.

Van Kammen, A. (1984) Plant Mol. Biol. Rep. 2, 43-45.

Vasse, J., de Billy, F., Camut, S. and Truchet, G. (1990) J. Bacteriol. 172, 4295-4306.

Vest, G. and Caldwell, B.E. (1972) Crop Science, 692-693.

Yang, W.-C, Horvath, B. Hontelez, J.,Van Kammen, A. and Bisseling, T. (1991) Mol. Plant-Microbe Interact. 4, 464-468.

Young, J.P.W. and Haukka, K.E. (1996) New Phytol. 133, 87-94.

Young, J.P.W. and Johnston, A.W.B. (1989) Trends Ecol. Evol. 4, 341-349.

Chapter 19

Genetic Organization and Transcriptional Regulation of Rhizobial Nodulation Genes

Helmi R.M. Schlaman, Donald A. Phillips and Eva Kondorosi

I. Introduction

I.A. NODULATION GENES

The ability of *Rhizobium* spp., *Mesorhizobium* spp., *Sinorhizobium* spp., *Bradyrhizobium* spp. and *Azorhizobium* spp., collectively called rhizobia, to nodulate plants is determined by both plant and bacterial genes. Nodulation genes are defined as those rhizobial genes which play a role in nodulation or which are coordinately regulated with such genes. In chronological order of discovery they were designated as *nod*, *nol* and *noe* genes followed by a subsequent letter of the alphabet. Thus, *nodA*, *nodB* and *nodC* were the first nodulation genes described (Rossen *et al.*, 1984; Török *et al.*, 1984), and *noeL* in *Rhizobium* sp. NGR234 was the one most recently identified (Freiberg *et al.*, 1997). Homologous nodulation genes in various rhizobia have identical names. Protein products of nodulation genes commonly are represented with a capitalized abbreviation (*e.g.* NodA is the protein encoded by the *nodA* gene). The *nodD*-homologue *syrM* was identified as a symbiotic regulator involved both in enhanced exopolysaccharide synthesis and in nodulation gene regulation (Mulligan and Long, 1989) and is therefore also discussed in this chapter. Initial studies on nodulation genes were carried out mainly in *Sinorhizobium meliloti* and in *Rhizobium leguminosarum* biovars *viciae* and *trifolii*. The identification of nodulation genes was advanced enormously by the discovery that in these rhizobia many of the nodulation genes are localized extra chromosomal on a large indigenous plasmid (Johnston and Beringer, 1977). These plasmids encoding symbiotic functions are the so-called Sym plasmids whose size can be even as large as one third of the chromosome in the case of *S. meliloti* (Bánfalvi *et al.*, 1981). When these rhizobial strains are cured of their Sym plasmid, they are unable to nodulate, whereas re-introduction of the homologous or a heterologous Sym plasmid restored nodulation (Bánfalvi *et al.*, 1981; Beynon *et al.*, 1980; Djordjevic *et al.*, 1983; Hooykaas *et al.*, 1981; Johnston *et al.*, 1978; Kondorosi *et al.*, 1982; van Brussel *et al.*, 1982). Transferring the rhizobial Sym plasmid into *Agrobacterium tumefaciens* and *Philobacterium myrsinacearum* also results in a capacity to induce root nodule formation (Hooykaas *et al.*, 1981, 1982; Rodriquez-Quinones *et al.*, 1989; van Veen *et al.*, 1988). The development of genetic tools, such as the availability of broad-host range cloning and cosmid vectors (Ditta *et al.*, 1980; Friedman *et al.*, 1982), construction of cosmid libraries, complementation and mapping of nodulation-deficient mutants, introduction of random and directed transposon mutagenesis techniques coupled to marker exchange (Beringer *et al.*, 1978; Buchanan-Wollaston,1979; Meade *et al.*, 1982) or used in combination with phage transduction methods (Pees *et al.*, 1986; Wijffelman *et al.*, 1985), led to the identification of regions involved in nodulation and, subsequently the first nodulation genes (Djordjevic *et al.*, 1985; Downie *et al.*, 1983,1985; Fisher *et al.*, 1985; Kondorosi *et al.*, 1984; Rossen *et al.*, 1984; Schofield *et al.*, 1984; Török *et al.*, 1984). DNA sequence analysis, as well as complementation studies, revealed that some nodulation genes are conserved in all rhizobia, whereas others are restricted to a few or a single species or strain. Thus, in early literature nodulation genes were designated as "common" or "host specific" (*hsn*), respectively (Horváth *et al.*, 1986; Kondorosi *et al.*, 1984; Putnoky and Kondorosi, 1986). The DNA sequence revealed also that promoters of many nodulation genes contain a conserved DNA sequence called the *nod* box that

allows coordinated regulation of *nod* genes (Rostas *et al.*, 1986).

Nodulation genes in rhizobia other than those mentioned above have been identified by functional complementation of strains deficient in nodulation on one or more host plants, by isolating homologs of known nodulation genes or by sequencing DNA regions linked to a *nod* box. In some cases biochemical activity known to be due to nodulation genes has been determined but still awaits proof of the physical presence of the gene. Recently powerful automatic nucleotide sequencing methods have made it possible to determine the entire nucleotide sequence of Sym plasmids. The 536 kb-sized p*Sym* of *Rhizobium* sp. NGR234 was shown to contain over 400 open reading frames (ORFs) and led to the identification of ORFs representing nodulation genes (Freiberg *et al.*, 1997).

I.B. HOST SPECIFICITY OF NODULATION

A certain rhizobial species or biovar can nodulate only a restricted number of host plants, a feature leading to the classification of rhizobial cross-inoculation groups (see Chapter 18). This so-called host specificity of nodulation is determined by the exchange of signal molecules between the plant and the bacterium. Such classification of rhizobial species is outdated in bacterial systematics (see Chapter 1), but it helps in understanding the nodulation process on the level of genetics and plant-microbe interaction as will be discussed in the next chapters and as we will describe below.

I.C. REGULATED NODULATION GENE EXPRESSION

Most nodulation genes, other than the regulatory ones, are expressed at (very) low levels when rhizobia are grown in the absence of host plants in the usual laboratory media (Djordjevic *et al.*, 1987; Mulligan and Long,1985; Spaink *et al.*, 1987b). Requirements for transcriptional activation could first be investigated when fusions of the *E. coli lacZ* reporter gene with the nodulation genes became available. These assays, where the β-galactosidase activity reflects the rate of transcription (Miller,1972), showed that transcriptional activation of the genes requires the rhizobial activator NodD protein and a plant factor often of flavonoid nature, as coinducer. Moreover, the *nod* box was found to be an essential part of the promoter region. In addition to the universal NodD-*nod* box system, alternative transcriptional activating circuits may operate in certain species. The best known example is the NodVW two-component regulatory system acting in *Bradyrhizobium japonicum*. *Sinorhizobium fredii* appears to employ also a second *nod*-regulatory system for subsets of nodulation genes that lack the *nod* box but still require NodD and flavonoids for expression. Nodulation gene expression is tightly controlled at the transcriptional level in most rhizobia as continuous high expression of nodulation genes may have an adverse effect on nodulation (Knight *et al.*, 1986; Kondorosi *et al.*, 1989). Indeed, in bacteroids the nodulation genes are transcriptionally silent. Fine tuning of nodulation gene expression can be achieved by different mechanisms including negative autoregulation of *nodD*, sensitivity of NodDs to different flavonoids and a transcriptional repressor such as NolR. In *B. japonicum*, NolA has an indirect negative effect on transcription of nodulation genes. Environmental factors like Ca^{2+}, ammonia, organic acids or pH, might contribute also to controlled nodulation gene expression.

In the following paragraphs the genetics of the nodulation genes and the different aspects of their regulated expression will be discussed in detail.

II. Genetic Organization of Nodulation Genes

In several rhizobial species, like *R. leguminosarum* bvs. *viciae* and *trifolii*, *S. meliloti*, *Rhizobium etli*, *Rhizobium galegae*, *Rhizobium tropici* and *Rhizobium* sp. NGR234, the nodulation genes are localized on a Sym plasmid, and only the regulatory *nolR* gene, found in *S. meliloti*, *R. leguminosarum* bv. *viciae* strain TOM and *Rhizobium* sp. NGR234 (Kondorosi *et al.*, 1989; Broughton, personal communication; Kiss *et al.*, in preparation), is located on the chromosome. In other rhizobia like *B. japonicum*, *Azorhizobium caulinodans*, *S. fredii* and *Mesorhizobium loti*, all nodulation genes are localized on the chromosome. Certain nodulation genes such as *nodA*, *nodB* and *nodC* are present in a single copy whereas others, like *nodM* (*glmS*), *nodP*, *nodQ* and *nodT*, have homologous sequences elsewhere in the genome (Baev *et al.*, 1991; Rivilla and Downie,1994; Schwedock and Long,1989; Surin and Downie,1988). In many cases the nodulation

genes are clustered in a relatively small region. Most of the nodulation genes are organized in different operons which are often conserved among various rhizobia. The operons can cover as many as 12 individual genes as illustrated by the *nodYABCSUIJnolMNOnodZ* operon in *B. japonicum* with an estimated length of approximately 12 kb. The *nodA*, *nodB*, *nodC*, *nodD*, *nodI* and *nodJ* genes are present in all rhizobia. Other nodulation genes are found in only some or a single rhizobial species or strain.

II.A. TRANSCRIPTION AND TRANSLATION

The direction of transcription of nodulation genes has been established most often by studying expression of fusions of nodulation genes with a reporter gene or by mapping RNA transcripts. Alternatively it was deduced from the gene organization and structure. Transcriptional start sites have been identified for several of the nodulation genes (indicated in Figure 2). The number of start sites may vary from one to four successive nucleotides. In *S. meliloti* the transcriptional start sites of *nodA* and *nodH* were found to be the same whether transcription was regulated by the coinducer-dependent NodD1 or by the constitutive NodD3 (Mulligan and Long,1989). Leader sequences longer than 100 nt were commonly found, but extraordinarily long leader sequences have been reported only for *S. meliloti nodD3* (659 nt) and *syrM* (499 nt) genes (Barnett *et al.*, 1996). No sequence conservation nor any function has been revealed for these long untranslated regions (Kalinowski and Long,1996).

Common to almost all bacterial genes, the ORFs of most of the nodulation genes begin with an ATG translational start codon, but aberrant start codons such as GTT (e.g. *M. loti nodS*) and TTG (e.g. *R. leguminosarum* bv. *viciae nodT*, *A. caulinodans nodD*) have been identified as well. The presence of a putative Shine and Dalgarno sequence located either by its presence just upstream from the open reading frames or by functional analysis suggests that those start codons are functional. In a few cases start and stop codons of successive genes within an operon overlap (e.g. *nodAB*, *nodFE* and *nodIJ* of *R. leguminosarum* bv. *viciae* and *nodCS* in *B. japonicum*) and translational coupling has been described for *nodAB* and *nodEF* in *S. meliloti* (Horváth *et al*, 1986; Török *et al.*, 1984).

II.B. PHYSICAL MAP

In Figure 1 an overview is given on the genetic organization of nodulation genes in the best-studied rhizobia. Relevant literature references can be found in the legend. Here, we will discuss only the most striking features of the nodulation genes in a chronological order.

The *nodA*, *nodB* and *nodC* genes are always present in the same operon except in *M. loti*, *R. etli* and *Rhizobium* sp. strain N33. In these species either the *nodAC* or *nodBC* operon contains a small ORF with homology to *nodB* and *nodA*, respectivily (Cloutier *et al.*, 1996b; Scott *et al.*, 1996; Vazquez *et al.*, 1991). Usually, *nodA* is the first gene of the operon, but in *Bradyrhizobium* spp. another nodulation gene is present upstream of *nodA* which is *nodY* in *B. japonicum* (Nieuwkoop *et al.*, 1987) and *nodK* in *Bradyrhizobium (Parasponia)* sp. (Scott, 1986).

Downstream of *nodC* other nodulation genes are always present in the same operon. Commonly found genes are *nodI*, *nodJ*, *nodS* and *nodU*. The genes *nodI* and *nodJ* as well as the genes *nodS* and *nodU* are always found pairwise with *nodJ* and *nodU* downstream of *nodI* and *nodS*, respectively. The order of the respective gene pairs may vary, however. Other nodulation genes being reported in the same operon as *nodC* are *nodT* in *R. leguminosarum* bv. *trifolii*, *nodX* in *R. legumino-sarum* bv. *viciae* strain TOM, *nolMNOnodZ* in *B. japonicum*, *nodZnoeCD* in *A. caulinodans* and *nolOnoeI* in *Rhizobium* sp. NGR234.

The *nodD* gene is present in a single copy in *R. leguminosarum* bvs. *viciae* and *trifolii* and *A. caulinodans*, but multiple copies have been identified in all other rhizobia. In most cases, *nodD* genes form a single transcriptional unit, but in *R. etli* *nolE* and *nodD1* form a two-gene operon. The *syrM* gene found in a single copy in *S. meliloti* and in two copies in *Rhizobium* sp. NGR234, has a strong phylogenetic relationship with *nodD* (Schlaman *et al.*, 1992b) but less sequence homology compared to the other *nodD* genes (Barnett and Long,1990; Kondorosi *et al.*, 1991a).

Many nodulation genes are restricted to certain species. In *S. meliloti* the genes *nodFEGPQ* and *nodH* (originally described as *hsnABC* for *nodFEG* and *hsnD* for *nodH* (Horváth *et al.*, 1986)) are organized in two operons. The genes *nodFE* are present also in *R. leguminosarum* bvs. *viciae* and *trifolii* and *nodE* homologous sequences were found in *R. tropici*. The genes *nodFE* form a single operon

with *nodL* and *nodRL* in *R. leguminosarum* bv. *viciae* and bv. *trifolii*, respectively. The *nodL* gene in *S. meliloti* is separated from *nodFE* and forms an operon together with *noeAB*. The *nodP* and *nodQ* genes, always found pairwise and together with the presence of *nodH*, were identified also in *R. tropici* (in one operon with *nodH*) and in *Rhizobium* sp. BR816. No *nodH* gene has been identified yet in *Rhizobium* BR816.

The *nodM* and *nodN* genes are combined with a variety of other genes in the same operon depending on the species. Those include *nodT* (*R. legumino-sarum* bv. *viciae*), *nodX* (*R. legumino-sarum* bv. *trifolii*) and *nolFGHI* (*S. meliloti*). Nodulation gene *nodO* has only been identified in *R. leguminosarum* bv. *viciae* and *Rhizobium* sp. BR816. The *nodR* gene has been described in *R. leguminosarum* bv. *trifolii* in the *nodFERL* operon, however, its function in nodulation has not been proven yet. As outlined above, the *nodSU* genes are found often downstream of *nodC* in the same operon. However, in *S. fredii*, *M. loti* and *Rhizobium* sp. NGR234 these genes are in a separate operon. The genes *nodV* and *nodW* were found sofar only in *B. japonicum* where they form one operon. The gene *nodZ* is either in a separate transcriptional unit (*R. etli* and *M. loti*), or in the same operon as *nodABC* (*B. japonicum* and *A. caulinodans*) or *noeLnolK* (*Rhizobium* sp. NGR234).

The nodulation genes designated as *nol* and *noe* generally are found in only a single rhizobial sp. except for *nolR* which occurs in *R. leguminosarum* bv. *viciae* strain TOM, *S. meliloti*, *S. fredii* and *Rhizobium* sp. NGR234. In several cases open reading frames are indicated in the same operon with nodulation genes or in their close vicinity. They are often small and in most cases a phenotype or a homologue is not known (see Chapter 20).

III. *nod* Box-Containing Promoters

Promoter sequences of nodulation genes show no homology with well-known bacterial promoters containing classical -10 and -35 or -70 boxes. Consistently, it was found that *E. coli* RNA polymerase does not recognize rhizobial nodulation gene promoters (Fisher *et al.*, 1987a). However, a consensus DNA sequence designated the *nod* box, is present upstream of many nodulation genes which show induced expression (Rostas *et al.*, 1986; Schofield and Watson,1986; Spaink *et al.*, 1987a).

The inducible nodulation gene expression which depends on the NodD protein and flavonoids derived from the host plant, is much better characterized than transcription from nodulation gene promoters lacking a *nod* box sequence. One exception would be the promoter analysis of *S. fredii nolW* using deletions (Gu *et al.*, 1997). Flavonoid-dependent nodulation gene expression is not necessarily indicative of the presence of a *nod* box-containing promoter as is illustrated by flavonoid-enhanced expression of *nolX* and *nolBTU* in *S. fredii* and *Rhizobium* sp. NGR234 which are not preceded by *nod* box sequences (Balatti *et al.*, 1995; Bellato *et al.*, 1996, 1997; Gu *et al.*, 1997; Meinhardt *et al.*, 1993).

In Figure 1 the position of *nod* boxes and homologous sequences are indicated and in Figure 2 an alignment is given for all known *nod* box sequences. Literature references can be found in the legends. It should be noted that the function of only a subset of these *nod* boxes has been shown. The relevance of the other *nod* boxes may be debated, especially in those cases where no downstream genes have been identified. The *nod* box consensus sequence consists of conserved modules of 7, 8 and 26 bp separated by 2 and 6 bp, respectively, of non-conserved sequences (the core consensus sequence). At a distance of generally 1 to 13 bp downstream of the 26-bp module a less well-conserved AT(T)AG sequence can often be recognized in which the G-residue is the most flexible nucleotide. Other DNA sequences adjacent the *nod* box are not conserved. The least well-conserved *nod* box is present in the *nodA* promoter of *A. caulinodans* (Goethals *et al.*, 1992). Sequences with poor homology to *nod* box sequences have been found in the promoter regions of some of the *nodD* genes. This is the case for *nodD1* in *B. japonicum* (Wang and Stacey, 1991), *S. meliloti syrM* (Barnett *et al.*, 1996) and *M. loti nodD3* (Scott *et al.*, 1996). No genes have been identified downstream of *nod* box sequences, for instance in *S. fredii* and *A. caulinodans* (Appelbaum *et al.*, 1988; Goethals *et al.*, 1990). A downstream gene of unknown function may also be present (*e.g.* gene *y4mC* downstream of N_{11} in *Rhizobium* sp. NGR234) (Freiberg *et al.*, 1997).

Several motifs have been designated in the *nod* box sequence (Goethals *et al.*, 1992) (Figure 2). (i) The LysR motif, characterised by a T-N_{11}-A motif is present three times in most cases. This motif was recognized in all promoters regulated by members of the LysR family to which the NodD protein belongs

(for a review see Schell (1993)). The conserved nucleotides of this motif are essential for coinducer-dependent transcription of *nodA* and *nodF* and for binding of NodD protein to the *nodA* promoter, which contains the overlapping *nodD* promoter, but not for NodD binding to the *nodF* promoter (Goethals *et al.*, 1992; Okker *et al.*, in preparation). These results suggest that these nucleotides are essential for binding of NodD protein to function in *nodD* autoregulation (see below) and not necessarily to induce transcription of *nodABCIJ* and *nodFEL*. (ii) The so-called NodD-box which is composed of an inverted repeat with the basic structure A-T-C-N_9-G-A-T. Two to three of these structures are present in the consensus *nod* box sequence.

Sequences essential for inducible expression have been determined by deletion mapping (Spaink *et al.*, 1987a) and point mutations (Goethals *et al.*, 1992; Okker *et al.*, in preparation). Those studies revealed that the entire region of 49 bp harbouring the core consensus sequence is essential as well as the 21 bp downstream sequence containing the less well-conserved sequence AT(T)AG. Moreover, the spacing between the individual conserved modules appeared to be essential as well as a conservation of the LysR motif. These requirements are also nicely illustrated in the case of the n6 *nod* box in *S. meliloti* where transcription of the downstream gene *nolQ* was not inducible with flavonoids (Cren 1994; Plazanet *et al.*, 1995) whereas NodD protein was able to bind to it (Kondorosi *et al.*, 1989). In this *nod* box a single nucleotide is deleted between the 2nd and 3rd modules, it lacks the *nod* box-ATTAG connecting region as well as the second putative LysR motif in the 26-bp-conserved module. Interestingly, two single base pair changes, one in the 7-bp and another in the 26-bp modules of the *nod* box directing *nodFEL* in *R. leguminosarum* bv. *viciae*, were shown both to result in constitutive expression (Okker *et al.*, in preparation). The presence of a *nod* box-like sequence does not predict inducible promoter activity. For example, in *S. fredii* the *nodSU* genes are preceded by a *nod* box but induced transcription could not be demonstrated (Krishnan *et al.*, 1992). The *Rhizobium* sp. NGR234 symbiotic plasmid (pNGR234a) contains 19 *nod* box-like sequences, however, it is yet unknown how many of them are functional (accession number U00090). Only 5 *nod* boxes are linked to nodulation genes and control probably the expression of the *nodABCIJnolOnoeI*, *nodSU* and *nodZnoeLnolK* operons and the *nolL* and *noeE* genes. Transcription analysis suggested that 11 of the putative *nod*

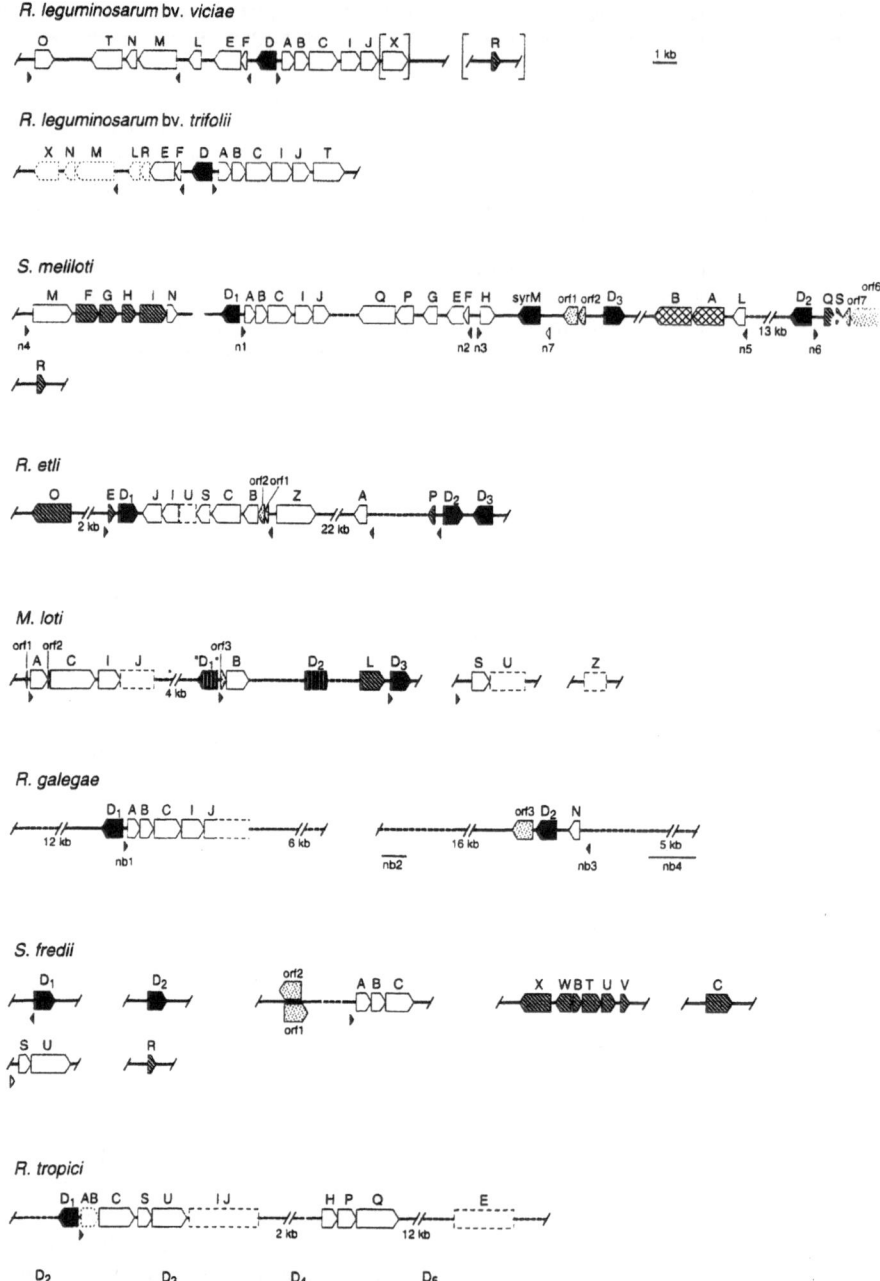

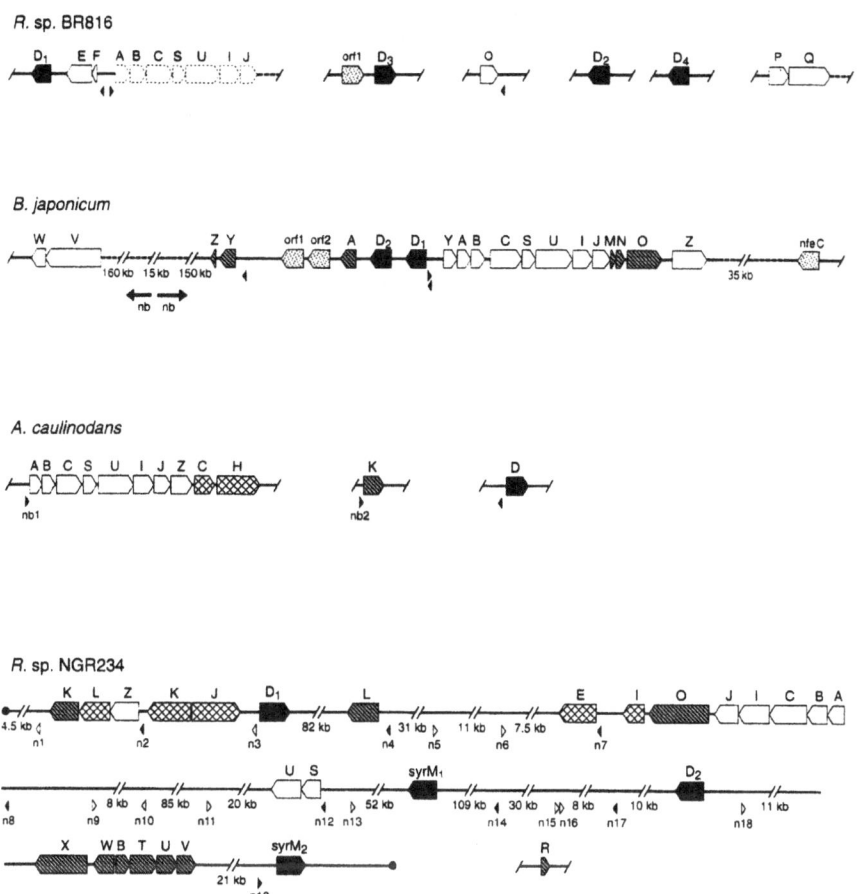

Figure 1. Genetic maps of the nodulation genes in various rhizobia.

The genes and the direction of their transcription are indicated by wide arrows; The genes with unknown direction of transcription are represented by blocks. Putative nodulation genes only known from hybridization or partial sequence data are indicated tentatively by broken lines. Open arrows, *nod* genes; black arrows, *nodD* and *syrM* genes; hatched arrows, *nol* genes; crossed arrows, *noe* genes; stipled arrows, other nodulation genes and unknown open reading frames (ORF). Positions of *nod* box sequences are marked with arrow heads below the sequence. Black arrow heads are functional and open arrow heads indicate probable non-functional *nod* boxes. Stretches of DNA hybridizing with *nod* box sequences are annotated with lines or arrows below the genes. In some bacterial species in which the *nod* boxes used to be named (for example, n1 in *S. meliloti* or nb1 in *R. galegae*) they are marked accordingly. References for the different rhizobia are the following: *R.leguminosarum* bv. *viciae*: Canter-Cremers *et al.*, 1989; de Maagd *et al.*, 1989; Economou *et al.*, 1990; Evans and Downie,1986; Rossen *et al.*, 1984; Shearman *et al.*, 1986; Spaink *et al.*, 1987a, 1989b; Surin and Downie,1988; Surin *et al.*, 1990; *R. leguminosarum* bv. *trifolii*: Innes *et al.*, 1985; Lewis-Henderson and Djordjevic,1991; McIver *et al.*, 1993; Schofield and Watson,1986; Spaink *et al.*, 1989b; Surin *et al.*, 1990; *S. meliloti*: Ardourel *et al.*, 1995; Baev *et al.*, 1991,1992; Baev and Kondorosi,1992; Barlier, 1995; Barnett and Long,1990; Cervantes *et al.*, 1989; Cren, 1994; Debellé and Sharma,1986; Egelhoff *et al.*, 1985; Fisher *et al.*, 1985, 1987b; Gerhold *et al.*, 1989; Göttfert *et al.*, 1986; Honma *et al.*, 1990; Horváth *et al.*, 1986; Jacobs *et al.*, 1985; Kondorosi *et al.*, 1991b; Plazanet *et al.*, 1995; Rostas *et al.*, 1986; Rushing *et al.*, 1991; Schwedock and Long,1989; Török *et al.*, 1984; *R. etli*: Cardenas *et al.*, 1996; Davis and Johnston,1990; Girard *et al.*, 1991; Vazquez *et al.*, 1991, 1993; Villalobos *et al.*, 1994; *M. loti*: Collins-Emmerson *et al.*, 1990; Scott *et al.*, 1996; Young *et al.*, 1990; D. Kafetzopoulos, unpublished results; *R. galegae*: Lindström *et al.*, 1995; Rasanen *et al.*, 1991; K. Lindström, unpublished results; *S. fredii*: Appelbaum *et al.*, 1988; Kovacs *et al.*, 1995; Krishnan and Pueppke,1991a,b; Meinhardt *et al.*, 1993; *R. tropici*: Folch-Mallol *et al.*, 1996; Sousa *et al.*, 1993; Van Rhijn *et al.*, 1993; Vargas *et al.*, 1990; Waelkens *et al.*, 1995; *Rhizobium* sp. BR816: Van Rhijn *et al.*, 1993, 1996; Van Rhijn, 1994; *B. japonicum*: Chun *et al.*, 1994; Deshmane and Stacey, 1989; Dockendorff *et al.*, 1994a,b; Göttfert *et al.*, 1989, 1990a,b, 1992; Luka *et al.*, 1993; Nieuwkoop *et al.*, 1987; Russell *et al.*, 1985; Sadowsky *et al.*, 1991; Sanjuan *et al.*, 1994; Stacey *et al.*, 1994; M. Göttfert, unpublished results; G. Stacey, unpublished results; *A. caulinodans*: Geelen *et al.*, 1993; Goethals *et al.*, 1989, 1990; Mergaert *et al.*, 1996; *Rhizobium* sp. NGR234 (isogenic to MPIK3030): Bachem *et al.*, 1985; Balatti *et al.*, 1995; Freiberg *et al.*, 1997; Horváth *et al.*, 1987; Lewin *et al.*, 1990; Perret *et al.*, 1991; Relic *et al.*, 1994; accession number U00090. Not presented are the data of less-well characterized rhizobia, like *Rhizobium* sp. strain N33 (Cloutier *et al.*, 1996a,b), *Bradyrhizobium elkanii* (Dobert *et al.*, 1994), *Bradyrhizobium* sp. (*Parasponia*) (Scott, 1986) and *Bradyrhizobium* (*Arachis*) sp. strain NC92 (Gilette and Elkan,1996).

boxes are symbiotically active (Freiberg *et al.*, 1997). Two *nod* box-like sequences were found together with putative NifA-dependent upstream activator sequences and a σ54-dependent promoter that allow flavonoid-induced, NodD-mediated gene expression in the nodule.

Transcriptional start sites of genes downstream of *nod* boxes have been determined for several operons, and they have been found 23 to 25 bp downstream of the conserved 26-bp module of the *nod* box (Figure 2).

IV. The Transcriptional Regulators NodD and SyrM

IV.A. EXPRESSION OF *NODD* AND *SYRM* GENES

In many rhizobia, the regulatory gene *nodD* is transcribed divergently from one of the nodulation gene operons regulated by this gene (Figure 1) and the promoters are often overlapping. For instance, the *nodD(1)* promoter overlaps with the *nod(Y)ABCIJ* promoter in *B. japonicum*, *R. leguminosarum* bv. *viciae* and *S. meliloti*. Consequently, the *nodD(1)* transcriptional start site falls within the conserved *nod* box sequence or just downstream of it (Figure 2). A possible role of divergently transcribed promoters is to increase local concentrations of RNA polymerase to elevate transcription levels from less active promoters. In *R. leguminosarum* bv. *trifolii* the *nod* box upstream of *nodA* overlaps with the ATG translational start codon of *nodD* suggesting a strong coupling in transcription of *nodABCIJT* and translation of *nodD*.

Expression levels of *nodD* are usually low in wild type rhizobia. For instance, it has been estimated that 20-80 molecules of NodD protein are present in *R. leguminosarum* bv. *viciae* (Schlaman *et al.*, 1989). In *R. leguminosarum* bvs. *viciae* and *trifolii* but not in *S. meliloti*, the expression of *nodD* is autoregulated negatively *ex planta* (Rossen *et al.*, 1985; Spaink *et al.*, 1987a). DNA sequence analysis revealed the presence of inverted repeats, designated as A-elements, in the *nodD* upstream region of *R. leguminosarum* bv. *viciae*. These A-elements, which are absent in the *S. meliloti* nodD1 promoter region, have slight homology to *nod* box sequences, and they were shown to be the target for autoregulation (Mao *et al.*, 1994). Interestingly, the *nodD1* genes of *R. etli* (Davis and Johnston, 1990),

S. fredii (Appelbaum *et al.*, 1988) and *B. japonicum* (Wang and Stacey,1991) and *syrM2* but not *syrM1* of *Rhizobium* sp. NGR234 (Freiberg *et al.*, 1997) are preceded by a *nod* box sequence and their transcription is enhanced in the presence of NodD1 and certain flavonoids (Davis and Johnston, 1990; Freiberg *et al.*, 1997; Wang and Stacey, 1991; Smit *et al.*, 1992). In *S. meliloti* the multiple *nodD* genes, including *syrM* form a complex autoregulatory network (Figure 6). The *nodD1* and *nodD2* transcription is under negative regulation of NolR. The expression of *nodD3* is positively controlled by its own product and by SyrM and NodD1 (Kondorosi *et al.*, 1991a; Maillet *et al.*, 1990; Mulligan and Long, 1989; Rushing *et al.*, 1991). Expression of *syrM* requires its own product, NodD3 and NodD2 proteins as well as a putative activator, indicated as CII in Figure 6 (Kondorosi *et al*, 1991a; Swanson *et al.* 1993). It should be noted that *S. meliloti syrM* is also preceded by a *nod* box-like sequence (Barnett *et al.*, 1996).

In vitro assays showed the formation of various complexes between proteins and the *nodD* and *syrM* promoter regions in *R. leguminosarum* bv. *viciae* and *S. meliloti* (Figure 6). Some of them could be ascribed to NodD and NolR proteins, respectively, but others were less clear (Burn *et al.*, 1987; Kondorosi *et al.*, 1989; Schlaman *et al.*, 1992a). The C1 complex in *R. leguminosarum* bv. *viciae* is probably involved in expression of *nodABCIJ* and not of *nodD* since it is also formed with the *nod* box in front of *nodMNT*. Except for CII which may be an activator (Kondorosi *et al.*, 1991a), it is unknown whether the proteins in the complexes indicated as C and CI in *S. meliloti*, play a role in *nodD c.q. syrM* transcription regulation.

IV.B. NODD AND SYRM PROTEINS AND THEIR FUNCTION

NodD and SyrM proteins function as transcriptional activators of nodulation genes downstream of a *nod* box-containing promoter. Not every NodD protein contributes equally to the expression of nodulation genes. For instance, the NodD2 proteins of *Rhizobium* spp. NGR234 and BR816, *B. japonicum* and *R. tropicii* seem not to play an important role in nodulation gene transcription (Broughton *et al.*, 1991, (Göttfert *et al.*, 1992; Van Rhijn *et al.*, 1993). On basis of homology, NodD and SyrM proteins have been classified as members of the LysR family (Henikoff *et al.*, 1988) (for a review see Schell

(1993)). This family consists of a still growing group of DNA-binding proteins which, in most cases, act as transcriptional activators. All members have a helix-turn-helix DNA-binding motif more or less at the same position in the N-terminus of the protein. Specific DNA-binding by NodD has been shown indeed, as will be discussed below. Although LysR family members require inducing factors, which are often environmental, they are not part of a two-component system of regulatory proteins (for reviews see Stock *et al.* (1989) and Parkinson (1993)). Instead, they seem to have a dual role both in DNA binding and in coinducer reception.

NodD proteins require plant signal molecules (*e.g.* flavonoids) as coinducers to regulate transcription. Exceptions to this rule are *S. meliloti* NodD3 and SyrM proteins, which act together in plant-independent nodulation gene activation (Kondorosi *et al.*, 1991a; Mulligan and Long, 1989). NodD proteins probably interact directly with plant signal molecules, though this has never been shown due to practical difficulties. However, this assumption is supported by several indirect evidences. Each rhizobial species responds to a characteristic set of flavonoids which gives good to high rates of transcription of the inducible nodulation genes and this feature is only determined by the origin of the NodD protein (Horváth *et al.*, 1987; Spaink *et al.*, 1987b; Zaat *et al.*, 1987). Therefore, NodD is a determinant of host-specific nodulation (Spaink *et al.*, 1987b). For instance, NodD1 of *Rhizobium* sp. NGR234 responds to a broad spectrum of flavonoids whereas 7,4'-dihydroxyflavone and luteolin are the only flavonoids which give good activation of *R. leguminosarum* bv. *trifolii* (Bassam *et al.*, 1988). These data are consistent with the wide and narrow host range, respectively, of these two bacterial species (see Chapter 18). There are a few examples that the transfer of the *nodD* gene from one *Rhizobium* species or biovar to another altered or extended the host range to that of the donor strain (Bender *et al.*, 1988; Horváth *et al.*, 1987; Spaink *et al.*, 1987b; Surin and Downie, 1988; Vargas *et al.*, 1990). Moreover, hybrid NodD proteins consisting of *S. meliloti* and *R. leguminosarum* bv. *trifolii* or *Rhizobium* sp. NGR234 sequences, have been described that were able to extend the range of nodulated host plants of the parental strains (Horváth *et al.*, 1987; Spaink *et al.*, 1989a). Furthermore, the different NodD proteins present within one cell of many rhizobia, exhibit partially overlapping as well as distinct specificity towards flavonoids (Györgypál *et al.*, 1988, 1991a,b; Honma *et al.*, 1990; Phillips

1992). Presumably, a conformational change is induced in the NodD protein upon activation by flavonoids. This notion is supported by the fact that mutant and hybrid NodD proteins have been isolated which activate transcription of inducible nodulation genes independent of flavonoids, the so-called FITA (for flavonoid independent transcription activation) NodD proteins (Burn *et al.*, 1987; McIver *et al.*, 1989; Spaink *et al.*, 1989a,c). The inducer molecules are thought to be sensed by the C-terminal part of the protein as suggested by mutant studies (Spaink *et al.*, 1989c). It should be noted that all LysR family members lack sequence homology in their C-terminal parts while the homology in the N-terminal part of the proteins is very high.

The presumed perception of flavonoids by NodD probably takes place at the inside of the cytoplasmic membrane in the bacterial cell. Support for this hypothesis comes from the observations that flavonoids accumulate in the inner membrane (Recourt *et al.*, 1989) and that NodD protein is localized exclusively or at least partially in the inner membrane in *R. leguminosarum* bv. *viciae* (Schlaman *et al.*, 1989) and *S. meliloti* (Kondorosi *et al.*, 1989), respectively. Interestingly, two regions in NodD proteins have significant homology with the ligand-binding site of estrogen receptors and one of these regions was predicted to be membrane associated (Györgypál and Kondorosi, 1991). It is not known whether the presumed interaction between flavonoids and NodD is temporary or permanent.

NodD does not separate simply into two functional domains, one for DNA binding and one for flavonoid sensing. Results with *nodD* double mutants (Burn *et al.*, 1989) and with hybrid *nodD*s (Spaink *et al.*, 1989c) suggested that aminoacids at various locations in the protein are required for signal perception. The structures of the NodD and SyrM proteins have not been determined. Protein analysis has been hampered by very low abundance of the protein, co-immunopurification of the chaperone GroEL-like proteins upon overproduction of NodD (Fisher and Long, 1989; Ogawa and Long, 1995; Fisher, personal communication) and the subcellular localization of NodD.

NodD does not only function as the transcriptional activator of other nodulation genes, it can act as a repressor of transcription as well. This is most clearly illustrated by the negative autoregulation of *nodD* in *R. leguminosarum* bvs. *viciae* and *trifolii* (Rossen *et al.*, 1985; Spaink *et al.*, 1989c). Furthermore, the expression of *rhiA*, localized on the

Rl bv viciae
Rl bv trif.
S. meliloti
R. etli
M. loti
S. fredii
Rhiz. BR816
Rhiz. N33
B. japonicum
A. caul.
Rhiz. NGR234

Seq	bp	to	gene
	173	bp to	nodA
	147	bp to	nodF
	69	bp to	nodM
	31	bp to	nodO
	152	bp to	nodA
	114	bp to	nodF
n1	182	bp to	nodA
n2	174	bp to	nodF
n3	219	bp to	nodH
n4	59	bp to	nodM
n5	117	bp to	nodL
n6	385	bp to	syrM
n7	496	bp to	syrM
	62	bp to	nodA
	146	bp to	nolE
	289	bp to	nolP
	20	bp to	orf1
	63	bp to	nodA
	26	bp to	orf3
	0	bp to	nodD3
nD1	10t	bp to	nodA
		...to	?
n1	292	bp to	nodS
n2	176	bp to	nodA
n3	283	bp to	nodF
n4	31	bp to	nodO
n5	298	bp to	nodB
n6	134	bp to	nodY
n7	55	bp to	nodD1
n8	53	bp to	nodA
nD		...to	?
n1	475	bp to	put. n1 regulated gene "fal"
n2	59	bp to	nodz
n3	2207	bp to	noeK
n4	249	bp to	nolL
n5	1309	bp to	insertion sequence homologue
n6	862	bp to	fixF
n7	174	bp to	noeE
n8	190	bp to	nodA
n9	1063	bp to	y4hM
n10	89	bp to	psiB
n11	301	bp to	y4mC
n12	125	bp to	nodS
n13	506	bp to	insertion sequence IS5b
n14	437	bp to	y4vC
n15	131	bp to	class-2 aminotransferase
n16	208	bp to	A. tumefaciens ORF2 homologue
n17	89	bp to	put. ABC-type transporter
n18	249	bp to	y4xI
n19	546	bp to	syrM2

LysR motif   Consensus nod box
YATCCAY..YRYRGATG......ATCYAAACAATCRATTTTACCAATCY   1-13 bp AT(T)AG
ATC.........GAT.......ATC........   GAT   NodD box
                                        ATC   .....AAT

Figure 2. nod box sequences in promoter regions of nodulation genes of various rhizobia. Sequences homologous to the consensus *nod* box (Rostas *et al.*, 1986) modified according to Spaink *et al.* (1987a) are in bold face. The LysR motif characterized as a T-N$_{11}$-A repeat (Goethals *et al.*, 1992) is marked by asterisks on top of the consensus sequence. The NodD-box motif which has the conserved A-T-C-N$_9$-G-A-T repeat (Goethals *et al.*, 1992) is indicated as well. The distance from the depicted DNA sequence towards the first approached gene may vary depending on the bacterial strain. Underlined nucleotides represent transcriptional start sites. In the case two transcriptional start sites are indicated in the *nod(Y)A* promoter region the most left one is of *nodD(1)*. Abbreviations: bp, base pairs; *caul., caulinodans*; nD, nodD; nD1, nodD1; R, purine; *Rhiz., Rhizobium* sp.; *Rl, R. leguminosarum*; *trif., trifolii*; Y, pyrimidine; ?, *nod* box-like sequence upstream of the *nodD(1)* gene and read in the opposite direction towards an unknown gene (compare Figure 1) in *S. fredii* (Appelbaum *et al.*, 1988), *A. caulinodans* (Goethals *et al.*, 1990) and *Rhizobium* sp. NGR234 (Freiberg *et al.*, 1997). Additional relevant references to the ones mentioned already in the legend of Figure 1: *R. leguminosarum* bv. *viciae*: Schlaman, 1992; Spaink *et al.*, 1987a; *R.meliloti*: Barnett *et al.*, 1996; Mulligan and Long, 1989; *R. etli*: Davis and Johnston, 1990; *S fredii*: Krishnan *et al.*, 1992; *Rhizobium* sp. (*Oxytropis arctobia*) strain N33 Cloutier *et al.*, 1996b; *B. japonicum* USDA110: Wang and Stacey, 1991.

Sym plasmid of *R. leguminosarum* bv. *viciae*, encoding an abundant 24 kDa protein, is under negative control of NodD (Economou *et al.*, 1989). Also the expression of *exo* genes appears to be repressed by NodD2 in *S. fredii* (Appelbaum *et al.*, 1988).

V. Nodulation Gene Inducing Compounds

V.A. IDENTIFICATION OF ACTIVE NODULATION GENE INDUCERS

Flavonoids, the largest class of nodulation gene coinducer molecules, are synthesized by many flowering plant species, conifers, ferns and bryophytes, but not by bacteria (Stafford, 1990). Flavonoids have been studied in numerous root extracts (Rao, 1990), but identification and quantification of flavonoids in root exudates was first reported for soybean seedlings (D'Arcy-Lameta, 1986). The first evidence that flavonoids function as transcriptional regulators of rhizobial nodulation genes was presented simultaneously by several research groups at a conference in May 1986 (Lugtenberg, 1986) and more detailed reports appeared soon thereafter (Firmin *et al.*, 1986; Kosslak *et al.*, 1987; Peters *et al.*, 1986; Redmond *et al.*, 1986; Spaink *et al.*, 1987b). Those contributions depended upon the availability of *nodABC-lacZ* reporter gene fusions and upon the knowledge that transcription of *nodABC* required an undefined factor from the host legume (Innes *et al.*, 1985; Mulligan and Long, 1985; Rossen *et al.*, 1985; Zaat *et al.*, 1987). Within a few years many different flavonoids had been identified as naturally occurring nodulation gene-inducers from various legumes (Table 1). Flavonoids are present in plants as glycosides with at least one sugar attached and as aglycones

unconjugated to any sugar. The flavonoid moiety generally is an active nodulation gene inducer in both glycoside and aglycone forms (Hungria *et al.*, 1991a; Smit *et al.*, 1992). In the case of luteolin-7-*O*-glucoside, however, this major component of the alfalfa seed coat is not itself an active nodulation gene inducer, but extracellular β-glucosidase activity exuded from the seed and by *S. meliloti* cells converts it to luteolin, an active inducer molecule (Hartwig and Phillips, 1991).

Two natural non-flavonoid nodulation gene inducers, which are active in *S. meliloti*, have also been identified (Phillips *et al.*, 1992) (Table 1). These molecules, stachydrine and trigonelline (Figure 4), are quaternary ammonium compounds known as betaines that are found in many plant tissues exposed to osmotic stress (Jones *et al.*, 1986). Stachydrine and trigonelline are present on ungerminated seeds of alfalfa and at least five other *Medicago* species (Phillips *et al.*, 1995). Attributes of these betaines, which are important for their function as nodulation gene inducers, include their high solubility in water and their abundance on dry legume seeds. The concentrations of betaines required for nodulation gene induction are much higher than flavonoids, but their prevalence on seed coats overcomes this potential problem during the early germination and formation of the first nodules. The relative importance of betaines and flavonoids as nodulation gene inducers cannot be settled on the basis of any single trait, but it is intriguing that stachydrine and trigonelline occur more widely than flavonoids on seeds of different *Medicago* species (Phillips *et al.*, 1995). Whether species such as *Medicago littoralis* and *Medicago truncatula*, which have few, if any, flavonoids on their seeds, exude flavonoids from roots is unknown.

A new report has just identified two aldonic acids, erythronic acid and tetronic acid (Table 1), as natural nodulation gene inducers (Gagnon and Ibrahim, 1998). These compounds, which were isolated from

Legume	Source	Compound	Reference
Aldonic Acids			
Lupin (*Lupinus albus*)	Seed	Erythronic acid Tetronic acid	Gagnon and Ibrahim, 1998
Betaines			
Alfalfa and five other *Medicago* species	Seed	Stachydrine Trigonelline	Phillips *et al.*, 1992, 1995
Flavonoids			
Alfalfa (*Medicago sativa*)	Seed	Luteolin (5,7,3',4'-tetrahydroxyflavone) Chrysoeriol (5,7,4'-trihydroxy-3'-methoxyflavone)	Peters *et al.*, 1986 Hartwig *et al.*, 1990a
	Root	4,4'-Dihydroxy-2'-methoxychalcone	Maxwell *et al.*, 1989
		7,4'-Dihydroxyflavone	Maxwell *et al.*, 1989
		Liquiritigenin (7,4'-dihydroxyflavanone)	Maxwell *et al.*, 1989
Clover (*Trifolium repens*)	Seedling	7,4'-Dihydroxyflavone	Redmond *et al.*, 1986
		Geraldone (7,4'-dihydroxy-3'-methoxyflavone)	Redmond *et al.*, 1986
		4'-Hydroxy-7-methoxyflavone	Redmond *et al.*, 1986
Common Bean (*Phaseolus vulgaris*)	Seed	Delphinidin (3,5,7,3',4',5'-hexahydroxyflavylium)	Hungria *et al.*, 1991a
		Kaempferol (3,5,7,4'-tetrahydroxyflavone)	Hungria *et al.*, 1991a
		Malvidin (3,5,7,4'-pentahydroxy-3',5'-dimethoxyflavylium)	Hungria *et al.*, 1991a
		Myricetin (3,5,7,3',4',5'-hexahydroxyflavone)	Hungria *et al.*, 1991a
		Petunidin (3,5,7,4',5'-pentahydroxy-3'-methoxyflavylium)	Hungria *et al.*, 1991a
		Quercetin (3,5,7,3',4'-pentahydroxyflavone)	Hungria *et al.*, 1991a
	Root	Eriodictyol (5,7,3',4'-tetrahydroxyflavanone)	Hungria *et al.*, 1991a
		Genistein (5,7,4'-trihydroxyisoflavone)	Hungria *et al.*, 1991a
		Naringenin (5,7,4'-trihydroxyflavanone)	Hungria *et al.*, 1991a
Pea (*Pisum sativum*)	Seed	Apigenin (5,7,4'-trihydroxyflavone)	Firmin *et al.*, 1986
		Eriodictyol (5,7,3',4'-tetrahydroxyflavanone)	Firmin *et al.*, 1986
Soybean (*Glycine max*)	Seedling	Daidzein (7,4'-dihydroxyisoflavone)	Kosslak *et al.*, 1987
		Genistein (5,7,4'-trihydroxyisoflavone	Kosslak *et al.*, 1987
	Root	Coumestrol	D'Arcy-Lameta, 1986 Kape *et al.*, 1992
Vetch (*Vicia sativa* subsp. *nigra*)	Root	3,5,7,3'-Tetrahydroxy-4'-methoxyflavanone	Zaat *et al.*, 1989
		7,3'-Dihydroxy-4'-methoxyflavanone	Zaat *et al.*, 1989 Recourt *et al.*, 1991
		4,2',4'-Trihydroxychalcone	Recourt *et al.*, 1991
		4,4'-Dihydroxy-2'-methoxychalcone	Recourt *et al.*, 1991
		Naringenin (5,7,4'-Trihydroxyflavanone)	Recourt *et al.*, 1991
		Liquiritigenin (7,4'-Dihydroxyflavanone)	Recourt *et al.*, 1991
		7,4'-Dihydroxy-3'-methoxyflavanone	Recourt *et al.*, 1991
		5,7,4'-Trihydroxy-3'-methoxyflavanone	Recourt *et al.*, 1991
		5,7,3'-Trihydroxy-4'-methoxyflavanone	Recourt *et al.*, 1991

Table 1. Natural nodulation gene-inducing aglycones produced by legumes. These compounds are released from roots or germinating seeds of legumes, and they induce transcription of nodulation genes in homologous rhizobia. In some cases the presence of rhizobia was required to elicit exudation of the compound.

lupin seed effusates, induce Nod factor formation in *R. lupini*. Concentrations in the low millimolar range are required for activity, similar to the situation with betaines. In addition, tetronic acid induces Nod factor formation by *M. loti* and *S. meliloti*. The structural similarities of these compounds to stachydrine (Figure 4) and their presence on lupin seeds reinforces the importance of non-flavonoid compounds as inducers of nodulation genes. It seems probable that additional nodulation gene inducers remain to be found on seeds as investigators extend the search to other leguminous species.

Some simple phenolics, which generally are found in plant lignin, also induce rhizobial nodulation genes (Kape *et al.*, 1991; Le Strange *et al.*, 1990). While these compounds can be extracted from many plants, there is no direct evidence that they are released by legumes from either roots or seeds. These compounds require much higher concentrations than flavonoids for half-maximum induction of nodulation gene transcription in *B. japonicum* (Kape *et al.*, 1991).

V.B. SYNTHESIS AND RELEASE OF NODULATION GENE INDUCERS

Diverse classes of flavonoids that induce *nod* genes in rhizobia, including chalcones, flavanones, flavones, flavonols, isoflavonoids, coumestans and anthocyanidin, (Table 1; Figure 3), all are derived from phenylpropanoid molecules that enter the flavonoid pathway through chalcone synthase (CHS) (Stafford, 1990). Exudation of flavonoid nodulation gene inducers is linked closely to their concurrent synthesis in sterile alfalfa roots (Maxwell and Phillips, 1990), but the presence of homologous rhizobia elicits quantitative increases and qualitative changes in flavonoid exudation in vetch (Recourt *et al.*, 1991; van Brussel *et al.*, 1990), alfalfa (Dakora *et al.*, 1993a), common bean (Dakora *et al.*, 1993b), soybean (Schmidt *et al.*, 1994) and clover (Lawson *et al.*, 1996). This process is accompanied by a rapid and transient increase in expression of CHS (Mathesius *et al.*, 1996; Recourt *et al.*, 1991). Experiments with soybean and vetch showed that Nod factor alone could elicit this response (Schmidt *et al.*, 1994; Spaink *et al.*, 1991). The nodulation gene-inducing flavonoids are released in greatest amounts near the root tip in both alfalfa (Peters and Long, 1988) and soybean (Graham, 1991), which may be related to sloughing of root cells (Hawes, 1990). While nitrogen limitation enhances flavonoid

production in alfalfa (Coronado *et al.*, 1995), there is no reason to postulate that this response necessarily is related to root nodule formation. Nitrogen stress also enhances flavonoids in tomato through transcriptional regulation (Bongue-Bartelsman and Phillips, 1995). Presumably, flavonoid synthesis is used as a disposal mechanism for extra carbon skeletons produced by deamination of phenylalanine under nitrogen-limited conditions. Alternatively, flavonoids could be intrinsic growth regulators of plants (Djordjevic *et al.*, 1997; Jacobs and Rubery,1988), which are affected by many factors.

Stachydrine and trigonelline are formed by two independent sets of biosynthetic reactions which are separate from the phenylpropanoid pathway (Phillips *et al.*, 1994). Thus, alfalfa has at least three pathways for synthesizing nodulation gene inducers. Simple phenolic nodulation gene inducers are products of phenylpropanoid metabolism.

V.C. STRUCTURE-ACTIVITY RELATIONSHIPS

The prevalence of hydroxylation at the 7- and 4'-carbon position in flavonoid *nod*-gene inducers (Table 1; Figure 3) was noted in early analyses of structure-activity relationships (Rolfe, 1988). For example, tests with *nodA-lacZ* fusions in rhizobia containing *nodD* from *R. leguminosarum* bv. *viciae* showed that flavone, which lacks any substituents on the three flavonoid rings, was inactive, while 7-hydroxyflavone produced about half the maximum induction obtained with luteolin, apigenin, naringenin or eriodictyol (Zaat *et al.*, 1987). In general, NodD proteins from broad host range rhizobia are responsive to a wide range of flavonoids (Györgypál *et al.*, 1991a, 1991b). For example, the presence of a hydroxyl moiety at the 7-carbon position in the flavonoid skeleton is sufficient for NodD1 activation in *Rhizobium* sp. NGR234. By contrast, NodD proteins from narrow host range rhizobia need more specific substitutions of the core flavonoid structure (including also at the C5 and C4' positions) for their activity (Hartwig *et al.*, 1989, 1990a; Peters and Long, 1988; Györgypál *et al.*, 1988, 1991a). Trials with more than 1,000 flavonoids or structural analogues supported the conclusion that hydroxylation at the position equivalent to the 7-carbon facilitated binding to some active site because such compounds had major effects on both induction and inhibition of nodulation genes in many different rhizobia (Cunningham *et al.*, 1991).

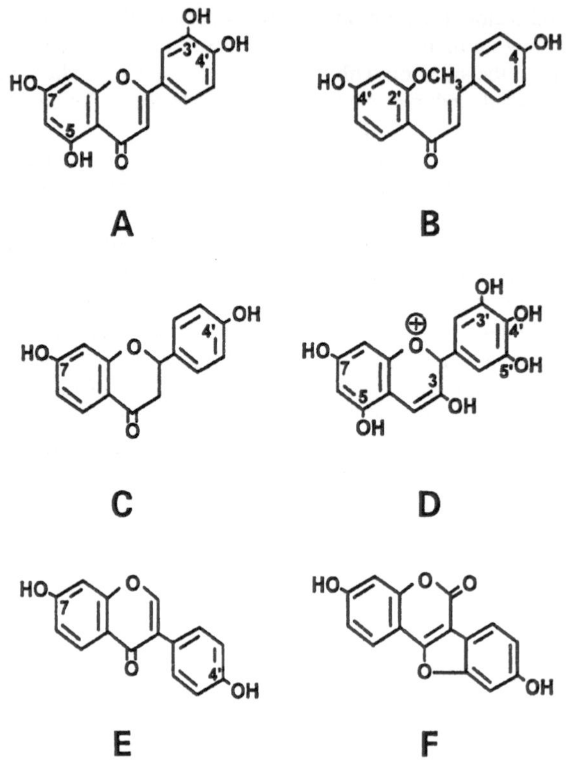

Figure 3. Representative flavonoid nodulation gene inducers. Carbon positions used in Table 1 are identified for the flavone luteolin (A), the chalcone 4,4'-dihydroxy-2'-methoxychalcone (B), the flavanone liquiritigenen (C), the anthocyanidin delphinidin (D), the isoflavone daidzein (E) and the coumestan coumestrol (F).

Not all inducer molecules regulate every nodulation gene by the same mechanism and NodD proteins exhibit significant variations in their ability to interact with flavonoids. In *B. japonicum* several glycosides of soybean seed isoflavones induce *nodD1* transcription but not the *nodYABCSUIJ* operon (Smit *et al.*, 1992). In *S. meliloti* luteolin induces *nodABCIJ* transcription only in strains containing extra *nodD1* genes, but 4,4'-dihydroxy-2'-methoxychalcone induces transcription in strains containing extra copies of either *nodD1* or *nodD2* (Hartwig *et al.*, 1990b). Stachydrine and trigonelline induce *S. meliloti* nodulation genes only in strains containing extra copies of *nodD2* (Phillips *et al.*, 1992). *R. etli* strains containing extra copies of three separate *nodD* genes showed no specificity for various bean flavonoids because all of them induced transcription of nodulation genes in the presence of each compound tested (Hungria *et al.*, 1992). While it is clear that the regulatory effects of some plant

signals differ, few rigorous conclusions about regulation of *nod*-gene transcription can be drawn from published studies because most experiments used rhizobial strains which contained at least one copy of every *nodD* gene in addition to extra copies of individual *nodD* genes.

Early studies suggesting that legumes may control nodulation gene-inducing activity in the rhizosphere by releasing inhibitors (Djordjevic *et al.*, 1987; Firmin *et al.*, 1986) were supported by the demonstration that the weak nodulation gene-inducers 7,4'-dihydroxyflavone and liquiritigenin exuded by alfalfa roots decreased activity of the very strong inducer 4,4'-dihydroxy-2'-methoxychalcone present in the same exudate (Hartwig *et al.*, 1989). It is tempting to impute some ecological importance to this phenomenon, but the levels of nodulation gene-inducing activity recovered from alfalfa rhizosphere soil are so low (León-Barrios *et al.*, 1993), that it appears doubtful that plants benefit from decreasing them further. In addition, not all strains of rhizobia within a single species respond identically to flavonoids (Kosslak *et al.*, 1990). For example, nodulation gene transcription in some *B. japonicum* strains was not inhibited by 7-hydroxy-5-methylflavone, a synthetic compound that strongly reduced nodulation gene induction in *B. japonicum* strain USDA110 (Cunningham *et al.*, 1991). Thus it seems doubtful that plants evolved certain flavonoid structures primarily to inhibit nodulation gene induction by other flavonoids.

These comments have emphasized natural nodulation gene inducers over synthetic compounds that produce the same effect. One danger in searching for natural molecules in root exudates is that bioassays often are more sensitive than the analytical methods required to identify the compound. For this reason, identifications of putative nodulation gene inducers should be confirmed with an authentic sample of the molecule. For example, traces of an inducer in alfalfa root exudate were tentatively identified as a formononetin-7-*O*-6"-*O*-malonylglycoside on the basis of ^{1}H-NMR and FAB-MS data, but no samples of such molecules were available for confirmation (Dakora *et al.*, 1993a). When other workers purified large amounts of formononetin-7-*O*-6"-*O*-malonylglucoside from alfalfa roots, they found that this compound did not induce nodulation genes in *S. meliloti* (Coronado *et al.*, 1995). It may therefore be safer to assay for nodulation gene induction by compounds known to be present in a host plant and then search for exudation of those molecules. For example, triterpenoid saponins, which comprise as much as 3% of alfalfa roots (Massiot *et al.*, 1988) and show some structural similarities to flavonoids have been studied as possible nodulation gene inducers. Four compounds were isolated and identified on the basis of published ^{1}H-NMR, ^{13}C-NMR and FAB-MS data (Oleszek *et al.*, 1990) as being a hederagenin-tetraglycoside, a medicagenic acid-hexaglycoside and the aglycones hederagenin and medicagenic acid. None of the compounds, however, induced nodulation genes in *S. meliloti* (D.A. Phillips and C.M. Joseph, unpublished results).

VI. Interaction Between NodD Proteins and *nod* Box Sequences

The mechanism of transcriptional activation of nodulation genes involves three components: the coinducer flavonoid, the *trans*-activator NodD protein and the conserved *cis* promoter element, the *nod* box providing co-regulation of nodulation gene operons. Binding of the NodD protein to the *nod* box has been demonstrated by gel retardation DNA binding assays (Fisher and Long, 1989; Goethals *et al.*, 1992; Hong *et al.*, 1987; Kondorosi *et al.*, 1989; Schlaman *et al.*, 1992a) and DNA footprinting analyses (Fisher and Long, 1989; Kondorosi *et al.*, 1989). The binding of NodD to the *nod* box is independent of flavonoids, in contrast to transcriptional activation of the nodulation genes, which requires an inducer. It cannot be excluded however, that differences in binding are only detectable *in vivo* and not *in vitro* as has been demonstrated for the NodD-homologue NahR (Huang and Schell, 1991). Interaction of NodD with the *nod* gene promoters protects approximately the -75 to -20 bp region from the transcriptional initiation sites comprising the entire *nod* box with the exception of the central region which is more accessible to DNase I (Fisher and Long, 1989; Kondorosi *et al.*, 1989). Large deletions of the *nod* box eliminate or strongly reduce complex formation indicating that NodD-*nod* box interaction requires all conserved modules of the *nod* box (Kondorosi *et al.*, 1989). In *B. japonicum* and *A. caulinodans* the *nod* box sequences diverge significantly from the consensus sequence (Figure 2) but still serve as a binding site for NodD (Goethals *et al.*, 1992; Wang and Stacey, 1991). The LysR-type T-N$_{11}$-A binding motif has been proposed as the binding target for NodD (Goethals *et al.*, 1992). Insertions, deletions or point mutations affecting the LysR motif abolished or impaired NodD-binding in the *nodA nod* box of *S. meliloti* and *A. caulinodans* (Fisher and Long, 1993; Goethals *et al.*, 1992). Single base pair substitutions along the *nodF nod* box of *R. leguminosarum* bv. *viciae* showed that mutations affecting the LysR motif, although abolishing the flavonoid-dependent promoter activity, unexpectedly did not alter binding of NodD to these sequences (Okker *et al.*, in preparation). This result indicates that flavonoid-dependent promoter activity is based on interaction of NodD with the entire *nod* box. Moreover, the LysR motif might be not the only sequence required for NodD binding.

NodD seems to be active in dimeric form, thus either one NodD dimer interacts with both modules of the *nod* box (Fisher and Long, 1993) or each target site binds a NodD dimer resulting in a tetrameric *nod* box-bound NodD (Goethals *et al.*, 1992; Wang and Stacey, 1991). DNA curving would be a logical

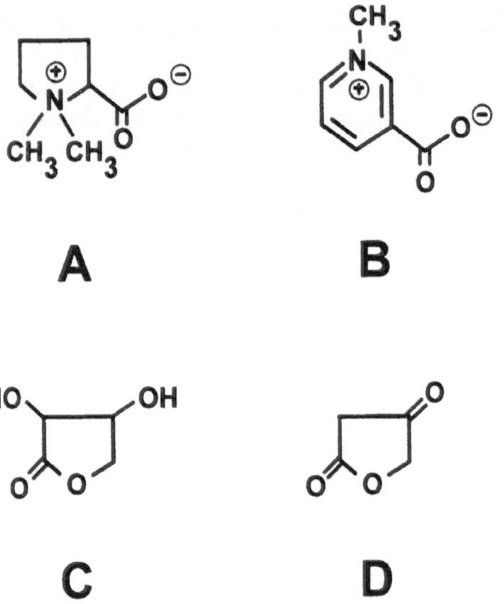

Figure 4. Non-flavonoid nodulation gene inducers. Stachydrine (A), trigonelline (B), erythronic acid (C) and tetronic acid (D).

result in the first model but could be evoked also by the second one. In sequence-specific protein-DNA interaction, protein-folding transitions are coupled to DNA bending (Harrington, 1992). This occurs on complementary surfaces in the DNA grooves comprising the binding site for precise steric matching of protein-DNA hydrogen-bonding interactions. DNA bending is often the result of binding by regulatory proteins with a helix-turn-helix motif. In accordance with this, it was shown that binding of *S. meliloti* NodD3 protein to *nod* promoters is at one site of the DNA helix and induces bending of the DNA (Fischer and Long, 1993). In addition, the chaperonins GroEL assist NodD binding to the *nod* box and might be involved in further interactions of the NodD protein with other components of the transcription apparatus.

Formation of the NodD-*nod* box complex, with the exception of flavonoid-independent systems (SyrM-NodD3, FITA-NodD), is insufficient for nodulation gene activation. Plant flavonoids probably are responsible for the formation of an active transcription complex probably by inducing further conformational changes of the NodD-*nod* box complexes. This is supported by differences in the DNase I footprint patterns between flavonoid-induced and uninduced protein extracts, by the strength of complex formation (Kondorosi *et al.,* 1989) and by enhanced *in vitro* NodD-*nod* box binding by flavonoids in some cases (Goethals *et al.,* 1992). This latter indicates direct interaction of

flavonoids with the binding complex. Alternatively, a temporary interaction of NodD with flavonoids may cause a irreversible conformational change in the protein. In the case of flavonoid-independent *nod* gene activation by the hybrid FITA NodD protein (Spaink *et al.,* 1989a,c) and by the *S. meliloti* NodD3-SyrM system, it is likely that the NodD-*nod* box complexes achieve their correct folding for binding of the transcription apparatus even in the absence of inducer molecules. This hypothesis is supported by differences in DNA footprinting of *S. meliloti* NodD3 and NodD1 proteins with the *nod* gene promoters (Fisher and Long,1989; Kondorosi *et al.,* 1989).

Transcriptional regulation of nodulation genes by NodD may show (subtle) differences. In *Rhizobium* species which harbour a single *nodD* gene, like *R. leguminasarum* bv. *viciae* (Figure 1 and 6), the unique NodD protein interacts with multiple, highly conserved but not identical *nod* boxes (Figure 2). Structural differences in the *nod* box sequences determining the DNA curving and steric matching of NodD protein and *nod* box DNA may cause a hierarchy in the NodD-mediated activation of nodulation gene operons. In *Rhizobium* species containing multiple NodD copies, of which *S. meliloti* is the best-studied example (Figure 1), eventually all NodD copies as well as SyrM contribute to nodulation gene activation (Figure 6). Concentration, stability and competition of these proteins for multiple binding sites and differences in binding efficiency are the major components determining promoter activity. In addition to formation of binding complexes, transcription of inducible nodulation genes depends on the additive, synergistic or inhibitory effect of inducers. Protein folding and complex formation of the individual NodD proteins and *nod* box-containing promoters might be affected differently by flavonoids leading to a range of well- and less well-fitted binding complexes with regards to transcriptional activation of nodulation gene operons.

The NodD-*nod* box interactions can be further complicated by binding of additional regulatory elements to the *nod* box or downstream sequences. For example, in *S. meliloti*, formation of the NodD-*nod* box complexes is strongly influenced by binding of the NolR repressor to a sequence adjacent to the *nod* box (Kondorosi *et al.*, 1989). Furthermore, additional complexes observed with *nod* boxes in *R. leguminosarum* bv. *viciae* indicated as C1 and C2 in Figure 6, contain proteins of which neither the identity nor their possible interference with the NodD-*nod* box interaction is known (Hong *et al.*, 1987; Schlaman *et al.*, 1992a).

VII. Positive Regulation of Nodulation Genes by Other Regulatory Systems

In addition to the regulation systems based on NodD and SyrM, rhizobia evolved control systems to modulate nodulation gene expression for optimal interaction with their host plants. The best known alternative regulatory system is based on NodVW found sofar only in *B. japonicum* (Figure 6) which activate the *nodD1* and *nodYABCSUIJ* promoters in an isoflavonoid-dependent manner (Göttfert *et al.*, 1992; Sanjuan *et al.*, 1994). NodVW belong to the large family of two-component regulatory systems comprising an inner membrane-localized sensor and a cytoplasm-localized transcriptional regulator protein. Upon perception of an extracellular coinducer molecule by the sensor, the regulator is activated *via* protein phosphorylation (for reviews see Parkinson (1993) and Stock *et al.* (1989)). NodV and NodW act as the sensor and the regulator, respectively (Göttfert *et al.*, 1990a). They have been shown to activate the nodulation genes also in the absence of NodD (Sanjuan *et al.*, 1994). NodV senses different sets of isoflavonoids than NodD1 thereby making the nodulation of other sets of host plants feasible. NodV and NodW proteins have a host-specific role in that they are required for nodulation of *Vigna* (cowpea) and *Macroptilium* (siratro) but not for *Glycine* (soybean). However, in the absence of NodD they are also essential for *Glycine* nodulation. It has been suggested therefore that NodV provides a second isoflavone receptor system in addition to NodD1 (Stacey, 1995). The interaction of NodW with the *nod* gene promoter or with NodD in controlling nodulation gene expression is not clear. Nor is it apparent why such a two-component regulatory circuit apparently does not exist in other rhizobia.

In *S. fredii* a second regulatory system is involved in regulation of the *nolJ*, *nolBTUV* and *nolX* transcriptional units (Bellato *et al.*, 1996; Meinhardt *et al.*, 1993). These genes, in spite of lacking the *nod* box sequence in their promoters, are flavonoid-inducible and NodD-dependent. Although the molecular mechanism involved in transcriptional activation of these genes has not been elucidated yet binding of a putative repressor was demonstrated to a 23 bp oligonucleotide 188 bp upstream of the transcriptional initiation site of *nolX*. This complex formation was dependent on both *nodD1* and *nodD2*, and the binding sequence exhibited homology to the A-element (Mao *et al.*, 1994).

VIII. Negative Regulation of Nodulation Gene Expression

VIII.A. THE NOLR REPRESSOR

A high level of nodulation gene expression apparently is detrimental for efficient nodulation in certain rhizobia-host plant combinations (Knight *et al.*, 1986). This would explain why nodulation gene expression is tightly regulated in most species and might involve negative regulatory factors. Downregulation of nodulation genes by a repressor protein called NolR has been demonstrated in *S. meliloti* strain 41 (Kondorosi *et al.*, 1989, 1991b). NolR is required for optimal nodulation (Kondorosi *et al.*, 1989), and it has a drastic effect on Nod factor production resulting in apparently optimal amounts of Nod factor for nodulation of the host plant (Cren *et al.*, 1995). The NolR protein is encoded by the single copy gene *nolR* located on the chromosome, and it is present in all *S. meliloti* strains. Curiously, the NolR function is missing from *S. meliloti* strain SU40 and its derivatives 1021 and 2011, whereas the gene is transcribed at the same level as in strain 41

R.leguminosarum bv. viciae	nodA	**TATAG**AAAACCCGG**AA**	(+)
	nodF	CA**TAG**CAGGGCAGCCG	(−)
	nodM	**ATTAG**CACGCGCTGG**A**	(−)
	nodO	GA**TA**AGGGGCACAGGC	(−)
strain TOM	nolR	**ATTAG**GAAGATAAC**AT**	+
R.leguminosarum bv. trifolii	nodA	**ATTAG**AAAGGCCGG**AA**	(+)
	nodF	CA**TAG**CACAAAACC**AG**	(−)
S. meliloti	nodA-nodD1* (n1)	**ATTAG**AGAACCCTG**AA**	+
	nodF (n2)	GA**TAT**GAGCACAAGC**T**	−
	nodH (n3)	**ATTA**AAACGCTAAGC**A**	−
	nodM (n4)	**TTAG**GAGACCCTG**AA**	+
	nodL (n5)	GA**TAG**GAGACAAGGCG	−
	nolQ-nodD2* (n6)	**ATTAG**AGAACCCTG**AA**	+
	syrM (n7)	**ATTA**CAAGAACGTT**AG**	(+)
	nolR	**ATTAG**CCGTGATGC**AT**	+
	nodD3	**ATTA**CACCGGTAGG**AA**	(+)
S. fredii	nodA	**ATTAG**AAGATGCTC**AC**	(+)
Rhizobium sp. NGR234	nodA (n8)	**ATTAG**AAGATGCTC**AC**	(+)
	nodZ (n2)	**ATTAG**GAAGCTCTG**AA**	(+)
	noeK (n3)	**ATTAG**TGGATGCG**AAT**	(+)
Consensus		(A/T)**TTAG**--N(9)---**A**(T/A)	

Figure 5. Alignment of NolR target sequences (Cren et al., 1995) in promoters of nodulation genes of various rhizobia. *, reverse complementary sequence; +, demonstrated NolR control; -, experimental evidence for NolR-independent regulation; (+), putative NolR-binding site; (-), probably NolR-independent control.

(E. Kondorosi, unpublished data). A single insertional point mutation in the C-terminal-coding sequence of nolR appeared to be responsible for the inactivation of the NolR protein in strain 1021 (Cren et al., 1994).

The NolR protein isolated from S. meliloti strain 41 has a molecular mass of 13 kDa, and it contains a helix-turn-helix DNA-binding motif (Kondorosi et al., 1991b). Mutations in this motif abolished or strongly reduced binding of NolR to its target sequence confirming its functionality, however, the C-terminal part of the protein contributed also to DNA binding (Cren et al., 1995). NolR shares an overall homology with members of the ArsR transcriptional regulators such as SmtB, MerR, AsrR or CadC (Kiss et al., in preparation). Both the size of these proteins and the location of the helix-turn-helix motif are similar to that of NolR. Moreover, all of them act as a repressor and their target sequence lies between the -35 and -10 promoter regions (Kondorosi et al., 1989; O'Halloran et al., 1989). Therefore NolR appears to be member of this repressor family.

NolR was shown to repress both S. meliloti nodD1 and the nodABCIJ genes by binding to the overlapping promoters of these genes as well as nodD2 and the nodMnolFGHInodN operon. The protein binds in dimeric form to the (A/T)TTAG-N(9)-A(T/A) target sequence present in the S. meliloti n1, n4 and n6 nod box promoters 2-12 bp downstream of the nod-boxes (Figure 2 and 5) (Cren et al., 1995). DNase I footprinting revealed protection by NolR in the -20 to +14 and in the -26 to +1 regions of the overlapping nodABCIJ and nodD1 promoters, respectively (Kondorosi et al., 1989). These NolR-protected regions are slightly overlapping (1-2 nucleotides) with the NodD-protected region in strain 41. The nodulation gene inducer luteolin has a negative effect on formation of the NolR-DNA complexes. Similarly, the presence of heavy metals resulted also in a diminished affinity of the AsrR-type repressors to the operator probably by undergoing an allosteric change (O'Halloran et al., 1989). Based on the altered strength of complex formation and footprint patterns in the presence of luteolin it is likely that the inducer promotes the NodD-nod box interactions and thereby decreases

R. leguminosarum bv. viciae

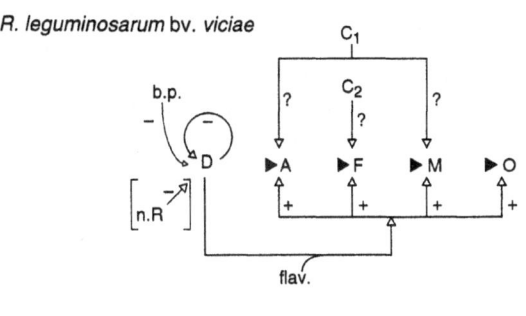

S. meliloti

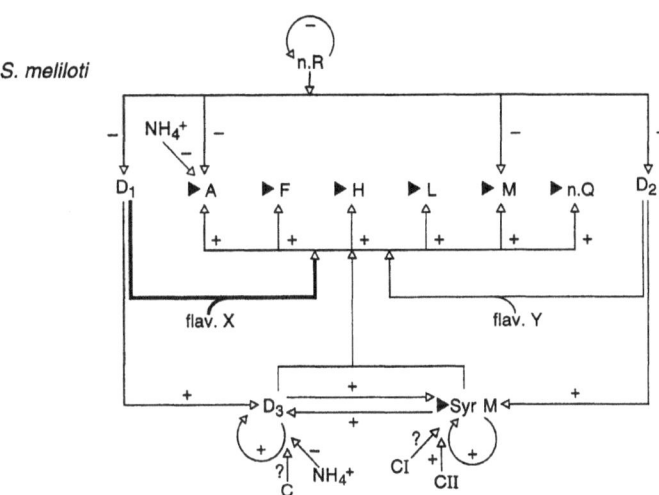

B. japonicum

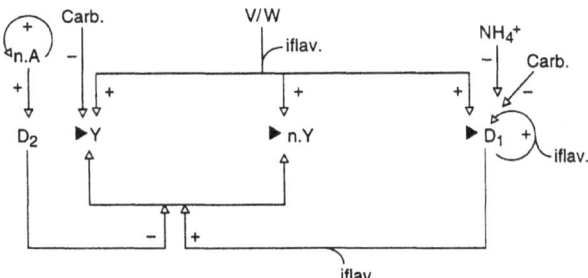

Figure 6. Summary of the regulation of nodulation gene expression in *R. leguminosarum* bv. *viciae*, *S. meliloti* and *B. japonicum*. Stimulation (+) and inhibition (-) of transcription are indicated. The involvement of flavonoids and isoflavonoids is depicted with flav. and iflav., respectively. Nodulation genes and operons are indicated by the letter of the first *nod* gene (*e.g.* A: *nodA* indicating the *nodABCIJ* operon, D: *nodD*) or by n. followed by a letter for *nol* genes (*e.g.* n.R: *nolR*). *Nod* boxes are indicated by a black arrowhead. b.p.: bacteroid-specific protein which binds *in vitro*. C, C1, C2, CI, CII: Proteins of unknown nature which bind specifically to *nod* box DNA *in vitro*. Environmental factors affecting nodulation gene transcription are also indicated. Carb.: carbohydrates. For more details, see text. Relevant literature references are as follows: *R. leguminosarum* bv. *viciae*: Hong *et al.*, 1987; Schlaman *et al.*, 1992a; Spaink *et al.*, 1987a; Zaat *et al.*, 1987; *S. meliloti*: Barnett *et al.*, 1996; Cren *et al.*, 1995; Dusha *et al.*, 1989; Fisher *et al.*, 1988; Kondorosi *et al.*, 1989, 1991a,b; Maillet *et al.*, 1990; Mulligan and Long,1989; *B. japonicum*: Dockendorff *et al.*, 1994a; Garcia *et al.*, 1996; Göttfert *et al.*, 1990; Sanjuan *et al.*, 1994; Stacey *et al.*, 1994.

binding of NolR to the operator and liberating the binding site for RNA polymerase (Kondorosi et al., 1989). NolR, in addition, mediates autorepression by binding to its own promoter. Expression of nolR compared to other nod genes is high both in free-living and symbiotic states (Cren et al., 1995).

NolR-mediated negative regulation of nod genes has been demonstrated recently also in R. legumino-sarum bv. viciae strain TOM (Kiss et al., in preparation) and in Rhizobium sp. NGR234 (Broughton, personal communication). Sequences homologous to nolR exist also in S. fredii and in R. leguminosarum bv. trifolii (Kiss et al., in preparation). Moreover, in these species as well as in various other rhizobia a putative NolR target sequence is present in the some of the nodulation gene promoters (Figure 5) suggesting NolR-medi-ated negative repression of downstream genes. Thus it appears that although NolR is not universal in rhizobia, at least in a subset of Rhizobium species it provides a common mechanism for negative regulation of nodulation genes.

VIII.B. THE NolA PROTEIN

The nolA gene has only been found in Bradyrhizobium spp. and it is involved in genotype specific nodulation (Gilette and Elkan,1996; Sadowsky et al., 1991). First, it was proposed to be a repressor of nodulation genes (Dockendorff et al., 1994a), however, recent studies contradict this hypothesis since no direct effect of NolA was found on the expression of nodD1 or the nodYABC operon (Garcia et al., 1996). On the other hand, it is required for its own expression as well as that of nodD2. Constitutive expression of nodD2 negatively affects transcription of nodulation genes and therefore it is possible that NolA via NodD2 or perhaps other proteins exerts indirect repression of nodulation genes. Surprisingly, nolA mutations have more impact on later stages of symbiosis than onnodulation by reducing the number of infected cells and altering bacteroid development in cowpea nodules.

Until now there is no direct proof for the regulatory function of NolA. The 238 amino acid long NolA protein with a predicted molecular weight of 27 kDa shows slight homology to the MerR regulatory protein (Sadowsky et al., 1991). This homology is restricted to the first 70 amino acid residues at the N-terminus and does not comprise the centrally located, conserved helix-turn-helix DNA-binding

domain of MerR and the other members of the ArsR regulatory family. Thus, both the size of the NolA protein and the position of the putative helix-turn-helix motif disfavour the proposition that NolA belongs to this family of bacterial regulators. NolA contains on the other hand a putative helix-turn-helix motif between amino acid positions 13 and 32 (Sadowsky et al., 1991). Although DNA-binding ability of the NolA protein has not been demonstrated yet the presence of this putative DNA-binding helix-turn-helix motif and stimulation of nolA and nodD2 expression support its regulatory function.

IX. Other Regulatory Factors in Nodulation Gene Expression

Establishment of symbiosis, formation of nitrogen fixing nodules are influenced by many environmental factors. Nodule formation is adversely affected by low pH and factors associated with soil acidity (Richardson et al., 1989). The acid sensitivity is especially pronounced during the first symbiotic events occurring before and at the time of root hair curling. One reason for acid sensitivity can be poor growth of rhizobia at low pH, which by extending the log phase decreases rapid inducibility of nodulation gene expression, lectin-binding ability and infectivity. Poor nodulation gene induction could be due also to the host plant which produces lower amounts of inducer flavonoids in acidic soil (Richardson et al., 1988). Moreover, accumulation of the flavonoid-inducer in the bacterium appears to be also pH-dependent that is another potential factor for controlling nodulation gene expression by the bacterial cell (Recourt et al., 1989). Sensitivity to low pH is further influenced by the Ca^{2+} concentration (Richardson et al., 1988). While low pH and low Ca^{2+} concentration inhibit nodulation, elevated concentration of Ca^{2+} at the same pH improves nodulation. Although the role of Ca^{2+} is unknown, it might be an important element of the Nod signal transduction pathway, or it could be involved also in stimulation of flavonoid production and thereby increasing nodulation gene expression.

Limitation of combined nitrogen in the soil is the prerequisite for symbiotic interactions between rhizobia and host plants. Elevated level of combined nitrogen, sensed by both partners, has a negative effect on the production of phenolic compounds in the root which, as in the case of low pH, can be a limiting factor for nodulation gene induction and for the establishment of symbiosis (Kapulnik et al., 1987). In *S. meliloti* and *B. japonicum* expression of several *nod* genes is controlled by combined nitrogen. In *S. meliloti* ammonia regulation of *nodABCIJ* genes and *nodD3*, but not *nodD1*, involves both the general nitrogen regulatory (*ntr*) system as well as the chromosomal *ntrR* gene (Dusha et al., 1989). The nitrogen level is sensed by the Ntr system and it is likely that a signal is transmitted to *syrM* and *nodD3* via the NtrA and NtrC regulatory proteins. The NtrC protein binds to the *nodD3* promoter, and the DNA regions protected by NtrC corresponded to the putative NtrC-binding sites (Dusha and Dixon, 1994). Repression of nodulation genes at high ammonia concentration is provided by NtrR (Dusha and Kondorosi, 1993). Maximal expression of the *S. meliloti nodABC* genes in *Rhizobium* sp. NGR234 requires also the *ntrA* gene (Stanley et al., 1989). In *B. japonicum* ammonia repression of the *nodD1* and *nodYABC* genes is independent of NtrC and NifA (Wang and Stacey, 1990). Ammonia repression of nodulation genes is, however, not general in rhizobia since neither *nodD* nor *nodABC* genes of *R. leguminosarum* bv. *viciae* are regulated in response to combined nitrogen (Baev and Kondorosi, 1992).

X. Nodulation Gene Expression at Later Stages of Symbiosis

In planta, expression of the inducible nodulation genes has been detected during the invasion process, in the infection threads in the root hairs and in the infection zone of the nodule but it then decreased rapidly during the process of differentiation of bacteria into bacteroids through the central nodule region and disappearing finally in the nodule senescence zone (Cren et al., 1995; Freiberg et al., 1997; Schlaman et al., 1991; Sharma and Signer, 1990). The expression of *R. leguminosarum* bv. *viciae nodD* is severely reduced in bacteroids probably due to an as yet unidentified repressor (Schlaman et al., 1991; 1992a). In contrast, the *S. meliloti syrM* gene expression is enhanced in bacteroids (Sharma and Signer, 1990).

The cause for repressed inducible nodulation gene expression is still not completely elucidated. Neither the reduced amounts of NodD nor that of the inducing flavonoids seems to be the limiting factor for transcription of nodulation genes (Schlaman et al., 1991). Gel retardation assay performed with protein extracts from bacteroids of *R. leguminosarum* bv. *viciae* and different *nod* gene promoters indicated that NodD probably has an altered form in the bacteroids that prevents or significantly reduces its binding to the *nod* boxes (Schlaman et al., 1992a). Despite its high expression in bacteroids, NolR appears not to be responsible for the suppression of nodulation gene expression in the bacteroids (Cren et al., 1995). Most recently, studies on flavonoid structures in *in vitro* bacterial cultures indicated that flavonoid catabolism by the rhizobia may influence nodulation gene inducing activity during symbiosis (Rao and Cooper, 1995). Organic acids are proposed to be the primary carbon sources for rhizobia *in planta* and might be involved in regulation of nodulation genes during symbiosis. Recently it was shown that specific dicarboxylic acids (*i.e.* acetate, fumarate, L-malate, succinate) at 1 mM concentration inhibited significantly the *nodD1* and *nodYABC* expression in *B. japonicum* (Yuen and Stacey, 1996). When L-malate and acetate, that are major carbon sources for bacteroids, were added together the inhibition increased up to 91%. Therefore, these compounds can be, at least partially, responsible for reduced nodulation gene expression in the bacteroids.

The function of NodD and SyrM proteins during later stages of symbiosis is still not clear. In *Rhizobium* sp. NGR234 NodD-regulated promoters have been found which are expressed at the later stages of the infection process suggesting a role of flavonoid-induced gene expression in the nodule (Freiberg et al., 1997). Several observations are consistent with each other in that they suggest a role of NodD in the process of nitrogen fixation. (i) Bacteria harbouring a hybrid *nodD* gene (FITA *nodD604*) give normal nodulation but the levels of nitrogen fixation are higher compared to the wild type control (Spaink et al., 1989a). This could not be explained by a continous expression of their inducible nodulation genes since this does not occur (Schlaman et al., 1991). (ii) A mutant *nodD* has been described which exhibits normal nodulation kinetics but the nodules formed are Fix⁻ (Burn et al., 1987). (iii) *S. meliloti syrM* gene is the only *nodD* allele which is highly expressed in nitrogen-fixing bacteroids and expressed at low levels in free-living bacteria whereas the expression levels of the other *S. meliloti nodD* alleles are just the opposite (Sharma and Signer, 1990). (iv) The general system for nitrogen-regulated gene expression (Ntr) appears to control *S. meliloti nodD3* transcription (Dusha and

Kondorosi,1993; Dusha and Dixon, 1994; Kondorosi *et al.*, 1991a). (v) The expression of several *nodD* genes in various rhizobia is regulated by ammonia (Dusha *et al.*, 1989; Stanley *et al.*, 1989; Wang and Stacey, 1990).

XI. Conclusions and Perspectives

Optimal nodulation requires a finely tuned expression of nodulation genes. This is understandable in view of the function of most nodulation genes being involved in the synthesis of Nod factors of which the concentration and composition are crucial to induce infection structures and nodule primordia on host plants (see chapter 20 and 21). A detailed study on the regulation of nodulation gene expression has mainly been made in *R. leguminosarum* bv. *viciae*, *S. meliloti* and *B. japonicum* which differ from each other with respect to several aspects of regulated transcription. In Figure 6 a summary is given of the current knowledge in this area. In free-living cells, activation of expression of *nod* box-containing nodulation genes is *via* NodD, present as one copy (*R. leguminosarum* bv. *viciae*) or multiple alleles (*S. meliloti* and *B. japonicum*) and plant-derived (iso)flavonoids. Various NodDs respond differently to plant signals, moreover they might be different in their affinity to various *nod* boxes. Thereby both the range and quantity of Nod factors can be modulated on a host-dependent manner to achieve optimal nodulation of definied host plants. Additionally, nodulation genes are also activated by the NodV/NodW system in *B. japonicum* making it possible to extend the set of host plants being nodulated. It should be noted that activation of NodD is required but not sufficient to achieve successful nodulation. This is clearly illustrated by the activation of *Rhizobium* sp. NGR234 NodD by extracts of the nonlegumes wheat, maize and rice which are not nodulated, however (Bender *et al.*, 1988).

Nod factor production must be finely adjusted to the host plants and this can be achieved by negative regulation of nodulation genes. It applies different mechnisms: (i) strong negative autoregulation of *nodD* (*R. leguminosam* bv. *viciae*), (ii) the NolR repressor downregulating *nodD* as well as the common nodulation genes (*S. meliloti*, *R. leguminosarum* bv. *viciae* strain TOM) and (iii) indirect negative control of common nodulation gene

expression (NolA *via* NodD2 which act as a repressor of nodulation genes in *B. japonicum*). Apart from these well-characterised *trans*-acting factors other proteins (C1 and C2 in *R. leguminosarum* bv. *viciae*, C, CI and CII in *S. meliloti*, Figure 6) interact with nodulation gene promoters. However, no role in transcriptional regulation has been observed for these proteins in most cases. In addition to regulatory proteins, environmental factors may play a role in regulated nodulation gene expression. In bacteroids, the nodulation genes are not expressed, but the mechanism by which their transcription is down-regulated is unknown. In *R. leguminosarum* bv. *viciae* a bacteroid-specific protein (indicated as b.p. in Figure 6) is suggested to be a repressor of *nodD* expression.

The precise mechanism of induced nodulation gene expression in biochemical terms is still not completely understood despite the fact that many components involved in it have been described fairly detailed. A major problem concerns the lack of any direct evidence for an interaction between flavonoids and NodD proteins. Remaining questions are whether membrane-bound NodD protein is able to induce transcription or whether it is released from the membrane after it is activated by flavonoids in order to stimulate expression of downstream genes as has been suggested (Kondorosi *et al.*, 1989; Schlaman, 1992). This hypothesis is supported by the observation that the chaperone GroEL proteins interacts strongly with NodD and these sort of proteins are believed to assist in the folding and translocation of other proteins.

Signal transduction in bacteria is primarily conducted through a two-component system involving a sensor and regulator protein. There is growing evidence, however, that one-component regulatory systems, like that of the LysR-type, play an important role in many regulatory pathways as well. At this point, the rhizobial NodD proteins remain the best-studied members of the LysR family of transcriptional regulators, and therefore they can be considered as a model protein for this kind of one-component regulatory proteins.

XII. Acknowledgements

We are very grateful to the people who provided us with data prior to publication and to those who helped us to construct the figure on the genetic organization of nodulation genes: B. Broughton, B.

Fisher, M. Göttfert, M. Holsters, D. Kafetzopoulos, E. Kiss, H. Krishnan, K. Lindström, P. Mergaert, D. Paelinck, C. Quinto, M. Schell, G. Stacey, J. Vanderleyden and K. Vlassak.

XIII. References

Appelbaum, E.R., Thompson, D.V., Idler, K. and Chartrain, N. (1988) J. Bacteriol. 170, 12-20.

Ardourel, M., Lortet, G., Maillet, F., Roche, P., Truchet, G., Promé, J-C. and Rosenberg, C. (1995) Mol. Microbiol. 17, 687-699.

Bachem, C.W.B., Kondorosi, E., Bánfalvi, Z., Horváth, B., Kondorosi, A. and Schell, J. (1985) Mol. Gen. Genet. 199, 271-278.

Baev, N., Endre, G., Petrovics, G., Bánfalvi, Z. and Kondorosi, A. (1991) Mol. Gen. Genet. 228, 113-124.

Baev, N. and Kondorosi, A. (1992) Plant Mol. Biol. 18, 643-846.

Baev, N., Schultze, M., Barlier, I., Cam Ha, D., Virelizier, H., Kondorosi, E. and Kondorosi A. (1992) J. Bacteriol. 174, 7555-7565.

Balatti, P.A., Kovacs, L.G., Krishnan, H.B. and Pueppke, S.G. (1995) Mol. Plant-Microbe Interact. 8, 693-699.

Bánfalvi, Z., Sakanyan, V., Konez, C., Kisc, A., Dusha, I. and Kondorosi, A. (1981) Mol. Gen. Genet. 184, 318-325.

Barlier, I. (1995) PhD. thesis, Analyse de génes et de signaux bactériens impliques dans les étapes precoces de la symbiose *Rhizobium meliloti-Medicago sativa*, Université Paris VI, France.

Barnett, M.J. and Long, S.R. (1990) J. Bacteriol. 172, 3695-3700.

Barnett, M.J., Rushing, B.G., Fisher, R.F. and Long, S.R. (1996) J. Bacteriol. 178, 1782-1787.

Bassam, B.J., Djordjevic, M.A., Redmond, J.W., Batley, M. and Rolfe, B.G. (1988) Mol. Plant-Microbe Interact. 1, 161-168.

Bellato, C.M., Balatti, P.A., Pueppke, S.G. and Krishnan, H.B. (1996) Mol. Plant-Microbe Interact. 9, 457-463.

Bellato, C., Krishnan, H.B., Cubo, T., Temprano, F. and Pueppke, S.G. (1997) Microbiol. 143, 1381-1388.

Bender, G.L., Nayudu, M., Le Strange, K.K. and Rolfe, B.G. (1988) Mol. Plant-Microbe Interact. 1, 259-266.

Beringer, J.E., Beynon, J.L., Buchanon-Wallaston, A.V. and Johnston, A.W.B. (1978) Nature 276, 633-634.

Beynon, J.L., Beringer, J.E. and Johnston, A.W.B. (1980) J. Gen. Microbiol. 10, 421-429.

Bongue-Bartelsman, M. and Phillips, D.A. (1995) Plant Physiol. Biochem. 33, 539-546.

Broughton, W.J., Krause, A., Lewin, A., Perret, X. and Price, N.P.J. (1991) pp. 162-167, in H. Hennecke and D.P.S. Verma (eds.) Advances in molecular genetics of plant microbe interactions, Vol.1, Kluwer Acad. Press, Dordrecht, The Netherlands.

Buchanan-Wollaston, A.V. (1979) J. Gen. Microbiol. 112, 135-142.

Burn, J., Rossen, L. and Johnston, A.W.B. (1987) Genes Dev. 1, 456-464.

Burn, J.E., Hamilton, W.D., Wootton, J.C. and Johnston, A.W.B. (1989) Mol. Microbiol. 3, 1567-1577.

Canter-Cremers, H.C.J., Spaink, H.P., Wijffjes, A.H.M., Pees, E., Wijffelman, C.A., Okker, R.J.H. and Lugtenberg, B.J.J. (1989) Plant Mol. Biol. 13, 163-174.

Cardenas, L., Dominguez, J., Santana, O. and Quinto, C. (1996) Gene 173, 183-187.

Cervantes, E., Sharma, S.B., Maillet, F., Vasse, J., Truchet, G. and Rosenberg, C. (1989) Mol. Microbiol. 3, 745-755.

Cloutier, J., Laberge, S., Castonguay, Y. and Antoun, H. (1996a) Mol. Plant-Microbe Interact. 9, 720-728.

Cloutier, J., Laberge, S., Prevost, D. and Antoun, H. (1996b) Mol. Plant-Microbe Interact. 9, 523-531.

Chun, J.Y. and Stacey, G. (1994) Mol. Plant-Microbe Interact. 7, 248-255.

Collins-Emmerson, J.M., Terzaghi, E.A. and Scott, D.B. (1990) Nucleic Acids Res. 18, 6690-6690.

Coronado, C., Zuanazzi, J.A.S., Sallaud, C., Quirion, J.C., Esnault, R., Husson, H.P., Kondorosi, A. and Ratet, P. (1995) Plant Physiol. 108, 533-542.

Cren, M. (1994) PhD. thesis, Interaction *Rhizobium meliloti-Medicago*: Caractérisation et rôle physiologique du répresseur NolR, Univ. Paris Sud, France.

Cren, M., Kondorosi, A. and Kondorosi, E. (1994) J. Bacteriol. 176, 518-519.

Cren, M., Kondorosi, A. and Kondorosi, E. (1995) Mol. Microbiol. 15, 733-747.

Cunningham, S., Kollmeyer, W.D. and Stacey, G. (1991) Appl. Environ. Microbiol. 57, 1886-1892.

Dakora, F.D., Joseph, C.M. and Phillips, D.A. (1993a) Plant Physiol. 101, 819-824.

Dakora, F.D., Joseph, C.M. and Phillips, D.A. (1993b) Mol. Plant-Microbe Interact. 6, 665-668.

D'Arcy-Lameta, A. (1986) Plant and Soil 92, 113-123.

Davis, E.O. and Johnston, A.W.B. (1990) Mol. Microbiol. 4, 933-941.

de Maagd, R.A., Wijfjes, A.H.M., Spaink, H.P., Ruiz-Sainz, J.E., Wijffelman, C.A., Okker, R.J.H. and Lugtenberg, B.J.J. (1989) J. Bacteriol. 171, 6764-6770.

Debellé, F. and Sharma, S.B. (1986) Nucleic Acids Res, 14, 7453-7472.

Deshmane, A. and Stacey, G. (1989) J. Bacteriol. 171, 3324-3330.

Ditta, G., Stanfield, S., Corbin, D., Helinski, D.R. (1980) Proc. Natl. Acad. Sci. USA 77, 7347-7351.

Djordjevic, M.A., Zurkowski, W., Shine, J. and Rolfe, B.G. (1983) J. Bacteriol. 156, 1035-1045.

Djordjevic, M.A., Schofield, P.R., Ridge, R.W., Morrison, N.A., Bassam, B.J., Planzinski, J., Watson, J.M. and Rolfe, B.G. (1985) Plant Mol. Biol. 4, 147-160.

Djordjevic, M.A., Redmond, J.W., Batley, M. and Rolfe, B.G. (1987) EMBO J. 6, 1173-1179.

Djordjevic, M.A., Mathesius, U., Arioli, T., Weinman, J.J. and Gartner, E. (1997) Aust. J. Plant Physiol. 24, 119-132.

Dobert, R.C., Brei, B.T. and Triplett, E.W. (1994) Mol. Plant-Microbe Interact. 7, 564-572.

Dockendorff, T.C., Sanjuan, J., Grob, P. and Stacey, G. (1994a) Mol. Plant-Microbe Interact. 7, 596-602.

Dockendorf, T.C., Sharma, A. and Stacey, G. (1994b) Mol. Plant-Microbe Interact. 7, 173-180.

Downie, J.A., Hombrecher, G., Ma, Q-S., Knight, C.D., Wells, B. and Johnston, A.W.B. (1983) Mol. Gen. Genet. 190, 359-365.

Downie, J.A., Knight, C.D., Johnston, A.W.B. and Rossen, L. (1985) Mol. Gen. Genet. 198, 255-262.

Dusha, I., Bakos, A., Kondorosi, A., de Bruijn, F. and Schell, J. (1989) Mol. Gen. Genet. 219, 89-96.

Dusha, I. and Kondorosi, A. (1993) Mol. Gen. Genet. 240, 435-444.

Dusha, I. and Dixon, R.A. (1994) p. 307, in G.B. Kiss and G. Endre (eds.), Proceedings of the 1st European Nitrogen Fixation Conference Officina P., Szeged, Hungary.

Economou, A., Hawkins, F.K.l., Downie, J.A. and Johnston, A.W.B. (1989) Mol. Microbiol. 3, 87-93.

Economou, A., Hamilton, W.D.O., Johnston, A.W.B. and Downie, J.A. (1990) EMBO J. 9, 349-354.

Egelhoff, T.T., Fisher, R.F., Jacobs, T.W., Mulligan, J.T. and Long, S.R. (1985) DNA 4, 241-248.

Evans, I.J. and Downie, J.A. (1986) Gene 43, 95-101.

Firmin, J.L., Wilson, K.E., Rossen, L. and Johnston, A.W.B. (1986) Nature 324, 90-92.

Fisher, R.F., Tu, J.K. and Long, S.R. (1985) Appl. Environm. Microbiol. 49, 1432-1435.

Fisher, R.F., Brierley, H.L., Mulligan, J.T. and Long, S.R. (1987a) J. Biol. Chem. 262, 6849-6855.

Fisher, R.F., Swanson, J.A., Mulligan, J.T. and Long, S.R. (1987b) Genetics 117, 191-201.

Fisher, R.F., Egelhoff, T., Mulligan, J.T. and Long, S.R. (1988) Genes Dev. 2, 282-293.

Fisher, R.F. and Long, S.R. (1989) J. Bacteriol. 171, 5492-5502.

Fisher, R.F. and Long, S.R. (1993) J. Mol. Biol. 233, 336-348.

Folch-Mallol, J.L., Marroqui, S., Sousa, C., Manyani, H., López-Lara, I.M., van der Drift, K.M.G.M., Haverkamp, J., Quinto, C., Gil-Serrano, A., Thomas-Oates, J., Spaink, H.P. and Megias, M. (1996) Mol. Plant-Microbe Interact. 9, 151-163.

Freiberg, C., Fellay, R., Bairoch, A., Broughton, W.J., Rosenthal, A. and Perret, X. (1997) Nature 387, 394-401.

Friedman, A.M., Long, S.R., Brown, S.E., Buikema, W.J. and Ausubel, F.M. (1982) Gene 18, 289-296.

Gagnon, H. and Ibrahim, R.K. (1998) Mol. Plant-Microbe Interact. In press.

Garcia, M., Dunlap, J., Loh, J. and Stacey, G. (1996) Mol. Plant-Microbe Interact. 9, 625-635.

Geelen, D., Mergaert, P., Geremia, R.A., Goormachtig, S., Van Montagu, M. and Holsters, M. (1993) Mol. Microbiol. 9, 145-154.

Gerhold, D., Stacey, G. and Kondorosi, A. (1989) Plant Mol. Biol. 12, 181-188.

Gilette, W.K. and Elkan, G.H. (1996) J. Bacteriol. 178, 2757-2766.

Girard, M.L., Flores, M., Brom, S., Romero, D., Palacios and R., Davila, G. (1991) J. Bacteriol. 173, 2411-2419.

Goethals, K., Gao, M., Tomekpe, K., Van Montagu, M., Holsters, M. (1989) Mol. Gen. Genet. 219, 289-298.

Goethals, K., Van den Eede, G., Van Montagu, M. and Holsters, M. (1990) J. Bacteriol. 172, 2658-2666.

Goethals, K., Van Montagu, M. and Holsters, M. (1992) Proc. Natl. Acad. Sci. USA 89, 1646-1650.

Göttfert, M., Horváth, B., Kondorosi, E., Rodriguez-Quinones, F. and Kondorosi, A. (1986) J. Mol. Biol. 191, 411-420.

Göttfert, M., Lamb, J.W., Gasser, R., Semenza, J. and Hennecke, H. (1989) Mol. Gen. Genet. 215, 407-415.

Göttfert, M., Grob, P. and Hennecke, H. (1990a) Proc. Natl. Acad. Sci. USA 87, 2680-2684.

Göttfert, M., Hitz, S. and Hennecke, H. (1990b) Mol. Plant-Microbe Interact. 3, 308-316.

Göttfert, M., Holzhäuser, D., Bäni, D. and Hennecke, H. (1992) Mol. Plant-Microbe Interact. 5, 257-265.

Graham, T.L. (1991) Plant Physiol. 95, 594-603.

Gu, J., Balatti, P.A., Krishnan, H.B. and Pueppke, S.G. (1997) Mol. Plant-Microbe Interact. 10, 138-141.

Györgypál, Z., Iyer, N. and Kondorosi, A. (1988) Mol. Gen. Genet. 212, 85-92.

Györgypál, Z. and Kondorosi, A. (1991) Mol. Gen. Genet. 226, 337-340.

Györgypál, Z., Kondorosi, E. and Kondorosi, A. (1991a) Mol. Plant-Microbe Interact. 4, 356-364.

Györgypál, Z., Kiss, G.B. and Kondorosi, A. (1991b) Bioessays 13, 575-581.

Harrington, R.E. (1992) Mol. Microbiol. 6, 2549-2555.

Hartwig, U.A., Maxwell, C.A., Joseph, C.M. and Phillips, D.A. (1989) Plant Physiol. 91, 1138-1142.

Hartwig, U.A., Maxwell, C.A., Joseph, C.M. and Phillips, D.A. (1990a) Plant Physiol. 92, 116-122.

Hartwig, U.A., Maxwell, C.A., Joseph, C.M. and Phillips, D.A. (1990b) J. Bacteriol. 172, 2769-2773.

Hartwig, U.A. and Phillips, D.A. (1991) Plant Physiol. 95, 804-807.

Hawes, M.C. (1990) Plant and Soil 129, 19-27.

Henikoff, S., Haughn, J.M., Calvo, J.M. and Wallace, J.C. (1988) Proc. Natl. Acad. Sci. USA 85, 6602-6606.

Hong, G.-F., Burn, J.E. and Johnston, A.W.B. (1987) Nucleic Acids Res. 15, 9677-9690.

Honma, M.A., Asomaning, M. and Ausubel, F.M. (1990) J. Bacteriol. 172, 901-911.

Hooykaas, P.J.J., van Brussel, A.A.N., den Dulk-Ras, A., van Slogteren, G.M.S. and Schilperoort, R.A. (1981) Nature 291, 351-353.

Hooykaas, P.J.J., Peerbolte, R., Regensburg-Tuink, A.J.G., de Vries, P. and Schilperoort, R.A. (1982) Mol. Gen. Genet. 188, 12-17.

Horváth, B., Kondorosi, E., John, M., Schmidt, J., Török, I., Györgypál, Z., Barabás, I., Wieneke, U., Schell, J. and Kondorosi, A. (1986) Cell, 46, 335-343.

Horváth, B., Bachem, C.W.B., Schell, J. and Kondorosi, A. (1987) EMBO J. 6, 841-848.

Huang, J. and Schell, M.A. (1991) J. Biol. Chem. 266, 10830-10838.

Hungria, M., Joseph, C.M. and Phillips, D.A. (1991a) Plant Physiol. 97, 751-758.

Hungria, M., Joseph, C.M. and Phillips, D.A. (1991b) Plant Physiol. 97, 759-764.

Hungria, M., Johnston, A.W.B. and Phillips, D.A. (1992) Mol. Plant-Microbe Interact. 5, 199-203.

Innes, R.W., Kuempel, P.L., Plazinski, J., Canter-Cremers, H., Rolfe, B.G. and Djordjevic, M.A. (1985) Mol. Gen. Genet. 201, 426-432.

Jacobs, T.W., Egelhoff, T.T. and Long, S.R. (1985) J. Bacteriol. 162, 469-476.

Jacobs, M. and Rubery, P.H. (1988) Science 241, 346-349.

Johnston, A.W.B. and Beringer, J.E. (1977) Nature 267, 611-613.

Johnston, A.W.B., Beynon, J.L., Buchanan-Wallaston, A.V., Setchell, S.M., Hirsch, P.R. and Beringer, J.E. (1978) Nature 276, 634-636.

Jones, G.P., Naidu, B.P., Starr, R.K. and Paleg, L.G. (1986) Aust. J. Plant Physiol. 13, 649-658.

Kalinowski, G. and Long, S.R. (1996) Mol. Plant-Microbe Interact. 9, 869-873.

Kape, R., Parniske, M. and Werner, D. (1991) Appl. Environ. Microbiol. 57, 316-319.

Kape, R., Wex, K., Parniske, M., Görge, E., Wetzel, A. and Werner, D. (1992) J. Plant Physiol. 141, 54-60.

Kapulnik, Y., Joseph, C.M. and Phillips, D.A. (1987) Plant Physiol. 84, 1193-1196.

Knight, C.D., Rossen, L., Robertson, J.G., Wells, B. and Downie, J.A. (1986) J. Bacteriol. 166, 552-558.

Kondorosi, A., Kondorosi, E., Pankhurst, C.E., Broughton, W.J. and Bánfalvi, Z. (1982) Mol. Gen. Genet. 188, 433-439.

Kondorosi, E., Bánfalvi, Z. and Kondorosi, A. (1984) Mol. Gen. Genet. 193, 445-452.

Kondorosi, E., Gyuris, J., Schmidt, J., John, M., Duda, E., Hoffmann, B., Schell, J. and Kondorosi, A. (1989) EMBO J. 8, 1331-1340.

Kondorosi, E., Buiré, M., Cren, M., Iyer, N., Hoffmann, B. and Kondorosi, A. (1991a) Mol. Microbiol. 5, 3035-3048.

Kondorosi, E., Pierre, M., Cren, M., Haumann, U., Buiré, M., Hoffmann, B., Schell, J. and Kondorosi, A. (1991b) J. Mol. Biol. 222, 885-896.

Kosslak, R.M., Bookland, R., Barkei, J., Paaren, H.E. and Appelbaum, E.R. (1987) Proc. Natl. Acad. Sci. USA 84, 7428-7432.

Kosslak, R.M., Joshi, R.S., Bowen, B.A., Paaren, H.E. and Appelbaum, E.R. (1990) Appl. Environ. Microbiol. 56, 1333-1341.

Kovacs, L.G., Balatti, P.A., Krishnan, H.B. and Pueppke, S.G. (1995) Mol. Microbiol. 17, 923-933.

Krishnan, H.B. and Pueppke, S.G. (1991a) Mol. Microbiol. 5, 737-745.

Krishnan, H.B. and Pueppke, S.G. (1991b) Mol. Plant-Microbe Interact. 4, 521-529.

Krishnan, H.B., Lewin, A., Fellay, R., Broughton, W.J. and Pueppke, S.G. (1992) Mol. Microbiol. 6, 3321-3330.

Lawson, C.G.R., Rolfe, B.G. and Djordjevic, M.A. (1996) Aust. J. Plant Physiol. 23, 93-101.

Lewin, A., Cervantes, E., Chee-Hoong, W. and Broughton, W.J. (1990) Mol. Plant-Microbe Interact. 3, 317-326.

Lewis-Henderson, W.R. and Djordjevic, M.A. (1991) Plant Mol. Biol. 16, 515-526.

Le Strange, K.K., Bender, G.L., Djordjevic, M.A., Rolfe, B.G. and Redmond, J.W. (1990) Mol. Plant-Microbe Interact. 3, 214-220.

León-Barrios, M., Dakora, F.D., Joseph, C.M. and Phillips, D.A. (1993) Appl. Environ. Microbiol. 59, 636-639.

Lindström, K., Paulin, L., Roos, C. and Suominen, L. (1995) pp.365-370, in I. Tikhonovich, V. Provorov, V. Romanov, and W.E. Newton (eds.), Nitrogen fixation: fundamentals and applications. Proceedings of the 10th International Symposium on Nitrogen Fixation., St. Petersburg, Russia.

Lugtenberg, B.J.J. (ed.) (1986) Recognition in Microbe-Plant Symbiotic and Pathogenic Interactions Springer-Verlag, Berlin, Germany.

Luka, S., Sanjuan, J., Carlson, R.W. and Stacey, G. (1993) J. Biol. Chem. 268, 27053-27059.

Maillet, F., Debellé, F. and Dénarié, J. (1990) Mol. Microbiol. 4, 1975-1984.

Mao, C., Downie, J.A. and Hong, G. (1994) Gene 144, 87-90.

Massiot, G., Lavaud, C., Guillaume, D. and Le Men-Olivier, L. (1988) J. Agric Food Chem. 36, 902-909.

Mathesius, U., Schlaman, H.R.M., Meijer, D., Lugtenberg, B.J.J., Spaink, H.P., Weinman, J.J., Roddam, L.F., Sautter, C., Rolfe, B.G. and Djordjevic, M.A. (1996) p. 353-358, in G. Stacey, B. Mullin and P.M. Gresshoff (eds.) Biology of plant-microbe interactions, Int. Society Mol. Plant-Microbe Interact., St. Paul, Mn, USA.

Maxwell, C.A., Hartwig, U.A., Joseph, C.M. and Phillips, D.A. (1989) Plant Physiol. 91, 842-847.

Maxwell, C.A. and Phillips, D.A. (1990) Plant Physiol. 93, 1552-1558.

McIver, J., Djordjevic, M.A., Weinman, J.J., Bender, G.L. and Rolfe, B.G. (1989) Mol. Plant-Microbe Interact. 2, 97-106.

McIver, J., Djordjevic, M.A., Weinman, J.J. and Rolfe, B.G. (1993) Protoplasma 172, 166-179.

Meade, H.M., Long, S.R., Ruvkun, G.B., Brown, S.E. and Ausubel, F.M. (1982) J. Bacteriol. 149, 114-122.

Meinhardt, L.W., Krishnan, H.B., Balatti, P.A. and Pueppke, S.G. (1993) Mol. Microbiol. 9, 17-29.

Mergaert, P., D'Haeze, W., Fernández-López, M., Geelen, D., Goethals, K., Promé, J-C., Van Montagu, M. and Holsters, M. (1996) Mol. Microbiol. 21, 409-419.

Miller, J.H. (1972) pp.352-355, Experiments in molecular genetics, Cold Spring Harbor Laboratory, Cold Spring Harbor, NY, USA.

Mulligan, J.T. and Long, S.R. (1985) Proc. Natl. Acad. Sci. USA 82, 6609-6613.

Mulligan, J.T. and Long, S.R. (1989) Genetics 122, 7-18.

Nieuwkoop, A.J., Bánfalvi, Z., Deshmane, N., Gerhold, D., Schell, M.G., Sirotkin, K.M. and Stacey, G. (1987) J. Bacteriol. 169, 2631-2638.

Ogawa, J. and Long, S.R. (1995) Genes & Dev. 9, 719-729.

O'Halloran, T.V., Frantz, B., Shin, M.K., Ralston, D.M. and Wright, J.G. (1989) Cell 56, 119-129.

Oleszek, W., Price, K.R., Colquhoun, I.J., Jurzysta, M., Ploszynski, M. and Fenwick, G.R. (1990) J. Agric. Food Chem. 38, 1810-1817.

Parkinson, J.S. (1993) Cell 73, 857-871.

Pees, E., Wijffelman, C.A., Mulders, I., van Brussel, A.A.N. and Lugtenberg, B.J.J. (1986) FEMS Microbiol. Lett. 33, 165-171.

Perret, X., Broughton, W.J. and Brenner, S. (1991) Proc. Natl. Acad. Sci. USA 88, 1923-1927.

Peters, N.K., Frost, J.W. and Long, S.R. (1986) Science 233, 977-980.

Peters, N.K. and Long, S.R. (1988) Plant Physiol. 88, 396-400.

Phillips, D.A. (1992) pp. 201-231, in H.A. Stafford and R.K. Ibrahim (eds.), Phenolic metabolism in plants, Plenum Press, New York, USA.

Phillips, D.A., Dakora, F.D., Sande, E., Joseph, C.M. and Zon, J. (1994) Plant and Soil 161, 69-80.

Phillips, D.A., Joseph, C.M. and Maxwell, C.A. (1992) Plant Physiol. 99, 1526-1531.

Phillips, D.A., Wery, J., Joseph, C.M., Jones, A.D. and Teuber, L.R. (1995) Crop Sci. 35, 805-808.

Plazanet, C., Refregier, G., Demont, N., Truchet, G. and Rosenberg, C. (1995) FEMS Microbiol. Lett. 133, 285-291.

Putnoky, P. and Kondorosi, A. (1986) J. Bacteriol. 167, 881-887.

Rao, A.S. (1990) Bot. Rev. 56, 1-84.

Rao, J.R. and Cooper, J.E. (1995) Mol. Plant-Microbe Interact. 8:855-862.

Rasanen, L.A., Heikkila-Kallio, U., Suominen, L., Lipsanen, P. and Lindström, K. (1991) Mol. Plant-Microbe Interact. 4, 535-544.

Recourt, K., van Brussel, A.A.N., Driessen, A.H.M. and Lugtenberg, B.J.J. (1989) J. Bacteriol. 171, 4370-4377.

Recourt, K., Schripsema, J., Kijne, J.W., van Brussel, A.A.N. and Lugtenberg, B.J.J. (1991) Plant Mol. Biol. 16, 841-852.

Redmond, J.W., Batley, M., Djordjevic, M.A., Innes, R.W., Kuempel, P.L. and Rolfe, B.G. (1986) Nature 323, 632-635.

Relic, B., Perret, X., Estrada-Garcia, M.T., Kopcinska, J., Golinowski, W., Krishnan, H.B., Pueppke, S.G. and Broughton, W.J. (1994) Mol. Microbiol. 13, 171-178.

Richardson, A.E., Djordjevic, M.A., Rolfe, B.G. and Simpson, R.J. (1988) Plant and Soil 109, 37-47.

Richardson, A.E., Djordjevic, M.A., Rolfe, B.G. and Simpson, R.J. (1989) Aust. J. Plant Physiol. 16, 117-129.

Rivilla, R. and Downie, J.A. (1994) Gene 144, 87-91.

Rolfe, B.G. (1988) Biofactors 1, 3-10.

Rodriguez-Quinones, F., Fernandez-Burriel, M., Bánfalvi, Z., Megias, M. and Kondorosi, A. (1989) Mol. Plant-Microbe Interact. 2, 75-83.

Rossen, L., Johnston, A.W.B. and Downie, J.A. (1984) Nucleic Acids Res. 12, 9497-9508.

Rossen, L., Shearman, C.A., Johnston, A.W.B. and Downie, J.A. (1985) EMBO J. 4, 3369-3373.

Rostas, K., Kondorosi, E., Horváth, B., Simoncsits, A. and Kondorosi, A. (1986) Proc. Natl. Acad. Sci. USA 83, 1757-1761.

Rushing, B.G., Yelton, M.M. and Long, S.R. (1991) Nucleic Acids Res. 19, 921-927.

Russell, P., Schell, M., Neslson, K., Halverson, L.J., Sirotkin, K. and Stacey, G. (1985) J. Bacteriol. 164, 1301-1308.

Sadowsky, M.J., Cregan, P.B., Göttfert, M., Sharma, A., Gerhold, D., Rodriguez-Quinones, F., Keyser, H.H., Hennecke, H. and Stacey, G. (1991) Proc. Natl. Acad. Sci. USA 88, 637-641.

Sanjuan, J., Grob, P., Göttfert, M., Hennecke, H. and Stacey, G. (1994) Mol. Plant-Microbe Interact. 7, 364-369.

Schell, M.A. (1993) Annu. Rev. Microbiol. 47, 597-626.

Schlaman, H.R.M. (1992) PhD. thesis, Regulation of nodulation gene expression in Rhizobium leguminosarum biovar viceae, Leiden University, The Netherlands.

Schlaman, H.R.M., Spaink, H.P., Okker, R.J.H. and Lugtenberg, B.J.J. (1989) J. Bacteriol. 171, 4686-4693.

Schlaman, H.R.M., Horváth, B., Vijgenboom, E., Okker, R.J.H. and Lugtenberg, B.J.J. (1991) J. Bacteriol. 173, 4277-4287.

Schlaman, H.R.M., Lugtenberg, B.J.J. and Okker, R.J.H. (1992a) J. Bacteriol. 174, 6109-6116.

Schlaman, H.R.M., Okker, R.J.H. and Lugtenberg, B.J.J. (1992b) J. Bacteriol. 174, 5177-5182.

Schmidt, P.E., Broughton, W.J. and Werner, D. (1994) Mol. Plant-Microbe Interact. 7, 384-390.

Schofield, P.R., Ridge, R.W., Rolfe, B.G., Shine, J. and Watson, J.M. (1984) Plant Mol. Biol. 3, 3-11.

Schofield, P.R. and Watson, J.M. (1986) Nucleic Acids Res. 14, 2891-2905.

Schwedock, J. and Long, S.R. (1989) Mol. Plant-Microbe Interact. 2, 181-194.

Scott, K.F. (1986) Nucleic Acids Res. 14, 2905-2910.

Scott, D.B., Young, C.A., Collins-Emmerson, J.M., Terzaghi, E.A., Rockman, E.S., Lewis, P.E. and Pankhurst, C.E. (1996) Mol. Plant-Microbe Interact. 9, 187-197.

Sharma, S.B. and Signer, E.R. (1990) Genes Dev. 4, 344-356.

Shearman, C.A., Rossen, L., Johnston, A.W.B. and Downie, J.A. (1986) EMBO J. 5, 647-652.

Smit, G., Puvanesarajah, V., Carlson, R.W., Barbour, W.M. and Stacey, G. (1992) J. Biol. Chem. 267, 310-318.

Sousa, C., Folch, J.L., Boloix, P., Megias, M., Nava, N. and Quinto, C. (1993) Mol. Microbiol. 9, 1157-1168.

Spaink, H.P., Okker, R.J.H., Wijffelman, C.A., Pees, E. and Lugtenberg, B.J.J. (1987a) Plant Mol. Biol. 9, 27-39.

Spaink, H.P., Wijffelman, C.A., Pees, E., Okker, R.J.H. and Lugtenberg, B.J.J. (1987b) Nature 328, 337-340.

Spaink, H.P., Okker, R.J.H., Wijffelman, C.A., Tak, T., Goosen-de Roo, L., Pees, E., van Brussel, A.A.N. and Lugtenberg, B.J.J. (1989a) J. Bacteriol. 171, 4045-4053.

Spaink, H.P., Weinman, J., Djordjevic, M.A., Wijffelman, C.A., Okker, R.J.H. and Lugtenberg, B.J.J. (1989b) EMBO J. 8, 2811-2818.

Spaink, H.P., Wijffelman, C.A., Okker, R.J.H. and Lugtenberg, B.J.J. (1989c) Plant Mol. Biol. 12, 59-73l.

Spaink, H.P., Sheeley, D.M., van Brussel, A.A.N., Glushka, J., York, W.S., Tak, T., Geiger, O., Kennedy, E.P., Reinhold, V.N. and Lugtenberg, B.J.J. (1991) Nature 354, 125-130.

Stacey, G. (1995) FEMS Microbiol. Lett. 127, 1-9.

Stacey, G., Luka, S., Sanjuan, J., Bánfalvi, Z., Nieuwkoop, A.J., Chun, J.Y., Forsberg, L.S. and Carlson, R. (1994) J. Bacteriol. 176, 620-633.

Stafford, H.A. (1990) Flavonoid Metabolism, CRC Press, Boca Raton, USA.

Stanley, J., van Slooten, J., Dowling, D.N., Finan, T. and Broughton, W.J. (1989) Mol. Gen. Genet. 217, 528-532.

Stock, J.B., Ninfa, A.J. and Stock, A.M. (1989) Microbiol. Rev. 53, 450-490.

Surin, B.P. and Downie, J.A. (1988) Mol. Microbiol. 2, 173-183.

Surin, B.P., Watson, J.M., Hamilton, W.D.O., Economou, A. and Downie, J.A. (1990) Mol. Microbiol. 4, 245-252.

Swanson, J.A., Mulligan, J.T. and Long, S.R. (1993) Genetics 134, 435-444.

Török, I., Kondorosi, E., Stepkowski, T., Pósfai, J. and Kondorosi, A. (1984) Nucleic Acids Res. 12, 9509-9524.

van Brussel, A.A.N., Tak, T., Pees, E. and Wijffelman, C.A. (1982) Plant Sci. Lett. 27, 317-325.

van Brussel, A.A.N., Recourt, K., Pees, E., Spaink, H.P., Tak, T., Wijffelman, C.A., Kijne, J.W. and Lugtenberg, B.J.J. (1990) J. Bacteriol. 172, 5394-5401.

Van Rhijn, P. (1994) PhD. thesis, Analysis of nodulation genes of two Rhizobium strains nodulating a wide range of leguminous plants, including Phaseolus vulgaris, KU Leuven, Belgium.

Van Rhijn, P.J.S., Feys, B., Verreth, C. and Vanderleyden, J. (1993) J. Bacteriol. 175, 438-447.

Van Rhijn, P., Luyten, E., Vlassak, K. and Vanderleyden, J. (1996) Mol. Plant-Microbe Interact. 9, 74-77.

van Veen, R.J.M., den Dulk-Ras, H., Bisseling, T., Schilperoort, R.A. and Hooykaas, P.J.J. (1988) Mol. Plant-Microbe Interact. 1, 231-234.

Vargas, C., Martinez, L.J., Megias, M. and Quinto, C. (1990) Mol. Microbiol. 4, 1899-1910.

Vazquez, M., Davalos, A., de las Penas, A., Sanchez, F. and Quinto, C. (1991) J. Bacteriol. 173, 1250-1258.

Vazquez, M., Santana, O. and Quinto, C. (1993) Mol. Microbiol. 8, 369-377.

Villalobos, M.A., Nava, N., Vazquez, M. and Quinto, C. (1994) Gene 150, 201-202.

Waelkens, F., Voets, T., Vlassak, K., Vanderleyden, J. and Van Rhijn, P. (1995) Mol. Plant-Microbe Interact. 8, 147-154.

Wang, S-P. and Stacey, G. (1990) Mol. Gen. Genet. 223, 329-331.

Wang, S-P. and Stacey, G. (1991) J. Bacteriol. 173, 3356-3365.

Wijffelman, C.A., Pees, E., van Brussel, A.A.N., Okker, R.J.H. and Lugtenberg, B.J.J. (1985) Arch. Microbiol. 143, 225-232.

Young, C., Collins-Emmerson, J.M., Terzaghi, E.A. and Scott, D.B. (1990) Nucleic Acids Res. 18, 6691.

Yuen, J.P. and Stacey, G. (1996) Mol. Plant-Microbe Interact. 9, 424-428.

Zaat, S.A.J., Wijffelman, C.A., Spaink, H.P., van Brussel, A.A.N., Okker, R.J.H. and Lugtenberg, B.J.J. (1987) J. Bacteriol. 169, 198-204.

Zaat, S.A.J., Schripsema, J., Wijffelman, C.A., van Brussel, A.A.N. and Lugtenberg, B.J.J. (1989) Plant Mol. Biol. 13, 175-188.

Chapter 20

Functions of Rhizobial Nodulation Genes

J. Allan Downie

I. Introduction

The formation of nodules on leguminous plants depends on a highly specific exchange of signals between plant and bacteria. These nodules are usually induced on roots, although in several aquatic legumes they can be formed on stems. This chapter focuses on those rhizobial "nodulation" genes that are involved in the biosynthesis and secretion of components that are involved in inducing nodule formation in legumes.

Over fifty different nodulation genes have been identified as being involved in legume nodulation. Initially they were called *nod* genes but as the numbers continued to increase other gene designations (*nol* and *noe*) had to be used. For convenience the term "*nod*" genes is used here to refer generically to all of the genes except when specific genes are being referred to. The *nod* genes can be divided into a number of groups according to their function. The regulatory genes (*nodD, nodV, nodW, nolA, nolD, nolR, and syrM*) all appear to be involved in regulation of the other *nod* genes: their role in

regulation is described by Schlaman *et al.*, in this volume (Chapter 19).

The definition of what is a *nod* gene is not always precise. This is because mutations in several of the *nod* genes may have little or no effect on nodulation. Sometimes this is due to duplication of gene function. Another explanation for lack of a nodulation phenotype with mutants, may be that the tests have not been done on the appropriate legumes or under the appropriate growth conditions. A useful definition of *nod* genes is those genes that are induced under the control of the various regulatory *nod* gene products. However, some host-specific *nod* genes have been isolated by selecting for strains containing cloned DNA that confers increased nodulation range (allowing nodulation of host plants other than those normally nodulated). Such genes are not always regulated in the same way as the other *nod* genes.

The various *nod* genes (other than regulatory genes) can be subdivided into a number of different categories: these include those that are (or may be) involved in (i) the biosynthesis and modification of lipo-chito-oligosaccharide Nod factors, (ii) Nod factor secretion (iii) protein secretion and (iv) genes of undefined function. Clearly, as we learn more, those genes in category (iv) will be redistributed into other groups, and may even establish other categories. Table 1 summarises the functions of the various *nod* gene products. In many cases these functions have been deduced on the basis of the effects of mutations and similarities to proteins of known function. Where there is such similarity but no good experimental evidence for function, a question mark is included in the table.

II. Nod Factor Biosynthesis

Before describing the specific Nod factors made by different rhizobia, I will outline the biosynthetic pathway that is common to all rhizobia and describe the structure of a generic "minimal" Nod factor. As shown in Figure 1, Nod factors are made up of a backbone of β 1,4-linked *N*-acetyl glucosamine residues. The precursor required for the formation of this oligosaccharide is UDP-*N*-acetyl-D-glucosamine (UDP-GlcNAc), which is also a precursor for the lipopolysaccharide and peptidoglycan components of the bacterial outer membrane (Carlson *et al.*, 1994; Inon de Iannino *et al.*, 1995).

Some strains have a nodulation-specific glucosamine synthase (encoded by *nodM*) which makes the glucosamine-6-P that is a precursor of UDP-GlcNAc (Baev *et al.*, 1991; Marie *et al.*, 1992). Other rhizobia rely on the normal housekeeping glucosamine synthase (GlmS), which appears to be sufficiently active to be able to supply the additional demand for glucosamine precursors required during Nod factor synthesis (Marie *et al.*, 1992, 1994).

NodC is a processive β-glycosyl transferase that forms chitin oligomers from UDP-GlcNAc (Geremia *et al.*, 1994; Mergaert *et al.*, 1995; Spaink *et al.*, 1994; Kamst *et al.*, 1995; Roche *et al.*, 1996). It is an integral inner membrane protein structurally similar to several other glycosyl transferases (Saxena *et al.*, 1995; Barny *et al.*, 1993, 1996) with the major catalytic domain on the cytoplasmic face of the inner membrane (Barny *et al.*, 1996). Some strains principally make Nod factors in which the backbone has either four or five sugars, whereas other strains such as *R.l. viciae* make equal amounts of tetrameric and pentameric Nod factors. This chain length is determined by the NodC protein, which therefore can impart a measure of host-specificity on the type of

Nod factor made (Kamst *et al.*, 1997; Roche *et al.*, 1996).

The next stage in "minimal" Nod factor biosynthesis is the attachment of an N-linked fatty acyl group to the terminal (non-reducing) glucosamine residue. This requires the sequential action of NodB and NodA (Figure 1). NodB deacetylates the glucosamine residue, leaving a free amino group (John *et al.*, 1993) which is then acylated in a reaction that requires NodA (Röhrig *et al.*, 1994; Atkinson *et al.*, 1994; Mergaert *et al.*, 1995). Although *nodA* genes are common to all rhizobia it is evident that different NodA proteins can preferentially add different acyl groups (Ritsema *et al.*, 1996; Debellé *et al.*, 1996). Therefore NodA can introduce some specificity into the Nod factor structure. The types of fatty acyl groups present often reflect the composition of the fatty acids found in the bacterial phospholipids (Geiger *et al.*, 1994; Cedergren *et al.*, 1995), although in some rhizobia the *nodFE* genes are involved in formation of specific fatty acyl groups (see below).

The Nod factors produced as a result of NodA, NodB and NodC activity have biological activity and can induce root hair deformation on vetch (Spaink *et al.*, 1991). The structure illustrated in Figure 1 might be considered to be a progenitor of all known Nod factors. However, at this stage no natural rhizobial isolate has been described as making only such a Nod factor.

III. Rhizobia-Specific Modifications to Nod Factors

It is beyond the scope of this review to detail the chemical structure of every Nod factor that has been published. Instead, the various types of modifications that are made will be outlined. As shown (Table 2) there are several sites on the chitin backbone that can carry substituents and it is these modifications that are the major determinants of host specificity in the rhizobia-legume interactions. There have been several excellent recent reviews in this area (Carlson *et al.*, 1994; Downie, 1994; Lerouge, 1994; Martinez-Romero, 1994; Relic *et al.*, 1994; Shultze *et al.*, 1994; Spaink & Lugtenberg, 1994; Fellay *et al.*, 1995; Mylona *et al.*, 1995; Spaink 1995, 1996; Stacey, 1995; van Rhijn and VanderLeyden, 1995; Dénarié *et al.*, 1996; Guerts and Franssen 1996; Long, 1996; Promé, 1996) which detail the function of Nod factors and their biosynthesis.

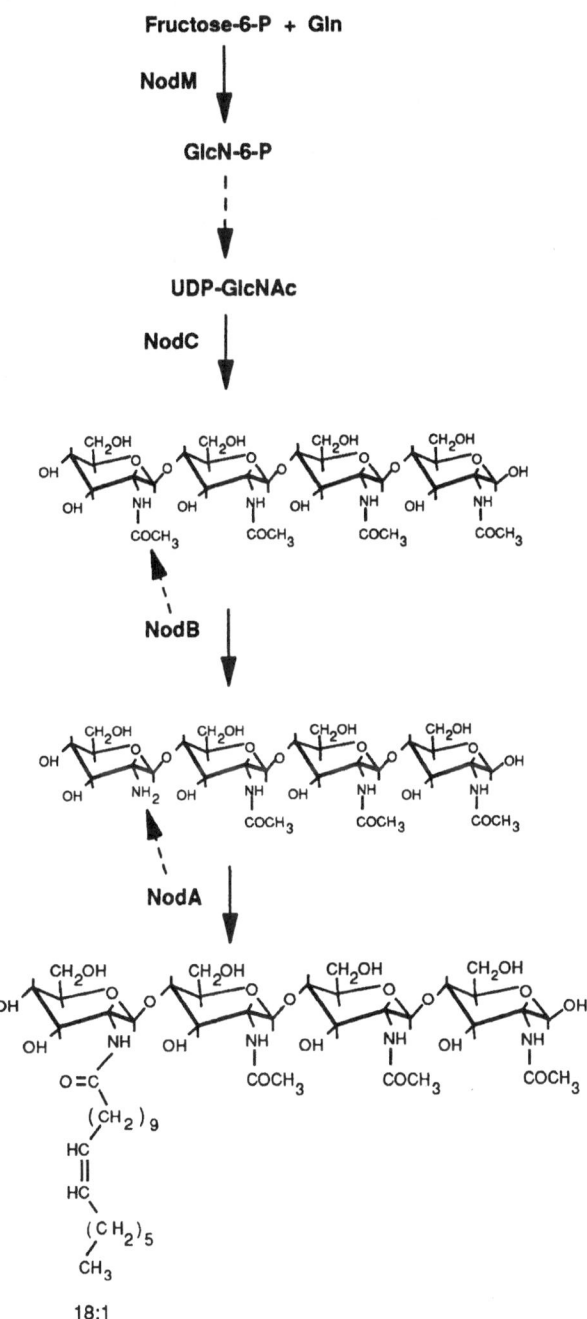

Figure 1. Biosynthetic pathway of a "minimal" Nod factor

III.A. N-LINKED MODIFICATIONS (GROUPS R1 AND R2)

The only N-linked substitutions found are on the terminal non-reducing glucosamine residue, which always carries a fatty acyl group and in some rhizobia is N-methylated. There is a strong correlation between the type of legume nodulated and the presence or absence of polyunsaturated N-linked fatty acyl groups on the cognate Nod factors. Thus, Nod factors with polyunsaturated acyl groups are made by *R.l. viciae*, *R.l. trifolii* and *S. meliloti* (Table 2) that nodulate legumes which form nodules with indeterminate meristems and long or broad infection threads (see chapter 22).

Conversely, polyunsaturated acyl groups are not found in those bacteria that normally nodulate those legumes forming determinate meristems and in which infection occurs via narrow infection threads or by infection pockets (Table 2). At this stage it may be premature to say that this correlation will hold for all legume-rhizobia combinations but it does suggest that there may be something characteristic about nodulation of legumes that form indeterminate nodules. The *nodFE* genes (Table 1) encode enzymes (an acyl carrier protein and β-keto acyl synthase) involved in fatty acid synthesis and are required to produce the polyunsaturated acyl groups in *R.l. viciae*, *R.l. trifolii* and *S. meliloti* Nod factors (Spaink *et al.*, 1991, 1995; Demont *et al.*, 1993; Bloemberg *et al.*, 1995a; Van der Drift *et al.*, 1996), although in one wild-type strain of *R.l. trifolii* polyunsaturated fatty acyl groups were not detected (Orgambide *et al.*, 1995; Philip-Hollingsworth *et al.*, 1995). Mutation of the *nodFE* genes in *S. meliloti*, *R.l. trifolii* and *R.l. viciae* leads to the formation of Nod factors carrying only a $C_{18:1}$ acyl group. It appears that in the absence of the appropriate acyl group, the NodA protein will use intermediates from the normal fatty-acid biosynthetic pathway, even though NodA in strains such as *R.l. viciae* and *S. meliloti* has a preference for addition of polyunsaturated acyl groups (Debellé *et al.*, 1996; Ritsema *et al.*, 1996). However, although mutation of the *nodFE* genes reduces nodulation efficiency (Spaink *et al.*, 1989, Surin and Downie, 1989; Demont *et al.*, 1993), nodulation is not totally blocked. This demonstrates that the appropriate acyl group is not necessarily essential for nodulation. In fact, *nodE* mutants of *R.l. trifolii* gain the ability to nodulate pea (Djordjevic *et al.*, 1985). In the absence of the *nodFE* genes, other nodulation genes can become more crucial for nodulation. Thus for example

in *R.l. viciae* the secreted nodulation protein NodO becomes essential for the residual nodulation of a *nodE* mutant (Downie and Surin, 1990; Economou *et al.*, 1994), and in *S. meliloti nodL* is essential for alfalfa nodulation in *nodE* mutants whereas in the presence of *nodE* it is not (Ardourel *et al.*, 1994).

On the basis of protein homology NodG is likely to be a dehydrogenase involved in fatty acyl synthesis and hence in synthesis of the Nod-factor fatty group. However, mutation of *nodG* had no observed effect on Nod factor biosynthesis, suggesting that it may have a homologue that can compensate for its absence (Demont *et al.*, 1993).

A second N-linked modification of the terminal non-reducing glucosamine residue is the attachment by NodS (an S-adenosyl methionine methyl transferase) of an N-linked methyl group (Geelen *et al.*, 1993, 1995; Jabbouri *et al.*, 1995). This has not been observed in Nod factors carrying polyunsaturated acyl groups and the *nodS* gene appears to be absent from those strains that have *nodF* and *nodE*. The *nodS* gene plays a role in host-specific nodulation (Lewin *et al.*, 1990). It is likely that NodS acts on deacetylated chitooligosaccharides prior to fatty acylation by NodA (Geelen *et al.*, 1995; Jabbouri, *et al.*, 1995; Mergaert *et al.*, 1995). Possibly the presence of the N-methyl group precludes attachment of the polyunsaturated fatty acyl groups containing double bonds conjugated to the ester linkage.

III.B. O-LINKED MODIFICATIONS TO THE TERMINAL NON-REDUCING RESIDUE (GROUP R3)

Two types of substituents are included in this group. One is an acetyl group introduced by NodL, an O-acetyl transferase (Downie, 1989; Bloemberg *et al.*, 1994, 1995b) that acetylates the 6-C position. This enzyme falls into a specific group of O-acetyl transferases, some of which have been extensively characterised. NodL is structurally similar to these proteins based on its sequence and on preliminary protein crystallographic data (Dunn *et al.*, 1996). It appears that NodL acts on Nod factors after the action of NodB (Figure 1), but before the acylation by NodA, since chitin oligomers deacetylated by NodB are a significantly better substrate than the unacetylated oligomers, and N-acylation strongly decreases the ability of NodL to carry out the O-acetylation (Bloemberg *et al.*, 1995b). Interestingly, there is a negative correlation between the presence of *nodL* and *nodS*. No strain has been described as carrying functional copies of both genes and so Nod factors are not found carrying an N-methyl group together with NodL-determined O-acetyl group.

The second type of O-linked substituent is a carbamoyl group, which has been found, O-linked to the C6, C5 and/or C4 positions. It appears that there are specific O-carbamoyl transferases, one of which, NodU, is involved in carbamoylation of the C6 position (Jabbouri et al., 1995; Geelen et al., 1993).

III.C. O-LINKED SUBSTITUTIONS ON THE REDUCING GLUCOSAMINE

The most frequently substituted position in Nod factors is the C-6 of the reducing glucosamine residue. These substituents are usually very important for legume nodulation specificity.

III.C.1. 6-O-Sulphation
Three nod gene products, NodH, NodP and NodQ, are involved in sulphation of Nod factors. This sulphation is crucial for host specificity: thus, nodH mutants of S. meliloti, which make unsulphated Nod factors, lose the ability to nodulate alfalfa but acquire the ability to nodulate vetch. (Faucher et al., 1988; Lerouge et al., 1990, Roche et al., 1991a). NodH is a sulphotransferase that catalyses the transfer of a sulphate group onto unsulphated Nod factors or chitin oligomers (Ehrhardt et al., 1995; Schultze et al., 1995; Bourdineaud et al., 1995). Fatty acylated Nod factors are the preferred substrate, indicating that sulphate addition occurs after acylation (Schulze et al., 1995).

As illustrated in Figure 2, the substrate used by NodH as a sulphate donor is PAPS (3'-phosphoadenosine 5' phosphosulphate); this is formed by the activity of NodP and NodQ which together form a complex with ATP sulphurylase and adenosine 5'-phosphosulphate (APS) kinase activities (Schwedock and Long, 1990, 1992; Leyh et al., 1992; Schwedock et al., 1994; Folch-Mallol et al., 1996; Laeremans et al., 1996). Mutation of nodP or nodQ in S. meliloti leads to the production of a mixture of sulphated and non-sulphated Nod factors and this is accompanied by the ability of the mutants to nodulate both vetch and alfalfa (Roche et al., 1991a). The reason for this is that S. meliloti has additional homologues of the nodPQ genes which allow for the formation of sufficient PAPS to allow some Nod factor modification (Schwedock and Long, 1989, 1992). The nodPQ homologues are probably involved in the formation of PAPS, that is used for reactions such as sulphation of the lipopolysaccharide (Cedergren et al., 1995). The

concentration of PAPS can be a limiting factor in sulphation of Nod factors in some rhizobia (Poupot et al., 1995b).

The presence of the sulphate group plays a key role in recognition of Nod factors by some legumes such as Medicago spp. (Roche et al., 1991a; Journet et al., 1994). However, it has also been noted that sulphation of Nod factors also confers a degree of resistance to degradation by plant chitinases (Staehelin et al., 1994a and b; 1995).

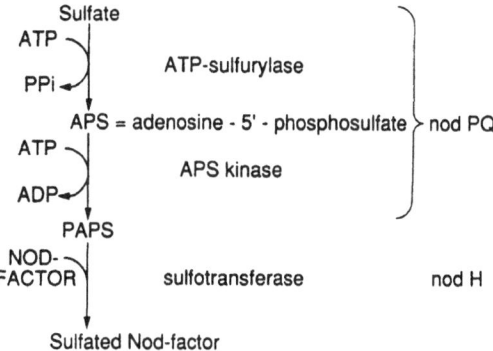

Figure 2. Sulfatation of Nod-factors. NodPQ correspond to E.coli cysD and cysC + cysN, respectively. These nod genes can substitute for ATP sulfuralyse (CysD and CysN as well as APS kinase (CysC))

III.C.2. 6-O-Acetylation
The nodX gene product is an O-acetyl transferase (Firmin et al., 1993) that is required for the nodulation of cv. Afghanistan peas that are resistant to nodulation by many R.l. viciae strains. This resistance of cv. Afghanistan peas to nodulation by some R.l. viciae strains segregates as a single gene and the fact that R.l. viciae strains carrying nodX overcome this nodulation resistance has been seen as circumstantial evidence for a modified Nod factor receptor in cv. Afghanistan peas (Kozik et al., 1995). The nodX gene was first isolated from a strain (TOM) that could nodulate cv. Afghanistan peas. Strains carrying nodX make a mixture of pentameric and tetrameric Nod factors, but only the pentameric factors are acetylated. Although both NodX and NodL are O-acetyl transferases that O-acetylate the C-6 of glucosamine, there is remarkably little similarity between these two proteins (Davis et al., 1988; Downie 1989) and NodX falls into a specific group of O-acetylases (Clark et al., 1991).

Gene	Function of gene product	Strains	Selected References
nodA	Acyl transferase (Figure 1)	All	Atkinson *et al.*, (1994), Röhrig *et al.*, (1994), Mergaert *et al.*, (1995), Ritsema *et al.*, (1996)
nodB	de N-acetylase (Figure 1)	All	Spaink *et al.*, (1994), Atkinson *et al.*, (1994), John *et al.*, (1993), Mergaert *et al.*, (1995)
nodC	Chitin synthase (Figure 1)	All	Geremia *et al.*, (1994), Spaink *et al.*, (1994), Inon de Iannino *et al.*, (1995), Bec-Ferté *et al.*, (1996), Kamst, *et al.*, (1997), Mergaert *et al.*, (1995), Barny, *et al.*, (1993, 1996)
nodD	Regulator	All	Schlaman, *et al.*, (Chapter 19)
nodE	β-ketoacyl synthase	Sm Rlv Rlt	Spaink *et al.*, (1989, 1991, 1995a), Demont *et al.*, (1993), Bloemberg *et al.*, (1995a), van der Drift *et al.*, (1996)
nodF	Acyl carrier protein	Sm Rlv Rlt	Demont *et al.*, (1993), Shearman, *et al.*, (1986), Geiger *et al.*, (1991), Bibb *et al.*, (1989), Ritsema *et al.*, (1994)
nodG	Dehydrogenase?	Sm	Demont *et al.*, (1993)
nodH	Sulphotransferase	Sm Rt Rn	Roche *et al.*,(1991a), Bourdineaud *et al.*, (1995), Schultze *et al.*, (1995), Ehrhardt *et al.*, (1995), Folch-Mallol *et al.*,(1996)
nodI *nodJ*	Transporter of Nod factors	All	Evans and Downie (1986), Vazquez *et al.*, (1993), McKay and Djordjevic *et al.*, (1993), Spaink, *et al.*, (1995b), Fernandez-Lopez *et al.*, (1996), Cardenas *et al.*, (1996)
nodK	Unknown; similar to *nodY*	Rn	Scott (1986)
nodL	O-acetyl transferase	Rlv Rlt Sm	Downie (1989), Baev and Kondorosi (1992) Ardourel *et al.*, (1995), Bloemberg *et al.*, (1994, 1995b)
nodM	Glucosamine synthase (Figure1)	Sm Rlt Rlv	Baev *et al.*, (1991, 1992), Marie *et al.*, (1992, 1994)
nodN	Unknown; (Nod factor synthesis)	Sm Rlv	Surin *et al.*, (1988, 1989), Baev *et al.*, (1991, 1992)
nodO	Secreted pore-forming protein	Rlv Rsp BR816	de Maagd *et al.*, (1989), Economou *et al.*, (1990, 1994), Sutton *et al.*, (1994), van Rhijn *et al.*, (1996)
nodP *nodQ*	ATP sulphurylase APS kinase	Sm Rt Rsp BR816	Schwedock and Long (1989, 1990, 1992), Schwedock *et al.*, (1994), Folch-Mallol *et al.*, (1996), Laeremans *et al.*, (1996)
nodR	Never defined in refereed literature		
nodS	N-methyl transferase	Rn Rt Ac Bj Sf	Mergaert *et al.*, (1995), Jabbouri *et al.*, (1995), Geelen *et al.*, (1993, 1995), Waelkens *et al.*, (1995)
nodT	Outer membrane protein. Transport of Nod factors?	Rlv Rlt	Surin *et al.*, (1990), Rivilla *et al.*, (1995)
nodU	6-O-carbamoyl transferase	Rn Rt Ac Bj Sf	Jabbouri *et al.*, (1995), Geelen *et al.*, (1995)
nodV *nodW*	2 component regulator	Bj	Schlaman *et al.*, (Chapter 19)
nodX	O-acetyl transferase	Rlv (Tom)	Firmin *et al.*, (1993)
nodY	Unknown	Bj	Nieuwkoop *et al.*, (1987)
nodZ	Fucosyl transferase	Ac Rn Re	Stacey *et al.*,(1994), Fellay *et al.*,(1995), Mergaert *et al.*,(1996), Lopez-Lara *et al.*, (1996), Quinto *et al.*,(1997), Quesada-Vincens *et al.*,(1997)
nolA	Regulator (by homology) Cultivar specificity	Bj	Schlaman *et al.*, (Chapter 19)
nolB	Secretion system?	Rn Sf	Balatti *et al.*,(1995), Meinhardt *et al.*,(1993)
nolC	Stress inducible (DnaJ) Mutation extends nodulation	Sf	Krishnan and Pueppke (1991a, 1992)

Gene	Function of gene product	Strains	Selected References
nolD	Regulator?		Schlaman *et al.*, (Chapter 19)
nolE	Periplasmic protein	Rlp	Davis and Johnston (1990)
nolF *nolG* *nolH* *nolI*	Transport?	Sm	Baev *et al.*,(1991, 1992), Saier (1994)
nolJ	Delayed nodulation; no known function	Sf	Boundy-Mills *et al.*,(1994)
nolK	GDP Fucose synthesis?	Ac	Goethals *et al.*, (1992), Mergaert *et al.*, (1996)
nolL	O-acetyl transferase (acetyl fucose)	Ml	Scott *et al.*, (1996), Freiberg *et al.*, (1997)
nolM	Function together with *nodZ*?	Bj	Luka *et al.*, (1993)
nolN	Unknown	Bj	Luka *et al.*, (1993)
nolO	2 Methyl fucose?	Bj	Luka *et al.*, (1993)
nolP	Unknown	Rlp	Davis and Johnston (1990)
nolQ	Unknown function	Sm	Plazanet *et al.*, (1995) and Kondorosi (cited therein)
nolR	*nod* gene repressor	Sm Rlv Tom (others)	Schlaman *et al.*, (Chapter 19)
nolS	Unknown; is similar to Thi1 thiamine biosynthetic enzyme	Sm	Plazanet *et al.*, (1995)
nolT	Secretion system (HrpI)?	Sf Rn	Balatti *et al.*, (1995); Meinhardt *et al.*, (1993); van Gijsem *et al.*, (1995); Kovacs *et al.*, (1995)
nolU	Secretion system?	Sf Rn	Balatti *et al.*, (1995); Meinhardt *et al.*, (1993); van Gijsem *et al.*, (1995); Kovacs *et al.*, (1995)
nolV	Secretion system?	Sf Rn	Balatti *et al.*, (1995); Meinhardt *et al.*, (1993); van Gijsem *et al.*, (1995); Kovacs *et al.*, (1995)
nolW	Secretion system (HrpA)?	Sf Rn	Balatti *et al.*, (1995); Meinhardt *et al.*, (1993); van Gijsem *et al.*, (1995); Kovacs *et al.*, (1995)
nolX	Secretion system?	Sf Rn	Balatti *et al.*, (1995); Meinhardt *et al.*, (1993); van Gijsem *et al.*, (1995); Kovacs *et al.*, (1995)
nolY	Unknown; affects mung bean	Bj	Dockendorf *et al.*, (1994)
nolZ	Normal nodulation; inducible	Sm	Dockendorf *et al.*, (1994)
noeA	Methyl transferase?	Sm	Ardourel *et al.*, (1994)
noeB	Unknown (inner membrane)	Sm	Ardourel *et al.*, (1994)
noeC	Arabinosylation	Ac	Mergaert *et al.*, (1996)
noeD	Unknown	Bj	Sadowsky *et al.*, (1996)
noeE	Fucose sulphation	Rn	Hanin *et al.*, (1997)
noeF	Unknown	Sf	Sadowsky (pers. comm.)
noeI	2-O methyl transferase	Rn	Freiberg *et al.*, (1997); Jabbouri and Broughton (pers. comm.)
noeJ	Mannose-1-P guanylyl transferase	Rn	Freiberg *et al.*, (1997)
noeL	GDP-Mannose 4, 6 dehydratase	Rn	Freiberg *et al.*, (1997); Jabbouri and Broughton (pers. comm.)

Table 1. Function of *nod* gene products

<u>Abbreviations</u>: Ac, *Azorhizobium caulinodans*; Be, *Bradyrhizobium elkanii*; Bj, *Bradyrhizobium japonicum*; Re, *Rhizobium etli*; Sf, *Sinorhizobium fredii*; Ml, *Mesorhizobium loti*; Rlp, *Rhizobium leguminosarum* biovar *phaseoli*; Rlt, *Rhizobium leguminosarum* biovar *trifolii*; Rlv, *Rhizobium leguminosarum* biovar *viciae*; Sm, *Sinorhizobium meliloti*; Rn, *Rhizobium* sp. NGR234; RspBR816, *Rhizobium* sp. BR816; Rt, *Rhizobium tropici*.

III.C.3. 6-O-Glycosylation

Different fucosyl groups have been identified at this position. Several derivatives of fucose (methyl fucose, acetyl methyl fucose, acetyl fucose and sulphated methyl fucose) have been identified in different rhizobial strains (Table 2). In the absence of the fucose group (due to mutation of *nodZ*), *B. japonicum* is unable to nodulate siratro (Stacey *et al.*, 1994). The *nodZ* and *nolK* genes have been shown to be involved in formation and attachment of the 3-methyl fucose group and *nolM* and *nolO* in *B. japonicum* are also thought to be involved in its formation (Luka *et al.*, 1993). It was shown that NolK is involved in the synthesis of GDP-fucose (Luka *et al.*, 1993; Mergaert *et al.*, 1996) and that NodZ is a fucosyl transferase (Stacey *et al.*, 1994; Fellay *et al.*, 1995a; Lopez-Lara *et al.*, 1996; Mergaert *et al.*, 1996; Quinto *et al.*, 1997). Transfer of *nodZ* to *R. l. viciae* resulted in the production of Nod factors carrying fucose on the C-6 position. This enabled *R.l. viciae* to nodulate various tropical legumes which it previously could not nodulate (Lopez-Lara *et al.*, 1996). Significantly *R.l. viciae* carrying *nodZ* also acquired the ability to nodulate cv. Afghanistan peas (Ovtsynia *et al.*, 1996). Thus this fucose substituent can functionally replace the acetyl group attached by NodX. Since these substituents are structurally very different, it is difficult to see how these modifications could be directly related to alterations in Nod factor recognition by a modified receptor in cv. Afghanistan peas. An alternative idea is that they may be related to increased resistance to Nod factor degradation or could be involved in avoiding activation of some defence related response (see chapter 21).

The formation of acetyl-fucose in *M. loti* is dependent upon the *nolL* gene. NolL shows similarity to NodX indicating that it is a fucose O-acetyl transferase (Scott *et al.*, 1996). A *nolL* gene has also been identified in *Rhizobium* sp. NGR234 (Freiberg *et al.*, 1997) where it has been shown to be involved in acetylation of fucose (Jabbouri and Broughton, personal communication). NoeE is involved in the sulphation of fucose in *Rhizobium* sp. NGR234 (Hanin *et al.*, 1997). Also present in *Rhizobium* sp. NGR234 are the *noeJ*, *noeK* and *noeL* genes (Freiberg *et al.*, 1997), that are involved in the biosynthesis of GDP-fucose, while *noeI* is an O-methyl transferase involved in 2-O-methylation of fucose (Jabbouri and Broughton, personal communication). Figure 3 summarises the proposed roles of several *nod* gene products in GDP-fucose biosynthesis.

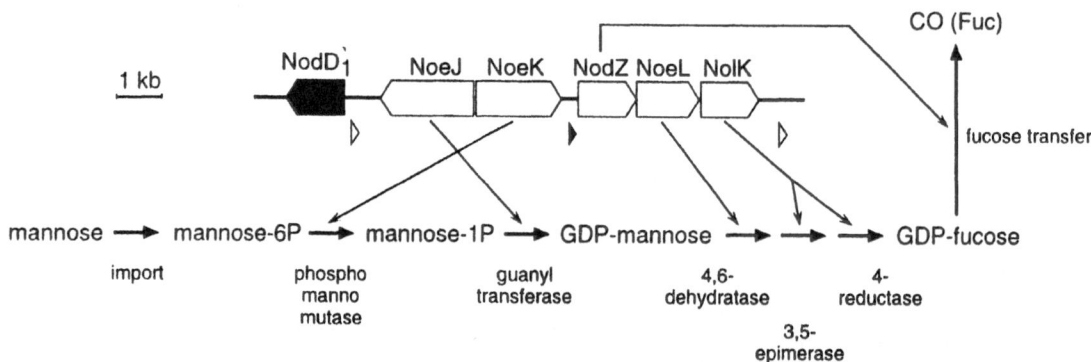

Figure 3. Roles of several *nod* gene products of NGR234 in GDP-fucose biosynthesis. For further details on the gene map see Chapter 19. This scheme is based on the work of Quesada-Vincens *et al.*, (1997) and is reproduced courtesy of R. Fellay and W.J. Broughton.

III.C.4. 3-O-Arabinose

In *A. caulinodans*, *S. saheli* and *S. teranga* bv. *sesbaniae* arabinose can be present linked to C-3 of the reducing glucosaminyl residue (Mergaert *et al.*, 1997; Lorquin *et al.*, 1997). It was initially thought that arabinose was present on C6 (rather than C5) of the reducing glucosaminyl residue of *A. caulinodans* Nod factors (Mergaert *et al.*, 1993, 1996) but this seems to have been an artefact of the genetically-modified strain used for Nod factor production (Mergaert *et al.*, 1997). The *noeC* gene encodes a potential integral membrane protein thought to be involved in arabinosylation, since mutation of *noeC* blocked attachment of arabinose. However, the possibility could not be excluded that *nolK* and additional gene(s) downstream of *noeC* might be involved in arabinosylation (Mergaert *et al.*, 1996).

III.C.5. 1-O-Glycerol

Some of the Nod factors from *Bradyrhizobium* strains have been found to carry glycerol linked to the reducing glucosamine residue (Carlson *et al.*, 1993; Luka *et al.*, 1993). These are usually minor components in wild-type bacteria and it has been suggested that they represent a biosynthetic intermediate (Carlson *et al.*, 1994). No *nod* gene has been identified as being involved in attachment of glycerol.

III.D. SUBSTITUTIONS TO NON-TERMINAL RESIDUES

Relatively few modifications of the non-terminal sugar residues of the Nod factors have been described. In some strains of *M. loti*, one of the minor Nod factor components was found to have a glucose residue replacing *N*-acetyl glucosamine as the central sugar in the pentameric backbone (Bec-Ferté *et al.*, 1996). In *R. galegae*, the *N*-acetyl-glucosamine adjacent to the acylated glucosamine residue is O-acetylated on the C-3 position (Yang *et al.*, 1997). Presumably a *nod* gene product with specific O-acetyl transferase activity is required for this addition, but the *nod* gene has not yet been identified.

In *Mesorhizobium loti* the sugar backbone of the Nod factor is substituted with a methyl fucose residue, which is α-1, 3-linked to the 3-C of the glucosamine residue adjacent to the acylated glucosamine, in effect creating a branched oligosaccharide. This indicates that an α-(1-3) fucosyl transferase must be present, but the gene encoding this activity has not yet been identified (Olsthoorn *et al.*, 1997).

IV. Nod Factor Secretion

Analysis of bacterial genes encoding transport systems is often difficult due to redundancy in transporters and this evidently holds true for Nod factor export. It is now clear that *nodI* and *nodJ* are important for secretion of Nod factors (McKay and Djordjevic 1993, Spaink *et al.*, 1995b, Cardenas *et al.*, 1996, Fernandez Lopez *et al.*, 1996) although it is also apparent that other transport system(s) can compensate for the absence of the *nodIJ*-encoded transporter. Based on protein sequence similarities, it has been evident for some time (Evans and Downie, 1985; Göttfert *et al.*, 1989; Vázquez *et al.*, 1993; Geelen *et al.*, 1993) that the NodIJ proteins were likely to be involved in transport. NodI contains an ATP-binding motif (Evans and Downie 1985), is found associated with the cytoplasmic membrane (Schlaman *et al.*, 1990) and contains a leucine zipper motif, (Göttfert 1993) suggesting it forms dimers or complexes with another protein. NodJ is an integral membrane protein (Surin *et al.*, 1990; Vázquez *et al.*, 1993) that has a similar hydrophobicity profile to other integral membrane transporters.

Mutations in the *nodIJ* genes caused relatively little effect on nodulation by *R.l. viciae* (Downie *et al.*, 1985, Spaink *et al.*, 1995b), *B. japonicum* (Göttfert *et al.*, 1989) or *A. caulinodans* (Geelen *et al.*, 1993). However, these genes proved to be important for subclover or bean nodulation by *R.l. trifolii* or *R. etli* (Djordjevic *et al.*, 1985., Canter-Cremers *et al.*, 1988, Fernandez-Lopez *et al.*, 1996). The NodI and NodJ proteins show most similarity to other bacterial genes encoding transporters involved in export of capsular polysaccharide (Vazquez *et al.*, 1993). The absence of a clear phenotype for *nodI* or *nodJ* mutants of several rhizobia might be explained if Nod factors are also exported by a transporter normally involved in *e.g.* polysaccharide secretion.

The *nodT* gene is immediately downstream of *nodIJ* in *R.l. trifolii* (Surin *et al.*, 1990) whereas it is in the *nodMNT* operon in *R.l. viciae*. Mutations in *nodT* in *R.l. viciae* or *R.l. trifolii* have little or no observed effect on nodulation (Surin *et al*, 1990, Lewis-Henderson , 1991) although transfer of *nodT* into a *R.l. trifolii* strain lacking *nodT* enabled that strain to nodulate cv Woolgenellup clover (Lewis-Henderson and Djordjevic 1991). NodT is found in the outer membrane and is related to a family of outer-membrane proteins (Rivilla *et al.*, 1995), that are involved in secretion of substrates across both the inner and outer bacterial membranes. By analogy with

Table 2. Structure of Nod factors from various rhizobial species[a]

Host plant[b]	Rhizobial species	Nod factor substituents[c]						n[d]	Ref.
		R1	R2[e]	R3	R4	R5	R6[fgh]		
Medicago	*S. meliloti*	H	C16:2.C16:3 C18:C26(w-1)OH	Ac(C-6).H	S	H		**1,2,3**	1,2,3,4
Vicia	*R.l.* bv. *viciae*	H	C18:1.C18:4	Ac(C-6)	H	H		**2,3**	5
Pisum cv. Afghanistan	*R.l.* bv. *viciae* TOM	H	C18:1.C18:4	Ac(C-6)	Ac	H		**2,3**	6
Trifolium	*R.l.* bv. *trifolii*	H	C18:1.C18:3 C20:3.C20:4	Ac(C-6)	H	H		**1,2,3**	7,8,9
Astragalus	*R. huakii*	H	C18:4	H	S	H		3	27
Galega	*R. galegae*	H	C18:2;C18:3	Cb(C-6)	H	H	Ac(C-3)	2	27
Lotus	*M. loti*	Me	C18:1	Cb(C-4)	AcFuc	H		3	10
		H	C18:1.C18	Cb(C-3)	Fuc AcFuc	H	Fuc	3	26
Phaseolus	*R. etli*	Me	C18:1	Cb(C-4).H	AcFuc	H		3	11,12
	R. tropici	Me	C18:1	H	S.H	H		3	13,14
	R.l. bv. *phaseoli*	Me	C18:1	H	H	H		3	15
Acacia	*R.* sp. GRH2	Me.H	C18:1	H	S.H	H		**2,3**,4	16
	S. teranga	Me	C16:0.C18:0.C18:1	Cb	S.H	H		3	17
Lablab	*R.* sp. NGR234	Me	C18:1	Cb(C-6 and C-3 or C-4) H	MeFuc AcMeFuc MeSFuc	H		3	18,19
Glycine	*S. fredii*	H	C18:1	H	MeFuc.Fuc	H	glc	**1,2,3**	20,21
	B. japonicum	H	C18:1	H	MeFuc	H		3	22,23
	B. elkanii	H.Me	C18:1	Ac(C-6).H,Cb	MeFuc.Fuc	H		**2,3**	23
Sesbania	*A. caulinodans*	Me	C18:1	Cb(C-6)	Fuc.H	D-Ara.H		**2,3**	24,25,28
	S. saheli	Me	C18:1;C16:0	Cb(C-3, C-4 or C-6)	Fuc.H	D-Ara.H		**2,3**	29
	S. teranga bv.sesbaniae	Me	C18:1;C16:0	Cb(C-3, C-4 or C-6)	Fuc.H	D-Ara.H		**2,3**	29

This table is based on Table 1 of Dénarié *et al.*,(1996) copyright © Annual Reviews Inc and is published here by kind permission of Annual Reviews Inc and J. Dénarié, F. Debellé and J.C. Promé. b. Plant genera from which rhizobial strains were isolated. c. In *B. elkanii*, C-1 of the terminal reducing glucosaminyl residue can be occasionally substituted with glycerol. d. The bold numbers indicate the number of glucosamine residues of the most abundant Nod factors. e. Selected fatty acyl substituents. f. In *M. loti* Fuc is linked α (1→3) to the glucosamine adjacent to the acylated glucosamine. g. In *R. galegae* the acetyl group is linked to the 3-C of the glucosamine adjacent to the reducing glucosamine. h. One of the minor Nod-factor components made by *S. fredii* contained a glucose residue instead of glucosamine.

Reference: 1, Lerouge *et al.*,(1990); 2, Roche *et al.*,(1991b); 3, Schultze *et al.*,(1992); 4, Demont *et al.*,(1993); 5, Spaink *et al.*,(1991); 6, Firmin *et al.*,(1993); 7, Spaink *et al.*,(1995a); 8; Orgambide *et al.*,(1995); 9, Van der Drift *et al.*,(1996); 10, Lopez-Lara *et al.*,(1995a); 11, Cárdenas *et al.*,(1995); 12, Poupot *et al.*,(1995a); 13, Poupot *et al.*,(1993); 14, Folch-Mallol *et al.*,(1996); 15, K.E. Wilson, R. Carlson, J. Firmin and J.A. Downie (unpublished); 16, Lopez-Lara *et al.*, (1995b); 17, Lorquin et al., (1997), 18, 19, Price *et al.*, (1992, 1996); 20 ,21, Bec-Ferté *et al.*,(1994, 1996); 22, Sanjuan *et al.*,(1992); 23, Carlson *et al.*,(1993); 24, 25, Mergaert *et al.*,(1993, 1996); 26, Olsthoorn *et al.*,(1997); 27, Yang *et al.*,(1997); 28, Mergaert *et al.*,(1997); 29, Lorquin *et al.*, (1997).

other secretion systems it was suggested that NodT might form part of a Nod factor secretion complex together with NodI and NodJ (Downie, 1994, Rivilla et al., 1995). However, there is no direct evidence for such a role for NodT.

The *nolFGHI* genes have only been found in *S. meliloti* where they are in the same operon as *nodM* and *nodN* (Baev et al., 1991). Mutations in *nolFGHI* delay and reduce nodulation on species of *Medicago*. It is unlikely that the delayed nodulation was simply caused by polar effects on the downstream *nodN* because mutation of *nolFGHI* had a much lower effect on Nod factor production (based on bioassays) than mutation of *nodN*. The *nolF* gene product was proposed to be similar to integral membrane proteins (Baev et al., 1991) and the *nolGHI* gene products can be aligned with single integral membrane proteins such as CnrA and CscA, that are involved in the efflux of cobalt, nickel and zinc. On the basis of this similarity it was suggested that the *nolFGHI* genes might be involved in Nod factor export (Saier et al., 1994). However, mutations in these genes appeared to have very little effect on root hair deformation activity (Baev et al., 1991) suggesting that their mutation did not have a significant effect on Nod factor export.

V. Protein Secretion

Relatively few *nod* genes appear to be involved in protein export. The best-characterised secreted protein is NodO from *R.l. viciae* (De Maagd et al., 1989; Economou et al., 1990). It is a Ca^{2+}-binding protein that forms cation-selective pores in membranes (Sutton et al., 1993) and a homologue has been found in a broad host range strain of *Rhizobium* (van Rhijn et al., 1996). NodO does not appear to be involved in Nod factor biosynthesis, export or binding (Spaink et al., 1991; Sutton et al., 1994), although it clearly stimulates nodulation in mutants that do not make a fully modified Nod factor. Thus in *nodE* mutants of *R.l. viciae*, *nodO* is needed for nodulation (Downie and Surin, 1990; Economou et al., 1994), even though mutation of *nodO* alone had little effect on nodulation. The cloned *nodO* gene can extend the nodulation range of some rhizobial strains into which it is introduced (Economou et al, 1994, Van Rhijn et al., 1996). Although its function is not precisely known, it is thought (Sutton et al., 1994) to stimulate nodulation by altering ion flow across the plasma membrane of root cells, possibly enhancing the effects of Nod factors on intracellular ion concentrations (Ehrhardt et

al., 1992, 1996; Felle et al., 1995,1996; Kurkdjian (1995). The genes required for NodO export are not *nod* genes (Scheu et al, 1992) and NodO is secreted via a Type I export pathway that secretes other (non nodulation) proteins (Finnie et al., 1996, 1997). Since it is likely that NodO will only directly affect plant cells that come into direct contact with the bacteria, NodO probably plays a role in the infection process.

A number of flavonoid-inducible proteins from other rhizobia have been identified in culture supernatants (Krishnan and Pueppke 1993), although these have not been extensively characterised. The *nolXWBTUV* gene region of *S. fredii* is involved in host specific nodulation; mutation of any of these genes enables a strain of *S. fredii* to nodulate a soybean cultivar that is normally resistant to nodulation by that strain (Meinhardt et al., 1993). A similar gene cluster is present in *Rhizobium* sp. NGR234 (Freiberg et al., 1997). NolW and NolT are homologous to proteins involved in protein secretion typical of those found in bacterial plant pathogens (Van Gijsegem et al., 1995). Expression of *nolBTUV* and *nolX* is induced by flavonoids whereas *nolW* is constitutively expressed (Kovacs et al., 1995). It is possible that these genes may be involved in the secretion of protein(s) that block nodulation of some soybean cultivars; this would explain the observation that mutation of any of these genes (encoding a putative export system) can extend nodulation.

VI. Genes of Unknown Function

The biochemical roles of several of the *nod* genes have not been defined. For several of these which show no homologies to other proteins in databases, it is difficult even to make an educated guess. Nevertheless, some have clear biological effects. For the sake of completeness I will mention these other genes and what is known of them.

The *nodK* gene was one of the earliest defined *nod* genes (Scott, 1986) and it was found in *Rhizobium* sp NGR234 which has a broad host range including *Parasponium*, a non legume. It is found at the start of the *nodABCIJ* operon but no function has been established. It is related to *nodY* from *B. japonicum*, which in this strain is also upstream of *nodA* (Nieuwkoop et al., 1987). Both *nodY* and *nodK* are likely to encode cytoplasmic proteins.

The function of *nodN* has not been established. It has been identified in the same operon as *nodM* in *R.l. viciae* and *S. meliloti* (Surin and Downie, 1988; Baev *et al.*, 1991) and mutation of *nodN* decreased the levels of Nod factors made by *S. meliloti* (Baev *et al.*, 1991). Therefore, like *nodM*, it may be involved in the production of precursors required for UDP-GlcNAc formation.

The *nolC* gene was identified as a gene in *S. fredii* OSDA257, which when mutated enables that strain to nodulate soybean cv. McCall (Krishnan and Pueppke 1991a, 1992). The *nolC* gene product is predicted to be a soluble protein with similarity to DnaJ, a heat shock gene found in a variety of bacteria (Krishman and Pueppke 1991a). Unlike DnaJ, NolC is not induced by heat shock and no evidence was presented for it being under NodD control.

The *nolJ* gene is one of two adjacent isoflavone inducible genes identified in *S. fredii* (Sadowsky *et al.*, 1988; Boundy-Mills *et al.*, 1994). Each of the genes had its own promoter and there was sequence similarity between the two promoters but little or no homology with recognised NodD-binding nod-box promoter sequences. Mutation of *nolJ* inhibited nodulation of soybean and mutation of the upstream gene (referred to as orf1) reduced competitiveness in nodulation tests.

The *nolP* and *nolE* genes were identified as genes downstream of inducible *nod*-box promoters in *R.l. phaseoli* (Davis and Johnston, 1990). The *nolE* gene precedes the *nodD1* gene in an isoflavonoid inducible operon. No function for NolE was found although it does appear to be located in the periplasm. NolP was predicted to be a relatively short hydrophilic protein. No sequence homologies to NolP or NolE were found and mutations in these genes had no observed effect on nodulation of *Phaseolus* beans. Michiels *et al.*, (1995) identified another (unnamed) *R.l. phaseoli* gene that is important for nodule competitiveness.

The *nolN* gene in *B. japonicum* is in the *nodYABCSUIJnolMNO* operon and analysis of NolN function is made complicated by the fact that the deletion mutation in *nolN* extended into *nolO*. However, there was no difference in Nod factor production when comparing a *nolNO* deletion mutant and a *nolO* mutant. The primary effect of these mutations was on the methyl fucosyl group of the Nod factor (Luka *et al.*, 1993).

The *nolQ* and *nolS* genes were identified downstream of a *nod* box promoter in *S. meliloti*, although there seemed to be very little flavanoid induction of these genes (Plazanet *et al.*, 1995 and Kondorosi cited therein). The predicted *nolQ* gene product from *S. meliloti* RM41 appears to be present as two ORFs (called *nolQa* and *nolQb*) in *S. meliloti* SU47.

Mutation of *nolQ* resulted in reduced efficiency of nodulation on *Medicago lupulina* and the effect was addative with regard to mutations of the *nodFE* genes. Mutants lacking *nolQ* seemed to be unaffected in Nod factor biosynthesis. The downstream *nolS* gene encodes a protein with similarity to the *S. meliloti* ThiI protein involved in thiamine synthesis, but since *nolS* insertional mutants could not be constructed, no phenotype has been described (Plazanet *et al.*, 1995).

In *B. japonicum* the *nolYZ* genes were identified downstream of a NodD (and NodW) -regulated *nod*-box (Dockendorf *et al.*, 1994). Mutation of *nolY* had an effect on nodulation of Mung bean but no effect was seen with a *nolZ* mutant. No effect on Nod factor production was observed and no homologies to other proteins were reported.

The *noeA* and *noeB* genes were found downstream of *nodL* in *S. meliloti* and mutation of either gene strongly inhibited nodulation of *Medicago lupulina* and *Medicago littoralis* (but not *Medicago truncatula*). Infection thread formation was strongly affected but no effects on Nod factor production were found (Ardourel *et al.*, 1995). NoeA showed limited homology to an *E. coli* methyl transferase around a domain involved in binding to S-adenosyl methionine, but it was not signficantly homologous to NodS. The *noeB* gene appears to encode an integral membrane protein that showed no homology to other proteins.

VII. Concluding Remarks

While Nod factor biosynthesis is clearly the role for many of the *nod* gene products, it is evident that there are many other roles for several of the *nod* gene products. Some of these proteins may play a role in infection, possibly of specific species of legume. Currently it is not obvious how some of these gene products function, but as more gene sequences are added to database libraries the roles of unknown *nod* gene products may well come to light. Genome sequencing has already had an impact on identification of *nod* genes with several novel genes (*noeIJKL*) being identified by sequencing the *Rhizobium* sp NGR234 symbiotic plasmid (Freiberg *et al.*, 1997). As additional more unusual rhizobial strains are studied (see e.g. Lortet *et al.*, 1996; Clutier *et al.*, 1996; Lorquin *et al.*, 1997) it is likely that other *nod* genes will be identified; such *nod* gene products are likely to be involved in enabling the rhizobia to efficiently nodulate their host legume and maximise their competitive nodulation ability. Currently, *nod* gene nomenclature is co-ordinated by Prof. G. Stacey, Department of Microbiology, University of Tennessee, Knoxville, TN 37996-0845, USA (Email: GSTACEY@utk.edu) to whom requests for new *nod* gene assignations should be addressed.

VIII. Acknowledgements

I am indebted to various colleagues for helping supply recent citations. In particular, I would like to thank F. Debelle, J. Dénarié and J.C. Promé for help with references and H. Spaink, A. Kondorosi, G. Stacey and R. Fisher for useful comments. I also thank K. E. Wilson, J. Firmin, R. Carlson, W. Broughton, S. Jabbouri, M. Sadowsky, H. Spaink and J. Denarié and their colleagues for providing information prior to publication. This article was written with the support of the BBSRC.

IX. References

Ardourel, M., Demont, N., Debellé, F., Maillet, F., De Billy, F., Promé, J.C., Dénarié, J. and Truchet, G. (1994) Plant Cell 6, 1357-1374.

Ardourel, M., Lortet, G., Maillet, F., Roche, P., Truchet, G., Promé, J.C. and Rosenberg, C. (1995) Mol. Microbiol. 17, 687-699.

Atkinson, E.M., Palcic, M.M., Hindsgaul, O. and Long, S.R. (1994) Proc. Natl. Acad. Sci. USA 91, 8418-8422.

Baev, N. and Kondorosi, A. (1992) Plant Mol. Biol. 18, 843-846.

Baev, N., Endre, G., Petrovics, G., Banfalvi, Z. and Kondorosi, A. (1991) Mol. Gen. Genet. 228, 113-124.

Baev, N., Schultze, M., Barlier, I., Ha, D.C., Virelizier, H., Kondorosi, E. and Kondorosi, A. (1992) J. Bacteriol. 174, 7555-7565.

Balatti, P.A., Kovacs, L.G., Krishnan, H.B. and Pueppke, S.G. (1995) Mol. Plant-Microbe Interact. 8, 693-699.

Barny, M.A. and Downie, J.A. (1993) Mol. Plant-Microbe Interact. 6, 669-672.

Barny, M.A., Schoonejans, E., Economou, A., Johnston, A.W.B. and Downie, J.A. (1996) Mol. Microbiol. 19, 443-453.

Bec-Ferté, M.P., Krishnan, H.B., Promé, D., Savagnac, A., Pueppke, S.G. and Promé, J.C. (1994) Biochemistry 33, 11782-11788.

Bec-Ferté, M.P., Krishnan, H.B., Savagnac, A., Pueppke, S.G. and Promé, J.C. (1996) FEBS Letters 393, 273-279.

Bibb, M.J., Biro, S., Motamedi, H., Collins, J.F. and Hutchinson, C.R. (1989) EMBO J. 9, 2727-2736.

Bloemberg, G.V., Thomas-Oates, J.E., Lugtenberg, B.J.J. and Spaink, H.P. (1994) Mol. Microbiol. 11, 793-804.

Bloemberg, G.V., Kamst, E., Harteveld, M., van der Drift, K.M.G.M., Haverkamp, J., Thomas-Oates, J.E., Lugtenberg, B.J.J. and Spaink, H.P. (1995a) Mol. Microbiol. 16, 1123-1136.

Bloemberg, G.V., Lagas, R.M., van Leeuwen, S., Van der Marel, G.A., Van Boom, J.H., Lugtenberg, B.J.J. and Spaink, H.P. (1995b) Biochemistry 39, 12712-12720.

Boundy-Mills, K.I., Kosslak, R.M., Tully, R.E., Pueppke, S.G., Lohrke, S. and Sadowsky, M.J. (1994) Mol. Plant-Microbe Interact. 7, 305-308.

Bourdineaud, J.P., Bono, J-J., Ranjeva, R. and Cullimore, J.V. (1995) Biochem. J. 306, 259-264.

Canter-Cremers, H.C.J., Wijffelman, C.A., Pees, E., Rolfe, B.G., Djordjevic, M.A. and Lugtenberg, B.J.J. (1988) J. Plant Physiol. 132, 398-404.

Cardenas, L., Dominguez, J., Quinto, C., Lopez-Lara, I.M., Lugtenberg, B.J.J., Spaink, H.P., Rademaker, G.J., Haverkamp, J. and Thomas-Oates, J.E. (1995) Plant Mol. Biol. 29, 453-464.

Cardenas, L., Dominguez, J., Santana, O. and Quinto, C. (1996) Gene 173, 183-187.

Carlson, R.W., Sanjuan, J., Bhat, U.R., Glushka, J., Spaink, H.P., Wijfjes, A.H.M., van Brussel, A.A.N., Stokkermans, T.J.W., Peters, N.K. and Stacey, G. (1993) J. Biol. Chem. 268, 18372-18381.

Carlson, R.W., Price, N.P.J. and Stacey, G. (1994) Mol. Plant-Microbe Interact. 7, 684-695.

Cedergren, R.A., Lee, J.G., Ross, K.L. and Hollingsworth, R.I. (1995) Biochemistry 34, 4467-4477.

Clark, C.A., Beltrame, J. and Manning, P.A. (1991) Gene 107, 43-52.

Cloutier, J., Laberge, S., Prevost, D. and Antoun, H. (1996) Mol. Plant-Microbe Interact. 9, 523-531.

Davis, E.O. and Johnston, A.W.B. (1990) Mol. Microbiol. 4, 921-932.

Davis, E.O., Evans, I.J. and Johnston, A.W.B. (1988) Mol. Gen. Genet. 212, 531-535.

De Lajudie, P., Willems, A., Pot, B., Dewettinck, D., Maestrojuan, G., Neyra, M., Collins, M.D., Dreyfus, B., Kersters, K. and Gillis, M. (1994) Int. J. Syst. Bact. 44, 715-733.

De Maagd, R.A., Wijfjes, A.H.M., Spaink, H.P., Ruiz-Sainz, J.E., Wijffelman, C.A., Okker, R.J.H. and Lugtenberg, B.J.J. (1989) J. Bacteriol. 171, 6764-6770.

Debellé, F., Plazanet, C., Roche, P., Pujol, C., Savagnac, A., Rosenberg, C., Promé, J.C. and Denarie, J. (1996) Mol. Microbiol. 22, 303-314.

Demont, N., Debellé, F., Aurelle, H., Dénarié, J. and Promé, J.C. (1993) J. Biol. Chem. 268, 20134-20142.

Dénarié, J., Debellé, F. and Promé, J.C. (1996) Annu. Rev. Biochem. 65, 503-535.

Djordjevic, M.A., Schofield, P.R. and Rolfe, B.G. (1985) Mol. Gen. Genet. 200, 463-471.

Dockendorff, T.C., Sharma, A.J. and Stacey, G. (1994) Mol. Plant-Microbe Interact. 7, 173-180.

Downie, J.A. (1989) Mol. Microbiol. 3, 1649-1651.

Downie, J.A. (1994) Trends Microbiol. 2, 318-324.

Downie, J.A. and Surin, B.P. (1990) Mol. Gen. Genet. 222, 81-86.

Downie, J.A., Knight, C.D. and Johnston, A.W.B. (1985) Mol. Gen. Genet. 198, 255-262.

Dunn, S.M., Moody, P.C.E., Downie, J.A. and Shaw, W.V. (1996) Prot. Sci. 5, 538-541.

Economou, A., Hamilton, W.D.O., Johnston, A.W.B. and Downie, J.A. (1990) EMBO J. 9, 349-354.

Economou, A., Davies, A.E., Johnston, A.W.B. and Downie, J.A. (1994) Microbiology 140, 2341-2347.

Ehrhardt, D.W., Atkinson, E.M. and Long, S.R. (1992) Science 256, 998-1000.

Ehrhardt, D.W., Atkinson, E.M., Faull, K.F., Freedberg, D.I., Sutherlin, D.P., Armstrong, R. and Long, S.R. (1995) J. Bacteriol. 21, 6237-6245.

Ehrhardt, D.W., Wais, R. and Long, S.R. (1996) Cell, 85, 673-681.

Evans, I.J. and Downie, J.A. (1986) Gene 43, 95-101.

Faucher, C., Maillet, F., Vasse, J., Rosenberg, C., van Brussel, A.A.N., Truchet, G. and Dénarié, J. (1988) J. Bacteriol. 170, 5489-5499.

Fellay, R., Perret, X., Viprey, V., Broughton, W.J. and Brenner, S. (1995) Mol. Microbiol. 16, 657-667.

Fellay, R., Rochepeau, P., Relic, B. and Broughton, W.J. (1995) Singh, U.S., Singh, R.P. and Kohmoto, K. (eds) Pathogenesis and host specificity in plant diseases. Histopathological, biochemical, genetic and molecular bases. Pergamon/Elsevier Science Ltd, Oxford, pp. 199-220.

Felle, H.H., Kondorosi, E., Kondorosi, A. and Schultze, M. (1995) Plant J. 7, 939-947.

Felle, H.H., Kondorosi, E., Kondorosi, A. and Schultze, M. (1996) Plant J., 10, 295-301.

Fernandez-Lopez, M., D'Haeze, W., Mergaert, P., Verplancke, C., Promé, J.C., Van Montagu, M. and Holsters, M. (1996) Mol. Microbiol., 20, 993-1000.

Finnie, C., Dean, G., Sutton, J.M., Gehlani, S. and Downie, J.A. (1996) Stacey, G., Mullin, B. and Gresshoff, P.M. (eds) Advances in Molecular Genetics of Plant Microbe Interactions. pp. 343-348.

Finnie, C., Hartley, N.M., Findlay, K.C. and Downie, J.A. (1997) Mol. Microbiol., 25, 135-146

Firmin, J.L., Wilson, K.E., Carlson, R.W., Davies, A.E. and Downie, J.A. (1993) Mol. Microbiol., 10, 351-360.

Folch-Mallol, J.L., Marroqui, S., Sousa, C., Manyani, H., Lopez-Lara, I.M., van der Drift, K.M.G.M., Haverkamp, J., Quinto, C., Gil-Serrano, A., Thomas-Oates, J., Spaink, H.P. and Megias, M. (1996) Mol. Plant-Microbe Interact., 9, 151-163.

Freiberg, C., Fellay, R., Bairoch, A., Broughton, W.J., Rosenthal, A. and Perret, X. (1997) Nature, 387, 394-401.

Geelen, D., Mergaert, P., Geremia, R.A., Goormachtig, S., Van Montagu, M. and Holsters, M. (1993) Mol. Microbiol., 9, 145-154.

Geelen, D., Leyman, B., Mergaert, P., Klarskov, K., Van Montagu, M., Geremia, R.A. and Holsters, M. (1995) Mol. Microbiol., 17, 387-379.

Geiger, O., Spaink, H.P. and Kennedy, E.P. (1991) J. Bacteriol., 173, 2872-2878.

Geiger, O., Thomas-Oates, J.E., Glushka, J., Spaink, H.P. and Lugtenberg, B.J.J. (1994) J. Biol. Chem., 269, 11090-11097.

Geremia, R.A., Mergaert, P., Geelen, D., Van Montagu, M. and Holsters, M. (1994) Proc. Natl. Acad. Sci. USA, 91, 2669-2673.

Geurts, R. and Franssen, H. (1996) Plant Physiol., 112, 447-453.

Goethals, K., Mergaert, P., Gao, M., Geelen, D., Van Montagu, M. and Holsters, M. (1992) Mol. Plant-Microbe Interact., 5, 405-411.

Göttfert, M. (1993) FEMS Microbiol. Rev., 104, 39-63.

Göttfert, M., Lamb, J.W., Gasser, R., Semenza, J. and Hennecke, H. (1989) Mol. Gen. Genet., 215, 407-415.

Hanin, M., Jabbouri, S., Quesada-Vincens, D., Freiberg, C., Perret, X., Prome, J-C., Broughton, W.J. and Fellay, R. (1997) Mol. Microbiol., 24, 1119-1129.

Inon de Iannino, N., Pueppke, S.G. and Ugalde, R.A. (1995) Mol. Plant-Microbe Interact., 8, 292-301.

Jabbouri, S., Fellay, R., Talmont, F., Kamalaprija, P., Burger, U., Relic, B., Promé, J.C. and Broughton, W.J. (1995) J. Biol. Chem., 270, 22968-22973.

John, M., Röhrig, H., Schmidt, J., Wieneke, U. and Schell, J. (1993) Proc. Natl. Acad. Sci. USA, 90, 625-629.

Journet, E.P., Pichon, M., Dedieu, A., De Billy, F., Truchet, G. and Barker, D. (1994) Plant J., 6, 241-249.

Kamst, E., Pilling, J., Raamsdonk, L.M., Lugtenberg, B.J.J. and Spaink, H.P. (1997) J. Bacteriol., 179, 2103-2108.

Kamst, E., van der Drift, K.M.G.M., Thomas-Oates, J.E., Lugtenberg, B.J.J. and Spaink, H.P. (1995) J. Bacteriol., 177, 6282-6285.

Kovacs, L.G., Balatti, P.A., Krishnan, H.B. and Pueppke, S.G. (1995) Mol. Microbiol., 17, 923-933.

Kozik, A., Heidstra, R., Horvath, B., Kulikova, O., Tikhonovich, I., Ellis, T.H.N., Vankammen, A., Lie, T.A. and Bisseling, T. (1995) Plant Sci, 108, 41-49.

Krishnan, H.B. and Pueppke, S.G. (1991a) Mol. Microbiol., 5, 737-745.

Krishnan, H.B. and Pueppke, S.G. (1991) Mol. Plant-Microbe Interact., 4, 512-520.

Krishnan, H.B. and Pueppke, S.G. (1992) Mol. Plant-Microbe Interact., 5, 14-21.

Krishnan, H.B. and Pueppke, S.G. (1993) Mol. Plant-Microbe Interact., 6, 107-113.

Kurkdjian, A.C. (1995) Plant Physiol., 107, 783-790.

Laeremans, T., Caluwaerts, I., Verreth, C., Rogel, M.A., Vanderleyden, J. and Martinez-Romero, E. (1996) Mol. Plant-Microbe Interact., 9, 492-500.

Lerouge, P. (1994) Glycobiology, 4, 127-134.

Lerouge, P., Roche, P., Faucher, C., Maillet, F., Truchet, G., Promé, J.C. and Dénarié, J. (1990) Nature, 344, 781-784.

Lewin, A., Cervantès, E., Wong, C.H. and Broughton, W.J. (1990) Mol. Plant-Microbe Interact., 3, 317.

Lewis-Henderson, W.R. and Djordjevic, M.A. (1991) Plant Mol. Biol., 16, 515-526.

Leyh, T.S., Vogt, T.F. and Suo, Y. (1992) J. Biol. Chem., 267, 10405-10410.

Long, S.R. (1996) Plant Cell, 8, 1885-1898.

Lopez-Lara, I.M., van den Berg, J.D.J., Thomas-Oates, J.E., Glushka, J., Lugtenberg, B.J.J. and Spaink, H.P. (1995) Mol. Microbiol., 15, 627-638.

Lopez-Lara, I.M., van der Drift, K.M.G.M., van Brussel, A.A.N., Haverkamp, J., Lugtenberg, B.J.J., Thomas-Oates, J.E. and Spaink, H.P. (1995a) Plant Mol. Biol., 29, 465-477.

Lopez-Lara, I.M., Blok-Tip, L., Quinto, C., Garcia, M.L., Stacey, G., Bloemberg, G.V., Lamers, G.E.M., Lugtenberg, B.J.J., Thomas-Oates, J.E. and Spaink, H.P. (1996b) Mol. Microbiol., 21, 397-408.

Lorquin, J., Lortet, G., Ferro, M., Mear, N., Dreyfos, B., Prome, J-C. and Boivin, C. (1997) Mol. Plant-Microbe Interact. 10, 879-890.

Lorquin, J., Lortet, G., Ferro, M., Mear, N., Promé, J-C. and Boivin, C. (1997) J. Bacteriol., 179, 3079-3083.

Lortet, G., Mear, N., Lorquin, J., Dreyfus, B., De Lajudie, P., Rosenberg, C. and Boivin, C. (1996) Mol. Plant-Microbe Interact., 9, 736-747.

Luka, S., Sanjuan, J., Carlson, R.W. and Stacey, G. (1993) J. Biol. Chem., 268, 27053-27059.

Marie, C., Barny, M.A. and Downie, J.A. (1992) Mol. Microbiol., 6, 843-851.

Marie, C., Plaskitt, K.A. and Downie, J.A. (1994) Mol. Plant-Microbe Interact., 7, 482-487.

Martinez-Romero, E. (1994) Plant and Soil, 161, 11-20.

Mckay, I.A. and Djordjevic, M.A. (1993) Appl. Environ. Microbiol., 59, 3385-3392.

Meinhardt, L.W., Krishnan, H.B., Balatti, P.A. and Pueppke, S.G. (1993) Mol. Microbiol., 9, 17-29.

Mergaert, P., Van Montagu, M., Promé, J.C. and Holsters, M. (1993) Proc. Natl. Acad. Sci. USA, 90, 1551-1555.

Mergaert, P., D'Haeze, W., Geelen, D., Promé, D., Van Montagu, M., Geremia, R.A., Promé, J.C. and Holsters, M. (1995) J. Biol. Chem., 270, 29217-29223.

Mergaert, P., D'Haeze, W., Fernandez-Lopez, M., Geelen, D., Goethals, K., Promé, J.C., Van Montagu, M. and Holsters, M. (1996) Mol. Microbiol., 21, 409-419.

Mergaert, P., Ferro, M., D'Haeze, W., Van Montagu, M., Holsters, M. and Promé, J.C. (1997) Mol. Plant-Microbe Interact., 10, 683-687.

Michiels, J., Pelemans, H., Vlassak, K., Verreth, C. and Vanderleyden, J. (1995) Mol. Plant-Microbe Interact., 8, 468-472.

Mylona, P., Pawlowski, K. and Bisseling, T. (1995) Plant Cell, 7, 869-885.

Nieuwkoop, A.J., Banfalvi, Z., Deshmane, N., Gerhold, D., Schell, M.G., Sirotkin, K.M. and Stacey, G. (1987) J. Bacteriol., 169, 2631-2638.

Olsthoorn, M.M.A., López-Lara, I.M., Petersen, B.O., Bock, K., Haverkamp, J., Spaink, H.P. and Thomas-Oates, J.E. (1997) Biochemistry, In press.

Orgambide, G.G., Lee, J., Hollingsworth, R.I. and Dazzo, F.B. (1995) Biochemistry, 34, 3832-3840.

Ovtsyna, A.O., Veldhuis, A., Lopez-Lara, I.M., Wijfjes, A.H.M., Quinto, C., Geurts, R., Bisseling, T., Scott, D.B., Tikhonovich, I.A., Lugtenberg, B.J.J. and Spaink, H.P. (1997) Legocki, A., Bothelt, Puhler, A. (eds.) Biological Fixation of Nitrogen for Ecology and Sustainable Agriculture pp. 25-28.

Philip-Hollingsworth, S., Orgambide, G.G., Bradford, J.J., Smith, D.K., Hollingsworth, R.I. and Dazzo, F.B. (1995) J. Biol. Chem., 270, 20968-20977.

Plazanet, C., Refregier, G., Demont, N., Truchet, G. and Rosenberg, C. (1995) FEMS Microbiol. Lett., 133, 285-291.

Poupot, R., Martinez-Romero, E. and Promé, J.C. (1993) Biochemistry, 32, 10430-10435.

Poupot, R., Martinez-Romero, E., Gautier, N. and Promé, J.C. (1995a) J. Biol. Chem., 270, 6050-6055.

Poupot, R., Martinez-Romero, E., Maillet, F. and Promé, J.C. (1995b) FEBS Letters, 368, 536-540.

Price, N.P.J., Relic, B., Talmont, F., Lewin, A., Promé, D., Pueppke, S.G., Maillet, F., Dénarié, J., Promé, J.C. and Broughton, W.J. (1992) Mol. Microbiol., 6, 3575-3584.

Price, N.P.J., Talmont, F., Wieruszeski, J.M., Promé, D. and Promé, J.C. (1996) Carbohydr. Res., 289, 115-136.

Promé, J.C. (1996) Curr. Opin. Struct. Biol., 6, 671-678.

Quesada-Vincens, D., Fellay, R., Nasim, T., Broughton, W.J. and Jabbouri, S. (1997) Abstract of the 11th International Congress on Nitrogen Fixation p. 106.

Quesada-Vincens, D., Fellay, R., Nasim, T., Viprey, V., Burger, U., Prome, J-C., Broughton, W.J. and Jabbouri, S. (1997) J. Bacteriol., 179, 5087-5093.

Quinto, C., Wijfjes, A.H.M., Bloemberg, G.V., Blok-Tip, L., Lopez-Lara, I.M., Lugtenberg, B.J.J., Thomas-Oates, J.E. and Spaink, H.P. (1997) Proc. Natl. Acad. Sci. USA, 94, 4336-4341.

Relic, B., Perret, X., Estrada-Garcia, M.T., Kopcinska, J., Golinowski, W., Krishnan, H.B., Pueppke, S.G. and Broughton, W.J. (1994) Mol. Microbiol., 13, 171-178.

Ritsema, T., Geiger, O., van Dillewijn, P., Lugtenberg, B.J.J. and Spaink, H.P. (1994) J. Bacteriol., 176, 7740-7743.

Ritsema, T., Wijfjes, A.H.M., Lugtenberg, B.J.J. and Spaink, H.P. (1996) Mol. Gen. Genet., 251, 44-51.

Rivilla, R., Sutton, J.M. and Downie, J.A. (1995) Gene, 161, 27-31.

Roche, P., Debellé, F., Maillet, F., Lerouge, P., Faucher, C., Truchet, G., Dénarié, J. and Promé, J.C. (1991a) Cell, 67, 1131-1143.

Roche, P., Lerouge, P., Ponthus, C. and Promé, J.C. (1991b) J. Biol. Chem., 266, 10933-10940.

Roche, P., Maillet, F., Plazanet, C., Debellé, F., Ferro, M., Truchet, G., Promé, J.C. and Denarie, J. (1996) Proc. Natl. Acad. Sci. USA, 93, 15305-15310.

Röhrig, H., Schmidt, J., Wieneke, U., Kondorosi, E., Barlier, I., Schell, J. and John, M. (1994) Proc. Natl. Acad. Sci. USA, 91, 3122-3126.

Sadowsky, M.J., Olson, E.R., Foster, V.E., Kosslak, R.M. and Verma, D.P.S. (1988) J. Bacteriol., 170, 171-178.

Sadowsky, M.J., Cregan, P.B., Göttfert, M., Sharma, A., Gerhold, D., Rodriguez-Quinones, F., Keyser, H.H., Hennecke, H. and Stacey, G. (1991) Proc. Natl. Acad. Sci. USA, 88, 637-641.

Sadowsky, M.J., Lohrke, S.M. and Orf, J. (1996) Legocki, A.G. (eds) Biological Fixation of Nitrogen for Ecology and Sustainable Agriculture. Proceedings of the 2nd European Nitrogen Fixation Conference and NATO Advanced Research Network. OWN, Polish Academy of Sciences, Poznan, Poland, pp. 41.

Saier, M.H., Tam, R., Reizer, A. and Reizer, J. (1994) Mol. Microbiol., 11, 841-847.

Sanjuan, J., Carlson, R.W., Spaink, H.P., Bhat, R.U., Barbour, M.W., Glushka, J. and Stacey, G. (1992) Proc. Natl. Acad. Sci. USA, 89, 8789-8793.

Saxena, I.M., Brown, R.M., Fevre, M., Geremia, R.A. and Henrissat, B. (1995) J. Bacteriol., 177, 1419-1424.

Scheu, A.K., Economou, A., Hong, G.F., Ghelani, S., Johnston, A.W.B. and Downie, J.A. (1992) Mol. Microbiol., 6, 231-238.

Schlaman, H.R.M., Okker, R.J.H. and Lugtenberg, B.J.J. (1990) J. Bacteriol., 172, 5486-5489.

Schultze, M., Kondorosi, E., Ratet, P., Buire, M. and Kondorosi, A. (1994) International Review of Cytology, 156, 1-75.

Schultze, M., Quiclet-Sire, B., Kondorosi, E., Virelizier, H., Glushka, J.N., Endre, G., Géro, S.D. and Kondorosi, A. (1992) Proc. Natl. Acad. Sci. USA, 89, 192-196.

Schultze, M., Staehelin, C., Röhrig, H., John, M., Schmidt, J., Kondorosi, E., Schell, J. and Kondorosi, A. (1995) Proc. Natl. Acad. Sci. USA, 92, 2706-2709.

Schwedock, J.S. and Long, S.R. (1989) Mol. Plant-Microbe Interact., 2, 181-194.

Schwedock, J.S. and Long, S.R. (1990) Nature, 348, 644-647.

Schwedock, J.S. and Long, S.R. (1992) Genetics, 132, 899-909.

Schwedock, J.S., Liu, C.X., Leyh, T.S. and Long, S.R. (1994) J. Bacteriol., 176, 7055-7064.

Scott, D.B., Young, C.A., Collins-Emerson, J.M., Terzaghi, E.A., Rockman, E.A., Lewis, P.E. and Pankhurst, C.E. (1996) Mol. Plant-Microbe Interact., 9, 187-197.

Scott, K.F. (1986) Nucl. Acids Res., 14, 2905-2919.

Semino, C.E. and Dankert, M.A. (1994) Cell Mol. Biol., 40, 1029-1037.

Shearman, C.A., Rossen, L., Johnston, A.W.B. and Downie, J.A. (1986) EMBO J., 5, 647-652.

Spaink, H.P. (1995) Annu. Rev. Phytopathol., 33, 345-368.

Spaink, H.P. (1996) Crit. Revs. Plant. Sci., 15, 559-582.

Spaink, H.P. and Lugtenberg, B.J.J. (1994) Plant Mol. Biol., 26, 1413-1422.

Spaink, H.P., Weinman, J., Djordjevic, M.A., Wijffelman, C.A., Okker, R.J.H. and Lugtenberg, B.J.J. (1989) EMBO J., 8, 2811-2818.

Spaink, H.P., Sheeley, D.M., van Brussel, A.A.N., Glushka, J., York, W.S., Tak, T., Geiger, O., Kennedy, E.P., Reinhold, V.N. and Lugtenberg, B.J.J. (1991) Nature, 354, 125-130.

Spaink, H.P., Wijfjes, A.H.M., van der Drift, K.M.G.M., Haverkamp, J., Thomas-Oates, J.E. and Lugtenberg, B.J.J. (1994) Mol. Microbiol., 13, 821-831.

Spaink, H.P., Bloemberg, G.V., van Brussel, A.A.N., Lugtenberg, B.J.J., van der Drift, K.M.G.M., Haverkamp, J. and Thomas-Oates, J.E. (1995a) Mol. Plant-Microbe Interact., 8, 155-164.

Spaink, H.P., Wijfjes, A.H.M. and Lugtenberg, B.J.J. (1995b) J. Bacteriol., 177, 6276-6281.

Stacey, G. (1995) FEMS Microbiol. Lett., 127, 1-9.

Stacey, G., Luka, S., Sanjuan, J., Banfalvi, Z., Nieuwkoop, A.J., Chun, J.Y., Forsberg, L.S. and Carlson, R. (1994) J. Bacteriol., 176, 620-633.

Staehelin, C., Granado, J., Muller, J., Wiemken, A., Mellor, R.B., Felix, G., Regenass, M., Broughton, W.J. and Boller, T. (1994a) Proc. Natl. Acad. Sci. USA, 91, 2196-2200.

Staehelin, C., Schultze, M., Kondorosi, E., Mellor, R.B., Boller, T. and Kondorosi, A. (1994) Plant J., 5, 319-330.

Staehelin, C., Schultze, M., Kondorosi, E. and Kondorosi, A. (1995) Plant Physiol., 108, 1607-1614.

Surin, B.P. and Downie, J.A. (1988) Mol. Microbiol., 2, 173-183.

Surin, B.P. and Downie, J.A. (1989) Plant Mol. Biol., 12, 19-29.

Surin, B.P., Watson, J.M., Hamilton, W.D.O., Economou, A. and Downie, J.A. (1990) Mol. Microbiol., 4, 245-252.

Sutton, J.M., Lea, E.J.A. and Downie, J.A. (1994) Proc. Natl. Acad. Sci. USA, 91, 9990-9994.

van der Drift, K.M.G.M., Spaink, H.P., Bloemberg, G.V., van Brussel, A.A.N., Lugtenberg, B.J.J., Haverkamp, J. and Thomas-Oates, J.E. (1996) J. Biol. Chem., 271, 22563-22569.

van Gijsegem, F., Gough, C., Zischek, C., Niqueux, E., Arlat, M., Genin, S., Barberis, P., German, S., Castello, P. and Boucher, C. (1995) Mol. Microbiol., 15, 1095-1114.

van Rhijn, P., Luyten, E., Vlassak, K. and Vanderleyden, J. (1996) Mol. Plant-Microbe Interact., 9, 74-77.

van Rhijn, P. and Vanderleyden, J. (1995) Microbiol. Rev, 59, 124-142.

Vazquez, M., Santana, O. and Quinto, C. (1993) Mol. Microbiol., 8, 369-377.

Waelkens, F., Voets, T., Vlassak, K., Vanderleyden, J. and van Rhijn, P. (1995) Mol. Plant-Microbe Interact., 8, 147-154.

Yang, G.P., Debellé, F., Ferro, M., Maillet, F., Schiltz, O., Vialas, C., Savagnac, A., Promé, J.C. and Dénarié, J. (1998) Kondorosi, A. and Newton, W.E. (Eds.) Biological Nitrogen Fixation for the 21st Century pp. 185-188.

Chapter 21

Responses of the Plant to Nod Factors

Az-Eddine Hadri and Ton Bisseling

I. Introduction

The rhizobial signal molecules that are involved in and in most cases are sufficient for the induction of the different steps of nodule formation are the Nod factors of which the structure and biosynthesis are described in more detail in chapter 20. Briefly, Nod factors are molecules that are composed of a chitin backbone attached to a fatty acyl chain. Depending on the rhizobial species and within the same species, different susbtitutions can occur on the chitin backbone. The length and the saturation level of the fatty acyl chain is also subject to variations as well.

The aim of this chapter is to summarize the different responses induced by Nod factors in the host plant, and to describe the structure-function relationship and possible perception mechanisms of Nod factors. Furthermore, we included in the last part of this chapter a brief description of some of the defense-related responses induced during the interaction between rhizobia and legumes.

II. Nod Factor-Induced Responses

Nod factors induce responses in three different root tissues of the host; epidermis, cortex, and pericycle (Figure 1). The various responses that are induced are described in the following paragraphs (see Table 1). In addition, the current knowledge on the activity of Nod factors on protoplast or suspension culture cells of legume as well as non-legume cells is discussed (see Table 1).

II.A. EPIDERMIS

When rhizobia colonize legume roots, they attach to the tips of root hairs where they induce deformation and curling. Curled root hairs form the so-called shepherd's crooks, in which rhizobia become entrapped in a small confinement formed by the curls. There, rhizobia enter the root by local hydrolysis of the root hair cell wall and invagination of the plasma membrane (Turgeon and Bauer 1985, van Spronsen *et al.*, 1994). Deposition of new plant

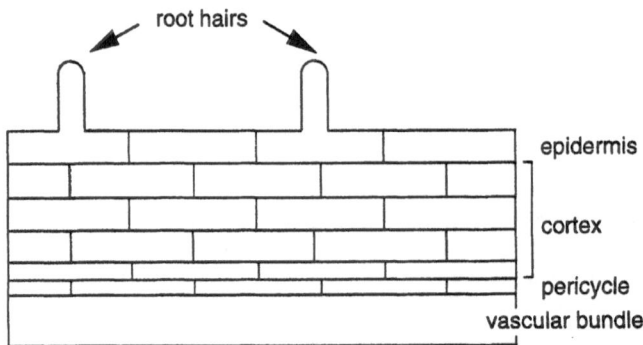

Figure 1. Schematic drawing of the different root tissues that respond to Nod factors. Nod factors can induce different responses in the root, namely in the epidermis, cortex and pericycle. In Table 1, a detailed description of these responses is presented.

cell wall material around the invaginating membrane results in the formation of a tubular structure called the infection thread. This is used by the bacteria to enter the plant root (Kijne *et al.*, 1992).

The molecular mechanisms underlying root hair curling and infection thread formation are not fully understood. However, our knowledge on molecular changes that are induced by Nod factors is rapidly increasing. In the following paragraphs the knowledge on Nod factor mediated changes is described.

II.A.1. Root Hair Deformation

Root hair deformation and curling are the first morphological changes induced by rhizobia and they are preceded by a rapid change of cytoplasmic streaming. Purified Nod factors, at concentrations as low as 10^{-12} M, are sufficient to induce root hair deformation (Lerouge *et al.*,1990; Spaink *et al.*, 1991; Price *et al.*, 1992; Sanjuan *et al.*, 1992; Schultze *et al.*, 1992; Mergeart *et al.*, 1993; Heidstra *et al.*, 1994) but in most cases curling is only observed when the bacteria are present (Relic *et al.*, 1993).

Root hair deformation is induced only in root hairs in a specific stage of development (Heidstra *et al.*, 1994, 1997; de Ruijter *et al.*, 1997). In the development of root hairs three successive stages are distinguished, represented by root hairs located along the root system in three adjacent zones (Figure 2). In zone I, root hairs are growing and have, like other tip

growing cells (review Sievers and Schnepf, 1981), a polarly organized cytoplasm. Since the apical region of zone I root hairs lacks big organelles, including vacuoles, it is called a clear zone (Figure 2B). Zone I root hairs show a reverse fountain type like cytoplasmic streaming that does not include the cytoplasm in the tip, and the direction of streaming is reversed below the clear zone. Only in root hairs of zone I, a Ca^{2+} gradient and a spectrin like-antigen is present at their tips. The spectrin-like antigen is recognised by a heterologous antiserum raised against chicken spectrin. This chicken antiserum recognises an antigen present at the tip of growing root hairs, pollen tubes and hyphae. Although the nature of the recognised antigen is unknown, its strict correlation with tip growth makes it a useful molecular marker.

In zone III, root hairs are full grown and are characterized by the presence of a large vacuole, filling most of the hair, surrounded by a thin layer of cortical cytoplasm (Figure 2B). There, the cytoplasmic streaming is of a rotation type. In zone II, the hairs are terminating growth and the clear zone is no longer visible, cytoplasmic streaming reaches into the tip but is still of the reverse fountain type. About 80% of root hairs in zone II deform upon application of Nod factors while the hairs of the other two zones do not respond (Figure 2C). Incubation for 5-10 min in the presence of 10^{-12} M NodRlv factor is sufficient to induce the deformation process (Heidstra *et al.*, 1994). This deformation starts with a swelling of the root hair tip, which is apparent within one hour, and followed one hour later by the formation of a new cylindrical outgrowth initiated at the swelling (Figure 2C) (Heidstra *et al.*, 1994). The new outgrowth has the characteristics of a growing root hair since a clear zone is found at the tip (Figure 2C), a cytosolic calcium gradient is built up, and the spectrin-like antigen accumulates. Therefore, it has been concluded that root hair deformation involves the reinitiation of tip growth in zone II hairs which is in agreement with a mathematical model proposed by Van Batenburg *et al.*, 1986. It was also shown that root hair deformation can be inhibited by actinomycin D and cyclohexamide suggesting that both DNA dependent RNA and protein synthesis are prerequisites for root hair deformation (Vijn *et al.*, 1995a).

Root hair deformation starts with a swelling of the tip. Since this swelling requires protein/RNA synthesis and a Ca^{2+} gradient is formed at its periphery and the spectrin-like antigens accumulate, it is probable that Nod factor-induced swelling of root hair involves growth. When pollen tube growth is blocked with a Ca^{2+} chelating agent, these tubes reinitiate growth after some time. Strikingly, these tubes first swell at their tip after which polar growth is established (Miller *et al.*, 1992).

Since ethylene promotes the formation of root hairs, which involves tip growth, in the root epidermis of *Arabidopsis* (Tanimoto *et al.*, 1995) and legumes (Heidstra *et al.*, 1997), it was investigated whether it would also be involved in the Nod factor-induced tip growth. It was shown that ethylene is not required for Nod factor-induced root hair tip growth. In addition, ethylene alone is neither sufficient to induce, nor to arrest root hair deformation. Ethylene is not only involved in the induction of root hair formation, but is also essential for maintenance of tip growth of normal root hairs. When ethylene blockers (*e.g.* Ag^+ and AVG) are used, growing hairs of zone I obtain a cytoarchitecture typical of zone II hairs (terminating growth). Such zone I hairs have now obtained the ability to deform upon treatment with Nod factor. This observation stresses the strict correlation of the growth phase of a hair and its ability to deform (Heidstra *et al.*, 1997).

In some cases, the application of Nod factors most likely results in the accumulation of ethylene in the host plant. Especially hosts of the genus *Vicia* appear to be sensitive to or produce higher amounts of ethylene. In *Vicia sativa* Nod factors can induce the formation of thick short roots (ethylene-related cell swelling, Table 1), a response that is also induced by

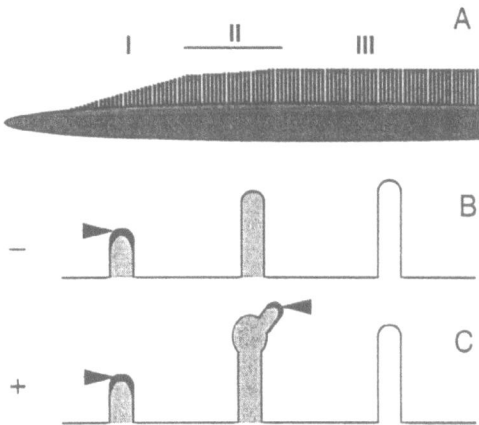

Figure 2. Relationship between developmental stage of root hairs and their susceptibility to Nod factors.
(A) Root hairs along the root representing 3 successive stages of root hairs development, zone I, zone II, and zone III. (B) Representative root hairs that belong to the different zones above-described in (A). In zone I, root hairs are growing and have a polarly organized cytoplasm. The apical region of zone I root hairs lacks big organelles and is therefore called a clear zone (black arrowhead). Cytoplasmic streaming is of reverse fountain type (gray). In zone II, root hairs lack the clear zone and cytoplasmic streaming reaches into the tip but is still of the reverse fountain type (gray). In zone III, root hairs are fully-grown and cytoplasmic streaming is of rotation type. (-) Indicates before nod factors application. (C) After Nod factors application (+), only root hairs in zone II deform. Deformation starts with a swelling of the root hair tip, followed by the formation of a new outgrowth initiated at the swelling. Cytoplasmic streaming is of reverse fountain type (gray), and the clear zone is present at the tip (arrowhead).

Effect and tissue type	Plant species	References
	Root epidermis	
Formation of new root hairs	*Vicia*	van Brussel *et al.*, 1992
	Sesbania	Mergaert *et al.*, 1993
	Medicago	Roche *et al.*, 1991b
	Lotus	Lopez-Lara *et al.*, 1995a
	Phaseolus	Lopez-Lara *et al.*, 1995b
Deformation, branching, swelling	Many legumes	See Schulze *et al.*, 1994
Formation of Shepherd's crooks	Macroptilium	Relic *et al.*, 1993; Relic *et al.*, 1994
Stimulation of cytoplasmic streaming	*Vicia*	Heidstra *et al.*, 1994
Alkalinization	*Medicago*	Felle *et al.*, 1996
Depolarization of membrane potential	*Medicago*	Ehrhardt *et al.*, 1992; Felle *et al.*, 1995; Kurkdjian, 1995
	Vicia	Kurkdjian, 1995
Modulation of proton and calcium ion fluxes	*Medicago*	Allen *et al.*, 1994
Calcium spiking	*Medicago*	Ehrhardt *et al.*, 1996
Induction of early nodulins	*Pisum, Medicago*	Horvath, *et al.*, 1993; Pichon *et al*, 1992; Cook *et al.*, 1995
	Root cortex	
Ethylene-related cell swelling	*Vicia*	van Spronsen *et al.*, 1995
Polar tip growth of outer cortical cells	*Vicia*	van Brussel *et al.*, 1992
Formation of preinfection threads	*Vicia*	van Brussel *et al.*, 1992
Formation of nodule primordia (inner cortex)	*Vicia*	Spaink *et al.*, 1991
	Medicago	Truchet *et al.*, 1991
	Trifolium	Bloemberg *et al.*, 1995a
	Sesbania	Mergaert *et al.*, 1993
Formation of nodule primordia (outer cortex)	*Lotus*	Lopez-Lara *et al.*, 1995a
	Phaseolus, Acacia	Lopez-Lara *et al.*, 1995b
	Glycine	Stokkermans and Peters, 1994
Formation of complete nodules[a]	*Medicago*	Truchet *et al.*, 1991
Local induction of cell cycle genes	*Vicia*	Yang *et al.*, 1994
Induction of early nodulins	*Pisum, Vicia*	Horvath *et al.*, 1993
	Medicago	Vijn *et al.*, 1995c; Bauer *et al.* 1994; Journet *et al.*, 1994
	Root pericycle	
Induction of enod40	*Vicia*	Kouchi *et al.*, 1993; Yang *et al.*, 1993; Matvienko *et al.*, 1994; Vijn *et al.*, 1995c
	Whole root[b]	
Production of additional flavonoids	*Vicia*	Spaink *et al.*, 1991
Induction of a Nod factor-hydrolase	*Medicago*	Staehelin *et al.*, 1995
	Suspension cell cultures	
Induction of flavonoid synthesis genes	*Medicago*	Savoure *et al.*, 1994
Induction of cell cycle genes	*Medicago*	Savoure *et al.*, 1994
Complementation of embryogenesis mutant	*Daucus*	de Jong *et al.*, 1993
Alkalinization	*Lycopersicum*	Staehelin *et al.*, 1994
	Protoplast cultures	
Induction of mitosis	*Nicotiana*	Roehrig *et al.*, 1995
Induction of an auxin-responsive gene	*Nicotiana*	Roehrig *et al.*, 1995

Table 1. Nod factors-induced responses in plants. [a]Complete nodules are distinguished from other externally visible nodule structures resulting from the formation of nodule primordia by the formation of a vascular system. Nodule structures that do not contain complete vascular bundles or that are not described as such are categorized as nodule primordia. [b]The particular effect has not been attributed to a particular cell type.

ethylene (Zaat *et al.*, 1989). In several cases (Table 1), it has been shown that Nod factors can induce the formation of root hairs. It has not been shown whether this response involves ethylene.

II.A.2. Electrophysiological Changes

Nod factors induce several rapid changes in the epidermis of the root like alkalinization and membrane depolarisation (Ehrhardt *et al.*, 1992, Kurkdjian, 1995, Felle et *al.*, 1995, 1996,) which are followed by calcium spiking (Ehrhardt *et al.*, 1996). These responses have especially been studied in alfalfa and are elicited by Nod factors at nanomolar concentration. Probably, these responses can also be induced in other legumes by the proper Nod factor, like the induction of calcium spiking in vetch by NodRlv (Ehrhardt *et al.*, 1996). However, the *R. meliloti* NodRm-IV(C16:2, S) was not able to trigger such responses in a non-legume like tomato (Ehrhardt *et al.*, 1992, Ehrhardt *et al.*, 1996).

R. meliloti Nod factors elicit an increase of the intracellular pH of alfalfa root hair tips of 0.2-0.3 pH unit with a delay time of about 15 sec after Nod factor addition (Felle *et al.*, 1995) and membrane depolarisation after 15 to 60 sec (Ehrhardt *et al.*, 1992, Felle *et al.*, 1995). The intensity of these two responses was shown to be Nod factor concentration-dependent (Felle *et al.*, 1995, 1996; Kurkdjian, 1995). Furthermore, measurements of both responses in a single cell showed that alkalinization persists as long as Nod factor is present in the medium while depolarization occurs transiently lasting for 15 to 30 min (Ehrhardt *et al.*, 1992, Kurkdjian, 1995, Felle *et al.*, 1995, 1996).

Whether membrane depolarization is induced in a root epidermal cell depends on its developmental stage (Kurkdjian, 1995). Depolarization is induced in growing root hairs as well as root hairs just bulging out from the epidermal cells (Ehrhardt *et al.*, 1992, kurkdjian, 1995, Felle *et al.*, 1995). The epidermal cells that are not yet differentiated into root hairs as well as the old mature root hairs are not able to respond (Kurkdjian, 1995). However, epidermal cells without root hairs, in the zone containing root hairs responded to Nod factors (Felle *et al.*, 1995). It is unclear in which epidermal cells and at which developmental stage calcium spiking and alkalinization can be induced.

Studies on changes in cytoplasmic calcium have shown that *R. meliloti* Nod factor induces regular oscillations in cytoplasmic free calcium in alfalfa root hairs especially in the region of the cytoplasm proximal to the cell nucleus (Ehrhardt *et al.*, 1996). This calcium spiking occurs after a delay period of about 9 min and persists for an average time period of about 60 min.

Although the kinetics of these different electrophysiological changes are not studied simultaneously in a single cell, it seems probable that alkalinization and membrane depolarisation precedes the calcium spiking. However, it is unknown whether alkalinization, depolarization and calcium spiking responses are causally related.

II.A.3. Gene Expression

Several genes that are induced by Nod factors in the root epidermis have been cloned and characterized. Examples are *ENOD5*, *ENOD12*, *Mtrip1* and *VsLb1* which are induced shortly (few hours) after addition of Nod factors to the root at nanomolar concentration (Horvath *et al.*, 1993, Journet *et al.*, 1994, Cook *et al.*, 1995, Heidstra *et al.*, 1997).

ENOD5 and *ENOD12* are proline-rich proteins that are probably located in the cell wall. Whether they are located in the infection thread wall is unknown (Scheres *et al.*, 1990a, 1990b).

Mtrip1 is a peroxidase with an unknown function (Cook *et al.*, 1995). In vetch *VsLb1* expression precedes the expression of *ENOD5* and *ENOD12* (Heidstra *et al.*, 1997). This gene encodes a leghemoglobin and it is also expressed, but at a markedly higher level in the nodule. Which function *VsLb1* has in epidermal cells is unclear.

The use of RNA display has recently resulted in the cloning of several new nodulin genes. It can be expected that among these, new genes will be found that can serve as valuable markers (Heidstra *et al.*, 1997). The expression of *ENOD12* and *Mtrip1* in the epidermis of alfalfa has been studied in most detail. These genes are activated in all epidermal cells of a rather broad zone of the root (Pichon *et al.*, 1992, Cook *et al.*, 1995). The expression starts in epidermal cells just above the root tip and extends to the area where root hairs are formed. Taken together these data and the studies on membrane depolarization clearly show that epidermal cells with and without root hairs are able to respond to Nod factors. Furthermore the gene expression studies show that even epidermal cells at a developmental stage preceeding root hair formation are susceptible to Nod factors.

The induction of the early nodulin genes like *ENOD12* requires *de novo* protein synthesis (Vijn *et al.*, 1995a). Hence the induction of such genes is a rather late step in Nod factor-activated epidermal responses and it is probable that the signal transduction cascade activated by Nod factors leading to the induction of such genes is rather complex. For this reason, genes induced very early after addition of Nod factors will be more useful as marker genes for studying the early signalling events elicited by Nod factors in the root epidermis.

II.B. CORTICAL CELLS

In addition to the responses induced in the root epidermis, Nod factors induce various responses in the root cortex (Table 1). These responses are thick short root (ethylene-related cell swelling) (van Spronsen et al., 1995) and cortical cell activation which result in starch granules accumulation (Ardourel et al., 1994). Because these latter responses do not appear to be essential for infection or primordium formation, they will not be described in further detail. The aspects that will be described are preinfection thread formation and cortical cell divisions.

II.B.1. Preinfection Threads Formation
In legumes, such as Medicago and pea, that form indeterminate nodules, nodule primordia are formed in the inner cortex. Hence, in the formation of this nodule type, the infection threads must traverse the outer cortex before they reach the nodule primordium. Prior to infection thread penetration, the outer cortical cells undergo morphological changes. The nuclei move to the center of the cells, and the microtubules and the cytoplasm rearrange to form a radially oriented conical structure, the cytoplasmic bridge, that resembles a phragmoplast (Kijne, 1992). The infection threads traverse the cortical cells through the radially aligned cytoplasmic bridges, which are therefore called preinfection threads (Van Brussel et al., 1992).

Although the preinfection thread-forming outer cortical cells never divide, the induced morphological changes resemble those in cells that are entering the cell cycle. In situ hybridization experiments (Yang et al., 1994) showed that narrow rows of outer cortical cells express the S phase-specific histone H4 gene (Figure 3A). However, a mitotic cyclin gene that is specifically expressed during the G2 to M phase transition, is not activated. Hence, the cells that form the preinfection thread reenter the cell cycle and most likely become arrested in the G2 phase, whereas the inner cortical

cells progress all the way through the cell cycle and form the primordia (Figure 3B). This shows that part of the infection process is derived from a general process, namely cell division. In some way, this process is modified and now exploited for a completely different purpose, the infection process.

Purified Nod factors induce preinfection thread formation, but infection threads are not formed unless rhizobia are present (van Brussel et al., 1992). These preinfection threads are formed in the outer cortex, a process that is preceded by cell polarization (van Brussel et al., 1992). It was also shown that these outer cortical cells can form tubular outgrowth as a result of polar tip growth preferentially opposite the cortical cells where nodule primordia are formed (van Brussel et al., 1992). Therefore, it was suggested that preinfection thread formation is related to the elicitation of polar tip growth (van Brussel et al., 1992, van Spronsen et al., 1994).

II.B.2. Cortical Cell Divisions
During mitotic reactivation of root cortical cells by rhizobia, genes that control the progression through the cell cycle, such as cdc2 and mitotic cyclins, are induced (Yang et al., 1994). In addition, several nodulin genes are expressed, allowing a distinction to be made between nodule primordia and root or shoot meristems. Examples of such nodulin genes are ENOD12 (Scheres et al., 1990a), Gm93 (Kouchi and Hata, 1993), ENOD40 (Kouchi and Hata, 1993; Yang et al., 1993; Asad et al., 1994; Matvienko et al., 1994), and MtPRP4 (Wilson et al., 1994). These genes are activated in all cells of the primordia. Furthermore, ENOD40 is also induced in the region of the pericycle opposite to the dividing cortical cells (Kouchi and Hata, 1993; Yang et al., 1993; Asad et al., 1994). Another early nodulin gene, ENOD5 (Scheres et al., 1990b), is transcribed only in primordial cells that contain rhizobia.

Nod factors are sufficient for mitotic reactivation of the cortical cells (Spaink et al., 1991; Truchet et al., 1991; Relic et al., 1993). The early nodulins ENOD12 and ENOD40 are activated in such primordia (Vijn et al., 1993). In some host plants, purified Nod factors even induce nodule formation (Truchet et al., 1991; Mergaert et al., 1993; Stokkermans and Peters, 1994).

Interestingly, only certain cortical cells are susceptible to Nod factors. In (sub)-tropical legumes, such as soybean and bean, the outer cortical cells are mitotically activated. In temperate legumes, such as pea, vetch, and alfalfa, the inner cortical cells, and

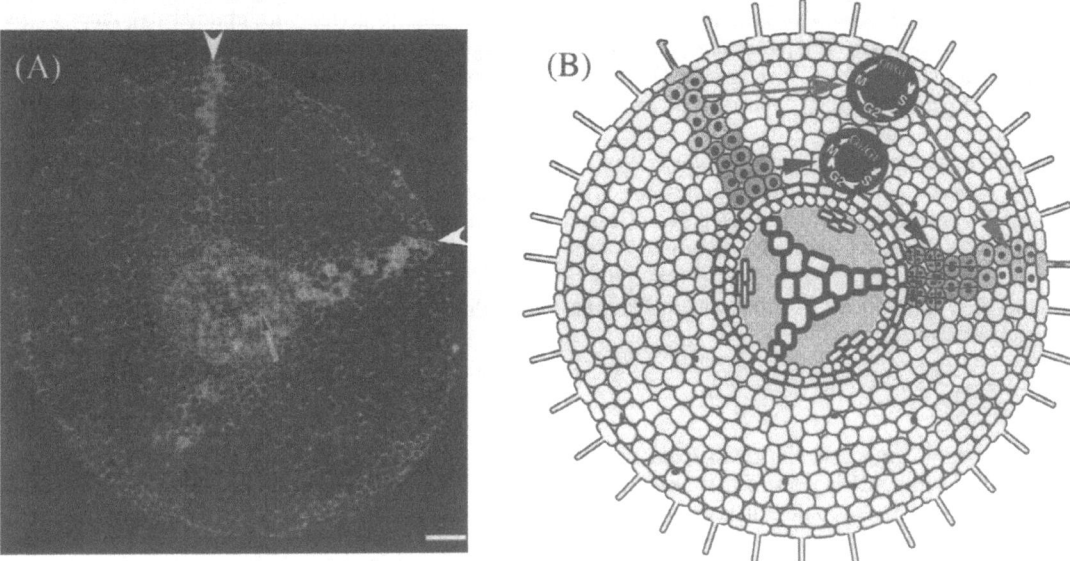

Figure 3. Activation of the cortex during the induction of an indeterminate nodule.
(A) Dark-field micrograph showing a pea root cross-section that originates from a pea root one day after inoculation with *R. l.* bv. *viciae*. This section was hybridized with a histone *H4* gene probe. In front of the infection sites indicated by arrowheads, *H4* transcripts are localized in narrow rows of cortical cells. Hybridization signal is represented by the silver grains. Infection sites are opposite to the protoxylem poles (arrow). Bar = 50 λm. (B) Schematic drawing of a pea root cross-section showing the reactivation of cortical cells after application of Nod factors or inoculation with rhizobia. The outer cortical cells shown in light grey, reenter the cell cycle, proceeding from the G0/G1 phase to the S phase, and become arrested in the G2, as indicated in the cell cycle in the light grey circle. However, inner cortical cells as shown in dark grey, progress all the way through the cell cycle, as indicated by the cell cycle in the dark grey circle, dividing and forming the nodule primordium. The cells that are activated are opposite the protoxylem poles of the root as shown in thick black.

especially those located opposite protoxylem poles, divide (Kijne, 1992). The mechanism that controls the susceptibility of cortical cells is unknown. It has been postulated for decades that the susceptibility of cortical cells is conferred by an arrest in the G2 phase (Wipf and Cooper, 1938; Verma, 1992). But the use of cell phase specific genes as probes in *in situ* hybridization experiments shows that this is not the case (Yang *et al.*, 1994). The pattern of responding cortical cells provides some hints about a possible mechanism. Figure 3A shows that only narrow rows of cortical cells are activated to express the histone *H4* gene by rhizobia. At this time the infection thread tips - the site where Nod factors are released - are still in the epidermis, indicating that Nod factors act at a distance. These rows of susceptible cells are located opposite protoxylem poles. About two decades ago, Libbenga *et al.*, (1973) found that an alcohol extract of the stele could induce cell divisions in explants of the pea root cortex in the presence of auxin and cytokinin. A substance responsible for this activity, the so-called

stele factor, has since been isolated and is thought to be released from the protoxylem poles. One active compound was shown to be uridine (Smit *et al.*, 1993). Such stele factors might confer susceptibility to the cortical cells located opposite the protoxylem poles (Smit *et al.*, 1993). Recently, it was shown that ethylene provides positional information for nodule primordium formation (Heidstra *et al.*, 1997). Ethylene is most likely formed in the pericycle opposite the phloem poles since the gene encoding the last step in the biosynthesis of ethylene, ACC oxidase, is specifically expressed in this part of the pericycle. Furthermore, when ethylene synthesis was inhibited, a higher proportion of nodules were found opposite a phloem pole. These results have suggested that ethylene may suppress cortical cell division, in the area opposite the phloem pole (Heidstra *et al.*, 1997).

Which Nod factors can induce mitotic reactivation of cortical cells depends on the host plant. Rhizobia that induce cell divisions in the inner cortical layers, *e.g. R. leguminosarum* bv. *viciae* and *R. meliloti*, produce

Nod factors with highly unsaturated fatty acyl groups, whereas rhizobia that mitotically reactivate outer cortical cells, *e.g. Bradyrhizobium japonicum*, generally contain a C18:1 acyl group. The highly unsaturated fatty acyl group appears to be important in inducing cell divisions in the inner cortex. For example, only those *R. leguminosarum* bv. *viciae* factors containing a C18:4 acyl group cause the formation of nodule primordia in vetch (Van Brussel *et al.*, 1992). Whether the highly unsaturated fatty acyl moiety is recognized by a specific receptor or whether it is required for *e.g.* transport to the inner layers, is unknown.

II.C. PERICYCLE

When plants are inoculated with rhizobia or Nod factors a rapid response is induced in the pericycle of the root since the *ENOD40* gene is activated within 3 hours in this tissue. This is rather surprising since in contrast to lateral root formation, nodules do not originate from this root tissue. Furthermore, Nod factors do not induce cytological changes in the pericycle cells in the early stages of nodule formation. What could be the function of the *ENOD40* gene during these early stages of nodule development?

ENOD40 has been cloned and identified in legumes as well as in non-legumes and it encodes a peptide of 10-13 amino acids (Figure 4) (Kouchi and Hata, 1993; Yang *et al.*, 1993; Crespi *et al.*, 1994; Matvienko *et al.*, 1994, and van de Sande *et al.*, 1996). Interestingly the studied *ENOD40s* also contain a conserved area in the 3' untranslated region, therefore it can not be excluded that *ENOD40* is active at the RNA level although experimental data supporting this hypothesis are still lacking (Crespi *et al.*, 1994, Matvienko *et al.*, 1994, van de Sande *et al.*, 1996).

The function of the ENOD40 peptide has been studied in the non-legume tobacco as well as in *Medicago*. Overexpression of soybean *ENOD40* in tobacco resulted in the formation of additional side shoots (van de Sande *et al.*, 1996), whereas ectopic expression in *Medicago* disturbed embryogenesis and regeneration (Crespi *et al.*, 1994). In addition tobacco protoplast division has been used to study the activity of the ENOD40 peptide. Wildtype protoplasts divide with an optimal frequency of about 50 % at 0.9 μM kinetin and 5.5 μM NAA. At higher NAA concentrations the efficiency of division markedly decreases. Protoplasts treated with the

ENOD40 peptide or expressing an *ENOD40* gene were able to divide with optimal frequency at such high NAA concentrations, thus the ENOD40 peptide confered tolerance of high auxin concentration to the protoplasts (van de Sande *et al.*, 1996).

Because *ENOD40* is activated markedly before cortical cells divide, it has been postulated that this gene might be involved in the induction of cortical cell division. Recently, Crespi *et al.*, (Knoxville)

PsENOD40	MKFLCWQKSIHGS
MsENOD40	MKLLCWQKSIHGS
MtENOD40	MKLLCWEKSIHGS
GmENOD40-1	MEL.CWQTSIHGS
GmENOD40-2	MEL.CWLTTIHGS
NtENOD40	M...QWDEAIHGS

Figure 4. Amino acid sequences of the ENOD40 encoded peptides from pea (Matvienko *et al.*, 1994), alfalfa (Asad *et al.*, 1994 and Crespi *et al.*, 1994), soybean (Yang *et al.*, 1993) and tobacco plants (van de Sande *et al.*, 1996).

have shown that overexpression of *MsENOD40* in *Medicago* is indeed sufficient to trigger cortical cell divisions. A chitin pentamer, which is unable to induce root hair deformation or cortical cell division, induces a transient accumulation of *ENOD40* mRNA (Minami *et al.*, 1996).

To unravel the mechanism, by which Nod factors elicit cortical cell divisions, studies with compounds that can mimic their mitogenic activity have been performed. Two lines of evidence strongly suggest that Nod factors cause a change in the auxin/cytokinin balance. Both cytokinin (Cooper and Long, 1994) and compounds (NPA, TIBA) that block polar auxin transport (Hirsch *et al.*, 1989) induce the formation of nodule-like structures in which early nodulin genes are expressed.

Because *ENOD40* expression is sufficient to cause a phytohormone-like effect in tobacco and it is sufficient to trigger root cortical cell division, it can be hypothesized that *ENOD40* expression can cause a change of the cytokinin/auxin ratio of the cortical cells.

The tobacco protoplast studies also revealed that exogenously applied peptide is recognised by protoplasts (van de Sande *et al.*, 1996), which strongly suggests that cells contain a perception or uptake mechanism at their surface. This makes the ENOD40 peptide an interesting candidate to be involved in cell-cell signalling. Since *ENOD40* transcription in the root pericycle precedes cortical cell divisions it is an attractive hypothesis that after production of the ENOD40 peptide in the pericycle it can diffuse to the inner cortex.

II.D. NOD FACTOR-INDUCED RESPONSES IN NON-LEGUMINOUS PLANTS

Rhizobial Nod factors are special molecules in that molecules with a similar sructure have not been identified in other organisms. This suggested that Nod factors would only be involved in the interaction of rhizobia and plants. However several research lines have shown that Nod factors are recognised by plants that are unable to establish an interaction with rhizobia. For example, a mutated carrot cell line that has lost the ability to form somatic embryos, can be rescued by Nod factors (De Jong et al., 1993). It was also shown that the expression of rhizobial nod genes, nodA and/or nodB, in tobacco affects the development of these plants (Schmidt et al., 1993). Moreover, purified and synthetic Nod factors are able to induce cell division of tobacco protoplasts, activate an auxin-responsive promoters and induce the expression of axi1, a gene implicated in auxin action (Röhrig et al., 1995). In tomato cell culture, Nod factors stimulate a rapid, transient alkalinization of the medium (Staehelin et al., 1994).

Such studies clearly show that non-legumes have the ability to recognise and respond to Nod factors. This could mean that Nod factors resemble endogenous plant signal molecules although endogenous plant Nod factors-like molecules have not been identified yet.

III. Structure-Function Relationship And Perception Mechanisms Of Nod Factors

As we have described above, Nod factors are able to trigger different responses in the epidermis, cortex and pericycle. Based on these responses, semi-quantitative assays have been developed. The use of these assays provided the means to address questions related to perception mechanisms of the Nod factors. The questions that will be discussed in more detail below are: 1/Do intact or processed Nod factor induce these responses?. 2/Is more than one Nod factor receptor involved?. 3/Do cortical cells directly interact with Nod factors or are second messengers produced in the epidermis?

III.A. NOD FACTOR PROCESSING

When Nod factors are added to roots, they are degraded and Nod factor derived molecules containing only two or three sugar residues are found in the growth medium and on the root. These molecules are probably generated by chitinases secreted by the plant. The fact that these degraded Nod factors are at least 1000-fold less active in the root hair deformation assay than Nod factors with four or five sugars (Heidstra et al., 1994; Staehelin et al., 1994a) indicates that these molecules are not involved in inducing root hair deformation.

Studies on degradation of Nod factors revealed remarkable differences in the speed by which different Nod factors are degraded. The degradation rate of NodRlv-V(C18:4) and NodRlv-IV(C18:4) showed that the tetramer is more stable than the pentamer (Heidstra et al., 1994). A purified endochitinase from alfalfa was able to hydrolase R. meliloti Nod factor NodRm-V, NodRm-V(S), NodRm-IV but not Nod-RmIV(S) which was stable (Staehelin et al., 1994). In vivo studies showed that Nod-RmIV(S) Nod factor was also degraded but at a slower rate than the non-sulfated NodRm factor (Staehelin et al., 1994). Thus substitutions at the reducing terminal sugar might provide some protection against degradation by chitinases or enzymes specifically degrading Nod factors, i.e. Nod factors hydrolases (Staehelin et al., 1994). However, responses like alkalinization (Felle et al., 1996), membrane potential depolarization (Felle et al., 1995) and root hair deformation (Heidstra et al., 1994) require only a short period of contact with Nod factors. Hence the degradation of Nod factors does not seem to play a role in the induction of these early responses. Therefore it is most probable that these responses are induced by intact Nod factors. Whether this is the case for steps induced at a later stage is unknown.

Using a ballistic microtargetting approach, Spaink et al., (1994) have shown that chitin fragments are sufficient to induce cell division. Furthermore, chitin fragments are also able to induce ENOD40 expression (Minami et al., 1996). This could mean that the chitin backbone is the part of the Nod factor active for eliciting cell division.

III.B. NOD FACTOR PERCEPTION IN THE EPIDERMIS

Nod factors are able to induce various responses in the epidermis at nanomolar concentrations. This suggests, although experimental evidence is lacking, that high affinity receptors are involved in Nod factor perception. The various responses that are induced by Nod factors in the epidermis have been used as assays to study Nod factor structure-function relationships. Such studies have been performed with several legumes, but only with *Medicago sativa* a rather complete picture has emerged. Therefore this

paragraph is focussed on the studies with this particular legume.

R. meliloti, the microsymbiont of *Medicago*, produces predominantly a Nod factor with a tetrameric N-acetyl-glucosamine backbone, a C16:2 acyl chain, an O-acetate substitution at the non-reducing sugar moiety and an O-sulphate substitution at the reducing sugar moiety. Therefore, it is named NodRm-IV(Ac, C16:2, S). By using *R. meliloti* mutants and NodRm factors the effect of the substitutions and the structure of the acyl chain on the induction of the various epidermal responses have been analysed in alfalfa (Roche *et al.*, 1991,

Responses	Number of glucosamine	O-substitutions at reducing end	N-acyl chain		O-substitution at non-reducing ends, O-acetate	Ref.
			Presence Sulphation	Structure		
Group1						
Alkalinization	+[a]	-	+	-	-	A
Group2						
Membrane potential depolarization	+	+	-	-	-	B
Root hair deformation	+	+	+	-	-	C
ENOD12 induction	nt	+	+	-	-	D
Group3						
infection thread formation	nt	+	+	+	+	E

Table 2. Structural requirements for Nod factor-induced responses in the epidermis
nt= not tested; [a] = play an important role in the induction of these responses; A= Felle *et al.*, 1996; B= Kurkdjian, 1995; Felle *et al.*, 1995; C= Lerouge *et al.*, 1990; D= Journet *et al.*, 1994; E= Ardourel *et al.*, 1994.

Truchet *et al.*, 1991, Schultze *et al.*, 1992, Ardourel *et al.*, 1994, Journet *et al.*, 1994, Kurkdjian 1995, Felle *et al.*, 1995, 1996). The Nod factor structure-function relationship has been determined for the induction of root hair deformation, *ENOD12* gene expression, root hair membrane potential depolarization, alkalinization and infection thread formation. These studies showed that the responses induced by *R. meliloti* Nod factor on alfalfa can roughly be categorised in three groups and these are summarized in Table 2.

The first group comprises alkalinization of the root hair cytoplasm. NodRm-IV(C16:2, S) is most active in increasing the cytoplasmic pH, whereas the unsulphated Nod factor is active but a higher

concentration is required (Felle *et al.*, 1996). The presence of an O-acetyl group at the non-reducing terminal residue of the Nod factor molecule is not required (Felle *et al.*, 1996). However, the acyl group is essential although its structure does not seem to be very important.

The second group of responses includes deformation of root hairs, membrane potential depolarization and *ENOD12* expression. Neither the O-acetyl substitution nor the structure of the acyl group are important for the induction of these responses, although the presence of a fatty acyl group is required (Heidstra *et al.*, 1994; Journet *et al.*, 1994). The sulphate substitution has a dramatic effect on the ability to induce root hair deformation, *ENOD12* expression and membrane potential depolarization. Desulfation of the NodRm factors reduces at least 1000-fold their activity to induce root hair deformation, *ENOD12* expression and membrane potential depolarization (Journet *et al.*, 1994; Felle *et al.*, 1995). Thus the major difference between group 1 and 2 responses is the necessity of the sulphate substitution in the induction of the latter responses. The unsulphated Nod factor does induce alkalinization while the sulphated Nod factor induces, in the same root hair, alkalinization as well as membrane potential depolarization (Felle *et al.*, 1996). The unsaturation level of the acyl chain might be important in inducing membrane potential depolarization but it can not be excluded that this in part is related to the solubility of the Nod factor (Felle *et al.*, 1995).

The third group includes infection thread formation. This response involves the most strict Nod factor structure-function relationship, since it requires both the sulphate and the acetyl substitutions and the unsaturated acyl chain. *R. meliloti* mutated in either *nodE* or *nodL* produce Nod factors that do not contain the appropriate unsaturated fatty acid or lack the O-acetyl group, respectively. These mutants initiate markedly fewer infection threads (Ardourel *et al.*, 1994). A double mutant (*nodE⁻/nodL⁻*) secreting Nod factors without the O-acetyl group and with a C18:1 fatty acid, have lost the ability to induce infection threads (Ardourel *et al.*, 1994). In addition *R. meliloti* strains mutated in *nodH* producing only non-sulphated NodRm factors are unable to induce infection thread formation (Roche *et al.*, 1991a). This difference was clearly demonstrated when membrane depolarization and alkalinization were studied in the same root hair. Strikingly, the plant responses for which the length of the chitin backbone was shown to be important also require the sulphate substitution. For example, the lenght of the chitin backbone is very important in the induction of the membrane potential depolarization and root hair deformation. The pentameric NodRm-V (C16:2, S) is active only at higher concentration than the tetrameric NodRm IV (C16:2, S) (Felle *et al.*, 1995). This indicates that the distance between the non-reducing end and the sulphate group at the reducing end is important, and suggests that a Nod factor receptor recognises both ends.

To explain the different Nod factor structural properties required to induce the various responses it has been proposed that more than one receptor might be involved in Nod factor perception in the root

epidermis (Ardourel *et al.*, 1994, Felle *et al.*, 1995, 1996). The three distinct groups of Nod factor responses, suggest that there might be even three receptors, at least one for non-sulphated and two for sulphated Nod factors. One of the presumed receptors for sulphated Nod factors only recognises Nod factors containing the O-sulphate substitution, the O-acetyl substitution and the specific acyl chain (C16:2). Since this receptor seems to be specifically involved in initiation of infection thread growth it was named the uptake receptor (Ardourel *et al.*, 1994). The strict regulation of the infection process is confirmed by studies on the pea *sym2* gene. The *sym2* allele originates from wild Afghanistan peas (Lie, 1984; Davis, Evan and Johnston, 1988; Kozik *et al.*, 1995). *R. leguminosarum* bv. *viciae* carrying an additional *nod* gene, *nodX*, is able to nodulate Afghanisatn pea and the *sym2* introgressed lines. NodX is an O-acetyl transferase that mediates the production of Nod factors with an additional O-acetyl group at the reducing sugar moiety of the pentameric NodRlv factor (Clark *et al.*, 1991, Firmin *et al.*, 1993). Rhizobia lacking *nodX* induce root hair deformation but infection thread growth is arrested (Geurts *et al.*, 1997).

The hypothesis that more than one Nod factor receptor is active in the root epidermis is supported by the work from Felle and co-workers (1995, 1996) who demonstrated that application of the sulphated as well as the non-sulphated Nod factor both resulted in alkalinisation of alfalfa root hairs. A second addition of the same Nod factor did not increase the alkalinisation response, which was explained as desensitisation of a receptor as a result of saturation. When first non-sulphated Nod factor was added at saturating concentration, followed by sulphated Nod factor or vice versa, alkalinisation took place at each addition of Nod factor. This indicates that the alkalinisation response is not just a consequence of differential binding of the Nod factors to the same receptor but of binding to different receptors (Felle *et al.*, 1996).

The hypothesis that more than one receptor is involved in Nod factor perception is a relatively simple way to explain the above described data. Only one Nod factor binding protein has been identified in roots of *Medicago truncatula*, binding both sulphated and non-sulphated Nod factors (Bono *et al.*, 1995). Recently two classes of binding sites for Nod factor NodRm (Ac, 35S, C16:2) have been found in the microsomal fraction of *M. varia* cell suspension culture extracts. This binding was reversible and saturable (Niebel *et al.*, 1997).

Binding studies suggested that these binding sites have a high and low affinity of binding to NodRm (Niebel *et al.*, 1997). The low affinity binding site is similar to the previously characterized in *Medicago truncatula* root extracts (Bono *et al.*, 1995). Whether a differential binding of sulphated and non-sulphated Nod factor occur with these two candidates and whether this binding occurs also in root extract is unknown.

III.C. NOD FACTOR PERCEPTION IN THE CORTEX AND PERICYCLE

Exogenously applied Nod factors induce responses in the epidermis, i.e. the tissue in direct contact with Nod factors, but also in the cortex and the pericycle. The first question that can be asked is whether Nod factors are translocated to and also recognized in the cortex and the pericycle, or whether secondary messengers generated in the epidermis are responsible for the induction of responses in these tissues.

Temperate legumes forming indeterminate nodules require Nod factors substituted with a highly unsaturated fatty acyl group for the induction of cell divisions in the inner cortex, whereas Nod factors of rhizobia nodulating tropical legumes forming determinate nodules generally contain a C18:1 acyl group, it is possible that the highly unsaturated acyl group might be required to transport the Nod factor to the inner layers. This hypothesis is consistent with the observation that chitin fragments ballistically targeted to the cortex are sufficient to induce cortical cell division (Spaink *et al.*, 1994).

Technical achievements by which labeled Nod factors can now be synthesized might be of help. The use of such molecules will make it possible to answer the question whether Nod factors are transported to the cortex or processed in order to induce cortical cell division and *ENOD40* expression in the pericycle.

IV. Defense Related-Responses Induced during Rhizobia-Legume Interaction

During the different steps of legume nodulation, rhizobia are in intimate contact with their host plant. Ultimately, this interaction culminates in the formation of a nodule where the bacteria are present intracellularly, in the so-called infected cells. Despite the fact that these cells are completely packed with

bacteria an obvious host defense response does not occur. The avoidance of a defense response is most likely controlled by surface determinants of the rhizobia as well as the host membrane that always separates the bacteria and the host cytoplasm. The rhizobial surface determinants that are essential to avoid a host defense response include most likely lipopolysaccharides and exopolysaccharides since mutants disturbed in the production of these compounds are unable or have a dramatically reduced ability to infect plants and trigger Hypersensitive Reaction-like responses (Niehaus *et al.*, 1993, Parniske *et al.*, 1994, and Perotto *et al.*, 1994).

Although rhizobia appear to be successful in avoiding a defense response in nodules it is striking that even in a wild type interaction defense responses are induced. In general only a small percentage of the rhizobium induced infection sites leads to the development of a nodule and the majority of the infection threads get aborted in the epidermis or cortex. In the *R. meliloti*-alfalfa interaction it has been shown that there are infection sites where complete necrosis of bacterial and plant cells occur (Vasse *et al.*, 1993). Furthermore, the host defense-related molecular, acidic chitinase was found to be specifically active in these necrotic cells. In contrast, phenylalanine ammonialyase (PAL) and chalcone synthase (CHS), enzymes that play a an important role in flavonoid biosynthesis, were found in both necrotic and non-necrotic cortical cells. It has been suggested that the host plant restricts the progression of a certain proportion of infection threads by eliciting a hypersensitive-like reaction. Since such HR-like responses are only induced by rhizobia producing Nod factors, it is possible that these compounds have a dual role since on one hand they induce the developmental programme leading to nodule development and on the other hand they might be involved in the activation of a HR-like response to control the number of infections. Why the HR-like response is only induced at certain infection sites is unclear.

The expression of PAL and CHS in the cortex might be related to a synthesis of specific flavonoids and does not necessarily mean that phytoalexins are synthesised. This is illustrated by studies on the interaction between a mycorrhizal fungus and alfalfa (Harrison *et al.*, 1994). In myccorrhiza colonized roots, elevated levels of PAL and CHS transcripts were detected specifically in the infected cortical cells containing arbuscules while isoflavone reductase transcripts, encoding the enzyme involved in the biosynthesis of the phytoalexin medicarpin, was only detected at relatively high levels in cortical cells of non-colonized roots.

Several studies have shown that CHS mRNA accumulates during nodule development, for example it was shown to be present in all cells of a nodule primordium (Yang *et al.*, 1992). However it is unclear whether it plays a role in a defense response or *e.g.* the synthesis of specific flavonoids that can act as inducers of the rhizobial nod genes or might even affect the polar transport of auxin (Hirsch *et al.*, 1989). The fact that Nod factors are composed of a chitin backbone, beside the fatty acid group, makes them prone to act as potential plant defense elicitors (Baureithel *et al.*, 1994, *Staehelin et al.*, 1994). In *Trifolium subterraneum*, Nod factors appear to play a role in the induction of CHS since this gene was only induced by rhizobia secreting Nod factors (Lawson *et al.*, 1994).

V. References

Allen, N.S., Bennett, M.N., Cox, D.N., Shipley, A., Ehrhardt, D.W., Long, S.R. (1994) In, Daniels, M.G., Downie J.A., Osbourne, A.E., eds. Advances in Molecular Genetics of Plant-Microbe Interactions, vol 3. Dordrecht: Kluwer Academic Publishers, 107-114.

Ardourel, M., Demont, N., Debellé, F., Maillet, F., de Billy, F., Promé, J-C., Dénarié, J., Truchet, G. (1994) Plant Cell 6, 1357-1374.

Asad, S., Fang, Y., Wycoff, K.L., Hirsch, A.M. (1994) Protoplasma 183, 10-23.

Bono, J-J., Riond, J., Nicolaou, K.C., Bockovich, N.J., Estevez, V.A., Cullimore, J.V., Ranjeva, R. (1995) Plant J. 7, 253-260.

Baureithel, K., Felix, G., and Boller, T. (1994) J. Biol. Chem. 269, 17931-17938.

Clark, C.A., Beltrame, J., Manning, P.A. (1991) Gene 106, 43-52.

Cook, D., Dreyer, D., Bonnet, D., Howell, M., Nony, E., VandenBosch, K. (1995) The Plant Cell 7, 43-55.

Cooper, J.B., Long, S.R. (1994) The Plant Cell 6, 215-225.

Crespi, M.D., Jurkevitch, E., Poiret, M., d'Aubenton-Carafa, Y., Petrovics, G., Kondorosi, E., Kondorosi, A. (1994) EMBO J. 13, 5099-5112.

Davis, E.O., Evans, I.J., Johnston, A.W.B. (1988) Mol. Gen. Genet. 212, 531-535.

De Jong, A.J., Heidstra, R., Spaink, H.P., Hartog, M.H., Meijier, E.A., Hendriks, T., Lo Schiavo, F., Terzi, M., Bisseling, T., Van Kammen, A., de Vries, S.C. (1993) The Plant Cell 5, 615-620.

De Ruijter, N., (1997) Plant J. in press.

Ehrhardt D.W., Atkinson E.M., Long S.R. (1992) Science 256, 998-1000.

Erhrhardt, D.W., Wais, R., and Long, S. R. (1996) Cell 85, 673-681.

Felle, H.H., Kondorosi, E., Kondorosi, A., Schultze, M. (1995) Plant J. 7, 939-947.

Felle, H.H., Kondorosi, E., Kondorosi, A., and Schultze, M. (1996) Plant J. 10, 295-301.

Firmin, J.L., Wilson, K.E., Carlson, R.W., Davies, A.E., Downie, J.A. (1993) Mol. Microbiol. 10, 351-360.

Geurts, R., Heidstra, R., Hadri, A-E., Downie, J.A., Franssen, H., Van Kammen, A., and Bisseling, T. (1997) Plant Physiol. 115, 351-359.

Harrisson, M.J., and Dixon R.A. (1994) Plant J. 6, 9-20.

Heidstra, R., Geurts, R., Franssen, H., Spaink, H.P., Van Kammen, A., Bisseling, T. (1994) P. Physiol. 105, 787-797.

Heidstra, R., Yang ,W.-C., Yalcin, Y., Peck, S., Emons, A., Van Kammen, A., and Bisseling, T. (1997) Development 124, 1781-1787.

Hirsh, A.M., Bhuvaneswari, T.V., Torrey, J.G., and Bisseling, T. (1989) Proc. Natl. Acad. Sci. USA. 86, 1244-1248.

Horvath, B., Heidstra, R., Lados, M., Moerman, M., Spaink, H.P., Promé, J-C., Van Kammen, A., Bisseling, T. (1993) Plant J. 4, 727-733.

Journet, E.P., Pichon, M., Dedieu, A., de Billy, F., Truchet, G., Barker, D.G. (1994) Plant J. 6, 241-249.

Kijne, J.W. (1992) In: Stacey G., Burris R.H., Evans H.J., eds. Biological Nitrogen Fixation. New York: Chapman and Hall, 349-398.

Kouchi, H., and Hata, S. (1993) Mol. Gen. Genet. 238, 106-119.

Kozik, A., Heidstra, R., Horvath, B., Kulikova, O., Tikhonovich, I., Noel, Ellis, T.H., Van Kammen, A., Lie, T.A., Bisseling, T. (1995) Plant Science 108, 41-49.

Kurkdjian, A.C. (1995) Plant Physiol. 107, 783-790.

Lawson, C.G.R., Djordjevic, M.A., Weinman, J.J. and Rolfe, B.G. (1994) Mol. Plant-Microbe Interact. 7, 498-507.

Lerouge, P., Roche, P., Faucher, C., Maillet, F., Truchet, G., Promé, J-C., Dénarié, J. (1990) Nature 344, 781-784.

Libbenga, K.R., Van Iren, F., Bogers, R.J., Schraag-Lammers, M.F. (1973) Planta 114, 29-39.

Lie, T.A. (1984) Plant and Soil 82, 415-425.

Lopez-Lara, I.M., Van den Berg, J.D.J., Thomas-Oates, J.E., Glushka, J., Lughtenberg, B.J.J., Spaink, H.P. (1995) Mol. Microbiol. 15, 627-638.

Matvienko, M., Van de Sande, K., Yang, W.C., Van Kammen, A., Bisseling, T., Franssen, H. (1994) Plant Mol. Biol. 26, 487-493.

Miller, D.D., Callaham, D.A., Gross, D.J., and Hepler, P.K. (1992) J. Cell Sci. 101, 7-12.

Minami, E., Kouchi, H., Cohn, J.R., Ogawa, T., and Stacey, J. (1996) Plant J. 10, 23-32.

Mergaert, P., Van Montagu, M., Promé, J-C., Holsters, M. (1993). Proc. Natl. Acad. Sci. USA 90, 1551-1555.

Niebel, A., Bono, J.J., Ranjeva, R. , and Cullimore, J.V. (1997) Mol. Plant-Microbe Interact. 10, 132-134.

Niehaus, K., Kapp, D., Pühler, A. (1993) Planta 190, 415-425.

Parniske, M., Schmidt, P.E., Kosch, K., and Müller, P. (1994) Mol. Plant-Microbe Interact. 7, 631-638.

Perotto S, Brewin, N.J., and Kanenberg, E.L. (1994) Mol. Plant-Microbe Interact. 7, 99-112.

Pichon, M., Journet, E-P., Dedieu, A., de Billy, F., Truchet, G., Barker, D.G. (1992) The Plant Cell 4, 1199-1211.

Price, N.P.J., Relic, B., Talmont, F., Lewin, A, Promé, D., Pueppke, S.G., Maillet, F., Dénarié, J., Promé, J-C., Broughton, W.J. (1992) Mol. Microbiol. 6 , 3575-3584.

Relic, B., Talmont, F., Kopcinska, J., Golinowski, W., Promé, J-C., Broughton, W.J. (1993) Mol. Plant-Microbe Interact. 6, 764-774.

Roche, P., Debellé, F., Maillet, F., Lerouge, P., Truchet, G., Dénarié, J., Promé, J-C. (1991a) Cell 67, 1131-1143.

Roche, P., Lerouge, P., Ponthus, C., Promé, J-C. (1991b) J. Biol. Chem. 266, 10933-10940.

Röhrig, H., Schmidt, J., Wieneke, U., Kondorosi, E., Barlier, I., Schell, J., John, M. (1994) Proc. Natl. Acad. Sci. USA 91, 3122-3126.

Sanjuan, J., Carlson, R.W., Spaink, H.P., Bhat, R., Mark Barbour W., Glushka J., Stacey J. (1992) Proc. Natl. Acad. Sci. USA 89, 8789-8793.

Scheres, B., Van de Wiel, C., Zalensky,A., Horvath, B., Spaink,H.P., Van Eck, H., Zwartkruis, F., Wolters, A-M., Gloudemans, T., Van Kammen, A., Bisseling, T. (1990a) Cell 60, 281-294.

Scheres, B., Van Engelen, F., Van der Knaap, E., Van de Wiel, C., Van Kammen, A., Bisseling, T. (1990b) The Plant Cell 8, 687-700.

Schmidt, J., Röhrig, H., John, M., Wieneke, U., Koncz, C., Schell, J. (1993) Plant J. 4, 651-658.

Schultze, M., Quiclet-Sire, B., Kondorosi, E., Virelizier, H., Glushka, J.N., Endre, G., Géro, S.D., Kondorosi, A. (1992) Proc. Natl. Acad. Sci. USA 89, 192-196.

Sievers, A., and Schnepf, E. (1981) In: Cytomorphogenesis in plants, ed. Kiermayer O., Cell Biology Monographs 8, (Springer verlag, Wien New York)

Smit, G., Van Brussel, A.A.N., Kijne, J.W. (1993) In: Palacios R, Mora J, Newton WE, eds. New Horizons in Nitrogen Fixation. Dordrecht: Kluwer Academic Publishers, 371

Spaink, H.P., Sheely, D.M., Van Brussel, A.A.N., Glushka, J., York, W.S., Tak, T., Geiger, O., Kennedy, E.P., Reinhold, V.N., Lugtenberg, B.J.J. (1991) Nature 354, 125-130.

Spaink, H.P., Bloemberg, G.V., Wijfjes, A.H.M., Ritsema, T., Geiger, O., López-Lara, I.M., Harteveld, M., Kafetzopoulos, D., Van Brussel, A.A.N., Kijne, J.W., Lugtenberg, B.J.J., Van de drift, K.M.G.M., Thomas-Oates, J.E., Potrykus, I., and Sautter, C. (1994a) In: Advances in Molecular Genetics of Plant-Microbe Interactions. 3.p.91-98. Daniels, M.J., Downie, J.A., and Osbourne, A.E., Eds., Kluwer Academic Publishers: Dordrecht.

Staehelin, C., Granado, J., Müller, J., Wiemken, A., Mellor ,R.B., Felix, G., Regenaas, M., Broughton, W.J., Boller, T. (1994a) Proc. Natl. Acad. Sci. USA 91, 2196-2200.

Staehelin, C., Schultze, M., Kondorosi, E., Mellor, R.B., Boller, T., Kondorosi, A. (1994b) Plant J. 5, 319-330.

Stokkermans, T.J.W., Peters, N.K. (1994) Planta 193, 413-420.

Tanimoto, M., Robert, K., and Dolan, L. (1995) Plant J. 8, 943-948.

Truchet, G., Roche, P., Lerouge, P., Vasse, J., Camut, S., de Billy, F., Promé, J-C., Dénarié, J. (1991) Nature 351, 670-673.

Turgeon, B.G., and Bauer, W.D. (1985) Planta 163, 328-349.

Van Batenburg, F. H. D., Jonker, R., and Kijne, J. W. (1986) Physiol. Plant. 66, 476-480.

Van Brussel, A.A.N., Bakhuizen, R., Van Spronsen, P.C., Spaink, H.P., Tak, T, Lugtenberg, BJJ, Kijne, JW. (1992) Science 257, 70-71.

Van de Sande, K., Pawlowski, K., Csaja, I., Wieneke, U., Schell, J., Schmidt, J., Walden, R., Matvienko, M., wellink, J., Van Kammen, A., Franssen, H., and Bisseling, T. (1996) Science 273, 370-373.

Van Spronsen, P.C., Bakhuizen, R., Van Brussel, A.A.N., Kijne, J.W. (1994) Eur. J. Cell Biol. 64, 88-94.

Van Spronsen, P.C., Van Brussel, A.A., and Kijne, J.W. (1995) Eur. J. Cell Biol. 68, 463-469.

Vasse, J., de Billy, F., Truchet, G. (1993) Plant J. 4, 555-566.

Verma, D.P.S. (1992) The Plant Cell 4, 373-382.

Vijn, I., das Neves, L., Van Kammen, A., Franssen, H., and Bisseling, T. (1993) Science, 260, 1764-1765.

Vijn, I., Martinez-Abarca, F., Yang, W.C., das Neves, L., Van Brussel, A., Van Kammen, A., Bisseling, T. (1995a) Plant J. 8, 111-119.

Vijn, I., Yang, W.C., Pallisgård, N., Østergaard Jensen, E., Van Kammen, A., Bisseling T. (1995b) Plant Mol. Biol. 28, 1111-1119.

Yang, W.C., Canter Cremers, H.C.J., Hogendijk, P., Katinakis, P., Wijfelman, C. A., Franssen, H., Van Kammen, A., Bisseling, T. (1992) Plant J. 2, 143-151

Wilson, C.W., Long, F., Maruoka, E.M., Cooper, J.B. (1994) The Plant Cell 6, 1265-1275.

Wipf, L., Cooper, D.C. (1938). Proc. Natl. Acad. Sci. USA 24, 87-91.

Yang, W.C., Katinakis, P., Hendriks, P., Smolders, A., de Vries, F., Spee, J., Van Kammen A., Bisseling T., Franssen H. (1993) Plant J. 3, 573-585.

Yang, W.C., de Blank, C., Meskiene, I., Hirt, H., Bakker, J., Van Kammen, A., Franssen, H., Bisseling, T. (1994) The Plant Cell 6, 1415-1426.

Zaat, S.A.J., Van Brussel, A.A.N., Tak, T., Lugtenburg, B.J.J., Kijne, J.W. (1989) Planta 177, 141-150.

Chapter 22

Tissue and Cell Invasion by *Rhizobium:* The Structure and Development of Infection Threads and Symbiosomes

Nicholas J. Brewin

I. Colonisation of the Intercellular Space: The Formation of Infection Threads

I.A. DEVELOPMENT

Elsewhere in this volume, the importance of *Rhizobium*-derived lipochitin oligosaccharide Nod factors has been described. While Nod factors appear to be sufficient to induce many of the early events in nodule morphogenesis (Long, 1996), additional host-symbiont recognition systems are apparently required to promote tissue and cell invasion (van Spronsen *et al.*, 1994; Vijn *et al.*, 1995). These processes have been most extensively studied for alfalfa (Benaben *et al.*, 1995), clover (de Boer and Djordjevic, 1995) and pea (Kijne, 1992; Rae *et al.*, 1992).

Prior to bacterial ingress, the curled appearance of root hairs which can be caused by Nod factor becomes further distorted into a 'shepherd's crook' (Dart, 1975). This final deformation cannot be mimicked by Nod factor and appears to depend on the presence of live rhizobia on the root hair surface. Within the pocket formed by the curled hair, bacterial intrusion takes place, apparently caused by a weakening of the root hair cell wall and subsequent invagination of the plant membrane (van Spronsen *et al.*, 1994; Kijne, 1992). Bacterial invasion is then facilitated by the development of an infection thread involving the localised deposition of plant cell wall materials. A tubular intrusion is subsequently formed, which is continuous with the plant cell wall and is apparently of similar composition (Figure 1). Within the lumen of the infection thread or the intercellular space, the invading bacteria are topologically extracellular (Van den Bosch *et al.*, 1989; Rae *et al.*, 1992) (Figure 2). These and other results suggest that a common mechanism may be responsible for the synthesis of the cell wall and the infection thread and that the invading rhizobia have merely modified an existing system of plant cell development. The wall of the thread prevents direct contact between the bacterial cells and the plant cell membrane.

Growth of the infection thread occurs only at the tip (Dart, 1975), and is apparently associated with a restructuring of the plant cell cytoskeleton. This is characterised by the uncoupling of the nucleus from its

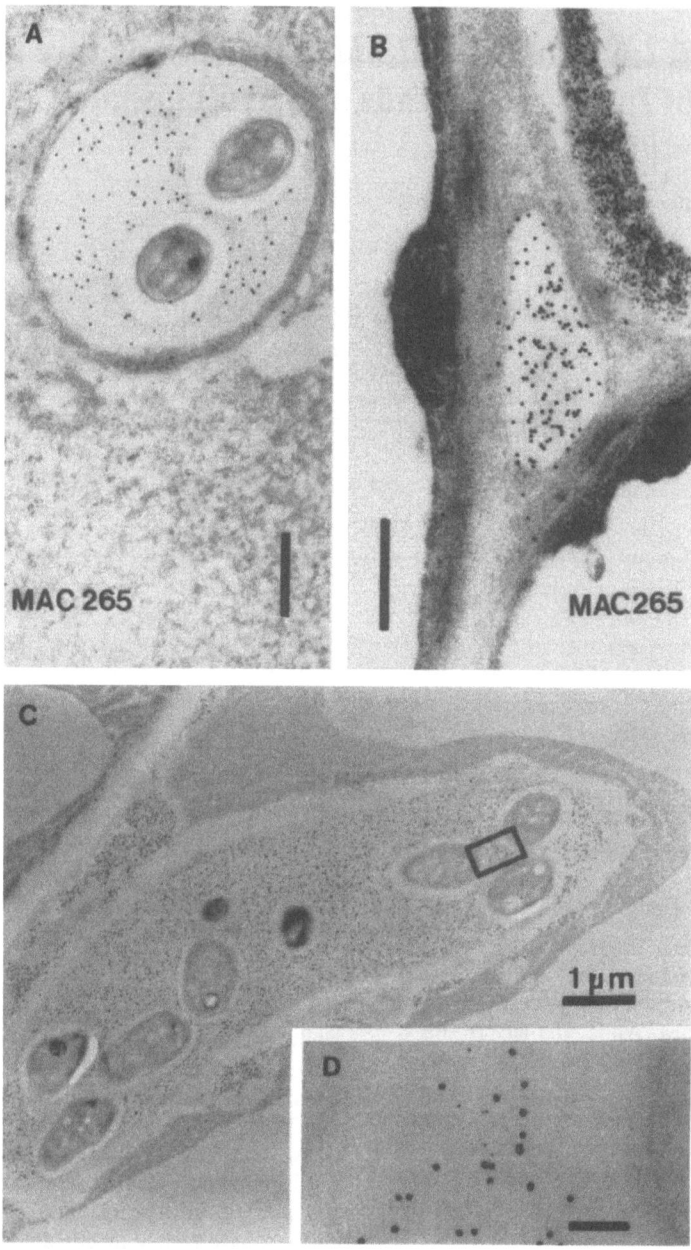

Figure 1. Electron micrographs of infection threads and intercellular spaces showing immunogold labelling of matrix material.
A. Transverse section through an infection thread from pea nodule tissue, showing immunogold labelling of matrix material. The halo surrounding each of the two rhizobial cells is thought to be due to the presence of capsular material. B. Section through a 3-way intercellular junction derived from uninfected nodule parenchyma, showing the presence of intercellular matrix glycoprotein revealed by immunogold labelling. C. Glancing longitudinal section through an intracellular infection thread with adjoining cell wall. D. Inset represents magnification of boxed area from micrograph (C) to show by dual immunogold labelling the presence of matrix glycoprotein and lipoxygenase antigens (20 and 5nm colloidal gold respectively). Scale bars represent 1μm (A,B,C) or 200nm (D). A and B reproduced from VandenBosch *et al.*, (1989) by kind permission of IRL Press; C and D courtesy of D.J. Sherrier.)

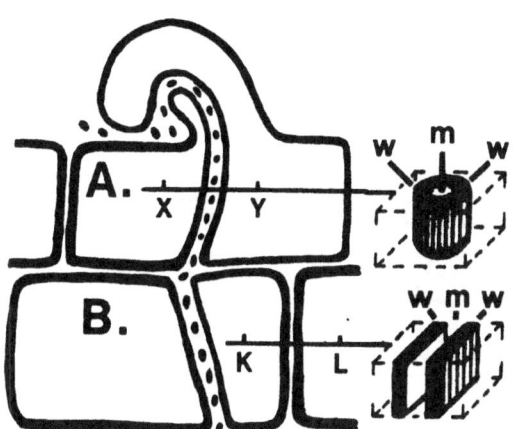

Figure 2. Topology of an infection thread relative to an intercellular space.

A. Root hair curling and infection thread initiation. B. Propagation of the infection thread as a transcellular tunnel through the root cortex and also in the subapical invasion zone of the nodule. The transects X - Y and K - L compare and contrast the topology of a transcellular infection thread (which is a cylinder) with a transcellular cell plate (which is planar): w=wall; m=extracellular matrix material for the infection thread lumen and for the contents of intercellular spaces and m=middle lamella for the cell plate. Reproduced from Rae *et al.*, (1992) by kind permission of Bios Scientific Publishers

normal position near the growing tip of the root hair (Lloyd *et al.*, 1987) and its reassociation with the inwardly growing tip of the infection thread (Figure 1a). The axis of growth of the infection thread across the host cell is apparently pre-determined by a transcellular cytoplasmic strand (Van Brussel *et al.*, 1992), the orientation of which is established as a 'pre-infection structure' following the re-activation of cell cycle processes by *Rhizobium*-released Nod factors (Yang *et al.*, 1994). Having grown across the cytoplasmic space, the infection thread then fuses with the mother cell wall by an unknown mechanism, releasing the contents of the lumen into the intercellular matrix on the proximal side of the epidermal cell (Brewin, 1991; Rae *et al.*, 1992). At this point, the adjacent cortical cell is stimulated to develop an infection thread and the process of transcellular infection repeats itself across the root cortex and into the post-meristematic cells derived from the nodule primordium. Thus, in mature indeterminate nodules, there is a sub-meristematic region, the invasion zone, in which branched infection threads pass through, and into, recently divided cortical cells and thereby

generate the invaded central tissue of the nodule in which nitrogen fixation can take place (Figure 3).

In other legumes, *e.g. Phaseolus vulgaris*, the importance of transcellular infection threads is diminished, either because the rhizobia spread through the intercellular matrix or because the bacteria are released from infection threads into the cytoplasm of hypodermal cortical cell which are themselves dividing as a result of cell cycle reactivation by Nod factor. In this case, the original intracellular infection threads become merely vestigial in the mature infected host cells. Thus, as discussed in Chapter 1, it seems that three rather different structural features have all been described as 'infection threads' in various legumes. There are the intercellular threads, proceeding by crack entry and/or by dissolution of the middle lamella, *e.g.* in *Andira* and *Mimosa*; there are the trans-cellular threads that promote cell-to-cell spread of rhizobia through non-mitotic tissue, *e.g.* in *Pisum*; and there are the intracellular intrusions, or 'infection pegs' (Rae *et al.*, 1992) which penetrate individual cells and release rhizobia into the cytoplasm as endocytotic symbiosomes, *e.g.* in *Phaseolus*. Furthermore, it is interesting to note that the transcellular infection threads observed in legumes with determinate nodules are generally narrower than those from legumes with indeterminate nodules. In the latter type, *e.g.* pea, more than half the volume of the central lumen is apparently filled with matrix material (Figure 1), whereas in the former type, *e.g. Phaseolus*, there is very little matrix material and the rhizobia are held tightly in single file within a tube of cell wall material (Rae *et al.*, 1992; Kijne, 1992).

I.B. PLANT COMPONENTS

Despite the inaccessibility of infection threads for biochemical analysis, some progress has been made towards understanding their structure and the mechanisms regulating their development. Using specific antibody probes for immunogold localisation in pea nodule sections, it has been demonstrated that both the infection thread wall and plant cell wall contain cellulose, xyloglucan, methyl-esterified pectin and non-esterified pectin. It has, however, been observed that the infection thread wall is more resistant to treatment with cell wall degrading enzymes (Higashi *et al.*, 1987), implying some further chemical modification. Studies of nodule-enhanced gene expression suggest that glycine rich proteins may be novel components of the invasion pathway (Küster *et al.*, 1995).

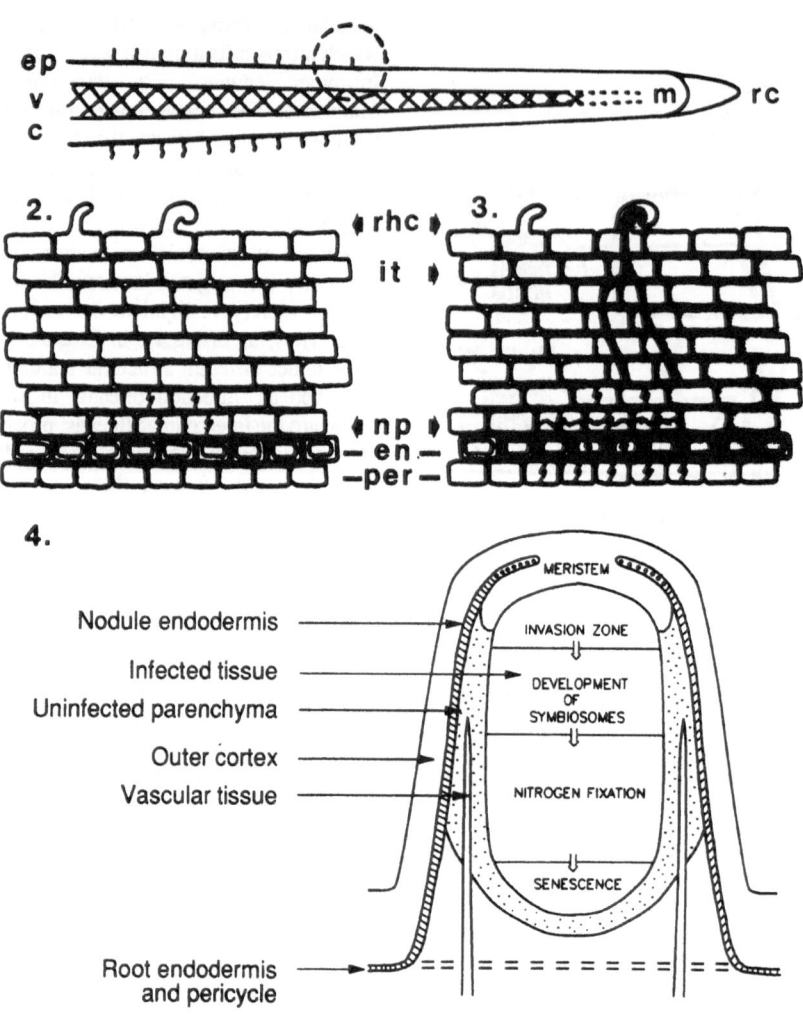

Figure 3. Stages in tissue and cell invasion by *Rhizobium*

1. Root tip, showing the infectible zone (circled) at the site of emergence of root hairs in the region proximal to the apical meristem (m) and root cap (rc). 2. Cross section of root cortex showing two early plant responses to diffusible signal molecules from rhizobia, root hair curling (rhc) and induction of cortical cell divisions, creating the nodule primordium (np) in the tissue adjacent to the endodermis (en) and pericycle (per). 3. A slightly later stage of development in which infection threads (it) develop as transcellular tunnels containing rhizobia and the nodule primordium has spread to include cells from the endodermis and pericycle and further differentiated into a fully meristematic region which remains uninfected by rhizobia. 4. At a later stage, this meristem has grown outwards from the root as an uninfected apical meristem, giving rise to the variety of differentiating cell types that comprise the nodule. Central tissues of the nodule become invaded by rhizobia and differentiate as a nitrogen-fixing tissue which functions under microaerobic (low oxygen) conditions. Outer uninfected tissues develop vascular strands, an endodermal sheath and an inner cortex of uninfected parenchymatous cells which may function as a gaseous diffusion barrier, regulating the supply of oxygen to the central nitrogen fixing tissues. Reproduced from Brewin (1991) by kind permission of Annual Reviews Inc.

The composition of the matrix material is probably of great significance in terms of the regulation of infection thread development. Van den Bosch *et al.*, (1989) used monoclonal antibody MAC 265 to identify a plant glycoprotein (95-110 kDa) as a major component of the luminal matrix of infection threads in pea nodules (Figure 1). The antigen is also detectable in intercellular spaces in the uninfected nodule parenchyma and in uninfected tissue in the vicinity of the apical meristem. Another study showed that proline-rich proteins (PRPs) are major components of the infection thread matrix in pea (Sherrier and van den Bosch, 1994). The authors also noted similarities between the locations of PRPs as revealed by these antibodies and the expression patterns of certain early nodulin genes, particularly ENOD12 which is known to encode a 12.5 kDa protein containing two proline-rich pentapeptide repeats (Scheres *et al.*, 1990a). Although this nodulin gene has proved very useful for studies of Nod factor induced transcriptional regulation, ENOD12 has been shown not to be essential for nodulation in *Medicago* spp (Csanadi *et al.*, 1994).

Studies of several plant pathogenic interactions have implicated glycoproteins recognised by MAC 265 and the anti-PRP antibodies detailed above in protein cross-linking reactions (Bradley *et al.*, 1992). Cross-linking of proteins, apparently driven by the release of hydrogen peroxide in the extracellular matrix, can be viewed as a rapid physical response, which might decrease pathogenic penetration of host tissue. Thus, the localisation of similar matrix glycoproteins in infection threads raises the possibility that these components are synthesised as part of a host defence response that could function to prevent or control the extent of tissue invasion by *Rhizobium*. Other indirect evidence for a possible role for active oxygen species in the regulation of infection thread development comes from the identification of cell peroxidases (Cook *et al.*, 1995; Salzwedel and Dazzo,

1993) and the localisation of a lipoxygenase antigen in the lumen of infection threads (Gardner *et al.*, 1996) (Figure 1D). Furthermore, diamine oxidase, which is an enzyme that could generate hydrogen peroxide and promote protein cross-linking in the intercellular matrix (Chiarello *et al.*, 1996; Tipping and McPherson, 1995) has been identified in pea nodules and is apparently present in the lumen of infection threads (C.D. Gardner and J.P. Wisniewski, personal communication).

The characterisation of other plant-derived components of the infection thread lumen is still somewhat anecdotal at the present time. There are indications from *in situ* hybridisation studies of pea nodules that the infection thread lumen may include an arabinogalactan protein encoded by PsENOD5 (Scheres *et al.*, 1990b) and a cysteine protease encoded by PsCYP15a (Kardailsky and Brewin, 1996). The activity of such a protease in the extracellular matrix could perhaps explain why most amino-acid auxotrophic mutants of *Rhizobium* are still able to nodulate normally (Beringer *et al.*, 1980). Enhanced expression of genes encoding a cysteine protease and a subtilisin-like protease have also been shown to be associated with invasion by *Frankia* in the actinorhizal symbiosis with *Alnus* (Goetting-Minesky and Mullin, 1994; Ribeiro *et al.*, 1995).

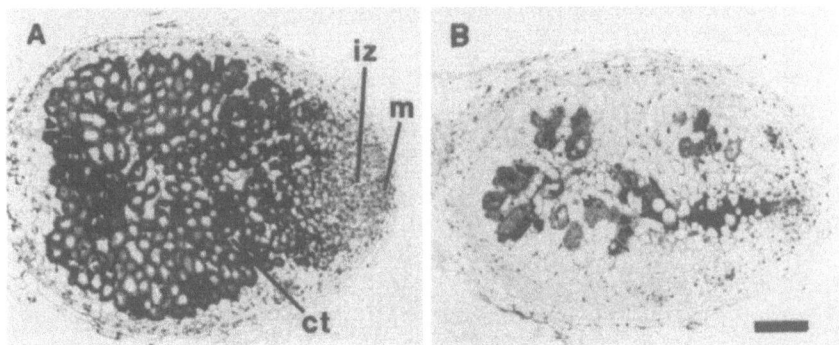

Figure 4. Pea nodules induced by *Rhizobium leguminosarum* bv. *viciae*.
(a) Longitudinal section of a pea root nodule containing a wild-type strain of *Rhizobium*: m, uninfected apical meristem; iz, invasion zone containing infection threads; ct, central tissue with host cells each containing several thousand endosymbiotic bacteroids. (b) Section of a pea nodule occupied by a mutant strain of *Rhizobium* with defective lipopolysaccharide (LPS) structure. The LPS-defective mutants show reduced tissue and host cell invasion, abnormal bacteroid morphology and evidence for the induction of host defence responses. Scale bar represents 0.5 mm. Reproduced from Perotto *et al.*, (1994) by kind permission of American Phytopathology Society.

I.C. BACTERIAL COMPONENTS

The identification of components of a potential system for restricting bacterial growth and division within the lumen of infection threads emphasises the fact that we still do not understand what is special about rhizobia that enables them to invade host plant cells without eliciting host defence responses. Localised defence reactions are often seen in cases where rhizobia associate with an inappropriate host legume (Kannenberg and Brewin, 1994; de Boer and Djordjevic, 1995). Even during colonisation of *Medicago* root hairs by a compatible strain, many invasions are apparently arrested by a localised defence reaction (Vasse *et al.*, 1993), emphasizing that, during nodule formation, the dividing line between symbiotic and pathogenic interactions is always delicately balanced.

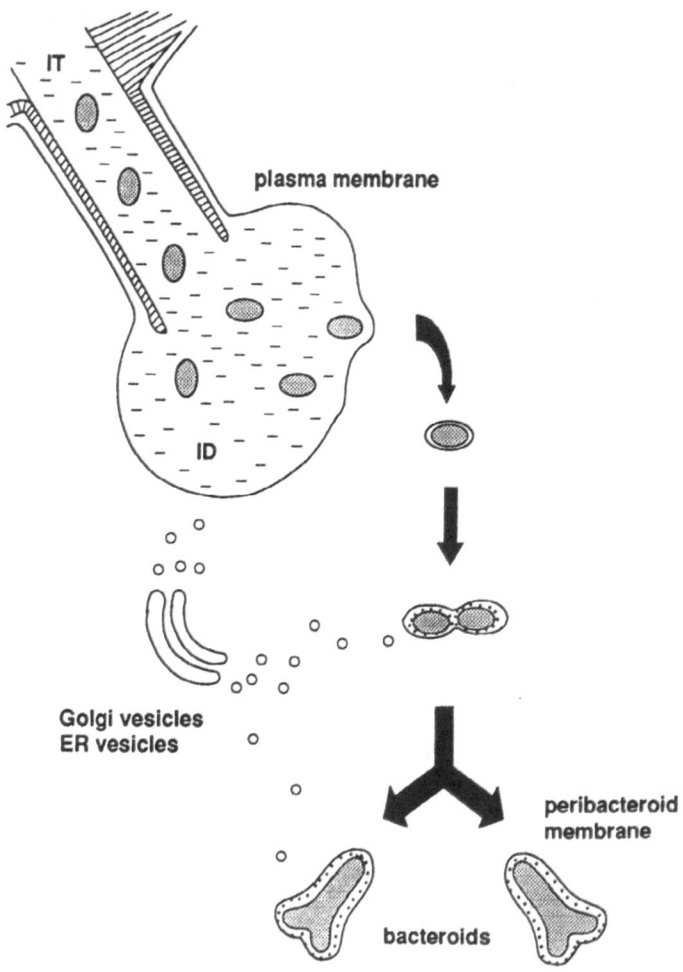

Figure 5. Biogenesis of the symbiosomal membrane.
Initially, rhizobial cells are taken into the host cytoplasm by endocytosis, i.e., by involution of the plasma membrane from the infection thread or infection droplet. Subsequent growth and division of intracellular bacteroids necessitates further extension of the symbiosomal membrane by inclusion of vesicles derived from the Golgi or endoplasmic reticulum (Roth and Stacey, 1989b; Roth and Stacey, 1989a). The nature of protein and vesicle targetting pathways to the symbiosome probably changes during the course of development as the structure and function of the symbiosomal membrane becomes more specialised. Verma and co-workers have suggested that membrane-anchored phosphatidyinositol-3-kinase and small GTP-binding proteins might be involved in the targetting of membrane vesicles to the symbiosome compartment (Cheon *et al.*, 1993; Verma *et al.*, 1994). Reproduced from Brewin *et al.*, (1992) by kind permission of Academic Press Inc.

The structure and function of bacterial cell surface components are discussed comprehensively in other chapters. Rhizobial cells within infection threads are apparently always coccooned within a capsule (Figure 1A). This is composed of extracellular polysaccharide (EPS) (Leigh and Walker, 1994) and a more tightly bound acidic capsular polysaccharide termed the K antigen (Rolfe *et al.*, 1996; Reuhs *et al.*, 1995). Lipopolysaccharide and divalent calcium ions may add further stability to this extracellular sheath, which may help the bacterium to avoid elicitation of host defence responses (Kannenberg and Brewin, 1994). In peas inoculated with lipopolysaccharide (LPS)-defective mutants (Figure 3), infection thread formation was grossly distorted, bacteria were only sporadically released into host cells, and there were many manifestations of a localised host defence response (Perotto *et al.*, 1994). Even more severe effects on nodule development were observed when LPS mutants were applied to *Phaseolus* beans or soybeans (Noel *et al.*, 1986; Stacey *et al.*, 1991): this resulted in infection threads that aborted in the root hair cells or in the underlying cortex. In addition to their structural role, there is increasing evidence that components of cell wall polysaccharides may also function as signal molecules promoting tissue invasion or suppressing a host defence response (Ozga *et al.*, 1994; Becker *et al.*, 1995)

II. Endocytosis of Bacteria and the Differentiation of Bacteroids

II.A. ENDOCYTOSIS

The process of tissue and cell invasion is accompanied by the differentiation of host cell types derived from the apical meristem (Figure 3). As nodule development proceeds, the differentiation of a nitrogen-fixation capability by *Rhizobium* is always preceded by the endocytosis of bacteria. In peas and other herbaceous legumes with indeterminate nodules, this occurs from unwalled infection droplets that extrude from infection threads (Figure 4). These droplets are bounded by a single membrane which is continuous with the plant plasma membrane (Dart, 1975). They contain a matrix which is apparently similar in composition to that of the infection thread; for example, the plant matrix glycoprotein recognised by MAC 265 is present (Rae *et al.*, 1992). In pea, the symbiosomal membrane (otherwise known as the peribacteroid membrane) undergoes division in synchrony with the bacteroids so

that each remains singly enclosed, except in the case of an interesting symbiotically defective mutant termed Sprint2 Fix⁻ (Borisov *et al.*, 1992). The released bacterial cells (bacteroids) eventually cease division and differentiate into the fully endosymbiotic, nitrogen-fixing form. This is associated with a progressive change in cell morphology to the characteristic 'Y' shape and involves alterations in bacterial gene expression, as described in chapter 23.

Little is known about what actually drives the internalisation of bacteria into the cytoplasmic compartment (Bassett *et al.*, 1977) but it is presumably significant that the symbiosomal membrane and the membrane surrounding infection droplets have no associated plant cell wall and therefore close contact with the bacterial cell surface is possible (Rae *et al.*, 1992). Using isolated membrane fragments from pea nodule symbiosomes, a physical interaction between symbiosomal and bacteroid outer membranes has been demonstrated *in vitro* (Bradley *et al.*, 1986). Such a direct interaction between plant and bacterial membrane surfaces could be involved in the uptake of rhizobial cells (endocytosis). The O-antigen of the bacterial LPS appears to be important in this process because mutants which are defective in its production fail to undergo normal endocytosis (Perotto *et al.*, 1994). Similarly, there is evidence for a role of boron at this stage of nodule development and a molecular model has been tentatively proposed implicating borate ions in the process of stabilising the interaction between the bacterial cell surface and the plant cell membrane at the point of endocytosis (Bolanos *et al.*, 1996). Similarly, surface interactions between these two membranes could explain how subsequent division of intracellular bacteroids is accompanied by concomitant division of the symbiosomal membrane.

II.B. BACTERIAL ADAPTATION

Clearly, there are significant changes in microenvironment for bacteria in infection threads compared to symbiosomes, as discussed in previous chapters. Several bacterial mutants have been described that show severely reduced survival of intracellular bacteroids (Glazebrook *et al.*, 1993; Morrison and Verma, 1987; Muller *et al.*, 1995b). For example, mutation in a bacterial gene encoding signal peptidase resulted in impaired stability of bacteroids within symbiosomes and consequently there were extensive changes in the expression of symbiosome-specific nodulins of infected soybean cells (Muller *et al.*, 1995a). Experiments involving the use of reporter genes fused to *exo* gene promoters indicate that many *exo* genes are not expressed in mature bacteroids but are actively transcribed in infection thread bacteria. Although some genetic regulatory systems have been identified, it is not known what physiological mechanisms control expression of *exo* genes in different regions of the nodule (Leigh and Walker, 1994). Similarly, studies with *nod* gene promoters fused to reporter genes (such as β-glucuronidase) and *in situ* hybridization experiments have revealed that *nod* genes are also actively transcribed by bacteria in the invasion zone of pea and alfalfa nodules, but that the genes are probably not transcribed in mature bacteroids (Sharma and Signer, 1990; Schlaman *et al.*, 1991).

As bacteroids differentiate, they must adapt to new physiological conditions. Cyclic β-linked glucans, present in the periplasmic space of *Rhizobium* and *Bradyrhizobium* spp, appear to be important in adapting these cells to endophytic conditions (Dunlap *et al.*, 1996). These polysaccharides are thought to confer osmotic stability but, in *S. meliloti*, an osmotically stable strain was constructed that lacked cyclic β-(1,2)-glucans (Quandt *et al.*, 1992): although this mutant was capable of host cell invasion through infection threads, the released bacteria did not multiply in host cells and were blocked in bacteroid development, suggesting special functions for the cell envelope in adapting bacterial cells to symbiosome conditions. Immunocytochemical studies with monoclonal antibodies have revealed that the LPS of bacteroids undergoes subtle variation in epitope structure (Kannenberg *et al.*, 1994): this is apparently regulated by the microaerobic conditions and by the developmental stage of the infected host cells in pea nodules. The abnormal morphology of bacteroids derived from LPS-defective mutants (Perotto *et al.*, 1994) also indicates that LPS structure may be important for the survival and viability of bacteroids. Normally, division of intracellular bacteroids is immediately followed by division of the enclosing symbiosomal membrane, such that no more than two bacteria are found in the same envelope (Figure 5), but with LPS mutants many bacteroids occupied a single symbiosome, suggesting that the modified LPS. may have less interaction with symbiosomal membrane. Interestingly, the LPS of *Rhizobium* and *Bradyrhizobium* carries long chain (28-C) fatty acyl substituents (Bhat *et al.*, 1991), which probably provide extra stability to the bacterial outer membrane and might be considered as an adaption to the physiological stresses of the endophytic environment.

III. Symbiosomes and the Role of the Symbiosomal Membrane

III.A. DEVELOPMENT

In a mature host legume cell containing several thousand bacteroids, it has been estimated that the total surface area of symbiosomal membrane (SM) exceeds that of the plasma membrane by 20- to 50-fold (Robertson *et al.*, 1985; Verma and Hong, 1996), depending on the legume species. Host cells containing differentiating bacteroids usually become enlarged: the cytoplasm becomes tightly packed with bacteroids and the tonoplast is partially or completely occluded. The intactness of the SM appears to be a precondition for bacteroid function and, conversely, degradation of the SM appears to be a very early consequence of bacteroid malfunction (Werner *et al.*, 1985). Despite its abundance in host infected cells, the biogenesis and differentiation of SM are not well understood (Verma and Hong, 1996). Initially, bacteria are taken into the host cytoplasm by endocytosis, i.e., by involution of the plasma membrane from the infection thread or infection droplet (Figure 5). Subsequent growth and division of intracellular bacteroids necessitates further extension of the SM by inclusion of additional vesicles derived from the Golgi or endoplasmic reticulum (Roth and Stacey, 1989b; Roth and Stacey, 1989a). Verma and co-workers have suggested that membrane-anchored phosphatidyinositol-3-kinase and small GTP-binding proteins might be involved in the targetting of membrane vesicles to the symbiosome compartment (Cheon *et al.*, 1993; Verma *et al.*, 1994; Hong and Verma, 1994).

The general concept of the symbiosome as a specialised compartment of the endomembrane system (Roth and Stacey, 1989b) is somewhat challenged by the discovery that, in many woody legumes, *e.g.* *Andira* (de Faria *et al.*, 1987) and also in the non-legume *Parasponia* (Trinick, 1979), rhizobia are not released into the cytoplasm as small membrane-enclosed units. Instead, rhizobia develop the capacity for nitrogen fixation within tubular threads, resembling thin-walled infection threads, but perhaps somewhat specialised in structural and metabolic capabilities (Becking, 1992). Thus, when examining the role of the symbiosomal membrane in evolutionarily advanced legumes, it is important to consider to what extent the symbiosomal membrane has become differentiated from the plasmamembrane or infection thread membrane (in terms of specialised functions lost or acquired). We should also perhaps reflect on how the systems of metabolic exchange that apparently control the activity of bacteroids within the symbiosomes of herbaceous legumes might have evolved from the still unknown processes associated with the fixation threads of more primitive legumes.

As it differentiates, the SM apparently loses some functions characteristic of the plasma membrane: for example, SM is not associated with the accumulation of cell wall material. At the same time, it acquires new functions that are more reminiscent of the tonoplast membrane or endoplasmic reticulum (Verma and Hong, 1996). The functional role of the SM can perhaps be rationalised from two points of view, depending on whether the intracellular bacteria are perceived as potential friend or as potential pathogen. Regarding rhizobia as potential friends, the bacteroids and their enclosing plant-derived symbiosomal membrane are almost equivalent to a prokaryotic organelle. The symbiosome is thus the under-evolved equivalent of a mitochondrion or a chloroplast, a specialist cytoplasmic organelle concerned with the metabolism of dinitrogen gas. One role of the SM is therefore to facilitate the exchange of metabolites between the bacteroid and the host cell cytoplasm in a way that will optimise the overall efficiency of biological nitrogen fixation. Regarding rhizobia as potential pathogens involves recognition of the fact that not all rhizobia capable of invading nodule tissue will necessarily be capable of fixing nitrogen. For example, the bacteria may be better adapted to function in an alternative legume host, or they may simply be mutant in some aspect of the nitrogen fixation process (Huang, 1986; Huang *et al.*, 1993; Perotto *et al.*, 1994). In such circumstances, intracellular rhizobia become pathogenic rather than symbiotic and therefore an important role of the SM is to retain those properties of the plasma membrane that are concerned with the capacity to respond to microbial attack and invasion by a potentially pathogenic endophyte.

III.B. COMPOSITION

Callose (glucan 1-3β) synthase, previously recognised as a plasma membrane marker, has also been identified in purified preparations of SM from soybean (Ahlborn and Werner, 1992). This enzyme is often considered to be an important component of the plant defence system responding to pathogen attack. However, in the context of root nodules, it seems that callose synthase is only activated when the normal balance of symbiosis is disturbed. Callose deposits could only be visualised in

thin sections of clover or soybean root nodules when these harboured symbiotically defective or incompatible strains of rhizobia (Ahlborn and Werner, 1992; Kumarasinghe and Nutman, 1977). In addition to callose synthase, a number of other inducible defence-related responses have been shown to operate in symbiosomes resulting from incompatible *Rhizobium*-legume associations. These include synthesis of antimicrobial phytoalexins (Werner *et al.*, 1985; Niehaus *et al.*, 1993), secretion and oxidation of lignins and phenolics (Perotto *et al.*, 1994), and accumulation of hydroxyproline-rich glycoproteins (Benhamou *et al.*, 1991). Analysis of nodule-specific cDNA libraries from pea and *Vicia faba* have indicated the presence of a group of small cysteine-rich proteins expressed at different stages during symbiosome development (Kardailsky *et al.*, 1993) and various glycine-rich proteins (Kuster *et al.*, 1995; Rice *et al.*, 1993).

In order to analyse the components of the symbiosomal membrane, symbiosomes with intact membranes can be isolated from nodule homogenates by density gradient centrifugation under conditions which protect against physical or osmotic rupture (Brewin *et al.*, 1985; Robertson *et al.*, 1978; Blumwald *et al.*, 1985; Garbers *et al.*, 1988; Herrada *et al.*, 1989; Price *et al.*, 1987; Christiansen *et al.*, 1995). In the case of pea nodules, the antigenic characteristics of the glycocalyx associated with the luminal face of the SM have been investigated by the isolation and use of a wide range of monoclonal antibodies as immunological probes (Perotto *et al.*, 1991). These glycocalyx antigens were also present on plasma membranes of infected or uninfected host cells and sometimes on Golgi membranes, but not on endoplasmic reticulum. Using monoclonal antibody JIM 18, a novel complex inositol-containing glycolipid of plasma membranes has recently been identified and tentatively characterised as a glycophospho-sphingolipid. Immunolocalisation studies on pea nodule sections revealed that the antigen was present on plasma membranes and juvenile (undifferentiated) symbiosomal membranes but was lost from the SM (but not the plasma membrane) of differentiating host cells. Loss of this antigen is the earliest known cytological marker for the differentiation of the SM and perhaps signals the advent of a specific targeting pathway for SM biogenesis (Perotto *et al.*, 1995). As discussed by Lugtenberg (Chapter 3) this developmental transition point is apparently accompanied by fundamental changes in bacteroid outer membrane proteins (Roest *et al.*, 1995). Following this differentiation step, the lipid composition of the mature SM appears to become more similar to that of microsomal membranes than to the plasma membrane. Compared to plasma membrane, SM is high in phosphatidylcholine and low in phosphatidylethanolamine (Hernandez and Cooke, 1996). Choline kinase II has been shown to be very active in infected soybeans nodule tissue (Mellor et al., 1986).

Bacterial components of the symbiosomal membrane and symbiosomal fluid have not been systematically described. In soybean nodules, a nodule-specific polysaccharide is synthesised by bacteroids of some strains of *B. japonicum* (Streeter et al., 1995). Similarly in the symbiosomes of *Neptunia natans* (Subba-Rao et al., 1995), bacteroids are embedded in a fibrillar matrix.

III.C. METABOLIC EXCHANGE

Because the SM has a specialised metabolic role and represents the major membrane component of infected plant cells, it is not surprising that it should harbour the products of several nodule-specific genes (nodulins) (Cheon et al., 1994; Verma and Hong, 1996). The best studied example is Nodulin-26 (N-26) from soybean, which encodes an integral membrane protein (Miao and Verma, 1993). N-26 is specifically expressed in root nodules of soybean, although a nodule-specific counterpart has not yet been identified in any other host legume. The protein is encoded by a member of an ancient gene family, conserved from bacteria to humans. A plant protein with homology to N-26 has also been identified as the tonoplast intrinsic protein of vacuolar membranes (Maurel et al., 1993) and at least two homologues of N-26 are expressed in vegetative tissues of soybean, their maximum level of expression being found in the root elongation zone (Miao and Verma, 1993). During the evolution of the legume symbiosis, it therefore seems probable that a pre-existing gene expressed in root tissues was recruited for symbiotic function and brought under nodule-specific developmental control. Based on its homology with several eukaryotic and prokaryotic channel-type membrane proteins, it is suggested that Nodulin-26 functions as a channel protein that facilitates the transport of small molecules or ions across the SM. Its activity is apparently regulated by a protein kinase located in the SM (Lee et al., 1995).

The exchange of metabolites across the symbiosomal membrane is discussed elsewhere in this volume. However, the exact role of the SM in initiating and regulating the process of biological nitrogen fixation by bacteroids is still a matter for conjecture. One of the curious and little-understood features of the SM is the mechanism by which the enclosed bacteroids are supplied with respiratory substrates and other materials necessary for nitrogen fixation, but the further growth and division of bacteroids is inhibited. The symbiosome system appears to be maintained in a state of dynamic equilibrium and if the process of nitrogen fixation does not take place, either because of a bacterial mutation, a plant mutation, or some other physiological constraint, the bacteroids rapidly become senescent and the host cell subsequently dies (Huang et al., 1993; Kneen et al., 1990; Novak et al., 1995). Several models have been proposed based on the concept of coupled metabolic exchanges (Udvardi and Kahn, 1993) or of a dynamic equilibrium regulating pH in the symbiosomal compartment (Brewin, 1991). However, the fact that the symbiosome is constantly receiving protein-filled vesicles derived from the Golgi apparatus has not received much attention hitherto, although this process could be important in relation to the overall regulation of bacteroid nutrition (Kardailsky et al., 1996; Dahiya et al., 1997).

III.D. AUTOLYSIS

Components of the symbiosomal fluid that have been identified suggest that the symbiosome has the capacity to develop into a lytic vesicle (Mellor, 1989). There is biochemical evidence for the existence of α-mannosidases (Kinnbach et al., 1987), proteases and protease inhibitors (Manen et al., 1991; Mellor et al., 1984). Expression of a nodule-specific cysteine protease gene is associated with senescent infected tissue within pea nodules (Kardailsky and Brewin, 1996). In addition, there is accumulating evidence that some proteins within the symbiosomal fluid may turnover very rapidly, at least under some conditions. For a pea nodule lectin (Kardailsky et al., 1996; Dahiya et al., 1997) and for a *Phaseolus* nodulin Npv30 (Campos et al., 1995), there appears to be discrepancy between a very high level of gene transcript detectable by *in situ* hybridisation and a very low level of protein antigen detectable by immunogold localisation. This suggests that certain proteins may turn over very rapidly in the symbiosomal fluid. Furthermore, in soybean nodules, artificially induced senescence has been shown to result in selective proteolytic degradation of proteins associated with the symbiosomal membrane (Jacobi et al., 1994).

It is interesting to note that the final stage in the developmental pathway for nearly all bacteroids in indeterminate nodules is senescence. Symbiosome degradation and host cell autolysis can generally be observed in the older tissue at the base of the nodule (Truchet and Coulomb, 1973) and is frequently induced precociously in symbiotically ineffective nodules (Novak *et al.*, 1995; Pladys and Vance, 1993). In herbaceous legumes with determinate nodules, *e.g.* soybean (Pfeiffer *et al.*, 1983), symbiosome senescence occurs more synchronously at the time of pod-fill. A possible cue for bacteroid senescence might be the acidification of the symbiosomal fluid which could allow the symbiosome to develop into a lytic vesicle, thus triggering the degradation of the enclosed bacteroids (Brewin, 1991). One early aspect of senescence is the breakdown of the symbiosomal membrane, which parallels the loss of nitrogen-fixing capacity (Herrada *et al.*, 1993). The mechanism of this degradation appears to be via the auto-oxidation of leghaemoglobin and the subsequent release of superoxide anions and hydrogen peroxide (Puppo *et al.*, 1991). From the point of view of the host plant, the perceived benefits from host cell autolysis are straightforward: this process serves to recycle fixed nitrogen to other parts of the plant, and at the same time it serves to limit the systemic spread of rhizobia. However, from the bacterial viewpoint, it may seem rather curious that most endosymbiotic bacteria within a nodule are eventually killed by host cell autolysis. How then can nitrogen fixation within a nodule be of evolutionary importance to *Rhizobium*? The nodule is of course a 'safe haven', a unique ecological niche, but what is the selective advantage of symbiosis when the majority of nitrogen-fixing bacteria eventually die? The answer to this enigma seems to lie in the concept of *Rhizobium* as a colonial endophyte (as discussed in relation to opine metabolism in Chapter 9). At the time of nodule initiation, it is well known that the progeny of only one or a few bacterial cells colonise the nodule tissue through the initiation of infection threads (Dart, 1975). Therefore, the entire population of bacteria within a nodule can be considered as a genetically homogeneous clone. Within the nodule, it is therefore possible to recognise in the specialised physiological role of bacteroids features that are somewhat reminiscent of the "altruistic" behaviour often observed in colonial insects. In other words, although nitrogen-fixing bacteroids may ultimately die within the nodule, their physiological activity creates a growth opportunity for other undifferentiated bacterial siblings that occupy intercellular spaces, infection threads or infection droplets in the same nodule. Thus, the final phase in the life cycle of root nodule bacteria apparently involves saprophytic growth by undifferentiated nodule bacteria on senescent host tissue.

IV. Conclusions

The hallmarks of the *Rhizobium* legume symbiosis are fairly typical of a highly evolved mutualistic symbiosis. It is characterised by progressive morphological and physiological specialisation and by the development of more intimate interactions between plant and microbial cell surfaces. At each stage in the process of tissue and cell invasion by *Rhizobium*, it is possible to identify adaptive features that promote the symbiotic interaction and other features related to host defence systems that could be activated to control the growth of a potential pathogen. Despite the fact that infection threads, symbiosomes, and the legume root nodule itself are all morphological structures unique to the symbiosis, much of nodule development can be viewed as a re-organisation of the normal processes of plant cell morphogenesis as a result of new morphogenetic cues and signals provided by *Rhizobium*. Although the products of some plant genes expressed during nodule development (the so-called nodulins) are either unique to nodules or are strongly up-regulated in nodule tissue, such genes are probably not "novel" in an evolutionary sense, and these genes generally have functional counterparts in other tissues and in the nodules of other legumes where they may not have acquired 'nodulin' status. Finally, as discussed by Quispel (Chapter 25), it is becoming clear that there may be many common features linking the morphology and morphogenesis of the *Rhizobium*-legume symbiosis with other major root symbioses of angiosperms, notably the actinorhizal and arbuscular mycorrhizal symbioses (Pawlowski and Bisseling, 1996; Gianinazzi-Pearson, 1996). With this in mind, it may be helpful to encourage the more widespread use of the terms 'symbiosome' and 'symbiosomal membrane' in preference to their more traditional counterparts, namely 'peribacteroid unit' and 'peribacteroid membrane'.

V. Acknowledgements

I acknowledge financial support from the UK Biotechnology and Biological Research Council and the EC-TMR programme 'Symbiosis and Defence' (ERB FMR XCT 960039). I am grateful to Janet Sprent for useful discussions on comparative legume anatomy.

VI. References

Ahlborn, B. and Werner, D. (1992) Physiol. Mol. Plant Pathol. 40, 299-314.

Appels, M.A. and Haaker, H. (1991) Plant Physiol. 95, 740-747.

Bassett, B., Goodman, R.N. and Novacky, A. (1977) Can. J. Microbiol. 23, 573-582.

Becker, A., Neihaus, K. and Pühler, A. (1995) Mol. Microbiol. 16, 191-203.

Benaben, V., Duc, G., Lefebvre, V. and Huguet, T. (1995) Plant Physiol. 107, 53-62.

Benhamou, N., Lafontaine, P.J., Mazau, D. and Esquerré-Tugayé, M.-T. (1991) Planta 184, 457-467.

Beringer, J.E., Brewin, N.J. and Johnston, A.W.B. (1980) Heredity 45, 161-186.

Bhat, U.R., Mayer, H., Yokota, A., Hollingsworth, R.I. and Carlson, R.W. (1991) J. Bacteriol. 173, 2155-2159.

Blumwald, E., Fortin, M.G., Rea, P.A. and Verma, D.P.S. (1985) Plant Physiol. 78, 665-672.

Bolanos, L., Brewin, N.J. and Bonilla, I. (1996) Plant Physiol. 110, 1249-1256.

Borisov, A.Y., Morzina, E.V., Kulikova, O.A., Tchetkova, S.A., Lebsky, V.K. and Tikhonovich, I.A. (1992) Symbiosis 14, 297-313.

Bradley, D.J., Butcher, G.W., Galfrè, G., Wood, E.A. and Brewin, N.J. (1986) J. Cell Sci. 85, 47-61.

Bradley, D.J., Kjellbom, P. and Lamb, C.J. (1992) Cell 70, 21-30.

Brewin, N.J., Robertson, J.G., Wood, E.A., Wells, B., Larkins, A.P., Galfrè, G. and Butcher, G.W. (1985) EMBO J. 4, 605-611.

Brewin, N.J. (1991) Ann. Rev. Cell Biol. 7, 191-226.

Campos, F., Carsolio, C., Kuin, H., Bisseling, T., Rocha-Sosa, M. and Sanchez, F. (1995) Plant Physiol. 109, 363-370.

Cheon, C.I., Lee, N.G., Siddique, A.B.M., Bal, A.K. and Verma, D.P.S. (1993) EMBO J. 12, 4125-4135.

Cheon, C.I., Hong, Z.L. and Verma, D.P.S. (1994) J. Biol. Chem. 269, 6598-6602.

Chiarello, M.D., Larré, C., Kedzior, Z.M. and Gueguen, J. (1996) J. Agric. Food Chem 44, 3723-3726.

Christiansen, J.H., Rosendahl, L. and Widell, S. (1995) J. Plant Physiol. 147, 175-181.

Chun, J.Y., Sexton, G.L., Roth, L.E. and Stacey, G. (1994) J. Bacteriol. 176, 6717-6729.

Cook, D., Dreyer, D., Bonnet, D., Howell, M., Nony, E. and VandenBosch, K.A. (1995) Plant Cell 7, 43-55.

Csanadi, G., Szecsi, J., Kalo, P., Kiss, P., Endre, G., Kondorosi, A., Kondorosi, E. and Kiss, G.B. (1994) Plant Cell 6, 201-213.

Dakiya, P., Kardailsky, I.V. and Brewin, N.J. (1997) Plant Physiol. 115, In Press.

Dart, P.J. (1975) J.G. Torrey and D.T. Clarkson (eds.) The Development and Function of Roots, Academic Press, London, N.Y., pp. 467-506.

de Boer, M.H. and Djordjevic, M.A. (1995) Protoplasma 185, 58-71.

Dunlap, J., Minami, E., Bhagwat, A.A., Keister, D.L. and Stacey, G. (1996) Mol. Plant-Microbe Interact. 9, 546-555.

Gamas, P., de Carvalho, N., Lescure, N. and Cullimore, J.V. (1996) Mol. Plant-Microbe Interact. 9, 233-242.

Garbers, C., Meckbach, R., Mellor, R.B. and Werner, D. (1988) J. Plant Physiol. 132, 442-445.

Gianinazzi-Pearson, V. (1996) Plant Cell 8, 1871-1883.

Glazebrook, J., Ichige, A. and Walker, G.C. (1993) Genes and Development 7, 1485-1497.

Goetting-Minesky, M.P. and Mullin, B.C. (1994) Proc. Natl. Acad. Sci. USA 91, 9891-9895.

Gu, X.J. and Verma, D.P.S. (1996) EMBO J. 15, 695-704.

Hernandez, L.E. and Cooke, D.T. (1996) Phytochemistry In Press,

Herrada, G., Puppo, A. and Rigaud, J. (1989) J. Gen. Microbiol. 135, 3165-3171.

Herrada, G., Puppo, A., Moreau, S., Day, D.A. and Rigaud, J. (1993) FEBS Lett. 326, 33-38.

Higashi, S., Kushiyama, K. and Abe, M. (1987) Can. J. Microbiol. 32, 947-952.

Hong, Z.G. and Verma, D.P.S. (1994) Proc. Natl. Acad. Sci. USA 91, 9617-9621.

Huang, J-S. (1986) Ann. Rev. Phytopathol. 24, 141-157.

Huang, S.S., Djordjevic, M.A. and Rolfe, B.G. (1993) Protoplasma 172, 180-190.

Jacobi, A., Katinakis, P. and Werner, D. (1994) J. Plant Physiol. 144, 533-540.

Jacobsen-Lyon, K., Jensen, E.O., Jorgensen, J-E., Marcker, K.A., Peacock, W.J. and Dennis, E.S. (1995) Plant Cell 7, 213-223.

James, E.K., Sprent, J.I., Minchin, F.R. and Brewin, N.J. (1991) Plant, Cell and Environment 14, 467-476.

Kannenberg, E.L., Perotto, S., Bianciotto, V., Rathbun, E.A. and Brewin, N.J. (1994) J. Bacteriol. 176, 2021-2032.

Kannenberg, E.L. and Brewin, N.J. (1994) Trends in Microbiology 2, 277-283.

Kardailsky, I.V., Yang, W.-C., Zalensky, A.O., Van Kammen, A. and Bisseling, T. (1993) Plant Mol. Biol. 23, 1029-1037.

Kardailsky, I.V., Sherrier, D.J. and Brewin, N.J. (1996) Plant Physiol. 111, 49-60.

Kardailsky, I.V. and Brewin, N.J. (1996) Molecular Plant-Microbe Interact. 9, 689-695.

Kijne, J.W. (1992) G. Stacey, R.H. Burris and H.J. Evans (eds.) Biological Nitrogen Fixation, Chapman and Hall, NY, London, pp. 349-398.

Kinnbach, A., Mellor, R.B. and Werner, D. (1987) J. Exp. Bot. 38, 1373-1377.

Kneen, B.E., LaRue, T.A., Hirsch, A.M., Smith, C.A. and Weeden, N.F. (1990) Plant Physiol. 94, 899-905.

Kouchi, H. and Hata, S. (1993) Mol. Gen. Genet. 238, 106-119.

Kumarasinghe, R.M.K. and Nutman, P.S. (1977) J. Exp. Bot. 28, 961-976.

Küster, H., Quandt, H.J., Broer, I., Perlick, A.M. and Puhler, A. (1995) Plant Mol. Biol. 29, 759-772.

Küster, H., Schröder, G., Frühling, M., Pich, U., Rieping, M., Schubert, I., Perlick, A.M. and Pühler, A. (1995) Plant Mol. Biol. 28, 405-421.

Lee, J.W., Zhang, Y.X., Weaver, C.D., Shomer, N.H., Louis, C.F. and Roberts, D.M. (1995) J. Biol. Chem. 270, 27051-27057.

Leigh, J.A. and Walker, G.C. (1994) Trends Genet. 10, 63-67.

Lloyd, C.W., Pearce, K.J., Rawlins, D.J., Ridge, R.W. and Shaw, P.J. (1987) Cell Motil. Cytoskeleton 8, 27-36.

Long, S.R. (1996) Plant Cell 8, 1885-1898.

Manen, J.F., Simon, P., Slooten, J.C., Osteras, M., Frutiger, S. and Hughes, G.J. (1991) Plant Cell 3, 259-270.

Maurel, C., Reizer, J., Schroeder, J.I. and Chrispeels, M.J. (1993) EMBO J. 12, 2241-2247.

Mellor, R.B., Morschel, E. and Werner, D. (1984) Z. Naturforsch. 39c, 123-125.

Mellor, R.B., Christensen, T.M.I.E. and Werner, D. (1986) Proc. Natl. Acad. Sci. USA 83, 659-663.

Mellor, R.B. (1989) J. Exp. Bot. 40, 831-839.

Miao, G.-H. and Verma, D.P.S. (1993) Plant Cell 5, 781-794.

Miller, I.M. and Baker, B.D. (1986) Protoplasma 131, 82-91.

Morrison, N. and Verma, D.P.S. (1987) Plant Mol. Biol. 9, 185-196.

Muller, P., Ahrens, K., Keller, T. and Klaucke, A. (1995a) Mol. Microbiol. 18, 831-840.

Muller, P., Klauke, A. and Wegel, P. (1995b) Planta 197, 163-175.

Niehaus, K., Kapp, D. and Puhler, A. (1993) Planta 190, 415-425.

Noel, K.D., VandenBosch, K.A. and Kulpaca, B. (1986) J. Bacteriol. 168, 1392-1401.

Novak, K., Pesina, K., Nebesarova, J., Skrdelta, V., Lisa, L. and Nasinec, V. (1995) Ann. Bot. 76, 303-315.

Ouyang, L.-J. and Day, D.A. (1992) Plant Physiol. Biochem. 30, 613-623.

Ozga, D.A., Lara, J.C. and Leigh, J.A. (1994) Mol. Plant-Microbe Interact. 7, 758-765.

Pawlowski, K. and Bisseling, T. (1996) Plant Cell 8, 1899-1913.

Perlick, A.M., Fruhling, M., Schroder, G., Frosch, S.C. and Puhler, A. (1996) Plant Physiol. 110, 147-154.

Perotto, S., VandenBosch, K.A., Butcher, G.W. and Brewin, N.J. (1991) Development 112, 763-773.

Perotto, S., Brewin, N.J. and Kannenberg, E.L. (1994) Mol. Plant-Microbe Interact. 7, 99-112.

Perotto, S., Donovan, N., Drobak, B.K. and Brewin, N.J. (1995) Mol. Plant-Microbe Interact. 8, 560-568.

Pfeiffer, N.E., Torres, C.M. and Wagner, F.W. (1983) Plant Physiol. 71, 797-802.

Pladys, D. and Vance, C.P. (1993) Plant Physiol. Oct 1993. 103, 379-384.

Price, G.D., Day, D.A. and Gresshoff, P.M. (1987) J. Plant Physiol. 130, 157-164.

Puppo, A., Herrada, G. and Rigaud, J. (1991) Plant Physiol. 96, 826-830.

Quandt, J., Hillemann, A., Niehaus, K., Arnold, W. and Pühler, A. (1992) Mol. Plant-Microbe Interact. 5, 420-427.

Rae, A.L., Perotto, S., Knox, J.P., Kannenberg, E.L. and Brewin, N.J. (1991) Mol Plant-Microbe Interact. 4, 563-570.

Rae, A.L., Bonfante-Fasolo, P. and Brewin, N.J. (1992) Plant J. 2, 385-395.

Reuhs, B.L., Williams, M.N.V., Kim, J.S., Carlson, R.W. and Cote.F., (1995) J. Bacteriol. 177, 4289-4296.

Ribeiro, A., Akkermans, A.D.L., Van Kammen, A., Bisseling, T. and Pawlowski, K. (1995) Plant Cell 7, 785-794.

Rice, S.J., Grant, M.R., Reynolds, P.H.S. and Farnden, K.J.F. (1993) Plant Science 90, 155-166.

Robertson, J.G., Lyttleton, P., Bullivant, S. and Grayston, G.F. (1978) J. Cell Sci. 30, 129-149.

Robertson, J.G., Wells, B., Brewin, N.J., Wood, E.A., Knight, C.D. and Downie, J.A. (1985) J. Cell Sci. Suppl. 2, 317-331.

Roest, H.P., Goosen-de-Roo, L., Wijffelman, C.A., de Maagd, R.A. and Lugtenberg, B.J.J (1995) Mol. Plant-Microbe Interact. 8, 14-22.

Rolfe, B.G., Carlson, R.W., Ridge, R.W., Dazzo, F.B., Mateos, P.F. and Pankhurst, C.E. (1996) Aus. J. Plant Physiol. 23, 285-303.

Rosendahl, L., Dilworth, M.J. and Glenn, A.R. (1992) J. Plant Physiol. 139, 635-638.

Roth, L.E. and Stacey, G. (1989a) Eur. J. Cell Biol. 49, 24-32.

Roth, L.E. and Stacey, G. (1989b) Eur. J. Cell Biol. 49, 13-23.

Salzwedel, J.L. and Dazzo, F.B. (1993) Mol. Plant-Microbe Interact. 6, 127-134.

Scheres, B., Van de Weil, C., Zalensky, A., Horvath, B., Spaink, H., Van Eck, H., Zwartkruis, F., Wolters, A.M., Gloudemans, T., Van Kammen, A. and Bisseling, T. (1990a) Cell 60, 281-294.

Scheres, B., van Engelen, F., van der Knaap, E., van de Wiel, C., Van Kammen, A. and Bisseling, T. (1990b) Plant Cell 2, 687-700.

Schlaman, H.R.M., Horvath, B., Vijgenboom, E., Okker, R.J.H. and Lugtenberg, B.J.J. (1991) J. Bacteriol. 173, 4277-4287.

Sharma, S.B. and Signer, E.R. (1990) Genes and Development 4, 344-356.

Sherrier, D.J. and VandenBosch, K.A. (1994) Protoplasma 183, 148-161.

Stacey, G., So, J.-S., Roth, L.E., Lakshmi S.K., B. and Carlson, R.W. (1991) Mol. Plant-Microbe Interact. 4, 332-340.

Streeter, J.G. (1991) Adv. Bot. Res. 18, 129-187.

Streeter, J.G. (1995) Symbiosis 19, 175-196.

Streeter, J.G., Peters, N.K., Salminen, S.O., Pladys, D. and Zhaohua, P. (1995) Plant Physiol. 107, 857-864.

Szafran, M.M. and Haaker, H. (1995) Plant Physiol. 108, 1227-1232.

Tipping, A.J. and McPherson, M.J. (1995) J. Biol. Chem. 270, 16939-16946.

Truchet, G.L. and Coulomb, P. (1973) J. Ultrastr. Res. 43, 36-57.

Turgeon, B.G. and Bauer, W.D. (1985) Planta 163, 328-349.

Tyerman, S.D., Whitehead, L.F. and Day, D.A. (1995) Nature 378, 629-632.

Udvardi, M.K., Yang, L-J.O., Young, S. and Day, D.A. (1990) Mol. Plant-Microbe Interact. 3, 334-340.

Udvardi, M.K. and Day, D.A. (1989) Plant Physiol. 90, 982-987.

Udvardi, M.K. and Kahn, M.L. (1993) Symbiosis 14, 87-101.

Van Brussel, A.A.N., Bakhuizen, R., van Spronsen, P.C., Spaink, H.P., Tak, T., Lugtenberg, B.J.J. and Kijne, J.W. (1992) Science 257, 70-72.

van Spronsen, P.C., Bakhuizen, R., Van Brussel, A.A.N. and Kijne, J.W. (1994) Eur. J. Cell Biol. 64, 88-94.

VandenBosch, K.A., Bradley, D.J., Knox, J.P., Perotto, S., Butcher, G.W. and Brewin, N.J. (1989) EMBO J. 8, 335-342.

Vasse, J., de Billy, F. and Truchet, G. (1993) Plant J. 4, 555-566.

Verma, D.P.S., Cheon, C.I. and Hong, Z.L. (1994) Plant Physiol. 106, 1-6.

Verma, D.P.S. and Hong, Z.L. (1996) Trends Microb. 4, 364-368.

Vijn, I., Martinez-Abarca, F., Yang, W.C., das Neves, L., Van Brussel, A.A.N., Van Kammen, A. and Bisseling, T. (1995) Plant Journal 8, 111-119.

Werner, D., Mellor, R.B., Hahn, M.G. and Grisebach, H. (1985) Z. Naturforsch. 40c, 179-181.

Wexler, M., Gordon, D. and Murphy, P.J. (1995) Soil Biol. Biochem. 27, 531-537.

Whitehead, L.F., Tyerman, S.D., Salom, C.L. and Day, D.A. (1995) Symbiosis 19, 141-154.

Yang, L-J.O., Udvardi, M.K. and Day, D.A. (1990) Planta 182, 437-444.

Chapter 23

A Survey of Symbiotic Nitrogen Fixation by Rhizobia

Pierre A. Kaminski, Jacques Batut and Pierre Boistard

I. Introduction

Nitrogen fixation, that is reduction of nitrogen gas into ammonia, is not restricted to rhizobia. In fact, this property, although unique to prokaryotic organisms, is widely spread among the prokaryotic phylogenetic tree (Young, 1992). In addition, the properties of the enzymatic complex which allows the saturation of the very stable triple bond of dinitrogen are remarkably conserved. A recent illustration has been provided by the resolution of the three dimensional structure of the two components of the nitrogenase complex in as different organisms as *Azotobacter vinelandii* and *Clostridium pasteurianum* (reviewed in Smith *et al.*, 1995).

The high conservation of the nitrogenase complex and of the DNA sequences encoding its different polypeptide components can be accounted for by vertical transmission, which would assume that the ability to fix nitrogen was a primitive property of the prokaryotic ancestor and that it was lost in the course of evolution. However, it is not excluded that nitrogen fixation could have been reacquired by horizontal transmission in some instances.

Because of this high conservation, one can wonder why we should study nitrogen fixation in symbiotic organisms where the symbiotic association inevitably complicates biochemical and genetic approaches. Indeed, most of the data concerning the basic process have been obtained in organisms which fix nitrogen in culture. However, there are at least two compelling reasons to overcome the methodological constraints which hinder the study of symbiotic nitrogen fixation: The first reason is the unquestionable agronomical and ecological importance of symbiotic nitrogen fixation (see Chapter 26). In this respect, any reasonable hope to improve nitrogen fixation depends on the ability to genetically control the process. For this purpose, identification of nitrogen fixation genes and of their regulatory pathways is an obligate step. Indeed, this task, although still under the way, has been facilitated by the knowledge acquired in the simpler non-symbiotic nitrogen fixers.

A second reason which argues in favour of the study of symbiotic nitrogen fixation is the possibility to discover how a basic process such as nitrogen fixation can adapt to such a highly specialized metabolic context as that resulting from the plant environment. As will be illustrated in the body of this chapter, this adaptation as well as the symbiotic coevolution have resulted in the adoption by rhizobia of environmental sensing mechanisms of general interest or of original biological processes such as bacteroid differentiation. In fact, symbiosomes, that is nitrogen-fixing bacteroids surrounded by the plant-derived peribacteroid membrane can be considered as incompletely evolved organelles of endocytic origin. Due to this incomplete evolution, the endosymbiotic partner has retained its capacity to multiply in culture.

The recent progress in molecular biology has led to several revisions in the taxonomy of rhizobia (see Chapter 1). Accordingly, there are presently five accepted genera, *Rhizobium, Sinorhizobium,* *Mesorhizobium, Bradyrhizobium* and *Azorhizobium*. Although it was recently proposed that *Rhizobium meliloti* should be considered as a member of the genus *Sinorhizobium*, this nomenclature is not followed in this chapter.

Because this book is devoted to the *Rhizobiaceae*, this chapter will deal essentially with the bacterial functions involved in nitrogen fixation although one should not underestimate the role of the plant, as exemplified by Figure 3, which shows the metabolic connections between both symbiotic partners.

In the first section of this chapter, we will summarize the biochemical features of the reduction of nitrogen into ammonia, the characteristics of the nitrogenase enzymatic complex which allows this reaction, and describe what is presently known about the genes which control nitrogenase structure and maturation in the different rhizobia. As already mentioned, the acquisition of most of the knowledge was made possible by the high degree of conservation of the proteins and the corresponding genes. In the same section, we will describe additional, non-conserved genes which are likely involved in the maturation and functioning of nitrogenase, such as *fixABCX* and *fdxN*. Besides their clustering together with nitrogenase structural genes, an argument that supports a direct role of these genes in nitrogen fixation is their coordinated expression with nitrogenase structural genes.

In the second section of this chapter, we will review the differentiation process of endosymbiotic nitrogen fixing bacteria, also called bacteroids, and its genetic control. The process of nitrogen reduction is closely connected to bacteroid carbon metabolism, and energy supply through respiration. We will discuss the distinctive traits which characterize these metabolic routes in the nitrogen fixing bacteroids as well as the changes in nitrogen metabolism which accompany nitrogen fixation. Some of the genes implicated in these metabolic pathways have been designated as *fix* because of the phenotype conferred by the mutant alleles.

In the last section, we will review the regulatory molecules and the regulatory circuits which control the expression of nitrogen fixation genes. This review will examplify the modular structure of many regulatory molecules as well as of the pathways that they constitute in the various rhizobial strains in which they have been studied. In addition, it will provide a working hypothesis to explain the coupling between nodule development and the construction of the nitrogen-fixing apparatus.

II. Nitrogenase Synthesis and Activity in Rhizobia

Reduction of N_2 into NH_3 is catalyzed in all nitrogen-fixing organisms by the same basic enzyme, nitrogenase. It was thought for a long time that Mo-nitrogenase was unique. However, certain strains of the soil bacterium *Azotobacter* harbor alternative nitrogenases in which Va or Fe replaces Mo (Bishop and Premakumar, 1992). As these alternative nitrogenases have not been found in rhizobia, they will not be discussed further in this chapter.

Mo-Nitrogenase is a complex enzyme consisting of two components. Component I or dinitrogenase or FeMo protein is an $\alpha 2\beta 2$ heterotetramer of about 220 kDa containing 4 (Fe_4S_4) clusters (the P-clusters), and a cofactor with Fe and Mo, which is the catalytic site for N_2 reduction. Component II or dinitrogenase reductase or Fe Protein is an homodimer of about 60 kDa that contains a single Fe_4S_4 cluster. A recent milestone in the field of nitrogen fixation was the elucidation of nitrogenase structure in *Azotobacter vinelandii* and *Clostridium pasteurianum* with the subsequent derivation of new models to account for nitrogenase activity (reviewed in Smith *et al.*, 1995).

The Fe protein (dinitrogenase reductase) is reduced *in vivo* by low-potential electron donors whose nature varies with the physiology of the organism. Electrons are then transferred one at a time to the FeMo protein in a process that involves MgATP hydrolysis. This cycle repeats until enough electrons have been provided for complete substrate reduction. Nitrogenase catalyzes the following reaction:

$$N_2 + 16\ MgATP + 8e^- + 8\ H^+ \Rightarrow 2\ NH_3 + 16MgADP + 16Pi + H_2$$

This chemical equation reflects one main feature of biological nitrogen fixation: its high energetic cost. This is, of course, no surprise given the high stability of the N_2 triple bond. Two ATP molecules are hydrolysed at every electron transferred from dinitrogenase reductase to nitrogenase. In addition, N_2 reduction is always accompanied by the concomitant reduction of two protons, thus adding to the energetic cost of the process. This high energy cost of nitrogen fixation explains why symbiotic nitrogen fixers, which benefit from plant photosynthesis, are, by far, the main contributors of N_2 fixed biologically on earth. A second, major, feature of nitrogenase is its oxygen-lability, mainly contributed by the Fe protein component.

Nitrogenase requires a strict anaerobic environment for activity and, as a consequence, aerobic nitrogen fixers such as rhizobia need to protect their nitrogenase from oxygen (Hill, 1992).

II.A. IDENTIFICATION AND FUNCTION OF *NIF* GENES IN RHIZOBIA

The characterization of nitrogen fixation genes in rhizobia was greatly facilitated by the thorough biochemical and genetic work conducted during the seventies on the free-living diazotroph prototype, *Klebsiella pneumoniae*. Twenty *nif* (for *ni*trogen *f*ixation) genes have been identified in *K. pneumoniae* by fine genetic mapping and, more recently, by DNA sequencing. These 20 genes span a 24kb-cluster near the *his* operon (Arnold *et al.*, 1988; Dean and Jacobson 1992; Gussin *et al.*, 1986; Merrick, 1992).

By using *nifH* and *nifD* specific fragments from *K. pneumoniae*, Ruvkun and Ausubel (1980) showed that these genes were well conserved among nitrogen-fixing organisms, including rhizobia (Ruvkun and Ausubel, 1980). *R. meliloti* was the first *Rhizobium* in which *nif* containing fragments could be cloned by virtue of homology. The subsequent development of the powerful reverse genetics used by Ruvkun and Ausubel in 1981 led to the discovery of a cluster of genes essential for symbiotic nitrogen fixation in this bacterium (Ruvkun *et al.*, 1982). This cluster comprises the classical *nif* genes but also *fix* genes for which no homologues have been found in *K. pneumoniae*.

II.A.1. Function of nif Genes

To date, 10 *nif* genes homologous to those of *K. pneumoniae* have been found in rhizobia. The α and β apo-subunits of component I are encoded by *nifD* and *nifK* respectively, whereas component II polypeptides are encoded by *nifH*. Full assembly of nitrogenase requires the product of other *nif* genes, in particular *nifB*, *nifE* and *nifN*, which are required for the biosynthesis of the FeMo-cofactor. The strong amino acid sequence similarity between NifE and NifD and between NifN and NifK, respectively, led to the suggestion that the *nifN* and *nifE* products might form an heterotetrameric complex structurally similar to the FeMo protein (Aguilar *et al.*, 1987; Aguilar *et al.*, 1990; Brigle *et al.*, 1987). This complex would serve as a scaffold for FeMo-cofactor biosynthesis whereas the product of *nifB* would be required for cofactor assembly into the complex (Paustian *et al.,* 1989). Since both nitrogenase components contain metallo clusters, it was conceivable that the acquisition of Fe or S involved the product of a *nif* gene. A NifS protein recently purified from *Azotobacter vinelandii* was shown to have a cysteine desulphurase activity that would release the sulphur necessary for metallocluster formation (Zheng *et al.,* 1993). Interestingly, *B. japonicum nifS* is partially dispensable for nitrogen fixation since Tn*5* insertion mutants retain 30% nitrogenase activity (Ebeling *et al.,* 1987). Possibly, another enzyme might substitute for NifS function in *B. japonicum.*

The function of NifW is unknown but it is absolutely required for both free-living and symbiotic nitrogen fixation in *A. caulinodans,* in contrast to the situation in *K. pneumoniae* and *A. vinelandii* where it is only required for full activity of the FeMo protein. Recently, Kim and Burgess (1996) have reported that NifW directly interacts with the FeMo protein to form a high molecular weight complex. This complex was observed only when the cell free extracts were exposed to O_2, suggesting that NifW could participate to the O_2 protection of the FeMo protein (Kim and Burgess, 1996). Finally, the last two *nif* genes are regulatory. *nifA* codes for the specific transcriptional activator of *nif* operons (see section IV). *nifX,* identified in *B. japonicum* (Hennecke, 1990), might play a role in negative regulation of nitrogenase gene expression, as proposed in *K. pneumoniae* (Gosink *et al.,* 1990).

Other *nif* genes such as *nifQ* or *nifZ* identified by the complete nucleotide sequence of a megaplasmid from *Rhizobium* sp. NGR234 (Freiberg *et al.,* 1997) may exist in rhizobia even though they have not been found until now. On the other hand, it is also very likely that the *nifL, nifF* and *nifJ* genes from *K. pneumoniae* do not exist in rhizobia. NifL is a flavoprotein that antagonizes the activity of the transcriptional activator NifA in response to oxygen and combined nitrogen (Hill *et al.,* 1996). Although NifA is inactivated by oxygen in rhizobia, NifA sensitivity to oxygen does not involve a NifL protein in rhizobia (see section IV). *nifF* and *nifJ* code respectively for a flavodoxin and a pyruvate-flavodoxin oxidoreductase that channel electrons from pyruvate to the Fe-protein in the anaerobic nitrogen-fixer *K. pneumoniae.* In rhizobia, which are aerobic, electrons for feeding nitrogenase do not originate from pyruvate.

II.A.2. Function of fix Genes

fixABCX. The *fixABCX* genes were initially identified in *R. meliloti* because of their close linkage to the *nif* cluster and because mutants in these genes induced Fix⁻ nodules (Dusha *et al.,* 1987; Earl *et al.,* 1987). *fixABCX* homologues have been found in *B. japonicum* (Fuhrmann *et al.,* 1985; Gubler and Hennecke, 1986), *R. leguminosarum* bv. *viciae* (Grönger *et al.,* 1987), bv. *trifolii* (Iismaa and Watson, 1987), bv. *phaseoli* (Michiels and Vanderleyden, 1993) and *A. caulinodans* (Arigoni *et al.,* 1991). Mutants in any of the four genes of *R. meliloti* and *B. japonicum* were defective in symbiotic nitrogen fixation and also in free-living nitrogen fixation in the case of *A. caulinodans* (Dusha *et al.,* 1987; Gubler and Hennecke, 1986; Gubler *et al.,* 1989; Kaminski *et al.,* 1988).

The full-length nucleotide sequence of the *fixABCX* operon has been established in *R. meliloti, A. caulinodans* and *B. japonicum* (Earl *et al.,* 1987; Arigoni *et al.,* 1991; Weidenhaupt *et al.,* 1996) whereas only part of it has been reported in *R. leguminosarum* (Grönger *et al.,* 1987; Michiels and Vanderleyden, 1993). Amino-acid sequence comparisons revealed a significant homology between the products of the *fixA* and *fixB* genes and the β and α-subunits of human electron transfer flavoprotein, respectively (Arigoni *et al.,* 1991; Finocchiaro *et al.,* 1993). FixC contains a consensus sequence found in proteins that bind NAD or FAD (Arigoni *et al.,* 1991) and shares sequence similarity with the human electron transfer flavoprotein-ubiquinone oxidoreductase (Goodman *et al.,* 1994). FixX is clearly homologous to ferredoxin-like proteins (Earl *et al.,* 1987; Gubler and Hennecke, 1986). No definitive biochemical functions have yet been assigned to the *fixABCX* gene products but the features described above suggest that, together, these genes might code for an electron tranport chain to nitrogenase. Thus FixABCX might be functional analogs of NifF and NifJ from *K. pneumoniae.* This hypothesis was challenged by the fact that dithionite, an artificial electron donor, did not restore nitrogen fixation to crude extracts from *A. caulinodans fixA, fixC* or *fixX* mutants (Arigoni, 1992; Kaminski *et al.,* 1988). However, elimination of the electron transport to nitrogenase might lead to instability of both components I and II which could explain the absence of nitrogenase activity.

It is very likely that FixABCX participate in a redox process in microaerobic or aerobic diazotrophs since homologues have been found in *Azospirillum brasilense* (Fogher *et al.*, 1985; Galimand *et al.*, 1989), *Azotobacter chroococum* and *Azotobacter vinelandii* (Evans *et al.*, 1988). The systematic sequencing of the *E. coli* chromosome revealed the presence of *fixABCX* homologues in the 0-2.4 min region (Yura *et al.*, 1992) that could be involved in carnitin metabolism (Eichler *et al.*, 1995). However, their role in this metabolism remains to be determined.

frxA and *fdxN*, two ferredoxins-like proteins have been found in *R. meliloti* (FdxN) and *B. japonicum* (FrxA) but only FdxN is required for symbiotic nitrogen fixation (Ebeling *et al.*, 1988; Klipp *et al.*, 1989). FdxN could be involved in electron transfer to nitrogenase since *in vitro* experiments have shown that the *R. meliloti* FdxN can donate electrons to *R. capsulatus* nitrogenase (Riedel *et al.*, 1995). Alternatively, based on *fdxN* localisation downstream *nifB*, it was proposed that FdxN could be involved in a redox step necessary for the maturation of the nitrogenase cofactor but no biochemical data are available yet to support this hypothesis (Klipp *et al.*, 1989).

II.B. GENETIC ORGANIZATION OF *NIF* AND *FIX* GENES IN RHIZOBIA

In *K. pneumoniae*, *nif* genes are tightly packed in a single cluster on the chromosome. Rhizobial *nif* and *fix* genes can be organized in more than one cluster whose location(s) differs in different rhizobia. Symbiotic functions (nodulation and nitrogen fixation) are carried by large (up to more than 1Mb) plasmids, termed pSym, in fast growing rhizobia (*R. meliloti*, *R. leguminosarum*, *R. etli*, and *R.* sp NGR234 (see Chapter 2) whereas they are located on the chromosome in *A. caulinodans* and in slow-growing rhizobia. The *R. meliloti nif* region resembles the *K. pneumoniae nif* cluster for it comprises all the *nif* genes identified so far. Interestingly however, a cluster of *nod* genes, including *nodABC*, is interspersed between *nifE* and *nifN*. This organization is well conserved in the other fast-growing rhizobia: *nif* and *fixABCX* genes are arranged in the same sequential order in *R. leguminosarum* bv. *trifolii* and bv. *viciae* (Iismaa and Watson, 1989) although the linkage with *nod* genes is different in the three species. *nif* and *fix* genes are located in at least two or three separate chromosomal clusters in *B. japonicum* and *A. caulinodans* (Figure 1).

Most *nif* and *fix* genes are organized in operons. However, the genes contributing to one operon vary from one species to the other. *R. meliloti nifH* and *nifD* are cotranscribed with *nifK*, *nifE*, ORF110 and *nifN* (Aguilar *et al.*, 1987; Aguilar *et al.*, 1990; Corbin *et al.*, 1982; Ruvkun *et al.*, 1982), whereas they form a transcriptional unit with only *nifK* and *nifE* in *A. caulinodans* (Denèfle *et al.*, 1987; Elmerich *et al.*, 1982). *nifH* is transcribed separately from *nifDK* which form an operon with *nifE*, *nifN*, and *nifX* (Hennecke, 1990) in *B. japonicum* and in other slow-growing rhizobia, such as *R.* sp. *parasponium* (Scott *et al.*, 1983) or *R.* sp. *cowpea* (Fischer and Hennecke, 1984; Kaluza *et al.*, 1983; Kaluza and Hennecke, 1984; Yun and Szalay, 1984). Similarly, *fixABCX* are transcribed as a single operon in all species but *B. japonicum* where *fixA* and *fixBCX* are two transcriptional units (Arigoni *et al.*, 1992; Earl *et al.*, 1987; Gubler and Hennecke, 1988; Gubler *et al.*, 1989). *nifA* can be transcribed as a single unit in *A. caulinodans* (Ratet *et al.*, 1989) or as part of an operon: *fixRnifA* in *B. japonicum* or *nifABfdxN* in *R. meliloti* (Buikema *et al.*, 1987; Klipp *et al.*, 1989).

A reiteration of the *nifH* gene has been reported in *R. etli* and *A. caulinodans* (Quinto *et al.*, 1982; Quinto *et al.*, 1985; Norel and Elmerich, 1987). *R. etli* has three identical *nifH* genes whereas *nifH1* differs from *nifH2* by 6 nucleotides in *A. caulinodans*. All these *nifH* copies code for functional Fe proteins and none of them is essential for nitrogen fixation. The high conservation of their nucleotide sequence indicates that these copies have been acquired by gene duplication rather than by separate lateral transfer events. These reiterations may confer an advantage to the strains having them such as better adaptation and rapid differential response to the environmental conditions or higher nitrogen fixation ability (Valderrama *et al.*, 1996). Another possibility would be that these extra *nifH* copies have a reductase activity unrelated to nitrogen fixation.

II.C. FREE-LIVING NITROGEN FIXATION

It was believed for a long time that rhizobia were able to fix N_2 only in symbiosis. However, several slow-growing strains of *B. japonicum* and *Bradyrhizobium* spp. associated with cowpea can display nitrogenase activity in the absence of the host plant (Keister, 1975; McComb *et al.*, 1975;

Rhizobium meliloti

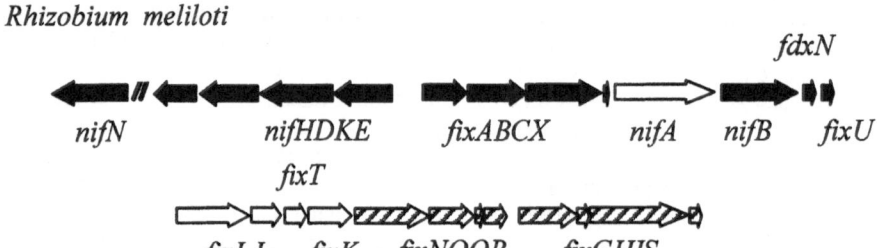

Bradyrhizobium japonicum

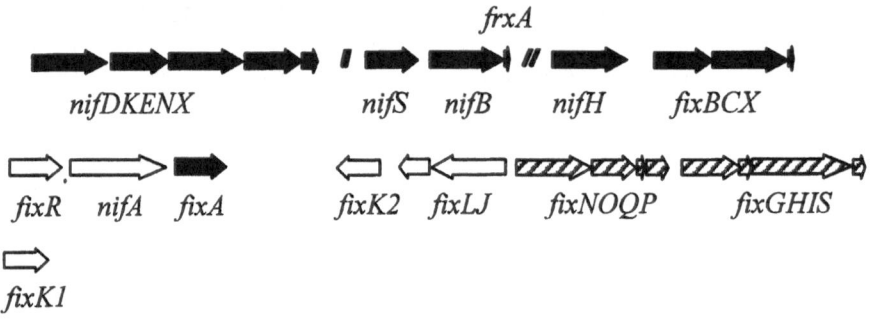

Azorhizobium caulinodans

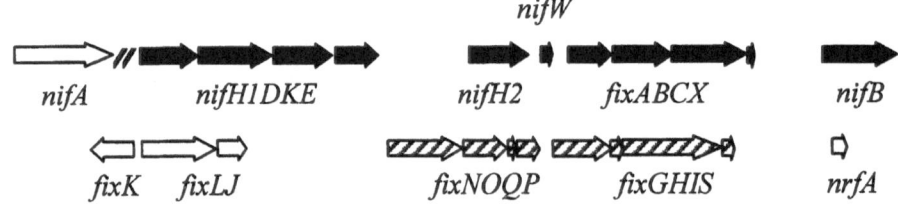

Rhizobium sp. NGR234

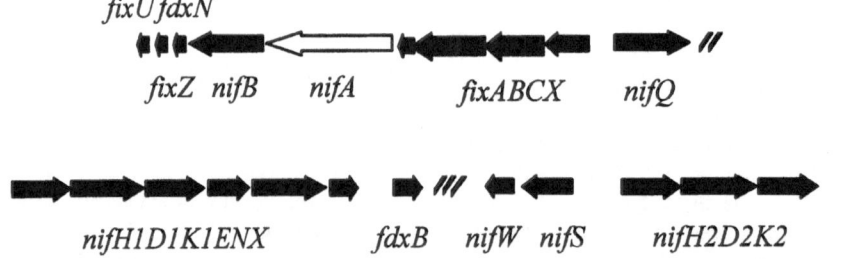

Figure 1. nif and *fix* clusters in rhizobia

Pagan *et al.*, 1975). Asymbiotic acetylene reduction has been reported also in some fast-growing *R. leguminosarum* strains TA101 (Kurz and LaRue, 1975) and 128C30 (Stam *et al.*, 1983) or *Bradyrhizobium* spp. associated with cowpea (Bender *et al.*, 1986) but only strain IHP100 reduces acetylene at a rate similar to *Bradyrhizobium* strains. This nitrogenase activity, as assayed by acetylene reduction could only be detected when cells were cultivated on defined medium. In all cases a very low O_2 tension was required (*ca.* 0,1% O_2) together with an appropriate nitrogen source (*e.g.* glutamate) that can sustain growth without inhibiting nitrogenase synthesis or activity. None of these strains were however able to grow at the expense of free molecular nitrogen indicating that either nitrogen fixation or assimilation was limiting for growth.

At the beginning of the 80's in Senegal, a major discovery in the field of nitrogen fixation was the isolation of a new *Rhizobium* strain, ORS571 (now known as *Azorhizobium caulinodans* ORS571) with unusual characteristics. First, it was isolated not from root but from stem nodules of a tropical legume, *Sesbania rostrata*. Second, ORS571 turned out to be a genuine diazotroph as it could grow at the expense of free N_2 in pure culture (Dreyfus *et al.*, 1983; Elmerich *et al.*, 1982). The acetylene reduction activity of *A. caulinodans* cultures (up to 2000 nmoles C_2H_4 produced per mg of dry weight per hour) was similar to those of other free-living diazotrophs and much higher than that of any *Bradyrhizobium* strain examined so far.

Why is *A. caulinodans* capable of diazotrophy? A factor that may contribute significantly to the ability of *A. caulinodans* to grow on N_2 is its high tolerance to oxygen (up to 12 μM as compared to <1μM in the case of *B. japonicum* CB576). This unusual tolerance to O_2 for a rhizobial strain might be explained by a high respiration rate conferred by the presence of several terminal oxidases in *A. caulinodans* (Kaminski *et al.*, 1996; Kitts and Ludwig, 1994). In addition to nitrogenase protection against O_2 damage, intense respiration may also favor nitrogenase activity *via* high ATP synthesis. Another relevant difference between *A. caulinodans* and *B. japonicum* may be that *B. japonicum* exports most (90%) of its fixed N_2 (which is then diluted out) whereas *A. caulinodans* does not (Gebhardt *et al.*, 1984; Ludwig, 1984). Finally, it should be noted that *A. caulinodans* ORS571, as the founding member of a new genus, *Azorhizobium*, may have an entirely

different physiology from *Rhizobium* and *Bradyrhizobium* from which it is distantly related (Dreyfus *et al.*, 1988; see Chapter 1). Whether *A. caulinodans* should be regarded as a free-living nitrogen fixer that has acquired recently, in evolutionary terms, the ability to form and colonize nodules remains to be elucidated.

III. Symbiotic Nitrogen Fixation by Rhizobia

In a first section we will review the process of nitrogen fixation in the nodule, with special emphasis on bacteroid differentiation. For the purpose of this review we will define bacteroid differentiation as a multi-step process leading to bacterial forms competent for symbiotic nitrogen fixation. However this definition by no means implies that all changes happening during bacteroid differentiation are instrumental for the ability to fix nitrogen. In a second section, the physiology of bacteroids will be reviewed in its three main aspects: oxygen, carbon and nitrogen metabolism. As the bacteroid and plant metabolisms are closely interconnected, it is not possible to describe bacteroid metabolism without referring, at least briefly, to the plant genes (nodulins) involved. In a last section, we will describe a selected set of regulatory mechanisms that have been shown to occur during bacteroid differentiation although, very often, their direct relevance to the process of nitrogen fixation *per se* is not known. Nevertheless this will illustrate the multiplicity of regulations operating inside the nodule and our large ignorance of the underlying physiological factors and genetic pathways.

III.A. FIXATION IN THE SYMBIOTIC STATE: A SURVEY OF BACTEROID DIFFERENTIATION

Since the early days of studies on the *Rhizobium*-legume symbiosis, more than a century ago, bacteroid differentiation has attracted a considerable interest and the corresponding literature describes in detail the morphological, cytological and biochemical changes occurring during bacteroid differentiation, in different biological systems. More recently, Vasse *et al.*, (1990) published a comprehensive description of bacteroid differentiation in the alfalfa-*R. meliloti* system (Vasse *et al.*, 1990). As this paper also established the currently accepted nomenclature for both nodule tissue organization and bacteroid types, it will be used as a guide for the following discussion.

A tremendous advantage of the alfalfa-*R. meliloti* association for cytological studies is that it forms indeterminate type of nodules, *i.e.* nodules with an apical persistent meristem. As a consequence, tissues and bacteria at successive stages of differentiation can be visualized along a longitudinal section of a single nodule. The youngest and poorly differentiated cells, are located at the distal (relative to the root) part of the nodule (*i.e.* near the apex) whereas highly differentiated cells are found in the more proximal part of the nodule. The nodule itself can be best described by a succession of four distinct zones, each identified by a roman number (Figure 2A).

At the apex of the nodule, zone I corresponds to the nodule meristem, which is free of bacteria. This meristem forms in the inner cortex of the root upon the action of Nod factors, the plant morphogens synthesized by rhizobia (see Chapter 20). In indeterminate nodules, such as alfalfa or pea nodules, the meristem keeps strongly active all along the life of the nodule, growing outward, giving rise to cells that are in turn infected by rhizobia.

Zone II corresponds to the region of the nodule where bacteria penetrate the plant cells *via* infection threads and quit the infection threads to invade the plant cell cytoplasm, from which they are separated by the peribacteroid membrane. Accordingly, zone II is called the infection zone of the nodule or, also, the pre-fixing zone to emphasize the fact that it has not yet reached competence for nitrogen fixation. Bacteria released from infection threads, called type 1 bacteroids, very much resemble free-living bacteria in size and cytoplasm appearance (Figure 2B). Type 2 bacteroids are most abundant in the proximal part of zone II and consist of elongated rod-shaped cells in which the nucleoid is no more visible (Figure 2C). Two important physiological changes are known to occur at these early stages of bacteroid differentiation. First, the very intense cell division inside the infection threads stops when the type 2 bacteroid stage is reached. Second, DNA replication stops after a few rounds when bacteria are freed into the plant cytoplasm. As a consequence, bacteroids have an increased nucleic content as compared to free-living cells, although the ratio to cell volume is not changed. In addition, it has been shown that *R. leguminosarum* bv. *viciae* bacteroids do have a

higher (7 fold) content of total proteins per cell but that the proportion of housekeeping proteins relative to the total protein content was unchanged (Schlaman *et al.*, 1991).

Interzone II-III is a very restricted zone of the nodule, 3 to 4 layers of cells in a wild-type mature nodule, that is easily identified under the microscope thanks to its richness in amyloplasts (Figure 2C). Interzone II-III in many respects is a landmark in nodule differentiation. First, it exclusively contains type 3 bacteroids which are fully elongated forms, about 7 fold larger than free-living bacteria, that have regained cytoplasmic heterogeneity with electron-dense areas. Second, a major event associated with bacteroid differentiation, the transcription of nitrogen fixation genes (*nif* and *fix*) starts there, although the cells do not fix nitrogen yet (Soupène *et al.*, 1995; Vasse *et al.*, 1990; Yang *et al.*, 1991). Third, expression of key late nodulins, such as leghaemoglobin or PsNod6, also starts at interzone II-III (de Billy *et al.*, 1991; Franssen *et al.*, 1992; Kardailsky *et al.*, 1993). We shall also detail later the example of a bacterial gene, *ropA*, that is turned off at the interzone II-III boundary. This is also the case for the early nodulin PsENod5 (Franssen *et al.*, 1992). Usually, these changes in gene expression occur very abruptly, sometimes from cell to cell, in the first layer of cells of interzone II-III. It is thus anticipated that key signals act at the level of interzone II-III, one of which has recently been identified as oxygen (Soupène *et al.*, 1995 see section IV).

Zone III is the region where type 4 bacteroids are found (Figure 4D). The study of both young immature (non-fixing) and N_2-fixing nodules supported the conclusion that zone III is the zone of the nodule where nitrogen fixation takes place and that type 4 bacteroids are thus fully differentiated forms, competent for N_2 fixation (Vasse *et al.*, 1990). They are of the same size as type 3 bacteroids but are best distinguished from the latter by the greater heterogeneity of their cytoplasms, characterized by a zonation of electron-dense and electron-transparent areas. In mature nitrogen-fixing nodules, type 4 bacteroids usually spread into 8 to 12 cell layers.

The last step in bacteroid differentiation occurs in the proximal zone III. Type 5 bacteroids vary in morphology and show a progressive loss of cytoplasmic heterogeneity. The number of type 5 bacteroids gradually decreases during the differentiation of the proximal degenerative zone IV of the nodule. Ghost membranes of plant and

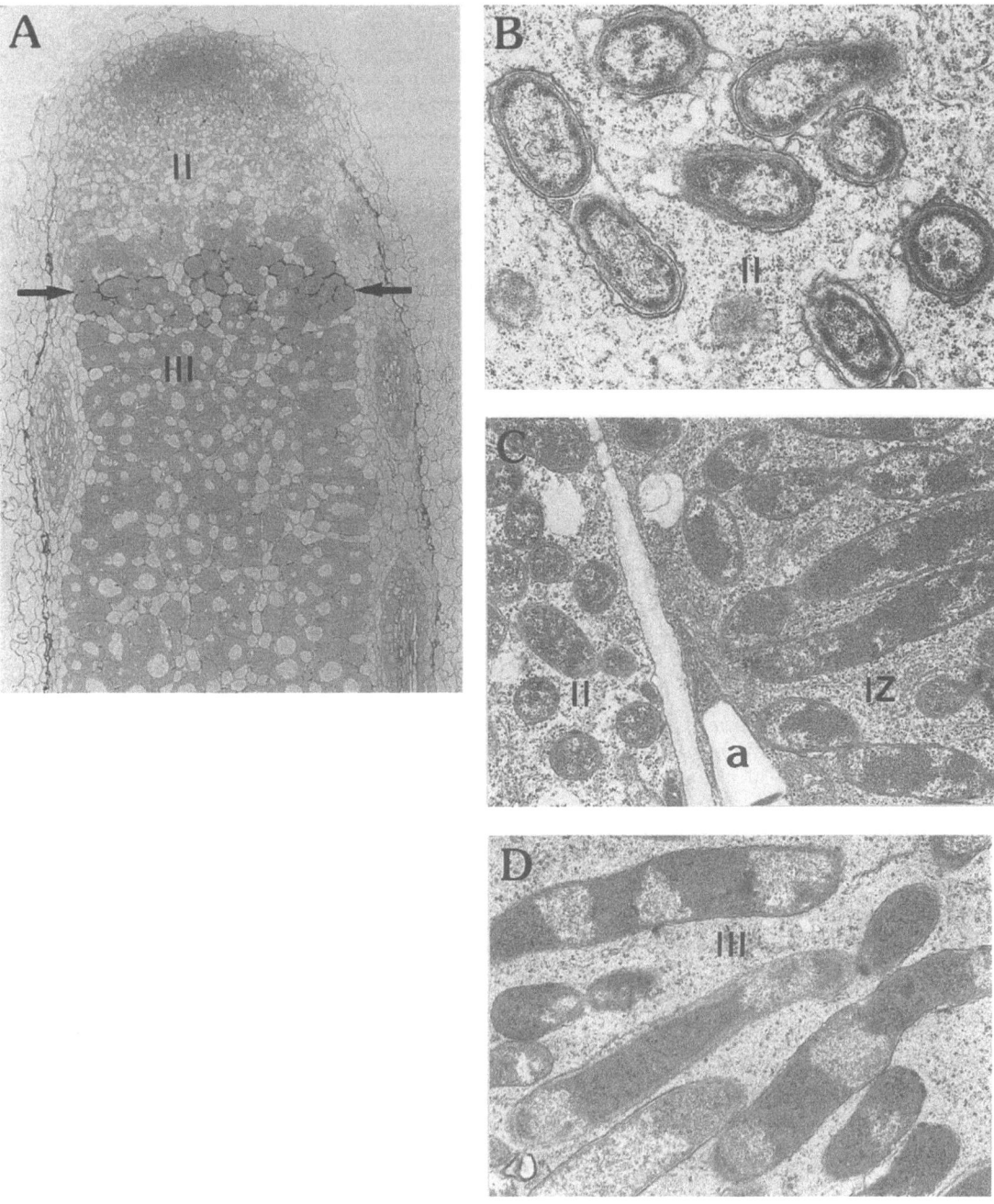

Figure 2. Nodule and bacteroid differentiation in the *R. meliloti*-alfalfa symbiosis (after Vasse *et al.*, 1990)
Panel A: Longitudinal section of a 4-week old alfalfa nodule. I: Meristem. II: Infection zone. III: Nitrogen-fixing zone. Arrows: Interzone II-III. Panels B to D: Successive stages of bacteroid differentiation. (D): Type 1 bacteroids in distal zone II. (C): Type 2 bacteroids in proximal zone II (top) and type 3 bacteroids at IZII-III (bottom); a: amyloplast. (D): Type 4 bacteroids in zone III. Panel B is enlarged 2 fold as compared to panels C and D. Photographs: Courtesy of Dr G. Truchet.

bacterial origin are the ultimate result of the senescing process.

III.B. PHYSIOLOGY OF THE NITROGEN-FIXING BACTEROID

Functionality of the nodule requires that the plant and microbe partners mutually adapt to meet the physiological requirements for nitrogen fixation and the assimilation of fixed nitrogen by the plant. The plant must provide the bacteroids with a microoxic environment compatible with nitrogenase activity and with photosynthates for bacteroid energy metabolism. In addition, the bacteroids must adapt to efficiently export fixed nitrogen to fuel the plant amino acid metabolism. These three complementary aspects of bacteroid metabolism will be reviewed separately below, although they are obviously intimately interconnected.

III.B.1. O_2 Metabolism in Bacteroids

Nitrogen fixation is an extremely energy-consuming process requiring at least 16 molecules of ATP per molecule of N_2 reduced. In rhizobia as in other aerobes, the production of ATP is generated by oxidative phosphorylation but, on the other hand, nitrogenase is rapidly inactivated by oxygen. Both the plant and the bacteroid contribute to solve this problem known as the oxygen paradox. It is now well established that the low O_2 concentration in the infected nodule cells (5-30 nM as compared with 250 µM under aerobiosis) is maintained by an oxygen diffusion barrier, by the buffering capacity of leghaemoglobin and also in part by bacteroid respiration.

Leghaemoglobin (Lb) is the predominant plant protein in nodules (up to 30% of the total soluble proteins) to which it confers a red colour. It has an extremely high affinity for O_2 with a fast kinetics of association and a slow dissociation rate. Lb is detectable in infected cells just before nitrogenase activity appears. Its role is to buffer free oxygen and facilitate O_2 diffusion to the actively respiring bacteroids (Appleby, 1984).

Existence of a specific bacteroid oxidase is supported by several biochemical observations: First, several oxidase activities were revealed by measuring the respiration of B. japonicum bacteroids (Bergersen et al., 1976), and the oxidase with the highest affinity for oxygen (Km = 5nM) was the most efficient in terms of nitrogenase activity. Second, the cytochrome oxidases aa_3 and o present

in aerated cells were absent in bacteroids while a new set of cytochromes appeared: P450, c_{550}, c_{552} and c_{554} (Appleby, 1984).

An important step in the characterization of the symbiotic terminal oxidase was the discovery of an insertion mutant in the fbcFH genes coding for cytochrome bc_1 complex (Thöny-Meyer et al., 1989). This mutant has a Fix⁻ phenotype whereas mutations in cycM coding for a cytochrome c or coxA coding for the subunit I of cytochrome aa_3 are Fix⁺ (Bott et al., 1990; Bott et al., 1991; O'Brian and Maier, 1987). These data are consistent with a branched respiratory system originating at the cytochrome bc_1 complex. By analogy with the other known cytochrome bc_1 dependent respiratory chains, it was thought that the symbiotic respiratory chain was constituted of a cytochrome c and a cytochrome c oxidase. However, neither cytochromes c_{550}, c_{552} and c_{554} encoded by cycAB and C respectively nor cox MNOP coding for an alternative cytochrome c oxidase were part of the bacteroid oxidase complex since insertion mutants formed effective nitrogen fixing nodules (Fix⁺) (Bott et al., 1992; Bott et al., 1995). The genes coding for the bacteroid oxidase were identified beacuse of their clustering with the fixLJ operon in both R. meliloti and B. japonicum (Kahn et al., 1993; Preisig et al., 1993).

The nucleotide sequence was established and amino acid comparisons suggested an oxidase function for the FixNOQP products. FixN has the characteristics of a heme b and copper oxidase subunit, FixO and FixP are thought to be mono and diheme cytochrome c respectively whereas FixQ does not present any sequence similarity to known proteins (Preisig et al., 1993). Genes homologous to fixNOQP have been identified in other Rhizobia such as R. leguminosarum bv. viciae and A. caulinodans (Schlüter et al., 1993; Mandon et al., 1994) but also in the non-symbiotic bacteria Rhodobacter capsulatus and Agrobacterium tumefaciens (Thöny-Meyer et al., 1994b; Schlüter et al., 1995).

fixN, fixO and fixP but not the fixQ B. japonicum mutants have reduced TMPD (Trimethylphenylene-diamine) and cytochrome oxidase activities and a Fix⁻ phenotype, features which are in agreement with the proposed role of bacteroid oxidase. This oxidase has been purified from B. japonicum membranes. Because of the presence of c and b types cytochromes and the absence of the subunit II carrying the Cu_A center found in the classical heme-copper oxidases, this oxidase was called cbb_3-type cytochrome oxidase. The B. japonicum cbb_3-oxidase is responsible for at least 85% of the cytochrome c

oxidase activity of the bacteroid. Moreover it has a Km of 7nM for O_2, a value compatible with the O_2 concentration inside the nodule (Preisig *et al.*, 1996a). In each bacterium mentionned above, the so-called cbb_3-oxidase is induced under microaerobic conditions with the exception of *R. capsulatus* where, surprisingly, the cbb_3-oxidase can support aerobic growth.

Contrary to the situation in *R. meliloti* and *B. japonicum*, *A. caulinodans* fixNOQP mutants were only slightly affected for both free-living and symbiotic nitrogen fixation (Mandon *et al.*, 1994). This discrepancy probably originates from the fact that *A. caulinodans* has an extended set of oxidases. Recently, it was shown that both cytochrome *bd* and cytochrome cbb_3-oxidases could be used under symbiotic conditions whereas others, such as the aa_3 oxidase, would be functional in the free-living state (Kaminski *et al.*, 1996; Kitts and Ludwig, 1994).

A role for other cytochromes is not excluded in rhizobia. Cytochrome *d* is synthesized at low oxygen concentrations in *R. leguminosarum* and *R. trifolii*. However, it is apparently absent from *R. leguminosarum* bv. *viciae* bacteroids (Vargas *et al.*, 1996) and thus its role could be restricted to free-living conditions. The presence of an *o*-like cytochrome has been reported in bacteroids of *R. leguminosarum*. These differences in cytochrome composition might account for the ability of bacteroids to cope with different O_2 concentrations. Isolated *R. leguminosarum* bacteroids have a maximum nitrogenase activity at 800nM O_2 as compared to 100nM for *B. japonicum*. Furthermore, *A. caulinodans* displays maximal nitrogenase activity at a dissolved O_2 concentration of 10μM whereas the O_2 concentration is in the range of 20nM within both root and stem nodules (Gebhardt *et al.*, 1984).

The fixNOQP operon, located downstream fixK in *R. meliloti* and upstream fixLJ in *B. japonicum*, is followed by the fixGHIS genes in all rhizobia studied until now (Kahn *et al.*, 1989; Mandon *et al.*, 1993; Preisig *et al.*, 1996b). fixGHIS and fixNOQP are both regulated by fixK2 in *B. japonicum* and mutations in either operon confers a Fix⁻ phenotype (Fischer, 1996; Fischer, 1994; Preisig *et al.*, 1996b). Analysis of the deduced translation products of fixGHIS led Kahn *et al.*, (1989) to propose that the operon would code for a membrane-located cation pump involved in symbiotic nitrogen fixation (Kahn *et al.*, 1989). This suggestion was based on the following observations: i) The four proteins contain potential transmembrane helices ii) FixG contains cysteine

rich clusters found in proteins containing iron-sulfur centers and, thus, might be involved in a redox process iii) FixI was homologous to eukaryotic and prokaryotic P-type ATPases that transport cations across membranes. More recently, Preisig *et al.*, (1996) have postulated that the membrane bound complex encoded by fixGHIS might play a role in uptake and metabolism of copper required for cbb_3-oxidase synthesis (Preisig *et al.*, 1996b). In support of this model, *B. japonicum* FixI shows 31% sequence identity with the copper-uptake protein, CopA from *Enterococcus hirae*. Furthermore, assembly of the cytochrome cbb_3 oxidase was found to depend on a functional fixGHIS region. Copper would be required for formation or stabilization of the FixNO core complex before FixP association (Preisig *et al.*, 1996b, Zufferey *et al.*, 1996).

C-type cytochromes, which comprise FixO and FixP, have their prosthetic haem group covalently attached to the apoprotein and require for the formation of the mature holoprotein the products of at least 8 genes (reviewed by Thöny-Meyer *et al.*, 1994a; Thöny-Meyer, 1997). Because of the importance of bacteroid respiration for nitrogen fixation, it was expected that mutants affected for cytochrome *c* maturation would have a Fix⁻ phenotype. Several *B. japonicum* and *R. phaseoli* mutants lacking cytochromes *c* have been characterized (Ramseier *et al.*, 1991; Ritz *et al.*, 1993; Soberon *et al.*, 1993). The corresponding genes cycHJKL identified in *B. japonicum*, *R. meliloti* and *R. leguminosarum* are required for a Fix⁺ phenotype and would code for a membrane bound complex involved in the attachment of the haem to FixO and/or FixP as well as cytochrome c_1 (Delgado *et al.*, 1995; Kereszt *et al.*, 1995; Ritz *et al.*, 1995; Thöny-Meyer, 1997).

III.B.2. Carbon Metabolism in Bacteroids

Symbiotic nitrogen fixation has a high demand for carbon sources. Photosynthates are first required to support nodule growth and bacterial proliferation at the early stages of the symbiotic interaction. In nitrogen-fixing bacteroids, carbon compounds are essential for the generation of ATP and reducing power needed for nitrogenase activity. Lastly, carbon skeletons are required for assimilation of fixed ammonia in the plant cells. Therefore carbon metabolism is central to the process of nitrogen fixation. We will only briefly summarize in this section the outlines of carbon metabolism in bacteroids as this is fully reviewed in chapter 24.

Emphasis will be put in this section on the genetic control of carbon import into bacteroids.

N₂-Fixing Bacteroids Use Dicarboxylic Acids as Carbon Source. Although sucrose is the major photosynthate translocated to the nodule (reviewed by Gordon, 1995) a large body of evidence indicates that bacteroids primarily take up dicarboxylic acids (that we shall abbreviate as dcA), to satisfy their carbon needs (reviewed in Chapter 24). Accordingly, *Rhizobium* mutants affected in sugar metabolism are generally fully effective (reviewed by Day and Copeland, 1991 and McDermott and Kahn, 1992) whereas the failure to import dcA results in the loss of the ability of bacteroids to fix N_2 (Bolton *et al.*, 1986; Engelke *et al.*, 1989; Finan *et al.*, 1983; Ronson *et al.*, 1981; van Slooten *et al.*, 1992; Watson *et al.*, 1988; Yarosh *et al.*, 1989). Once internalized, dcA are metabolized through the TCA cycle as rhizobia are strict aerobes (see Chapter 24).

Besides the compelling evidence that dcA are the obligate carbon source for N_2-fixing bacteroids, other carbon sources can apparently support bacteroid proliferation and differentiation in some rhizobia. Whereas dcA import is essential for *Rhizobium* NGR234 infection of *Macroptilium*, a tropical legume (van Slooten *et al.*, 1992), nodules induced by dcA transport mutants of both *R. meliloti* and *R. leguminosarum* contain differentiated bacteroids (Engelke *et al.*, 1989; Ronson *et al.*, 1981). It is thus clear in these latter cases that other carbon sources than dcA can support bacteroid division and differentiation.

dcA Supply to Bacteroids. Photosynthates enter the nodule as sucrose *via* the phloem. The conversion of sucrose into organic acids probably follows a fermentation pathway (Figure 3). Sucrose is primarily degraded by sucrose synthase that cleaves sucrose phosphorylytically. UDP-glucose is degraded to pyruvate which undergoes carboxylation by PEP-carboxylase resulting in oxaloacetate. Oxaloacetate is converted to malate which is then taken up by bacteroids (see Gordon, 1995 for a recent review).

dcA import into bacteroids involves two different active transporters: one is of plant origin and is located in the peribacteroid membrane of the symbiosome (Day *et al.*, 1989; Udvardi *et al.*, 1988) The second transporter, the *dctA* gene product of rhizobia, is a single-protein permease embedded in

the inner membrane of the bacteroid (Engelke *et al.*, 1989; Jording and Pühler, 1993; Ronson *et al.*, 1984).

Genetics of dcA Transport to Bacteroids. In free-living cells of *R. leguminosarum* and *R. meliloti*, dcA are transported *via* a common transport system which is active and inducible. Ronson and coworkers first analysed the dcA transport system in *R. leguminosarum* and have identified three genes, organized in two divergently transcribed operons, *dctA* and *dctBD* (Ronson, 1988; Ronson *et al.*, 1984). A similar genetic organization was found in *R. meliloti* (Engelke *et al.*, 1989; Jiang *et al.*, 1989; Watson, 1990; Yarosh *et al.*, 1989). The *dctB* and *dctD* genes are regulatory and activate expression of *dctA*, the gene for the permease, in the presence of dcA in the growth medium. DctB and DctD belong to the two-component family of regulatory proteins (Nixon *et al.*, 1986; Ronson *et al.*, 1987; see section IV). DctD, the transcriptional activator of *dctA*, belongs to the NtrC subfamily (class I) of response regulators and, as such, requires *ntrA*, the gene encoding sigma 54, for activity. DctB is homologous to sensors of the two-component family of regulatory proteins and, thus, was proposed to sense the presence of dcA in the periplasm. In addition, *dctA* gene expression was found to be constitutive in a *dctA* mutant thus suggesting that the DctA permease is involved in the regulation of its own synthesis, possibly *via* a direct interaction with the membrane-located DctB sensor protein (Ronson and Astwood, 1985; Yarosh *et al.*, 1989).

The genetics of dcA transport is less understood in other *Rhizobium* spp. Recent data suggest that some rhizobia may use more than one system for dcA transport, at least under free-living conditions. In *Rhizobium* NGR234, a second permease allows growth of a *dctA* mutant on succinate (van Slooten *et al.*, 1992). However, this permease does not operate under symbiotic conditions as a *dctA* mutant was clearly Fix⁻. Similarly, two succinate uptake systems have been detected in *B. japonicum*, one of which is *ntrA* (*rpoN*)-independent (Humbeck and Werner, 1987; Kullik *et al.*, 1991).

Whereas the DctA permease is required for both free-living growth on dcA and nitrogen fixation by bacteroids, the regulatory operon *dctBD* appeared not to be essential under symbiotic conditions in *R. meliloti* and *R. leguminosarum*. Null mutations in *dctD* and mutations in *dctB* that were polar on *dctD* had a limited effect on *dctA* expression, dcA

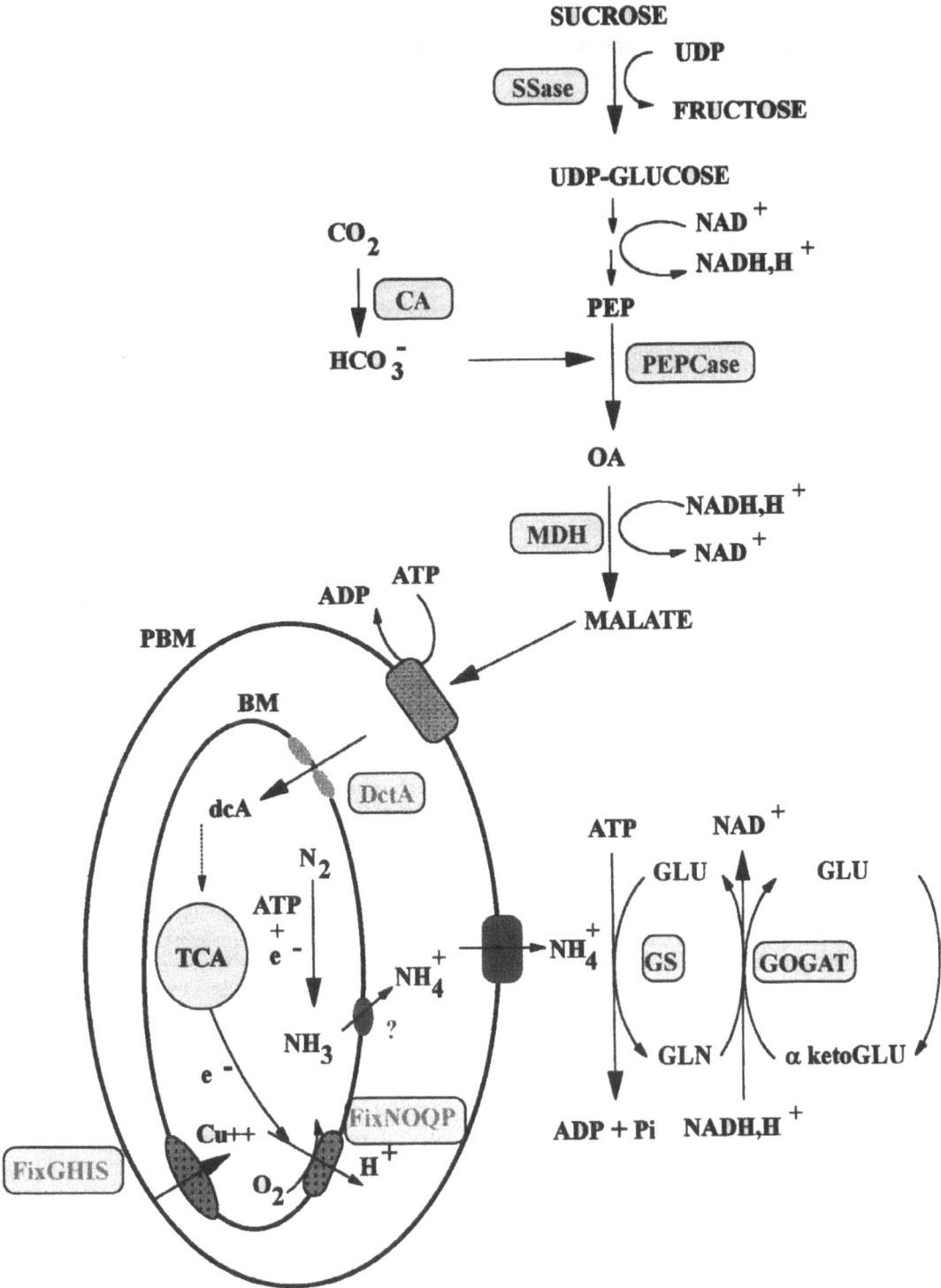

Figure 3. Schematic representation of metabolite exchange between a nitrogen-fixing bacteroid and the plant cell. PBM: peribacteroid membrane. BM: bacteroid membrane.

transport to bacteroids and nitrogen fixation (*ca.* 50 to 70% of wild-type activities) (Engelke *et al.*, 1987; Ronson, 1988; Wang *et al.*, 1989; Yarosh *et al.*, 1989). This suggested the existence of an alternative system operating on *dctA* in bacteroids, subsequently called Alternative Symbiotic Activator (Ronson, 1988). However, the nature of ASA could not be elucidated so far. One way toward ASA identification consisted in looking for extragenic suppressors of *dctD* mutations in pure culture. This led to the isolation of mutations in the DctD-homolog NtrC that, however, was not ASA (Labes *et al.*, 1993). Ronson (1988) suggested, on the basis of promoter sequence analysis, that NifA might be the ASA. Indeed, *nifA* inactivation was found to affect *dctA* expression in alfalfa nodules. However, *dctA* expression could not be demonstrated *ex planta* under conditions where *nifH*, a known NifA-target, was (Wang *et al.*, 1989). Thus the nature of ASA remains elusive.

The physiological relevance of ASA is also a matter of debate. One possibility is that symbiotic expression of *dctA* is indeed taken over, or at least contributed for, by ASA. Alternatively, activation of *dctA* expression by ASA may correspond to a cross-talk mechanism that only takes place in *dctBD* mutants, thus being physiologically irrelevant. Identifying the gene encoding ASA is clearly instrumental to solve this question. Recently, new *ntrA*-dependent activators have been isolated from *R. meliloti* by PCR means, that offer new candidates for ASA (Kaufman and Nixon, 1996).

III.B.3. Nitrogen Metabolism in Symbiosomes

In free-living diazotrophs, nitrogen fixation serves to support bacterial growth under nitrogen starvation. Symbiotic nitrogen fixation instead aims at feeding the host legume with fixed nitrogen, thus allowing plant growth on nitrogen-depleted soils. We shall describe in this section the two ways rhizobia may have adapted to this function. First, nitrogenase synthesis in rhizobia, with the exception of *A. caulinodans*, is uncoupled from the nitrogen status of bacteria. In such a way, rhizobia can fix nitrogen in amounts well behind those covering their own needs. Second, ammonia assimilation is essentially turned off in bacteroids which permanently pump out fixed ammonia to the cytoplasm of the plant cells where it is assimilated. A full description of nitrogen metabolism in rhizobia can be found in Chapter 24.

nif Gene Expression in Rhizobia is Uncoupled From the ntrBC System. A central player of the genetic circuitry governing nitrogen metabolism in rhizobia as in enterics is the NtrBC two-component system. In enterics, *ntrBC* primarily activates glutamine synthetase (*glnA*) gene expression, thus promoting ammonia assimilation under nitrogen-limiting conditions. Szeto *et al.*, (1987) identified *ntrBC* homologs in *R. meliloti* and demonstrated that symbiotic *nif* gene expression was not under *ntrBC* control (Szeto *et al.*, 1987). Moreover, *ntrBC* mutants were essentially Fix$^+$, indicating that the *ntrBC* system in *R. meliloti* played no (essential) role in symbiotic nitrogen fixation (Szeto *et al.*, 1987). Accordingly, it was demonstrated that the *ntrBC* genes are not required for the expression of the *fixK/fixNOQP* respiratory pathway (David *et al.*, 1988). Instead, it was found that *nif* and *fix* gene expression in *R. meliloti* is under oxygen control (David *et al.*, 1988; Ditta *et al.*, 1987) (see section IV). Thus the primary signal that elicits *nif/fix* gene expression in *R. meliloti* is the low oxygen concentration of the nodule that signals the bacteria that they have entered an environment appropriate for nitrogen fixation. Hence, in *R. meliloti*, a two-component system responsive to oxygen, *fixLJ*, substitutes for the nitrogen-sensing *ntrBC* two-component system of *K. pneumoniae*.

Oxygen-dependence and nitrogen-independence of nitrogen fixation gene expression has also been established in *B. japonicum*, although the genetic circuit that operates in this case differs of that described in *R. meliloti* (see section IV).

The situation is markedly different in *A. caulinodans* in which *nif* expression is controlled both by oxygen via *fixLJ*, and nitrogen *via* two different -albeit closely related- two-component systems, *ntrBC* and *ntrXY* (see section IV). Whereas the nitrogen control of *A. caulinodans nif* gene expression in the free-living state is easy to understand, the physiological significance of nitrogen control in the symbiotic state is not clear. Nevertheless, the genetic control of nitrogen fixation in *A. caulinodans*, involving *fixLJ* as well as *ntrBC*-like systems, perfectly illustrates the dualistic nature of this bacterium which is presently unique in being both a genuine diazotroph and a very efficient symbiotic nitrogen fixer.

Ammonia Assimilation is Repressed in Bacteroids. Whereas *A. caulinodans*, as enterics, has a single GS (Donald and Ludwig, 1984), rhizobia and bradyrhizobia possess at least two different glutamine synthetases, encoded by non-homologous genes. GSI, the homolog of the enterics GS is encoded by *glnA* (Carlson *et al.*, 1985; Colonna *et al.*, 1987; Moreno *et al.*, 1991; Somerville and Kahn, 1983). The regulation of *glnA* is presently unknown besides the fact that it is only slightly affected by nitrogen (Arcondéguy *et al.*, 1996; Carlson *et al.*, 1987). (Brady)Rhizobia have a second GS, GSII, encoded by the *glnII* gene, that more resembles eukaryotic GSs and is under *ntrBC* control (Carlson and Chelm, 1986; de Bruijn *et al.*, 1989). A third GSIII has been described in *R. meliloti* and *R. etli* (Chiurazzi *et al.*, 1992; de Bruijn *et al.*, 1989; Espin *et al.*, 1990; Shatters *et al.*, 1993). However its physiological relevance is unclear as it could only be detected in double *glnAglnII* mutants and has a poor biosynthetic activity (Shatters *et al.*, 1993). Rhizobial GSI, GSIII and possibly GSII are regulated post-translationally (Arondéguy *et al.*, 1996; Liu and Kahn, 1995; Ludwig, 1978; Manco *et al.*, 1992; Rossi *et al.*, 1989).

Upon entering the symbiotic state, bacteroid metabolism shifts from ammonium assimilation to ammonium export. How this is achieved has not been elucidated yet. The emerging picture is that ammonia assimilation is turned off during bacteroid differentiation, primarily at the level of GS synthesis. It was recently shown that GSI is severely down-regulated in *R. meliloti* bacteroids and, furthermore, that GSI adenylylation was not essential for symbiotic nitrogen fixation (Arondéguy *et al.*, 1996). GSII could not be detected in *R. meliloti* bacteroids using either a *glnII-lacZ* fusion (de Bruijn *et al.*, 1989) or an anti-GSII specific antiserum (Shatters and Kahn, 1989). Similarly, GSII activity was not detected in *R. etli* bacteroids (Moreno *et al.*, 1991). One interesting observation that may rationalize the down regulation of GSII expression in bacteroids is that *ntrBC* expression in *R. etli* is severely down-regulated during bacteroid differentiation (Patriarca *et al.*, 1996). Repression takes place before the onset of *nif* gene expression in *Vicia hirsuta* nodules (Patriarca *et al.*, 1996).

However a low level of GS activity could be required for symbiosis. This is suggested by the observation that *B. japonicum glnAglnII* mutants are severely affected in nodule infection and nitrogen fixation (Carlson *et al.*, 1987). Similarly, a *glnA* mutant of *R. etli* is impaired (by 50%) in nitrogen fixation (Moreno *et al.*, 1991). A GS⁻ mutant of *A. caulinodans* is also Fix⁻ (Donald and Ludwig, 1984). In contrast, *R. meliloti glnAglnII* mutants are Nod⁺Fix⁺ on alfalfa (de Bruijn *et al.*, 1989) possibly,

because these mutants are rescued by *glnT* (Shatters *et al.*, 1993). Hence a low level of ammonia assimilation could be required for proper infection of the nodules or, later on, for bacteroid differentiation and/or metabolism.

Bacteroids Excrete NH_4^+ which is Assimilated by the Plant Cells. Early evidence indicated that symbiotic nitrogen fixing bacteroids exported ammonia to the surrounding plant cell cytosol where it is assimilated (Bergersen and Turner, 1967). Congruently, no ^{14}C-labeled compounds were excreted from isolated bacteroids fed with ^{14}C succinate (Miller *et al.*, 1991). In support of the plant assimilation of the ammonia fixed by *Rhizobium*, it has been shown that the plant Glutamine synthetase activity is strongly enhanced in nodule tissues as compared to roots and that induction of specific GS genes occurs in nodules (see Cullimore and Bennett, 1992 for a review).

The mechanism by which ammonia is excreted from the bacteroids into the plant cytoplasm is not completely clear yet. A recent milestone has been the identification of an ammonium channel in the peribacteroid membrane of soybean (Tyerman *et al.*, 1995). The existence of a matching ammonium carrier in the bacteroid membrane, however, remains to be demonstrated. The alternative possibility would be that NH_3 diffuses out through the bacteroid membrane to the peribacteroid space where it is protonated.

III.C. THE GENETICS OF BACTEROID DIFFERENTIATION

Many modifications and regulations take place during the process of bacteroid differentiation. Whether these modifications/regulations are instrumental for the ability of bacteroids to fix nitrogen remains an open question in most cases. However elucidating their genetic control is of interest as this may give insights into the physiological conditions that prevail inside the nodule and into the plant signals that coordinate plant and bacteroid differentiation. We review below a selected set of bacterial genes that seem to be particularly interesting.

III.C.1. Genetic Control of DNA Replication and Cell Division in Bacteroids.

As indicated above, endosymbiotic bacteria cease division and DNA replication shortly after release into the plant cells. In order to understand how these regulations operate, S. Long and coworkers have isolated candidate genes that may provide control points in bacteroid proliferation.

The earliest known step in *Escherichia coli* cell division is the formation of a ring consisting of the FtsZ protein at the division site. Whereas *ftsZ* is unique in *E. coli*, two *ftsZ* homologs have been identified in *R. meliloti* which encode significantly different protein products (Margolin *et al.*, 1991; Margolin and Long, 1994). Future studies on the regulation and function of these two *ftsZ* genes will probably allow a better understanding of the control of cell division in bacteroids. As proper DNA replication is normally a prerequisite for cell septation, a homolog of *dnaA*, a gene essential for initiating DNA replication in *E. coli*, has been cloned recently from *R. meliloti* (Margolin *et al.*, 1995).

III.C.2. Suppression of Nodulation Gene Expression in Bacteroids

In *R. leguminosarum* bv. *viciae*, it has been demonstrated that the structural *nod* genes are turned off, at the level of transcription, before the bacteria leave the infection threads (Schlaman *et al.*, 1991). This negative regulation of *nod* gene expression seems to be important biologically since artificial constitutive expression of *nod* genes *in planta* led to a Fix⁻ phenotype (Knight *et al.*, 1986). The molecular mechanism underneath this down-regulation of inducible *nod* genes in bacteroids is still unknown. However exciting clues have been obtained recently. The silencing of *nod* gene expression is not due to a paucity in NodD regulator protein nor to the absence of flavonoids, the NodD-coinducers, nor to the presence of anti-inducers, thus pointing to an as yet unknown negative regulation mechanism (Schlaman *et al.*, 1991; see Chapter 19). Recently, direct biochemical evidence for this repressor was obtained (Schlaman *et al.*, 1992). In addition, two papers suggest a link between carbon metabolism and *nod* gene repression. First, Mavridou *et al.*, (1995) isolated a *R. leguminosarum* bv. *viciae* mutant with reduced *nodC* and *nodD* expression *ex planta* (Mavridou *et al.*, 1995). The mutant turned out to contain an IS50 insertion in *dctB* that caused constitutive transport of C₄-dicarboxylic acids. It was independently found that organic acids inhibited *nod* gene expression in *B. japonicum* (Yuen and Stacey, 1996). Thus inhibition of *nod* gene expression in bacteroids might be mediated by the presence of dicarboxylic acids inside the nodule.

III.C.3. Down Regulation of ropA

Outer membrane proteins of free-living *R. leguminosarum* bv. *viciae* are divided into four major antigen groups, two of which are severely reduced in bacteroids (de Maagd *et al.*, 1989; de Maagd *et al.*, 1994; see Chapter 3). De Maagd *et al.* (1992) isolated a gene, *ropA*, that encodes (part of) the antigen group III proteins (de Maagd *et al.*, 1992). *In situ* hybridization experiments showed that transcription of this gene is down-regulated at the transition from zone II to interzone II-III (de Maagd *et al.*, 1994). This down regulation thus operates at a stage posterior to the *nod* gene repression described above and is concomitant to *nif/fix* gene induction. However it is clearly independent of nitrogen fixation gene expression and nitrogen fixation (de Maagd *et al.*, 1994; Roest *et al.*, 1995a) Interestingly, high calcium concentrations were found to repress *ropA* expression *ex planta* (de Maagd *et al.*, 1994). Whether calcium is the relevant signal operating *in planta* remains to be demonstrated. A gene, *ropB*, coding for the antigen group II protein was recently isolated (Roest *et al.*, 1995 b).

III.C.4. BacA, a Putative Bacteroid Developmental Sensor

If bacteroid differentiation is essential for nitrogen fixation, one would expect to find mutants affected in its genetic control among Fix⁻ mutants. Unfortunately, many Fix⁻ mutations induce premature senescence of their bacteroids thus making *bona fide* bacteroid differentiation mutants difficult to isolate. There are exceptions to this general observation. Some *nif* and *fix* mutants of *R. meliloti* differentiate into type 4 and 5 bacteroids (Hirsch *et al.*, 1983; Hirsch and Smith, 1987) which clearly indicates that bacteroid differentiation is independent of the nitrogen-fixing activity of the nodule. *R. meliloti fixLJ* mutants are much more affected in bacteroid development since nodules only contain type 2 bacteroids (Vasse *et al.*, 1990; see section IV). Nevertheless, this indicates that the onset of bacteroid differentiation is not *fixLJ*-dependent in this bacterium. In *B. japonicum*, *nifA* mutants are markedly affected in bacteroid differentiation or viability thus suggesting that *nifA* may have a general role in bacteroid differentiation (Fischer *et al.*, 1986).

The best candidate for being a genuine developmental sensor is the *bacA* gene isolated from *R. meliloti* (Glazebrook *et al.*, 1993). The key observation is that *bacA* mutants seem to be blocked very early, at the onset of bacteroid differentiation. *bacA* mutants are released from infection threads and penetrate the plant cells, but then rapidly senesce so that type 2 and even type 1 bacteroids could not be found. Other suggestive, albeit indirect, evidence is that the *bacA* gene expresses maximally in the region of the nodule where bacteroid differentiation starts (*i.e.* the proximal part of zone II, interzone II-III and distal zone III). Sequence analysis revealed that *bacA* is homologous to *sbmA* from *E. coli*, whose physiological role is presently unknown but which is likely involved in (modified) peptide transport. Sequence analysis was also consistent with an inner-membrane location for BacA. Thus it was speculated that BacA might be involved in the uptake of a plant-derived molecule (perhaps a peptide) that would somehow signal the bacteria to begin differentiation (Glazebrook *et al.*, 1993; Yorgey and Kolter, 1993).

IV. Regulation of Nitrogen Fixation Genes

As illustrated in the preceding sections, the nitrogen fixing nodule is an organ of mixed origin whose full morphogenesis and differentiation results from the dialogue between the plant and the prokaryotic symbiotic partners. Expression of nitrogen fixation genes reflects the success of this dialogue which uses metabolic as well as developmental signals. The fact that one of the partners is a prokaryotic organism which can be grown in culture independently of the plant partner, has greatly facilitated the identification of its regulatory pathways. This in turn has allowed the development of genetic tools for the study of the symbiotic interplay and of the signals involved.

In the last fifteen years, there has been considerable progress in the identification of the regulatory pathways controlling nitrogen fixation gene expression (see for example the review by Fischer, 1994). As discussed in the first section of this chapter, there are two categories of rhizobia, according to their ability to express nitrogenase activity exclusively in association with their symbiotic partner, or in culture as well. Oxygen concentration appeared as a major regulatory factor of the expression of nitrogenase activity for cultures of *B. japonicum* (Bergersen *et al.*, 1976). On the other hand, it was thought that a specific symbiotic signal was required for the expression of nitrogen fixation genes in *R. meliloti* (Gussin *et al.*, 1986). It

was not before 1987 that it was realized that it was possible to induce the expression of *R. meliloti* nitrogen fixation genes in microooxic conditions (Ditta *et al.*, 1987). In addition to identifying oxygen concentration as a candidate for symbiotic regulation of nitrogen fixation gene expression, this finding facilitated the genetic analysis of the regulatory pathways and in particular the individual role of their constituents.

In this section we will first describe the various regulatory proteins which are components of the regulatory pathways controlling the expression of genes involved in nitrogen fixation. A remarkable characteristic of these pathways in different rhizobia is their modular architecture which uses homologous regulatory proteins in different combinations. Therefore we will describe first the prototypic regulatory proteins and the approach which allowed their identification in the organism where they were first identified, then their variations in different rhizobia. It is interesting to note that identification of several regulatory genes was facilitated by the fact that they are clustered together with some of the structural genes whose expression they control. In a second part we will compare the different regulatory cascades. Finally, we will review the recent data on regulation in the symbiotic state.

IV.A. COMPONENTS OF THE REGULATORY PATHWAYS OF NITROGEN FIXATION GENE EXPRESSION

IV.A.1. The NifA Transcriptional Activator
The *nifA* regulatory gene was identified as part of a cluster of genes homologous to *K. pneumoniae nif* genes because it was required for the symbiotic transcription of *R. meliloti nifHDK* structural genes (Szeto *et al.*, 1984; Zimmerman *et al.*, 1983). Sequence data indicated that *R. meliloti* NifA is homologous to *K. pneumoniae* NifA and to *K. pneumoniae* NtrC, the protein which activates genes of nitrogen metabolism in response to a deprivation of ammonium (Drummond *et al.*, 1986). Because NifA proteins from various rhizobia show extensive homology and results obtained in different species have illuminated the properties of NifA as a generic protein, we will gather the data concerning *Rhizobium* NifA from different species in the same section (see Figure 4).

Both NifA and NtrC activate transcription of genes in conjunction with sigma 54-RNA polymerase which recognizes particular -12/-24 DNA sequences (Kustu *et al.*, 1989). Activation by NtrC involves the hydrolysis of an ATP molecule bound in the central domain of the NtrC molecule which allows the transition from closed to open RNA polymerase-DNA complexes (Austin and Dixon, 1992; Weiss *et al.*, 1991). A homologous nucleotide-binding motif is present in NifA and it has been shown recently that ATP or another nucleoside triphosphate is needed for NifA-mediated open complex formation (Berger *et al.*, 1994). The C-terminal domain of NifA shows homology to NtrC and contains a helix-turn-helix motif which is responsible for NifA binding to regulatory sequences about 100 bp upstream of the transcription start site. Contrary to NtrC, NifA does not contain a N-terminal receiver domain and thus does not belong to two-component systems family (see below). Nevertheless it has been shown that *Rhizobium* NifA activity is regulated by an environmental signal which in this case is oxygen. This has been possible by studying the activation of *nifH* in *E. coli* by a constitutively expressed NifA protein under different aeration conditions (Beynon *et al.*, 1988; Fischer and Hennecke, 1987; Klipp *et al.*, 1989). Sensitivity of *Rhizobium* NifA to oxygen has been correlated with the presence of cysteine residues which could be part

of a metal (Fe) binding motif located between the central and the C-terminal domains in the so-called interdomain linker which is absent in *K. pneumoniae* NifA (Fischer *et al.*, 1988). It is interesting to note that in the case of *K. pneumoniae* NifA, which does not show intrinsic sensitivity to oxygen, NifL mediates inactivation of NifA under aerobic conditions (Berger *et al.*, 1994; Hill *et al.*, 1996).

A role for a bound metal in the activity of NifA is indicated by the effects of a chelating agent on the induction of NifA-dependent gene expression (Fischer *et al.*, 1988). The isolation of NifA mutants able to activate *nifH* expression in *E.coli* in the presence of oxygen, provides an approach to understanding the mechanism of the oxygen-mediated inhibition of NifA activity (Krey *et al.*, 1992). All the resistant mutants isolated so far have a methionine to isoleucine change in the immediate vicinity of the ATP binding site (Figure 4). Therefore it has been proposed that a change in the redox state of the metal linked to the cysteine cluster in the interdomain linker may alter the conformation of the NifA protein around the ATP binding site. The M to I mutation would maintain this site in its active configuration for either ATP binding or hydrolysis irrespective of the redox state of the NifA-bound metal. An alternative view proposes that the M to I mutation would render the protein more stable in the presence of oxygen. *In*

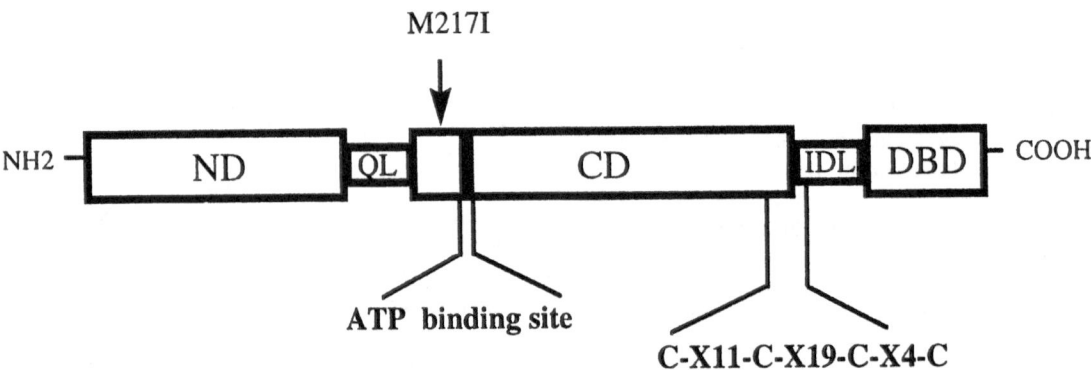

Figure 4. R. meliloti NifA protein. ND : N-terminal domain ; CD:Central domain; DBD:DNA-binding domain; QL:Q linker; IDL:Interdomain linker present in O_2-sensitive NIFAs

vivo experiments have indicated that oxygen could modify both activity and stability of NifA. As evidenced by *in vivo* footprinting experiments, *B. japonicum* NifA protein was no longer able to protect upstream regulatory sequences in the *nifH* promoter region after a short exposure of *B. japonicum* cultures to air. At longer exposure times, NifA protein was degraded (Morett *et al.*, 1991).

All *Rhizobium* NifA proteins characterized show a high degree of identity in their central and C-terminal domain which are involved in the activation process and binding to regulatory sequences, respectively. On the other hand, the N-terminal domain is much less conserved and in particular, it is entirely lacking in *R. leguminosarum* bv. *trifolii* (Iismaa and Watson, 1989). Therefore it has been hypothesized that this domain could play a regulatory role (Huala and Ausubel, 1989). Finally, *fixR*, that is cotranscribed with *nifA* in *B. japonicum* is not essential for nitrogen fixation (Thöny *et al.*, 1987). The similarity of its translation product with NodG and other dehydrogenases has suggested that FixR could be implied in the activation-inactivation of NifA in response to the redox status.

IV.A.2. The Alternative Sigma 54

The alternative sigma 54 component of RNA polymerase holoenzyme is encoded by *ntrA* (*rpoN*). The *ntrA* gene was identified in *R. meliloti* because the existence of -12/-24 cis regulatory sequences predicted the need of the sigma 54 RNA polymerase holoenzyme for the expression of the *dctA* gene which encodes the dicarboxylic acid transporter (Ronson *et al.*, 1987). In addition to their inability to express *dctA*, *ntrA* mutants are unable to express *nif* genes, which confirmed the absolute necessity of the sigma 54 RNA polymerase holoenzyme for *nif* gene expression.

Homologs of *ntrA* have been identified in *B. japonicum*, *A. caulinodans* and the broad host-range strain NGR234 (Kullik *et al.*, 1991, Stigter *et al.*, 1993, van Slooten *et al.*, 1990). Two copies of *ntrA* are present in the genome of *B. japonicum* and appear to be differently regulated (see below). Similarly there are indications that several *ntrA* homologs are present in *A. caulinodans* (Loroch *et al.*, 1995).

IV.A.3. FixL/J, a Two-Component Regulatory System

The *fixL/J* regulatory genes were first identified as part of a *R. meliloti* *fix* cluster (David *et al.*, 1987;

Renalier *et al.*, 1987). The FixLJ regulatory system was further characterized on the basis of its necessity for the expression of *nif* and *fix* genes and particularly of the regulatory gene *nifA* (David *et al.*, 1988). FixL and FixJ proteins are members of the two-component family of regulatory proteins (David *et al.*, 1988). These regulatory systems allow the cell to respond to an environmental stimulus through a transfer of phosphoryl groups from a sensor kinase to a response regulator, which is generally a transcriptional activator. (Albright *et al.*, 1989; Nixon *et al.*, 1986; Parkinson and Kofoid, 1992 for a recent review). FixL belongs to the sensor subfamily. Because FixLJ-dependent genes are expressed in microoxic conditions, it was hypothesized that the stimulus perceived by FixL was related to O_2 concentration (David *et al.*, 1988). Characterization of FixL as a haemoprotein (Gilles-Gonzalez *et al.*, 1991) and subsequent biochemical studies confirmed this hypothesis. FixL is a 505 amino acid long protein made up of three structural entities, only one of which is conserved among sensor proteins. The conserved C-terminal or transmitter domain carries an ATP-binding site and is the substrate of an autophosphorylation reaction very likely on a conserved histidine residue (Monson *et al.*, 1992). The central non-conserved region of FixL is also cytoplasmic and carries a non-covalently attached haeme group, which is likely responsible for the modification of the catalytic properties of FixL in response to oxygen (de Philip *et al.*, 1992; Monson *et al.*, 1995; Monson *et al.*, 1992). It is worth noting that FixL is the first example of a haemoprotein with the property of an oxygen sensor. An amino-terminal hydrophobic region probably anchors FixL to the membrane (Lois *et al.*, 1993). It has been shown that FixL proteins devoid of this region are still able to mediate oxygen response both *in vivo* and *in vitro* (de Philip *et al.*, 1992; Gilles-Gonzalez *et al.*, 1991).

FixJ, a 204 amino acid long protein, belongs to the subfamily of regulator components. Its activity as a transcriptional activator has been demonstrated first by using *E. coli* for reconstituting the regulatory pathway (de Philip *et al.*, 1990; Hertig *et al.*, 1989), then directly by *in vitro* transcription experiments. The FixJ amino-terminal domain is homologous to the receiver domain of other members of the family notably NtrC, VirgG and the CheY protein which is composed of only the receiver domain (David *et al.*, 1988; Kahn and Ditta, 1991; Parkinson and Kofoid, 1992). In these proteins, a conserved aspartate residue corresponding to aspartate 54 of FixJ is phosphorylated in the presence of the cognate phosphorylated sensor protein (Parkinson and Kofoid, 1992; Sanders *et al.*, 1992). FixJ as well as other regulator proteins can also be phosphorylated using some highly reactive phosphate donors such as acetyl-phosphate (Lukat *et al.*, 1992; Reyrat *et al.*, 1993).

As a first example of a complete *in vitro* reconstitution of a signal transduction pathway, it has been possible recently to reconstitute the complete pathway of signal transduction from oxygen sensing to transcriptional activation *in vitro* with purified FixL and FixJ components (Agron *et al.*, 1993; Reyrat *et al.*, 1993). This allowed the following model to be proposed. Under anoxic conditions, FixL autophosphorylation is enhanced (Gilles-Gonzalez and Gonzalez, 1993; Lois *et al.*, 1993) and FixJ is phosphorylated using FixL-Phosphate as a donor in a reaction which is independent of O_2 concentration (Lois *et al.*, 1993). As a result, FixJ transcriptional activity is enhanced at least 100 fold (Reyrat *et al.*, 1993). On the other hand, FixL possesses a phosphatase activity which might be repressed under anoxic conditions (Lois *et al.*, 1993; Monson *et al.*, 1995). Therefore FixL-mediated oxygen regulation of FixJ transcriptional activity relies on the antagonist effects of oxygen on FixL kinase and phosphatase activities. Phosphorylation of FixJ allows it to bind to regulatory sequences present in the upstream regions of *fixK*, one of the genes activated by *fixJ* (Galinier *et al.*, 1994). One high affinity site is located between positions -69 and -44 relative to the transcription start site and a lower affinity site located between -57 and -31 overlaps with the -35 region of the promoter. The C-terminal domain of FixJ, FixJC, is able to bind to these sequences and to activate transcription, indicating that the N-terminal phosphorylatable domain has a regulatory role (Da Re *et al.*, 1994; Galinier *et al.*, 1994). In addition, sequence data indicate that the C-terminal domain of FixJ, FixJC, is homologous to region 4 of sigma factors which is involved in recognition of the -35 region of the promoters (Kahn and Ditta, 1991). However sigma 70 is required in addition to *E. coli* core RNA polymerase to initiate transcription at the *nifA* and *fixK* promoters (Batut *et al.*, 1991; Da Re *et al.*, 1994; Reyrat *et al.*, 1993).

Hybridization techniques and systematic sequencing in the vicinity of identified *fix* genes have allowed the identification and the cloning of *fixJ* and/or *fixL*-homologs in *B. japonicum*, *Rhizobium leguminosarum* bv. *viciae* and *phaseoli* and *A. caulinodans* (Anthamatten and Hennecke, 1991; D'hooghe *et al.*, 1995; Kaminski and Elmerich,

1991; Patschkowski *et al.*, 1996). All FixL homologs possess the C-terminal 250 amino acid highly conserved transmitter domain. In addition all of them share homology in the central cytoplasmic 115 amino acid domain responsible for oxygen sensing in *R. meliloti* FixL. The existence of this heme-binding domain is reinforced by its presence in an *A. caulinodans* ORF in a completely different combination (Kahn, 1993). The *R. leguminosarum* bv. *phaseoli* FixL central domain, is also homologous to these heme-binding domains. However, it lacks histidine 194 and the surrounding conserved residues which are assumed to be involved in the formation of the heme binding site (D'hooghe *et al.*, 1995). The N-terminal domain is less well conserved. A hydrophobic region able to anchor FixL to the cytoplasmic membrane is present in *A. caulinodans* and *R. meliloti* protein but not in *B. japonicum* FixL. It is presently not known whether these predicted differences in localisation might be associated with differences in the process of oxygen sensing, or whether they reflect the fact that in addition to oxygen, *A. caulinodans* and *R. meliloti* FixL may respond to another environmental signal. The use of chimaeric proteins or the reconstitution of the regulatory pathway in a heterologous host might shed some light on the significance of this anchoring of *A. caulinodans* and *R. meliloti* FixL to the membrane. *R. leguminosarum* bv. *viciae* possesses an additional C-terminal domain homologous to the receiver domain of the response regulators of the two component regulatory systems. This is remarkable because *R. leguminosarum* bv. *viciae*, contrary to the other rhizobia, does not carry a FixLJ cluster.

FixJ homologs have been identified in all *Rhizobium* species where they have been looked for, except *R. leguminosarum* bv. *viciae*. The N-terminal domain is homologous to the phosphorylatable "receiver" domain of regulators of the two-component family. In addition FixJ homologs show similarity in their "output" C-terminal domain which contains the helix-turn-helix motif responsible for DNA binding. Conservation of this motif is in agreement with the presence of a highly conserved sequence in the -35 region of each of the FixJ-activated genes (Waelkens *et al.*, 1992).

IV.A.4. FixK, a Close Relative of Fnr

The 211 amino acid long FixK protein was identified by sequencing the *fix* region of *R. meliloti* which also carries *fixL/J* and was predicted to be a regulatory protein due to its homology to Fnr and Crp (Batut *et al.*, 1989; Shaw *et al.*, 1983). Two functionally equivalent FixK proteins are encoded by highly conserved sequences (Renalier *et al.*, 1987) (J. Fourment and A.M. Garnerone, personal communication). FixK acts as a positive regulator of the *fixNOQP* operon which codes for a respiratory chain (see section III). FixK controls negatively its own expression as well as that of *nifA*. The C-terminal end of FixK contains a helix-turn-helix motif highly similar to that of Fnr, thus suggesting that FixK may recognize DNA sequences identical to the Fnr sites (Cherfils *et al.*, 1989). Indeed purified FixK binds such a sequence upstream of *fixN* (Foussard *et al.*, 1997). A similar sequence located 480 bp upstream of the *fixK* transcription start site

has been shown to be involved in the negative control of *fixK* expression (Waelkens *et al.*, 1992) and to be a binding site for FixK (Foussard *et al.*, 1997). Although FixK and Fnr are homologous over almost their entire sequence, the cysteine residues at the N-terminal end which are thought to be responsible for Fnr response to oxygen are not found in FixK. Indeed, constitutive expression of *fixK* under the control of an oxygen-independent mutant of FixJ allowed Soupène *et al.* (1995) to show that FixK activity is not subject to oxygen control.

FixK homologs have been identified in various rhizobia by three different methods: 1) hybridization with a *R. meliloti fixK* probe allowed the identification of a *fixK* homolog in *B. japonicum* (Anthamatten *et al.*, 1992) 2) heterologous

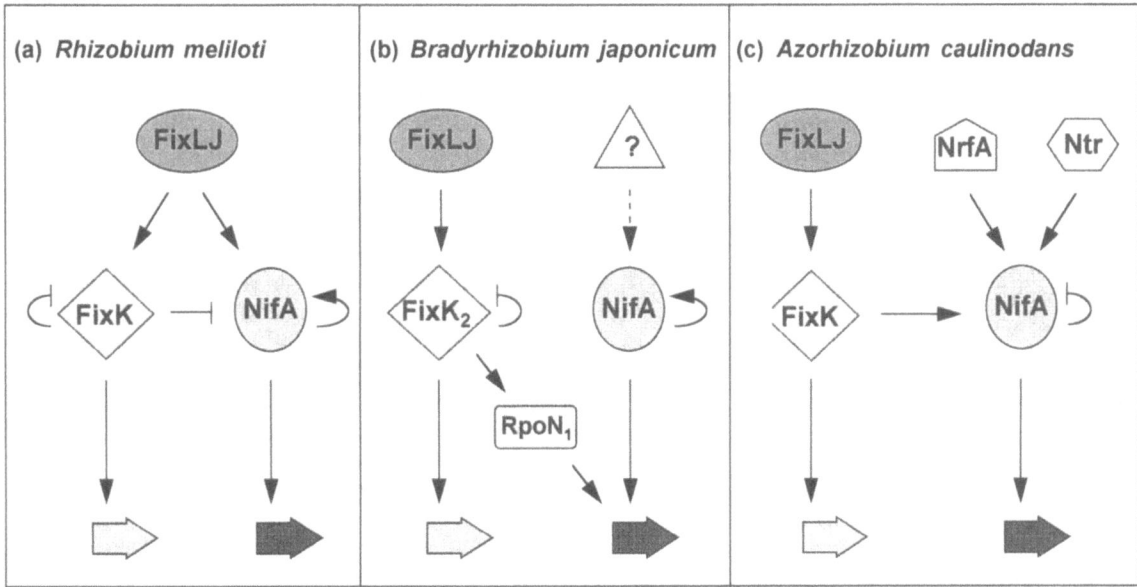

Figure 5. Regulators and regulatory patways in rhizobia (Courtesy of Dr H.M. Fischer, reproduced with permission from reference Fischer, 1996)

complementation of a *R. meliloti* mutant devoid of FixK activity led to the identification of FnrN of *R. leguminosarum* bv. *viciae* (Colonna *et al.*, 1990; Schlüter *et al.*, 1992) 3) sequencing of *fix* clusters led to the identification of a *fixK*-like gene in *A.caulinodans* and of a second FixK homolog in *R. leguminosarum* bv. *viciae* and in *B. japonicum* (Fischer, 1994; Kaminski *et al.*, 1991; Patschkowski *et al.*, 1996).

All FixK homologs share two highly conserved structural elements: first, in the central domain, six glycine residues are thought to be involved in the formation of a beta-roll structure and, secondly, the helix turn helix motif at the C-terminal end, due to its high similarity to DNA binding motifs of CRP and FNR, is likely to interact with regulatory sequences of the target genes.

Oxygen sensing by Fnr is due to the presence of an iron-sulfur cluster bound by essential cysteine residues. Cysteine residues with similar spacing are present in the N-terminal part and central domain of *R. leguminosarum* bv. *viciae* FnrN and *B. japonicum* FixK1 which have been shown to respond to oxygen. By contrast, these cysteine residues are absent from *R. meliloti* FixK, from *A. caulinodans* FixK and from *B. japonicum* FixK2. It is therefore likely that these two latter FixK homologs do not respond to oxygen, as has been shown to be the case for *R. meliloti* FixK.

IV.A.5. NtrB/C, a Regulatory System Sensing the Nitrogen Nutritional Status

NtrB and NtrC constitute a two-component regulatory system which, in concert with the P_{II} protein, allows the adaptation of nitrogen metabolism to the availability of nitrogen sources in *E. coli* and *K. pneumoniae* (see for instance the review by Reitzer and Magasanik, 1987). In particular *K. pneumoniae* NtrB/C regulates the expression of *nifA* in response to the nitrogen status (Gussin *et al.*, 1986). Homologs of NtrB and NtrC have been found in rhizobia and have been shown to be involved in the control of nitrogen metabolism in culture (see for example Szeto *et al.*, 1987). In *A. caulinodans*, a second two-component regulatory system homologous to NtrB/C, NtrY/X, has been identified and NtrY characterized as a putative transmembrane sensor. In addition to controlling nitrogen metabolism, these two systems are involved in the control of symbiotic nitrogen fixation in *A. caulinodans* (Pawlowski *et al.*, 1991).

IV.A.6. NrfA, New Functions for an Old Gene

The regulatory gene *nrfA* was identified following EMS mutagenesis of *A. caulinodans* and characterization of mutants impaired in the expression of a set of proteins normally present in nitrogen fixation conditions (Denèfle *et al.*, 1987). Mutation of *nrfA* prevented *nifA* expression and as a consequence, that of *nifA*-dependent genes (Kaminski *et al.*, 1994). NrfA was found to be homologous to *E.coli* host factor I (HF-I), an RNA binding protein required for replication of RNA phage Qβ. The process by which *nrfA* allows *A. caulinodans nifA* expression has not been elucidated yet. However, recent results indicate that HF-I is essential for *rpoS* translation in *E. coli* (Muffler *et al.*, 1996) and it was suggested that NrfA might be required for *nifA* mRNA translation.

IV.B. COMPARISON OF *NIF* AND *FIX* GENE REGULATORY CIRCUITS

The regulatory circuits of *R. meliloti*, *B. japonicum* and *A. caulinodans nif* and *fix* genes are presented in Figure 5. One obvious characteristic of these different regulatory circuits is that they use homologous regulatory proteins. However, these regulatory proteins are assembled in different combinations. One major difference resulting from these various combinations is that whereas in *R. meliloti* and *A. caulinodans*, *nifA* transcription is under the dependence of FixL/J, ensuring a first level of oxygen control of *nif* gene expression, in *B. japonicum*, oxygen control of *nifA* gene expression is achieved mainly via an autoregulatory circuit. In microaerobic conditions, *B. japonicum nifA* transcription is activated by NifA itself at a sigma 54-dependent promoter. In aerobic conditions, in which NifA is not in its active state, a significant although lower level of *nifA* transcription results from the activation of an alternative promoter (Barrios *et al.*, 1995). Although not involved in *nifA* expression in *B.japonicum*, FixL/J could nevertheless control nitrogen fixation gene expression by modulating the concentration of the alternative sigma factor NtrA needed for *nif* gene transcription (Kullik *et al.*, 1991). This modulation is achieved through activation by the FixL/J system of one of the two *rpoN (ntrA)* copies.

Studies on *nif* and *fix* regulatory circuits have also begun in *R. leguminosarum* bv. *viciae* and bv. *phaseoli*. NifA is expressed independently of *fixL* and *fixK* in *R. leguminosarum* bv. *viciae* in which no FixJ homoloque has been identified so far (Patschkowski *et al.*, 1996). In *R.leguminosarum* bv. *phaseoli*, *nifA* expression is independent of FixL/J (D'hooghe *et al.*, 1995).

An illustration of the shuffling of homologous regulatory proteins in the establisment of the various regulatory pathways is provided by the fact that FixK is a repressor of *nifA* in *R. meliloti* and an activator of *nifA* in *A. caulinodans* (Batut *et al.*, 1989; Kaminski *et al.*, 1991; Loroch *et al.*, 1995).

In addition to the variation in the architecture of the regulatory pathways caused by the different combinations of homologous regulatory proteins, another type of complexity is introduced by the combination of several regulatory pathways responding to different signals. As already mentioned, in *A. caulinodans*, whereas oxygen response is mediated by FixL/J and FixK, an additional control of *nifA* expression, by nitrogen availability, is exerted through the regulatory proteins NtrB/C and NtrY/X.

IV.C. SYMBIOTIC CONTROL OF NITROGEN FIXATION GENE EXPRESSION

As has been reviewed above, nodule development on one hand and bacteroid differentiation and expression of the nitrogen fixing apparatus on the other hand are tightly coupled. This has been shown essentially in indeterminate nodules where the expression of *nif* and *fix* genes appears to be activated precisely at the interzone II-III (see section III) (Soupène *et al.*, 1995; Yang *et al.*, 1991). The genetic and physiological knowledge acquired on the regulatory cascade responsible for the expression of the *R. meliloti* nitrogen fixation genes, provided tools for addressing the question of the mechanism coupling nodule development and bacterial nitrogen fixation gene expression. Two complementary approaches have shown that oxygen distribution along the longitudinal axis of the nodule is a major determinant of the expression pattern of nitrogen fixation genes in alfalfa nodules elicited by *R. meliloti* (Soupène *et al.*, 1995) First, by modifying the concentration of oxygen in nodule environment, it has been possible to induce the ectopic expression of *nifA*, *fixK* and *fixN* genes in the invasion zone II, that is distally to the interzone II-III where they normally start being expressed. A second approach made use of a mutation in the transcriptional

activator FixJ which resulted in the constitutive oxygen-independent expression of *nifA*, *fixK*, and *fixN*. The mutant strains elicited nodules where *nifA*, *fixK* and *fixN* expression could be observed in the invasion zone II, similarly to wild type nodules under microoxic environment. The existence of a longitudinal gradient of oxygen in the nodule was confirmed by the use of oxygen sensitive microelectrodes. Oxygen concentration decreased from 250 to 1 µM from the apical end to the interzone II-III of the nodule. Therefore, whereas the oxygen barrier surrounding the central tissues, together with the high bacteroid respiration rate and leghaemoglobin, produces an environment necessary for nitrogenase function, the longitudinal gradient can be considered as a morphogen gradient since it determines the pattern of gene expression. The fact that, contrary to *nifA*, *fixK* and *fixN*, the *nifA*-dependent *nif* genes are not ectopically expressed in anoxiated nodules could be due to the oxygen sensitivity of NifA itself, introducing a second level of control of the spatial pattern of *nif* genes expression, or to the existence of an additional signal controlling the activity of *nifA* (Soupène *et al.*, 1995).

The question remains whether the same oxygen signal operates in *Rhizobium* strains where the expression of nitrogen fixation genes depends on a different regulatory cascade from *R. meliloti*. In determinate nodules, where nodule differentiation is essentially temporally programmed, it will be interesting to determine whether there is a time course decrease of oxygen concentration during nodule development.

V. Conclusions and Perspectives

V.A. SYMBIOTIC FUNCTIONS OF BACTEROIDS

Clearly, during these last 15 years, a lot of effort has been devoted to the elucidation of the symbiotic regulation of the expression of the already identified *nif* genes rather than to the identification of additional *nif* genes. How many *nif* genes, homologous to *K. pneumoniae nif* genes, are present in rhizobia in addition to the 10 already identified, remains to be determined. Three *Klebsiella nif* genes, *nifL*, *nifF* and *nifJ*, have not been found and are probably absent from rhizobia. *nifL*, that mediates oxygen and nitrogen control of NifA activity in

Klebsiella, probably has no counterpart in rhizobia. NifF and NifJ, that transfer electrons from pyruvate to nitrogenase in anaerobic *K. pneumoniae*, may have been replaced by FixABCX in rhizobia as well as in other aerobic nitrogen fixers.

In the long lasting search for genes involved in bacteroid functions, a major achievement has been the identification of the *fixNOQP* genes coding for a bacteroid terminal oxidase with high affinity for oxygen (Preisig *et al.*, 1993). Furthermore, *B. japonicum* is now one of the best known bacteria with respect to cytochrome biosynthesis (Thöny-Meyer *et al.*, 1994a; Thöny-Meyer, 1997). Identification of genes involved in carbon and nitrogen metabolism has begun and will likely be pursued actively in the near future. Questions such as the nature of the alternative symbiotic activator (ASA), which is able to activate genes involved in carbon metabolism in bacteroids, and the mechanism by which fixed NH_4^+ is excreted need to be answered.

Have we identified all bacterial functions that are essential for symbiotic nitrogen fixation? This sounds unlikely given the great number of *fix* loci evidenced by random transposon mutagenesis (Forrai *et al.*, 1983; Putnoky *et al.*, 1988). Identification and characterization of these genes may reveal new aspects of the biology of nitrogen-fixing bacteroids and of the physiological conditions in which they operate. Random transposon mutagenesis indeed sounds as the method of choice for identifying new *fix* genes. However its usefulness is limited by the fact that duplications are quite frequent in rhizobia, and its efficiency, somewhat paradoxically, by the fact that it is random. Thus, more specific screens are preferred in some instances to identify new genes in a specific pathway, for example genes under the control of *fixLJ*, *nifA* or *fixK*. Biochemical and genetic approaches could also be set up to search for genes which are preferentially expressed under symbiotic conditions. Finally, large scale sequencing of rhizobial genomes, that has just started with the symbiotic plasmid of *Rhizobium* sp. NGR234 (Freiberg *et al.*, 1997; Freiberg *et al.*, 1996), will undoubtedly open a new age for the analysis of symbiotic nitrogen fixation.

The genetic control of bacteroid differentiation is another area in which a great deal remains to be done. The central issue here is the causal relationship between bacteroid differentiation and the capacity to fix nitrogen. Indeed, apart from a few key events, such as the induction of *nif* and *fix* gene expression, we still do not know whether the numerous modifications that are known to happen during the course of bacteroid differentiation are instrumental for the capacity to fix nitrogen symbiotically, or whether most of them are simply concomitant or just the consequence of metabolic changes resulting from nitrogen fixation.

V.B. DEVELOPMENTAL AND METABOLIC SIGNALS OPERATING INSIDE THE NODULE

Establishment of symbiosis requires coordinated differentiation of the two partners, that includes both induction and repression of a great number of bacterial and plant genes and, at a later stage, permanent adjustement of the two metabolisms. It is tempting to speculate that the same physiological signals may apply to both bacterial and plant genes, eventhough they are sensed and transduced by different signaling pathways. As bacteria are readily amenable to genetic manipulation, they are very convenient tools for probing the nodule environment and identifying such signals. Hence, the study of plant gene regulation could be facilitated.

Work on the regulation of *nif* and *fix* genes led to the identification of "low oxygen" as a developmental signal operating inside the nodule (Soupène *et al.*, 1995; Witty *et al.*, 1986). Oxygen concentration inside the nodule not only controls *nif* and *fix* gene expression on a quantitative ground but also determines where the genes are expressed. Thus oxygen has a prominent role in symbiosis being a respiratory substrate, an inhibitor of NifA and nitrogenase activity and a signalling molecule.

A burning question for the future is whether "low oxygen" is the only regulatory signal operating on symbiotic *nif* and *fix* gene expression or, as one may anticipate from the high energy cost of nitrogen fixation, whether the carbon and/or nitrogen status of the plant influence nitrogen fixation gene expression in the bacteroid. *A. caulinodans* is the only example so far in which *nifA* expression is under nitrogen control *via* Ntr-like proteins. However the significance of this regulation under symbiotic conditions is not established. Obviously, elucidating the connection(s) between oxygen, nitrogen and carbon control is crucial for our understanding of symbiotic nitrogen fixation.

Elucidation of the genetic control of molecular events taking place during the course of bacteroid differentiation is a good strategy for the identification of signals operating inside the nodule, besides oxygen. Recent data indicate that dicarboxylic acids, which are the source of energy for bacteroids, also may block *nod* gene expression in bacteroids. Calcium was suggested to affect *ropA* repression. Thus a set of different signals may operate inside the nodule either independently or in combination with each other. For example, some of the changes in surface components observed in bacteroids can be mimicked by growing bacteria under microoxic conditions in the presence of succinate (Sindhu *et al.*, 1990). Similarly, it was shown recently that the combination of a low pH and the presence of acetate had a bacteriostatic effect on rhizobia (Perez-Goldano and Kahn, 1994). Thus, bacteroid differentiation, which is indeed a complex process, is likely under multi-signalling control.

V.C. REGULATORS AND REGULATORY PATHWAYS

The study and comparison of regulatory pathways in different rhizobia has been a very active and successful area of research in the last ten years.
In spite of their diversity, all pathways share common features that we shall briefly discuss below: (1) they are highly integrated, (2) they are built in cascades, (3) they make use of homologous proteins (FixLJ, FixK, NifA) in different combinations.
Let us consider the *nifA*-regulon. As in *K. pneumoniae*, rhizobial *nif* genes are under *nifA* control. In addition, the *fixABCX* genes, that likely perform *nif*-like functions but are not *nif* homologs, are under *nifA* control in all rhizobia studied so far. This suggests a selective pressure in nitrogen fixers for all nitrogenase-related genes being co-regulated. Furthermore, it has been discovered in the recent years that rhizobial genes that are not essential for nitrogen fixation under laboratory conditions were under *nifA* control. For example, *nifA* controls genes involved in rhizopine synthesis or genes controlling nodulation efficiency (Saint *et al.*, 1993; Sanjuan and Olivares, 1991; reviewed by Batut and Boistard, 1994). The reason for having these ancillary genes co-regulated with functions central to N_2 fixation is not understood yet. One attractive possibility would be that these genes would be beneficial in the field, possibly in the long term, and their integration into the *nifA* regulon may represent a selective advantage. Why are the regulatory pathways organized in cascades? Most often, regulators respond to physiological signals. Thus, the clearest biological significance for having two regulators in a cascade is

the possibility to integrate two different signals. How do the data in rhizobia fit with this model? If we consider the *nif* cascade of *R.meliloti*, *nif* expression is under both *fixLJ* and *nifA* control. Both FixL and NifA are known to be oxygen responsive proteins. Then the need for two regulators sensitive to the same signal is not obvious. However there is the possibility that FixL and NifA actually respond to different levels of microaerobiosis, thus allowing the onset of respiration, *via fixK*, before *nif* induction, *via nifA*. In addition, oxygen exerts its effect on NifA at the protein level, which would allow a rapid block in nitrogenase synthesis under unfavorable oxygen conditions. Another possibility, which cannot be formally excluded, is that either NifA or FixL senses another signal than oxygen. A similar analysis can be conducted on the *fixLJ/fixK* cascade. *fixNOQP* expression involves both *fixLJ* and *fixK* in all three rhizobia examined (Figure 4), thus suggesting a biological significance for the conservation of these two levels of regulation. Nevertheless, FixK activity is not controlled by oxygen (Soupène *et al.*, 1995) and no physiological signal associated with FixK could be evidenced so far.

Why have rhizobia combined differently the same basic regulators? The diversity of the pathways is most easily envisioned by considering the way two different cascades, the *nifA*-dependent cascade and a respiratory *fixNOQP* operon, regulated by *fixLJ/K*, are connected. One can distinguish three different strategies (see Figure 4). In the first two strategies, *nifA* expression is under *fixLJ*-mediated oxygen control either directly (*R. meliloti*) or *via fixK* (*A. caulinodans*). In *R. meliloti*, the connection is tight as there is absolutely no *nifA* expression in a *fixLJ* mutant. In *A. caulinodans*, a *fixJ* or *fixL* mutant is leaky for both *nifA* expression and nitrogenase activity (*ca.* 10% of wild-type). In the third strategy, illustrated by *B. japonicum*, *nifA* expresses independently of the *fixLJ/fixK* regulators. The connection is at the level of *rpoN₁*, a gene encoding a sigma factor indispensable for NifA activity, as well as for the activity of many regulators of the same family. However the connection between respiration and nitrogen fixation is loose in *B. japonicum* because, in addition to *rpoN₁* which is under the control of *fixLJ*, a functionally homologous *rpoN₂* gene is expressed independently of oxygen.
It should be emphasized however that the generality of the *fixLJ/fixK/fixNOQP* cascade has not been established. Instead, recent data in *R. leguminosarum* bv. *viciae* indicate that *fixN* expression may not depend on *fixL* (Patschkowski *et al.*, 1996). Further

characterization of *fixN* regulation in other rhizobia, in *Agrobacterium* and in *Rhodobacter capsulatus* should allow to clarify the issue of whether *fixNOQP* is constantly under *fixLJ/K* control or whether, as yet another evidence for the plasticity of regulatory pathways, different regulators account for *fixN* expression in diverse bacteria. These results should also shed some light on whether the FixLJ regulators, that have been only described in rhizobia so far, have a special commitment with symbiotic nitrogen fixation or, alternatively, are more likely related to life under microoxic conditions. Hence, understanding the modularity of these networks should have interesting implications in both functional and evolutionary terms.

VI. Acknowledgements

The authors are very grateful to Dr H-M. Fischer and Dr G. Truchet for Figures 2 and 5 and Prof. T. Finan and Dr H-M. Fischer for their constructive criticisms of the manuscript. The authors apologize to all colleagues whose work could not be referred to in this review.

VII. References

Agron, P.G., Ditta, G.S. and Helinski, D.R. (1993) Proc. Natl. Acad. Sci. USA 90, 3506-3510.

Aguilar, O.M., Reilander, H., Arnold, W. and Pühler, A. (1987) J. Bacteriol. 169, 5393-5400.

Aguilar, O.M., Taormino, J., Thony, B., Ramseier, T., Hennecke, H. and Szalay, A.A. (1990) Mol. Gen. Genet. 224, 413-420.

Albright, C.M., Huala, E. and Ausubel. F.M. (1989) Annu. Rev. Genet. 23, 311-336.

Anthamatten, D. and Hennecke, H. (1991) Mol. Gen. Genet. 225, 38-48.

Anthamatten, D., Scherb, B. and Hennecke, H. (1992) J. Bacteriol. 174, 2111-20.

Appleby, C. A. (1984) Ann. Rev. Plant. Physiol. 35, 443-478.

Arcondéguy, T., Huez, I., Fourment, J. and Kahn, D. (1996) FEMS Lett. 145, 33-40.

Arigoni, F. (1992) PhD Thesis. Swiss Federal Institute of Technology Zürich.

Arigoni, F., Kaminski, P.A., Celli, J. and Elmerich, C. (1992) Mol. Gen. Genet. 235, 422-431.

Arigoni, F., Kaminski, P.A., Hennecke, H. and Elmerich, C. (1991) Mol. Gen. Genet. 225, 514-520.

Arnold, W., Rump, A., Klipp, W., Priefer, U.B. and Pühler, A. (1988) J. Mol. Biol. 203, 715-738.

Austin, S., and Dixon R. (1992) EMBO J. 11, 2219-2228.

Barrios, H., Fischer, H-M., Hennecke, H. and Morett, E. (1995) J. Bacteriol. 177, 1760-1765.

Batut, J. and Boistard, P. (1994) Antonie van Leeuwenhoek 66, 129-150.

Batut, J., Daveran, M.M., David, M., Jacobs, J., Garnerone, A.M. and Kahn, D. (1989) EMBO J. 8, 1279-1286.

Batut, J., Santero, E. and Kustu, S. (1991) J. Bacteriol.173, 5914-5917.

Bender, G.L., Plazinski, J. and Rolfe, B.G. (1986) Appl. Environ. Microbiol. 51, 868-871.

Berger, D.K., Narberhaus, F. and Kustu, S. (1994) Proc. Natl. Acad. Sci. USA 91, 103-107.

Bergersen, F.J. and Turner, G.L. (1967) Biochim. Biophys. Acta. 141, 507-515.

Bergersen, F.J., Turner, G.L., Gibson, A.H. and Dudman, W.E. (1976) Biochim. Biophys. Acta. 444, 164-174.

Beynon, J.L., Williams, M.K. and Cannon F.C. (1988) EMBO J. 7, 7-14.

Bishop, P.E. and Premakumar R. (1992). pp. 736-762. In Stacey, G., Burris, R.H. and Evans, H.J. (eds), Biological nitrogen fixation. Chapman and Hall, New York.

Bolton, E., Higgisson, B., Harrington, A. and O'Gara, F. (1986) Arch. Microbiol. 144, 142-146.

Bott, M., Bolliger, M.and Hennecke, H. (1990) Mol. Microbiol. 4, 2147-2157.

Bott, M., Preisig, O. and Hennecke, H. (1992) Arch. Microbiol. 158, 335-343.

Bott, M., Ritz, D. and Hennecke, H. (1991) J. Bacteriol. 173, 6766-6772.

Bott, M., Thöny-Meyer, L., Loferer, H., Rossbach, S., Tully, R.E., Keister, D., Appleby, C.A. and Hennecke, H. (1995) J. Bacteriol. 177, 2214-2217.

Brigle, K.E., Weiss, M.C., Newton, W.E. and Dean, D.E. (1987) J. Bacteriol. 169, 1574-1553.

Buikema, W.J., Klingensmith, J.A., Gibbons, S.L. and Ausubel, F.M. (1987) J. Bacteriol. 169, 1120-1126.

Carlson, T.A. and Chelm, B.K. (1986) Nature (London) 322, 568-570.

Carlson, T.A., Guerinot, M.L. and Chelm, B.K. (1985) J. Bacteriol. 162, 698-703.

Carlson, T.A., Martin, G.B. and Chelm, B.K. (1987) J. Bacteriol. 169, 5861-5866.

Cherfils, J., Gibrat, J.F., Levin, F., Batut, J. and Kahn, D. (1989) J. Mol. Recognition 2, 114-121.

Chiurazzi, M., Meza, R., Lara, M., Lahm, A., Defez, R., Iaccarino, M. and Espin, G. (1992) Gene 119, 1-8.

Colonna, R.S., Arnold, W., Schluter, A., Boistard, P., Pühler, A. and Priefer, U.B. (1990) Mol. Gen. Genet. 223, 138-147.

Colonna, R.S., Riccio, A., Guida, M., Defez, R., Lamberti, A., Iaccarino, M., Arnold, W., Priefer, U. and Pühler, A. (1987) Nucleic. Acids. Res. 15, 1951-1964.

Corbin, D., Ditta, G. and Helinski, D.R. (1982) J. Bacteriol. 149, 221-228.

Cullimore, J.V. and Bennett, M.J. (1992) Can. J. Microbiol. 38, 461-466.

D'hooghe, I., Michiels, J., Vlassak, K., Verreth, C., Waelkens, F. and Vanderleyden, J. (1995) Mol. Gen. Genet. 249, 117-126.

Da Re, S., Bertagnoli, S., Fourment, J., Reyrat, J.M. and Kahn, D. (1994) Nucleic. Acids. Res. 22, 1555-1561.

David, M., Daveran, M.L., Batut, J., Dedieu, A., Domergue, O., Ghai, J., Hertig, C., Boistard, P. and Kahn, D. (1988) Cell 54, 671-683.

David, M., Domergue, O., Pognonec, P. and Kahn, D. (1987) J. Bacteriol. 169, 2239-2244.

Day, D.A. and Copeland, L. (1991) Plant. Physiol. Biochem. 29, 185-201.

Day, D.A., Price, G.D. and Udvardi, M.K. (1989) Austral. J. Plant. Physiol. 16, 69-84.

de Billy, F., Barker, D.G., Gallusci, P. and Truchet, G. (1991) The Plant J. 1, 27-35.

de Bruijn, F.J., Rossbach, S., Schneider, M., Ratet, P., Messmer, S., Szeto, W.W., Ausubel, F.M. and Schell, J. (1989) J. Bacteriol. 171, 1673-1682.

de Maagd, R., de Rijk, R. Mulders, I.H. and Lugtenberg, B.J.J. (1989) J. Bacteriol. 171, 1136-1142.

de Maagd, R., Mulders, I.H., Canter, C.H. and Lugtenberg, B.J.J. (1992) J. Bacteriol. 174, 214-221.

de Maagd, R., Yang, W.-C., Goosen-de Roo, L., Mulders, I.H., Roest, H.P., Spaink, H.P., Bisseling, T. and Lugtenberg, B.J.J. (1994) Mol. Plant-Microbe Interact. 7, 276-281.

de Philip, P., Batut, J. and Boistard, P. (1990) J. Bacteriol. 172, 4255-4262.

de Philip, P., Soupene, E., Batut, J. and Boistard, P. (1992) Mol. Gen. Genet. 235, 49-54.

Dean, D.R. and Jacobson, M.R. (1992). pp. 763-834. In Stacey, G., Burris, R.H. and Evans, H.J. (eds), Biological Nitrogen Fixation. Chapman and Hall, New York.

Delgado, M.J., Yeoman, K.H., Wu, G., Vargas, C., Davies, A.E., Poole, R.K., Johnston, A.W. and Downie, J.A. (1995) J. Bacteriol. 177, 4927-4934.

Denèfle, P., Kush, A., Norel, F., Paquelin, A. and Elmerich, C. (1987) Mol. Gen. Genet. 207, 280-287.

Ditta, G., Virts, E., Palomares, A. and Kim, C.H. (1987) J. Bacteriol. 169, 3217-3223.

Donald, R.G. and Ludwig, R.A. (1984) J. Bacteriol. 158, 1144-1151.

Dreyfus, B., Garcia, J.L. and Gillis, M. (1988) Int. J. Syst. Bacteriol. 38, 89-98.

Dreyfus, B.L., Elmerich, C. and Dommergues, Y.R. (1983) Appl. Environ. Microbiol. 45, 711-713.

Drummond, M., Whitty, P. and Wootton, J. (1986) EMBO J. 5, 441-447.

Dusha, I., Kovalenko, S., Banfalvi, Z. and Kondorosi, A. (1987) J. Bacteriol. 169, 1403-1409.

Earl, C.D., Ronson, C.W. and Ausubel, F.M. (1987) J. Bacteriol. 169, 1127-1136.

Ebeling, S., Hahn, M., Fischer, H-M. and Hennecke, H. (1987) Mol. Gen. Genet. 207, 503-508.

Ebeling, S., Noti, J.D. and Hennecke, H. (1988) J. Bacteriol. 170, 1999-2001.

Eichler, K., Buchet, A., Bourgis, F., Kleber, H.P. and Mandrand-Berthelot, M.A. (1995) J. Basic. Microbiol 35, 217-227.

Elmerich, C., Dreyfus, B.L., Reysset, G. and Aubert, J.P. (1982) EMBO J. 1, 499-503.

Engelke, T.H., Jording, D., Kapp, D. and Pühler, A. (1989) J. Bacteriol. 171, 5551-5560.

Engelke, T.H., Jagadish, M.N. and Pühler, A. (1987) J. Gen. Microbiol. 133, 3019-3029.

Espin, G., Moreno, S., Wild, M., Meza, R. and Iaccarino, M. (1990) Mol. Gen. Genet. 223, 513-516.

Evans, D., Jones, R., Woodley, P. and Robson, R. (1988) J. Gen. Microbiol. 134, 931-942.

Finan, T.M., Wood, J.M. and Jordan, D.C. (1983) J. Bacteriol. 154, 1403-1413.

Finocchiaro, G., Colombo, I., Garavaglia, B., Gellera, C., Valdemari, G., Garbuglio, N. and Didonato, S. (1993) Eur. J. Biochem. 213, 1003-1008.

Fischer, H-M. (1996) Trends in Microbiol. 4, 317-320.

Fischer, H-M., Alvarez-Morales, A. and Hennecke, H. (1986) EMBO J. 5, 1165-1173.

Fischer, H-M., and Hennecke, H. (1987) Mol. Gen. Genet. 209, 621-626.

Fischer, H-M. (1994) Microbiol. Rev. 58, 352-386.

Fischer, H-M., Bruderer, T. and Hennecke, H. (1988) Nucleic. Acids. Res. 16, 2207-2224.

Fischer, H-M. and Hennecke, H. (1984) Mol. Gen. Genet. 196, 537-540.

Fogher, C., Dusha, I., Barbot, P. and Elmerich, C. (1985) FEMS Microbiol. Lett. 30, 245-249.

Forrai, T., Vincze, E., Banfalvi, Z., Kiss, G.B., Randhawa, G.S. and Kondorosi, A. (1983) J. Bacteriol. 153, 635-643.

Foussard, M., Garnerone, A.M., Ni, F., Soupene, E., Boistard, P. and Batut, J. (1997) Mol. Microbiol. 25, 27-37.

Franssen, H.J., Vijn, I., Yang, W.C. and Bisseling, T. (1992) Plant. Mol. Biol. 19, 89-107.

Freiberg, C., Fellay, R., Bairoch, A., Broughton, W.J., Rosenthal, A. and Perret, X. (1997) Nature 387, 394-401.

Freiberg, C., Perret, X., Broughton, W.J. and Rosenthal, A. (1996) Genome Research. 6, 590-600.

Fuhrmann, M., Fischer, H-M. and Hennecke, H. (1985) Mol. Gen. Genet. 199, 315-322.

Galimand, M., Perroud, B., Delorme, F., Paquelin, A., Vieille, C., Bozouklian, H. and Elmerich, C. (1989) J. Gen. Microbiol. 135, 1047-1059.

Galinier, A., Garnerone, A.M., Reyrat, J.M., Kahn, D., Batut, J. and Boistard, P. (1994) J. Biol. Chem. 269, 23784-23789.

Gebhardt, C., Turner, G.L., Gibson, A.H., Dreyfus, B.L. and Bergersen, F.J. (1984) J. Gen. Microbiol. 130, 843-848.

Gilles-Gonzalez, M., Ditta, G.S. and Helinski, D.R. (1991) Nature 150, 170-172.

Gilles-Gonzalez, M.A., and Gonzalez, G. (1993) J. Biol. Chem. 268, 16293-16297.

Glazebrook, J., Ichige, A. and Walker, G.C. (1993) Genes & Dev. 7, 1485-1497.

Goodman, S.I., Axtell, K.M. Bindoff, L.A., Beard, S.E., Gill, R.E. and Frerman, F.E. (1994) Eur. J. Biochem. 219, 277-286.

Gordon, A.J. (1995). pp. 533-538. In Palacios, R., Mora, J. and Newton, W.E. (eds), Nitrogen Fixation: Fundamentals and Applications. Kluwer Academic Publishers, Dordrecht Boston London.

Gosink, M.M., Franklin, N.M. and Roberts, G.P. (1990) J. Bacteriol. 172, 1441-1447.

Grönger, P., Manian, S.S., Reiländer, H., O'Connell, M., Priefer, U.B. and Pühler, A. (1987) Nucleic. Acids. Res. 15, 31-49.

Gubler, M. and Hennecke, H. (1986) FEBS lett. 200, 186-192.

Gubler, M. and Hennecke, H. (1988) J. Bacteriol. 170, 1205-1214.

Gubler, M., Zurcher, T. and Hennecke, H. (1989) Mol. Microbiol. 3, 141-148.

Gussin, G.N., Ronson, C.W. and Ausubel, F.M. (1986) Annu. Rev. Genet. 20, 567-591.

Hennecke, H. (1990) FEBS Lett. 268, 422-426.

Hertig, C., Li, R.Y., Louarn, A.M., Garnerone, A.M., David, M., Batut, J., Kahn, D. and Boistard, P. (1989) J. Bacteriol. 171, 1736-1738.

Hill, S. (1992). p. 87-134. In G. Stacey, R. H. Burris and H. J. Evans (eds), Biological Nitrogen Fixation. Chapman and Hall, New York, London.

Hill, S., Austin, S., Eydmann, T., Jones, T. and Dixon, R. (1996) Proc. Natl. Acad. Sci. USA 93, 2143-2148.

Hirsch, A.M., Bang, M. and Ausubel, F.M. (1983) J. Bacteriol. 155, 367-380.

Hirsch, A.M. and Smith, C.A. (1987) J. Bacteriol. 169, 1137-1146.

Huala, E. and Ausubel, F.M. (1989) J. Bacteriol. 171, 3354-3365.

Humbeck, C. and Werner, D. (1987) Curr. Microbiol. 14, 259-262.

Iismaa, S.E., Ealing, P.M., Scott, K.F. and Watson, J.M. (1989) Mol. Microbiol. 3, 1753-1764.

Iismaa, S.E. and Watson, J.M. (1987) Nucleic. Acids. Res. 15, 318.

Iismaa, S.E. and Watson, J.M. (1989) Mol. Microbiol. 3, 943-955.

Jiang, J., Gu, B., Albright, L.M. and Nixon, T.B. (1989) J. Bacteriol. 171, 5244-5253.

Jording, D. and Pühler, A. (1993) Mol. Gen. Genet. 241, 106-114.

Kahn, D. (1993) Mol. Microbiol. 8, 786-787.

Kahn, D., Batut, J., Daveran, M-L. and Fourment, J. (1993). pp. 474. In Palacios, R., Mora, J. and Newton, W.E. (eds), New Horizons in Nitrogen Fixation. Kluwer Academic Publishers, Dordrecht.

Kahn, D., David, M., Domergue, O., Daveran, M.L., Ghai, J., Hirsch, P.R. and Batut, J. (1989) J. Bacteriol. 171, 929-939.

Kahn, D., and Ditta, G. (1991) Mol. Microbiol. 5, 987-997.

Kaluza, K., Fuhrmann, M., Hahn, M., Regensburger, B.and Hennecke, H. (1983) J. Bacteriol. 155, 915-918.

Kaluza, K. and Hennecke, H. (1984) Mol. Gen. Genet. 196, 35-42.

Kaminski, P.A., Desnoues, N. and Elmerich, C. (1994) Proc. Natl. Acad. Sci. U S A 91, 4663-4667.

Kaminski, P.A., and Elmerich, C. (1991) Mol. Microbiol. 5, 665-673.

Kaminski, P.A., Kitts, C.L., Zimmerman, Z. and Ludwig, R.A. (1996) J. Bacteriol. 178, 5989-5994.

Kaminski, P.A., Mandon, K., Arigoni, F., Desnoues, N. and Elmerich, C. (1991) Mol. Microbiol. 5, 1983-1991.

Kaminski, P.A., Norel, F., Desnoues, N., Kush, A., Salzano, G. and Elmerich, C. (1988) Mol. Gen. Genet. 214, 496-502.

Kardailsky, I., Yang, W.C., Zalensky, A., van Kammen, A. and Bisseling, T. (1993) Plant. Mol. Biol. 23, 1029-1037.

Kaufman, R.I. and Nixon, T.B. (1996) J. Bacteriol. 178, 3967-3970.

Keister, D.L. (1975) J. Bacteriol. 123, 1265-1268.

Kereszt, A., Slaska, K.K., Putnoky, P., Banfalvi, Z. and Kondorosi, A. (1995) Mol. Gen. Genet. 247, 39-47.

Kim, S. and Burgess, B.K. (1996) J. Biol. Chem. 271, 9764-9770.

Kitts, C. L., and R. A. Ludwig. (1994) J. Bacteriol. 176, 886-895.

Klipp, W., Reiländer, H., Schlüter, A., Krey, R. and Pühler, A. (1989) Mol. Gen. Genet. 216, 293-302.

Knight, C.D., Rossen, L., Robertson, J.G., Wells, B. and Downie, A.J. (1986) J. Bacteriol. 166, 552-558.

Krey, R., Pühler, A. and Klipp, W. (1992) Mol. Gen. Genet. 234, 433-441.

Kullik, I., Fritsche, S., Knobel, H., Sanjuan, J., Hennecke, H. and Fischer, H-M. (1991) J. Bacteriol. 173, 1125-1138.

Kurz, W.G.W. and LaRue, T.A. (1975) Nature 256, 407-409.

Kustu, S., Santero, E., Keener, J., Popham, D. and Weiss, D. (1989) Microbiol. Rev. 53, 367-376.

Labes, M., Rastogi, V., Watson, R. and Finan, T.M. (1993) J. Bacteriol. 175, 2662-2673.

Liu, Y. and Kahn, M.L. (1995) J. Biol. Chem. 270, 1624-1628.

Lois, A.F., Ditta, G.S. and Helinski, D.R. (1993) J. Bacteriol. 175, 1103-1109.

Lois, A.F., Weinstein, M., Ditta, G.S. and Helinski, D.R. (1993) J. Biol. Chem. 268, 4370-4375.

Loroch, A., Nguyen, B.G. and Ludwig, R.A. (1995) J. Bacteriol. 177, 7210-7221.

Ludwig, R.A. (1978) J. Bacteriol. 135, 114-123.

Ludwig, R.A. (1984) Proc. Natl. Acad. Sci. USA 81, 1566-1569.

Lukat, G.S., McCleary, W.R., Stock, A.M. and Stock, J.B. (1992) Proc. Natl. Acad. Sci. USA 89, 718-722.

Manco, G., Rossi, M., Defez, R., Lamberti, A., Percuoco, G. and Iaccarino, M. (1992) J. Gen. Microbiol. 138, 1453-1460.

Mandon, K., Kaminski, P.A. and Elmerich, C. (1994) J. Bacteriol. 176, 2560-2568.

Mandon, K., Kaminski, P.A., Mougel, C., Desnoues, N., Dreyfus, B. and Elmerich, C. (1993) FEMS Microbiol. Lett. 114, 185-189.

Margolin, W., Bramhill, D. and Long, S.R. (1995) J. Bacteriol. 177, 2892-2900.

Margolin, W., Corbo, J.C. and Long, S.R. (1991) J. Bacteriol. 173, 5822-5830.

Margolin, W. and Long, S.R. (1994) J. Bacteriol. 176, 2033-2043.

Mavridou, A., Barny, M.A., Poole, P., Plaskitt, K., Davies, A.E., Johnston, A.W. and Downie, J.A. (1995) Microbiology 141, 103-111.

McComb, J.A., Elliott, J. and Dilworth, M.J. (1975) Nature 256, 409-410.

McDermott, T.R. and Kahn, M.L. (1992) J. Bacteriol. 174, 4790-4797.

Merrick, M.J. (1992). p. 835-876. In G. Stacey, R. H. Burris and H. J. Evans (eds), Biological Nitrogen Fixation. Chapman & Hall, New York.

Michiels, J. and Vanderleyden, J. (1993) Biochim. Biophys. Acta. 1144, 232-233.

Miller, R.W., McRae, D.G. and Joy, K. (1991) Mol. Plant-Microbe. Interact. 4, 37-45.

Monson, E.K., Ditta, G.S. and Helinski, D.R. (1995) J. Biol. Chem. 270, 5243-5250.

Monson, E.K., Weinstein, M., Ditta, G.S. and Helinski, D.R. (1992) Proc. Natl. Acad. Sci. USA 89, 4280-4284.

Moreno, S., Meza, R., Guzman, J., Carabez, A. and Espin, G. (1991) Mol. Plant-Microbe Interact. 4, 619-622.

Morett, E., Fischer, H-M. and Hennecke, H. (1991) J. Bacteriol. 173, 3478-3487.

Muffler, A., Fischer, D. and Hengge-Aronis, R. (1996) Genes & Dev. 10, 1143-1151.

Nixon, B.T., Ronson, C.W. and Ausubel, F.M. (1986) Proc. Natl. Acad. Sci. USA 83, 7850-7854.

Norel, F., Desnoues, N. and Elmerich, C. (1985) Mol. Gen. Genet. 199, 352-356.

Norel, F. and Elmerich, C. (1987) J. Gen. Microbiol. 133, 1563-1576.

O'Brian, M.R., and Maier, R.J. (1987) Proc. Natl. Acad. Sci. USA 84, 3219-3223.

Pagan, J.D., Child, J.J., Scowcroft, W.R. and Gibson, A.H. (1975) Nature (London) 256, 406-407.

Parkinson, J.S. and Kofoid, E.C. (1992) Annu. Rev. Genet. 26, 71-112.

Patriarca, E.J., Tate, R., Fedorova, E., Riccio, A., Defez, R. and Iaccarino, M. (1996) Mol. Plant-Microbe Interact. 9, 243-251.

Patschkowski, T., Schluter, A. and Priefer, U.B. (1996) Mol. Microbiol. 21, 267-280.

Paustian, T., Shah, V.K. and Roberts, G.P. (1989) Proc. Natl. Acad. Sci. USA 86, 6082-6086.

Pawlowski, K., Klosse, U. and de Bruijn, F.J. (1991) Mol. Gen. Genet. 231, 124-138.

Perez-Goldano, R. and Kahn, M.L. (1994) Microbiology 140, 1231-1235.

Preisig, O., Anthamatten, D. and Hennecke, H. (1993) Proc. Natl. Acad. Sci. USA 90, 3309-3313.

Preisig, O., Zufferey, R., Thöny-Meyer, L., Appleby, C.A. and Hennecke, H. (1996a) J. Bacteriol. 178, 1532-1538.

Preisig, O., Zufferey, R. and Hennecke, H. (1996b) Arch. Microbiol. 165, 297-305.

Putnoky, P., Grosskopf, E., Ha, D.T.,. Kiss, G.B. and Kondorosi, A. (1988) J. Cell. Biol. 106, 597-607.

Quinto, C., De la Vega, H., Flores, M., Fernandez, L., Ballado, T., Soberon, G. and Palacios, R. (1982) Nature 299, 724-726.

Quinto, C., De la Vega, H., Flores, M., Leemans, J., Cevallas, M.A., Pardo, M.A., Azpiroz, R., Girard, M.L., Calva, E. and Palacios, R. (1985) Proc. Natl. Acad. Sci. USA 82, 1170-1174.

Ramseier, T.M., Winteler, H.V. and Hennecke, H. (1991) J. Biol. Chem. 266, 7793-7803.

Ratet, P., Pawlowski, K., Schell, J. and de Bruijn, F.J. (1989) Mol. Microbiol. 3, 825-838.

Reitzer, L.J. and Magasanik, B. (1987). pp. 302-320. In Neidhardt, F.C., Ingisham, J.L., Lav, K.B., Mayasonik, B., Schoechter, M., Umbarger, H.E. (eds), *Escherichia coli* and *Salmonella typhimurium*: cellular and molecular biology. American Society for Microbiology., Washington, D. C.

Renalier, M.H., Batut, J., Ghai, J., Terzaghi, B., Gherardi, M., David, M., Garnerone, A.M., Vasse, J., Truchet, G., Huguet, T. and Boistard, P. (1987) J. Bacteriol. 169, 2231-2238.

Reyrat, J.M., David, M., Blonski, C., Boistard, P. and Batut, J. (1993) J. Bacteriol. 175, 6867-6872.

Riedel, K.U., Jouanneau, Y., Masepohl, B., Pühler, A. and Klipp, W. (1995) Eur. J. Biochem. 231, 742-746.

Ritz, D., Bott, M. and Hennecke, H. (1993) Mol. Microbiol. 9, 729-740.

Ritz, D., Thöny-Meyer, L., and Hennecke, H. (1995) Mol. Gen. Genet. 247, 27-38.

Roest, H.P., Goosen-de Roo, L., Wijffelman, C.A., de Maagd, R.A. and Lugtenberg, B.J.J. (1995 a) Mol. Plant-Microbe Interact. 8, 14-22.

Roest, H.P., Mulders, I.H., Wijffelman, C.A. and Lugtenberg, B.J.J. (1995 b) Mol. Plant-Microbe Interact. 8, 576-583.

Ronson, C.W. (1988). pp. 547-557. In Bothe, H., de Bruijn, F.J. and Newton, W.E. (eds), Nitrogen Fixation: Hundred Years After. Gustav Fischer, Stuttgart New York.

Ronson, C.W., and Astwood, P.M. (1985). pp. 201-207. In Evans, H.J., Bottomley, P.J. and Newton, W.E. (eds), Nitrogen Fixation Research Progress. Martinus Nijhoff Publishers, Dordrecht.

Ronson, C.W., Astwood, P.M. and Downie, J.A. (1984) J. Bacteriol. 160, 903-909.

Ronson, C.W., Astwood, P.M., Nixon, B.T. and Ausubel, F.M. (1987b) Nucleic. Acids. Res. 15, 7921-7934.

Ronson, C.W., Lyttleton, P. and Robertson, J.G. (1981) Proc. Natl. Acad. Sci. USA 78, 4284-4288.

Rossi, M., Defez, R., Chiurazzi, M., Lamberti, A., Fuggi, A. and Iaccarino, M. (1989) J. Gen. Microbiol. 135, 629-637.

Ruvkun, G.B. and Ausubel, F.M. (1980) Proc. Natl. Acad. Sci. USA. 77, 191-195.

Ruvkun, G.B. and Ausubel, F.M. (1981) Nature 289, 85-89.

Ruvkun, G.B., Sundaresan, V. and Ausubel, F.M. (1982) Cell 29, 551-9.

Saint, C.P., Wexler, M., Murphy, P.J., Tempé, J., Tate, M.E. and Murphy, P.J. (1993) J. Bacteriol. 175, 5205-5215.

Sanders, D.A., Gillece, C.B., Burlingame, A.L. and Koshland, D.J. (1992) J. Bacteriol. 174, 5117-5122.

Sanjuan, J. and Olivares, J. (1991) Arch. Microbiol. 155, 543-548.

Schlaman, H.R., Horvath, B., Vijgenboom, E., Okker, R.J. and Lugtenberg, B.J.J. (1991) J. Bacteriol. 173, 4277-4287.

Schlaman, H.R., Lugtenberg, B.J.J. and Okker, R.J. (1992) J. Bacteriol. 174, 6109-6116.

Schlüter, A., Patschkowski, T., Unden, G. and Priefer, U.B. (1992) Mol. Microbiol. 6, 3395-3404.

Schlüter, A., Patschkowski, T., Weidner, S., Unden, G., Hynes, M.F. and Priefer, U.B. (1993). pp. 493. In Palacios, R., Mora, J. and Newton, W.E. (eds), New Horizons in Nitrogen fixation. Kluwer Academic Publishers, Dordrecht.

Schlüter, A., Ruberg, S., Kramer, M., Weidner, S. and Priefer, U.B. (1995) Mol. Gen. Genet. 247, 206-215.

Scott, K.F., Rolfe, B.G. and Shine, J. (1983) DNA 2, 141-148.

Shatters, R.G. and Kahn, M.L. (1989) J. Mol. Evol. 29, 422-428.

Shatters, R.G., Liu, Y. and Kahn, M.L. (1993) J. Biol. Chem. 268, 469-475.

Shaw, D.J., Rice, D.W. and Guest, J.R. (1983) J. Mol. Biol. 166, 241-7.

Sindhu, S.S., Brewin, N.J. and Kannenberg, E.L. (1990) J. Bacteriol. 172, 1804-1813.

Smith, B.E., Roe, S.M. and Yousafzai, F.K. (1995). pp. 19-27. In Tikhonovitch, I.A., Provorov, N.A., Romanov, V.I. and Newton, W.E. (eds), Nitrogen fixation: fundamentals and applications. Kluwer Academic Publisher, Dordrecht.

Soberon, M., Aguilar, G.R. and Sanchez, F. (1993) Mol. Microbiol. 8, 159-166.

Somerville, J.E. and Kahn, M.L. (1983) J. Bacteriol. 156, 168-176.

Soupène, E., Foussard, M., Boistard, P., Truchet, G. and Batut, J. (1995) Proc. Natl. Acad. Sci. USA 92, 3759-3763.

Stam, H., van Verseveld, H.W. and Stouthamer, A.H. (1983) Arch. Microbiol. 135, 199-204.

Stigter, J., Schneider, M. and de Bruijn, F.J. (1993) Mol. Plant-Microbe Interact. 6, 238-252.

Szeto, W.W., Nixon, B.T., Ronson, C.W. and Ausubel, F.M. (1987) J. Bacteriol. 169, 1423-32.

Szeto, W.W., Zimmerman, J.L., Sundaresan, V. and Ausubel, F.M. (1984) Cell 36, 1035-43.

Thöny, B., Fischer, H-M., Anthamatten, D., Bruderer, T. and Hennecke, H. (1987) Nucleic. Acids. Res. 15, 8479-8499.

Thöny-Meyer, L. (1997) Microbiol. Mol. Biol. Rev. 61, 337-376.

Thöny-Meyer, L., Ritz, D. and Hennecke, H. (1994a) Mol. Microbiol. 12, 1-9.

Thöny-Meyer, L., Beck, C., Preisig, O. and Hennecke, H. (1994b) Mol. Microbiol. 14, 705-716.

Thöny-Meyer, L., Stax, D. and Hennecke, H. (1989) Cell 57, 683-97.

Tyerman, S.D., Whitehead, L.F. and Day, D.A. (1995) Nature 378, 629-632.

Udvardi, M.K., Price, D.G., Gresshoff, P.M. and Day, D.A. (1988) FEBS Lett. 231, 36-40.

Valderrama, B., Davalos, A.G., Morett, L.E. and Mora, J. (1996) J. Bacteriol. 178, 3119-3126.

van Slooten, J.C., Bhuvanasvari, T.V., Bardin, S. and Stanley, J. (1992) Mol. Plant-Microbe Interact. 5, 179-186.

van Slooten, J.C., Cervantes, E., Broughton, W.J., Wong, C.H. and Stanley, J. (1990) J. Bacteriol. 172, 5563-5574.

Vargas, C., Wu, G.H., Delgado, M.J., Poole, R.K. and Downie, J.A. (1996) Microbiology 142, 41-46.

Vasse, J., de Bruijn, F.J., Camut, S. and Truchet, G. (1990) J. Bacteriol. 172, 4295-4306.

Waelkens, F., Foglia, A., Morel, J.B., Fourment, J., Batut, J. and Boistard, P. (1992) Mol. Microbiol. 6, 1447-1456.

Wang, Y.P., Birkenhead, K., Boesten, B., Manian, S. and O'Gara, F. (1989) Gene 85, 135-144.

Watson, R.J. (1990) Mol. Plant-Microbe Interact. 3, 174-181.

Watson, R.J., Chan, Y.K., Wheatcroft, R., Yang, A.F. and Han, S.H. (1988) J. Bacteriol. 170, 927-934.

Weidenhaupt, M., Rossi, P., Beck, C., Fischer, H-M and Hennecke, H. (1996) Arch. Microbiol. 165, 169-178.

Weiss, D.S., Batut, J., Klose, K.E., Keener, J. and Kustu, S. (1991) Cell 67, 155-167.

Witty, J.F., Minchin, F.R., Skot, L. and Sheehy, J.E. (1986). pp. 275-314. In Oxford Surveys of Plant Molecular and Cell Biology. Oxford University Press, Oxford.

Yang, W.C., Horvath, B., Hontelez, J., Van Kammen, A. and Bisseling, T. (1991) Mol. Plant-Microbe Interact. 4, 464-468.

Yarosh, O.K., Charles, T.C. and Finan, T.M. (1989) Mol. Microbiol. 3, 813-823.

Yorgey, P. and Kolter, P. (1993) Trends in Genetics 9, 374-375.

Young, J.P.W. (1992). pp. 43-86. In Stacey, G., Burris, R.H. and Evans, H.J. (eds), Biological Nitrogen Fixation. Chapman and Hall, New York.

Yuen, J. and Stacey, G. (1996) Mol. Plant-Microbe Interact. 9, 424-428.

Yun, A.C. and Szalay, A.A. (1984) Proc. Natl. Acad. Sci. USA. 81, 7358-7362.

Yura, T., Mori, H., Nagai, H., Nagata, T., Ishihama, A., Fujita, N., Isono, K., Mizobuchi, K. and Nakata, A. (1992) Nucleic. Acids. Res. 20, 3305-3308.

Zheng, L., White, R.H., Cash, V.L., Jack, R.F. and Dean, D.R. (1993) Proc. Natl. Acad. Sci. USA 90, 2754-2758.

Zimmerman, J.L., Szeto, W.W. and Ausubel, F.M. (1983) J. Bacteriol. 156, 1025-1034.

Zufferey, R., Preisig, O., Hennecke, H. and Thöny-Meyer, L. (1996) J. Biol. Chem. 271, 9114-9119.

Chapter 24

Carbon and Nitrogen Metabolism in Rhizobia

Michael L. Kahn, Tim R. McDermott and Michael K. Udvardi

I. Introduction

The *Rhizobiaceae* are a family of aerobic soil bacteria that can establish mutualistic or parasitic associations with higher plants. Life in the soil is challenging because the soil environment can vary widely from place to place, and from time to time in any one place. The *Rhizobiaceae*, like other soil bacteria, possess diverse metabolic pathways that enable them to grow and multiply under many different conditions. For example, the *Rhizobiaceae* can grow on a range of inorganic and organic nitrogen compounds as sole nitrogen source, including molecular nitrogen, nitrate, nitrite, ammonium, and amino acids. Likewise, they can utilize a wide range of carbon compounds, including sugars, organic acids, amino acids, and phenolics. Metabolism of carbon and nitrogen compounds involves a number of metabolic pathways that are common to many soil microorganisms, as well as some that are restricted to just a few. The *Rhizobiaceae* possess many of the common primary metabolic pathways, although the regulation of these pathways is sometimes unique in this family. Also unique are a number of secondary metabolic pathways that produce signaling or cell surface compounds that are involved in establishing or maintaining specific associations with host plants (see elsewhere in this book).

Higher plants are the principal photoautotrophs in terrestrial environments and are the primary source of reduced carbon for most other terrestrial organisms, including soil microorganisms. Plants are also a potential source of most or all of the other nutrients required by the soil biota. Therefore, plants provide a relatively stable and rich environment for any microorganism that can establish and maintain intimate contact with them. However, symbiosis requires integration of microbial metabolism with that of the host, and this has shaped bacterial evolution in interesting ways. Within the Rhizobiaceae, the challenge to integrate metabolically with the host has been answered in two very different ways. To understand this, it is most instructive to consider the interplay of carbon and nitrogen metabolism between plants and the Rhizobiaceae. In the nitrogen fixing symbioses that involve legumes (and a few non-legumes) and members of the genera *Rhizobium, Bradyrhizobium,* and *Azorhizobium* (collectively referred to as rhizobia), the central metabolic interaction is the exchange of reduced carbon from the plant for reduced nitrogen from the bacteria. Underlying this apparently simple exchange are quite complex and often subtle changes in carbon and nitrogen metabolism in both organisms. In contrast, members of the genus *Agrobacterium* are much less subtle in the manner in which they obtain carbon and nitrogen from their hosts. The agrobacteria transfer part of their DNA (the T-DNA) to the plant cell nucleus. The T-DNA encodes enzymes that redirect plant carbon and nitrogen flow into compounds, called rhizopines, that only the bacteria can use. From the perspective of higher plants, the Rhizobiaceae can be categorized into two classes: the 'good guys' and the 'bad guys'. This chapter focuses on the organization and regulation of carbon and nitrogen metabolism in the 'good guys', or the rhizobia. The metabolism of both free-living and symbiotic bacteria will be considered, in part to illustrate the metabolic flexibility of these bacteria, and in part to highlight the metabolic specialization that enables the integration of bacterial and plant metabolism during symbiosis.

II. Carbon and Nitrogen Metabolism in Free-Living Rhizobia

II.A. NITROGEN METABOLISM

Rhizobia are often prototrophic or require one or more vitamins or amino acids as supplements (see Chakrabarti *et al.*, 1981 for an example of the kinds of variation found) and are therefore commonly grown using yeast extract or other complex media. However, much of our knowledge of nitrogen metabolism in rhizobia comes from *in vitro* studies on the physiology and biochemistry of pure strains grown in defined minimal media. A common feature of rhizobia is their ability to grow on a variety of inorganic nitrogen compounds as the sole source of nitrogen. These include nitrate, nitrite, and ammonium. Despite the fact that rhizobia have the potential to fix nitrogen, few species express this potential *in vitro*, and only *Azorhizobium* species are able to grow on molecular dinitrogen (N_2) as the sole source of nitrogen. In nature, it appears that nitrogen fixation in rhizobia is confined largely to the symbiotic state. Nonetheless, we will consider some aspects of nitrogen fixation in this section.

II.A.1. Nitrate Assimilation

Nitrate assimilation is a three step process that begins with nitrate uptake. A number of bacteria, including *Klebsiella pneumoniae* (Thayer and Huffaker, 1982) and *Pseudomonas fluorescens* (Betlach *et al.*, 1981) possess nitrate transporters that enable them to accumulate nitrate in an energy-dependent manner. Similar systems probably exist in rhizobia, although they remain to be characterized in detail. Following nitrate uptake into the cell, nitrate is reduced to ammonium in two steps by the enzymes nitrate reductase and nitrite reductase. Rhizobia possess two nitrate reductases, an assimilatory nitrate reductase that is required for growth on nitrate under aerobic conditions (Sik and Barabas, 1977; Kiss *et al.*, 1979) and a dissimilatory nitrate reductase that uses nitrate as the terminal electron acceptor during anaerobic respiration (Cheniae and Evans 1960; Daniel and Appleby, 1972; Manhart and Wong, 1979; Streeter and deVine, 1983; Gollop and Avissar, 1984).

Assimilatory nitrate reductase is regulated by nitrate (Manhart and Wong, 1979; Howitt *et al.*, 1988; Monza *et al.*, 1992) via a global nitrogen regulatory system (see below), while the dissimilatory nitrate reductase is regulated by oxygen (Daniel and Appleby, 1972, O'Hara *et al.*, 1983). Assimilatory nitrate reductase reduces nitrate to nitrite, which is then reduced to ammonium by nitrite reductase. Little is known about nitrite reductases in rhizobia, but it has been suggested that different enzymes may be used for the assimilation of nitrite and its dissimilation to nitrogen-containing gases (Giannakis *et al.*, 1988).

II.A.2. Nitrogen Fixation

Rhizobia can reduce molecular nitrogen (N_2) to ammonium using the enzyme nitrogenase (see Chapter 23). Reduction of molecular nitrogen to ammonium requires a large amount of metabolic energy to break the triple bond between the two nitrogen atoms. Consequently, expression of nitrogenase is tightly regulated at a number of different levels. Most free-living nitrogen-fixing bacteria mobilize the nitrogenase machinery only after other nitrogen sources are exhausted, and only if certain physical conditions are met (Burris and Roberts, 1993). Paramount amongst these is the requirement for very low oxygen concentrations, since nitrogenase is very oxygen labile. A number of different adaptive mechanisms that provide the low oxygen concentrations required for nitrogenase activity have evolved in these organisms (Robson and Postgate, 1980). As a result of their ability to utilize molecular nitrogen as the sole source of nitrogen for growth, free-living diazotrophs are able to inhabit niches that are inaccessible to other organisms. Unlike in the strictly free-living diazotrophs, however, molecular nitrogen apparently cannot serve as the sole source of nitrogen for growth of any of the rhizobia except *Azorhizobium*. In fact, nitrogen fixation in rhizobia appears to be largely confined to the symbiotic state where the bacteria appear not to be nitrogen limited at all. This interesting paradox is considered in more detail in Section III.

II.A.3. Ammonia Assimilation

Rhizobia differ in the nitrogen sources they prefer. Some strains grow best with ammonium as nitrogen source while others prefer glutamate or other amino acids and, when grown on amino acids with a high N:C ratio, excess ammonia can be found in the media. In many bacteria, ammonium can be assimilated into glutamate by glutamate dehydrogenase (GDH) when the ammonium concentration is relatively high but at low concentrations ammonium is incorporated first into glutamine and then glutamate via the enzymes glutamine synthetase (GS) and glutamate synthase (GOGAT). This GS-GOGAT pathway is thought to be the major pathway for ammonia assimilation in rhizobia. These bacteria are unusual in possessing two or even three different GS enzymes, which were reviewed by Espin *et al.* (1994). The two major forms, GSI and GSII, have similar affinities for substrates and inhibitors (Darrow, 1980) and both are often found in growing cultures. To quickly estimate the fraction of GS that each protein contributes, the heat stable fraction of the activity is often assigned to GSI and the heat labile fraction to GSII and, although GSI is not completely unaffected by heat, this distinction is sufficiently precise for most purposes.

GSI is similar to the GS found in other Gram-negative bacteria (Rhee *et al.*, 1985). The behavior of GSI in enzyme assays in the presence of Mn^{2+} and Mg^{2+} (Arcondeguy *et al.*, 1996, Noguez *et al.*, 1994) and the regulation of GSI by *glnB* (Amar *et al.*, 1994), which encodes the PII adenyltransferase, suggest that like the enteric enzyme, GSI is post-translationally inhibited by adenylylation in the presence of ammonium. Unlike GS regulation in the enteric bacteria (Reitzer, 1996), rhizobial *glnA* transcription is relatively independent of the nitrogen status of the cell or the NtrA or NtrC proteins (Martin *et al.*, 1988; DeBruijn *et al.*, 1989; Shatters *et al.*, 1989). Thus, rhizobia appear to contain GSI protein under most growth conditions although GSI specific activity can be altered posttranslationally as appropriate for the growth conditions. This situation may be useful in allowing the bacteria to respond immediately under conditions where glutamine synthesis is needed but protein synthesis is not fully active, conditions that may be common in marginal soil environments.

In contrast to GSI, the major regulation of GSII appears to be transcriptional rather than posttranslational. GSII expression requires the nitrogen stress response regulator NtrC, acting together with the NtrA subunit of RNA polymerase (Martin *et al.*, 1988; DeBruijn *et al.*, 1989; Shatters *et al.*, 1989). GSII synthesis and activity are generally greatly depressed by the presence of ammonium in the growth medium. Under these conditions GSII transcription appears to stop immediately and GSII activity in the culture decreases rapidly, suggesting that the protein is unstable or specifically inactivated (Manco et al., 1992). The amino acid sequence of GSII is similar to those of eukaryotic GS enzymes (Carlson and Chelm, 1986) but the presence of GSII in rhizobia can be explained without invoking prokaryotic-eukaryotic gene exchange (Pesole et al., 1995; Shatters and Kahn, 1989). Taboada et al. (1996) have found that GSII is a good taxonomic marker for rapid classification of Rhizobium strains.

The third GS, GSIII, is not found in free-living cultures grown on normal laboratory medium but is found in S. meliloti mutants that lack GSI and GSII (Shatters et al., 1993) and has also been characterized as the product of the R. phaseoli glnT gene expressed in Klebsiella pneumoniae (Espin et al., 1990). The S. meliloti enzyme has unusual kinetic properties, with very high K_m values for glutamate (13.3 mM) and ammonium (33 mM) (Shatters et al., 1993) and a very low GS transferase activity. The K_m values are consistent with the requirement for both ammonium and glutamate or aspartate for glutamine independent growth of a S. meliloti double mutant (Somerville et al., 1989). Because GSIII is about 4% of soluble protein in the S. meliloti double mutant, it can plausibly synthesize enough glutamine to support growth despite its low affinity for substrates. A distinct function for the three GS enzymes is not yet clear, especially in symbiosis (see Section III).

There are three potential routes for glutamate synthesis, by GOGAT (using glutamine, an oxidant and α–ketoglutarate as substrates), by GDH (using ammonium, an oxidant and α–ketoglutarate) and by transamination of α–ketoglutarate. Both NAD and NADP-dependent forms of GOGAT are found in Azorhizobium caulinodans (Donald and Ludwig, 1984). Mutants producing only one of these enzymes are unable to grow on ammonium, although since some mutants lacking both enzymes appear to be due to a single mutation, at least some of the mutations studied were likely to have been in regulatory proteins. In S. meliloti 102F34, Lewis et al. (1990) cloned a gene encoding the single NADP-dependent GOGAT and constructed transposon mutants lacking the enzyme. The mutants were unable to grow on media that contained 10 mM ammonium and grew poorly when some other amino acids were used as nitrogen sources. Second site revertants partially restored growth on some of these amino acids but did not restore growth on ammonium, showing that

in this strain none of the amino acid dehydrogenases were able to assimilate nitrogen and contribute it to the anabolic pool. GDH increases about ten-fold in S. meliloti 102F34 grown in high concentrations of ammonium (Gonzalez-Gonzalez et al., 1990) but the assimilatory Km for NAD-GDH is 20 mM, a high value indicating that the enzyme is not optimum for biosynthesis.

Once incorporated into glutamate, the amino acid nitrogen can be transferred to other amino acids by various aminotransferases. The major aspartate aminotransferase, AAT-A, uses only oxaloacetate as an acceptor of nitrogen from glutamate (Bai and Kahn, unpublished). In S. meliloti JJc10, AAT-A is induced above its basal level of expression by growth on aspartate as carbon source (Rastogi and Watson, 1991), strongly suggesting that aspartate catabolism in this case first involves transamination. A second, cryptic aspartate aminotransferase gene, aatB, also is found in S. meliloti (Alfano and Kahn, 1993). aatB encodes an enzyme with a higher specific activity than AAT-A, but no phenotype has yet been detected in an aatB mutant. A major branched chain aminotransferase is also present in S. meliloti (Gonzales-Gonzales et al., 1990) and a gene encoding this activity has been cloned (Alfano and Kahn, 1993).

II.A.4. Amino Acid and Nucleotide Metabolism

Studies investigating the metabolism of amino acids and nucleotides suggest that the biosynthesis of most of these molecules is not unusual. Much of the work in this area has been directed at understanding the role of amino acid and nucleotide metabolism in the symbiosis. For this reason, further discussion of this topic is left to Section III.

II.B. REGULATION OF NITROGEN METABOLISM

In enteric bacteria such as Escherichia coli and Klebsiella pneumoniae, nitrogen metabolism is genetically regulated by a central nitrogen regulation (Ntr) system (Magasanik, 1993). Three genes, ntrA, ntrB, and ntrC are responsible for overall nitrogen regulation in these bacteria. NtrB and NtrC constitute a two-component regulatory system that responds to the nitrogen status of the cell. Under conditions of low nitrogen availability, the sensor protein NtrB phosphorylates and thereby activates NtrC, which in turn activates the transcription of several genes involved in nitrogen metabolism, including itself (Magasanik, 1993; Hirschman et al., 1985; Hunt and Magasanik, 1985). The NtrA protein is a unique sigma factor, σ54, that combines with the RNA polymerase core enzyme, and the regulatory protein NtrC to activate gene transcription (Hirsham et al., 1985; Hunt and Magasanik, 1985; Merrick and Gibbons, 1985). NtrC has sequence specific DNA-binding activity (Ames and Nikaido, 1985) which has been implicated in transcriptional activation (Kustu et al., 1991). Genes that are under Ntr control include the genes encoding glutamine synthetase (GS), nitrate reductase (NR), an ammonium transporter (Amt), and amino acid uptake and utilization functions (Magasanik, 1993). The nitrogen fixation (nif) genes in K. pneumoniae are also under Ntr control. (Ausubel, 1984; Gussin et al., 1986) The Ntr system induces nif gene expression by promoting the synthesis of NifA, which then activates the transcription of the other nif genes.

An Ntr system also regulates nitrogen metabolism in rhizobia. Homologues of ntrA, ntrB and ntrC have been identified in rhizobia, and both ntrA and ntrC have been shown to be necessary for expression of several genes involved in nitrogen assimilation, including nitrate reductase, GSII, and an ammonium transporter (Ausubel et al., 1985; Nixon et al., 1986; Pawlowski et al., 1987; Ronson et al., 1987; Szeto et al., 1987; Martin et al., 1988; Shatters et al., 1989; Stanley et al., 1989; Udvardi et al., 1992). However, there are differences between rhizobia and enteric bacteria in the pattern of Ntr control. For example, while glnA is under the control of the Ntr system in enteric bacteria (Magasanik, 1993), its homologue in Azorhizobium and Bradyrhizobium species is not (Pawlowski et al., 1987; Carlson et al., 1987). Whilst transcription of the nitrogen fixation (nif) genes is absolutely dependent on the activator NtrC in the free-living bacterium Klebsiella pneumoniae, this is not the case in Rhizobium, Azorhizobium, and Bradyrhizobium, at least during symbiotic nitrogen fixation (Gussin et al., 1986; Szeto et al., 1987; Pawlowski et al., 1987; Udvardi et al, 1992). Furthermore, NtrC is not necessary for the utilization of amino acids in rhizobia (Szeto et al., 1987; Pawlowski et al., 1987; Martin et al., 1988; Udvardi et al., 1992).

Regulation of nif genes in rhizobia represents an important difference between these bacteria and other nitrogen fixing bacteria. Low oxygen concentrations, rather than nitrogen starvation, appears to be the principal physiological signal that induces nitrogen fixation in rhizobia. What this means in effect is that nitrogen fixation can be uncoupled from nitrogen metabolism, and this appears to be exactly what happens during symbiosis, as we shall see in Section III.

II.C. CARBON METABOLISM

Rhizobia can use a wide range of carbon compounds for biosynthesis and energy production, although there are a few compounds that are not metabolized for lack of key enzymes. In the following paragraphs we survey the metabolism of several different classes of carbon compounds.

II.C.1. C-1, C-2, C-3 Compounds

Many rhizobia, including Bradyrhizobium japonicum, S. meliloti, and R. leguminosarum bvs. viciae, trifolii, and phaseoli, have an absolute requirement for CO_2, presumably for anapleurotic metabolism (Lowe and Evans, 1962). However, strains of B. japonicum that possess uptake hydrogenase can also utilize CO_2 as a sole source of carbon (Lepo et al., 1980) by employing the Calvin cycle (Simpson et al., 1979) [see Tabita, (1988) for a review of autotrophic growth]. B. japonicum can also use the Calvin cycle to grow on formate after CO_2 is first released by formate dehydrogenase (Manian and O'Gara, 1982). Ribulose bisphosphate carboxylase and phosphoribulokinase are the key enzymes of the Calvin cycle and both have been found in B. japonicum, although not in R. leguminosarum bvs. viciae and trifolii (Malik and Schlegel, 1981).

The metabolism of the C-2 compound acetate is important because acetyl-CoA is the precursor for a number of important anabolic pathways and for the generation of energy. Acetyl-CoA serves as fuel for the tricarboxylic acid (TCA) cycle, is found as a substituent of extracellular polysaccharide molecules, and it is the starting material for the poly-β-hydroxybutyrate pathway which plays an important role in many species during symbiosis (see Section III). In Escherichia coli, growth on acetate or fatty acids as the sole carbon source induces the glyoxylate bypass (reviewed by Clark and Cronan, 1996). This sequence of reactions serves an anapleurotic function that results in carbon being diverted from the TCA cycle at isocitrate, and avoids quantitative loss of carbon as CO_2. Enzymes involved in the glyoxylate bypass include isocitrate lyase and malate synthase. Malate synthase has been found to be constitutively expressed in S. meliloti, and *R. leguminosarum* bvs. *phaseoli, viciae,* and *trifolii.* Isocitrate lyase is not expressed in cells grown on a variety of C-3, C-4, and C-6 carbon sources (Johnson *et al* 1966), but is inducible by growth on the fatty acid oleate (Johnson *et al.,* 1966), which is metabolized via β-oxidation to produce acetyl-CoA. Duncan and Fraenkel (1979), showed that isocitrate lyase is induced when *S. meliloti* is grown on acetate. Malate synthase has also been shown to be constitutive in *B. japonicum,* but this organism will not grow on oleate (Johnson *et al* 1966).

Growth of rhizobia on acetate is usually not good (Johnson *et al* 1966; Perez-Galdona and Kahn, 1994; Preston *et al.,* 1989; Summers and McDermott, unpublished data) and is influenced by culture pH (Perez-Galdona and Kahn, 1994). In *B. japonicum,* acetate can be activated for metabolism by either the acetate kinase-phosphotransacetylase pathway (Smith *et al.,* 1994), or via acetyl-CoA synthetase (Preston *et al* , 1989). The highest levels of acetyl-CoA synthetase and acetate kinase are found in cells grown on acetate or pyruvate (Preston *et al* , 1989). A *S. meliloti ackA* mutant will grow on acetate (Summers and McDermott, unpublished data), and acetyl-CoA synthetase activity can be measured in the *S. meliloti* wild-type and *ackA* mutant. However, acetyl-CoA synthetase activity is low in cells grown on minimal mannitol medium and growth of the wild-type on acetate does not lead to increased levels of this enzyme (Summers and McDermott, unpublished). Curiously, *S. meliloti ackA* is highly induced by phosphate stress, but not by growth on acetate (Summers and McDermott, unpublished).

Aldehyde dehydrogenase has been observed in *B. japonicum* (Peterson and LaRue, 1982). This enzyme is active with a variety of aldehyde substrates, with Kms ranging from 4.0 µM for butyraldehyde to 430 µM and 454 µM for succinic semialdehyde and acetaldehyde, respectively.

Glycerol is utilized as a sole carbon source by a variety of rhizobia, and its presence induces the enzymes required for its metabolism (Glenn and Dilworth, 1981a; Arias and Martinez-Drets, 1976). In *B. japonicum* and *R. leguminosarum* bv. *trifolii,* glycerol metabolism is initiated by glycerol kinase which produces glycerolphosphate.

Glycerolphosphate is then converted to dihydroxyacetone phosphate by glycerolphosphate dehydrogenase (Arias and Martinez-Drets, 1976).

The dihydroxyacetone is further metabolized to pyruvate, which can then enter the TCA cycle.

Based on oxygen consumption rates, malonate apparently also induces at least some of the enzymes required for its metabolism in *R. leguminosarum* bv. *viciae* (Glenn and Dilworth, 1981a). Malonamidases, which have been proposed to play an important role in bacteroid nitrogen metabolism (Kim and Chae, 1990), have been purified to homogeneity from *B. japonicum* and characterized (Kim and Kang, 1994). These enzymes, as well as malonyl-CoA synthetase, are expressed constitutively in *B. japonicum* (Kim and Chae, 1991).

II.C.2. C-4 Compounds

The metabolism of fumarate, succinate, and malate has received considerable attention. Rhizobia take up these compounds via a high affinity C_4-dicarboxylate transporter, encoded by the *dctA* gene (Bolton *et al.,* 1986; Engelke *et al.,* 1989; Ronson *et al.,* 1984). The DctA protein has a Km of 2-15 µM and a Vmax of 10-80 nmol/min/mg protein in different genera (Engelke *et al.,* 1987; Finan *et al.,* 1981; McAllister and Lepo, 1983; San Francisco and Jacobson, 1985; Udvardi *et al.,* 1988a). Transcription of the *dctA* gene is induced by C_4-dicarboxylates in the external medium via a two component regulatory system consisting of the DctB and DctD proteins (Jording *et al.,* 1992; 1994). The dicarboxylate transport system will also transport aspartate (Watson *et al.,* 1993), and can be induced by this compound (Batista *et al* 1992).

Fumarate, succinate, and malate are all intermediates of the TCA cycle, and thus their metabolism proceeds directly through the TCA cycle. However, growth on C_4-dicarboxylates as the sole carbon source requires an anapleurotic sequence in order to provide acetyl-CoA for the condensation reaction with oxaloacetate, and also for the synthesis of phosphoenolpyruvate (PEP) for gluconeogenesis. The pathway that provides acetyl-CoA in *R. leguminosarum* bv. *viciae* includes malic enzyme and pyruvate dehydrogenase, and PEP is formed from oxaloacetate via PEP carboxykinase (McKay *et al.,* 1985; 1988). PEP carboxykinase is required for gluconeogenesis in the broad host-range *Rhizobium* NGR234 (Osteras *et al.,* 1991), and Finan *et al.* (1988) described several *S. meliloti* mutants that were unable to grow on succinate, most of which were defective in gluconeogenesis. One of these was a PEP carboxykinase mutant. Even though the PEP carboxykinase mutant showed no PEP carboxykinase activity, it could grow very slowly on succinate. This suggests that either the mutant retained a small, unmeasurable amount of PEP carboxykinase, or an alternative route, such as malic enzyme (ME) and PEP synthetase, was available to make PEP. Two MEs have been identified in *R. leguminosarum* bv. *viciae* (McKay *et al.*, 1988) and *S. meliloti* (Driscoll and Finan, 1993). One of the *S. meliloti* MEs specifically uses $NADP^+$ as a cofactor, whereas the other ME can use both $NADP^+$ and NAD^+, but has a preference for the latter (Driscoll and Finan, 1993). NAD^+-ME mutants have been isolated and have no apparent carbon utilization defect, suggesting that both MEs are expressed in free-living cells. However, the NAD^+-ME mutant was Fix⁻ in symbiosis (Driscoll and Finan, 1993) (see section III).

II.C.3. C-5 Compounds

Arabinose metabolism proceeds via 2-oxoglutarate semialdehyde, 2-oxoglutarate, and TCA cycle in *S. meliloti* and *R. leguminosarum* bvs. *trifolii*, *phaseoli*, and *viciae* (Duncan and Fraenkel, 1979; Dilworth *et al.*, 1986) and in the broad-host range *Rhizobium* NGR234 (Dilworth *et al.*, 1986). Cellular levels of some of the enzymes involved in arabinose metabolism are coordinately increased in response to growth on arabinose (Dilworth *et al.*, 1986). Resting cells of an *S. meliloti* 2-oxoglutarate dehydrogenase mutant were able to metabolize arabinose to CO_2, suggesting an alternative pathway for arabinose metabolism. Possible routes include conversion of 2-oxoglutarate to glutamate which could re-enter the TCA cycle via the γ-aminobutyrate (GABA) by-pass (discussed below), or conversion of 2-oxoglutarate to pyruvate and acetaldehyde via 2-keto-3-deoxy-L-arabinonate aldolase (KDA aldolase). Low levels of KDA aldolase have been observed in *S. meliloti* and *R. leguminosarum* bv. *viciae*, but not in *R. leguminosarum* bvs. *trifolii* and *phaseoli* (Duncan, 1979). Arabinose metabolism in *B. japonicum* proceeds primarily through the KDA aldolase branch (Pedrosa and Zancan, 1974), but GABA enzymes have also been found in *B. japonicum* and *Bradyrhizobium* sp. (cowpea) (Duncan, 1979) and in a fast-growing cowpea isolate (Jin *et al.*, 1990).

Ribose metabolism is initiated by ribokinase to form ribose-5-phosphate, which can then feed directly into nucleotide biosynthesis, or be isomerised to ribulose-5-phosphate and then epimerised to xylulose-5-phosphate (Dilworth *et al.*, 1986). Ribose-5-phosphate and xylulose-5-phosphate are then converted to fructose-6-phosphate and glyceraldehyde-3-phos-

phate via transketolase and transaldolase. Fructose-6-phosphate is then isomerised to glucose-6-phosphate (Martinez De Drets and Arias, 1970) to enter the Entner-Doudoroff pathway and glyceraldehyde-3-phosphate can be metabolized to pyruvate. Metabolism of ribose in *S. meliloti* apparently is not limited to the above scenario, as Duncan (1981) reported that an *S. meliloti* mutant lacking ribokinase could take up and partially metabolize this pentose. The behavior of a *S. meliloti* xylose isomerase mutant was similar (Duncan, 1981).

II.C.4. C-6 Compounds

Enzymes of the Entner-Doudoroff pathway have been detected in *B. japonicum* (Martinez De Drets and Arias, 1972; Katznelson and Zagallo, 1957; Keele *et al.*, 1969), *Bradyrhizobium* sp (Lupin) (Martinez De Drets and Arias, 1972), *Bradyrhizobium* sp (Cowpea) (Martinez De Drets and Arias, 1972), a slow-growing rhizobia isolated from *Desmodium* (Martinez De Drets and Arias, 1972), *S. meliloti* (Irigoyen *et al.*, 1990; Martinez De Drets and Arias, 1972), *R. leguminosarum* bvs. *phaseoli*, *trifolii*, and *viciae* (Glenn *et al.*, 1984; Martinez De Drets and Arias, 1972), *R. tropici* (Romanov *et al.*, 1994), and in fast-growing rhizobia isolated from *Lotus* (Martinez De Drets and Arias, 1972). Low levels of Embden-Meyerhof-Parnas (EMP) pathway enzymes have been found in *Bradyrhizobium* and in some strains of *Rhizobium* (Mulangoy and Elkan, 1977; Stowers and Elkan, 1983). In general, however, the EMP pathway is not a major route of carbon metabolism.

The oxidative pentose phosphate pathway has been found in fast-growing species, including *R. tropici* (Romanov *et al.*, 1994), *R. leguminosarum* bv. *viciae* (Martinez De Drets and Arias, 1972; Glenn *et al.*, 1984), bv. *trifolii* (Martinez De Drets and Arias, 1972; Ronson and Primrose, 1979), and *phaseoli* (Martinez De Drets and Arias, 1972). This pathway is also operational in snake bean rhizobia (Saroso *et al.*, 1986), and in one strain of *S. meliloti* it was found at higher levels than the Entner-Doudoroff pathway (Irigoyen *et al.*, 1990). Gluconate-6-phosphate dehydrogenase is absent in bradyrhizobia isolates from soybean, lupine, cowpea and *Desmodium* (Martinez De Drets and Arias, 1972; Mulongoy and Elkan, 1977; Stowers and Elkan, 1983) and thus the pentose phosphate pathway appears incomplete in slow-growers. However, transaldolase and transketolase have been detected in *B. japonicum* (Mulongoy and Elkan, 1977) and

Bradyrhizobium (cowpea) (Stowers and Elkan, 1983), which demonstrates a mechanism for generating precursors of nucleotide synthesis.

II.C.5. Disaccharides

Surveys of various rhizobia have shown consistent differences between the so-called fast-growers and slow-growers in their abilities to utilize disaccharides. Sucrose, lactose, and maltose are imported poorly, if at all, by *B. japonicum* and slow-growing lupine and cowpea rhizobia (Glenn and Dilworth, 1981b), and they lack appropriate disacharidases for initial hydrolysis (Martinez De Drets *et al.*, 1974; Glenn and Dilworth, 1981b). In contrast, *S. meliloti*, *R. leguminosarum* bvs. *viciae* and *trifolii*, and fast-growing isolates from cowpea, bean, and *Lotus* grow well on these disacharides (Martinez De Drets *et al.*, 1974; Glenn and Dilworth, 1981b). Sucrose and maltose uptake is constitutive in *R. leguminosarum* bvs. *viciae* and *trifolii*, but is inducible 10- to 20-fold by sucrose in *S. meliloti* and a fast-growing cowpea isolate (Glenn and Dilworth, 1981b). Transport of lactose in these *R. leguminosarum* biovars is by a lactose-inducible permease.

S. meliloti strain Rm2011 has two distinct β-galactosidases (Niel *et al.*, 1977); one is expressed constitutively at low levels and the other is inducible by lactose. In *S. meliloti* strain Rm1021, the inducible β-galactosidase is required for growth on lactose (Charles *et al.*, 1990). The gene coding for this enzyme is located on megaplasmid pRmeSU47b (Charles and Finan, 1990; 1991) and has recently been further characterized (Jelesko and Leigh, 1994). In *S. meliloti*, utilization of mannose and lactose is inhibited by the presence of succinate (Arias *et al.*, 1982; Ucker and Signer, 1978; Jelesko and Leigh, 1994).

II.C.6. Aromatic Compounds

In bacteria, a major pathway for the degradation of aromatic compounds is the β-ketoadipate pathway. Various aromatic substrates are converted to either protocatechuate or catechol, which are branch points for further degradation defined as the *meta* and *ortho* pathways. The *ortho* pathways from both protocatechuate and catechol lead to β-ketoadipate and ultimately succinate and acetyl-CoA (reviewed by Stanier and Ornston, 1973), which can be further oxidized in the TCA cycle.

A variety of rhizobia are capable of growing on aromatic compounds (Hussein *et al* 1974; Parke and Ornston, 1984). Muthukumar *et al.* (1982) reported

that in all biovars of *R. leguminosarum*, *p*-hydroxybenzoate is metabolized to protocatechuate. Glenn and Dilworth (1981a) initiated a series of studies on the metabolism of aromatic compounds in *R. leguminosarum* bv. *viciae*. Compounds studied have included phenol, benzoate, *p*-hydroxybenzoate, catechol, protocatechuate, salicylate, and a variety of hydroxy- and methoxy-substituted benzoates. Some of these aromatics are oxidized immediately, whereas the use of others involves prolonged lag periods (Glenn and Dilworth, 1981a; Rohm and Weiner, 1985). Most of the enzymes in this pathway are constitutively expressed in some strains of *Bradyrhizobium* (Parke and Ornston, 1986), but are inducible in *Rhizobium* (Chen *et al.*, 1984a; Gajendiran and Mahadevan, 1988; Hussein *et al.*, 1974; Muthukumar *et al.*, 1982; Parke and Ornston, 1986; Wong *et al.*, 1991). One strain of *R. leguminosarum* bv. *trifolii* possesses both the catechol and protocatechuate branches of the β-ketoadipate pathway, whereas another strain of bv. *trifolii* uses only the protocatechuate branch (Chen *et al.*, 1984b). The genes encoding protocatechuate 3,4-dioxygenase have been cloned from *B. japonicum* (Podila *et al.*, 1993) and the gene encoding 4-hydroxybenzoate hydroxylase in *R. leguminosarum* bv. *viciae* has been cloned and characterized (Wong *et al.*, 1994).

III. Integration of Bacteroid and Plant Carbon and Nitrogen Metabolism in Nodules

"When the living plant cell must live with another organism which is actually part of its protoplasm, it is then necessary that a subtle balance must exist between the growth of the plant and the growth of the bacterium"
Beijerinck, 1888 [translated by T. Brock, 1961]

Nitrogen fixing rhizobia (called bacteroids) in legumes nodules inhabit the cytoplasm of the host cells. However, like many other endocytobionts, they are separated from the cytoplasm *per se* by a host-derived membrane called the symbiosome membrane or peribacteroid membrane (PBM). The PBM plays a major role in determining the type and flux of nutrients that are exchanged by the plant and the bacteroids (Udvardi and Day, 1997). In fact, as we shall see in this section, the PBM, together with the bacteroid inner membrane, helps to establish the metabolic coupling that occurs between the plant and the bacteroids.

The study of bacteroid metabolism has contributed greatly to our understanding of how bacteroid and plant metabolism is integrated during symbiosis. Adaptation of rhizobia to their symbiotic niche requires activation of a limited set of metabolic pathways and repression of others. The physiological and biochemical conditions imposed by the plant appear to play a major role in orchestrating the changes in bacterial metabolism that occur during bacteroid differentiation. Plant metabolism also undergoes specialization during nodule development, and it is clear that bacteroids play an important role in bringing this about. In the introduction, we pointed out that the exchange of reduced carbon from the plant for reduced nitrogen from bacteroids forms the basis of the symbiotic interaction. In this section, we review plant and bacteroid carbon metabolism within nodules, then bacteroid and plant nitrogen metabolism. To begin, however, we first review briefly some aspects of nodule structure and physiology, particularly as they relate to oxygen.

III.A. NODULES PROVIDE THE MICROAEROBIC ENVIRONMENT ESSENTIAL FOR NITROGEN FIXATION

Nodule development and ultrastructure are described in detail in Chapter 18. There are two types of legume nodules: determinate and indeterminate. Determinate nodules are generally spherical and have of layer of uninfected cells that surrounds the central region of bacteroid-infected cells. Indeterminate nodules are generally cylindrical (and often branched) and, like determinate nodules, have a layer of uninfected cells that surround an inner core of infected cells. The outer layers of uninfected cells in both nodule types play an important role in nodule physiology: they restrict the flux of oxygen (and other gases) into the infected zone of the nodule and contain the peripheral vascularization through which nutrients are exchanged with the plant (Brown and Walsh, 1994). The resistance to oxygen diffusion, together with high rates of oxygen consumption by bacteroids and mitochondria inside the nodule, result in microaerobic conditions within the infected cells of nodules. Leghemoglobin, an oxygen binding protein found in millimolar concentrations in the plant cytoplasm (Appleby, 1992), increases the transport of oxygen through the cytoplasm by

increasing the amount of oxygen in the cell without increasing the amount of free oxygen. A low oxygen concentration is essential for biological nitrogen fixation because nitrogenase is extremely oxygen-labile. Leghemoglobin also attenuates changes in oxygen concentration that result from changes in the permeability of the diffusion barrier or the respiration rate. However, low concentrations of oxygen severely constrain oxidative metabolism and low levels of oxygen are a major, if not the principal, regulator of both plant and bacteroid metabolism in the nodule.

III.B. PLANT CARBON METABOLISM

Phillips (1980) estimated that for each gram of N_2 fixed, 5-10 grams of carbon are required. This carbon is used as a source of energy and reductant for N_2 fixation, carbon skeletons for assimilation of the fixed N_2 into amino acids, and for nodule and bacteroid growth and maintenance. Labeling studies have demonstrated that the primary photosynthate translocated from the shoot to the nodule is sucrose (Reibach and Streeter, 1983; Kouchi and Nakaji, 1985; Gordon et al., 1985). It is metabolized via the action of sucrose synthetase and glycolytic enzymes, mainly in the plant cytoplasm of the infected zone of the nodule (Day and Copeland, 1991). Both invertase and sucrose synthase are present at high levels in soybean nodules (Streeter, 1982; Kouchi et al., 1988; Copeland et al., 1989b) and levels of host sucrose synthase correlate with nitrogenase activity (Anthon and Emerich, 1990). The inventory of glycolytic enzymes present in the host cytosol (Reibach and Streeter, 1983; Copeland et al., 1989b; Hong and Copeland, 1990) suggests sufficient capacity exists to convert hexoses to trioses and is consistent with data that demonstrates rapid conversion of sucrose to organic acids (discussed below). The synthesis of organic acids does not appear to be through significant host TCA cycle activity as mitochondrial respiration is probably limited by the low oxygen concentrations in nodules (Rawsthorne and LaRue, 1986). Further, TCA cycle enzymes in the nodule cytosol are low and may limit catabolism through this pathway. For example, in soybean nodule cytosol samples, Suganuma and Yamamoto (1987) could not detect 2-oxoglutarate dehydrogenase and pyruvate dehydrogenase activities, and found only low levels of isocitrate dehydrogenase activity.

Several studies point to the trioses being routed through phosphoenolpyruvate (PEP) carboxylase and malate dehydrogenase. Relative to roots, PEP carboxylase levels in pea, faba bean, and lupin nodules are high (Lawrie and Wheeler, 1975; Christeller *et al.*, 1977; DeVries *et al.*, 1980), and $^{14}CO_2$-labeling studies with intact nodules have demonstrated very rapid and significant labeling of organic acids-particularly malate (Christeller *et al.*, 1977; Coker and Shubert, 1981; Vance *et al.*, 1983; Snapp and Vance, 1986). Also consistent with the incorporation of $^{14}CO_2$ into malate are the extremely high levels of malate dehydrogenase found in the nodule cytosol fraction (Lawrie and Wheeler, 1975; Christeller *et al.*, 1977; DeVries *et al.*, 1980). In summary, recent photosynthates appear to be translocated to the nodule in the form of sucrose, and much of this is metabolized to malate, via PEP and oxaloacetate.

In addition to sucrose and glucose, Streeter and Bosler (1976) identified *myo*-inositol, (+)-*chiro*-inositol, α,α-trehalose, and (+)-pinitol in soybean nodules. Concentrations of these compounds are significant, often surpassing sucrose as the prominent carbohydrate (Streeter, 1987). Maltose was also identified in soybean nodules (Streeter, 1980), and levels of it along with *chiro*-inositol and *myo*-inositol increase in nodules with the onset of nitrogen fixation (Streeter, 1980).

Given the very low concentrations of oxygen in the nodule, it is perhaps not surprising that evidence for fermentation-like pathways in nodules has been found. GABA, which preferentially accumulates in plant tissue cultures incubated anaerobically (Wickremasinghe *et al.*, 1963), has been found in relative abundance in nodules of clover (Butler and Bathurst, 1958; Freney and Gibson, 1975), alfalfa (Larher *et al.*, 1983), snake bean (Jin *et al.*, 1990) and soybean (Streeter, 1987). Acetaldehyde and ethanol are volatile exports from soybean nodules (Van Straten and Schmidt, 1974) and the levels of pyruvate decarboxylase and alcohol dehydrogenase parallel nitrogenase activity in soybean nodules (Tajima and LaRue, 1982). The concentrations of such compounds are relatively low compared to sugars or organic acids (Streeter, 1987), but nevertheless their presence, along with the accumulation of other partially metabolized compounds (e.g.. malate and succinate) serves to illustrate that the host lacks the capacity to fully oxidize all carbon entering the nodule.

To summarize, there are many different carbon compounds in the plant fraction of legume nodules and their relative concentrations change during nodule development. A major research focus has been to discover which of these compounds are supplied to the bacteroids to support nitrogen fixation. As we shall see in the next section, the study of bacteroid carbon metabolism has contributed significantly to answering this question.

III.C. BACTEROID CARBON METABOLISM

III.C.1. What Carbon Compounds are Supplied to Bacteroids for Nitrogen Fixation?

Sugars. As noted above, invertase is lacking in *Bradyrhizobium*, and sucrose is not used as an energy source by bacteroids of this species. Even for those rhizobia that can hydrolyze sucrose, evidence suggests that neither sucrose nor the hexoses released by its hydrolysis are important as direct sources of energy for the bacteroids. Uptake and metabolism of sucrose, glucose, or fructose by isolated bacteroids is minimal (*e.g.* Glenn and Dilworth, 1981b; Reibach and Streeter, 1983; Salminen and Streeter, 1987a; Copeland *et al.*, 1989a). Sugar-metabolizing enzymes induced in free-living rhizobia are not induced in bacteroids, implying that bacteroids are exposed to very low concentrations of these sugars (Saraso *et al.*, 1986; McKay *et al.*, 1988). In addition, several mutants of *S. meliloti* (El Guezzar *et al.*, 1988; Hornez *et al.*, 1994; Cervenansky and Arias, 1984), *R. leguminosarum* bv. *trifolii* (Ronson and Primrose, 1979), and bv. *viciae* (Glenn *et al.*, 1984; Dilworth *et al.*, 1986) with defined defects in transport or metabolism of various sugars are Fix[+]. Direct measurement of sugar transport across the soybean PBM indicates that it is relatively impermeable to these compounds (Udvardi *et al.*, 1990). These results are consistent with the conclusion that the plant does not supply bacteroids with significant amounts of sucrose, glucose or fructose. One exception to the above generalization is in french bean, where glucose is transported across the PBM at appreciable rates (Herrada *et al.*, 1989) and stimulates nitrogenase activity in isolated bean bacteroids (Trinchant *et al.*, 1981). Further, a comparison of glycolytic enzymes in *R. tropici* free-living cells and bacteroids isolated from bean nodules shows that bacteroids are similar to free-living cells cultured on sucrose (Romanov *et al.*, 1994).

Cyclitols are concentrated primarily in the infected tissue, and have been found at significant concentrations in pea (Skot and Egsgaard, 1984), alfalfa (Fougere et al., 1991) and soybean (Streeter, 1987) bacteroids. Initial steps of myo-inositol metabolism in R. leguminosarum bv. viciae have recently been characterized (Poole et al., 1994). Catabolism of myo-inositol is analogous to that demonstrated for Klebsiella aerogenes (Berman and Magasanik, 1966a; 1966b; Anderson and Magasanik, 1971a; 1971b). myo-inositol uptake is constitutive, and catabolism is initiated by myo-inositol dehydrogenase followed by 2-keto-myo-inositol dehydratase; both enzymes are inducible by growth on myo-inositol (Poole et al., 1994). While the rest of the catabolic sequence of myo-inositol awaits characterization, R. leguminosarum bv. viciae mutants that cannot use myo-inositol as a sole carbon source are Nod$^+$ and Fix$^+$ (Poole et al., 1994).

Dicarboxylic acids. In contrast to sugars, C_4-dicarboxylic acids from the plant appear to be essential for bacteroid nitrogen fixation. There are four lines of evidence that consistently point to the importance of succinate and/or malate as essential substrates:

1). $^{14}CO_2$ fed to legumes quickly accumulates in organic acids in the nodule cytosol fraction (Streeter, 1987);

2). The PBM has a dicarboxylate transporter that catalyses rapid transport of C_4-dicarboxylates from the nodule cytosol to the bacteroids (Udvardi et al., 1988a; Herrada et al., 1989);

3). C_4-dicarboxylic acids are actively taken up by bacteroids (San Francisco and Jacobson, 1985; Kouchi et al., 1988; Udvardi et al., 1988a) at rates 30- to 50-fold that of sugars (Salminen and Streeter, 1987b), and rapidly oxidized by bacteroids to CO_2 (Stovall and Cole, 1978; Salminen and Streeter, 1987a). C_4-dicarboxylates are also effective substrates for nitrogen fixation (Bergersen, 1977; Bergersen and Turner, 1967; Miller et al., 1988; Peterson and LaRue, 1982);

4). Mutant rhizobia that are impaired in C_4-dicarboxylic acid transport form ineffective nodules (Ronson et al., 1981; Finan et al., 1983; Bolton et al., 1986; Watson et al., 1988; El Din, 1992), and S. meliloti mutants that lack TCA cycle enzymes like citrate synthase (M. Mortimer, T.R. McDermott, and M.L. Kahn, unpublished), isocitrate dehydrogenase (McDermott and Kahn, 1992), α-ketoglutarate

dehydrogenase (Duncan and Fraenkel, 1979), and succinate dehydrogenase (Gardiol et al., 1987) are all ineffective.

As discussed above for free-living cells, for cells relying on C_4-dicarboxylates as the sole or principal source of reduced carbon, metabolism via the TCA cycle requires anapleurotic metabolism to provide acetyl-CoA as an oxaloacetate acceptor. Several routes from malate or oxaloacetate to pyruvate are possible, including pyruvate carboxylase, PEP carboxykinase, and malic enzyme. Mutants of R. tropici and R. etli (Dunn et al., 1996), and R. leguminosarum bv. trifolii (Ronson and Primrose, 1979) and viciae (Arwas et al., 1986) that lack pyruvate carboxylase are all Fix$^+$. PEP carboxykinase is normally viewed as part of gluconeogenesis and is highly induced in S. meliloti grown on succinate (Osteras et al., 1995). PEP carboxykinase is important to carbon flow and metabolism during symbiotic interactions, but the relative degree of importance depends on host genotype. An R. leguminosarum bv. viciae PEP carboxykinase mutant was Fix$^+$ (McKay et al., 1985), whereas the symbiotic competence of a PEP carboxykinase mutant of the broad-host-range strain NGR234 varied when inoculated onto Leucaena leucocephala and Macroptilium atropurpureum, and it formed only a few Fix$^-$ nodules on Vigna unguiculata (Osteras et al., 1991). Osteras et al. (1995) suggested that PEP carboxykinase is involved in carbon metabolism during infection, with the availability of different carbon compounds varying between hosts and cited several supporting lines of evidence. First, PEP carboxykinase is not found in mature wild-type S. meliloti bacteroids (Finan et al., 1991). Second, PEP carboxykinase is detected in only low levels in mature pea bacteroids and is not required for nodulation and symbiotic nitrogen fixation (McKay et al., 1985). Third, ultrastructural analysis of nodules in the variably-affected symbioses noted above shows that nodule development is interrupted (Osteras et al., 1991). In contrast, the activity of soybean bacteroid PEP carboxykinase increases in parallel with nodule nitrogenase activity (Smith et al., 1994). The latter observation does not prove PEP carboxykinase plays a major role in soybean bacteroid carbon metabolism, but it is consistent with such a conclusion.

Both NAD- and NADP-dependent malic enzymes (NAD-ME, NADP-ME) have been identified in rhizobia (Copeland et al., 1989a; Kouchi et al., 1988; McKay et al., 1988; Osteras et al., 1991), and both enzymes have been partially purified and characterized (Kimura and Tajima, 1989; Copeland et al., 1989a; Trinchant and Rigaud, 1990). The B. japonicum NADP-ME has a much greater affinity for malate (Km = 0.1 mM) than the NAD-ME (Km ~ 2.0 mM). Driscoll and Finan (1993; 1996) isolated S. meliloti mutants for both enzymes and found that only the mutant lacking NAD-ME is symbiotically defective.

Amino acids. There has been speculation that amino acids may also play a role in carbon and/or energy supply to the bacteroids (Kahn et al., 1985; Kohl et al., 1994). Rhizobia will grow on a number of amino acids as sole carbon or nitrogen sources and catabolic activities are often induced in bacteroids (Kahn et al., 1985; Duran and Calderon, 1995). No mutants that affect only the catabolism of an amino acid lead to an ineffective symbiosis, with the exception of the dct mutants, which affect both aspartate and dicarboxylate catabolism. Glutamate and proline are able to enhance respiration and support nitrogen fixation in isolated bacteroids (Bergersen and Turner, 1990b; Zhu et al., 1992) although this support is apparently a consequence of prior deamination of the amino acid (Bergersen and Turner, 1990b).

Proline has been proposed to play an important role as a carbon and energy source to support nitrogen fixation (Kohl et al., 1988; Kohl et al., 1994). However, although proline dehydrogenase activity was found in soybean bacteroids (Kohl et al., 1988), its activity was low (~5 nmol/min/mg-protein). Proline is taken up by intact soybean symbiosomes (Udvardi et al., 1990) and cowpea bacteroids (Glenn et al., 1991), but at substantially slower rates than dicarboxylic acids. $^{14}CO_2$ labelling of intact soybean nodules showed that relatively little bacteroid label was found in proline (Salminen and Streeter, 1992), and other labelling experiments failed to detect labelled proline (Kouchi et al., 1991). A R. leguminosarum bv. viciae mutant that is defective in proline catabolism (Chien et al., 1991) and a proline dehydrogenase (putA) mutant of S. meliloti (Jimenez-Zurdo et al., 1995) are both Fix[+]. Taken together, these experiments suggest that proline is not used as a carbon source to the extent of other possible substrates. However, the putA mutant is less competitive, suggesting that proline may be used during infection (Jimenez-Zurdo et al., 1997). Experiments with putA::lacZ fusions suggest that although proline is not an important carbon source

during nitrogen fixation it is available earlier, during infection of the plant cells. Under drought stress conditions, proline may play a more important role in establishing solute balance between bacteroids and host. Proline typically accumulates in water-stressed bacteria (see Harris, 1981) and in the nodule cytosol and bacteroids of salt-stressed soybeans (Kohl et al., 1991) and alfalfa (Fougere, et al, 1991).

GABA has been found in relative abundance in nodules of clover (Butler and Bathurst, 1958; Freney and Gibson, 1975), alfalfa (Larher et al., 1983), snake bean (Jin et al., 1990) and soybean (Streeter, 1987). However, the evidence to date suggests that host nodule GABA is not translocated to the bacteroid (Jin et al., 1990). Even though it appears that routing of carbon flow through the GABA shunt occurs in alfalfa and soybean bacteroids and plays a role in bacteroid metabolism in these symbioses, GABA shunt enzyme activities reported in bacteroids are relatively low (Miller et al., 1991; Freney and Gibson, 1975; Jin et al., 1990); much lower than those in fast-growing cowpea rhizobia growing on GABA as a sole carbon and nitrogen source (Jin et al., 1990).

Other carbon sources. Bacteroids have at least some of the enzymes required to metabolize other carbon compounds that may be made in nodules. As discussed above, nodules of some legume species produce acetaldehyde and ethanol (Van Straten and Schmidt, 1974) and contain pyruvate decarboxylase (Tajima and LaRue, 1982) and alcohol dehydrogenase (Tajima and LaRue, 1982; Suganuma and Yamamoto, 1987). A large number of aldehydes and alcohols increased acetylene reduction rates in isolated soybean bacteroids (Peterson and LaRue, 1981), and aldehyde dehydrogenase and alcohol dehydrogenase are present in soybean bacteroids (Triplett and Blevins, 1979; Wong et al., 1971, Peterson and LaRue, 1982). The aldehyde dehydrogenase partially purified from soybean bacteroids was active with several substrates, including succinic semialdehyde (Peterson and LaRue, 1982). If acetaldehyde is the substrate, the resulting acetate could be metabolized via conversion to its CoA ester by either acetyl-CoA synthetase or the combination of acetate kinase and phosphotransacetylase. All of these enzymes have been reported in soybean bacteroids (Preston et al., 1989) and enzymes of the second pathway increase in soybean bacteroids in parallel with acetylene reduction activity (Smith et al., 1994), suggesting that acetate is metabolized by these bacteroids. However, Miller et al., (1988) reported that while isolated alfalfa bacteroids could use succinate, fumarate, malate, and oxaloacetate to support nitrogenase activity, they could not use acetate. Thus, the significance of acetate as a carbon source for bacteroids may vary in different symbioses.

Nodule malonate concentrations are also considerable (Stumpf and Burris, 1979; Kouchi and Yoneyama, 1986; Streeter, 1987) and malonyl-CoA synthetase has been reported in soybean bacteroids (Kim and Chae, 1991). Three novel malonamidases have been characterized in B. japonicum (Kim and Kang, 1994), and Kim and Chae (1990) proposed that malonate taken up by bacteroids is part of a "malonamate shuttle" that would function to export ammonia from the bacteroid to the host. The evidence in support of such a shuttle is lacking. Rates of malonate uptake in soybean bacteroids (Reibach and Streeter, 1984) and across the symbiosome membrane (Humbeck and Werner, 1987; Ou Yang et al., 1990) are low. Furthermore, Werner et al. (1982) showed that malonate oxidation is severely inhibited by succinate.

In summary, C_4-dicarboxylates, particularly malate, appear to be the principal source of carbon supplied by the plant for bacteroid metabolism in most, if not all symbioses. However, other carbon compounds are undoubtedly provided by the plant and the type, amount, and importance of these probably changes during nodule development, and differs between different symbioses.

III.C.2. Regulation of the TCA Cycle and Diversion of Carbon to Other Pathways

Because C_4-dicarboxylates appear to be the major source of carbon transferred from the plant to the bacteroids, it is important to understand the regulation of the bacteroid TCA cycle and the pathways/enzymes that remove or divert cycle substrates from it. While a significant amount of the carbon that enters the TCA cycle is oxidized to CO_2, a large fraction is not, at least immediately. It has been suggested that diversion of carbon from the TCA cycle is due, in part, to regulation of TCA cycle enzymes (McDermott et al., 1989). Such regulation could occur at the transcriptional level, such as in E. coli where genes that code for TCA cycle enzymes are repressed when the organism is shifted from aerobic to anaerobic growth (reviewed by Lynch and Linn, 1996). Post-transcriptional controls may also be important. For instance, under the

microaerobic conditions of the nodule, high NAD(P)H/NAD(P) ratios may inhibit dehydrogenases such as isocitrate dehydrogenase or 2-oxoglutarate dehydrogenase (McDermott et al., 1989). Salminen and Streeter (1990) found that the B. japonicum 2-oxoglutarate dehydrogenase complex is inhibited when NADH/NAD ratios are as high as those found in soybean bacteroids (Tajima and Kouzai, 1989).

Flow of TCA Cycle Carbon to Amino Acids. Stovall and Cole (1978) fed differentially labeled substrates to isolated soybean bacteroids and showed that approximately 30% more $^{14}CO_2$ was evolved from [1,4-^{14}C]succinate than from [2,3-^{14}C]succinate. One of the conclusions of this study was that since the methylene carbons were evolved as $^{14}CO_2$, bacteroids have a complete TCA cycle (Stovall and Cole, 1978). However, the quantitative difference in $^{14}CO_2$ evolution between [1,4-^{14}C]succinate and [2,3-^{14}C]succinate showed that a significant amount of carbon is not fully oxidized and is diverted from the TCA cycle. In similar experiments with anaerobic bacteroids, Salminen and Streeter (1987b) observed that four-times as much $^{14}CO_2$ was evolved from bacteroids fed [1,4-^{14}C]succinate than [2,3-^{14}C]succinate even though uptake of these substrates was similar, and that part of this difference was due to a greater fraction of the [2,3-^{14}C]succinate label accumulating in glutamate. [1,4-^{14}C]succinate was taken up more rapidly than [U-^{14}C]malate (see also Reibach and Streeter, 1984) but malate respiration was nearly twice as rapid. Considered together, this data suggested that dicarboxylates were taken up very rapidly and entered the TCA cycle, but that some of the carbon was diverted from the TCA cycle to glutamate (and to a lesser extent aspartate and alanine). Furthermore, their results showed that succinate appeared to have multiple fates as compared to malate. In a later study, Kouchi et al. (1991) identified a relatively active aspartase in bacteroid extracts and this might account for the quantitative differences between malate and succinate utilization observed by Salminen and Streeter (1987b). Salminen and Streeter (1992) addressed the same issue in soybean and pea bacteroids with intact nodules fed with $^{14}CO_2$ and using bacteroid isolation techniques that significantly reduced processing times. They again found large amounts of label in glutamate, although the rate of label accumulation into glutamate in pea bacteroids was less than in soybean bacteroids. Although *S. meliloti* mutants lacking GOGAT activity are Fix$^+$ (Osborne and Signer, 1980; Lewis *et al.*, 1990), a *B. japonicum* GOGAT$^-$ mutant is Fix$^-$ (O'Gara *et al.*, 1984). Manipulation of the interconversion of glutamate and 2-oxoglutarate in *R. etli* by adding a gene for the *E. coli* GDH also leads to a Fix$^-$ phenotype (Mendoza *et al.*, 1995).

In addition to glutamate, other amino acids are rapidly synthesized from dicarboxylic acids taken up by bacteroids. In $^{14}CO_2$ feeding to intact nodules, label in bacteroids very steadily and appreciably accumulated in alanine and aspartate (Salminen and Streeter, 1992), and with intact pea symbiosomes fed [^{14}C]malate, Rosendahl *et al.* (1992) reported that the addition of glutamate to the incubation mixture resulted in a 3-fold increase in labeled amino acids accumulating in the incubation medium. This may simply be a result of the added glutamate enhancing the probability of oxaloacetate or pyruvate being transaminated by basal level expression of cognate aminotransferases (Rosendahl *et al.*, 1992). Aminotransferases that would yield aspartate and alanine are present in soybean (Ryan *et al.*, 1972; Stripf and Werner, 1978, Werner and Stripf, 1978; Kouchi *et al.*, 1991), snakebean (Jin *et al.*, 1990), and alfalfa (Miller *et al.*, 1991) bacteroids, although levels vary between reports. Alanine may also be generated by alanine dehydrogenase which has also been documented in bacteroids of the soybean (Muller and Werner, 1982; Smith and Emerich, 1993), snake bean (Jin *et al.*, 1990), and alfalfa (Miller *et al.*, 1991) symbioses. The significance of bacteroid amino acid metabolism is considered further in Section III.D.

III.C.3. Acetyl-CoA Metabolism

Acetyl-CoA occupies a central position in bacteroid carbon metabolism. As discussed above, operation of the TCA cycle requires that half of the malate/succinate be diverted to acetyl-CoA to serve as an oxaloacetate acceptor during the formation of citrate. An additional source of acetyl-CoA might also be derived from the oxidation of nicotinate (Kitts *et al.*, 1992), although the affect of this pathway on acetyl-CoA pool size during symbiosis may be limited. Possible sinks for acetyl-CoA include the TCA cycle, PHB synthesis (see below), or malonyl-CoA synthesis. Acetyl-CoA carboxylase has been observed in *R. etli* free-living cells (Encarnacion *et al.*, 1995).

III.C.4. Poly-β-hydroxybutyrate Metabolism

The massive accumulation of poly-β-hydroxybutyrate (PHB) in bacteroids of some species of rhizobia but the complete absence (or nearly so) in others is perplexing. Explaining PHB accumulation is likely to be important in understanding bacteroid metabolic responses to the microaerobic conditions in the nodules and how these responses differ between species of rhizobia (McDermott *et al.*, 1989). Early studies of PHB metabolism in *Azotobacter beijerinckii* (Senior and Dawes, 1971; Jackson and Dawes, 1976; Ritchie *et al*, 1971, Senior *et al.*, 1972) demonstrated separate pathways for PHB synthesis and degradation (Figure 1). PHB synthesis often occurs when carbon availability exceeds that of other nutrients such as nitrogen, phosphorous or sulfur, or when electron acceptors are limiting. These factors may also play a role in bacteroid PHB synthesis. PHB synthesis under different growth conditions, including salt stress, and a variety of carbon, nitrogen, and oxygen regimes, has been examined in free-living rhizobia (*e.g.* Stam *et al.*, 1986; De Vries *et al.*, 1986; Bonartseva *et al.*, 1994; Bonartseva *et al.*, 1995; Encarnacion *et al.*, 1995; Natarajan *et al.*, 1995). From these studies, it seems clear that PHB synthesis in the rhizobia is similar to that described in other Gram-negative bacteria and that essentially any growth condition that does not interrupt carbon supply but reduces carbon flow into biosynthetic reactions or respiration will increase cellular PHB content.

In *Azotobacter*, oxygen-limiting conditions inhibit carbon flow through the TCA cycle by inhibiting isocitrate dehydrogenase (Jackson and Dawes, 1976). It was proposed that under these circumstances, PHB synthesis acts as a sink for excess carbon and reducing equivalents (Senior and Dawes, 1971; Senior *et al.*, 1972). This would result in less carbon entering the TCA cycle and would reduce the inhibitory effects of excess reductant on the isocitrate dehydrogenase and 2-oxoglutarate dehydrogenase reactions of the TCA cycle. The role of PHB synthesis in *B. japonicum* bacteroids was proposed to be similar (McDermott *et al.*, 1989), and both 2-oxoglutarate dehydrogenase and isocitrate dehydrogenase in *B. japonicum* have been shown to be sensitive to their reduced nicotinamide pyridine nucleotide cofactors (McDermott *et al.*, 1989; Salminen and Streeter, 1990). However, the factors that increase PHB accumulation probably act at several levels. In a study where the effects of salt stress on PHB content were examined in a *Sesbania* root nodule isolate (Natarajan *et al.*, 1995), the specific activities of β-ketothiolase and β-hydroxybutyrate dehydrogenase (Figure 1) were inversely correlated, with levels of β-ketothiolase increasing in salt-stressed cells. This implies that at least in this strain of *Rhizobium*, PHB metabolism is regulated so that enzymes used in PHB synthesis are induced under circumstances that repress enzymes used in PHB degradation.

Karr *et al.* (1984) and Wong and Evans (1971) showed that soybean bacteroid PHB content and β-ketothiolase specific activity increased in parallel with nitrogenase activity. Using ^{14}C-labeling and flow-through chambers to control oxygen and carbon flow, Bergersen and Turner (1990a) studied PHB cycling in isolated soybean bacteroids. Bacteroids readily took up labeled malate or succinate but most of the label accumulated as labile pools. About half of the radioactivity was found in PHB, and ^{14}C incorporation into PHB continued after the supply of exogenous substrates was eliminated, apparently by taking label from other metabolically active carbon pools. Curiously, nitrogen fixation was suppressed while 10 mM dicarboxylic acids were supplied, but when the carbon supply was cut off and the carbon concentration in the chamber effluent dropped below 0.5 mM, nitrogen fixation rates increased sharply (see also Bergersen and Turner 1990b). While nitrogen fixation rates were increasing, ^{14}C in PHB also increased, suggesting PHB synthesis is rapid and readily consumes pools of acetyl-CoA. As with earlier studies (Stovall and Cole, 1978; Salminen and Streeter, 1987b), $^{14}CO_2$ evolution was significantly greater with [1,4-^{14}C]succinate than with [2,3-^{14}C]succinate, indicating that the terminal carbons were removed in decarboxylation reactions at malic enzyme and pyruvate dehydrogenase, as suggested by Bergersen and Turner (1990a), and/or at isocitrate dehydrogenase and 2-oxoglutarate dehydrogenase in the TCA cycle. However, during the PHB synthesis phase, label from [U-^{14}C]malate fed to bacteroids accumulated at twice the rate than when [2,3-C^{14}]succinate was used and $^{14}CO_2$ was released from [U-^{14}C]malate at ten times the rate as from [2,3-C^{14}]succinate. These results are similar to those discussed above for Salminen and Streeter (1987b) and Kouchi *et al.* (1991), and indicate that some succinate taken up by bacteroids is converted to aspartate via the enzyme aspartase.

Wild-type *S. meliloti* bacteroids do not contain PHB, and *S. meliloti* PHB$^-$ mutants have a normal symbiosis phenotype (Povolo *et al.*, 1994). These observations suggest that carbon in alfalfa nodules does not cycle through PHB. In contrast, *R. etli* accumulates large amounts of PHB under a variety of growth conditions, including symbiosis (Encarnacion *et al.*, 1995). *R. etli* PHB$^-$ mutants form nodules on bean that have prolonged capacity for nitrogen fixation (Cevallos *et al.*, 1996), indicating that PHB synthesis and nitrogen fixation may represent alternative fates for reductant. Free-living wild-type *R. etli* cultures display a "fermentation-like" growth behavior (Encarnacion *et al.*, 1995) in that growth rates and cell yield decrease dramatically after successive sub-cultures in minimal succinate medium. This non-growing state is characterized by reduced activity of TCA cycle enzymes, accumulation of PHB, and excretion of organic acids and amino acids. The phenomenon can be eliminated by high levels of oxygen or if specific supplements, such as biotin or thiamine, are added to the culture. With added biotin, TCA cycle enzymes remained at high levels, β-ketothiolase activity was reduced, and pyruvate carboxylase activity increased, although other biotin-requiring enzymes measured were not affected. The fermentative response of a PHB$^-$ mutant is more pronounced than the wild-type (Cevallos *et al.*, 1996). In general, organic acid accumulation in the medium of the PHB$^-$ mutant was several-fold higher and the NADH:NAD ratio was 2- to ~20-fold higher than wild-type. This suggests that carbon otherwise destined for PHB synthesis is exported from the cell as organic acids and amino acids, and supports the proposed role of PHB synthesis in controlling cellular redox levels.

III.C.5. Synthesis of other bacteroid carbon compounds

Soybean nodule trehalose concentrations increase with nitrogen fixation, but then decline (Streeter, 1980). Trehalose synthesis by bacteroids (Reibach and Streeter, 1983, Phillips *et al.*, 1984; Salminen and Streeter, 1986) probably results from a combination of bacteroid responses to the anaerobic conditions of the nodule (Hoelzle and Streeter, 1990) and host activities. UDP-glucose is an intermediate in trehalose metabolism (see Salminen and Streeter, 1986), and enzymes involved in its synthesis are present at high levels in host and bacteroid fractions (Anthon and Emerich, 1990; Salminen and Streeter, 1986; Copeland *et al.*, 1989b). UDP-glucose is taken up by soybean bacteroids at rates 5-fold that of uncharged glucose (Salminen and Streeter, 1987a) and other enzymes involved in trehalose biosynthesis are found in bacteroids (Salminen and Streeter,

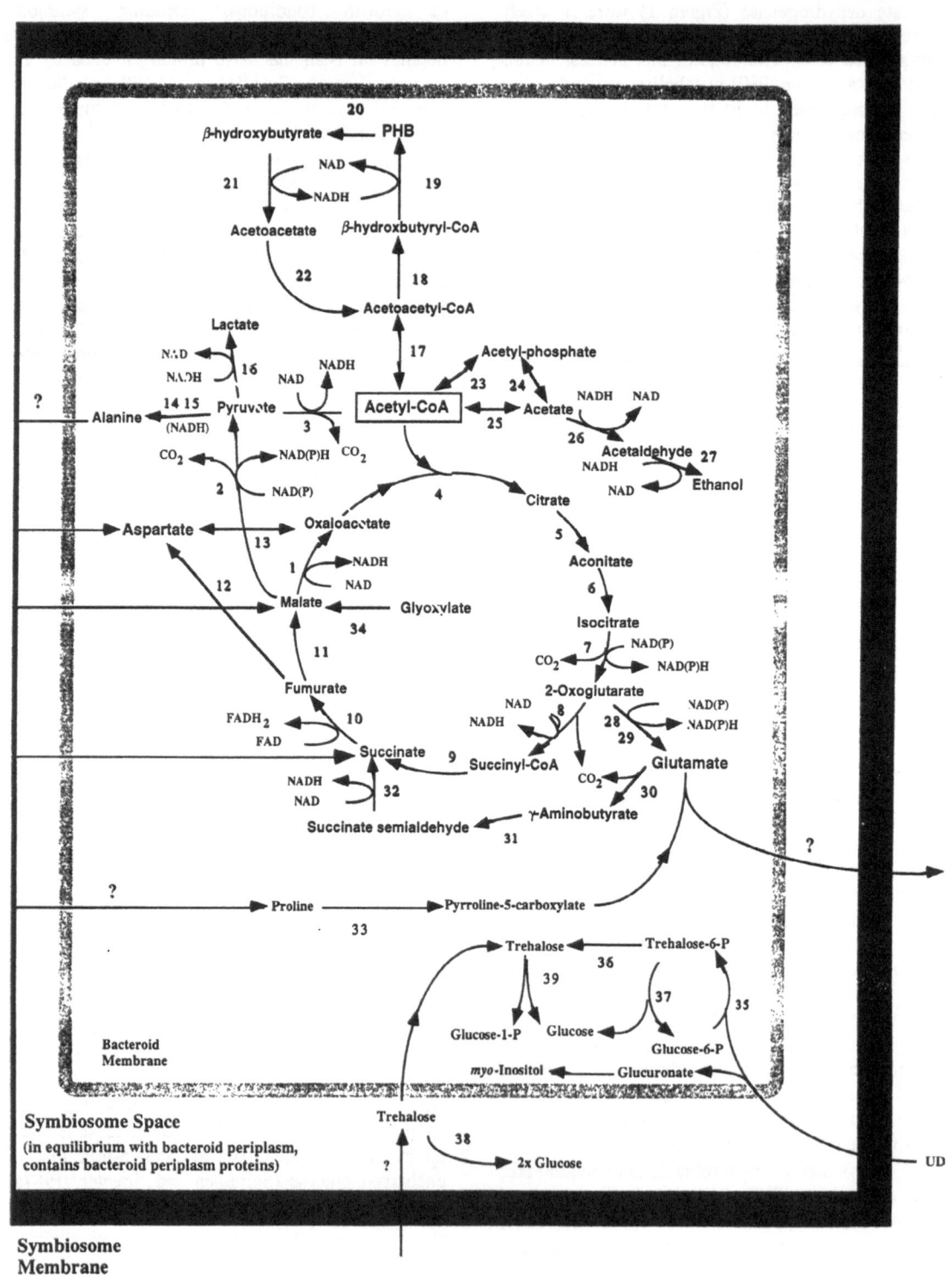

Figure 1. Generalized routes of carbon flow within bacteroids. All reactions shown have been documented. However, not all reactions have been demonstrated (yet) for species within the genus *Rhizobium*. For example in bacteroids of *R meliloti*, which is probably the best characterized of the rhizobia that form indeterminate nodules, enzymes/pathways that are thought not to occur, are at very low levels, or have not yet been reported, include the PHB cycle, acetate energizing reactions, malate synthase, aldehyde and alcohol dehydrogenases, lactate dehydrogenase, and trehalose metabolism. Numbers refer to the following enzymes: (1) malate dehydrogenase; (2) malic enzyme; (3) pyruvate dehydrogenase; (4) citrate synthase, (5) and (6) aconitase; (7) isocitrate dehydrogenase; (8) 2-oxoglutarate dehydrogenase; (9) succinate thiokinase; (10) succinate dehydrogenase; (11) fumarase; (12) aspartase; (13) aspartate aminotransferase; (14) alanine aminotransferase; (15) alanine dehydrogenase; (16) lactate dehydrogenase; (17) β-ketothiolase; (18) acetoacetyl-CoA reductase; (19) β-hydroxybutyryl-CoA polymerase; (20) PHB depolymerase; (21) β-hydroxybutyrate dehydrogenase; (22) acetoacetate:succinyl-CoA transferase; (23) phosphotransacetylase; (24) acetate kinase; (25) acetyl-CoA synthetase; (26) aldehyde dehydrogenase; (27) alcohol dehydrogenase; (28) glutamate dehydrogenase; (29) glutamine:oxoglutarate aminotransferase; (30) glutamate decarboxylase; (31) g-aminobutyric-glutamic transaminase; (32) succinate semialdehyde dehydrogenase; (33) proline dehydrogenase; (34) malate synthase; (35) UDPG pyrophosphorylase; (36) trehalose-6-P synthetase; (37) phosphotrehalase; (38) trehalase; (39) trehalose phosphorylase. The trehalose "cycle" is modified from that provided by Mellor (1988), whereas the rest of the information in this figure is an update from that provided from McDermott *et al.* (1989).

1986). The role of trehalose in symbiosis is not understood (see Mellor, 1988; Streeter, 1991), but it has been suggested that trehalose may be part of a futile cycle in which glucose moves in and out of the bacteroids/symbiosomes (Streeter, 1991).

Streeter *et al.* (1992) observed that some strains of soybean rhizobia accumulate polysaccharide in the symbiosome space. The basis for nodule polysaccharide synthesis is unknown. It is also not known why some strains produce it while others do not. However, the presence of polysaccharide has little affect on nitrogen fixation (Streeter and Salminen, 1993), and mutants that fail to produce it have a normal symbiotic phenotype (Streeter *et al.*, 1995).

III.D. BACTEROID NITROGEN METABOLISM AND EXPORT

III.D.1. Nitrogen Fixation

Bacteroid nitrogen fixation and its regulation are covered in detail in Chapter 23. As we mentioned in Section II, nitrogen fixation in rhizobia is regulated principally by oxygen. In response to low oxygen concentration, the FixLJ two-component regulatory system activates transcription of *nifA* and/or *fixK*, the products of which control transcription of other *nif* and *fix* genes that are required for nitrogen fixation (Fischer, 1994). In the case of *B. japonicum*, the NifA protein is activated by low oxygen independently of FixJ. Induction of nitrogen fixation in bacteroids is generally insensitive to the nitrogen status of the cells, or at least independent of the Ntr system. Thus, the microaerobic environment of the nodule not only allows but also promotes, independently, nitrogen fixation by the bacteroids. Of course nitrogen fixation requires that other conditions be met, including especially an adequate

supply of reductant and ATP. As we have seen, the TCA cycle plays a central role in bacteroid energy metabolism but operation of the cycle under the microaerobic conditions inside the nodule requires the deployment of one or more terminal oxidases that have very high affinity for oxygen. The *fixNOQP* operon is essential for symbiotic nitrogen fixation in rhizobia and encodes a high affinity oxidase complex (Fischer, 1994). Transcription of the *fixNOQP* operon is induced by FixK in response to low oxygen concentrations in nodules. The electrons that are required by nitrogenase for the reduction of nitrogen may arrive via an electron transport chain that includes one or more of the products of the *fixABCX* operon (Fischer, 1994). Transcription of this operon is induced by NifA in response to low oxygen concentrations in nodules. Thus, low oxygen concentrations in legume nodules induce the expression of the nitrogenase proteins as well as other proteins that facilitate ATP and electron supply to this enzyme. Interestingly, much of the nitrogen fixed by bacteroids is lost to the plant cytoplasm, as will be explained below.

III.D.2. Nitrogen Assimilation

The nitrate and ammonia assimilation pathways are down-regulated in bacteroids. The activities of nitrate reductase, the high-affinity ammonium transporter, GS, and GOGAT are all low relative to nitrogenase activity (Howitt and Gresshoff, 1985; Howitt *et al.*, 1986; Carlson *et al.*, 1987; Shatters *et al.*, 1989; DeBruijn *et al.*, 1989; Brown and Dilworth, 1975; Darrow, 1980). Because these enzymes are normally induced under conditions of nitrogen stress, it appears that bacteroids are not nitrogen-starved in nodules. Indeed, mutations in *ntrC*, which is required for the synthesis of these proteins, have little or no effect on symbiotic nitrogen fixation by rhizobia (Gussin *et al.*, 1986; Szeto *et al.*, 1987; Pawlowski *et al.*, 1987; Udvardi *et al*, 1992). As a result of the down-regulation of ammonia assimilation enzymes, much of the ammonia produced by nitrogenase is lost from bacteroids as NH_3 (Bergersen and Turner, 1967) by simple diffusion across the bacteroid membranes. Because bacteroids do not express the high-affinity ammonium transporter, this ammonia cannot be recovered. Instead, it is transported from the peribacteroid space to the plant cell cytoplasm probably via an ammonium transporter on the PBM (Tyerman *et al.*, 1995).

III.D.3. Amino Acid Metabolism

As one probe of symbiotic metabolism, a number of auxotrophs (references in Kerpolla and Kahn, 1988; Rossbach and Hennecke, 1991) have been isolated and tested for symbiotic competence. Few types of auxotroph are Fix⁻ in all systems, suggesting that an adequate supply of the required nutrient is available during bacteroid development at least in some symbioses. Exceptions to this are adenosine auxotrophs (see Djordjevic *et al.*, 1996) and leucine auxotrophs (see Aguilar and Grasso, 1991). However, in some of the work with auxotrophs, nodulated plants were not always checked for reversion of the original, usually chemically induced, mutations and nodule formation by revertants might underestimate the severity of the original mutation. Although the effectiveness of some well characterized auxotrophs might seem to contradict the information summarized above concerning the apparent impermeability of the symbiosome membrane to amino acids, it is possible that the transport rate required to transfer small but sufficient amounts of nutrients across the peribacteroid membrane may be very low. The fact that some auxotrophs are Fix- might also be due to insufficient uptake by the bacteroids of the required nutrient. *hemA* mutants of *B. japonicum* are effective but those of several other rhizobia are not, a difference that O'Brian (1996) attributes to the greater ability of *B. japonicum* to import ᴀ-amino levulinic acid.

The Role of Bacteroid GSs.

A distinct function for the three GS enzymes in bacteroids is not yet clear. As mentioned above, GS is down regulated in bacteroids (Brown and Dilworth, 1975; Darrow, 1980) and, in fact, glutamine degradative enzymes are found in *R. etli* bacteroids at high levels (Duran and Calderon, 1995). Mutants in any one GS and

GSI⁻GSIII⁻ and GSII⁻GSIII⁻ mutants are able to grow well and form nitrogen fixing nodules. A GSI⁻GSII⁻ mutant of *S. meliloti* is also effective, although it requires high concentrations of ammonium and amino acids to grow on minimal media (De Bruijn *et al.*, 1989; Somerville *et al.*, 1989). *B. japonicum* mutants lacking GSI or GSII are effective but a GSI⁻GSII⁻ mutant is not infective unless the medium is supplemented with glutamine and is ineffective under these conditions (Carlson *et al.*, 1987). A mutant of *Azorhizobium caulinodans* lacking the only GS found in this species is still infective but is not effective (Donald and Ludwig, 1984). *R. tropici* GSI mutants form fewer nodules than wild type, but *B. japonicum* mutants in GSI or GSII formed a higher number of nodules with greater specific activity (Carlson *et al.*, 1987). GSI protein is found in bacteroids but based on relative activities with Mg and Mn appears to be adenylylated (Darrow, 1980). Recently, it has been shown that a mutant lacking the probable adenylylation site in the GSI protein is still effective, suggesting that "proper" symbiotic behavior does not require this posttranslational modification (Arcondeguy *et al.*, 1996). GSII protein was not detected in soybean or alfalfa nodules formed by wild-type rhizobia (Shatters *et al.*, 1989) but GSII mRNA was detected in *B. japonicum* isolated from soybean nodules (Martin *et al.*, 1988). GSIII protein was found in nodules formed by the *S. meliloti* double mutant (Shatters *et al.*, 1993).

The Role of Bacteroid Aspartate Aminotransferase.

S. meliloti mutants lacking AAT-A are ineffective (Rastogi and Watson, 1991) but can be complemented either by the *aatA* gene or by *tatA*, which codes for a tyrosine aminotransferase that has substantial aspartate aminotransferase activity. Since the *aatA* mutant is not an aspartate auxotroph, the source of the mutant's symbiotic defect seems to be its inability to generate a high level of AAT activity.

The dependence of the *S. meliloti*–alfalfa symbiosis on AAT-A has refocussed attention on the possibility that an exchange of organic acids and amino acids similar to the malate-aspartate shuttle used in mitochondria (Meijer and VanDam, 1974) might play a role in symbiotic metabolism. Work in both *R. leguminosarum* (Appels and Haaker, 1991) and *B. japonicum* (Streeter and Salminen, 1990) suggests that both bacteroids and free-living bacteria would be willing players in such a scheme since both release aspartate and α-ketoglutarate when incubated in the presence of malate and glutamate. Since dicarboxylates are thought to be provided to bacteroids (see above), the major argument against the operation of a shuttle is the lack of transport of glutamate through the symbiosome membrane (Udvardi et al., 1988b). However, there may be more glutamate transport through the symbiosome membrane at higher but physiological glutamate concentrations (Kouchi et al., 1991). Recently, Haaker et al. (1995) suggested on the basis of a kinetic analysis of the plant AAT enzymes, that aspartate synthesis in the plant was not great enough to generate precursors for the asparagine exported from pea nodules. They suggest that the classic shuttle may be modified in nodules by transporting malate from the plant cytoplasm into the symbiosomes then taking the aspartate generated there and amidating it for transport out of the nodule.

Tryptophan Metabolism. Tryptophan metabolism has been of interest both for comparison to other bacterial pathways (Bae and Crawford, 1990) and because of tryptophan's role as a precursor of the plant hormone indole acetic acid (IAA) or auxin. In *Agrobacterium*, the conversion of tryptophan to IAA via indoleacetamide using the tryptophan monooxygenase pathway is directly involved in tumor formation (Thomashow et al., 1986). The second step in this pathway has been found in *R. leguminosarum* (Kawaguchi et al., 1990). The first committed step in *S. meliloti* aromatic amino acid synthesis, DHAP synthetase, is necessary for effective symbiosis (Jelesko et al., 1993). Mutants blocked late in tryptophan biosynthesis form effective nodules but those unable to make anthranillate are either partially effective or ineffective (Barsomian et al., 1992) It is possible that anthranillate may be needed as a siderophore or that excretion of a biosynthetic intermediate may interfere with plant auxin metabolism. Four aromatic aminotransferases were found in *S. meliloti* that can use tryptophan as substrate (Kittell et al., 1989). Mutations in the two major enzymes did not affect nodule function although both mutations lowered the level of IAA produced when tryptophan was added at high concentrations to the growth medium.

III.E. PLANT NITROGEN METABOLISM

The ammonium that is exported from symbiosomes is rapidly incorporated into the amino acids glutamine and glutamate by glutamine synthetase (GS) in the cytoplasm and glutamate synthase (GOGAT) in the plastids, respectively (Atkins, 1991). The ammonia assimilation pathway is induced during nodule development, and this involves expression of nodule-specific isoforms of GS and GOGAT in some symbioses (Atkins, 1991). Ammonia produced by the bacteroids appears to play a role in regulating the expression of some of the enzymes required in its assimilation (Atkins, 1991; Hirel et al., 1987). Glutamine is transported from nodules for use in other parts of the plant, at least in some legumes (Atkins, 1991). However, it is rarely the major transport form of nitrogen exported from nodules. In general, temperate legumes such as pea and alfalfa export nitrogen from nodules in the form of amides (particularly asparagine, but glutamine also) whilst tropical legumes such soybean export nitrogen as ureides (allantoin and allantoic acid) (Atkins, 1991). Continual synthesis and export of asparagine from nodules requires a constant supply of precursors of aspartate (oxaloacetate or malate). As mentioned in Section III.B., PEP carboxylase and malate dehydrogenase activities are high in nodules, and a number of studies have concluded that CO_2 fixation by PEP carboxylase is linked to amide synthesis. Interestingly, addition of ammonia to anaerobic cells of the green alga, *Selenastrum minutum* resulted in an eight-fold increase in CO_2 fixation via PEP carboxylase which presumably was required to produce intermediates for amino acid biosynthesis in the absence of substantial oxidative TCA cycle metabolism (Vanlerberghe and Turpin, 1990). It is tempting to speculate that ammonium exported by bacteroids may induce plant PEP carboxylase activity in the microaerobic environment of the nodule. Given the high activity of plant malate dehydrogenase in nodules, induction of PEP carboxylase activity would accelerate the rate of oxaloacetate and malate production for both amino acid synthesis by the plant and C_4-dicarboxylate supply to the bacteroid.

Synthesis of ureides involves *de novo* purine synthesis and this increases sharply with nodule development, apparently in response to ammonium exported from bacteroids (Atkins, 1991). Thus, ammonium exported by the bacteroids appears to exert a positive regulatory effect not only on ammonium assimilation, but also on both ureide and amide production in different legumes.

IV. Summary and Future Directions

In the first paper written on the rhizobia, Beijerinck (1888) studied the properties of the free-living bacteria and wrote: "Because of the simple nutritional requirements of the organism, the symbiotic relationship with the leguminous plants seems all the more surprising, since, except for sugar, its nutrient requirements are just as simple as the plant itself, and we know of many non symbiotic organisms, such as *Bacillus [Pseudomonas] fluorescens*, which have just as simple requirements". Beijerinck also observed that "...it was not possible to demonstrate the fixation of free nitrogen gas (in culture)" and concluded that "It seems to me hardly likely that a bacterium which has strong chemical abilities, such as the ability to fix nitrogen...would be suitable in the development of such a delicate equilibrium. Therefore, only an organism like *Bacillus radicicola (Rhizobium)*, which is similar in its chemical properties to that of the protoplasm of the plant cell, would be suitable for such a symbiotic relationship...". It should be apparent from this chapter that we have learned much about rhizobia and the symbiosis since 1888. For example, there is no question that the rhizobia have strong enough chemical abilities to fix nitrogen in the symbiosis, despite their properties in culture. Indeed it is the differences, more than the similarities, between the metabolic capabilities of rhizobia and legumes that enables each to benefit from living together. Thus, nitrogen fixation by rhizobia complements carbon fixation by plants. As a result, the plants are able to grow in soils devoid of fixed nitrogen and rhizobia are treated to much higher concentrations of fixed carbon than are available in the soil.

In the past century, we have accumulated a large amount of data on metabolism in rhizobia and their plant hosts and a basic picture of the interaction between carbon and nitrogen metabolism in these organisms has emerged. We now know that for a successful interaction to occur at all, symbiotic metabolism must be conducted under microaerobic conditions. This is principally because the bacteroid enzyme nitrogenase is extremely oxygen labile. However, microaerobic conditions in legume nodules have profound effects on other aspects of plant and bacteroid metabolism, especially carbon metabolism. Sucrose that is imported into nodules from the plant shoot is metabolised in the cytoplasm to produce malate and other dicarboxylic acids that are then transported to bacteroids for further metabolism. The bacteroid TCA cycle provides a major sink for this carbon and the deployment of very high affinity terminal oxidases enables bacteroids to compete effectively with mitochondria for nodule oxygen. Bacteroid respiration and oxidative phosphorylation provide reductant and ATP for nitrogen fixation, although the precise manner by which this is achieved remains unknown. A substantial amount of the malate/succinate that enters the bacteroid TCA cycle is only partially oxidized, at least initially, and the fate of this carbon remains an important question. Is it returned to the host carrying with it reduced nitrogen? If so, is any of this exported carbon recycled back to the bacteroid? Likewise, the fate of amino acids synthesized in bacteroids is unclear, but is an important area for future research.

Despite some confusion over the role of bacteroid amino acid metabolism in symbiosis, many basic aspects of nitrogen metabolism in nodules are now quite well understood. Bacteroids are essentially ammonia-exporting organelles because ammonia assimilation is repressed in bacteroids despite high rates of nitrogen fixation. Ammonia is transported to the plant cytoplasm where it is incorporated into glutamine, then into other amides or ureides prior to export from the nodules. Ammonia has a positive regulatory effect on the expression of enzymes necessary for its assimilation by the plant, including those that provide the carbon skeletons necessary for nitrogen assimilation and transport.

The symbiosis appears to provide a selective advantage for both partners, but why do bacteroids devote such a substantial amount of energy to nitrogen fixation when they clearly have little inclination to assimilate the ammonia that is produced? An answer to this question may be forthcoming when we define the conditions that are necessary to achieve routinely high rates of nitrogen fixation *in vitro* in all rhizobia, not just *Azorhizobium*. Our inability to achieve this at present underscores our lack of knowledge about the bacteroid environment. Routine induction of nitrogen fixation *ex planta* may also provide us with insight into another physiological property of bacteroids, namely inhibition of cell division and apparent loss of viability observed in highly differentiated bacteroids in some symbioses. Interestingly, high concentrations of dicarboxylic acids together with low pH inhibit rhizobial growth *ex planta* (Perez-Galdona and Kahn, 1994). Because such conditions may prevail inside the symbiosomes of nodules, it follows that they may have similar effects on bacteroids. It will be interesting to learn more about the chemical and physical factors that limit bacteroid populations inside nodules.

A nitrogen fixing symbiosis is a solution to the problem of nitrogen availability for the plant and probably of carbon availability for the bacteria. This solution requires the integration of plant and bacterial carbon and nitrogen metabolism and it is obvious perhaps that the activities of each organism affect those of the other at many levels. The study of carbon and nitrogen metabolism in rhizobia has enriched our understanding and appreciation of this complex interaction and will, no doubt, continue to do so in the future.

V. References

Aguilar, O.M. and Grasso, D.H. (1991) J.Bacteriol. 173, 7756-7764.

Alfano, J.R. and Kahn, M.L. (1993) J. Bacteriol. 175, 4186-4196.

Amar, M., Patriarca, E.J., Manco, G., Bernard,P., Riccio, A., Lamberti, A., Defez, R. and Iaccarino, M. (1994) Molec. Microbiol. 11, 685-693.

Ames, G. F.-L. and Nikaido, K. (1985) EMBO J. 4, 539-547.

Anderson, W.A. and Magasanik, B. (1971a) J. Biol. Chem. 246, 5653-5661.

Anderson, W.A. and Magasanik, B. (1971b) J. Biol. Chem. 246, 5662-5675.

Anthon, G.E. and Emerich, D.W. (1990) Plant Physiol. 92, 346-351.

Appels, M. A. and Haaker, H. (1991) Plant Physiol. 95, 740-747.

Appleby, C.A. (1992) Sci. Progress 76, 365-398.

Arcondeguy, T., Huez, I., Fourment, J. and Kahn, D. (1996) FEMS Microbiol. Letters. 145, 33-40.

Arias, A. and Martinez-Drets, G. (1976) Can. J. Microbiol. 22, 150-153.

Arias, A., Gardiol, A. and Martinez-Drets, G. (1982) J. Bacteriol. 151, 1069-1072.

Arwas, R., Glenn, A.R., McKay, I.A. and Dilworth, M.J. (1986) J. Gen. Microbiol. 132, 2743-2747.

Atkins, C.A. (1991) In M.J. Dilworth and A.R. Glenn (eds.) Biology and Biochemistry of Nitrogen Fixation, Elsevier, Amsterdam, pp.293-319.

Ausubel, F.M. (1984) Cell 37, 5-6.

Ausubel, F.M., Buikema, W.J., Earl, C.D., Klingensmith, B.T., Nixon, B.T., Szeto, W.W. (1985) In HJ Evans, PJ Bottomley, WE Newton (eds.) Nitrogen Fixation Research Progress, Martinus Nijhoff, Dordrecht, pp. 167-179.

Bae, Y.M. and Crawford, I.P. (1990) J. Bacteriol. 172, 3318-3327.

Barsomian, G. D., Urzainqui, A., Lohman, K. and Walker, G. C. (1992) J. Bacteriol. 174, 4416-4426.

Batista, S., Castro, S., Aguilar, O.M. and Martinez-Drets, G. (1992) Can. J. Microbiol. 38, 51-55.

Beijerinck, M.W. (1888) Botanische Zeitung 46, 725-804.

Bergersen F.J. (1977) In RWF Hardy, 5W Silver (eds.) A Treatise on Dinitrgen Fixation Section III, Biology, Wiley, New York, pp. 519-56.

Bergersen F.J. and Turner, G.L. (1990a) Proc. Roy. Soc. (Lond.) B 240, 39-59.

Bergersen F.J. and Turner, G.L. (1990b) Proc. R. Soc. Lond. B 238, 295-320.

Bergersen F.J. and Turner, G.L. (1967) Biochim. Biophys. Acta. 141, 507-15.

Berman, T. and Magasanik, B. (1966a) J. Biol. Chem. 241, 800-806.

Berman, T. and Magasanik, B. (1966b) J. Biol. Chem. 241, 807-813.

Betlach, M.R., Tiedje, J.M. and Firestone, R.B. (1981) Arch. Microbiol. 129, 135-140.

Bolton, E., Higgisson, B., Harrington, A. and O'Gara, F. (1986) Arch. Microbiol. 144, 142-146.

Bonartseva, G.A., Myshkina, V.L. and Zagreba, E.D. (1994) Microbiology 63, 45-48.

Bonartseva, G.A., Myshkina, V.L. and Zagreba, E.D. (1995) Microbiology 64, 30-33.

Brock, T.D. (1961) Milestones of Microbiology, ASM Press, Washington, D.C. pp 220-224.

Brown, C. M. and Dilworth, M. J. (1975) J. Gen. Microbiol. 86, 39-48.

Brown, S.M. and Walsh, K.B. (1994) Aust. J. Plant Physiol. 21, 49-68.

Burris, R.H. and Roberts, G.P. (1993) Annu. Rev. Nutr. 13, 317-335.

Butler, G.W. and Bathurst, N.O. (1958) Aust. J. Biol. Sci. 11, 529-537.

Carlson, T.A. and Chelm, B. (1986) Nature 322, 568- 570.

Carlson, T.A., Martin, G.B. and Chelm, B.K. (1987) J. Bacteriol. 169, 5861-5866

Cervenansky, C. and Arias, A. (1984) J. Bacteriol. 160, 1027-1030.

Cevallos, M.A., Encarnacion, S., Leija, A., Mora, Y. and Mora, J. (1996) J. Bacteriol. 178, 1646-1654.

Chakrabarti, S.M., Lee, S. and Gibson, A. H. (1981) Soil Biol. Biochem. 13, 349-354.

Charles, T.C. and Finan, T.M. (1990) J. Bacteriol. 172, 2469-2476.

Charles, T.C. and Finan, T.M. (1991) Genetics 127, 5-20.

Charles, T.C., Singh, R.S. and Finan, T.M. (1990) J. Gen. Microbiol. 136, 2497-2502.

Chen, Y.P., Dilworth, M.J. and Glenn, A.R. (1984a) Arch. Microbiol. 138, 187-190.

Chen, Y.P., Glenn, A.R. and Dilworth, M.J. (1984b) FEMS Microbiol. Lett 21, 201-205.

Cheniae, G. and Evans, H.J. (1960) Plant Physiol. 35, 454-562.

Chien, C. T., Rupp, R., Beck, S. and Orser, C. S. (1991) FEMS Microbiol. Lett. 77, 299-302.

Christeller, J.T., Laing, W.A. and Sutton, W.D. (1977) Plant Physiol. 60, 47-50.

Clark, D.P. and Cronan, Jr., J.E. (1996) In F.C. Neidhardt, R. Curtiss III, J.L. Ingraham, E.C.C. Lin, K. B. Low, B. Magasanik, W.S. Reznikoff, M. Riley, M. Schaechter, and H.E. Umbarger (eds.) Escherichia coli and Salmonella, Cellular and molecular biology, ASM Press, American Society for Microbiology, Washington, DC, pp. 343-357.

Coker, G.T. and Shubert, K.R. (1981) Plant Physiol. 67, 691-696.

Copeland, L., Quinnell, R.G. and Day, D.A. (1989a) J. Gen. Microbiol. 135, 2005-2011.

Copeland, L., Vella, J. and Hong, Z. (1989b) Phytochemistry. 28, 57-61.

Daniel, R.M. and Appleby, C.A. (1972) Bioohim. Biophys. Acta. 275, 347-354.

Darrow, R. A. (1980) In J. Mora and R. Palacios (eds.), Glutamine Synthetase, Metabolism, Enzymology and Regulation. Academic Press, pp. 139-166.

Day D.A. and Copeland L. (1991) Plant Physiol. Biochem. 29, 185-201.

DeBruijn, F. J., S. Rossbach, M. Schneider, P. Ratet, S. Messmer, W. W. Szeto, F. M. Ausubel, and J. Schell. (1989) J. Bacteriol. 171, 1673-1682.

deVries, G.E., In't Veld, P. and Kijne, J.W. (1980) Plant Sci. Lett. 20, 115-123.

deVries, W., Stam, H., Duys, J.G., Ligtenberg, A.J.M., Simons, L.H. and Stouthamer, A.H. (1986) Ant. van Leeuwenhoek 52, 85-96.

Dilworth, M.J., Arwas, R., McKay, I.A., Saroso, S. and Glenn, A.R. (1986) J. Gen. Microbiol. 132, 2733-2742.

Djordjevic, S. P., Weinman, J. J. L., Redmond, J. W., Djordjevic, M. A. and Rolfe, B. G. (1996) Mol. Plant-Microbe Interact. 9, 114-124.

Donald, R. G. and Ludwig, R. A. (1984) J. Bacteriol. 158, 1144-1151.

Driscoll, B.T. and Finan, T.M. (1993) Mol. Microbiol. 7, 865-873.

Driscoll, B.T. and Finan, T.M. (1996) J. Bacteriol. 178, 2224-2231.

Duncan, M.J. (1979) J. Gen. Microbiol.113, 177-179.

Duncan, M.J. (1981) J. Gen. Microbiol. 122, 61-67.

Duncan, M.J. and Fraenkel, D.G. (1979) J. Bacteriol. 37, 415-419.

Dunn, M.F., Encarnacion, S., Araiza, G., Vargas, M.C., Davalos, A., Peralta, H., Mora, Y. and Mora, J. (1996) J. Bacteriol. 178, 5960-5970.

Duran, S. and Calderon, J. (1995) Microbiol. 141, 589-595.

El Din, A.K.Y.G. (1992) Can. J. Microbiol. 38, 230-234.

El Guezzar, M., Hornez, J.P., Courtois, B. and Deriex, J.C. (1988) FEMS Microbiol. Lett. 49, 429-434.

Encarnacion, S., Dunn, M., Willms, K. and Mora, J. (1995) J. Bacteriol. 177, 3058-3066.

Engelke, T., Jagadish, M.N. and Puhler, A. (1987) J. Gen. Microbiol. 133, 3019-3029.

Engelke, T., Jording, D., Kapp, D. and Puhler, A. (1989) J. Bacteriol. 171, 5551-5560.

Espin, G., Moreno, S. and Guzman, J. (1994) Crit. Rev. Microbiol. 20, 117-123.

Espin, G., Moreno, S., Wild, M., Meza, R. and Iaccarino, M. (1990) Mol. Gen. Genet. 223, 513-516.

Finan, T.M., McWhinnie, E., Driscoll, B. and Watson, R.J. (1991) Mol. Plant-Microbe Interact. 4, 386-392.

Finan, T.M., Oresnik, I. and Bottacin, A. (1988) J. Bacteriol., 170,3396-3403.

Finan, T.M., Wood, J.M, and Jordan, D.C. (1981) J. Bacteriol. 148, 193-202.

Finan, T.M., Wood, J.M, and Jordan, D.C. (1983) J. Bacteriol. 154, 1403-1413.

Fischer, H.M. (1994) Microbiol. Rev. 58, 352-386.

Fougère, F., Rudulier, D. L. and Streeter, J. G. (1991) Plant Physiol. 96, 1228-1236.

Freney, J.R. and Gibson, A.H. (1975) Aust. J. Plant Physiol. 2, 663-668.

Gajendiran, N. and Mahadevan, A. (1988) Plant and Soil 108, 263-266.

Gardiol, A.E., Truchet, G.L. and Dazzo, F.B. (1987) Appl. Environ. Microbiol. 53, 1947-1950.

Giannakis, C., Nicholas, D.J.D. and Wallace, W. (1988) Planta 174, 51-58.

Glenn, A.R. and Dilworth, M.J. (1981a) J. Gen. Microbiol. 126, 243-247.

Glenn, A.R. and Dilworth, M.J. (1981b) Arch. Microbiol. 129, 233-239.

Glenn, A.R., Holliday, S. and Dilworth, M.J. (1991) FEMS Microbiol. Lett. 82, 307-312.

Glenn, A.R., McKay, I.A., Arwas, R. and Dilworth, M.J. (1984) J. Gen. Microbiol. 130, 239-245.

Gollop, R. and Avissar, Y.J. (1984) Can. J. Microbiol. 30, 890-893.

Gonzalez-Gonzalez, R., Botsford, J. L. and Lewis, T. (1990) Can. J. Microbiol.36, 469-474.

Gordon, A.J., Ryle, G.J.A., Mitchell, D.F. and Powell, C.E. (1985) J. Exp. Bot. 36, 756-769.

Gussin, G.N., Ronson, C.W., Ausubel, F.M. (1986) Annu. Rev. Genet. 20, 567-591.

Haaker, H. , Szafrau, M. M., Wassink, H. J. and Appels, M. A. (1995) In I.A. Tikhonovich, N.A. Provorov, V.I. Romanov, and W.E. Newton (eds.) Nitrogen fixation, Fundamentals and Applications, Kluwer Academic Publishers, Dordrecht, 564-572.

Harris, R.F. (1981) In J.F. Parr, W.R. Gardner, and L.F. Elliot (eds.) Water potential relations in soil microbiology, Soil Science Society of America.Madison, WI, pp. 23-95.

Herrada, G., Puppo, A. and Rigaud, J. (1989) J. Gen. Microbiol. 135, 3165-3177.

Hirel, B., Bouet, C., King, B., Layzell, B., Jacobs, F. and Verma, D.P.S. (1987) EMBO J. 6, 1167-1171.

Hirschman, J., Wong, P.K., Sei, K., Keener, J., Kustu, S. (1985) Proc. Natl. Acad. Sci. USA 82, 7525-7529.

Hoelzle, I. and Streeter, J.G. (1990) Appl. Environ. Microbiol. 56, 3213-3215.

Hong, Z.Q. and Copeland, L. (1990) Phytochem. 8, 2437-2440.

Hornez, J.P., Timinouni, M., Defives, C. and Derieux, J.C. (1994) Curr. Microbiol. 28, 225-229.

Howitt, S.M. and Gresshoff, P.M. (1985) J. Gen. Microbiol. 132, 257-261.

Howitt, S.M., Day, D.A., Scott, K.F. and Gresshoff, P.M. (1988) J. Plant Physiol. 132, 5-9.

Howitt, S.M., Udvardi, M.K., Day, D.A. and Gresshoff, P.M. (1986) J. Gen. Microbiol. 132, 257-261.

Humbeck, C. and Werner, D. (1987) Endcyt. Cell Research 4, 185-196.

Hunt, T.P. and Magasanik, B. (1985) Proc. Natl. Acad. Sci. USA 83, 8453-8457.

Hussein, Y.A., Tewfik, M.S. and Hamdi, Y.A. (1974) Soil Biol. Biochem. 6, 377-381.

Irigoyen, J.J., Sanchez-Diaz, M. and Emerich, D.W. (1990) Appl. Envion. Microbiol. 56, 2587-2589.

Jackson, F.A. and Dawes, E.A. (1976) J. Gen. Microbiol. 97, 303-312.

Jelesko, J. G., Lara, J. C. and Leigh., J. A. (1993) Mol. Plant-Microbe Interact. 6, 135-143.

Jelesko, J.G. and Leigh, J.A. (1994) Mol. Microbiol. 11, 165-173.

Jimenez-Zurdo, J. I.,Dillewijn, P. V., Soto, M. J., DeFelipe, M. R., Olivares, J. and Toro, N. (1995) Mol. Plant-Microbe Interact. 8, 492-498.

Jimenez-Zurdo, J.I., Garcia-Rodriguez, F.M. and Toro, N. (1997) Mol. Microbiol. 23, 85-93.

Jin, H.N., Dilworth, M.J. and Glenn, A.R. (1990) Arch. Microbiol. 153, 455-462.

Johnson, G.V., Evans, H.J. and Ching, T.M. (1966) Plant Physiol. 41, 1330-1336.

Jording, D., Sharma, P.K., Scmidt, R., Engelke, T., Uhde, C. and Puhler, A. (1992) J. Plant Physiol. 141, 18-27.

Jording, D., Uhde, C., Schmidt, R. and Puhler, A. (1994) Experientia 50, 874-883.

Kahn, M.L., Kraus, J. and Sommerville, J.E. (1985) In H.J. Evans, P.J. Bottomley, and W.E. Newton (eds.) Nitrogen Fixation Research Progress, Martinus Nijhoff, Dordrecht, pp. 193-199.

Karr, D.B., Waters, J.K., Suzuki, F. and Emerich, D.W. (1984) Plant Physiol. 75, 1158-1162.

Katznelson, H. and Zagallo, A.C. (1957) Can. J. Microbiol. 3, 879-884.

Kawaguchi, M., Sekine, M. and Syono, K. (1990) Plant Cell Physiol. 31, 449-455.

Keele, B.B., Hamilton, Jr., P.B. and Elkan, G.H. (1969) J. Bacteriol. 97, 1184-1191.

Kerpolla, T.K. and Kahn, M.L. (1988) J. Gen. Microbiol. 134, 913-919.

Kim, Y.S. and Chae, H.Z. (1990) Biochem. Biophys. Res. Comm. 169, 692-699.

Kim, Y.S. and Chae, H.Z. (1991) Biochem J. 273, 511-516.

Kim, Y.S. and Kang, S.W. (1994) J. Biol. Chem. 269, 8014-8021.

Kimura, I. and Tajima, S. (1989) Soil Sci. Plant Nutr. 35, 271-279.

Kiss, G.B., Vincze, E., Kalman, Z., Forrai, T. and Kondorosi, A. (1979) J. Gen. Microbiol. 113, 105-118.

Kittell, B. L., Helinski, D. R. and Ditta, G. S. (1989) J. Bacteriol. 171, 5458-5466.

Kitts, C.L, LaPointe, J.P., Lam, V.T. and Ludwig, R.A. (1992) J. Bacteriol. 174, 7791-7797.

Kohl, D.H., Kennedy, E.J., Zhu, Y-X, Shubert, K.R. and Shearer, G. (1991) J. Exp. Bot. 42, 831-837.

Kohl, D.H., Schubert, K.R., Carter, M.B., Hagedorn, C.H. and Shearer, G. (1988) Proc. Natl. Acad. Sci. USA 85, 2036-2040.

Kohl, D.H., Straub, P.F. and Shearer, G. (1994) Plant Cell Environ. 17, 1257-1262.

Kouchi, H. and Nakaji, K. (1985) Soil Sci. Plant Nutr 31, 323-334.

Kouchi, H. and Yoneyama, T. (1986) Physiol. Plant. 68, 238-244.

Kouchi, H., Fukai, K., Katagiri, H., Minamisawa, K. and Tajima, S. (1988) Physiol. Plant. 73, 327-334.

Kouchi, H., Fukai. K. and Kihara, A. (1991) J. Gen. Microbiol. 137, 2901-2910.

Kustu, S., North, A. and Weiss, D.A. (1991) Trends Biochem. Sci. 16, 397-402.

Larher, F., Goas, G., Le Rudulier, D., Gerard, J. and Hamelin, J. (1983) Plant Sci. Lett. 29, 315-326.

Lawrie, A.C. and Wheeler, C.T. (1975) New Phytol.74, 437-445.

Lepo, J.E., Hanus, F.J. and Evans, H.J. (1980) J. Bacteriol. 141, 661-670.

Lewis, T.A., Gonzalez, R. and Botsford, J. L. (1990) J. Bacteriol. 172, 2413-2420.

Lowe, R.H. and Evans, H.J. (1962) Soil Sci. 94,351-356.

Lynch, A.S. and Lin, E.C.C. (1996) In F.C. Neidhardt, R. Curtiss III, J.L. Ingraham, E.C.C. Lin, K. B. Low, B. Magasanik, W.S. Reznikoff, M. Riley, M. Schaechter, and H.E. Umbarger (eds.) Escherichia coli and Salmonella Cellular and molecular biology, ASM Press, American Society for Microbiology, Washington, DC, pp. 1526-1538.

Magasanik, B. 1993. J. Cell. Biochem. 51, 34-40.

Malik, K.A. and Schlegel, H.G. (1981) FEMS Microbiol. Letts. 11, 63-67.

Manco, G., Rossi, M., Defez, R., Lamberti, A., Percuoco, G. and Iaccarino, M. (1992) J. Gen. Microbiol. 138, 1453-1460.

Manhart, J. R. and Wong, P. P. (1979) Can. J. Microbiol. 25, 1169-1174.

Manian, S.S. and O'Gara, F. (1982) Arch. Microbiol. 131, 51-54.

Martin, G.B., Chapman , K.A. and Chelm, B.K. (1988) J. Bacteriol. 170, 5452-5459.

Martinez De Drets, G. and Arias, A. (1970) J. Bacteriol. 103, 97-103.

Martinez De Drets, G. and Arias, A. (1972) J. Bacteriol. 109, 467-470.

Martinez De Drets, G., Arias, A. and Rovira de Cutinella, M. (1974) Can. J. Microbiol. 20, 605-609.

McAllister, C.F. and Lepo, J.E. (1983) J. Bacteriol. 153, 1155-1162.

McDermott, T.R., Griffith, S.M., Vance, C.P. and Graham, P.H. (1989) FEMS Microbiol. Rev. 63, 327-340.

McDermott, T.R.and Kahn, M.L. (1992) J. Bacteriol. 174, 4790-4797.

McKay, I.A., Dilworth, M.J. and Glenn, A.R. (1988) J. Gen. Microbiol. 134, 1433-1440.

McKay, I.A., Glenn, A.R. and Dilworth, M.J. (1985) J. Gen. Microbiol. 131, 2067-2073.

Mendoza, A., Leija, A., Martínez-Romero, E., Hernández, G. and Mora J.. 1995. Mol. Plant-Microbe Int. 8, 584-592.

Meijer, A. J. and VanDam, K. (1974) Biochem. Biophys. Acta. 346, 213-244.

Mellor, R.B. (1988) J. Plant Physiol. 133, 173-177.

Merrick, M.J. and Gibbins, J.R. (1985) Nucleic Acids Res. 13, 7607-7619

Miller, R.W., McRae, D.G., Al Jobore, A. and Berndt, W.B. (1988) J. Cell. Biochem. 38, 35-49

Miller, R.W., McRae, D.G. and Joy, K. (1991) Mol. Plant-Microbe Int. 4, 37-45.

Monza, J., Delgado, M.J. and Bedmar, E.J. (1992) Plant and Soil 139, 203-207.

Muller, P. and Werner, D. (1982) Z. Naturforsch. 37c, 927-936.

Mulongoy, K. and Elkan, G.H. (1977) J. Bacteriol. 131, 179-187.

Muthukumar, G., Arunakumari, A. and Mahadevan, A. (1982) Plant and Soil. 69, 163-169.

Natarajan, K., Kishore, L. and Babu, C.R. (1995) Microbios 82, 95-107.

Niel, C., Guillaume, J.B. and Bechet, M. (1977) Can. J. Microbiol. 23, 1178-1181.

Nixon, B.T., Ronson, C.W. and Ausubel, F.M. (1986) Proc. Natl. Acad. Sci. USA 83, 7850-7854.

Noguez, R., Moreno, S., Guzman, J. and Espin, G. (1994) Can. J. Microbiol. 40, 965-968.

O'Brian, M. (1996) J. Bacteriol. 178, 2471-2478.

O'Gara, F., Manian, S. and Meade, J. (1984) FEMS Microbiol. Lett. 24, 241-245.

O'Hara, G.W., Daniel, R.M. and Steele, K.W. (1983) J. Gen. Microbiol. 129, 2405-2412.

Osburne, M.S. and Signer, E.R. 1980. J. Bacteriol. 143, 1234-1240.

Osteras, M., Driscoll, B.T. and Finan, T.M. (1995) J. Bacteriol. 177, 1452-1460.

Osteras, M., Finan, T.M. and Stanley, J. (1991) Mol. Gen. Genet. 230, 257-269.

Ou Yang, L.-J., Udvardi, M.K. and Day, D.A. (1990) Planta 182, 437-444.

Parke, D. and Ornston, L.N. (1984) J. Gen. Microbiol. 130, 1743-1750.

Parke, D. and Ornston, L.N. (1986) J. Bacteriol. 165, 288-292.

Pawlowski, K., Ratet, P., Schell, J. and de Bruijn, F.J. (1987) Mol. Gen. Genet. 206, 207-19.

Pedrosa, F.O. and Zancan, G.T. (1974) J. Bacteriol. 119, 336-338.

Perez-Galdona, R. and Kahn, M.L. (1994) Microbiol. 140, 1231-1235.

Pesole, G., Gissi, C., Lanave, C. and Saccone, C. (1995) Mol. Biol. Evolution. 12, 189-197.

Peterson, J.B. and LaRue, T.A. (1981) Plant Physiol.68, 489-493.

Peterson, J.B. and LaRue, T.A. (1982) J. Bacteriol. 151, 1473-1484.

Phillips, D.A. (1980) Annu. Rev. Plant Physiol. 31, 29-49.

Phillips, D.V., Wilson, D.O. and Dougherty, D.E. (1984) J. Agric. Food Chem. 32, 1289-1291.

Podila, G.K., Kotagiri, S. and Shantharam, S. (1993) Appl. Environ. Microbiol. 59, 2717-2719.

Poole, P.S., Blyth, A., Reid, C.J. and Walters, K. (1994) Microbiol. 140, 2787-2795.

Povolo, S., Tombolini, R., Morea, A., Anderson, A.J., Casella, S. and Nuti, M.P. (1994) Can. J. Microbiol. 40, 823-829.

Preston, G.G., Zeiher, C., Wall, J.D. and Emerich, D.W. (1989) Appl. Environ. Microbiol. 55, 165-170.

Rastogi, V.K. and Watson, R.J. (1991) J. Bacteriol. 173, 2879-2887.

Rawsthorne, S. and LaRue, T.A. (1986) Plant Physiol. 81, 1092-1096.

Reibach, P.H. and Streeter, J.G. (1983) Plant Physiol. 72, 634-640.

Reibach, P.H. and Streeter, J.G. (1984) J. Bacteriol. 159, 47-52.

Reitzer, L. (1996) In F.C. Neidhardt, R. Curtiss III, J.L. Ingraham, E.C.C. Lin, K. B. Low, B. Magasanik, W.S. Reznikoff, M. Riley, M. Schaechter, and H.E. Umbarger (eds.) *Escherichia coli* and *Salmonella*, Cellular and molecular biology (2nd edition), ASM Press, American Society for Microbiology, Washington, DC, p. 380-407.

Rhee, S. G., Chock, P. B. and Stadtman, E. R. (1985) Meth. Enzymol. 113, 213-241.

Ritchie, G.A.F., Senior, P.J. and Dawes, E.A. (1971) Biochem. J. 121, 309-316.

Robson, R.L. and Postgate, J.R. (1980) Ann. Rev. Microbiol. 34, 183-207.

Rohm, M. and Weiner, D. (1985) Arch. Microbiol. 140, 375-379.

Romanov, V.I., Hernandez-Lucas, I. and Martinex-Romero, E. (1994) Appl. Environ. Microbiol. 60, 2339-2342.

Ronson, C. W., Astwood, P. M. and Downie, J. A. (1984) J. Bacteriol. 160, 903-909.

Ronson, C.W. and Primrose, S.B. (1979) J. Gen. Microbiol. 112, 77-88.

Ronson, C.W., Lyttleton, P. and Robertson, J.G. (1981) Proc. Natl. Acad. Sci. USA 78, 4284-4288.

Ronson, C.W., Nixon, B.T., Albright, L.M. and Ausubel, F.M. (1987) J. Bacteriol. 169, 2424-2431.

Rosendahl, L., Dilworth, M.J. and Glenn, A.R. (1992) J. Plant Physiol. 139, 635-638.

Rossbach, S. and Hennecke, H. (1991) Mol. Microbiol. 5, 39-47.

Ryan, E., Bodley, F. and Fottrell, P.F. (1972) Phytochem. 11, 957-963.

Salminen, S. O. and Streeter, J. G. (1987b) J. Bacteriol. 169, 495-499.

Salminen, S. O. and Streeter, J. G. (1990) J. Gen. Microbiol. 136, 2119-2126.

Salminen, S.O. and Streeter, J.G. (1986) Plant Physiol. 81, 538-541.

Salminen, S.O. and Streeter, J.G. (1987a) Plant Physiol. 83, 535-540.

Salminen, S.O. and Streeter, J.G. (1992) Plant Physiol. 100, 597-604.

SanFransisco, M.J.D. and Jacobson, G.R. (1985) J. Gen. Microbiol.131, 765-773.

Saroso, S., Dilworth, M.J. and Glenn, A.R. (1986) J. Gen. Microbiol. 132, 243-249.

Senior, P.J. and Dawes, E.A. (1971) Biochem J. 125, 55-66.

Senior, P.J., Beech, G.A., Ritchie, G.A.F. and Dawes, E.A. (1972) Biochem. J. 128, 1193-1201.

Shatters, R.G. and Kahn, M.L. (1989) J. Mol. Evolution 29, 422-428.

Shatters, R.G., Liu, Y. and Kahn, M.L. (1993) J. Biol. Chem. 268, 1-7.

Shatters, R.G., Somerville, J.E. and Kahn, M.L. (1989) J.Bacteriol. 171, 5087-5094.

Sik, T. and Barabas, I. (1977) In W. Newton, J.R. Postgate and C. Rodriguez Barrueco (eds.), Recent Developments in Nitrogen Fixation, Academic Press, London, pp. 365-373.

Simpson, F.B., Maier, R.J. and Evans, H.J. (1979) Arch. Microbiol. 123, 1-8.

Skot, L. and Egsgaard, H. (1984) Planta 161, 32-36.

Smith, M.T. and Emerich, D.W. (1993) Arch. Biochem. Biophys. 304, 379-385.

Smith, M.T., Preston, G.G. and Emerich, D.W. (1994) Symbiosis 17, 33-42.

Snapp, S.S. and Vance, C.P. (1986) Plant Physiol. 82, 390-395.

Somerville, J. E., Shatters, R. G. and Kahn, M. L. (1989) J. Bacteriol. 171, 5079-5086.

Stam, H., van Verseveld, H.W., de Vries, W. and Stouthamer, A.H. (1986) FEMS Microbiol. Lett. 35, 215-220.

Stanier, R.Y. and Ornston, L.N. (1973) Adv. Microbiol. Physiol. 9, 89-151.

Stanley, J., van Slooten, J., Dowling, D.N., Finan, T., Broughton, W.J. (1989) Mol. Gen. Genet. 217, 528-32

Stovall, I. and Cole, M. (1978) Plant Physiol. 61, 787-790.

Stowers, M.D. and Elkan, G.H. (1983) Can. J. Microbiol. 29, 398-406.

Streeter, J. G. and Salminen, S. O. (1990) Biochim. Biophys. Acta Gen.Subj. 1035, 257-265.

Streeter, J.G. (1980) Plant Physiol. 66, 471-476.

Streeter, J.G. (1982) Planta. 155, 112-115.

Streeter, J.G. (1987) Plant Physiol. 85, 768-773.

Streeter, J.G. (1991) Adv. Bot. Res. 18, 129-187.

Streeter, J.G. and Bosler, M.E. (1976) Plant Sci. Letts. 7, 321-329.

Streeter, J.G. and DeVine, P.J. (1983) Appl. Env. Microbiol. 46, 521-524.

Streeter, J.G. and Salminen, S.O. (1993) Plant Physiol. Biochem. 31, 73-79.

Streeter, J.G., Peters, N.K., Salminen, S.O., Pladys, D. and Zhaohua, P. (1995) Plant Physiol. 107, 857-864.

Streeter, J.G., Salminen, S.O., Whitmoyer, R.E. and Carlson, R.W. (1992) Appl. Environ. Microbiol. 58, 607-613.

Stripf, R. and Werner, D. (1978) Z. Naturforsch. 33c, 373-381.

Stumpf, D.K. and Burris, R.H. (1979) Anal. Biochem. 95, 311-315.

Suganuma, N. and Yamamoto, Y. (1987) Soil Sci. Plant Nutr. 33, 79-91.

Szeto, W.W., Nixon, B.T., Ronson, C.W. and Ausubel, F.M. (1987) J. Bacteriol. 169, 1423-1432.

Tabita, F.R. (1988) Microbiol. Rev. 52, 155-189.

Taboada, H., Encarnacion, S., Vargas, M. C., Mora, Y., Romero-Martinez, E. and Mora, J. (1996) Int. J. Sys Bacteriol. 46, 485-491

Tajima, S. and Kouzai, K. (1989) Plant Cell Physiol. 30, 589-593.

Tajima, S. and LaRue, T.A. (1982) Plant Physiol. 70, 388-392.

Thayer, J.R. and Huffaker, R.C. (1982) J. Bacteriol. 149, 198-202.

Thomashow, M. F., Hugly, S., Buchholz, W. G. and Thomashow, L. S. (1986) Science 231, 616-618.

Trinchant, J.C. and Rigaud, J. (1990) Plant Physiol. 94, 1002-1008.

Trinchant, J.C., Birot, A.M. and Rigaud, J. (1981) J. Gen. Microbiol. 125, 159-165.

Triplett, E.W. and Blevins, D.G. (1979) Plant Physiol. 63(Suppl), 113.

Tyerman, S.D., Whitehead, L.F., Day, D.A. (1995) Nature 378, 629-32.

Ucker, D.S. and Signer, E. (1978) J. Bacteriol. 136, 1197-1200.

Udvardi MK, Price GD, Gresshoff PM, Day DA. (1988a) FEBS Lett. 231, 36-40.

Udvardi, M. K., Salom, C. L. and Day, D. A. (1988b) Mol. Plant-Microbe Interact. 1, 250-254.

Udvardi, M.K. and Day, D.A. (1997) Annu. Rev. Plant Physiol. Plant Mol. Biol. 48, 493-523.

Udvardi, M.K., Lister, D.L. and Day, D.A. (1992) J. Gen. Microbiol. 138, 1019-25.

Udvardi, M.K., Ou Yang, L-J., Young, S. and Day, D.A. (1990) Molec. Plant-Microbe Interact. 3, 334-340.

Van Straten, J. and Schmidt, E.L. (1974). Soil Biol. Biochem. 6, 347-351.

Vance, C.P., Stade, S. and Maxwell, C.A. (1983) Plant Physiol. 72, 469-473.

Vanlerberghe, G.C. and Turpin, D.H. (1990) Plant Physiol. 94(3), 1124-1130.

Watson, R.J., Chan, Y.-K., Wheatcroft, R., Yang, A.-F. and Han, S. (1988) J. Bacteriol. 170, 927-934.

Watson, R.J., Rastogi, V.K. and Chan, Y.-K. (1993) J. Gen. Microbiol. 139, 1315-1323.

Werner, D. and Stripf, R. (1978) Z. Naturforsch. 33c, 245-252.

Werner, D., Dittrich, W. and Thierfelder, H. (1982) Z. Naturforsch 37c, 921-926.

Wickremasinghe, R.L., Swain, T. and Goldstein, J.L. (1963) Nature 199, 1302-1303.

Wong, C.M., Dilworth, M.J. and Glenn, A.R. (1991) Arch. Microbiol. 156, 385-391.

Wong, C.M., Dilworth, M.J. and Glenn, A.R. (1994) Microbiol. 140, 2775-2786.

Wong, P., Evans, H.J., Klucas, R. and Russell, S. (1971) Plant Soil special volume, 525-543.

Wong, P.P. and Evans, H.J. (1971) Plant Physiol. 47, 750-755.

Zhu, Y.X., Shearer, G. and Kohl, D.H. (1992) Plant Physiol. 98, 1020-28.

Chapter 25

Evolutionary Aspects of Symbiotic Adaptations
Rhizobium's Contribution to Evolution by Association

Anton Quispel

I. Introduction

The symbioses between plants and diazotrophic bacteria and especially those in which representatives of the *Rhizobiaceae* are involved are the best studied of all symbiotic associations. This is mainly caused by their great practical, agronomical, importance and stimulated by expectations for improved future applications. On the other hand these symbioses are most interesting because of their highly specialized mutual relations and corresponding physiological adaptations.Their study therefore may lead to important conclusions about fundamental aspects of plant-microbe associations and their contributions to the evolution of life on earth.

Since the concept of "symbiosis" has been introduced by the Swiss botanist de Bary in 1879 as "Die Erscheinungen des Zusammenlebens ungleichnamiger Organismen" (the phenomena of the living together of organisms with different names) this concept has been used with different meanings. While De Bary did not speak about the nature of this "living together" and thus included forms of parasitism as well as forms of mutualism, many later authors restricted the concept to the mutualistic associations. However, the distinction between parasitic and mutualistic interactions is, if ever, only possible after thorough investigations.

Only gradually it was recognized that symbiotic associations with a more or less mutualistic character are not incidental situations with only some anecdotical interest, but must be considered as one of the most important aspects of the evolution of life. In a recent historical book Sapp (1994) described the different types of reluctance to accept an important role for symbiosis in the evolution of life. Though a few scientists speculated about a role of symbiotic associations in the evolution of higher organisms their arguments could not yet be impressive. However, recent studies on molecular evolution leave little doubt that the mitochondria and plastids of eukaryotic cells arose through a process of symbiogenesis from symbiotic associations with α-purple bacteria and cyanobacteria respectively. More far-reaching concepts about the symbiogenetic origin of eukaryotic cells still are disputed (Margulis 1993).

The progress in the study of molecular aspects of extant symbiotic associations has shown that symbiosis may function as a vehicle by which many eukaryotic organisms gained access to complex metabolic capabilities of prokaryotes (Douglas 1994). While the symbiogenesis of mitochondria and chloroplasts introduced the benefits of the metabolic capabilities of respiration and fotosynthesis respectively, the extant symbioses of the *Rhizobiaceae* enable the leguminous plants to use the benefits of the complex metabolic capabilities of dinitrogen fixation. For mitochondria and chloroplasts this happened long ago, for rhizobial symbioses their development can be studied today.

II. Evolution of Symbiosis: from Loose Associations to Endocytosis

While studying the putative evolution of symbiotic associations we have to realize that we can only study and compare such associations in extant organisms. Only in few cases, like some mycorrhizal symbioses, fossil remains are found which indicate their presence already in periods about 350-450 million years ago, indicating a correlation with the origin of land plants (Remy *et al.*, 1994) and long before the origin of the leguminosae in the late Cretaceous period (discussed in Chapter 22). Baker and Miller (1980) described fossil actinorhizal root nodules with the characteristic clusters of hyphae and vesicles from pleistocenic sediments. No such fossils are as yet available for leguminous root nodule symbioses. Further data about the evolutionary history of leguminous plants and their symbionts are given by Sprent and Raven (1985), Sprent (1994) and in

Chapter 1. A recent phylogenetic tree for the *Rhizobiaceae* is given by Young (1996).

Even if sufficient fossil remains were available they could tell us something about morphological developments, but hardly or not about physiological and molecular aspects. Studies on the sequence of steps in the ontogenetic development of highly evoluated symbiotic systems, like those in root nodule symbioses, show the present results of this physiological and molecular evolution. It is tempting to conclude that such steps are representative for their phylogenetic history as well, but ontogenetic steps do not need to be identical with phylogenetic steps. More convincing conclusions are possible by comparisons with other, less evoluated, symbiotic associations. Such studies on "comparative symbiontology" may help us in providing data for the evaluation of putative evolutionary trends like: association ⇨ ectosymbiosis ⇨ endosymbiosis, broad host range ⇨ specific associations, parasitic ⇨ mutualistic relations and facultative ⇨ obligatory dependence upon the partner.

II.A. ASSOCIATION⇨ECTOSYMBIOSIS ⇨ENDOSYMBIOSIS

II.A.1. Root Colonization

Bacteria belonging to the *Rhizobiaceae* are present in the microbial soil population (Laguerre *et al.*, 1993), where they are subject to the different biotic and abiotic influences which affect their population dynamics and even their symbiotic capacities (Sullivan 1996). Plant roots exert marked effects on the microbial populations as described in Chapter 8. They may affect the pH, absorb oxygen, use nutrient ions, excrete different organic subtances and gases like CO_2 and C_2H_4. These effects create an ecological niche, the rhizosphere (Lynch 1990, de Weger *et al.*, 1995).The existing populations of micro-organisms will be selected according to their reactions on the plant-root environment in relation to the environment elsewhere in the soil. As a secondary effect the changed populations will affect the interactions between the different organisms e.g. by competition for nutrients and production of antibiotics and therewith further influence the constitution of the rhizosphere microbial populations.

Some of the organic factors in the root exudates can exert chemotactic effects on certain soil bacteria. Studies on chemotaxis in *Rhizobiaceae* have demonstrated the stimulating effect of dicarboxylic acids, glutamate and aspartate. Flavonoids, which play such important roles as signal substances in later step of infection, either are inactive or attract the bacteria through other types of receptors (Barbour *et al.,* 1991, Dharmatilake and Bauer 1992). This is an indication that the genetic regulation of chemotaxis is not related to regulation of infection. Mutants, which are defective in their motility are normally infective, but are impaired in their competition with other strains. This could have some selective value under natural conditions.

For most bacteria which are attracted and selected in the ecological niche of the rhizosphere there are no further steps in the plant-microbe interactions. For the *Rhizobiaceae* and for other plant endosymbionts, like *Frankia* and mycorrhizal fungi, it is the first step in the sequence of interactions, which lead to endosymbiosis. These later steps may have different kinds of regulatory feed-back on the rhizosphere. If the symbionts have some positive effect on the host plants, e.g. because of their dinitrogen fixation, the resulting better growth of the plants will lead to an increase of root exudates. It has been suggested that those bacteria in the root nodules, which do not differentiate into bacteroids, proliferate in such a way that they finally increase the rhizobial populations in the rhizosphere (Jimenez and Casadesus 1989, Olivieri and Frank 1994).

As a conclusion we may consider the rhizosphere as an environmental niche in which many types of micro-organisms, including diazotrophic bacteria, are selected. Once selected, evolutionary adaptations may increase their chances for survival. However, the rhizosphere is an ecological niche, which hardly protects the participating microorganisms against harmful environmental disturbances. They may be easily washed away, affected by antibiotics from other bacteria and are subject to predation by protists. Better ways of protection therefore will have a positive selective influence.

II.A.2. Adhesion at Plant Surfaces

A better protection against removal from the rhizosphere is obtained by adhesion to the cell surface of plant root cells. Moreover adhesion may concentrate the bacteria at sites from where further steps of the infection can be initiated. Adhesion of bacteria is a general phenomenon, not specific for plant-bacterium associations, since similar binding processes are observed in less specific associations or even to soil or other non-biological particles.

All such types of adhesion depend on surface proteins and polysaccharides. These multiple

polymers explain the differences in attachment properties, their variability and flexibility (Williams and Fletcher 1996). There are numerous examples of two-step binding processes in which the binding by a first, often less-specific, step is strengthened by a second step. Such a two-step binding process already is observed in the binding of *Azospirillum* to wheat roots (Michiels *et al.,* 1991) and the intensively studied binding of *Agrobacterium* and (*Brady*)-*Rhizobium* to plant cells (Kijne 1992, and Chapter 11). There seems to be no reason to consider such binding processes with their great variability in surface compounds as steps in an evolution. They could be regarded as different selections from the multitude of possibilities of binding between plants and micro-organisms, in which bacterial proteins like rhicadhesin, bacterial surface polysaccharides like LPS and EPS, cellulose fibrills, fimbriae and plant lectins can be involved. However, it is especially the second step, which may have far-reaching consequences for further steps in the plant-microbe interactions. In parasitic infections the type of binding to the surface of host cells may determine the compatibility or the hypersensitive reaction (HR) leading to incompatibility (Sequiera *et al.,* 1977). Mutations in genes for surface polysaccharides in (*Brady*)*Rhizobium* may lead to defects in the further development of the infection and nodulation process (see section II.A.4.). A surface binding of lectins to such polysaccharides might interfere with the subsequent infection process (Diaz *et al.,* 1995, van Eysden *et al.,* 1995). These are aspects, which certainly will have had selective importance for further evolution of the association. In associations, where the development does not proceed further than surface adhesion, the type of adhesion may not have had much selective importance. However, in those infection processes where the type of binding affects essential later steps their evolutionary role may have been considerable. This may be called an example of "feed-back selection" in evolutionary selection.

II.A.3. The Infection Process

A better form of protection from outside influences is obtained by further penetration into the host plant. Moreover an endophytic way of life ensures a more direct provision with nutritive plant metabolites, though in an essentially hostile host-environment in which different types of defence systems have to be overcome.

The entrance of bacteria from the rhizosphere into the intercellulars, or even inside the cells of the outer cortex, is not restricted to the real ecto- and endosymbioses (or parasitisms) but is already of common occurrence in rhizosphere associations. Already the classical study by Döbereiner and Day (1976) on the role of diazotrophic bacteria in the rhizosphere described how these bacteria could be present in cells of the outer cortex. In these and many subsequent descriptions of e.g. *Azospirillum* species in the rhizosphere the bacteria were present in dead cells, but the presence of pectolytic enzymes may enable a penetration into some living cells as well. Very interesting in this respect are the associations on the roots of Kallar grass (*Leptochoa fusca* L.Kunth) with bacteria of the newly described genus *Azoarcus*, where agglomerations of bacteria in the cortex are not only ectocellular, but even endocellular (Hurek *et al.*, 1991).

Root nodule bacteria belonging to the *Rhizobiaceae* can enter the roots in different ways (Sprent and de Faria 1987, Kijne 1992). In Chapter 22, Table 1 gives a clear survey of the different possibilities, from the most simple "crack" infection to the most studied infection threads. While simple types of infection might be explained by the activity of pectolytic enzymes, either from the invading organisms or after induction in the host cells, the formation of the infection threads is a far more organized morphogenetic process. Recent concepts of infection thread initiation in root hair tips, as based on computer models by van Batenburg *et al.*, (1986), explain their formation as a reversed inside-directed growth of the root-hair tip (Kijne 1992, and Chapter 11). There seem to be few arguments to suggest that the more highly organized types of infection represent a higher step in their evolution since all types may lead to equally effective dinitrogen-fixing root nodules without obvious differences in their selective value.

Why do some bacteria enter the root cells, while other strains remain outside the roots in the rhizosphere? This is an aspect belonging to a general problem of plant and animal pathogenesis. It has been demonstrated that the invasion of animal tissues by bacterial pathogens, e.g. enteric bacteria, depends on a machinery by which certain proteins are secreted that promote their entry into the host cells. Such proteins bind receptors that are linked to the cytoskeleton or initiate a signal transduction cascade that promotes cytoskeletal rearrangements. DNA sequences, which are related to the loci of the invasion gene clusters, appear to be present in other animal and plant pathogens as well (Ochman and Groisman 1995). Though structure and organization

of such gene complexes appear to be broadly conserved, their phylogenetic distribution suggests their independent origin in different animal and plant pathogens with indications for horizontal gene transfer.

nod Genes and Nod Factors. The sequence of steps, which is needed for the formation of infection threads and further nodule initiation, already starts in the rhizosphere since it is here that the *nod* genes of the bacterial cells are induced to synthesize Nod factors. These Nod factors are needed as signals for different aspects of the infection and nodulation as reviewed by e.g. Dénarié and Cullimore 1993, Spaink 1995 and the Chapters 19, 20 and 21. After external application of a specific Nod factor to non-infected roots the first detectable reaction consists of the induction of a membrane depolarisation in the root hairs (Ehrhardt *et al.*, 1992). Membrane depolarisation as such is neither host specific nor specific for the rhizobia, since e.g. inoculation of wheat roots by *Azospirillum* reduces the membrane potential in every part of the root surface, accompanied by proton efflux (Bashan and Levanony 1991). The Nod factors, however, have more far-reaching effects like curling of the root hairs, formation of the infection threads, induction of nodule meristems and induction of some nodule-specific proteins, the early nodulins (ENOD). The induction of the *nod* genes to synthesize Nod factors thus can be regarded as a most crucial step in the selection from the rhizosphere populations of those species, which have the genetic potentialities to prepare the host cells for further penetration.

This leads to a first question with regard to the evolution of the *nod* genes: why do these bacteria contain such *nod* genes, whose functions appear to be only involved as a preparation for infection and are only induced under the specific influence of the roots? There are not yet indications that the Nod factors, which are synthesized by a cooperative and specific action of the induced *nod* gene products, play any role during the life of the free-living bacteria. This does not mean that the products of the individual *nod* genes are not homologous with other gene products with a well-known function. The oligosaccharide backbone of the Nod factors is biochemically similar to components of the well-known polysaccharide structure of bacterial cell walls. The formation and attachment of different substituents, which are essential for their specificity, is achieved by enzymes with functions already known from the synthesis of essential membrane constituents like phospholipids (Geiger *et al.*, 1994, and Chapter 4).

Unfortunately there are at this moment no data for comparisons with related endosymbionts. Nothing is as yet known about the genetic system for the infection of the most closely comparable actinorhizal symbioses. No relevant data enable comparisons with infections by mycorrhizal fungi. The role of different types of oligosaccharide signals as elicitors for plant defence reactions to pathogens is well-known (Ryan and Farmer 1991), chitin oligomers have been described as elicitors for phytoalexin production in rice (Ren and West 1992). In the diazotrophic rhizosphere bacterium *Azospirillum*, a few *nod* genes, *nod* P, Q en G, have been identified (Elmerich *et al.*, 1991). The products NodP and NodQ encode subunits for ATP sulphurylase, NodG is homologous with dehydrogenases and could have a general function as β-ketoacylreductase. These enzymes can have functions in other biosynthetic pathways than the synthesis of a Nod factor-like molecule.

We may expect that before their evolution as symbionts certain genes of the associated bacteria already encoded enzymatic functions for the synthesis of e.g. chitin oligomeric elicitors or of phospholipids. Such enzymes could be gradually adapted to other related functions like the synthesis of Nod signals. Another important evolutionary aspect may have been the reorganization of genes from their original gene clusters into the new structural units of the *nod*-operons, either as plasmids or as parts of the bacterial chromosomes. Plasmids could have played important roles in gene transfers. For many strains of *Rhizobiaceae* we now know the genetic organization of the *nod* genes in operons, each with their *nod* boxes. Comparisons between different species demonstrated their conserved character as well as their specific arrangements (see e.g. figure 1 in Schlaman *et al.*, 1992 and Chapter 19). Such comparisons illustrate the common origin of the system of *nod* genes as well as the specific evolutionary divergences in organization.

Regulation of nod-Genes. The synthesis of Nod factors could never have played any role in the evolution of these plant-microbe interactions if there had not been a regulatory system developed for the induction of the *nod* genes. Most of the structural *nod* genes are under the regulatory control of the

products of the *nodD* genes, which are present in one to three alleles. In *R. meliloti* the regulation of the *nod* genes involves in addition to the gene products NodD1, NodD2 and NodD3 another transcriptional activator SyrM and the chaperone-like protein GroEL (Dénarié *et al.*, 1993 and Chapter 19).

Evolutionary suggestions for the origin of such complicated systems can be based on the comparisons of the different types of NodD's. The amino acid sequences of all NodD proteins indicate that they all belong, as well as SyrM, to the LysR family of transcriptional regulators (Schell 1993). Unrooted phylogenetic trees as constructed for the NodD's by Rushing *et al.*, (1991), and in relation to other members of this family by Schlaman *et al.*, (1992) show that the NodD group forms a separate subfamily within the LysR family, while the SyrM proteins are more closely related to another part of this family. The separation of NodD must have occurred at an early date in the evolutionary history of the genes for the LysR family. The present distinction between the different alleles of *nodD* shows a more recent evolution at work. According to our present knowledge of the central role of the regulatory system by NodD's it is tempting to conclude that the origin of this system must have been one of the crucial steps in the evolution of the rhizobial symbioses. It may have marked a most decisive step in the transgression from the superficial surface associations to the ecto- or even endosymbiotic state. It marked a selection from other bacteria, which had to enter in a less coordinated way or simply had to remain outside. Of course we have to realize that the evolution of the genes for the NodD regulatory proteins was only effective after co-evolution with genes for the specific recognition DNA sequences (Goethals *et al.*, 1992).

Plant Flavonoid and Other Regulatory Signals. In most cases the regulatory function of the NodD products depends on the interaction with specific plant flavonoids (Schlaman *et al.*, 1992 and Chapter 19). Depending on the specific flavonoid the regulation is inducing, sometimes repressing. When there are more alleles of *nodD* the reactions of their NodD's to flavonoids are different. In few cases the binding of the NodD to the *nod* box of the *nod* genes is possible without further activation, as in the case for *nod* D3 in *R. meliloti* (Maillet *et al.*, 1990) and the hybrid NodD's studied by Spaink *et al.*, (1989). The structural requirements for these specific effects indicate a high degree of specificity between the plant flavonoids and the bacterial NodD. However, this is not a simple gene for gene relation, but consists of a reaction towards a complicated mixture of flavonoids.

A biological effect on other organisms by plant flavonoids is not at all exceptional. Phenolic compounds are well-known as regulators of gene expression in plant-microbe interactions as well as in many other aspects of plant metabolism (Peters and Verma 1990, Koes *et al.,* 1994). The chemotactic attraction of soil bacteria to the rhizosphere was already discussed above, several flavonoids or their glucosides were identified as factors responsable for the stimulation of spore germination and early hyphal growth in arbuscular mycorrhizae (AM) (Siquiera *et al.,* 1991, Bécard *et al.,* 1992). Kape *et al.,* (1993) made direct comparisons between the effect of flavonoids on *nod* gene induction in *Bradyrhizobium* and their effect in the chemotaxis of hyphae from spores of AM mycorrhizal fungi (*Glomus* spp.). While cinnamic acid and coumaric acid had a high chemotactic activity on *Glomus* spp., hardly any effect was observed on the *nod* genes of *B. japonicum*. The reverse was observed with genistein and daidzein. This might be explained by specific differences in flavonoid receptors in the different reactions. Comparisons between the *nod* gene inducing flavonoids and phytoalexins, which are formed by plant cells after infection with pathogens, were made by Dakora *et al.,* (1993a, 1996). In *Medicago sativa* inoculation with *Rhizobium meliloti* induced the formation of isoflavonoids, identified as the aglucons and glucosides of the phytoalexins medicarpin and a conjugated form of its precursor formononetin. It appeared that this precursor induced the *nod* genes, while the medicarpin was induced as a phytoalexin. Similar results were obtained in experiments with *Phaseolus vulgaris*, where *Rhizobium* infection induced the synthesis of the phytoalexin coumestrol and its isoflavin precursor daidzein (Dakora *et al.,* 1993b).

Though far more data have to be obtained about the role of flavonoids in other plant-microbe interactions a general picture emerges in which specific binding of flavonoids affects different regulatory systems, with stimulatory or inhibitory effects on the microorganisms involved. From an evolutionary point of view it is extremely interesting that these flavonoid-mediated regulatory systems in plant-microbe interactions show remarkable points of similarity with regulatory systems in higher organisms. Györgypal and Kondorosi (1991) compared the ligand-binding regions of the symbiotic regulatory protein NodD with other regulatory proteins with a dual sensor-activator function in vertebrates under the influence of steroid and thyroid hormones. Some flavonoids, like coumestrol and estradiol, bind as well to mammalian hormone-receptor proteins as to NodD proteins. Two regions of NodD (with 63 and 37 amino acids) had a homology of 45 and 36% respectively with modules of the estrogen-receptor. These regions thus coincided with conserved parts of the ligand-binding domains of the nuclear receptors. Moreover, the NodD overlapping module is 46% homologous with the membrane-spanning sensory segments of the VirA protein of *Agrobacterium tumefaciens*. These homologous modules suggest a common ancestor protein. Independently the similarities between flavonoid-dependent legume-rhizobium interactions, and steroid-mediated intercellular communication in vertebrates was discussed by Baker (1991,1994). He too stressed the related effects of steroids in the endocrine system of vertebrates and the induction of *nod* gene synthesis by flavonoids. This relation between the effects of steroids and flavonoids is well illustrated by those observations where flavonoids were able to bind to estrogen receptors (e.g.Martin *et al.,* 1978). Steroid plant hormones are known (e.g. brassinolides). Related structures, like different steroles,which play important roles in cell membranes of plants, are present in the peribacteroid membranes (PBM) (Hernandez and Cooke, 1996) while the bacterial hopanoids are especially abundant in the endophytic vesicles of *Frankia* (Berry and Kleemann 1993). Yet there are only few reports in plant-microbe interactions about signal molecules with related structures. The only signal molecule described up till now in actinorhizal root nodules, exerting an effect on the viability of endophytic clusters of a particular strain of *Frankia* in nodules of *Alnus glutinosa*, was identified as the triterpenoid dipterocarpol (Quispel *et al.,* 1989). The parasitic phase of *Ustilago violaceae* is induced by α-tocopherol (Castle and Day 1984). Phenylpropanoids and triterpenoids were identified as inducers of haustoria in parasitic angiosperms (Riopel and Timko 1992).

Though still many more observations have to be made, the present results indicate the existence of a vast amount of evolutionary related regulatory systems in which different signal molecules like flavonoids, steroids and other molecules bind to receptor proteins in a more or less specific way and affect their regulatory actions. Such regulatory actions are involved in biosynthetic, developmental and morphogenetic biochemical pathways, as well as in interactions between different organisms with parasitic or mutualistic aspects (Ryan and Farmer 1991, Boller 1995). The great variability of the chemical structures of the signal molecules, the mutational variability of their receptor proteins, the binding of these regulatory proteins to the DNA sequences of specific operator genes in operons for different biochemical pathways must have enabled an overwhelming amount of flexible possibilities for adaptations during evolution of plant-microbe associations.

II.A.4. From Infection to Bacterial Release.

Though Nod factors can "mimic" most of the early steps in the process of infection and nodulation, this does not necessarily mean that all steps react in a similar way to the specific Nod factors, nor that no other factors play a role. Ardourel et al., (1994) made a detailed analysis of the structural requirements of the Nod factor of R.meliloti for the initiation of the different steps in the infection process. By comparing the effects of Nod factors from different nod gene mutants the structural demands for the first formation of infection threads appeared to be more stringent than for the initiation and elicitation of plant developmental responses like the root hair tip growth and the differentiation of cortex cells. Many experiments have shown that mutants of (Brady)Rhizobium strains differed in the progress of the infection process. This sometimes could be explained by specific differences in the chemical structures of their Nod factors, but other aspects too may play a role in which different plant and bacterial components are involved (discussed in Chapter 22). In most if not all interactions of bacteria with plants and animals the complex oligosaccharide structures at their surface play a decisive role in the recognition of other cells, host micro-environments and immunological reactions. In (Brady)Rhizobium EPS and LPS mutants interruptions of the infection process have been observed (Kijne 1992, and the Chapters 6 and 7). Such interruptions may already occur during the early phases of infection thread growth up till the ultimate step of nodulation: the bacterial release from the infection threads into the cells of the initiated nodule.

The initiation of meristems for nodule formation in the root cortex is another result of the activity of Nod factors. Cell divisions in the root cortex of peas can be explained from the interactions of auxins, cytokinins and an originally unidentified C-factor from the vascular bundle (Libbenga et al., 1973).

Recent studies have indicated that Nod factors affect auxin transport, while the C-factor now has been identified as uridin (Smit et al., 1995). Nod factors induce similar cortical cell divisions as cytokinins (Bauer et al., 1996). Röhrig et al., (1996) suggested convergent pathways for lipo-chitin oligosaccharides (LCO's) and auxins as signals for cell divisions in tobacco cells, with cytokinins as common effector. The effects of chitin-oligosaccharides is especially interesting since there are ample indications that oligosaccharides are important factors in normal morphogenesis of plants (Ryan and Farmer 1991). Spaink (1996) postulated that LCO's, like the Nod factors, are representatives of a general class of signal molecules involved in plant and animal morphogenesis.

The resulting morphogenetic differentiation in the root cortex cells, in which not only the changed patterns of phytohormones, but certainly other factors like the reduced oxygen concentration play a role, is especially important. It prepares the cells and especially their cell membranes for the bacterial release after contact with infection threads. This bacterial release is extensively described in Chapter 22 so that here only those aspects need to be summarized, which may be important for evolutionary conclusions. The bacterial release is the final stage in the process from ecto- to endosymbiosis, a process of endocytosis during which the bacteria remain surrounded by a membrane, the peribacteroid membrane (PBM). This PBM at first is evidently a continuation of the plant cytoplasmic membrane (Kijne 1975), though during their further development contact with host cell debris and ER- and Golgi-derived substances may play a role (Kijne and Planqué 1979, Mellor 1989). Some of the PBM-specific proteins are nodule-specific nodulins, which are already induced during a step in the infection process preceding the bacterial release (Morrison and Verma 1987).

After some cell divisions, the bacteria stop dividing and are transformed into the bacteroids which, still surrounded by the PBM, form the dinitrogen-fixing "symbiosomes". At this step in nodule development the nod genes are no longer induced (Schlaman 1991).

Comparisons with other endophytic micro-organisms show that all are surrounded by a PBM-like cytoplasmic membrane of the host plant (Smith and Smith 1990). The PBM is just one example of a perisymbiont membrane (PSM). Such membranes are essential for the transport between the two symbiotic partners. In biotrophic parasitic associations the transport in these membranes is uni-directional, while mutualistic associations are characterized by surrounding membranes around the endophytic symbionts with bi-directional transport systems. In most plant endosymbioses, however, these membranes remain surrounded by cell-wall deposits of plant origin and as such they still must be called ectosymbiotic. In mycorrhizae and actinorhizae this is the general situation, though in AM mycorrhizae the cell-wall deposits decrease around the arbuscles. Even in the rhizobial symbioses there are host-bacteria combinations where no bacterial release is observed and the bacteria remain separated from the cytoplasm of the host plants by cell-wall deposits e.g. in the non-leguminous host *Parasponia*. Though we might expect that these surrounding cell-wall deposits, due to the absence of bacterial release, will reduce the efficiency of the dinitrogen fixation and thus are evolutionary selective, such expectations are not supported by the existing data.

II.B. BROAD RANGE ⇨ SPECIFIC ASSOCIATIONS

Some bacterial strains infect and nodulate a broad range of host plants, like the *Bradyrhizobium* strain 32H1, which even nodulates the non-leguminous *Parasponia* (*Ulmaceae*). Other strains are highly specific for related plant species or even particular strains. Do such differences represent different degrees of evolutionary adaptations?

The formation of all endosymbioses can be divided into a series of steps as summarized above for those of *Rhizobiaceae*. In the sequence of interactive steps, all with their own degree of mutual specificity, the specific restrictions of every next step will narrow the final range of specificity. Important restrictive steps are based on the structure of the Nod factors, their backbone structure and especially the different substituents, the specific types of NodD and the recognition of the available flavonoids, and the bacterial types of EPS and LPS with at the plant side the specific lectins (Diaz *et al.*, 1989, 1995). Mutations in one of the participating genes of a specificity-limiting step, either at the bacterial or the plant side, may broaden or narrow the specificity. But why should a broad host range be less succesful in evolution than a narrow range of specificity? The opposite may look more plausible.

For answering such questions comparisons with related symbioses are interesting (Gianinazzi-Pearson 1996, Pawlowski and Bisseling 1996). The AM mycorrhizae are generally cited as examples of a highly ubiquitous distribution among the great majority of Angiosperms though, if once their fungi can be cultivated, their specificity might appear to be more restricted than originally expected. In symbioses with Cyanobacteria in most cases the endosymbiotic *Nostoc* or *Anabaena* strains are not specific since plants may be infected by strains, which were isolated from other hosts or even by originally non-symbiotic strains. However, the non-infectivity of all strains isolated from *Azolla* shows how careful we must be in concluding that the isolated strains are really the endosymbiotic forms. The real endophytic strains may be far more specific (Zimmerman *et al.*, 1989). Comparable problems are encountered in *Frankia* strains from *Alnus glutinosa* (Burggraaf and Valstar 1984, Quispel 1992). The endophytic populations appear to be highly heterogenous since isolated strains, even from one nodule, may differ in their infectivity, their effectivity and their spore-formation *in situ*. An interesting comparison can be made with the mycorrhizal symbiosis in monotrope species with different mycorrhizal fungi. Comparisons between the phylogenetic lineages of host plants and their fungi resulted in the conclusion that evolution had led to a narrowing from generalists to specialisation on one fungus only (Cullings *et al.*, 1996). However, it was concluded that specificity resulted from a one-sided selection on these parasitic plants to increase their fitness by optimizing carbon-allocation, rather than from co-evolution in the sense of reciprocal selection. In a comparable sense it is conceivable that in the rhizobial symbioses the efficiency of the resulting dinitrogen fixation had a greater selective value than the occasional plant-microbe specificity. The broad or narrow host range might be merely a "frozen accident", rather than a selective factor itself.

II.C. PARASITIC ⇨ MUTUALISTIC ASSOCIATION

There have been many discussions about the differences between parasitic and mutualistic associations. Are mutualistic associations higher steps in the co-evolution? The difficulties in distinguishing these associations is evident from recent reviews in which rhizobial symbionts are indicated as examples of "prokaryotic plant parasites" (Long and Staskawicz 1993),"refined parasites of legumes" (Djordjevic *et al.*, 1987) or "sympathogenesis" (Spaink 1995). Definitions of parasitism or mutualism depend on our knowledge of the different types of interactions between the associated organisms. These interactions always consist of mixtures of synergistic and antagonistic effects (Quispel 1951). Actions, which are synergistic in one direction mostly, are antagonistic in the other way. Even then the comparisons appear to be complicated since many biochemically closely comparable interactions depend on signals which are rejecting the "parasite" in one type of association and inviting the "symbiont" in another one.

When comparing reactions between rhizobial bacteria and their host plants with those in incompatible combinations of phytopathogenic fungi or bacteria with host plants there appear to be remarkable analogies. These are especially interesting with respect to the induction of the hypersensitive reactions (HR), the synthesis of phytoalexins and their elicitors (Darvill and Albersheim 1984, Kuc 1995, Boller 1995). Comparisons between the NodD-activating or inactivating flavonoids with phytoalexins and the chitin-oligosaccharide Nod factors with elicitors in phytopathogenic infections were discussed above (section II.A.3.).

It is especially interesting that in Leguminosae the dominating phytoalexins are *iso*-flavonoids like glyceolin in soybeans and pisatin in peas. Infections by *Bradyrhizobium* induce only low amounts of glyceolin, but far higher amounts may be induced by certain non-infective mutants e.g. mutants in EPS synthesis (Parniske *et al.*, 1994). Comparable results were given for *R.meliloti* in alfalfa by Niehaus *et al.*, (1994). In *Trifolium pratense* absence of EPS synthesis triggered plant defense responses after release from the infection thread (Skorupska *et al.*, 1995). In peas a role of pisatin was postulated during senescence (van Iren *et al.*, 1983). A *fix⁻* mutant of *R.meliloti* induced far higher levels of chalcone synthase, a key enzyme in the phenylpropanoid pathway, and a more defence-like response (Grosskopf *et al.*, 1993). Vasse *et al.*, (1993) described a mutant of *R. meliloti* which after infection of alfalfa plants induced abortion of infection threads and necrotic reactions reminding of a HR. On the other hand a strain of *R. trifolii* with inhibited formation of infection threads did not show any indications of HR (de Boer *et al.*, 1995). Werner *et al.*, (1985) described an ineffective type of *Bradyrhizobium*-soybean interaction in which an early loss of the PBM was correlated with accumulation of glyceollin. This was explained by

assuming that the PBM protects the bacteroids against defense reactions by the host plant (Bassarab and Werner 1987). Mellor and Collinge (1995) proposed a model based on the elicitor-like activities of Nod factors e.g. the induction of a plant chitolytic enzyme which "down" regulates further Nod factor synthesis. Many aspects of nodule development could be described from the interactions of inducing and repressing reactions inside the symbiosomes up till the final plant defense reactions leading to their senescence.

A general conclusion can be that the difference between parasitic and mutualistic interactions is far from absolute. In the multifunctional network of possibilities, e.g. in the regulation of the synthesis of elicitors or the pathways in flavonoid biosynthesis, there are ample possibilities that small changes lead to important effects on the equilibrium between beneficial and antagonistic interactions. This opens continuous possibilities for adaptations in the parasitic as well as in the mutualistic direction.

II.D. FACULTATIVE ⇨ OBLIGATE DEPENDENCE ON THE ASSOCIATION

In studies on phytopathogenic organisms the terms facultative versus obligate have been rather popular in distinguishing between organisms which can be cultivated or are found as saprophytes in natural habitats and those which are only observed in or on their hosts and can not be isolated or cultivated. The latter case could be easily explained by the assumption that these organisms were so adapted to the environment inside the host that no growth under any other conditions was possible. This might be a consequence of highly specific adaptations during evolution. In practice it might simply mean that we were unable to prepare the right nutrient media. A more senseful definition of "obligate" is only possible if we could make sure that the needed constituents are, under natural conditions, only found in the endophytic situation and nowhere else.

If indeed the comparison between facultative and obligate parasites or symbionts might represent the extremes in a trend of evolution we might expect that facultative relations are found in superficial associations, while obligate symbionts will be found in highly adapted endosymbioses. This is not supported by the experimental data. In soil populations and rhizosphere associations it has been claimed that only 1-10% of the microorganisms can be cultivated on the usual laboratory nutrient media

(Campbell and Greaves 1990). On the other hand highly evoluated endosymbionts like most (Brady)Rhizobium strains can be isolated from the root nodules and cultivated without any problems.

Problems in isolating endosymbionts do not necessarily mean that their growth requirements have become very complicated. Another explanation is that their vitality has been impaired as a consequence of host defense-activities. It has been often observed that so-called obligate parasites can be easily cultivated on simple nutrient media once their primary isolation has succeeded. The endophytes from actinorhizal root nodules now can be easily isolated and cultivated, but some strains withstand isolation and cultivation (Lechevalier and Lechevalier 1990). One strain from *Alnus glutinosa* root nodules could be only isolated from the endophytic hyphal clusters after addition of a plant factor, which could be identified as the triterpenoid dipterocarpol. For further cultivation this factor can be omitted, though fatty acids still have a stimulating effect (Quispel *et al.*, 1992). Such effects, and the general growth-promoting effects of fatty acids (e.g. Selim *et al.*, 1996), indicate a role in the establishment of membrane integrity as essential structures for endosymbiotic systems.

The question, whether endophytic organisms may be so adapted to the endophytic way of life that they can no longer be cultivated on the usual nutrient media, has led to some controversies with regard to the viability of the bacteroid state of *(Brady)Rhizobium* in root nodules. In a classical paper from 1933 Almon showed by single-cell isolates of bacteroids from peas, alfalfa and red clover that they rarely or not multiplied on nutrient media, on which the original bacteria had been easily cultivated (Bergersen 1974).

These results could be critisized since the bacteroids might have been damaged by the crushing of the nodules during the isolation. A more careful procedure of isolation from macerated protoplasts yielded successful cultures from 80-90% of the isolated bacteroids in subterraneum clover and 60-70% in red clover (Greshoff and Rolfe 1978). Comparisons of total and viable counts from soybean nodule bacteroids led MacDermott *et al.*, (1987) to the conclusion that the viability of the bacteroids was highly variable, dependent on the bacterial strains and the age of the bacteroids. Such results must warn us not to confuse a non-viability of endophytic forms of a symbiotic organism with their obligatory dependence on host plants. The absolute obligate dependence, based on the genetic impossibility to

multiply elsewhere, can never be expected in parasitic or symbiotic organisms, which need a free-living period in their life cyles from where new infections are initiated. Only in those endosymbioses where all developmental phases are endophytic (or endozoic) is an obligate host-dependence conceivable. Difficulties in isolation of micro-symbionts can be explained by highly adapted nutritional requirements or by reduced viability in the endophytic state. This must make us cautious to make conclusions about possible evolutionary trends, though both aspects certainly have important evolutionary consequences.

III. Evolution by Symbiosis: Symbiosomes and Organelles.

In the preceding section the evolution of the symbiotic association could be understood from the point of view that the micro-symbionts invade the host in a way which is comparable with the infection of a parasite. Evolution in such an infective system will be possible for instance if mutations in the infective micro-organism yield better infectious possibilities. At the other hand the selective advantages of the host depend on its success to defend itself against its invader. New situations arise when the host may benefit from a new nutritional contribution by the pathogen. In mycorrhizae this may be the supply of nutrients taken up by the fungus, in symbioses with photo- or chemo-autotrophic microsymbionts the provision with organic assimilation-products and in symbioses with diazotrophic bacteria the use of the products of atmospheric dinitrogen fixation. As soon as an infecting micro-organism has some nutritional value for the host, there is the paradoxical situation that the host plant not only has to defend itself against the invader, but as well benefits from its protection or even its exploitation. Certainly every stimulation of growth of the host plant may increase the provision of nutrients for endophytic endosymbionts. Jimenez and Casadesus (1989) made an interesting "altruistic model" in which the dinitrogen-fixing bacteroids finally loose their vitality, but provide the possibilities for better growth of other non-differentiated bacteria (see also Olivieri and Frank 1994). The increased dependence of both symbiotic partners on each other leads to a new situation. While initially selection operated on the levels of the individual partners, now selection starts to operate on

Aspects	Cultivated Bacteria	Bacteroids
Forms	Small Rods	Swollen, Branched
Cell Walls	LPS and Proteins	Less LPS with loose structure [a] New epitopes in outer membrane [b]
Inner Membrane	(Phospho)Lipids and (Transport)Proteins	Phospholipids and Proteins changed transport functions [c]
C-Metabolism	Entner-Doudoroff (main pathway) Krebs cycle active	E.D.enzymes present [d] (no glucose used by cells)[e] high activity of dicarboxylic acid enzymes [f]
Respiration	Cytochromes, Cyt. Oxidase Membrane Energetization	id. Spec. Cytochromes [g] (adapted to low oxygen content) Membrane Energetization coupled to nitrogen fixation [h]
N-Metabolism	No Nitrogenase activity GS-GOGAT activity	High Nitrogenase activity[i] No GS-GOGAT activity [k]
Transport Systems	Active glucose transport Active transport of dicarboxylic acids Active uptake ammonium	No active glucose transport [l] Very active transport of dicarboxylic acids[m] No active uptake ammonium [n] Passive efflux ammonium [o]

Table 1 Some general comparisons between cultivated bacteria and bacteroids
[a]van Brussel *et al.*, 1977, Brewin *et al.*, this volume; [b]VandenBosch *et al.*, 1989; [c]Glazebrook *et al.*, 1993, Roest *et al.*, 1995; [d]Stowers 1985, Day and Copeland 1991, Kahn this volume; [e]de Vries *et al.*, 1982; [f]Bergersen and Turner 1967, Finan *et al.*, 1991, MacKay *et al.*, 1988; [g]Appleby, 1974, 1984; [h]Laane *et al.*, 1980; [i]Long 1989, Boistard and Batut this volume; [k]Brown and Dilworth 1975, Planque *et al.*, 1978; [l]de Vries *et al.*, 1982, Udvardi *et al.*, 1990; [m]Quispel *et al.*, 1984, Ronson 1988; [n]O'Hara *et al.*, 1985; [o]Laane *et al.*, 1980.

the efficiency of the combined system. Long ago the lichenologist Tobler (1925) stated with respect to the lichen symbiosis: "Die duale Natur ist heute nicht mehr das Wesen der Flechten, ihr Wesen ist die neue Einheit" ("the dual nature no longer is the essence of lichens, their essence is the new unity"). This may be applied to all higher evoluated symbiotic associations. In rhizobial root nodule symbioses this new unity is most manifest in the structure and function of the symbiosome: the dinitrogen-fixing bacteroid in the micro-environment of the peribacteroid space (PBS) surrounded by the peribacteroid membrane (PBM). The symbiosome is the result of the evolution of the symbiosis, but its efficient dinitrogen-fixation now enables new directions in evolution by the symbiosis.

III.A. CHARACTERISTICS AND FUNCTION OF THE SYMBIOSOME

Some of the most important aspects of the dinitrogen-fixing bacteroids as compared to free-living and cultivated bacteria are summarized in

Table 1. All these aspects are extensively discussed in other chapters *e.g.* Chapter 22,23 and 24, so that here, with reference to these chapters, the discussion can be restricted to a short summary of those aspects which are essential for our understanding of the mutual adaptations and the evolutionary significance as centres of symbiotic dinitrogen-fixation.

Comparisons like those of Table 1 can be no more than superficial approximations. The characteristics of cultivated bacteria of course depend on the nutrient and other environmental conditions. It is well-known that under "stress" conditions bacteroid-like forms can be formed. Particularly interesting were the observations that in the broad host range strain 32H1 low oxygen concentration induce bacteroid-like cells in which nitrogenase activity could be demonstrated (Pankhurst *et al.*, 1978, van Brussel *et al.*, 1979). Yet we must be careful in calling such cells "artificial bacteroids" and therewith suggest an identity. Data on isolated bacteroids may be averages since nodules contain bacteroids in different phases of development (Vasse *et al.*, 1990). Moreover they may be damaged during the isolation from the root nodule tissue and further fractionation.

While many aspects of metabolism are alike in free bacteria and bacteroids, important differences are found. Since there are few reasons to doubt the genetic identity of cultivated bacteria and their symbiotic bacteroids, these differences must be explained from regulatory effects of the micro-environment in the PBS. These regulatory aspects are especially important with regard to the reaction which is involved in the symbiotic introduction to the host plant of the new metabolic capacity: dinitrogen-fixation.

III.A.1. Dinitrogen-Fixation.

In all diazotrophic bacteria dinitrogen fixation depends on the induction of the structural genes *nif* D, K and H for the two components nitrogenase and nitrogenase dehydrogenase and the function of different regulatory and supporting *nif* and *fix* genes (Long 1989, Merrick 1993, Boistard and Batut this volume). The genes *nif*H, D and K are highly conserved in taxonomically unrelated organisms and must have a common and old evolutionary origin. The same may be concluded for regulatory systems like the two-component regulatory system NifL-NifA and the global N-regulatory system NtrB-NtrC. The most important external factors for the regulation of the transcription of the *nif* genes are the nitrogen status and the oxygen concentration. In rhizobia the regulation by the oxygen concentration is mediated by another two-component sensor-regulator system FixL-FixJ. Moreover O_2 has a post-transcriptional inhibiting effect on NifA (Fischer and Hennecke 1987).

Though much is known about the complicated regulatory cascades (de Bruyn 1990) the reason why *Rhizobium* only is able to fix dinitrogen in the bacteroid state is not yet fully understood though without any doubt the regulation of oxygen is a dominant factor. All aerobic diazotrophic bacteria had to find a solution for the problem of protecting nitrogenase synthesis and activity against oxygen, while this same oxygen was needed for the production of energy for the nitrogenase reaction. During their evolution different solutions have been found, either of a biochemical or a structural nature. In all symbiotic associations with diazotrophic bacteria the plants stimulate their nitrogenase activity, e.g. by reduction of oxygen supply. Most diazotrophic rhizosphere bacteria, like *Azospirillum*, and the symbiotic Cyanobacteria, *Frankia* and the

stem nodule bacterium *Azorhizobium* can fix dinitrogen under free-living, though preferably N-limited, conditions. *Bradyrhizobium* species may fix dinitrogen in special media under micro-aerobic culture conditions, but are unable to use the fixed N for their protein synthesis. In *Rhizobium* species all signs of dinitrogen fixation in cultures are absent or at best negligible, while the benefit of dinitrogen fixation for the plant is obvious after the formation of the bacteroids inside the symbiosomes. We do not know how these remarkable differences between different symbiotic bacteria arose during evolution and did have selective importance. Were rhizobia from the beginning bacteria with *nif* genes, which remained repressed until a symbiosis with host plants was realized? Or were they initially quite normal diazotrophic bacteria that restricted this capacity to the endosymbiotic situation in the course of evolution? Did the functions for nodulation and dinitrogen fixation arise by the uptake of *sym* plasmids? What was the original situation in evolution: the localization of *sym* genes on plasmids as in *Rhizobium* species or on the bacterial chromosomes as in *Bradyrhizobium*? Did the *sym* plasmid evoluate from plasmids with genes for infective functions, like those of the related agrobacteria? From an evolutionary point of view it can be stated beforehand that it is difficult to understand why restriction of the capacity for dinitrogen fixation to an endosymbiotic situation could have been of any selective advantage. Might the repression of dinitrogen-fixation in the soil and rhizosphere have been a stimulant to infect N-containing roots under N-limiting conditions? Or are such regulatory differences between *Rhizobium* and *Azorhizobium* merely incidental correlations with non-selective mutations, examples of "frozen accidents"?

III.A.2. Ammonium Assimilation.

In all diazotrophic organisms the first product of the nitrogenase reaction is ammonium. In free-living bacteria, like *Klebsiella*, this ammonium is assimilated through the GS (glutamine synthetase)-GOGAT (glutamate synthase) cycle. The regulation is dependent on the general N-status in the environment by the Ntr system on the level of transcription and by the post-translational adenylylation-deadenylylation cascade. All *Rhizobiaceae* are characterized by the presence of at least two types of GS (Darrow *et al.*, 1981). GS 1 belongs to the normal bacterial type, GS II is immunologically related to the plant types of GS (de Vries *et al.*, 1983). This relationship and the remarkable occurrence in bacteria, which have close interactions with plants, like *Agrobacterium* and *Frankia* (Edmons *et al.*, 1987, Tsai and Benson 1989) has led to the suggestion of a horizontal gene transfer during evolution from plant to micro-symbiont. However, Cullimore and Miflin (1983) did not find any homology with the cDNA of GS from *Phaseolus*.

In all endosymbiontic diazotrophs the activity of these enzymes is reduced concomitant with the increase of nitrogenase. This is in contrast with the situation in all free-living diazotrophic bacteria, but is as well observed in those cases where *Bradyrhizobium* species were induced for nitrogenase activity under microaerobic conditions on special media e.g. with glutamate (O'Gara and Shanmugam 1976). The regulatory system for the two genes: *gln A* for GS1 and *glnII* for GSII is connected with the general Ntr regulatory system (Amar *et al.*, 1994). However, it is doubtful whether the Ntr system is active in the bacteroids. The GS1 may be still present in the bacteroids (Moreno *et al.*, 1991), though no longer in an active state. The oxygen concentration regulates the gene *glnII* (Adams and Chelm 1988).

The micro-environmental factors that induce the activity of the nitrogenase in the bacteroids thus lead to a reduction and even annihilation of the activity of the enzymes for the assimilation of the fixation product ammonium. This means that the bacteria, though induced to reduce dinitrogen to ammonia, are not able to use themselves this fixation product for their nitrogen nutrition.

Cultivated bacteria of *(Brady)Rhizobium* possess an active ammonium permease which is normally active in the uptake of ammonium from the environment. This has been demonstrated by different authors with methylammonium as analogue and by using a selective NH_4^+ electrode (O'Hara *et al.*, 1985). In the bacteroids this system could not be demonstrated. There is instead a passive export of ammonium according to a concentration and pH gradient which is produced by membrane energetization (Laane *et al.*, 1980).

III.A.3. The Two-Component Regulatory Systems.

As mentioned in Table 1 there is a general consensus that glucose (and other carbohydrates) are bad substrates for respiration of bacteroids, while there is a dominating role for dicarboxylic acids like succinic and malic acid. This is mainly explained from differences in the active transport systems in the bacteroid cytoplasmic membranes (and the PBM which will be discussed later). While mutants in glucose transport may form normally effective nodules, mutants in the dicarboxylic acid transport system are still infective, but are not effective (Glenn and Brewin 1981, Ronson *et al.*, 1981). The supply of dicarboxylic acids, like malic acid and succinic acid, from the host cells is mainly responsable for the induction of their transport system and the repression of glucose transport (de Vries *et al.*, 1982).

The genes for the synthesis and the regulation of the dicarboxylate transport system have been extensively studied, first in enteric bacteria, later in *Rhizobiaceae* as well (Ronson 1988). The original concept explains the transport by the role of a specific transport protein encoded by a gene *dctA*, which is regulated by two genes *dctB* and *dctD*. The gene products DctB and DctD form again an example of a two-component sensor-regulator system in which DctB is the sensor for the inducing dicarboxylic acids, while DctD is the actual regulator of *dctA*. They bind, just like the two-component systems NtrB-NtrC, NifL-NifA and FixL-FixJ, at operons with the required consensus sequence for the σ^{54} factor. Mutations, which change the sequence in the N-terminal regions of DctD, either prevent the activation of *dctA* or make this gene constitutive. In strains with *dctA-lacZ* fusions in wild-type and mutant backgrounds it was shown that in their bacteroids *dctA* was expressed in *dctB* and *dctD* mutants, though these mutants did not express *dctA* in the free-living state. This led to the conclusion that an alternative regulatory system might operate in bacteroids. The structural similarity of DctD and NifA suggested that NifA might be the regulator, not only of the *nif* genes, but of *dctA* as well. This hypothesis is the more likely since in enteric bacteria an interference with NtrA has been found. The details of these regulatory systems and their interactions are still but partially understood. Interpretations are difficult because the related structures of their regulatory proteins lead to the theoretical possibility of a "cross-talk" between related regulator proteins (Birkenhead *et al.*, 1990, Alloway *et al.*, 1995, Labes *et al.*, 1993, Labes and Finan 1993). Direct evidence for such "cross-talks" has been obtained for the two-component regulatory systems NodV-NodW and NwsA-NwsB (Grob *et al.*, 1994).

It is tempting to suggest that the phylogenetic relations between these two-component regulatory systems, with closely related domains (Ronson 1988), have played a decisive role in the evolution of the complicated host-microbe interactions leading to the functional symbiosome. The interactions between the related regulatory systems and their possibility for cross-talks leads to the conclusion that "a fine-tuned regulation of gene expression in nodules may represent an evolutionary optimization of symbiosis" (Labes and Finan 1993). Initially *Rhizobium* and *Bradyrhizobium* lived saprophytically in the soil or in the rhizosphere. Here the Ntr system directed the regulation of the energy-consuming dinitrogen-fixation in relation to the availability of bound forms of nitrogen. Already in the rhizosphere adaptations, which enabled a better uptake of dicarboxylic acids, may have been selected. After the evolution of the potentialities for an endophytic way of life, which asked for regulatory systems to a micro-aerobic environment. Further adaptive steps had to be selected. It must have been of utmost importance that such adaptations could be obtained by relatively small mutations in already existing regulatory systems for other biochemical functions.

The adaptations to a micro-aerobic environment were as well directed towards the respiratory e-transport chain. Alternative cytochromes are induced in bacteroids as an adaptation to the lower oxygen content. Reducing the aeration in cultures affects the synthesis of the key-enzyme for heme synthesis *hemA*, but this effect is not observed with mutants in the regulatory genes *fixL-fixJ* (Page and Guérinet 1995). This is again an example of the same two-component regulatory system involved in different processes with functional connections.

III.A.4. The Micro-Environment of the Bacteroids in the Symbiosome.

The direct micro-environment of the bacteroids is the PBS surrounded by the PBM. Bassarab and Werner (1987) concluded from their studies in soybean nodules that the PBM has two main functions: 1. Protection of the bacteroids against defense reactions from the host cell and 2. Communications between the plant cells and their microsymbionts. The first function was already discussed above (sections III.C. and II.D.). The presence of different lytic enzymes, as studied by Mellor (1989), is important for the understanding of the special characteristics of the bacteroids since in studying the difference between bacteroids and free-

living bacteria we must acknowledge the possible effects from lytic enzymes *e.g.* on the bacteroid membranes.

For the communications between bacteroids and host cells two PBM enzymes are especially important: the H^+ ATPase and the already discussed transport system for dicarboxylic acids (Blumwald *et al.*, 1985, Day and Udvardi 1992). The H^+ATPase not only functions as a proton pump, lowering the internal pH of the PBS, but may activate the dicarboxylic acid and other transport systems. The ammonium, which is secreted from the symbiosomes, will affect the pH of the surrounding plant cells. This may lead to an increased production of plant dicarboxylic acids, like malic acid, since this is a well-known phenomenon in plant cells (LaRue *et al.*, 1983).

Besides the effects on metabolic activities in the understanding of the mutual adaptations of both partners, important effects at the transcriptional level play a role. The synthesis of many plant proteins is increased. This may be explained by a type of substrate induction like the increase of GS by the produced ammonium (Sengupta-Gopal *et al.*, 1986, Hirel *et al.*, 1987). Other proteins, like the leghemoglobins and the PBM-specific protein nodulin 26, are supposed to be nodule-specific. Since oxygen appears to be the most important regulatory factor for nitrogenase activity in the symbiosomes, the induction of leghemoglobins has obtained most attention (Appleby 1974, 1984). The theory that apo-leghemoglobin is formed by the plant and the heme by the bacteroids is no longer plausible because the induction of plant heme synthesis in the root nodule plant cells appears to be sufficient for the leghemoglobin formation (O'Brian 1996).

III.A.5. Comparisons with Other Symbiotic Associations.

The general conclusion of the preceding discussions must be that the functional "symbiosome" of the effective legume-(*Brady*)*Rhizobium* symbiosis is the result of a highly complex series of mutual interactions between plants and bacteria based on many structural and regulatory genes. Did these genes already have a function in the non-symbiotic situation? Or did they arise during evolution of the association? A possible answer may once be obtained if we not only possess more knowledge about the symbiotic interactions in the legume-(*Brady*)*Rhizobium* symbioses, but if we can compare them with related symbioses, since such comparisons might tell us whether they belong to general aspects and consequences of plant-microbe interactions.

This certainly is the case with regard to the development of a peribacteroid membrane (PBM). In their fascinating review Smith and Smith (1990) demonstrated that in most if not all mutualistic symbiotic systems the endophytes are surrounded by a cytoplasmic membrane of predominantly host origin. All such membranes possess a $H^+ATPase$, which is absent in membranes around endoparasites. This enables the operation of bi-directional transport systems, which determine the composition of the micro-environment around the endophytes at one side and the provision of nutrients to the host at the other side. Induction of genes by infecting micro-symbionts is not restricted to the nodulins. In analogy with the nodulins of leguminous root nodules the induced proteins in mycorrhizae have been called 'mycorrhizins' (Wyss et al., 1990). A mycorrhizin in soybeans is homologous with the PBM-incorporated nodulin 26. Further studies must show how far other mycorrhizins and nodulins are comparable. In any case they show that the effects of infecting microsymbionts on gene regulation of host plants is a general phenomenon. The induction of leghemoglobins too is not restricted to leguminosae. The presence of leghemoglobin in the root nodules of Parasponia (Appleby et al., 1983), demonstrates that Rhizobium may induce leghemoglobin in a plant belonging to a quite different family (Ulmaceae). In actinorhizal root nodules leghemoglobins were first detected in Casuarina and Myrica, but after application of detergents during the homogenisation procedure their presence could be demonstrated in other species, like Alnus glutinosa, as well (Suharjo and Tjepkema 1995). The higher amount of leghemoglobins in Casuarina and Myrica might be related to the absence of real vesicles whose thick cell walls protect the nitrogenase against oygen diffusion. In nodules with typical vesicles a regulation of oxygen by leghemoglobins therefore will be of less importance. It is doubtful whether in plants like Alnus glutinosa the small amount of leghemoglobins really plays a role in the regulation of oxygen. The original suggestions that the occurrence of leghemoglobins outside the leguminosae could be explained by horizontal gene-transfer in evolution is therefore less likely. The occurrence of leghemoglobins even in cereals (Tjepkema and Asa 1987) makes it evident that leghemoglobins are of far more general occurrence than originally expected and may have other functions than oxygen transport.

The most important aspect of the rhizobial symbiosome is the induction of the nitrogenase activity in correlation with the repression and inactivation of the ammonium assimilation. Comparisons with other symbioses show this to be an example of a general aspect of endosymbioses: the induction or stimulation of a primary assimilatory reaction and the repression or inactivation of later steps. In all symbioses with diazotrophic bacteria, Cyanobacteria or Frankia, the nitrogenase in the endophytic cells is stimulated, while the assimilation by GS and GOGAT is repressed so that all ammonium is transported to the surrounding host cells (Bergman et al., 1992, Silvester et al., 1996, Huss-Danell, 1990). In symbioses where photosynthetic Cyanobacteria or eukaryotic algae are endosymbionts in fungi or animals there are some indications that the symbiotic situation stimulates the photosynthetic activity. In lichens the algae of the species Trebouxia hardly are able to live autotrophically after isolation in pure cultures, though their photosynthesis in the lichen thallus is considerable (Quispel 1959). Excretion of the first products of photosynthesis to the symbiotic partner has been amply demonstrated in lichens (Richardson 1968, Galun and Bubrick (1984), Hydra (Mews 1980) and marine corals (Muscatine et al., 1972). Like in the diazotrophic symbioses this excretion is stimulated by a low pH produced by a $H^+ATPase$ in the surrounding perisymbiont membranes, which act as a proton-pump towards the micro-environment of the endosymbionts.

Notwithstanding the specific differences and adaptations the general aspects of metabolic and transport regulation in endophytic and endozooic symbioses appear to be based on general principles, comparable regulatory systems and transport systems in the participating membranes. Further evolutionary adaptations of course will play a role, but these are only later modifications within these general patterns.

III.B. FROM SYMBIOSOME TO ORGANELLE

The term "symbiosome", was first introduced by Roth et al., (1988) as "a membrane-bound compartment containing one or more symbionts and certain metabolic components and located in the cytoplasm of eukaryote cells". The word symbiosome implies a comparison to organelles. Just like in mitochondria originally endosymbiotic bacteria introduced the metabolic capacities of respiration and in chloroplasts originally endosymbiotic cyanobacteria introduced the capacities of photosynthesis, the endosymbiotic diazotrophic rhizobia introduce the capacities of dinitrogen fixation. Are we allowed to regard symbiosomes as intermediate steps on the symbiogenetic way to real organelles?

III.B.1. Symbiogenetic Organelles

The long debated idea that chloroplasts and mitochondria originally were endosymbiotic microorganisms (Sapp 1994) is now established without any reasonable doubt. The similarities between chloroplasts and mitochondria with (cyano)bacteria were already extensively reviewed by Gray and Doolittle in 1979. According to the theory of symbiogenesis the original prokaryotic bacteria were taken up by eukaryotic cells in a process of endocytosis in such a way that the surrounding outer membrane was derived from the eukaryotic cytoplasmic membrane (Taylor 1974) and this was a PSM. The inner membrane then should be derived from the original cytoplasmic membrane of the symbiotic bacteria, entirely comparable with the situation in symbiosomes. All doubts about the symbiogenetic origin of chloroplasts and mitochondria have been removed by studies on their molecular evolution (Bonen and Doolittle 1976, Giovannoni et al., 1988, Gray 1985, Sankoff et al., 1992). Comparisons between amino-sequences of common proteins like cytochrome c, base sequences in 5S, 16S and 28S rRNA and some tRNA's in different organisms, in chloroplasts, mitochondria, the nuclear encoded products of their "host"cells and representatives of different groups of prokaryotes, showed that chloroplasts (in these respects) belong to the Cyanobacteriaea, while mitochondria appear to belong to the group of the α-purple bacteria, and both are not related to their host cells.

It is extremely fascinating that the putative original endosymbionts of chloroplasts and mitochondria belonged to those groups of prokaryotes, whose representatives are found today as endophytes of plants: the Cyanobacteriae in liverworts, Azolla, Cycads and Gunnera, the Rhizobiaceae as heterotrophic representatives of the α purple bacteria in leguminous root nodules. However, we must realize that their origins as symbionts belong to very different periods of evolution. There are no direct indications for the first establishment of rhizobial infections, but it is conceivable that this has

happened in a precursor of the Leguminosae in the same period that the first mycorrhizal infections developed, around the appearance of the first land plants. The infection of primitive eukaryotic cells by the ancestors of the mitochondria must have occurred before the division of animals and plants and most probably will have been the direct cause of the splendid evolution of the eukaryotes.The infection by the cyanobacterial ancestors of the plastids must have taken place somewhat later at the diversification of plants from animals (Gray 1985, Giovannoni et al., 1988).

III.B.2. Gene Organization in Symbiosomes and Organelles.

Though the molecular evidence, based on comparisons of sequences in proteins and RNA's, indicates the evolutionary relationships between extant cell organelles and endosymbionts, the differences in gene arrangements and organization between organelles and free-living bacteria are considerable. For many chloroplasts complete DNA sequences have been published (Palmer 1985). Though there are important specific differences there is a general tendency that most genes for RNA synthesis are present as well as some genes for the synthesis of enzymatic proteins. Other original chloroplast genes appear to have been transferred to the nuclear genome of the plant cell. In the mitochondrial genomes the same phenomenon is observed, but still more complicated. The arrangement of the genes in the mitochondrial genomes is highly different in different groups of organisms. Extremes are the highly efficient gene organization in the mitochondria of animals, including humans (Attardi 1985), and the complex mixture of circular and linear types of DNA in higher plants (Leaves and Gray 1982). Still the about 80 identified plant mitochondrial genes, together with the originally mitochondrial genes, that during their evolution have been transferred to the nucleus or to the chloroplasts, are functionally operative thanks to a complex system of regulation at the transcriptional, translational and post-translational level (Mullet 1988). Moreover, the situation in mitochondrial genomes is complicated by some deviations from the universal genetic code and special types of RNA editing.

The remarkable situations in plant mitochondria form an extreme example of an endosymbiotic situation in which the endosymbiont as such no longer is important, if only the essential genes are present. Selection pressure is operative on the genes because of their function, not on their organization provided that a regulatory system is operative to coordinate the different gene activities. Chloroplast or mitochondrial genes could be transferred to the nucleus or to the other organelle as promiscuous genes (Timmis and Scott 1984, Schuster and Brennecke 1987, Palmer 1990, Breiman and Galun 1990). This is the ultimate result of endosymbiosis, where evolutionary selection no longer affects the interests of the individual symbionts but of the symbiotic complex as a whole. There is no explanation for the great difference between plant and human mitochondria. Is it caused by different types of host cells with different evolutionary adaptations, or merely the result of "frozen accidents"?

Such conclusions and questions lead us back to the problem as formulated in the beginning: are we right in considering the rhizobial symbiosome as an intermediate step in the evolution to organelles? Up till now there are no observations which indicate any transfer of rhizobial genes to the plant nuclei, though such transfers are well-known in the related genus *Agrobacterium*. On the other side there are some indications that there did occur some transfer of plant DNA to the bacteria. The homology of bacterial GSII with plant GS in *Rhizobium*, *Agrobacterium* and *Frankia* was already discussed (section III.A.2.). There are some fascinating indications of DNA insertions of eukaryotic character in the plasmids of *Rhizobium* (Yeoman 1996) and *Agrobacterium* (Magrelli *et al.*, 1994).

If bacteroids might ever have lost some DNA, this remains restricted to non-viable cells and thus can not develop into a genetic defect. In the symbiosomes the bacteroids developed from normal bacteria, which entered the plant roots from the rhizosphere. Periods of saprophytic growth therefore are an essential part of the rhizobial life-cycle. These saprophytic periods in the soil or the rhizosphere ask for the activity of many essential genes. Bacterial cells, which lost some of these essential genes, will die unless they are immediately taken up by host cells. This is impossible when a complicated infection process, in which many bacterial cell multiplications play a role, is needed before an endocytotic stage is obtained. Evolution to real organelles can only be expected if endophytic rhizobial cells could enter reproductive cells of their hosts and take part in the normal reproductive cycle of their host plants.

The original symbiogenesis of the chloroplasts and mitochondria started very early in evolution, most probably when the eukaryotic host cells still were unicellular so that all cell divisions could transfer their endosymbionts to new host cells. The rhizobial symbioses started far later in already multicellular hosts and always depended on new infections. They therefore are not (or not yet?) intermediary steps on the way to real organelles, but different examples of symbiogenetic evolution.

IV. General Conclusions and Final Remarks

The aim of this chapter was to study the special place of the rhizobial symbioses in the great variety of plant-microbe interactions. This variety consists of surface interactions, like the rhizosphere, entrance in pre-existing cavities, like in many cyanobacterial symbioses, infections from surface cells up till the endosymbiotic forms. They varied from adaptations selected for the benefit of the individual partners to those highly evoluated forms in which selection was directed to the new unity of the symbiotic combination. The situation in the rhizobial symbiosomes is complicated because the demands of the endosymbiotic system had to be combined with the demands of the saprophytic and infectious phases. The situation is unique, in relation to all other symbioses with diazotrophic bacteria, because the induction of nitrogenase as the most important selective aspect for the symbiotic evolution only is manifest at a late step in the development of the symbiotic interactions. Its evolutionary success therefore could only affect the earlier steps in a kind of "feed-back selection".

When trying to draw conclusions about the evolution of symbiotic systems the first question must be whether the evolution of the symbionts is comparable to the evolution of free-living relatives. The special aspects of the endophytic way of life, the population structure of the endophytic populations and the special characteristics of the endophytic environment with relatively ineffective selective constraints may lead to a faster evolution as described for the insect endosymbionts of the genus *Buchnera* (Moran 1996). The indications of a transfer of genes between endosymbiont and host, and the possibilities for gene transfer when plasmids are involved, like in *Rhizobium* and *Agrobacterium*, can be important aspects for our understanding of symbiotic evolution.

The success of the evolution of such highly complicated symbiotic systems, in which a multitude of interactions is involved, was possible because of relatively small adaptations in the structure and activities of some already existing signal factors of normal metabolism and some multi-functional and flexible regulatory systems. For the bacterial side these are e.g. the regulatory proteins of the Lys R family,of the two-component systems and the role of different surface polysaccharides, at the plant side the receptors to these substances, of lectins, components in the plant hormone regulatory systems, the production of flavonoids and other signal substances. At both sides the functions of adaptive membranes may be the most important aspects of symbiotic interactions (Assman 1995). Comparative studies on related symbioses and pathogenic interactions certainly will give inspiration for the discovery of more signals and regulatory aspects and their possible roles in evolution. The results of the studies of other comparable symbioses, like the actinorhizae and the mycorrhizae, still lags behind the progress in rhizobial symbioses, and progress in these fields thus will profit from such comparisons. On the other side results in other symbioses might be stimulating for comparitive studies of rhizobial symbioses e.g. the recent studies on the role of small, cysteine-rich proteins like elicitins and hydrophobins in pathogenic and symbiotic fungi (Templeton *et al.*, 1994, Wessels 1996).

Not all specific aspects of the individual plant-microbe interactions need to be considered as evolutionary adaptations. In many cases we have to conclude that they could be considered as "frozen accidents". The quite different function attributed to comparable signal molecules and their regulations and the many examples of adaptations in systems, which are already active for other functions in normal metabolism of the individual partners, indicate that even in highly evoluated symbioses evolution did not so much depend on real inventions. Most selective adaptations appear to be excellent illustrations of the principle of "evolutionary tinkering" as postulated by Jacob (1982).

V. References

Adams, T.H. and Chelm, B.K. (1988) J. Gen. Microbiol. 134, 611-618.

Alloway, D., Voesten, B. and O'Gara (1995) FEMS Microbiol. Lett. 128, 241-245.

Amar, M., Patriarca, E.J., Manco, G. and Bernard, P. (1994) Mol. Microbiol. 11, 685-693.

Appleby, C.A. (1974) in The biology of nitrogen fixation (Quispel, A. ed.), pp. 521-554, North Holland/Amer.Elsevier, Amsterdam/New York.

Appleby, C.A. (1984) Annu. Rev. Plant Physiol. 35, 443-478.

Appleby, C.A., Tjepkema, J.D. and Trinick, M.D. (1983) Science 220, 951-953.

Ardourel, M., Demont, N., Debellé, F., Maillot, F., de Billy, F., Promé, J.C., Dénarié, J. and Truchet, G. (1994) The Plant Cell 6, 1357-1374.

Assmann, S.M. (1995) Proc. Natl. Acad. Sci. USA 92, 1795-1796.

Attardi, G. (1985) Int. Rev. Cytol. 93, 93-145.

Baker, D. and Miller, N.G. (1980) Can. J. Bot. 58, 1612-1620.

Baker, M.E. (1991) Can. J. Microbiol. 38, 541-547.

Baker, M.E. (1994) Steroids 59, 248-258.

Barbour, M., Hattermann, D.R. and Stacey, G. (1991) Appl. Environm. Microbiol. 57, 2635-2639.

Bashan, Y. and Levanony, H. (1991) Plant Soil 137, 175-179

Bauer,P.,Ratet,P.,Crespi,M.D.,Schultze,M.   and Kondorosi,A.(1996) The Plant J. 10, 91-105

Bassarab, S. and Werner, D. (1987) J. Plant Physiol. 130, 233-241.

Bergersen, F.J. (1974) in The biology of nitrogen fixation, (Quispel, A. ed.), pp. 473-498, North Holland,Amer.Elsevier, Amsterdam/New York.

Bergersen, F.J. and Turner, G.L. (1967) Biochem. Biophys. Acta 141, 507-515.

Bergman, B., Rai, A.N. and Johansson, C. (1993) Symbiosis 14, 61-81.

Berry, A.M. and Kleemann, G. (1993) in Nitrogen fixation with non-legumes; (Hegazi, N.A., Fayez, M. and Monib, M. eds.), pp. 191-199, Amer.Univ.Cairo Press, Cairo.

Bécard, G., Douds, D.D. and Pfeffer, P.E. (1992) Appl. Environm. Microbiol. 58, 821-825.

Birkenhead, K., Noonan, B., Reville, W.J., Boesten, B., Manian, S.S. and O'Gara.F. (1990) Mol. Plant-Microbe Interact. 3, 167-173.

Blumwald, E.B., Fortin, M.G., Phillips, A.R. and Poole, R.J. (1985) Plant Physiol. 78, 665-672.

Boller, T. (1995) Annu. Rev. Plant Physiol. Plant Mol. Biol. 46, 189-214.

Bonen, L. and Doolittle, W.F. (1976) Nature (London) 261, 669-673.

Breiman, A. and Galun, E. (1990) Plant Science 71, 3-19.

Brown, C.M. and Dilworth, M.J. (1979) J. gen. microbiol. 101, 51-56.

Burggraaf, A.J.P. and Valstar, J. (1984) Plant Soil 78, 29-43.

Campbell, R. and Greaves, M.P. (1990) in The rhizosphere, (Lynch, J.M. ed.), pp. 11-35, Wiley, Chichester.

Castle, A.J. and Day, A.W. (1984) Phytopathology 74, 1194-1984

Cullimore, J.V. and Miflin, B.J. (1983) FEBS Lett. 158, 107-112.

Cullings, K.W., Szaro, T.M. and Bruns, T.D. (1996) Nature 379, 63-66.

Dakora, F.D., Joseph, C.M. and Phillips, D.A. (1993) Plant Physiol. 101, 819-824.

Dakora, F.D., Joseph, C.M. and Phillips, D.A. (1993) Mol. Plant-Microbe Interact. 6, 665-668.

Dakora,F.D. and Phillips.D.A. (1996) Physiol.Mol.Plant Pathol. 49, 1-20

Darrow, R.A., Crist, D., Evans, W.R., Jones, B.L., Keister, D.L. and Knotts, R.R. (1981) in Current perspectives in nitrogen fixation (Gibson, A.H. and Newton, W.E. eds.), pp. 182-185, Canberra, Austral.Acad.Sci.

Darvill, A.G. and Albersheim, P. (1984) Annu. Rev. Plant Physiol. 35, 243-275.

Day, D.A. and Copeland, L. (1991) Plant Physiol. Biochem. 29, 185-201.

Day, D.A. and Udvardi, M.K. (1992) Symbiosis 14, 175-189.

de Boer, M.H. and Djordjevic.M.A. (1995) Protoplasma 185, 58-71.

de Bruijn, F.J., Hilgert, U., Stigter, J., Schneider, M., Meyer, H., Klosse, U. and Pawlowski, K. (1990) in Nitrogen fixation achievements and objectives,(Gresshoff, P.M., Stacey, G. and Newton, W.E. eds.), pp. 33-44, Chapman and Hall, New York.

de Vries, G.E., Oosterwijk, E. and Kijne, J.W. (1983) Plant Science Lett. 32, 333-341.

de Vries, G.E., van Brussel, A.A.N. and Quispel, A. (1982) J. Bacteriol. 149, 872-879.

de Weger, L.A., van der Bij, Dekkers, L.C., Simons, M., Wijffelman.C.A. and Lugtenberg, B.J.J. (1995) FEMS Microbiol. Lett. 17, 221-228.

Dénarié, J. and Cullimore, J. (1993) Cell 74, 951-954.

Dénarié, J., Debellé, Truchet, G. and Promé, J.C. (1993) in New horizons in nitrogen fixation (Palacios, R., Mora, J. and Newton, W.E. eds.), pp. 19-30, Kluwer Acad.Publ. Dordrecht.

Dharmatilake, A.J. and Bauer, W.D. (1992) Appl. Environm. Microbiol. 58, 1153-1158.

Diaz, C.L., Melchers, L.S., Hooykaas, P.J.J., Lugtenberg, B.J.J. and Kijne, J.W. (1989) Nature (London) 338, 579-581.

Diaz, C.L.,Spaink, H.P., Wijffelman, C.A. and Kijne, J.W.(1995) Mol.Plant-Microbe Interact. 8, 348-356

Djordjevic, M.A., Gabriel, D.W. and Rolfe, B.G. (1987) Annu. Rev. Phytopathology 25, 145-168.

Douglas, A.E. (1994) Symbiotic interactions, Oxford Science Publ.Oxford Univ.Press, Oxford.

Döbereiner, L. and Day, J.M. (1976) in Nitrogen fixation by free-living organisms (Newton, W.E. and Nyman, C.J. eds.), pp. 518-537, Washington State Univ.Press, Pullman,Wash.

Edmonds, J., Noridge, N.A. and Benson, D.R. (1987) Proc. Natl. Acad. Sci. USA 84, 6126-6130.

Ehrhardt, D.W., Atkinson, E.M. and Long, S.R. (1992) Science 256, 998-1000.

Elmerich, C., de Zamaroczy, M., Vieille, C., Delorme, F., Onyeocha, Y.Y. and Zimmer, W. (1991) in Nitrogen fixation (Polsinelli, M., Materassi, R. and Vincenzini, M. eds.), pp. 79-87, Kluwer Acad.Publ. Dordrecht.

Finan, T.M., McWhinnie, E., Driscoll, B. and Watson, R. (1991) Mol. Plant-Microbe Interact. 4, 386-392.

Fischer, H.M. and Hennecke, H. (1987) Mol. Gen. Genet. 209, 621-626.

Galun, M. and Bubrick, P. (1984) in Encyclopedia Plant Physiology NS 17, (Linskens, H.F. ed.), pp. 362-406, Springer, Berlin.

Geiger, O., Thomas-Oates, J.E., Glushka, J., Spaink, H.P. and Lugtenberg, B.J.J. (1994) J. Biol. Chem. 269, 11090-11097

Gianinazzi-Pearson,V. (1996) The Plant Cell 8, 1871-1883.

Giovannoni, S.J., Turner, S., Olsen, G.J., Barns, S., Lane, D.J. and Pace, N.R. (1988) J. Bacteriol. 170, 3584-3592.

Glazebrook, J., Ichige, A. and Walker, G.C. (1993) Genes Development 7, 1485-1497.

Glenn, A.R. and Brewin, N.J. (1981) J. Gen. Microbiol. 126, 237-241.

Goethals, K., van Montagu, M. and Holsters, M. (1992) Proc. Natl. Acad. Sci. USA 89, 1646-1650.

Gray, M.W. (1985) Trends in Genet. 5, 294-299.

Gray, M.W. and Doolittle, W.F. (1982) Microbiol. Rev. 46, 1-42.

Greshoff, P.M. and Rolfe, B.G. (1978) Planta 142, 329-332.

Grob, P., Hennecke, H. and Göttfert, M. (1994) FEMS Microbiol. Lett. 120, 349-354.

Grosskopf, E., Ha, D.T.C., Wingerder, R., Röhrig, H., Szecsi, J., Kondorosi, E., Schell, J. and Kondorosi, A. (1993) Mol. Plant-Microbe Interact. 6, 173-181.

Györgypal, Z. and Kondorosi, A. (1991) Mol. Gen. Genet. 226, 337-340.

Hirel, B., Bouet, C., King, B., Layzell, D., Jacobs, F. and Verma, D.P.S. (1987) EMBO J. 6, 11671171

Hurek, T., Reinhold-Hurek, B., van Montagu, M. and Kellenberger, E. (1991) in Nitrogen fixation (Polsinelli, M., Materassi, R. and Vincenzini, M. eds.), pp. 235-242, Kluwer Acad.Publ. Dordrecht.

Huss-Danell, K. (1990) in The biology of Frankia and actinorhizal plants (Schwintzer, C. and Tjepkema, J.D. eds.), pp. 129-156, Acad.Press, San Diego.

Jacob, F. (1982) The possible and the actual, Pantheon Books, New York.

Jimenez, J. and Casadesus, J. (1989) J. Heredity 80, 335-337.

Kape, R., Wex, K., Parniske, M., Görge, E., Wetzel, A. and Werner, D. (1993) J. Plant Physiol. 141, 54-60.

Kijne, J.W. (1975) Physiol. Plant Pathol. 5, 75-79.

Kijne, J.W. (1992) in Biological nitrogen fixation (Stacey, G., Burris, R.H. and Evans, H.J. eds.), pp. 349-399, Chapman and Hall, New York.

Kijne, J.W. and Planqué, K. (1979) Physiol. Plant Pathol. 14, 339-345.

Koes, R., Quattrocchio, F. and Mol, J.N.M. (1994) BioEssays 16, 123-132.

Kuc, J. (1995) Annu. Rev. Phytopathol. 33, 275-297.

Laane, C., Krone, W., Konings, W., Haaker, H. and Veeger, C. (1980) Eur. J. Biochem. 103, 39-46.

Labes, M. and Finan, T.M. (1993) J. Bacteriol. 175, 2674-2681.

Labes, M., Rastogi, V., Watson, R. and Finan, T.M. (1993) J. Bacteriol. 175, 2662-2673.

Laguerre, G., Bardin, M. and Amarger, N. (1993) Can. J. Microbiol. 39, 1142-1149.

LaRue, T.A., Peterson, J.B. and Tajima, S. (1989) in Advances in nitrogen fixation research (Veeger, C. and Newton, W.E. eds.), pp. 437-443, Nijhoff/Junk, The Hague/Wageningen.

Leaves, C.J. and Gray, N.W. (1982) Annu. Rev. Plant Physiol. 33, 373-402.

Lechevalier, M.P. and Lechevalier, H.A. (1990) in The biology of Frankia and actinorhizal plants. (Schwintzer, C.R. and Tjepkema, J.D. eds.), pp. 35-60, Acad.Press, San Diego.

Libbenga, K.R., van Iren, F., Bogers, R.J. and Schraag-Lamers, M.F. (1973) Planta 114, 29-39

Long, S.R. (1989) Annu. Rev. Genet. 23, 483-506.

Long, S.R. and Staskawicz, B.J. (1993) Cell 73, 921-935.

Lynch, J.M. (1990) The Rhizosphere, Wiley, Chichester.

MacDermott, T.R., Graham, P.H. and Brandwein, D.H. (1987) Arch. Microbiol. 148, 100-106.

MacKay, I.A., Dilworth, M.J. and Glenn, A.R. (1988) J. Gen. Microbiol. 134, 1433-1440.

Maillt, F., Debellé, J. and Dénarié, J. (1990) Mol. Microbiol. 4, 1975-1984.

Margulis, L. (1993) Symbiosis in cell evolution, Freeman, New York.

Martin, P.M., Hortwitz, K.B., Ryan, D.S. and NcGuire, W.L. (1978) Endocrinology 103, 1860-1867.

Mellor, R.B. (1989) J. exp. bot. 40, 831-839.

Mellor, R.B. and Collinge, D.B. (1995) J. exp. bot. 46, 1-18.

Merrick, M.J. (1993) in New horizons in nitrogen fixation, (Palacios, R., Mora, J. and Newton, W.E. eds.), pp. 43-54, Kluwer Acad.Publ. Dordrecht.

Mews, L.K. (1980) Proc. Roy. Soc. B 173, 397-413.

Michiels, K., Croes, C. and VanderLeyden, J. (1991) J. Gen. Microbiol. 137, 2241-2246.

Moran, N.A. (1996) Proc.Natl. Acad.Sci.USA 93, 2873-2878

Moreno, V.S., Meza, R., Guzman, J., Carabez, A. and Espin, G. (1991) Mol. Plant-Microbe. Int. 4, 619-622.

Morrison, N. and Verma, D.P.S. (1987) Plant Mol. Biol. 9, 185-196.

Mullet, J.E. (1988) Annu. Rev. Plant Physiol. 39, 475-502.

Muscatine, L., Pool, R.R. and Cernichiari, E. (1972) Marine Biol. 13, 298-308.

Niehaus, K., Baier, R., Kapp, D., Lagares, A., Meyer-Gatterman, P., Sieben, S. and Pühler, A. (1994) in Proc. First Eur.Nitrogen Fixation Conf FEMS, (Kiss, G.B. and Endre, G. eds.), pp. 205-209, Officina Press, Szeged Hungary.

O'Brian, M.R. (1996) J. Bacteriol. . 178, 2471-2478.

O'Gara, F. and Shanmugam, K.T. (1976) Biochem. Biophys. Acta 437, 313-332.

O'Hara, G.W., Riley, L.T., Glewnn, A.R. and Dilworth, M.J. (1985) J. Gen. Microbiol. 131, 757-764.

Ochman, H. and Groisman, E.A. (1995) Can. J. Microbiol. 41, 555-561.

Olivieri, I. and Frank, S.A. (1994) J. Heredity 85, 46-47.

Page, K.M. and Guérinet, M.L. (1995) J. Bacteriol. 477, 3979-3984.

Pankhurst, C.E., Schwinghmer, E.A. and Bergersen, F.J. (1972) J. Gen. Microbiol. 70, 161-177

Parniske, M., Schmidt, P.E., Kosch, K. and Müller, P. (1994) Mol. Plant-Microbe Interact. 7, 631-638.

Pawlowski, K. and Bisseling, T. (1996), The Plant Cell 8, 1894-1913

Peters, N.K. and Verma, D.P.S. (1990) Mol. Plant-Microbe Interact. 3, 4-8.

Planqué, K., de Vries, G.E. and Kijne, J.W. (1978) J. Gen. Microbiol. 106, 173-178.

Quispel, A. (1951) Ant. Leeuwenhoek, J. Microbiol. Serol. 17, 69-80.

Quispel, A. (1959) in Handbuch der Pflanzenphysiologie XI Heterotrophie (Mothes, K. ed.), pp. 577-604, Springer, Berlin.

Quispel, A. (1992) in Molecular signals in plant-microbe communications (Verma,D.P.S. ed.), pp. 471-491, CRC Press, Inc. Boca Raton, Florida.

Quispel, A., Kijne, J.W., van Brussel, A.A.N., Pees, E., Wijffelman, C.A. and Burggraaf, A.J.P. (1984) in Advances in Nitrogen Fixation Research (Veeger, C. and Newton, W.E. eds.), pp. 381-388, Nijhoff,Junk,Pudoc, The Hague/Wageningen.

Quispel, A., Svendsen, A.B., Schripsema, J., Baas, W.J., Erkelens, C. and Lugtenburg, J. (1989) Mol. Plant-Microbe Int. 2, 107-112.

Quispel, A., van Brussel, A.A.N., Hooymans, J.J.M., Priem, W.J.J. and Staal, H.J.M. (1984) in Nitrogen fixation and CO_2 metabolism, (Ludden, P.J. and Burris, J.W. eds.), pp. 193-202, Elsevier, New York.

Remy, W., Taylor, T.N., Hass, H. and Kerp, H. (1994) Proc. Natl. Acad. Sci. USA 91, 11841-11843.

Ren, Y.Y. and West, C.A. (1992) Plant Physiol 99, 1169-1178.

Richardson, D.H.S., Hill, D.J. and Smith, D.C. (1968) New Phytol. 67, 469-486.

Riopel, J.L. and Timko, M.P. (1992) in Molecular signals in plant-microbe interactions (Verma, D.P.S. ed.), pp. 494-507, CRC Press, Boca Raton.

Roest, H.P., Goosen-de Roo, L., Wijffelman, C.A., de Maagd, R.A. and Lugtenberg, B.J.J. (1995) Mol. Plant-Microbe Int. 8, 14-22.

Röhrig, H., Schmidt, J., Walden, R., Czaja, L., Lubenow, H., Wencke,U., Schell, J. and John, M. (1996) Proc.Natl.Acad.Sci.USA 93, 13389-13392.

Rolfe, B.G., Djordjevic, M., Scott, K.F., Hughes, J.E., Badenoch Jones, J., Greshoff, P.M., Cen, Y., Dudman, W.F., Zurkowski, W. and Shine, J. in Current perspectives in nitrogen fixation (Gibson, A.H. and Newton, W.E. eds.), pp. 142-145, Austral. Acad. Sci. Canberra.

Ronson, C.W. (1988) in Nitrogen fixation, hundred years after (Bothe, H., de Bruijn, F.J. and Newton, W.E. eds.), pp. 547-551, G.Fischer, Stuttgart.

Ronson, C.W., Lyttleton, P. and Robertson, J.G. (1981) Proc. Natl. Acad. Sci. USA 78, 4248-4288.

Roth, L.E., Jeon, K. and Stacey, G. (1989) in Molecular genetics of plant-microbe interactions (Palacios, R. and Verma, D.P.S.eds.),pp 220-225,Amer.Phytopathol.. Soc.Press, St.Paul.

Rushing, B.G., Yelton, M.M. and Long, S.R. (1991) NucleicAcid Res. 19, 921-927.

Ryan, C.A. and Farmer, E.E. (1991) Annu. Rev. Plant Physiol. Plant Mol. Biol. 42, 651-674.

Sapp, J. (1994) Evolution by association, Oxford Univ. Press, New York, Oxford.

Schell, M.A. (1993) Annu. Rev. Microbiol. 47, 597-626.

Schlaman, H.R.M., Horvath, B., Vijgenboom, E., Okker, R.J.H. and Lugtenberg, B.J.J. (1991) J. Bacteriol. 173, 4277-4287.

Schlaman, H.R.M., Okker, R.J.H. and Lugtenberg, B.J.J. (1992) J. Bacteriol. 174, 5177-5182.

Schuster, W. and Brennicke, A. (1988) Plant Science 54, 1-10.

Selim, S., Delacour, S. and Schwencke, J. (1996) Arch.Microbiol. 165, 252-257.

Sengupta-Gopalan, C. and Pitas, J.W. (1986) Plant Mol. Biol. 7, 189-200.

Sequeira, L., Gaard, G. and de Zoeten, G.A. (1977) Physiol. Plant Pathol. 10, 43-50.

Silvester, W.B., Parsons, R. and Watt, P.W. (1996) New-Phytol. `132, 617-625.

Siquiera, J.O., Safir, G.R. and Nair, N.G. (1991) New Phytol. 118, 87-93.

Skorupska, A., Bialek, U., Urbanik-Sypniewska, T. and van Lammeren, A. (1995) J. Plant Physiol. 147, 93-100.

Smit, G., de Koster, C.C., Schripsema, J., Spaink, H.P., van Brussel, A.A.N. and Kijne, J.W. (1995) Plant Mol. Biol. 29, 869-873.

Smith, S.E. and Smith, F.A. (1990) New Phytol. 114, 1-38.

Spaink, H.P. (1995) Annu. Rev. Phytopathol. 33, 345-368

Spaink, H. (1996) Crit.Rev.Plant Sci. 15, 559-582.

Spaink, H.P., Okker, R.J.H., Wijffelman, C.A., Tak, T., Goosen-de Roo, L., Pees, E. and van Brussel, A.A.N. (1989) J. Bacteriol. 171, 4045-4053

Sprent, J.I. (1994) Plant Soil 161, 1-10.

Sprent, J.I. and de Faria, M. (1987) in Nitrogen fixation with non-legumes (Skinner, F.A., Boddey, R.M. and Fendrik, I. eds.), pp. 3-11, Kluwer Acad.Publ. Dordrecht.

Sprent, J.I. and Raven, J.A. (1985) Proc. Roy. Soc. Edinburgh 85 B, 215-237.

Stowers, M.D. (1985) Annu. Rev. Microbiol. 39, 89-108.

Suharjo, U.K.J. and Tjepkema, J.D. (1995) Physiol. Plant. 95, 247-252

Sullivan, J.T., Eardly, B.D., van Berkum, P. and Ronson, C.W. (1996) Appl. Environm. Microbiol. 62, 2818-2825.

Taylor, F.J.R. (1974) Taxon 23, 229-258.

Templeton, M.D., Rikkerink, E.H.A. and Beever, R.E. (1994) Mol.Plant-Microbe Int. 7, 320-325

Timmis, J.N. and Scott, N.S. (1984) Trends Biochem. Sci. 9, 271-273.

Tjepkema, J.D. and Asa, D.J. (1987) Plant Soil 100, 225-236.

Tobler, F. (1925) Biologie der Flechten, Bornträgaer, Berlin.

Tsai, Y.L. and Benson, D.R. (1989) Arch. Microbiol. 152, 382-386.

Udvardi, M.K., Ou Yang, L.J., Young, S. and Day, D.A. (1990) Mol. Plant-Microbe Int. 3, 334-340.

van Batenburg, F.H.D., Jonker, R. and Kijne, J.W. (1986) Physiol. Plant. 66, 476-480.

van Brussel, A.A.N., Costerton, J.W. and Child, J.J. (1979) Can. J. Microbiol. 25, 352-361

van Brussel, A.A.N., Planqué, K. and Quispel, A. (1977) J. Gen. Microbiol. 101, 51-56.

van Eijsden, R.R., Diaz, C.L., de Pater, B.S. and Kijne, J.W. (1995) Plant Mol. Biol. 29, 431-439.

van Iren, F., van der Knaap, M., van den Heuvel, J. and Kijne, J.W. (1983) in Advances in nitrogen fixation research

(Veeger, C. and Newton, W.E. eds.), pp.433, Nijhoff,Junk, The Hague,Wageningen.

VandenBosch, K.A., Brewin, N.J. and Kannenberg, E.L. (1989) J. Bacteriol. 171, 4537-4542.

Vasse, J., de Billy, F., Camut, S. and Truchet, G. (1990) J. Bacteriol. 172, 4295-4306.

Vasse, J., de Billy, F. and Truchet, G. (1993) The Plant J. 4, 555-566.

Werner, D., Mellor, R.B., Hahn, M.G. and Grisebach, H. (1985) Z. Naturforsch. 40 c, 171-181.

Wessels, J.G.H. (1996) Trends Plant Sci 1, 9-15.

Williams, V. and Fletcher, M. (1996) Appl. Environm. Microbiol. 62, 100-104.

Wyss, P., Mellor, R.B. and Wiemken, A. (1990) Planta 182, 22-26

Young, P.W. and Haukka, K.E. (1996) New Phytol. 133, 87-94.

Zimmerman, W.J., Rosen, B.H. and Lumpkin, T.A. (1989) New Phytol. 113, 497-503.

Chapter 26

Legume Symbiotic Nitrogen Fixation: Agronomic Aspects

Carroll P. Vance

I. Importance of Legume Symbiotic N_2 Fixation

The Earth's population, increasing exponentially (Figure 1), is expected to reach 10 billion, nearly double its present status, by 2035 (Bockman et al., 1990). Of this projected population, 90% is expected to reside in tropical and subtropical regions of the developing countries in Asia, Africa, and Latin America (Waggoner, 1994). Plant sources currently provide 80% of the caloric and dietary protein needs for tropical countries, and this is not expected to change in the near future. In 1910 human beings used about 10% of the total carbon fixed through photosynthesis (Golley et al., 1992). Currently humans use 40% of that carbon, and it is estimated that by 2030 humans will require 80%. Individual protein and caloric consumption of the Earth's current 5.7 million people averages 70 g·protein·d^{-1} and 2,400 calories·d^{-1}, respectively (Waggoner, 1994). The range of protein consumption varies from 38 to 125 g protein·d^{-1}, while the range of caloric intake varies from 1,800 to 3,500·d^{-1}, with the low range values associated with developing countries (Bongaarts, 1994). The anticipated doubling of Earth's population will exacerbate the current inequalities in nutritional intake. Clearly, to maintain the current level of protein and caloric intake over the next 40 years will necessitate unprecedented increases in crop production. This enhanced production will need to be achieved despite a significant deterioration of much prime agricultural land and will require the utilization of large areas now considered marginal.

Nitrogen is the major limiting nutrient for most crop species. Acquisition and assimilation of N_2 is second in importance only to photosynthesis for plant

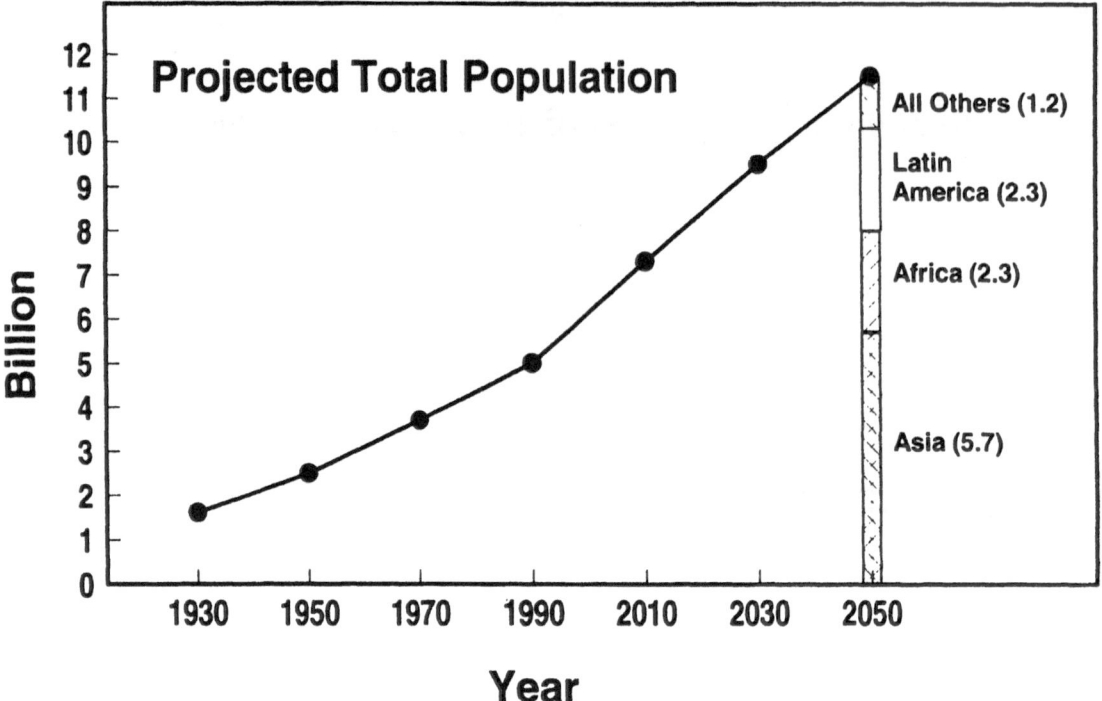

Figure 1. Projected world population growth through the year 2050 and geographical distribution of that population. Adapted from Bockman *et al.*, 1990, and Waggoner, 1994.

growth and development (Newbould, 1989). Production of high-quality, protein-rich food is extremely dependent upon the availability of necessary N. The striking rise in cereal grain yields in developed countries between 1950 and 1990 is directly attributable to a 10-fold increase in N_2 fertilizer use (Figure 2). The "Green Revolution" has been spurred by the development of cereal crops that respond favorably to high N_2 fertilization rates. A typical cereal yield of $7T \cdot ha^{-1}$ requires the uptake of 200 to 300 kg $N \cdot ha^{-1}$ (Bockman *et al.*, 1990; Waggoner, 1994). Concomitant with high application of N_2 fertilizer in developed countries are volatilization of N_2 oxides (greenhouse gases) into the atmosphere, depletion of nonrenewable resources, an imbalance in the global N_2 cycle, and leaching of NO_3^- into groundwater (Kinzig and Socolow, 1994). By contrast, in developing countries the high cost of N_2 fertilizer, the energy requirements for production, and the suboptimal transportation capabilities limit its use, especially for small farms. Sustainable agriculture is broadly defined as agriculture that is managed toward greater resource efficiency and conservation while maintaining an environment favorable for evolution of all species

(Bohlool *et al.*, 1992; Golley *et al.*, 1992). More simply, it is meeting the needs of the present without compromising the needs of the future. One of the driving forces behind agricultural sustainability is effective management of N_2 in the environment. Moreover, judicious management of N_2 inputs into cropping systems is a prerequisite for land stewardship. Successful manipulation of N_2 inputs through the use of biologically fixed N_2 results in farming practices that are economically viable and environmentally prudent (Bohlool *et al.*, 1992; Vance and Graham, 1995). For example, use of N_2-fixing species in cropping systems reduces the need for N_2 fertilizers and increases soil tilth. Additionally, biologically fixed N_2 is bound in soil organic matter and thus is much less susceptible to soil chemical transformations and physical factors that lead to volatilization and leaching. Although many diverse associations (Table 1) contribute to biological N_2 fixation (BNF) (Sprent, 1984), in most agricultural settings the primary source (80%) of biologically fixed N_2 is through the soil bacteria *Rhizobium, Bradyrhizobium, Sinorhizobium, and Azorhizobium*-legume symbiosis (Vance, 1996). Legumes provide 25-35% of the worldwide protein

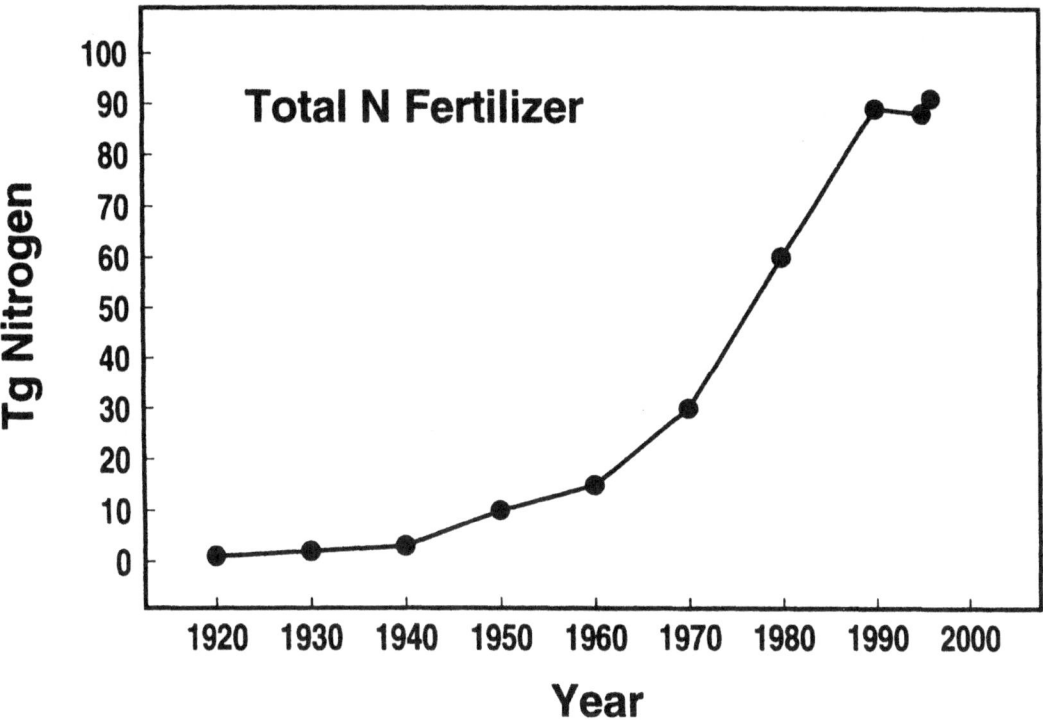

Figure 2. Total nitrogen (N) fertilizer use from 1920-1996. Adapted from Newbould, 1989 and Waggoner, 1994.

intake. Approximately 250 million hectares of legumes are grown worldwide and they fix about 90 Tg N2·yr[-1] (Kinzig and Socolow, 1994). The amount of N_2 fixed by legumes is quite amazing since the total amount of nitrogenase in the world amounts to only a few kilograms (Delwiche, 1970). To replace the N_2 fixed by legumes with anhydrous ammonia produced by the Haber-Bosch process would require 288 Tg of fuel and cost approximately $30 billion annually. Obviously, important goals for agriculture are enhancing the use of and improving the management of legume biologically fixed N_2 for both humanitarian and economic reasons.

II. Symbiotic N_2 Fixation

II.A. NODULES AND NITROGENASE

The ecological niche for the *Rhizobium*-legume symbiosis is usually the root nodule (Figure 3). However, *Azorhizobium* can form N_2-fixing nodules on the stems of *Sesbania rostrata*. Root nodules are

highly organized, hyperplastic tissue masses derived from root cortical cells (Hirsch, 1992; Vance, 1996). Nodules are generally divided into two major groupings characterized by shape, meristematic activity, and fixed N_2 transport products (Figure 3 and see chapter 24): 1) nodules that are elongate-cylindrical with indeterminate apical meristematic activity that transport fixed N_2 as amides such as alfalfa (*Medicago sativa* L.), pea (*Pisum sativum* L.), and clover (*Trifolium* spp.); and 2) nodules that are spherical with determinate internal meristematic activity that transport fixed N_2 as ureides, such as soybean (*Glycine max* L. Merr.) and common bean (*Phaseolus vulgaris* L.). In general, perennial species tend to have indeterminate nodules, while those of annual species are determinate. However, exceptions do occur, for example *Lotus* is a perennial forage legume with spherical determinate nodules yet its major N_2 transport products are amides (Maxwell *et al.*, 1984). *Pisum*, *Lens*, and *Vicia* are annual species with elongate indeterminate nodules that transport fixed N_2 as amides (Rosendahl *et al.*, 1990). This author is not aware of any species with

Plant type	Genus	Microbe	Location	Range of N_2 fixed (kg N_2 ha^{-1}season^{-1})
Leguminosae[c]	Pisum Glycine Medicago (etc.)	Rhizobium, Bradyrhizobium, and Azorhizobium	root nodules	10-350
Ulmaceae[d]	Parasponia	Bradyrhizobium	root nodules	20-70
Betulaceae	Alnus	Frankia[e]	root nodules	15-300
Casuarinacea[e]	Casuarina	(Actinomycete)	root nodules	10-50
Eleagnaceae	Eleagnus	(Actinomycete)		ND[b]
Rosaceae	Rubus	(Actinomycete)		ND
Pteridophytes	Azolla	Anabaena	heterocysts in cavities of dorsal leaf lobes	40-120
Cycads	Ceratozamia	Nostoc	modified corraloid shaped roots	19-60
Lichens	Collema	Nostoc[f]	interspersed between fungal hyphae	ND

Table1. Diversity of symbiotic N_2 fixing associations.[a]
[a]Table adapted from Sprent (1984) and Vance (1996). [b]ND, not determined. [c]Numerous (over 3,000) species within the Leguminoseae form symbiotic associations with Rhizobium and Bradyrhizobium (Allen and Allen, 1981). [d]The only non-legume known to form root nodules with Rhizobium. [e]Frankia, an actinomycete, forms root nodules on species in eight non-legume families of Angiosperms. [f]Nostoc and other blue-green algae form symbiotic associations with 35 species of lichens.

indeterminate nodules that transport symbiotically fixed N_2 as ureides.

The actual reduction of N_2 to NH_4^+ is catalyzed by the nitrogenase enzyme complex within rhizobial cells. For every N_2 reduced, 10 protons are required, 8 are used for ammonia formation and 2 for generation of H_2 (Newton, 1993). The formation of H_2 by nitrogenase is considered a wasteful reaction (see reaction below), thus reducing the efficiency of the enzyme by some 25% (Phillips, 1980). Nitrogenase can also reduce other triple-bond substrates including acetylene (Hardy et al., 1968). As evidenced from the nitrogenase equation, the reaction involves a large expenditure of energy with approximately 6g C required for each g N_2 reduced. Carbon use for symbiotic N_2 fixation theoretically may necessitate 10 to 20% of the total plant photosynthesis (Phillips, 1980; Schubert, 1981). The physical characteristics of nitrogenase are quite similar, irrespective of the source (Newton, 1993). The enzyme is comprised of two easily separable proteins designated the iron (Fe) protein compo-

nent II and the molybdenum-iron (MoFe) protein or component I (Dean and Jacobsen, 1992; Merrick, 1992). The Fe-protein is a homodimer with a native M_r of 60 to 64 kD and a subunit mass of 30 to 32 kD. The Fe-protein contains approximately 4 g-atoms of Fe and S per mole of preparation. The Fe and S form a single [4Fe-4S] cluster, which is bound between the subunits of component II. The Fe protein has two Mg·ATP binding sites and as ATP binds to these sites, the potential of electrons at the [4Fe-4S] cluster is reduced, allowing the Fe protein to donate electrons to the MoFe protein.

The MoFe protein is a tetramer ($\alpha 2\beta 2$) of 220 kD molecular mass. The subunit molecular mass of subunit α is 56 kD while the β subunit has a molecular mass of about 60 kD. The MoFe protein contains 2 atoms of Mo and 24 to 32 atoms of Fe and S per molecule. There appear to be two to four [4Fe-4S] clusters and two [MoFe^{6}S^8] clusters. The two [MoFe^{6}S^8] clusters comprise the MoFe cofactor. The role of the MoFe protein is to transfer electrons to N_2 and H^+.

$$N_2 + 16ATP + 8e^- + 10H^+ \xrightarrow{\text{MgH}} 2NH_4 + H_2 + 16\ ADP + 16\ Pi \qquad (1)$$

Figure 3. Root nodules on (a) alfalfa (*Medicago sativa* L.) and (b) birdsfoot trefoil (*Lotus corniculatus* L.)

II.B. PHOTOSYNTHESIS AND NITRO-GENASE SUBSTRATE

Leaf photosynthate is the ultimate source of energy and of carbon skeletons for nodule growth and maintenance, bacteroid respiration, N_2 fixation, and N_2 assimilation (Phillips, 1980; Vance and Heichel, 1991; Hunt and Layzell, 1993). The interdependence of photosynthesis and N_2 fixation in nodulated legumes is clearly shown by the similarity of patterns in plant growth, carbon assimilation, and N_2 assimilation (Figures 4 and 5). Long term experiments that increase photosynthesis e.g. elevated CO_2 (Table 2) and increased photosynthetically active radiation, enhance nodule mass, N_2 fixation, and N_2 accumulation (Zanetti et al., 1996).

In short term experiments it has been more difficult to establish a relationship between current photosynthate supply and N_2 fixation. Stem girdling, shoot removal, and placing shoots into darkness rapidly inhibit N_2 fixation in perennial legumes (Figure 4, Vance and Heichel, 1991). This response was originally attributed to a reduction in current photosyn-

thate. More recent studies, however, indicate that the immediate effect of those treatments is to alter nodule oxygen diffusion (Denison et al., 1992; Hunt and Layzell, 1993; Hartwig and Nosberger, 1994) rather than a direct effect on N_2 fixation through photosynthate availability. The concept that current photosynthate is non-limiting for N_2 fixation is also supported by experiments showing little diurnal variation in nitrogenase activity in nodulated roots maintained at a constant temperature and short term stimulation (1.5-36 h) of photosynthesis by CO_2 enrichment had no effect on N_2 fixation (Vance and Heichel, 1991). The lack of dependence of N_2 fixation on current photosynthate was shown in mature alfalfa plants by continued unreduced nitrogenase activity for 12.5 h after complete shoot removal. Low temperatures prolonged activity while high temperatures shortened it.

The question of whether energy supply limits N_2 fixation remains perplexing. Labeling studies show that nodules are weak sinks for photosynthate (Vance and Heichel, 1991). Moreover, starch accumulates in both infected and uninfected nodule cells even under conditions of maximum fixation.

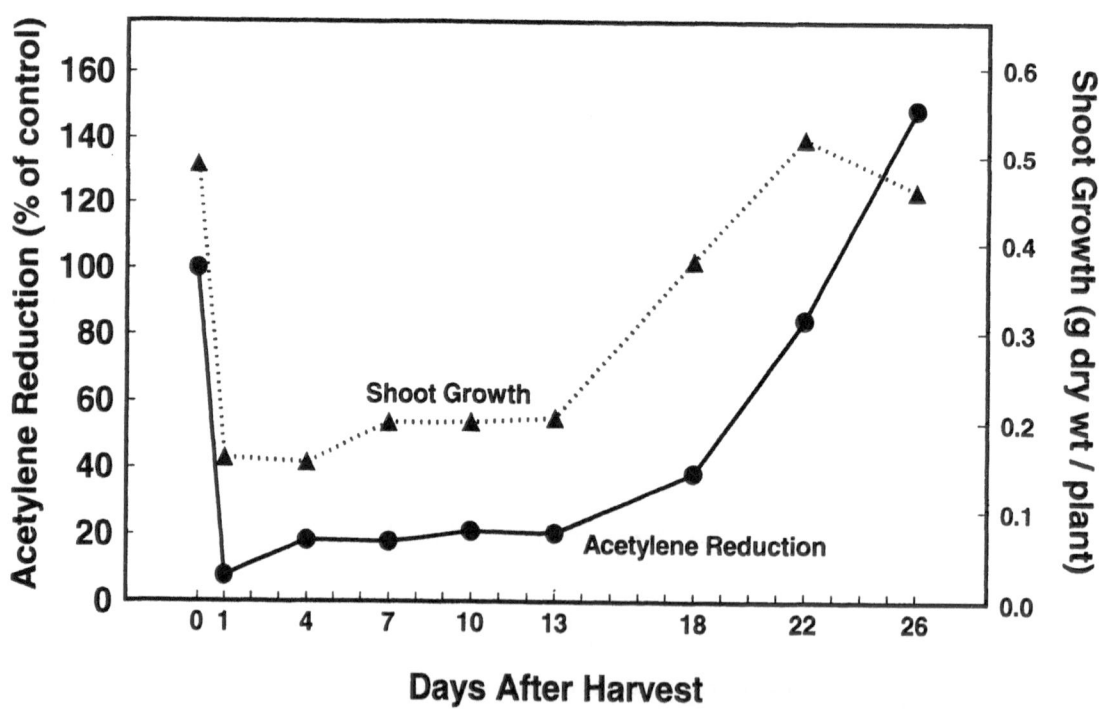

Figure 4. Shoot regrowth and nitrogen fixation as measured by acetylene reduction of alfalfa after shoot harvest. Shoots were removed on day 0. Adapted from Vance et al., 1978.

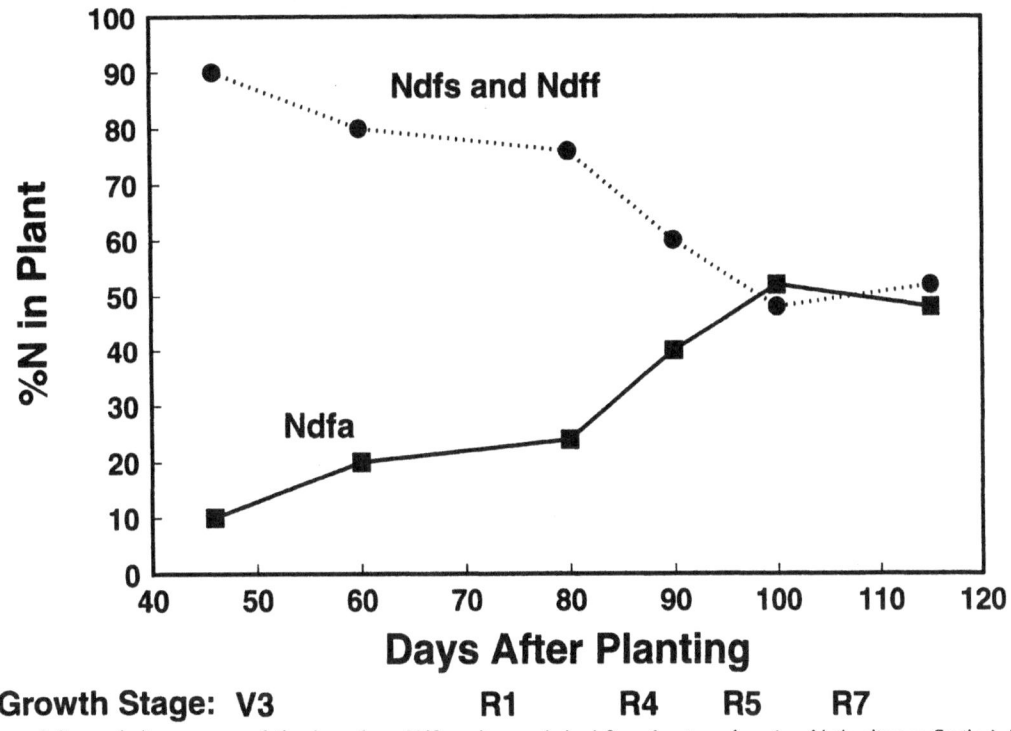

Figure 5. Seasonal nitrogen accumulation in soybean. Ndfa = nitrogen derived from the atmosphere (symbiotic nitrogen fixation); Ndfs and Ndff = nitrogen derived from soil and fertilizer, respectively. Adapted from Zapata *et al.*, 1987.

Nodules generally account for less than 4% of legume dry matter. However, in bean and pea mutants selected for super nodulation nodule mass has been increased 200% with little accompanying loss of shoot dry matter (Rosendahl *et al.*, 1989; Hansen *et al.*, 1993). Taken together, the data to date show nodules require photosynthate, but that may not be a primary factor limiting N$_2$ fixation.

While photosynthesis may not be the primary factor limiting N$_2$ fixation, O$_2$-limited carbon metabolism may well do so (Vance and Heichel, 1991; Hunt and Layzell, 1993; Hartwig and Nosberger, 1994). Glycolysis in the O$_2$-limited cells of the nodule interior is directed towards malate rather than pyruvate as normally expected. High concentrations of malate and succinate are synthesized in nodules and these dicarboxylic acids are the primary carbon sources used by the bacterial for N$_2$ fixation (Rosendahl *et al.*, 1990; Day and Copeland, 1991; Vance, 1996). They also provide the carbon skeleton for N$_2$ assimilation. Whether the anaerobic adaptation

of glycolysis going to malate is as efficient as glycolysis to pyruvate is not known. However we do know that glycolysis to ethanol is much less efficient than that to pyruvate. Conditions which result in reduced synthesis of dicarboxylic acids in nodules cause decreased N$_2$ fixation.

II.C. NITROGENASE: O$_2$ PROTECTION

The nitrogenase enzyme complex is rapidly and irreversibly denatured by O$_2$, and yet the large ATP requirement for nitrogenase is derived from O$_2$ dependent oxidative phosphorylation within the rhizobial cell. Thus the enzyme is functional only in low O$_2$ environments. This paradox is resolved by several exquisite adaptive features during the development of root nodule symbiosis (Vance and Heichel, 1991; Hunt and Layzell, 1993), the first of which involves O$_2$ diffusion. The O$_2$ concentration of nodule outer cortical cells is very close to atmospheric levels, while the interior infected and

Plant	Treatment	Total dry weight (g)	N (%)	Nodule		Nitrogenase	
				Mass (mg)	Number	Specific (nmol·mg nodule h^{-1})	Total (µmol·plant^{-1})
Alfalfa	CO_2:1000 ppm	3.1[a]	2.52[a]	46[a]	124[a]	344	16[a]
	Control	2.0	2.95	28	77	356	10
Clover	CO_2:1000 ppm	3.8[a]	2.47[a]	85[a]	344[a]	284	24[a]
	Control	2.3	3.01	66	270	252	16
Pea	CO_2:1000 ppm	3.6[a]	2.48[a]	27[a]	131[a]	226	37[a]
	Control	3.0	2.28	14	87	193	18

Table 2. Effect of enriched atmospheric CO_2 on nitrogenase activity of yield of alfalfa, clover, and pea.
[a] Denotes significant difference between CO_2 enriched and control. Data adapted from Masterson and Sherwood (1978), Murphy (1986), and Phillips *et al.* (1976).

uninfected cells have an O_2 concentration of 10-30 nM (nearly anaerobic). Separating the outer cortex from the interior cells of the nodules is a compact layer of two to four cells, which have few intercellular air spaces. This cell layer comprises the O_2 diffusion barrier (Tjepkema and Yocum, 1974; Sheehy, 1987; Hunt and Layzell, 1993). This diffusion barrier is variable displaying increased resistance to O_2 diffusion under adverse environmental conditions (providing greater protection for nitrogenase). By contrast, under favorable conditions diffusion decreases allowing increased nitrogenase activity. Although control of the diffusion barrier is not well understood, it apparently involves shrinking and swelling of diffusion barrier cells somewhat similar to the changes that occur in stomata. The model also invokes increases and/or decreases in intercellular water and protein content.

Another important contributor to solving the O_2 dilemma is leghemoglobin (Lb) an O_2 binding protein found within nodule infected cells. This plant protein which is very similar to animal hemoglobin gives nodules their pink color. Leghemoglobin facilitates the diffusion of available O_2 through the plant cell cytoplasm to the bacterial cells at concentrations which allow oxidative phosphorylation to occur without inactivation of nitrogenase activity (Appleby, 1992). Leghemoglobin can comprise as much as 35% of the total plant nodule soluble protein in most legume species. It appears to be the most highly expressed plant protein in nodules (Egli *et al.*, 1991). Proteins and genes homologous to Lb have been identified in nodules of the non-legume *Parasponia* and in non-legume actinorhizal nodules (Jacobsen-Lyon *et al.*, 1995). Moreover, hemoglobin genes

have been isolated from the non-nodulating, non-legume barley and maize (Taylor *et al.*, 1994). Thus Lb or Lb equivalents are found in many non-legume plants, suggesting genetic elements contributing to nodulation and O_2 regulation may be widespread in plants.

II.D. HYDROGEN EVOLUTION

Nitrogenase allocates a significant fraction of reductant to protons to form H_2 during the process of N_2-fixation (Evans *et al.*, 1987). This reduction of protons is an ATP-dependent process, which seemingly represents an energy loss from the nodule. For every $2NH_4^+$ produced by nitrogenase, 4 ATPs and 2e$^-$ are used for H_2 formation (Arp, 1992). Some strains of *Rhizobium* have evolved a separate uptake hydrogenase system, which can oxidize H_2 to water and, in some cases, couple that oxidation to ATP formation. Although the uptake hydrogenase system confers several potential biochemical advantages to *Rhizobium*, including energy conservation and additional protection of nitrogenase from inhibition by H_2 and O_2, rigorous attempts to demonstrate such advantages for soybean production have not been reproducibly successful. Further complexity is added to this poorly understood process in that the plant genotype can suppress or enhance the rhizobial uptake hydrogenase reaction.

II.E. METHODS FOR MEASUREMENT OF N_2 FIXATION

Field grown legumes obtain variable amounts of N_2 from both the soil solution as NO_3^- or NH_4^+ and the soil as N_2. The N_2 derived from any of these sources is a function of soil inorganic N_2 availability, efficiency of symbiosis, environmental constraints, and genetic variability (Hardarson and Danso, 1993). Thus, when simple Kjeldahl N_2 is measured, the proportion of plant total N_2 derived from symbiosis versus that derived from inorganic soil N_2 is unknown. Three methods: (i) acetylene reduction; (ii) N_2-difference; (iii) ^{15}N isotope dilution and natural abundance, have been developed to measure/estimate the quantity of N_2 derived from symbiotic N_2 fixation (Nutman, 1976; Heichel, 1987; Hardarson and Danso, 1993; Herridge and Danso, 1995). Compositional analysis of xylem sap N_2 content has also been advocated as a method for assessing N_2 fixation in beans (Herridge and Danso, 1995).

The acetylene reduction assay (ARA) is based on the principle that N_2 fixing enzyme nitrogenase not only reduces N_2 to NH_3, but also reduces acetylene (C_2H_2) to ethylene (C_2H_4) (Hardy et al., 1968). Thus, nodulated roots can be incubated in C_2H_2 and the C_2H_4 quantitated by gas chromatography. The technique is extremely sensitive, rapid, simple, and relatively inexpensive to perform.

The C_2H_2 reduction assay, however, has numerous shortcomings (Sheehy, 1987; Hunt and Layzell, 1993). It is so highly variable that significant differences between treatments are very difficult to detect unless they are quite large. The theoretical stoichiometry of the reaction three moles of C_2H_2 to one mole N_2 reduced is far from accurate in biological systems and varies with rhizobial strain, plant genotype, and plant development (Vessey, 1994). Acetylene reduction is a stopped time assay and is generally made once or twice a week over the growing season. Therefore, integration of seasonal N_2 fixation values with plant performance is frequently inaccurate. Lastly, exposure of nodules to C_2H_2 has detrimental effects on nodule physiology, particularly O_2 diffusion (Sheehy, 1987; Hunt and Layzell, 1993; Minchin et al., 1994). However, if one is cautious, useful data may still be obtained with this assay. Evaluation of intact plants with an open flow-through system for short periods (5-10 min) allows C_2H_4 analysis without inhibition of the nitrogenase enzyme (Minchin et al., 1994; Vessey, 1994). The C_2H_2 reduction assay is also quite useful for simply measuring for the presence of nitrogenase activity.

The N_2-difference method is based on determining the difference in the amount of reduced N_2 between a N_2 fixing crop and an appropriate non-N_2 fixing crop (Heichel, 1987; Hardarson and Danso, 1993; Herridge and Danso, 1995). The only parameters that need to be measured are plant N_2 by Kjeldahl analysis and total plant dry matter. The method is accurate, simple, and requires no expensive equipment. It integrates N_2 fixation and plant growth over the season. Difficulties with the N_2-difference method primarily involve obtaining an appropriate control. Non-N_2-fixing genotypes are available for Medicago, Glycine, Cicer, Trifolium, Pisum, and Vicia (Vance, 1996). For the numerous other species of legumes, either uninoculated controls or controls inoculated with ineffective rhizobia must be maintained contamination free. Contamination free plants are difficult to achieve since most soils are readily infested with both effective and ineffective rhizobia. An alternative control may be an annual or perennial grass with a growth habit similar to the legume. However, differences in N_2 uptake and other physiological parameters between the legume and grass open this choice of control to considerable criticism.

The ^{15}N isotope dilution methods involves labeling the soil N_2 pool with the stable isotope $^{15}NO_3^-$ and/or $^{15}NH_4^+$ and determine the ratio of plant N_2 occurring as ^{15}N versus that occurring as ^{14}N (the form most predominant in nature 99.65%) (Hardarson and Danso, 1993; Heichel, 1987; Herridge and Danso, 1995). A legume fixing N_2 will accumulate most of its N_2 as ^{14}N from the atmosphere, thus diluting the ^{15}N that may come from the soil. The greater the fixation, the more the dilution. Conversely, in plants deriving more N_2 from soil the ^{15}N pool will be less diluted. This method is accurate and integrates N_2 fixation and plant growth over the season. The isotope dilution method is effective for determining partitioning patterns. It can also be used to measure the contribution of legumes to the N_2 budget of a companion crop and/or a subsequent crop grown in rotation (Peoples and Craswell, 1992; Herridge and Danso, 1995).

The primary criticisms of the isotope dilution techniques are the high cost of the stable isotopes and the expensive instrumentation required to measure ^{15}N. This procedure also requires a non-N_2-fixing control and thus encounters problems similar to those described for the N-difference method.

II.F. QUANTITIES OF N_2 FIXED BY LEGUMES

The amount of N_2 fixed symbiotically and the % N_2 derived from symbiosis is a function of numerous genetic and environmental effectors and thus, can

Species	N_2 fixed ha^{-1} per season[a] median value (kg)	Plant N_2 from atmosphere median value (%)	Equivalent cost of fertilizer N_2 production[b] (U.S. dollars/ha)
Forage Legumes			
Trifolium pratense	170	59	119.00
Lotus corniculatus	92	55	64.40
Medicago sativa	180	70	126.00
Vicia sativa	130	70	91.00
Trifolium repens	172	75	120.40
Desmodium sp	200	85	140.00
Pulse Legumes			
Pisum sativum	72	35	50.40
Glycine max	120	53	84.00
Arachis hypogaea	114	57	79.80
Phaseolus vulgaris	65	40	45.50
Vigna angularis	80	70	56.00
Vicia faba	151	80	105.70
Lupinus angustifolius	170	65	119.00
Lens culinaris	100	63	70.00

Table3. Nitrogen-fixation characteristics of selected legumes and the cost of an equivalent of N_2 fertilizer.
[a]Calculated or adapted from Heichel (1987), Nutman (1976), Peoples and Craswell (1992), and Peoples *et al*. (1995a and 1995b). [b]Cost for fertilizer N, 0.70 dollar kg^{-1}.

vary substantially within and across species. An exhaustive listing of quantities of N_2 fixed and % N_2 derived from symbiosis can be found in Ladha and Peoples (1995). Median values of N_2 fixed per season and % N_2 derived from symbiosis calculated fromseveral sources are given in Table 3 and 4. A noticeable feature of these tables is that irrespective of the method of assessment perennial forage legumes tend to fix more N_2 symbiotically than do annual pulse legumes. The limited data available for N_2 fixing trees suggest that they also fix substantially more N_2 than pulse legumes. The fact that several harvests of forage legume herbage are taken in a single season and that trees have substantially more biomass than annual legumes contribute to these differences in N_2 fixation. Another feature that may also contribute to increased fixation by forage and tree legumes is that these species have indeterminate nodules that can function for long periods (Vance, 1996).

The typical pattern for N_2 fixation and shoot regrowth after herbage harvests in the perennial forage legume alfalfa is shown in Figure 4. The pattern shows that N_2 fixation declines immediately after herbage harvest, remains low for 10 to 14 days, and then recovers rapidly. Studies with Lotus and *Trifolium* show similar patterns (Denison *et al.*, 1992; Hartwig and Nosberger, 1994). Depending upon the species and location this pattern repeats from two to eight times a year for as long as the crop remains in the field, resulting in large quantities of N_2 being fixed. Indeterminate nodules formed in one herbage regrowth cycle may function through several subsequent cycles (Vance *et al.*, 1979).

By comparison, the dry matter accumulation and N_2 fixation data for soybean (Figure 5) and adzuki bean (*Vigna angularis* L.) (Table 5) typifies the seasonal profile for pulse legumes. For most pulses the majority of N_2 fixation occurs after flowering during pod development (Hardarson, 1993). In soybean, which has determinate nodules, crown nodules formed during early growth account for 100% of the N_2 fixed at 20 d after planting but less than 20% at 76 d after planting (McDermott and Graham, 1989). Later nodules formed on lateral roots account for most of N_2 fixed in soybean and lupin. Thus, the amount of N_2 fixed shifts according to the nodule population throughout the season.

Species	N$_2$ fixed ha^{-1} per season[a] median value (kg)	Plant N$_2$ from atmosphere[a] median value (%)	Equivalent cost of fertilizer N$_2$ production[b] (U.S. dollars/ha)
Casurina	180	47	126.00
Alnus	188	55	131.60
Gliricidia	170	53	119.00
Acacia	30	25	21.00
Leucaena	300	60	210.00
Sesbania	160	63	112.00

Table 4. Nitrogen fixation characteristics of selected trees and cost of an equivalent of N$_2$ fertilizer.
[a]Calculated or adapted from Danso *et al.*, 1992; Peoples and Craswell, 1992; Peoples *et al.*, 1995a, and Sanginga *et al.*, 1995. [b]Cost for fertilizer N, 0.70 dollar kg^{-1}

Since N$_2$ fixed by legumes is in essence a gain in reduced N$_2$ without the high cost associated with fertilizer N, it is worthwhile to calculate how much energy would be required for production of an equivalent amount of fertilizer N$_2$ as that produced through fixation (Table 3 and 4). Depending upon the species, the amount of N$_2$ gained as potential fertilizer would be equivalent to a saving of 45 to 210 dollars per ha. Currently 32 million ha of maize are grown in the U.S.A modest replacement of 25 kg of fertilizer N$_2$ with that derived from legumesymbiotic N$_2$ fixation would result in a yearly savings of \$560 million (Peterson and Russelle, 1991). This savings would equate to about \$2,600/year for a typical 150 ha farm. These values reflect an almost doubling of the cost of N$_2$ fertilizer in the last ten years.

III. Host Selection for Improved N$_2$ Fixation

III.A. GENETIC VARIABILITY

Most legume species studied to date have exhibited genetic variability in traits related to biological nitrogen fixation (BNF). These have included differences in nodule number and mass, seasonal accumulation of N$_2$, % N$_2$ derived from symbiosis, acetylene reduction activity (ARA), host restriction of nodulation, and enzyme function associated with N$_2$ assimilation (Herridge *et al.* 1994). More recent studies have shown single gene loci regulating nodule number, host strain compatibility, and sensitivity to nitrate (Caetano-Anolles and Gresshoff,

1991). When properly managed, such variation is more than sufficient to improve current rates of BNF in the major legume species (Buttery *et al.*, 1992; Herridge *et al.*, 1994), to ensure enhanced yields under N$_2$-deficient conditions, and in specified conditions to provide significant N$_2$ for succeeding or companion crops. Layzell and Moloney (1994) have suggested that increases in BNF of up to 300% could be achieved through host plant selection and effective crop management, providing a major incentive for a continued focus in this area of research.

BNF in legumes can be enhanced through either indirect selection for traits such as biological yield and N$_2$ concentration under low soil N$_2$ conditions, or by direct selection using traits associated with BNF including nodule mass, xylem ureide content (Herridge *et al.* 1994), ARA (Bliss, 1993; Heichel *et al.*, 1989), and ^{15}N isotope enrichment (Heichel *et al.*, 1989; Herridge *et al.*, 1994). Breeding activities with soybean, common bean, and alfalfa typify differences in the balance of direct and indirect approaches used to date.

III.B. SOYBEAN

Although increased BNF has not been a consistent breeding objective in soybean selection in the U.S., improved BNF fixation has resulted from indirect selection for yield and protein content. Comparisons of modern soybean lines with both ancestral cultivars and agronomically superior plant introductions show that enhanced N$_2$ accumulation is characteristic of newer soybean genotypes (Cregan and Yaklick, 1986). Moreover, field studies of the soybean

Harvest date	Dry matter $(g \cdot plant^{-1})$	Total N_2 $(mg \cdot plant^{-1})$	N_2 fixed (mg)	N_2 fixed $(\%)$
7-2-80	0.96	29.50	11.47	38
8-4-80	4.06	96.14	58.08	60
9-22-80	19.63	438.40	328.78	75

Table 5. Field assessment of Adzuki bean (*Vigna angularis*) nitrogen fixation. Data from C. P. Vance and L. L. Hardman (unpublished)

cultivars 'Lincoln,' 'Shelby,' and 'Williams' which represent serial enhancement of the 'Lincoln' line, show that selection for increased yield resulted in higher N_2 concentration, total N_2, and N_2 fixed (Coale *et al.*, 1985). Duc *et al.* (1988) have shown that improvement of BNF by selection for yield is not limited to soybean. Their evaluation of 21 faba bean genotypes selected for increased yield demonstrated that the highest yielding genotypes also fixed the greatest amount of N_2. It should be noted that improvement of U.S. soybean cultivars came through selection for high yielding varieties grown in well fertilized soil with little to no control of rhizobial strain interactions. This selection strategy has resulted in a troublesome ecological situation in which agronomically superior lines of soybean frequently are nodulated by native *Bradyrhizobium japonicum* strains that are inferior in N_2 fixation to strains currently available for inoculum (Buttery *et al.*, 1992). Thus maximum yields of U.S. soybean cultivars rarely can be achieved under solely symbiotic growth conditions. This is evidenced by the fact that U.S. varieties

seldom derive more than 60% of their N_2 from symbiosis. By comparison, in Brazil soybean selection has been conducted in the presence of highly effective *B. japonicum* without N_2 fertilizer (Döbereiner, 1987). This strategy has resulted in cultivars that require no added N_2 for maximum yield. Also, a marked increase in the production of soybean in the Brazilian cerrado was achieved by selection of antibiotic resistant strains of *B. elkanii* (Rumjanek *et al.*, 1993) because of the high levels of antibiotic *Streptomyces* sp. in these soils.

III.C. BEAN

The limited BNF ability of most varieties of *Phaseolus vulgaris* has been recognized for many years (Graham, 1981). Bliss and coworkers were the first to apply the backcross inbred method of population development of the transfer of superior BNF ability from "Puebla 152" into otherwise agronomically acceptable bush bean cultivars (Rosas and Bliss, 1986; Bliss, 1993). Improvement

Line	Data taken at R7 mid-podfill			
	Ndfa $(\%)$	N_2 fixed $(mg \cdot plant^{-1})$	Shoot N_2 $(mg \cdot plant^{-1})$	Shoot dry wt $(g \cdot plant^{-1})$
Puebla 152 (donor)	50	589	1128	47
Sanilac (RC)*	19	143	729	32
(17 x 48) - 22	61	1028	1680	59
(17 x 65) - 25	43	707	1511	63
24 x 17	46	573	1267	48
24 x 48	37	426	1146	43
24 x 65	36	504	1348	60
LSD 0.05	13	336	487	16

Table 6. Improvement in N_2 fixation in common bean (IBC)
IBC = inbred backcross line methodology. Source: Bliss, 1993.

in bean was initially based upon both ARA and ^{15}N analysis then in subsequent generations upon seed yield and N_2 accumulation data for plants grown under N_2 limited conditions (Table 6).

III.D. ALFALFA

Several different approaches have been used in attempts to enhance BNF by alfalfa. Teuber et al. (1984) selected within two unimproved alfalfa germplasms first by growth on low N_2 and then by transfer to a high N_2 environment. Selection at low N_2 was to emphasize symbiotic effectiveness, whereas growth on high N_2 reflected differences in N_2 uptake and assimilation. This approach improved herbage N_2 concentration and dry matter production under controlled conditions, but not in the field (Teuber and Phillips, 1988). A similar approach with the cultivar Moapa 69 (Table 7) enhanced total forage N_2 and % N_2 derived from symbiosis, but did not affect yield (Phillips and Teuber, 1992). Increased nodulation obtained in this study may have resulted from enhanced release of flavonoid nod-gene inducers (Kapulnik et al. 1987). A multiple trait bidirectional recurrent selection program for alfalfa (Figure 6) at the University of Minnesota selected for ARA, nodule and root mass, and shoot dry weight (Barnes et al., 1984). Reproducible positive results were obtained in the glasshouse, but results were equivocal in field grown plants (Heichel et al., 1989). During the evaluation of this material, root nodule enzymes associated with N_2 assimilation were also assayed (Groat et al., 1984). Phosphoenolpyruvate carboxylase (PEPC) and NADH-glutamate synthase (GOGAT) activity were highly correlated with BNF and shoot growth. Subsequently, alfalfa germplasms previously selected for nitrogenase activity and agronomic performance were selected

for differences in PEPC and GOGAT activities (Jessen et al., 1987, 1988). Crosses between high enzyme selections of the experimental line MnPl-11 and three other germplasm sources showed increased N_2 yield, dry matter accumulation, and % N_2 derived from fixation (Table 8). It is not yet clear whether these gains were due solely to selection or whether they resulted in part from release of inbreeding depression in this outcrossing tetraploid species. Interestingly, plants selected for reduced PEPC and GOGAT activity showed decreased yield and N_2 accumulation (Jessen et al., 1988) suggesting that threshold levels of these two enzymes are necessary for normal growth.

Although few breeding lines can lay claim to having been selected directly for improved N_2 fixation, it is imperative that selection for this trait play an important role in the cultivars of the future. Plant improvement strategies must ensure that new varieties fix as much N_2 as the best N_2-fixing germplasm currently available. This can be achieved in both developed and developing countries using established methods of direct and indirect selection. Improvement of N_2 fixation and % N_2 derived from symbiosis will depend upon recurrent selection programs that pyramid genetic traits desirable for this process including: an adequate balance of nodule mass and number; duration of activity; improved N_2 assimilation; nitrate tolerance; and superior strain selection. Other activities that could be important in enhancing the use of fixed N_2 in agriculture may be the introgression of supernodulation into agronomically acceptable cultivars, and the development of marker assisted protocols to select for improved BNF. Moreover, it will be increasingly important to obtain accurate quantification of N_2 fixed by trees, and select for improved yield and BNF in these species (Table 4).

Selected alfalfa population	Yield (kg·ha^{-1})	Nitrogen concentration (g·kg^{-1})	Nitrogen yield (kg·ha^{-1})	N_2 from symbiosis	
				Difference (kg·ha^{-1})	^{15}N dilution (kg·ha^{-1})
MOAPA 69	3,330	32.4	107	37.3	58.1
MOAPA 69-33	3,330	34.7	115	45.3	68.2
CUF 101	3,400	32.1	109	39.3	63.3
LSD (0.05)	141	1.4	6	5.5	5.7

Table 7. Selection for alfalfa forage yield and N_2 concentration as a strategy to increase N_2 fixation. Data represent mean of six field harvest in first year of production. Adapted from Phillips and Teuber, 1992.

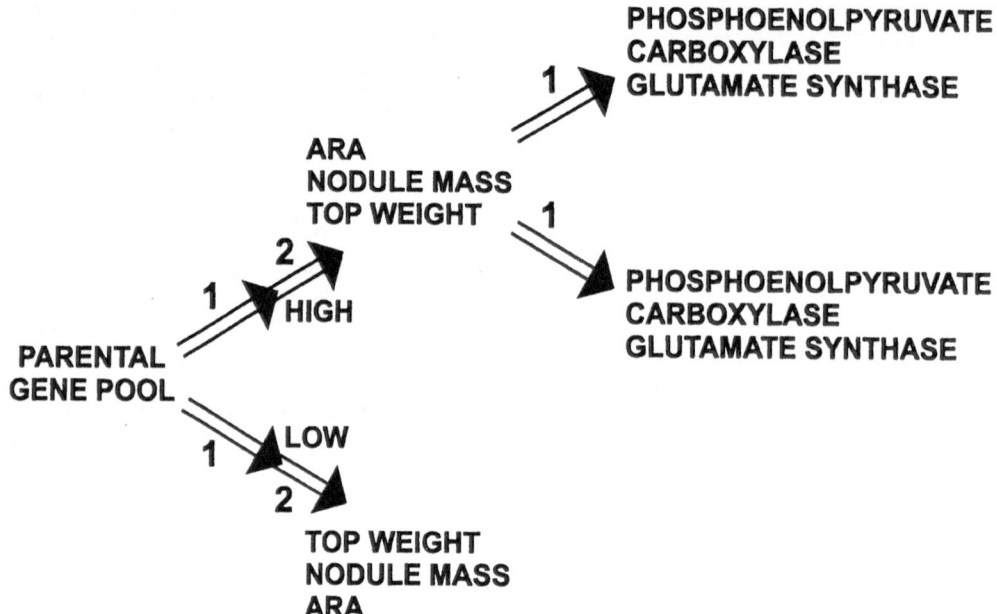

Figure 6. Bidirectional recurrent selection scheme for selection of traits associated with nitrogen fixation in alfalfa. Adapted from Heichel *et al.*, 1989.

IV. Legume Fixed N₂: Benefits to Cropping Systems

Nitrogen fixing species have played an integral role in cropping systems since the domestication of plants and have been prominently featured in rotations and intercropping, as alley crops, in pasture systems, as green manures, in agroforestry, and cover crops (Heichel, 1987; Peoples and Craswell, 1992; West and Mallarino, 1996). Mixed cropping with N_2 fixing species as a major component was the accepted norm prior to the development of the Haber Bosch process (Fujiata *et al.*, 1992). In the U.S. and Europe the use of legumes to provide N_2 for subsequent crops reached a maximum in the 1940s (Heichel, 1987). Since then implementation of legumes into cropping systems has declined due to the low cost of N_2 fertilizer, high yielding cereal crops, and other economic and political forces (Heichel, 1987; Bohlool *et al.*, 1992). A return to the widespread use of N_2 fixing plants in cropping systems of developed countries will require convincing evidence of the economic and environmental advantages accrued due to their use. Alternatively, social and/or political pressures may force changes favorable for the enhanced use of N_2 fixing species in management schemes. By contrast, more than 50% of the crops grown in Africa, India, and Latin America (Table 9) are either intercropped or rotated with N_2 fixing species (Fujiata *et al.*, 1992). However, new and improved strategies need to be developed and transmitted to farmers to more effectively use biological N_2 fixation in developing countries.

Symbiotically fixed N_2 may become available to an intercrop or subsequent crop through several avenues including: (i) release from root exudates; (ii) vesicular arbuscule mycorrhizal (VAM) mediated transfer between species; (iii) N_2 leaching from leaves and leaf litter decomposition; (iv) plow down of green manures; (v) decomposition of roots and nodules, and (vi) release of animal waste (Peoples *et al.*, 1995a). The quantity and availability of symbiotically fixed N_2 available to a companion or subsequent crop are governed by genotypic, environmental, and management factors. Accurate assessment of the legume contribution almost invariably requires the ^{15}N methodologies (Heichel, 1987). However, with the use of new non-N_2 fixing genotypes representative estimates of symbiotically fixed N_2 available for other species can be obtained.

Alfalfa germplasm		Dry matter (m tons)	N$_2$ yield (kg·ha^{-1})	N$_2$ fixed (kg·ha^{-1})	Ndfa (%)
I	Blazer	1.7	75	57	76
	MNPL-11	1.7	72	55	77
	Hi Enz Cross	2.0	85	67	79
II	Citation	1.5	63	47	74
	MNPL-11	1.7	72	55	77
	Hi Enz Cross	1.7	79	62	79
III	Saranac AR	1.9	86	69	80
	MNPL-11	1.7	72	55	77
	Hi Enz Cross	2.2	98	81	83
	LSD (0.10)	.26	13	13	--

Table 8. Field performance of high nodule enzyme crosses and parental lines in three alfalfa germplasm sources. High enz cross = cross between high PEPC and NADH-GOGAT enzyme activity parents (C. P. Vance, J. F. Lamb, and D. K. Barnes, unpublished)

The long-held assumption is that planting legumes with companion crops, either interspersed or between rows provides additional N$_2$ to and improves the performance of the non-N$_2$ fixing species. However, most evidence supporting N$_2$ transfer to a companion crop is indirect (Heichel, 1987; Russelle, 1996; West and Mallarino, 1996). In more exact studies where ^{15}N has been used, the amount of N$_2$ in the companion crop that is derived from the N$_2$-fixing legumes varies from 0 to 70% (Heichel, 1987; Ledgard and Steele, 1992; West and Mallarino, 1996). Although there is a wide range in the amount of N$_2$ transferred and the % of companion crop N$_2$ that is derived from the legume (Table 10), these figures are generally less than 40 kg N$_2$·ha^{-1} and 30%, respectively (Ledgard and Steele, 1992; Peoples *et al.*, 1995b; West and Mallarino, 1996). Because N$_2$ transfer is a function of plant growth and development, soil N$_2$ content, and plant proximity, the most reliable data and maximum amounts of N$_2$ transfer are obtained when the growth habit of the legume and companion species is similar, the experiment is conducted in low soil N, and plant spacing is optimal (Brophy *et al.*, 1987; Ledgard and Steele, 1992; Russelle, 1996; Thomas, 1995). Studies conducted in the presence of high soil N$_2$ frequently fail to show transfer of fixed N$_2$ due to the inhibitory effect of soil N$_2$ on N$_2$ fixation and the greater competitiveness of the non-legume species (Heichel, 1987; Fujiata *et al.*, 1992; West and Mallarino, 1996). Brophy *et al.* (1987) elegant study with alfalfa and trefoil clearly demonstrated that at suboptimal

densities and populations N$_2$ transfer was hard to detect, while at optimal populations and spacing striking amounts of N$_2$ transfer could occur.

Although most studies of N$_2$ transfer do not differentiate what proportion of above and below ground components contribute to transfer, experiments with alfalfa and trefoil provide insight into this question (Ta and Faris, 1987; Lory *et al.*, 1992; Dubach and Russelle, 1994). The quantity of symbiotically fixed N$_2$ deposited in the rhizosphere of alfalfa was less than 3 kg·ha^{-1}. Additionally, the amount of N$_2$ potentially available for a companion crop from the turnover of roots and nodules was approximately 13 kg N$_2$·ha^{-1} and 2 kg N$_2$·ha^{-1}, respectively. Heichel (1987) estimated that nodule turnover would contribute less than 6 kg N$_2$·ha^{-1}. These data indicate that above ground N$_2$ contributions are of similar if not greater importance than below ground parts toward N$_2$ transfer because total N$_2$ transfer is usually around 30 to 40 kg N$_2$·ha.

Legume N$_2$ as a replacement for fertilizer N$_2$ derives from the practice of comparing the yield of a non-legume grown after a legume to that of a non-legume grown with fertilizer. This assessment provides an estimate of the amount of N$_2$ in the non-legume that is derived from the legume. Replacement N$_2$ is usually thought of in terms of rotational systems in which either the legume leaf debris and roots or the entire plant is returned to the soil for the subsequent crop. In most instances the replacement N$_2$ value represents the response of the non-legume to the total available soil N$_2$ and does not distinguish

Species	Geographical region (% intercropped)			
	Nigeria	Uganda	India	Latin America
Maize	76	84		50
Peanut	95	56		
Pigeon pea	99	76	90+	
Bean		81		80-90

Table 9. Proportion of species intercropped Adapted from Fujiata et al., 1992.
+ Used in almost all cropping systems

between symbiotically fixed N_2 and N_2 from other sources (Heichel, 1987; Peoples et al., 1995a; Wani et al., 1995). Because only the amount the N_2 derived from fixation and acquired by the subsequent crop is gain of N_2 to be credited against fertilizer use, the fertilizer N_2 replacement value frequently is an overestimate of that contributed by fixation.

The fertilizer replacement value for legumes when grown preceding a grain crop varies from 0 to 110 kg $N_2 \cdot ha^{-1}$ (Table 11). This range represents composite data taken from several sources. As expected, there is substantial variation between species for N_2 replacement values. Such variation reflects the impact of: (i) environment; (ii) moisture; (iii) plant composition, particularly C:N and phenolic content; (iv) soil texture and organic content; and (v) amount of legume incorporated into soil. Heichel (1987) and Hesterman et al. (1987) have suggested that about 40 to 70% of the fertilizer N_2 replacement value of legumes is due to legume N_2. The remainder of the beneficial effect of legumes is due to increased soil health, improved disease and pest control, reduced allelopathic compounds, and enhanced water retention (Peoples and Craswell, 1992).

With the advent of increased population pressures, farming of marginal land, and degradation of the environment, research on N_2 fixation must include renewed efforts to deliver biologically fixed N_2 to cropping systems (Becker et al., 1995). Research must identify the best species to incorporate into cropping systems, show the amount of N_2 fixed by these species, and accurately predict the availability of fixed N_2 as replacement for fertilizer N_2 within intercrops and in rotations. Extension must be able to show on-farm success with enhanced use of biologically fixed N_2 in practical management schemes. Lastly, we must develop new and/or value-added uses for N_2 fixing crops that make them attractive for production agriculture in developed countries.

V. New Uses for N_2-Fixing Legumes

V.A. SIGNIFICANCE

Earlier sections of this article have documented the current agronomic significance of N_2 fixing species,

Legume-grass mixture	N_2 transferred amount ($kg \cdot ha^{-1}$)	% of legume N_2 fixation (%)	Grass N_2 derived from legume (%)
Alfalfa-reed canarygrass	9	13	68
Alfalfa-orchardgrass	13	7	22
Alfalfa-timothy	10	5	24
White clover-ryegrass	78	26	27
White clover-tall fescue	30	21	37
Birdsfoot trefoil-reed canarygrass	14	10	28
Red clover-Italian ryegrass	30	23	39
Red clover-orchardgrass	38	19	58

Table 10. Transfer of fixed N_2 from forage legume to accompanying grass. Data adapted from Ledgard and Steele, 1992; Peoples et al., 1995a; and West and Mallarino, 1996.

Preceding legume	Fertilizer N_2 replacement value $(kg\ N_2 \cdot ha^{-1})$
Forage legumes	
Alfalfa	110
Sweet Clover	115
Red Clover	88
Birdsfoot Trefoil	94
Pulse Legumes	
Chickpea	70
Cowpea	50
Pigeonpea	44
Winged Bean	70
Pea	26

Table 11. Fertilizer N_2 replacement value of preceding legume for subsequent year grain yield of non-N_2 fixing species. Data adapted from Heichel, 1987; Wani *et al.*, 1995; Giller and Cadisch, 1995

but what about the future? Undoubtedly the use of legumes for food and fiber sources will continue at current to increased levels in the foreseeable future. New uses, however, may alter how we perceive N_2-fixing plants in agriculture and their importance to sustainable systems. Visions for the use of biological N_2 fixation in the future are numerous. Adoption of new uses could result in substantial increases in land sown to N_2-fixing species and increased diversity for cropping systems. Three projected new uses exemplify realistic possibilities: (i) develop N_2-fixing plants to decontaminate soil and water (phytoremediation); (ii) grow legumes for the generation of electrical energy; and (iii) produce industrial, pharmaceutical, and natural products in N_2 fixing plants. Nitrogen-fixing crops are ideally suited for these purposes because they use the sun's energy to achieve the desired outcome, they fix N_2 and require less fertilizer than other crops, and lastly many can be transformed with and express foreign genes. A description of the use of plants in phytoremediation will provide insight into projected new roles for legumes in sustainable agriculture.

V.B. PHYTOREMEDIATION

While the use of biological agents to restore contaminated soil and water has been recognized for years (Salt *et al.*, 1995) and plants have been used to reclaim mine spills and purify water moving into wetlands, the concept of phytoremediation of specific toxic compounds is a fairly recent development (Anderson *et al.*, 1993; Schnoor *et al.*, 1995). Plants may facilitate detoxification of compounds indirectly by stimulating the metabolism of rhizosphere microorganisms and through the release of enzymes into the rhizosphere. Alternatively, direct uptake of contaminants and subsequent sequestration and/or degradation can contribute to remediation (Shimp *et al.*, 1993; Stomp *et al.* 1994).

The rhizosphere is characterized as the zone of soil under the immediate influence of plant roots (Anderson *et al.* 1993). In the rhizosphere, plant root exudates can affect soil pH, O_2 concentration, redox potential, C:N ratio, and microbial growth (Marschner and Romheld, 1996). Release of root exudates into the rhizosphere has been shown to increase microbial biomass and metabolic activity. Plant enzymes released from roots are also important components of rhizosphere biochemistry (Schnoor *et al.*, 1995; Anderson *et al.*, 1993). Thus, the rhizosphere has striking metabolic diversity, and this diversity is mediated through plant genotype by microbe interaction.

Legumes have been shown to facilitate the degradation (mineralization) of pesticides and other soil contaminants (Table 12). Walton and Anderson (1990) have shown that rhizosphere soil of bush clover (*Lespedeza cuneata* L.) and soybean have greater rate of trichloroethylene degration than does bulk soil. Mineralization of diazinon and parathion

Plant	Contaminant
Pea	Diazinon
Bushclover	Trichloroethylene
Edible bean	Parathion
	Diazinon
Soybean	Trichloroethylene
	Diethylcarbamates
Broadbean	Sulphonamides

Table 12. Legume species shown to enhance degradation of soil contaminants by rhizosphere bacteria. Data adapted from Anderson *et al.*, 1993, and Shimp *et al.*, 1995

was 18% in the rhizosphere of edible bean as compared to about 7% in root-free soil (Hsu and Bartha, 1979). Rhizosphere soil of pea containing diazinon supports substantially higher microbial populations than soil lacking vegetation (Anderson *et al.*, 1993). Legume exudates stimulate growth and expression of novel genes in *Rhizobium* and *Bradyrhizobium* (see Chapter 19 and 20). Moreover, these bacteria degrade numerous phenolics (Parke and Ornston, 1984; Tepfer *et al.*, 1988) and are amenable to genetic modification (Fischer and Long, 1992). Thus, rhizobia might be engineered to degrade any number of toxic compounds in soils planted with legumes. An alternative strategy might be to screen for and select legume genotypes that release enzymes from roots that degrade contaminants. Plant roots are known to exude a wide range of enzymes including esterases, peroxidases, hydroxylases, laccases, nitrilases, and dehalogenases (Schnoor *et al.*, 1995; Waisel *et al.*, 1996).

Not only can plants affect decontamination through the rhizosphere but also they can either sequester or degrade soil and water pollutants through direct uptake in the xylem stream. Salt *et al.* (1995) have demonstrated that *Brassica juncea* and *Thalspi caerulescens* can remove substantial quantities of heavy metals from soils. They also showed that sunflower (*Helianthus annulus* L.) roots rapidly absorbed and precipitated heavy metals from solution. Stomp *et al.* (1994) reported that several laboratories have transformed plants with the genes encoding metallothioneins, metal binding proteins and the resultant transgenic plants grew normally in the presence of heavy metals. They suggest that the development of plants for bioremediation of heavy metals through biotechnology will occur in the near future.

Although it is well known that plants can take up and degrade herbicides and insecticides, they can also accumulate other organic aromatic compounds (Table 13) that pose as hazardous waste such as trinitrotoluene, dioxin, nitrobenzene, and pentachlorophenol (Paterson *et al.*, 1990; Schnoor *et al.*, 1995). Studies of herbicide resistance show that legumes and other species can develop the capacity to degrade or inactivate organic toxins (Harrison, 1992). This may occur through natural selection, mutagenesis, or genetic engineering. In collaboration with M. Sadowsky we have recently initiated a project to bioremediate atrazine and enhance the use of alfalfa in sustainable agriculture (Sadowsky and Smith, 1996). Alfalfa will be transformed with the bacterial *atz*A gene, atrazine halidohydrolase, under the control of the cauliflower mosaic virus 35S (CaMV 35S) promoter to obtain high levels of *atz*A gene expression throughout the plant. The transgenic alfalfa should be able to detoxify atrazine. Ground water contaminated with atrazine, which is not uncommon in the U.S., will be irrigated onto transgenic alfalfa and atrazine degradation monitored.

In the U.S. more than 32,000 toxic waste sites have been targeted for remediation, and this does not include rural areas showing contaminants such as pesticides or herbicides in ground water. Cleanup of these sites by conventional strategies could cost more than $200 billion (Salt *et al.*, 1995). On a worldwide basis these figures are substantially higher. Legumes in phytoremediation offer a sustainable and less costly approach for cleanup of these sites.

V.C. ELECTRICAL ENERGY AND INDUSTRIAL PRODUCTS

Electricity generation from renewable sources of biomass is becoming an attractive economic and

Plant	Contaminant
Alfalfa	Polychlorinated Biphenyls Aroclors DDT, DDE, DDD
Edible Bean	Polynuclear Aromatic Hydrocarbons Chlorobenzenes Polychlorinated Biphenyls Hexachlorocyclohexanes Phenapronil
Lentils	Polybrominated Biphenyls Metsalfulron
Peanut	Polynuclear Aromatic Hydrocarbons Methazole Dinitroanilines
Soybean	Aldrin Chlordane Polychlorinated Biphenyls DDT, DDE, DDD Hexachlorocyclohexanes Chlorinated Phenols, Nitreophenols Carbamates and Organophosphates

Table 13. Legume uptake of organic contaminants. Data adapted from Paterson *et al.*, 1990.

environmental alternative to the use of nonrenewable natural resources such as oil, coal, and gas. In the U.S. more than 1,000 power plants use biomass fuels to produce electricity (DOE Report). The U.S. Department of Energy's (DOE) strategic plan advocates the production of 50,000 megawatts (MW) of electric energy through biomass by the year 2010. "More than any other energy technology, biomass power is capable of decoupling energy production from environmental degradation" (DeLong *et al.*, 1995). With newly developed biomass gasification systems, the amount of electricity produced from biomass can be increased by 50%. Moreover, this new technology coupled to high efficiency gas cleaning systems removes a substantial portion of ash, particulates and alkaline materials plus reduces carbon monoxide and nitrogen oxides. These advances in technology now allow biomass material previously unsuitable for direct burning to be used with high efficiency.

Plant components used as biomass fuel for electricity have included branches pruned from orchards, sawdust from wood processing, nut shells and husks, residue from processing sugarcane and other herbaceous crops, and wood pieces or chips remaining from trees harvested for paper and lumber

production (DeLong *et al.*, 1995). The major drawback for the use of such plant products as biofuels has been proximity to the power plant and long-term availability of the fuel supply (US DOE Report, 1995). A dedicated feed stock supply system (DFSS) would guarantee a consistent long-term fuel supply (DeLong *et al.*, 1995) to the power plant.

Recently, USDA-ARS, DOE, the University of Minnesota, and Minnesota Valley alfalfa producers have initiated a collaborative project to develop alfalfa as a feed stock for the biomass to generate electricity (DeLong *et al.*, 1995). Alfalfa, 100,000 ha grown by over 2,000 farmers, will be harvested and separated into stems and leaves. The stems will be gasified and used to produce electricity, while the leaf meal will be used as high quality animal feed.

This approach for producing biomass for electricity has several attractive features. Generation of electricity from alfalfa biomass will contribute little to global greenhouse gases. Alfalfa grown in the area currently being proposed would introduce a N_2 fixing legume into a cropping system that previously was dedicated to corn and soybeans. Introduction of alfalfa would require fewer inputs of fertilizer, reduce erosion and runoff of chemicals, while providing a habitat for wildlife. Alfalfa meal as a by-product would provide value-added benefits. Demonstration of the feasibility of using alfalfa to produce electricity from biomass will open the door for the use of other legumes, such as fast growing trees and clovers.

Other value-added qualities than simply leaf meal for animal feed could be developed in this project. Alfalfa is easily transformed with foreign genes through *Agrobacterium tumefaciens* (Austin *et al.*, 1995, Vance *et al.*, 1995) and industrial products like α-amylase and Mn-dependent peroxidase have been produced in alfalfa (Austin *et al.*, 1995). Valuable products have also been produced in a number of species through transgenic technology including polyhydroxy-alkanoates in *Brassica* sp. (Poirer *et al.*, 1995), oral immunogens in tobacco and potato (Haq *et al.*, 1995) and antibodies in tobacco (Ma *et al.*, 1995). Transformation of alfalfa with such genes and production of either industrial or pharmaceutical products in alfalfa grown for electricity production would therefore result in three useful products: (i) stems for electricity; (ii) pharmaceuticals processed from leaves; (iii) and leaf meal for animal feed remaining after processing. The production of value added traits through biotechnology is not limited to alfalfa. Many legumes can now be transformed and regenerated (Nisbet and Webb, 1990). The use of N_2-fixing legumes for making industrial and pharmaceuticals could reduce production costs substantially for many items because both the carbon and N_2 required for synthesis would be derived from a sustainable renewable system.

VI. Conclusions

Although striking advances have been made in understanding the molecular and biochemical components regulating symbiotic N_2 fixation, this has yet to be translated into applied improvements. In fact, over the last 40 years the importance of legume symbiotic N_2 fixation to agriculture has been overlooked, if not forgotten. The doubling of Earth's population expected early in the new millennium necessitates that the significance of legume N_2 fixation in cropping systems be reaffirmed. Moreover, scientists must redefine the economic and social benefits of N_2 fixing crops in sustainable agriculture. This will require accurate documentation of reduced fertilizer N expenses, improved water quality and soil health, new uses and value-added components, and reduced pesticide inputs accruing

due to the use of N_2 fixing species. Nontraditional and novel roles for legumes in cropping systems will contribute to their enhanced use in sustainable agriculture. Both traditional and biotechnological approaches will be required to increase yield and maintain high rates of N_2 fixation in crops targeted for use in sustainable agriculture. Biological N_2 fixation is perhaps the single-most important factor impacting the development of more sustainable farming systems.

VII. Acknowledgements

This article is a joint contribution of the Plant Science Research Unit, U.S. Department of Agriculture, Agricultural Research Service, and the Minnesota Agricultural Experiment Station, Paper No. 971130030, Scientific Journal Series.

VIII. References

Allen, O.N. and Allen, E.K. (1981) The Leguminoseae, University of Wisconsin Press, Madison.

Anderson, T.A., Guthrie, E.A. and Walton, B.T. (1993) Environ. Sci. Technol. 27, 2629-2636.

Appleby, C.A. (1992) Sci. Prog. (Oxford) 76, 365-398.

Arp, D. J. (1992) Hydrogen cycling in symbiotic bacteria, in G. Stacey, R.H. Burris, H.J. Evans (eds.), Biological Nitrogen Fixation, Chapman and Hall, New York, pp. 432-460.

Austin, S.E., Bingham, T., Matthews, D.E., Shahan, M.N., Will, J. and Burgess, R.R. (1995) Euphytica 85, 381-393.

Barnes, D.K., Heichel, G.H., Vance, C.P. and Ellis, W.R. (1984) Plant Soil 82, 303-314.

Becker, M., Ladha, J.K. and Ali, M. (1995) Plant Soil 174, 181-194.

Bliss, F.A. (1993) Plant Soil 152, 71-79.

Bockman, O.-C., Kaarstad, O., Lie, O.H. and Richards, I. (1990) Agriculture and Fertilizers: Fertilizers in Perspective, Norsk Hydro, Drammen, Norway.

Bohlool, B.B., Ladha, J.K., Garrity, D.P. and George, T. (1992) Plant Soil 141, 1-11.

Bongaarts, J. (1994) Sci. Amer. 237, 36-42.

Brophy, L.S., Heichel, G.H. and Russelle, M.P. (1987) Crop Sci. 27, 753-758.

Buttery, B. R., Park, S. J. and Hume, D. J. (1992) Can. J. Plant Sci. 72, 323-349.

Caetano-Anolles, G. and Gresshoff, P.M. (1991) Annu. Rev. Microbiol. 45, 345-382.

Coale, F.J., Meisinger, J.J. and Wiebold, W.J. (1985) Plant Soil 86, 357-367.

Cregan, P.B. and Yaklich, R.W. (1986) Theor. Appl. Genet. 72, 782-786.

Danso, S.K.A., Bowen, G.D. and Sanginga, N. (1992) Plant Soil 141, 171-196.

Day, D.A. and Copeland, L. (1991) Plant Physiol. Biochem. 29, 185-201.

Dean, D.R. and Jacobsen, M.R. (1992) Biochemical genetics of nitrogenase, in G. Stacey, R.H. Burris, H.J. Evans (eds.), Biological Nitrogen Fixation, Chapman-Hall, New York, pp. 763-831.

DeLong, M.M., Oelka, E.A., Onischak, M,. Schmid, M.R. and Wiant, B.C. (1995) Sustainable biomass energy production and rural economic development using alfalfa as a feedstock, in Proc. Second Biomass Conf. Americas, pp. 1582-1592, National Renewable Energy Laboratory, Department of Energy, Publication NREL/CP-200-8098.

Delwiche, C.C. (1970) Sci. Amer. 223, 136-146.

Denarie, J. and Cullimore, J. (1993) Cell 74, 951-954.

Denison, R.F., Hunt, S. and Layzell, D. (1992) Plant Physiol. 98, 894-900.

Döbereiner, J. (1987) Biotechnologies using dinitrogen fixation as an alternative to traditional agrochemicals, in G.B. Marini Bettolo (ed.), Toward A Second Green Revolution, Elsevier, Oxford, pp. 351-365.

Dubach, M. and Russelle, M.P. (1994) Agron. J. 86, 259-266.

Duc, G., Mariotti, A. and Amarger, N. (1988) Plant Soil 106, 269-276.

Egli, M.A., Larson, R.J., Hruschka, W.R. and Vance, C.P. (1991) J. Exp. Bot. 42, 969-977.

Evans, H.J., Harker, A.R., Papen, H., Russell, S.A., Hanus, F.J. and Zuber, M. (1987) Annu. Rev. Microbiol. 41, 335-361.

Fischer, R.F. and Long, S.R. (1992) Nature 357, 655-660.

Fujiata, K., Ofosu-Budu, K.G. and Ogata, S. (1992) Plant Soil 141, 155-175.

Giller, K.E. and Cadisch, G. (1995) Plant Soil 174, 255-277.

Golley, F., Baudry, J., Berry, R., Bornkamm, R., Dahlberg, K., Jansson, M., King, V., Lee, J., Lenz, R., Sharitz, R. and Svedin, U. (1992) INTECOL Bull. 20, 15-20.

Graham, P.H. (1981) Field Crops Res. 4, 93-112.

Groat, R.G., Vance, C.P. and Barnes, D.K. (1984) Crop Sci. 24, 895-898.

Hansen, A.P., Yoneyama, T., Kouchi, H. and Martin, P. (1993) Planta 189, 538-545.

Haq, T.A., Mason, H.S., Clements, J.D. and Arntzen, C.J. (1995) Science 268, 714-716.

Hardarson, G. (1993) Plant Soil 152, 1-17.

Hardarson, G. and Danso, S.K.A. (1993) Plant Soil 152, 19-23.

Hardy, R.W.F., Holsten, R.O., Jackson, E.K. and Burns, R.C. (1968) Plant Physiol. 43, 1185-1207.

Harrison, H.F. (1992) Weed Technol. 6, 613-614.

Hartwig, U.A. and Nosberger, J. (1994) Plant Physiol. 161, 109-114.

Heichel, G.H. (1987) Legume nitrogen: symbiotic fixation and recovery by subsequent crops, in Z.R. Helsel (ed.), Energy in Plant Nutrition and Pest Control, Elsevier Science, Amsterdam, pp. 63-80.

Heichel, G.H., Barnes, D.K., Vance, C.P. and Sheaffer, C.C. (1989) J. Prod. Agric. 2, 24-32.

Herridge, D.F. and Danso, S.K.A. (1995) Plant Soil 174, 51-82.

Herridge, D.F., Rupela, O.P., Serraj, R. and Beck, D.P. (1994) Euphytica 73, 95-108.

Hesterman, O.B., Russelle, M.P., Sheaffer, C.C. and Heichel, G.H. (1987) Agron J. 79, 726-731.

Hirsch, A.M. (1992) New Phytol. 122, 211-237.

Hsu, T.S. and Bartha, R. (1979) Appl. Environ. Microbiol. 37, 36-41.

Hunt, S. and Layzell, D.B. (1993) Annu. Rev. Plant Physiol. Plant Mol. Biol. 44, 483-511.

Jacobsen-Lyon, K., Jensen, E.O., Jorgensen, J.-E., Marcker, K.A., Peacock, W.J. and Dennis, E.S. (1995) Plant Cell 7, 213-223.

Jessen, D.L., Barnes, D.K., Vance, C.P. and Heichel, G.H. (1987) Crop Sci. 27, 627-631.

Jessen, D.L., Barnes, D.K. and Vance, C.P. (1988) Crop Sci. 28, 18-22.

Kapulnik, Y., Joseph, C.M. and Phillips, D.A. (1987) Plant Physiol. 84, 1193-1196.

Kinzig, A.P. and Socolow, R.H. (1994) Physics Today 47, 24-35.

Ladha, J.K. and Peoples, M.B. (1995) Management of Biological Nitrogen Fixation for the Development of More Productive and Sustainable Agricultural Systems, Kluwer Academic Publishers, Dordrecht.

Layzell, D.B. and Moloney, A.H.M. (1994) Dinitrogen fixation, in K.J. Boote, J.M. Bennett, T.R. Sinclair, G.M. Paulsen (eds.), Physiology and Determination of Crop Yield, American Society of Agronomy, Madison, pp. 311-334.

Ledgard, S.F. and Steele, K.W. (1992) Plant Soil 141, 137-152.

Lory, J.A., Russelle, M.P. and Heichel, G.H. (1992) Agron. J. 84, 1023-1040.

McDermott, T.R. and Graham, P.H. (1989) Appl. Environ. Microbio. 55, 2493-2498.

Ma, J.K.-C., Hiatt, A., Hein, M., Vine, N.D., Wang, F., Stabila, P., van Dolleweerd, C., Mortov, K. and Lehner, T. (1995) Science 268, 716-719.

Marschner, H. and Romheld, V. (1996) Root-induced changes in the availability of micronutrients in the rhizosphere, in Y. Waisel, A. Eschel, and U. Kafkafi (eds.), Plant Roots: The Hidden Half, Marcel Dekker, Inc., New York, pp. 581-606.

Masterson, C.L. and Sherwood, M.T. (1978) Plant Soil 49, 421-426.

Maxwell, C.A., Vance, C.P., Heichel, G.H. and Stade, S. (1984) Crop Sci. 24, 257-264.

Merrick, M.J. (1992) Regulation of nitrogen fixation genes in free-living and symbiotic bacteria, in G. Stacey, R.H. Burris, H.J. Evans (eds.), Biological Nitrogen Fixation, Chapman-Hall, New York, pp. 835-876.

Minchin, F.R., Witty, J.F. and Mytton, L.R. (1994) Plant Soil 158, 163-167.

Murphy, P.M. (1986) Plant Soil 95, 399-409.

Newbould, P. (1989) Plant Soil 115, 297-311.

Newton, W.E. (1993) Nitrogenases: distribution, composition, structure and function, in R. Palacios, J. Mora, and W. Newton (eds.), New Horizons in Nitrogen Fixation, Kluwer, Dordrecht, pp. 5-17.

Nisbet, G.S. and Webb, K.J. (1990) Transformation in legumes, in Y.P.S. Bajej (ed.), Biotechnology in Agriculture and Forestry, Springer-Verlag, Berlin, pp. 38-48.

Nutman, P.S. (1976) IBP field experiments on nitrogen fixation by nodulated legumes, in P.S. Nutman (ed.), Symbiotic Nitrogen Fixation in Plants, Cambridge University Press, Cambridge, pp. 211-237.

Parke, D. and Ornston, L.N. (1984) J. Gen. Microbiol. 130, 1743-1747.

Paterson, S., Mackay, D., Tam, D. and Shiu, W.Y. (1990) Chemosphere 21, 297-331.

Peoples, M.B. and Craswell, E.T. (1992) Plant Soil 141, 13-39.

Peoples, M.B., Herridge, D.F. and Ladha, J.K. (1995a) Plant Soil 174, 3-28.

Peoples, M.B., Ladha, J.K. and Herridge, D.F. (1995b) Plant Soil 174, 83-102.

Peterson, T.A. and Russelle, M.P. (1991) J. Soil Water Conserv. 46, 229-233.

Phillips, D.A. (1980) Annu. Rev. Plant Physiol. 11, 29-49.

Phillips, D.A., Newell, D.A., Hassell, S.A. and Felling, C.E. (1976) Am. J. Bot. 63, 356-362.

Phillips, D.A. and Teuber, L.R. (1992) Plant genetics of symbiotic nitrogen fixation, in G. Stacy, R.H. Burris, H.J. Evans (eds.), Biological Nitrogen Fixation, Chapman-Hall, New York, pp. 625-645.

Poirer, Y., Nwrath, C. and Somerville, C. (1995) Bio/Technology 13, 142-149.

Rosas, J.C. and Bliss, F.A. (1986) Hort. Sci. 21, 287-289.

Rosendahl, L., Vance, C.P., Miller, S.S. and Jacobsen, E. (1989) Physiol. Plant 77, 606-612.

Rosendahl, L., Vance, C.P. and Pedersen, W.B. (1990) Plant Physiol. 93, 12-19.

Rumjanek, N.G., Dobert, R.C., Vanberkum, P. and Triplett, E.W. (1993) Appl. Environ. Microbiol. 59, 4371-4373.

Russelle, M.P. (1996) Nitrogen cycling in pasture systems, in R.E. Joost and C.A. Roberts (eds.), Nutrient Cycling in Forage Systems, Potash and Phosphorous Institute and Foundation for Agronomic Research, Manhattan, Kansas, pp. 125-166.

Sadowsky, M.J. and Smith, D.R. (1996) Use of phytoremediation strategies to bioremediate contaminated soils and water, in G. Stacy, et al. (eds.), Proceedings of 8th International Congress on Molecular Plant-Microbe Interactions, Kluwer Academic Publishers, Dordrecht (in press).

Salt, D.E., Blaylock, M., Kumar, N.P.B.A., Dushenkov, V., Ensley, B.D., Chet, I. and Raskin, I. (1995) Biotechnology 13, 468-474.

Sanginga, N., Vanlauwe, B. and Danso, S.K.A. (1995) Plant Soil 174, 119-141.

Schnoor, J.L., Licht, L.A., McCutcheon, S.C., Urolfe, N.L. and Carreira, L.H. (1995) Environ. Sci. Technol. 7, 318-323A.

Schubert, K.R. (1981) American Society Plant Physiologist, Rockville, MD.

Sheehy, J.E. (1987) Crit. Rev. Plant Sci. 5, 121-159.

Shimp, J.F., Tracy, J.E., Davis, L.C., Lee, E., Huang, W., Erickson, L.E. and Schnoor, J.L. (1993) Crit. Rev. Envir. Sci. Technol. 23, 41-57.

Sprent, J.I. (1984) Nitrogen fixation, in M. B. Wilkins (ed.), Advances in Plant Physiology, Pitman, London, pp. 249-276.

Stomp, A.M., Han, K.H., Wilbert, S., Gordon, M.P. and Cunningham, S.D. (1994) Ann. N.Y. Acad. Sci. 721, 481-491.

Ta, T.C. and Faris, M.A. (1987) Agron J. 79, 820-824.

Taylor, E.R., Nie, X.Z., MacGregor, A.W. and Hill, R.D. (1994) Plant Mol. Biol. 24, 853-862.

Tepfer, D., Goldmann, A., Pamboukdijian, N., Maille, M., Lepingle, A., Chevalier, D., Denarie, J. and Rosenberg, C. (1988) J. Bacteriol. 170, 1153-1157.

Teuber, L.R., Levin, R.P., Sweeney, T.C. and Phillips, D.A. (1984) Crop Sci. 24, 553-558.

Teuber, L.R. and Phillips, D.A. (1988) Crop Science 28, 599-604.

Thomas, R.J. (1995) Plant Soil 174, 103-118.

Tjepkema, J.D. and Yocum, C.S. (1974) Planta 119, 351-360.

United States Department of Energy (1995) Biomass Power: Program Overview, DOE/GO-10095-089, pp. 12.

Vance, C.P. (1996) Root bacteria interactions: symbiotic nitrogen fixation, in Y. Waisel, A. Eschel, and U. Kafkafi (eds.), Plant Roots: The Hidden Half, Marcel Dekker, Inc., New York, pp. 723-756.

Vance, C.P. and Graham, P.H. (1995) Nitrogen fixation in agriculture: application and perspective, in I.A. Tikhonovich, N.A. Provorov, V.I. Romanov, and W.E. Newton (eds.), Nitrogen Fixation: Fundamentals and Applications, Kluwer Academic Publishers, Dordrecht, pp. 77-86.

Vance, C.P. and Heichel, G.H. (1991) Annu. Rev. Plant Physiol. Plant Mol. Biol. 42, 373-392.

Vance, C.P., Heichel, G.H., Barnes, D.K., Bryan, J.W. and Johnson, L.E. (1979) Plant Physiol. 64, 1-8.

Vance, C.P., Miller, S.S., Gregerson, R.G., Samac, D.A., Robinson, D.L. and Gantt, J.S. (1995) Plant J. 8, 345-358.

Vessey, J.K. (1994) Plant Soil 158, 151-162.

Waggoner, P.E. (1994) Council for Agricultural Science and Technology Task Force Report 121, Ames, Iowa.

Waisel, Y., Eschel, A. and Kafkafi, U. (1996) Plant Roots: The Hidden Half, Marcel Dekker, Inc., New York.

Walton, B.T. and Anderson, T.A. (1990) Appl. Envir. Microbiol. 56, 1012-1016.

Wani, S.P., Rupela, O.P. and Lee, K.K. (1995) Plant Soil 174, 29-50.

West, C.P. and Mallarino, A.P. (1996) Nitrogen transfer from legumes to grasses, in R.E. Joost and C.A. Roberts (eds.), Nutrient Cycling in Forage Systems, Potash and Phosphate Institute and Foundation for Agronomic Research, Manhattan, Kansas, pp. 167-175.

Zanetti, S., Hartwig, U.A., Luscher, A., Hebeisen, T., Hendrey, G.R., Blum, H. and Nosberger, J. (1996) Plant Physiol., in press.

Contributors

Batut, J. (Chapter 23)
Laboratoire de Biologie Moléculaire des Relations Plantes-Microorganismes, CNRS INRA, BP 27 31326 Castanet-Tolosan Cedex, France.

Becker, A. (Chapter 6)
Lehrstuhl für Genetik, Fakultät für Biologie, Universität Bielefeld, Postfach 100131, 33501 Bielefeld, Germany.

Binns, A.N. (Chapter 12)
Department of Biology, University of Pennsylvania, Philadelphia, PA, 19104-6018, USA.

Bisseling, T. (Chapter 18 and 21)
Department of Molecular Biology, Agricultural University, Dreijenlaan 3, 6703 HA, Wageningen, The Netherlands.

Boistard, P. (Chapter 23)
Laboratoire de Biologie Moléculaire des Relations Plantes-Microorganismes, CNRS INRA, BP 27 31326 Castanet-Tolosan Cedex, France.

Botterman, J. (Chapter 17)
Plant Genetic Systems, Jozef Plateaustraat 22, B-9000 Gent Belgium.

Breedveld, M.W. (Chapter 5)
Groningen University, Centre for Biology, Department Microbiology, Kerklaan 30, 9751 NN Haren, The Netherlands.

Brewin, N.J. (Chapter 18 and 22)
Department of Genetics, John Innes Centre, Colney Lane, Norwich NR4 7UH, Great Britain.

Carlson, R.W. (Chapter 7)
Complex Carbohydrate Research Center, University of Georgia, 220 Riverbend Rd., Athens, GA 30602, USA.

Costantino, P. (Chapter 12)
Dip. Genetica e Biologia Molecolare, Università "La Sapienza" Rome, P. le A. Moro 5, 00185 Roma, Italy.

D'Halluin, K. (Chapter 17)
Plant Genetic Systems, Jozef Plateaustraat 22, B-9000 Gent Belgium.

Das, A. (Chapter 13)
Department of Biochemistry and Plant Molecular Genetics Institute, University of Minnesota, 1479 Gortner Avenue, St. Paul, MN 55108, USA.

de la Cruz, F. (Chapter 14)
Departamento de Biología Molecular, Universidad de Cantabria, C. Herrera Oria s/n, 39011 Santander, Spain.

Deng, W. (Chapter 16)
Department of Microbiology, University of Washington, Box 357242, Seattle, WA 98195-7242, USA.

Dessaux, Y. (Chapter 9)
CNRS, Institut des Sciences Végétales, Bâtiment 23, Gif-sur-Yvette, France.

Downie, J.A. (Chapter 20)
John Innes Centre, Colney Lane, Norwich, NR4 7UH, Great Britain.

Eardly, B.D. (Chapter 1)
Department of Biology, The Pennsylvania State University, University Park, Pennsylvania, USA.

Farrand, S.K. (Chapter 9 and 10)
Departments of Microbiology and Plant Pathology, University of Illinois at Urbana-Champaign, Urbana, IL. 61801, USA.

Finan, T.M. (Chapter 2)
Department of Biology, McMaster University, 1280 Main Street West, Hamilton ON L8S 4K1, Canada.

Forsberg, S. (Chapter 7)

Complex Carbohydrate Research Center, University of Georgia, 220 Riverbend Rd., Athens, GA 30602, USA.

Geiger, O. (Chapter 4)

Technische Universität Berlin, Institut für Biotechnologie, FG Technische Biochemie, Seestraße 13, D-13353 Berlin, Germany.

Graham, P.H. (Chapter 8)

Department of Soil, Water, and Climate, University of Minnesota, St. Paul, MN, 55108, USA.

Hadri, AZ-E. (chapter 18 and 21)

Department of Molecular Biology, Agricultural University, Dreijenlaan 3, 6703 HA, Wageningen, The Netherlands.

Hohn, B. (Chapter 15)

Friedrich Miescher-Institut, P.O.Box 2543, CH4002 Basel, Switzerland.

Hooykaas, P.J.J. (Editor)

Leiden University, Institute of Molecular Plant Sciences, Clusius Laboratory, Wassenaarseweg 64, 2333 AL Leiden, The Netherlands.

Hynes, M.F.(Chapter 2)

Department of Biological Sciences, University of Calgary, 2500 University Drive NW, Calgary, Alberta, T2N 1N4, Canada.

Johnson, T.M. (Chapter 13)

Department of Biochemistry and Plant Molecular Genetics Institute, University of Minnesota, 1479 Gortner Avenue, St. Paul, MN 55108, USA.

Kahn, M.L. (Chapter 24)

Department of Microbiology, Institute of Biological Chemistry, Washington State University, Pullman WA 99164-6340, USA.

Kaminski, P.A. (Chapter 23)

Unité de Physiologie Cellulaire, Département des Biotechnologies URA1300. Institut Pasteur. 28, rue du Dr Roux. 75724 Paris Cedex 15, France.

Kannenberg, E.L. (Chapter 7)

Complex Carbohydrate Research Center, University of Georgia, 220 Riverbend Rd., Athens, GA 30602, USA.

Kijne, J.W. (Chapter 11)

Leiden University, Institute of Molecular Plant Sciences, Clusius Laboratory, Wassenaarseweg 64, 2333 AL Leiden, The Netherlands.

Kondorosi, A. (Editor)

CNRS, Institute des Sciences Végétales, Avenue de la Terasse, 91198 Gif sur Yvette, CEDEX, France.

Kondorosi, E. (Chapter 19)

CNRS, Institute des Sciences Végétales, Avenue de la Terasse, 91198 Gif sur Yvette, CEDEX, France.

Lanka, E. (Chapter 14)

Max-Planck-Institut für Molekulare Genetik, Dahlem, D-14195 Berlin, Germany.

Lugtenberg, B.J.J. (Chapter 3)

Leiden University, Institute of Molecular Plant Sciences, Clusius Laboratory, Wassenaarseweg 64, 2333 AL Leiden, The Netherlands.

Matthysse, A.G. (Chapter 11)

Department of Biology, University of North Carolina, Chapel Hill, NC 27599-3280, USA.

McDermott, T.R. (Chapter 24)

Department of Plant, Soil and Environmental Science, Montana State University, Bozeman, Montana 59717-0312, USA.

Miller, K.J. (Chapter 5)

Department of Food Science, The Pennsylvania State University, University Park, Pennsylvania 16802, USA.

Murphy, P.J. (Chapter 9)

Department of Crop Protection, University of Adelaide, Glen Osmond, South-Australia.

Nester, E.W. (Chapter 16)
    Department of Microbiology, University of Washington, Box 357242, Seattle, WA 98195-7242, USA.
Petit, A. (Chapter 9)
    Institut des Sciences Végétales, Bâtiment 23, CNRS, Gif-sur-Yvette, France.
Phillips, D.A. (Chapter 19)
    Department of Agronomy, University of California, Davis, CA 95616, USA.
Pühler, A. (Chapter 6)
    Lehrstuhl für Genetik, Fakultät für Biologie, Universität Bielefeld, Postfach 100131, 33501 Bielefeld, Germany.
Quispel, A. (Chapter 25)
    Leiden University, Institute of Molecular Plant Sciences, Clusius Laboratory, Wassenaarseweg 64, 2333 AL Leiden, The Netherlands.
Reuhs, B.L. (Chapter 7)
    Complex Carbohydrate Research Center, University of Georgia, 220 Riverbend Rd., Athens, GA 30602, USA.
Rossi, L. (Chapter 15)
    Friedrich Miescher-Institut, P.O.Box 2543, CH4002 Basel, Switzerland.
Sadowsky, M.J. (Chapter 8)
    Department of Soil, Water, and Climate, University of Minnesota, St. Paul, MN, 55108, USA.
Schlaman, H.R.M. (Chapter 19)
    Leiden University, Institute of Molecular Plant Sciences, Clusius Laboratory, Wassenaarseweg 64, 2333 AL Leiden, The Netherlands.
Spaink, H.P. (Editor and Chapter 18)
    Leiden University, Institute of Molecular Plant Sciences, Clusius Laboratory, Wassenaarseweg 64, 2333 AL Leiden, The Netherlands.
Tinland, B. (Chapter 15)
    Swiss Federal Institute of Technology, Institute of Plant Sciences, ETH-Zürich CH-8092, Zürich, Switzerland.
Udvardi, M.K. (Chapter 24)
    Division of Biochemistry and Molecular Biology, Australian National University, Faculty of Science, Canberra ACT 0200, Australia.
van Berkum, P. (Chapter 1)
    Soybean and Alfalfa Research Laboratory, USDA, ARS, HH-19, Bldg. 011, BARC-West, 10300 Baltimore Blvd., Beltsville, Maryland 20705, USA.
Vance, C.P. (Chapter 26)
    United States Department of Agriculture, Agricultural Research Service, Plant Science Research Unit, Department of Agronomy and Plant Genetics, University of Minnesota, St. Paul, MN, 55108, USA.

Abbreviations

11-Me-VA	11-methyl-*cis*-vaccenic acid		*cfa*	structural gene for CFA synthase
3-O-MSI	3-O-methyl-*scyllo*-inosamine		*cgm*	cyclic glucan modification
AAI	*Agrobacterium* autoinducer		CHR	chromosome
Aas	acyl transferase, acyl-ACP synthase		CHS	chalcone synthase
ABC	ATP binding cassette		*chv*	chromosomal virulence gene
Ac	acetate		CL	cardiolipin
acc	catabolism of agrocinopines A+B		CLD	chain length determining
ACC	1-aminocyclopropane-1-carboxylate		Cls	cardiolipin synthase
ACC	acetyl-CoA carboxylase		*cls*	structural gene for cardiolipin synthase
AccB	biotin carboxyl carrier protein		CM	cytoplasmic membrane
AccC	biotin carboxylase		CoA	coenzyme A
AcFuc	acetyl fucose		COMT	3-O-methyltransferase
AcMeFuc	acetyl methyl fucose		CPS	capsular polysaccharide
ACP	acyl carrier protein		CRP	cAMP binding protein
AcpP	constitutive acyl carrier protein		C-terminal	carboxy-terminal
acpP	structural gene for constitutive acyl carrier protein		*cvi*	chromosomal virulence
			CY	cytoplasm
AcpS	holo-ACP synthase		D-Ara	*D*-arabinose
AcpXL	acyl carrier protein involved in transfer of 27-hydroxyoctacosanoic acid		Dct	dicarboxylic acid
			dctA	dicarboxylic acid transport gene
acs	gene involved in agrocinopine synthesis		DegT	pleiotropic regulatory protein from *Bacillus Stearothermophilus*
Ag	agrocin producing plasmid			
agc	gene(s) involved in agropine catabolism		dfg	deoxyfructosyl-glutamine
ags	agropine synthesis gene, determining the anabolic amnopine cyclase		dfop	deoxyfructosyl-5-oxo-proline
			DFSS	dedicated feed stock supply system
AM	arbuscular mycorrhiza		DG	diacylglycerol, diglyceride
Amt	ammonium transporter		DHAP	dihydroxyacetone phosphate
APS	adenosine 5'-phosphosulphate		DMPE	dimethyl phosphatidylethanolamine
Ar	*Agrobacterium rhizogenes* accessory plasmid		DNA	desoxyribonucleic acid
			DnrJ	regulatory protein from *Streptomyces peucetius*
ARA	acetylene reduction activity			
arc	gene involved in arginine catabolism		DP	degree of polymerization
Arg	proteins responsible for arginine uptake and transport		dsDNA	double stranded DNA
			dso	double stranded origin
At	*Agrobacterium* spp. accessory plasmid		Dtr	DNA transfer and replication functions
ATA	acetyl-CoA:ACP transacylase		EC	extracellular
ATP	adenosine triphosphate		ED	Entner-Doudoroff pathway
att	attachment		Eex	entry exclusion
BCCP	biotin carboxyl carrier protein		EMP	Embden-Meyerhof-Parnas
BF-7	1,2-diacyl-3-*O*-(α-D-glucopyranosyl-(1-3)-*O*-α-D-mannopyranosyl)-glycerol		ENOD	early nodulins
			EPS	extracellular polysaccharide
Bj38	adhesin from *Bradyrhizobium japonicum*		ER	endoplasmatic reticulum
BNF	biological N_2 fixation		ERIC	sequences for enterobacterial repetitive intergenic consensus sequences
bp	base pair			
bv.	biovar		ET	electrophoretic types
c.q.	*casu quo*		*exo*	exopolysaccharide biosynthesis or regulation genes
CaMV	cauliflower mosaic virus			
Cb	carbamoyl		*exoC*	phosphoglucomutase
CdsA	CDP-diacylglycerol synthetase		*exp*	exopolysaccharide biosynthesis or regulation genes
cdsA	structural gene for CDP-diacylglycerol synthetase			
			exs	exopolysaccharide biosynthesis or regulation genes
cel	cellulose synthesis			
CFA	cyclopropane fatty acid		FAB	fatty acid biosynthesis
Cfa	CFA synthase		*FAB*	locus involved in synthesis or regulation of fatty acids

FAO	food and agriculture organization
FITA	flavonoid independent transcription activation
fix	nitrogen fixation gene locus
fixA	electron transport for nitrogen fixation in *Rhizobium*
fixU	unknown function in *Rhizobium*
g	gram
G3P	glycerol-3-phosphate
Gal	galactose
GalA	galacturonic acid
GDH	glutamate dehydrogenase
Glc	glucose
GlcA	glucuronic acid
GlcNAc	*N*-acetyl glucosamine
gln	glutamine
glp	glycerol catabolic operon
GntR	class of bacteria; regulatory proteins
GOGAT	glutamate synthase
GpsA	biosynthetic glycerol-3-phosphate dehydrogenase
gpsA	structural gene for glycerol-3-phosphate dehydrogenase
GS	glutamine synthase
gus	β-glucuronidase
ha	hectare
hem	heme synthesis
HIS	proteins responsible for histidine uptake and transport
HMW	high molecular weight
HR	hypersensitive response
IBC	inbred backcross line methodology
IFR	isoflavone reductase
IHF	integration host factor
Inc	incompatibility group
ipt	isopentenyl transferase
IV	intervening sequences
kb	kilobase
kD	kilodaltons
KDO	keto deoxy octonate
Kdo	3-deoxy-D-*manno*-2-octulosonic acid
kps	capsular polysaccharide synthesis gene
LB	leghemoglobin
LCO	lipo-chitin oligosaccharide
LHR	limited host range
LMW	low molecular weight
lps	lipopolysaccharide synthesis gene
LPS	lipopolysaccharide
LpxA	UDP-*N*-acetyl glucosamine-3-*O*-acyltransferase
LSU rRNA	large subunit ribosomal RNA
lux	bioluminescence
Mabs	monoclonal antibodies
Man	mannose
mas	genes involved in the synthesis of the mannityl opines mannopine and mannopinic acid
MCP	methyl accepting chemotaxis protein
MDO	membrane-derived oligosaccharide
MdoH	glucosyltransferase
Me	methyl
MeFuc	methyl fucose
MeSFuc	methyl sulpho fucose
Ml	*Mesorhizobium loti*
MLEE	multilocus enzyme electrophoresis
MMPE	monomethyl phosphatidylethanolamine
Mob	locus conferring ability to be mobilised
moc	rhizopine catabolism (*Rhizobium*), mannityl opine catabolism (*Agrobacterium*)
modA	synthesis membrane-derived oligosaccharides
mos	genes involved in rhizopine biosynthesis
MP	megaplasmid
Mpf	mating pair formation functions
MS medium	Murashige and Skoog medium
muc	(mucoid), exopolysaccharide biosynthesis or regulation genes
ndv	locus involved in nodule development
nfe	nodule formation efficiency locus
Ngrol	*Nicotiana glauca rol*
nic	specific cleavage site in *oriT*
nif	nitrogen fixation gene locus
NLS	nuclear localization signal
NMR	nuclear magnetic resonance
noc	genes involved in nopaline catabolism
nod	nodulation gene locus
Nod Factor	nodulation factor (lipochitin oligosaccharide)
noe	nodulation gene locus
nol	nodulation gene locus
nos	gene involved in nopaline biosynthesis, determining the nopaline synthase
nox	genes determining the nopaline oxidase
NPS	nodule polysaccharide
NR	nitrate reductase
NtBBF1	*Nicotiana tabacum* domain B bindingfactor 1
N-terminal	amino-terminal
ntr	genes involved in the regulation of the nitrogen metabolism of the bacterial cell
occ	genes involved in octopine catabolism
ocd	gene determining the OCDase
OCDase	ornithine cyclodeaminase
ocs	gene involved in octopine synthesis, determining the octopine synthase
OM	outer membrane
omp	outer membrane protein
onc	oncogenic
oox	genes determining the octopine oxydase
orf	open reading frame
oriT	origin of conjugal transfer
oriT/tra	origin of transfer/genes involved in conjugal transfer of the plasmid
oriV	origin of vegetative replication

oriV/rep	origin of replication/genes involved in replication of the plasmid	Re	*Rhizobium etli*
P5C	delta-1-pyrroline-5-carboxylate	*rep*	replication
PA	phosphatidic acid	REP	sequences for repetitive extragenic palindromic sequences
PAL	phenylalanine ammonialyase	Rf	*Rhizobium fredii*
PAPS	3'-phosphoadenosine 5' phosphosulphate	*rfa*	lipopolysaccharide core oligosaccharide synthesis gene in enteric bacteria
PBM	peri-bacteroid membrane		
PBS	peri-bacteroid space	*rfb*	lipopolysaccharide O-chain synthesis gene in enteric bacteria
PC	phosphatidylcholine, lecithin		
PCR	polymerase chain reaction	RFLP	restriction fragment length polymorphism
PE	phosphatidylethanolamine	RGD	arginine-glycine-aspartate
PEPC	phosphoenolpyruvate carboxylase	RGE	arginine-glycine-glutamate
PG	phosphatidylglycerol	R-Gene	resistance gene
PGP	phosphatidylglycerol phosphate	Rha	rhamnose
PgpA	PGP phosphatase I	*rhiABC*	rhizosphere-expressed nodulation genes
pgpA	structural gene for PGP phosphatase I	RhiR	regulator of *rhiABC* operon
PgpB	PGP phosphatase II	Ri	root inducing
pgpB	structural gene for PGP phosphatase II	*rkp*	rhizobial K-antigen synthesis gene
PgsA	phosphatidylglycerol phosphate synthase	RkpF	RkpF protein, an ACP
pgsA	structural gene for phosphatidylglycerol phosphate synthase	*rkpF*	structural gene for RkpF protein, involved in formation of rhizobial capsular polysaccharide
PHB	polyhydroxybutyrate		
PHB	poly-β-hydroxybutyrate	Rlp	*Rhizobium leguminosarum* biovar *phaseoli*
pho	genes involved in phosphate uptake		
PI	phosphatidylinositol	Rlt	*Rhizobium leguminosarum* biovar *trifolii*
pin	plant inducible locus	Rlv	*Rhizobium leguminosarum* biovar *viciae*
PL	phospholipid	Rm	*Rhizobium meliloti*
PlsB	glycerol-3-phosphate acyltransferase	Rn	*Rhizobium* species NGR234
plsB	structural gene for glycerol-3-phosphate acyltransferase	RNA	ribonucleic acid
		rol	(A, B, C, D), root locus (A, B, C, D)
PlsC	1-acyl-glycerol-3-phosphate acyltransferase	*rop*	*Rhizobium* outer membrane protein
		rrn	ribosomal genes
plsC	structural gene for 1-acyl-glycerol-3-phosphate acyltransferase	rRNA	ribosomal RNA
		Rsp	*Rhizobium* species
pmtA	structural gene for phospholipid *N*-methyltransferase	Rt	*Rhizobium tropici*
		RT-PCR	reverse transcriptase polymerase chain reaction
PP	periplasm		
PRA	peanut root agglutinin	*sacB*	levansucrase gene from *B. subtilis*
PRP	proline-rich proteins	SAH	*S*-adenosylhomocysteine
PS	phosphatidylserine	SAM	*S*-adenosylmethionine
psc	polysaccharide synthesis	SDS-PAAGE	sodium dodecyl sulphate gel electrophoresis
Psd	phosphatidylserine decarboxylase		
psd	phosphatidylserine decarboxylase structural gene	SecA	ATP-binding protein of Sec-dependent protein secretion
psi	polysaccharide inhibition genes	Shc	squalene-hopene cyclase
PSL	*Pisum sativum* lectin	*shc*	squalene-hopene cyclase structural gene
PSM	peri-symbiont membrane	SI	*scyllo*-inosamine
psr	polysaccharide regulation genes	SM	symbiosome membrane
pss	polysaccharide synthesis genes	spp	species
Pss	phosphatidylserine synthase	*sqd*	genes involved in sulfolipid biosynthesis
pssA	structural gene for phosphatidylserine synthase	ssDNA	single stranded DNA
		ss-T-DNA	single stranded transferred DNA
put	catabolism of proline	SSU rRNA	small subunit ribosomal RNA
pv.	pathovar	Suc	succinate
Pyr	pyruvate	Sym	symbiotic gene-containing plasmid
R/M	restriction/modification	Syr	symbiotic regulation
RAPD	random amplified polymorphic DNA	*syr*	symbiotic regulator gene
RB,LB	right border and left border sequences	TCA	tricarboxylic acid

T-DNA	transferred DNA	tRNA	transfer RNA
Ti	tumor inducing	TY	trypton yeast extract
TIP	tumor inducing principle	*tzs*	*trans*-zeatin synthesis
T$_L$	T-Left	UDP-GlcNAc	uridine-diphosphate-*N*-Acetyl Glucosamine
tml	tumor morphology large		
tmr	tumor morphology root	*Ugp*	utilization of glycerol-phosphodiesters
tms	tumor morphology shoot	VA	*cis*-vaccenic acid
T$_R$	T-right	VAI	*Vibrio* autoinducer
Tr	tartrate-utilization plasmid	VAM	vesicular arbuscule mycorrhizal
Tra	transfer phenotype	*vir*	virulence genes
tra	conjugal transfer	WHR	wide host range
TraI	conjugative transfer of Ti plasmid	YM	yeast extract mannitol
trb	conjugal mating bridge		
trl	*traR*-like		

Subject Index

C

F

G

I

LPS   32, 45, 46, 47, 48, 49, 50, 59, 67, 89, 109, 119, 120, 121, 122, 123, 124, 125, 126, 127, 129, 130, 131, 133, 134, 136, 137, 138, 139, 140, 141, 142, 143, 144, 145, 146, 147, 148, 149, 237, 241, 243, 245, 423, 424, 489, 493, 494, 497, 536

role in infection   423, 424
lps genes   131, 133, 536
lps region   131, 133
Lpx proteins
  LpxA   64, 536
LSU   1, 4, 5, 7, 21, 536
lumen of infection thread   417, 421, 422
lupin   159, 373, 470, 519
Lupinus   19, 76, 127, 129, 159, 348, 372, 518
  angustifolius   518
  densiflorus   19
lux genes
  luxI   30, 216
Lux proteins
  LuxR   214, 216, 222
LysR family   365, 368, 369, 382, 491
LysR motif   365, 371, 375
lytic enzyme   143, 500

M

MAbs   129, 133, 149
Macroptilium   113, 161, 168, 348, 377, 406, 442, 471
  atropurpureum   113, 471
Maize   334, 524
malate   442, 465, 466, 470-475, 479, 480, 515
malate dehydrogenase   470, 479
malate synthase   465
malic enzyme   466, 471, 475
malonyl-ACP   61, 62
malonyl-CoA   61, 466, 473, 474
mannopine   175, 176, 182, 187, 188, 190, 192, 202, 222, 254, 255, 258, 325, 536
mannopine-type strains   323
mannose   98, 100, 124, 125, 127, 130, 147, 187, 359, 468, 536
mannose 4, 6 dehydratase   393
mannose-1-P guanylyl transferase   393
mannosidases   426
  α-mannosidases   426
Mar proteins
  MarR   107
mas genes   536
mating   34, 35, 200, 206-209, 224-226, 228, 230, 288, 291, 293, 297, 536, 538
mating bridge   206-209, 225, 226, 538
mating-pair-formation   283, 290, 291, 536
matrix glycoprotein   418, 421, 423
MCP   182, 536
McpA   218
MDO   64, 73, 88, 89, 536
  biosynthesis   89

mdo genes
  mdoA   88
MdoH   64, 536
Medicago   18, 19, 20, 113, 114, 138, 148, 152, 159, 160, 161, 178, 348, 352, 359, 371, 372, 391, 396, 397, 398, 406, 408, 410, 412, 413, 421, 422, 492, 511, 512, 517, 518
  cerulea   114
  littoralis   371, 398
  lupulina   398
  polymorpha   159, 161
  sativa   113, 138, 148, 152, 372, 412, 492, 511, 518
  truncatula   114, 371, 398, 413
membrane   45-52, 56, 59, 67, 69, 70, 73, 75-78, 85-89, 98, 99, 104, 108, 111, 120, 121, 131, 133, 138, 140-145, 148, 149, 161, 163, 179, 190, 206, 242, 268, 290, 293, 295, 297, 299, 306, 312, 352, 355, 356, 357, 369, 382, 388, 392, 395, 397, 404, 406, 407, 411-414, 417, 423-427, 441, 445, 446, 449, 450, 468, 473, 478, 479, 490, 491, 493, 496, 499, 502, 537
  biogenesis   423, 424
  depolarization   407, 413, 490
  energetization   499
  integrity   496
  lipid   56, 76, 77, 78
  potential depolarization   411, 412, 413
  separation   47
  transport   395
membrane-derived oligosaccharide   64, 73, 536
meristem   113, 135, 137, 262, 349, 352, 353, 356, 420, 423, 438
meristematic tissue   140, 263, 334, 352, 356, 511
Mesorhizobium   16
  ciceri   15, 16
  huakuii   9, 15, 16
  loti   9, 15, 16, 58, 113, 127, 129, 130, 363, 364, 365, 367, 373, 394, 395, 396
    Nod factors   396
  mediterraneum   9, 15, 16
  thianshanense   9, 15, 16
metabolism   12, 27, 57, 59, 78, 79, 161, 163, 173, 187, 204, 209, 218, 262, 264, 373, 385, 425, 427, 435, 437, 440, 441, 442, 445, 452, 462, 464, 465, 466, 467, 468, 469, 470, 471, 472, 473, 474, 475, 477, 478, 479, 480, 481, 492, 498, 504, 506, 525
metallothioneins   526
11-methyl-*cis*-vaccenic acid   58, 535
methyl fucose   393, 394, 395, 535, 536
methyl fucosyl group   398
methyl transferase   390, 393, 398
methylation   63, 71, 82, 136, 179, 341
methyl-esterified pectin   419
11-Me-VA   58, 535
Mg^{2+} 86, 282, 286, 297, 463
mia genes
  miaA   330
microaerobic environment   477, 479

N

O

S

T

W

X

566